SYMBOLS AND UNITS OF PHYSICAL QUANTITIES

QUANTITY	SYMBOL*	UNITS	QUANTITY	SYMBOL*	UNITS
Acceleration	$\mathbf{a}$	m/s^2	Entropy	S	J/K
Angle	θ, ϕ	rad	Force	$\mathbf{F}$	N
Angular acceleration	$\boldsymbol{\alpha}$	$rad/s^2, s^{-2}$	Frequency	f	Hz
Angular frequency, angular			Heat	Q	J
speed, angular velocity	$\omega, \boldsymbol{\omega}$	$rad/s, s^{-1}$	Heat flow	H	J/s
Angular momentum	$\mathbf{L}$	$kg \cdot m^2/s$	Inductance	L	H
Atomic number	Z		Intensity	$I, S, \mathbf{S}$	W/m^2
Capacitance	C	F	Magnetic field	$\mathbf{B}$	T
Charge	q, Q, e	C	Magnetic flux	ϕ_B	$T \cdot m^2$
Charge density			Mass	m, M	kg
Line	λ	C/m	Mass number	A	
Surface	σ	C/m^2	Molar specific heat	C_V, C_P	$J/mol \cdot K$
Volume	ρ	C/m^3	Momentum	$\mathbf{p}$	$kg \cdot m/s$
Conductivity	σ	$\Omega^{-1} \cdot m^{-1}$	Period	T	s
Current	I	A	Power	P	W
Current density	$\mathbf{J}$	A/m^2	Pressure	P	Pa
Density	ρ	kg/m^3	Resistance	R	Ω
Dielectric constant	κ		Resistivity	ρ	$\Omega \cdot m$
Dipole moment, electric	$\mathbf{p}$	$C \cdot m$	Rotational inertia	I	$kg \cdot m^2$
Dipole moment, magnetic	$\boldsymbol{\mu}$	$A \cdot m^2$	Temperature	T	K
Distance, displacement,	$x, y, z, s, d,$		Time	t	s
length, position	$\ell, w, h, \mathbf{r}$	m	Torque	τ	$N \cdot m$
Electric field	$\mathbf{E}$	N/C, V/m	Specific heat	c	$J/kg \cdot k$
Electric flux	ϕ, ϕ_E	$N \cdot m^2/C$	Speed, velocity	$v, \mathbf{v}$	m/s
Electric potential	V	V	Volume	V	m^3
Electromotive force	$\mathscr{E}$	V	Wavelength	λ	m
Energy	E, U, K	J	Wavenumber	k	m^{-1}
			Work	W	J

*Boldface indicates vector quantities.

TRIGONOMETRY

Definition of angle (in radians): $\theta = \dfrac{s}{r}$

2π radians in complete circle
1 radian $\simeq 57.3° = 2.06$ arc sec

TRIGONOMETRIC FUNCTIONS

$\sin \theta = \dfrac{y}{r}$

$\cos \theta = \dfrac{x}{r}$

$\tan \theta = \dfrac{\sin \theta}{\cos \theta} = \dfrac{y}{x}$

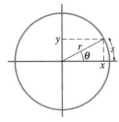

THE GREEK ALPHABET

	UPPERCASE	LOWERCASE
Alpha	A	α
Beta	B	β
Gamma	Γ	γ
Delta	Δ	δ
Epsilon	E	ε
Zeta	Z	ζ
Eta	H	η
Theta	Θ	θ
Iota	I	ι
Kappa	K	κ
Lambda	Λ	λ
Mu	M	μ
Nu	N	ν
Xi	Ξ	ξ
Omicron	O	o
Pi	Π	π
Rho	P	ρ
Sigma	Σ	σ
Tau	T	τ
Upsilon	Υ	υ
Phi	Φ	ϕ
Chi	X	χ
Psi	Ψ	ψ
Omega	Ω	ω

PHYSICS

FOR SCIENTISTS AND ENGINEERS

Second Edition

Richard Wolfson
Middlebury College

Jay M. Pasachoff
Williams College

HarperCollins*CollegePublishers*

Executive Editor: Doug Humphrey

Senior Development Editor: Kathy Richmond

Project Editor: Carol Zombo

Design Administrator: Jess Schaal

Text and Cover Design: Ellen Pettengell

Cover Photos: Hang gliding: Bill Ross/All Stock;
   Cloud chamber: Photo Researchers, Inc.

Photo Researchers: Lynn Mooney, Roberta Knauf

Production Administrator: Randee Wire

Project Coordination: Elm Street Publishing Services, Inc.

Compositor: Interactive Composition Corporation

Art Studio: Precision Graphics

Printer and Binder: R. R. Donnelley & Sons Company

For permission to use copyrighted material, grateful acknowledgment is made to the copyright holders on pp. A-36 to A-40, which are hereby made part of this copyright page.

Physics for Scientists and Engineers, Second Edition
Copyright © 1995 by HarperCollins College Publishers

Library of Congress Cataloging-in-Publication Data

Wolfson, Richard.
    Physics : for scientists and engineers / Richard Wolfson, Jay M.
  Pasachoff. — 2nd ed.
        p.   cm.
    Includes index.
    ISBN 0-673-52041-2
    1. Physics.   I. Pasachoff, Jay M.   II. Title.
  QC21.2.W653   1995
  530—dc20   93-29622

94 95 96 97 9 8 7 6 5 4 3 2 1

BRIEF CONTENTS

CONTENTS

PART 5 895

OPTICS

PART 2

OSCILLATIONS, WAVES, AND FLUIDS

EXAMPLES AND APPLICATIONS

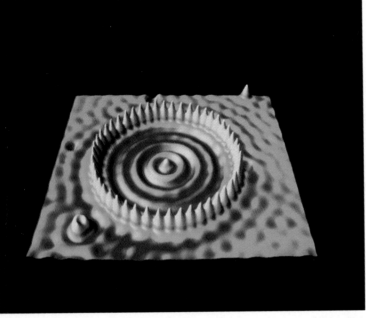

PART 3
THERMODYNAMICS

PART 4
ELECTROMAGNETISM

PREFACE

PHYSICS: CHALLENGE AND SIMPLICITY

Physics is fundamental. To understand physics is to understand how the world works, both at the everyday level and on scales of time and space so small and so large they defy intuition.

To the student, physics can be at once fascinating, challenging, subtle, and yet simple. Physics fascinates with its applications throughout science and engineering and in its revelation of unexpected phenomena like superconductivity, black holes, and chaos. It challenges with its need for precise thinking and skillful application of mathematics. It can be subtle in its interpretation, especially in describing phenomena at odds with everyday intuition. Most importantly, physics is simple. Its few fundamental laws are stated in the simplest of terms, yet they encompass a universe of natural phenomena and technological devices. Students who recognize the simplicity of physics develop confidence that stems from understanding the physical universe at a fundamental level.

This text is for science and engineering students. The standard version covers a full sequence of calculus-based university physics, and the extended version adds seven chapters on modern physics. The extended version is also available as a two-volume set.

Coverage The book is organized into six parts. Part 1 (Chapters 2 through 14) covers the basics of mechanics. Part 2 (Chapters 15 through 18) studies motion in terms of oscillations, waves, and fluids. Part 3 (Chapters 19 through 22) is on thermodynamics. Part 4 (Chapters 23 through 34) deals with electricity and magnetism. Part 5 (Chapters 35 through 37) treats optics. Part 6 (Chapters 38 and 39) briefly introduces modern physics. Each part ends with Cumulative Problems that help students synthesize concepts from several chapters. Part 6 of the extended version (Chapters 38 through 45) begins with relativity and continues with quantum mechanics and its applications to atoms, molecules, and the solid state; nuclear physics and its applications; and particle physics and cosmology.

DISTINGUISHING FEATURES OF THE TEXT

In the second edition, like the first, we emphasize careful and thorough explanations. We have pared down wordiness without sacrificing clarity of explanation. And we've added many new features to help you learn.

Contents This revision includes many substantial changes and improvements, most of which were suggested by instructors who used the text or reviewed the manuscript. Here are the most important changes that were implemented in this edition:

- We added a separate chapter (Chapter 3) on vectors.
- We reorganized Chapter 9 on gravitation so that it is more focused than in the first edition.
- We added a second chapter on wave motion (Chapter 17) that includes extensive new material on sound.
- We reorganized the material on electricity and magnetism (Part 4) for a shorter, clearer presentation.
- The optics chapters in Part 5 are completely rewritten, and we added a chapter to this edition for more thorough coverage.

- Relativity (Chapter 38) is now treated after optics, for an introduction to modern physics at the end of the text.
- The number of problems in the second edition is double that of the previous edition.
- A complete package of supplements is offered for this edition.

Applications We include a rich array of practical applications—from the workings of the compact disc through skyscraper engineering, biomedical technology, antilock brakes, space exploration, global warming, microelectronics, lasers, and much more. We chose to integrate many shorter applications into the text, where they are more likely to be read, rather than presenting a few applications as guest essays. An index of examples and applications appears at the beginning of the text. High-quality color photographs and figures enhance the applications.

Questions These follow the chapter synopsis and can be used for class discussion or to get students thinking about concepts in the chapter before they start on the problem sets.

Problems Science and engineering students learn physics best by working physics problems. The second edition contains nearly *3,000* end-of-chapter problems—double the number from the first edition. Problems range from simple confidence builders, to more complex and realistic problems involving the application of multiple concepts, to difficult **Supplemental Problems** that will challenge the best students. A section of **Paired Problems** for each chapter lets students practice problem-solving techniques on a pair of problems whose solutions involve closely related physical concepts or mathematical approaches. **Cumulative Problems** at the end of each part of the text integrate the material from several chapters.

In-Text Exercises Reinforce the Examples Worked examples in the text are generally followed by an exercise to reinforce concepts or processes. A **Similar Problems** line after each exercise points out problems at the end of the chapter that relate to the in-text exercises.

Tip Boxes Tip boxes point out useful problem-solving techniques and warn against common pitfalls. Frequent text references to specific problems link text and problems in the common purpose of enhancing the student's understanding of physics.

Pedagogical Use of Color The book's design enhances its readability; from the carefully planned use of color in figures and to highlight important equations and definitions to the selection of photographs, design is an essential pedagogical feature. Physical elements are coded in color to make them more apparent. A list of the elements and the colors used for them follow this preface.

Chapter Synopses Chapter summaries emphasize key concepts and remind students of new terms and mathematical symbols. Even in a field as fundamental as physics, many theories and their equations have limited applicability, and each chapter concludes with a reminder of limitations students should keep in mind.

Appendices and Endpapers The book's appendices and endpapers contain a mathematical review and a wealth of up-to-date physical data, conversion factors, and information on measurement systems.

PHYSICS AND MATHEMATICS

For many students, the university physics course is their first contact with practical applications of calculus. We recognize that our readers bring a range of mathematical abilities, from those taking their first calculus course concurrently with physics to those fluent in both differential and integral calculus. The former will find tip boxes and figures to build understanding of and confidence in their new mathematical skills; the latter will find a selection of challenging calculus-based problems.

For all students, we refuse to let mathematical derivations dangle. We ask frequently after the derivation of an equation or an example solution: "Does this result make sense?" We show that it does by examining easily understood special cases, thus building both physical intuition and confidence in the application of mathematics. We explore the meaning of equations verbally and through figures, ensuring that concepts are clear as students begin to use the material qualitatively.

THE SUPPLEMENT PACKAGE TO ACCOMPANY THIS TEXT

Professors and students alike often find it useful to have supplements designed to complement their text. For the second edition, we provide a new expanded package.

For the Instructor

The *Solution Manual*, prepared by Edward S. Ginsberg, University of Massachusetts-Boston, includes worked solutions to all problems in the text.

TestMaster software, available in IBM and Macintosh formats, enables instructors to select problems for any chapter, scramble them as desired, or create problems. A print version of the Test Generator is also available.

Overhead transparencies are available to instructors who adopt the text. This set of 150 color acetates of figures from the text will be useful in classroom discussions.

For the Student

The *Student Study Guide*, by Jeffrey J. Braun, University of Evansville, briefly summarizes the text discussion and important equations in each chapter. It also gives some common pitfalls for students to avoid and includes plenty of practice problems (with solutions). To order, use ISBN 0-673-52369-1.

The *Student Solution Manual*, by Edward S. Ginsberg, University of Massachusetts-Boston, includes complete, worked-out solutions to some of the odd-numbered problems in the text. To order, use ISBN 0-06-501873-7.

PhysiCad Explorer, by Tara C. Woods, consists of approximately 200 examples from the text that have been adapted for use with Mathcad® for Windows software (version 4.0 or higher). The student can call up representative examples from the text and substitute new variables to see how the results differ. Mathcad® will perform all the calculations, generating numerical solutions and graphics where appropriate. Besides offering additional problem-solving drills, *PhysiCad Explorer* gives the student hands-on practice using one of the most powerful numerical tools available, Mathcad®. *PhysiCad Explorer* is self-contained and includes the Mathcad Reader, so students need not own the full Mathcad® software package.

A Calculus Laboratory Workshop with Applications to Physics, by Joan R. Hundhausen and F. Richard Yeatts, both from the Colorado School of Mines, contains 30 self-contained projects that can be used to strengthen the calculus skills of introductory physics students. To order, use ISBN 0-06-501719-6.

ACKNOWLEDGMENTS

A project of this magnitude is not the work of its authors alone. Here we acknowledge the many people whose contributions made this book possible. Colleagues using the first edition at universities and colleges throughout the world have volunteered suggestions, and most will find those incorporated here. We are especially grateful to Bob Gould at Middlebury College and Al Bartlett at the University of Colorado for their many thoughtful comments. Students in Middlebury's Physics 109–110 courses have made significant contributions to the book's accuracy and readability. Middlebury's laboratory supervisor, Cris Butler, skillfully prepared lab demonstrations illustrated in the book, and photographer Erik Borg captured them on film. Linda Eisenhart, Ann Broughton, and Tucky Ceballos at Middlebury and Susan Kaufman at Williams College provided invaluable assistance with many details. Willie S. M. Yong of Singapore, a physics educator and publisher with extensive involvement in the International Physics Olympiad, suggested new problems and helped keep the authors aware of our international readership.

It is frustrating for students and professors to find numerical errors in a textbook, especially in answers to problems. We have gone to considerable lengths to make this book as free from error as possible, and we credit the people who helped achieve this goal. Edward Ginsberg, University of Massachusetts-Boston, meticulously checked all numerical results in examples, exercises, and end-of-chapter problems. Alan Goldman of Iowa State University, Sven Rudin of Ohio State University, Kent Scheller of the University of Evansville, and Thomas Suleski of the Georgia Institute of Technology rechecked these numerical results. Claire D. Dewberry, Florida Community College at Jacksonville, checked the art proofs and made numerous suggestions. Naomi Pasachoff provided expert proofreading.

Every chapter of the second edition was reviewed by physics professors who examined part or all of the manuscript. Their comments were incorporated into the final product. We are very grateful to them all:

Edward Adelson, *Ohio State University*
Vijendra Agarwal, *Moorhead State University*
William Anderson, *Phoenix College*
Gordon J. Aubrecht, *Ohio State University-Marion*
Paul Avery, *University of Florida*
John Bartelt, *Vanderbilt University*
Marvin Blecher, *Virginia Polytechnic Institute and State University*
John Brient, *University of Texas at El Paso*
James H. Burgess, *Washington University*
Bernard Chasan, *Boston University*
Roger Clapp, *University of South Florida*

Claire D. Dewberry, *Florida Community College at Jacksonville*
Robert J. Endorf, *University of Cincinnati*
Heidi Fearn, *California State University-Fullerton*
Shechao Feng, *University of California, Los Angeles*
Albert L. Ford, *Texas A & M University*
Ian Gatland, *Georgia Institute of Technology*
James Goff, *Pima Community College*
Alan I. Goldman, *Iowa State University*
Philip Goode, *New Jersey Institute of Technology*
Denise S. Graves, *Clark Atlanta University*
Donald Greenberg, *University of Alaska*
Phillip Gutierrez, *University of Oklahoma*
Stephen Hanzely, *Youngstown State University*
Randy Harris, *University of California, Davis*
Warren Hein, *South Dakota State University*
Gerald Harmon, *University of Maine at Orono*
Roger Herman, *The Pennsylvania State University*
Francis L. Howell, *University of North Dakota*
J. N. Huffaker, *University of Oklahoma*
Wayne James, *University of South Dakota*
Karen Johnston, *North Carolina State University*
Evan Jones, *Sierra College*
Dean Langely, *St. John's University*
Chew-Lean Lee, *Florida Community College*
Brian Logan, *University of Ottawa*
Peter Loly, *University of Manitoba*
Hilliard K. Macomber, *University of Northern Iowa*
David Markowitz, *University of Connecticut*
Daniel Marlow, *Princeton University*
Nolan Massey, *University of Texas at Arlington*
Ralph McGrew, *Broome Community College*
Victor Michalk, *Southwest Texas State University*
P. James Moser, *Bloomsburg University*
Vinod Nangia, *University of Minnesota*
Robert Osborne, *Yakima Valley Community College*
Michael O'Shea, *Kansas State University*
George Parker, *North Carolina State University*
R. J. Peterson, *University of Colorado*
Joseph F. Polen, *Shasta College*
Talat Rahman, *Kansas State University*
Dennis Roark, *The King's College*
Roger F. Sipson, *Moorhead State University*

John Sperry, *Sierra College*
Lon D. Spight, *University of Nevada at Las Vegas*
Robert Sprague, *Foothill College*
J. C. Sprott, *University of Wisconsin-Madison*
Konrad M. Stein, *Golden West College*
Bryan H. Suits, *Michigan Technological University*
Leo Takahashi, *The Pennsylvania State University*-Beaver Campus
Frank Tanherlini, *College of the Holy Cross*
Karl Trappe, *University of Texas*
Michael Trinkala, *Hudson Valley Community College*
Loren Vian, *Centralia College*
Clarence Wagener, *Creighton University*
Robert Weidman, *Michigan Technological University*
George Williams, *University of Utah*
Robert J. Wilson, *Colorado State University*
Arthur Winston, *Gordon Institute*
David Yee, *City College of San Francisco*
John Yelton, *University of Florida*
Dale Yoder-Short, *Michigan Technological University*

We also thank editors Kathy Richmond, Jane Piro, and Doug Humphrey at HarperCollins for their vigorous support of this project, and Sue Nodine of Elm Street Publishing Services for her skillful efforts in bringing it to fruition. Finally, we thank our families for their patience during the intense process of revising this book.

We invite suggestions from readers—students, professors, and others—for improvements to our text. We promise a speedy reply to each correspondent and an effort to incorporate appropriate suggestions in subsequent printings. Please write us at our respective institutions or contact us by electronic mail.

Richard Wolfson
Middlebury College
Middlebury, Vermont 05753
wolfson@middlebury.edu

Jay M. Pasachoff
Williams College
Williamstown, Massachusetts 01267
pasachoff@williams.edu

A VISUAL GUIDE TO THE BOOK

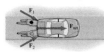

FIGURE 5-13 When two people push a stalled car, the force they exert is the vector sum of their individual forces.

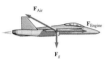

FIGURE 5-14 Forces on a jet include the force of the engines, the force of the air that provides both lift and drag, and the force of gravity. When the plane moves with constant velocity, these forces sum to zero.

5-7 ADDING FORCES

So far we've considered only situations where a single force acts on an object. But often there are several forces acting, as suggested in Fig. 5-13. Airplane flight involves the forces associated with engine thrust, air flowing over the wings and body of the plane, and the force of gravity (Fig. 5-14).

How are we to apply Newton's law in these multiple-force cases? The answer is given ultimately by experiment: We add vectorially the individual forces to find the **net force** on an object. Newton's second law then relates the object's mass and acceleration to this net force. As long as that net force is not zero, the object must be accelerating.

The *net* force is all-important! There may be all sorts of forces acting on a body, but it's their vector sum alone that determines the net force and therefore the acceleration. We've written Newton's law $\mathbf{F}_{net} = m\mathbf{a}$ to emphasize this point; some authors write $\Sigma\mathbf{F} = m\mathbf{a}$ for the same reason.

TIP   Trust Newton   Newton's second law really does relate acceleration to the net force in all inertial frames. In particular, if you see an object at rest or moving with constant velocity, with respect to an inertial frame, then $\mathbf{a} = \mathbf{0}$ and Newton says the net force on that object must be zero. If you find a nonzero net force, look again; there *must* be additional forces acting (see Fig. 5-15).

Tips point out strategies for helping students to solve problems.

In this introductory chapter on Newton's laws, we'll consider the addition of forces in one dimension only. But forces are vectors, and vector addition is generally necessary to get the net force. In the next chapter we'll deal with Newton's laws in two and three dim...

FIGURE 5-15 The gravitational force F... acceleration is zero, the net force must be ze... case, it is the upward contact force **F**, provid... electrical force involving interactions between...

● EXAMPLE 5-7   AN ELEVATOR

A 740-kg elevator is accelerating upward at 1.1 m/s^2, pulled by a cable of negligible mass, as shown in Fig. 5-16. What is the tension force in the elevator cable?

Solution

This is an important example, and understanding it thoroughly will help you apply Newton's second law correctly in more complicated situations. We're given the acceleration and mass and asked for the force in the cable. Can't we just write $F = ma$ for that force? No! $F = ma$ is true only for the *net force*, and here the cable tension is *not* the only force acting. There is also the force of gravity, which has magnitude mg and points downward. We've shown both forces in Fig. 5-16b.

TIP   Identify the Forces   A key step in solving any Newton's law problem is to identify the forces acting on the object or objects of interest. There's no sense proceeding unless you know the *net force*– and that requires identifying *all* the individual forces acting.

Calling the tension force **T** and the gravitational force $\mathbf{F}_g$, Newton's second law becomes

$$\mathbf{F} = \mathbf{T} + \mathbf{F}_g = m\mathbf{a}. \qquad (5\text{-}5)$$

TIP   Vectors Tell it All   You might be tempted to put minus signs in Equation 5-5. But not yet! This is a *vector* equation, and information about signs is built into the directional nature of the vectors. Skipping the step of writing Newton's law in vector form will often get you into trouble. And in writing the vector equation, you never need to worry about signs.

Having written the vector form of Newton's second law for our problem, we next choose a coordinate system and write the components of Newton's law in that system. Here all the forces are vertical, so we choose our y axis pointing vertically upward. Now we write the components of our vector equation 5-5. In this one-dimensional problem only the y equation is interesting; formally, it reads

$$T_y + F_{gy} = ma_y. \qquad (5\text{-}6)$$

Now the tension force **T** points upward, or in the +y direction; its vertical component T_y is positive and is equal to the magnitude T of **T**. The gravitational force $\mathbf{F}_g$ has magnitude mg and points vertically downward, in the −y direction. Its vertical component F_{gy} is therefore $-mg$. Then Equation 5-6 is

$$T - mg = ma_y,$$

so $$T = ma_y + mg = m(a_y + g). \qquad (5\text{-}7)$$

Before putting in specific numbers, let's see if this result makes sense. Suppose the acceleration a_y were zero. Then the

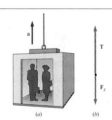

FIGURE 5-16 Forces on the elevator include the cable tension **T** and the gravitational force $\mathbf{F}_g$; the net force is their vector sum. Since the elevator is accelerating upward, this net force must be upward.

net force on the elevator would have to be zero. In this case, Equation 5-7 tells us that $T = mg$, showing that the tension force and gravity have the same magnitude. Since they point in opposite directions, they indeed sum to zero.

If, on the other hand, the elevator is accelerating upward, so a_y is positive, then Equation 5-7 requires that the magnitude of the tension force exceed that of the gravitational force by an amount ma_y. Why is this? Because in this case the cable not only supports the elevator against gravity, but also provides the upward acceleration. For the numbers of this example, we have

$$T = m(a_y + g) = (740 \text{ kg})(1.1 \text{ m/s}^2 + 9.8 \text{ m/s}^2)$$

$$= 8100 \text{ N}.$$

What if the elevator were accelerating downward? Then a_y is negative and Equation 5-7 shows that the tension force is less than mg. Does this make sense? Yes. For a downward acceleration, the net force must be downward so that the magnitude of the downward gravitational force exceeds that of the upward tension force. Were the elevator in free fall, so $a_y = -g$, the tension force would be exactly zero.

You probably could have reasoned out the answer to this problem in your head. But by applying the steps illustrated here– specifically, writing Newton's second law as a vector equation and then breaking it into components– you will be able to handle more complicated situations without confusion. We will outline the steps in solving a Newton's law problem more formally in the next chapter.

◆ **EXERCISE**   A 270-kg rocket accelerates straight upward from Earth at 51 m/s^2. What is the thrust (force) provided by the rocket's engine?

Answer: 1.6×10^4 N

Some problems similar to Example 5-7:   27±30   ●

Worked examples are followed by exercises to reinforce concepts. A **Similar Problems** line after the exercise indicates the relevant end-of-chapter problems.

A rich array of practical applications is presented—from the workings of a compact disc to skyscraper engineering, biomedical technology, antilock brakes, global warming, and microelectronics.

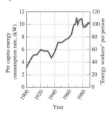

---

166 CHAPTER 7 WORK, ENERGY, AND POWER

● **EXAMPLE 7-12** CLIMBING MOUNT WASHINGTON

A 55-kg hiker ascends New Hampshire's Mount Washington, a vertical rise of 1300 m from the base elevation. The hike takes 2 hours. A 1500-kg car drives up the Mount Washington Auto Road, the same vertical rise, in half an hour. What is the average power output in each case? Assume the hiker and car maintain constant speed and neglect friction, so each does work only against the gravitational force.

Solution

In Example 7-9, we found that the work done by gravity depends only on the overall change in vertical position. Here the hiker and car do positive work $\Delta W = mgh$ to overcome the negative work $-mgh$ done by gravity. So the average power outputs are

$$\bar{P}_{hiker} = \frac{\Delta W}{\Delta t} = \frac{(55\ kg)(9.8\ m/s^2)(1300\ m)}{(2.0\ h)(3600\ s/h)} = 97\ W$$

and

$$\bar{P}_{car} = \frac{\Delta W}{\Delta t} = \frac{(1500\ kg)(9.8\ m/s^2)(1300\ m)}{(0.50\ h)(3600\ s/h)} = 1.1 \times 10^4\ W.$$

These values correspond to 0.13 hp and 14 hp, respectively. The figure of 97 W is typical of the sustained long-term power output of the human body. Remember that next time you leave a 100-W light bulb burning! The power plant supplying the electricity is doing work at about the rate your body can (actually, it's doing work about three times this rate, for reasons we will examine in Chapter 22). You may be surprised at the low output of the car, given that it probably has an engine rated at several hundred horsepower. Actually, only a small fraction of a car engine's rated horsepower is available in mechanical power to the wheels. The rest is lost in friction and heating.

Some problems similar to Example 7-12: 58, 60, 62
●

When power is constant, so the average and instantaneous power are the same, then Equation 7-17 shows that the amount of work W done in a time Δt is just

$$W = P\Delta t. \tag{7-19}$$

When the power is not constant, we can consider small amounts of work ΔW, each taken over so small a time interval Δt that the power is nearly constant. Adding all these amounts of work, and taking the limit as Δt becomes arbitrarily small, we have

$$W = \lim_{\Delta t \to 0} \sum P\Delta t = \int_{t_1}^{t_2} P\,dt, \tag{7-20}$$

where t_1 and t_2 are the beginning and end of the time interval over which we calculate the power.

● **EXAMPLE 7-13** AN ELECTRIC BILL: YANKEE STADIUM

Each of the 500 floodlights at Yankee Stadium uses electrical energy at the rate of 1.0 kW. How much does it cost to run these lights during a 4-hour night game, if electricity costs 9.5¢/kWh (see Fig. 7-19)?

Solution

The total power consumption of the 500 lights is 500 kW. Since the power is constant, the total work done to run the lamps is given by Equation 7-19:

$$W = P\Delta t = (500\ kW)(4.0\ h) = 2000\ kWh.$$

The cost is then (2000 kWh)(9.5¢/kWh) = $190.

Note in this example that energy in kilowatt-hours is simply the product of power in kilowatts and time in hours.

EXERCISE New Hampshire's Seabrook nuclear power plant produces electrical energy at the rate of 1150 MW. (a) How much energy does it produce in a year? (b) At a wholesale price of 2¢/kWh, how much is this energy worth?

Answers: (a) 10^{10} kWh; (b) $200 million

Some problems similar to Example 7-13: 67, 70, 74

◆ ■ **APPLICATION** ENERGY AND SOCIETY

We hear a great deal about today's energy-intensive industrial societies and about environmental and other negative consequences of our energy use. Just how much energy do we use, and what does it mean for a society to be energy intensive?

Example 7-12 showed that the average power output of the human body is about 100 W. Before our species harnessed Æe and domesticated animals, that 100 W was all the power available to the average human being. At what rate, on the average, do *you* use energy? There's the 100 W associated with your own body. Then there's that 100-W light bulb you keep on several hours a day. Maybe your car burns an average of a gallon of gasoline a day; that gallon is equivalent to about 40 kWh (Appendix C lists this and many other fuel equivalents). On average, that's 40 kWh/24 h or about 1.7 kW. In the winter, your share of the heat may account for several kilowatts. The energy to cook

your meals adds further. And what about the energy used to plow the fields that grew your food or to produce the fertilizers and pesticides used in agriculture? Or to manufacture all the goods you consume? In all, you use energy at a substantial rate. In the United States at the end of the twentieth century, in fact, the average energy consumption rate is just about 10 kW per person (see Fig. 7-20). Other industrial nations are within a factor of two of this value, most of them lower.

This quantity, 10 kW, is 10,000 W, or 100 times the average power output of the human body. If our energy were supplied by human labor, instead of gasoline engines and nuclear power plants, each of us would need about 100 laborers working around the clock. What do our 100 "energy workers" do? Figure 7-21 shows that 36 of them work in the industries that supply us with goods; 27 are in transportation, moving us and our goods

FIGURE 7-20 Per-capita energy consumption rate in the United States through the twentieth century. One "energy worker" is 100 W, about the average power output of the human body.

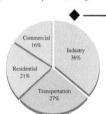

FIGURE 7-21 Energy use in the United States for the late Twentieth Century.

We chose to integrate many shorter applications into the text, where they are more likely to be read, rather than present a few as guest essays.

An index of examples and applications appears at the beginning of the text.

---

330   CHAPTER 14   STATIC EQUILIBRIUM

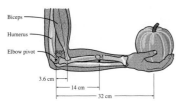

Biceps

Humerus

Elbow pivot

CM

3.6 cm

14 cm

32 cm

FIGURE 14-13   A human arm.

T

$\theta = 80°$

$\mathbf{F}_c$ $m\mathbf{g}$ $M\mathbf{g}$

FIGURE 14-14   Forces on the forearm include its own weight $m\mathbf{g}$, the pumpkin weight $M\mathbf{g}$, the muscle tension $\mathbf{T}$, and the elbow contact force $\mathbf{F}_c$.

tension $\mathbf{T}$, and the contact force $\mathbf{F}_c$. We can read the horizontal and vertical components of force balance from the diagram:

horizontal: $F_{cx} - T\cos\theta = 0$

vertical: $T\sin\theta - F_{cy} - mg - Mg = 0$.

TIP   You Can Avoid Solving for Trig Functions
Here we've chosen to treat the two components of the contact force as unknowns rather than its magnitude and direction as we did with the hinge force in Example 14-2. That way we avoid solving for trig functions, but in the end we'll have to calculate the magnitude of the force from its components. It doesn't make much difference in these examples, but sometimes it's more straightforward if you can avoid solving for an unknown angle.

Choosing the pivot at the elbow gives a torque equation in which the contact force doesn't appear:

$$d_1 T\sin\theta - d_2 mg - d_3 Mg = 0,$$

where the d's are the distances from the elbow to the force application points. Solving for the biceps tension T gives

$$T = \frac{(d_2 m + d_3 M)g}{d_1 \sin\theta}$$

$$= \frac{[(0.14\ \mathrm{m})(2.7\ \mathrm{kg}) + (0.32\ \mathrm{m})(4.5\ \mathrm{kg})](9.8\ \mathrm{N/kg})}{(0.036\ \mathrm{m})(\sin 80°)}$$

$$= 500\ \mathrm{N}.$$

The horizontal and vertical force equations then give the components of the elbow contact force:

$$F_{cx} = T\cos\theta = (500\ \mathrm{N})(\cos 80°) = 87\ \mathrm{N},$$

and

$$F_{cy} = T\sin\theta - (m + M)g$$
$$= (500\ \mathrm{N})(\sin 80°) - (2.7\ \mathrm{kg} + 4.5\ \mathrm{kg})(9.8\ \mathrm{N/kg})$$
$$= 420\ \mathrm{N},$$

making the total force on the elbow $F_c = \sqrt{(87\ \mathrm{N})^2 + (420\ \mathrm{N})^2} = 430\ \mathrm{N}$. This example shows how the human body often sustains forces far larger than the weights of objects it may be lifting.

EXERCISE   A 95-kg horizontal tree branch has its center of mass 2.2 m from the tree trunk. A rope helps support the branch, attached as shown in Fig. 14-15. A 4.5-kg swing hangs from a point 7.3 m out along the branch. How massive a child can sit stationary on the swing if the rope tension is not to exceed 1750 N? Neglect any supporting torque where the branch joins the trunk.

Answer: 33.4 kg.

Some problems similar to Example 14-5: 19, 39, 40

FIGURE 14-1

14-3   EXAMPLES OF STATIC EQUILIBRIUM   **331**

■ **APPLICATION**   RESTORING THE STATUE OF LIBERTY

In 1986, workers completed major renovation of the Statue of Liberty, France's famous gift to the United States that was first dedicated in 1886. Although its designer, French sculptor Frédéric-Auguste Bartholdi, suggested that his creation should last as long as Egypt's pyramids, many factors conspired to make major renovation necessary after only one century. These include corrosion from air pollution and from a chemical reaction between the iron framework and the copper skin, as well as an assembly change that resulted in excess torques on the statue's structural members.

Sculptor Bartholdi was no engineer, and without the work of the French engineer Eiffel (of tower fame) the statue could not have maintained itself in static equilibrium. Eiffel designed an inner skeleton of iron to provide the forces necessary to counteract the weights and torques associated with components of the statue, as well as with the force exerted by wind (Fig. 14-16). The statue was constructed in France in 300 separate pieces, then shipped to New York. During assembly, probably as a conscious aesthetic decision, Liberty's head and upraised arm were mounted 2 feet from their locations on Eiffel's plan,

with the arm making a greater angle to the vertical than planned (Fig 14-17). This made the arm exert a much greater torque about the shoulder than planned, resulting in greater forces on structural components. For historical integrity, renovators did not return to Eiffel's original design; instead, they reinforced the support structure to withstand better the excess forces and torques.

FIGURE 14-16   The Statue of Liberty's interior skeleton counteracts forces and torques acting on the statue. Computer-drawn images of Liberty's skeleton helped engineers and architects plan its renovation.

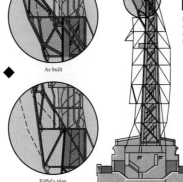

As built

Eiffel's plan

FIGURE 14-17   The statue's head and arm are offset from their planned positions, resulting in greater than anticipated forces and torques on structural members.

High-quality color figures and photographs enhance the text. For a full explanation of how color is used to show physical quantities, see page xxxii.

Chapter Synopses emphasize key concepts and remind students of new terms and mathematical symbols.

CHAPTER SYNOPSIS ◆

Summary

1. **Fluid** is matter that readily deforms and flows under the influence of forces. Fluids are characterized by **density,** or mass per unit volume, and **pressure,** or force per unit area. Liquids are nearly **incompressible,** meaning liquid density hardly changes. Gases are **compressible,** capable of large density changes, but such changes generally occur only when flow speeds approach or exceed the sound speed.

2. In **hydrostatic equilibrium** there is no net force on any element of a fluid. In the absence of external forces, hydrostatic equilibrium implies uniform pressure throughout the fluid. In the presence of gravity, fluid pressure increases with depth so pressure forces balance the gravitational force; in a liquid, that increase is described by

$$P = P_0 + \rho g h,$$

where P_0 is the surface pressure and h the depth.

3. In hydrostatic equilibrium, a pressure increase at any point is transmitted throughout the fluid, a fact known as **Pascal's law.**

4. **Buoyancy** is the upward pressure force on an object wholly or partly immersed in a fluid. The buoyancy force is equal to the weight of fluid displaced by the object– a fact known as **Archimedes's principle.** If an object is less dense than the fluid, the buoyancy force exceeds the gravitational force, and the object rises.

5. A moving fluid is characterized by its flow velocity at each point in space and time. In **steady flow** the velocity is always the same at a given point; such flow is represented by **streamlines** that mark the paths of the fluid elements. In **unsteady flow** the flow velocity varies with time as well as position.

6. The laws of conservation of mass and conservation of energy provide a simplified description of a fluid in steady flow. Both laws are applied to a narrow **flow tube**– a volume bounded by nearby streamlines. Conservation of mass results in the **continuity equation:**

$$\rho v A = \text{constant along a flow tube,}$$

where A is the tube area and $\rho v A$ the **mass flow rate.** In a liquid, or a gas with flow speed well below the sound speed, the density ρ is constant and therefore the **volume flow rate** $v A$ remains constant along a flow tube.

In steady, incompressible flow in the absence of viscosity or other forms of energy loss or addition, conservation of energy yields **Bernoulli's equation:**

$$P + \frac{1}{2}\rho v^2 + \rho g h = \text{constant along a flow tube.}$$

The continuity equation and Bernoulli's equation together help explain a great many fluid phenomena; other phenom-

ena, like airplane flight, are more simply explained through the action-reaction forces of Newton's third law.

7. Fluid friction, or **viscosity,** is especially important near fluid boundaries, particularly in narrowly confined flows. Viscosity exerts a stabilizing influence on flows that would otherwise become **turbulent,** or chaotically unstable.

Terms You Should Understand

(Pairs are closely related terms whose distinction is important; number in parentheses is chapter section where term first appears.)
fluid (introduction)
density (18-1)
pressure (18-1)
pascal (18-1)
hydrostatic equilibrium (18-2)
barometer, manometer (18-2)
gauge pressure (18-2)
Pascal's law (18-2)
buoyancy force (18-3)
Archimedes's principle (18-3)
neutral buoyancy (18-3)
streamlines (18-4)
steady flow, unsteady flow (18-4)
flow tube (18-4)
continuity equation (18-4)
mass flow rate, volume flow rate (18-4)
Bernoulli's equation (18-4)
venturi (18-5)
Bernoulli effe
lift (18-5)
viscosity (18-
turbulence (1

Symbols Yo

P (18-1)
ρ (18-1)
Pa (18-1)

Problems Y

calculating pr
calculating pr
analyzing sim
determining d
  objects (18
determining fl
  using the cont
  (18-4)
using the cont
  gether to s

Even in a field as fundamental as physics, many theories and their equations have limited applicability. Each chapter concludes with a reminder of limitations that students should keep in mind.

◆ Limitations to Keep in Mind

Treating matter as a continuous fluid is an approximation valid only when the spacing between molecules is much smaller than any length of interest– including the wavelength of any significant wave motion.

Bernoulli's equation applies only to incompressible flows– that is, to liquids or to gases moving at much less than the sound speed.

QUESTIONS

1. Explain the difference between hydrostatic equilibrium, steady flow, and unsteady flow.
2. Why do your ears "pop" when you drive up a mountain?
3. The cabins of commercial jet aircraft are usually pressurized to the pressure of the atmosphere at about 2 km above sea level. Why don't you feel the lower pressure on your entire body?
4. Water pressure at the bottom of the ocean arises from the weight of the overlying water. Does this mean that the water exerts pressure only in the downward direction? Explain.
5. The three containers in Fig. 18-37 are filled to the same level and are open to the atmosphere. How do the pressures at the bottoms of the three containers compare?

FIGURE 18-37   Question 5.

6. Municipal water systems often include tanks or reservoirs mounted on hills or towers. Besides water storage, what function might these reservoirs have?
7. Why is it easier to float in the ocean than in fresh water?
8. Figure 18-38 shows a cork suspended from the bottom of a sealed container of water. The container is on a turntable rotating about a vertical axis, as shown. Explain the position of the cork.

FIGURE 18-38   Question 8.

FIGURE 18-39   Why don't snorkels work at more than a meter or so of depth (Question 13)?

9. An ice cube is floating in a cup of water. Will the water level rise, fall, or remain the same when the cube melts?
10. Meteorologists in the United States usually report barometer readings in "inches." What are they talking about?
11. A mountain stream, frothy with entrained air bubbles, presents a serious hazard to hikers who fall into it, for they may sink in the stream where they would float in calm water. Why?
12. Why are dams thicker at the bottom than at the top?
13. It's not possible to breathe through a snorkel from a depth greater than a meter or so (Fig. 18-39). Why not?
14. Most humans float naturally in fresh water. Yet the body of a drowning victim generally sinks, often rising several days later after bodily decomposition has set in. What might explain this sequence of floating, sinking, and floating again?
15. A helium-filled balloon stops rising long before it reaches the "top" of the atmosphere, while a cork released from the bottom of a lake rises all the way to the surface of the water. Explain the difference between these two behaviors.
16. A barge filled with steel beams overturns in a lake, spilling its cargo. Does the water level in the lake rise, fall, or remain the same?
17. Imagine a vertical cylinder filled with water and set rotating about its axis. If pieces of wood and stone are introduced into the cylinder, where will each end up?
18. When gas in steady, subsonic flow through a tube encounters a constriction, its flow speed increases. When it flows supersonically through the same situation, flow speed decreases in the constriction. What must be happening to the gas density at the constriction in the supersonic case?
19. A ball moves horizontally through the air without spinning. Where on the ball's surface is the air pressure greatest?

13. Can there be a true "line drive" in baseball? What is required for the ball to travel in a nearly straight, horizontal trajectory?
14. What is the vertical component of a projectile's velocity at the peak of its trajectory?
15. In terms of its initial velocity, what is the horizontal component of a projectile's velocity at the peak of its trajectory?
16. Is there any point on a projectile's trajectory where the velocity and acceleration are perpendicular?
17. How is the up-and-down motion of an object thrown straight up consistent with our conclusion that projectiles follow parabolic trajectories?
18. Projectiles launched at 30° and 60° have the same range. Does this mean they stay in the air the same amount of time?
19. Medieval invaders are attacking a walled village on a flat-topped hill, as shown in Fig 4-22. Is the horizontal range of the rocks they're catapulting given by Equation 4-10? Explain.
20. Earth's curvature renders Equation 4-10 inexact for projectiles with very long ranges. Do you expect the actual range of such a projectile to be longer or shorter than Equation 4-10 would imply? Explain.
21. A 45° launch angle maximizes the horizontal range of a projectile. Is this still true for the situation shown in Fig. 4-22?
22. A friend who's not taking physics insists that you can't be accelerating when you drive around a curve since the speedometer reading remains steady. Refute this argument.
23. You're driving around a curve. Is there any way of stepping on the gas or brake such that your tangential acceleration cancels your radial acceleration, giving a net acceleration of zero? How or why not?
24. An object starts from rest, accelerating with constant tangential acceleration on a circular path. Which is greater when it starts, its tangential or its radial acceleration? Which is greater after a long time? Explain.
25. An object moves outward along the spiral path shown in Fig. 4-23. Must its speed increase, decrease, or remain the same if the magnitude of its radial acceleration is to remain constant?

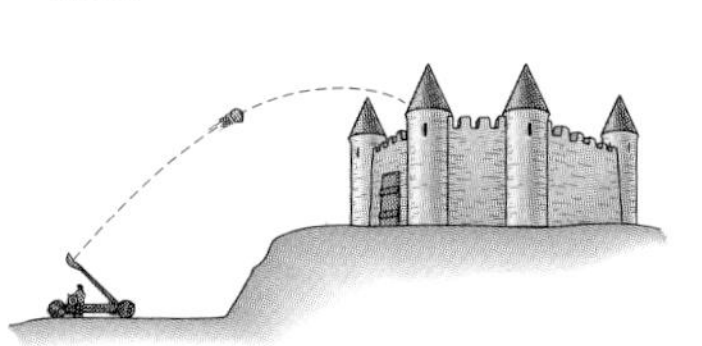

FIGURE 4-22 Questions 19, 21.

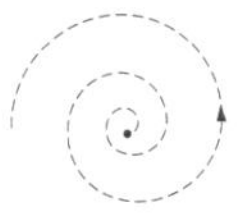

FIGURE 4-23 Question 25.

Nearly 3000 new, interesting, and diverse problems will help your students understand concepts and develop problem-solving skills.

PROBLEMS

Section 4-1 Velocity and Acceleration

1. A skater is gliding along the ice at 2.8 m/s, when she undergoes an acceleration of magnitude 1.1 m/s² for 2.0 s. At the end of that time she is moving at 5.0 m/s. What must be the angle between the acceleration vector and the initial velocity vector?
2. In the preceding problem, what would have been the magnitude of the skater's final velocity if the acceleration had been perpendicular to her initial velocity?
3. An object is moving in the x direction at 1.3 m/s when it is subjected to an acceleration given by $\mathbf{a} = 0.52\,\hat{\mathbf{j}}$ m/s². What is its velocity vector after 4.4 s of acceleration?
4. An airliner is flying at a velocity of $260\hat{\mathbf{i}}$ m/s, when a wind gust gives it an acceleration of $0.38\hat{\mathbf{i}} + 0.72\hat{\mathbf{j}}$ m/s² for a period of 24 s. (a) What is its velocity at the end of that time? (b) By what angle has it been deflected from its original course?

Section 4-2 Constant Acceleration

5. The position of an object as a function of time is given by $\mathbf{r} = (2.4t + 1.2t^2)\hat{\mathbf{i}} + (0.89t - 1.9t^2)\hat{\mathbf{j}}$ m, where t is the time in seconds. What are the magnitude and direction of the acceleration?
6. An airplane heads northeastward down a runway, accelerating from rest at the rate of 2.1 m/s². Express the plane's velocity and position at $t = 30$ s in unit vector notation, using a coordinate system with x axis eastward and y axis northward, and with origin at the start of the plane's takeoff roll.

...ady 21 km/s. To ... rocket to the as... 5 km/s² at right ... firing lasts 250 s, ... asteroid's motion ... the firing? ... ction at 4.5 m/s, ... y direction for a ... s in the x and y ... magnitude of its

... n a stick imparts ... 0.0° angle to the ... cceleration lasts ... ng this time? ... now the acceler... inal direction of

... l velocity $\mathbf{v}_0 =$ ... cceleration given by $\mathbf{a} = -1.2t + 0.26\hat{\mathbf{j}}$ m/s². (a) When does the particle cross the y axis? (b) What is its y coordinate at the time? (c) How fast is it moving, and in what direction, at that time?

12. A particle starts from the origin with initial velocity $\mathbf{v}_0 = v_0\hat{\mathbf{i}}$ and constant acceleration $\mathbf{a} = a\hat{\mathbf{j}}$. Show that the particle's distance from the origin and its direction relative to the x axis are given by $d = t\sqrt{v_0^2 + \frac{1}{4}a^2t^2}$ and $\theta = \tan^{-1}(at/2v_0)$.
13. Figure 4-24 shows a cathode-ray tube, used to display electrical signals in oscilloscopes and other scientific instruments. Electrons are accelerated by the electron gun, then move down the center of the tube at 2.0×10^9 cm/s. In the 4.2-cm-long deflecting region they undergo an acceleration directed perpendicular to the long axis of the tube. The acceleration "steers" them to a particular spot on the screen, where they produce a visible glow. (a) What acceleration is needed to deflect the electrons through 15°, as shown in the figure? (b) What is the shape of an electron's path in the deflecting region?

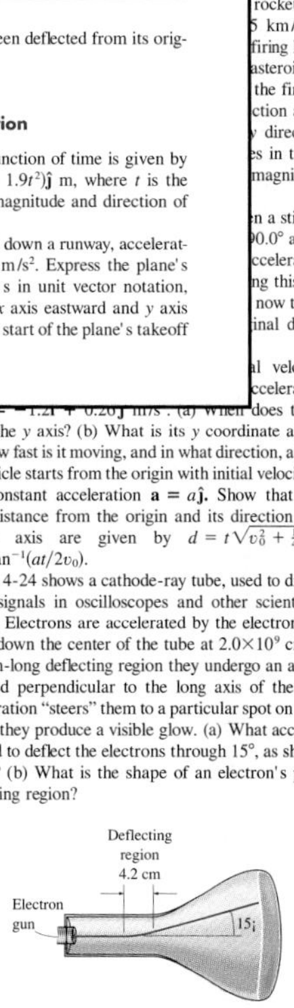

FIGURE 4-24 A cathode-ray tube (Problem 13).

Section 4-3 Projectile Motion

14. You toss an apple horizontally at 9.5 m/s from a height of 1.8 m. Simultaneously, you drop a peach from the same height. How long does each take to reach the ground?
15. A carpenter tosses a shingle off a 9.4-m-high roof, giving it an initially horizontal velocity of 7.2 m/s. (a) How long does it take to reach the ground? (b) How far does it move horizontally in this time?
16. An arrow fired horizontally at 41 m/s travels 23 m horizontally before it hits the ground. From what height was it fired?
17. A kid fires a blob of water horizontally from a squirt gun held 1.6 m above the ground. It hits another kid 2.1 m away square in the back, at a point 0.93 m above the ground (see Fig. 4-25). What was the initial speed of the blob?

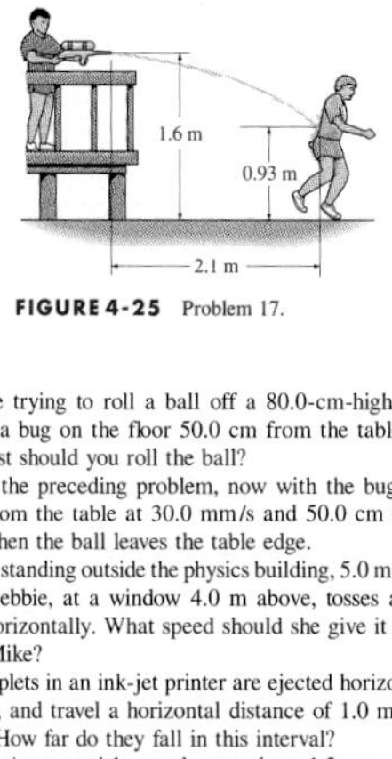

FIGURE 4-25 Problem 17.

18. You are trying to roll a ball off a 80.0-cm-high table to squash a bug on the floor 50.0 cm from the table's edge. How fast should you roll the ball?
19. Repeat the preceding problem, now with the bug moving away from the table at 30.0 mm/s and 50.0 cm from the table when the ball leaves the table edge.
20. Mike is standing outside the physics building, 5.0 m from the wall. Debbie, at a window 4.0 m above, tosses a physics book horizontally. What speed should she give it if it is to reach Mike?
21. Ink droplets in an ink-jet printer are ejected horizontally at 12 m/s, and travel a horizontal distance of 1.0 mm to the paper. How far do they fall in this interval?
22. Protons in a particle accelerator drop 1.2 μm over the 1.7-km length of the accelerator. What is their approximate average speed?
23. You're on the ground 3.0 m from the wall of a building, and want to throw a package from your 1.5-m shoulder level to someone in a second floor window 4.2 m above the ground. At what speed and angle should you throw it so it just barely reaches the window?
24. Derive a general formula for the horizontal distance covered by a projectile launched horizontally at speed v_0 from a height h.
25. A car moving at 40 km/h strikes a pedestrian a glancing blow, breaking both the car's front signal light lens and the pedestrian's hip. Pieces of the lens are found 4.0 m down the road from the center of a 1.2-m wide crosswalk, and a lawsuit hinges on whether or not the pedestrian was in the crosswalk at the time of the accident. Assuming that the lens

Section-referenced problems allow students to refer back to the corresponding section for help using a worked example.

452 CHAPTER 18 FLUID MOTION

52. A drinking straw 20 cm long and 3.0 mm in diameter stands vertically in a cup of juice 8.0 cm in diameter. A section of straw 6.5 cm long extends above the juice. A child sucks on the straw, and the level of juice in the glass begins dropping at 0.20 cm/s. (a) By how much does the pressure in the child's mouth differ from atmospheric pressure? (b) What is the greatest height from which the child could drink, assuming this same mouth pressure?

◆ **Paired Problems**

(Both problems in a pair involve the same principles and techniques. If you can get the first problem, you should be able to solve the second one.)

53. A steel drum has volume 0.23 m³ and mass 16 kg. Will it float in water when filled with (a) water or (b) gasoline (density 860 kg/m³)? Neglect the thickness of the steel.
54. A 260-g circular pan 20 cm in diameter has straight sides 6.0 cm high and is made from metal of negligible thickness. To what maximum depth can the pan be filled with water and still float on water?
55. A spherical rubber balloon with mass 0.85 g and diameter 30 cm is filled with helium (density 0.18 kg/m³). How many 1.0-g paper clips can you hang from the balloon before it loses its buoyancy?
56. A string of negligible diameter has mass per unit length 1.4 g/m. You tie a 3-m-long piece of the string to a spherical helium balloon 23 cm in diameter and find that the balloon floats with 1.8 m of string off the floor. Find the combined mass of the balloon and helium.
57. Water at a pressure of 230 kPa is flowing at 1.5 m/s through a pipe, when it encounters an obstruction where the pressure drops by 5%. What fraction of the pipe's area is obstructed?
58. A venturi flowmeter in an oil pipeline has a radius half that of the pipe. The flow speed in the unconstricted flow is 1.9 m/s. If the pressure difference between the unconstricted flow and the venturi is 16 kPa, what is the density of the oil?
59. Find an expression for the volume flow rate from the siphon shown in Fig. 18-51, assuming the siphon area A is much less than the tank area.
60. (a) Find the initial siphon flow speed in Fig. 18-51 if the tank is sealed, with its top at only one-fourth of atmospheric pressure. Answer in terms of atmospheric pressure P_a, liquid density ρ, height h, and g. (b) What is the maximum distance between the bend at the top of the siphon and the liquid level in the tank for which the siphon will work under these conditions, assuming the liquid is water? (Give a numerical value here.)

◆ **Supplementary Problems**

61. A 1.0-m-diameter tank is filled with water to a depth of 2.0 m and is open to the atmosphere at the top. The water

FIGURE 18-51 Problems 59, 60.

drains th ough a 1.0-cm-diameter pipe at the bottom; that pipe then joins a 1.5-cm-diameter pipe open to the atmosphere, as shown in Fig. 18-52. Find (a) the flow speed in the narrow section and (b) the water height in the *sealed* vertical tube shown.

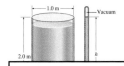

454 PART 2 CUMULATIVE PROBLEMS

PART 2 CUMULATIVE PROBLEMS

1. A cylindrical log of total mass M and uniform diameter d has an uneven mass distribution that causes it to float in a vertical position, as shown in Figure 1. (a) Find an expression for the length ℓ of the submerged portion of the log when it is floating in equilibrium, in terms of M, d, and the water density ρ. (b) If the log is displaced vertically from its equilibrium position and released, it will undergo simple harmonic motion. Find an expression for the period of this motion, neglecting viscosity and other frictional effects.

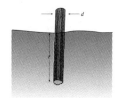

FIGURE 1 Cumulative Problem 1.

2. A cable of total mass m and length ℓ hangs vertically, with a mass M attached to its bottom end, as shown in Fig 2. The mass is given a sudden sideways blow that starts a low-amplitude pulse propagating up the cable. Show that the time it takes the pulse to reach the top of the cable is

$$t = 2\left(\sqrt{\frac{(m + M)\ell}{mg}} - \sqrt{\frac{M\ell}{mg}}\right).$$

FIGURE 2 Cumulative Problem 2.

3. Let P_0 and ρ_0 be the atmospheric pressure and density at Earth's surface. Assume that the ratio P/ρ is the same throughout the atmosphere (this implies that the tempera-

ture is uniform). Show that the pressure a vertical height z above the surface is given by $P(z) = P_0 e^{-\rho_0 gz/P_0}$, for z much less than Earth's radius (this amounts to neglecting Earth's curvature, and thus taking g to be constant).

4. A piece of rope of length ℓ and mass m has its two ends spliced together to form a continuous loop. The loop is set spinning at so high a rate that it forms a circle with essentially uniform tension. It is then placed in contact with the ground, where it rolls, without slipping, like a rigid hoop. The loop is rolling on level ground when it rolls over a stick that produces a small distortion (see Fig. 3). As a result, two pulses, initially coinciding, propagate along the loop in opposite directions. (a) Where will they again coincide? (b) Through what angle will the loop have rotated while the pulses are separated?

FIGURE 3 Cumulative Problem 4.

5. A U-shaped tube containing liquid is mounted on a table that tilts back and forth through a slight angle, as shown in Fig. 4. The diameter of the tube is much less than either the height of its arms or their separation. When the table is rocked very slowly or very rapidly, nothing particularly dramatic happens. But when the rocking takes place at a few times per second, the liquid level in the tube oscillates violently, with maximum amplitude at a rocking frequency of 1.7 Hz. Explain what is going on, and find the total length of the liquid including both vertical and horizontal portions.

FIGURE 4 Cumulative Problem 5.

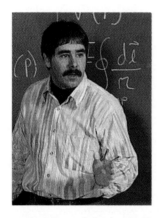

Richard Wolfson is Professor of Physics and George Adams Ellis Professor of the Liberal Arts at Middlebury College, where he has taught since 1976. He did undergraduate work at the Massachusetts Institute of Technology and Swarthmore College and holds the M.S. degree from the University of Michigan and Ph.D. from Dartmouth. He has published widely in scientific journals, including works ranging from medical physics research to experimental plasma physics, electronic circuit design, solar energy engineering, and theoretical astrophysics. He is also an interpreter of science for the nonspecialist, a contributor to *Scientific American*, and author of the book *Nuclear Choices: A Citizen's Guide to Nuclear Technology*. Wolfson has spent sabbatical years as Visiting Scientist at the National Center for Atmospheric Research in Boulder, Colorado, and in 1993 was Visiting Scientist at St. Andrews University in Scotland.

Jay M. Pasachoff is Field Memorial Professor of Astronomy and Director of the Hopkins Observatory at Williams College. He was born and brought up in New York City. After attending the Bronx High School of Science, he received his A.B. degree from Harvard College and his A.M. and Ph.D. from Harvard University. He then held postdoctoral fellowships at Harvard and at the California Institute of Technology before going to Williams in 1972. His research has dealt mainly with solar physics and nuclear astrophysics, namely, the solar atmosphere, and with the abundances of the light elements and their formation in the first minutes of the universe. Pasachoff has spent sabbatical leaves at the University of Hawaii, at l'Institut d'Astrophysique in Paris, at the Institute for Advanced Study in Princeton, and at the Harvard-Smithsonian Center for Astrophysics. He is also author or co-author of major texts in physics, calculus, physical science, and astronomy.

The artwork for this second edition is carefully designed to make effective use of color printing as a learning aid. In particular, we have assigned colors to the vector quantities that are so important in physics, and we have used those colors consistently throughout the book. The table below lists some important physical quantities, along with their text and graphic symbols and color assignments. We also include electric circuit symbols, which, to be consistent with usage in engineering, are printed in black.

Vector	Text Symbol	Graphic Symbol
Displacement	$\mathbf{r}, \boldsymbol{\ell}$	
Velocity	$\mathbf{v}$	
Acceleration	$\mathbf{a}$	
Force	$\mathbf{F}$	
Linear momentum	$\mathbf{p}$	
Angular velocity	ω	
Torque	τ	
Angular momentum	$\mathbf{L}$	
Electric field	$\mathbf{E}$	
Magnetic field	$\mathbf{B}$	
Electric dipole moment	$\mathbf{p}$	
Magnetic dipole moment	$\boldsymbol{\mu}$	
Electric charge		
Positive charge	q, Q	⊕
Negative charge	q, Q	⊖
Circuit symbols		
Battery, emf	$\mathcal{E}$	
Resistor	R	
Capacitor	C	
Inductor	L	
Switch	S	

CHAPTER 1

DOING PHYSICS

The laws of physics determine whether the Golden Gate Bridge will stand or fall, the fuel efficiency of the cars crossing the bridge, the dynamics of the ocean currents that flow beneath the bridge, and the behavior of the moisture-laden air that swirls around the bridge.

When you ride a bicycle, fly in a jet plane, or watch a thunderstorm, you experience physical laws. **Physics** is the science that tells us how and why things work. Knowledge of physics will help you understand both natural phenomena and the technologies that increasingly pervade our lives. In this book, we set forth the basic laws of physics and show how they apply to a myriad of phenomena from atoms and molecules to cars and airplanes, rockets and computers, stars and galaxies.

1-1 FIELDS OF PHYSICS

The branch of physics governing the motion of bodies is called **mechanics.** You use the laws of mechanics when you drive a car, ride a skateboard, or build a skyscraper. Through most of this book, we deal with **classical mechanics,** which applies to objects from molecules to galaxies that are moving at speeds small compared with the speed of light. In later chapters, we will see that classical mechanics is but an approximation to a more comprehensive set of physical laws that includes Einstein's theory of relativity and the theory of quantum mechanics.

In the first nine chapters we deal with the motion of single objects, often called particles. Then, in Chapters 10 to 15, we generalize to the motion of many-particle systems. The wave of standing people that sweeps through a football stadium, ocean waves that pound the coastlines, and the sound waves by which we communicate are examples of a special kind of many-particle motion called **wave motion,** which we examine in Chapters 16 and 17.

Our Earth is five-eighths covered by liquid water and is surrounded entirely by a gaseous atmosphere. Even the continents on which we live float around on an underlying fluid layer. The cells of our bodies are nurtured by the motion of fluid through our circulatory systems. And the bulk of the matter in the universe—stars and interstellar gas—is in the fluid state. In Chapter 18, we apply the laws of mechanics to fluids.

Nearly all life on Earth thrives on a constant supply of energy from the Sun, and a delicate balance of energy-transfer processes keeps our planet at a habitable temperature; we worry increasingly that human activity may upset that balance and bring severe climatic changes. For our technological society, we have developed nuclear power plants, gasoline engines, solar collectors, oil and gas furnaces, and a host of other energy sources. Worries about our planet's energy balance and the operation of our energy technologies are both concerns of **thermodynamics**—the study of heat and its interaction with matter. We follow our study of mechanics with four chapters on thermodynamics.

Following, we deal extensively with another major branch of physics: **electromagnetism.** The Greeks knew something of the electric and magnetic properties of matter; the words *electricity* and *magnetism* themselves originate in Greek words for natural materials displaying electric and magnetic effects. The Chinese invention of the magnetic compass resulted by the twelfth century A.D. in widespread use of Earth's natural magnetism to aid navigation. In 1800, Alessandro Volta invented the electric battery, paving the way for a series of experiments involving electric current and its magnetic effects. By the late 1800s, application of electromagnetism led to practical technologies including telegraph, telephone, and electric power distribution. In one of the most sweeping syntheses in the history of science, James Clerk Maxwell in the 1860s formulated a complete description of electromagnetism, showing that electricity and magnetism are intimately related. A startling prediction of Maxwell's theory was that light is an electromagnetic wave and that other electromagnetic waves should be possible as well. Radio and television follow directly from that prediction. Today, electromagnetic technology dominates our civilization, as evidenced by our digital watches, video recorders, microwave ovens, and especially computers (Fig. 1-1).

The behavior of light is intimately tied to electromagnetism. In Chapters 35 through 37 we explore the branch of physics called **optics,** building an understanding of light and of optical devices like lenses, prisms, diffraction gratings, and the instruments we build from them—microscopes, telescopes, spectrometers, and the like (Fig. 1-2).

Electromagnetic theory led directly to many practical technologies, but it also raised bafflingly deep questions about the nature of light, motion, space, and

FIGURE 1-1   This tiny silicon chip, containing over a million electronic components, forms the heart of a personal computer. The chip is small enough to fit through the eye of a needle.

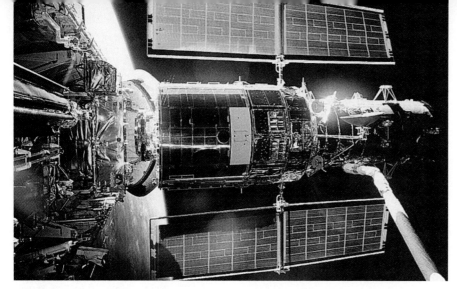

FIGURE 1-2 The Hubble Space Telescope is an example of an optical system. Here the telescope undergoes repairs in the cargo bay of the space shuttle Endeavor in 1993.

time. With a radically simple statement whose implications seem to violate our common sense, Albert Einstein in 1905 confronted these questions (Fig. 1-3). His **special theory of relativity** gives a surprisingly simple picture of reality, in the context of radically altered notions of space and time. Although relativity provides a more correct description of physical reality than does classical mechanics, the two theories differ significantly only at speeds approaching that of light—seven times around the Earth in one second. In our everyday existence we need not worry about relativity. But in particle accelerators probing the basic structure of matter; in studies of exploding supernovas, neutron stars, and the early universe; and even in designing a color television, we must take account of relativity. We explore the theory of relativity in Chapter 38.

But what *is* light? This question leads to one of the major developments of twentieth-century thought: **quantum mechanics.** Dealing with the smallest objects in the universe, quantum mechanics blurs the distinction between light and matter, giving a description of matter on the atomic scale for which our macroscopic intuition is simply inadequate. We give a brief introduction to quantum mechanics and its philosophical implications in Chapter 39, where we also discuss our evolving knowledge of the elementary particles that are the fundamental constituents of matter and the implications of that knowledge for the origin of the universe. The extended version of this text explores quantum mechanics in four chapters including quantum applications in atomic, molecular, and solid-state physics as well as technological devices like lasers. Two chapters on nuclear physics and its applications follow, and the extended version ends with a look at elementary particle physics, cosmology, and future directions in physics.

FIGURE 1-3 Albert Einstein at age 26, when he developed the special theory of relativity.

1-2 THE SIMPLICITY OF PHYSICS

Physics is the fundamental science. Its laws and theories describe the workings of the universe at the most basic level. For that reason physics is immensely powerful; the same laws describe the behavior of molecules, of airplanes, and of galaxies. Although no one has done so, most scientists believe in principle that it would be possible to describe the operation of a living cell or organism using only the fundamental laws of physics.

Applying the laws of physics can give rise to challenging problems whose solutions call for clever insight and mathematical agility. The challenge of problem solving is what gives physics some of its intellectual interest and also its reputation as a difficult subject. But if you approach this course thinking that physics presents you with numerous difficult things to learn, you are missing the point. Because it is so fundamental, physics is inherently simple. There are only a few basic laws to learn; if you really understand those laws, you can apply them in a wide variety of situations. We wrote this book in a spirit that emphasizes the underlying simplicity of physics by reminding you how diverse examples are really manifestations of the *same* underlying physical laws. You should come to understand the basic laws thoroughly so you can apply them confidently in new situations. As you read the text and work the problems, remember the simplicity of the underlying physical principles. Ask yourself how each problem you approach is really similar to other problems and to the text examples. And you will find that similarity, because the many problems and examples really do involve only a few underlying laws. So physics is simple—challenging, too—but with an underlying simplicity that reflects the scope and power of this fundamental science.

1-3 MEASUREMENT SYSTEMS

"A long way" means different things to a sedentary person, to a marathon runner, to a pilot, and to an astronaut. We need to quantify our measurements. Today, nearly every country in the world uses the **metric system,** in which the fundamental quantities length, mass, and time are measured in meters, kilograms, and seconds, respectively. The United States remains the only industrialized country that has not "gone metric," although economic pressures and new legislation are gradually moving the United States toward adoption of the metric system. The modern version of the metric system is known as **SI,** for Système International d'Unités (International System of Units). The SI is a universal language for science, and it incorporates scientifically precise definitions of the fundamental quantities.

Length

The **meter** was first defined at the time of the French Revolution to be one ten-millionth of the distance from the equator to the north pole through Paris. This distance, of course, is not trivial to determine—surveyors sent out at the time to measure a portion of the path were nearly guillotined—and more readily accessible measuring standards proved necessary.

In 1889, a standard meter was carefully made; it is the distance between two lines on a certain bar made of a platinum-iridium alloy when examined at the temperature of the melting point of ice. It is kept in a temperature-controlled room at the Bureau International des Poids et Mesures in Sèvres, France, where the international treaty on measurement signed in 1875 is monitored. Thirty secondary standards were also made from the same ingot of alloy; the United States Bureau of Standards in Washington, D.C., has numbers 21 and 27. (Comparisons over 15 years showed that the standard meters remained the same to within 2 ten-millionths of a meter.) But as new methods developed, even the standard meter was not sufficiently accurate. In 1960, the meter was redefined as 1,650,763.73 wavelengths of orange-red light emitted by the isotope krypton-86.

In recent years, even this standard became insufficient. The speed of light is now one of the most accurately determined quantities. In 1983, the meter was redefined in terms of the speed of light, which, in vacuum, has been measured at 299,792,458 meters per second (Fig. 1-4). The 1983 Conférence Général des Poids et Mesures inverted this result to make the current definition of the meter:

> *The meter is the length of the path traveled by light in vacuum during a time interval of 1/299,792,458 of a second.*

This definition is more useful than a meter stick locked in a vault in Paris since the standard meter can be reproduced in any laboratory with a sufficiently accurate clock.

Time

The division of time into hours, minutes, and seconds goes back to the Babylonians. The **second** used to be defined by the Earth's rotation, as $1/(24\times60\times60)$ of a mean solar day, until it was recognized that the Earth's rotation is slowing slightly. The second was then redefined as a specific fraction of the year 1900. In 1967 the second was again redefined, now in terms of the vibrations of a specific atom, to give our current definition:

> *The second is the duration of 9,192,631,770 periods of the radiation corresponding to the transition between the two hyperfine levels of the ground state of the cesium-133 atom.*

The device that implements this definition—which will seem a lot less obscure once you've studied some atomic physics—is called an atomic clock (Fig. 1-5). Like the definition of the meter, the atomic-clock definition of the second is reproducible in any laboratory with the appropriate equipment.

Recently, devices called hydrogen masers have been used as even more stable time standards. And some scientists anticipate that even more accuracy will be available using pulsars—celestial objects that emit radio signals at extremely steady rates. So the definition of the second may someday change again.

FIGURE 1-4 A surveyor using laser light for accurate distance determination. The standard measure of length—the meter—is defined in terms of the speed of light.

FIGURE 1-5 This cesium atomic clock at the National Institute of Standards and Technology (NIST) keeps time to an accuracy of about 3 μs/y. It is part of the NIST Atomic Time System, which established the United States' standard for time measurement.

FIGURE 1-6 The international prototype kilogram in its vault at the International Bureau of Weights and Measures in Sèvres, France.

It's a good thing that we have better standards of time than the rotation of the Earth: Recent research shows that Earth's rotation rate changes slightly as the wind blows! To keep atomic time and Earth-spin time in agreement, "leap seconds" are added to our terrestrial clocks when necessary, usually about once a year.

Mass

The standard of mass is now the least satisfactory. Unlike the standards of length and time, which are based on measurement procedures that can be repeated by scientists anywhere, the unit of mass is still defined in terms of a particular object—the international prototype **kilogram** kept at the International Bureau of Weights and Measures at Sèvres, France (Fig. 1-6):

> **The kilogram is the unit of mass; it is equal to the mass of the international prototype of the kilogram.**

The prototype kilogram is made out of a special platinum-iridium alloy that is very hard, not subject to corrosion, and very dense. Nevertheless, it could conceivably change, and in any event comparison with such a standard is less convenient than a procedural standard that can be checked in a laboratory. So scientists are now working on techniques to measure the spacing of atoms in a crystal of silicon, essentially counting the atoms in a given volume, to scale up from the mass of a single silicon atom to a new definition of the kilogram.

Other SI Units

SI includes seven independent base units: In addition to the three we've just defined, there are the ampere (A) for electric current, the kelvin (K) for temperature, the mole (mol) for the amount of substance, and the candela (cd) for luminosity. Two supplementary units are used to measure angle—the radian (rad) for ordinary angles and the steradian (sr) for solid angles (Fig. 1-7). Units for all other quantities are derived from these base units. In mechanics, we use only the first three of the base units—meter, kilogram, and second—along with measures of angle.

The metric system is a decimal system. Its base units are used with the prefixes listed in Table 1-1 to indicate multiplication by powers of 10. For example, k is the SI symbol for the prefix "kilo-," which means "times 1000;" 1 km is thus 1000 m. Table 1-1 also appears inside the front cover.

When two units are used together, a hyphen appears between them: newton-meter. Each unit has a symbol, such as m for meter or km for kilometer. Since these are symbols, rather than abbreviations, they are not followed by periods. Neither are the plural forms followed by an "s." Symbols are ordinarily lowercase; only those named after people are uppercase. Thus, the symbol for gram is a lowercase g; the unit "newton" is written with a small "n" but its symbol is a capital N. The only exception is for the volume unit, the liter, defined as the volume of a cube one-tenth of a meter on each side. Since the lowercase "l" is easily confused with the number one, the symbol for liter is written with a capital L. When two units are multiplied, their symbols are separated by a

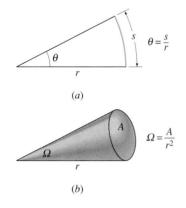

$\theta = \dfrac{s}{r}$

(a)

$\Omega = \dfrac{A}{r^2}$

(b)

FIGURE 1-7 The SI units of angle. (a) The radian measure of an angle is defined as the ratio of the subtended arc length to the radius. Since the circumference of a circle is $2\pi r$, there are 2π radians in a complete circle. One radian is then equal to $360°/2\pi$ or $57.3°$. (b) The measure of a solid angle is defined as the ratio of the subtended area to the radius squared.

▲ TABLE 1-1 SI PREFIXES

PREFIX	SYMBOL	POWER	
yetta	Y	10^{24} =	1 000 000 000 000 000 000 000 000
zetta	Z	10^{21} =	1 000 000 000 000 000 000 000
exa	E	10^{18} =	1 000 000 000 000 000 000
peta	P	10^{15} =	1 000 000 000 000 000
tera	T	10^{12} =	1 000 000 000 000
giga	G	10^{9} =	1 000 000 000
mega	M	10^{6} =	1 000 000
kilo	k	10^{3} =	1 000
hecto	h	10^{2} =	100
deca	da	10^{1} =	10
—	—	10^{0} =	1
deci	d	10^{-1} =	0.1
centi	c	10^{-2} =	0.01
milli	m	10^{-3} =	0.001
micro	μ	10^{-6} =	0.000 001
nano	n	10^{-9} =	0.000 000 001
pico	p	10^{-12} =	0.000 000 000 001
femto	f	10^{-15} =	0.000 000 000 000 001
atto	a	10^{-18} =	0.000 000 000 000 000 001
zepto	z	10^{-21} =	0.000 000 000 000 000 000 001
yocto	y	10^{-24} =	0.000 000 000 000 000 000 000 001

centered dot: N·m for newton-meter. Division of units is expressed using the slash (/) or by writing the symbol for the denominator unit raised to the −1 power. Thus the SI unit of speed is the meter per second, written m/s or m·s^{-1}. When four or more digits are used together, SI recommends the use of spaces to separate groups of three: 34 365.456 32. We tend to use commas instead of spaces in the United States, though some countries use a comma as we use a period to indicate a decimal point.

Other Unit Systems

The inches, feet, yards, miles, and pounds of the so-called English system still dominate measurement in the United States. Other non-SI units like the hour for time are often mixed with English or SI units, as when speed limits are given in miles per hour or kilometers per hour. In some specialized fields of physics there are good reasons for using non-SI units. We'll discuss these as the need arises and will occasionally use non-SI units in examples and problems. We'll also often find it convenient to use degrees rather than radians in measuring angles. The vast majority of examples and problems in this book, however, use strictly SI units.

1-4 CHANGING UNITS

Sometimes we need to change from one unit system to another—for example, from English to SI units. Appendix C contains extensive tables for converting among unit systems; you should familiarize yourself with this and the other appendices and refer to them often.

FIGURE 1-8 With a height of 533 m, Toronto's CN Tower is the world's tallest free-standing structure.

In changing units, it is helpful to treat the units as algebraic quantities. For example, Appendix C shows that 1 ft = 0.3048 m. Then the 1815-ft CN Tower in Toronto—the world's tallest free-standing structure—is

$$(1815 \text{ ft})\left(\frac{0.3048 \text{ m}}{1 \text{ ft}}\right) = 553.2 \text{ m}$$

high (Fig. 1-8). Had we wanted to convert meters to feet, we would have used the ratio 1 ft/0.3048 m.

Often, you'll need to change several units in the same expression. Keeping track of the units through a chain of multiplications helps prevent you from carelessly inverting any of the conversion factors.

● **EXAMPLE 1-1** CHANGING UNITS

Express a 55 mi/h speed limit in meters per second.

Solution

In Appendix C, we find that 1 mi = 1609 m, so we can multiply miles by the ratio 1609 m/mi to get meters. Similarly, we use the conversion factor 3600 s/h to convert hours to seconds. So we have

$$55 \text{ mi/h} = \frac{(55 \text{ mi})(1609 \text{ m/mi})}{(1 \text{ h})(3600 \text{ s/h})} = 24.6 \text{ m/s}.$$

EXERCISE Aeronautical engineers in the United States often express aircraft speeds in knots (nautical miles per hour). For a 767 jetliner, touchdown speed is 145 knots. Use information from Appendix C to express this speed in meters per second.

Answer: 74.6 m/s

Some problems similar to Example 1-1: 10, 15, 17, 20 ●

1-5 DIMENSIONAL ANALYSIS

Whether we measure length in meters or feet, we're still talking about the same physical quantity—length. And meters per second, miles per hour, or furlongs per fortnight all measure the same thing—speed. No matter what the unit system, speed is always expressed in units of length per unit of time. We call these underlying terms the **dimensions** of a physical quantity. Symbolically, we use the letters L, T, and M to signify the dimensions length, time, and mass; thus the dimensions of speed are L/T.

In **dimensional analysis,** we examine physical quantities and equations with regard to their dimensions only. For example, both sides of an equation must have the same dimensions for it to make sense. (You can't equate apples and oranges!) Thus a dimensional analysis of the familiar formula "distance equals speed times time" is

$$L = \left(\frac{L}{T}\right)T.$$

The T's cancel on the right-hand side, giving $L = L$ and showing that this equation makes dimensional sense. Dimensional analysis is an important check on the answer to a physics problem since an answer with the wrong dimensions can't be right.

Dimensional analysis also gives scientists and engineers hints as to the form equations must take. For example, you've surely heard of Einstein's famous formula $E = mc^2$ relating energy E, mass m, and the speed of light c. (You'll meet this equation in Chapter 8 and again in Chapter 38.) But why not $E = mc^3$ (Fig. 1-9)? It doesn't take an Einstein to see why not. The dimensions of the energy E happen to be ML^2/T^2. So a dimensional analysis of $E = mc^3$ gives

$$M\left(\frac{L^2}{T^2}\right) = M\left(\frac{L}{T}\right)^3.$$

Dividing both sides by $M(L^2/T^2)$ and multiplying by T then yields $T = L$, which is *not* true. So $E = mc^3$ is dimensionally incorrect. Dimensional analysis of $E = mc^2$ shows that this equation *is* dimensionally correct. What dimensional analysis won't tell us is whether the right equation is $E = mc^2$, $E = \frac{1}{2}mc^2$, $E = 5mc^2$, or $E = \pi mc^2$. What it will tell us is that an equation relating energy to mass and the speed of light must involve the term mc^2.

FIGURE 1-9 Dimensional analysis could help Einstein with his problem.

● **EXAMPLE 1-2** WATER WAVES

The speed of waves in shallow water depends only on the acceleration of gravity g, a quantity with dimensions L/T^2, and on the water depth h. Which of the following formulas for the wave speed v could be correct? (a) $v = \frac{1}{2}gh^2$; (b) $v = \sqrt{gh}$

Solution
Water depth clearly has the dimension L, and speed has the dimensions L/T, so a dimensional analysis of formula (a) gives $(L/T) = (L/T^2)(L^2) = L^3/T^2$. You can probably see that this can't be an identity, but if that isn't obvious, then a little algebra reduces it to $T = L^2$, which is patently false. Note that we ignore the factor $\frac{1}{2}$ since in dimensional analysis we're concerned only with dimensions and not with numerical size.

Analysis of formula (b), on the other hand, gives $L/T = \sqrt{(L/T^2)(L)} = \sqrt{L^2/T^2} = L/T$. This is an identity, so (b) could be the right formula.

Again, we only know that (b) has the right dimensions; it might be that $v = 2\sqrt{gh}$, for example. In this case, though, (b) is, in fact, the correct expression for the wave speed.

EXERCISE The frequency of a pendulum (number of back-and-forth swings it executes in a unit time) depends only on the length ℓ of the pendulum and on the acceleration of gravity g. Which of these formulas for the frequency f of a pendulum could be correct?

(a) $f = \dfrac{1}{2\pi}\sqrt{\dfrac{g}{\ell}}$; (b) $f = 2\pi\sqrt{g\ell^2}$

Answer: $f = \dfrac{1}{2\pi}\sqrt{\dfrac{g}{\ell}}$

Some problems similar to Example 1-2: 27, 28, 61

●

1-6 SCIENTIFIC NOTATION

The range of measured quantities in the universe is enormous; lengths alone range from about 1/1,000,000,000,000,000 m for the radius of the proton to 10,000,000,000,000,000,000,000 m for the size of a galaxy; our telescopes see

10,000 times farther still (see Tables 1-2, 1-3, and 1-4 and Figs. 1-10, 1-11, and 1-12). Therefore, we frequently express numbers in **scientific notation,** with a number of reasonable size multiplying some power of ten. For example, 4,185 is 4.185×10^3, and 0.00012 is 1.2×10^{-4}.

▲ **TABLE 1-2** SOME TYPICAL DISTANCES

Distance to farthest quasar detected	1×10^{26} m
Distance to nearest large galaxy (Andromeda)	2×10^{22} m
Diameter of our galaxy (Milky Way)	8×10^{20} m
Distance to nearest star (Proxima Centauri)	4×10^{16} m
Radius of Pluto's orbit	6×10^{12} m
Radius of Earth's orbit	1.5×10^{11} m
Diameter of Earth	1.3×10^{7} m
Deepest ocean trench (Marianas Trench)	1.1×10^{4} m
Skyscraper (Sears Tower, Chicago)	4.4×10^{2} m
Height of person	2 m
Smallest object visible to naked eye	1×10^{-4} m
Diameter of red blood cell	8×10^{-6} m
Diameter of virus	1×10^{-7} m
Atomic diameter (hydrogen)	1×10^{-10} m
Diameter of proton	2×10^{-15} m

▲ **TABLE 1-3** SOME TYPICAL TIMES

Time since Big Bang	5×10^{17} s
Age of Earth	1.5×10^{17} s
Time for Sun to orbit galactic center	8×10^{15} s
Existence of human species	6×10^{13} s
Half-life of plutonium	8×10^{11} s
Human lifespan	2×10^{9} s
Orbital period of Earth (1 year)	3×10^{7} s
Rotation period of Earth (1 day)	9×10^{4} s
Flight time of intercontinental missile	1×10^{3} s
Human heartbeat	1 s
Reaction time of human nervous system	1×10^{-1} s
Period of highest audible sound	5×10^{-5} s
Period of AM radio wave	1×10^{-6} s
Time for fast computer to add two numbers	1×10^{-8} s
Period of rotation of typical molecule	1×10^{-12} s
Shortest light pulse produced in laboratory	1×10^{-15} s
Half-life of neutral pion	2×10^{-16} s
Time for light to cross a proton	7×10^{-24} s
Earliest time after Big Bang to which known laws of physics apply	10^{-43} s

▲ **TABLE 1-4** SOME TYPICAL MASSES

Galaxy (Milky Way)	4×10^{41} kg
Sun	2×10^{30} kg
Earth	6×10^{24} kg
Mountain	2×10^{18} kg
747 jetliner	4×10^{5} kg
Compact car	1×10^{3} kg
Human	65 kg
Raisin	1×10^{-3} kg
Raindrop	1×10^{-6} kg
Red blood cell	1×10^{-13} kg
DNA molecule	2×10^{-18} kg
Uranium atom	4×10^{-25} kg
Proton	1.7×10^{-27} kg
Electron	9.1×10^{-31} kg

FIGURE 1-10 A galaxy can be 10^{21} m across.

Scientific calculators handle numbers in scientific notation. But straightforward rules allow you to manipulate scientific notation if you don't have such a calculator handy:

STRATEGY **Addition/Subtraction** To add (or subtract) numbers in scientific notation, first give them the same exponent, then add (or subtract):

$$3.75 \times 10^6 + 5.2 \times 10^5 = 3.75 \times 10^6 + 0.52 \times 10^6 = 4.27 \times 10^6.$$

Multiplication/Division To multiply (or divide) numbers in scientific notation, multiply (or divide) the digits and add (or subtract) the exponents:

$$(3.0 \times 10^8 \text{ m/s})(2.1 \times 10^{-10} \text{ s}) = (3.0)(2.1) \times 10^{8+(-10)} \text{ m}$$
$$= 6.3 \times 10^{-2} \text{ m}.$$

Powers/Roots To raise numbers in scientific notation to any power, raise the digits to the given power and multiply the exponent by the power:

$$\sqrt{(3.61 \times 10^4)^3} = \sqrt{3.61^3 \times 10^{(3)(4)}} = (47.04 \times 10^{12})^{1/2}$$
$$= \sqrt{47.04} \times 10^{(12)(1/2)} = 6.86 \times 10^6.$$

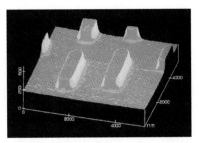

FIGURE 1-11 Among the smallest commercially manufactured structures are the pits that encode information on compact discs (CDs). This photo, taken with a scanning-tunneling microscope, shows part of a stamper used to produce a CD. The region shown measures 5 μm × 5 μm, and the projections that produce the pits are only about 200 nm high.

FIGURE 1-12 A 747 jetliner has a mass of 4×10^5 kg.

"Powers" include roots since these can be expressed as fractional powers. Example 1-3 illustrates the application of all these rules.

● EXAMPLE 1-3 SCIENTIFIC NOTATION

As we'll see in Chapter 9, the altitude necessary to put a satellite in geosynchronous orbit—where it appears fixed in the sky, like the satellites at which TV dish antennas are pointed—is given by

$$h = \sqrt[3]{\frac{GM_E T^2}{4\pi^2}} - R_E,$$

where $M_E = 5.97 \times 10^{24}$ kg is the mass of the Earth, T is the orbital period (24 hours or 8.64×10^4 s for a geosynchronous satellite), $R_E = 6.37 \times 10^6$ m is Earth's radius, and $G = 6.67 \times 10^{-11}$ m³/kg·s² is a constant. Find the value of h.

Solution
Writing the cube root as its argument to the one-third power, we have

$$h = \left[\frac{(6.67\times10^{-11} \text{ m}^3/\text{kg·s}^2)(5.97\times10^{24} \text{ kg})(8.64\times10^4 \text{ s})^2}{4\pi^2}\right]^{1/3}$$
$$- 6.37\times10^6 \text{ m}$$

$$= \left[\frac{(6.67)(5.97)(8.64)^2}{4\pi^2}\right]^{1/3}[(10^{-11})(10^{24})(10^4)^2]^{1/3}$$
$$\times \left[\left(\frac{\text{m}^3}{\text{kg·s}^2}\right)(\text{kg})(\text{s})^2\right]^{1/3} - 6.37\times10^6 \text{ m}$$

$$= (75.30)^{1/3}(10^{-11+24+(2)(4)})^{1/3}(\text{m}^3)^{1/3} - 6.37\times10^6 \text{ m}$$

$$= (4.222)(10^{21})^{1/3} \text{ m} - 6.37\times10^6 \text{ m}$$

$$= 4.222\times10^7 \text{ m} - 0.637\times10^7 \text{ m} = 3.59\times10^7 \text{ m}.$$

This is about 36,000 km or 22,000 miles above Earth's surface. Note again how the units worked out. In this case, the exponent (21) turned out to be a multiple of three, so taking the one-third power was easy. Had this not been the case, we would have had to rewrite the exponential term with an appropriate exponent; for example, $(10^{20})^{1/3} = (100\times10^{18})^{1/3} = (100)^{1/3}(10^{18})^{1/3} = 4.64\times10^6$. A scientific calculator would handle all this automatically.

EXERCISE The speed of sound in air is given by $v = \sqrt{\gamma kT/m}$, where γ is a dimensionless constant equal to 1.4, $k = 1.38\times10^{-23}$ kg·m²/K·s² is another constant, T is the temperature in kelvins (K), and $m = 4.84\times10^{-26}$ kg is the mean mass of an air molecule. Find the speed of sound in air at room temperature (300 K).

Answer: 346 m/s

Some problems similar to Example 1-3: 30, 35
●

1-7 ACCURACY AND SIGNIFICANT FIGURES

How accurate are the numbers we have just calculated? In Example 1-3, we were given the value 6.37×10^6 m for the radius of the Earth. The three **significant figures** in this number imply that the radius lies closer to this value than to 6.36×10^6 m or 6.38×10^6 m. The fewer significant figures, the less accurately we can claim to know a given quantity.

In Example 1-3 we were, in fact, given three significant figures for all quantities. The mere act of calculating cannot add accuracy, so we rounded our answer to three significant figures as well. Calculators and computers often give us numbers with many figures, but some or even most of those are usually meaningless.

What is the circumference of the Earth? It's $2\pi R_E$, of course. And π is approximately 3.14159. . . . But if you only know Earth's radius as 6.37×10^6 m, knowing π to more significant figures doesn't mean you can claim to know the diameter any more accurately. This example suggests a rule for handling calculations involving numbers with different accuracies:

> **When multiplying or dividing, the answer should have the same number of significant figures as the *least accurate* of the quantities entering the calculation.**

You're engineering an access ramp to a bridge whose main span is 1.248 km long. The ramp will be 65.4 m long. What will be the overall length of the structure? A simple calculation gives 1.248 km + 0.0654 km = 1.3134 km. How should you round this? Here, you know the bridge length to ±0.001 km, so even the addition of an amount this small is significant. Thus, your answer should have three digits to the right of the decimal point, giving 1.313 km as the appropriate answer. This example suggests the rule handling significant figures when adding or subtracting:

> **When adding or subtracting, the number of digits to the right of the decimal point should equal that of the term in the sum or difference that has the smallest number of digits to the right of the decimal point.**

When subtracting, this rule can quickly lead to loss of accuracy, as Example 1-4 illustrates.

● **EXAMPLE 1-4** NUCLEAR FUEL

A uranium fuel rod is 3.241 m long before insertion in a nuclear reactor. After insertion, heat from the nuclear reaction has raised its length to 3.249 m. What is the increase in its length?

Solution
Simple subtraction gives 3.249 m − 3.241 m = 0.008 m or 8 mm. Should this be 8 mm or 8.000 mm? Just 8 mm. Subtraction affected only the last digit of our four-significant-figure lengths, leaving only one significant figure in the answer.

EXERCISE Roof 1 measures 8.22 m by 6.51 m; roof 2 measures 8.46 m by 6.11 m. Find the area of each roof, and determine how much more area of roofing material is needed for the larger roof.

Answers: $A_1 = 53.5$ m²; $A_2 = 51.7$ m²; $A_1 − A_2 = 1.8$ m²

Some problems similar to Example 1-4: 36, 49, 50
●

Numbers with zeroes before the decimal point can be ambiguous. When we say the speed of a jetliner is 950 km/h, do we mean 9.50×10^2 km/h or 9.5×10^2 km/h? Strictly speaking, it is best to write the former if we really know the number to three significant figures. Sometimes, though, we will opt for the convenience of a number like 950 km/h; usually in such cases other numbers present will set the accuracy for the calculation.

TIP **Intermediate Results** Although it's important that your final answer reflect the accuracy of the numbers that went into it, any intermediate results you get on the way to that answer should be expressed with at least one more significant figure. Otherwise rounding of intermediate results could alter the rightmost significant figure in your answer.

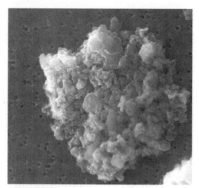

FIGURE 1-13 A scanning-electron-microscope image of a micrometeorite.

1-8 ESTIMATION

Some problems in physics and engineering call for precise numerical answers. We need to know exactly how long to fire a rocket to put a space probe on course for a distant planet or exactly what size to cut the tiny quartz crystal whose vibrations set the pulse of a digital watch. But for many other purposes, we need only a rough idea of the size of a given physical effect. Often a rough estimate will tell us that we can ignore a certain effect before we go to the trouble of calculating it exactly. In other cases, an estimate—even though it may be off by a factor of ten or more—may help decide among competing theories. And we can often make a rough estimate to check whether the results of more difficult calculations make sense.

● EXAMPLE 1-5 ESTIMATION: IS EARTH GROWING?

(a) Estimate the present mass of the Earth. (b) Earth is about 5 billion years old. Micrometeorites, roughly the size of dust grains, rain down on Earth at the rate of about 10^6 kg per day (see Fig. 1-13). Has this rain of dust contributed significantly to Earth's present mass?

Solution

(a) How big is the Earth? Maybe you know its diameter, but if you don't, consider how the United States looks on a globe. In its east-west dimension, the United States spans perhaps one-sixth of the globe. From the Atlantic to the Pacific coast is about 3000 miles, a little less than 5000 km. So Earth is about 5000 km × 6 = 3×10^4 km around at the latitude of the United States. At the equator, it is farther around—say about 4×10^4 km. The circumference of a circle is $2\pi r$, so Earth's radius is 4×10^4 km/2π, or about 6×10^3 km. (The actual value, given in Appendix E, is 6.37×10^3 km, so our estimate is quite close.) Now the volume of a sphere is $\frac{4}{3}\pi r^3$, or, since π is close to 3, about $4r^3$. So Earth's volume is

$$V_E \sim (4)(6 \times 10^6 \text{ m})^3 \sim 8 \times 10^{20} \text{ m}^3,$$

where we use the sign $\sim$ to indicate a rough approximation. How dense is Earth? It is made of rock, which is certainly denser than water. How much denser? Significantly denser, but on the other hand not ten times as dense. Say about twice as dense. Now how dense is water? Maybe you know that 1 cm³ of water has a mass of 1 g. But if not, maybe you've heard that "a pint's a pound," or maybe you can guess from experience that a gallon of water is around 10 lb, or 5 kg. Let's use the latter estimate. A gallon contains 4 quarts, and a quart is about a liter (look at the label on a 1-quart soda bottle to confirm this!) So a quart of water has a mass of about 5 kg/4, or about 1 kg. A liter is 1000 cm³, or 10^{-3} m³, so the density of water is about 1 kg/10^{-3} m³, or 1×10^3 kg/m³. If rock is twice as dense, the density of rock is 2×10^3 kg/m³. Then, using our value for Earth's volume we can estimate the mass of the Earth as

$$M_E = \text{(volume)(density)} = (8 \times 10^{20} \text{ m}^3)$$
$$\times (2 \times 10^3 \text{ kg/m}^3) = 1.6 \times 10^{24} \text{ kg}.$$

Appendix E gives the correct value as 5.97×10^{24} kg; our answer is low by a factor of about 4, but is still quite adequate to our purpose of determining whether the rain of micrometeorites is significant. (Part of the reason for our low estimate is the fact that material in the core of the Earth is compressed to much higher densities than the rocks with which we are familiar.)

(b) Coming down at the rate of 10^6 kg/day, the total mass of micrometeorites landing on Earth in 5×10^9 years is

$$m = (10^6 \text{ kg/day})(365 \text{ days/year})(5 \times 10^9 \text{ years})$$
$$= 2 \times 10^{18} \text{ kg}.$$

Since this is far less than our estimate of Earth's mass (only about $10^{18}/10^{24}$, or one-millionth), the rain of micrometeorites has made an insignificant contribution to Earth's mass.

Our conclusion rests on an important assumption: that micrometeorites have fallen at roughly the same rate for 5 billion years. Actually, the rate was probably higher when the solar system was young. But except for the earliest times, when Earth itself was forming from the matter of the primordial solar system, extraterrestrial matter probably has not contributed significantly to Earth's mass. ●

● **EXAMPLE 1-6** ESTIMATION: NUCLEAR-ELECTRIC CARS?

(a) Estimate the annual consumption of gasoline by cars in the United States. (b) Burned in a car, a gallon of gasoline is roughly equivalent to 30 kilowatt-hours (kWh) of electricity. If a large nuclear power plant produces electricity at the rate of 10^6 kWh per hour, how many such plants would have to be built if the United States converted to electric cars recharged with nuclear-generated electricity?

Solution
How far is your car driven each year? Mine goes about 12,000 miles, or roughly 10^4 mi. A typical car might get 20 miles per gallon, so at 10^4 miles/year, it would use $(10^4 \text{ mi/y})/(20 \text{ mi/gal}) = 500$ gal/y. How many cars are there in the United States? There are over 200 million people, but certainly not everyone has a car. My family has four people and two cars, so perhaps there are about 100 million, or 10^8 cars. The annual gasoline consumption of all these cars is then $(500 \text{ gal/y/car}) \times (10^8 \text{ cars}) = 5 \times 10^{10}$ gal/y. (Figures from the U.S. Department of Energy show total gasoline consumption of somewhat over 10^{11} gal/y, so our estimate is low by only a factor of about 2.)

At 30 kWh/gal, our estimated gasoline consumption is equivalent to $(30 \text{ kWh/gal})(5 \times 10^{10} \text{ gal/y}) \sim 10^{12}$ kWh/y. How many power plants would this require if we converted to electric cars? Each plant produces electricity at the rate of 10^6 kWh/h. But there are 24×365 or about 10^4 hours in a year, so each plant produces $(10^6 \text{ kWh/h})(10^4 \text{ h/y})$ or 10^{10} kWh/y. We need 10^{12} kWh/y; this requires $(10^{12} \text{ kWh})/(10^{10} \text{ kWh/plant})$ or 100 new nuclear plants. Since there are just over 100 nuclear power plants in the United States, this would represent substantial new construction.

EXERCISE Estimate the thickness of tread worn off in one revolution of an average car tire.

Possible Answer: 10^{-7} cm, with reasonable assumptions about tire diameter, total distance, and initial tread thickness. There's no single right answer to an estimation problem!

Some problems similar to Examples 1-5 and 1-6: 41–50 ●

1-9 ANALYTIC AND NUMERICAL ANALYSIS

Most of the examples and problems in this book—and in most introductory physics courses—involve calculations you can do with pencil and paper and a simple calculator. In particular, we arrive at most equations through familiar techniques of algebra, trigonometry, and calculus—an approach known as **analytic analysis.** We then solve for numerical answers using simple arithmetic.

Increasingly, though, working scientists and engineers resort to computers to handle their problems. That's not just because computers make calculations easier; more importantly, it's because they allow us to solve problems that would have been either unsolvable or would have taken literally lifetimes to solve. Learning the detailed techniques for such **numerical analysis** is beyond the scope of this book. However, sophisticated calculators now have built-in numerical analysis capabilities, and ever more powerful computer software allows almost automatic solution of problems requiring numerical analysis. An added benefit is the ability to display solutions in graphical form (Fig. 1-14). Although the problems in this book do not require the use of computers or advanced calculators, you may nevertheless find these tools helpful.

FIGURE 1-14 Today's calculators can solve equations numerically and display the results graphically.

1-10 PROBLEM SOLVING

Most physics problems are quantitative, and for many the answer is a number. That's why we've reviewed topics like scientific notation and significant figures.

But if you view physics as a quest for numerical answers, you're missing the point. The excitement of physics lies not in plugging numbers into formulas but in asking broad questions about how the natural world works and about how we forge the technologies at the heart of modern civilization. The power of physics lies in its answering those questions with quantitatively precise laws, and the challenge of physics—and of engineering based on physical principles—lies in applying those laws in specific situations.

The best way to learn physics is to work problems because that will both hone your skills and give you a deeper understanding of the underlying physical principles. This book emphasizes problem-solving skills in several ways. Each chapter contains numerous **worked examples** that illustrate the application of specific principles and techniques. Most examples are followed by **exercises** that are similar, with answers supplied. If you really understand the example you should have no trouble working the exercise. After each example is a list of some end-of-chapter problems whose solutions are similar to that of the example. The **end-of-chapter problems** themselves range from simple "confidence builders" to multistep problems; some have **hints** to get you started. Most of the problems are grouped according to the text section to which they most directly relate; within a group, problems appear in order of increasing difficulty. A separate section contains **paired problems** that are about different situations but are the same in their mathematical essentials. An answer to the first problem in each pair is given in the back of the book; if you can get that one, you should also be able to solve the second problem of the pair. A **supplementary problems** section contains additional problems not linked to specific text sections; some of these relate material from different chapters, are especially challenging, or go beyond the text material. Each of the book's six major parts is also followed by a brief set of problems designed to draw on several chapters' material or to present special challenges. **Answers** to all odd-numbered problems are in the back of the book.

At the end of each chapter, we give a **synopsis** including a brief summary of key concepts and lists of terms and symbols you should recognize. We also list the kinds of problems you should be able to solve after studying the chapter and state any limitations that apply to the material. Finally, within the text are numerous **tips,** set off from the main text, that will help your problem-solving efforts.

TIP **Read!** Before you start work on a problem, *read* the problem statement thoroughly. Be sure you understand what information is given, what's being asked for, and the meaning of the terms used in stating the problem. If you don't know what something means, reread the appropriate text sections or look in the index. If some information seems missing, check the appendices. Although physics has a reputation of being hard because it's so mathematical, experienced physics teachers agree that one of the greatest difficulties students have comes from not reading thoroughly!

CHAPTER SYNOPSIS

Summary

1. **Physics** is the fundamental science, dealing with the most basic laws governing the behavior of the physical universe. This book deals mainly with **classical physics,** an approximation to the more accurate description provided by **quantum mechanics** and **relativity.** Classical physics is a valid approximation for dealing with all but the smallest systems and the most rapid motion.
2. Physics is simple, in that a few fundamental principles describe a host of diverse phenomena.
3. Physics is a quantitative science. In this book and throughout the scientific community, we use the **SI** system of measurement units. Although seven SI base units are defined, the three most central to mechanics are those of length, time, and mass.
   a. The SI unit of length is the **meter,** defined in terms of the distance light travels in a fixed time.
   b. The SI unit of time is the **second,** defined in terms of the vibration period of the cesium-133 atom.
   c. The SI unit of mass is the **kilogram,** defined by the international prototype kilogram at the International Bureau of Weights and Measures in France.
4. **Dimensional analysis** is used to check that the units of a physical quantity are correct and can also help reveal the correct forms for equations.
5. Numbers are generally expressed in **scientific notation,** with a number typically between one and ten multiplying some power of ten. Rules governing the manipulation of scientific notation are given in the text.
6. **Significant figures** convey the accuracy of a number. Results of a calculation should not convey more accuracy than was in the numbers used in the calculation.
7. In some applications, **estimation** may be all that is required to grasp the significance of a physical phenomenon.
8. Most problems in introductory physics involve **analytic analysis**—the use of simple algebra, trigonometry, and calculus. **Numerical analysis** using computers or sophisticated calculators extends the range of problems that scientists and engineers can tackle.
9. **Problem-solving skills** are crucial to your success in physics. This book provides many ways of building your problem-solving skills.

Terms You Should Understand

(Number in parentheses is the chapter section where term first appears.)
metric system (1-3)
SI (1-3)
meter, kilogram, second (1-3)
SI prefixes (Table 1-1)
dimensional analysis (1-5)
scientific notation (1-6)
significant figures (1-7)

Symbols You Should Recognize

kg, m, s; other SI symbols as they arise (1-3)
M, L, T in dimensional analysis (1-5)
scientific notation (e.g., 4.85×10^6) (1-6)

Problems You Should Be Able to Solve

changing units (1-4)
checking answers and equations using dimensional analysis (1-5)
adding, subtracting, multiplying, and dividing numbers in scientific notation (1-6)
rounding answers to the appropriate number of significant figures (1-7)
estimating answers (1-8)
using the problem-solving help throughout this book (1-10)

QUESTIONS

1. Explain why measurement standards based on laboratory procedures are preferable to those based on specific objects like the international prototype kilogram.
2. Which measurement standards are now defined procedurally? Which are not?
3. Why is it important that the international prototype kilogram be made from corrosion-resistant material?
4. Given present-day definitions of the fundamental units, is it meaningful to attempt a more accurate measurement of the speed of light?
5. Why are "leap seconds" necessary?
6. When a computer carrying 7 significant figures adds 1.000000 and 2.5×10^{-15}, what answer does it display? Why?
7. In what way are the sciences of biology and chemistry based in physics?
8. Why does Earth's rotation not provide a suitable standard of time?
9. For which of the two coordinates—latitude or longitude—do navigators need a precise measure of time? Explain.

10. Astronomers distinguish two different measures of the length of a day—solar and sidereal. A solar day is the interval between times when the Sun is at its highest position in the sky. A sidereal day is the interval between times when any star other than the Sun is at its highest position in the sky. Why are they different? Which is longer?

11. To raise a power of ten to another power, you multiply the exponent by the power. Explain why this works.

12. A scientist and a creationist are arguing about the age of the Earth. What facts might the scientist use in estimating this age?

13. How would you determine the length of a curved line?

14. How would you measure the thickness of a sheet of paper?

15. Write $1/\sqrt{x}$ as x to some power.

16. Why can't dimensional analysis tell if answers are numerically correct?

17. How might you define the kilogram so that the standard kilogram could be reproduced anywhere?

PROBLEMS

Section 1-3 Measurement Systems

1. What is your mass in (a) kg; (b) g; (c) Gg; (d) fg?

2. The power output of a typical large power plant is 1,000 megawatts (MW). Express this result in (a) W, (b) kW, (c) GW.

3. The diameter of a hydrogen atom is about 0.1 nm, and the diameter of a proton is about 1 fm. How much bigger is a hydrogen atom than a proton?

4. Use the definition of the meter to determine how far light travels in 1 ns.

5. How long, in nanoseconds, is the period of the cesium radiation used to define the second?

6. Lake Baikal in Siberia holds the world's largest quantity of freshwater, about 14 Eg. How many kilograms is that?

7. A hydrogen atom is about 0.1 nm in diameter. How many hydrogen atoms lined up side-by-side would make a line 1 cm long?

8. How long a piece of wire would you need to form a circular arc subtending an angle of 1.4 rad, if the radius of the arc is 8.1 cm?

9. Making a turn, a jetliner flies 1.9 km on a circular path of radius 2.4 km. Through what angle does it turn?

Section 1-4 Changing Units

10. A car is moving at 35.0 mi/h. Express its speed in (a) m/s and (b) ft/s.

11. I have enough postage for a 1-oz letter but only a metric scale. What's the maximum mass for my letter, in grams?

12. A year is very nearly $\pi \times 10^7$ s. By what percentage is this figure in error?

13. How many cubic centimeters (cm^3) are there in a cubic meter (m^3)?

14. I need a 14-mm wrench to remove the spark plugs in my car, but I have only an English set with sizes that are increments of one-sixteenth of an inch. What size should I use?

15. By what percentage do the 1500-m and 1-mile foot races differ?

16. A gallon of paint covers 350 ft². What is its coverage in m²/L?

17. Superhighways in Canada have speed limits of 100 km/h. Does this exceed the 55 mi/h speed limit common in the United States? If so, by how much?

18. One m/s is how many km/h?

19. Express Neptune's distance from the Sun in astronomical units (see Fig 1-15). *Hint*: Consult Appendices C and E.

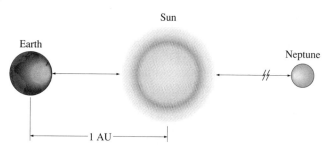

FIGURE 1-15 Problem 19 (sketch is not to scale).

20. The United States uses about 550 billion barrels of oil a year. Express this rate in (a) gallons/s and (b) L/s.

21. An aircraft carrier goes 17 feet on a gallon of fuel. Express this result in (a) miles per gallon (mpg) and (b) kilometers per liter (km/L).

22. A radian is how many degrees?

23. A degree is how many radians?

24. Engineers sometimes use an angular measure called the *grad,* defined as one one-hundredth of a right angle. What is one grad in (a) degrees and (b) radians? (Your calculator may have a switch labelled "DRG" to select degrees, radians, or grads as its angular measure.)

Section 1-5 Dimensional Analysis

25. An equation you'll encounter in the next chapter is $x = \frac{1}{2}at^2$, where x is distance and t is time. Use dimensional analysis to find the dimensions of a.

26. Show that the quantity $\sqrt{2GM/r}$ has the dimensions of speed (i.e., L/T), where M is mass, r a length, and where G has the dimensions given in Example 1-3.

27. The speed of waves in deep water depends only on the gravitational acceleration g, with dimensions L/T^2 and on the wavelength λ, with dimension L. Which of these formulas *could* be the right one for the wave speed? (a) $v = \sqrt{g/\lambda}$; (b) $v = \lambda g^2$; (c) $v = \sqrt{\lambda g}$.

28. Energy has the dimensions $M \cdot L^2/T^2$. The energy U stored in a stretched spring is given by an equation of the form $U = \frac{1}{2}kx^\alpha$, where k is a constant with the dimensions M/T^2, and x is the stretch. What should be the value of the exponent α?

Section 1-6 Scientific Notation

29. Add 3.6×10^5 m and 2.1×10^3 km.

30. The volume of the moon is 2.21×10^{19} m^3. What is its radius?

31. Divide 4.2×10^3 m/s by 0.57 ms, and express your answer in m/s^2.

32. Add 5.1×10^{-2} cm and 6.8×10^3 μm, and multiply the result by 1.8×10^4 N (1 N is the SI unit of force).

33. If there are 100,000 electronic components on a semiconductor chip that measures 5.0 mm by 5.0 mm, (a) how much area does each component occupy? (b) If the individual components are square, how long is each on a side?

34. What is the cube root of 2.7×10^{13}?

35. In Chapter 19, we will show that the average temperature of the Earth should be given roughly by $T = \sqrt[4]{S/4\sigma}$, where T is the temperature in kelvins, $S = 1.4 \times 10^3$ kg·s^{-3} is the intensity of sunlight, and $\sigma = 5.7 \times 10^{-8}$ kg·s^{-3}·K^{-4} is a constant. Find the value of T.

Section 1-7 Accuracy and Significant Figures

36. Add 1.46 m and 2.3 cm.

37. A 3.6-cm long radio antenna is added to the front of an airplane 41 m long (Fig. 1-16). What is the overall length?

FIGURE 1-16 Problem 37.

38. Repeat the preceding problem, now given that the airplane's length is 41.05 m.

39. "Machine epsilon" is a computer term describing the minimum quantity that can be added to 1 to give a different number. If machine epsilon for a certain computer is 0.000001, roughly how many significant figures does the computer carry in its calculations?

40. To see that it is important to carry more digits in intermediate calculations, determine $(\sqrt{3})^3$ to three significant figures in two ways: (a) find $\sqrt{3}$ and round to three significant figures, then cube and again round. (b) find $\sqrt{3}$ to four significant figures, then cube and round to three significant figures.

Section 1-8 Estimation

41. Paper is made from wood pulp. Estimate the number of trees that must be cut to make one day's run of a big city's daily newspaper. Assume no recycling.

42. Estimate the number of characters (letters and numbers) in this book.

43. How many Earths would fit inside the Sun?

44. Estimate the number of people, standing with outstretched arms touching, needed to form a line from New York to Los Angeles (see Fig. 1-17).

FIGURE 1-17 Problem 44.

45. If you set your watch to be exactly correct now, and if it runs slow by 1 s per month, in how many years would it again read exactly the right time?

46. (a) Estimate the volume of water going over Niagara Falls each second. (b) The falls provides the outlet for Lake Erie; if the falls were shut off, estimate how long it would take Lake Erie to rise 1 m.

47. (a) Estimate the volume of water in Earth's oceans. (b) If scientists succeed in harnessing nuclear fusion as an energy source, each gallon of sea water will be equivalent to 340 gallons of gasoline. At our present rate of gasoline consumption (see Example 1-6), how long would the oceans supply our fuel needs?

48. Estimate the number of air molecules in your dormitory room.

49. The density of interstellar space is about 1 atom per cubic cm. Stars in our galaxy are typically a few light-years apart (one light-year is the distance light travels in one year) and have typical masses of 10^{30} kg. Estimate whether there is more matter in the stars or in the interstellar gas.

50. Solar power roughly equivalent to the light from ten 100-watt light bulbs falls on each square meter of Earth's surface. Using the fact that Earth is 150 million km from the Sun, and that the Sun radiates energy equally in all directions, estimate how many 100-watt light bulbs are equivalent to the Sun's total power output.

Paired Problems

(Both problems in a pair involve the same principles and techniques. If you can get the first problem, you should be able to solve the second one.)

51. A biologist studying the effect of diet measures a rat's tail and finds it to be 18.7 cm long. A more precise measurement a week later gives 19.21 cm. How many significant figures should the biologist use in reporting the tail's growth?

52. A computer that carries seven significant figures subtracts 5.879312×10^6 from 5.879401×10^6. How many significant figures should you use in expressing the answer?

53. Find the dimensions of F in the formula $F = Gm_1 m_2 / r^2$, where G is given in Example 1-3, both m's are masses, and r is a length.

54. The quantity ρv^2 arises in the study of fluids (Chapter 18), where ρ is density (mass per volume) and v is a speed. What are the dimensions of ρv^2?

Supplementary Problems

55. A human hair is about 100 μm across. Estimate the number of hairs in a typical braid.

56. The density of bubble gum is about 1 g/cm^3. You blow an 8-g wad of gum into a bubble 10 cm in diameter. What is the thickness of the bubble? *Hint:* think about unrolling the bubble into a flat sheet. The surface area of a sphere is $4\pi r^2$.

57. The moon barely covers the Sun at a solar eclipse. Given that the moon is 4×10^5 km from Earth and that the Sun is 1.5×10^8 km from Earth, determine how much bigger the Sun's diameter is than the moon's. If the moon's radius is 1800 km, how big is the Sun?

58. The semiconductor chip at the heart of a personal computer is a square 4 mm on a side, and contains 10^6 electronic components. (a) If each component is a square, what is the distance across each component? (b) If a calculation requires that electrical impulses traverse 10^4 elements on the chip, each a million times, how many such calculations can the computer perform each second? The maximum speed of an electrical impulse is close to the speed of light, 3×10^8 m/s.

59. Estimate the number of (a) atoms and (b) cells in your body.

60. When we write the number 3.6 as typical of a number with two significant figures, we are saying that the actual value is closer to 3.6 than to 3.5 or 3.7. That is, the actual value lies between 3.55 and 3.65. Show that the percent uncertainty implied by two-significant-figure accuracy varies with the value of the number, being the least for numbers beginning with 9 and most for numbers beginning with 1. In particular, what is the percent uncertainty implied by the numbers (a) 1.1, (b) 5.0, and (c) 9.9?

61. A good-size nuclear weapon has an explosive yield equivalent to one million tons (1 megaton) of the chemical explosive TNT. Estimate the length of a train of boxcars needed to carry 1 megaton of TNT. (A 1-megaton nuclear weapon, in contrast, is on the order of 1 m long and may have a mass of a few hundred kg.)

62. Continental drift occurs at about the rate at which your fingernails grow. Estimate the age of the Atlantic Ocean, assuming eastern and western hemispheres have been drifting apart.

63. As you will see in Chapter 16, the speed of waves on a stretched string depends only on the tension, F_0, in the string and on the mass per unit length of string μ. Tension has the dimensions $M \cdot L/T^2$ and mass per unit length obviously has dimensions M/L. Find a combination of F_0 and μ that has the units of speed; this combination must play a prominent role in the expression for the speed of waves on the string.

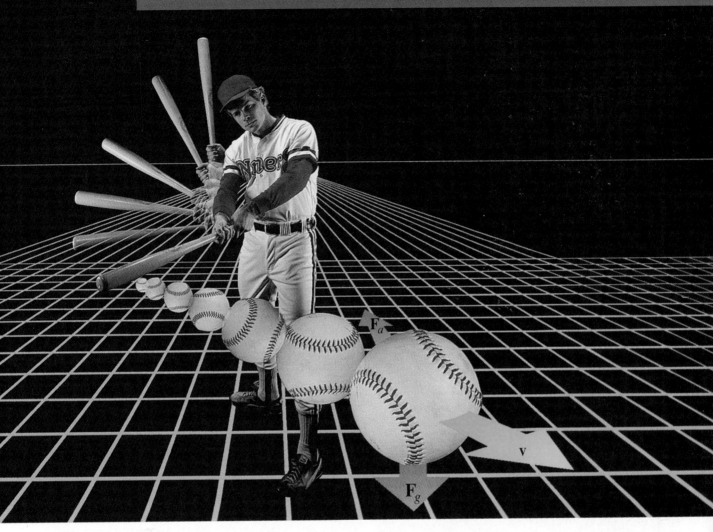

Mechanics is the branch of physics that deals with motion—whether it be the motion of atoms, baseballs, or planets. In this strobe photo, the bat undergoes a complex three-dimensional motion, imparting energy to the ball. After it leaves the bat, the ball's subsequent motion, characterized by its velocity **v** (blue arrow), is governed by the force of gravity $\mathbf{F}_g$ and the aerodynamic force $\mathbf{F}_a$.

CHAPTER 2

MOTION IN A STRAIGHT LINE

"How fast?" is a fundamental question about motion. The cheetah is the fastest land animal, with a top speed of 110 kilometers per hour (about 30 meters per second). Here **v** indicates the cheetah's velocity.

Electrons swarming around atomic nuclei, cars speeding along a highway, the swirling winds of a hurricane, the galaxies rushing apart in the expanding universe—all these are examples of matter in motion. In physics, the study of motion is called **kinematics** (from the Greek "kinema," motion, as in motion pictures). In this chapter, we deal with the simplest case of motion: a single particle moving along a straight line. Later, we generalize to motion in more dimensions and with more complicated objects. But the basic concepts we develop here remain fundamental to these more complex situations.

We define a **particle** as an isolated point of mass, having no size or structure, and incapable of rotation, vibration, or other internal motions. Whether we can treat a real object as a particle depends on the circumstances. To an astronomer describing Earth's orbit, our planet is essentially a particle; to a geologist it is a complex system exhibiting an array of internal motions. In later chapters, we will see how the behavior of such complex objects follows from the laws we develop for their constituent particles.

2-1 DISTANCE, TIME, SPEED, AND VELOCITY

"How fast?" is a basic question about motion. We define the **average speed** of an object as the distance it travels divided by the travel time. The units of speed are therefore length per time, such as m/s (read "meters per second"), km/h, or mi/h (often written mph). For example, a car that goes 7.2 km in 12 min has a speed of

$$\frac{7.2 \text{ km}}{12 \text{ min}} = 0.60 \text{ km/min} = \frac{0.60 \text{ km}}{(1 \text{ min})(1 \text{ h/60 min})} = 36 \text{ km/h}.$$

At this speed, the car would go $(36 \text{ km/h})(2.5 \text{ h}) = 90$ km in 2.5 h.

> **TIP** **Watch Your Units** Note how in these calculations we canceled units just as we multiplied and divided numbers. Had our answers not come out with the correct units we would have had a sure indication of an error. Checking the units of your answers is always a good practice and is easier if you carry units along with the numbers throughout your calculations.

Distance and Displacement

Suppose you drive 15 minutes to a hamburger stand 10 miles away, grab your food at a drive-through window, and take another 15 minutes to drive home. You've gone a total of 20 miles in half an hour, giving an average speed of 40 mi/h. You certainly have been traveling, as the hamburger proves. But in another sense, you have gone nowhere at all since you end up back at your starting point.

We distinguish between the distance actually traveled and the overall change in position. This overall change—between the position at which you started and the position at which you finished—is the **displacement.** In our hamburger example, the displacement is zero even though the overall distance traveled is 20 miles.

When motion is confined to one dimension, there are still two possible ways to go: forward or backward, east or west, up or down, left or right. In specifying displacement, we distinguish these possibilities using positive and negative numbers. For vertical motion, we might call the upward direction positive and the downward direction negative; on Earth's surface, north might be positive and south negative. For convenience, we specify an origin and mark coordinates designating positive and negative displacements from that origin (Fig. 2-1). The choices of direction and origin are arbitrary, but once we've made them, we must stick with them to provide a consistent mathematical description of our physical situation.

Velocity

In describing motion, we want to know not only the displacement an object undergoes, but also how rapidly that displacement occurs. Accordingly, we

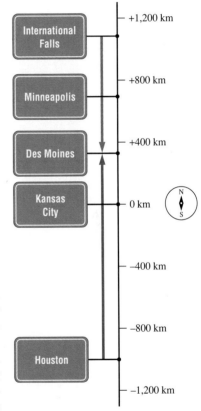

FIGURE 2-1 Describing north-south motion in the central United States. Kansas City has been arbitrarily chosen as the origin and the northward direction as positive. Motion from Houston to Des Moines (red arrow) represents approximately a displacement of +1,300 km, from International Falls to Des Moines (blue arrow) an approximate displacement of −750 km. How would you describe the overall displacement for a trip from Kansas City to International Falls and then on to Houston?

define **average velocity** as displacement divided by time interval. In our hamburger example, the average velocity was zero because the displacement was zero. If the trip from Houston to Des Moines in Fig. 2-1 takes 2.6 hours by plane, then the average velocity is (1,300 km)/(2.6 hours) = 500 km/h. If the 750-km trip from International Falls to Des Moines takes 10 hours by car, then the average velocity is −75 km/h. This number is negative because the direction of motion is from north to south, which we have defined as the negative direction.

In calculating the average velocity, all that matters is the overall displacement and the time. Maybe that trip from Houston to Des Moines was in a single plane going steadily at 500 km/h. Or maybe it involved a faster plane that stopped for a while in Minneapolis; as long as the overall displacement is 1,300 km and the time 2.6 hours, the average velocity is still 500 km/h.

Using symbols, we can write the average velocity as

$$\bar{v} = \frac{\Delta x}{\Delta t}, \tag{2-1}$$

where Δx is the displacement in the time interval Δt. The bar over the v indicates an average quantity (and is read "v bar"). We have here introduced the symbol Δ (the Greek capital delta) for "the change in . . . ."

● **EXAMPLE 2-1** FLOWING LAVA

Geologists studying the flow of lava underground often drop a branch of wood in at one hole and watch downstream a measurable distance away to see how long it takes for the wood to travel that known distance. At a recent volcanic eruption on the island of Hawaii, a branch traveled 100 meters in 12 seconds. What was its average velocity?

Solution

Applying Equation 2-1, we have

$$\bar{v} = \frac{\Delta x}{\Delta t} = \frac{100 \text{ m}}{12 \text{ s}} = 8.3 \text{ m/s}.$$

In writing Δx and the answer as positive numbers, we are implicitly taking the downstream direction as positive.

Some problems similar to Example 2-1: 1–4, 85 ●

2-2 MOTION IN SHORT TIME INTERVALS

Although our geologists' experiment told them the average velocity of the lava flow, it did not give them all the details of the motion. For example, did the lava move faster at the beginning of the interval? Was there a region where the flow slowed significantly? The geologists could answer these questions with a series of observations measuring the lava flow velocity over smaller intervals of time and distance (Fig. 2-2).

As the size of the intervals shrinks, and their number grows, an ever-more-detailed picture of the velocity emerges. In the limit of arbitrarily small time intervals, we achieve a description that gives a value for the velocity at essentially each instant. We call the velocity obtained through this limiting process the **instantaneous velocity,** v. The magnitude of the instantaneous velocity is called the **instantaneous speed.**

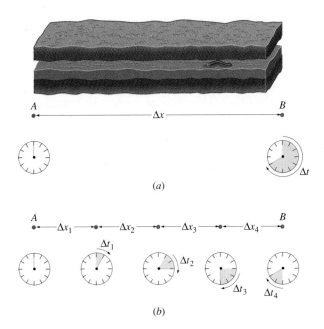

FIGURE 2-2 Determining the velocity of a lava flow. (*a*) Measuring the time Δt for a branch to go the distance Δx from A to B gives the average flow velocity $\bar{v} = \Delta x/\Delta t$. (*b*) A series of measurements over shorter intervals gives a more detailed picture of the flow, by providing values for the velocities $\bar{v}_1 = \Delta x_1/\Delta t_1$, $\bar{v}_2 = \Delta x_2/\Delta t_2$, etc. Can you tell from the clock readings when the flow is fastest and when it is slowest?

(*a*)

(*b*)

Physically, we can determine an approximate value for the instantaneous velocity by measuring the average velocity $\Delta x/\Delta t$ over a very small time interval; we will soon see how we can calculate the instantaneous velocity mathematically given displacement as a function of time.

● **EXAMPLE 2-2**   AVERAGE AND INSTANTANEOUS VELOCITY

You arrive at Boston's Logan Airport at 2:00 P.M. to check in for a trip to Detroit. Your plane departs at 3:00 P.M. and flies westward 1400 km to Chicago at 900 km/h. There, you catch a 5:30 flight to Detroit, which lies 390 km east of Chicago. Your plane lands in Detroit at 6:10. Assuming the planes maintain constant velocity while in the air, what are your instantaneous velocities at 2:30 P.M., 4:00 P.M., and 6:00 P.M.? What is your average velocity for the entire trip?

Solution
At 2:30, you are still waiting in Boston, so your instantaneous velocity is zero. At 900 km/h, the 1400-km trip to Chicago will take well over an hour, so at 4:00 P.M. you are still en route to Chicago, with instantaneous velocity 900 km/h. (Here we're taking the westward direction as positive.) At 6:00 P.M., you are en route to Detroit. Since the 390-km trip from Chicago to Detroit takes 40 min or 0.67 h, your velocity is then −390 km/0.67 h = −580 km/h (negative because Detroit is east of Chicago). Because the plane's velocity is constant, your

instantaneous velocity at 6:00 P.M. is also −580 km/h. The entire trip takes 4 h 10 min, or 4.17 h. Your net displacement is 1400 km + (−390 km) = 1010 km. Then your average velocity is 1010 km/4.17 h = 242 km/h.

EXERCISE   You get in your car at 7 P.M. and head toward a movie theater 5.0 km from home. You drive the first 2.0 km in 2.0 min. Then you're stuck at a red light for 3.0 min. The light turns green and you continue at a steady speed, reaching a parking garage 1.0 km beyond the theater at 7:13 P.M. You walk back, arriving at the theater at 7:28 P.M. What are your instantaneous velocities at (a) 7:01 P.M., (b) 7:03 P.M., (c) 7:10 P.M., and (d) 7:20 P.M.? (e) What is your average velocity for the entire trip? Take the direction from home toward the theater as positive.

Answers: (a) 60 km/h; (b) 0 km/h; (c) 30 km/h; (d) −4.0 km/h; (e) 10.7 km/h

Some problems similar to Example 2-2: 5, 7, 9, 15

●

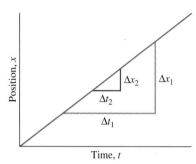

FIGURE 2-3 A graph of position versus time for an object moving at constant velocity. Shown are two time intervals Δt and the corresponding displacements Δx. The average velocity over an interval is the ratio $\Delta x/\Delta t$, or the slope of a line connecting the ends of the interval. With constant velocity this slope is the same for all intervals.

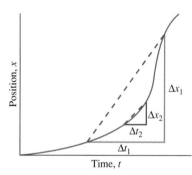

FIGURE 2-4 A graph of position versus time for an object whose velocity is continually changing. Dashed lines connect the endpoints of two intervals; since the lines have different slopes, the average velocities $\Delta x/\Delta t$ have different values over the two intervals.

To make the notion of instantaneous velocity more precise, consider first motion at constant velocity, as shown in Fig. 2-3. The graph is drawn in the usual format for a graph describing motion: time along the horizontal axis and distance along the vertical axis. Note that if we shorten the length of the time interval by any factor, the distance is shortened by the same factor. We get the same value for velocity no matter how short the time interval; thus for this constant-velocity situation, the average and instantaneous velocities are equal. Both velocities are, indeed, given by the slope of the line on the graph—that is, by distance divided by time.

In Fig. 2-4, we plot the motion of an object whose velocity is continually changing. Now as we consider different time intervals, we get different slopes. But if we examine shorter and shorter intervals around one specific time, we find that part of the curve gets closer and closer to a straight line (Fig. 2-5). The average velocity we compute then approaches a single value, which in the limit of arbitrarily small intervals is just the slope of the curve at a single point.

Thus, to compute the instantaneous velocity we apply Equation 2-1 as the time interval Δt becomes arbitrarily small. In doing so, we are carrying out the mathematical equivalent of the physical procedure our geologists could have used to discover the detailed motion of their lava flow in Example 2-1. Of course, we cannot take Δt all the way to zero, for then $\Delta x = 0$, and the velocity would be undefined. But the velocity remains perfectly well defined for all nonzero values of Δt, no matter how small. So as we consider ever smaller time intervals, the average velocity we calculate comes ever closer to the instantaneous velocity. We can say this by writing

$$v = \lim_{\Delta t \to 0} \frac{\Delta x}{\Delta t}. \qquad (2\text{-}2a)$$

Read this equation as "v is the limit, as Δt approaches zero, of Δx divided by Δt." This statement is true even though the value of $\Delta x/\Delta t$ would be undefined if we evaluated it at $\Delta t = 0$. And it is true even for nonconstant velocity since a sufficiently short segment of a curve comes arbitrarily close to being straight (Fig. 2-5).

The limiting process we have just described occurs so frequently in physics and mathematics that we give the result a special name and symbol. We define the **derivative** of x with respect to t (symbol dx/dt) as

$$\frac{dx}{dt} = \lim_{\Delta t \to 0} \frac{\Delta x}{\Delta t}.$$

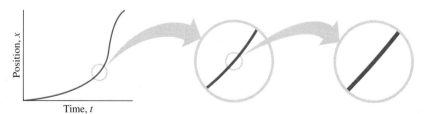

FIGURE 2-5 As we examine an ever-smaller segment of the curve, it begins to resemble a straight line. The slope of this line is the instantaneous velocity.

The quantities dx and dt are called **infinitesimals;** they represent arbitrarily small quantities associated with the limiting process that defines the derivative. We can then write Equation 2-2a for the velocity in the form

$$v = \frac{dx}{dt}. \tag{2-2b}$$

In general, the derivative is the rate of change of one quantity with respect to another. Thus Equation 2-2 says that velocity is the rate of change of position with respect to time.

Given position as a function of time, we could carry out the limiting process leading to Equation 2-2 by computing average velocities over ever smaller time intervals (see Problem 22). But that's not necessary. In calculus you either have or will derive formulas for the derivatives of common functions. For example, any function of the form

$$x = bt^n,$$

where b and n are constants, has derivative given by

$$\frac{dx}{dt} = bnt^{n-1}. \tag{2-3}$$

● EXAMPLE 2-3   A SPACE SHUTTLE ASCENDS

The altitude of a space shuttle in the first half minute of its ascent is given approximately by $x = bt^2$, where the constant b is 2.90 m/s^2, x is in meters, and t is in seconds. Find a general expression for the shuttle's velocity as a function of time and from it the instantaneous velocity at 20 seconds after liftoff. Determine also the average velocity for the first 20 seconds.

Solution
The instantaneous velocity v is the derivative of position with respect to time. Applying Equations 2-2 and 2-3, we have

$$v = \frac{dx}{dt} = \frac{d(bt^2)}{dt} = 2bt.$$

This is our expression for the instantaneous velocity at any time t; evaluating it at $t = 20$ s gives $v = (2)(2.9 \text{ m/s}^2)(20 \text{ s}) = 116$ m/s. Note how the units work out; multiplying meters per second squared (m/s^2) by seconds gives meters per second (m/s), the correct unit for velocity.

What about the *average* velocity for the first 20 s? Here we need the total displacement—given by the expression for x itself: $x = bt^2$. So the average velocity from liftoff to some time t is given by

$$\bar{v} = \frac{\Delta x}{\Delta t} = \frac{x}{t} = \frac{bt^2}{t} = bt,$$

where we've used x for Δx and t for Δt since both position and time are taken to be zero at liftoff. Comparing with our result for the instantaneous velocity, we see that the average velocity from liftoff to any time is just half the instantaneous velocity at that time. For the first 20 s, then, the average velocity is 58 m/s.

> **TIP   Watch Your Language**   The language of physics often holds clues to the meaning and use of physical concepts. In this example, for instance, we speak of the *instantaneous* velocity *at* a particular time. That wording should remind you of the limiting process that focuses on a single instant. But we talk about the *average* velocity *over an interval* of time since averaging explicitly involves more than one time.

EXERCISE   At time $t = 0$ you slam on your car's brakes. From then until the car stops, your position is given by $x = ct - bt^2$, where $c = 20$ m/s, $b = 1.2$ m/s^2, and x is the displacement in meters from the point where you applied the brakes. (a) What is your speed 5.0 s after you start braking? (b) What is the average velocity during the first 5.0 s of braking? Remember that the derivative of a sum of terms is the sum of the derivatives of those terms.

Answers: (a) 8.0 m/s; (b) 14 m/s

Some problems similar to Example 2-3:   23, 24, 25, 91

●

Derivatives of other common functions, including exponentials, logarithms, and the trigonometric functions, are given in Appendix A. If a function is made up of a sum of terms, its derivative is simply the sum of the derivatives of the individual terms.

2-3 ACCELERATION

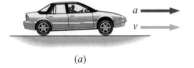

(a)

(b)

FIGURE 2-6 In one-dimensional motion, acceleration implies speeding up or slowing down. In (*a*), velocity and acceleration are both to the right, and the car speeds up. In (*b*), velocity and acceleration are in opposite directions, and the car slows down. Algebraically, the two quantities have the same sign in (*a*) and opposite signs in (*b*).

In our space shuttle example, the velocity of the shuttle was changing. When velocity changes, an object is said to undergo **acceleration.** Quantitatively, we define acceleration as the rate of change of velocity, just as we defined velocity as the rate of change of position. The average acceleration over a time interval Δt is

$$\bar{a} = \frac{\Delta v}{\Delta t}, \qquad (2\text{-}4)$$

where Δv is the change in velocity and where the bar over the a indicates that this is an average value. Just as we defined instantaneous velocity through a limiting procedure, so we define instantaneous acceleration as

$$a = \lim_{\Delta t \to 0} \frac{\Delta v}{\Delta t} = \frac{\Delta v}{dt}. \qquad (2\text{-}5)$$

In one-dimensional motion, the acceleration must be either in the direction of the velocity or directly opposite it. In the latter case, the magnitude of the velocity decreases, and we sometimes use the term *deceleration.* In solving problems, though, it's easier to use only the term *acceleration* and let the algebra indicate whether the velocity change is a speeding up or a slowing down (Fig. 2-6). When we deal with two-dimensional motion, we'll find a much richer range of relationships between the directions of velocity and acceleration.

Since acceleration is the rate of change of velocity, the units of acceleration must be (unit of velocity) per (unit of time). Since the units of velocity are distance per time, the units of acceleration are (distance/time) per time, which can be written distance/time². For example, if you accelerate in such a way that your velocity changes by 5 m/s during each second, your acceleration is 5 (m/s)/s

● **EXAMPLE 2-4** AVERAGE ACCELERATION

A jetliner rolls down the runway with constant acceleration. From rest, it reaches its takeoff speed of 250 km/h in 1 min. What is its acceleration? Express in km/h².

Solution
The plane undergoes a velocity change Δv of 250 km/h in time $\Delta t = 1$ min or 1/60 h. The average acceleration is therefore

$$\bar{a} = \frac{\Delta v}{\Delta t} = \frac{250 \text{ km/h}}{(1/60 \text{ h})} = 15{,}000 \text{ km/h}^2.$$

Does this huge result make sense? Yes—the jet reaches 250 km/h in only 1 min, so an increase of 15,000 km/h in 1 h is not unreasonable. Of course, the plane never reaches such a high speed because it does not accelerate for a full hour. ●

● **EXAMPLE 2-5** ACCELERATION AND DECELERATION

Two cars are moving eastward at 20 m/s. At time $t = 0$ both undergo accelerations of magnitude 2 m/s², with one car's acceleration eastward and the other westward. How fast is each car going 5 s later?

Solution
Each car's velocity changes at the rate of 2 m/s², or 2 (m/s)/s, so in 5 s, each velocity has changed by (2 m/s²)(5 s) = 10 m/s. (Note how the units work out.) Both cars are *moving* eastward. For the car that is also *accelerating* eastward, this change represents a speeding up to 30 m/s. But for the car that

is accelerating westward—opposite to the direction of its velocity—the change represents a slowing down, to 10 m/s.

EXERCISE A car moving at 90 km/h brakes and, 5.0 s later, is moving at 40 km/h. (a) Find the average acceleration. (b) Find the additional time it would take for the car to stop, assuming the same average acceleration.

Answers: (a) 2.78 m/s²; (b) 4.0 s

Some problems similar to Examples 2-4 and 2-5: 28–32 ●

---

or 5 m/s². Sometimes acceleration is specified in mixed units; for example, a car going from 0 to 60 mi/h in 10 seconds has an average acceleration of 6 mi/h/s.

Just as velocity is the slope of the position-versus-time graph, so acceleration is the slope of the velocity-versus-time graph. Figure 2-7 shows graphs of position, velocity, and acceleration versus time for a particle undergoing one-dimensional motion with position x given by $x = bt^2 - ct^3$, where $b = 4$ m/s² and $c = 1$ m/s³. We obtained the velocity by differentiating the position: $v = d(bt^2 - ct^3)/dt = 2bt - 3ct^2$; a second differentiation gives the acceleration: $a = d(2bt - 3ct^2)/dt = 2b - 6ct$. Note that where the curve of Fig. 2-7a is steep—where the position x changes rapidly—the magnitude of the velocity in Fig. 2-7b is large. When x is decreasing with time, the velocity is negative. Similarly, where the slope of the velocity graph is greatest, the acceleration has its greatest magnitude. And when the velocity is decreasing—regardless of whether its value is positive or negative—the acceleration is negative. Finally, we have marked on Fig. 2-7 the points where velocity and acceleration are zero. The zero-velocity points appear where the *slope* of the position graph is zero—where that graph is flat. Similarly, the acceleration is zero where the *slope* of the velocity curve is zero.

Notice in Fig. 2-7 that the *value* of the velocity is unrelated to the *value* of the position. The velocity is the *slope* of the position curve—and the slope depends on how the position is changing, not on its actual value. Similarly, the value of the acceleration depends only on the rate of change of velocity, not on the velocity itself. In particular, we can have zero velocity and still be accelerating! You can see this in Fig. 2-7, where the point of zero velocity does not correspond to zero acceleration.

How can something be accelerating if it's not moving? Consider a ball thrown straight up. Just before it reaches the peak of its flight, it's moving upward. Just after, it's moving downward. Right at the peak its velocity is instantaneously zero. No matter how small a time interval you consider about the peak, there's always a change in velocity—and therefore the ball is accelerating, even right at the instant its velocity is zero (see Fig. 2-8).

Acceleration is the rate of change—that is, the derivative with respect to time—of velocity, and velocity is the rate of change of position. That makes acceleration the rate of change of the rate of change of position. This sounds less awkward mathematically: We say that the acceleration is the **second derivative**

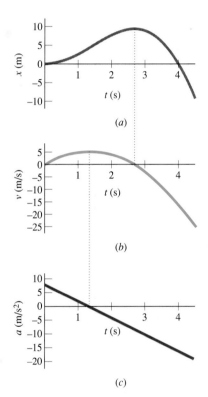

FIGURE 2-7 Position, velocity, and acceleration versus time for one-dimensional motion with position given by $x = bt^2 - ct^3$, where $b = 4$ m/s² and $c = 1$ m/s³. Note that zero velocity occurs where the *slope* of the position graph is zero, and zero acceleration where the *slope* of the velocity graph is zero. In general, each graph gives the *slope* of the one above it; there is no particular relation between the actual *values* on one graph and those on a lower graph.

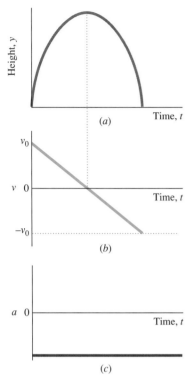

FIGURE 2-8 Position, velocity, and acceleration versus time for a ball thrown straight up. At the peak of its flight, the ball is instantaneously at rest ($v = 0$). But even at that point, the velocity is in the process of changing—here from upward to downward—so the acceleration is nonzero. In this case the acceleration is, in fact, constant and downward (i.e., negative).

of the position with respect to time. Symbolically, we write the second derivative as d^2x/dt^2; this is just a symbol and doesn't mean that anything is actually squared. Then the relationship among acceleration, velocity, and position can be written

$$a = \frac{dv}{dt} = \frac{d}{dt}\left(\frac{dx}{dt}\right) = \frac{d^2x}{dt^2}. \tag{2-6}$$

2-4 CONSTANT ACCELERATION

When acceleration is constant, the description of motion takes an especially simple form. Suppose an object starts at time $t = 0$ with some initial velocity v_0 and constant acceleration a. Later, at some time t, it has velocity v. We can write the average acceleration as the change in velocity divided by the time interval. But because the acceleration doesn't change, its average and instantaneous values are identical, so we have

$$a = \bar{a} = \frac{\Delta v}{\Delta t} = \frac{v - v_0}{t - 0},$$

or, rearranging,

$$v = v_0 + at. \tag{2-7}$$

This equation says simply that the velocity changes from its initial value by an amount that is the product of acceleration and time.

TIP **Know Your Limits** Many equations we develop are special cases of more general laws. Understanding how they are derived will help you to recognize their limitations and to know when it is appropriate to use them. Equation 2-7 is a case in point: It applies *only when acceleration is constant*. Only then are instantaneous and average accelerations identical, and it was by identifying a and $\bar{a}$ that we were led to Equation 2-7.

Having determined velocity as a function of time, we now consider position. Since the velocity is not constant, we can't simply multiply velocity by time to get the change in position. But in this case of constant acceleration, Fig. 2-9a shows that the average velocity over some interval is just the average of the velocities at the beginning and end of the interval—a result that is, in general, *not true* when the acceleration varies (Fig. 2-9b). So we can write

$$\bar{v} = \tfrac{1}{2}(v_0 + v) \tag{2-8}$$

for the average velocity over the interval from $t = 0$ to some later time when the velocity is v. We can also write the average velocity as the change in position

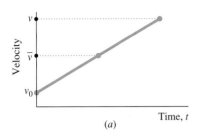

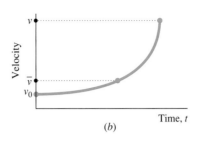

FIGURE 2-9 (*a*) With constant acceleration, the velocity rises steadily, with slope equal to the acceleration. Because the rise is steady, the average velocity lies midway between the initial velocity v_0 and the final velocity v. (*b*) When the acceleration is not constant, the velocity-versus-time curve is not a straight line, and the average velocity is not midway between the initial and final velocities.

divided by the time interval. Suppose that at time $t = 0$ our object was at some position x_0. Then its average velocity over a time interval from 0 to some time t is

$$\bar{v} = \frac{\Delta x}{\Delta t} = \frac{x - x_0}{t - 0},$$

where x is the object's position at time t. Equating this expression for $\bar{v}$ with that of Equation 2-8, and rearranging, gives

$$x = x_0 + \bar{v}t = x_0 + \tfrac{1}{2}(v_0 + v)t. \tag{2-9}$$

But we already derived an expression for the instantaneous velocity v appearing in this expression; it's given by Equation 2-7. Substituting from that equation and simplifying the result then gives an expression for position as a function of time:

$$x = x_0 + v_0 t + \tfrac{1}{2}at^2. \tag{2-10}$$

Does this equation make sense? In the absence of any acceleration ($a = 0$), the position would increase linearly with time, at a rate given by the initial velocity v_0. With constant acceleration, the additional term $\frac{1}{2}at^2$ describes the effect of the ever-changing velocity. That time is squared in this term makes sense, too: With constant acceleration, the longer an object travels, the faster it moves, so the more distance it covers in a given time. Figure 2-10 shows the meaning of the terms in Equation 2-10.

How much runway do I need to land a jetliner, given the touchdown speed and assuming constant deceleration? Sometimes we're interested in questions like this, where we want to relate position, velocity, and acceleration directly without explicit mention of time. We can solve Equation 2-7 for time:

$$t = \frac{v - v_0}{a}.$$

Substituting this expression for t in Equation 2-10, we can write

$$x - x_0 = \frac{1}{2}\frac{(v_0 + v)(v - v_0)}{a},$$

or, since $(a + b)(a - b) = a^2 - b^2$,

$$v^2 = v_0^2 + 2a(x - x_0). \tag{2-11}$$

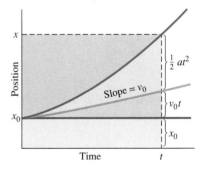

FIGURE 2-10 Meaning of the terms in Equation 2-10. An object starts at position x_0. Were there no velocity or acceleration, it would stay at x_0, and its position-versus-time curve would be the horizontal straight line. If it started with an initial velocity v_0 but no acceleration, its position would change linearly with time, as shown in the slanted straight line. A constant acceleration increases the velocity, causing the actual position-versus-time to bend upward in a parabola (the upper curve). How would the curves differ if the acceleration and velocity had opposite signs?

Equations 2-7 through 2-11 link all possible combinations of position, velocity, and acceleration for motion with constant acceleration. We summarize these equations in Table 2-1.

We derived the equations of Table 2-1 starting from the assumption of constant acceleration, and we emphasize that they apply *only* in that case. It's also possible to go backwards, differentiating Equation 2-10 once to get Equation 2-7 and again to get the acceleration. Problem 38 explores this approach.

2-5 USING THE EQUATIONS OF MOTION

The equations in Table 2-1 provide a full description of motion under constant acceleration. Don't regard these equations as a set of separate laws, but instead recognize them as complementary descriptions of a single underlying phenomenon—one-dimensional motion with *constant acceleration*. The advantage in having several equations lies in their providing convenient starting points for problem solving. Don't memorize these equations; instead, you should grow familiar with them as you work problems.

● **EXAMPLE 2-6** LANDING A JETLINER

A jetliner touches down at a speed of 270 km/h (Fig. 2-11). On landing, the plane decelerates at 4.5 m/s². What is the minimum runway length on which the aircraft can land?

Solution

Here we're given an initial velocity and acceleration and asked for a distance. Equation 2-11 provides the link we need. Solving for x, we have

$$x = x_0 + \frac{v^2 - v_0^2}{2a}.$$

We can take $x_0 = 0$ at the beginning of the runway, where the speed is v_0; at the unknown position x, we want the plane to be stopped, so $v = 0$. Then we have

$$x = \frac{-v_0^2}{2a} = \frac{-[(270 \text{ km/h})(1{,}000 \text{ m/km})/(3{,}600 \text{ s/h})]^2}{(2)(-4.5 \text{ m/s}^2)}$$

$$= 625 \text{ m}.$$

Note that we used a negative value for the acceleration since the plane decelerates after landing. We also converted the touchdown speed to meters per second for compatibility with the units given for the acceleration. A required safety margin brings the minimum legal runway for this aircraft to about one half mile.

TIP Be Careful with Mixed Units Frequently problems are stated in units other than those of the SI system. Although it's sometimes possible to work consistently in other units, when in doubt it's always best to convert to SI units. In this problem, a sure indication of the need for conversion is the fact that the acceleration is expressed in SI (m/s²), while the velocity is not.

EXERCISE On a dry road, a car with conventional brakes can achieve a maximum deceleration of 7.3 m/s². Assuming constant deceleration at this value, what is the stopping distance for a car moving initially at 100 km/h?

Answer: 53 m

Some problems similar to Example 2-6: 45, 48, 52

FIGURE 2-11 A 767 touches down. How much runway does it need to stop safely (Example 2-6)?

●

▲ TABLE 2-1 EQUATIONS OF MOTION FOR CONSTANT ACCELERATION

EQUATION	CONTAINS	NUMBER
$v = v_0 + at$	v, a, t; no x	2-7
$\bar{v} = \frac{1}{2}(v_0 + v)$	Average v	2-8
$x = x_0 + \frac{1}{2}(v_0 + v)t$	x, v, t; no a	2-9
$x = x_0 + v_0 t + \frac{1}{2}at^2$	x, a, t; no v	2-10
$v^2 = v_0^2 + 2a(x - x_0)$	x, v, a; no t	2-11

● **EXAMPLE 2-7** ESCAPING A DOG

I am riding my bicycle at a leisurely 10 km/h, then speed up at a steady rate for 10 s to escape a dog. At the end of that time, I am going at 30 km/h. How far did I go during the 10-s interval?

Solution
We want to link v and t, which we know, with x, which we don't know. Equation 2-9 does this. Our answer is the distance $x - x_0$ traveled during the time t:

$$x - x_0 = \frac{1}{2}(v_0 + v)t$$

$$= \frac{1}{2}(10 \text{ km/h} + 30 \text{ km/h})(10 \text{ s})(1 \text{ h}/3600 \text{ s})$$

$$= 0.056 \text{ km} = 56 \text{ m}.$$

Again, with time in seconds and velocity in kilometers per hour, we had to convert seconds to hours for consistency of units.

EXERCISE Electrons in a particle accelerator are moving at 9.0×10^5 m/s when they enter a tube where they are accelerated to 6.5×10^6 m/s. If they spend 0.61 μs in the tube, what is the tube's length?

Answer: 2.26 m

Some problems similar to Example 2-7: 43, 46, 56
●

Sometimes, we're interested in comparing the one-dimensional motions of two different objects, as the following example illustrates.

● **EXAMPLE 2-8** SPEED TRAP

A speeding motorist zooms through a 50 km/h zone at 75 km/h, without noticing a stationary police car by the roadside. The police officer immediately heads after the speeder, accelerating at 9.0 km/h/s. When the officer pulls alongside the speeder, how far down the road are they, and how fast is the police car going?

Solution
In this example, we deal with two different one-dimensional motions, both described by Equation 2-10. We want to know when the two cars are in the same place—that is, when equations for the positions of the two cars give the same value. In working such a problem, it is important to define the initial positions and times consistently for both motions; otherwise a comparison between the two is meaningless.

Let $t = 0$ be the time the speeder first passes the police car, and let $x = 0$ be the position at which this occurs. Then $x_0 = 0$ for both cars. For the speeder, we have $v_{s0} = 75$ km/h =

21 m/s and $a_s = 0$. For the police car, $v_{p0} = 0$ but $a_p = 9.0$ km/h/s = 2.5 m/s². (Here we use subscripts s and p for quantities associated with the speeder and police, respectively.) Then for the two cars, Equation 2-10 becomes

$$x_s = v_{s0}t \quad \text{(speeder)}$$

and

$$x_p = \frac{1}{2}a_p t^2 \quad \text{(police)}$$

Equating these expressions tells when the speeder and police car are at the same place:

$$v_{s0}t = \frac{1}{2}a_p t^2,$$

so $t = 0$ or $t = 2v_{s0}/a_p$. Why two answers? We asked the equations for *any* times when the two cars were in the same place. Our first answer is just the initial encounter as the speeder passes the stationary police car. The second answer is the one we want—the time when the officer catches up with the speeder. *Where* does this occur? Using the time $2v_{s0}/a_p$ in the speeder equation, we have

$$x = v_{s0}t = v_{s0}\frac{2v_{s0}}{a_p} = \frac{2v_{s0}^2}{a_p} = \frac{(2)(21 \text{ m/s})^2}{2.5 \text{ m/s}^2} = 350 \text{ m}.$$

We can calculate the police car's speed at this time from Equation 2-7, with $v_{p0} = 0$:

$$v_p = a_p t = a_p\frac{2v_{s0}}{a_p} = 2v_{s0} = 150 \text{ km/h}.$$

Does this result make sense? Yes—the police car and the speeder cover the same distance in the same time. Therefore, their average velocities must be the same. The speeder maintains constant velocity, but the police car accelerates from rest. Since the average velocity with constant acceleration is half the sum of the initial and final velocities, the police car must be going twice as fast as the speeder when they meet again.

Figure 2-12 shows both motions plotted on the same graph. Sketching such a graph is often helpful in understanding problems involving two or more one-dimensional motions.

EXERCISE An astronaut making repairs on a space station accidentally lets go of a wrench, which drifts off at a speed of 1.1 m/s. The astronaut immediately fires a rocket-powered

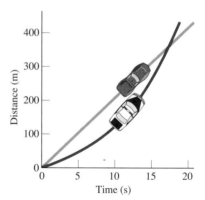

FIGURE 2-12 Position versus time for the speeder and police car of Example 2-8.

backpack and heads after the wrench with a constant acceleration of 0.64 m/s². (a) How far does the wrench travel before it's retrieved? (b) How fast is the astronaut going on reaching the wrench?

Answers: (a) 3.8 m; (b) 2.2 m/s

Some problems similar to Example 2-8: 58–61, 96

●

2-6 THE CONSTANT ACCELERATION OF GRAVITY

When you drop an object, it falls at an ever-increasing rate (Fig. 2-13), accelerating because of gravity—a phenomenon we will study further in Chapter 9. Experimentally, we find that the acceleration remains constant for objects falling near the surface of the Earth, and furthermore that this constant acceleration has the same value for all objects, independent of mass or other properties. The rate at which the speed of a falling object increases—the **acceleration of gravity**—is approximately 9.8 m/s² near Earth's surface. (A more exact value is given in Appendix E, along with other properties of Earth and other planets.)

The acceleration of gravity applies strictly only to objects in **free fall**—that is, objects under the influence of gravity alone. The presence of other forces, in particular air resistance, may alter dramatically the statement that all objects fall with the same constant acceleration. Try dropping a flat piece of paper and a rock, for example. Their behavior is radically different. But if you crumple the paper, lessening its air resistance, it falls with an acceleration closer to that of the rock. An experiment illustrating that all objects fall with the same acceleration was carried out by astronauts on the moon in 1971, when they showed that a feather and a hammer fell together at the airless lunar surface. The acceleration of gravity on the moon is different from that on Earth, but the crucial concept is that all objects at the same location experience the same gravitational acceleration.

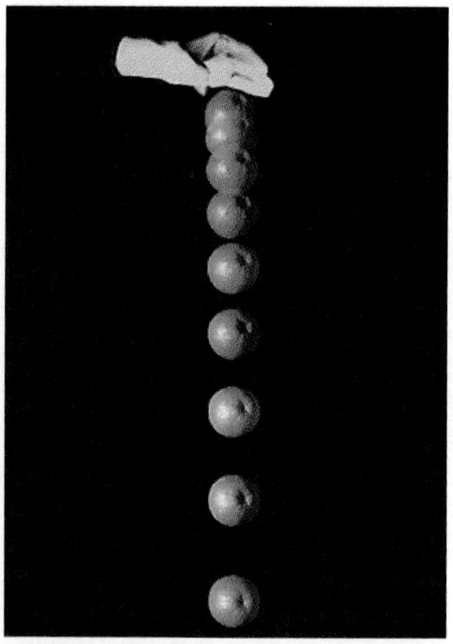

FIGURE 2-13 Strobe photo of a falling orange. Successive images are farther apart, showing that the orange is accelerating.

Galileo is reputed to have shown that all objects fall with the same acceleration by dropping objects off the Leaning Tower of Pisa in about 1600. There is no evidence that the Tower of Pisa played a role, but it is known that Galileo carried out careful experiments by rolling balls down inclined planes. The gradual inclines "diluted" the effect of gravity, allowing Galileo to make more precise measurements. We now study vertically falling bodies with strobe lights or video cameras, equipment that was not available to Galileo.

That the acceleration of gravity near Earth's surface is a constant 9.8 m/s^2 means that objects moving in vertical paths change velocity at the rate of 9.8 m/s in each second. Moving upward, an object slows down until it stops and reverses direction; moving downward, its speed increases. These changes in velocity, for an object initially moving upward, were shown earlier in Fig. 2-8.

The magnitude of the gravitational acceleration is given the symbol g. Remember that g is relatively constant near Earth's surface but is not fundamentally a constant quantity. If we define the positive y direction as vertically upward, then the downward acceleration enters Equations 2-7, 2-10, and 2-11 with a minus sign, and these equations become

$$v = v_0 - gt \tag{2-12}$$

$$y = y_0 + v_0 t - \tfrac{1}{2} g t^2 \tag{2-13}$$

$$v^2 = v_0^2 - 2g(y - y_0), \tag{2-14}$$

where we have replaced x by y since we are describing motion along the vertical direction.

● **EXAMPLE 2-9** CLIFF DIVER

A diver drops from a 10-m-high cliff (Fig. 2-14). At what speed does he enter the water? Give the answer in both meters per second and miles per hour. How long is he in the air? If a photographer's motor drive can take three frames per second, how many frames cover the dive?

Solution
To find the speed, given the distance, we use Equation 2-14. Since the diver starts from rest, $v_0 = 0$, and the equation becomes

$$v^2 = -2g(y - y_0).$$

Calling the water $y = 0$, the initial height is $y_0 = 10$ m. Then we have

$$v^2 = (-2)(9.8 \text{ m/s}^2)(0 \text{ m} - 10 \text{ m}) = 196 \text{ m}^2/\text{s}^2.$$

Taking the square root, we find that the speed—the magnitude of the velocity—at impact with the water is 14 m/s. (The actual velocity is -14 m/s, with the negative sign indicating the diver's downward motion.) Converting this result to miles per hour gives

$$|v| = (14 \text{ m/s})(3600 \text{ s/h})(0.001 \text{ km/m})(0.62 \text{ mi/km})$$

$$= 31 \text{ mi/h}.$$

FIGURE 2-14 A cliff dive (Example 2-9).

Note how the units cancel. Appendix C is useful for conversions such as this one.

Knowing the initial and final velocities, we can use Equation 2-12 to find out how long the diver is in the air. Solving that equation for time t gives

$$t = \frac{v_0 - v}{g} = \frac{0 \text{ m/s} - (-14 \text{ m/s})}{9.8 \text{ m/s}^2} = 1.4 \text{ s}.$$

Note here that we were careful to use the negative sign with v, indicating a downward velocity. Alternatively, we could have obtained the time from Equation 2-13, knowing the distance and acceleration. Solving for t^2 and setting $y_0 = 10$ m (top of cliff), $y = 0$ (water), and $v_0 = 0$, we have

$$t = \sqrt{\frac{2(y_0 - y)}{g}} = \sqrt{\frac{2(10 \text{ m} - 0 \text{ m})}{9.8 \text{ m/s}^2}} = 1.4 \text{ s}.$$

At three frames per second, the diver will appear on $(1.4)(3)$, or four complete frames after the dive starts.

●

● **EXAMPLE 2-10** A VOLCANIC ERUPTION

The fountains of lava at the volcanic eruption shown in Fig. 2-15 are so hot that geologists can't get close. But they can watch chunks of lava that have been thrown upward, and time their fall. If lava chunks at the peak of the fountain take 5.0 s to fall, how high are the fountains?

Solution
Here we know the time and the acceleration, so we can calculate the distance from Equation 2-13. Let $y_0 = h$, our unknown height at the peak of the fountain, and take $y = 0$ at the ground. With $v_0 = 0$ at the peak, Equation 2-13 becomes

$$0 = h - \tfrac{1}{2}gt^2,$$

so

$$h = \tfrac{1}{2}gt^2 = \tfrac{1}{2}(9.8 \text{ m/s}^2)(5.0 \text{ s})^2 = 120 \text{ m}.$$

EXERCISE Dropped by a seagull, a clamshell crashes down on a seaside rock at 8.3 m/s. (a) From what height did the gull drop the rock? (b) How long was it falling?

Answers: (a) 3.5 m; (b) 0.85 s

Some problems similar to Examples 2-9 and 2-10: 62, 67, 72, 93

FIGURE 2-15 A lava fountain at the 1983 eruption of Kilauea volcano in Hawaii. ●

So far, we have treated only examples of objects that are actually moving downward. In that case, the magnitude of the velocity increases because the acceleration is in the same direction as the velocity. But the acceleration of an object subject only to gravity is 9.8 m/s² *downward* no matter what the direction of its motion. When you throw a ball straight up, for example, it is accelerating *downward* even while moving *upward*. Why? Because its upward velocity is *decreasing*—that is, the *change* in velocity is negative. Equations 2-12 to 2-14 continue to describe the situation no matter what the direction of this initial velocity.

● EXAMPLE 2-11   UP AND DOWN

You toss a ball straight up with an initial speed of 7.3 m/s. When it leaves your hand, it is 1.5 m above the floor. Use Equation 2-13 to determine when it hits the floor. Find also the maximum height it reaches and its speed when it passes your hand on the way down.

Solution

If we choose $y = 0$ at the floor, then $y_0 = 1.5$ m and $v_0 = +7.3$ m/s. To find out when the ball hits the floor, we must ask of Equation 2-13: At what time t is the ball at $y = 0$? That is, we want t such that

$$0 = y_0 + v_0 t - \tfrac{1}{2} g t^2 .$$

Using the quadratic formula (see Appendix A), we have

$$t = \frac{v_0 \pm \sqrt{v_0^2 + 2 y_0 g}}{g} .$$

The quantity under the square root is

$$v_0^2 + 2 y_0 g = (7.3 \text{ m/s})^2 + 2(1.5 \text{ m})(9.8 \text{ m/s}^2)$$

$$= 82.7 \text{ m}^2/\text{s}^2 ,$$

so

$$t = \frac{7.3 \text{ m/s} \pm \sqrt{82.7 \text{ m}^2/\text{s}^2}}{9.8 \text{ m/s}^2} = 1.7 \text{ s or } -0.18 \text{ s} .$$

The first answer, 1.7 s, is reasonable, but what about the second answer, -0.18 s? Remember that Equations 2-12 to 2-14 "assume" that gravity alone provides the acceleration. The equations don't "know" anything about the upward acceleration of your hand or about the ball stopping when it hits the floor; they "think" the ball always undergoes a downward acceleration g. Our negative answer means simply this: had our hand not been involved, a ball traveling upward at 7.3 m/s when it was 1.5 m off the floor would have been on the floor 0.18 s *earlier*. The positive answer, the one we expected, says that the ball will again hit the floor 1.7 s *later*.

At the peak of the ball's flight, its instantaneous velocity is zero since it is moving neither up nor down at that point. Setting $v^2 = 0$ in Equation 2-14, we can then solve for the peak height y:

$$0 = v_0^2 - 2g(y - y_0) ,$$

so

$$y = \frac{2 g y_0 + v_0^2}{2g} = y_0 + \frac{v_0^2}{2g} = 1.5 \text{ m} + \frac{(7.3 \text{ m/s})^2}{(2)(9.8 \text{ m/s}^2)}$$

$$= 4.2 \text{ m}$$

above the floor.

To determine the speed when the ball reaches 1.5 m on the way down, we can set $y = 1.5$ m in Equation 2-14 and solve for v. Since $y_0 = 1.5$ m, the term $y - y_0 = 0$, and Equation 2-14

becomes simply $v^2 = v_0^2$. There are two possibilities: $v = v_0$ and $v = -v_0$. Again, our formulation of the problem does not really ask only about the downward flight; it simply asks for the velocity *whenever* the height is 1.5 m. The positive answer corresponds to the initial upward velocity and the negative answer to the velocity on the downward flight. In deriving this result, we have demonstrated the important point that an object thrown upward returns to its initial height with the same speed with which it left that height. (This result is exactly true only in the absence of air resistance.) Figure 2-16 is a plot of the motion in this example.

TIP   Think About Multiple Answers

Frequently you'll encounter physics problems where the mathematics gives you more than one answer. Before you discard one of the answers, think about what each means. Sometimes an extraneous answer won't be consistent with assumptions you made in setting up the problem. But often, as in this example, both answers have physical meaning—even though both may not be what you're looking for. Often you'll gain more insight into the physics of your problem if you consider the meaning of multiple answers.

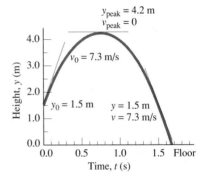

FIGURE 2-16   Height-versus-time curve for the ball of Example 2-11. The ball's trajectory in space is straight up and down; the curve shown is not the trajectory but rather the ball's height as a function of time. Tangents to the curve give the instantaneous velocities: ± 7.3 m/s when the ball is at 1.5 m height and zero at the peak.

EXERCISE   A student tosses a book straight up and just outside her dormitory window, with an initial velocity of 3.1 m/s. How much time passes before she has to catch it? If she misses, how long after it was thrown will it hit the ground, 8.6 m below the window?

Answers: (a) 0.63 s; (b) 1.68 s

Some problems similar to Example 2-11:   72, 73, 83, 84   ●

CHAPTER SYNOPSIS

Summary

1. **Kinematics** is the study of motion. In this chapter we limit ourselves to the one-dimensional motion of **particles**—objects that have no extent and therefore cannot rotate or vibrate. Real objects approximate particles to the extent we can ignore their rotation, vibration, or other internal motions.
2. **Displacement** is the change in an object's position. In one dimension, displacement is characterized by a number whose sign describes which of the two possible directions is involved.
3. **Velocity** is the rate of change of position. The **average velocity** $\bar{v}$ is the displacement Δx divided by the time interval Δt over which it occurred:

$$\bar{v} = \frac{\Delta x}{\Delta t}.$$

4. To find the **instantaneous velocity**—the velocity at a specific instant of time—we take the limit of the average velocity as the time interval becomes arbitrarily small. Mathematically, this process defines the **derivative** dx/dt:

$$v = \lim_{\Delta t \to 0} \frac{\Delta x}{\Delta t} = \frac{dx}{dt}.$$

When position can be written in the form $x = bt^n$, the derivative dx/dt, or v, is

$$v = \frac{dx}{dt} = bnt^{n-1}.$$

Graphically, the instantaneous velocity—the derivative of position with respect to time—is the slope of the position-versus-time curve.

5. **Acceleration** is the rate of change of velocity. If Δv is the change in velocity in a time interval Δt, then the **average acceleration** over that interval is

$$\bar{a} = \frac{\Delta v}{\Delta t}$$

Like the instantaneous velocity, the **instantaneous acceleration** is defined as the average acceleration taken in the limit of arbitrarily small time intervals Δt:

$$a = \lim_{\Delta t \to 0} \frac{\Delta v}{\Delta t} = \frac{\Delta v}{dt}.$$

The direction of the acceleration—indicated by algebraic sign—is the same as that of the velocity when speed is increasing and opposite when speed is decreasing.

6. For **constant acceleration**, position is a quadratic function of time. The following equations relate displacement, time,

velocity, and acceleration in one-dimensional motion with constant acceleration:

$$v = v_0 + at$$
$$x = x_0 + v_0 t + \tfrac{1}{2}at^2$$
$$\bar{v} = \tfrac{1}{2}(v_0 + v)$$
$$x - x_0 = \tfrac{1}{2}(v_0 + v)t$$
$$v^2 = v_0^2 + 2a(x - x_0).$$

7. The **acceleration of gravity** is the same for all objects at the same location. Near Earth's surface, the acceleration of gravity is directed downward and is essentially constant at 9.8 m/s².

Terms You Should Understand

(Pairs are closely related terms whose distinction is important; number in parentheses is chapter section where term first appears.)

kinematics (introduction)
particle (introduction)
distance, displacement (2-1)
speed, velocity (2-1)
average, instantaneous (2-2)
derivative (2-2)
acceleration (2-3)

Symbols You Should Recognize

$\bar{v}$ (2-1)
$\Delta, \Delta x, \Delta t$ (2-1)
$\frac{dx}{dt}$ (2-2)
$\bar{a}$ (2-3)
Δv (2-3)
$\frac{dv}{dt}$ (2-3)

Problems You Should Be Able to Solve

calculating time, displacement, or average velocity from the other two (2-1)
taking a derivative to get instantaneous velocity, given position as a function of time (2-2, possibly in consultation with Appendix A)
calculating time, velocity change, or average acceleration from the other two

taking derivatives to get instantaneous acceleration, given either position or velocity as a function of time (2-3, possibly in consultation with Appendix A)

constructing plots of velocity and acceleration versus from position-versus-time curves (2-3)

solving any problem involving the kinematics of one-dimensional motion (2-5)

solving problems involving vertical motion under the influence of uniform gravity (2-6)

Limitations to Keep in Mind

This chapter covers only motion in one dimension; changes of direction other than reversals along a single line have not yet been treated.

Many of the equations derived apply only to cases of constant acceleration.

Equations for motion under the influence of gravity apply only near Earth's surface (for heights much less than Earth's radius), and do not account for air resistance.

QUESTIONS

1. Why can we consider Earth to be a particle in some applications but not in others?

2. Give three common examples of velocity units.

3. What is the difference between your displacement and the total distance you travel between the time you wake up and the time you go back to bed on a typical day?

4. You are driving straight at a steady 80 km/h, but stop a while for a picnic lunch. How does the stop affect your average velocity?

5. Does a speedometer measure speed or velocity?

6. You check your odometer at the beginning of a day's driving, and again at the end. Under what conditions would the difference between the two readings represent your displacement?

7. In the previous question, under what conditions would the difference in odometer readings, divided by the total time, equal the magnitude of your average velocity?

8. Consider two possible definitions of average speed: (a) average speed is the average of the values of the instantaneous speed over a time interval; (b) average speed is the magnitude of the average velocity. Are these definitions equivalent? Give examples to demonstrate your conclusion.

9. Is it possible to be at the position $x = 0$ and still be moving?

10. Is it possible to have zero velocity and still be accelerating?

11. Is it possible to have zero average velocity over a 10-s interval and still be accelerating during the interval? If so, give an example. If not, why not?

12. Is it possible to have zero instantaneous velocity at all times in a 10-s interval and still be accelerating during the interval? If so, give an example. If not, why not?

13. Suppose your car's brakes provide nearly constant deceleration, independent of speed. If you double your initial speed, what does this do to the time it will take to stop your car? What does it do to the distance the car travels while stopping?

14. If you know the initial velocity v_0 and the initial and final heights y_0 and y, you can use Equation 2-13 to solve for the time t when the object will be at height y. But the equation is quadratic in t, so you'll get two answers. Physically, why is this?

15. Starting from rest, an object undergoes an acceleration given by $a = bt$, where t is time and b is a constant. Can you use the expression bt for a in Equation 2-10 to predict the object's position as a function of time? Why or why not?

16. In which of the velocity-versus-time graphs shown in Fig. 2-17 would the average velocity over the interval shown equal the average of the velocities at the ends of the interval?

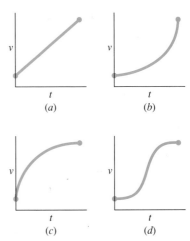

FIGURE 2-17 Question 16.

17. What would the third derivative of position signify? (The third derivative is sometimes called the "jerk." It's what can make you sick on an amusement park ride.)

18. Standing on a roof, you simultaneously throw one ball straight up and drop another from rest. Which hits the ground first? Which hits the ground moving faster?

19. Can the height of the lava fountains in Fig. 2-14 be used to estimate the masses of the lava chunks? Why or why not?

20. If you travel in a straight line at 50 km/h for 1 h and at 100 km/h for another hour, is your average velocity 75 km/h? If not, is it more or less?

21. If you travel in a straight line at 50 km/h for 50 km, and then at 100 km/h for another 50 km, is your average velocity 75 km/h? If not, is it more or less?

22. If you drop one rock from rest, and throw another down with velocity v_0, is the difference in their velocities when they hit the ground more than, less than, or equal to v_0? Don't apply any equations here; just think about how long each rock experiences the acceleration of gravity.

PROBLEMS

Section 2-1 Distance, Time, Speed, and Velocity

1. In 1994 Leroy Burrell of the United States set a world record in the 100-m dash, with a time of 9.84 s. What was his average speed?

2. When races in a track meet are timed manually, timers start their watches when they see smoke from the starting gun, rather than when they hear the gun. How much error is introduced in timing a 200-m dash over a straight track if the watch is started on the sound rather than the smoke? The speed of sound is about 340 m/s.

3. In 1992 Willie Mtolo of South Africa won the New York City Marathon, completing the 26-mi, 385-yd course in 2 h 9 min 29 s. What was Mtolo's average speed, in meters per second?

4. Human nerve impulses travel at about 10^2 m/s. Estimate the minimum time that must elapse between the time you perceive a stalled car in front of you and the time you can activate the muscles in your leg to brake your car. (Your actual "reaction time" is much longer than this estimate.) Moving at 90 km/h, how far would your car travel in this time?

5. Starting from home, you bicycle 24 km north in 2.5 h, then turn around and pedal straight home in 1.5 h. What are your (a) displacement at the end of the first 2.5 h, (b) average velocity over the first 2.5 h, (c) average velocity for the homeward leg of the trip, (d) displacement for the entire trip, and (e) average velocity for the entire trip?

6. The European Space Agency's Giotto spacecraft encountered Halley's Comet in March 1986, when the comet was 93 million km from Earth. How long did it take radio signals (travelling at the speed of light) to reach Earth from Giotto?

7. The triathlon is a grueling event in which contestants first swim 2.4 mi, then bicycle 112 mi, then complete a 26-mi, 385-yd marathon on foot. Mark Allen won the 1990 Ironman Triathlon World Championship in a time of 8 h 28 min 17 s. What was Allen's average speed?

8. (a) Find a value, good to one significant figure, for the speed of light in feet per nanosecond (ft/ns) (1 ns = 10^{-9} s). (b) Electrical signals in wires travel at about half the speed of light. What is the maximum possible separation between a computer's central processing unit and its memory if the central processor is to be able to get a signal to memory requesting data, and have the data return, all in 8 ns?

9. You allow yourself 40 min to drive 25 mi to the airport, but are caught in heavy traffic and average only 20 mi/h for the first 15 min. What must your average speed be on the rest of the trip if you are to get there on time?

10. Taking Earth's orbit to be a circle of radius 1.5×10^8 km, determine the speed of Earth's orbital motion in (a) meters per second and (b) miles per second.

11. What is the conversion factor from meters per second to miles per hour?

12. If the average American driver goes 5000 mi each year on interstate highways, how much more time did the average driver spend on interstate highways each year as a result of the 1974 drop in the speed limit from 70 mi/h to 55 mi/h?

13. A fast base runner can get from first to second base in 3.4 s. If he leaves first base as the pitcher throws a 90 mi/h fastball the 61-ft distance to the catcher, and if the catcher takes 0.45 s to catch and rethrow the ball, how fast does the catcher have to throw the ball to second base to make an out? Home plate to second base is the diagonal of a square 90 ft on a side.

14. Despite the fact that jet airplanes fly at about 1000 km/h, plane schedules and connections are such that the 4800-km trip from Burlington, Vermont, to San Francisco ends up taking about 11 h. (a) What is the average speed of such a trip? (b) How much time is spent on the ground, assuming that the actual distance covered by the several aircraft involved in connecting flights is 6700-km and that the planes maintain a steady 960 km/h in flight?

15. If you drove the 4600 km from coast to coast of the United States at 55 mi/h (88 km/h), stopping an average of 30 min for rest and refueling after every 2 h of driving, (a) how long would it take? (b) What would be your average velocity for the entire trip?

16. I can run 9.0 m/s, 20% faster than my kid brother. How much head start should I give him in order to have a tie race over 100 m?

17. A jetliner leaves San Francisco for New York, 4600 km away. With a strong tailwind, its speed is 1100 km/h. At the same time, a second jet leaves New York for San Francisco. Flying into the wind, it makes only 700 km/h. When and where do the two planes pass each other?

18. Figure 2-18 shows the position of an object as a function of time. Determine the average velocity for (a) the first 2 s; (b) the first 4 s; (c) the first 6 s; (d) the interval from 3 s to 4 s.

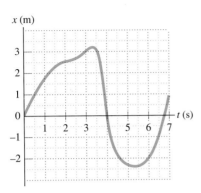

x (m)

FIGURE 2-18   Problem 18.

Section 2-2 Motion in Short Time Intervals

19. On a single graph, plot distance versus time for the two trips from Houston to Des Moines described on page 24. For each trip, identify graphically the average velocity and, for each segment of the trip, the instantaneous velocity.

20. For the motion plotted in Fig. 2-19, estimate (a) the greatest velocity in the positive x direction; (b) the greatest velocity in the negative x direction; (c) any times when the object is instantaneously at rest; and (d) the average velocity over the interval shown.

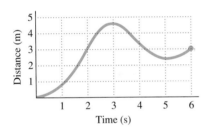

FIGURE 2-19   Problem 20.

21. Figure 2-20 shows the position of an object as a function of time. From the graph, determine the instantaneous velocity at (a) 1.0 s; (b) 2.0 s; (c) 3.0 s; and (d) 4.5 s. (e) What is the average velocity over the interval shown?

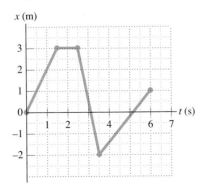

x (m)

FIGURE 2-20   Problem 21.

22. The position of an object as a function of time is given by $x = bt + ct^3$, where $b = 1.50$ m/s and $c = 0.640$ m/s³. To study the limiting process leading to the definition of instantaneous velocity, calculate the average velocity of the object over time intervals from (a) 1.00 s to 3.00 s; (b) 1.50 s to 2.50 s; (c) 1.95 s to 2.05 s. (d) Obtain the instantaneous velocity as a function of time by differentiating, and compare its value at 2 s with your average velocities.

23. A model rocket is launched straight upward; its altitude y as a function of time is given by $y = bt - ct^2$, where $b = 68$ m/s, $c = 4.9$ m/s², t is the time in seconds, and y is in meters. (a) Use differentiation to find a general expression for the rocket's velocity as a function of time. (b) When is the velocity zero?

24. The position of an object as a function of time is given by $x = bt^3$, where b is a constant. Find an expression for the instantaneous velocity as a function of time, and show that the average velocity over the interval from $t = 0$ to any time t is one-third of the instantaneous velocity at t.

25. The position of an object is given by $x = bt^3 - ct^2 + dt$, with x in meters and t in seconds. The constants b, c, and d are $b = 3.0$ m/s³, $c = 8.0$ m/s², and $d = 1.0$ m/s. (a) Find all times when the object is at position $x = 0$. (b) Determine a general expression for the instantaneous velocity as a function of time, and from it find (c) the initial velocity and (d) all times when the object is instantaneously at rest. (e) Graph the object's position as a function of time, and identify on the graph the quantities you found in (a) to (d).

26. In a drag race, the position of a car as a function of time is given by $x = bt^2$, with $b = 2.000$ m/s². In an attempt to determine the car's velocity midway down a 400-m track, two observers stand 20 m on either side of the 200-m mark and note the time when the car passes them. (a) What value do the two observers compute for the car's velocity? Give your answer to four significant figures. (b) By what percentage does this observed value differ from the actual instantaneous value at $x = 200$ m?

Section 2-3 Acceleration

27. A car roars away from a red light. Half a minute later, it screeches to a halt at a second light, one-fourth mile from the first. What is its average acceleration between the times it is stopped at the two lights?

28. Starting from rest, a subway train first accelerates to 25 m/s, then begins to brake. Forty-eight seconds after starting, it is moving at 17 m/s. What is its average acceleration in this 48-s interval?

29. The 1986 explosion of the space shuttle Challenger occurred 74 s after liftoff. At that time, mission control reported a shuttle speed of 2900 ft/s (880 m/s). What was the Challenger's average acceleration during its brief flight? Compare with the acceleration of gravity.

30. An egg drops from a second-story window, taking 1.12 s to fall and reaching a speed of 11.0 m/s just before hitting the ground. On contact with the ground, the egg stops completely in 0.131 s. Calculate the average magnitudes of its acceleration while falling and of its deceleration while stopping.

31. An airplane's takeoff speed is 320 km/h. If its average acceleration is 2.9 m/s^2, how long is it on the runway after starting its takeoff roll?

32. A car can accelerate from rest to 60 miles per hour in 7.1 s. What is its average acceleration, in m/s^2?

33. Your plane reaches its takeoff runway and then holds for 4.0 min because of air-traffic congestion. The plane then heads down the runway with an average acceleration of 3.6 m/s^2. It is airborne 35 s later. What are (a) its takeoff speed and (b) its average acceleration from the time it reaches the takeoff runway until it's airborne?

34. Under the influence of a radio wave, an electron in an antenna undergoes back-and-forth motion whose velocity as a function of time is described by Fig. 2-21. From the graph, estimate the electron's maximum acceleration.

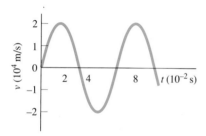

FIGURE 2-21 Problem 34.

35. Determine the instantaneous acceleration as a function of time for the motion in Problem 25.

36. The position of an object is given by $x = bt^3$, where x is in meters, t in seconds, and where the constant b is 1.5 m/s^3. Determine (a) the instantaneous velocity and (b) the instantaneous acceleration at the end of 2.5 s. Find (c) the average velocity and (d) the average acceleration during the first 2.5 s.

Section 2-4 Constant Acceleration

37. A car accelerates from rest to 25 m/s in 8.0 s. Determine the distance it travels in two ways: (a) by multiplying the average velocity given in Equation 2-8 by the time and (b) by calculating the acceleration from Equation 2-7 and using the result in Equation 2-10.

38. Differentiate both sides of Equation 2-10, and show that you get Equation 2-7.

39. If you square Equation 2-7, you'll have an expression for v^2. Equation 2-11 also gives an expression for v^2. Equate

the two expressions for v^2, and show that the resulting equation reduces to Equation 2-10.

Section 2-5 Using the Equations of Motion

40. In 1982, drag racer Gary Beck set a new record, traveling a quarter mile from rest in 5.484 s. Assuming constant acceleration, determine (a) Beck's acceleration, (b) the maximum speed achieved, and (c) the time to reach 55 mi/h.

41. A rocket rises with constant acceleration to an altitude of 85 km, at which point its speed is 2.8 km/s. (a) What is its acceleration? (b) How long does the ascent take?

42. On packed snow, use of computerized anti-lock brakes can reduce the stopping distance for a car by 55%. By what percentage is the stopping *time* reduced?

43. Starting from rest, a car accelerates at a constant rate, reaching 88 km/h in 12 s. (a) What is its acceleration? (b) How far does it go in this time?

44. If you are running at 6.0 m/s and decelerate at 2.0 m/s^2, (a) how long will it take you to stop? (b) How far will you go while decelerating?

45. A car moving initially at 50 mi/h begins decelerating at a constant rate 100 ft short of a stoplight. If the car comes to a full stop just at the light, what is the magnitude of its deceleration?

46. In an x-ray tube, electrons are accelerated to a velocity of 10^8 m/s, then slammed into a tungsten target. When they collide with the tungsten atoms, the electrons undergo rapid deceleration, producing x rays in the process. If the stopping time for an electron is on the order of 10^{-19} s, approximately how far does an electron move while decelerating? Assume constant deceleration.

47. A particle leaves its initial position x_0 at time $t = 0$, moving in the positive x direction with speed v_0, but undergoing acceleration of magnitude a in the negative x direction. Find expressions for (a) the time when it returns to the position x_0 and (b) its speed when it passes that point.

48. The Barringer meteor crater in northern Arizona is 180 m deep and 1.2 km in diameter. The fragments of the meteor lie just below the bottom of the crater. If these fragments decelerated at a constant rate of 4×10^5 m/s^2 as they ploughed through the Earth in forming the crater, what was the speed of the meteor's impact at Earth's surface?

49. A gazelle accelerates from rest at 4.1 m/s^2 over a distance of 60 m to outrun a predator. What is its final speed?

50. A frog's tongue flicks out 10 cm in 0.1 s to catch a fly. If it accelerates at a constant rate during this time, what are (a) its average velocity and (b) its acceleration?

51. Amtrak's 20th-Century Limited is en route from Chicago to New York at 110 km/h, when the engineer spots a cow on the track. The train brakes to a halt in 1.2 min with constant deceleration, stopping just in front of the cow. (a) What is the magnitude of the train's acceleration? (b) What is the direction of the acceleration? (c) How far was

the train from the cow when the engineer first applied the brakes?

52. A jetliner touches down at 220 km/h, reverses its engines to provide braking, and comes to a halt 29 s later. What is the shortest runway on which this aircraft can land, assuming constant deceleration starting at touchdown?

53. A flea extends its 0.80-mm-long legs, propelling itself 10 cm into the air. Assuming that the legs provide constant acceleration as they extend from completely folded to completely extended, find the magnitude of that acceleration. Compare with the acceleration of gravity at Earth's surface.

54. A motorist suddenly notices a stalled car and slams on the brakes, decelerating at the rate of 6.3 m/s^2. Unfortunately this isn't good enough, and a collision ensues. From the damage sustained, police estimate that the car was moving at 18 km/h at the time of the collision. They also measure skid marks 34 m long. (a) How fast was the motorist going when the brakes were first applied? (b) How much time elapsed from the initial braking to the collision?

55. The maximum acceleration that a human being can survive even for a short time is about 200g. In a highway accident, a car moving at 88 km/h slams into a stalled truck. The front end of the car is squashed by 80 cm on impact. If the deceleration during the collision is constant, will a passenger wearing a seatbelt survive?

56. A racing car undergoing constant acceleration covers 120 m in 2.7 s. (a) If it is moving at 53 m/s at the end of this interval, what was its speed at the beginning of the interval? (b) How far did it travel *from rest* to the end of the 120-m distance?

57. The maximum deceleration of a car on a dry road is about 8 m/s^2. If two cars are moving head-on toward each other at 88 km/h (55 mi/h), and their drivers apply their brakes when they are 85 m apart, will they collide? If so, at what relative speed? If not, how far apart will they be when they stop? On the same graph, plot distance versus time for both cars.

58. George, a physics student, leaves his dormitory at a speed of 1.2 m/s, heading for the physics building 95 m away. Just as he leaves his dorm, Amy, another physics student, leaves the physics building and heads toward George at a steady 1.6 m/s. George immediately spots her and begins accelerating at 0.075 m/s^2. Where and when do the two meet? Plot position-versus-time curves for both students on a single graph. *Hint:* Be sure to set the algebraic signs of positions, velocities, and accelerations consistently.

59. After 35 minutes of running, at the 9-km point in a 10-km race, you find yourself 100 m behind the leader and moving at the same speed. What should your acceleration be if you are to catch up by the finish line? Assume that the leader maintains constant speed throughout the entire race.

60. You're speeding at 85 km/h when you notice that you're only 10 m behind the car in front of you, which is moving at the legal speed limit of 60 km/h. You slam on your brakes, and your car decelerates at 4.2 m/s^2. Assuming the car in front of you continues at constant speed, will you collide? If so, at what relative speed? If not, what will be the distance between the cars at their closest approach?

61. Repeat the preceding problem, now assuming your initial speed is 95 km/h.

Section 2-6 The Constant Acceleration of Gravity

62. You drop a rock into a deep well and 2.7 s later hear the splash. How far down is the water? Neglect the travel time of the sound.

63. Your friend is sitting 6.5 m above you in a tree branch. How fast should you throw an apple so that it just reaches her?

64. A model rocket leaves the ground, heading straight up at 49 m/s. (a) What is its maximum altitude? What are its speed and altitude at (b) 1 s; (c) 4 s; (d) 7 s?

65. A foul ball leaves the bat going straight upward at 23 m/s. (a) How high does it rise? (b) How long is it in the air? Neglect the distance between the bat and the ground.

66. A Frisbee is lodged in a tree branch, 6.5 m above the ground. A rock thrown from below must be going at least 3 m/s to dislodge the Frisbee. How fast must such a rock be thrown upward, if it leaves the thrower's hand 1.3 m above the ground?

67. Space pirates kidnap an earthling and hold him imprisoned on one of the planets of the solar system. With nothing else to do, the prisoner amuses himself by dropping his watch from eye level (170 cm) to the floor. He observes that the watch takes 0.62 s to fall. On what planet is he being held? *Hint:* Consult Appendix E.

68. The earliest attempt to land instruments on the moon involved Ranger spacecraft that were to release instrument capsules about 11 km above the lunar surface. After a retrorocket firing, the capsules were to fall freely to the moon. Surrounded by a balsawood sphere, the instrument capsule was designed to withstand an impact speed equivalent to a free fall from 150 m above Earth's surface. From what height above the lunar surface could the capsule drop, assuming the retrorocket slowed it to zero speed at that height? (In fact, none of the Ranger instrument packages ever made the intended cushioned landing.)

69. A falling object travels one-fourth of its total distance in the last second of its fall. From what height was it dropped?

70. The defenders of a castle throw rocks down on their attackers from a 15-m-high wall. If the rocks are thrown with an initial speed of 10 m/s, how much faster are they moving when they hit the ground than if they were simply dropped?

71. A kingfisher is 30 m above a lake when it accidentally drops the fish it is carrying. A second kingfisher 5 m above the first dives toward the falling fish. What initial speed should it have if it is to reach the fish before the fish hits the water?

72. Two divers jump from a 3.00-m platform. One jumps upward at 1.80 m/s, and the second steps off the platform as the first passes it on the way down. (a) What are their speeds as they hit the water? (b) Which hits the water first and by how much?

73. A balloon is rising at 10 m/s when its passenger throws up a ball at 12 m/s. How much later does the passenger catch the ball?

74. A conveyer belt moves horizontally at 80 cm/s, carrying empty shoe boxes. Every 3 s, a pair of shoes is dropped from a chute 1.7 m above the belt. (a) How far apart should the boxes be spaced? (b) At the instant a pair of shoes drops, where should a box be in relation to a point directly below the chute?

Paired Problems

(Both problems in a pair involve the same principles and techniques. If you can get the first problem, you should be able to solve the second one.)

75. You drive 14 km to the next town, maintaining a speed of 50 km/h except for a stop lasting 4.1 min at a red light. You shop for 20 min, then head back toward your starting point at a steady 70 km/h. You stop at a gas station 4.4 km beyond the town. What are (a) your average speed and (b) the magnitude of your average velocity between your starting point and the gas station?

76. The itsy-bitsy spider climbed 3.7 m up the water spout, starting at the bottom, in 6.2 minutes. She paused at the top for a 5.0-min rest. Then down came the rain, and washed the spider out—all the way to the ground, 0.41 m below the bottom of the spout, in 2.8 s. What were the spider's (a) average speed and (b) magnitude of her average velocity for the entire adventure?

77. A skier starts from rest, and heads downslope with a constant acceleration of 2.3 m/s². How long does it take her to go 15 m, and what is her speed at that point?

78. Landing on the moon, a spacecraft fires its retrorockets and comes to a complete stop just 12 m above the lunar surface. It then drops freely to the surface. How long does it take to fall, and what is its impact speed? (Consult Appendix E for the moon's gravitational acceleration.)

79. A frustrated student drops a book out of his dormitory window, releasing it from rest. After falling 2.3 m, it passes the top of a 1.5-m high window on a lower floor. How long does it take to cross the window?

80. Launched from the ground, a rocket accelerates vertically upward at 4.6 m/s². It passes through a band of clouds 5.3 km thick, extending upward from an altitude of 1.9 km. How long is it in the clouds?

81. A subway train is traveling at 80 km/h when it approaches a slower train 50 m ahead traveling in the same direction at 25 km/h. If the faster train begins decelerating at 2.1 m/s², while the slower train continues at constant speed, how soon and at what relative speed will they collide?

82. A parachutist is drifting vertically downward at a constant 11 m/s. An airplane passes a mere 8.4 m directly above the parachutist, and the pilot throws an orange straight downward at 2.2 m/s. How much later do the orange and parachutist meet, and what is their relative speed?

83. You toss a hammer over the 3.7-m high wall of a construction site, starting your throw at a height of 1.2 m above the sidewalk. On the other side of the wall, the hammer falls to the bottom of an excavation 7.9 m below the sidewalk (see Fig. 2-22). (a) What is the minimum speed at which you must throw the hammer for it to clear the wall? (b) Assuming it's thrown with the speed given in part (a), when will it hit the bottom of the excavation?

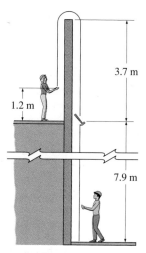

3.7 m

1.2 m

7.9 m

FIGURE 2-22 Problem 83.

84. You toss a book into your dorm room, just clearing a windowsill 4.2 m above the ground. (a) If the book leaves your hand 1.5 m above the ground, how fast must it be going to clear the sill? (b) How long after it leaves your hand will it hit the floor, 0.87 m below the windowsill?

Supplementary Problems

85. Compute and rank order the average speeds, in m/s, associated with the following sporting events: (a) the 1990 Tour de France, a 2121-mi bicycle race won by Greg Lemond in 90 h 43 min 20 s; (b) the 1987 world indoor record women's 200-m dash, by Heike Dreschler in 22.27 s; (c) the 1990 world record men's indoor 400-m run, by Danny Everett in 45.04 s; (d) the 1986 world record women's 100-m freestyle, swum by Kristin Otto in 54.73 s.

86. A penny dropped from a 90-m-high building embeds itself 3.6 cm into the ground. Compare its average acceleration on stopping with the acceleration of gravity. *Hint:* You need not compute any velocities or times.

87. Consider an object traversing a distance ℓ, part of the way at speed v_1 and the rest of the way at speed v_2. Find expressions for the average speeds (a) when the object moves at each of the two speeds for half the total *time* and (b) when it moves at each of the two speeds for half the *distance.*

88. Referring to the preceding problem, find an expression for the difference between the times it takes the object to move the distance ℓ. Which approach takes more time?

89. You see the traffic light ahead of you is about to turn from red to green, so you slow to a steady speed of 10 km/h and cruise to the light, reaching it just as it turns green. You then accelerate to 60 km/h in the next 12 s, then maintain constant speed. At the light, you pass a Corvette that has stopped for the red light. Just as you pass (and the light turns green) the Corvette begins accelerating, reaching 65 km/h in 6.9 s, then maintaining constant speed. (a) Plot the motions of both cars on a graph showing the 10-s period after the light turns green. (b) How long after the light turns green does the Corvette pass you? (c) How far are you from the light when the Corvette passes you? (d) How far ahead of you is the Corvette 1 min after the light turns green?

90. In the accident of Problem 55, calculate the relative speed with which a passenger not wearing a seatbelt collides with the dashboard. Assume the passenger undergoes no deceleration before striking the dashboard, and that the passenger is initially 1 m from the dashboard.

91. The position of a particle as a function of time is given by $x = x_0 \sin \omega t$, where x_0 and ω are constants. (a) Take derivatives to find expressions for the velocity and acceleration. (b) What are the maximum values of velocity and acceleration? *Hint:* Consult the table of derivatives in Appendix A.

92. A basketball player runs 12 m down court at 4.6 m/s, dribbling the ball so it has an initial downward velocity of 75 cm/s at 90 cm above the floor. How many times does the ball hit the floor during the 12-m run (see Fig. 2-23)? Neglect any up-and-down motion of the player's hand. *Hint:* The vertical motion of the ball is independent of its horizontal motion.

93. You drop a rock into a well. 2.7 s later, you hear the splash. (a) How far down is the water? The speed of sound is 340 m/s. (b) What percentage error would be introduced by neglecting the travel time for the sound?

94. The depth of a well is such that an object dropped into the well hits the water going far slower than the speed of

FIGURE 2-23 Problem 92.

sound. Use the binomial theorem (see Appendix A) to show that, under these conditions, the depth of the well is given approximately by

$$d = \frac{1}{2}gt^2\left(1 - \frac{gt}{v_s}\right).$$

where t is the time from when you drop the object until you hear the splash, and v_s is the speed of sound.

95. A student is staring idly out her dormitory window when she sees a water balloon fall past. If the balloon takes 0.22 s to cross the 130-cm-high window, from what height above the top of the window was it dropped?

96. A police radar has an effective range of 1.0 km, while a motorist's radar detector has a range of 1.8 km. The motorist is going 100 km/h in a 70 km/h zone when the radar detector beeps. At what rate must the motorist decelerate to avoid a speeding ticket?

THE VECTOR DESCRIPTION OF MOTION

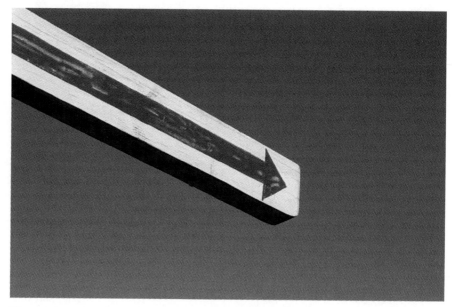

Direction is an essential attribute of motion and of many other physical quantities, as suggested by this highway sign. The mathematical language of vectors helps characterize such directional quantities.

Motion is not limited to a single straight road or to purely vertical flight. As we walk, as we drive, as we sail, we move freely about Earth's surface—that is, in two dimensions. As we fly from city to city, scale a mountain, or are launched into space, we move in three dimensions. Chapter 2's use of positive and negative numbers to designate the direction of motion in one dimension is no longer adequate in two and three dimensions. In this chapter we develop mathematical tools for the full description of motion.

3-1 VECTORS AND SCALARS

Motion in more than one dimension offers rich possibilities: I can head north, east, south, or west—or any direction in between. I can throw a ball straight up, or horizontally, or at any angle I choose. I can go in a circle or any other curved path. Describing such motions involves not only questions like "how far?" and "how fast?" but also "which direction?"

In Chapter 2 we measured position by specifying a single number representing distance from some origin, with a positive or negative sign to indicate either of the two possible directions in one dimension. In two dimensions we can still give the distance from the origin, but now we need another number—namely, the angle relative to some axis—to specify direction. A quantity that requires two numbers—a size and a direction—for its full specification is called a **vector.** We illustrate vectors graphically using arrows; the length of the arrow represents the **magnitude,** or size, of the vector quantity, while the orientation of the arrow represents the vector's direction. Position in two or three dimensions is clearly a vector; so are velocity and acceleration and a number of other quantities we will encounter in physics.

In contrast, a **scalar** is a quantity that can be represented by a single number. Examples of scalars include time, mass, and temperature. We distinguish vectors from scalars by using boldface type for vector symbols and italics for scalars.

3-2 ADDING VECTORS

Figure 3-1 shows the position vector $\mathbf{r}_1$ for an object located initially 2 m from the origin, at an angle of 30° to the horizontal. What if we move the object 1 m to the right, as suggested in Fig. 3-2? The new position is described by a vector $\mathbf{r}_2$; we obtain $\mathbf{r}_2$ by adding to $\mathbf{r}_1$ a vector $\Delta\mathbf{r}$ that is 1 m long and points to the right. To accomplish the addition, we simply place $\Delta\mathbf{r}$ with its tail at the head of the initial position vector $\mathbf{r}_1$; the head of $\Delta\mathbf{r}$ then lies at the head of the new position vector $\mathbf{r}_2$. This procedure defines the general rule for adding any two vectors: place the vectors to be added head-to-tail; their vector sum is then a vector from the tail of the first vector to the head of the second. Symbolically, we write

$$\mathbf{r}_2 = \mathbf{r}_1 + \Delta\mathbf{r}$$

for the vectors of Fig. 3-2. Remember that a vector is specified fully by its magnitude and direction. Where the vector starts doesn't matter, so we can slide vectors around—as long as we preserve length and direction—to form vector sums. Note that we've written the **+** and **=** signs in boldface to emphasize that these are operations performed on vectors. We'll use this convention throughout the text to help remind you that vectors are different mathematical objects than scalars. For the same reason, we boldface the **0** that results when vectors sum to zero.

In Fig. 3-2, we formed the vector sum $\mathbf{r}_1 + \Delta\mathbf{r}$ graphically; to specify the resultant vector $\mathbf{r}_2$, we could measure its angle and length on our graph. Or we could use the law of cosines to calculate the length of $\mathbf{r}_2$ and the law of sines to determine its angle, as in Example 3-1. (The laws of cosines and of sines are described in Appendix A.)

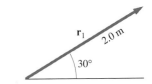

FIGURE 3-1   The position vector $\mathbf{r}_1$ for an object located 2 m from the origin at 30° to the horizontal.

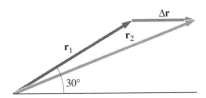

FIGURE 3-2   Vector addition. The new position vector $\mathbf{r}_2$ defines the vector sum of the old position vector $\mathbf{r}_1$ and the change $\Delta\mathbf{r}$. Graphically, we sum the vectors by placing them head-to-tail.

● **EXAMPLE 3-1** VECTOR ADDITION

Determine the vector $\mathbf{r}_2$ in Fig. 3-3 using the laws of cosines and sines.

Solution
The law of cosines (see Appendix A) is like a modified Pythagorean theorem, giving the third side, C, of a triangle in terms of the other two sides A and B and the angle θ between them:

$$C^2 = A^2 + B^2 - 2AB \cos\theta.$$

Applying the law of cosines to the lengths of the vectors in Fig. 3-2, we have

$$r_2^2 = r_1^2 + (\Delta r)^2 - 2r_1\Delta r \cos\theta.$$

We indicate the lengths—magnitudes—of the vectors by writing their symbols in italic type rather than boldface. Figure 3-3 shows the angle θ is 150°; therefore

$$r_2 = \sqrt{(2.0 \text{ m})^2 + (1.0 \text{ m})^2 - (2)(2.0 \text{ m})(1.0 \text{ m}) \cos 150°}$$

$$= 2.9 \text{ m}.$$

This value is the length of $\mathbf{r}_2$; what about its direction? Applying the law of sines (see Appendix A) to the triangle formed by the vectors $\mathbf{r}_1$, $\Delta\mathbf{r}$, and $\mathbf{r}_2$, we have

$$\frac{\sin\alpha}{\Delta r} = \frac{\sin 150°}{r_2},$$

where α is the angle between $\mathbf{r}_1$ and $\mathbf{r}_2$ (see Fig. 3-3). Then

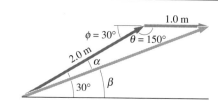

FIGURE 3-3 We are given that the angle between $\mathbf{r}_1$ and the horizontal axis is 30°. Therefore, the angle ϕ is also 30°, so $\theta = 180° - 30° = 150°$. Knowing θ, we use the law of sines to get the angle α between $\mathbf{r}_1$ and $\mathbf{r}_2$; the angle β between $\mathbf{r}_2$ and the horizontal axis is $30° - \alpha$.

$$\alpha = \sin^{-1}\left(\frac{\Delta r \sin 150°}{r_2}\right) = \sin^{-1}\left(\frac{(1.0 \text{ m})(0.50)}{2.9 \text{ m}}\right) = 9.9°.$$

Figure 3-3 then shows that $\mathbf{r}_2$ makes an angle $\beta = 30° - \alpha$, or 20°, with the x axis. The full specification of the vector $\mathbf{r}_2$ includes both its 2.9-m magnitude and its 20° direction.

EXERCISE Vector **A** is 3.0 units long and makes an angle of 45° above the horizontal. Vector **B** is 2.0 units long and makes an angle of 30° *below* the horizontal. Determine the magnitude and direction of the vector sum **A** + **B**.

Answer: magnitude 4.0 units, at 16° above the horizontal

Some problems similar to Example 3-1: 2, 3, 6 ●

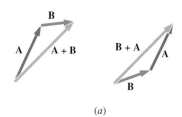

(a)

(b)

FIGURE 3-4 (a) Vector addition is commutative; whether we form the sum **A** + **B** or the sum **B** + **A**, we get the same result. (b) Vector addition is also associative; that is, (**A** + **B**) + **C** = **A** + (**B** + **C**).

Adding three or more vectors is a simple extension of the two-vector addition we've just described: to form the sum **A** + **B** + **C,** for example, first add the vectors **A** and **B,** then add **C** to the sum **A** + **B.** Figure 3-4 shows that vector addition is both commutative and associative.

3-3 SUBTRACTING VECTORS

What if we had been given $\mathbf{r}_1$ and $\mathbf{r}_2$ of Example 3-1 and were asked to find $\Delta\mathbf{r}$? Since $\mathbf{r}_2 = \mathbf{r}_1 + \Delta\mathbf{r}$, we can write

$$\Delta\mathbf{r} = \mathbf{r}_2 - \mathbf{r}_1.$$

But what does this subtraction mean? It means the same thing as subtraction with real numbers: To find the difference **A** − **B**, we add **A** and the opposite of **B.** And what is the opposite of a vector? Simply this: another vector with the same length but opposite direction (Fig. 3-5). Then a graphical determination of the displacement vector $\Delta\mathbf{r}$ of Fig. 3-3 involves adding the vectors $\mathbf{r}_2$ and $-\mathbf{r}_1$, as shown in Fig. 3-6.

TIP **The Starting Point Doesn't Matter**   Does it bother you that Fig. 3-6 doesn't look much like Fig. 3-3? That's ok: The vector $-\mathbf{r}_1$ shown in Fig. 3-6 is the same length as $\mathbf{r}_1$, and it points in the opposite direction. Length and direction are all that matter; we've moved $-\mathbf{r}_1$ into the right position to add it to $\mathbf{r}_2$, and that's all right as long as we preserve length and direction. Get used to sliding vectors around as needed when you add them.

FIGURE 3-5   A vector **A** and its opposite, $-\mathbf{A}$.

3-4 MULTIPLYING VECTORS BY SCALARS

You and I both set out jogging toward the northeast, but you go twice as far as I do. Both our displacements are described by northeast-pointing vectors, and your displacement vector—call it **B**—is twice as long as mine, which we'll call **A**. Mathematically, $\mathbf{B} = 2\mathbf{A}$. That is, multiplying a vector by a scalar simply changes the length of the vector. If the scalar is positive, the direction of the product vector is unchanged; if the scalar is negative, the direction reverses. Figure 3-7 shows the effects of multiplying a vector by different scalars.

Multiplication by a scalar obeys the familiar distributive property; that is, $c(\mathbf{A} + \mathbf{B}) = c\mathbf{A} + c\mathbf{B}$, as shown in Fig. 3-8.

Is it possible to multiply one vector by another? In fact, there are several types of vector multiplication. We will introduce them in later chapters as the need arises.

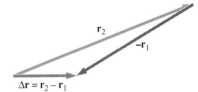

FIGURE 3-6   To perform the subtraction $\mathbf{r}_2 - \mathbf{r}_1$, we add to $\mathbf{r}_2$ the vector $-\mathbf{r}_1$—that is, a vector the same length as $\mathbf{r}_1$ but pointing in the opposite direction. The result is the vector difference $\Delta \mathbf{r}$.

3-5 COORDINATE SYSTEMS, VECTOR COMPONENTS, AND UNIT VECTORS

We will be adding lots of vectors throughout this book, sometimes infinitely many in a single sum! Graphical methods or the laws of sines and cosines will prove tedious; instead we'll find it more convenient to use the methods of analytic geometry, describing our vectors in reference to coordinate systems that we establish.

Coordinate Systems

A **coordinate system** is a framework for the quantitative description of positions in space. A familiar example in two dimensions is the grid of latitude and longitude that locates points on Earth's surface; the pair of numbers specifying a given position are its **coordinates.** Adding an altitude coordinate gives three-dimensional position information.

You're probably familiar with the **Cartesian,** or **rectangular coordinate system,** in which points in a plane are represented by pairs of numbers (x, y) corresponding to distances along two perpendicular axes (Fig. 3-9). We might equally well think of these points as marking the heads of vectors—vectors that describe the position of each point relative to the origin. Then the coordinates x and y fully specify each vector. In this context, the coordinates are called the **components** of a vector. Although the points on our graph could represent actual positions, they might equally well be velocities, forces, or other vector

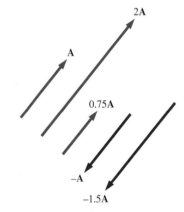

FIGURE 3-7   Multiplying a vector by a scalar. A positive scalar leaves the direction unchanged, while a negative scalar reverses the direction. Note again that where a vector starts doesn't matter; its direction and magnitude fully characterize the vector.

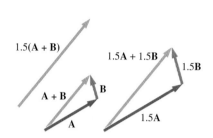

FIGURE 3-8 Multiplication of a vector by a scalar is distributive.

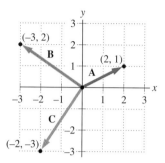

FIGURE 3-9 Several points, labeled by their coordinates on a rectangular coordinate system. We can describe the points equally well with position vectors. The components of the position vectors are the coordinates of the points, for example, $A_x = 2$ and $A_y = 1$.

quantities. We therefore consider an arbitrary vector **A,** and its x and y components A_x and A_y. If we're dealing with a position vector **r,** then the components r_x and r_y are simply the coordinates x and y.

Previously, we specified a vector by giving its length and direction; now we have the option of giving its components. How are these specifications of the same vector related? Figure 3-10 shows the relation. Using the Pythagorean theorem and the definition of the tangent function, we can relate the magnitude A and direction θ to the components:

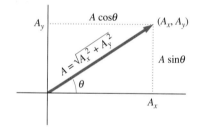

FIGURE 3-10 The Pythagorean theorem and simple trigonometry give the relation between the components and the length and magnitude of a vector.

$$A = \sqrt{A_x^2 + A_y^2} \tag{3-1}$$

and

$$\tan \theta = \frac{A_y}{A_x}. \tag{3-2}$$

Going the other way, we can use the definitions of sine and cosine to write

$$A_x = A \cos \theta \tag{3-3}$$

and

$$A_y = A \sin \theta. \tag{3-4}$$

We stress that giving a pair of x-y components or the magnitude A and direction θ are perfectly equivalent ways of describing the same vector.

Unit Vector Notation

It's cumbersome to say "a vector of length 2 m at 30° to the x axis," or, equivalently, "a vector whose x component is 1.73 m and whose y component is 1.0 m." To express vectors in compact form, we define three **unit vectors, î** and **ĵ,** and **k̂,** which have length 1, no units, and which point in the x, y, and z directions, respectively. To make a vector of any magnitude pointing in the x direction, we multiply the magnitude (a scalar) by the unit vector **î.** Similarly,

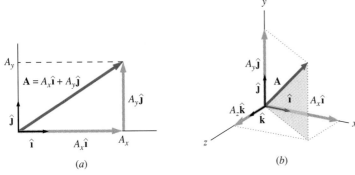

FIGURE 3-11 (a) Any vector in the x-y plane can be written as a combination of the two unit vectors $\hat{\imath}$ and $\hat{\jmath}$, each multiplied by the appropriate components. (b) Similarly, any vector in space can be written in terms of the three unit vectors $\hat{\imath}$, $\hat{\jmath}$, and $\hat{k}$.

any vector pointing in the y direction can be expressed as a scalar multiplying $\hat{\jmath}$. Suppose we have an arbitrary vector $\mathbf{A}$ in the x-y plane, with components A_x and A_y. As Fig. 3-11a shows, we can think of this vector as the sum of two vectors, one of length A_x pointing in the x direction and the other of length A_y pointing in the y direction. In unit vector notation, those two vectors are $A_x\hat{\imath}$ and $A_y\hat{\jmath}$, respectively. So the vector $\mathbf{A}$ itself is given by $\mathbf{A} = A_x\hat{\imath} + A_y\hat{\jmath}$, as suggested in Fig. 3-11. And in general, we can write *any* vector in the x-y plane as a sum of the unit vectors $\hat{\imath}$ and $\hat{\jmath}$ with suitable scalars multiplying each. Analogously, any vector in space can be written in terms of the three unit vectors $\hat{\imath}$, $\hat{\jmath}$, and $\hat{k}$ (Fig. 3-11b).

Note that vector components themselves are scalars; the directionality and therefore the vector character of the vector $\mathbf{A}$, for example, is contained entirely in the unit vectors $\hat{\imath}$ and $\hat{\jmath}$. On the other hand, the unit vectors themselves are dimensionless; the dimensions of the vector $\mathbf{A}$ are associated with its components A_x and A_y.

● **EXAMPLE 3-2** UNIT VECTORS

You drive to a city 160 km distant, in a direction 35° north of east. Express your displacement vector $\Delta\mathbf{r}$ in unit vector notation, using a coordinate system with the x direction eastward and the y direction northward.

Solution
Using Equations 3-3 and 3-4 we have

$$x = r\cos\theta = (160\text{ km})(\cos 35°) = 131\text{ km}$$

and

$$y = r\sin\theta = (160\text{ km})(\sin 35°) = 91.8\text{ km}.$$

Then the displacement vector $\Delta\mathbf{r}$ is

$$\Delta\mathbf{r} = 131\hat{\imath} + 91.8\hat{\jmath}\text{ km}.$$

Note that we treat the entire $131\hat{\imath} + 91.8\hat{\jmath}$ as a single vector quantity, labeling it at the end with the appropriate unit. It's

just as important to keep track of units with vectors as it is with scalars.

TIP **Be Consistent with Vectors** Vectors and scalars are different, so it's impossible for a scalar to equal a vector. That means any equation you write with a vector on one side *must* have a vector on the other side. In the answer to this example, for instance, the boldface $\mathbf{r}$ on the left tells you this is a vector equation, so the right-hand side must also be a vector. And it is, as shown by the presence of the unit vectors $\hat{\imath}$ and $\hat{\jmath}$. On the other hand, the magnitude and the components of a vector are simply scalars, so they can appear in scalar equations.

Some problems similar to Example 3-2: 16–21 ●

Vector Addition and Subtraction with Unit Vectors

Unit vector notation makes vector addition and subtraction straightforward extensions of their scalar counterparts. Suppose we have two vectors **A** and **B**, expressed in terms of unit vectors as $\mathbf{A} = A_x\mathbf{\hat{i}} + A_y\mathbf{\hat{j}}$ and $\mathbf{B} = B_x\mathbf{\hat{i}} + B_y\mathbf{\hat{j}}$. Their sum **A** + **B** is then

$$\mathbf{A} + \mathbf{B} = (A_x\mathbf{\hat{i}} + A_y\mathbf{\hat{j}}) + (B_x\mathbf{\hat{i}} + B_y\mathbf{\hat{j}}) = (A_x + B_x)\mathbf{\hat{i}} + (A_y + B_y)\mathbf{\hat{j}}.$$

In other words, the components of the sum are just the sums of the components of the individual vectors. Had we subtracted **B** from **A,** we would have subtracted the components. To find the length and direction of the sum or difference vector, we can apply Equations 3-1 and 3-2.

● **EXAMPLE 3-3** VECTOR ADDITION WITH UNIT VECTORS

Repeat Example 3-1, this time using unit vector notation.

Solution

Referring to Fig. 3-3, we see that the components x_1 and y_1 of the vector $\mathbf{r}_1$ are given by

$$x_1 = r_1 \cos\theta_1 = (2.0 \text{ m})\cos 30° = 1.73 \text{ m}$$

and

$$y_1 = r_1 \sin\theta_1 = (2.0 \text{ m})\sin 30° = 1.00 \text{ m}.$$

Similarly, the 1.0-m-long horizontal vector $\Delta\mathbf{r}$ obviously has components $\Delta x = 1.0$ m and $\Delta y = 0$. Then $\mathbf{r}_1 = 1.73\mathbf{\hat{i}} + 1.00\mathbf{\hat{j}}$ m and $\Delta\mathbf{r} = 1.0\mathbf{\hat{i}}$ m, so their sum is

$$\mathbf{r}_2 = \mathbf{r}_1 + \Delta\mathbf{r} = (1.73\mathbf{\hat{i}} + 1.00\mathbf{\hat{j}} \text{ m}) + (1.0\mathbf{\hat{i}} \text{ m})$$

$$= 2.73\mathbf{\hat{i}} + 1.00\mathbf{\hat{j}} \text{ m}.$$

The final expression here is a complete answer for $\mathbf{r}_2$ expressed compactly in unit vector notation. If we want the magnitude and direction of $\mathbf{r}_2$, we apply Equations 3-1 and 3-2:

$$r_2 = \sqrt{x_2^2 + y_2^2} = \sqrt{(2.73 \text{ m})^2 + (1.00 \text{ m})^2} = 2.9 \text{ m}$$

and

$$\theta_2 = \tan^{-1}\left(\frac{y_2}{x_2}\right) = \tan^{-1}\left(\frac{1.00}{2.73}\right) = 20°,$$

in agreement with the graphical approach of Example 3-1.

EXERCISE Express the vectors of the exercise following Example 3-1 in unit vector notation and form their sum. Show that the magnitude and direction of the sum agree with the answers to that exercise.

Answers: $\mathbf{A} = 2.12\mathbf{\hat{i}} + 2.12\mathbf{\hat{j}}$; $\mathbf{B} = 1.73\mathbf{\hat{i}} - 1.00\mathbf{\hat{j}}$; $\mathbf{A} + \mathbf{B} = 3.85\mathbf{\hat{i}} + 1.12\mathbf{\hat{j}}$

Some problems similar to Example 3-3: 22, 23, 25

●

Coordinate Systems and Reality

Vectors represent real physical quantities like the displacement between two cities, the velocity of an airplane, or the acceleration of a rocket. Coordinate systems, in contrast, are mathematical constructs we use for convenience in doing calculations. Coordinate systems have no intrinsic reality, and the choice of coordinate system is ours to make. Figure 3-12 shows two different coordinate systems for treating the same vector. Although its components are different in the two systems, the vector's overall length and its orientation in space are the same.

How do we choose a coordinate system? Usually the geometry suggests a choice. Often convention dictates: for example, coordinate systems describing motion on Earth's surface often use east-west and north-south oriented coordinate axes. But in a city with a rectangular grid of streets that don't coincide with the compass directions, it would be more convenient to use a coordinate system aligned to the street grid. In problems including vertical motion, the vertical and horizontal directions often provide the most convenient choice of coordinate axes. But not always; analyzing a skier's motion, for example, may be easier in a coordinate system with axes parallel and perpendicular to the ski slope.

We stress again that the choice of coordinate system is purely for convenience, with no physical content. You should get the right answer no matter what coordinate system you choose, but a sensible choice may make your work a lot easier (and less prone to error).

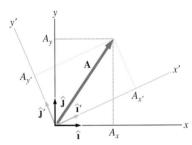

FIGURE 3-12 A vector **A** analyzed into components in two different coordinate systems. Although the components A_x and A_y depend on the choice of coordinates, the actual physical properties of the vector—its length and orientation in space—do not. Here, for example, $A_x^2 + A_y^2 = A_x'^2 + A_y'^2$, showing that the two different pairs of components do indeed represent a vector with the same length in both coordinate systems.

> **TIP** **Choose Your Coordinates Wisely** Figure 3-13 shows the path a climber takes to the top of a mountain. Suppose you were asked to find the magnitude of the overall displacement on this trip. You could treat each of the three distinct segments as a separate vector, and sum them to get the overall vector displacement, then take its magnitude. But think before you automatically draw horizontal and vertical axes! Two of the vectors are in the same direction, so why not make that the direction of one coordinate axis? Then you'll have two fewer components to calculate and the problem becomes that much easier. Problem 25 explores the two coordinate systems suggested for Fig. 3-13.

3-6 VELOCITY VECTORS

If I'm in Boston and I want to go the 400 km to Montreal in 4 hours, it's not enough that I go at 100 km/h. I've also got to go in the right direction, namely toward the north-northwest. In other words, velocity is a vector; I must specify both magnitude and direction in order to give a full description of how something is moving.

In one dimension, we defined velocity as the rate of change of position. Here we do the same, except that now the change of position—the displacement—is itself a vector. So we write

$$\overline{\mathbf{v}} = \frac{\Delta \mathbf{r}}{\Delta t} \qquad (3\text{-}5)$$

for the average velocity, in analogy with Equation 2-1. Here division of the vector $\Delta \mathbf{r}$ by the scalar Δt means multiplying $\Delta \mathbf{r}$ by the reciprocal of Δt; we defined such multiplication in Section 3-4.

We can apply the limiting process defined in Chapter 2 to get the instantaneous velocity:

$$\mathbf{v} = \lim_{\Delta t \to 0} \frac{\Delta \mathbf{r}}{\Delta t} = \frac{d\mathbf{r}}{dt}. \qquad (3\text{-}6)$$

What does this vector derivative, $d\mathbf{r}/dt$, mean? Again, it's just a shorthand way of expressing the result of the limiting process, taking ever smaller times and the

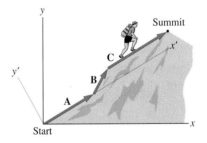

FIGURE 3-13 A climber climbs to the summit of a mountain on a path described by the displacement vectors **A**, **B**, and **C**. Describing this climb quantitatively would be easier in coordinate system x'-y' than in the horizontal-vertical system x-y, since the vectors **A** and **C** would then have only one nonzero component each.

associated displacements **Δr.** Another way of looking at Equation 3-6 is in terms of components; if **r** = $x\hat{\mathbf{i}} + y\hat{\mathbf{j}}$, then Equation 3-6 becomes

$$v_x\hat{\mathbf{i}} + v_y\hat{\mathbf{j}} = \frac{dx}{dt}\hat{\mathbf{i}} + \frac{dy}{dt}\hat{\mathbf{j}}. \tag{3-7}$$

In other words, the vector equation 3-6 incorporates the two scalar equations $v_x = dx/dt$ and $v_y = dy/dt$. For instance, we found that the trip of Example 3-2 was described by a displacement vector **Δr** = $131\hat{\mathbf{i}} + 91.8\hat{\mathbf{j}}$ km. Suppose this displacement took place in 1.43 h; then the average velocity would be

$$\bar{\mathbf{v}} = \frac{\Delta\mathbf{r}}{\Delta t} = \frac{131\hat{\mathbf{i}} + 91.8\hat{\mathbf{j}}\ \text{m}}{1.43\ \text{h}} = 91.6\hat{\mathbf{i}} + 64.2\hat{\mathbf{j}}\ \text{km/h}.$$

● EXAMPLE 3-4   LA TO DENVER VIA SALT LAKE CITY

You leave Los Angeles on a flight to Denver, with a stop in Salt Lake City. Flying 55° north of east, the plane makes the 970-km trip to Salt Lake City in 72 min. After 40 min on the ground, the plane heads for Denver, which is 600 km away and in a direction 10° south of east. The plane reaches Denver 55 min after leaving Salt Lake City. Determine the average velocity of the plane between Los Angeles and Denver.

Solution
Figure 3-14 shows the situation graphically. The vector displacement **Δr**$_{LD}$ from Los Angeles to Denver is the sum of the displacement **Δr**$_{LS}$ from Los Angeles to Salt Lake City and the displacement **Δr**$_{SD}$ from Salt Lake City to Denver. Choosing coordinate axes x and y running east-west and north-south, respectively, we can compute the components needed to write these vectors in unit vector notation:

$$\Delta\mathbf{r}_{LS} = (970\ \text{km})(\cos 55°)\hat{\mathbf{i}} + (970\ \text{km})(\sin 55°)\hat{\mathbf{j}}$$

$$= 556\hat{\mathbf{i}} + 795\hat{\mathbf{j}}\ \text{km}$$

$$\Delta\mathbf{r}_{SD} = (600\ \text{km})[\cos(-10°)]\hat{\mathbf{i}} + (600\ \text{km})[\sin(-10°)]\hat{\mathbf{j}}$$

$$= 591\hat{\mathbf{i}} - 104\hat{\mathbf{j}}\ \text{km}.$$

Note that the angle associated with the Salt Lake City to Denver displacement is negative since the direction is south of east. Adding these vectors gives the displacement vector **Δr**$_{LD}$ from Los Angeles to Denver:

$$\Delta\mathbf{r}_{LD} = \Delta\mathbf{r}_{LS} + \Delta\mathbf{r}_{SD} = (556\hat{\mathbf{i}} + 795\hat{\mathbf{j}}\ \text{km})$$

$$+ (591\hat{\mathbf{i}} - 104\hat{\mathbf{j}}\ \text{km}) = 1147\hat{\mathbf{i}} + 691\hat{\mathbf{j}}\ \text{km}.$$

The total elapsed time for this trip includes the two flight times and the 40-min ground time, so Δt = 72 min + 40 min + 55 min = 167 min or 2.78 h. The average velocity is then

$$\bar{\mathbf{v}} = \frac{\Delta\mathbf{r}_{LD}}{\Delta t} = \frac{1147\hat{\mathbf{i}} + 691\hat{\mathbf{j}}\ \text{km}}{2.78\ \text{h}} = 413\hat{\mathbf{i}} + 249\hat{\mathbf{j}}\ \text{km/h}.$$

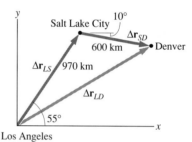

FIGURE 3-14   Displacement vectors for a flight from Los Angeles to Denver via Salt Like City (Example 3-4).

The magnitude of this average velocity is given by Equation 3-1:

$$\bar{v} = \sqrt{\bar{v}_x^2 + \bar{v}_y^2} = \sqrt{(413\ \text{km/h})^2 + (249\ \text{km/h})^2}$$

$$= 482\ \text{km/h}.$$

Since the actual path is not straight and is therefore longer than the magnitude of the vector **Δr**$_{LD}$, the average *speed* would be greater than this.

EXERCISE   Oceanographers on a research voyage sail 340 km in a direction 10° east of north, at a steady 12 km/h. They pause for 5.2 h to take a sediment sample from beneath the ocean floor, then proceed 115 km directly east. The eastward leg takes 12.1 h. After a pause of 3.6 h, the ship heads southwest for 190 km; this takes 18.4 h. Determine the average velocity vector for this trip, and calculate its magnitude and compass direction. Use a coordinate system with the x direction eastward and the y direction northward.

Answers: $\bar{\mathbf{v}} = 0.587\hat{\mathbf{i}} + 2.96\hat{\mathbf{j}}$ km/h; $v = 3.02$ km/h; $\theta = 11.2°$ east of north

Some problems similar to Example 3-4:   27, 28, 31

3-7 ACCELERATION VECTORS

Just as velocity is the rate of change of displacement, so is acceleration the rate of change of velocity. And just as where you end up depends on which direction you go, so the change in your velocity depends not only on the magnitude of your acceleration, but also on its direction. That is, acceleration is a vector quantity. Mathematically, we write

$$\overline{\mathbf{a}} = \frac{\Delta \mathbf{v}}{\Delta t} \tag{3-8}$$

for the average acceleration and

$$\mathbf{a} = \lim_{\Delta t \to 0} \frac{\Delta \mathbf{v}}{\Delta t} = \frac{d\mathbf{v}}{dt} \tag{3-9}$$

for the instantaneous acceleration. We could also write the vector equation 3-9 in terms of components, just as we did with Equation 3-7 for the velocity.

● **EXAMPLE 3-5** COMET HALLEY

Figure 3-15 shows part of the orbit of Comet Halley as it swung through the inner solar system in 1985–1986. In mid-November of 1985, the comet was moving at approximately 31 km/s, at an angle of 34° to the orbital axis, as shown in the figure. The comet made its closest approach to the Sun in February, and by mid-March it was moving at approximately 43 km/s, at 129° to the orbital axis, as shown. Determine the average acceleration vector for the comet during the November–March interval.

Solution
In a coordinate system with the y axis along the orbital axis, we can write the components of the November velocity $\mathbf{v}_1$ as

$$v_{x1} = v_1 \cos\theta = (31 \text{ km/s})\cos(90° - 34°) = 17.3 \text{ km/s}$$

and

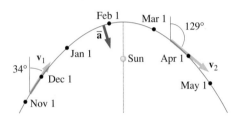

FIGURE 3-15 A portion of the orbit of Comet Halley as it rounded the Sun in 1985–1986. Example 3-5 uses velocities in mid-November and mid-March to calculate the average acceleration for that interval; the average acceleration vector points approximately toward the Sun. A calculation of the comet's instantaneous acceleration would show it always pointing exactly sunward.

$$v_{y1} = v_1 \sin\theta = (31 \text{ km/s})\sin(90° - 34°) = 25.7 \text{ km/s},$$

giving $\mathbf{v}_1 = 17.3\hat{\mathbf{i}} + 25.7\hat{\mathbf{j}}$ km/s. Note that we used the complement of the 34° angle given, since θ in Equations 3-3 and 3-4 is the angle with the x axis. A similar calculation gives $\mathbf{v}_2 = 33.4\hat{\mathbf{i}} - 27.1\hat{\mathbf{j}}$ km/s. The change in velocity, $\Delta\mathbf{v}$, is then given by

$$\Delta\mathbf{v} = \mathbf{v}_2 - \mathbf{v}_1 = (33.4\hat{\mathbf{i}} - 27.1\hat{\mathbf{j}} \text{ km/s})$$
$$- (17.3\hat{\mathbf{i}} + 25.7\hat{\mathbf{j}} \text{ km/s})$$
$$= 16.1\hat{\mathbf{i}} - 52.8\hat{\mathbf{j}} \text{ km/s}.$$

The interval from mid-November to mid-March comprises 120 days, or $(120 \text{ d})(24 \text{ h/d})(3600 \text{ s/h}) = 1.04 \times 10^7$ s, so Equation 3-8 gives

$$\overline{\mathbf{a}} = \frac{\Delta\mathbf{v}}{\Delta t} = \frac{16.1\hat{\mathbf{i}} - 52.8\hat{\mathbf{j}} \text{ km/s}}{1.04 \times 10^7 \text{ s}}$$
$$= 1.5 \times 10^{-6}\hat{\mathbf{i}} - 5.1 \times 10^{-6}\hat{\mathbf{j}} \text{ km/s}^2.$$

As Equation 3-9 suggests, the average acceleration we've calculated is a rough approximation to the instantaneous acceleration during the middle of the November–March interval. In Fig. 3-15 we've sketched the average acceleration vector $\overline{\mathbf{a}}$ at the middle of that interval. Note that the direction of the acceleration is quite different from that of either velocity; again, as we emphasized in Chapter 2, acceleration depends not on velocity itself but on *change* in velocity (Fig. 3-16). In two dimensions that applies to direction as well as magnitude. Here, in fact, the acceleration vector points approximately sunward. A more accurate calculation, approaching the limit in Equation

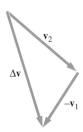

FIGURE 3-16   Graphical representation of the vector subtraction used in determining the average acceleration of Comet Halley. The direction of the acceleration is that of the velocity change $\Delta \mathbf{v}$, not of the velocity itself.

3-9, would show that the acceleration points *exactly* toward the Sun. This is not surprising since the Sun's gravity is what causes the comet to accelerate. We will explore gravitational acceleration and orbits further in Chapter 9.

EXERCISE   A proton enters a particle accelerator moving in the x direction at 3.8×10^5 m/s. It emerges 0.94 μs later, moving at 7.6×10^6 m/s at an angle of 23° to the x axis. Determine the average acceleration vector, as well as its magnitude and direction.

Answers: $\bar{\mathbf{a}} = 7.0 \times 10^{12}\hat{\mathbf{i}} + 3.2 \times 10^{12}\hat{\mathbf{j}}$ m/s², with the y axis in the plane defined by the initial and final velocities; $a = 7.7$ m/s²; $\theta = 24°$

Some problems similar to Example 3-5:   33–35, 51, 52   ●

● **EXAMPLE 3-6**   DIFFERENTIATING VECTORS

The position of a particle as a function of time t is given by $\mathbf{r} = (bt^3 + ct)\hat{\mathbf{i}} + (dt^2 + e)\hat{\mathbf{j}}$, where b, c, d, and e are constants. Find expressions for the velocity and acceleration of the particle.

Solution
Velocity is the rate of change, or derivative, of position—as expressed in Equation 3-6. So:

$$\mathbf{v} = \frac{d\mathbf{r}}{dt} = \frac{d}{dt}[(bt^3 + ct)\hat{\mathbf{i}} + (dt^2 + e)\hat{\mathbf{j}}]$$

$$= \frac{d}{dt}(bt^2 + ct)\hat{\mathbf{i}} + \frac{d}{dt}(dt^2 + e)\hat{\mathbf{j}},$$

where we made use of the fact that the derivative of a sum is the sum of the derivatives. Now the unit vectors $\hat{\mathbf{i}}$ and $\hat{\mathbf{j}}$ are *constants;* they don't change with time. So, as with the constants b, c, d, and e, they aren't affected by differentiation. Then, applying Equation 2-3 to differentiate the various powers of t, we have

$$\mathbf{v} = \frac{d}{dt}(bt^3 + ct)\hat{\mathbf{i}} + \frac{d}{dt}(dt^2 + e)\hat{\mathbf{j}} = (3bt^2 + c)\hat{\mathbf{i}} + 2dt\hat{\mathbf{j}}.$$

We get the acceleration by differentiating once again:

$$\mathbf{a} = \frac{d\mathbf{v}}{dt} = \frac{d}{dt}[(3bt^2 + c)\hat{\mathbf{i}} + 2dt\hat{\mathbf{j}}] = 6bt\hat{\mathbf{i}} + 2d\hat{\mathbf{j}}.$$

Thus we have an acceleration that initially points in the y direction, with magnitude $2d$, but whose x component then grows linearly with time while the y component remains constant.

EXERCISE   Find the acceleration as a function of time for an object whose position is given by $\mathbf{r} = (v_0t + bt^2)\hat{\mathbf{i}} + (y_0 + ct^4)\hat{\mathbf{j}}$, where all quantities except $\mathbf{r}$ and t are constants.

Answer: $\mathbf{a} = 2b\hat{\mathbf{i}} + 12ct^2\hat{\mathbf{j}}$

Some problems similar to Example 3-6:   39, 40   ●

3-8 RELATIVE MOTION

What does it mean to say that a car is moving at 80 km/h? Usually, it means 80 km/h relative to Earth. But Earth is moving at 30 km/s relative to the Sun, so the car's speed relative to the Sun is much greater. Statements about velocity and speed are meaningful only when we have an answer to the question "velocity relative to what?" The object or system with respect to which velocity is measured is called a **frame of reference;** in most of the examples we have considered, Earth is our frame of reference. But playing tennis on the deck of a moving ship, you would find the ship's frame of reference more appropriate for describing the motion of the tennis ball. Similarly, the motion of a comet or interplanetary space probe is described most simply in the reference frame of the Sun.

Suppose a car is moving at 80 km/h. Then its speed relative to another car moving in the same direction at 50 km/h is 80 km/h − 50 km/h = 30 km/h. More generally, if an object is moving at velocity **v** relative to some frame of reference S, and if another frame of reference S′ moves with velocity **V** relative to S, then the velocity **v′** of the object with respect to the frame S′ is given by

$$\mathbf{v}' = \mathbf{v} - \mathbf{V}. \tag{3-10}$$

In our simple example, frame S is the Earth, frame S′ is the car moving at 50 km/h, $v = 80$ km/h, $V = 50$ km/h, and $v' = 30$ km/h. Here all the velocities are in the same direction, so we deal only with their magnitudes. More generally, questions of relative velocity involve vectors, as in the following example.

● EXAMPLE 3-7   NAVIGATING A JETLINER

A jetliner has airspeed 960 km/h. It embarks on a flight from Houston to Omaha, a distance of 1290 km northward. In what direction should it fly if there is a steady 190 km/h wind from the west at its cruising altitude? What will be its groundspeed? How long will the trip take?

Solution
Let S be the frame of reference of the ground, and choose x and y axes pointing east and north, respectively. Let S′ be the frame of reference of the air, so the velocity **V** is 190**î** km/h. The velocity **v′** of the plane relative to the air has magnitude 960 km/h, but we don't know its direction. We can write this velocity as

$$\mathbf{v}' = v'\cos\theta\,\hat{\mathbf{i}} + v'\sin\theta\,\hat{\mathbf{j}},$$

where v' is 960 km/h and θ is the angle **v′** makes with the positive x axis (see Fig. 3-17). The velocity **v** points northward, so it can be expressed in the form

$$\mathbf{v} = v\hat{\mathbf{j}},$$

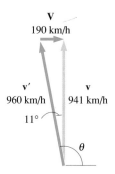

FIGURE 3-17   Vector diagram for Example 3-7. The plane's velocity **v′** with respect to the air sums with the wind velocity **V** to give the velocity **v** with respect to the ground.

where we don't know the magnitude v. We can now write the two components of Equation 3-10:

$$v'_x = v_x - V_x$$
$$v'_y = v_y - V_y.$$

Using our expressions for the various velocities, these equations become

$$v'\cos\theta = 0 - V,$$

where $V = 190$ km/h is the magnitude of **V**, and

$$v'\sin\theta = v - 0.$$

We can solve the first of these equations for θ:

$$\theta = \cos^{-1}\left(\frac{-V}{v'}\right) = \cos^{-1}\left(\frac{-190 \text{ km/h}}{960 \text{ km/h}}\right) = 101.4°.$$

Using this result in the second equation gives

$$v = v'\sin\theta = (960 \text{ km/h})(\sin 101.4°) = 941 \text{ km/h}.$$

So the plane should fly at 101° to the x axis, or 11° west of north, and its groundspeed will be 941 km/h. The 1290-km trip will then take (1290 km)/(941 km/h) = 1.37 h. Figure 3-17 is a vector diagram relating the velocities **v′**, **v**, and **V**. Before the advent of sophisticated navigation equipment, pilots routinely carried out such calculations, often using graphical techniques.

EXERCISE   A boat sails at 14 km/h relative to the water. Its captain wants to sail from one island to another 83 km directly northward. However, a current flows at 3.6 km/h from west to east between the islands. In what direction should the captain head the boat, and how long will the trip take?

Answers: 15° west of north; 6.1 h

Some problems similar to Example 3-7:   42, 44, 46, 55, 56, 65

What about the acceleration of an object viewed in two different reference frames? Suppose **a** is the acceleration in some frame S; that is, $\mathbf{a} = d\mathbf{v}/dt$, where **v** is the velocity relative to frame S. Equation 3-10 gives the velocity **v′** in a frame S' moving with velocity **V** relative to S. Differentiating this equation then gives the acceleration **a′**:

$$\mathbf{a'} = \frac{d\mathbf{v'}}{dt} = \frac{d\mathbf{v}}{dt} - \frac{d\mathbf{V}}{dt}.$$

We generally consider only reference frames moving with *constant* relative velocity **V**, so $d\mathbf{V}/dt = \mathbf{0}$. Furthermore, $d\mathbf{v}/dt$ is just the acceleration **a** in the frame S, so

$$\mathbf{a'} = \mathbf{a}. \tag{3-11}$$

That is, the acceleration of an object is the same in all frames of reference moving at constant velocity relative to one another. This simple result is true because acceleration depends on *changes* in velocity, and the addition of any constant velocity does not alter those changes.

That acceleration does not change from one uniformly moving reference frame to another has a deep significance in physics. Consider the simple experiment of tossing a ball into the air. The experiment will have exactly the same outcome—the ball will go straight up and down in accordance with our equations for one-dimensional motion—both on an airplane moving at a constant 1000 km/h and on the ground. Why? Because in both cases the ball experiences a downward acceleration of magnitude g, which is unaffected by the uniform motion of the plane relative to the Earth.

If you were on a plane flying through perfectly smooth air, there is in fact no experiment involving motion that you could do to tell that the plane was moving. All experiments would come out exactly the same as they would on the ground. This fact is expressed in a simple statement, called the **principle of Galilean relativity:**

The laws of motion are the same in all frames of reference in uniform motion.

What this means is that there is no way of using the laws of motion—the only laws of physics we will consider until Chapter 23—to answer the question "am I moving?" With respect to the laws of motion, the question is meaningless; only *relative motion* matters.

Although our discussion of relative motion and the principle of Galilean relativity may seem almost obvious and straightforward, in fact that discussion rests on deep-seated notions about the nature of time and space. In Chapter 38, after we have completed our study of electromagnetism, we will see how Albert Einstein was able to extend the principle of Galilean relativity to all of physics; the result is Einstein's special theory of relativity. In the process, Einstein showed that our commonsense notions of space and time are not quite right. As a result, Equations 3-10 and 3-11—and indeed all the equations we develop in Chapters 1 through 22—are really only approximately correct; they work well for our everyday experience, and even for spacecraft probing the solar system, but they break down when relative speeds approach the speed of light.

CHAPTER SYNOPSIS

Summary

1. **Vectors** are used to describe motion in two or three dimensions. Vector quantities have magnitude and direction; we visualize vectors using arrows whose length represents magnitude and whose direction is that of the vector. A vector is described fully by giving its magnitude and direction or, equivalently, its components on a chosen set of coordinate axes.

2. **Unit vectors** are dimensionless vectors of length 1 that lie along the coordinate axes. They allow vectors to be written in compact mathematical notation; if a vector $\mathbf{A}$ has componets A_x and A_y, then $\mathbf{A} = A_x\hat{\mathbf{i}} + A_y\hat{\mathbf{j}}$, where $\hat{\mathbf{i}}$ and $\hat{\mathbf{j}}$ are the unit vectors in the x and y directions, respectively.

3. Vectors may be added: the vector sum $\mathbf{A} + \mathbf{B}$ is formed by placing the tail of vector $\mathbf{B}$ at the head of $\mathbf{A}$; the sum is then a vector from the tail of $\mathbf{A}$ to the head of $\mathbf{B}$. Alternatively, vectors may be added by adding their components to get the components of the sum: the x component of the vector $\mathbf{A} + \mathbf{B}$, for example, is just $A_x + B_x$. Vector addition is commutative: $\mathbf{A} + \mathbf{B} = \mathbf{B} + \mathbf{A}$. It is also associative: $(\mathbf{A} + \mathbf{B}) + \mathbf{C} = \mathbf{A} + (\mathbf{B} + \mathbf{C})$.

4. Vector subtraction $\mathbf{A} - \mathbf{B}$ means adding to $\mathbf{A}$ a vector whose magnitude is that of $\mathbf{B}$, but which points in the opposite direction.

5. A vector may be multiplied by a scalar. If c is a scalar and $\mathbf{A}$ a vector of length A, the vector $c\mathbf{A}$ is a vector of length cA having the same direction as $\mathbf{A}$. Multiplication by a scalar is distributive: $c(\mathbf{A} + \mathbf{B}) = c\mathbf{A} + c\mathbf{B}$.

6. **Velocity** is a vector describing the rate of change of position. The average velocity over a time interval Δt is given by the change in position divided by the time interval:

$$\bar{\mathbf{v}} = \frac{\Delta \mathbf{r}}{\Delta t},$$

where $\Delta \mathbf{r} = \mathbf{r}_2 - \mathbf{r}_1$, with $\mathbf{r}_1$ and $\mathbf{r}_2$ the positions at the beginning and end of the time interval Δt. The **instantaneous velocity** is obtained by taking the limit of arbitrarily small time intervals:

$$\mathbf{v} = \lim_{\Delta t \to 0} \frac{\Delta \mathbf{r}}{\Delta t} = \frac{d\mathbf{r}}{dt}.$$

7. **Acceleration** is a vector describing the rate of change of velocity; the average acceleration over a time interval Δt is

$$\bar{\mathbf{a}} = \frac{\Delta \mathbf{v}}{\Delta t},$$

while the instantaneous acceleration is

$$\mathbf{a} = \lim_{\Delta t \to 0} \frac{\Delta \mathbf{v}}{\Delta t} = \frac{d\mathbf{v}}{dt}.$$

Since velocity is a vector, it can change in ... or direction or both; any change in velocit... acceleration.

8. The concept of velocity is meaningful only in r.... ...on to a particular frame of reference; the statement "I am moving" is incomplete unless the answer to the question "moving relative to what?" is understood. Only **relative motion** matters. This idea is summarized in the **principle of Galilean relativity:**

> **The laws of motion are the same in all frames of reference in uniform motion.**

In the context of Galilean relativity, we can determine the velocity $\mathbf{v}'$ of an object relative to some frame of reference S' moving with constant velocity $\mathbf{V}$ relative to a second frame S:

$$\mathbf{v}' = \mathbf{v} - \mathbf{V},$$

where $\mathbf{v}$ is the velocity of the object with respect to the frame S. The acceleration of the object is the same in either frame of reference: $\mathbf{a}' = \mathbf{a}$.

Terms You Should Understand

(Pairs are closely related terms whose distinction is important; number in parentheses is chapter section where term first appears.)

vector, scalar (3-1)
magnitude (3-1)
coordinate system (3-5)
component (3-5)
unit vector (3-5)
velocity vector (3-6)
acceleration vector (3-7)

Symbols You Should Recognize

$\mathbf{r}$ (3-2)
$\mathbf{A} + \mathbf{B}$ (3-2)
$\mathbf{A} - \mathbf{B}$ (3-3)
$c\mathbf{A}$ (3-4)
x, y, r, θ (3-5)
$\hat{\mathbf{i}}, \hat{\mathbf{j}}, \hat{\mathbf{k}}$ (3-5)
$\mathbf{A} = A_x\hat{\mathbf{i}} + A_y\hat{\mathbf{j}}$ (3-5)
$\dfrac{d\mathbf{r}}{dt}$ (3-6)
$\dfrac{d\mathbf{v}}{dt}$ (3-7)

Problems You Should Be Able to Solve

adding and subtracting vectors graphically (3-2, 3-3)
multiplying vectors by scalars (3-4)
determining vector components (3-5)
writing vectors in unit vector notation (3-5)
adding and subtracting vectors using unit vector notation (3-5)
determining average and instantaneous velocity vectors (3-6)
determining average and instantaneous acceleration vectors (3-7)
evaluating vectors in different frames of reference (3-8)

Limitations to Keep in Mind

Coordinate systems are artificial constructs established for convenience. When solving a problem, you are free to pick any coordinate system you choose. A wise choice may make your work much easier.

The description of motion developed in the first half of this book is valid only for relative speeds much less than the speed of light.

QUESTIONS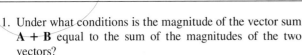

1. Under what conditions is the magnitude of the vector sum **A** + **B** equal to the sum of the magnitudes of the two vectors?

2. Which of the following are valid expressions? Explain. (a) **C** = **A** − **B**; (b) a = **B** + c; (c) **B** = c**A**; (d) **B** = **A**/c; (e) **B** = c/**A**.

3. Can two vectors of equal magnitude sum to zero? Can two vectors of unequal magnitude?

4. Repeat the preceding question for three vectors.

5. One way of adding vectors is to place them tail-to-tail, and then form a parallelogram as shown in Fig. 3-18. The sum of the two vectors is then a vector along the diagonal of the parallelogram. Why is this method equivalent to the head-to-tail method described in the text?

6. Three vectors sum to zero. Placed head-to-tail, what geometrical figure must they form? Explain.

7. Is vector subtraction commutative?

8. Two vectors **A** and **B** have lengths A and B, respectively. What is the maximum possible length for the vector **A** + **B**? How must **A** and **B** be oriented for their sum to have this maximum possible length?

9. A fly heads west 5 m, then north 3 m, then straight up 6 m. A mosquito starts from the same place, flies straight up 6 m, west 5 m, and north 3 m. Compare their final positions.

10. Is it possible to form the unit vector $\hat{\mathbf{k}}$ from a sum of $\hat{\mathbf{i}}$ and $\hat{\mathbf{j}}$, each multiplied by appropriate scalars? Explain.

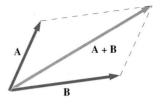

FIGURE 3-18 Parallelogram method of adding vectors (Question 5).

11. Is it meaningful to talk about vectors without mention of coordinate systems or components?

12. Can an object have a southward acceleration while moving northward? A westward acceleration while moving northward?

13. Can an object move northward for 10 minutes, all the while having a southward acceleration?

14. Can an object move northward for 10 minutes, all the while having a westward acceleration?

15. Why can we usually use Cartesian coordinates on Earth's surface even though Earth is round?

16. A space shuttle completes one full orbit in 89 min. What are its average velocity and average acceleration over this interval? Is its average speed equal to its average velocity?

17. Why is the statement "I am moving" meaningless?

PROBLEMS

Sections 3-2 and 3-3 Adding and Subtracting Vectors

1. You walk west 220 m, then turn 45° toward the north and walk another 50 m. How far and in what direction from your starting point do you end up?

2. For the vectors shown in Fig. 3-19, evaluate the vectors **A** + **B**, **A** − **B**, **A** + **C**, **A** + **B** + **C**.

3. A plane going from Washington, DC, to Burlington, Vermont, first flies 360 km due northeast to a point over central Long Island. It then turns due north and flies 400 km to its destination. Determine the magnitude and direction of its displacement vector.

4. Three vectors **A**, **B**, and **C** have the same length L and form an equilateral triangle, as shown in Fig. 3-20. Find the magnitude and direction of the vectors (a) **A** + **B**, (b) **A** − **B**, (c) **A** + **B** + **C**, (d) **A** + **B** − **C**.

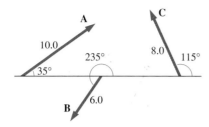

FIGURE 3-19 Vectors for Problem 2. Lengths are in arbitrary units.

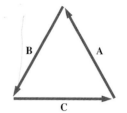

FIGURE 3-20 Problem 4.

5. The vector **A** is 12 units long and points 30° north of east. The vector **B** is 18 units long and points 45° west of north. Find a vector **C** such that **A** + **B** + **C** = 0.

6. In a Hawaiian foot race, runners travel a twisting 58-km path, climbing from the ocean shore to an altitude of 3300 m at the summit of Haleakala volcano. They wind up a horizontal distance 15 km northwest of their starting point. (a) At what angle must someone at the starting line beam a laser to send a signal to someone at the finishing line? (b) Moving at the speed of light, how long does it take the laser beam to go from start to finish line?

7. An ion in a mass spectrometer (a device that sorts atomic-size particles) follows a semicircular path of radius 15.2 cm. What are (a) the distance it travels and (b) the magnitude of its displacement?

8. Three vectors, **A, B,** and **C,** sum to zero. Vector **C** points horizontally to the right and has length 2.00 units. Vector **A** makes an angle of 110° with the horizontal, and has length 2.92 units. (a) How do the lengths of **A** and **B** compare? (b) What is the direction of vector **B**?

9. A direct flight from Orlando, Florida, to Atlanta, Georgia, covers 660 km and heads at 29° west of north. Your flight, however, stops at Charleston, South Carolina, on the way to Atlanta. Charleston is 510 km from Orlando, in a direction 9.3° east of north. What are the magnitude and direction of the Charleston-to-Atlanta leg of your flight?

10. Vector **A** has length 2.50 units, and makes an angle of 30° to a rightward-pointing horizontal axis. Vector **B** makes an angle of 135° to the same axis, and the sum **A** + **B** has a length of 3.41 units. What is the length of vector **B**?

11. Two vectors have the same length a and their sum has magnitude $0.845a$. What is the angle between the vectors? *Hint:* Remember the law of cosines! Here "angle between the vectors" means the angle between their actual directions; that is, the angle between them when they're tail to tail.

Section 3-4 Multiplying Vectors by Scalars

12. Two vectors **A** and **B** have the same length A and are at right angles. What is the length of the vector **A** + 2**B**?

13. Two vectors **A** and **B** have the same length A. **A** points horizontally to the right, while **B** makes a 30° angle with the horizontal. What are the length and direction of the vector **A** − 2**B**?

14. For the vectors of Fig. 3-19, determine the magnitude and direction of the vector 1.5**A** + 3.0**B** − **C**

15. For the vectors of Fig. 3-20, find 2 values for the scalar c such that **A** + c**B** has length 2.18L.

Section 3-5 Coordinate Systems, Vector Components, and Unit Vectors

16. Express each of the vectors of Fig. 3-19 in unit vector notation, with the x axis horizontally to the right and the y axis vertically upward.

17. Repeat Problem 2, using unit vector notation.

18. Express the vectors of Fig. 3-20 in unit vector notation, taking the x axis horizontal and the y axis vertical. Each vector has length A.

19. In Fig. 3-12 the angle between the x and x' axes is 21°, the angle between the vector **A** and the x axis is 54°, and **A**'s length is 10 units. (a) Find the components of **A** in both coordinate systems shown. (b) Verify that the length of **A**, computed using Equation 3-1, is the same in both coordinate systems.

20. A proton travels in a circular path around the 2.0-km-diameter accelerator at Fermilab, near Chicago. Write expressions for the proton's displacement vector from the center of the circle when it is at (a) 0°; (b) 30°; (c) 45°; (d) 60°; (e) 90°; (f) 180°.

21. A vector **A** is 10 units long and points 30° counterclockwise (CCW) from horizontal. What are the x and y components on a coordinate system (a) with the x axis horizontal and the y axis vertical; (b) with the x axis at 45° CCW from horizontal and the y axis 45° CCW from vertical; and (c) with the x axis at 30° CCW from horizontal and the y axis 90° CCW from the x axis?

22. Express the sum of the unit vectors $\hat{\imath}, \hat{\jmath},$ and $\hat{k}$ in unit vector notation, and determine its magnitude.

23. Let **A** = $15\hat{\imath} - 40\hat{\jmath}$ and **B** = $31\hat{\jmath} + 18\hat{k}$. Find a vector **C** such that **A** + **B** + **C** = 0.

24. A mountain expedition starts a base camp at an altitude of 5500 m. Four climbers then establish an advance camp at an altitude of 7400 m; the advance camp is southeast of the

base camp, at a horizontal distance of 8.2 km. From the advance camp, two climbers head directly north to a 8900-m summit, a horizontal distance of 2.1 km. Using a coordinate system with the x axis eastward, the y axis northward, and the z axis upward, and with origin at the base camp, express the positions of the advance camp and summit in unit vector notation, and determine the straight-line distance from base camp to summit.

25. In Fig. 3-13, suppose that vectors **A** and **C** both make 30° angles with the horizontal while **B** makes a 60° angle, and that $A = 2.3$ km, $B = 1.0$ km, and $C = 2.9$ km. (a) Express the displacement vector $\Delta\mathbf{r}$ from start to summit in each of the coordinate systems shown, and (b) determine its length.

Section 3-6 Velocity Vectors

26. An object is moving at 18 m/s at an angle of 220° counterclockwise from the x axis. What are the x and y components of its velocity?

27. A car drives north at 40 mi/h for 10 min, then turns east and goes 5.0 mi at 60 mi/h. Finally, it goes southwest at 30 mi/h for 6.0 min. Draw a vector diagram and determine (a) the car's displacement and (b) its average velocity for this trip.

28. A biologist studying the motion of bacteria notes a bacterium at position $\mathbf{r}_1 = 2.2\hat{\mathbf{i}} + 3.7\hat{\mathbf{j}} - 1.2\hat{\mathbf{k}}$ μm (1 μm = 10^{-6} m). After 6.2 s the bacterium is at $\mathbf{r}_2 = 4.6\hat{\mathbf{i}} + 1.9\hat{\mathbf{k}}$ μm. What is its average velocity? Express in unit vector notation, and calculate the magnitude.

29. The Orlando-to-Atlanta flight described in Problem 9 takes 2.5 h. What is the average velocity vector? Express (a) as a magnitude and direction, and (b) in unit vector notation with the x axis east and the y axis north.

30. The minute hand of a clock is 5.5 cm long. What is the average velocity vector for the tip of the hand during the interval from the hour to 20 minutes past the hour, expressed in a coordinate system with the y axis toward noon and the x axis toward 3 o'clock?

31. A hot-air balloon rises vertically 800 m over a period of 10 min, then drifts eastward 14 km in 27 min. Then the wind shifts, and the balloon moves northeastward for 15 min, at a speed of 24 km/h. Finally, it drops vertically in 5 min until it is 250 m above the ground. Express the balloon's average velocity in unit vector notation, using a coordinate system with the x axis eastward, the y axis northward, and the z axis upward.

32. Figure 3-21 shows the path of a bug as it crawls around a tabletop. Dots mark the position of the bug at each second. Determine the average velocity of the bug over the interval (a) from 1.0 s to 2.0 s; (b) from 2.0 s to 4.0 s; (c) 0 to 6.0 s.

Section 3-7 Acceleration Vectors

33. A supersonic aircraft is traveling east at 2100 km/h. It then begins to turn southward, emerging from the turn 2.5 min

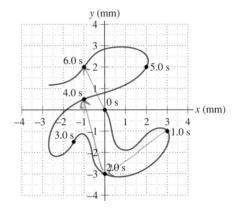

FIGURE 3-21 Problem 32.

later heading due south at 1800 km/h. What are the magnitude and direction of its average acceleration during the turn?

34. A car, initially going eastward, rounds a 90° bend and ends up heading southward. If the speedometer reading remains constant, what is the direction of the car's average acceleration vector?

35. Earth moves in a nearly circular orbit about the Sun. What is the magnitude of its average acceleration over the intervals (a) from January 1 to July 1 and (b) from January 1 to April 1? (c) What is the angle between the two average acceleration vectors for the intervals given? Consult Appendix E for Earth's orbital speed.

36. What are (a) the average velocity and (b) the average acceleration of the tip of the 2.4-cm-long hour hand of a clock in the interval from 12 P.M. to 6 P.M.? Express in unit vector notation, with the x axis pointing toward 3 P.M. and the y axis toward 12 P.M.

37. Attempting to stop on a slippery road, a car moving at 80 km/h skids across the road at a 30° angle to its initial motion, coming to a stop in 3.9 s. Determine the average acceleration in m/s², using a coordinate system with the x axis in the direction of the car's original motion and y axis toward the side of the road to which the car skids.

38. An object undergoes acceleration of $2.3\hat{\mathbf{i}} + 3.6\hat{\mathbf{j}}$ m/s² over a 10-s interval. At the end of this time, its velocity is $33\hat{\mathbf{i}} + 15\hat{\mathbf{j}}$ m/s. (a) What was its velocity at the beginning of the 10-s interval? (b) By how much did its speed change? (c) By how much did its direction change? (d) Show that the speed change is *not* given by the magnitude of the acceleration times the time. Why not?

39. An object's position as a function of time is given by $\mathbf{r} = (bt^3 + ct)\hat{\mathbf{i}} + dt^2\hat{\mathbf{j}} + (et + f)\hat{\mathbf{k}}$, where b, c, d, e, and f are constants. Determine the velocity and acceleration as functions of time.

40. The position of an object is given by $\mathbf{r} = (ct - bt^3)\hat{\mathbf{i}} + dt^2\hat{\mathbf{j}}$, with constants $c = 6.7$ m/s, $b = 0.81$ m/s³, and $d = 4.5$ m/s². (a) Determine the object's velocity at time $t = 0$. (b) How long does it take for the direction of motion

to change by 90°? (c) By how much does the speed change during this time?

Section 3-8 Relative Motion

41. A dog paces around the perimeter of a rectangular barge that is headed up a river at 14 km/h relative to the river-bank. The current in the water is at 3.0 km/h. If the dog walks at 4.0 km/h, what are its speeds relative to (a) the shore and (b) the water as it walks around the barge?

42. A jetliner with an airspeed of 1000 km/h sets out on a 1500-km flight due south. To maintain a southward direction, however, the plane must be pointed 15° west of south. If the flight takes 100 min, what is the wind velocity?

43. A spacecraft is launched toward Mars at the instant Earth is moving in the $+x$ direction at its orbital speed of 30 km/s, in the Sun's frame of reference. Initially the space-craft is moving at 40 km/s relative to Earth, in the $+y$ direction. At the launch time, Mars is moving in the $-y$ direction at its orbital speed of 24 km/s. Find the space-craft's velocity relative to Mars.

44. You wish to row straight across a 63-m-wide river. If you can row at a steady 1.3 m/s relative to the water, and the river flows at 0.57 m/s, (a) in what direction should you head? (b) How long will it take you to cross the river?

45. You're on an airport "people mover," a conveyor belt going at 2.2 m/s through a level section of the terminal. A button falls off your coat and drops freely 1.6 m, hitting the belt 0.57 s later. What are the magnitude and direction of the button's displacement and average velocity during its fall in (a) the frame of reference of the "people mover" and (b) the frame of reference of the airport terminal? (c) As it falls, what is its acceleration in each frame of reference?

46. An airplane with airspeed of 370 km/h flies perpendicu-larly across the jet stream. To achieve this flight, the plane must be pointed into the jet stream at an angle of 32° from the perpendicular direction of its flight. What is the speed of the jet stream?

Paired Problems

(Both problems in a pair involve the same principles and tech-niques. If you can get the first problem, you should be able to solve the second one.)

47. A rabbit scurries across a field, going eastward 21.0 m. It then turns and darts southwestward for 8.50 m. Then it pops down a rabbit hole, 1.10 m vertically downward. What is the magnitude of the displacement from its start-ing point?

48. A cosmic ray plows into Earth's upper atmosphere, liberat-ing a shower of subatomic particles. One of these particles moves downward 3.2 km, then undergoes a collision after which it moves 1.6 km at 27° northward of the vertical. It then undergoes another collision that sends it moving hori-

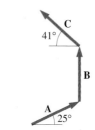

FIGURE 3-22 Problem 49.

zontally eastward for 2.1 km before it annihilates with another particle. What is the magnitude of its overall displacement?

49. The three displacement vectors shown in Fig. 3-22 are each 10.0 m long. (a) Write each in unit vector notation. (b) Find a vector **D** such that $\mathbf{A} + \mathbf{B} + \mathbf{C} + \mathbf{D} = \mathbf{0}$. (c) Find the length of **D**.

50. The three displacement vectors shown in Fig. 3-23 are each 10.0 m long. (a) Write each in unit vector notation. (b) Find a vector **D** such that $\mathbf{A} + \mathbf{B} + \mathbf{C} + \mathbf{D} = \mathbf{0}$. (c) Find the length of **D**.

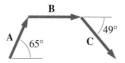

FIGURE 3-23 Problem 50.

51. A car is heading into a turn at 85 km/h. It enters the turn, slows to 55 km/h, and emerges 28 s later at 35° to its original direction, still moving at 55 km/h. What are (a) the magnitude and (b) the direction of its average acceler-ation, the latter measured with respect to the car's original direction?

52. The Galileo space probe was originally to be launched directly toward its destination, Jupiter. But the 1986 explo-sion of the space shuttle Challenger led to a decision that Galileo's liquid-fueled booster rocket was unsafe to fly on the shuttle. As a result, Galileo's trajectory became a com-plicated path through the inner solar system, picking up speed through a so-called "gravity assist" maneuver in-volving close encounters with the planets Venus and Earth. The first Earth encounter occurred on December 8, 1990, when Galileo, outbound from Venus, passed 200 miles from Earth. Thirty days before the encounter, Galileo was approaching Earth at 2.99×10^4 m/s. Thirty days after the encounter, Galileo was moving at 54° to its pre-encounter direction, at a speed of 3.5×10^4 m/s. What were (a) the magnitude and (b) the direction of Galileo's average accel-eration during this interval, the latter measured with re-spect to its pre-encounter direction?

53. The sweep-second hand of a clock is 3.1 cm long. What are the magnitude of (a) the average velocity and (b) the average acceleration of the hand's tip over a 5.0-s interval? (c) What is the angle between the average velocity and acceleration vectors?

54. A proton in a cyclotron follows a circular path 23 cm in diameter, completing one revolution in 0.17 μs. What are the magnitude of (a) the average velocity and (b) the average acceleration as the proton sweeps through one-twelfth of the full circle? (c) What is the angle between the average velocity and acceleration vectors?

55. A ferryboat sails between two towns directly opposite one another on a river. If the boat sails at 15 km/h relative to the water, and if the current flows at 6.3 km/h, at what angle should the boat head?

56. A flock of geese is attempting to migrate due south, but the wind is blowing from the west at 5.1 m/s. If the birds can fly at 7.5 m/s relative to the air, in what direction should they head?

Supplementary Problems

57. A satellite is in a circular orbit 240 km above Earth's surface, moving at a constant speed of 7.80 km/s. A tracking station picks up the satellite when it is 5.0° above the horizon, as shown in Fig. 3-24. The satellite is tracked until it is directly overhead. What are the magnitudes of (a) its displacement, (b) its average velocity, and (c) its average acceleration during the tracking interval? Is the value of the average acceleration approximately familiar?

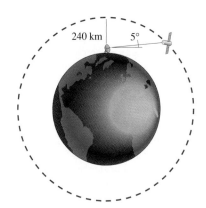

FIGURE 3-24   Problem 57 (figure is not to scale).

58. The sum, **A + B,** of two vectors is perpendicular to the difference, **A − B.** How do the magnitudes of the two vectors compare?

59. Find two vectors in the x-y plane that are perpendicular to the vector $a\hat{\imath} + b\hat{\jmath}$.

60. Find the angle between the vectors $3.0\hat{\imath} + 1.7\hat{\jmath}$ and $6.1\hat{\imath} − 4.2\hat{\jmath}$.

61. Write an expression for a unit vector that lies at 45° between the positive x and y axes.

62. A vector **A** has components A_x and A_y in a coordinate system with axes x and y. Find its components A'_x and A'_y in a coordinate system whose axes x' and y' are rotated counterclockwise through an angle θ from the x and y axes. Test your result for the cases $\theta = 0$ and $\theta = 90°$.

63. Figure 3-25 shows two arbitrary vectors **A** and **B** that sum to a third vector **C.** By working with components, prove the law of cosines: $C^2 = A^2 + B^2 − 2AB\cos\gamma$.

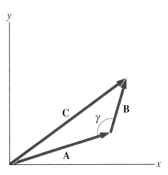

FIGURE 3-25   Problem 63.

64. You wish to paddle a canoe perpendicularly across a river of width w and back. If the river's flow speed is c, and you can paddle at speed v relative to the water, show that your round-trip travel time is given by $\Delta t = 2w/\sqrt{v^2 − c^2}$.

65. Town B is located across the river from town A and at a 40.0° angle upstream from A, as shown in Fig. 3-26. A ferryboat travels from A to B; it sails at 18.0 km/h relative to the water. If the current in the river flows at 5.60 km/h, at what angle should the boat head? What will be its speed

FIGURE 3-26   Problem 65.

relative to the ground? *Hint:* Set up Equation 3-10 for this situation. Each component of Equation 3-10 yields two equations in the unknowns v, the magnitude of the boat's velocity relative to the ground, and ϕ, the unknown angle. Solve the x equation for $\cos \phi$ and substitute into the second equation, using the relation $\sin \phi = \sqrt{1 - \cos^2 \phi}$. You can then solve for v, then go back and get ϕ from your first equation.

66. A space shuttle orbits the Earth at 27,000 km/h, while at the equator Earth rotates at 1300 km/h. The two motions are in roughly the same direction (west to east), but the shuttle orbit is inclined at 25° to the equator. What is the shuttle's velocity relative to scientists tracking it from the equator?

CHAPTER 4

MOTION IN MORE THAN ONE DIMENSION

Landing a jetliner involves a complex sequence of motions in three dimensions.

What is the speed of an orbiting satellite? How should I leap for best results in a long-jump competition? How much advance warning does my country have against a nuclear missile fired from a submarine offshore? How should I throw this ball so it gets over the roof? How should I engineer the curve in this road for safe driving? How wide a stage does a dancer need to execute a grand jeté? Chapter 3's introduction of the displacement, velocity, and acceleration vectors has prepared us to answer these and many other questions about motion in two and three dimensions.

4-1 VELOCITY AND ACCELERATION

Motion in a straight line may or may not involve acceleration. But motion in two or more dimensions *must* be accelerated motion. Why? Because moving in two dimensions requires changing direction—and that means at least the direction of the velocity vector must change. We emphasize again that *any* change in velocity—in magnitude, direction, or both—involves acceleration.

In Chapter 2 we stressed that the numerical values of velocity and acceleration in one dimension are completely unrelated; acceleration depends not on velocity but on *changes* in velocity. The same is true in two dimensions, where both velocity and acceleration are *vectors*. That means the acceleration vector can have not only any magnitude, but also any direction relative to the velocity vector. Mathematically, the relation between velocity and acceleration is described completely by Equations 3-8 and 3-9, which we repeat here:.

average acceleration:
$$\overline{\mathbf{a}} = \frac{\Delta \mathbf{v}}{\Delta t} \qquad (4\text{-}1)$$

instantaneous acceleration:
$$\mathbf{a} = \lim_{\Delta t \to 0} \frac{\Delta \mathbf{v}}{\Delta t} = \frac{d\mathbf{v}}{dt}. \qquad (4\text{-}2)$$

If the acceleration is constant—in *both* magnitude and direction—then the average and instantaneous accelerations are the same. To understand further the relation between acceleration and velocity, it's helpful to consider several special cases.

Colinear Velocity and Acceleration

Suppose you're driving along a straight road at speed v_0, and you step on the gas to give a constant acceleration a for a time Δt. Then Equation 4-1 says that the magnitude of your change in velocity is $\Delta v = a\Delta t$. But what about direction? Stepping on the gas while driving on a straight road doesn't change your direction at all; it just makes you go faster. That is, the acceleration vector for this case is in the same direction as the velocity vector. So, too, is the velocity change $\Delta \mathbf{v}$, since it's just the acceleration vector $\mathbf{a}$ multiplied by Δt. Figure 4-1a shows the vector addition for this case.

What if you had stepped on the brake instead of the gas? Then you would have slowed while continuing to move forward. In that case the acceleration vector points opposite to the velocity vector, and the result is a decrease in the magnitude of the velocity but, again, no change in direction (Fig. 4-1b).

We summarize the situation when velocity and acceleration are colinear, either parallel or antiparallel:

| **Acceleration colinear to velocity changes only the magnitude of the velocity, not its direction.**

Use of vectors here is overkill; we could have handled this case using the one-dimensional methodology of Chapter 2. But we use the full vector description to emphasize both the similarities and differences between this and our next special case.

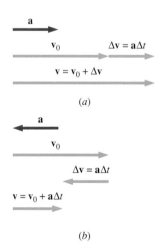

FIGURE 4-1 With acceleration and velocity in the same direction, the velocity vector changes in magnitude but not direction. (*a*) With velocity and acceleration parallel, the magnitude of the velocity increases. (*b*) With velocity and acceleration antiparallel, the magnitude of the velocity decreases.

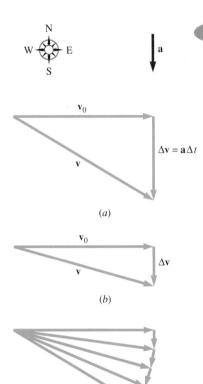

FIGURE 4-2 (*a*) A jetliner flying eastward accelerates southward during a brief wind gust. Its new velocity is changed considerably in direction but only slightly in magnitude. (*b*) If the perpendicular velocity change $\Delta\mathbf{v}$ is smaller, the initial and final velocities are more nearly equal in magnitude. (*c*) If the acceleration direction changes, so the velocity and acceleration are *always* perpendicular, then the velocity changes in direction only.

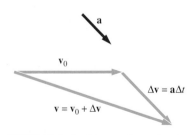

FIGURE 4-3 In general, acceleration changes both the magnitude and direction of the velocity vector.

TIP **Vectors Tell It All** Are you thinking there should be a minus sign somewhere in Fig. 4-1*b*, because the magnitude of the velocity is decreasing? No, there shouldn't! The vector symbols stand for quantities with magnitude *and* direction. Full specification of **a** in Fig. 4-1*b* includes the fact that it's pointing to the left—and that's enough to show that this acceleration vector will decrease the magnitude of the right-pointing velocity. The vector addition $\mathbf{v} = \mathbf{v}_0 + \mathbf{a}\Delta t$ really tells it all. It's only when you write this in components that a negative sign will appear; in a coordinate system with its *x* axis to the right, for example, a_x of Fig. 4-1*b* is a negative number and the addition of the components v_{x0} and $\Delta v_x = a_x\Delta t$ becomes a subtraction. In general, you shouldn't worry about signs when adding vectors, even those in opposite directions; only when you write components in your chosen coordinate system do signs become important.

Perpendicular Velocity and Acceleration

A jetliner is flying eastward at speed v_0, when a gust of wind blows from the north, giving the plane a constant acceleration *a* for a short time Δt. Again, the magnitude of the change in velocity is $\Delta v = a\,\Delta t$, but now the direction of the acceleration is perpendicular to that of the initial velocity. Figure 4-2*a* shows the vector addition in this case. We could calculate the magnitude and direction of the new velocity vector, but the important point to make here is a qualitative one: the new velocity differs in direction from the initial velocity and is not much different in magnitude. Why? Because the acceleration vector is perpendicular to the initial velocity vector, so the dominant effect of the acceleration is to change the direction of the velocity. When it first acted, in fact, the acceleration changed *only* the direction; Fig 4-2*b* shows that for very short times, the initial and final velocities have essentially the same magnitude. As the velocity changed, it was no longer exactly perpendicular to the acceleration, and the magnitude began to change as well. Had the acceleration direction changed to keep it perpendicular to the velocity, only the direction would have changed (Fig. 4-2*c*). We conclude:

> **Acceleration that is always perpendicular to velocity changes only the direction of the velocity, not its magnitude.**

In a case like that of Fig. 4-2*a*, where the acceleration is perpendicular to the velocity and produces a change $\Delta\mathbf{v}$ considerably smaller in magnitude than the initial velocity, this statement is approximately true. When the acceleration and velocity are *always* perpendicular, as in Fig. 4-2*c*, it becomes exactly true.

The General Case

In general, velocity and acceleration vectors can be at any angle to each other. Then both the magnitude and direction of the velocity will change, as suggested in Fig. 4-3. Which changes most depends on the orientation of the vectors and on the magnitude of the change $\Delta\mathbf{v}$.

● EXAMPLE 4-1   BLOWN OFF COURSE

A plane is flying eastward at 250 m/s when it encounters a crosswind that gives it a southward acceleration of 1.2 m/s². What are the magnitude and direction of its velocity at the end of 1 minute? Repeat for the case of an eastward acceleration.

Solution

Choosing a coordinate system with the x axis eastward and the y axis northward, the initial eastward velocity is $\mathbf{v}_0 = 250\hat{\imath}$ m/s. Similarly, the southward acceleration is $\mathbf{a} = -1.2\hat{\jmath}$ m/s². Using this acceleration in Equation 4-1 gives the change in velocity over the 1-min interval:

$$\Delta\mathbf{v} = \mathbf{a}\Delta t = (-1.2\hat{\jmath} \text{ m/s})(60 \text{ s}) = -72\hat{\jmath} \text{ m/s}.$$

Then the velocity at the end of 1 min is given by

$$\mathbf{v} = \mathbf{v}_0 + \Delta\mathbf{v} = 250\hat{\imath} \text{ m/s} - 72\hat{\jmath} \text{ m/s}.$$

The magnitude of the vector $\mathbf{v}$ is given by Equation 3-1:

$$v = \sqrt{v_x^2 + v_y^2} = \sqrt{(250 \text{ m/s})^2 + (-72 \text{ m/s})^2} = 260 \text{ m/s},$$

while the tangent of the angle θ that $\mathbf{v}$ makes with the x-axis is the ratio of v_y to v_x:

$$\theta = \tan^{-1}\left(\frac{v_y}{v_x}\right) = \tan^{-1}\left(\frac{-72 \text{ m/s}}{250 \text{ m/s}}\right) = -16°,$$

or 16° south of east. Note that this acceleration—directed initially at right angles to the velocity—has relatively little effect on the magnitude of the velocity. The latter increases by only 10 m/s, even though the acceleration produces a 72 m/s change in the southward component of the velocity. This is essentially the situation shown in Fig. 4-2a.

When the acceleration acts eastward, the situation is reduced to a one-dimensional problem. In vector notation, we still have $\mathbf{v}_0 = 250\hat{\imath}$ m/s, but now $\mathbf{a} = 1.2\hat{\imath}$ m/s², so $\Delta\mathbf{v} = 72\hat{\imath}$ m/s. Then the velocity after 1 min is

$$\mathbf{v} = \mathbf{v}_0 + \Delta\mathbf{v} = 250\hat{\imath} \text{ m/s} + 72\hat{\imath} \text{ m/s} = 322\hat{\imath} \text{ m/s}.$$

Here the acceleration—now in the same direction as the velocity—results in a substantial change in speed but no change in direction. This case is essentially that of Fig. 4-1a.

EXERCISE   A hockey puck is gliding across the ice at 14 m/s, when a player's stick gives it an acceleration of 81 m/s² at an angle of 72° to its initial velocity. If the acceleration is applied for 0.23 s, what are the direction and magnitude of the final velocity?

Answer: 27 m/s at 42°

Some problems similar to Example 4-1: 3, 4   ●

4-2 CONSTANT ACCELERATION

When acceleration is constant—no change in either magnitude or direction—then the individual components of the acceleration vector must themselves be constants. Experimentally, we find that the component of acceleration in one direction has no effect on the motion in a perpendicular direction (Fig. 4-4). Then with constant acceleration, the separate components of the motion must obey the constant-acceleration formulas we developed in Chapter 2 for one-dimensional motion. Using vector notation, we can then generalize Equations 2-7 and 2-10 to read

$$\mathbf{v} = \mathbf{v}_0 + \mathbf{a}t \tag{4-3}$$

$$\mathbf{r} = \mathbf{r}_0 + \mathbf{v}_0 t + \tfrac{1}{2}\mathbf{a}t^2. \tag{4-4}$$

In two dimensions, each of these vector equations represents a pair of scalar equations describing constant acceleration in two mutually perpendicular directions; Equation 4-4, for example, contains the pair $x = x_0 + v_{x0}t + \tfrac{1}{2}a_x t^2$ and $y = y_0 + v_{y0}t + \tfrac{1}{2}a_y t^2$. In three dimensions there would be a third equation for the z component of the motion. The other one-dimensional equations of Table 2-1 also hold for the individual components of multidimensional motion.

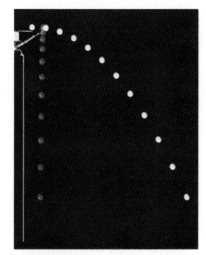

FIGURE 4-4   Strobe photo showing two marbles, one dropped and the other projected horizontally at the same time. At each instant their vertical positions are identical, showing that the vertical and horizontal motions are independent. Time interval between images is 0.025 s.

● **EXAMPLE 4-2** A SPACESHIP ACCELERATES

A spaceship is traveling in a straight line at 6.2 km/s, when a rocket is fired that gives it an acceleration of 0.36 km/s^2 in a direction 60° to its initial velocity. If the rocket firing lasts 22 s, what is its net displacement during the firing?

Solution
Equation 4-4 gives the position of an object undergoing constant acceleration. Since we're working in two dimensions, we need an x-y coordinate system. A reasonable choice is to take the x axis along the direction of the spaceship's initial velocity; then we have $\mathbf{v}_0 = 6.2\hat{\mathbf{i}}$ km/s and, as Fig. 4-5 shows, $\mathbf{a} = 0.18\hat{\mathbf{i}} + 0.31\hat{\mathbf{j}}$ km/s^2. (Another reasonable choice would be to take the x axis along the acceleration direction; then the acceleration would have only one nonzero component, while the initial velocity would have two.) If we choose the origin to be the point where the rocket is first fired, we have $x_0 = y_0 = 0$, and the two components of Equation 4-4 become

$$x = v_{x0}t + \tfrac{1}{2}a_x t^2 = (6.2 \text{ km/s})(22 \text{ s})$$
$$+ \tfrac{1}{2}(0.18 \text{ km/s}^2)(22 \text{ s})^2 = 180 \text{ km}$$
$$y = \tfrac{1}{2}a_y t^2 = \tfrac{1}{2}(0.31 \text{ m/s}^2)(22 \text{ s})^2 = 75 \text{ km},$$

where in writing the y equation, we used the fact that $v_{y0} = 0$. The new position vector is then given by $\mathbf{r} = x\hat{\mathbf{i}} + y\hat{\mathbf{j}} = 180\hat{\mathbf{i}} + 75\hat{\mathbf{j}}$ km, so the net displacement is

$$r = \sqrt{x^2 + y^2} = \sqrt{(180 \text{ km})^2 + (75 \text{ km})^2} = 195 \text{ km}.$$

Figure 4-6 shows the position vector $\mathbf{r}$ formed from the sum of component vectors $x\hat{\mathbf{i}}$ and $y\hat{\mathbf{j}}$. Although this vector gives the overall displacement of the spaceship during the rocket firing, it is *not* the path of the ship; rather, since the ship's velocity is constantly changing, the actual path is the curve shown in Fig. 4-6.

EXERCISE A sailboard is sailing at 7.3 m/s when a gust of wind hits, causing it to accelerate at 0.82 m/s^2 at a 75° angle to its original direction of motion. If the acceleration lasts 8.7 s, what is the board's net displacement during the wind gust?

Answer: 77.6 m

Some problems similar to Example 4-2: 7, 9, 10

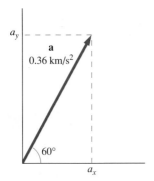

FIGURE 4-5 The components of the acceleration vector are $a_x = a \cos 60° = 0.18$ km/s^2 and $a_y = a \sin 60° = 0.31$ km/s^2, where $a = 0.36$ km/s^2 is the magnitude of the acceleration vector.

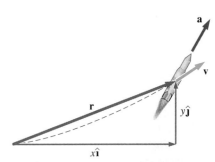

FIGURE 4-6 Displacement vector for the accelerating spaceship. Because the ship's velocity is constantly changing, its actual trajectory is the curved path shown. Note that displacement, velocity, and acceleration vectors all point in different directions. ●

As Example 4-2 shows, we handle problems of multidimensional motion by breaking them into two or three one-dimensional problems. We solve each of these using the methods of Chapter 2, then combine the solutions to give a full vector description of the motion.

4-3 PROJECTILE MOTION

A **projectile** is an object that is launched into the air and then moves predominantly under the influence of gravity. Examples are numerous: baseballs, streams of water, fireworks, missiles, ejecta from volcanoes, drops of ink in an ink-jet computer printer, and leaping dolphins are all projectiles (Fig. 4-7).

To treat projectile motion, we make two simplifying assumptions: (1) we neglect any variation in the direction or magnitude of the gravitational acceleration and (2) we neglect air resistance. Assumption 1 is equivalent to neglecting the curvature of the Earth, and is valid for projectiles whose displacements are small compared with Earth's radius. Air resistance has a more variable effect; for dense, compact objects it is often negligible, but for objects whose ratio of surface area to mass is large—like ping-pong balls and parachutes—air resistance may dramatically alter the motion (Fig 4-8). And the more subtle motions of a baseball—its curves and wobbles—can only be explained by invoking air resistance.

To describe projectile motion, it is usually convenient to choose a coordinate system with the y axis vertically upward, and the x axis horizontal and in the direction of the horizontal component of the projectile's initial velocity. Then there is neither velocity nor acceleration in the z direction, and we have purely two-dimensional motion in the x-y plane. Furthermore, with the only acceleration provided by gravity, $a_x = 0$ and $a_y = -g$, so the components of Equations 4-3 and 4-4 become

FIGURE 4-7   Sparks fly in a welding operation. Each spark describes a parabolic trajectory.

FIGURE 4-8   Air resistance has a dominant effect on a parachutist's motion, so the trajectory is not parabolic. Here the air resistance force $\mathbf{F}_a$ and the gravitational force $\mathbf{F}_g$ balance, so the parachutist floats down at constant speed.

$$v_x = v_{x0} \tag{4-5}$$

$$v_y = v_{y0} - gt \tag{4-6}$$

$$x = x_0 + v_{x0}t \tag{4-7}$$

$$y = y_0 + v_{y0}t - \tfrac{1}{2}gt^2. \tag{4-8}$$

In writing these equations, we take g to be a positive number, as we did in Chapter 2, and account for the downward direction using minus signs. Equations 4-5 to 4-8 tell us mathematically what Fig. 4-4 tells us physically: that the horizontal motion of a projectile is not affected by gravity, and that the vertical motion is not affected by the horizontal motion.

● **EXAMPLE 4-3** OVER THE EDGE

Standing on the edge of a 120-m-high cliff, you throw a rock horizontally at 14 m/s. How far from the bottom of the cliff does it land? Neglect air resistance (although this assumption is not entirely justified for such a long fall).

Solution

Since there is no component of acceleration in the horizontal direction, the horizontal component of the velocity remains constant at $v_x = 14$ m/s. The answer we seek is the distance the rock travels horizontally; knowing the horizontal component of velocity, we could find this distance if we knew the flight time.

We can find the time by analyzing the vertical motion. Since the rock is thrown horizontally, $v_{y0} = 0$. If we choose our origin at the cliff bottom, then $y_0 = 120$ m. We want to find the time when the rock hits the ground—that is, when $y = 0$. Setting y and v_{y0} to zero in Equation 4-8, and solving for t, we have

$$t = \sqrt{\frac{2y_0}{g}} = \sqrt{\frac{(2)(120 \text{ m})}{9.8 \text{ m/s}^2}} = 4.95 \text{ s}.$$

During this time, the rock drops the 120-m distance to the bottom of the cliff. The horizontal displacement during this time is given by Equation 4-7:

$$x - x_0 = v_{x0}t = (14 \text{ m/s})(4.95 \text{ s}) = 69 \text{ m}.$$

Instead of solving for a numerical value of t, we could have used the expression $t = \sqrt{2y_0/g}$ in Equation 4-7 before substituting numerical values. It is usually best to carry calculations as far as possible in symbolic form.

Some problems similar to Example 4-3: 15, 24, 25

●

TIP **Multistep Problems** In Example 4-3, we were asked for the horizontal position of the rock. But to find that, we needed to know the time of flight—which we weren't given. This is a common situation in all but the simplest "plug-in" physics problems: It isn't immediately obvious how to get from the given information to the final answer. In such cases it's often best to work backwards, asking "What would I need to know to get the answer?" and then figuring out how to get that quantity from the given information. Here the missing quantity was the time, which we got by analyzing the vertical motion. In essence, we had to formulate a second problem: "Given a vertical drop of 120 m, how long does it take an object to fall?" Solving that problem then gave the information we needed to solve the original problem.

When faced with a multistep problem, don't hesitate to assign a symbol to any unknown quantity you need—in this case the time, t—even if the problem doesn't mention it. That quantity will not appear in your final answer, but it may be essential in doing the intermediate calculations.

In Example 4-3 we were asked about the rock's horizontal position when it hit the ground. We didn't really care about the time of its flight, although we needed it to solve the problem. More generally, we are often interested in describing the path, or **trajectory,** of a projectile without the details of where it is at each instant of time. A look at Fig. 4-7 suggests that objects undergoing projectile motion have similarly shaped trajectories. What is that shape?

Mathematically, we could specify a projectile's trajectory by giving its height y as a function of horizontal position x. To do so, consider a projectile launched from the origin ($x_0 = y_0 = 0$) at some angle θ_0 to the horizontal, with initial speed v_0. As Fig. 4-9 suggests, the components of the initial velocity are $v_{x0} = v_0 \cos \theta_0$ and $v_{y0} = v_0 \sin \theta_0$. Then Equations 4-7 and 4-8 become

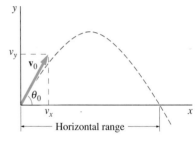

FIGURE 4-9 Parabolic trajectory of a projectile.

$$x = v_0 \cos \theta_0 t$$

and

$$y = v_0 \sin \theta_0 t - \tfrac{1}{2} g t^2.$$

Solving the x equation for the time t gives

$$t = \frac{x}{v_0 \cos \theta_0};$$

using this result in the y equation, we have

$$y = v_0 \sin \theta_0 \left(\frac{x}{v_0 \cos \theta_0} \right) - \frac{1}{2} g \left(\frac{x}{v_0 \cos \theta_0} \right)^2,$$

or

$$y = x \tan \theta_0 - \frac{g}{2v_0^2 \cos^2 \theta_0} x^2. \quad \text{(projectile trajectory)} \qquad (4\text{-}9)$$

In Equation 4-9 we have the mathematical description of the projectile's trajectory. Since y is a quadratic function of x, the trajectory is a parabola.

● EXAMPLE 4-4 OUT OF THE HOLE

A construction worker is standing in a 2.6-m-deep cellar hole, 3.1 m from the side of the hole. He tosses a hammer to a companion outside the hole. If the hammer leaves his hand 1.0 m above the bottom of the hole, at an angle of 35° to the horizontal, what is the minimum speed it must have to clear the edge of the hole, as shown in Fig. 4-10? How far from the edge of the hole does it land?

Solution
If it just clears the edge of the hole, the hammer must be on a trajectory that has $y = 1.6$ m when $x = 3.1$ m, where we take the origin at the point where the hammer leaves his hand, 1.0 m

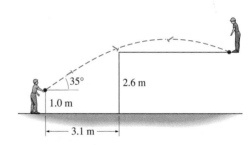

FIGURE 4-10 Example 4-4.

above the bottom of the 2.6-m-deep hole. Solving Equation 4-9 for v_0 and using these x and y values gives

$$v_0 = \sqrt{\frac{gx^2}{2\cos^2\theta_0(x\tan\theta_0 - y)}}$$

$$= \sqrt{\frac{(9.8 \text{ m/s}^2)(3.1 \text{ m})^2}{(2)(\cos^2 35°)[(3.1 \text{ m})(\tan 35°) - 1.6 \text{ m}]}} = 11 \text{ m/s}.$$

To find where the hammer lands on the level ground outside the hole, we need to know the horizontal position when $y = 1.6$ m. Rearranging Equation 4-9 into the standard form $ax^2 + bx + c = 0$ for a quadratic equation gives

$$\frac{g}{2v_0^2\cos^2\theta_0}x^2 - (\tan\theta_0)x + y = 0,$$

so the coefficients a, b, and c are $a = g/2v_0^2\cos^2\theta_0 = 0.0594 \text{ m}^{-1}$, $b = -\tan\theta_0 = -0.700$, and $c = y = 1.6$ m. Applying the quadratic formula (see Appendix A) then gives

$$x = \frac{-b \pm \sqrt{b^2 - 4ac}}{2a}$$

$$= \frac{0.700 \pm \sqrt{0.700^2 - (4)(0.0594)(1.6)}}{(2)(0.0594)}$$

$$= 3.1 \text{ m or } 8.7 \text{ m},$$

where for clarity we have left off units in the calculation. These two answers give the two horizontal positions where the hammer is at ground level. The first provides a useful check on our work: it is just the horizontal position when the hammer clears the edge of the hole. The second answer is the one we want; it shows that the hammer lands 8.7 m − 3.1 m = 5.6 m to the right of the hole's edge.

EXERCISE A daredevil motorcyclist attempts to leap a 48-m-wide gorge, as shown in Fig. 4-11. At the side where the cyclist starts, the ground slopes upward at 15° to the gorge rim, as shown in Fig. 4-11. Beyond the far rim the ground is level and is 5.9 m below the near rim, as shown. (a) What is the minimum speed necessary for the cyclist to clear the gorge? (b) If the cyclist exceeds that minimum by 50%, how far from the far rim will he land?

Answers: (a) 25 m/s; (b) 44 m

Some problems similar to Example 4-4: 31, 32, 36, 59, 60

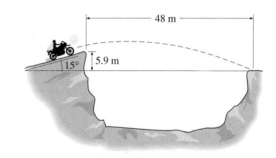

FIGURE 4-11 A daredevil cyclist.

How far will a soccer ball go if I kick it at 12 m/s at 50° to the horizontal? If I can throw a rock at 15 m/s, can I get it across a 30-m-wide pond? At what angle should I throw it for maximum distance? How far off vertical can a rocket's trajectory be and still have it land within 50 km of its launch point? As in these examples, we are frequently interested in the **horizontal range** of a projectile—that is, in how far it moves horizontally over level ground.

Equation 4-9 describes the trajectory—height y versus horizontal position x—for a projectile. For a projectile launched on level ground, we can ask when the projectile will return to the ground by setting $y = 0$ in Equation 4-9:

$$0 = x\tan\theta_0 - \frac{g}{2v_0^2\cos^2\theta_0}x^2.$$

Factoring this equation gives

$$x\left(\tan\theta_0 - \frac{gx}{2v_0^2\cos^2\theta_0}\right) = 0,$$

so either $x = 0$, corresponding to the launch point, or

$$x = \frac{2v_0^2}{g} \cos^2 \theta_0 \tan \theta_0 = \frac{2v_0^2}{g} \sin \theta_0 \cos \theta_0.$$

Here we have used $\tan \theta_0 = \sin \theta_0 / \cos \theta_0$; recalling further that $\sin 2\theta_0 = 2 \sin \theta_0 \cos \theta_0$ (see Appendix A), we can write

$$x = \frac{v_0^2}{g} \sin 2\theta_0. \qquad \text{(horizontal range)} \qquad (4\text{-}10)$$

TIP   **Know Your Limits**   We emphasize that Equation 4-10 gives the *horizontal* range—the distance a projectile travels horizontally before returning *to its starting height*. From the way it was derived—setting $y = 0$—you can see that it does *not* give the horizontal distance when the projectile returns to a different height (Fig. 4-12).

Does Equation 4-10 make sense? When $\theta_0 = 0$, the range is zero: a projectile launched horizontally on level ground immediately hits the ground. When $\theta_0 = 90°$, $\sin 2\theta_0 = \sin 180° = 0$, and again the range is zero. Here the projectile is launched vertically upward, so of course it returns to the same point. When is the range a maximum? The largest value of $\sin 2\theta_0$ is 1, which occurs when $\theta_0 = 45°$. For a given initial speed v_0, then, the maximum range is attained by launching at 45°. Figure 4-13 shows the trajectories of projectiles launched at different angles with the same initial speed. At angles less than 45° the projectile has a greater horizontal component of velocity, but it doesn't get as high and therefore isn't in the air as long as the projectile launched at 45°; the net effect is a shorter range. At angles greater than 45°, the projectile rises higher and is therefore in the air

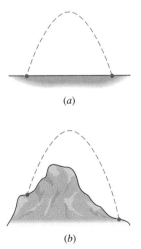

(a)

(b)

FIGURE 4-12   (*a*) Equation 4-10 gives the horizontal range of a particle that returns to its starting height. (*b*) It does *not* apply otherwise.

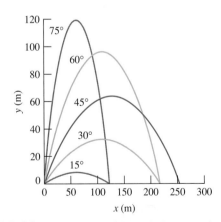

FIGURE 4-13   For a given initial speed, the range of a projectile is maximum at a launch angle of 45°. For launch angles equally spaced above and below 45°, the range is the same. Curves are for an initial speed of 50 m/s, with launch angles indicated.

longer, but its horizontal velocity is lower and the net effect is again a shorter range. In fact, as you can prove in Problem 41, the range is the same for angles equally spaced on either side of 45°. (These results are valid only in the absence of air resistance; when air resistance is taken into account, the maximum range angle turns out to be less than 45°.)

● EXAMPLE 4-5   A SOUNDING ROCKET

A sounding rocket is launched from a remote desert site to probe the atmosphere. With a short firing of its engine, the rocket quickly reaches a speed of 4.6 km/s. If the rocket is to land within 50 km of its launch site, what is the maximum allowable deviation from a vertical trajectory? Neglect air resistance.

Solution

Here we neglect the short distance over which the rocket accelerates, considering that it left the ground with its initial speed of 4.6 km/s. Then the answer to the problem is the angle θ_0 for which the range x is 50 km. Solving Equation 4-10 for $\sin 2\theta_0$ gives

$$\sin 2\theta_0 = \frac{gx}{v_0^2} = \frac{(9.8 \text{ m/s}^2)(50 \times 10^3 \text{ m})}{(4.6 \times 10^3 \text{ m/s})^2} = 0.0232.$$

There are two solutions to this equation, corresponding to $2\theta_0 = \sin^{-1}(0.0232) = 1.33°$ and $2\theta_0 = 180° - 1.33°$. The second solution is the one we want; it represents a nearly vertical launch angle $\theta_0 = 90° - 0.67°$, so the launch angle must be within 0.67° of vertical.

EXERCISE   As an engineer, you're asked to design a fountain for a public park. Water droplets squirt from holes arranged around a circle of negligible size, and the water is supposed to land in a circular trough of radius 1.7 m. If the water emerges from the holes at 4.3 m/s, at what angle should the holes be oriented? Choose the answer that gives the maximum fountain height consistent with the given radius and initial speed.

Answer: 58°

Some problems similar to Example 4-5:   35, 37   ●

● EXAMPLE 4-6   LIGHTING THE OLYMPIC FLAME

The Olympic Flame at the 1992 Barcelona Olympics was lit by a flaming arrow shot from a point located 60 m horizontally and 24 m vertically from the flame (Fig. 4-14). Find the initial speed necessary for the arrow to reach the flame at the peak of its trajectory. Ignore aerodynamic effects (actually significant for an arrow).

Solution

Let $\ell = 60$ m and $h = 24$ m be the horizontal and vertical distances, respectively. If the trajectory is to peak at $x = \ell$, the range must be 2ℓ and Equation 4-10 then gives

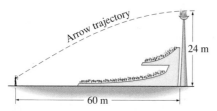

Arrow trajectory

24 m

60 m

FIGURE 4-14   Cross section of the Olympic stadium in Barcelona, Spain, showing trajectory of arrow used to light the Olympic flame (Example 4-6).

$$2\ell = \frac{v_0^2}{g} \sin 2\theta_0,$$

or

$$v_0^2 = \frac{2g\ell}{\sin 2\theta_0} = \frac{g\ell}{\sin \theta_0 \cos \theta_0}.$$

The trajectory must have $y = h$ when $x = \ell$; using our expression for v_0^2 in the trajectory equation 4-9 then gives

$$h = \ell \tan \theta_0 - \frac{g}{2 \dfrac{g\ell}{\sin \theta_0 \cos \theta_0} \cos^2 \theta_0} \ell^2$$

$$= \ell \tan \theta_0 - \tfrac{1}{2}\ell \tan \theta_0 = \tfrac{1}{2}\ell \tan \theta_0.$$

Thus the launch angle is

$$\theta_0 = \tan^{-1}\left(\frac{2h}{\ell}\right) = \tan^{-1}\left(\frac{(2)(24 \text{ m})}{60 \text{ m}}\right) = 38.66°,$$

which then gives an initial speed of

$$v_0 = \sqrt{\frac{2g\ell}{\sin 2\theta_0}} = \sqrt{\frac{(2)(9.8 \text{ m/s}^2)(60 \text{ m})}{\sin[(2)(38.66°)]}} = 34.7 \text{ m/s}.$$

TIP **Trust Your Algebra** This example was a bit more complicated, in that it wasn't at all obvious at first how to get the answer. But the two physical constraints—the given range and height—could be expressed by two independent algebraic equations, each containing the two unknowns θ_0 and v_0. And that's enough to ensure that if a solution exists, it can be found—usually by the methods of algebra.

EXERCISE A zany astronaut takes a golf ball and club to the moon and wants to hit the ball over a 380-m high hill on an otherwise level plain. If the astronaut is 870 m horizontally from the hilltop, at what speed and angle should the ball be hit so it just clears the hill at the peak of its trajectory? Would either or both of your answers differ if this were Earth rather than the moon?

Answer: 53 m/s at 41°; only speed would be different on Earth

Some problems similar to Example 4-6: 36, 74, 75

●

■ APPLICATION BALLISTIC MISSILE DEFENSE

Since the development of long-range nuclear-armed missiles in the late 1950s, it's been the case that the United States could be destroyed in half an hour—and there's nothing we can do to prevent it. Although the threat of nuclear attack by the former Soviet Union has diminished since the end of the Cold War, the proliferation of missile technology raises the danger that others could launch a missile attack. And already ballistic missiles are playing a role in regional conflicts, like the 1991 Persian Gulf war.

Is there a defense against ballistic missiles? That question is wrapped in technological and political controversy, but the general parameters in which it must be answered are determined by the physics of projectile motion. A ballistic missile is one that is given an initial rocket boost, then allowed to coast under the influence of gravity alone; as soon as the rockets stop firing, the missile is simply a projectile. (So-called cruise missiles, in contrast, are like airplanes, using aerodynamic lift and engine power to get to their targets.)

The range of intercontinental missiles is so great that Earth's curvature becomes important, rendering invalid our simplifying assumption that the gravitational acceleration is constant. But for short- and intermediate-range missiles, treating the missile like a projectile is a good approximation. Consider, for example, a missile launched from a submarine 1,000 km off the coast (Fig. 4-15). How long does the defense have to deal with the attack?

For the optimum launch angle of 45°, Equation 4-10 shows that a 1,000-km range requires a launch speed of 3.1 km/s; the horizontal component of the velocity is then $v_x = (3.1 \text{ km/s})(\cos 45°) = 2.2 \text{ km/s}$. At this rate, the missile flight lasts only (1,000 km)/(2.2 km/s) = 450 s or 7.5 minutes! It might take a minute or so to detect the launch, determine the missile's trajectory, confirm that the attack was real, and notify defensive systems. That leaves very little time to intercept the incoming missile. Many of today's missiles are armed with multiple warheads that get released on separate trajectories early in the flight, giving the defense even less time to destroy a single missile before it becomes a "threat cloud" of multiple dangers.

Since the early 1980s, the United States has spent billions of dollars on schemes for defense against ballistic missiles. Although many scientists and engineers are convinced that an

FIGURE 4-15 The flight time for a submarine-launched ballistic missile can be under 10 minutes.

effective defense is technologically impossible, the search continues. Some of the most exotic proposals involve high-power lasers and particle beams, which would function most effectively in the vacuum of space. For submarine-launched missiles, a simple expedient to foil such systems is to launch at less than the optimum angle. At 45°, for example, our 1,000-km-range missile reaches a peak altitude given by solving Equation 2-14 for y, with $v_y = 0$ at the peak:

$$y = \frac{v_{y0}^2}{2g} = \frac{[(3100 \text{ m/s})(\sin 45°)]^2}{(2)(9.8 \text{ m/s}^2)} = 245 \text{ km},$$

well above the atmosphere. But a launch angle of 15°, for example, gives a maximum altitude of 67 km, well within the atmosphere. There's a cost to the attacker: The launch speed now rises to 4.4 km/s, requiring a more powerful rocket. But there's also an advantage: the missile's flight time drops to just 3.9 minutes.

Problems 29 and 42 explore further the physics of ballistic missiles.

4-4 UNIFORM CIRCULAR MOTION

An important case of accelerated motion in two dimensions is **uniform circular motion**—the motion of an object describing a circular path at constant speed. Although the speed is constant, the motion is accelerated because the *direction* of the velocity is changing.

Examples of uniform circular motion are numerous. Many spacecraft are in circular orbits, and the orbits of the planets and their natural satellites are approximately circular. The Earth's daily rotation carries you around in uniform circular motion. Pieces of rotating machinery describe uniform circular motion. In laboratory apparatus and astrophysical objects, electrons undergo uniform circular motion when they encounter magnetic fields. Even with the atom itself, we can gain some insight by picturing electrons in uniform circular motion about the nucleus—although this view is not fully consistent with modern quantum mechanics. Other situations involve uniform circular motion over a limited path. Driving around a curve at constant speed, you are temporarily in uniform circular motion. As you swing a baseball bat, golf club, or hockey stick, you produce motion that is circular and that may be approximately uniform.

Here we derive an important relation among the acceleration, speed, and radius of uniform circular motion. Figure 4-16 shows several velocity vectors for an object moving with speed v around a circle of radius r. Note that the velocity vectors are tangent to the circle, indicating the instantaneous direction of motion. In Fig. 4-17a we focus on two nearby points described by position vectors $\mathbf{r}_1$ and $\mathbf{r}_2$, showing also the velocity vectors $\mathbf{v}_1$ and $\mathbf{v}_2$. Figure 4-17b shows the displacement $\Delta\mathbf{r}$ that represents the difference between $\mathbf{r}_1$ and $\mathbf{r}_2$; similarly, Fig. 4-17c shows the velocity difference $\Delta\mathbf{v}$.

Because $\mathbf{v}_1$ is perpendicular to $\mathbf{r}_1$, and $\mathbf{v}_2$ is perpendicular to $\mathbf{r}_2$, the angles θ shown in all three parts of Fig. 4-17 are the same angle. Therefore the triangles in Fig. 4-17b and c are similar, and we can write

$$\frac{\Delta v}{v} = \frac{\Delta r}{r}.$$

Now suppose the angle θ is small, corresponding to a short time interval Δt for motion from position $\mathbf{r}_1$ to $\mathbf{r}_2$. Then the length of the vector $\Delta\mathbf{r}$ is approximately that of the circular arc joining the endpoints of the position vectors, as suggested in Fig. 4-17b. The length of this arc is just the distance traveled by the object in

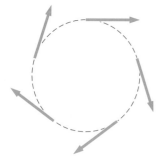

FIGURE 4-16 Velocity vectors for an object in uniform circular motion are tangent to the circular path.

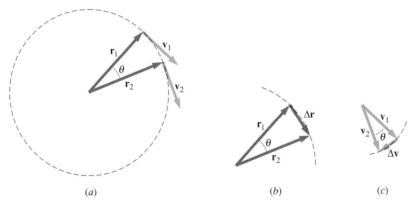

FIGURE 4-17 (a) Position and velocity vectors for two nearby points on the circular path. (b) The vectors $\mathbf{r}_1$ and $\mathbf{r}_2$ differ by $\Delta\mathbf{r}$. For small angles θ, the length of $\Delta\mathbf{r}$ is very nearly equal to the arc connecting the ends of $\mathbf{r}_1$ and $\mathbf{r}_2$. (c) Similarly, vectors $\mathbf{v}_1$ and $\mathbf{v}_2$ differ by $\Delta\mathbf{v}$. The triangles in (b) and (c) are similar, so the angles θ are the same.

the time Δt, or $v\Delta t$, so we can write $\Delta r \simeq v\Delta t$. Then our relation between similar triangles becomes

$$\frac{\Delta v}{v} \simeq \frac{v\Delta t}{r}.$$

Rearranging this equation gives an approximate expression for the magnitude of the average acceleration:

$$\bar{a} = \frac{\Delta v}{\Delta t} \simeq \frac{v^2}{r}.$$

Taking the limit $\Delta t \to 0$ gives the instantaneous acceleration; in this limit the angle θ approaches 0, the circular arc and the vector $\Delta\mathbf{r}$ become indistinguishable, and the relation $\Delta r \simeq v\Delta t$ becomes exact. So we have

$$a = \frac{v^2}{r} \text{ (uniform circular motion)} \qquad (4\text{-}11)$$

for the magnitude of the instantaneous acceleration of an object moving in a circle of radius r at constant speed v. What about the direction? As Fig. 4-17c suggests, $\Delta\mathbf{v}$ is very nearly perpendicular to both velocity vectors; in the limit $\Delta t \to 0$, $\Delta\mathbf{v}$ and therefore the acceleration $\Delta\mathbf{v}/\Delta t$ becomes exactly perpendicular to the velocity. The direction of the acceleration vector is therefore toward the center of the circle.

Clearly, our geometrical argument would have applied to any point on the circle, so we conclude that the acceleration in uniform circular motion has constant magnitude v^2/r, and that the acceleration vector always points toward the center of the circle. Isaac Newton coined the term *centripetal* to describe this center-pointing acceleration. However, we'll use that term sparingly because we want to emphasize that centripetal acceleration is not fundamentally different from any other acceleration: It's simply a vector describing the rate of change of velocity.

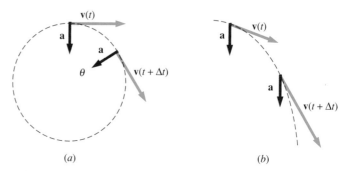

FIGURE 4-18 (*a*) In uniform circular motion, the acceleration vector is always perpendicular to the velocity, and only the direction of the velocity changes. Although the acceleration vector is constant in magnitude, its direction continually changes. (*b*) With truly constant acceleration, the velocity and acceleration cannot stay perpendicular, and the velocity changes in both direction and magnitude unless it happens to be colinear with the acceleration.

Does Equation 4-11 make sense? Yes. An increase in speed v means the time Δt for a given change in direction of the velocity becomes shorter. Not only that, but the associated change $\Delta \mathbf{v}$ in velocity is larger. These two effects combine to give an acceleration that depends on the *square* of the speed. On the other hand, an increase in radius with a fixed speed simply increases the time Δt associated with a given change in velocity, so the acceleration is inversely proportional to the radius. Although the form of the acceleration in Equation 4-11 does not look familiar from our work with constant acceleration, you can verify that its SI units are indeed those of acceleration—m/s^2.

Although we derived Equation 4-11 geometrically, we could have obtained the same result by writing the position in unit vector notation and then differentiating twice to get the acceleration. Problem 82 explores this approach.

> **TIP**   **Circular Motion and Constant Acceleration**   The direction toward the center changes as an object moves around a circular path, so the acceleration vector is *not constant,* even though its magnitude is. Uniform circular motion is *not* motion with constant acceleration, and our constant-acceleration equations *do not apply.* In fact, we know that constant acceleration in two dimensions implies a parabolic trajectory, not a circle. Figure 4-18 clarifies this difference.

● EXAMPLE 4-7   A SPACE SHUTTLE ORBIT

A space shuttle is in a circular orbit at a height of 250 km, where the acceleration of Earth's gravity is 93% of its surface value. What is the period of its orbit? ("Period" means the time to complete one orbit.)

Solution
Here we are given the acceleration and radius and asked to find the period. Our equations for uniform circular motion relate acceleration, radius, and speed. But we can write the speed as the distance traveled in one orbit—the orbital circumference $2\pi r$—divided by the orbital period T:

$$v = \frac{2\pi r}{T}.$$

Using this result in Equation 4-11, we have

$$a = \frac{v^2}{r} = \frac{(2\pi r/T)^2}{r} = \frac{4\pi^2 r}{T^2}. \qquad (4\text{-}12)$$

This equation is equivalent to Equation 4-11, but relates a, r, and T rather than a, r, and v. Here we are given that $a = 0.93g$ and that the shuttle is 250 km above Earth's surface. But the r in Equation 4-12 is the distance to the center of the circular path—

that is, to the center of the Earth. So $r = R_{Earth} + 250$ km $= 6.62 \times 10^6$ m, where we used $R_{Earth} = 6.37 \times 10^6$ m from Appendix E. Solving Equation 4-12 for the period T then gives

$$T = \left(\frac{4\pi^2 r}{a}\right)^{1/2} = \left(\frac{(4\pi^2)(6.62 \times 10^6 \text{ m})}{(0.93)(9.8 \text{ m/s}^2)}\right)^{1/2}$$

$$= 5355 \text{ s} = 89 \text{ min},$$

or about an hour and a half. This value is approximately the orbital period of any object orbiting the Earth at altitudes that are small compared with Earth's radius. Scientists and engineers have no choice here; the orbital period is fixed by the size and mass of the Earth. (At higher altitudes the decrease in the strength of gravity becomes more noticeable, and orbital periods lengthen. The moon, 390,000 km distant, orbits Earth with a period of 27 days. We will study gravity and orbits further in Chapter 9.)

EXERCISE An airplane on its landing approach executes a 180° turn by describing a semicircular path, at constant speed, with radius 5.3 km. An accelerometer in the cockpit reads 1.7 m/s² during the turn. How long will it take to complete the turn?

Answer: 2.92 min

Some problems similar to Example 4-7: 47, 51–53

● **EXAMPLE 4-8** ENGINEERING A ROAD

A flat, horizontal road is being designed for an 80-km/h speed limit. If the maximum acceleration of a car traveling this road at the speed limit is to be 1.5 m/s², what is the minimum radius for curves in the road?

Solution
Here we are given speed and acceleration, so we can solve Equation 4-11 for the radius r:

$$r = \frac{v^2}{a} = \frac{[(80 \times 10^3 \text{ m/h})(1 \text{ h}/3600 \text{ s})]^2}{1.5 \text{ m/s}^2} = 330 \text{ m}.$$

A limit on acceleration is necessary to keep the car on the road, as we will see more clearly in Chapters 5 and 6.

EXERCISE An engineer is designing a circular saw blade. For cutting efficiency, the teeth of the blade are to move at 40 m/s. The bonds that hold the carbide cutting tips to the blade can withstand a maximum acceleration of 2.0×10^4 m/s². What is the maximum diameter of the blade?

Answer: 16 cm

Some problems similar to Example 4-8: 49, 50 ●

4-5 NONUNIFORM CIRCULAR MOTION

What about circular motion whose speed changes, like a ball whirling around in a vertical circle (Fig 4-19)? Since both the direction and magnitude of the velocity are changing, the acceleration cannot be either parallel or perpendicular to the velocity. But we can break the acceleration vector into two parts, one perpendicular to the velocity and one parallel. We call the former the **radial acceleration** a_r, since it points radially inward toward the center of the circular path; the latter is the **tangential acceleration** a_t, since it is tangent to the path. The radial acceleration changes only the direction of the velocity; it acts just like the acceleration in uniform circular motion, and therefore has magnitude $a_r = v^2/r$. The tangential acceleration changes only the magnitude of the velocity; it acts just like one-dimensional acceleration, and therefore has magnitude $a_t = dv/dt$. The net acceleration vector can be found by combining these two components, as Example 4-9 illustrates.

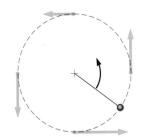

FIGURE 4-19 A ball whirled about on a string in a vertical circle undergoes circular motion with nonconstant speed; it moves faster near the bottom and more slowly near the top.

● **EXAMPLE 4-9** TAKING THAT CURVE TOO FAST

A road makes a 90° bend with a radius of 190 m. A car enters the bend moving at 20 m/s. Finding this too fast, the driver decelerates at 0.92 m/s². Determine the acceleration of the car when its speed rounding the bend has dropped to 15 m/s.

Solution
Since it is rounding a curve, the car has a radial acceleration associated with its changing direction, in addition to the tangential deceleration that changes its speed. We are given that $a_t =$

0.92 m/s^2; since the car is slowing down, the tangential acceleration is directed opposite the velocity. The radial acceleration is given by Equation 4-11:

$$a_r = \frac{v^2}{r} = \frac{(15 \text{ m/s})^2}{190 \text{ m}} = 1.2 \text{ m/s}^2.$$

Figure 4-20 shows both components of the acceleration and their vector sum, which is the net acceleration. From the figure, we see that this sum has magnitude

$$a = \sqrt{a_r^2 + a_t^2} = \sqrt{(1.2 \text{ m/s})^2 + (0.92 \text{ m/s})^2} = 1.5 \text{ m/s}^2$$

and points at an angle

$$\theta = \tan^{-1}\left(\frac{a_r}{a_t}\right) = \tan^{-1}\left(\frac{1.2 \text{ m/s}^2}{0.92 \text{ m/s}^2}\right) = 53°$$

relative to the tangent line to the circle, as shown.

EXERCISE You slip a compact disk into a CD player, and the disk rapidly spins up toward its operating speed. During the spin-up, what are the direction and magnitude of the accelera-

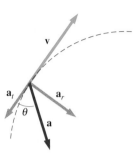

FIGURE 4-20 The radial and tangential accelerations add vectorially to give the net acceleration.

tion vector for a point on the rim of the 12.7-cm-diameter disk when the rim is moving at 31.2 cm/s, with its speed increasing at 91.4 cm/s^2?

Answer: 178 cm/s^2 at 59.2° to the direction of motion

Some problems similar to Example 4-9: 58, 69, 70

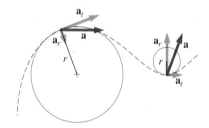

FIGURE 4-21 Each point on an arbitrary path can be characterized by the radius of curvature r at that point. An object moving with speed v has radial acceleration of magnitude v^2/r and tangential acceleration of magnitude dv/dt. In general, both v and r may change as the particle moves.

Although we've discussed radial and tangential acceleration in the context of a circular path, those concepts apply even when the path has an arbitrary shape. At any point on the path, we can define a **radius of curvature** that describes how sharply the path is bending (Fig. 4-21). (Mathematically, the radius of curvature is that of a circle tangent to the path and whose second derivative of one rectangular coordinate with respect to the other is equal to that of the given curve.) If a particle is moving at speed v at a point where the curvature radius is r, then its radial acceleration is still v^2/r; if its speed is also changing, then its tangential acceleration is still dv/dt. Now, however, both v and r may change as the particle moves along the path.

CHAPTER SYNOPSIS

Summary

1. The acceleration vector describes the rate of change of the velocity vector. In two or three dimensions, the acceleration can have any direction relative to the velocity vector. When the acceleration lies along the same line as the velocity, then only the magnitude of the velocity changes. When the acceleration is strictly perpendicular to the velocity, only the direction of the velocity changes. An acceleration at an arbitrary angle generally changes both the direction and magnitude of the velocity.

2. When an object undergoes **constant acceleration,** its motion is described by the vector equations

$$\mathbf{v} = \mathbf{v}_0 + \mathbf{a}t$$
$$\mathbf{r} = \mathbf{r}_0 + \mathbf{v}_0 t + \tfrac{1}{2}\mathbf{a}t^2.$$

These equations may be broken into components along any chosen coordinate axes; the result is a set of equations describing independently the components of the motion along the different axes.

3. **Projectile motion** is an important case of motion with constant acceleration. Launched from the origin ($x = y = 0$) with speed v_0 at an angle θ_0 to the horizontal, a projectile follows a parabolic trajectory described by

$$y = x \tan \theta_0 - \frac{g}{2v_0^2 \cos^2 \theta_0} x^2.$$

4. An object moving around a circle of radius r at constant speed v is in **uniform circular motion.** Since the direction of its velocity is changing, it is accelerating. The magnitude of its acceleration is

$$a = \frac{v^2}{r}, \quad \text{(uniform circular motion)}$$

and the acceleration vector always points toward the center of the circle.

5. When an object moves in a curved path with varying speed, the acceleration vector is the sum of the **radial acceleration** v^2/r, where r is the local radius of curvature, and the **tangential acceleration** dv/dt that is tangent to the circle. The radial acceleration is associated with changes in the direction of motion, while the tangential acceleration is associated with changes in speed.

Terms You Should Understand

constant acceleration (4-2)
projectile (4-3)
projectile motion (4-3)
trajectory (4-3)
range (of a projectile) (4-3)
uniform circular motion (4-4)
radial acceleration (4-5)
tangential acceleration (4-5)
radius of curvature (4-5)

Symbols You Should Recognize

vector equations, e.g., $\mathbf{r} = \mathbf{r}_0 + \mathbf{v}_0 t + \frac{1}{2}\mathbf{a}t^2$ (4-2)
$\mathbf{a}_r$, $\mathbf{a}_t$ (4-5)

Problems You Should Be Able to Solve

Use vector analysis to show the effects of accelerations at different angles to the velocity. (4-1)
Given initial position and velocity for objects undergoing constant acceleration, determine position and velocity as functions of time. (4-2)
Determine the motion of projectiles, including position and velocity as functions of time, trajectories, ranges, and initial conditions needed to achieve stated trajectories. (4-3)
Relate velocity, radius, and acceleration in circular motion. (4-4)
Determine radial and tangential acceleration for objects moving on curved trajectories. (4-5)

Limitations to Keep in Mind

Equations for constant acceleration apply *only* when both the magnitude and direction of the acceleration are unchanged.
Equation 4-10 for the horizontal range of a projectile applies *only* when the launch and landing points are at the same height.
Uniform circular motion is **accelerated** motion since the direction of the velocity vector changes.
Uniform circular motion is *not* motion with *constant* acceleration since the direction of the acceleration vector changes.

QUESTIONS

1. Moving along a road with the wheels pointed straight ahead, you step on the brake of your car. Describe the direction of the resulting acceleration in relation to your car's velocity.
2. How is it possible for an object to be moving in one direction but accelerating in another?
3. Is there any way to negotiate a curved path without accelerating?
4. A satellite glides through space at a steady 29,000 km/h in a circular orbit around Earth. Is the satellite accelerating? If so, in what direction? If not, why not?
5. You're moving northward when you accelerate briefly toward the east. Is your subsequent motion strictly eastward? Explain.
6. Why is the case of Fig. 4-2a not strictly a case of acceleration always perpendicular to velocity? After all, the velocity and acceleration vectors shown are perpendicular.
7. An object is moving at some modest initial speed, when a large acceleration is applied perpendicular to the direction of the initial velocity. What can you say, approximately,

about the relative directions of velocity and acceleration after the acceleration has been applied for a long time? Explain with a simple vector diagram.

8. In what sense is Equation 4-4 really two (or three) equations?
9. In a problem involving constant acceleration, why might it be sensible to choose a coordinate system with one axis along the direction of the acceleration?
10. Is the speed of a projectile constant throughout its parabolic trajectory?
11. Three objects are released from the same height, one from rest, one with horizontal velocity component v_x, the other with horizontal velocity component $2v_x$. Compare the times each is in the air. Compare the horizontal distances traveled by each.
12. A hunter's bullet emerges from a gun, moving horizontally above level ground. At the same instant, the hunter drops another bullet from the height of the gun muzzle. Which hits the ground first? Explain.

13. Can there be a true "line drive" in baseball? What is required for the ball to travel in a nearly straight, horizontal trajectory?

14. What is the vertical component of a projectile's velocity at the peak of its trajectory?

15. In terms of its initial velocity, what is the horizontal component of a projectile's velocity at the peak of its trajectory?

16. Is there any point on a projectile's trajectory where the velocity and acceleration are perpendicular?

17. How is the up-and-down motion of an object thrown straight up consistent with our conclusion that projectiles follow parabolic trajectories?

18. Projectiles launched at 30° and 60° have the same range. Does this mean they stay in the air the same amount of time?

19. Medieval invaders are attacking a walled village on a flat-topped hill, as shown in Fig 4-22. Is the horizontal range of the rocks they're catapulting given by Equation 4-10? Explain.

20. Earth's curvature renders Equation 4-10 inexact for projectiles with very long ranges. Do you expect the actual range of such a projectile to be longer or shorter than Equation 4-10 would imply? Explain.

21. A 45° launch angle maximizes the horizontal range of a projectile. Is this still true for the situation shown in Fig. 4-22?

22. A friend who's not taking physics insists that you can't be accelerating when you drive around a curve since the speedometer reading remains steady. Refute this argument.

23. You're driving around a curve. Is there any way of stepping on the gas or brake such that your tangential acceleration cancels your radial acceleration, giving a net acceleration of zero? How or why not?

24. An object starts from rest, accelerating with constant tangential acceleration on a circular path. Which is greater when it starts, its tangential or its radial acceleration? Which is greater after a long time? Explain.

25. An object moves outward along the spiral path shown in Fig. 4-23. Must its speed increase, decrease, or remain the same if the magnitude of its radial acceleration is to remain constant?

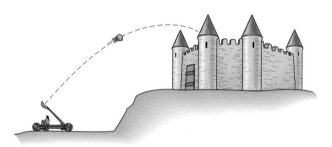

FIGURE 4-22 Questions 19, 21.

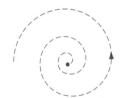

FIGURE 4-23 Question 25.

PROBLEMS

Section 4-1 Velocity and Acceleration

1. A skater is gliding along the ice at 2.8 m/s, when she undergoes an acceleration of magnitude 1.1 m/s² for 2.0 s. At the end of that time she is moving at 5.0 m/s. What must be the angle between the acceleration vector and the initial velocity vector?

2. In the preceding problem, what would have been the magnitude of the skater's final velocity if the acceleration had been perpendicular to her initial velocity?

3. An object is moving in the x direction at 1.3 m/s when it is subjected to an acceleration given by $\mathbf{a} = 0.52\,\hat{\mathbf{j}}$ m/s². What is its velocity vector after 4.4 s of acceleration?

4. An airliner is flying at a velocity of $260\hat{\mathbf{i}}$ m/s, when a wind gust gives it an acceleration of $0.38\hat{\mathbf{i}} + 0.72\,\hat{\mathbf{j}}$ m/s² for a period of 24 s. (a) What is its velocity at the end of that time? (b) By what angle has it been deflected from its original course?

Section 4-2 Constant Acceleration

5. The position of an object as a function of time is given by $\mathbf{r} = (2.4t + 1.2t^2)\hat{\mathbf{i}} + (0.89t - 1.9t^2)\hat{\mathbf{j}}$ m, where t is the time in seconds. What are the magnitude and direction of the acceleration?

6. An airplane heads northeastward down a runway, accelerating from rest at the rate of 2.1 m/s². Express the plane's velocity and position at $t = 30$ s in unit vector notation, using a coordinate system with x axis eastward and y axis northward, and with origin at the start of the plane's takeoff roll.

7. An asteroid is heading toward Earth at a steady 21 km/s. To save their planet, astronauts strap a giant rocket to the asteroid, giving it an acceleration of 0.035 km/s² at right angles to its original motion. If the rocket firing lasts 250 s, (a) by what angle does the direction of the asteroid's motion change? (b) How far does it move during the firing?

8. An object is moving initially in the x direction at 4.5 m/s, when an acceleration is applied in the y direction for a period of 18 s. If it moves equal distances in the x and y directions during this time, what is the magnitude of its acceleration?

9. A hockey puck is moving at 7.15 m/s when a stick imparts a constant acceleration of 63.5 m/s² at a 90.0° angle to the original direction of motion. If the acceleration lasts 0.132 s, how far does the puck move during this time?

10. Repeat the preceding problem, except that now the acceleration makes a 78.0° angle with the original direction of motion.

11. A particle leaves the origin with initial velocity $\mathbf{v}_0 = 11\hat{\imath} + 14\hat{\jmath}$ m/s. It undergoes a constant acceleration given by $\mathbf{a} = -1.2\hat{\imath} + 0.26\hat{\jmath}$ m/s². (a) When does the particle cross the y axis? (b) What is its y coordinate at the time? (c) How fast is it moving, and in what direction, at that time?

12. A particle starts from the origin with initial velocity $\mathbf{v}_0 = v_0\hat{\imath}$ and constant acceleration $\mathbf{a} = a\hat{\jmath}$. Show that the particle's distance from the origin and its direction relative to the x axis are given by $d = t\sqrt{v_0^2 + \frac{1}{4}a^2t^2}$ and $\theta = \tan^{-1}(at/2v_0)$.

13. Figure 4-24 shows a cathode-ray tube, used to display electrical signals in oscilloscopes and other scientific instruments. Electrons are accelerated by the electron gun, then move down the center of the tube at 2.0×10^9 cm/s. In the 4.2-cm-long deflecting region they undergo an acceleration directed perpendicular to the long axis of the tube. The acceleration "steers" them to a particular spot on the screen, where they produce a visible glow. (a) What acceleration is needed to deflect the electrons through 15°, as shown in the figure? (b) What is the shape of an electron's path in the deflecting region?

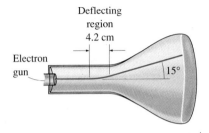

FIGURE 4-24 A cathode-ray tube (Problem 13).

Section 4-3 Projectile Motion

14. You toss an apple horizontally at 9.5 m/s from a height of 1.8 m. Simultaneously, you drop a peach from the same height. How long does each take to reach the ground?

15. A carpenter tosses a shingle off a 9.4-m-high roof, giving it an initially horizontal velocity of 7.2 m/s. (a) How long does it take to reach the ground? (b) How far does it move horizontally in this time?

16. An arrow fired horizontally at 41 m/s travels 23 m horizontally before it hits the ground. From what height was it fired?

17. A kid fires a blob of water horizontally from a squirt gun held 1.6 m above the ground. It hits another kid 2.1 m away square in the back, at a point 0.93 m above the ground (see Fig. 4-25). What was the initial speed of the blob?

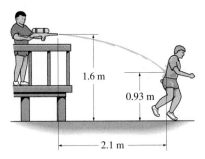

FIGURE 4-25 Problem 17.

18. You are trying to roll a ball off a 80.0-cm-high table to squash a bug on the floor 50.0 cm from the table's edge. How fast should you roll the ball?

19. Repeat the preceding problem, now with the bug moving away from the table at 30.0 mm/s and 50.0 cm from the table when the ball leaves the table edge.

20. Mike is standing outside the physics building, 5.0 m from the wall. Debbie, at a window 4.0 m above, tosses a physics book horizontally. What speed should she give it if it is to reach Mike?

21. Ink droplets in an ink-jet printer are ejected horizontally at 12 m/s, and travel a horizontal distance of 1.0 mm to the paper. How far do they fall in this interval?

22. Protons in a particle accelerator drop 1.2 μm over the 1.7-km length of the accelerator. What is their approximate average speed?

23. You're on the ground 3.0 m from the wall of a building, and want to throw a package from your 1.5-m shoulder level to someone in a second floor window 4.2 m above the ground. At what speed and angle should you throw it so it just barely reaches the window?

24. Derive a general formula for the horizontal distance covered by a projectile launched horizontally at speed v_0 from a height h.

25. A car moving at 40 km/h strikes a pedestrian a glancing blow, breaking both the car's front signal light lens and the pedestrian's hip. Pieces of the lens are found 4.0 m down the road from the center of a 1.2-m wide crosswalk, and a lawsuit hinges on whether or not the pedestrian was in the crosswalk at the time of the accident. Assuming that the lens

was initially 63 cm off the ground, and that the lens pieces continued moving horizontally with the car's speed at the time of the impact, was the pedestrian in the crosswalk?

26. Repeat Problem 15 for the case when the shingle is thrown with a speed of 7.2 m/s at 12° below the horizontal.

27. In part (b) of the exercise following Example 4-4, what is the vertical component of the velocity with which the cyclist strikes the ground?

28. Compare the travel times for the projectiles launched at 30° and 60° in Fig. 4-13, both of which have the same starting and ending points.

29. A submarine-launched missile has a range of 4500 km. (a) What launch speed is needed for this range when the launch angle is 45°? (Neglect the distance over which the missile accelerates.) (b) What is the total flight time? (c) What would be the minimum launch speed at a 20° launch angle, used to "depress" the trajectory so as to foil a space-based antimissile defense?

30. A rescue airplane is flying horizontally at speed v_0 at an altitude h above the ocean, attempting to drop a package of medical supplies to a shipwreck victim in a lifeboat. At what line-of-sight angle α (Fig. 4-26) should the pilot release the package?

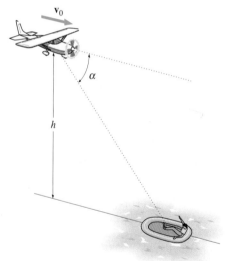

FIGURE 4-26 Problem 30.

31. At a circus, a human cannonball is shot from a cannon at 35 km/h at an angle of 40°. If he leaves the cannon 1.0 m off the ground, and lands in a net 2.0 m off the ground, how long is he in the air?

32. As you stand on the rim of the Grand Canyon, a friend 10 m back from the rim throws you a baseball at 60 km/h and an initial angle of 10°. You miss the ball, and it winds up dropping 500 m below the rim before it hits a trail descending into the canyon. Neglecting air resistance, (a) what horizontal distance from the rim would it travel and (b) at what speed would it hit the trail? (Neglect of air resistance here is

actually not justified; with air resistance, your answers would be quite different.)

33. If you can hit a golf ball 180 m on Earth, how far can you hit it on the moon? (Your answer is an underestimate, because the distance on Earth is restricted by air resistance as well as by a larger g.)

34. Prove that a projectile launched on level ground reaches its maximum height midway along its trajectory.

35. A projectile launched at an angle θ to the horizontal reaches a maximum height h. Show that its horizontal range is $4h/\tan\theta$.

36. You're 5.0 m from the left-hand wall of the house shown in Fig. 4-27, and you want to throw a ball to a friend 5.0 m from the right-hand wall. (a) What is the minimum speed that will allow the ball to clear the roof? (b) At what angle should it be thrown? Assume the throw and catch both occur 1.0 m above the ground.

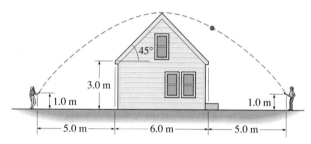

FIGURE 4-27 Problem 36.

37. A circular fountain has jets of water directed from the circumference inward at an angle of 45°. Each jet reaches a maximum height of 2.2 m. (a) If all the jets converge in the center of the circle and at their initial height, what is the radius of the fountain? (b) If one of the jets is aimed at 10° too low, how far short of the center does it fall?

38. When the Olympics were held in Mexico City in 1968, many sports fans feared that the high altitude would result in poor athletic performances due to reduced oxygen. To their surprise, new records were set in track and field events, probably as a result of lowered air resistance and a decrease in g to 9.786 m/s², both ultimately associated with the high altitude. In particular, Robert Beamon set a new world record of 8.90 m in the long jump. Photographs suggest that Beamon started his jump at a 25° angle to the horizontal. If he had jumped at sea level, where $g = 9.81$ m/s², at the same angle and initial speed as in Mexico City, how far would Beamon have gone? Neglect air resistance in both cases (although its effect is actually more significant than the change in g).

39. In 1991 Mike Powell shattered Bob Beamon's 1968 world long jump record with a leap of 8.95 m (see Fig. 4-28). Studies show that Powell jumps at 22° to the vertical. Treating him as a projectile, at what speed did Powell begin his jump?

FIGURE 4-28   Problem 39.

40. A motorcyclist driving in a 50-km/h zone hits a stopped car on a level road. The cyclist is thrown from his bike and lands 32 m down the road. Was the cyclist speeding? To answer, find the minimum speed he could have been going just before the accident.

41. Show that, for a given initial speed, the horizontal range of a projectile is the same for launch angles $45° + \alpha$ and $45° - \alpha$, where α is between $0°$ and $45°$.

42. One model of the Scud missile used in the 1991 Persian Gulf war has a range of 630 km. (a) What is its launch speed, given a $40°$ launch angle? (b) What is the missile flight time?

43. A basketball player is 15 ft horizontally from the center of the basket, which is 10 ft off the ground. At what angle should the player aim the ball if it is thrown from a height of 8.2 ft with a speed of 26 ft/s?

Section 4-4 Circular Motion

44. How fast would a car have to round a turn 50 m in radius in order for its acceleration to be numerically equal to that of gravity?

45. Estimate the acceleration of the moon, which completes a nearly circular orbit of 385,000 km radius in 27 days.

46. An object is in uniform circular motion. Make a graph of its average acceleration, measured in units of v^2/r, versus angular separation for points on the circular path spaced $40°$, $30°$, $20°$, and $10°$ apart. Your graph should show the average acceleration approaching the instantaneous value v^2/r as the angular separation decreases.

47. When Apollo astronauts landed on the moon, they left one astronaut behind in a circular orbit around the moon. For the half of the orbit spent over the far side of the moon, that individual was completely cut off from communication with the rest of humanity. How long did this lonely state last?

Assume a sufficiently low orbit that you can use the moon's surface gravitational acceleration (see Appendix E) for the spacecraft.

48. A 10-in-diameter circular saw blade rotates at 3500 revolutions per minute. What is the acceleration of one of the saw teeth? Compare with the acceleration of gravity.

49. A beetle can cling to a phonograph record only if the acceleration is less than $0.25g$. How far from the center of a $33\frac{1}{3}$-rpm record can the beetle stand?

50. A jet is diving vertically downward at 1200 km/h (see Fig. 4-29). If the pilot can withstand a maximum acceleration of $5g$ (i.e., 5 times Earth's gravitational acceleration) before losing consciousness, at what height must the plane start a quarter turn to pull out of the dive? Assume that the plane's speed remains constant.

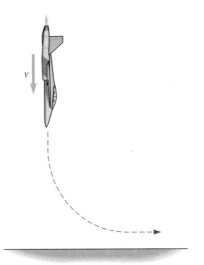

FIGURE 4-29   Problem 50.

51. Electrons in a TV tube are deflected through a $55°$ angle, as shown in Fig. 4-30. During the deflection they move at constant speed in a circular path of radius 4.30 cm. If they experience an acceleration of 3.35×10^{17} m/s², how long does the deflection take?

FIGURE 4-30   Problem 51.

52. How long would a day last if Earth were rotating so fast that the acceleration of an object on the equator were equal to g?
53. A runner rounds the semicircular end of a track whose curvature radius is 35 m. The runner moves at constant speed, with an acceleration of 1.8 m/s². How long does it take to complete the turn?

Section 4-5 Nonuniform Circular Motion

54. A space station 120 m in diameter is set rotating in order to give its occupants "artificial gravity." Over a period of 5.0 min, small rockets bring the station steadily to its final rotation rate of 1 revolution every 20 s. What are the radial and tangential accelerations of a point on the rim of the station 2.0 min after the rockets start firing?
55. A plane is heading northward when it begins to turn eastward on a circular path of radius 9.10 km. At the instant it begins to turn, its acceleration vector points 22.0° north of east and has magnitude 2.60 m/s². (a) What is the plane's speed? (b) At what rate is its speed increasing?
56. A horse runs at a steady 6.8 m/s around an oval track whose maximum and minimum radii of curvature are 50 m and 25 m. What are the maximum and minimum values for the magnitude of the horse's acceleration?
57. An object is set into motion on a circular path of radius r by giving it a constant tangential acceleration a_t. Derive an expression for the time t when the acceleration vector points at 45° to the direction of motion.
58. A car moving at 65 km/h enters a curve that describes a quarter turn of radius 120 m. The driver gently applies the brakes, giving a constant tangential deceleration of magnitude 0.65 m/s². Just before emerging from the turn, what are (a) the magnitude of the car's acceleration and (b) the angle between the acceleration vector and the direction of motion?

Paired Problems

(Both problems in a pair involve the same principles and techniques. If you can get the first problem, you should be able to solve the second one.)

59. An alpine rescue team is using a slingshot to send an emergency medical packet to climbers stranded on a ledge, as shown in Fig. 4-31. What should be the launch speed from the slingshot?
60. A cat leaps onto a counter 90 cm off the floor, starting 65 cm from the edge of the counter. It leaps at an initial angle of 79° to the horizontal and lands on the counter 22 cm from the edge. What was its initial speed?
61. If you can throw a stone straight up to a height of 16 m, how far could you throw it horizontally over level ground? Assume the same throwing speed and optimum launch angle.
62. In a conversion from military to peacetime use, a missile with a maximum horizontal range of 180 km is being adapted for studying the upper atmosphere. What is the maximum altitude it can achieve, if launched vertically?
63. I can kick a soccer ball 28 m on level ground, giving it an initial velocity at 40° to the horizontal. At the same initial speed and angle to the horizontal, what horizontal distance can I kick the ball on a 15° upward slope?
64. A model rocket has a horizontal range of 280 m on level ground, when given a 45° launch angle. What horizontal distance will the rocket cover when launched at 45° to the horizontal from the top of a hill whose sides slope down at 21°?
65. A fireworks rocket is 73 m above the ground when it explodes. Immediately after the explosion, one piece is moving at 51 m/s at 23° to the upward vertical direction. A second piece is moving at 38 m/s at 11° below the horizontal direction. At what horizontal distance from the explosion site does each piece land?
66. A hose nozzle sprays water drops in a fan-shaped pattern as suggested in Fig. 4-32. The drops leave the nozzle moving at 4.6 m/s. With the hose aimed as shown, the directions of the emerging drops range from 10° below the horizontal to 25° above. If the nozzle is 1.7 m above the ground, how wide is the region (marked w in Fig. 4-32) that gets wet?

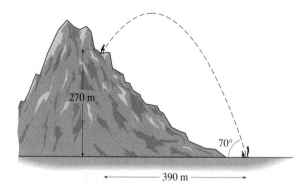

FIGURE 4-31 Problem 59.

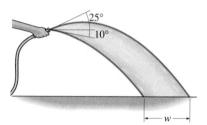

FIGURE 4-32 Problem 66.

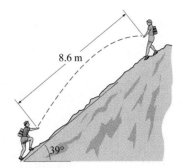

FIGURE 4-33   Problem 67.

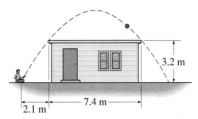

FIGURE 4-34   Problem 74.

67. You toss a chocolate bar to your hiking companion located 8.6 m up a 39° slope, as shown in Fig. 4-33. Determine the initial velocity vector so that the chocolate bar will reach your friend moving horizontally.

68. A circus lion prepares to leap through a flaming hoop. A line from the lion to the hoop is 2.2 m long and makes a 26° angle with the floor. With what initial velocity should the lion leap so as to pass through the hoop moving horizontally?

69. After takeoff, a plane makes a three-quarter circle turn of radius 7.1 km, maintaining constant altitude but steadily increasing its speed from 390 to 740 km/h. Midway through the turn, what is the angle between the plane's velocity and acceleration vectors?

70. On its landing approach a plane makes a semicircular turn of radius 6.7 km, remaining at constant altitude while its speed drops steadily from 840 km/h to 290 km/h. At the midpoint of the turn, what is the angle between the plane's velocity and acceleration vectors?

Supplementary Problems

71. Verify the maximum altitude and flight time for the 15° launch angle trajectory of the missile described at the end of the Application: Ballistic Missile Defense (page 77).

72. A juggler's hands are 80 cm apart, and the balls being juggled reach a maximum height of 100 cm above the juggler's hands. (a) At what velocity do the balls leave the juggler's hands? (b) If four balls are being juggled, how often must the juggler catch a ball?

73. A monkey is hanging from a branch a height h above the ground. A naturalist stands a horizontal distance d from a point directly below the monkey. The naturalist aims a tranquilizer dart directly at the monkey, but just as he fires the monkey lets go. Show that the dart will nevertheless hit the monkey, provided its initial speed exceeds $\sqrt{(d^2 + h^2)g/2h}$.

74. A child tosses a ball over a flat-roofed house 3.2 m high and 7.4 m wide, so it just clears the corners on both sides, as shown in Fig. 4-34. If the child stands 2.1 m from the wall, what are the ball's initial speed and launch angle? Assume the ball is launched essentially from ground level.

75. A diver leaves a 3-m board on a trajectory that takes her 2.5 m above the board, and then into the water a horizontal distance of 2.8 m from the end of the board. At what speed and angle did she leave the board?

76. Use the trigonometric identity $\cos^2\theta = 1/(1 + \tan^2\theta)$ to write Equation 4-9 for a projectile's trajectory in terms of $\tan\theta_0$. Given the initial speed v_0 and a point (x, y) through which you want the trajectory to go, Equation 4-9 can be rearranged as a quadratic in $\tan\theta_0$. Solve to show that there are two possible launch angles, given by

$$\tan\theta_0 = \frac{v_0^2}{gx} \pm \sqrt{\frac{v_0^4}{g^2x^2} - \frac{2v_0^2y}{gx^2} - 1}.$$

77. (a) Use the result of the preceding problem to find an expression for the launch angles on level ground (i.e., $y = 0$). (b) Taking an approximate value for the range for the 30° and 60° launch angles from the graph in Fig. 4-13, and the initial speed from the caption of that figure, show that your result yields approximately those angles.

78. Two projectiles are launched with the same speed v_0, at angles 45° + α and 45° − α. As Fig. 4-13 shows, they have the same horizontal range. Derive an expression for the difference in their flight times.

79. A well-engineered ski jump is less dangerous than it looks because skiers hit the ground with very small velocity components perpendicular to the ground. Skiers leave the Olympic ski jump in Lake Placid, New York, at 28 m/s, at an angle of 9.5° below the horizontal. Their landing zone is a horizontal distance of 55 m from the end of the jump. The ground at that point is contoured so skiers' trajectories make an angle of only 3.0° with the ground on landing, as suggested in Fig. 4-35. What is the slope of the ground in the landing zone?

80. A particle is moving along the x axis with velocity $\mathbf{v}_0$ in the positive x direction. As it passes the origin, it begins to

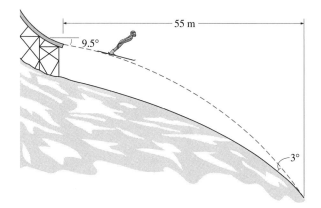

FIGURE 4-35 Problem 79.

experience a constant acceleration **a** in the y direction. Show that, at any subsequent time, the tangent of the angle its velocity makes with the x axis is twice the tangent of the angle its displacement vector makes with the x axis.

81. At the peak of its trajectory, a projectile is moving horizontally with speed v_x, and its acceleration (g, downward) is obviously perpendicular to its velocity. Use these facts to find an expression for the radius of curvature at the peak of the parabolic trajectory.

82. An object moves at constant speed v in the x-y plane, describing a circle of radius r centered at the origin. It is on the positive x axis at time $t = 0$. Show that the position of the object as a function of time can be written $\mathbf{r} = r[\cos(vt/r)\hat{\mathbf{i}} + \sin(vt/r)\hat{\mathbf{j}}]$, where the argument of the sine and cosine is in radians. Differentiate this expression once to obtain an expression for the velocity and again for the acceleration. Show that the acceleration has magnitude v^2/r and is directed radially inward (that is, opposite to **r**).

83. In the Olympic hammer throw, contestants whirl a 7.3-kg ball on the end of a 1.2-m-long steel wire before releasing it. In a particular throw, the hammer is released from a height of 1.3 m while moving in a direction 24° above the horizontal. If it travels 84 m horizontally, what is its radial acceleration just before release (see Fig. 4-36)?

84. A projectile is launched at an angle θ to the horizontal, with sufficient speed to give it a horizontal range x. Show that the radius of curvature at the top of its trajectory is given by $r = (x/2)\tan\theta$.

85. While increasing its speed, a train enters a 90° circular turn of radius r with speed v_0 and with tangential acceleration equal to half its radial acceleration. If its tangential acceleration remains constant, show that when the train leaves the turn its tangential and radial accelerations are related by $a_t = a_r/(2 + \pi)$.

FIGURE 4-36 The Olympic hammer throw (Problem 83).

86. A convertible is speeding down the highway at 130 km/h, when the driver spots a police airplane 600 m back at an altitude of 250 m. The driver decelerates at 2.0 km/h/s. (a) If the plane is flying horizontally at a steady 210 km/h, where should the plane be in relation to the car for the police officer to drop a speeding ticket into the car? Assume the ticket is dropped with no initial vertical motion, and it is in a heavy capsule that experiences negligible air resistance. (b) How fast will the car be moving when the ticket reaches it?

87. A projectile is launched with initial speed v_0 at an angle θ_0 to the horizontal. Find expressions for the angle the trajectory makes with the horizontal (a) as a function of time and (b) as a function of position.

88. A projectile is launched with speed v_0 at an angle θ_0 to the horizontal. By resolving the gravitational acceleration into components parallel and perpendicular to the trajectory, and recognizing that the latter is the radial acceleration, determine the curvature radius of the trajectory just after launch.

CHAPTER 5

DYNAMICS: WHY DO THINGS MOVE?

Forces of wind, water, gravity, and human muscle determine the motion of a sailboard.

A spacecraft moves effortlessly through space. No rocket firing is necessary to maintain its motion. But then what keeps it moving?

A baseball is heading toward the batter at 150 km/h. The batter swings, connects, and suddenly the ball is heading toward left field. What caused the change in its motion?

In the preceding chapters we studied **kinematics**—the description of motion without reference to its causes. Now we're asking *why* things move. The study of motion and its causes is called **dynamics,** from the Greek "dynamikos," meaning force or power. In this chapter we'll see how Isaac Newton, over 300 years ago, laid down the basic laws of motion that we apply today. Here we introduce Newton's laws and their immediate consequences; in the following chapter, we'll apply those laws in a variety of practical situations.

5-1 THE WRONG QUESTION

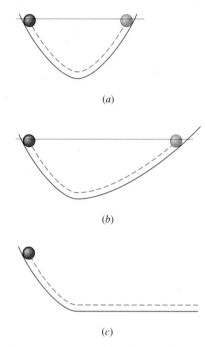

FIGURE 5-1   (*a*) Galileo recognized that a ball rolling down an incline would rise to nearly its starting height on a second incline. (*b*) Making the second incline more gradual would result in the ball's traveling farther in the horizontal direction. (*c*) If the second incline were reduced to a horizontal surface, Galileo reasoned, the ball's motion should continue forever.

What makes things move? For nearly 2000 years following the work of Aristotle (384–322 B.C.), most people believed that a force—a push or a pull—was needed to keep something moving. This idea made a lot of sense: when an ox stopped pulling an ox-cart, for example, the cart quickly came to a stop. Aristotelians extended their thinking to projectiles by saying that an arrow, for instance, could continue moving through the air because air rushed from the front to the back of the arrow to push it forward.

Actually, "What makes things move?" is the wrong question. In the early 1600s, Galileo Galilei carried out a series of experiments that convinced him that a moving object has an intrinsic "quantity of motion" that needs no cause to maintain itself (Fig. 5-1). Instead of answering the question "what makes things move?" Galileo asserted that the question needs no answer. In doing so, he set the stage for centuries of progress in physics, beginning with the achievements of Isaac Newton and culminating—so far—in the work of Albert Einstein.

The Right Question

So the question that forms the title of this chapter is the wrong question. Wrong, too, is our question about what keeps the spacecraft moving; according to Galileo, the fact of its motion is sufficient answer. Then what's the right question? An example of the right question is our question about the baseball—what caused its motion to *change*? The study of dynamics isn't really about what causes motion; rather, it's about what causes *changes* in motion. Such changes include an object at rest starting to move, or a moving object coming to a stop, or an object's changing its direction of motion. Any *change* in motion requires a cause, but motion itself does not.

It was Galileo who identified the right question about motion. But it was left to Isaac Newton (Fig. 5-2) to formulate quantitative laws describing exactly how motion changes. We use those laws today to design safer automobile braking systems, to guide spacecraft to distant planets, to build skyscrapers that won't fall down, and even to understand some aspects of atomic behavior.

5-2 NEWTON'S LAWS OF MOTION

Newton's laws of motion are the basis of mechanics. They involve three important concepts: **mass, force,** and **momentum.** Newton thought of mass as a measure of the quantity of matter in an object. Although this description is a valid one, we will see that it is more practical to give an operational definition of mass that arises from Newton's laws. We have an intuitive sense of **force** as a push or a pull, a sense that is consistent with Newton's concept of force. Since pushes and pulls have direction, force is a vector quantity. **Momentum** is a modern word for what Newton called "quantity of motion." Seeking a quantity that reflected both the velocity of a moving object as well as the amount of matter in motion, Newton hit on the product of mass and velocity as the best measure of the "quantity of motion" or momentum. Thus the momentum of an object of mass m moving with velocity $\mathbf{v}$ is $m\mathbf{v}$; we use the symbol $\mathbf{p}$ for momentum:

$$\mathbf{p} = m\mathbf{v}. \quad \text{(momentum)} \quad (5\text{-}1)$$

Since velocity is a vector, so is momentum.

We now state Newton's three laws, and then discuss each in detail:

Newton's first law of motion: A body in motion tends to remain in motion, or to remain stopped if stopped, except insofar as acted on by an outside force.

Newton's second law of motion: The rate at which a body's momentum changes is equal to the net force acting on the body; mathematically

$$\mathbf{F}_{net} = \frac{d\mathbf{p}}{dt}. \quad \text{(Newton's 2nd law)} \quad (5\text{-}2)$$

When the mass of a body is constant, we can use the definition of momentum, $\mathbf{p} = m\mathbf{v}$, to write

$$\mathbf{F}_{net} = \frac{d(m\mathbf{v})}{dt} = m\frac{d\mathbf{v}}{dt},$$

But $d\mathbf{v}/dt$ is the acceleration, $\mathbf{a}$, so

$$\mathbf{F}_{net} = m\mathbf{a}. \quad \text{(Newton's 2nd law, constant mass)} \quad (5\text{-}3)$$

Although Newton originally wrote his second law in the form 5-2, which remains the most general form, the form 5-3 is more widely recognized.

Newton's third law of motion: For every action there is always an equal and opposite reaction.

FIGURE 5-2 Isaac Newton at age 59. When Newton was in his twenties the plague had swept across England, shutting down the universities. After two years' seclusion in the English countryside, Newton had laid the groundwork for his laws of motion and gravity and had begun the development of calculus.

5-3 NEWTON'S FIRST LAW AND INERTIA

Newton's first law is a more precise statement of what Galileo had earlier realized—that, in the absence of friction or other forces, a moving body will keep moving steadily forward. In other words, its velocity (a vector quantity) will not change in either direction or magnitude. We use the term **inertia** to describe a body's resistance to change in motion. So Newton's first law is equivalent to saying that a body has inertia; for this reason, the first law is also known as the law of inertia.

A body at rest has zero velocity. A stationary body's tendency to stay at rest is thus the same thing as a moving body's tendency to keep moving; both are manifestations of Newton's first law.

Inertial Frames of Reference

Why don't flight attendants serve beverages when an airplane is accelerating down the runway? For one thing, their beverage cart wouldn't stay put; instead, it would go careering toward the back of the plane—even in the absence of any forces acting on it. So is Newton's law of inertia wrong? No. It's just that the law of inertia doesn't apply in a frame of reference that's accelerating. Neither do Newton's other laws. Try throwing a ball on a whirling merry-go-round; you'll have trouble getting it to go where you want, again because Newton's laws don't hold in the rotating—and therefore accelerating—frame of reference of the merry-go-round.

A frame of reference in which Newton's laws are valid is called an **inertial frame.** In the absence of external forces the acceleration of an object in an inertial frame is zero, and therefore its motion continues unchanged. We saw in Section 3-8 that an object with acceleration **a** in one frame of reference has the same acceleration in another frame of reference moving with constant velocity relative to the first frame. So if an object has zero acceleration in one frame of reference, it will have zero acceleration in any other frame moving uniformly relative to the first frame. If Newton's laws are satisfied in the first frame, then they will be satisfied in the other frame as well since an object's acceleration will remain zero in the absence of external forces. If we have one inertial frame of reference, we therefore know that any other frame of reference moving uniformly relative to our inertial frame is itself inertial. The laws of motion—Newton's laws—are therefore valid in all frames of reference in uniform motion. This is the principle of Galilean relativity that we introduced in Chapter 3.

Strictly speaking, our rotating Earth is not an inertial frame, and therefore Newton's laws aren't quite valid in an Earth-based frame of reference. But Earth's rotation has an insignificant effect on most motions we're interested in, so we are usually justified in treating Earth as an inertial frame. An exception comes in considering the large-scale motions of the atmosphere and oceans; here, oceanographers and atmospheric scientists must take account of Earth's rotation.

If Earth isn't exactly an inertial frame, what is? That's a surprisingly difficult question, and it pointed Einstein the way to his general theory of relativity. The law of inertia and the principle of relativity are intimately related, and both are tied to deep questions about the nature of space, time, and gravity—questions whose answers lie in Einstein's theory. We'll look briefly at those answers later in the book; for now, we'll be content with the understanding that Newton's laws hold in any frame of reference in uniform motion.

5-4 FORCE

For Aristotle, the natural state of motion was to be at rest. That's why the question "what makes things move" was meaningful in ancient times. For Galileo and Newton, the natural state of motion is uniform—that is, unchanging—motion. So the important question is not "what makes things move?" but "what causes changes in motion?" The answer is embodied in Newton's first law: "A body in motion tends to remain in motion . . . *except insofar as acted on by an outside force."* In the italicized phrase Newton identifies **force** as the cause of changes in motion.

What is force? We're familiar with forces in our everyday lives, and our common sense notion of force is pretty close to the scientific meaning. We push, pull, or lift objects and thus exert forces on them. We experience the pull of **gravity.** We see objects slow down, and infer the existence of **frictional forces.** We use ropes and cables to provide **tension forces** that support heavy objects (Fig. 5-3). Where two surfaces contact, as in a book resting on a table or a hand pushing a grocery cart, we must have **contact forces** that act between the surfaces. We know that structures like buildings or stadium seating can support heavy loads; they do so with **compression forces** that develop when material is compressed. We pull clothes from the dryer and find socks clinging to them; thus we experience **electrical forces.** We stick messages to refrigerators using **magnetic forces.** Probing the innermost realm of the atom, we infer the presence of a very strong **nuclear force** that binds the atomic nucleus.

The Fundamental Forces

We've just mentioned a plethora of forces—gravitational forces, contact forces, frictional forces, electrical forces, tension forces, magnetic forces, compression forces, nuclear forces. How many kinds of forces are there? Actually, physicists believe there is probably only one fundamental force, although we do not yet know how to describe it. At present, physicists identify three basic forces, each subsuming other forces once considered distinct. The unification of forces has been a major theme in the history of physics and is perhaps the central problem of contemporary physics (Fig. 5-4). Later we will explore one aspect of this unification in detail, as we come to understand how electricity and magnetism are aspects of a single phenomenon. In the final chapter, we consider in more detail the known fundamental forces and the prospects for their unification; here we simply introduce the three forces and give each a brief description.

Familiar though it is, the **gravitational force** is the weakest and perhaps the least understood of the fundamental forces. Although weak, gravity is cumulative; there is no such thing as antigravity. That means massive objects exert large gravitational forces, and for that reason gravity is the dominant force on the large-scale structure of the universe.

The **electroweak** force subsumes the force of electromagnetism and the so-called weak nuclear force. The theory that explains the unification of these two forces was developed only in the 1970s and confirmed in the 1980s; before then physicists listed four fundamental forces. Earlier, in the 1860s, a similar theoretical unification of electricity and magnetism had occurred. Virtually all the nongravitational forces we encounter in everyday life are manifestations of the electroweak force. Contact forces, friction, tension forces, the forces in your muscles, and the forces that hold ordinary matter together are all essentially electrical forces. Magnetism plays a somewhat lesser role in everyday life, and the weak force is seemingly unimportant. But a closer look at atomic-scale phenomena shows that these forces, too, are significant; in particular, processes involving the weak force occur in the core of the Sun and are partially responsible for the generation of solar energy that keeps us alive.

FIGURE 5-3 This crane provides many examples of forces. Tension forces (**T**) in the cables support the load against the force of gravity (**F**$_g$). The ground exerts a contact force that supports the entire crane. Unseen are electrical forces that hold matter together at the atomic and molecular scale and the strong or color force that binds individual particles into atomic nuclei.

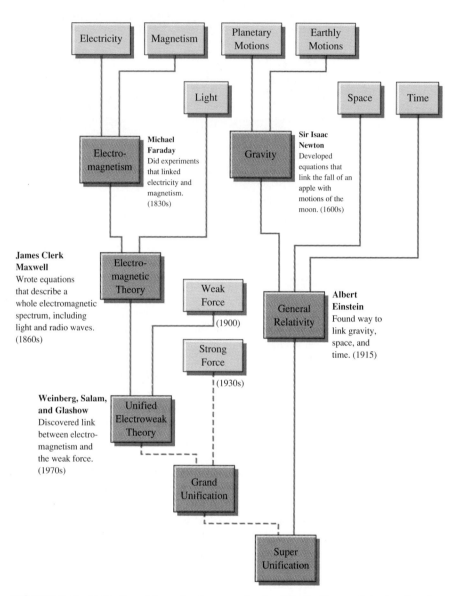

FIGURE 5-4 Unification of forces has been a major theme in the history of physics. Grand unification of the electroweak and strong, or color, forces may be achieved soon, but super unification including gravity is a more distant goal.

The third basic force, originally called the **strong force,** was advanced in the 1930s to explain how atomic nuclei stay together. Today we know that the protons and neutrons comprising the nucleus are themselves composites of simpler particles called quarks (Fig. 5-5). The force acting between quarks is the **color force,** and the force binding nuclei is just a residual manifestation of the very strong color force. Although the color force is not obvious in our everyday lives, its behavior determines the structure of matter at the most fundamental level. Without it the universe would be an entirely different place.

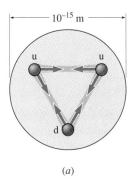

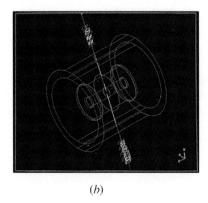

(a)

(b)

FIGURE 5-5   (a) A proton consists of three quarks, held together by a 50,000-N (10,000 pound) color force. Although the color force is not obvious in everyday life, it is essential in determining the basic structure of matter. (b) This image from CERN, the European Laboratory for Particle Physics, shows the decay of the Z^0 particle into a quark-antiquark pair. The quarks do not appear in isolation, but instead as jets of secondary particles moving in opposite directions. This experiment helped confirm the theory unifying the electromagnetic and weak forces.

5-5 NEWTON'S SECOND LAW

Why are we so interested in forces? Because forces cause changes in motion: if we know the force acting on a body, we can use Newton's second law to calculate its acceleration, and thus determine its subsequent motion.

We introduced Newton's second law in Section 5.2; for an object of constant mass, the law takes the form

$$\mathbf{F}_{net} = m\mathbf{a}. \tag{5-3}$$

Just what does this equation mean? We can think of it in several ways.

First, Equation 5-3 provides a measure of force. If we observe a body of mass m undergoing an acceleration of magnitude a, we know that there is a net force of magnitude ma acting on the body. The direction of the force is the same as that of the acceleration. Equation 5-3 shows that the units of force are $kg \cdot m/s^2$. If we observe a 2-kg object accelerating at 3 m/s², then we know that a force of $(2 \text{ kg})(3 \text{ m/s}^2)$ or $6 \text{ kg} \cdot \text{m/s}^2$ is acting on the object. To honor Isaac Newton, we call one $kg \cdot m/s^2$ a newton (symbol N). From Equation 5-3, we see that a force of 1 N will cause a 1-kg mass to accelerate at 1 m/s². Similarly, our 2-kg object accelerating at 3 m/s² experiences a force of 6 N. A 1-N force is about a quarter pound—a fairly small force on the human scale. You can readily exert forces of tens to hundreds of newtons.

Newton's law makes good sense in light of our everyday notions of force. We expect it should take a larger force to produce a given acceleration in a more massive object (Fig. 5-6a). And with a given mass, a larger force should be needed for a larger acceleration (Fig. 5-6b). Newton's second law, with force

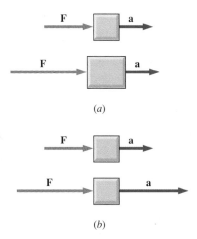

FIGURE 5-6   (a) For two objects to have the same acceleration, a greater force must be exerted on the more massive object. (b) When two objects have the same mass, the one with the greater force experiences the greater acceleration.

proportional to the product of mass and acceleration, reflects our intuitive under-standings suggested in Fig. 5-6.

● EXAMPLE 5-1   FORCE, MASS, AND ACCELERATION

A 1200-kg car accelerates from rest to 20 m/s in 7.8 s, moving in a straight line. What is the net force acting on it? If the car then rounds a bend 85 m in radius at a steady 20 m/s, what net force acts on it?

Solution

To calculate the net force from Newton's second law, we need to know the acceleration and the mass being accelerated. When the car accelerates from rest to 20 m/s in 7.8 s, it undergoes an acceleration of magnitude

$$a = \frac{\Delta v}{\Delta t} = \frac{20 \text{ m/s}}{7.8 \text{ s}} = 2.6 \text{ m/s}^2.$$

The magnitude of the force causing this acceleration is then

$$F_{net} = ma = (1200 \text{ kg})(2.6 \text{ m/s}^2) = 3100 \text{ kg·m/s}^2$$

$$= 3100 \text{ N}.$$

Force is a vector, and the full vector form of Newton's second law—$\mathbf{F}_{net} = m\mathbf{a}$—shows that the force is in the same direction as the acceleration, which in this case also happens to be in the direction of motion. What is the physical origin of the force? Newton's law does not tell us that; it only tells us the magnitude and direction of the force. *Any* physical mechanism providing the same net force on the car would produce the same acceleration. In the case of the accelerating car, the force is in fact provided by friction between tires and road—a force that is ultimately elec-trical in origin. If the driver tried to accelerate on frictionless ice, the wheels would simply spin and the car would not accelerate.

When the car rounds a circular bend at constant speed, we know from Section 4-4 that it undergoes an acceleration of magnitude v^2/r directed toward the center of the circle. New-ton's second law therefore requires a net force acting toward the center of the circle; the magnitude of this force is

$$F = ma = \frac{mv^2}{r} = \frac{(1200 \text{ kg})(20 \text{ m/s})^2}{85 \text{ m}} = 5600 \text{ N}.$$

Note that Newton's law does not distinguish between forces that result in a change in speed (car accelerating in a straight line) and those that result in a change in direction (car accelerating in a circular path at constant speed). Newton's second law relates force, mass, and acceleration in *all* instances; there is nothing special about circular or any other form of motion in the eyes of Newton's second law.

What provides the accelerating force in the car's circular motion? Again, it is the force of friction between tires and road. On an icy road, the car may skid off the road in a straight line, since the frictional force is virtually absent. We'll study circular motion and frictional forces in detail in the next chapter.

Figure 5-7 shows the force vectors for the two cases in this example.

EXERCISE   A 2.1×10^6-kg freight train takes 3.8 min to accelerate from rest to 75 km/h. What force does its locomotive exert?

Answer: 1.9×10^5 N

Some problems similar to Example 5-1:   4, 5, 8

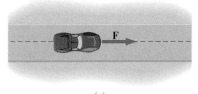

(a)

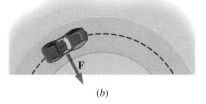

(b)

FIGURE 5-7   Forces on the car of Example 5-1. (*a*) Accelerating in a straight line. (*b*) Round-ing a curve. (Only forces in the horizontal plane are shown; there are also vertical forces from gravity and the road that sum to zero.)

● EXAMPLE 5-2   ELECTRONS IN A TV TUBE

Electrons in a TV tube are accelerated essentially from rest to a speed of 8.4×10^7 m/s in the "electron gun" at the back of the tube. If the electron gun is 8.2 cm long and if the acceleration in the gun is constant, what force is exerted on each electron?

Solution

To calculate the force, we need the acceleration and mass of the electron. Equation 2-11 relates acceleration, velocity, and dis-tance in one-dimensional motion. With $v_0 = 0$, this equation becomes

$$v^2 = 2a(x - x_0),$$

so

$$a = \frac{v^2}{2(x - x_0)}.$$

Here $x - x_0$ is the 8.2-cm length of the electron gun and v is the 8.4×10^7 m/s velocity of electrons emerging from the gun. Then Newton's second law gives

$$F = ma = \frac{mv^2}{2(x - x_0)} = \frac{(9.11 \times 10^{-31} \text{ kg})(8.4 \times 10^7 \text{ m/s})^2}{(2)(0.082 \text{ m})}$$

$$- 3.9 \times 10^{-14} \text{ N}.$$

That such a small force can accelerate electrons to such high speed in only a few centimeters is possible because of their minuscule mass. (Our answer in this case is only approximate because effects of relativity become important at this high speed.)

TIP Remember What You've Learned In this example we were given a velocity change and a distance and asked for a force. We can get force from Newton's second law if we know the acceleration, which we can get from the given information only by drawing on material from Chapter 2. You may not have remembered the relevant equation—it's not important that you do—but you should remember that you've studied equations relating the quantities you need to work with.

EXERCISE The merging lane on which cars enter a highway is 180 m long. A 1.2×10^3-kg car enters the merging lane moving at 40 km/h. What force must its drive wheels exert to get it safely to the 100 km/h speed limit before it merges with traffic on the main highway?

Answer: 2.2×10^3 N ●

TIP **Get To Know Your Book** Example 5-2 didn't give the mass of the electron, yet that mass was needed to get the answer. Inside the front cover of the book is a table listing common physical constants, including the electron mass. The appendices contain much more detailed information. Get used to consulting the inside covers and the appendices!

We just used Newton's second law as a prescription for measuring forces. Conversely, given the force on an object, we can calculate the object's acceleration and thus predict its motion. It is this use of Newton's law that lets us send a spacecraft to Jupiter, design engines appropriate to a new aircraft, or determine the safe distance between cars on a highway. It also helps us analyze a skyscraper's response to gale-force winds, predict the positions of the planets and the timing of eclipses, and develop better tennis rackets. It is only a slight exaggeration to say that the equation $\mathbf{F}_{net} = m\mathbf{a}$ covers all of classical physics.

● **EXAMPLE 5-3** AN AIRPORT RUNWAY

Fully loaded, a 747 jetliner has a mass of 3.6×10^5 kg. Its four engines provide a total thrust (force) of 7.7×10^5 N. How long a runway is needed for a 747 to achieve takeoff speed of 310 km/h (86 m/s)?

Solution
We are given force and mass, from which we can calculate acceleration using Newton's second law:

$$a = \frac{F}{m}.$$

We need then to relate acceleration to speed and distance; Equation 2-11 provides this relation:

$$v^2 = 2a(x - x_0).$$

Solving for the runway length $x - x_0$ and using $a = F/m$, we have

$$x - x_0 = \frac{v^2}{2a} = \frac{v^2}{2(F/m)} = \frac{mv^2}{2F}$$

$$= \frac{(3.6 \times 10^5 \text{ kg})(86 \text{ m/s})^2}{(2)(7.7 \times 10^5 \text{ N})} = 1700 \text{ m}.$$

EXERCISE A 50-kg skater can exert a horizontal force of 940 N against the ice. If she's gliding toward a wall of the rink at 6.7 m/s, how close can she get before attempting to stop and still avoid colliding with the wall?

Answer: 1.2 m

Some problems similar to Examples 5-2 and 5-3: 12, 13, 15, 16, 57, 58 ●

● **EXAMPLE 5-4** PREDICTING MOTION USING NEWTON'S SECOND LAW

A 1.2-kg object initially at rest at the origin is acted on by a force $\mathbf{F} = 2.4\hat{\mathbf{i}} + 1.7\hat{\mathbf{j}}$ N. What is the object's acceleration? Where is the object and how fast is it moving 3.5 s after the force is first applied?

Solution
Solving Newton's second law for the acceleration gives

$$\mathbf{a} = \frac{\mathbf{F}}{m} = \frac{2.4\hat{\mathbf{i}} + 1.7\hat{\mathbf{j}} \text{ N}}{1.2 \text{ kg}} = 2.0\hat{\mathbf{i}} + 1.4\hat{\mathbf{j}} \text{ m/s}^2.$$

Given the acceleration, we use methods of the previous chapters to find the position and speed. Equation 4-6 gives the position of an object undergoing constant acceleration:

$$\mathbf{r} = \mathbf{r}_0 + \mathbf{v}_0 t + \tfrac{1}{2}\mathbf{a}t^2 = \tfrac{1}{2}(2.0\hat{\mathbf{i}} + 1.4\hat{\mathbf{j}} \text{ m/s}^2)(3.5 \text{ s})^2$$
$$= 12\hat{\mathbf{i}} + 8.7\hat{\mathbf{j}} \text{ m}.$$

Here we have set $\mathbf{r}_0$ and $\mathbf{v}_0$ to zero since the object is initially at rest at the origin. The velocity after 3.5 s is given by Equation 4-3:

$$\mathbf{v} = \mathbf{v}_0 + \mathbf{a}t = (2.0\hat{\mathbf{i}} + 1.4\hat{\mathbf{j}} \text{ m/s})(3.5 \text{ s}) = 7.0\hat{\mathbf{i}} + 5.0\hat{\mathbf{j}} \text{ m/s},$$

so the speed is

$$v = \sqrt{v_x^2 + v_y^2} = [(7.0 \text{ m/s})^2 + (5.0 \text{ m/s})^2]^{1/2} = 8.6 \text{ m/s}.$$

EXERCISE A 940-kg spacecraft is moving uniformly at 4.8 km/s when it fires a small rocket that exerts a 4.5×10^3-N force at 67° to the initial direction of motion, as suggested in Fig. 5-8. If the rocket fires for 120 s, (a) how far does the ship move during the rocket firing? (b) How fast and in what direction is it moving after the firing?

FIGURE 5-8 A rocket thrust.

Answers: (a) 590 km; (b) 5.1 km/s, at 6.0° to its original direction

Some problems similar to Example 5-4: 17 ●

A third use of Newton's law is to determine unknown masses from measured forces and accelerations. This approach is actually used in spacecraft, where weightlessness (see Section 5-6) precludes more conventional measurements.

● **EXAMPLE 5-5** "MASSING" AN ASTRONAUT

To study physiological effects of spaceflight, astronauts routinely measure their masses using a movable chair subject to the force exerted by a spring (Fig. 5-9). If the 15-kg chair undergoes an acceleration of 24 mm/s² when the spring force is 1.8 N, what is the astronaut's mass?

Solution
Solving Newton's second law for mass gives

$$m = \frac{F}{a} = \frac{1.8 \text{ N}}{24\times10^{-3} \text{ m/s}^2} = 75 \text{ kg}.$$

This is the combined mass of chair and astronaut; subtracting the 15-kg chair then gives 60 kg for the astronaut's mass.

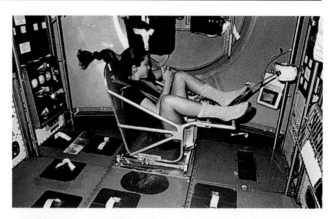

FIGURE 5-9 Astronaut Tamara Jernigan using a mass-measuring device aboard the space shuttle. ●

5-6 MASS AND WEIGHT: THE FORCE OF GRAVITY

Newton's second law shows that mass is a measure of a body's resistance to changes in motion—that is, of its inertia. A body's mass is an intrinsic property; it doesn't depend on location or any other circumstances. If my mass is 65 kg, it is 65 kg whether I am on Earth, in an orbiting spacecraft, on the moon, or in the remote reaches of intergalactic space. That means no matter where I am, a force of 65 N is needed to give me an acceleration of 1 m/s².

We commonly use the term "weight" to mean the same thing as mass. In physics, though, **weight** is the name for a force—the force that gravity exerts on a body. Near the surface of the Earth, a body allowed to fall freely accelerates downward at 9.8 m/s²; we designate this acceleration vector by **g.** Newton's second law, **F = ma,** then says that the force of gravity on a body of mass m is m**g;** this force is the body's weight:

$$\mathbf{W} = m\mathbf{g}. \qquad \text{(weight)} \qquad (5\text{-}4)$$

With my 65-kg mass, my weight near Earth's surface is then (65 kg)(9.8 m/s²) or 640 N. On the moon, where the acceleration of gravity is only 1.6 m/s², I would weigh only 100 N. And in the remote reaches of intergalactic space, far from any gravitating object, my weight would be essentially zero.

● EXAMPLE 5-6 MASS AND WEIGHT

The two Viking spacecraft that landed on Mars had weights of 5880 N on Earth. What were their masses on Earth and Mars, and their weights on Mars?

Solution
Weight is the force of gravity; Equation 5-4 shows that its magnitude is given by $W = mg$. Solving for m and using the weight and acceleration of gravity on Earth, we have

$$m = \frac{W}{g} = \frac{5880 \text{ N}}{9.8 \text{ m/s}^2} = 600 \text{ kg}.$$

The mass is the same everywhere; on Mars the weight is then given by

$$W = mg_{\text{Mars}} = (600 \text{ kg})(3.74 \text{ m/s}^2) = 2240 \text{ N},$$

where we obtained the acceleration of gravity on Mars from Appendix E.

EXERCISE A rope can support a maximum weight of 350 N. Could it support a 70-kg astronaut on (a) Earth and (b) Mercury? *Hint:* Consult Appendix E.

Answers: (a) no; (b) yes

Some problems similar to Example 5-6: 19, 20, 21, 22, 23, 25 ●

One reason we confuse mass and weight stems from the common use of the SI unit kilogram to describe "weight." At the doctor's office you may be told that you "weigh" 55 kg. You don't; you have a mass of 55 kg, so your weight is (55 kg)(9.8 m/s²) or 540 N. The unit of force in the English system is the pound, so giving your weight in pounds is correct. The English unit of mass is the slug, which is rarely used.

That we confuse mass and weight at all results from the remarkable fact that the gravitational acceleration of all objects at a given location is the same. This makes the *weight* of a body—a property defined in terms of gravity—proportional to its *mass*—a property that describes its inertia in terms that have nothing

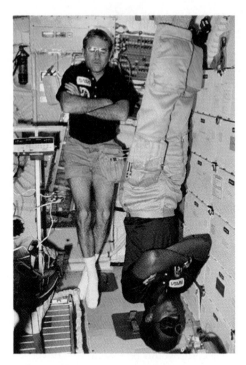

FIGURE 5-10 Astronauts Richard Truly and Guy Bluford floating about the cabin of a space shuttle.

to do with gravity. First inferred by Galileo from his experiments with falling bodies, this relation between gravitation and inertia seemed a coincidence until the early twentieth century. Finally, in his general theory of relativity, Albert Einstein showed how that simple relation reflects the underlying geometry of space and time in a way that links intimately the phenomena of gravitation and acceleration.

Although we defined weight in terms of the acceleration of gravity, a body need not be accelerating to have weight. When you're sitting in a chair, for example, gravity is still exerting a downward force $m\mathbf{g}$ on you. That force $m\mathbf{g}$ is your weight; it's just that the chair is exerting an upward force that keeps you from accelerating. In fact, it's the forces acting between you and the chair, and the forces of your muscles supporting the rest of your body, that make you conscious of your weight.

Weightlessness

What about astronauts aboard a space shuttle? Aren't they "weightless?" Not according to our definition. At the shuttle's altitude, the acceleration of gravity has about 93% of its surface value, so the gravitational forces $m\mathbf{g}$ on the shuttle and on its occupants are nearly the same as they would be at Earth's surface. But the astronauts *seem* weightless, and indeed they *feel* weightless (Fig. 5-10). What's going on?

Imagine yourself in an elevator whose cable has broken and is dropping freely downward with the gravitational acceleration g. In other words, the elevator and its occupant are in **free fall,** with only the force of gravity acting. If you let go of a book you're holding, it too falls freely with acceleration g. But so does every-

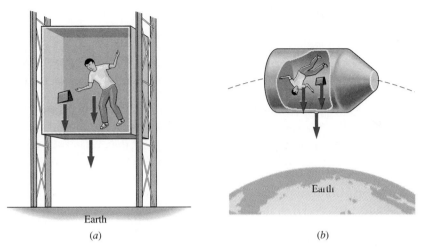

(a)   (b)

FIGURE 5-11 Objects in free fall appear weightless because they experience the same acceleration as their reference frame. Weightlessness would occur in a freely falling elevator near Earth's surface just as it does in an orbiting spacecraft. In both cases, the person, book, and elevator or spaceship all have the same acceleration toward Earth so none accelerates relative to its surroundings.

thing else around it—and therefore the book stays put relative to you and to the elevator (Fig. 5-11a). As far as you're concerned, the book is "weightless"; it doesn't seem to fall when you let go of it. And you're "weightless" too; if you jump off the floor of the elevator, you drift gradually to the ceiling rather than falling back down. Of course you, the book, and the elevator are *all* falling, but that isn't obvious from your perspective. It seems to you like everything in the freely falling elevator is weightless. But gravity is still acting; it's after all what's making you fall. So you really do have weight, and your condition is **apparent weightlessness.** Your **apparent weight** is zero even though your weight itself—the force gravity exerts on you—is not.

A falling elevator is a dangerous place, but the same apparent weightlessness occurs temporarily in aircraft flying parabolic trajectories that mimic projectile motion (Fig. 5-12). And it occurs permanently in a state of free fall that doesn't happen to intersect Earth—like in an orbiting spacecraft (Fig. 5-11b). Apparent weightlessness is not a condition peculiar to outer space; it occurs in any frame of reference in free fall. (Don't be misled by the word "fall." Free fall just means that only gravity is acting, so a spacecraft in orbit is just as much in free fall as an elevator with a broken cable.) The condition of apparent weightlessness in orbiting spacecraft is sometimes called "microgravity."

Many people think astronauts are weightless because they're in space, where gravity is weak or nonexistent. Nothing could be further from the truth. If there were no gravity in space, a spacecraft and its occupants would happily obey Newton's first law and go off in a straight line, leaving Earth forever behind. But there *is* gravity in space. Indeed, it's the gravitational force that keeps a spacecraft and its occupants in orbit, causing deviations from straight-line motion in accordance with Newton's second law. In Chapter 9 we'll take a detailed look at orbital motion.

FIGURE 5-12 Astronaut trainees experience apparent weightlessness as their NASA training aircraft executes a parabolic trajectory.

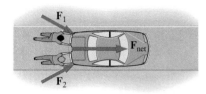

FIGURE 5-13   When two people push a stalled car, the force they exert is the vector sum of their individual forces.

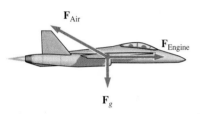

FIGURE 5-14   Forces on a jet include the force of the engines, the force of the air that provides both lift and drag, and the force of gravity. When the plane moves with constant velocity, these forces sum to zero.

5-7 ADDING FORCES

So far we've considered only situations where a single force acts on an object. But often there are several forces acting, as suggested in Fig. 5-13. Airplane flight involves the forces associated with engine thrust, air flowing over the wings and body of the plane, and the force of gravity (Fig. 5-14).

How are we to apply Newton's law in these multiple-force cases? The answer is given ultimately by experiment: We add vectorially the individual forces to find the **net force** on an object. Newton's second law then relates the object's mass and acceleration to this net force. As long as that net force is not zero, the object must be accelerating.

The *net* force is all-important! There may be all sorts of forces acting on a body, but it's their vector sum alone that determines the net force and therefore the acceleration. We've written Newton's law $\mathbf{F}_{net} = m\mathbf{a}$ to emphasize this point; some authors write $\Sigma\mathbf{F} = m\mathbf{a}$ for the same reason.

> **TIP**   **Trust Newton**   Newton's second law really does relate acceleration to the net force in all inertial frames. In particular, if you see an object at rest or moving with constant velocity, with respect to an inertial frame, then $\mathbf{a} = \mathbf{0}$ and Newton says the net force on that object must be zero. If you find a nonzero net force, look again; there *must* be additional forces acting (see Fig. 5-15).

In this introductory chapter on Newton's laws, we'll consider the addition of forces in one dimension only. But forces are vectors, and vector addition is generally necessary to get the net force. In the next chapter we'll deal with Newton's laws in two and three dimensions.

FIGURE 5-15   The gravitational force $\mathbf{F}_g$ acts on a seated person. But since the person's acceleration is zero, the net force must be zero. So there must be another force acting; in this case, it is the upward contact force $\mathbf{F}_c$ provided by the chair. (The contact force is ultimately an electrical force involving interactions between the molecules of the person and the chair.)

● EXAMPLE 5-7 AN ELEVATOR

A 740-kg elevator is accelerating upward at 1.1 m/s^2, pulled by a cable of negligible mass, as shown in Fig. 5-16. What is the tension force in the elevator cable?

Solution

This is an important example, and understanding it thoroughly will help you apply Newton's second law correctly in more complicated situations. We're given the acceleration and mass and asked for the force in the cable. Can't we just write $\mathbf{F} = m\mathbf{a}$ for that force? No! $\mathbf{F} = m\mathbf{a}$ is true only for the *net force,* and here the cable tension is *not* the only force acting. There is also the force of gravity, which has magnitude mg and points downward. We've shown both forces in Fig. 5-16*b*.

> **TIP Identify the Forces** A key step in solving any Newton's law problem is to identify the forces acting on the object or objects of interest. There's no sense proceeding unless you know the *net* force—and that requires identifying *all* the individual forces acting.

Calling the tension force $\mathbf{T}$ and the gravitational force $\mathbf{F}_g$, Newton's second law becomes

$$\mathbf{F} = \mathbf{T} + \mathbf{F}_g = m\mathbf{a}. \tag{5-5}$$

> **TIP Vectors Tell it All** You might be tempted to put minus signs in Equation 5-5. But not yet! This is a *vector* equation, and information about signs is built into the directional nature of the vectors. Skipping the step of writing Newton's law in vector form will often get you into trouble. And in writing the vector equation, you never need to worry about signs.

Having written the vector form of Newton's second law for our problem, we next choose a coordinate system and write the components of Newton's law in that system. Here all the forces are vertical, so we choose our y axis pointing vertically upward. Now we write the components of our vector equation 5-5. In this one-dimensional problem only the y equation is interesting; formally, it reads

$$T_y + F_{gy} = ma_y. \tag{5-6}$$

Now the tension force $\mathbf{T}$ points upward, or in the $+y$ direction; its vertical component T_y is positive and is equal to the magnitude T of $\mathbf{T}$. The gravitational force $\mathbf{F}_g$ has magnitude mg and points vertically downward, in the $-y$ direction. Its vertical component F_{gy} is therefore $-mg$. Then Equation 5-6 is

$$T - mg = ma_y,$$

so

$$T = ma_y + mg = m(a_y + g). \tag{5-7}$$

Before putting in specific numbers, let's see if this result makes sense. Suppose the acceleration a_y were zero. Then the

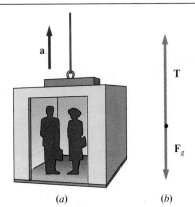

FIGURE 5-16 Forces on the elevator include the cable tension $\mathbf{T}$ and the gravitational force $\mathbf{F}_g$; the net force is their vector sum. Since the elevator is accelerating upward, this net force must be upward.

net force on the elevator would have to be zero. In this case, Equation 5-7 tells us that $T = mg$, showing that the tension force and gravity have the same magnitude. Since they point in opposite directions, they indeed sum to zero.

If, on the other hand, the elevator is accelerating upward, so a_y is positive, then Equation 5-7 requires that the magnitude of the tension force exceed that of the gravitational force by an amount ma_y. Why is this? Because in this case the cable not only supports the elevator against gravity, but also provides the upward acceleration. For the numbers of this example, we have

$$T = m(a_y + g) = (740 \text{ kg})(1.1 \text{ m/s}^2 + 9.8 \text{ m/s}^2)$$

$$= 8100 \text{ N}.$$

What if the elevator were accelerating downward? Then a_y is negative and Equation 5-7 shows that the tension force is less than mg. Does this make sense? Yes. For a downward acceleration, the net force must be downward so that the magnitude of the downward gravitational force exceeds that of the upward tension force. Were the elevator in free fall, so $a_y = -g$, the tension force would be exactly zero.

You probably could have reasoned out the answer to this problem in your head. But by applying the steps illustrated here—specifically, writing Newton's second law as a vector equation and then breaking it into components—you will be able to handle more complicated situations without confusion. We will outline the steps in solving a Newton's law problem more formally in the next chapter.

EXERCISE A 270-kg rocket accelerates straight upward from Earth at 51 m/s^2. What is the thrust (force) provided by the rocket's engine?

Answer: 1.6×10^4 N

Some problems similar to Example 5-7: 27–30 ●

5-8 NEWTON'S THIRD LAW

FIGURE 5-17 When you push on a book, the book pushes back on you; otherwise you wouldn't feel its presence. The forces **F**$_1$ and **F**$_2$ have equal magnitudes and constitute an action-reaction pair.

Push your book across your desk; you feel the book pushing back on your hand (Fig. 5-17). Kick a ball with bare feet, and your toes hurt. Why? You've exerted a force on the ball, and the ball has exerted a force back on you. Or imagine a rocket engine, exerting forces that expel hot gases out of its nozzle. Something else happens, too: the hot gases exert a force on the rocket, accelerating it forward (Fig. 5-18).

Experimentally, we find that whenever one object exerts a force on another, the second object also exerts a force on the first. The two forces are in opposite directions, but they have equal magnitudes. The familiar expression that "for every action there is an equal and opposite reaction" is Newton's statement of this fact in seventeenth-century language. By "action" Newton means an applied force. "Reaction" is the second force, equal in magnitude but opposite in direction. Technically, it never makes any difference which force we call the action and which the reaction; they are both always present as an **action-reaction pair.** Newton's statement about action and reaction constitutes his **third law of motion.** In more modern language, we express the third law by stating

> **If object A exerts a force on object B, then object B exerts an oppositely directed force of equal magnitude on object A.**

Notice that Newton's third law deals with *two forces* and also with *two objects*. The two forces of the action-reaction pair act on different objects. Therefore they do not cancel, even though they are of equal magnitude but opposite direction. Misunderstanding this point leads to a contradiction, embodied in the famous horse and cart dilemma described in Fig. 5-19.

FIGURE 5-19 The horse exerts a force on the cart, and the cart exerts an equal but opposite force on the horse. So how can the two start moving? The answer comes from looking at the *net* force on the horse. This includes not only the backward-pointing force of the cart, but also a forward-pointing force that the road exerts on the horse in reaction to the horse's pushing against the road. Those two aren't part of an action-reaction pair, so they need not have equal magnitudes.

FIGURE 5-18 The combustion chamber of a rocket engine exerts a force **F**$_1$ on the hot gases, expelling them through the nozzle. In return, the gases exert a force **F**$_2$ of equal magnitude on the rocket, accelerating it forward.

● EXAMPLE 5-8 NEWTON'S THIRD LAW

On a surface with negligible friction, you push with force $\mathbf{F}$ on a book of mass m_1 that in turn pushes on a book of mass m_2 (Fig. 5-20). What is the force exerted by the second book on the first?

Solution
Newton's *third* law tells us that the force of the second book on the first is equal in magnitude to that of the first book on the second. But what is that force? We can find it by applying Newton's *second* law to the combination of two books, then individually to the second book.

The total mass m of the two books is $m_1 + m_2$, and the net force applied to this combination is $\mathbf{F}$ (actually, this is only the horizontal force; the downward force of gravity and an upward force from the table also act, but they cancel). Then Newton's second law, $\mathbf{F} = m\mathbf{a}$, gives

$$\mathbf{a} = \frac{\mathbf{F}}{m} = \frac{\mathbf{F}}{m_1 + m_2}$$

for the acceleration of the two books as they move together. Now the only (horizontal) force on the second book is the force $\mathbf{F}_{12}$ exerted on it by the first book. So we can apply Newton's second law again to find this force:

$$\mathbf{F}_{12} = m_2\mathbf{a} = m_2 \frac{\mathbf{F}}{m_1 + m_2} = \frac{m_2}{m_1 + m_2} \mathbf{F}.$$

The answer we want is $\mathbf{F}_{21}$, the force book 2 exerts on book 1. Newton's third law tells us this force has the same magnitude as $\mathbf{F}_{12}$ we've just calculated, but points in the opposite direction:

$$\mathbf{F}_{21} = -\mathbf{F}_{12} = -\frac{m_2}{m_1 + m_2} \mathbf{F}.$$

Does all this make sense? We can see it does by considering the motion of the first book. It too undergoes an acceleration $\mathbf{a} = \mathbf{F}/(m_1 + m_2)$. And what is the *net* force on it? For the first book, there are *two* forces acting in the horizontal plane: the applied force $\mathbf{F}$ and the reaction force $\mathbf{F}_{21}$ exerted by the second book. So the net force on the first book is

$$\mathbf{F}_1 = \mathbf{F} + \mathbf{F}_{21}.$$

Using our expression of $\mathbf{F}_{21}$ then gives

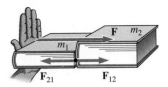

FIGURE 5-20 Example 5-8.

$$\mathbf{F}_1 = \mathbf{F} - \frac{m_2}{m_1 + m_2} \mathbf{F} = \frac{\mathbf{F}(m_1 + m_2) - m_2\mathbf{F}}{m_1 + m_2}$$

$$= \frac{m_1}{m_1 + m_2} \mathbf{F} = m_1\mathbf{a}.$$

This is consistent with Newton's second law: we know that both books have the same acceleration $\mathbf{a} = \mathbf{F}/(m_1 + m_2)$, and indeed we find that the force on the first book is the product of that acceleration with the book's mass. Our result shows how Newton's second and third laws are both necessary for a fully consistent description of the motion.

Understanding this example thoroughly will give you confidence in applying Newton's laws of motion correctly.

TIP Forces and Objects Newton's second law relates the *net force* on an object to the mass and acceleration *of that object*. In this example, the force $\mathbf{F}$ acts on the first book, but the acceleration of the book is *not* $\mathbf{F}/m_1$, becasue $\mathbf{F}$ is not the *net* force on the book. The reaction force $\mathbf{F}_{21}$ also acts on the first book, so its acceleration is $(\mathbf{F} + \mathbf{F}_{21})/m_1$, as we showed at the end of the example.

EXERCISE A railroad switching engine with a mass of 7.2×10^4 kg exerts a force of 8.5×10^4 N against the tracks, while pulling a boxcar with a mass of 5.5×10^4 kg. What force does the boxcar exert on the engine?

Answer: 3.7×10^4 N

Some problems similar to Example 5-8: 38, 39, 43, 44 ●

A contact force like the force between the books in Example 5-8 is called a **normal force** because it acts at right angles to the surfaces in contact. ("Normal" comes from the Latin for "according to the carpenter's square.") Other examples of normal forces include the upward force that a table or bridge exerts on objects it's supporting or the force perpendicular to a sloping surface supporting an object

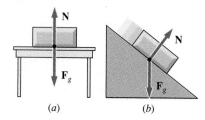

FIGURE 5-21 Normal forces. (*a*) The upward force of a table on a block. (*b*) The normal force of a sloping surface on a block. Also shown in each case is the gravitational force on the block; since both the normal force and the gravitational force act on the same object, they do not constitute an action-reaction pair. In (*a*) the two forces balance, but in (*b*) they add to give a net force that accelerates the block down the slope.

Calvin and Hobbes

by Bill Watterson

HOW DO THEY KNOW THE LOAD LIMIT ON BRIDGES, DAD?

LOAD LIMIT 10 TONS

THEY DRIVE BIGGER AND BIGGER TRUCKS OVER THE BRIDGE UNTIL IT BREAKS.

THEN THEY WEIGH THE LAST TRUCK AND REBUILD THE BRIDGE.

OH. I SHOULD'VE GUESSED.

DEAR, IF YOU DON'T KNOW THE ANSWER, JUST TELL HIM!

FIGURE 5-22 A bridge must be capable of providing normal forces sufficient to support the weights of vehicles crossing it.

(Figs. 5-21 and 5-22). Normal forces ultimately involve electrical interactions among atoms in the contacting surfaces.

Newton's third law also applies to a force like gravity that doesn't involve direct contact. Since Earth exerts a downward force on you, Newton's third law says you must exert an upward force on Earth. When you're standing on the ground, both those forces are cancelled by normal forces, as shown in Fig. 5-23. But if you're in free fall, then the forces are unbalanced. Following Newton's second law, you accelerate toward Earth—and Earth accelerates toward you (Fig. 5-24). Does this really happen? It does—but Newton's third law shows that the forces on you and on Earth have the same magnitude, and since Earth's mass is some 10^{23} times yours, Newton's second law implies its acceleration is negligibly small.

When we first listed Newton's laws in Section 5-2, we noted that Newton himself wrote his second law as a relation between force and the rate of change of "quantity of motion," or momentum, defined by $\mathbf{p} = m\mathbf{v}$. In Chapter 10, we'll see how Newton's second law in this form combines with the third law to make a powerful statement about the momentum of systems consisting of more than one particle.

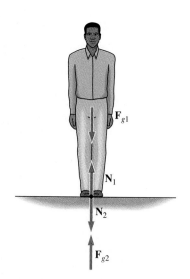

FIGURE 5-23 Standing on the ground involves two action-reaction pairs: gravitational forces that the Earth exerts on you and that you exert on the Earth, and normal forces between your feet and the ground. The net force on each object is zero and therefore neither accelerates.

5-9 MEASURING FORCE

We've talked a lot about force in this chapter, but have not said much about how to measure this important physical quantity. Newton's second law provides us a direct approach: apply an unknown force to a known mass and measure the resulting acceleration; the force is then the product of mass and acceleration.

Often, though, it is more convenient to use elastic objects like springs to measure forces. A spring, when stretched or compressed, exerts forces on whatever is deforming it (Fig. 5-25). In the case of a stretched spring, these are **tension forces;** for a compressed spring, they are **compression forces.** In either case, Newton's third law requires that the forces be oppositely directed but of equal magnitude to the external forces acting on the ends of the spring to cause the stretching or compression.

Over a limited range of stretching and compression, the force exerted by a spring is directly proportional to the distance stretched or compressed. When this direct proportionality holds, the spring is said to obey **Hooke's law.** We then write

$$F = -kx, \qquad \text{(Hooke's law; ideal spring)} \qquad (5\text{-}8)$$

where x is the amount by which the spring is stretched or compressed from its normal state, and where the minus sign indicates that the spring force is directed oppositely to the stretching or compression. The constant k is called the **spring constant;** its units are obviously N/m. A spring with $k = 200$ N/m (typical of a spring a few inches long that you might hold in your hand) would exert a force of 2 N when stretched or compressed 1 cm. If it were an **ideal spring**—one that obeyed Hooke's law (Equation 5-8) no matter how much it was deformed—it would exert a force of 200 N when stretched 1 m. A real spring a few inches long would break before reaching 1 m in length, and at any rate would cease to obey Hooke's law well before that point (Fig. 5-26).

We can use a spring to make a **spring scale** by attaching an indicator and a scale calibrated in force units (Fig. 5-27). Common bathroom scales, hanging scales in supermarkets, and laboratory spring scales are all examples of such scales. Even electronic scales widely used at supermarket checkouts are spring scales, with their "spring" a material that produces an electrical signal when it is deformed by an applied force.

We can use a spring scale to weigh objects by attaching one end of the scale to a fixed point and hanging the unknown weight from the other end (Fig. 5-28). When the object hangs at rest on the scale, the scale force balances the gravitational force mg, and is therefore equal in magnitude to the object's weight. Since mass and weight are proportional at a given location, the spring scale also provides a measure of the object's mass.

FIGURE 5-24 Forces on you and on Earth as you fall freely toward one another. The forces constitute an action-reaction pair, but since each acts on a different object and is the only force acting on that object, both objects accelerate. By Newton's third law, the force on each object has the same magnitude, but Earth is so much more massive that its acceleration is negligibly small.

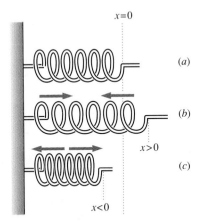

FIGURE 5-25 (*a*) An unstretched spring attached to a wall. (*b*) When stretched, the spring exerts **tension forces** that oppose the stretching. (*c*) When compressed, it exerts **compression forces** that oppose the compression. The distance x in Equation 5-8 is measured from the unstretched position, here marked $x = 0$. Compression corresponds to a negative value of x and therefore a positive direction of force on the free end of the spring, and vice versa for stretching.

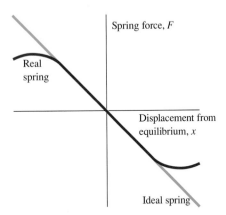

FIGURE 5-26 Force curves for ideal and real springs. The slopes are negative, indicating that the force is opposite to the deformation. Negative values of x mean compression; positive values mean stretching.

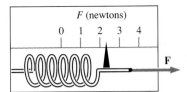

FIGURE 5-27 A simple spring scale, made from a spring of known spring constant k, an indicating needle, and a scale calibrated in force units.

● EXAMPLE 5-9 WEIGHING A FISH

When a fish is suspended from a spring of spring constant $k = 320$ N/m, the spring stretches 15 cm. What are the weight and mass of the fish?

Solution

From Equation 5-8, a 15-cm stretch corresponds to a spring force of magnitude

$$F = kx = (320 \text{ N/m})(0.15 \text{ m}) = 48 \text{ N}.$$

Since the net force on the fish is zero, the magnitude of the fish's weight is also 48 N. But weight is given by $W = mg$, so $m = W/g = (48 \text{ N})/(9.8 \text{ m/s}^2) = 4.9$ kg.

●

FIGURE 5-28 Weighing nectarines with a spring scale.

Apparent Weight

When a spring scale is in an accelerating reference frame, its reading—called the **apparent weight**—differs from the gravitational force mg. Why? Because the object it's weighing is also accelerating, so Newton's laws require a net force on that object. If the only forces acting are gravity and the spring force of the scale, then those two forces can't balance. We've already seen a special case of apparent weight in Section 5-6, where we found that the apparent weight of an object in free fall is zero. More generally the situation is like that of the elevator in Example 5-7, with the scale playing the role of the elevator cable. The scale reading will be more or less than the actual weight depending on whether the acceleration is upward or downward.

● EXAMPLE 5-10 A HELICOPTER RIDE

A helicopter is rising vertically, carrying a load of concrete to make the foundation for a ski lift. A 35-kg bag of concrete is resting on a scale in the helicopter. A construction worker in the copter notes that the scale reads 280 N. What is the acceleration of the copter?

Solution

This problem is similar to Example 5-7. The two forces acting on the bag of concrete are the force of gravity $\mathbf{F}_g$ and the force of the scale $\mathbf{F}_s$ (Fig. 5-29). Newton's second law tells us that the net force is the product of mass and acceleration:

$$\mathbf{F} = \mathbf{F}_g + \mathbf{F}_s = m\mathbf{a}.$$

Since the motion is entirely vertical, we need work only with the vertical component of this equation. With the positive y direction upward, the vertical component of the downward gravitational force is $-mg$; that of the scale force is positive and is equal to the magnitude F_s of the scale force. Then the y component of Newton's second law for this case becomes

$$-mg + F_s = ma_y,$$

so the vertical acceleration is

$$a_y = \frac{-mg + F_s}{m} = -g + \frac{F_s}{m} = -9.8 \text{ m/s}^2 + \frac{280 \text{ N}}{35 \text{ kg}}$$
$$= -1.8 \text{ m/s}^2.$$

What does this negative result mean? It means that the direction of the helicopter's *acceleration* is downward—even though the

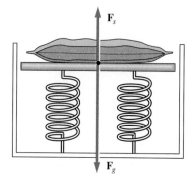

FIGURE 5-29 Forces on the concrete of Example 5-10.

copter is *moving* upward. That is, its speed is decreasing at the rate of 1.8 m/s² even as it rises. The result of this example is consistent with that of Example 5-7. There, a downward acceleration would result in a cable tension lower than the weight *mg*. Here, a downward acceleration results in a scale reading lower than the weight, as suggested by the length of the force vectors in Fig. 5-28.

EXERCISE A 65-kg person steps on a spring scale in the elevator of Example 5-7. Even though the scale measures force, it's calibrated in kilograms under the assumption that $g = 9.8$ m/s². What does it read?

Answer: 72 kg

Some problems similar to Example 5-10: 36, 37, 70

●

> **TIP** **Don't Confuse Acceleration and Velocity** As early as Chapter 2 we emphasized that acceleration and velocity are *independent* quantities; acceleration depends only on the *rate of change* of velocity, not on its value or even its direction. This example, with an upward velocity but downward acceleration, is a good case in point. The magnitude and direction of the helicopter's velocity are completely irrelevant.

CHAPTER SYNOPSIS

Summary

1. Newton's **first law of motion** states that a body continues in uniform motion unless acted on by a force. Philosophically, the first law represents a shift from the Aristotelian notion that the natural state of motion for objects on Earth is rest. In Newton's view, uniform motion is a natural state; the study of motion then emphasizes not the cause of motion itself, but the cause of *changes* in motion.

2. Newton's **second law of motion** quantifies the change in motion brought about by a force. The law states that the rate of change of a body's momentum (product of mass and velocity; symbol **p**) is equal to the net force on the body:

$$\mathbf{F}_{net} = \frac{d\mathbf{p}}{dt}.$$

As long as the mass of the body does not change, Newton's second law can also be written in the form

$$\mathbf{F}_{net} = m\mathbf{a}.$$

Newton's second law is valid only when **F** is the **net force** on a body—that is, the vector sum of all forces acting on the body. The SI unit of force is the **newton,** defined as the force that gives a 1-kg mass an acceleration of 1 m/s².

3. **Weight** is the force of gravity on an object: **W** = *m***g.** In an accelerated frame of reference, the **apparent weight** differs from the gravitational force; in particular, the apparent weight of an object in free fall is zero.

4. Newton's **third law of motion** states that forces always come in **action-reaction pairs.** When one object exerts a force on another, the second object exerts an oppositely directed force of equal magnitude back on the first. Newton's second and third laws together permit a consistent description of the motions of interacting objects.

5. Elastic objects like springs provide a practical way of measuring forces. When the force exerted by a spring is directly proportional to the amount of stretch or compression, the spring is said to obey **Hooke's law;** the force is then given by

$$F = -kx, \quad \text{(Hooke's law)}$$

where k is the **spring constant,** with SI units of N/m.

Terms You Should Understand

(Pairs are closely related terms whose distinction is important; number in parentheses is chapter section where term first appears.)

kinematics, dynamics (introduction)
Aristotle, Galileo, Newton (5-1)
force (5-2)
momentum (5-2)
Newton's laws of motion (5-2)
inertia (5-3)
inertial frame (5-3)
gravitational, electroweak, strong/color forces (5-4)
mass, weight (5-6)
apparent weight (5-6, 5-9)
weightlessness (5-6)
net force (5-7)
action-reaction pair (5-8)
normal force (5-8)
tension force, compression force (5-9)
Hooke's law (5-9)
spring constant (5-9)

Symbols You Should Recognize

$\mathbf{F}$ (5-2)
$\mathbf{p}$ (5-2)
N (5-5)
$\mathbf{g}$ (5-6)
$\mathbf{W}$ (5-6)
k (5-9)

Problems You Should Be Able to Solve

calculating force, mass, or acceleration from the other two quantities (5-5)
calculating weight or mass given the other (5-6)
solving for forces or accelerations in one-dimensional situations (5-7)
applying Newton's second and third laws together in one-dimensional situations (5-8)

determining spring force, extension/compression, or spring constant given the other two quantities (5-9)
determining apparent weight in accelerating frames of reference (5-9)

Limitations to Keep in Mind

Newton's laws apply only when relative velocities are much less than the speed of light; aside from that restriction, the basic material of this chapter is universally applicable.
Newton's laws apply to inertial frames of reference; correct treatment of accelerating frames requires consideration of the forces giving rise to the acceleration.
Real springs deviate from ideal Hooke's law behavior when stretched or compressed substantially.

QUESTIONS

1. Distinguish between the Aristotelian and Galilean/Newtonian view of the natural state of motion.
2. Use Newton's laws of motion to explain the function of an airbag in a car.
3. A ball bounces off a wall with the same speed it had before it hit the wall. Has its momentum changed? Has a force acted on the ball? Has a force acted on the wall? Relate your answers to Newton's laws of motion.
4. Give several examples from the history of physics in which seemingly unrelated forces were found to be related.
5. We often use the term "inertia" to describe human sluggishness. How is this usage related to the meaning of inertia in physics?
6. Why do subway riders lurch?
7. My high school physics teacher defined mass as "inverse pushability aroundness." Comment.
8. Which of the fundamental forces do you deal with in everyday life? Give several examples.
9. Does a body necessarily move in the direction of the net force on it?
10. Given that $\mathbf{F} = m\mathbf{a}$, is it possible to have a nonzero net force acting on a body that is at rest? Explain.
11. Can a motorcycle and a truck have the same momentum? Explain.
12. A barefooted astronaut kicks a ball across the recreation area of a space station. Does the ball's apparent weightlessness mean the astronaut's toes don't get hurt? Explain.
13. Does the length of time that a force is applied to an object affect the change in the object's momentum? Explain.
14. Does the length of time that a force is applied to an object affect the object's acceleration? Explain.
15. An elevator in the physics building at the University of Colorado includes a built-in physics demonstration consisting of a 10-kg mass suspended from a spring scale. Is the scale reading more than, less than, or equal to 98 N when the elevator (a) is stationary; (b) first starts moving upward; (c) moves steadily upward; (d) slows down on its way up; (e) starts moving downward; (f) moves steadily downward; (g) slows down on its way down?
16. The surface gravity of Jupiter's moon Io is one-fifth that of Earth. What would happen to your weight and to your mass if you were to travel to Io?
17. What does it mean to say that someone "weighs" 50 kilograms?
18. Consider two teams in a tug-of-war (Fig. 5-30). Newton's third law says that the force team A exerts on team B is equal in magnitude to the force team B exerts on team A. How, then, can either team win?

A   B

FIGURE 5-30 A tug-of-war (Question 18).

19. Describe the action and reaction forces involved in paddling a canoe.
20. When you step on a car's brake pedal, a force develops between the tires and the road that eventually stops the car. Are the force you exert on the pedal and the force of the road on the tires an action-reaction pair? Explain.
21. If you move from bow to stern of a canoe, the canoe moves in the opposite direction. Why?

22. As you sit in your chair, the chair exerts an upward force on you. Why don't you accelerate upward?
23. As you sit in a car, what direction is the force exerted on you by the car seat (a) when the car moves at constant speed; (b) when the car accelerates in the forward direction; (c) when the car brakes while moving in a straight line; (d) while the car rounds a corner at constant speed?
24. As your plane accelerates down the runway, you take your keys from your pocket and suspend them by a thread. Do they hang vertically? Explain.
25. Since every force has an oppositely directed reaction force of equal magnitude, how can there ever be a net force on an object?
26. A friend thinks that astronauts are weightless because there is no gravity in outer space. Criticize this statement, and supply your friend with a correct explanation for weightlessness.

27. Why does your stomach seem to head for your throat when a high-speed elevator stops its upward journey?
28. A tractor-trailer truck is stopped at a red light. How can it ever get started, given that Newton's third law requires that the force of the trailer on the tractor be equal and opposite that of the tractor on the trailer?
29. I tell passengers in my car to buckle their seatbelts, explaining that I believe in the law of inertia. What's that got to do with seatbelts?
30. Compare the forces needed to increase the speed of an object by 1 m/s (a) starting from rest and (b) if it's moving initially at 10 m/s. Neglect friction and air resistance.
31. What would happen to passengers in an elevator accelerating downward with acceleration greater than g?

PROBLEMS

Section 5-5 Newton's Second Law

1. A subway train has a mass of 1.5×10^6 kg. What force is required to accelerate the train at 2.5 m/s^2?
2. A railroad locomotive with a mass of 6.1×10^4 kg can exert a force of 1.2×10^5 N. At what rate can it accelerate (a) by itself and (b) when pulling a 1.4×10^6 -kg train?
3. A small plane starts down the runway with acceleration 7.2 m/s^2. If the force provided by its engine is 1.1×10^4 N, what is the plane's mass?
4. A car leaves the road traveling at 95 km/h and hits a tree, coming to a complete stop in 0.16 s. What average force does a seatbelt exert on a 55-kg passenger during this collision?
5. In an x-ray tube, electrons are accelerated to speeds on the order of 10^8 m/s, then slammed into a target where they come to a stop in about 10^{-18} s. Estimate the average stopping force on each electron.
6. A 280-kg crate is secured in a truck with ropes running horizontally forward and backward, as shown in Fig. 5-31. What is the maximum tension force these ropes must be able to withstand if the most rapid deceleration anticipated is 6.5 m/s^2? Assume the tension is initially zero.

FIGURE 5-31 Problem 6.

7. The magnitude of the force on an electron in a hydrogen atom is about 82 nN. What is the magnitude of the electron's acceleration?

8. A hockey player strikes a 170-g puck, accelerating it from rest to 50 m/s. If the hockey stick is in contact with the puck for 2.5 ms, what is the average force applied by the stick?
9. Object A accelerates at 8.1 m/s^2 when a 3.3-N force is applied. Object B accelerates at 2.7 m/s^2 when the same force is applied. (a) How do the masses of the two objects compare? (b) If A and B were stuck together and accelerated by the 3.3-N force, what would be the acceleration of the composite object?
10. By how much does the force required to stop a car increase if (a) the stopping time is halved and (b) the stopping distance is halved? Assume the same initial speed in both cases.
11. By how much does the force required to stop a car increase if the initial speed is doubled and the stopping distance remains the same?
12. A car moving at 70 km/h collides with a concrete bridge support. The bridge support is unaffected, but the front of the car is compressed by 0.94 m. What average force must a seatbelt exert in order to restrain a 75-kg passenger during this collision?
13. The maximum braking force of a 1400-kg car is about 8000 N. Estimate the stopping distance when the car is traveling (a) 40 km/h; (b) 60 km/h; (c) 80 km/h; (d) 55 mi/h.
14. As a function of time, the velocity of an object of mass m is given by $\mathbf{v} = bt^2\hat{\mathbf{i}} + (ct + d)\hat{\mathbf{j}}$, where b, c, and d are constants with appropriate units. What is the force acting on the object, as a function of time?
15. A car moving at 50 km/h collides with a truck, and the front of the car is crushed 1.1 m as it comes to a complete stop. The driver is wearing a seatbelt, but the passenger is not.

The passenger, obeying Newton's first law, keeps moving and slams into the dashboard after the car has stopped. If the dashboard compresses 5.0 cm on impact, find and compare the forces exerted on the driver by the seatbelt and on the passenger by the dashboard. Assume the two have the same 65-kg mass.

16. A 3800-kg jet touches down at 240 km/h on the deck of an aircraft carrier, and immediately deploys a parachute to slow itself down. If the plane comes to a stop in 170 m, what is the average force of air on the parachute? Assume the parachute provides essentially all the stopping force.

17. A 1.25-kg object is moving in the x direction at 17.4 m/s. 3.41 s later, it is moving at 26.8 m/s at 34.0° to the x axis. What are the magnitude and direction of the force applied during this time?

Section 5-6 Mass and Weight: The Force of Gravity

18. Show that the units of acceleration can be written N/kg. Why would it make sense to state g as 9.8 N/kg when talking about mass and weight?

19. My spaceship crashes on one of the Sun's nine planets. Fortunately, the ship's scales are intact, and show that my weight is 532 N. If I know my mass to be 60 kg, where am I? *Hint:* Consult Appendix E.

20. If I can barely lift a 50-kg concrete block on Earth, how massive a block can I lift on the moon?

21. A cereal box says "net weight 340 grams." What is the actual weight (a) in SI units? (b) In ounces?

22. I weigh 640 N. What's my mass?

23. A bridge specifies a maximum load of 10 tons. What's the maximum mass, in kilograms, that the bridge can carry?

24. The English unit of force is the pound; 1 pound equals 4.45 N. The English unit of mass is the slug; 1 slug equals 14.6 kg. Use these conversions to find the acceleration of gravity in English units. *Hint:* You don't need any length conversions. See Problem 18.

25. A neutron star is a fantastically dense object with the mass of a star crushed into a region about 10 km in diameter. If my mass is 65 kg, and if I would weigh 5.4×10^{14} N on a certain neutron star, what is the acceleration of gravity on the neutron star?

Section 5-7 Adding Forces

26. A 50-kg parachute jumper descends at a steady 40 km/h. What is the force of air on the parachute?

27. Two children are fighting over a 680-g toy. One pulls on it with a 50-N force; the other pulls in the opposite direction with a 54-N force. What is the magnitude of the toy's acceleration?

28. A 930-kg motorboat accelerates away from a dock at 2.3 m/s². Its propeller provides a thrust force of 3.9 kN. What is the drag force exerted by the water on the boat?

29. An elevator accelerates downward at 2.4 m/s². What force does the floor of the elevator exert on a 52-kg passenger?

30. What is the vertical lifting force on a 747 jetliner when the plane is (a) flying at constant altitude and (b) accelerating upward at 1.1 m/s²? The aircraft's mass is 4.5×10^5 kg.

31. An airplane encounters sudden turbulence, and you feel momentarily lighter. If your apparent weight seems to be about 70% of your normal weight, what are the magnitude and direction of the plane's acceleration?

32. A 74-kg tree surgeon rides a "cherry picker" lift (Fig. 5-32) to reach the upper branches of a tree. What force does the bucket of the lift exert on the surgeon when the bucket is (a) at rest; (b) moving upward at a steady 2.4 m/s; (c) moving downward at a steady 2.4 m/s; (d) accelerating upward at 1.7 m/s²; (e) accelerating downward at 1.7 m/s²?

FIGURE 5-32 Problem 32.

33. At liftoff, a space shuttle with 2.0×10^6-kg total mass undergoes an upward acceleration of $0.60g$. (a) What is the total thrust force developed by its engines? (b) What force does the seat back exert on a 60-kg astronaut during liftoff?

34. Ballet star Mikhail Baryshnikov (Fig. 5-33) executes a vertical jump during which the floor pushes up on his feet with a force 50% greater than his weight. What is his upward acceleration?

35. You're eating dinner on an airplane when the plane encounters sudden turbulence and drops abruptly downward. You've got your seatbelt on, but to your amazement your

FIGURE 5-33 Mikhail Baryshnikov jumping (Problem 34).

dinner rises momentarily off the seatback tray. What can you conclude about the magnitude of the downward acceleration caused by the turbulence?

36. A poorly designed elevator begins its descent with a downward acceleration of 11.4 m/s². To keep yourself on the floor, you grab a handrail in the elevator. If your mass is 69 kg, what upward force must you exert on the rail?

37. An elevator moves upward at 5.2 m/s. What is the minimum stopping time it can have if the passengers are to remain on the floor?

38. A 2.50-kg object is moving along the x axis at 1.60 m/s. As it passes the origin, two forces $\mathbf{F}_1$ and $\mathbf{F}_2$ are applied, both at right angles to the x axis. $\mathbf{F}_1 = 15\hat{\jmath}$ N. The forces are applied for 3.00 s, after which the object is at the point $x = 4.80$ m, $y = 10.8$ m. Find $\mathbf{F}_2$.

Section 5-8 Newton's Third Law

39. What upward force does a 5600-kg elephant exert on the Earth?

40. Blocks of 1.0, 2.0, and 3.0 kg are lined up from left to right in that order on a table so each block is touching the next one. A rightward-pointing 12-N force is applied to the leftmost block. What force does the middle block exert on the rightmost one?

41. Repeat the preceding problem, now with the left-right order of the blocks reversed. (That is, find the force on the rightmost mass, now 1.0 kg.)

42. A 68-kg astronaut pushes off a 420-kg satellite, exerting a force of 120 N for the 0.89 s they're in contact. (a) Find the speed of each after they've separated. (b) How far apart are they after 1.0 min?

43. I have a mass of 65 kg. If I jump off a 120-cm-high table, how far toward me does Earth move during the time I fall?

44. A child pulls an 11-kg wagon with a horizontal handle whose mass is 1.8 kg, giving the wagon and handle an acceleration of 2.3 m/s². (a) The tension at each end of the handle is different. Why? (b) Find the tension at each end of the handle.

45. A 2200-kg airplane is pulling two gliders, the first of mass 310 kg and the second of mass 260 kg, down the runway with an acceleration of 1.9 m/s² (Fig. 5-34). Neglecting the mass of the two ropes and any frictional forces, determine (a) the horizontal thrust of the plane's propeller; (b) the tension force in the first rope; (c) the tension force in the second rope; and (d) the net force on the first glider.

FIGURE 5-34 Problem 45.

46. In a tractor pulling contest, a 2300-kg tractor pulls a 4900-kg sledge with an acceleration of 0.61 m/s². If the tractor exerts a horizontal force of 7700 N on the ground, determine the magnitudes of (a) the force of the tractor on the sledge, (b) the force of the sledge on the tractor, and (c) the frictional force exerted on the sledge by the ground.

Section 5-9 Measuring Force

47. What force is necessary to stretch a spring 48 cm, if the spring constant is 270 N/m?

48. A 35-N force is applied to a spring with spring constant $k = 220$ N/m. How much does the spring stretch?

49. A spring stretches 22 cm when a 40-N force is applied. If a 6.1-kg mass is suspended from the spring, how much will it stretch?

50. A spring with spring constant $k = 340$ N/m is used to weigh a 6.7-kg fish. How far does the spring stretch?

51. A father pulls his 27-kg daughter across frictionless ice, using a horizontal spring with spring constant 160 N/m. If the spring is stretched 32 cm from its equilibrium position, what is the child's acceleration?

52. You want to make a spring scale that reads directly in mass units. If a change of 1 cm in the position of the scale's indicating needle is to correspond to a mass difference of 100 g, what should be the spring constant?

53. A biologist is studying the growth of rats in an orbiting space station. To determine a rat's mass, she puts it in a 320-g cage, attaches a spring scale, and pulls so the scale reads 0.46 N. If the resulting acceleration of the rat and cage is 0.40 m/s², what is the rat's mass?

54. An elastic tow rope has a spring constant of 1300 N/m. It is connected between a truck and a 1900-kg car. As the truck tows the car, the rope stretches 55 cm. Starting from rest, how far do the truck and car move in 1 min? Assume the car experiences negligible friction.

55. A 7.2-kg mass is hanging from the ceiling of an elevator by a spring of spring constant 150 N/m whose unstretched length is 80 cm. What is the overall length of the spring when the elevator (a) starts moving upward with acceleration 0.95 m/s²; (b) moves upward at a steady 14 m/s; (c) comes to a stop while moving upward at 14 m/s, taking 9.0 s to do so? (d) If the elevator measures 3.2 m from floor to ceiling, what is the maximum acceleration it could undergo without the 7.2-kg mass hitting the floor?

56. An accelerometer consists of a spring of spring constant $k = 1.25$ N/m and unstretched length $\ell_0 = 10.0$ cm fastened to a frictionless surface by a pivot that allows it to swivel in any direction in a horizontal plane. A 50.0-g mass is attached to the other end of the spring, as shown in Fig. 5-35. The whole system is mounted securely in an automobile. When the vehicle accelerates, the spring provides a force to keep the 50-g mass accelerating with the vehicle; by measuring the stretch of the spring, the acceleration can then be determined. To calibrate the accelerometer, circles marked with values of acceleration can be drawn on its frictionless surface. (a) How far apart should the circles be if each represents an acceleration of 0.250 m/s² larger than the next smaller circle? (b) What should be the radius of the circle marked 2.0 m/s²? (c) How do you read the direction of the car's acceleration from this device?

Paired Problems

(Both problems in a pair involve the same principles and techniques. If you can get the first problem, you should be able to solve the second one.)

57. Starting from rest, a 940-kg racing car covers 400 m in 4.95 s. What is the average force acting on the car?

58. Starting from rest, a 200-g hockey puck moves 1.1 m during the 75 ms it's in contact with the stick. What average force does the stick exert on the puck?

59. In an egg-dropping contest, a student encases an 95-g egg in a styrofoam block. If the force on the egg is not to exceed 2.0 N, and if the block hits the ground at 1.1 m/s, by how much must the styrofoam crush?

60. In a front-end collision, a 1300-kg car with shock-absorbing bumpers can withstand a maximum force of 65 kN before damage occurs. If the maximum speed for a nondamaging collision is 10 km/h, by how much must the bumper be able to move relative to the car?

61. You step into an elevator, and it accelerates to a downward speed of 9.2 m/s in 2.1 s. How does your apparent weight during this acceleration time compare with your actual weight?

62. A spacecraft blasts off from the moon's surface, achieving an upward speed of 2.5 km/s at the end of its rocket firing. During the firing, an astronaut on board feels she weighs exactly the same as she would on Earth. How long does the rocket firing last?

63. A 20-kg fish at the end of a fishing line is being yanked vertically into a boat. When its acceleration reaches 2.2 m/s², the line breaks and the fish goes free. What is the maximum tension the line can tolerate without breaking?

64. Tarzan (mass 73 kg) slides down a vine that can withstand a maximum tension of 620 N before breaking. What minimum acceleration must Tarzan have?

65. A 2.0-kg mass and a 3.0-kg mass are on a horizontal frictionless surface, connected by a massless spring with spring constant $k = 140$ N/m. A 15-N force is applied to the larger mass, as shown in Fig. 5-36. How much does the spring stretch from its equilibrium length?

FIGURE 5-36 Problem 65.

66. Two large crates, with masses 640 kg and 490 kg, are connected by a stiff, massless spring ($k = 8.1$ kN/m) and propelled along an essentially frictionless, level factory floor by a force applied horizontally to the more massive crate. If the spring compresses 5.1 cm from its equilibrium length, what is the applied force?

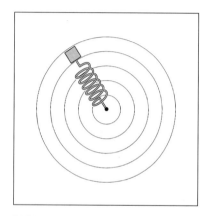

FIGURE 5-35 Problem 56.

Supplementary Problems

67. To escape a dog, a 1.3-kg squirrel runs up a vertical tree trunk, accelerating at 0.82 m/s^2. What is the average vertical force the tree exerts on the squirrel?

68. The second floor of a house can safely carry a load of 3500 N for each square meter of floor. Can it support a waterbed measuring 1.4 m wide by 1.8 m long by 25 cm thick? (The density of water is 1 g/cm^3.)

69. In throwing a 200-g ball, your hand exerts a constant upward force of 9.4 N for 0.32 s. How high does the ball rise after leaving your hand?

70. What downward force is exerted on the air by the blades of a 4300-kg helicopter when it is (a) hovering at constant altitude; (b) dropping at 21 m/s with speed decreasing at 3.2 m/s^2; (c) rising at 17 m/s with speed increasing at 3.2 m/s^2; (d) rising at a steady 15 m/s; (e) rising at 15 m/s with speed decreasing at 3.2 m/s^2?

71. What engine thrust (force) is needed to accelerate a rocket of mass m (a) downward at $1.40g$ near Earth's surface: (b) upward at $1.40g$ near Earth's surface; (c) at $1.40g$ in interstellar space far from any star or planet?

72. An elevator cable can withstand a maximum tension of 19,500 N before breaking. The elevator has a mass of 490 kg, and a maximum acceleration of 2.24 m/s^2. Engineering safety standards require that the cable tension never exceed two-thirds of the breaking tension. How many 65-kg people can the elevator safely accommodate?

73. You have a mass of 60 kg, and you jump from a 78-cm-high table onto a hard floor. (a) If you keep your legs rigid, you come to a stop in a distance of 2.9 cm, as your body tissues compress slightly. What force does the floor exert on you? (b) If you bend your knees when you land, the bulk of your body comes to a stop over a distance of 0.54 m. Now estimate the force exerted on you by the floor. Neglect the fact that your legs stop in a shorter distance than the rest of you.

74. An F-14 jet fighter has a mass of 1.6×10^4 kg and an engine thrust of 2.7×10^5 N. A 747 jumbo jet has a mass of 3.6×10^5 kg and a total engine thrust of 7.7×10^5 N. Is it possible for either plane to climb vertically, with no lift from its wings? If so, what vertical acceleration could it achieve?

75. A spider of mass m_S drapes a silk thread of negligible mass over a stick with its far end a distance h off the ground as shown in Fig 5-37. The stick is lubricated by a drop of dew, so that there is essentially no friction between silk and stick. The spider waits on the ground until a fly of mass $m_f (m_f > m_s)$ lands on the other end of the silk and sticks to it. The spider immediately begins to climb her end of the silk. (a) With what acceleration must she climb to keep the fly from falling? (b) If she climbs with acceleration a_S, at what height y above the ground will she encounter the fly?

FIGURE 5-37 Problem 75.

76. Each of the Viking spacecraft that landed on Mars used retrorockets to begin its descent to the Martian surface. At the beginning of the rocket firing, the craft had a mass of 1070 kg, and its rocket engine developed a thrust of 2840 N. If it dropped vertically with its retrorocket firing downward to slow the fall, what was its acceleration? Assume the craft was close enough to the Martian surface to neglect any variation in the local value of g.

77. Three identical massless springs of unstretched length ℓ and spring constant k are connected to three equal masses m as shown in Fig. 5-38. A force is applied at the top of the upper spring to give the whole system the same acceleration a. Determine the length of each spring.

FIGURE 5-38 Problem 77.

78. A 90-kg weightlifter is standing on a scale calibrated in newtons, holding a 140-kg barbell. (a) What does the scale read? (b) The lifter then gives the barbell a constant upward acceleration. If the scale reads 2400 N, what is the acceleration of the barbell? (c) Suppose the lifter were in an orbiting spaceship with his feet contacting an identical scale. If he gave the barbell the same acceleration as in (b), what would the scale read? Assume in (b) and (c) that the lifter

keeps his body rigid, bending only his arms, and neglect the mass of the arms and any downward acceleration of the lifter.

79. Two springs have the same unstretched length but different spring constants k_1 and k_2. (a) If they are connected side-by-side and stretched a distance x, as shown in Fig. 5-39a, show that the force exerted by the combination is $(k_1 + k_2)x$. (b) If they are now connected end-to-end and the combination is stretched a distance x (Fig. 5-39b), show that they exert a force $k_1 k_2 x / (k_1 + k_2)$.

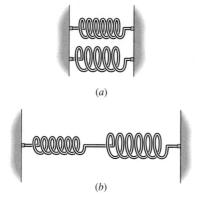

(a)

(b)

FIGURE 5-39 Problem 79.

80. A tow truck exerts a 750-N force on 54.4-kg tow chain. The other end of the chain is attached to a car, and the whole system accelerates at 0.436 m/s^2. Find (a) the force exerted on the truck by the cable, (b) the force exerted on the car by the cable, and (c) the mass of the car. Assume there are no horizontal forces acting on the car except for the cable. Why are your answers to (a) and (b) different?

81. Although we usually write Newton's law for one-dimensional motion in the form $F = ma$, the most basic version of the law reads $F = \dfrac{d(mv)}{dt}$. The simpler form holds only when the mass is constant. (a) Consider an object whose mass may be changing, and show that the rule for the derivative of a product (see Appendix A) can be used to write Newton's law in the form $F = ma + v\dfrac{dm}{dt}$. (b) A railroad car is being pulled beneath a grain elevator that dumps in grain at the rate of 450 kg/s. What force must be applied to keep the car moving at a constant 2.0 m/s?

82. The force of air resistance is roughly proportional to the square of an object's speed and is directed oppositely to the velocity: $F_{air} = -cv^2$, where c is a constant that depends largely on the shape of the falling body. As a result of air resistance, a falling object eventually reaches a constant **terminal speed,** at which it ceases to accelerate. The constant c for a 70-kg skydiver is approximately $0.30 \text{ N·s}^2/\text{m}^2$ (this depends on how the diver's body is oriented). Find (a) the skydiver's acceleration while falling at 25 m/s and (b) the terminal speed.

CHAPTER 6

USING NEWTON'S LAWS

This roller coaster's path is anything but a straight line at constant speed, and therefore it's subject to a net force. Can you identify the individual forces that sum to provide that net force?

In the preceding chapter we introduced Newton's three laws of motion and used them in simple one-dimensional applications. Now we apply those laws in a wider variety of physical situations, most involving two dimensions. The methods we develop are at the heart of all applications of Newton's laws, from the simplest textbook problems to the most complicated systems that guide spacecraft to distant planets.

> **TIP**   **Look for the Underlying Principles**   This chapter contains a wide array of examples. As you study them, don't look on each as an entirely new thing to learn; rather, think about how the different examples all represent applications of the same underlying physical principles—namely Newton's laws. That way you'll gain confidence in applying those laws in new situations—which is always the challenge for the scientist or engineer.

6-1 USING NEWTON'S SECOND LAW

Newton's second law, $\mathbf{F}_{net} = m\mathbf{a}$, is the cornerstone of mechanics. That law relates the *net force* on an object to the acceleration *of that object*. To apply Newton's law, we must identify the object whose motion interests us and all the forces that together provide the net force acting on that object. Only then can we write Newton's law and solve the equations it provides. In working Newton's law problems, you will find the following steps helpful in correctly applying the law and developing from it a mathematical description of each problem:

> **STRATEGY**   **Solving Newton's Law Problems**
>
> 1. Identify the object of interest.
> 2. Identify all the forces acting on the object, and draw them on a vector diagram. This diagram is often called a **free-body diagram.**
> 3. Write Newton's second law, $\mathbf{F}_{net} = m\mathbf{a}$, *in vector form*, using for $\mathbf{F}_{net}$ the vector sum of the forces in your diagram.
> 4. Choose a convenient coordinate system, and draw the coordinate axes on your force diagram.
> 5. Write the components of your vector equation in your chosen coordinate system. Be guided in this by the geometry of the force diagram, and by any constraints on the motion (for example, the constraint that an object move along a given surface).
> 6. Solve the equations symbolically for whatever the problem asks you to find.
> 7. Ask whether your results make sense. In particular, ask about extreme cases such as purely horizontal or purely vertical motion to see if they lead to results that you have seen before or that are physically obvious.
> 8. Finally, insert numerical values and work the arithmetic, if numerical answers are called for.

Don't memorize these steps. Rather, you should refer to them as you begin to work problems and will gradually come to realize that they simply reflect the full meaning of Newton's second law. Later, as you develop confidence in problem solving, you can take shortcuts. But whenever you deal with a complicated problem, following these steps will help avoid confusion.

● **EXAMPLE 6-1** AN INCLINED SURFACE

A block of mass m slides without friction on a surface tilted at an angle θ to the horizontal. Find its acceleration and the magnitude of the force the block exerts on the surface.

Solution
Here we outline the solution in the context of our eight steps:

1. The object of interest is the block.
2. There are two forces on the block: the gravitational force $\mathbf{F}_g$ and the normal force $\mathbf{N}$ from the surface. Both are shown in Fig. 6-1.
3. In vector form, Newton's second law becomes

$$\mathbf{F}_g + \mathbf{N} = m\mathbf{a}. \qquad (6\text{-}1)$$

4. Now we choose a coordinate system. Since gravity points vertically downward, you might pick a system with horizontal and vertical axes. But a better choice has axes parallel and perpendicular to the slope, for then two of the three vectors in Equation 6-1—the acceleration and the normal force—have only one nonzero component. The individual component equations will be simpler in such a coordinate system, which we show in Fig. 6-1. (In Example 6-2 we repeat this example using a horizontal/vertical coordinate system to show that the choice of coordinate system affects only the ease of computation, not the results.)
5. We now write the components of Equation 6-1. In our tilted coordinated system, the normal force has no x component, but the gravitational force does. From Fig. 6-2, we see that the x component of the gravitational force is $F_g \sin\theta$; since the magnitude F_g is mg, we have $F_{gx} = mg\sin\theta$. Are the signs right here? Yes. The gravitational force points vertically downward, and our positive x direction is downslope, so the x component of the gravitational force must be positive. What about the acceleration? The block is constrained to move along the slope, so there is no y component of acceleration. Since the x component of force is positive, so is the x component of

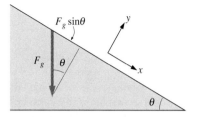

FIGURE 6-2 The two triangles shown are similar, so the angles marked θ are equal. Since the sine of an angle is the ratio of opposite side to hypotenuse, the x component of the gravitational force $\mathbf{F}_g$ is $F_g \sin\theta$.

acceleration, which is therefore just the magnitude a. Then the x component of Newton's law is

$$x \text{ component: } mg\sin\theta = ma.$$

Similarly, the y component of the gravitational force is $-mg\cos\theta$; with no acceleration perpendicular to the slope, the y component of Newton's law reads

$$y \text{ component: } N - mg\cos\theta = 0.$$

6. Now we solve these equations for the quantities of interest. The x equation gives immediately

$$a = g\sin\theta,$$

while the y equation gives $N = mg\cos\theta$.

This is the magnitude of the force that the surface exerts on the block; by Newton's third law, it is also the magnitude of the force that the block exerts on the surface.

7. To check our results, we can consider extreme values for the angle θ. When $\theta = 0$, corresponding to a horizontal surface, we have $\sin\theta = 0$ so $a = 0$, as we expect. In this case $N = mg$, and the surface supports the entire weight of the block. When $\theta = 90°$, $\sin\theta = 1$ and $a = g$. In this case the block falls freely; the normal force is $mg\cos 90° = 0$, so the presence of the surface is irrelevant.
8. There are no numerical values in this example.

Our result in this example shows that a frictionless, sloping surface "dilutes" the acceleration of gravity. It was through the use of such "diluted" gravity on inclined planes that Galileo first probed the laws of motion.

EXERCISE A ski slope is inclined at 34°. (a) What is the acceleration of a skier on this slope? (b) What force does a 70-kg skier exert on the slope?

Answer: (a) 5.5 m/s²; (b) 570 N

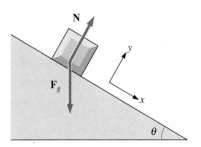

FIGURE 6-1 Force diagram for block sliding on incline.

Some problems similar to Example 6-1: 4–7 ●

● EXAMPLE 6-2 ANOTHER COORDINATE SYSTEM

Rework Example 6-1, solving for the acceleration using a coordinate system with x and y axes horizontal and vertical, respectively.

Solution

Steps 1 to 3 of our solution procedure are independent of the choice of coordinate system, so the forces are still as shown in Fig. 6-1 and Newton's second law is still given by Equation 6-1. The essential physics is summed up in that equation; the rest of the problem is algebra.

In Fig. 6-3 we show the forces on a horizontal/vertical coordinate system. The gravitational force is simpler in this system; it has only a vertical component given by $F_{gy} = -mg$. But now the normal force has two components; from Fig. 6-3, we see that they are $N_x = N\sin\theta$ and $N_y = N\cos\theta$. What about the acceleration? The block is constrained to accelerate down the slope, so from Fig. 6-4 we then have $a_x = a\cos\theta$ and $a_y = -a\sin\theta$, where a is the magnitude of the acceleration. Having resolved the forces and acceleration into components, we rewrite Equation 6-1 and then give its components in our coordinate system:

$$\mathbf{F}_g + \mathbf{N} = m\mathbf{a} \qquad (6\text{-}1)$$

x component: $\qquad N\sin\theta = ma\cos\theta \qquad (6\text{-}2)$

y component: $\qquad -mg + N\cos\theta = -ma\sin\theta. \qquad (6\text{-}3)$

With our sensible choice of coordinate system in Example 6-1, we needed only one of the component equations to get the acceleration; here, with a less appropriate coordinate system, we must deal with both equations. Solving Equation 6-2 for the unknown normal force N gives

$$N = \frac{ma\cos\theta}{\sin\theta};$$

using this result in Equation 6-3, we have

$$-mg + \frac{ma\cos^2\theta}{\sin\theta} = -ma\sin\theta.$$

Multiplying through by $\sin\theta$ and rearranging the equation gives

$$ma(\cos^2\theta + \sin^2\theta) = mg\sin\theta.$$

But $\cos^2\theta + \sin^2\theta = 1$, so we recover our previous result,

$$a = g\sin\theta.$$

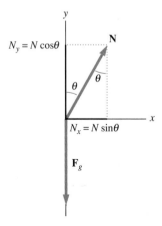

FIGURE 6-3 Forces of Example 6-1 and 6-2 on a horizontal/vertical coordinate system. Also shown is the breakdown of the normal force **N** into components $N_x = N\sin\theta$ and $N_y = N\cos\theta$.

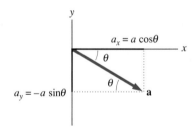

FIGURE 6-4 The acceleration vector points downslope, so its components are $a_x = a\cos\theta$ and $a_y = -a\sin\theta$.

Of course we *had* to get the same result, because we described the same physical situation. But our poor choice of coordinate system resulted in more complicated algebra.

TIP Choose Your Coordinate System Wisely
Comparison of Examples 6-1 and 6-2 shows that a little thought about coordinate systems can save a good deal of mathematical effort.

● EXAMPLE 6-3 BEAR PRECAUTIONS

To protect it from bears, a camper hangs her 17-kg food pack from a rope strung between two trees (Fig. 6-5). What is the tension force in the rope?

Solution

We're asked for the force in the rope, and the rope is supporting the pack, so the object of interest is the pack. The forces acting

FIGURE 6-5   Bear precautions (Example 6-3).

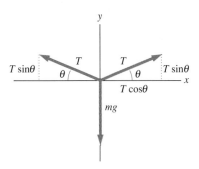

FIGURE 6-7   Vector components on a horizontal-vertical coordinate system.

on the pack are gravity and the tension forces from the two halves of the rope. Figure 6-6 is a free-body diagram showing these forces.

> **TIP   Draw Your Free-Body Diagrams**   In very simple cases, like Example 6-1, it's reasonable to draw the force vectors right on the physical diagram. But when things get at all complicated, it's best to draw a separate free-body diagram showing only the object of interest and the forces on it.

The pack isn't accelerating, so the forces in Fig. 6-6 sum to zero. Thus Newton's law reads

$$\mathbf{T}_1 + \mathbf{T}_2 + \mathbf{F}_g = \mathbf{0}.$$

Since the tension forces are in different directions, there's no sense orienting a coordinate axis along either rope; instead, a horizontal-vertical coordinate system is a reasonable choice. The symmetry of the situation—the pack hangs in the middle of the rope, and the angles made by the two halves of the rope are the same—ensures that the magnitudes of the two tension forces are

equal. We'll call that magnitude T. Then Fig. 6-7 shows that the components of the various forces are $T_{1x} = T\cos\theta$, $T_{2x} = -T_{1x} = -T\cos\theta$, $T_{1y} = T_{2y} = T\sin\theta$, $F_{gx} = 0$, and $F_{gy} = -mg$. Thus the components of Newton's law become

x component: $\qquad T\cos\theta + (-T\cos\theta) + 0 = 0$
y component: $\qquad T\sin\theta + T\sin\theta + (-mg) = 0.$

The x component isn't particularly interesting; it reduces to $0 = 0$. But the y component gives what we want; it reduces to $2T\sin\theta = mg$, or

$$T = \frac{mg}{2\sin\theta}.$$

Now we're through step 6 of our problem-solving strategy. Before working out the numerical value for T, let's see if our expression makes sense. Suppose θ were 90°, so the ropes hung vertically as shown in Fig. 6-8a. Then $T = \frac{1}{2}mg$, showing that each piece of rope supports half the weight of the pack. That

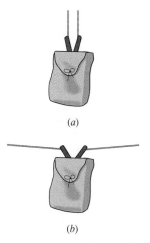

(a)

(b)

FIGURE 6-8   (a) When the ropes hang vertically, the tension in each is half the pack's weight. (b) The tension is enormous when the ropes are nearly horizontal.

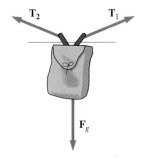

FIGURE 6-6   Free-body diagram showing forces on the pack.

makes sense. But what if θ approaches 0°? Then T becomes arbitrarily large. That, too, makes sense: With θ close to 0° the rope is nearly horizontal, and the vertical component of the tension is a tiny fraction of the horizontal component (Fig. 6-8b). Since the vertical component still has to support half the pack weight, the horizontal component must be enormous, and so must the total tension T.

Finally, using the numbers given in this example, we have

$$T = \frac{mg}{2 \sin \theta} = \frac{(18 \text{ kg})(9.8 \text{ m/s}^2)}{(2)(\sin 22°)} = 220 \text{ N}.$$

This is considerably more than the pack's weight.

EXERCISE You hang a wet blanket with a mass of 14 kg in the middle of a clothesline. The clothesline sags until its two halves make 27° angles with the horizontal. (a) What is the tension in the clothesline? (b) You decide the blanket is too close to the ground, so you tighten up the clothesline. If the line can withstand a maximum tension of 250 N before breaking, what is the minimum angle at which the clothesline can be allowed to sag?

Answer: (a) 150 N; (b) 16°

Some problems similar to Example 6-3: 9, 11, 13

● **EXAMPLE 6-4** SKI RACER

A 60-kg skier stands on a frictionless 30° slope at the starting gate of a ski race. What magnitude of force, applied horizontally, is necessary to restrain the skier before the start of the race?

Solution
Figure 6-9 shows the situation, which is similar to Examples 6-1 and 6-2 except that now there are three forces acting—gravity, the normal force, and the horizontal restraining force $\mathbf{F}_h$. We want to restrain the skier from moving, so $\mathbf{a} = 0$ and Newton's law becomes

$$\mathbf{F}_g + \mathbf{N} + \mathbf{F}_h = m\mathbf{a} = 0,$$

Now two of the three force vectors point in the horizontal or vertical direction, so we're best off with the horizontal/vertical coordinate system of Example 6-2. In this system, the gravitational force has only a vertical component $-mg$, and the normal force breaks down as in Example 6-2 into components $N_x = N \sin \theta$ and $N_y = N \cos \theta$. What about $\mathbf{F}_h$? It points in the negative x direction, so it has only an x component $-F_h$. Then the components of Newton's second law become

$$x \text{ component:} \quad N \sin \theta - F_h = 0$$

and

$$y \text{ component:} \quad -mg + N \cos \theta = 0.$$

Solving the y equation for the normal force gives

$$N = \frac{mg}{\cos \theta}.$$

Using this result in the x equation, we have

$$F_h = \frac{mg}{\cos \theta} \sin \theta = mg \tan \theta.$$

Does this answer make sense? When $\theta = 0$, corresponding to a horizontal surface, then $\tan \theta = 0$ and therefore $F_h = 0$. Of course: it takes no force to prevent acceleration on a horizontal surface. As the slope approaches vertical, however, $\tan \theta \to \infty$, and it becomes impossible for a purely horizontal force $\mathbf{F}_h$ to prevent the acceleration.

Finally, we determine F_h from the numbers given:

$$F_h - mg \tan \theta - (60 \text{ kg})(9.8 \text{ m/s}^2)(\tan 30°) = 340 \text{ N}.$$

EXERCISE A roofer secures a 28-kg toolbox while working on an essentially frictionless metal roof pitched at a 40° angle. To do so he uses a rope running horizontally to the roof ridge, as shown in Fig. 6-10. What is the tension in the rope?

Answer: 230 N

Some problems similar to Example 6-4: 12, 20

FIGURE 6-9 Restraining a skier. (Example 6-4).

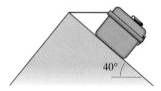

FIGURE 6-10 Securing a toolbox.

6-2 MULTIPLE OBJECTS

In each of the examples just given there is a single object whose motion (or prevention thereof) is of interest. But sometimes we have several objects whose motion is linked. The strategy outlined in the preceding section still applies, with several modifications:

> **STRATEGY**   **Solving Problems with Multiple Objects**
>
> 1. Draw a separate free-body diagram for each object.
> 2. Write Newton's second law, first in vector form, separately for each object.
> 3. Choose a coordinate system appropriate to each object. These systems need not have the same orientation.
> 4. Write each of the separate vector equations in components.
> 5. Use the physical linkage between the objects to relate quantities appearing in the equations for the different objects.
> 6. Solve your multiple equations for the quantities of interest.

● EXAMPLE 6-5 AN ALPINE RESCUE

An unfortunate 70-kg climber finds himself dangling over the edge of an ice cliff, as shown in Fig. 6-11. Fortunately, he's roped to a 940-kg rock located 51 m from the edge, and fortunately help is on the way. Unfortunately the ice is frictionless, and the climber starts to accelerate downward. How long does he have before the rock goes over the edge? Neglect the mass of the rope.

Solution

Here the two objects of interest are the climber and the rock, and they're linked by the rope. Gravity and the rope tension act on the climber, while gravity, the normal force of the ice, and the

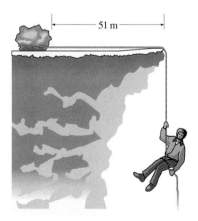

FIGURE 6-11 A climber in trouble. (Example 6-5).

rope tension act on the rock. Figure 6-12 shows free-body diagrams for both objects. Referring to the free-body diagrams, we write Newton's law for each object:

$$\text{climber:} \qquad \mathbf{T}_c + \mathbf{F}_{gc} = m_c \mathbf{a}_c$$

$$\text{rock:} \qquad \mathbf{T}_r + \mathbf{F}_{gr} + \mathbf{N} = m_r \mathbf{a}_r,$$

where the subscripts c and r stand for climber and rock, respectively. Here the motions are clearly either horizontal or vertical, so there's no need to adopt separate coordinate systems. In a standard horizontal-vertical x-y system we then have two component equations for each object.

$$\text{climber, } x: \qquad 0 = 0$$

$$\text{climber, } y: \qquad T_{cy} + (-m_c g) = m_c a_{cy}$$

$$\text{rock, } x: \qquad T_{rx} = m_r a_{rx}$$

$$\text{rock, } y: \qquad -m_r g + N = 0.$$

The climber's x equation is uninteresting because neither of the forces has a horizontal component, and therefore neither does the climber's acceleration. The rock's y equation says that the rock has no vertical acceleration, so the normal force balances the rock's weight; since this equation doesn't contain tension or acceleration, it won't help us further.

We're left with two interesting equations: the climber's y equation and the rock's x equation. And we seem to have four unknowns: T_c, a_c, T_r, and a_r. But we still haven't used the fact that climber and rock are physically linked by the rope. If the

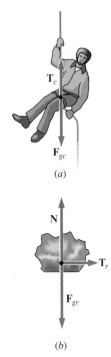

(a)

(b)

FIGURE 6-12 Free-body diagrams for (a) the climber and (b) the rock.

climber accelerates downward, the rock must accelerate rightward at the same rate. Mathematically, this means

$$a_{rx} = -a_{cy}.$$

> **TIP Watch Your Signs** Why the minus sign in this equation? Because a downward acceleration for the climber is in the *negative y* direction, while a rightward acceleration for the rock is in the *positive x* direction. There's no physics in this; it simply follows from our choice of coordinate systems. But once we've made that choice, we have to stick with it for consistency.

Now, the tension in the rope must have the same magnitude at both ends, for two reasons. First, the only things pulling on it are at the ends. This would not be true if there were friction where the rope rounds the cliff edge, since that would introduce another force acting *along* the rope. Second, we're neglecting the mass of the rope, so no force is needed to accelerate it. Had the rope's mass been important, the tension forces at the ends would differ by $m_{rope}a$ in order to satisfy Newton's law.

An *upward* tension on the climber corresponds to a *rightward* tension on the rock; since we've just argued that the two tensions have the same magnitude, we have

$$T_{rx} = T_{cy}.$$

Using these acceleration and tension links in our two "interesting" equations then gives

$$T_{cy} - m_c g = m_c a_{cy}$$
$$T_{cy} = -m_r a_{cy}.$$

Eliminating T_{cy} between these equations, we have

$$-m_r a_{cy} - m_c g = m_c a_{cy},$$

so $(m_c + m_r)a_{cy} = -m_c g$, or

$$a_{cy} = -\frac{m_c g}{m_c + m_r} = \frac{-(70 \text{ kg})(9.8 \text{ m/s}^2)}{(70 \text{ kg} + 940 \text{ kg})} = -0.679 \text{ m/s}^2.$$

Thus the climber accelerates downward at a small fraction of g, and the rock follows.

To find out how long his rescuers have before the rock goes over the edge, we use the expression $x = \frac{1}{2}a_{rx}t^2$ from Chapter 2. Solving for t gives

$$t = \sqrt{\frac{2x}{a_{rx}}} = \sqrt{\frac{(2)(51 \text{ m})}{0.679 \text{ m/s}^2}} = 12 \text{ s},$$

where $a_{rx} = -a_{cx}$.

EXERCISE Suppose the climber in Example 6-5 was on a steep, frictionless slope rather than over a cliff, as shown in Fig. 6-13. Show that the magnitude of the climber's acceleration is now given by $a_{cy} = -\dfrac{m_c g \sin\theta}{m_c + m_r}$. *Hint:* Here's a case where it makes sense to adopt different coordinate systems for the two objects. See Example 6-1.

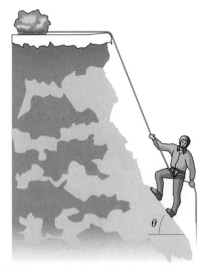

FIGURE 6-13 A climber in a little less trouble.

Some problems similar to Example 6-5: 15, 17–20

6-3 CIRCULAR MOTION

A sports car rounds a curve. A satellite circles Earth. A proton whirls around the giant particle accelerator at Fermilab (Fig. 6-14). Since they're not going in straight lines, Newton's first law tells us that a nonzero net force acts on each. Newton's second law gives us the magnitude of that force: $F = ma$. For an object moving at constant speed v in circular motion, we know from Section 4-4 that the object accelerates—because the *direction* of its motion is changing—with acceleration of magnitude $a = v^2/r$. So for an object of mass m to be in uniform circular motion, a net force of magnitude

$$F = ma = \frac{mv^2}{r} \qquad \text{(force in uniform circular motion)} \qquad (6\text{-}4)$$

must act on the object. Like the acceleration, this force is directed toward the center of the circular path; for this reason, it's often called the **centripetal force.**

FIGURE 6-14 Protons whirl around the main ring of the particle accelerator at Fermilab, held in their circular paths by magnetic forces. In this time exposure, lights of a vehicle driving around the Fermilab service road outline the accelerator's 2-km circumference.

> **TIP** **Look for Real Forces** "Centripetal force" is not a new kind of force; it's just a name for whatever force acts to keep an object in circular motion. The centripetal force is always a real, physical force, and to solve problems you need to identify that force or combination of forces. For our sports car, the centripetal force is the friction between tires and road; for the satellite, it's gravity; for the proton, it's a magnetic force.

Newton's second law describes circular motion in exactly the same way that it does any other motion: by relating the net force, the mass, and the acceleration. We can analyze circular motion using exactly the same steps we outlined in Section 6-1, for we are dealing with exactly the same law of motion.

● **EXAMPLE 6-6** A MASS ON A STRING

A ball of mass m is whirled around in a horizontal circle at the end of a string of length ℓ. The string makes an angle θ with the horizontal. Determine (a) the speed of the ball and (b) the tension in the string.

Solution

To see that this circular motion problem is like any other Newton's law problem, we apply rigorously the steps outlined in Section 6-1:

1. The object of interest is the ball.
2. There are two forces acting on the ball: the string tension and the gravitational force, as shown in Fig. 6-15.

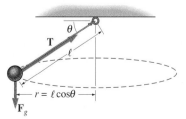

FIGURE 6-15 Forces on a ball being whirled about at the end of a string.

TIP Don't Introduce Forces That Aren't There! Were you tempted to draw a third force in Fig. 6-15, perhaps pointing outward to balance the other two? **DON'T!** The tension and gravitational forces **DO NOT BALANCE.** Why not? Because the ball is accelerating, so there must be a net force acting on it.

Or maybe you're tempted to draw a third force pointing inward, the force mv^2/r. **DON'T!** The quantity mv^2/r is *not* another physical force; it's just the magnitude of the net physical force that acts to give the ball its acceleration.

Students often complicate Newton's law problems by introducing forces that aren't there. That makes physics seem harder than it is. Circular motion is really quite simple, and basically no different from any other accelerated motion. An object is accelerating, so there must be a net force on it. Period. Here the net force is provided by gravity and the string tension. Period.

3. Newton's law tells us that $\mathbf{F} = m\mathbf{a}$; here we have

$$\mathbf{T} + \mathbf{F}_g = m\mathbf{a}.$$

4. Now we choose a coordinate system. The tension force is at some angle θ to the horizontal. The gravitational force is purely vertical. The acceleration points toward the center of the circular path—that is, horizontally. So if we choose a horizontal/vertical coordinate system, then two of the three terms in Newton's law will have only one nonzero component. Figure 6-16 shows the forces with appropriate coordinate axes.

5. The downward-pointing gravitational force $\mathbf{F}_g$ has a vertical component $F_{gy} = -mg$; from Fig. 6-16 we see that the tension force $\mathbf{T}$ has a vertical component $T_y = T\sin\theta$, where T is the magnitude of $\mathbf{T}$. The acceleration has no component in the vertical direction, so the y component of Newton's law becomes

$$y \text{ component:} \qquad T\sin\theta - mg = 0.$$

The gravitational force has no horizontal component, while Fig 6-16 shows that the horizontal component of the tension force is $T_x = T\cos\theta$. At the point where we show the ball in Fig. 6-15, the acceleration points entirely in the x direction. The constraint that the ball move in a circle tells us that the magnitude of the acceleration is mv^2/r,

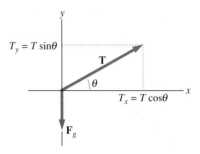

FIGURE 6-16 Force components on a horizontal/vertical coordinate system.

where r is the radius of the circular path. From Fig. 6-15, we see that this radius is $\ell\cos\theta$. Then the x component of the acceleration is $mv^2/\ell\cos\theta$, so the x component of Newton's law becomes

$$x \text{ component:} \qquad T\cos\theta = \frac{mv^2}{\ell\cos\theta}.$$

6. Now we're ready to solve for the tension and speed. The vertical equation gives the tension:

$$T = \frac{mg}{\sin\theta}.$$

Using this result in the horizontal equation, we have

$$\frac{mg}{\sin\theta}\cos\theta = \frac{mv^2}{\ell\cos\theta}.$$

or

$$v = \sqrt{\frac{g\ell\cos^2\theta}{\sin\theta}}.$$

7. Do these results make sense? When $\theta = 90°$, so the string hangs vertically downward, $\cos\theta = 0$, and therefore $v = 0$. In this case there is no acceleration, and the tension is vertically upward and equal in magnitude to the weight. But as $\theta \to 0°$, $\sin\theta \to 0$ and the speed v becomes very large. This tells us that it is not possible to whirl the ball around with the string exactly horizontal. Why not? Because there must always be a vertical component of string tension to balance the ball's weight. As the speed increases, so does the string tension, and therefore the vertical component can be a smaller fraction of the total tension. But it can never be zero.

8. There are no numerical values in this example. ●

● EXAMPLE 6-7 ENGINEERING A ROAD

Curves in roads designed for safe, high-speed travel are banked so that the normal force of the road has a component toward the center of the curve. This allows cars to turn without relying on any force between tires and road. At what angle should a road with a 150-m curvature radius be banked for travel at 75 km/h (21 m/s)?

Solution

There are two forces acting on the car—gravity and the normal force, as shown in Fig. 6-17. (Here we ignore forces between tires and road since the car is supposed to be able to turn even in the absence of such forces). Newton's second law then reads

$$\mathbf{F}_g + \mathbf{N} = m\mathbf{a}.$$

The gravitational force points vertically downward, and the vertical component of the normal force is $N\cos\theta$. There is no acceleration in the vertical direction, so the vertical component of Newton's second law is

vertical component: $\qquad N\cos\theta - mg = 0.$

The only force component in the horizontal direction is $N\sin\theta$; since the car describes a circular path, the horizontal component of Newton's law becomes

horizontal component: $\qquad N\sin\theta = \dfrac{mv^2}{r}.$

Substituting $N = mg/\cos\theta$ from the vertical equation gives

$$\frac{mg}{\cos\theta}\sin\theta = \frac{mv^2}{r},$$

so

$$\tan\theta = \frac{v^2}{gr}.$$

Does this make sense? At very low speeds, banking is hardly needed. But at high speeds, the force needed to maintain circular motion increases, and so does the banking angle.

For the numbers of this example,

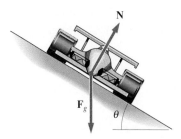

FIGURE 6-17 Forces on a car rounding a banked turn. The horizontal component of the normal force provides the acceleration.

$$\theta = \tan^{-1}\left(\frac{v^2}{gr}\right) = \tan^{-1}\left[\frac{(21\ \text{m/s})^2}{(9.8\ \text{m/s}^2)(150\ \text{m})}\right] = 17°.$$

EXERCISE When an airplane turns, it banks as shown in Fig. 6-18 in order to give the lifting force of the wings a horizontal component that turns the plane. If a plane is flying level at 950 km/h and the banking angle is not to exceed 40°, what is the minimum curvature radius for the turn?

Answer: 8.5 km

Some problems similar to Examples 6-6 and 6-7: 27, 30, 33–35

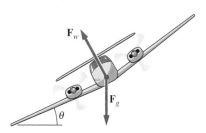

FIGURE 6-18 When a plane turns, the horizontal component of the wing force $\mathbf{F}_w$ is the centripetal force that turns the plane. ●

● EXAMPLE 6-8 LOOP-THE-LOOP

The "Great American Revolution" roller coaster at Valencia, California, includes a loop-the-loop section whose radius is 6.3 m (Fig. 6-19). What is the minimum speed of a train at the top of the loop if it is not to leave the track?

Solution

What does it mean for the train to stay on the track? It means that there must be a nonzero normal force between train and track; if that force goes to zero, then the train and track are no longer

in contact. The only forces acting on the train are gravity and the normal force of the track; Fig. 6-20 shows these forces at two points on the loop. Newton's second law relates the net force to the acceleration:

$$\mathbf{F}_g + \mathbf{N} = m\mathbf{a}.$$

At the top of the loop, both forces point downward. In a coordinate system with the positive direction downward, the vertical component of Newton's law then becomes

FIGURE 6-19 A loop-the-loop roller coaster. At the top of the loop, net force on the train and its passengers is downward; that force provides the acceleration that keeps them in their circular path. In general, the net force is the vector sum of the vertical gravitational force $\mathbf{F}_g$ and the normal force $\mathbf{N}$ from the track, as shown.

$$mg + N = ma = \frac{mv^2}{r},$$

so

$$v^2 = gr + \frac{N}{m}.$$

At the minimum speed, the normal force barely reaches zero as the train passes the top of the loop; at that instant the gravitational force alone provides the acceleration keeping the train in its circular path. The minimum speed is then given by setting $N = 0$ at the top of the loop:

FIGURE 6-20 Forces on the train include gravity and the normal force. At the top of the loop, both point downward.

FIGURE 6-21 If the train's speed is too low, it leaves the track on a parabolic trajectory, subject only to the gravitational force.

$$v_{min} = \sqrt{gr}.$$

For speeds lower than v_{min}, the normal force goes to zero before the train reaches the top, and the train leaves the track traveling in the parabolic trajectory of a projectile (Fig. 6-21). For speeds greater than v_{min}, the normal force is everywhere nonzero, and the train remains in contact with the track. For the "Great American Revolution," $r = 6.3$ m so that $v_{min} = \sqrt{gr} = \sqrt{(9.8 \text{ m/s}^2)(6.3 \text{ m})} = 7.9$ m/s. The actual speed of the train at the top of the loop is 9.7 m/s to maintain a margin of safety.

EXERCISE A stunt plane executes a vertical loop-the-loop. If the loop radius is 320 m and the plane maintains a steady 270 km/h, what force does the seat exert on a 75-kg pilot (a) at the top of the loop and (b) at the bottom of the loop?

Answers: (a) 580 N; (b) 2100 N

Some problems similar to Example 6-8: 28, 31, 36

●

6-4 FRICTION

Our everyday experience of motion does not seem consistent with Newton's first law. Roll a ball, and it soon stops. Take your foot off the gas pedal, and you coast to a stop. But Newton's law is right, so these examples tell us that some force must be acting. That force is **friction.**

Friction acts between two surfaces to oppose their relative motion. Even the smoothest of surfaces is highly irregular on a microscopic scale (Fig. 6-22). When two surfaces contact, their irregularities adhere because of electrical forces between their molecules (Fig. 6-23). The result is a force that opposes relative motion of the surfaces; in order to move them, an external force must be applied to break the microscopic electrical bonds.

On Earth, friction can rarely be ignored. As much as 20% of the gasoline burned in a car is used to overcome friction in the engine. But frictional forces are also beneficial; without friction between tires and road, the car could not stop, start, or turn corners. And it is friction between your feet and the floor that lets you walk.

We distinguish two types of friction: (1) kinetic, or sliding, friction and (2) static friction.

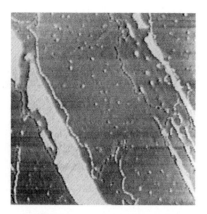

FIGURE 6-22 New studies using a so-called friction-force microscope show that friction is associated not only with surface irregularities but also with defects in the underlying molecular order. Photo shows frictional force encountered in dragging a microscopic probe across a sample measuring 2 μm by 2 μm. Lighter regions represent higher frictional forces, and are associated largely with surface irregularities up to 24 nm high.

Kinetic Friction

Kinetic friction is the friction between surfaces in relative motion. It is kinetic friction that slows down a book you shove across a table or makes it hard to push a heavy trunk even at constant speed. Experimentally, we find that the force of kinetic friction is independent of the surface areas that seem to be in contact, but is proportional to the normal force acting between the surfaces. Intuitively, you might have expected the frictional force to increase with increasing contact area. On a microscopic scale, your intuition is correct: the greater the area *in actual contact,* the greater the friction. But microscopically, only a small fraction of the area you measure macroscopically is actually in contact with the other surface (Fig. 6-23*a*). As the normal force between the surfaces increases, the surface irregularities are crushed together and the actual contact area increases, so the frictional force increases (Fig. 6-23*b*).

We can characterize the force of kinetic friction mathematically by writing

$$F_k = \mu_k N, \qquad (6\text{-}5)$$

where F_k is the force of kinetic friction, N the normal force between the two surfaces, and μ_k the **coefficient of kinetic friction,** a quantity that depends on the properties of the two surfaces. Equation 6-5 relates only the magnitudes; the direction of the frictional force is opposite to the relative motion of the surfaces, and is therefore perpendicular to the normal force (Fig. 6-24).

Since both F_k and N have the units of force, the coefficient of kinetic friction is a dimensionless quantity; it has no units. The coefficient μ_k ranges from about 0.01 for very smooth surfaces to 1.5 for the roughest surfaces. That is, the force of friction on a smooth surface is about 1% of the normal force; to push an object at constant speed on a smooth horizontal surface, you must apply a force about 1% of the object's weight. On a rough surface, on the other hand, you may have to push an object with a force greater than its weight. No wonder Aristotle was confused about the laws of motion!

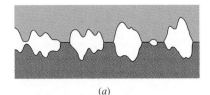

(a)

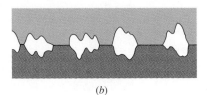

(b)

FIGURE 6-23 (a) When two surfaces contact, microscopic irregularities in the two surfaces adhere, giving rise to a force that opposes their relative motion. Only the microscopic irregularities actually touch, so the actual contact area is much less than the measured area. (b) When the normal force between the surfaces increases, the microscopic contact area increases as the irregularities are crushed together. This increases the frictional force.

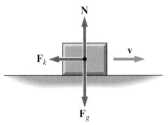

FIGURE 6-24 The force of kinetic friction $\mathbf{F}_k$ acts to oppose relative motion $\mathbf{v}$, and is therefore parallel to the surfaces in contact and perpendicular to the normal force. In the absence of an applied force to the right, the block shown here would be decelerated by the frictional force.

● **EXAMPLE 6-9** PULLING A TRUNK

You pull a trunk of mass m across a level floor using a massless rope that makes an angle θ with the horizontal. The coefficient of friction between trunk and floor is μ_k. What rope tension is required to move the trunk at constant velocity?

Solution
The trunk is our object of interest; Fig. 6-25 shows that the forces include the rope tension $\mathbf{T}$, the force of kinetic friction $\mathbf{F}_k$, the normal force $\mathbf{N}$, and the gravitational force $\mathbf{F}_g$. Since the trunk moves with constant velocity, its acceleration is zero and therefore the net force is zero:

$$\mathbf{T} + \mathbf{F}_k + \mathbf{N} + \mathbf{F}_g = 0.$$

The frictional force acts only in the horizontal direction, while the normal and gravitational forces act only in the vertical direction. So we choose a coordinate system with x axis horizontal and in the direction of the trunk's motion, and with y axis pointing vertically upward. From Fig. 6-25, we see that the tension force has horizontal and vertical components $T_x = T\cos\theta$ and $T_y = T\sin\theta$, respectively. Equation 6-5 shows that the magnitude of the frictional force is $\mu_k N$; this force points in the negative x direction to oppose the trunk's motion. Finally, the normal and gravitational forces are entirely vertical, with the vertical component of the gravitational force given by $-mg$ since it points downward. Then the two components of Newton's law are

x component: $\qquad T\cos\theta - \mu_k N = 0$
y component: $\qquad T\sin\theta + N - mg = 0.$

Solving the y equation for the normal force N then gives

$N = mg - T\sin\theta$; using this result in the x equation, we have

$$T\cos\theta - \mu_k(mg - T\sin\theta) = 0,$$

or

$$T(\cos\theta + \mu_k \sin\theta) = \mu_k mg.$$

Then the rope tension T is

$$T = \frac{\mu_k mg}{\cos\theta + \mu_k \sin\theta}.$$

Does this result make sense? As $\mu_k \to 0$, the frictional force vanishes and the force needed to pull the trunk at constant velocity becomes zero, in accordance with Newton's first law. And as the frictional force increases, it becomes more difficult

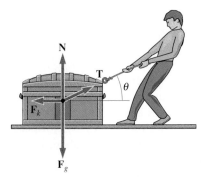

FIGURE 6-25 Forces on a trunk being pulled across a level floor include the rope tension $\mathbf{T}$, the force of kinetic friction $\mathbf{F}_k$, the normal force $\mathbf{N}$, and the gravitational force $\mathbf{F}_g$.

to pull the trunk. We can also look at extreme possibilities for the angle θ: for $\theta = 0$, we have $\cos\theta = 1$ and $\sin\theta = 0$, so $T = \mu_k mg$. When we pull with a horizontal force on a horizontal surface, the force we must provide is just the coefficient of friction times the object's weight. But in the more general case, the upward component of the tension force helps balance the gravitational force, so the normal force and therefore the frictional force are lowered. Finally, when $\theta = 90°$, our result gives $T = mg$, independent of μ_k. And this makes sense, too: here we are lifting the trunk vertically; with no acceleration, we provide a force equal to the trunk's weight. Friction plays no role because the trunk is no longer sliding along the surface.

EXERCISE A logging vehicle pulls a 2.1×10^6-kg redwood log at constant speed, using a chain that angles downward at 15°, as shown in Fig. 6-26. The coefficient of friction between

the log and ground is 1.3. What is the tension in the chain? Why is this not a sensible orientation for the towing chain?

FIGURE 6-26 Pulling a redwood log.

Answer: 4.3×10^7 N; downward pull increases the normal force, hence the friction

Some problems similar to Example 6-9: 39–42 ●

● **EXAMPLE 6-10** SLEDDING

A child sleds down a 20° snow-covered slope. If the coefficient of kinetic friction is $\mu_k = 0.085$, what is the child's acceleration? At what angle would the child slide with constant speed?

Solution
Here the forces on the child and sled together include only gravity, the normal force, and the frictional force, as shown in Fig. 6-27. Newton's second law then reads

$$\mathbf{F}_g + \mathbf{N} + \mathbf{F}_k = m\mathbf{a}.$$

As in Example 6-1, the simplest coordinate system is one with axes parallel and perpendicular to the slope. In this system the gravitational force has an x component $mg\sin\theta$ (see Fig. 6-28), while the frictional force points upslope, or in the negative x direction; $F_{kx} = -\mu_k N$. So the x component of Newton's law becomes

x component: $\qquad mg\sin\theta - \mu_k N = ma.$

There is no acceleration in the y direction, perpendicular to the slope; forces in the y direction include the normal force N and the y component $-mg\cos\theta$ of the gravitational force (see Fig. 6-28), so the y component of Newton's law is

y component: $\qquad N - mg\cos\theta = 0.$

Solving the y equation gives $N = mg\cos\theta$; using this result in the x equation, we have

$$a = g\sin\theta - \mu_k g\cos\theta,$$

where the mass m cancels from all terms. Note that if $\mu_k = 0$, we recover the result $a = g\sin\theta$ for acceleration down a frictionless slope.

For the sled to slide with constant velocity, its acceleration must be zero; then

$$g\sin\theta = \mu_k g\cos\theta$$

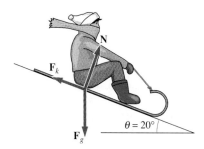

FIGURE 6-27 Forces on the child and sled.

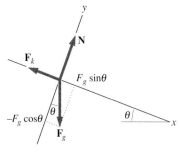

FIGURE 6-28 Resolving the forces into components. The magnitude F_g of the gravitational force is mg, so the components become $F_{gx} = mg\sin\theta$ and $F_{gy} = -mg\cos\theta$.

so

$$\mu_k = \frac{\sin\theta}{\cos\theta} = \tan\theta.$$

Does this make sense? As the slope becomes very steep, $\tan\theta$ increases without bound and it takes a very large frictional force to counteract gravity. For a horizontal slope, on the other hand, $\tan\theta = 0$ and the sled slides with constant velocity only in the absence of friction: $\mu_k = 0$.

For the numbers of this example, we have

$$a = g(\sin\theta - \mu_k\cos\theta) = (9.8)[\sin 20° - (0.085)(\cos 20°)]$$
$$= 2.6 \text{ m/s}^2$$

for the acceleration on a 20° slope, and

$$\theta = \tan^{-1}\mu_k = \tan^{-1}(0.085) = 4.9°$$

when the sled undergoes no acceleration.

Note that in working this problem we could not write the frictional force as μ_k times the weight *mg*. Why not? Because the frictional force depends on the *normal force*, whose magnitude is not equal to the weight when the object is on a sloping surface.

TIP You Don't Always Need Numbers
This example doesn't give the mass of a child and sled. Yet mass appears in Newton's second law, so you might think some vital information is missing. It isn't—as the canceling of mass from the equations shows. In cases like this it's best to assign a symbol to the information you think might be missing, and proceed to solve the problem. If the information really isn't necessary, the symbol you've assigned—in this case, *m*—will eventually drop from the mathematics.

EXERCISE Two skiers start down a 28° slope. Skier 1 accelerates at 2.7 m/s², skier 2 at half this rate. How do their coefficients of friction between skis and ground compare?

Answer: $\mu_{k2} = 1.7\mu_{k1}$

Some problems similar to Example 6-10: 44, 55, 56, 61, 79, 80 ●

● **EXAMPLE 6-11** LOSING A PIANO

Workers are pulling a 210-kg piano up a building using a massless rope over a frictionless, massless pulley as shown in Fig. 6-29. When lunchtime arrives they tie the rope to a 260-kg desk, leaving the piano dangling 3.1 m above the ground. Unfortunately the desk begins to slide; the coefficient of kinetic friction is $\mu_k = 0.71$. What is the piano's acceleration? At what speed does it hit the ground?

Solution
This is like Example 6-5, except now there's a frictional force involved. Figure 6-30 shows free-body diagrams giving the forces on piano and desk. Having identified those forces, we write Newton's law for both objects:

piano: $$\mathbf{F}_{gp} + \mathbf{T}_p = m_p\mathbf{a}_p$$
desk: $$\mathbf{F}_{gd} + \mathbf{N} + \mathbf{T}_d + \mathbf{F}_k = m_d\mathbf{a}_d.$$

Given the geometry, it's appropriate to use the same horizontal-vertical coordinate system for each object. There are no horizontal forces on the piano, and the remaining three components of our two Newton's law equations become:

piano, vertical: $$-m_p g + T_p = m_p a_p \qquad (6\text{-}6)$$
desk, horizontal: $$T_d - \mu_k N = m_d a_d \qquad (6\text{-}7)$$
desk, vertical: $$-m_d g + N = 0, \qquad (6\text{-}8)$$

where we use the fact that the frictional force has magnitude μ_k times the normal force and acts against the rightward motion of the desk.

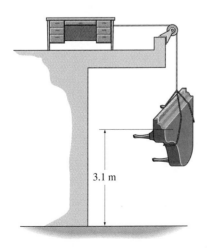

FIGURE 6-29 Example 6-11.

3.1 m

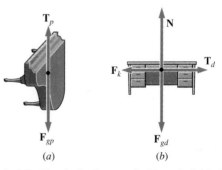

(a) $\qquad$ (b)

FIGURE 6-30 Free-body diagrams for Example 6-11. (*a*) Forces on the piano include gravity and the rope tension. (*b*) Forces on the desk include the normal force **N** and gravity in the vertical direction, and the rope tension and frictional force **F**$_k$ in the horizontal direction.

Solving Equation 6-8 for N and using the result in Equation 6-7 gives

$$T_d - \mu_k m_d g = m_d a_d. \tag{6-9}$$

Now piano and desk are linked by the rope. Since the pulley and rope are massless, and the pulley frictionless, the argument given in Example 6-5 shows that the tension forces at the two ends of the rope have equal magnitudes. Thus $T_d = T_p$. Furthermore, the accelerations of desk and piano are also equal in magnitude, and a rightward (positive) acceleration for the desk corresponds to a downward (negative) acceleration for the piano. So $a_d = -a_p$. These results allow us to write Equation 6-9 in terms of the piano quantities T_p and a_p:

$$T_p - \mu_k m_d g = -m_d a_p.$$

Solving for T_p and using the result in Equation 6-6 then gives

$$-m_p g + \mu_k m_d g - m_d a_p = m_p a_p.$$

Finally, we solve for the piano's acceleration:

$$a_p = -\frac{(m_p - \mu_k m_d)g}{m_p + m_d}.$$

Does this result make sense? In the frictionless case $\mu_k = 0$ we recover the result of Example 6-5. In the case $m_d = 0$, there's no friction or tension and the piano falls freely with acceleration g. Finally, if $u_k m_d = m_p$, there's no acceleration and the pair remain at rest. What about $\mu_k m_d > m_p$? That would seem to imply an upward acceleration for the piano! But that condition violates our assumption that the frictional force points to the left, so it's not actually covered by our result.

> **TIP Know Your Assumptions** Sometimes you'll get a result that doesn't seem to make sense under certain conditions. Check the conditions against the assumptions you made in setting up the problem; if they violate your assumptions, then the nonsensical result must be rejected.

Finally, the numbers of this example give

$$a_p = -\frac{(m_p - \mu_k m_d)g}{m_p + m_d}$$

$$= -\frac{[210 \text{ kg} - (0.71)(260 \text{ kg})](9.8 \text{ m/s}^2)}{210 \text{ kg} + 260 \text{ kg}} = -0.530 \text{ m/s}^2,$$

where the minus sign indicates a downward acceleration. The piano falls 3.1 m; Equation 2-11 then shows that it hits the ground with speed

$$v = \sqrt{2a\Delta y} = \sqrt{(2)(-0.530 \text{ m/s}^2)(-3.1 \text{ m})} = 1.8 \text{ m/s}.$$

Smash!

> **TIP Shortcuts** We've emphasized the importance of following rigorous steps in solving Newton's law problems. But as you get more skilled in problem solving, you may develop shortcuts. Here, for example, the piano's weight $m_p g$ pulls downward and, through the rope, to the right. It's opposed by the friction force $\mu_k m_d g$. So the net force on the piano and desk together is $(\mu_k m_d - m_p)g$; since this force acts on a total mass $m_p + m_g$, the acceleration must be $-(m_p - \mu_k m_d)g/(m_p + m_d)$, in agreement with our lengthier analysis. Use such shortcuts if you're sure you understand them; if at all in doubt, though, follow the rigorous procedure.

EXERCISE The coefficient of kinetic friction between a 2.5-kg block and a tabletop is 0.57. The block is connected by massless ropes over massless pulleys to a 5.0 kg block and a 3.0 kg block, as shown in Fig. 6-31. Find the acceleration of the block on the table.

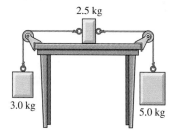

FIGURE 6-31 What's the acceleration?

Answer: 0.54 m/s², to the right.

Some problems similar to Example 6-11: 20, 46 ●

Static Friction

Kinetic or sliding friction is associated with microscopic bonds that continually form and break as surfaces slide past each other. In contrast, **static friction** describes the frictional force between two surfaces at rest with respect to each other. In that case more bonds have time to form, so a greater force is needed to initiate relative motion.

FIGURE 6-32   A book on a table. In the vertical direction, the normal force balances the gravitational force. (a) With no externally applied horizontal force, the force of static friction is also zero. (b) When a horizontal force $\mathbf{F}_h$ of magnitude less then $\mu_s N$ is applied, a static frictional force of the same magnitude acts to oppose the applied force. (c) An applied force of magnitude $\mu_s N$ is the largest that the static frictional force can balance. (d) For applied forces of magnitude greater than $\mu_s N$, the book begins to accelerate; its subsequent motion is determined by the applied force and the force of kinetic friction.

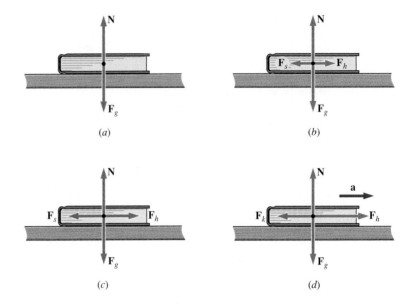

The static friction between two surfaces is characterized by the **coefficient of static friction**, μ_s, which relates the maximum frictional force to the normal force. Consider a book resting on a table (Fig. 6-32). With no externally applied forces in the horizontal direction, the frictional force must be zero; otherwise there would be an unbalanced frictional force on the book and it would accelerate. If we apply a small force in the horizontal direction, we find that the book still does not move; evidently the force of static friction has the same magnitude as the applied force. As the applied force is increased, there comes a point when the force of static friction can no longer balance the applied force; then the book begins to slide and its subsequent motion is governed by the applied force and the force of kinetic friction. Experimentally, we find that the maximum value for the static frictional force is proportional to the normal force between the surfaces. Thus we describe static friction by writing

$$F_s \le \mu_s N. \qquad (6\text{-}10)$$

Because bonds between stationary surfaces are harder to break, the coefficient of static friction μ_s is generally higher than the coefficient of kinetic friction μ_k.

● **EXAMPLE 6-12**   STATIC FRICTION

A 2.5-kg block is placed on a horizontal board, and the board is gradually tilted. The block remains at rest until the board makes an angle of 38° to the horizontal, at which point it begins to slide. What is the coefficient of static friction?

Solution
Figure 6-33 shows the situation, which is identical to that of Example 6-10 except that static friction replaces kinetic friction until the block begins to slide. Following Example 6-10, we can immediately write

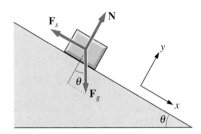

FIGURE 6-33 Example 6-12.

$$\mathbf{F}_g + \mathbf{N} + \mathbf{F}_s = 0$$

for Newton's second law before the block starts to slide. Then, using the tilted coordinate system of Example 6-10, and setting the frictional force to its maximum value $\mu_s N$, the components of Newton's law become

x component: $\qquad mg \sin \theta - \mu_s N = 0$
y component: $\qquad -mg \cos \theta + N = 0,$

where θ is the maximum angle at which the block remains at rest. Solving the y equation for N and using the result in the x equation gives

$$mg \sin \theta - \mu_s mg \cos \theta = 0,$$

so

$$\mu_s = \frac{\sin \theta}{\cos \theta} = \tan \theta.$$

Does this make sense? When $\theta = 0$, μ_s must be zero, showing that the block can slide freely without friction. As $\theta \to 90°$, the coefficient of friction must become very large. This is partly because the downslope component of the gravitational force becomes larger, but more importantly it is because the normal force gets smaller, so a much larger coefficient of friction is needed to counteract the increasing downslope component of gravity. For the numbers of this example, we have

$$\mu_s = \tan 38° = 0.78.$$

EXERCISE A cross-country skier heads up a mountain, using a ski wax whose coefficient of static friction with the snow is 0.27. What is the maximum slope on which the skier can stand without slipping backwards?

Answer: 15°

Some problems similar to Example 6-12: 47, 60

●

Static friction plays a vital role in everyday activities like walking and driving. When you walk, the foot contacting the ground is temporarily at rest, pushing back against the ground. In reaction, the frictional force of the ground on your foot accelerates you forward (Fig. 6-34). On a nearly frictionless surface, walking becomes difficult (Fig. 6-35). Similarly, the tires of an accelerating car push back against the road. If they aren't slipping, the bottom of each tire is momentarily at rest, and the force is static friction. In reaction, the road pushes forward to accelerate the car. When the car brakes, the tires now push forward and, in reaction, the road pushes backwards to decelerate the car (Fig. 6-36). Note that the brakes stop only the wheels; it's friction that stops the car. You know this if you've ever applied your brakes on ice!

FIGURE 6-34 Forces involved in walking. The foot in contact with the ground is temporarily at rest. It pushes back against the ground, exerting the static frictional force $\mathbf{F}_1$ on the ground. In reaction, the ground pushes forward with the equal but opposite frictional force $\mathbf{F}_2$. Transmitted through muscle and bone, this is the force that accelerates your body forward.

FIGURE 6-35 Walking is difficult on a frictionless surface!

FIGURE 6-36 The frictional force between tires and road stops the car. (Here we show only the frictional forces on the tires; each tire exerts an equal but opposite frictional force on the road.)

■ APPLICATION ANTI-LOCK BRAKES

That the coefficient of static friction is higher than that of kinetic friction has important implications for automotive safety. When you slam on your brakes so hard that the wheels lock and slide without turning, the force between tires and road is *kinetic* friction. But if you can manage to keep the wheels from skidding, then the bottoms of the tires are at rest and the force is *static* friction (Fig. 6-37). Since $\mu_s > \mu_k$, the deceleration is greater and therefore the stopping distance is shorter.

To maintain static friction conditions, drivers are often advised to "pump" their brakes when making emergency stops. Actually, a host of complicating factors limit the

(a)   (b)

FIGURE 6-37 (a) In a wheel that is locked and skidding, all points move with the same velocity. Kinetic friction acts between the wheel and road. (b) When the brakes are pumped so that the wheel keeps rolling, the bottom of the tire is instantaneously at rest so the force is static friction. Because the coefficient of static friction is greater, the stopping distance is reduced when the wheels don't slip.

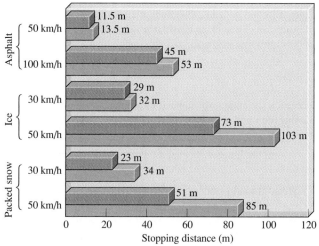

FIGURE 6-38 Stopping distances for computerized anti-lock braking system (brown) and for conventional brakes (green).

success of this strategy—including the fact that the driver can't know what each individual wheel is doing.* But electronic sensors *can* tell what's happening at each wheel and can feed that information to a computer that controls the braking system to keep each wheel just on the verge of slipping. Such anti-lock braking systems are increasingly common on new cars. Figure 6-38 shows that they can achieve substantial reductions in stopping distance.

● EXAMPLE 6-13 STOPPING A CAR

The kinetic and static coefficients of friction between a car's tires and a dry road are 0.61 and 0.89, respectively. If the car is traveling at 88 km/h when the brakes are applied, determine (a) the minimum stopping distance and (b) the stopping distance with the wheels fully locked.

Solution
Figure 6-39 shows the forces on the car; they include gravity, the normal force, and a frictional force $\mathbf{F}_f$. We have worked enough examples to see that the normal force in this case of horizontal motion is equal in magnitude to the weight mg. Then the horizontal component of Newton's law reads

---

*See Jearl Walker, "The Amateur Scientist," *Scientific American,* February 1989, p. 104, for a discussion of braking strategies.

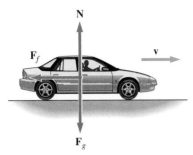

FIGURE 6-39 Forces on a braking car.

or

$$-\mu mg = ma.$$

$$a = -\mu g,$$

where the negative sign appears because we've taken the positive x direction in the direction of the car's motion. Equation 2-11

relates distance and acceleration: $v^2 = v_0^2 + 2a\Delta x$. Putting $v^2 = 0$ for the stopped car and using $a = -\mu g$, the stopping distance Δx becomes

$$\Delta x = \frac{v_0^2}{-2a_x} = \frac{v_0^2}{2\mu g}.$$

The car's initial speed is 88 km/h or 24 m/s; using the appropriate coefficients for μ, the stopping distances are

$$\Delta x_{\text{no skid}} = \frac{v_0^2}{2\mu_s g} = \frac{(24 \text{ m/s})^2}{(2)(0.89)(9.8 \text{ m/s}^2)} = 33 \text{ m}$$

and

$$\Delta x_{\text{skid}} = \frac{v_0^2}{2\mu_k g} = \frac{(24 \text{ m/s})^2}{(2)(0.61)(9.8 \text{ m/s}^2)} = 48 \text{ m}.$$

The 15-m difference could well be enough to prevent an accident. ●

Finally, static friction plays a role in steering a car, as Example 6-14 shows:

● **EXAMPLE 6-14** STEERING A CAR

A level road makes a 90° turn with a 73-m radius of curvature. What is the maximum speed with which a car can negotiate this turn (a) when the road is dry and the coefficient of static friction is 0.88 and (b) when the road is snow-covered and the coefficient of static friction is 0.21?

Solution
Figure 6-40 shows the forces on the car as it rounds the corner. The only force acting in the horizontal direction is the force of static friction between tires and road. Why *static* friction? Because there is no motion of the tire in the radial direction, per-

pendicular to the car (Fig. 6-41). This frictional force is what provides the car's acceleration; with the car in a circular path of radius r, we can write

$$\mu_s N = \frac{mv^2}{r}$$

FIGURE 6-41 Rounding a curve at high speed. The force of static friction acts on the bottom of the tire and is directed toward the center of the curvature to keep the car in its circular path.

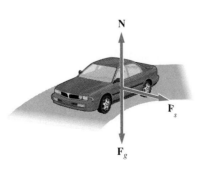

FIGURE 6-40 Forces on a car rounding a corner include gravity, the normal force, and the force of static friction between tires and road.

for the horizontal component of Newton's second law. Here we have used the maximum possible value, $\mu_s N$, for the frictional force because we want the maximum possible speed v. In the vertical direction there is no acceleration, so the normal force balances gravity: $N = mg$. Substituting this result in the horizontal equation, we have

$$\mu_s mg = \frac{mv^2}{r},$$

so

$$v = \sqrt{\mu_s gr}.$$

Does this make sense? On a frictionless road, $v = 0$: you can't corner without a force to deflect you from a straight-line path. With friction, the maximum safe speed increases as the curve becomes more gradual—that is, with increasing r.

For the numbers of this example, we have

$$v_{\text{dry road}} = \sqrt{(0.88)(9.8\ \text{m/s}^2)(73\ \text{m})} = 25\ \text{m/s} = 90\ \text{km/h}$$

and

$$v_{\text{snow}} = \sqrt{(0.21)(9.8\ \text{m/s}^2)(73\ \text{m})} = 12\ \text{m/s} = 44\ \text{km/h}.$$

If you exceed these speeds, your car must inevitably move in a path of greater radius—and that means going off the road!

EXERCISE   A car moving at 75 km/h enters a turn on an icy road where the coefficient of static friction is 0.17. If the radius of the turn is 150 m, will the car make the turn?

Answer: no

Some problems similar to Example 6-14:   52, 53

●

6-5 ACCELERATED REFERENCE FRAMES

FIGURE 6-42   As the car rounds a turn, a force is necessary to keep the passenger on the same circular path. If friction is insufficient to provide the full force necessary, the passenger moves on a path of larger radius. In the car's frame of reference, the passenger slides across the seat, or toward the outside of the curve.

Sometimes the most obvious choice for a frame of reference is itself accelerating. One example is a car rounding a turn; from your perspective as a passenger, you're at rest in an accelerating reference frame. Yet you seem to experience a force that pushes you toward the outside of the turn, and you may actually slide across your seat. What's going on here?

What's going on is nothing unusual: Passenger and car are simply obeying Newton's laws. A frictional force acts to keep the car accelerating in its circular path. If you're at rest in the car, some force must be acting on you, too. That force is friction of the seat against your bottom, and it's directed toward the center of the curved path that you and the car are following. There's no outward force whatsoever; the inward force is unbalanced, as it needs to be to keep you accelerating. If the frictional force isn't big enough to provide the force mv^2/r needed to keep you in circular motion with radius r, then you'll describe a path of larger radius and you'll therefore move with respect to the car, as Fig. 6-42 suggests.

There are two ways to look at this situation. One, which we'll go so far as to call the right way, is to apply Newton's laws in an *unaccelerated*—i.e., inertial—frame of reference. Another is to insist on using the car as your reference frame, to ignore the fact of its acceleration, and nevertheless to insist on applying Newton's laws. Since you slide toward the outside of the turn, those laws suggest that there must be a force in that direction. There isn't—but if you insist on applying Newton's laws in this accelerated reference frame where they don't apply, then you have to invent a **fictitious force** to account for what's going on. You may have heard the term "centrifugal force" to describe the fictitious force associated with circular motion. We urge you to forget that term and to work your problems whenever possible in inertial reference frames—those that aren't accelerating.

There are, however, occasions where scientists and engineers find it convenient to work in accelerating reference frames. For an aeronautical engineer studying air flow around a propeller, it's easier to work in the reference frame of the rotating propeller than in an inertial frame in which a propeller blade comes whizzing by every few milliseconds. An atmospheric scientist studying global weather patterns needs to account for the effect of Earth's rotation on atmospheric circulation, and this is more conveniently done in Earth's rotating frame of reference. In these cases fictitious force terms occur in the equations of motion. You may sometime have occasion to work on similar problems; when you do, be aware that the fictitious forces are simply artifacts of your choosing an accelerated reference frame. For now, though, you should always work in unaccelerated reference frames; then the procedures we've outlined in this chapter are sure to work.

● **EXAMPLE 6-15**   ROUNDING A TURN

You're a passenger in a car rounding a turn whose curvature radius is 180 m. You take your keys from your pocket and dangle them from your key chain. The chain makes an angle of 18° with the vertical. How fast is the car going?

Solution

Inside the car it seems like there must be some force on your keys causing them to deflect the chain. Actually, it's the other way round: The keys, along with everything else in the car, are accelerating because they're in circular motion. The only force capable of providing this horizontal acceleration is the horizontal component of the tension in the chain. So of course it's deflected. Figure 6-43 shows the forces on the keys; Newton's law then reads

$$\mathbf{T} + \mathbf{F}_g = m\mathbf{a},$$

or, in components on a horizontal-vertical coordinate system,

vertical component: $T\cos\theta - mg = 0$

horizontal component: $T\sin\theta = ma_x = \dfrac{mv^2}{r}.$

Eliminating T from these equations and solving for v gives

$$v = \sqrt{rg\tan\theta} = \sqrt{(180 \text{ m})(9.8 \text{ m/s}^2)(\tan 18°)}$$

$$= 24 \text{ m/s} = 86 \text{ km/h}.$$

This example is essentially identical to Example 6-6, and we handled it in exactly the same way. The important point was to apply Newton's laws not in the car's frame of reference but in an inertial frame in which the car was explicitly accelerating.

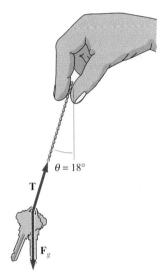

$\theta = 18°$

T

F$_g$

FIGURE 6-43   Forces on the keys as the car rounds the curve (Example 6-15). Note that the keys themselves also hang at an angle.

EXERCISE   A 15.0-N weight hangs from a spring scale attached to the ceiling of an airplane. Flying at 840 km/h, the plane executes a circular turn with curvature radius 7.50 km. What does the scale read? *Hint:* The scale reading is the magnitude of the tension force the scale exerts on the weight.

Answer: 18.7 N

Some problems similar to Example 6-15:   29, 73 ●

CHAPTER SYNOPSIS

Summary

1. Newton's second law provides a universal description of motion in the realm of classical physics. The following steps will help you apply Newton's law correctly in a variety of situations:

1. Identify the object of interest.
2. Identify all forces acting on the object, and draw the force vectors on a single diagram.
3. Write Newton's law in vector form: $\mathbf{F}_{net} = m\mathbf{a}$, where $\mathbf{F}_{net}$, the net force, is the vector sum of the forces you have identified.

4. Choose a suitable coordinate system.
5. Write the components of Newton's law in your coordinate system.
6. Solve algebraically for the quantities of interest.
7. Check some special cases to see if your answers make sense.
8. Insert numerical values into the equations to obtain numerical answers.

2. When two or more objects are linked, it's necessary to write Newton's law for each object, then use the physical linkage to relate quantities appearing in the separate equations.
3. When an object undergoes circular motion at speed v in a circle of radius r, it accelerates at the rate v^2/r in the direction toward the center of the circle. Newton's second law shows that a force of magnitude mv^2/r must then be acting in the same direction.
4. **Friction** is a force associated with microscopic bonds between two contacting surfaces. The frictional force always acts to oppose relative motion of the surfaces. Its magnitude depends on the normal force between the two surfaces. When the surfaces are moving relative to one another, the frictional force is the force of **kinetic friction, F_k** (also called **sliding friction**); in that case the relation between friction and normal forces is quantified by the **coefficient of kinetic friction, μ_k**:

$$F_k = \mu_k N.$$

When the surfaces are at rest, the force of **static friction** opposes an applied force:

$$F_s \leq \mu_s N,$$

where μ_s is the **coefficient of static friction.** In general, $\mu_s > \mu_k$. Once an applied force exceeds the maximum static frictional force $\mu_s N$, the surfaces start moving relative to one another and the subsequent motion is governed by the applied force and the force of kinetic friction.
5. **Fictitious forces** seem to appear in accelerated reference frames. But those forces are artifacts arising from the choice of a reference frame in which Newton's laws don't apply. A better description, conceptually clearer and consistent with Newton's laws, is given by working in unaccelerated reference frames.

Terms You Should Understand

(Pairs are closely related terms whose distinction is important; number in parentheses is chapter section where term first appears.)
free-body diagram (6-1)
centripetal force (6-3)
static friction, kinetic friction (6-4)
coeffiecient of (static, kinetic) friction (6-4)
fictitious force (6-5)

Symbols You Should Recognize

mv^2/r (6-3)
μ_k, μ_s (6-4)

Problems You Should Be Able to Solve

Setting up Newton's second law for a single object and solving for any unknown quantities in two dimensions (6-1)
Newton's law problems for two or more linked objects (6-2)
Problems involving forces in circular motion (6-3)
Problems involving kinetic and static friction (6-4)
Avoiding accelerated frames of reference (6-5)

Limitations to Keep in Mind

Newton's second law works only if you take into account *all* the forces acting on an object.
Newton's second law does not apply in accelerated reference frames; to use the law in such frames, unreal fictitious forces must be introduced. It's almost always best to work from the perspective of unaccelerated frames.

QUESTIONS

1. Compare the net force on a heavy trunk when it is (a) at rest on the floor; (b) being slid across the floor at constant speed; (c) being pulled upward in an elevator whose cable tension equals the combined weight of the elevator and trunk; (d) sliding down a frictionless ramp.
2. Can an object move if there is no net force on it? Can it start moving if there is no net force?
3. If you take your foot off the gas pedal while driving on a horizontal road, your car soon coasts to a stop. Is there a net force on the car during this time? If not, why not? If so, what forces might contribute to it?
4. The force of static friction acts only between surfaces at rest. Yet that force is essential in walking and in accelerating or braking a car. Explain.
5. A jet plane flies at constant speed in a vertical circular loop (Fig. 6-44). At what point in the loop does the seat exert the greatest force on the pilot? The least force?
6. In cross-country skiing, skis should easily glide forward but should remain at rest when the skier pushes back against the snow. What frictional properties should the ski wax have to achieve this goal?

FIGURE 6-44 Question 5.

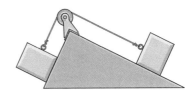

FIGURE 6-46 Question 15, Problems 17, 19.

7. Why is it easier for a child to stand nearer the inside of a rotating merry-go-round?

8. Can a coefficient of friction exceed 1?

9. By pushing horizontally with your hand, you can hold a book at rest on a perfectly vertical wall (Fig. 6-45). How is this possible?

10. In general, a mass suspended from a string will not point exactly toward the center of the Earth. Explain in terms of Earth's rotation. Are there places on Earth where the string will point toward the center?

11. Earth's gravity pulls a satellite toward the center of the Earth. So why doesn't the satellite actually fall to Earth?

12. Why can front-wheel-drive cars corner better on slippery roads than do rear-wheel-drive cars? Explain in terms of the forces acting on the wheels.

13. Explain why stepping as hard as you can on the brake pedal does not necessarily result in the shortest stopping distance for your car.

14. A fishing line has a breaking strength of 20 lb. Is it possible to break the line while reeling in a 15-lb fish? Explain.

15. Two blocks rest on slopes of unequal angle, connected by a rope passing over a pulley (Fig. 6-46). If the blocks have equal masses, will they remain at rest? Why? Neglect friction.

16. Why do a car's front brakes usually wear out before the rear brakes?

17. The dominant aerodynamic force on an airplane is the force of air approximately perpendicular to the wings. Why do airplanes bank when turning?

18. Someone once tried to describe the orbital motion of a satellite by saying that "the gravitational force balances the force mv^2/r." Criticize this statement.

19. In what sense is the force of friction between two surfaces independent of their contacting surface areas? In what sense does the force depend on surface area?

20. On landing, jet aircraft sometimes reverse their engine thrust to assist in stopping. Is this braking mechanism dependent on friction with the runway? Explain.

21. If surfaces are very highly polished, friction actually increases. Why might this be?

22. A car rounds a banked turn at a speed lower than the design speed. Which way does the frictional force point?

23. For safety, automobile racetracks are banked as shown in Fig. 6-47. Explain why this improves the safety of the track.

24. In contrast to the automobile racetrack of the previous question, well-designed running tracks are sometimes banked as shown in Fig. 6-48. Explain why, neglecting safety considerations, this is a more appropriate banking scheme.

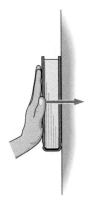

FIGURE 6-45 Question 9.

FIGURE 6-47 Banking of an automobile racetrack. Inside edge of turn is to left (Question 23).

FIGURE 6-48 Banking of a running track. Inside edge of turn is to left (Question 24).

PROBLEMS

1. Two forces, both in the *x-y* plane, act on a 1.5-kg mass, which accelerates at 7.3 m/s² in a direction 30° counterclockwise from the *x* axis. One force has magnitude 6.8 N and points in the +*x* direction. Find the other force.

2. Two forces act on a 3.1-kg mass, which undergoes acceleration $\mathbf{a} = 0.91\hat{\imath} - 0.27\hat{\jmath}$ m/s². If one of the forces is $\mathbf{F}_1 = -1.2\hat{\imath} - 2.5\hat{\jmath}$ N, what is the other force?

3. A 3700-kg barge is being pulled along a canal by two mules, as shown in Fig. 6-49. The tension in each tow rope is 1100 N, and the ropes make 25° angles with the forward direction. What force does the water exert on the barge (a) if it moves with constant velocity and (b) if it accelerates forward at 0.16 m/s²?

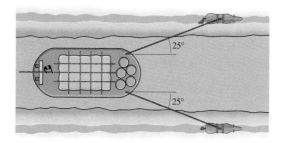

FIGURE 6-49 Problem 3.

4. At what angle should you tilt an air table to simulate motion on the moon's surface, where $g = 1.6$ m/s²?

5. A block of mass *m* slides with acceleration *a* down a frictionless slope that makes an angle θ to the horizontal; the only forces acting on it are the force of gravity $\mathbf{F}_g$ and the normal force **N** of the slope. Show that the magnitude of the normal force is given by $N = m\sqrt{g^2 - a^2}$.

6. A skier starts from rest at the top of a 24° slope 1.3 km long. Neglecting friction, how long does it take to reach the bottom?

7. A block is launched up a frictionless ramp that makes an angle of 35° to the horizontal. If the block's initial speed is 2.2 m/s, how far up the ramp does it slide?

8. At the start of a race, a 70-kg swimmer pushes off the starting block with a force of 950 N directed at 15° below the horizontal. (a) What is the swimmer's horizontal acceleration? (b) If the swimmer is in contact with the starting block for 0.29 s, what is the horizontal component of his velocity when he hits the water?

9. A 15-kg monkey hangs from the middle of a massless rope as shown in Fig. 6-50. What is the tension in the rope? Compare with the monkey's weight.

10. A tow truck is connected to a 1400-kg car by a cable that makes a 25° angle to the horizontal, as shown in Fig. 6-51. If the truck accelerates at 0.57 m/s², what is the magnitude of the cable tension? Neglect friction and the mass of the cable.

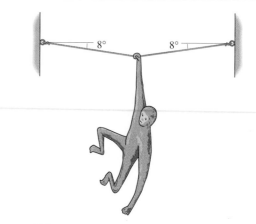

FIGURE 6-50 Problem 9.

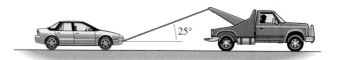

FIGURE 6-51 Problem 10.

11. A 10-kg mass is suspended at rest by two strings attached to walls, as shown in Fig. 6-52. What are the tension forces in the two strings?

12. A 1,100-kg car goes off the road and plunges down a 23° embankment, coming to rest against a tree. The contact between tree and car is such that the force exerted on the car by the tree is purely horizontal, as suggested in Fig. 6-53. Find the magnitude of that force once the car is fully stopped.

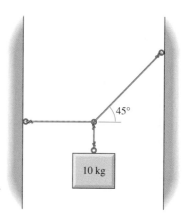

FIGURE 6-52 Problem 11.

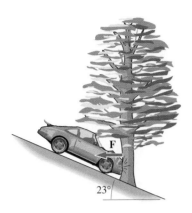

FIGURE 6-53 Problem 12.

13. A camper hangs a 26-kg pack between two trees, using two separate pieces of rope of different lengths, as shown in Fig. 6-54. What is the tension in each rope?

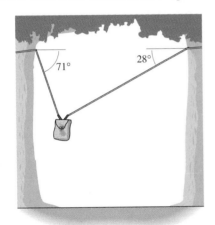

FIGURE 6-54 Problem 13.

14. A construction worker is lifting a 92-kg bundle of plywood onto an upper floor using the arrangement shown in

FIGURE 6-55 Problem 14.

Fig. 6-55. What force must the worker apply to lift the bundle at constant speed? Assume the pulley is massless and frictionless.

Section 6-2 Multiple Objects

15. Your 12-kg baby sister is hanging on the bottom of the tablecloth with all her weight. In the middle of the table, 60 cm from each edge, is a 6.8-kg roast turkey. (a) What is the acceleration of the turkey? (b) From the time she starts pulling, how long do you have to intervene before the turkey goes over the edge of the table?

16. Find expressions for the acceleration of the blocks in Fig. 6-56, where the string is fastened securely to the ceiling. Neglect friction and assume that the masses of pulley and string are negligible.

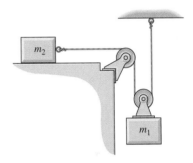

FIGURE 6-56 Problem 16.

17. If the left-hand slope in Fig. 6-46 makes a 60° angle with the horizontal, and the right-hand slope makes a 20° angle, how should the masses compare if the objects are not to slide along the frictionless slopes?

18. In a setup like that shown in Fig. 6-31, but with different masses, a 4.34-kg block starts from rest on the left edge of a frictionless tabletop 1.25 m wide. It accelerates to the right, and reaches the right edge in 2.84 s. If the mass of the block hanging from the left side is 3.56 kg, what is the mass hanging from the right side?

19. Suppose the angles shown in Fig. 6-46 are 60° and 20°. If the left-hand mass is 2.1 kg, what should be the right-hand mass in order that (a) it accelerates downslope at 0.64 m/s^2; (b) it accelerates upslope at 0.76 m/s^2?

20. Two unfortunate climbers, roped together, are sliding freely down an icy mountainside. The upper climber (mass 75 kg) is on a slope at 12° to the horizontal, but the lower climber (mass 63 kg) has gone over the edge to a steeper slope at 38°. (a) Assuming frictionless ice and a massless rope, what is the acceleration of the pair? (b) The upper climber manages to stop the slide with an ice ax. Once the climbers have come to a complete stop, what force must the ax exert against the ice?

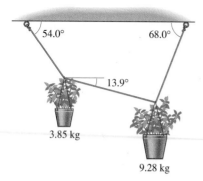

FIGURE 6-57   Problem 21.

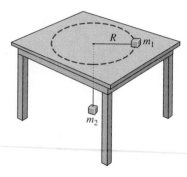

FIGURE 6-59   Problem 26.

21. In a florist's display, hanging plants of mass 3.85 kg and 9.28 kg are suspended from an essentially massless wire, as shown in Fig. 6-57. Find the tension in each section of the wire.

22. A rectangular block of mass m_1 rests on a wedge-shaped block of mass m_2, as shown in Fig. 6-58. All contact surfaces are frictionless. Find an expression for the magnitude of the horizontal force **F** that must be applied to the wedge in order that the rectangular block not slide along the wedge.

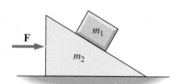

FIGURE 6-58   Problem 22.

Section 6-3 Circular Motion

23. A simplistic model for the hydrogen atom pictures its single electron in a circular orbit of radius 0.0529 nm about the fixed proton. If the electron's orbital speed is 2.18×10^6 m/s, what is the magnitude of the force between the electron and the proton?

24. Suppose the moon were held in its orbit not by gravity but by the tension in a massless cable. Estimate the magnitude of the cable tension. (See Appendix E for relevant data.)

25. Show that the force needed to keep a mass m in a circular path of radius r with period T is $4\pi^2 mr/T^2$.

26. A mass m_1 undergoes circular motion of radius R on a horizontal frictionless table, connected by a massless string through a hole in the table to a second mass m_2 (Fig. 6-59). If m_2 is stationary, find (a) the tension in the string and (b) the period of the circular motion.

27. A 940-g rock is whirled in a horizontal circle at the end of a 1.3-m-long string. (a) If the breaking strength of the string is 120 N, what is the maximum allowable speed of the rock? (b) At this maximum speed, what angle does the string make with the horizontal?

28. If the rock of the previous problem is whirled in a vertical circle, what is the minimum speed needed at the top of the circle in order that the string remain taut?

29. A subway train rounds an unbanked curve at 67 km/h. A passenger hanging onto a strap notices that an adjacent unused strap makes an angle of 15° to the vertical. What is the radius of the turn?

30. An Olympic hammer thrower whirls a 7.3-kg hammer on the end of a 120-cm chain. If the chain makes a 10° angle with the horizontal, what is the speed of the hammer?

31. Riders on the "Great American Revolution" loop-the-loop roller coaster of Example 6-8 wear seatbelts as the roller coaster negotiates its 6.3-m-radius loop with a speed of 9.7 m/s. At the top of the loop, what are the magnitude and direction of the force exerted on a 60-kg rider (a) by the roller-coaster seat and (b) by the seatbelt? (c) What would happen if the rider unbuckled at this point?

32. A 45-kg skater rounds a 5.0-m-radius turn at 6.3 m/s. (a) What are the horizontal and vertical components of the force the ice exerts on her skate blades? (b) At what angle can she lean without falling over?

33. An indoor running track is square-shaped with rounded corners; each corner has a radius of 6.5 m on its inside edge. The track includes six 1.0-m-wide lanes. What should be the banking angles on (a) the innermost and (b) the outermost lanes if the design speed of the track is 24 km/h?

34. A jetliner flying horizontally at 850 km/h banks at 32° to make a turn. What is the radius of the turn? *Note:* the main aerodynamic force on an airplane is normal to the wing surfaces.

35. A 550-g rock is whirled in a horizontal circle at the end of a string 80 cm long. If the period of the circular motion is 1.6 s, what angle does the string make with the vertical?

36. A bucket of water is whirled in a vertical circle of radius 85 cm. What is the minimum speed that will keep the water from falling out?

37. A 1200-kg car drives on the country road shown in Fig. 6-60. The radius of curvature of the crests and dips is 31 m. What is the maximum speed at which the car can maintain road contact at the crests?

FIGURE 6-60 Problems 37, 83.

38. The Tethered Satellite System (TSS) is a NASA experiment consisting of a 500-kg satellite connected to the space shuttle by a 20-km-long cable of negligible mass. Suppose the shuttle is in a 250-km-high circular orbit, where the acceleration of gravity is 0.926 times its value at Earth's surface. The TSS hangs vertically on its tether (Fig. 6-61), and at its 230-km altitude the acceleration of gravity is 0.932 times its surface value. What is the tension in the cable?

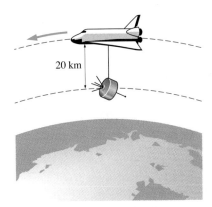

20 km

FIGURE 6-61 NASA's Tethered Satellite System (Problem 38).

Section 6-4 Friction

39. Movers slide a file cabinet along a floor. The mass of the cabinet is 68 kg, and the coefficient of kinetic friction between cabinet and floor is 0.74. What is the frictional force on the cabinet?

40. You make a huge snowball with a mass of 33 kg. If the coefficient of friction between the ball and an ice-covered pond is 0.16, with what force must you push the ball to move it (a) at constant velocity and (b) with an acceleration of 0.84 m/s²?

41. A 380-N force is required to push a table at constant velocity across a floor where the coefficient of sliding friction is 0.86. What is the mass of the table?

42. Eight 80-kg rugby players climb on a 70-kg "scrum machine," and their teammates proceed to push them with constant velocity across a field. If the coefficient of kinetic friction between scrum machine and field is 0.78, with what force must they push?

43. A hockey puck is given an initial speed of 14 m/s. If it comes to rest in 56 m, what is the coefficient of kinetic friction?

44. A child sleds down a 12° slope at constant speed. What is the coefficient of friction between slope and sled?

45. The handle of a 22-kg lawnmower makes a 35° angle with the horizontal. If the coefficient of friction between lawnmower and ground is 0.68, what magnitude of force is required to push the mower at constant velocity? Assume the force is applied in the direction of the handle. Compare with the mower's weight.

46. Repeat Example 6-5, now assuming that the coefficient of kinetic friction between rock and ice is 0.057.

47. During an ice storm, the coefficients of friction between car tires and road are reduced to $\mu_k = 0.088$ and $\mu_s = 0.14$. (a) What is the maximum slope on which a car can be parked without sliding? (b) On a slope just steeper than this maximum, with what acceleration will a car slide down the slope?

48. A bat crashes into the vertical front of an accelerating subway train. If the coefficient of friction between bat and train is 0.86, what is the minimum acceleration of the train that will allow the bat to remain in place?

49. In a factory, boxes drop vertically onto a conveyor belt moving horizontally at 1.7 m/s. If the coefficient of kinetic friction is 0.46, how long does it take each box to come to rest with respect to the belt?

50. The coefficient of static friction between steel train wheels and steel rails is 0.58. The engineer of a train moving at 140 km/h spots a stalled car on the tracks 150 m ahead. If he applies the brakes so that the wheels do not slip, will the train stop in time?

51. If you neglect to fasten your seatbelt, and if the coefficient of friction between you and your car seat is 0.42, what is the maximum deceleration for which you can remain in your seat? Compare with the deceleration in an accident that brings a 60-km/h car to rest in a distance of 1.6 m.

52. A Volvo model 240 sedan has a minimum turning radius of 9.8 m. If the coefficient of static friction between tires and road is 0.81, what is the maximum speed the car can have without skidding if the steering wheel is turned fully to the right?

53. A bug walks outward from the center of a turntable rotating at $33\frac{1}{3}$ revolutions per minute. If the coefficient of friction is 0.15, how far does the bug get before slipping?

54. A 310-g paperback book rests on a 1.2-kg textbook. A force is applied to the textbook, and the two books accelerate together from rest to 96 cm/s in 0.42 s. The textbook is then brought to a stop in 0.33 s, during which time the paperback slides off. Within what range does the coefficient of static friction between the two books lie?

55. A 2.5-kg block and a 3.1-kg block slide down a 30° incline as shown in Fig. 6-62. The coefficient of kinetic friction between the 2.5-kg block and the slope is 0.23; between the 3.1-kg block and the slope it is 0.51. Determine the (a) acceleration of the pair and (b) force the lighter block exerts on the heavier one.

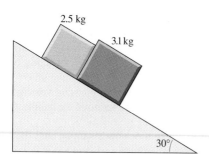

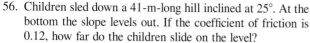

FIGURE 6-62 Problem 55.

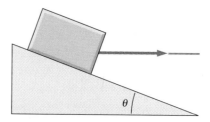

FIGURE 6-64 Problem 64.

56. Children sled down a 41-m-long hill inclined at 25°. At the bottom the slope levels out. If the coefficient of friction is 0.12, how far do the children slide on the level?

57. In a typical front-wheel-drive car, 70% of the car's weight rides on the front wheels. If the coefficient of friction between tires and road is 0.61, what is the maximum acceleration of the car?

58. Repeat the previous problem for a rear-wheel-drive car with the same portion of its weight over the front wheels.

59. A police officer investigating an accident estimates from the damage done that a moving car hit a stationary car at 25 km/h. If the moving car left skid marks 47 m long, and if the coefficient of kinetic friction is 0.71, what was the initial speed of the moving car?

60. A skier finds she must give herself a push to get started on slopes of less than 8°. What is the coefficient of static friction?

61. Starting from rest, the skier of the previous problem traverses a 1.8-km-long trail in 65 s. If the trail is inclined at 12°, what is the coefficient of kinetic friction?

62. A slide inclined at 35° takes bathers into a swimming pool. With water sprayed onto the slide to make it essentially frictionless, a bather spends only one-third as much time on the slide as when it is dry. What is the coefficient of friction on the dry slide?

63. You try to push a heavy trunk, exerting a force at an angle of 50° below the horizontal (Fig. 6-63). Show that, no matter how hard you try to push, it is impossible to budge the trunk if the coefficient of static friction exceeds 0.84.

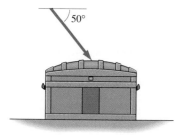

FIGURE 6-63 Problem 63.

64. A block of mass m is being pulled at constant speed v down a slope that makes an angle θ with the horizontal. The pulling force is applied through a horizontal rope, as shown in Fig. 6-64. If the coefficient of kinetic friction is μ_k, find an expression for the rope tension.

65. A block is shoved down a 22° slope with an initial speed of 1.4 m/s. If it slides 34 cm before stopping, what is the coefficient of friction?

66. If the block in the previous problem were shoved up the slope with the same initial speed, (a) how far would it go? (b) Once it stopped, would it slide back down?

Section 6-5 Accelerated Reference Frames

TIP   Avoid Accelerated Frames   Remember that these problems aren't really any different from those of previous sections. They're just phrased in terms of accelerated reference frames. You can apply Newton's laws and treat them like the other problems of this chapter provided you work in unaccelerated frames.

67. A plane starts its takeoff roll before flight attendants have had time to stow a 115-kg beverage cart. A passenger manages to keep the cart stationary by exerting a 790-N force on it. What is the plane's acceleration?

68. You weigh 638 N on a spring scale at the North Pole. Assuming Earth is perfectly round, what would be your apparent weight on a spring scale at the equator? *Hint:* There's a net force on you at the equator but not at the pole. Why?

69. A space station is in the shape of a hollow ring with an outer diameter of 150 m. How fast should it rotate to simulate Earth's surface gravity—that is, so that the force exerted on an object by the outside wall is equal to the object's weight on Earth's surface?

70. A huge Ferris wheel at the Tsukuba Exposition in Japan was 100 m in diameter and rotated once every 15 min. By what percentage would your apparent weight differ from your actual weight (a) at the top and (b) at the bottom of the rotating wheel?

71. An amusement-park ride consists of a vertical cylinder that is set rotating. Riders stand on a floor that falls away when the cylinder reaches its full rotation rate. If the cylinder's

radius is 2.3 m and it rotates at 4.0 rev/min, what is the minimum coefficient of friction that will keep riders from sliding down the cylinder walls? Riders argue that they were held to the wall by the "centrifugal force." What's really going on?

72. A heavy piece of machinery rests on a railroad flatcar. If the coefficient of friction between the machinery and car is 0.94, what is the maximum deceleration with which the train can brake if the machinery is to stay in place?

73. A spring scale hangs from the ceiling of a railroad car, and a 10.0-N weight is attached to the scale. What is the scale reading when the train is (a) moving in a straight line at a constant 90 km/h; (b) rounding a 215-m-radius turn on level tracks, at a constant 90 km/h; (c) rounding a banked 215-m-radius turn at a constant 90 km/h, assuming the banking angle is optimum for this speed? (d) What angle does the scale make with the ceiling of the car in each case?

74. A spring of spring constant $k = 44$ N/m and unstretched length 25 cm is attached to the center of a horizontal rotating turntable. A 1.5-kg mass is attached to the other end of the spring. The spring stretches to a total length of 30 cm, and then the mass remains at rest with respect to the turntable. How long does it take the turntable to complete one revolution?

Paired Problems

(Both problems in a pair involve the same principles and techniques. If you can get the first problem, you should be able to solve the second one.)

75. In Fig. 6-65, suppose $m_1 = 5.0$ kg and $m_2 = 2.0$ kg, and that the surface and pulley are frictionless. Determine the magnitude and direction of m_2's acceleration.

76. Repeat the preceding problem, now taking $m_1 = 3.0$ kg with m_2 still 2.0 kg.

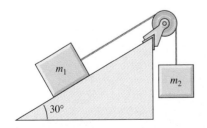

FIGURE 6-65   Problems 75, 76.

77. A tetherball on a 1.7-m rope is struck so it goes into circular motion in a horizontal plane, with the rope making a 15° angle to the horizontal. What is the ball's speed?

78. An airplane goes into a turn 3.6 km in radius. If the banking angle required is 28° from the horizontal, what is the plane's speed?

79. Starting from rest, a skier slides 100 m down a 28° slope.

How much longer does the run take if the coefficient of kinetic friction is 0.17 instead of 0?

80. At the end of a factory production line, boxes start from rest and slide down a 30° ramp 5.4 m long. If the slide is to take no more than 3.3 s, what is the maximum frictional coefficient that can be tolerated?

81. A car moving at 40 km/h negotiates a 130-m radius banked turn designed for 60 km/h. (a) What coefficient of friction is needed to keep the car on the road? (b) To which side of the curve would it move if it hit an essentially frictionless icy patch?

82. A passenger sets a coffee cup on the seatback tray of an airplane flying at 580 km/h. The plane goes into a 2.6-km-radius turn, getting part of its turning force from its rudder and part from banking at 25° (i.e., it's banking at a lower angle than required to give the full turning force). (a) What coefficient of friction is needed to keep the coffee cup on the tray? (b) If there were insufficient friction, which way would the cup slide?

Supplementary Problems

83. In Problem 37, suppose the maximum rated load for the car is 450 kg, and that it carries four 65-kg people. What is the maximum speed for which it will not exceed its rated load at the bottom of the dips shown in Fig. 6-60?

84. Figure 6-66 shows a 0.84-kg ball attached to a vertical post by strings of length 1.2 m and 1.6 m. If the ball is set whirling in a horizontal circle, find (a) the minimum speed necessary for the lower string to be taut and (b) the tension in each string if the ball's speed is 5.0 m/s.

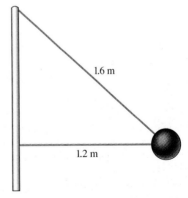

FIGURE 6-66   Problem 84.

85. In the loop-the-loop track of Fig. 6-21, show that the car leaves the track at an angle ϕ given by $\cos \phi = v^2/rg$, where ϕ is the angle made by a vertical line through the center of the circular track and a line from the center to the point where the car leaves the track.

86. An astronaut is training in a centrifuge that consists of a small chamber whirled around horizontally at the end of a

5.1-m-long shaft. The astronaut places a notebook on the vertical wall of the chamber and it stays in place. If the coefficient of static friction is 0.62, what is the minimum rate at which the centrifuge must be revolving?

87. A block of mass m is given an initial speed v_0 on a horizontal table of height h. It slides a distance x_1 across the table and lands on the floor a horizontal distance x_2 from the edge of the table. Find an expression for the coefficient of friction between block and table.

88. Driving in thick fog on a horizontal road, a driver spots a tractor-trailer truck jackknifed across the road, as in Fig. 6-67. To avert a collision, the driver could brake to a stop or swerve in a circular arc, as suggested in Fig. 6-67. Which offers the greater margin of safety? Assume that the same coefficient of static friction is operative in both cases, and that the car maintains constant speed if it swerves.

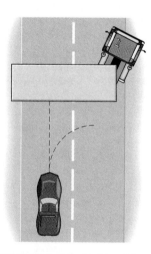

FIGURE 6-67 Problem 88.

89. A highway turn of radius R is banked for a design speed v_d. If a car enters the turn at speed $v = v_d + \Delta v$, where Δv can be positive or negative, show that the minimum coefficient of static friction needed to prevent slipping is

$$\mu_s = \frac{|\Delta v|}{gR} \frac{(2v_d + \Delta v)}{[1 + (v_d v/gR)^2]} .$$

90. Suppose the coefficient of friction between a block and a horizontal surface is proportional to the block's speed: $\mu = \mu_1 v/v_1$, where μ_1 and v_1 are constants. If the block is given an initial speed v_0, show that it comes to rest in a distance $x = v_0 v_1/\mu_1 g$.

91. A block is projected up an incline making an angle θ with the horizontal. It returns to its initial position with half its initial speed. Show that the coefficient of kinetic friction is $\mu_k = \frac{3}{5}\tan\theta$.

92. A conical pendulum consists of a mass m whirled in a horizontal circle at the end of a massless string of length ℓ. Show that in the limit when the string is nearly vertical, the period of the circular motion is $2\pi\sqrt{\ell/g}$.

93. A 2.1-kg mass is connected to a spring of spring constant $k = 150$ N/m and unstretched length 18 cm. The pair are mounted on a frictionless air table, with the free end of the spring attached to a frictionless pivot. The mass is set into circular motion at 1.4 m/s. Find the radius of its path.

94. The victim of a political kidnapping is forced into a north-facing car and then blindfolded. The car pulls into traffic and, from the sound of the surrounding traffic, the victim knows that the car is moving at about the legal speed limit of 85 km/h. The car then turns to the right; the victim estimates that the force the seat exerts on him is one-fifth of his weight. The victim experiences this force for 28 s. At the end of that time, in what direction can the victim conclude that he is heading?

WORK, ENERGY, AND POWER

Lifting a heavy object requires that work be done against the gravitational force $\mathbf{F}_g$. Here the cable tension forces $\mathbf{T}$ do work on the object.

Figure 7-1a shows a skier starting from rest at the top of a uniform slope. What is the skier's speed at the bottom? You could easily solve this problem by applying Newton's second law to find the acceleration, then using the appropriate constant acceleration equation to find the speed. But what about the skier in Fig. 7-1b? Here the slope is continuously changing and so is the acceleration. Constant acceleration equations do not apply, so solving for the details of the skier's motion would prove very difficult. Nevertheless, by introducing two important new concepts—work and energy—we will see how to "shortcut" the detailed application of Newton's law to arrive very easily at the answers to this and many other practical problems.

(a)

(b)

FIGURE 7-1 Skier on (a) a uniform slope and (b) a nonuniform slope.

151

FIGURE 7-2 When pushing in the direction of the car's motion, the person does work equal to the force he applies times the distance the car moves.

FIGURE 7-3 Only the component of force in the direction of motion contributes to the work.

7-1 WORK

We all have an intuitive sense of the term "work." Carrying a piece of furniture upstairs involves work. The heavier the furniture, or the higher the stairs, the greater the work we do. Pushing a stalled car involves work. Again, the heavier the car or the farther we push it, the more work we do. Our precise definition of work reflects this intuition:

For an object moving in one dimension, the work W done on the object by constant applied force F is

$$W = F_x \Delta x,$$ (7-1)

where F_x is the component of the force in the direction of the object's motion and where Δx is the distance moved.

The force **F** need not be the net force; if we are interested, for example, in how much work *we* must do to drag a heavy box across the floor, then **F** is the force *we apply* and *W* is the work *we do*.

Equation 7-1 shows that the SI unit of work is the newton-meter (N·m). One newton-meter is given the name **joule,** in honor of the nineteenth-century British physicist and brewer James Joule.

The SI unit of work is the joule (J), defined as 1 newton-meter (N·m).

Equation 7-1 shows that the person pushing the car in Fig. 7-2 does work equal to the force he applies times the distance the car moves. But the person pulling the suitcase in Fig. 7-3 does work equal only to the horizontal component of the force she applies times the distance the suitcase moves. Furthermore, by our definition, the waiter of Fig. 7-4 does no work on the tray. Why not? Because the direction of the force he applies is perpendicular to the horizontal motion of the tray; there is no component of force along the direction of motion. Similarly, the weightlifter of Fig. 7-5 does no work while he holds the barbell steady over his head.

FIGURE 7-5 The weightlifter, strain though he may, is not doing work on the barbell while he holds it stationary.

FIGURE 7-4 The waiter applies a vertical force $\mathbf{F}_{up}$ to the tray; since this is perpendicular to the tray's horizontal motion, the waiter does no work on the tray.

● EXAMPLE 7-1 PUSHING A TRUNK

You push a 75-kg trunk 6.4 m across a floor, exerting a horizontal force of 540 N. How much work do you do?

Solution
Since the force is in the direction of motion, Equation 7-1 gives simply

$$W = F\Delta x = (540 \text{ N})(6.4 \text{ m}) = 3.5\times10^3 \text{ N·m} = 3.5 \text{ kJ}.$$

If the trunk is pushed at constant speed, then this work is expended against the frictional force.

● EXAMPLE 7-2 A GARAGE LIFT

A garage lift raises a 1300-kg car a vertical distance of 2.3 m at constant speed. How much work does it do?

Solution
To find the work, we need the force the lift exerts on the car. Since the car is raised at constant speed, it is not accelerating and therefore the *net* force on it is zero. The two forces acting are the upward lift force and the downward force of gravity; with no acceleration, they have the same magnitude, *mg*. The lift force is in the same direction as the motion, so the work done by the lift is

$$W = F\Delta x = mg\Delta x = (1300 \text{ kg})(9.8 \text{ m/s}^2)(2.3 \text{ m})$$
$$= 2.9\times10^4 \text{ kg·m}^2/\text{s}^2 = 2.9\times10^4 \text{ J}.$$

EXERCISE The total mass of an elevator and its passengers is 1400 kg. How much work does the lifting mechanism do on the elevator and passengers in raising them 90 m at constant speed?

Answer: 1.23 MJ

Some problems similar to Examples 7-1 and 7-2: 1–4

In the previous two examples, the force we were interested in was in the same direction as the motion. When that is not the case, we must first determine the component of force along the direction of motion.

● EXAMPLE 7-3 PULLING A SUITCASE

An airline passenger pulls a wheeled suitcase, exerting a 60-N force at a 35° angle to the horizontal. How much work does she do in pulling the suitcase 45 m from a taxi to the check-in counter?

Solution
Figure 7-6 shows that the horizontal component of the force is $F_x = F\cos\theta$, so the work is

$$W = F_x\Delta x = F\cos\theta\Delta x = (60 \text{ N})(\cos 35°)(45 \text{ m})$$
$$= 2.2 \text{ kJ}.$$

EXERCISE You push a lawnmower, exerting a 95-N force along the handle, at 40° to the horizontal. How much work do you do in mowing a 22-m long stretch of grass?

Answer: 1.6 kJ

Some problems similar to Example 7-3: 8, 75, 76

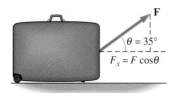

FIGURE 7-6 Example 7-3.

Work can be positive or negative. When a force acts in the same general direction as the motion, the work it does is positive. Acting at 90° to the direction of motion, a force does no work. And when a force acts to oppose motion, it does negative work (Fig. 7-7).

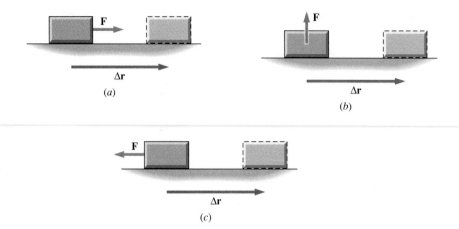

FIGURE 7-7   (*a*) A force applied in the same general direction as the motion does positive work. (*b*) A force acting at right angles to the motion—like the normal force of the floor—does no work. (*c*) A force acting to oppose the motion—like the frictional force—does negative work.

● **EXAMPLE 7-4**   STOPPING A CAR

The coefficient of kinetic friction between a wet, horizontal road and the wheels of a skidding car is 0.21. The mass of the car is 1500 kg. How much work is done on the car by the frictional force if the car comes to a stop in 54 m?

Solution

To find the work, we need to know the frictional force. On a horizontal road, the normal force balances gravity, so that $N = mg$. Then the magnitude of the frictional force is $\mu_k N = \mu_k mg$. The direction of this force is opposite the car's

motion. If we take the positive x direction in the direction of the car's motion, then the component of force along this direction is negative: $F_x = -\mu_k mg$. So the work done by friction is

$$W = F_x \Delta x = (-\mu_k mg)(\Delta x)$$

$$= -(0.21)(1500 \text{ kg})(9.8 \text{ m/s}^2)(54 \text{ m})$$

$$= -1.7 \times 10^5 \text{ kg·m}^2/\text{s}^2 = -1.7 \times 10^5 \text{ J}.$$

Some problems similar to Example 7-4:   6, 11   ●

7-2 WORK AND THE SCALAR PRODUCT

Work is a *scalar* quantity; it is specified completely by a single number and has no direction. But Fig. 7-7 shows clearly that work involves a relation between two *vectors*—the force **F** and the displacement **Δr**. If θ is the angle between these two vectors, then the component of the force along the direction of motion is $F\cos\theta$, and the work is

$$W = (F\cos\theta)(\Delta r) = F\Delta r \cos\theta. \tag{7-2}$$

(Equation 7-2 is a generalization of our definition 7-1; choose the x axis along **Δr**; then $\Delta r = \Delta x$ and $F\cos\theta = F_x$, so we recover Equation 7-1.)

Equation 7-2 shows that work is the product of the magnitudes of the vectors **F** and **Δr** and the cosine of the angle between them. This combination—a product of vector magnitudes with the cosine of the angle between the vectors—occurs so often that it's given a special name: the **scalar product** of two vectors.

The scalar product of any two vectors A and B is defined as

$$\mathbf{A} \cdot \mathbf{B} = AB\cos\theta, \qquad (7\text{-}3)$$

where A and B are the magnitudes of the vectors and θ the angle between them.

Note that a centered dot is used to designate the scalar product; for this reason, it's often called the **dot product.** Figure 7-8 shows a geometrical interpretation of the dot product, while Box 7-1 lists several other mathematical properties, including its expression in unit vector notation.

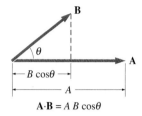

FIGURE 7-8 The scalar product of two vectors is the product of the vector lengths and the cosine of the angle between them. Geometrically, it is the length of vector **A** multiplied by the component of vector **B** in the direction of **A**.

BOX 7-1 THE SCALAR PRODUCT

Equation 7-3 defines the scalar product of two vectors:

$$\mathbf{A} \cdot \mathbf{B} = AB\cos\theta$$

Since the product of scalars is commutative, this defining equation shows that the scalar product is also commutative: $\mathbf{A} \cdot \mathbf{B} = \mathbf{B} \cdot \mathbf{A}$. Problems 13 and 14 ask you to confirm that the scalar product is also distributive:

$$\mathbf{A} \cdot (\mathbf{B} + \mathbf{C}) = \mathbf{A} \cdot \mathbf{B} + \mathbf{A} \cdot \mathbf{C}$$

and that it has the property

$$(c\mathbf{A}) \cdot \mathbf{B} = \mathbf{A} \cdot (c\mathbf{B}),$$

where c is a scalar.

The defining Equation 7-3 is written in terms of vector magnitudes and the angle between two vectors. With vectors expressed in unit vector notation, Problem 15 shows how the distributive law gives a simple form for the scalar product; if $\mathbf{A} = A_x\hat{\mathbf{i}} + A_y\hat{\mathbf{j}} + A_z\hat{\mathbf{k}}$ and $\mathbf{B} = B_x\hat{\mathbf{i}} + B_y\hat{\mathbf{j}} + B_z\hat{\mathbf{k}}$, then

$$\mathbf{A} \cdot \mathbf{B} = A_xB_x + A_yB_y + A_zB_z. \qquad (7\text{-}4)$$

Comparing Equation 7-2 with Equation 7-3 shows that the work done by a constant force **F** moving an object through a straight-line displacement $\Delta\mathbf{r}$ is just

$$W = \mathbf{F} \cdot \Delta\mathbf{r}. \qquad (7\text{-}5)$$

As the examples below show, either Equation 7-3 or Equation 7-4 can be used in evaluating the dot product in this expression for work.

● **EXAMPLE 7-5** A STALLED CAR: FINDING THE WORK

You push on a stalled car with a force $\mathbf{F} = 490\hat{\mathbf{i}} + 230\hat{\mathbf{j}}$ N, giving the car a displacement $\Delta\mathbf{r} = 1.77\hat{\mathbf{i}} + 1.90\hat{\mathbf{j}}$ m. How much work do you do?

Solution
We're given the force and displacement in unit-vector notation, so we use Equation 7-4 for the dot product in Equation 7-5:

$$\begin{aligned} W = \mathbf{F} \cdot \Delta\mathbf{r} &= F_x\Delta x + F_y\Delta y \\ &= (490 \text{ N})(1.77 \text{ m}) + (230 \text{ N})(1.90 \text{ m}) \\ &= 1.3 \text{ kJ}. \end{aligned}$$

● **EXAMPLE 7-6** A TUGBOAT: FINDING THE ANGLE

A tugboat pushes an ocean liner with a force of 830 kN, moving it 0.38 km and doing 290 MJ of work. What is the angle between the direction of the tugboat's push and the motion of the ship?

Solution

Here we know the magnitudes of the force and displacement, and their dot product $W = \mathbf{F} \cdot \Delta\mathbf{r}$, so we can solve Equation 7-3 for the angle:

$$\theta = \cos^{-1}\left(\frac{\mathbf{F} \cdot \Delta\mathbf{r}}{F\Delta r}\right) = \left(\frac{290\times10^6 \text{ J}}{(830\times10^3 \text{ N})(380 \text{ m})}\right) = 23°.$$

Is this answer reasonable? Yes: 830 kN multiplied by 0.38 km is a little more than 290 MJ, so we expect a small but nonzero angle between force and displacement.

EXERCISE A force $\mathbf{F} = 34\hat{\imath} + 61\hat{\jmath}$ N acts to move an object along a straight path given by $\Delta\mathbf{r} = 4.6\hat{\imath} + 5.1\hat{\jmath}$ m. Find (a) the work done by the force and (b) the angle between force and displacement. *Hint:* This exercise combines the two preceding examples.

Answers: (a) 470 J; (b) 13°

Some problems similar to Examples 7-5 and 7-6: 19, 22–24 ●

> **TIP** **Don't Confuse Vectors and Scalars** Vectors and scalars are different, and that means an equation with a scalar on one side must have a scalar on the other side. Be especially careful with this when you're working with dot products; the quantity $\mathbf{A} \cdot \mathbf{B}$, despite being written in terms of vectors, is itself a *scalar*. Thus an equation like $c = \mathbf{A} \cdot \mathbf{B}$ makes mathematical sense, while $\mathbf{C} = \mathbf{A} \cdot \mathbf{B}$ does not. Later we'll encounter a way of multiplying two vectors to make another vector.

7-3 A VARYING FORCE

Often the force applied to an object varies with position. Especially important examples include the fundamental forces of nature; the electrical and gravitational forces, for example, vary as the inverse square of the distance between interacting objects. The force of a spring that we encountered in Chapter 5 provides another example; as the spring stretches, the force increases.

How are we to calculate the work done by a varying force? We consider first the case of one-dimensional motion, with the force in the same direction as the motion. Figure 7-9 is a plot of a force F that varies with position x. We would like to find the work done by the force as an object moves from x_1 to x_2. We can't simply write $F(x_2 - x_1)$; since the force varies, there's no single value for **F.** What we can do, though, is to divide the region into a series of small rectangles of width Δx, as shown in Fig. 7-10a. If we make the width Δx small enough, the force will be nearly constant over the width of each rectangle. Then the work ΔW done in moving the width Δx of one such rectangle is approximately $F(x)\Delta x$, where $F(x)$ is the magnitude of the force at the midpoint x of that rectangle; we write $F(x)$ to show explicitly that the force is a function of position. Note that the quantity $F(x)\Delta x$ is simply the area of the rectangle, expressed in the appropriate units (N·m, or J). Suppose there are N rectangles. Let x_i be the midpoint of the ith rectangle. Then the total work done in moving from x_1 to x_2 is given approximately by the sum of the individual amounts of work ΔW_i associated with each rectangle, or

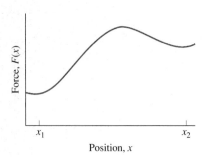

FIGURE 7-9 A force $F(x)$ that varies with position x.

$$W \simeq \sum_{i=1}^{N} W_i = \sum_{i=1}^{N} F(x_i)\Delta x. \qquad (7\text{-}6)$$

How good is this approximation? That depends on how small we make the rectangles (Fig. 7-10b). Suppose we let them get arbitrarily small. Then the number of rectangles must grow arbitrarily large. In the limit of infinitely many infinitesimally small rectangles, the approximation in Equation 7-6 becomes exact (Fig. 7-10c). Then we have

$$W = \lim_{\Delta x \to 0} \sum_i F(x_i)\Delta x, \qquad (7\text{-}7)$$

where the sum is over all the infinitesimal rectangles between x_1 and x_2. The sum of infinitely many infinitesimal quantities is a **definite integral.** The quantity on the right-hand side of Equation 7-7 is the definite integral of the function $F(x)$ over the interval from x_1 to x_2. We introduce a special symbolism for the limiting process of Equation 7-7, writing

$$W = \int_{x_1}^{x_2} F(x)\,dx. \qquad \left(\begin{array}{l}\text{work done by a varying}\\ \text{force in one dimension}\end{array}\right) \qquad (7\text{-}8)$$

What does expression 7-8 mean? It means, by definition, exactly the same thing as Equation 7-7: it tells us to break the interval from x_1 to x_2 into many small rectangles of width Δx, to multiply the value of the function $F(x)$ at each rectangle by the width Δx, and to sum those products. As we take arbitrarily many arbitrarily small rectangles, the result of this process gives us the value of the definite integral. You can think of the symbol $\int$ in Equation 7-8 as standing for "sum," and of the symbol dx as a limiting case of arbitrarily small Δx. The definite integral has a simple geometrical interpretation: as the sum of the areas of the many rectangles, it is the area under the curve $F(x)$ between the limits x_1 and x_2.

How are we to evaluate the infinite sum that is implied in Equation 7-8? One way is to back off and take a sum over a large but finite number of small, but not infinitesimally small, rectangles. This method gives an approximation that can be very accurate if the rectangle width Δx is small enough. It is widely used in computer calculations and for integrals that cannot be handled in any other way.

Calculus provides a powerfully simple way to evaluate many integrals. In your calculus course you've either learned, or will soon learn, that integrals and derivatives are inverses of each other. In Chapter 2 we saw that the derivative of x^n is nx^{n-1}; therefore, the integral of x^n involves the expression $(x^{n+1})/(n+1)$, as you can verify by differentiating. We determine the value of a definite integral by evaluating this expression at upper and lower limits and subtracting:

$$\int_{x_1}^{x_2} x^n\,dx = \left.\frac{x^{n+1}}{n+1}\right|_{x_1}^{x_2} = \frac{x_2^{n+1}}{n+1} - \frac{x_1^{n+1}}{n+1}, \qquad (7\text{-}9)$$

where the middle term in this expression is a shorthand notation for the difference given in the rightmost term. A review of integration and a table of common integrals are given in Appendix A.

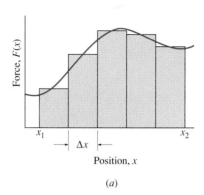

(a)

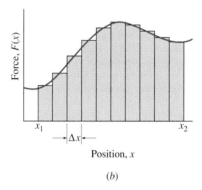

(b)

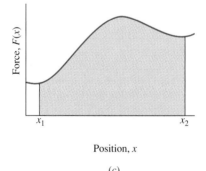

(c)

FIGURE 7-10 Work is the area under the force-versus-position curve. (a) It can be approximated by dividing the region into rectangles and summing their areas. (b) The more rectangles, the more accurate the approximation. (c) The limit of infinitely many infinitesimally small rectangles gives the exact area.

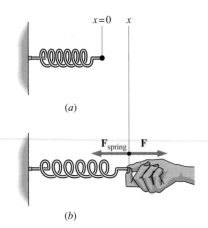

(a)

(b)

FIGURE 7-11 Stretching a spring. The left end of the spring is fixed, and in (a) the equilibrium position of the right end defines $x = 0$. (b) In stretching the spring, an external agent applies a force **F** to oppose the spring force. The work done by the external agent is $\frac{1}{2}kx^2$.

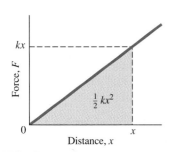

FIGURE 7-12 The force required to stretch the spring increases linearly with distance x. The work done in stretching the spring is the triangular area under the force-versus-distance curve, given by $W = (\frac{1}{2}x)(kx) = \frac{1}{2}kx^2$.

Stretching a Spring

A stretching spring provides an important example of a force that varies with position. We introduced springs in Section 5-9 and found that an ideal spring exerts a force proportional to its displacement from equilibrium

$$F = -kx,$$

where k is the spring constant and where the minus sign shows that the spring force is opposite to the direction of the displacement. It's not just coiled springs that we're interested in here; many physical systems, from molecules to skyscrapers to stars, behave as though they contain springs. The work and energy considerations we develop here and in subsequent chapters apply to those systems as well as to actual springs.

Suppose we stretch a spring from its equilibrium length. The force exerted *by* the spring is $-kx$, so the force exerted *on* the spring *by* the external stretching force is $+kx$. If we let $x = 0$ be one end of the spring at equilibrium and if we hold the other end fixed and pull the spring until its free end is at a new position x, as shown in Fig. 7-11, then Equation 7-8 shows that the work done on the spring by the external force is

$$W_{\text{done on spring}} = \int_0^x F(x)\,dx = \int_0^x kx\,dx = \frac{1}{2}kx^2 \Big|_0^x = \frac{1}{2}kx^2 - \frac{1}{2}k(0)^2,$$

or

$$W_{\text{done on spring}} = \frac{1}{2}kx^2, \tag{7-10}$$

where we've used Equation 7-9 to evaluate the integral. Does this result make sense? The more we stretch the spring, the greater the force we must apply—and that means we must do more work for a given amount of additional stretch. Figure 7-12 shows graphically why the work depends quadratically on the displacement. Although we used the word "stretch" in developing Equation 7-10, the result applies equally to compressing a spring a distance x from equilibrium.

● **EXAMPLE 7-7** BUNGEE JUMPING

An elastic cord used in bungee jumping (Fig. 7-13) is 20 m long and has a spring constant $k = 11$ N/m. At the lowest point in the jump, the cord has stretched to double its length. (a) How much work has been done on it? (b) Compare the amount of work involved in the first meter of stretch with that involved in the last meter.

Solution

(a) Equation 7-10 gives the work needed to stretch the cord from equilibrium:

$$W = \frac{1}{2}kx^2 = \left(\frac{1}{2}\right)(11 \text{ N/m})(20 \text{ m})^2 = 2.2 \text{ kJ},$$

where we use $x = 20$ m since the amount of stretch is equal to the initial length of the cord.

(b) For the first meter of stretch, we repeat the calculation with $x = 1$ m: $W_1 = (\frac{1}{2})(11 \text{ N/m})(1 \text{ m})^2 = 5.5$ J. For the last meter, we want the work needed to go from $x = 19$ m to $x = 20$ m. We can repeat the calculation leading to Equation 7-10, this time using 19 m as the lower limit instead of zero. Then we have

FIGURE 7-13 A bungee jumper leaps from a crane, suspended by a strong elastic cord. At its maximum extension, the cord length will double.

$$W_{19\text{-}20} = \left(\frac{1}{2}\right)(11 \text{ N/m})(20 \text{ m})^2 - \left(\frac{1}{2}\right)(11 \text{ N/m})(19 \text{ m})^2$$

$$= 214 \text{ J}.$$

Why is this number so much larger than the work required for the first meter? Because the bungee cord exerts a large force by the time it's been stretched 19 m, and stretching it further requires that a lot of work be done against that force (Fig. 7-14).

EXERCISE Uncompressed, the spring for an automobile suspension is 45 cm long. It needs to be fitted into a space 32 cm long. If the spring constant is 3.8 kN/m, how much work does a mechanic have to do to fit the spring?

TIP Don't Just Multiply Whenever a force depends on position, you can't just multiply force by distance to get work. You can see that because there's no unique value to use for the force—but it's still a common mistake. Instead you need to integrate, or use an equation resulting from integration, as with Equation 7-10. As you continue with physics, you'll find other times when you're essentially summing over a quantity that varies continuously. In all such cases you can't just multiply; rather, your answer must be the result of integration.

Answer: 32 J

Some problems similar to Example 7-7: 30, 31, 81, 82

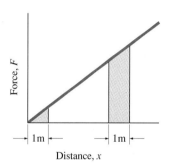

FIGURE 7-14 Work is the area under the force-distance curve. It takes a lot more work to stretch the last meter than it does the first meter.

● **EXAMPLE 7-8** ROUGH SLIDING

Workers pushing a 180-kg trunk across a level floor encounter a 10-m-long region where the floor becomes increasingly rough; the coefficient of kinetic friction increases from 0.17 to 0.79 according to the expression

$$\mu_k = \mu_0 + ax^2,$$

where $\mu_0 = 0.17$, $a = 0.0062 \text{ m}^{-2}$, and x is the distance in meters from the beginning of the rough region. How much work does it take to push the trunk at constant speed across the region?

Solution

Pushing at constant speed on a level floor requires a force equal in magnitude to the frictional force $\mu_k mg = (\mu_0 + ax^2)mg$. This force varies with position, so we must integrate—using Equation 7-8—to get the work:

$$W = \int_{x_1}^{x_2} F\,dx = \int_0^{10} (\mu_0 + ax^2)mg\,dx$$

$$= mg\left(\mu_0 x + \frac{1}{3}ax^3\right)\Bigg|_{x=0}^{x=10\,\text{m}}$$

$$= (180 \text{ kg})(9.8 \text{ m/s}^2)\Bigg[(0.17)(10 \text{ m})$$

$$+ \left(\frac{1}{3}\right)(0.0062 \text{ m}^{-2})(10 \text{ m})^3 - 0\Bigg] = 6.64 \text{ kJ}.$$

Does this result make sense? It is well above the 3.0 kJ we get by calculating $\mu_k mg\Delta x$, with μ_k equal to its value 0.17 at the beginning of the rough interval. But it's much less than the 14 kJ we get using the value $\mu_k = 0.79$ appropriate to the end of the interval. Since the actual coefficient varies between these limits, we expect the work to lie between these extremes.

Figure 7-15 shows the force $\mu_k mg$ as a function of position. Unlike the previous example, we cannot calculate the work using a simple geometrical formula; to get the exact area under the

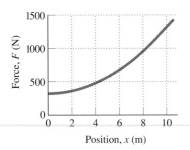

FIGURE 7-15 Force versus position for Example 7-8.

curve we must integrate. However, we can approximate the work by estimating this area graphically; Problem 32 explores this approach.

EXERCISE As a 7.6-cm-long nail is driven into a block of wood, the frictional force on the nail increases according to $F = ax^2$, where $a = 16$ N/cm². (a) Find the work needed to drive the entire nail into the wood. (b) What per cent of the total work is involved in driving the last centimeter?

Answers: (a) 23 J; (b) 35%

Some problems similar to Example 7-8: 29, 35, 36, 77, 78 ●

7-4 FORCE AND WORK IN THREE DIMENSIONS

What if a force varies not only in magnitude but also in direction? Or what if an object's path is not straight but curved? Then how do we calculate work? Our basic definition $W = \mathbf{F} \cdot \Delta \mathbf{r}$ provides the answer. Figure 7-16a shows the curved path of an object and several vectors of an applied force for which we would like to know the work. In this most general case, both the direction of the path and the direction and magnitude of the force vary. But if we magnify a small section $\Delta \mathbf{r}$ of the path (Fig. 7-16b), it will appear essentially straight and the force will not vary significantly over the segment. Over this short segment, with the force a nearly constant quantity $\mathbf{F}$, the work ΔW is approximately

$$\Delta W = \mathbf{F} \cdot \Delta \mathbf{r}.$$

If we determine the work ΔW associated with each small segment of the path from some point $\mathbf{r}_1$ to another point $\mathbf{r}_2$, then the total work is given by summing all the ΔW's:

$$W \simeq \sum \Delta W = \sum \mathbf{F} \cdot \Delta \mathbf{r}.$$

FIGURE 7-16 (a) An object moves on a curved path. We want to know the work done by a force that may vary in direction and/or magnitude along the path. (b) A very small segment $\Delta \mathbf{r}$ of the path appears straight and the force does not vary significantly over that segment. The work involved in traversing the short segment is $\Delta W = \mathbf{F} \cdot \Delta \mathbf{r}$.

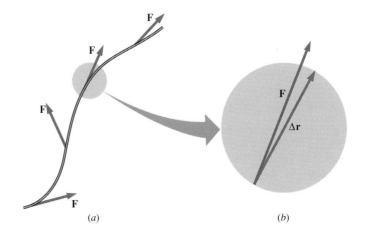

(a)   (b)

If we let the size of the segments grow arbitrarily small and their number increase without bound, this approximation becomes increasingly accurate. In the limit $\Delta\mathbf{r} \rightarrow 0$, the sum becomes an integral and we write

$$W = \int_{\mathbf{r}_1}^{\mathbf{r}_2} \mathbf{F} \cdot d\mathbf{r}, \tag{7-11}$$

where the limits $\mathbf{r}_1$ and $\mathbf{r}_2$ imply that the integral is taken over the specified path from point $\mathbf{r}_1$ to $\mathbf{r}_2$. An integral like that in Equation 7-11, involving the dot product of a vector with a particular path, is called a **line integral.** You will encounter line integrals again when you study electromagnetism.

What does the line integral 7-11 mean? Its meaning comes from the procedure through which we arrived at it: form the many scalar quantities $\mathbf{F} \cdot \Delta\mathbf{r}$ over the path of interest and sum them, taking the limit as $\Delta\mathbf{r}$ becomes arbitrarily small. Example 7-9 illustrates the use of a line integral.

● EXAMPLE 7-9   WORK AND THE GRAVITATIONAL FORCE

A car drives down the hill shown in Fig. 7-17a. How much work does the force of gravity do on the car as it drives from the top of the hill $(y = h)$ to the bottom $(y = 0)$?

Solution
Figure 7-17b shows a small segment $\Delta\mathbf{r}$ of the path, which can be written $\Delta\mathbf{r} = \Delta x\hat{\imath} + \Delta y\hat{\jmath}$ in a coordinate system with x axis horizontal and y axis vertically upward. The force of gravity is always downward: $\mathbf{F}_g = -mg\hat{\jmath}$. Then the amount of work ΔW done by gravity as the car undergoes the displacement $\Delta\mathbf{r}$ is

$$\Delta W = \mathbf{F} \cdot \Delta\mathbf{r} = -mg\hat{\jmath} \cdot (\Delta x\hat{\imath} + \Delta y\hat{\jmath}) = -mg\,\Delta y,$$

where $\hat{\jmath} \cdot \hat{\imath} = 0$ since the two unit vectors are perpendicular and $\hat{\jmath} \cdot \hat{\jmath} = 1$ since $\hat{\jmath}$ has length 1 and is parallel to itself. Summing over the entire path and taking the limit as $\Delta y \rightarrow 0$ gives

$$W = \int_{y=h}^{y=0} (-mg\,dy) = -mgy\,\Big|_h^0$$

$$= -mg(0) - (-mgh) = mgh.$$

In this case the details of the path don't matter at all; the work depends only on the vertical distance h. This illustrates an important property of the gravitational force that we will explore further in the next chapter.

EXERCISE   The irregular hill shown in Fig. 7-1b has a vertical drop of 44 m. How much work does the gravitational force do on a 62-kg skier traversing this run?

Answer: 27 kJ

Some problems similar to Example 7-9: 39, 41, 44

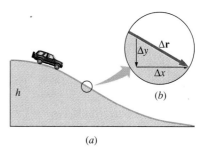

FIGURE 7-17   (a) Hill of Example 7–9; (b) shows a small portion of the hill where the car undergoes displacement $\Delta\mathbf{r} = \Delta x\hat{\imath} + \Delta y\hat{\jmath}$.

TIP   Don't Leave Out *d(something)*   An integral is a sum over infinitely many infinitesimally small quantities. It's the *d(something)* that makes the integrand—the quantity to the right of the integral sign—infinitesimally small. Any integral must include this infinitesimal, or its value is infinite! In this example, for instance, it would be incorrect to write $\int(-mg)$ for the work; not only would the value be infinite, but it would be dimensionally incorrect as well. The correct expression includes the infinitesimal dy: $W = \int(-mg)\,dy$. Leaving out the infinitesimal is a common error among introductory physics students; learn to avoid it as you start working with integrals.

An important case involving work as a line integral is provided by uniform circular motion. Here the force **F** is always perpendicular to the displacement vector $\Delta\mathbf{r}$; the product $\mathbf{F} \cdot \Delta\mathbf{r}$ is therefore always zero and no work is done, no matter how far the object moves in its circular path. Thus it takes no work to whirl a ball around on the end of a horizontal string; similarly, Earth's gravity does no work on a satellite in a circular orbit.

7-5 KINETIC ENERGY

So far we've discussed only the work done by a particular force acting on an object. We didn't distinguish between the case when that was the only force acting—in which case the object must be accelerating—and the case where other forces act as well, perhaps giving zero net force. We did not need to make a distinction because the work done by the one force of interest is $\int \mathbf{F} \cdot d\mathbf{r}$ regardless of what other forces may be acting.

We now turn our attention to the *net* force on an object. The *net* work done *on* an object is given by summing the work associated with each individual force. Since the sum of the forces is the net force, this amounts to using the net force in our expression for work:

$$W_{\text{net}} = \int \mathbf{F}_{\text{net}} \cdot d\mathbf{r}. \tag{7-12}$$

For example, if you push a heavy crate across the floor at constant speed, you do work on the crate. But the frictional force does negative work of equal magnitude; the net force is zero and so is the total work done on the crate.

Considering the net work leads to the important concept of **energy**—a concept that will become increasingly vital as your understanding of physics develops. Here we consider the case of one-dimensional motion; the general case is readily treated by retaining all three terms that form the dot product in Equation 7-12. In one dimension, Equation 7-12 becomes

$$W_{\text{net}} = \int F_{\text{net}} \, dx.$$

But the net force can be written in terms of Newton's second law: $F_{\text{net}} = ma$, or $F_{\text{net}} = m\,dv/dt$, so

$$W_{\text{net}} = \int m \frac{dv}{dt} dx. \tag{7-13}$$

The quantities dv, dt, and dx arose as the limits of small numbers Δv, Δt, and Δx. In your calculus class, you have shown by now that the limit of a product or quotient is just the product or quotient of the individual terms involved. For these reasons, we can rearrange the symbols dv, dt, and dx to rewrite Equation 7-13 in the form

$$W_{\text{net}} = \int m \, dv \frac{dx}{dt}.$$

But $dx/dt = v$, so we have

$$W_{\text{net}} = \int mv \, dv.$$

The integral is simple; it's just like the form $\int x \, dx$, which we integrate by raising the exponent and dividing by the new exponent. What about the limits? Suppose our object starts out at some speed v_1 and ends up at v_2. Then we have

$$W_{\text{net}} = \int_{v_1}^{v_2} mv \, dv = \frac{1}{2}mv^2 \Big|_{v_1}^{v_2} = \frac{1}{2}mv_2^2 - \frac{1}{2}mv_1^2. \tag{7-14}$$

Equation 7-14 shows that an object has associated with it a quantity $\frac{1}{2}mv^2$ that changes when, and only when, net work is done on the object. This quantity plays a vital role in physics and is called the object's **kinetic energy:**

> **The kinetic energy K of an object of mass m moving at speed v is given by**
>
> $$K = \frac{1}{2}mv^2. \tag{7-15}$$

Like velocity, kinetic energy is a relative term; its value depends on the reference frame in which it's measured. But unlike velocity, kinetic energy is a *scalar*. And since it depends on the *square* of the velocity, kinetic energy is never negative.

Equation 7-14 equates the change in an object's kinetic energy with the net work done on the object, a result known as the **work-energy theorem:**

> **Work-energy theorem: The change in an object's kinetic energy is equal to the net work done on the object:**
>
> $$\Delta K = W_{\text{net}}. \tag{7-16}$$

Equations 7-14 and 7-16 are equivalent statements of the work-energy theorem.

We've already seen that work can be positive or negative; so, therefore, can *changes* in kinetic energy. If I stop a moving object, for example, I must reduce its kinetic energy from $\frac{1}{2}mv^2$ to zero—a change $\Delta K = -\frac{1}{2}mv^2$. So I must do negative work, by applying a force directed oppositely to the motion. By Newton's third law, the object exerts an equal but oppositely directed force on me, therefore doing positive work $\frac{1}{2}mv^2$ on me. So an object of mass m moving at speed v can do work equal to its initial kinetic energy, $\frac{1}{2}mv^2$, if it's brought to rest.

● EXAMPLE 7-10   PASSING ZONE

A 1400-kg car enters a passing zone and accelerates from 70 to 95 km/h to pass a slower car. How much work must be done on the car? If it then brakes to a stop, how much work is done on it?

Solution

The work-energy theorem tells us that the work is given by the change in kinetic energy. To accelerate the car, we have

$$W_{net} = \frac{1}{2}mv_2^2 - \frac{1}{2}mv_1^2 = \frac{1}{2}m(v_2^2 - v_1^2)$$

$$= \left(\frac{1}{2}\right)(1400 \text{ kg})[(26.4 \text{ m/s})^2 - (19.4 \text{ m/s})^2]$$

$$= 2.2 \times 10^5 \text{ J},$$

where we converted speeds to meters per second (m/s) before doing the calculation.

The work-energy theorem applies equally to braking the car; now we have $v_1 = 26.4$ m/s and $v_2 = 0$, so

$$W_{net} = \frac{1}{2}m(v_2^2 - v_1^2) = -\frac{1}{2}mv_1^2$$

$$= -\left(\frac{1}{2}\right)(1400 \text{ kg})(26.4 \text{ m/s})^2$$

$$= -4.9 \times 10^5 \text{ J}.$$

Here the work done *on* the car is negative, implying that the car does positive work on the external world. In both cases the forces that do the work—either positive or negative—are ultimately the frictional forces between tires and road.

EXERCISE   A cyclotron accelerates protons from rest to a speed of 2.1×10^7 m/s. How much work does it do on each proton?

Answer: 3.7×10^{-13} J

Some problems similar to Example 7-10:   46, 49

● EXAMPLE 7-11   ROUGH SLIDING, AGAIN

Movers are pushing a 78.0-kg trunk at 0.710 m/s when they encounter a 2.25-m-long stretch of floor where the coefficient of kinetic friction is 0.295. If they push with a steady force $F = 220$ N, what is the speed of the trunk at the end of the 2.25-m stretch (see Fig. 7-18)?

Solution

The frictional force has magnitude $\mu_k mg$ and points opposite the trunk's motion. Taking the positive x direction in the direction of the trunk's motion, the x component of the net force becomes

$$F_{net} = F - \mu_k mg$$

$$= 220 \text{ N} - (0.295)(78.0 \text{ kg})(9.8 \text{ m/s}^2) = -5.498 \text{ N},$$

so the work done on the trunk is

$$W_{net} = F_{net}\Delta x = (-5.498 \text{ N})(2.25 \text{ m}) = -12.37 \text{ J}.$$

The work is negative because the frictional force opposing the motion is greater than the force F with which the movers push; thus, we expect the trunk will slow down. The work-energy theorem equates the work to the change in kinetic energy: $\Delta K = W_{net}$, or

$$\frac{1}{2}mv_2^2 - \frac{1}{2}mv_1^2 = W_{net}.$$

Solving for the final velocity v_2, we have

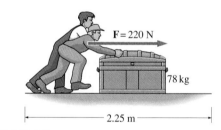

FIGURE 7-18   Rough sliding (Example 7-11).

$$v_2 = \sqrt{v_1^2 + \frac{2W_{net}}{m}} = \sqrt{(0.710 \text{ m/s})^2 + \frac{(2)(-12.65 \text{ J})}{78.0 \text{ kg}}}$$

$$= 0.432 \text{ m/s}.$$

As we expected, the trunk has slowed down since the net work done on it was negative.

EXERCISE   A block is sliding on a horizontal surface at an initial speed of 2.8 m/s. The coefficient of kinetic friction is 0.56. Use the work-energy theorem to determine how far the block will slide before it stops.

Answer: 71 cm

Some problems similar to Example 7-11:   52, 54

Energy Units

Since work is equal to the change in kinetic energy, the units of energy are the same as those of work. In SI, the unit of energy is therefore the joule, equal to 1 newton-meter. In science, engineering, and everyday life, though, you will probably encounter other energy units. Scientific units include the **erg,** used in the centimeter-gram-second system of units and equal to 10^{-7} J; the **electron-volt** used in nuclear, atomic, and molecular physics; and the **calorie** used in thermodynamics and to describe the energies of chemical reactions. English units include the **foot-pound** and the **British thermal unit** (Btu); the latter is commonly used in engineering of heating and cooling systems. Your electric company charges you for energy use in **kilowatt-hours** (kW·h or kWh); we'll see in the next section how this unit relates to the SI joule. Appendix C contains an extensive table of energy units and conversion factors, as well as the energy contents of common fuels.

7-6 POWER

Climbing a flight of stairs requires the same amount of work no matter how fast you go. Why? As long as you maintain constant speed, the upward force you apply is equal to your weight—and you apply that force over a fixed distance. Since work is the product of force and distance, the work is independent of speed.

But it's much harder to *run* up the stairs than to walk. Harder in what sense? In the sense that you do the same work in a shorter time; the *rate* at which you do work is greater. We define **power** as the rate of doing work:

> **If an amount of work ΔW is done in a time Δt, then the average power P̄ is**

$$\overline{P} = \frac{\Delta W}{\Delta t}. \tag{7-17}$$

Often the rate of doing work varies with time. Then we define the **instantaneous power** as the average power taken in the limit of arbitrarily small time interval Δt:

$$P = \lim_{\Delta t \to 0} \frac{\Delta W}{\Delta t} = \frac{dW}{dt}. \tag{7-18}$$

Equations 7-17 and 7-18 both show that the units of power are joules/second. One J/s is given the name **watt** (W) in honor of James Watt, a Scottish engineer and inventor who was instrumental in developing the steam engine as a practical power source.

> **The watt (W) is the SI unit of power: 1 watt is equal to 1 joule/second.**

Watt himself defined another unit, the horsepower. One horsepower (hp) is about 746 J/s or 746 W.

● **EXAMPLE 7-12** CLIMBING MOUNT WASHINGTON

A 55-kg hiker ascends New Hampshire's Mount Washington, a vertical rise of 1300 m from the base elevation. The hike takes 2 hours. A 1500-kg car drives up the Mount Washington Auto Road, the same vertical rise, in half an hour. What is the average power output in each case? Assume the hiker and car maintain constant speed and neglect friction, so each does work only against the gravitational force.

Solution

In Example 7-9, we found that the work done by gravity depends only on the overall change in vertical position. Here the hiker and car do positive work $\Delta W = mgh$ to overcome the negative work $-mgh$ done by gravity. So the average power outputs are

$$\overline{P}_{\text{hiker}} = \frac{\Delta W}{\Delta t} = \frac{(55 \text{ kg})(9.8 \text{ m/s}^2)(1300 \text{ m})}{(2.0 \text{ h})(3600 \text{ s/h})} = 97 \text{ W}$$

and

$$\overline{P}_{\text{car}} = \frac{\Delta W}{\Delta t} = \frac{(1500 \text{ kg})(9.8 \text{ m/s}^2)(1300 \text{ m})}{(0.50 \text{ h})(3600 \text{ s/h})} = 1.1 \times 10^4 \text{ W}.$$

These values correspond to 0.13 hp and 14 hp, respectively. The figure of 97 W is typical of the sustained long-term power output of the human body. Remember that next time you leave a 100-W light bulb burning! The power plant supplying the electricity is doing work at about the rate your body can (actually, it's doing work about three times this rate, for reasons we will examine in Chapter 22). You may be surprised at the low output of the car, given that it probably has an engine rated at several hundred horsepower. Actually, only a small fraction of a car engine's rated horsepower is available in mechanical power to the wheels. The rest is lost in friction and heating.

Some problems similar to Example 7-12: 58, 60, 62

●

When power is constant, so the average and instantaneous power are the same, then Equation 7-17 shows that the amount of work W done in a time Δt is just

$$W = P\Delta t. \tag{7-19}$$

When the power is not constant, we can consider small amounts of work ΔW, each taken over so small a time interval Δt that the power is nearly constant. Adding all these amounts of work, and taking the limit as Δt becomes arbitrarily small, we have

$$W = \lim_{\Delta t \to 0} \sum P\Delta t = \int_{t_1}^{t_2} P \, dt, \tag{7-20}$$

where t_1 and t_2 are the beginning and end of the time interval over which we calculate the power.

● **EXAMPLE 7-13** AN ELECTRIC BILL: YANKEE STADIUM

Each of the 500 floodlights at Yankee Stadium uses electrical energy at the rate of 1.0 kW. How much does it cost to run these lights during a 4-hour night game, if electricity costs 9.5¢/kWh (see Fig. 7-19)?

Solution

The total power consumption of the 500 lights is 500 kW. Since the power is constant, the total work done to run the lamps is given by Equation 7-19:

$$W = P\Delta t = (500 \text{ kW})(4.0 \text{ h}) = 2000 \text{ kWh}.$$

The cost is then $(2000 \text{ kWh})(9.5¢/\text{kWh}) = \$190$.

Note in this example that energy in kilowatt-hours is simply the product of power in kilowatts and time in hours.

EXERCISE New Hampshire's Seabrook nuclear power plant produces electrical energy at the rate of 1150 MW. (a) How much energy does it produce in a year? (b) At a wholesale price of 2¢/kWh, how much is this energy worth?

Answers: (a) 10^{10} kWh; (b) $200 million

Some problems similar to Example 7-13: 67, 70, 74
1

FIGURE 7-19 How much does it cost to run Yankee Stadium's lights for a night game? (Example 7-13)

■ APPLICATION ENERGY AND SOCIETY

We hear a great deal about today's energy-intensive industrial societies and about environmental and other negative consequences of our energy use. Just how much energy do we use, and what does it mean for a society to be energy intensive?

Example 7-12 showed that the average power output of the human body is about 100 W. Before our species harnessed fire and domesticated animals, that 100 W was all the power available to the average human being. At what rate, on the average, do *you* use energy? There's the 100 W associated with your own body. Then there's that 100-W light bulb you keep on several hours a day. Maybe your car burns an average of a gallon of gasoline a day; that gallon is equivalent to about 40 kWh (Appendix C lists this and many other fuel equivalents). On average, that's 40 kWh/24 h or about 1.7 kW. In the winter, your share of the heat may account for several kilowatts. The energy to cook your meals adds further. And what about the energy used to plow the fields that grew your food or to produce the fertilizers and pesticides used in agriculture? Or to manufacture all the goods you consume? In all, you use energy at a substantial rate. In the United States at the end of the twentieth century, in fact, the average energy consumption rate is just about 10 kW per person (see Fig. 7-20). Other industrial nations are within a factor of two of this value, most of them lower.

This quantity, 10 kW, is 10,000 W, or 100 times the average power output of the human body. If our energy were supplied by human labor, instead of gasoline engines and nuclear power plants, each of us would need about 100 laborers working around the clock. What do our 100 "energy workers" do? Figure 7-21 shows that 36 of them work in the industries that supply us with goods; 27 are in transportation, moving us and our goods

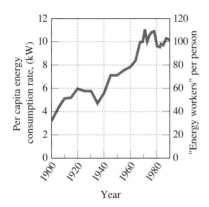

FIGURE 7-20 Per-capita energy consumption rate in the United States through the twentieth century. One "energy worker" is 100 W, about the average power output of the human body.

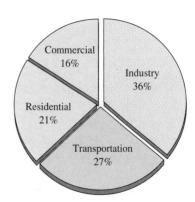

FIGURE 7-21 Energy use in the United States for the late Twentieth Century.

around; 21 are domestic workers, heating our homes and running refrigerators, lights, stereos, and computers. Finally, 16 are in the commercial sector, powering things like the vast banks of lights and air conditioners in shopping malls or the aisles of open freezer units in supermarkets.

Today, over 90% of our "energy workers" derive their power from the combustion of fossil fuels—coal, oil, or natural gas.

Carbon dioxide released in that combustion is thought to be a major contributor to global warming. Both our rate of energy consumption and the sources of that energy are therefore coming under intense scrutiny. Much of what you learn in an introductory physics course is directly relevant to contemporary problems like energy and global warming, and we'll focus on this again elsewhere in the book.

> **TIP** **Don't Confuse Energy and Power** Is that 10 kW per year, per day, or what? That question reflects a common confusion of energy and power. Power is the *rate* of energy use; it doesn't need any "per time" attached to it. It's just 10 kW—meaning energy is used at the average rate of 10,000 W, or 10,000 J/s, or, equivalently, 10 kWh per hour.

Power and Velocity

We can derive an expression relating power, applied force, and velocity by noting that the work ΔW done by a force $\mathbf{F}$ acting on an object that undergoes a small displacement $\Delta \mathbf{r}$ is given by Equation 7-5:

$$\Delta W = \mathbf{F} \cdot \Delta \mathbf{r}.$$

Dividing both sides by the time interval Δt in which the displacement $\Delta \mathbf{r}$ occurs, and taking the limit as $\Delta t \to 0$, we have

$$\lim_{\Delta t \to 0} \frac{\Delta W}{\Delta t} = \lim_{\Delta t \to 0} \mathbf{F} \cdot \frac{\Delta \mathbf{r}}{\Delta t} = \mathbf{F} \cdot \frac{d\mathbf{r}}{dt}.$$

But the limit of $\Delta W / \Delta t$ is just the instantaneous power, P, and $d\mathbf{r}/dt$ is the velocity, $\mathbf{v}$, so

$$P = \mathbf{F} \cdot \mathbf{v}. \tag{7-21}$$

● EXAMPLE 7-14 BICYCLING

As you ride your 14-kg bicycle on level ground, you do work against a 30-N force of air resistance (Fig. 7-22). If your mass is 65 kg, what is the power you must supply to maintain a steady 25 km/h (a) on level ground and (b) going up a 5° incline?

Solution
Air resistance opposes your motion; to overcome it, you must apply a force of equal magnitude in the direction of motion. The 25 km/h speed is equivalent to $(25 \times 10^3 \text{ m})/(3600 \text{ s}) = 6.9$ m/s, so the power you supply is

$$P = \mathbf{F} \cdot \mathbf{v} = Fv = (30 \text{ N})(6.9 \text{ m/s}) = 210 \text{ W},$$

or a little over one-fourth horsepower. (In writing $\mathbf{F} \cdot \mathbf{v} = Fv$, we used the fact that the force is in the same direction as the velocity, so the cosine in the dot product is 1.)

In climbing the hill, you must overcome not only air resistance but also the downslope component of the gravitational force. From Example 6-1, this downslope component is $mg \sin \theta$, here equal to $(14 \text{ kg} + 65 \text{ kg})(9.8 \text{ m/s}^2)(\sin 5°) = 67$ N. So the power you must supply is

$$P = Fv = (30 \text{ N} + 67 \text{ N})(6.9 \text{ m/s}) = 670 \text{ W},$$

or nearly one horsepower. Here again we wrote $P = Fv$ since the applied force is in the direction of the velocity.

Both our answers are considerably greater than the 100 W we've cited as the average human power output. That 100 W is only an average; many athletes can produce several hundred watts for sustained periods and several kilowatts in brief bursts.

EXERCISE A 1750-kg car delivers energy to its drive wheels at the rate of 35 kW. Neglecting air resistance, what is the greatest speed with which it can climb a 4.5° slope?

Answer: 94 km/h

Some problems similar to Example 7-14 73, 83, 84, 92

FIGURE 7-22 How much power does it take to overcome air resistance ($\mathbf{F}_a$) and the gravitational force ($\mathbf{F}_g$) as a bicycle climbs a hill?

CHAPTER SYNOPSIS

Summary

This chapter introduces the important concepts of work and energy, whose meanings in physics are similar, but not identical, to their meanings in everyday language.

1. **Work** is the product of the component of force in the direction an object moves and the distance through which the object moves:

$$W = F_x \Delta x.$$

The SI unit of work is the newton-meter, given the name **joule** (J).

2. The **scalar product** or **dot product** of two vectors provides a convenient shorthand for writing work in terms of the force and displacement vectors. The scalar product of any two vectors **A** and **B** is the product of their magnitudes with the cosine of the angle between them:

$$\mathbf{A} \cdot \mathbf{B} = AB \cos \theta.$$

In terms of the dot product, the work done by a force **F** acting on an object undergoing displacement $\Delta \mathbf{r}$ is

$$W = \mathbf{F} \cdot \Delta \mathbf{r}.$$

3. When force varies with distance, work must be calculated by summing the terms $\mathbf{F} \cdot \Delta \mathbf{r}$ in the limit of arbitrarily small $\Delta \mathbf{r}$:

$$W = \lim_{\Delta \mathbf{r} \to 0} \sum \mathbf{F} \cdot \Delta \mathbf{r}.$$

This limiting process defines the **definite integral:**

$$W = \int_{r_1}^{r_2} \mathbf{F} \cdot d\mathbf{r}.$$

Written in terms of a dot product of force with the infinitesimal displacement $d\mathbf{r},$ this form of the definite integral is called a **line integral.** In the case of one-dimensional motion, the line integral reduces to

$$W = \int_{x_1}^{x_2} F \, dx.$$

Integration is the inverse of differentiation; the methods of calculus show how to evaluate integrals. In particular, the integral of a power of a variable is obtained by raising the exponent by one and dividing by the new exponent:

$$\int_{x_1}^{x_2} x^n \, dx = \frac{x^{n+1}}{n+1} \Big|_{x_1}^{x_2},$$

where the $\Big|_{x_1}^{x_2}$ means to evaluate the expression at x_2 and subtract from it the same expression evaluated at x_1.

4. Work done by a nonzero net force results in a change in an object's **kinetic energy,** a scalar quantity given by $K = \frac{1}{2}mv^2$. The **work-energy theorem** shows that the change in kinetic energy is exactly equal to the work done by the net force:

$$W_{\text{net}} = \Delta K = \tfrac{1}{2}mv_2^2 - \tfrac{1}{2}mv_1^2.$$

The units of energy are the same as those of work.

5. **Power** is the rate at which work is done. In SI units, power is measured in joules/second, or **watts;** 1 W = 1 J/s. The average power $\overline{P}$ is given by

$$\overline{P} = \frac{\Delta W}{\Delta t},$$

while the instantaneous power is

$$P = \frac{dW}{dt}.$$

Inverting this last expression allows us to calculate work as the integral of the power:

$$W = \int_{t_1}^{t_2} P \, dt.$$

When a force **F** acts on an object moving with velocity **v,** the instantaneous power supplied by that force is

$$P = \mathbf{F} \cdot \mathbf{v}.$$

Terms You Should Understand

(Pairs are closely related terms whose distinction is important; number in parentheses is chapter section where term first appears.)
work (7-1)
joule (7-1)
scalar or dot product (7-2)
definite integral (7-3)
line integral (7-4)
kinetic energy (7-5)
work-energy theorem (7-5)
power, energy (7-6)
watt (7-6)

Symbols You Should Recognize

W (7-1)
J (7-1)
$\mathbf{A} \cdot \mathbf{B}$ (7-2)
$\int_{x_1}^{x_2} F(x)dx$ (7-3)

$\int_{r_1}^{r_2} \mathbf{F} \cdot d\mathbf{r}$ (7-4)
K (7-5)
P (7-6)
W (7-6)

Problems You Should Be Able to Solve

calculating work done when a constant force acts on an object moving in one dimension (7-1)
evaluating scalar products for vectors specified either by magnitude and direction or in unit vector notation (7-2)
using the scalar product to evaluate work (7-2)
setting up and evaluating integrals for the work done by a force that varies with position in one dimension (7-3)
solving problems involving the work done in stretching and compressings springs (7-3)
setting up and evaluating line integrals for the work done by a force that varies with position in two dimensions (7-4)
calculating kinetic energy (7-5)
applying the work-energy theorem to relate kinetic energy changes to the total work done on an object (7-5)
converting among common energy units (7-5)
evaluating average and instantaneous power (7-6)
calculating work from power and time, using integration if the power varies with time (7-6)
calculating power from force and velocity (7-6)

Limitations to Keep in Mind

Simple multiplicative formulas for work given force and displacement apply only when the force is constant; otherwise integration must be used to determine the work. The same holds for work determined from power and time, where integration must be used unless the power is constant.
Like much of Newtonian mechanics, the expression $K = \frac{1}{2}mv^2$ for kinetic energy holds only for speeds much less than the speed of light.

QUESTIONS

1. Give two examples of situations in which you might think you are doing work but in which, in the technical sense, you do no work.
2. If the scalar product of two nonzero vectors is zero, what can you conclude about their relative directions?
3. Must you do work to whirl a ball around on the end of a string? Explain.
4. If you pick up a suitcase and put it down, how much total work have *you* done on the suitcase? Does your answer change if you pick up the suitcase and drop it?
5. You lift a book from your desk to a bookshelf. It is initially at rest and ends up at rest, so its kinetic energy has not changed. Yet you have certainly done work on the book. Explain why this is not a violation of the work-energy theorem.
6. Would Equation 7-4 hold in a coordinate system whose axes were not perpendicular?

7. A given force F is applied to move a crate a fixed distance across the floor. Does the amount of work done by the force depend on the coefficient of friction? Does the total work done on the crate depend on the coefficient of friction?
8. Discuss the relation between a sum and an integral.
9. Two cross-country skiers race up opposite sides of the mountain shown in Fig. 7-23. If the skier coming up the steeper side wins, compare the work and the average power associated with each skier. Assume that no work is done by friction (although static friction, which does no work, is what makes the climb possible).
10. You want to raise a piano a given height, using a ramp. With a fixed, nonzero coefficient of friction, will you have to do more work if the ramp is steeper or more gradual? Explain.
11. Does the gravitational force of the Sun do work on a planet in a circular orbit? On a comet in an elliptical orbit? Explain.

FIGURE 7-23   Question 9.

12. How does the dependence of a car's kinetic energy on speed affect its stopping distance, assuming a constant braking force?
13. Give an example showing that kinetic energy depends on frame of reference.
14. Can kinetic energy ever be negative? Explain.
15. A pendulum bob swings back and forth on the end of a string, describing a circular arc. Does the tension force in the string do any work?

16. In the preceding question, does the kinetic energy of the bob change as it swings back and forth? Why?
17. Does your car's kinetic energy change if you drive at constant speed for 1 hour?
18. A watt-second is a unit of what quantity? Relate it to a more standard SI unit.
19. A body collides with a spring and compresses it an amount Δx. How will Δx change if you double the impact velocity? Half the mass of the body? Triple the spring constant?
20. Two particles of different mass have the same momentum. Compare their kinetic energies.
21. A truck is moving northward at 55 mph. Later, it is moving eastward at the same speed. Has its kinetic energy changed? Has its momentum changed? Has work been done on the truck? Has a force acted on the truck? Explain.
22. In a recent article, *The New York Times* described a new power plant by saying it "could yield 50 megawatts of electricity an hour. . ." Criticize this phrase.

PROBLEMS

Section 7-1 Work

1. How much work do you do as you exert a 95-N force to push a 30-kg shopping cart through a 14-m-long supermarket aisle?
2. If the coefficient of kinetic friction is 0.21, how much work do you do when you slide a 50-kg box at constant speed across a 4.8-m-wide room?
3. A crane lifts a 500-kg beam vertically upward 12 m, then swings it eastward 6.0 m. How much work does the crane do? Neglect friction, and assume the beam moves with constant speed.
4. You lift a 45-kg barbell from the ground to a height of 2.5 m. (a) How much work do you do on the barbell? (b) You hold the barbell aloft for 2.0 min. How much work do you do on the barbell during this time? (c) You lower the barbell to the ground. Now how much work do you do on it?
5. The world's highest waterfall, the Cherun-Meru in Venezuela, has a total drop of 980 m. How much work does gravity do on a cubic meter of water dropping down the Cherun-Meru?
6. A meteorite plunges to Earth, embedding itself 75 cm in the ground. If it does 140 MJ of work in the process, what average force does the meteorite exert on the ground?
7. You slide a box of books at constant speed up a 30° ramp, applying a force of 200 N directed up the slope. The coefficient of sliding friction is 0.18. (a) How much work have you done when the box has risen 1 m vertically? (b) What is the mass of the box?
8. Two people push a stalled car at its front doors, each applying a 330-N force at 25° to the forward direction, as shown in Fig. 7-24. How much work does each do in pushing the car 6.2 m?

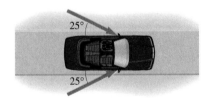

FIGURE 7-24   Problem 8.

9. A locomotive does 8.8×10^{11} J of work in pulling a 2×10^6-kg train 150 km. What is the average force in the coupling between the locomotive and the rest of the train?
10. An elevator of mass m rises a distance h up a vertical shaft with upward acceleration equal to one-tenth g. How much work does the elevator cable do on the elevator?
11. A 20-kg child lies on her back and scoots 7.4 m across the floor at constant speed by applying a 50-N force with the soles of her feet (Fig. 7-25). (a) How much work does she do? (b) What is the coefficient of friction between the child's back and the floor?
12. The mass of a hydrogen atom is 1.67×10^{-27} kg. How much work does gravity do on the atom when it falls a distance of 10 cm?

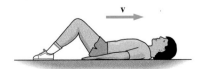

FIGURE 7-25   Problem 11.

Section 7-2 Work and the Scalar Product

13. Show that the scalar product obeys the distributive law:
$\mathbf{A} \cdot (\mathbf{B} + \mathbf{C}) = \mathbf{A} \cdot \mathbf{B} + \mathbf{A} \cdot \mathbf{C}$.

14. Show that the scalar product has the property that $(c\mathbf{A}) \cdot \mathbf{B} = \mathbf{A} \cdot (c\mathbf{B})$, where c is a scalar.

15. (a) Find the scalar products $\hat{\imath} \cdot \hat{\imath}$, $\hat{\jmath} \cdot \hat{\jmath}$, and $\hat{k} \cdot \hat{k}$. (b) Find the scalar products $\hat{\imath} \cdot \hat{\jmath}$, $\hat{\jmath} \cdot \hat{k}$, $\hat{k} \cdot \hat{\imath}$. (c) Use the distributive law to multiply out the scalar product of two arbitrary vectors $\mathbf{A} = A_x\hat{\imath} + A_y\hat{\jmath} + A_z\hat{k}$ and $\mathbf{B} = B_x\hat{\imath} + B_y\hat{\jmath} + B_z\hat{k}$. Then use the results of parts (a) and (b) to verify Equation 7-4.

16. One vector is 15 units long, and another is 6.5 units long. Find their scalar product if the angle between them is (a) 27° and (b) 78°.

17. Given the following vectors:
    A has length 10 and points 30° above the x axis
    B has length 4.0 and points 10° to the left of the y axis
    $\mathbf{C} = 5.6\hat{\imath} - 3.1\hat{\jmath}$
    $\mathbf{D} = 1.9\hat{\imath} + 7.2\hat{\jmath}$,
    compute the scalar products (a) $\mathbf{A} \cdot \mathbf{B}$; (b) $\mathbf{C} \cdot \mathbf{D}$; (c) $\mathbf{B} \cdot \mathbf{C}$.

18. (a) Find the scalar product of the vectors $a\hat{\imath} + b\hat{\jmath}$ and $b\hat{\imath} - a\hat{\jmath}$, and (b) determine the angle between them. (Here a and b are arbitrary constants.)

19. Rework Example 7-5 using Equation 7-3 instead of Equation 7-4.

20. Use Equations 7-3 and 7-4 to show that the angle between the vectors $\mathbf{A} = a_x\hat{\imath} + a_y\hat{\jmath}$ and $\mathbf{B} = b_x\hat{\imath} + b_y\hat{\jmath}$ is

$$\theta = \cos^{-1}\left\{ \frac{a_x b_x + a_y b_y}{[(a_x^2 + a_y^2)(b_x^2 + b_y^2)]^{1/2}} \right\}$$

21. Given that $\mathbf{A} = 2\hat{\imath} + 2\hat{\jmath}$, $\mathbf{B} = 5\hat{\imath}$, and $\mathbf{C} = \sqrt{2}\hat{\imath} - \pi\hat{\jmath}$, find the angles between (a) **A** and **B**; (b) **A** and **C**; (c) **B** and **C**. *Hint:* See previous problem.

22. Find the work done by a force $\mathbf{F} = 2.5\hat{\imath} + 3.7\hat{\jmath}$ N as it acts on an object moving from the origin to the point $\mathbf{r} = 6.1\hat{\imath} + 2.3\hat{\jmath}$ m.

23. A force $\mathbf{F} = 14\hat{\imath} + 11\hat{\jmath}$ N acts on an object. Find the work done by the force if the object moves from the origin to the point (a) $28\hat{\imath} + 22\hat{\jmath}$ m and (b) $22\hat{\imath} - 28\hat{\jmath}$ m.

24. A force $\mathbf{F} = 67\hat{\imath} + 23\hat{\jmath} + 55\hat{k}$ N is applied to a body as it moves in a straight line from $\mathbf{r}_1 = 16\hat{\imath} + 31\hat{\jmath}$ to $\mathbf{r}_2 = 21\hat{\imath} + 10\hat{\jmath} + 14\hat{k}$ m. How much work is done by the force?

25. A rope pulls a box a horizontal distance of 23 m, as shown in Fig. 7-26. If the rope tension is 120 N, and if the rope does 2500 J of work on the box, what angle does it make with the horizontal?

Section 7-3 A Varying Force

26. Find the total work done by the force shown in Fig. 7-27 as the object on which it acts moves from $x = 0$ to $x = 6.0$ m.

27. Find the total work done by the force shown in Fig. 7-28 as

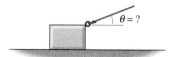

FIGURE 7-26 Problem 25.

the object on which it acts moves (a) from $x = 0$ to $x = 3.0$ km; (b) from $x = 3.0$ km to $x = 4.0$ km; (c) from $x = 0$ to $x = 4.0$ km.

28. Find the total work done by the force shown in Fig. 7-29, as the object on which it acts moves from $x = 0$ to $x = 5.0$ m.

29. A force F acts in the x direction, its magnitude given by $F = ax^2$, where x is in meters and a is exactly 5 N/m². (a) Find an exact value for the work done by this force as it acts on a particle moving from $x = 0$ to $x = 6$ m. Now find approximate values for the work by dividing the area under the force curve into rectangles of width (b) $\Delta x = 2$ m; (c) $\Delta x = 1$ m; (d) $\Delta x = \frac{1}{2}$ m with height equal to the magnitude of the force in the center of the interval. Calculate the per cent error in each case.

30. A spring has spring constant $k = 200$ N/m. How much work does it take to stretch the spring (a) 10 cm from equilibrium and (b) from 10 cm to 20 cm from equilibrium?

31. A certain amount of work is required to stretch spring A a certain distance. Twice as much work is required to stretch spring B half that distance. Compare the spring constants of the two springs.

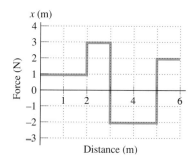

FIGURE 7-27 Problem 26.

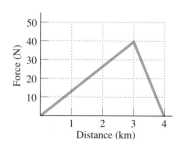

FIGURE 7-28 Problem 27.

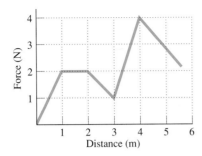

FIGURE 7-29 Problem 28.

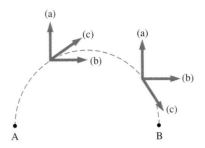

FIGURE 7-30 Problem 39. Arrows are force vectors, labeled to correspond to parts (a), (b), and (c) of the problem.

32. On graph paper, draw an accurate force-distance curve for the force of Example 7-8, and obtain a solution to that example by determining graphically the area under the curve.

33. A force $\mathbf{F}$ acts in the x direction, its x component given by $F = F_0 \cos(x/x_0)$, where $F_0 = 51$ N and $x_0 = 13$ m. Calculate the work done by this force acting on an object as it moves from $x = 0$ to $x = 37$ m. *Hint:* Consult Appendix A for the integral of the cosine function and treat the argument of the cosine as a quantity in radians.

34. Work the preceding problem graphically, by making an accurate plot of the force versus distance curve on graph paper, and determining the area (in units of work) under the curve.

35. A force given by $F = a\sqrt{x}$ acts in the x direction, where $a = 14$ N/m$^{1/2}$. Calculate the work done by this force acting on an object as it moves (a) from $x = 0$ to $x = 2.2$ m; (b) from $x = 2.2$ m to $x = 4.4$ m; (c) from 4.4 m to 6.6 m.

36. A force given by $F = b/\sqrt{x}$ acts in the x direction, where b is a constant with the units N·m$^{1/2}$. Show that even though the force becomes arbitrarily large as x approaches 0, the work done in moving from x_1 to x_2 remains finite even as x_1 approaches zero. Find an expression for that work in the limit $x_1 \to 0$.

37. The force exerted by a rubber band is given approximately by

$$F = F_0 \left[\frac{\ell_0 + x}{\ell_0} - \frac{\ell_0^2}{(\ell_0 + x)^2} \right],$$

where ℓ_0 is the unstretched length, x the stretch, and F_0 is a constant (although F_0 varies with temperature). Find the work needed to stretch the rubber band a distance x.

Section 7-4 Force and Work in Three Dimensions

38. A car drives 3.1 km southward, propelled by the 6.3-kN force of friction between its tires and the road. It then turns eastward and goes another 1.8 km; during this stretch the frictional force is 5.1 kN. Finally, the car turns toward the southeast and goes 2.6 km while the frictional force is 6.8 kN. Find the total work done on the car by the frictional force, assuming the force always acts in the direction of the car's motion.

39. A particle moves from point A to point B along the semicircular path of radius R, as shown in Fig. 7-30. It is subject to a force of constant magnitude F. Find the work done by the force (a) if the force always points upward in Fig. 7-30, (b) if the force always points to the right in Fig. 7-30, and (c) if the force always points in the direction of the particle's motion.

40. A force $\mathbf{F}$ acts on an object as it moves along the x axis from $x = 0$ to $x = 75$ m. The magnitude of the force is 2.5 N, and the angle θ that it makes with the x axis varies such that $\cos\theta$ is numerically equal to $x/100$ m. Find the total work done by the force.

41. A cylindrical log of radius R lies half buried in the ground, as shown in Fig 7-31. An ant of mass m climbs to the top of the log. Show that the work done by gravity on the ant is $-mgR$.

FIGURE 7-31 Problem 41.

42. A particle of mass m moves from the origin to the point $x = 3$ m, $y = 6$ m along the curve $y = ax^2 - bx$, where $a = 2$ m^{-1} and $b = 4$. It is subject to a force $\mathbf{F} = cxy\hat{\mathbf{i}} + d\hat{\mathbf{j}}$, where $c = 10$ N/m^2 and $d = 15$ N. Calculate the work done by the force.

43. Repeat the preceding problem for the case when the particle moves first along the x axis from the origin to the point $(3,0)$, then parallel to the y axis until it reaches $(3,6)$.

44. You put your little sister (mass m) on a swing whose chains have length ℓ, and pull slowly back until the swing makes an angle ϕ with the vertical. Show that the work you do is $mg\ell(1 - \cos\phi)$.

Section 7-5 Kinetic Energy

45. A 3.5×10^5-kg jumbo jet is cruising at 1000 km/h. (a) What is its kinetic energy relative to the ground? A 85-kg passenger strolls down the aisle at 2.9 km/h. What is the passenger's kinetic energy (b) relative to the ground and (c) relative to the plane?

46. Electrons in a color TV tube are accelerated to 25% of the speed of light. How much work does the TV tube do on each electron? (At this speed, relativity theory introduces small but measurable corrections; here you neglect these effects.) See inside front cover for the electron mass.

47. In a cyclotron used to produce radioactive isotopes for medical research, deuterium nuclei (mass 3.3×10^{-27} kg) are given kinetic energies of 60×10^{-12} J. What is their speed? Compare with the speed of light.

48. At what speed must a 950-kg subcompact car be moving to have the same kinetic energy as a 3.2×10^4-kg truck going 20 km/h?

49. A 60-kg skateboarder comes over the top of a hill at 5.0 m/s, and reaches 10 m/s at the bottom of the hill. Find the total work done on the skateboarder between the top and bottom of the hill.

50. Two unknown elementary particles pass through a detection chamber. If they have the same kinetic energy and their mass ratio is 4:1, what is the ratio of their speeds?

51. You do 14 J of work to stretch a spring of spring constant $k = 160$ N/m, starting with the spring unstretched. How far does the spring stretch?

52. After a tornado, a 0.50-g drinking straw was found embedded 4.5 cm in a tree. Subsequent measurements showed that the tree would exert a stopping force of 70 N on the straw. What was the straw's speed when it hit the tree?

53. You drop a 150-g baseball from a sixth-story window 16 m above the ground. What are (a) its kinetic energy and (b) its speed when it hits the ground? Neglect air resistance.

54. A hospital patient was being wheeled in for x ray when his leg slipped off the stretcher and his heel hit the concrete floor. As a physicist, you are called to testify about the forces involved in this accident. You estimate that the foot and leg had an effective mass of 8 kg, that they dropped freely a distance of 70 cm, and that the stopping distance was 2 cm. What force can you claim was exerted on the foot by the floor? Give your answer in pounds so the jury will have a feel for the size of the force.

55. From what height would you have to drop a car for its impact to be equivalent to a collision at 20 mph?

56. Catapults run by high-pressure steam from the ship's nuclear reactor are used on the aircraft carrier Enterprise to launch jet aircraft to takeoff speed in only 76 m of deck space. A catapult exerts a 1.1×10^6 N force on a 3.3×10^4 kg aircraft. What are (a) the kinetic energy and (b) the speed of the aircraft as it leaves the catapult? (c) How long does the catapulting operation take? (d) What is the acceleration of the aircraft?

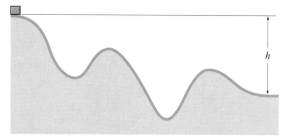

FIGURE 7-32 Problem 57.

57. A block of mass m slides from rest without friction down the slope shown in Fig. 7-32. (a) How much work is done on the block by the normal force of the slope? (b) Show that the final speed is $\sqrt{2gh}$ regardless of the details of the slope.

Section 7-6 Power

58. A horse plows a 200-m-long furrow in 5.0 min, exerting a force of 750 N. What is its power output, measured in watts and in horsepower?

59. A typical car battery stores about 1 kWh of energy. What is its power output if it is drained completely in (a) 1 minute; (b) 1 hour; (c) 1 day?

60. A sprinter completes a 100-m dash in 10.6 s, doing 22.4 kJ of work. What is her average power output?

61. How much work can a 3.5-hp lawnmower engine do in 1 h?

62. Water drops over 49-m-high Niagara Falls at the rate of 6.0×10^6 kg/s. If all the energy of the falling water could be harnessed by a hydroelectric power plant, what would be the plant's power output?

63. A "mass driver" is designed to launch raw materials mined on the moon to a factory in lunar orbit. The driver can accelerate a 1000-kg package to 2.0 km/s (just under lunar escape speed) in 55 s. (a) What is its power output during a launch? (b) If the driver makes one launch every 30 min, what is its average power consumption?

64. An 85-kg long-jumper takes 3.0 s to reach a prejump speed of 10 m/s. What is his power output?

65. Estimate your power output as you do deep knee bends at the rate of one per second.

66. At what rate can a one-half horsepower well pump deliver water to a tank 60 m above the water level in the well? Give your answer in kg/s and gal/min.

67. In midday sunshine, solar energy strikes Earth at the rate of about 1 kW/m². How long would it take a perfectly efficient solar collector of 15 m² area to collect 40 kWh of energy? (This is roughly the energy content of a gallon of gasoline.)

68. It takes about 20 kJ to melt an ice cube. A typical microwave oven produces 625 W of microwave power. How long will it take to melt the ice cube in this oven?

69. The rate at which the United States imports oil, expressed in terms of the energy content of the imported oil, is very nearly

500 GW. Using the "Energy Content of Fuels" table in Appendix C, convert this figure to gallons per day.

70. Which consumes more energy, a 1.2-kW hair dryer used for 10 min or a 7-W night light left on for 24 h?

71. By measuring oxygen uptake, sports physiologists have found that the power output of long-distance runners is given approximately by $P = m(bv - c)$, where m and v are the runner's mass and speed, respectively, and where b and c are constants given by $b = 4.27$ J/kg·m and $c = 1.83$ W/kg. Determine the work done by a 54-kg runner who runs a 10-km race at a speed of 5.2 m/s.

72. A 65-kg runner running at $v_0 = 4.8$ m/s accelerates to 6.1 m/s over a 25-s interval. (a) By writing $v = v_0 + at$, where a is the runner's acceleration, use the formula in the previous problem to express the runner's power output as a function of time. (b) How much work does the runner do during the acceleration period?

73. A 1400-kg car ascends a mountain road at a steady 60 km/h. The force of air resistance on the car is 450 N. If the car's engine supplies energy to the drive wheels at the rate of 38 kW, what is the slope angle of the road?

74. A machine does work at a rate given by $P = ct^2$, where $c = 18$ W/s^2 and t is time. Find an expression for the work done by the machine between $t = 10$ s and $t = 20$ s.

Paired Problems

(Both problems in a pair involve the same principles and techniques. If you can get the first problem, you should be able to solve the second one.)

75. You apply a 650-N force to push a stalled car at an 18° angle to its direction of motion, doing 990 J of work in the process. How far do you push the car?

76. A tractor tows a jumbo jet from its airport gate, doing 8.7 MJ of work. The link from the plane to the tractor makes a 22° angle with the direction of the plane's motion, and the tension in the link is 4.1×10^5 N. How far does the tractor move the plane?

77. A force pointing in the x direction is given by $F = F_0(x/x_0)$, where F_0 and x_0 are constants and x is the position. Find an expression for the work done by this force as it acts on an object moving from $x = 0$ to $x = x_0$.

78. A force pointing in the x direction is given by $F = ax^{3/2}$, where $a = 0.75$ N/m$^{3/2}$. Find the work done by this force as it acts on an object moving from $x = 0$ to $x = 14$ m.

79. Two vectors have equal length, and their scalar product is one-third of the square of their length. Find the angle between them.

80. Vector **A** has length A, vector **B** has length $2A$, and $\mathbf{A} \cdot \mathbf{B} = A^2$. Find the angle between **A** and **B**.

81. A 460-kg piano is pushed at constant speed up a ramp, raising it a vertical distance of 1.9 m (see Fig. 7-33). If the coefficient of friction between ramp and piano is 0.62, find the work done by the agent pushing the piano if the ramp

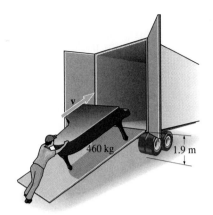

FIGURE 7-33 Problem 81.

angle is (a) 15° and (b) 30°. Assume the force is applied parallel to the ramp.

82. You have to do 3.7 kJ of work to push a 75-kg trunk 4.2 m along a slope inclined upward at 30°, pushing parallel to the slope. What is the coefficient of friction between trunk and slope?

83. (a) How much power is needed to push a 95-kg chest at 0.62 m/s along a horizontal floor where the coefficient of friction is 0.78? (b) How much work is done in pushing the chest 11 m?

84. You mix flour into a thick bread dough, exerting a 45-N force on the stirring spoon. If you move the spoon at 0.29 m/s, (a) what power do you supply? (b) How much work do you do if you stir for 1.0 min?

Supplementary Problems

85. The power output of a machine of mass m increases linearly with time, according to the formula $P = bt$, where b is a constant. (a) Find an expression for the work done between $t = 0$ and some arbitrary time t. (b) Suppose the machine is initially at rest and all the work it supplies goes into increasing its own speed. Use the work-energy theorem to show that the speed increases linearly with time, and find an expression for the acceleration.

86. You're trying to decide whether to buy an energy-efficient, 225-W refrigerator for $1150 or a standard, 425-W model for $850. The standard model will run 20% of the time, while its better insulation means the energy-efficient model will run 11% of the time. If electricity costs 9.5¢/kWh, how long would you have to own the energy-efficient model to make up the difference in cost? Neglect any interest you might earn on your money.

87. The per-capita energy consumption rate plotted in Fig. 7-18 can be approximated by the expression $P = P_0 + at + bt^2 + ct^3$, where $P_0 = 4.4$ kW, $a = -5.57\times10^{-2}$ kW/y, $b = 3.84\times10^{-3}$ kW/y^2, $c = -2.79\times10^{-5}$ kW/y^3, and t is the time in years since 1900 (i.e., 1960 is $t = 60$). Integrate

this expression to find approximate values for the energy used per capita during the decades (a) from 1940 to 1950 and (b) from 1960 to 1970. It's easiest to give your answers in kilowatt-years.

88. A spring of spring constant k is attached to a mass m and the other end of the spring is pulled vertically in order to lift the mass. Find an expression for the amount of work that must be done on the spring *before* the mass begins to leave the ground.

89. If the ant of Problem 41 climbs with constant speed v along the log surface, (a) what is its power output as a function of time? (b) Integrate your expression for power to show that the total work to climb the log is the same as in Problem 41.

90. A machine delivers power at a decreasing rate $P = P_0 t_0^2/(t + t_0)^2$, where P_0 and t_0 are constants. The machine starts at $t = 0$ and runs forever. Show that it nevertheless does only a finite amount of work, equal to $P_0 t_0$.

91. An unusual spring has the force-distance curve shown in Fig. 7-34 and described by $F = 100x^2$ for $0 \le x \le 1$ and $F = 100(4x - x^2 - 2)$ for $1 \le x \le 2$, where x is the displacement in meters from the spring's unstretched length and F is in newtons. Find the work done in stretching this

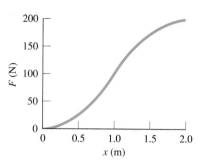

FIGURE 7-34 Problem 91.

spring (a) from $x = 0$ to $x = 1$ m and (b) from $x = 1$ m to $x = 2$ m.

92. A locomotive accelerates a freight train of total mass M from rest, applying constant power P. Determine the speed and position of the train as functions of time, assuming all the power goes to increasing the train's kinetic energy.

93. A 2.3-kg particle's position as a function of time is given by $x = bt^3 - ct$, where $b = 0.41$ m/s^3 and $c = 1.9$ m/s. Find the work done on the particle between the times $t = 0$ and $t = 2.0$ s.

CHAPTER 8

CONSERVATION OF ENERGY

This back flip on stilts involves multiple interchanges among different forms of energy.

The rock climber of Fig. 8-1a does work as she ascends the vertical cliff. So does the mover of Fig. 8-1b, as he pushes a heavy chest across the floor. But there's a difference. If the rock climber lets go, down she goes; the work she put into the climb comes back as kinetic energy of her fall. If the mover lets go of the chest, though, it just sits there.

The contrast between these two situations illustrates an important distinction between two types of forces. From that distinction we will now develop one of the most important principles in all of physics: conservation of energy. In later chapters, we will expand the conservation of energy principle to include new forms of energy we encounter. Ultimately, in Chapter 38, we will see how Einstein's theory of relativity subsumes two great principles—conservation of energy and conservation of matter—into a single statement.

(a)   (b)

FIGURE 8-1   Both the rock climber (a) and the mover (b) do work, but only the climber can recover that work as kinetic energy.

8-1 CONSERVATIVE AND NONCONSERVATIVE FORCES

Both the climber and the mover of Fig. 8-1 are working against external forces—gravity for the climber and friction for the mover. The difference between the two situations is this: if the climber lets go, the gravitational force "gives back" the work that she did; that work manifests itself as kinetic energy. But the frictional force does not "give back" the work that the mover did; that work cannot be recovered as kinetic energy.

We use the term **conservative force** to describe a force that, like gravity, "gives back" work that has been done against it. A force like friction is **nonconservative.** We can give a more precise mathematical sense of this distinction by considering the work involved in moving an object over a closed path—that is, a path that ends up at the same point where it started. Suppose our rock climber ascends a cliff of height h and then returns to her starting point. How much work has the gravitational force done on her? That force has magnitude mg, and points downward. Going up, the force is directed opposite to her motion, so the work done by gravity is $-mgh$. Coming down, the force is in the same direction as her motion, so the work done by gravity is $+mgh$. The total work that gravity does on the climber as she traverses the closed path from the bottom to the top of the cliff and back again is therefore zero.

Now suppose the mover of Fig. 8-1b pushes the chest a distance ℓ across a room, discovers it's the wrong room, and pushes it back to the door. Like the climber, the chest describes a closed path. How much work is done by the frictional force acting over this path? That force has magnitude μN, where the normal force in this case of horizontal motion is just mg. But the frictional force *always* acts to oppose the motion, so it *always* does negative work. With frictional force μmg opposing the motion, the work done in crossing the width ℓ of the room is $-\mu mg\ell$. Coming back, it is also $-\mu mg\ell$, so the total work done by the frictional force is $-2\mu mg\ell$.

The difference between our answers for the total work done by the gravitational and frictional forces acting over closed paths provides one precise definition of the distinction between conservative and nonconservative forces:

> **When the total work done by a force F acting as an object moves over any closed path is zero, then the force is conservative. Mathematically,**

$$\oint \mathbf{F} \cdot d\mathbf{r} = 0. \qquad \text{(conservative force)} \qquad (8\text{-}1)$$

The circle on the integral sign indicates that the integral is taken over a *closed* path. A force for which the integral is nonzero, like the frictional force, is nonconservative.

You can see why the mathematical definition given by Equation 8-1 is equivalent to our more physical statement that a conservative force "gives back" work that was done against it. When we do work against a conservative force, it simultaneously does negative work on us. But for the total work done by the conservative force to be zero, it must subsequently do positive work on us; during this time it "gives back" the work we did earlier.

Equation 8-1 suggests a related property of conservative forces. Suppose we move an object along the straight path between points A and B shown in Fig. 8-2, along which a conservative force acts; let the work done by the conservative force be W_{AB}. Since the work done by a conservative force over any closed path is zero, the work W_{BA} done in moving back from B to A must be $-W_{AB}$ whether we return along the straight path or take the curved path shown or any other path. So going from A to B involves work W_{AB}, regardless of the path taken. In other words:

> **The work done by a conservative force in moving between two points is independent of the path taken; mathematically, $\int_A^B \mathbf{F} \cdot d\mathbf{r}$ depends only on the endpoints A and B, not on the path between them.**

We saw this in Example 7-9, where we found that the work done by gravity does not depend on the details of the path taken (Fig. 8-3). In contrast, the work done by a nonconservative force is *not* path-independent. On a frictional surface, for example, the least work is done over a straight-line path; any other path involves more work (Fig. 8-4).

Important examples of conservative forces include gravity and the static electric force. The force of an ideal spring—fundamentally an electric force—is also conservative. Nonconservative forces include friction and the electric force in the presence of time-varying magnetic effects.

8-2 POTENTIAL ENERGY

Work done against a conservative force is somehow "stored," in the sense that we can get it back again in the form of kinetic energy. The climber of Fig. 8-1a is acutely aware of that "stored work"; it gives her the potential for a dangerous fall.

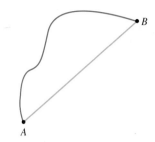

FIGURE 8-2 Since the work done by a conservative force over a closed path is zero, the work done in moving from point A to point B is the same for the straight path, the curved path, or any other path.

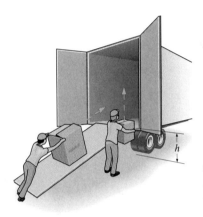

FIGURE 8-3 Movers raise identical boxes a vertical distance h. The work done by gravity is the same for the box lifted vertically and for the box pushed up the ramp. (The horizontal location of the starting point doesn't matter here since gravity does no work on an object moved horizontally.)

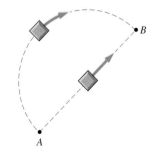

FIGURE 8-4 Top view of boxes being pushed along a horizontal floor. The work done by friction is least for the straight-line path.

FIGURE 8-5 The archer's taut bow stores potential energy that can become kinetic energy of the arrow.

So is the archer of Fig. 8-5; the work he did against the springiness of his bow has the potential to propel the arrow at high speed. "Potential" is an appropriate word here: we can consider the "stored work" as **potential energy,** potential in the sense that it can become actualized as kinetic energy.

We define potential energy formally in terms of the work done by a conservative force. Specifically:

The change ΔU$_{AB}$ in potential energy associated with moving an object from point A to point B is the negative of the work done by the conservative force F acting on that object:

$$\Delta U_{A \to B} = -\int_{A}^{B} \mathbf{F} \cdot d\mathbf{r}. \quad \text{(potential energy)} \qquad (8\text{-}2)$$

Why the minus sign? Because potential energy represents stored work. If a conservative force does *positive* work, its potential energy must decrease accordingly—and that means ΔU must be *negative*. Conversely, if a conservative force does *negative* work—as does gravity on a weight being lifted—then energy is stored, and ΔU must be positive. The minus sign in Equation 8-2 handles both these cases. We will often drop the subscript $A \to B$ and write simply ΔU for the potential energy change. Keeping the subscript is important, though, when we need to be very clear about whether we're going from A to B or from B to A.

Our definition describes only *changes* in potential energy. Such changes are all that ever matter physically; the actual value of potential energy is physically meaningless. Often, though, it is convenient to establish a reference point at which the potential energy is defined to be zero. When we say "the potential energy U," we really mean the potential energy difference ΔU between that reference point and whatever other point we're considering. Our rock climber, for example, might find it convenient to take the zero of potential energy at the base of the cliff. But the choice is purely for convenience; only potential energy *differences* really matter.

Equation 8-2 is a completely general definition of potential energy, applicable in all circumstances. Often, though, we can consider a path where force and displacement are parallel (or antiparallel). Then Equation 8-2 simplifies to

$$\Delta U = -\int_{x_1}^{x_2} F(x)\, dx, \tag{8-2a}$$

where x_1 and x_2 are the starting and ending points on the x axis, taken to coincide with the path. When the force is constant, this equation simplifies further to

$$\Delta U = -F(x_2 - x_1). \tag{8-2b}$$

TIP **Understand Your Equations** Equation 8-2b provides a very simple expression for potential energy changes, but it applies *only* when the force is constant. Equation 8-2b is a special case of Equation 8-2a that results when the force can be taken outside the integral sign—and that's allowed only when the force doesn't vary with position. Similarly, Equation 8-2a is a special case of Equation 8-2 that applies only when the force and path are parallel (or antiparallel).

Gravitational Potential Energy

We're frequently moving things up and down. Because of the gravitational force, these movements cause changes in potential energy. Figure 8-6 shows two possible paths for a book that's lifted from the floor to a shelf of height h. Which path should we use to calculate the potential energy change? Since the gravitational force is conservative, it doesn't matter. It's easiest to use the path consisting of straight horizontal and vertical segments. There's no work or potential energy change associated with the horizontal motion since the gravitational force is perpendicular to the motion. For the vertical lift, the force of gravity is constant and Equation 8-2b gives immediately $\Delta U = mgh$, where the minus sign in Equation 8-2b cancels with the minus sign associated with the *downward* direction of gravity. This result is quite general; when a mass m undergoes a vertical displacement Δy near Earth's surface, its gravitational potential energy changes by

$$\Delta U = mg\,\Delta y. \quad \text{(gravitational potential energy)} \tag{8-3}$$

The quantity Δy can be positive or negative, depending on whether the object moves up or down; correspondingly, the potential energy can either increase or decrease.

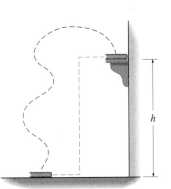

FIGURE 8-6 The work done by gravity as the book moves from floor to shelf is the same for both paths shown, but is more easily calculated for the path composed of horizontal and vertical sections.

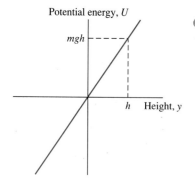

FIGURE 8-7 Since the gravitational force is constant, the potential energy increases linearly with height.

> **TIP** **Know Your Limitations** We emphasize that Equation 8-3 applies *near Earth's surface*—that is, for distances small compared with Earth's radius. That assumption allowed us to treat the gravitational force as constant over the path. For larger distances the gravitational force varies, and we must use the integral of Equation 8-2 to calculate the potential energy change. We'll explore this more general case in the next chapter.

We've found the *change* in the book's potential energy, but what about the potential energy itself? That depends on where we define the zero of potential energy. If we choose $U = 0$ at the floor, then $U = mgh$ on the shelf. But we could just as well take $U = 0$ at the shelf; then the book's potential energy on the floor would be $-mgh$. Negative potential energies arise frequently, and they're nothing to be concerned about because it's only *differences* in potential energy that really matter. Figure 8-7 shows a plot of potential energy versus height with $U = 0$ taken at the floor. The *linear* increase in potential energy with height reflects the *constant* gravitational force.

● EXAMPLE 8-1 GRAVITATIONAL POTENTIAL ENERGY

A 55-kg engineer leaves her office on the 33rd floor of a skyscraper and takes an elevator to the 59th floor. Later she descends to street level. If the engineer takes her office as the zero of potential energy and if the distance from one floor to the next is 3.5 m, what is the engineer's potential energy (a) in her office, (b) on the 59th floor, and (c) at street level?

Solution

In her office, the engineer's potential energy is zero since she defined it that way. The 59th floor is $59 - 33 = 26$ floors higher, so the potential energy there is

$$U_{59} = mg\Delta y = (55 \text{ kg})(9.8 \text{ m/s}^2)(26 \text{ floors})(3.5 \text{ m/floor})$$
$$- 49 \text{ kJ}.$$

Note that we can write U rather than ΔU since we're calculating the potential energy change from the place where $U = 0$. Similarly, street level is 32 floors *below* the engineer's office, so

$$U_{\text{street}} = mg\Delta y = (55 \text{ kg})(9.8 \text{ m/s}^2)(-32 \text{ floors})(3.5 \text{ m/floor})$$
$$= -60 \text{ kJ}.$$

EXERCISE You fly from Boston's Logan Airport, at sea level, to Denver (altitude 1.6 km). Taking your mass as 65 kg and the zero of potential energy at Boston, what is your gravitational potential energy (a) at the plane's 11-km cruising altitude and (b) in Denver?

Answers: (a) 7.0 MJ; (b) 1.0 MJ

Some problems similar to Example 8-1: 4, 5, 8, 10

●

■ APPLICATION PUMPED STORAGE

Electricity, which we'll study in detail in later chapters, is a wonderfully versatile form of energy. But it has a serious drawback: It's difficult to store electrical energy. Large electric power plants operate most efficiently on a continuous basis, yet electric power demand fluctuates considerably. Energy storage can match generating capability more closely to demand for electricity, thus reducing the need for new power plants.

Gravitational potential energy provides a simple storage medium. In so-called pumped storage facilities, surplus electric power is used to pump water from a lower reservoir to a higher

one, storing energy as gravitational potential energy of the water. When demand for electricity is high, water from the upper reservoir flows back down, turning electric generators to make electricity. (For reasons we'll see in Chapter 31, the same motors that run the pumps can also function as generators.)

Figure 8-8 shows an aerial view of the Northfield Mountain Pumped Storage Project in Massachusetts. The upper reservoir at this facility is 270 m above the pump/generators, and holds 1.6×10^{10} kg of water. According to Equation 8-3, the total energy stored is $U = mg\Delta y = (1.6 \times 10^{10} \text{ kg})(9.8 \text{ m/s}^2)$

$\times$ (270 m) = 4.2$\times$10^{13} J. The generators at Northfield Mountain have a total power output of 1080 MW. Since 1 MW = 10^6 J/s, it would take (4.2$\times$10^{13} J)/(1080$\times$10^6 J/s) = 3.9$\times$10^4 s, or about 11 hours to drain the upper reservoir. In practice, the facility can generate electricity for almost this long before recharging the reservoir.

The Northfield facility is quite efficient: 86% of the energy used to pump the water is recovered as electricity.

FIGURE 8-8 The Northfield Mountain Pumped Storage Project in Massachusetts stores surplus electrical energy as gravitational potential energy of water in this mountain top reservoir, 270 m above the pump/generators.

The Potential Energy of a Spring

An ideal spring exerts a force given by $F = -kx$, where x is the distance the spring is stretched from equilibrium, and where the minus sign shows that the force opposes the stretching (or compression for $x < 0$). Since the force varies with position, we use Equation 8-2b to evaluate the potential energy:

$$\Delta U = -\int_{x_1}^{x_2} F(x)\,dx = -\int_{x_1}^{x_2} (-kx)\,dx = \tfrac{1}{2}kx_2^2 - \tfrac{1}{2}kx_1^2,$$

where x_1 and x_2 are the initial and final values of the stretch. If we take $U = 0$ when $x = 0$—that is, when the spring is neither stretched nor compressed—then we can use this result to write the potential energy at an arbitrary stretch (or compression) x as

$$U = \tfrac{1}{2}kx^2. \qquad \text{(potential energy of a spring)} \qquad (8\text{-}4)$$

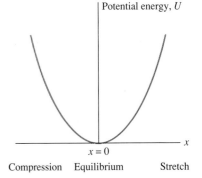

FIGURE 8-9 The potential energy curve for a spring is a parabola.

Comparison with Equation 7-10 shows that this is equal to the work done in stretching the spring. Of course: That work gets stored as potential energy of the spring. Figure 8-9 shows the potential energy as a function of the stretch or compression of a spring. The *parabolic* shape of the potential energy curve reflects the *linear* change of the spring force with stretch or compression.

● EXAMPLE 8-2 ENERGY STORAGE: SPRINGS VERSUS GASOLINE

How far would you have to stretch a spring of constant $k = 8500$ N/m in order to store the same amount of energy as is contained in 1.0 g of gasoline?

Solution

The "Energy Content of Fuels" table in Appendix C shows that 1 g of gasoline contains 44 kJ of energy. Solving Equation 8-4 for x then gives

$$x = \sqrt{\frac{2U}{k}} = \sqrt{\frac{(2)(44 \times 10^3 \text{ J})}{8500 \text{ N/m}}} = 3.2 \text{ m}.$$

Since $k = 8500$ N/m implies an extraordinarily stiff spring, this is an unrealistically large stretch. Furthermore, 1 gram isn't much gasoline. This example shows that springs, while useful energy storage devices, cannot compete with chemical fuels.

EXERCISE A hydrogen chloride (HCl) molecule can be modeled as a hydrogen and a chlorine atom connected by a spring of spring constant $k = 480$ N/m. If the molecule is stretched from equilibrium until its potential energy is 5.0×10^{-22} J, by how much does the interatomic separation increase? (This model holds only for very small deviations from equilibrium; beyond that the "spring" ceases to obey Hooke's law.)

Answer: 1.4 pm $= 1.4 \times 10^{-12}$ m

Some problems similar to Example 8-2: 11–13 ●

Other Forces

In Equations 8-3 and 8-4 we have the potential energy associated with two common forces—gravity and springs. In the gravitational case, calculation of the potential energy was easy because the gravitational force doesn't change with position. The spring force does change with position, but we've accounted for that by evaluating the integral in Equation 8-2a.

You will encounter other forces that vary with position. Unless you've already done an integral to evaluate the associated work or potential energy, it's necessary to integrate—according to Equation 8-2 or Equation 8-2a—to evaluate the potential energy.

● EXAMPLE 8-3 A CLIMBING ROPE

Ropes used in rock climbing are "springy" so they cushion a climber's fall (Fig. 8-10). A particular rope has a force law given by $F = -kx + bx^2$, where $k = 223$ N/m, $b = 4.1$ N/m^2, and x is the amount of stretch. Find the potential energy stored in this rope when it's been stretched 2.62 m, taking $U = 0$ at $x = 0$.

Solution

For this one-dimensional case we use Equation 8-2a:

$$U = -\int_{x_1}^{x_2} F(x)\,dx = -\int_0^x (-kx + bx^2)\,dx$$
$$= \frac{1}{2}kx^2 - \frac{1}{3}bx^3 \Big|_0^x = \frac{1}{2}kx^2 - \frac{1}{3}bx^3$$
$$= (\tfrac{1}{2})(223 \text{ N/m})(2.62 \text{ m})^2 - (\tfrac{1}{3})(4.1 \text{ N/m}^2)(2.62 \text{ m})^3$$
$$= 741 \text{ J}.$$

This is about 3% less than the potential energy $U = \frac{1}{2}kx^2$ of an ideal spring with the same spring constant.

EXERCISE An elementary-school student stretches a rubber band between thumb and forefinger and takes aim at the teacher. The force law for this particular rubber band is

FIGURE 8-10 The right degree of "springiness" in the climbing rope is essential for safety (Example 8-3).

$F = -1.0x + 0.12x^2$, where x is the stretch in centimeters and F is in newtons. How much energy is stored in the rubber band when the student has stretched it 4.3 cm?

Answer: 0.061 J

Some problems similar to Example 8-3: 14–17 ●

8-3 CONSERVATION OF MECHANICAL ENERGY

The work-energy theorem, introduced in the preceding chapter, shows that the change ΔK in a body's kinetic energy is equal to the net work done on it:

$$\Delta K = W_{\text{net}}.$$

Consider separately the work W_c done by conservative forces and the work W_{nc} done by nonconservative forces, so

$$\Delta K = W_c + W_{nc}.$$

We have defined the change in potential energy, ΔU, as the negative of the work done by conservative forces. So we can write

$$\Delta K = -\Delta U + W_{nc},$$

or

$$\Delta K + \Delta U = W_{nc}. \tag{8-5}$$

We define the sum of the kinetic and potential energy as the **mechanical energy.** Then Equation 8-5 can be written

$$\Delta(K + U) = W_{nc}, \tag{8-6}$$

showing that the change in mechanical energy is equal to the work done by nonconservative forces.

In the absence of nonconservative forces (or when those forces are perpendicular to displacement, so they do no work), then Equations 8-5 and 8-6 show that the mechanical energy is unchanged:

$$\Delta K + \Delta U = 0 \tag{8-7}$$

and, equivalently,

$$K + U = \text{constant} = K_0 + U_0, \tag{8-8}$$

where K_0 and U_0 are the values of kinetic and potential energy of a body at some point and K and U their values when the body is at any other point. Equations 8-7 and 8-8 express the **law of conservation of mechanical energy.** They show that, in the absence of nonconservative forces, the mechanical energy $K + U$ remains always the same. The kinetic energy K may change, but that change is always compensated by an equal but opposite change in potential energy.

Conservation of mechanical energy is a powerful principle. Throughout physics, from the subatomic realm through practical problems in engineering and on to astrophysics, the principle of energy conservation is widely used in solving problems that would be intractable without it.

> **TIP** **Applying Conservation of Energy** Applying the conservation of energy principle is as simple as writing $K + U = K_0 + U_0$. In classical physics, the kinetic energy is always given by $K = \frac{1}{2}mv^2$. But the form of the potential energy U depends on the particular conservative force; Equations 8-3 and 8-4 give expressions for the common forces of gravity and springs. You may be given other potential energy functions, or you may need to determine the form of the potential energy from a force law, as in Example 8-3. What about the constants K_0 and U_0? You can determine them from the conditions given in the problem; their sum is the mechanical energy that remains unchanged.

● EXAMPLE 8-4 TRANQUILIZING AN ELEPHANT

A biologist uses a spring-loaded gun to shoot tranquilizer darts into an elephant. The gun's spring has spring constant $k = 940$ N/m and is compressed a distance $x_0 = 25$ cm before firing a 38-g dart. At what speed does the dart leave the gun?

Solution
Here we have a system in which energy is initially stored in the compressed spring; later, that energy gets transformed into the kinetic energy of the dart. Initially the dart is at rest, so $K_0 = 0$. The spring is compressed a distance x_0, so the initial potential energy is $U_0 = \frac{1}{2}kx_0^2$ from Equation 8-4. When the dart leaves the gun the spring is back in equilibrium, so the potential energy is $U = 0$, while the kinetic energy is that of the dart: $K = \frac{1}{2}mv^2$. So the statement that mechanical energy is conserved—Equation 8-8—becomes $U_0 = K$, or

$$\tfrac{1}{2}kx_0^2 = \tfrac{1}{2}mv^2.$$

Solving for the dart speed v then gives

$$v = \sqrt{\frac{k}{m}}\, x = \left(\sqrt{\frac{940 \text{ N/m}}{0.038 \text{ kg}}}\right)(0.25 \text{ m}) = 39 \text{ m/s}.$$

Does this result make sense? Yes. In algebraic form, our answer $v = \sqrt{k/m}\, x$ shows that the stiffer the spring, the higher the speed; the greater the mass, the lower the speed, as we expect from Newton's second law. And the further we compress the spring, the greater the speed as well.

Notice how easy conservation of energy made this problem. Had we tried to calculate the speed using Newton's second law, we would have had a hard time because the force, and therefore the acceleration, is not constant.

EXERCISE In a railroad yard, a 35,000-kg boxcar moving at 7.5 m/s is brought to a stop by a spring-loaded bumper mounted at the end of the level track. If the spring has constant $k = 2.8$ MN/m, how far does it compress in stopping the boxcar?

Answer: 84 cm

Some problems similar to Example 8-4: 21, 22, 29
●

● EXAMPLE 8-5 CONSERVATION OF ENERGY: A SPRING AND GRAVITY

The spring in Fig. 8-11 has spring constant $k = 140$ N/m. A 50-g block is placed against the spring, which is then compressed 11 cm. When the block is released, how high up the slope does it rise? Both the horizontal surface and the slope are frictionless.

Solution
To solve this problem, we equate the total energy $K_0 + U_0$ in the initial position with the total energy $K_1 + U_1$ at the maximum height:

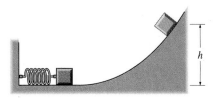

FIGURE 8-11 Example 8-5.

$$K_1 + U_1 = K_0 + U_0.$$

Initially, the block is at rest, so $K_0 = 0$. At maximum height on the slope, it is also at rest, so $K_1 = 0$. The initial potential energy is that of the compressed spring, or $\frac{1}{2}kx^2$. The final potential energy is the gravitational potential energy mgh. So our statement of energy conservation reads simply

$$\tfrac{1}{2}kx^2 = mgh.$$

Solving for the height h, we have

$$h = \frac{kx^2}{2mg} = \frac{(140 \text{ N/m})(0.11 \text{ m})^2}{(2)(0.050 \text{ kg})(9.8 \text{ m/s}^2)} = 1.7 \text{ m}.$$

TIP Save Steps You might have been tempted in working this example to solve for the speed after the block left the spring and then equate $\frac{1}{2}mv^2$ to mgh to find the height. And you certainly could have done that. But you don't have to! Conservation of energy shortcuts all details of the motion and allows you to equate directly the total energy at two points of interest. Here the speed is zero at both these points, so you need not be concerned with kinetic energy.

EXERCISE You're designing a toy rocket to be launched by a spring. The launching apparatus has room for a spring that can be compressed 14 cm, and the rocket's mass is 65 g. If the rocket is to reach 35 m altitude, what should be the spring constant?

Answer: 2.3 kN/m

Some problems similar to Example 8-5: 23, 26, 30

● **EXAMPLE 8-6** A PENDULUM

The pendulum of Fig. 8-12 is pulled back until its string makes an angle θ_0 with the vertical, then released from rest. (a) How fast is it going when it reaches its lowest point? (b) At its lowest point, the string catches on a nail located a distance a above the bottom of the string. Assuming the string remains taut, what is the maximum angle θ_1 that the string makes with the vertical?

Solution
Here two forces are acting: the conservative force of gravity and the string tension. But the string tension is perpendicular to the motion, so it does no work. So the total energy remains constant, and the potential energy is associated with the gravitational force alone:

$$K + U = K_0 + U_0.$$

To write out the potential energy terms, we need an expression for the potential energy as a function of angle. Figure 8-13 shows that an angle θ corresponds to a height $h = \ell(1 - \cos\theta)$. Setting the zero of potential energy at the lowest point, we can then write $U_0 = mgh = mg\ell(1 - \cos\theta_0)$, where m is the mass of the pendulum bob (we assume the string has negligible mass). Since the pendulum is released from rest, $K_0 = 0$, so $K_0 + U_0 = mg\ell(1 - \cos\theta_0)$.

(a) To find the speed at the bottom, we equate the energy at the bottom to our total energy $K_0 + U_0$. At the bottom, we have defined $U = 0$, so $K + U = \frac{1}{2}mv^2$ at the bottom. Then

$$\tfrac{1}{2}mv^2 = mg\ell(1 - \cos\theta_0),$$

so

$$v = \sqrt{2g\ell(1 - \cos\theta_0)}.$$

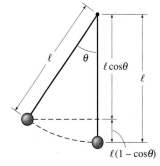

FIGURE 8-13 When the pendulum makes an angle θ with the vertical, it is a distance $h = \ell(1 - \cos\theta)$ above its lowest position.

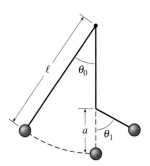

FIGURE 8-12 Example 8-6.

(b) To find the maximum angle as the pendulum swings to the right, we equate the energy in the rightmost position to the total energy $K_0 + U_0$. On the right, the geometry is the same as in Fig. 8-13 but with the length a replacing ℓ. So the potential energy in the rightmost position is $mga(1 - \cos\theta_1)$. Since the pendulum is instantaneously at rest in this position, $K = 0$, and conservation of energy becomes

$$mga(1 - \cos\theta_1) = mg\ell(1 - \cos\theta_0).$$

Solving for θ_1 gives

$$\theta_1 = \cos^{-1}\left(\frac{\ell}{a}\cos\theta_0 + \frac{a - \ell}{a}\right).$$

Does this result make sense? When $a = \ell$, so the nail is the pivot for the pendulum, then $\theta_1 = \theta_0$, and the pendulum swings symmetrically about its lowest position. And no matter where the nail, our conservation of energy statement shows that $a(1 - \cos\theta_1) = \ell(1 - \cos\theta_0)$; from Fig. 8-13, we see that this means the pendulum rises to the same height on either side, regardless of the presence of the nail. Of course it must! At its extreme positions, it has only gravitational potential energy. Since gravitational potential energy is mgh regardless of the details of the path taken to rise a distance h, the heights must be the same on both sides of the pendulum's arc.

Will the string really remain taut, as we've assumed? Not always! Problems 33 and 80 explore this question.

How long does it take the pendulum to complete its swing? That's a much harder question, one that energy conservation cannot answer. We'll solve Newton's second law for the pendulum in Chapter 15; only then can we determine the time dependence of the motion.

EXERCISE   A wrecking ball is used to demolish a building. The 600-kg ball starts from rest, with its 37-m-long cable making a 22° angle with the vertical. It strikes the building when the cable is vertical. What is the speed of the ball on impact?

Answer: 7.3 m/s

Some problems similar to Example 8-6:   24, 27, 33, 80   ●

In the examples of this section we've neglected friction and other nonconservative forces like air resistance. Under those circumstances the principle of mechanical energy conservation is exactly true. But in reality nonconservative forces are nearly always present, and in Section 8-6 we show how to deal with them. Often, though, the energy lost to nonconservative forces is small enough that it's a reasonable approximation to assume energy is strictly conserved.

8-4 POTENTIAL ENERGY CURVES

Figure 8-14 shows a frictionless roller-coaster track. How fast must a car be coasting at point A if it's to reach point D? Conservation of energy provides the answer. To get to D, the car must clear peak C. Clearing C requires that its total energy exceed its potential energy at C; that is, $\frac{1}{2}mv_A^2 + mgh_A > mgh_C$, where we've taken the zero of potential energy at the bottom of the track. Solving for v_A gives $v_A > \sqrt{2g(h_C - h_A)}$. If v_A satisfies this inequality, the car will reach C with some kinetic energy remaining and will coast over the peak.

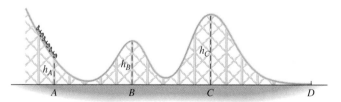

FIGURE 8-14   A roller-coaster track.

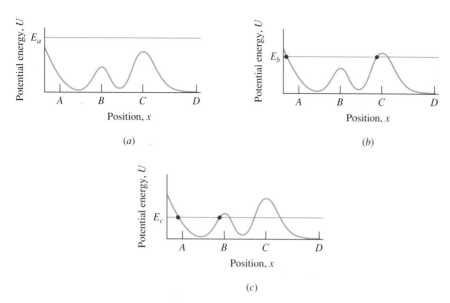

FIGURE 8-15   Potential and total energy for a car on a roller coaster. (*a*) The total energy E_a exceeds the potential energy at peak *C*, and the car can move over the entire track. (*b*) The total energy E_b is lower than the potential energy at *C*, and the car's motion is confined between the turning points (red dots) where the total and potential energy curves intersect. (*c*) The total energy E_c is lower than the potential energy at *B*, further limiting the car's motion.

What if the car is moving a little more slowly at *A*? Then it will stop at some point to the left of *C* where its energy is entirely potential—that is, at a height *h* such that $\frac{1}{2}mv_A^2 + mgh_A = mgh$. Then it will head back, clearing peak *B* and again coming to a stop at the height *h*. So it will run back and forth between two **turning points** set by the value of its total energy. Lowering the speed still further would confine the car to the valley between points *A* and *B*.

Figure 8-14 is a drawing of the actual roller-coaster track. But because gravitational potential energy is directly proportional to height, it's also a plot of potential energy versus position: a **potential energy curve.** We can study the car's motion by plotting its total energy on the same graph as the potential energy curve. Since total energy is constant, the total energy plot is a straight line. Figure 8-15 shows the potential energy and total energy for several values of the total energy. In Fig. 8-15*a*, the total energy exceeds the potential energy at *C*; therefore the car will reach *C* with kinetic energy to spare and will make it all the way to *D*. In Fig. 8-15*b*, the total energy is less than the potential energy at *C*, and the motion is confined between two turning points where the total energy and potential energy curves intersect. In Fig. 8-15*c* the energy is still lower, and now the car is confined to the leftmost valley.

Even though the car in Figs. 8-15*b* and *c* cannot get to *D*, its total energy still exceeds the potential energy at *D*. But it's blocked from reaching *D* by the **potential barrier** of peak *C*. We say that it's **trapped** in a **potential well** between its turning points.

Potential energy curves are useful even with nongravitational forces where there is no direct correspondence with hills and valleys. The terminology used here—potential barriers, wells, and trapping—remains appropriate in such cases and indeed is widely used throughout physics.

■ **APPLICATION**   KEEPING A MOLECULE TOGETHER

Figure 8-16 shows the potential energy of a pair of hydrogen atoms as a function of their separation. This energy is associated with attractive and repulsive electrical forces involving the electrons and the nuclei of the two atoms. The potential energy curve exhibits a potential well, showing that the atoms can form a **bound system** in which they are unable to separate fully. That bound system is a hydrogen molecule (H_2). The minimum energy, -7.6×10^{-19} J, corresponds to a separation of 0.074 nm; this is the equilibrium separation for the H_2 molecule. It's convenient to define the zero of potential energy when the atoms are infinitely far apart; Fig. 8-16 then shows that any total energy less than zero results in a bound system. At the energy marked E_1, for example, the atomic separation is limited to a narrow range defined by the interval where the $E = E_1$ line is above the potential energy curve.

What if the total energy is greater than zero? Then, as the line marked E_2 shows, the energy lies above the top of the potential well, and therefore, the atoms can be arbitrarily far apart. At positive energy the atoms are no longer bound together to form a molecule. So this potential energy curve allows a pair of hydro-

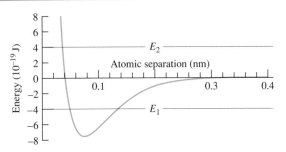

FIGURE 8-16   Potential energy curve for two hydrogen atoms. At total energy E_1 the atoms are trapped in the potential well, forming a hydrogen molecule (H_2). For energy E_2 and any other total energy above zero, the two atoms can move to arbitrarily large separations, so they don't constitute a molecule.

gen atoms two possibilities: Total energy less than zero means they're trapped in the potential well and form a molecule, while total energy above zero gives two separate, unbound atoms.

Readers of the extended version of this text will encounter many more details of molecular energy in Chapter 41.

● **EXAMPLE 8-7**   MOLECULAR ENERGY

Very near the bottom of the potential well in Fig. 8-16, the potential energy of the two hydrogen atoms is given approximately by $U = U_0 + a(x - x_e)^2$, with $U_0 = -7.6 \times 10^{-19}$ J, $a = 2.86 \times 10^{-16}$ J/nm^2 and $x_e = 0.0741$ nm. What range of atomic separations is allowed if the total energy is -7.17×10^{-19} J?

Figure 8-17 shows graphically what we want, namely, the points where the line $E = 0.40 \times 10^{-19}$ J intersects the curve $U = U_0 + a(x - x_e)^2$. We can find those points algebraically by writing $E = U_0 + a(x - x_e)^2$ and solving to get

$$x - x_e = \pm \sqrt{\frac{E - U_0}{a}}$$

$$= \pm \sqrt{\frac{-7.17 \times 10^{-19} \text{ J} - (-7.6 \times 10^{-19} \text{ J})}{2.86 \times 10^{-16} \text{ J/nm}^2}}$$

$$= \pm 0.0123 \text{ nm}.$$

With $x_e = 0.0741$ nm, this gives

$$x = x_e \pm 0.0123 \text{ nm} = 0.086 \text{ nm or } 0.062 \text{ nm}.$$

In working this example we've assumed the atoms behave like classical particles. In fact, as we'll see in later chapters, quantum mechanics constrains the energies of the system to certain fixed values. Our choice of $E = 7.17 \times 10^{-19}$ J happens to be one of those values.

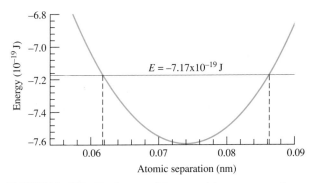

FIGURE 8-17   The bottom of the potential energy curve for two hydrogen atoms approximates a parabola. With total energy -7.17×10^{-19} J, the atoms are constrained to separations between 0.62 nm and 0.86 nm.

EXERCISE   A particle is trapped in a potential well described by $U = a|x|$, where x is meters and $a = 3.8$ J/m. If the particle has total energy 5.0 J, find the turning points for its motion.

Answer: $x = \pm 1.3$ m

Some problems similar to Example 8-7:   38–41   ●

8-5 FORCE AND POTENTIAL ENERGY

Figure 8-16 describes the potential energy of a hydrogen molecule as a function of atomic separation. But it also tells us about the force acting between the atoms. As they get very close—to the left of the minimum energy point in Fig 8-16—they experience a repulsive force. This is evident because the potential energy increases with decreasing distance, showing that you have to do work to move the atoms closer. How strong is that force? That depends on *how rapidly* the potential energy changes with position. If we change the separation by a small distance Δx—so small that the force F_x is essentially constant over this distance—then we can use Equation 8-2b to write $\Delta U = -F_x \Delta x$, or $F_x = -\Delta U/\Delta x$. In the limit $\Delta x \to 0$, this equation becomes exact, and $\Delta U/\Delta x$ becomes the derivative, and we have

$$F_x = -\frac{dU}{dx}. \tag{8-9}$$

This equation makes mathematical as well as physical sense; we've already written potential energy as the *integral* of force over distance, so it's no surprise that force is the *derivative* of potential energy.

We were careful to write F_x in Equation 8-9 because the rate at which the potential energy changes with position in the x direction tells only about the force component in that direction. The y and z components of force are, in analogy with Equation 8-9, the derivatives of the potential energy with respect to y and z. Taking derivatives with respect to all three coordinates thus determines the entire force vector.

In Fig. 8-16 we see that dU/dx is negative to the left of the energy minimum; Equation 8-9 then shows that the force is positive, or to the right—indicating a repulsion of the two atoms that tends to increase their separation. To the right of the minimum dU/dx is positive, and the force is to the left, showing that the atoms attract each other. What about the minimum itself? Here the slope dU/dx is zero, so there's no force on the atoms. That's the one place where they can remain at rest. We'll explore such **equilibrium points** further in Chapter 14.

● **EXAMPLE 8-8** INTERATOMIC FORCES

Find the force on the hydrogen atoms at the turning points given in Example 8-7.

Solution

Example 8-7 gives $U = a(x - x_e)^2$ for the potential energy near the bottom of the potential well. We then find the force from Equation 8-9:

$$F_x = -\frac{dU}{dx} = -\frac{d}{dx}[U_0 + a(x - x_e)^2] = -2a(x - x_e).$$

In Example 8-7 we found $x - x_e = \pm 0.0123 \times 10^{-9}$ m; then we have

$$F_x = (-2)(2.86 \times 10^{-16} \text{ J/nm}^2)(\pm 0.0123 \text{ nm}) = \mp 7.0 \text{ nN}.$$

The minus sign goes with the point to the right of the energy minimum, indicating a leftward (attractive) force, and vice versa.

EXERCISE The energy stored in a rubber band is given approximately by $U(x) = ax^2 - bx^3$, where $a = 90$ J/m^2, $b = 750$ J/m^3, and x is the stretch in meters. Find an expression for the force exerted by the rubber band as a function of stretch, and evaluate your result for a stretch of 2.5 cm.

Answers: $F = -2ax + 3bx^2$; $F(x = 2.5 \text{ cm}) = -3.1$ N (The minus sign indicates that the force opposes the stretch.)

Some problems similar to Example 8-8: 45, 47, 48

●

8-6 NONCONSERVATIVE FORCES

We developed the conservation of energy principle and the notion of potential energy on the assumption that only conservative forces were acting. But with nonconservative forces like friction, mechanical energy is not conserved and we must go back to Equation 8-6:

$$\Delta(K + U) = W_{nc}, \qquad (8\text{-}6)$$

where K is the kinetic energy, U the potential energy associated with any conservative forces acting, and W_{nc} the work done by nonconservative forces. It makes no sense to define a potential energy associated with nonconservative forces since these forces do not "give back" work done against them.

If we have an expression for W_{nc} in Equation 8-6, we can use that equation as we did the conservation of energy principle to provide a shortcut to the solution of many practical problems.

● **EXAMPLE 8-9**   LOUSY SKIING
_____

A cross-country skier moving at 4.8 m/s on level ground encounters a nearly frictionless downward slope 6.1 m high. On the level ground below, the snow has been worn thin, and consequently, the coefficient of friction is 0.27. After coasting down the hill, how far will the skier glide across the level stretch?

Solution

The skier starts with mechanical energy $K + U = \frac{1}{2}mv_0^2 + mgh$, where v_0 is the speed at the top of the slope and h the height of the hill and where we've taken the zero of potential energy at the bottom of the slope. When the skier has stopped on the level stretch, both the kinetic and potential energy are zero; the total change in mechanical energy is therefore $\Delta(K + U) = -(\frac{1}{2}mv_0^2 + mgh)$. The work done by friction is the frictional force times the displacement. On level ground, the frictional force has magnitude μmg; if we take the x direction as the direction of motion, the work is then $-\mu mg \Delta x$, where the minus sign arises because the force is opposite the direction of motion. Then Equation 8-6, $\Delta(K + U) = W_{nc}$, becomes

$$-(\tfrac{1}{2}mv_0^2 + mgh) = -\mu mg \Delta x.$$

so

$$\Delta x = \frac{\frac{1}{2}v_0^2}{\mu g} + \frac{h}{\mu} = \frac{(\frac{1}{2})(4.8 \text{ m/s})^2}{(0.27)(9.8 \text{ m/s}^2)} + \frac{6.1 \text{ m}}{0.27} = 27 \text{ m}.$$

Does this result make sense? Yes. As μ gets very small, Δx grows arbitrarily large, showing that the skier would coast forever in the absence of friction.

EXERCISE   At the end of a factory production line, boxes are placed on a 30° ramp of height 2.7 m, as shown in Fig. 8-18. They slide down the ramp and drop into a waiting truck, whose bed lies 2.4 m below the end of the ramp. If a 10-kg box has kinetic energy 380 J when it lands in the truck, what is the coefficient of friction on the ramp?

Answer: 0.26

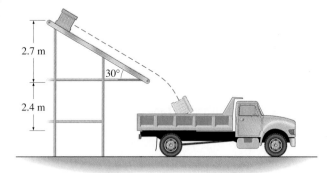

FIGURE 8-18   What is the coefficient of friction on the ramp?

Some problems similar to Example 8-9:   50, 52, 59, 73, 74   ●

8-7 CONSERVATION OF ENERGY AND MASS-ENERGY

When conservative forces act, we've seen that mechanical energy is conserved—although it may change from kinetic to potential energy and vice versa. Those are only two of many forms energy can take. With nonconservative forces, the energy that is apparently "lost" changes into forms we don't usually associate with mechanical energy, like the random molecular motions called, loosely, heat. Fuels like gasoline store energy in the chemical bonds of their molecules, energy ultimately associated with electric forces. The even stronger forces binding atomic nuclei result in correspondingly greater energy storage. Electric and magnetic energy together make electromagnetic waves that, like the light from the Sun, carry energy over great distances.

Much of physics and engineering involves identifying or controlling the interchanges of energy among its different forms. In classical physics, one overriding principle governs those interchanges: the principle of **conservation of energy.** If we keep track of all forms, we find that energy is never lost in any interchange.

Alongside conservation of energy is the principle of conservation of mass. But a closer look shows that neither principle stands by itself. If we measure the mass of a system before it emits energy, and again afterwards, in principle we will find the mass has decreased. Einstein's famous equation $E = mc^2$ describes this effect, saying that a quantity of mass m is equivalent to an amount of energy mc^2, where c is the speed of light.

Conversion of matter to energy is popularly associated with nuclear energy. That misconception does injustice to Einstein's universal statement of mass-energy equivalence. *Any* time a system loses energy E, its mass drops by E/c^2. This is as true in burning gasoline or in releasing the tension on a stretched spring as it is in fissioning uranium; it's just that nuclear reactions convert a far greater portion of their mass to energy, and so the effect is readily measurable only for nuclear reactions. Under extremely high-energy conditions it's even more obvious, as particles of matter form out of pure energy and annihilate to give again pure energy (Fig. 8-19). If we could extract all the energy mc^2 in a single raisin, it would be enough to supply the energy needs of a major city for a day!

Einstein's mass-energy equivalence replaces the two separate conservation statements for mass and energy with a single statement: the **conservation of mass-energy.** Interchanges occur not only among different forms of energy but also among mass and energy, conserving only the total mass-energy. In this sense matter and energy are aspects of the same basic "stuff" that makes up our universe. We will see how mass-energy equivalence arises when we study the theory of relativity in Chapter 38.

FIGURE 8-19 The creation of matter out of energy. In this image from the Fermi National Accelerator Laboratory, a high-energy proton and antiproton have collided head-on. Each of the colored tracks represents a new particle created from the energy of original proton-antiproton pair.

CHAPTER SYNOPSIS

Summary

1. A **conservative force** "stores" work done against it. Mathematically, a conservative force is one that does zero work on an object moved around any closed path:

$$\oint \mathbf{F} \cdot d\mathbf{r} = 0. \quad \text{(conservative force)}$$

A corollary is that the work done by a conservative force as an object is moved between two points is independent of the path taken. Gravity is a familiar example of a conservative force.

2. With a **nonconservative force,** like friction, the work done by the force is not path-independent, and the total work done on an object describing a closed path need not be zero.

3. **Potential energy** describes the "stored work" associated with a conservative force. Mathematically, the potential energy difference as an object moves from point A to point B is defined as the negative of the work done on the object by the conservative force:

$$\Delta U_{A \to B} = -\int_A^B \mathbf{F} \cdot d\mathbf{r}. \qquad \text{(potential energy)}$$

Only potential energy differences have physical significance. We are free to assign zero potential energy to any point we choose; when we then speak of the potential energy U at some other point, we really mean the potential energy difference between that point and the point where $U = 0$.

The form of the potential energy function depends on the specific conservative force involved. Two commonly encountered cases include gravitational potential energy near Earth's surface:

$$U = mgh, \qquad \text{(gravitational potential energy)}$$

and the potential energy of an ideal spring:

$$U = \tfrac{1}{2}kx^2. \qquad \text{(ideal spring)}$$

4. When only conservative forces act, the **mechanical energy**—the sum $K + U$ of the kinetic and potential energies—remains constant. This statement is the principle of **conservation of mechanical energy.** The conservation of energy principle allows us to solve easily problems that would be difficult to solve using Newton's second law.

5. Graphs of potential energy as a function of position—**potential energy curves**—reveal features of an object's motion. In particular, from such curves we can identify **turning points** where kinetic energy goes to zero; when there are two such points, the object is confined to a **potential well** by **potential barriers** on either side.

6. Potential energy is related to the integral of force over distance; inversely, force is the negative of the derivative of potential energy:

$$F_x = -\frac{dU}{dx}.$$

7. Energy takes many forms besides kinetic and potential. The **conservation of energy** principle states that energy is neither created nor destroyed as it undergoes changes among the various forms.

8. Einstein showed that energy and matter are interchangeable according to the equation $E = mc^2$. His theory of relativity therefore replaces the classical principles of conservation of energy and conservation of matter with the single **conservation of mass-energy** principle.

Terms You Should Understand

(Pairs are closely related terms whose distinction is important; number in parentheses is chapter section where term first appears.)

conservative force, nonconservative force (8-1)
potential energy (8-2)
mechanical energy (8-3)
conservation of mechanical energy (8-3)
potential energy curve (8-4)
turning point (8-4)
potential well (8-4)
potential barrier (8-4)
conservation of energy (8-7)
conservation of mass-energy (8-7)

Symbols You Should Recognize

ΔU, U (8-2)
$\Delta(K + U)$ (8-3)

Problems You Should Be Able to Solve

calculating gravitational potential energy near Earth's surface (8-2)
calculating the potential energy of a spring (8-2)
calculating potential energy given simple force laws (8-2)
applying conservation of mechanical energy in situations involving gravity, springs, or other simple forces alone or in combination (8-3)
evaluating motion given potential energy curves and total energy; determining turning points (8-4)
calculating force given potential energy as a function of position (8-5)
solving problems involving both nonconservative and conservative forces (8-6)

Limitations to Keep in Mind

Potential energy can be computed as the negative of the product of force and distance only when the force does not vary with position. Otherwise, integration must be used to evaluate the potential energy.

The simple expression $\Delta U = mg\,\Delta y$ for gravitational potential energy changes applies only near Earth's surface, for Δy small compared with Earth's radius.

The expression $U = \tfrac{1}{2}kx^2$ applies to ideal springs whose force-distance curve is linear; real springs deviate from this law, especially when stretched or compressed substantially.

Mechanical energy is conserved only in the absence of non-conservative forces like friction or air resistance. In reality these forces are almost always present, so mechanical energy generally decreases.

Conservation of energy is itself only approximately true; accurate treatment of highly energetic processes, especially, requires the principle of conservation of mass-energy.

QUESTIONS

1. Figure 8-20 shows force vectors at different points in space for two forces. Which is conservative and which nonconservative? Explain.

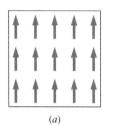 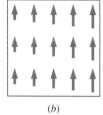

(a)   (b)

FIGURE 8-20 Question 1, Problem 3.

2. Is the conservation of energy principle related to Newton's laws, or is it an entirely separate physical principle? Discuss.
3. Why can't we define a potential energy associated with friction?
4. Why are you free to choose the zero of potential energy?
5. Can potential energy be negative? Can kinetic energy? Can total mechanical energy? Explain.
6. If the potential energy is zero at a point, must the force also be zero at that point? Give an example.
7. If the force is zero at a point, must the potential energy also be zero at that point? Give an example.
8. Explain how path independence of the work done by a conservative force follows from the fact that the work done over a closed path is zero.
9. If the difference in potential energy between two points is zero, does that necessarily mean that an object moving between those points experiences no force?
10. A tightrope walker follows an essentially horizontal rope between two mountain peaks of equal altitude. A climber descends from one peak and climbs the other. Compare the work done by the gravitational force on the tightrope walker and on the climber.
11. A bowling ball is tied to the end of a long rope and suspended from the ceiling. A student climbs a stool at one side of the room and holds the ball to her nose, then releases it from rest (Fig. 8-21a). Should she duck as it swings back (Fig. 8-21b)? Argue from conservation of energy.
12. An avalanche thunders down a mountainside. Discuss the various energy transfers that resulted in water from the ocean eventually ending up as snow in the avalanche.

(a)   (b)

FIGURE 8-21 Question 11.

13. Could you define a potential energy function for a velocity-dependent force?
14. A block of mass m is held against a spring of constant k while the spring is compressed a distance x. Discuss how the speed of the block when released scales with k, x, and m.
15. Figure 8-22 shows a potential energy curve along with total energies for three particles. If the particles are all initially at x_0 and moving to the right, discuss qualitatively their subsequent motions.

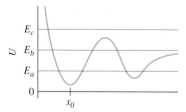

FIGURE 8-22 Question 15.

16. If conservation of energy is a law of nature, why do we have programs—like mileage requirements for cars or insulation standards for buildings—designed to encourage energy conservation?

17. If potential energy is an even function, meaning that $U(x) = U(-x)$, what can you conclude about the force acting at the origin?

18. Why is it a misconception to apply $E = mc^2$ only to nuclear reactions?

19. High-energy physicists often give the mass of elementary particles in energy units. Explain.

20. You use a microwave oven to boil a cup of coffee. Assuming that your electricity comes from a hydroelectric power plant, trace the energy of your hot coffee as far back as you can.

21. In terms of potential energy, how do a new and a rundown battery differ?

PROBLEMS

Section 8-1 Conservative and Nonconservative Forces

1. Determine the work done by the frictional force in moving a block of mass m from point 1 to point 2 over the two paths shown in Fig. 8-23. The coefficient of friction has the constant value μ over the surface. (The diagram lies in a horizontal plane.)

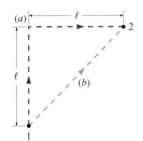

FIGURE 8-23   Problem 1.

2. Now take Fig. 8-23 to lie in a vertical plane, and find the work done by the gravitational force as an object moves from point 1 to point 2 over each of the paths shown.

3. The force in Fig. 8-20a is given by $\mathbf{F} = F_0\hat{\mathbf{j}}$, where F_0 is a constant. The force in Fig. 8-20b is given by $\mathbf{F} = F_0(x/a)\hat{\mathbf{j}}$, where the origin is taken at the lower left corner of the box, a is the width of the square box, and the distance x increases horizontally to the right. Determine the work done in going counterclockwise around each box, starting at the lower left corner.

Section 8-2 Potential Energy

4. Rework Example 8-1, now taking the zero of potential energy at street level.

5. Find the potential energy of a 9000-kg camper when it's (a) atop New Hampshire's Mount Washington, 1900 m above sea level, and (b) in Death Valley, California, 86 m below sea level. Take the zero of potential energy at sea level.

6. An incline makes an angle θ with the horizontal. Find the gravitational potential energy associated with a mass m located a distance x measured along the incline. Take the zero of potential energy at the bottom of the incline.

7. Show using Equation 8-2b that the potential energy difference between the ground and a distance h above the ground is mgh regardless of whether you choose the y axis upward or downward.

8. The top of the volcano Haleakala on Maui, Hawaii, is 3050 m above sea level and 18 km inland from the sea. By how much does your gravitational potential energy change as you come down from the mountain-top observatory to swim in the ocean? Assume your mass is 75 kg.

9. A 1.50-kg brick measures 20.0 cm × 8.00 cm × 5.50 cm. Taking the zero of potential energy when the brick lies on its broadest face, what is the potential energy (a) when the brick is standing on end and (b) when it is balanced on its 8-cm edge, with its center directly above that edge? *Note:* You can treat the brick as though all its mass is concentrated at its center.

10. A 60-kg hiker ascending 1250-m-high Camel's Hump mountain in Vermont has potential energy -2.4×10^5 J; the zero of potential energy is taken at the mountain top. What is her altitude?

11. How much energy can be stored in a spring with $k = 340$ N/m if the maximum allowed stretch is 21 cm?

12. A carbon monoxide molecule can be modeled as a carbon atom and an oxygen atom connected by a spring. If a displacement of the carbon by 1.6×10^{-12} m from its equilibrium position relative to the oxygen increases the molecule's potential energy by 0.015 eV, what is the spring constant?

13. How far would you have to stretch a spring of spring constant $k = 1700$ N/m until it stored 250 J of energy?

14. A more accurate expression for the force law of the rope in Example 8-3 is $F = -kx + bx^2 - cx^3$, where k and b have the values given in Example 8-3, and $c = 3.1$ N/m³. Find the energy stored in stretching the rope 2.62 m. By what percentage does your result differ from that of Example 8-3?

15. The force exerted by an unusual spring when it's compressed a distance x from equilibrium is given by $F = -kx - cx^3$, where $k = 130$ N/m and $c = 2.9\times10^3$ N/m³. Find the energy stored in this spring when it's been compressed 8.7 cm.

16. The force on a particle is given by $\mathbf{F} = A\hat{\mathbf{i}}/x^2$, where A is a positive constant. (a) Find the potential energy difference

between two points x_1 and x_2, where $x_1 > x_2$. (b) Show that the potential energy difference remains finite even when $x_1 \rightarrow \infty$.

17. A particle moves along the x axis under the influence of a force $F = ax^2 + b$, where a and b are constants. Find its potential energy as a function of position, taking $U = 0$ at $x = 0$.

18. A 3.0-kg fish is hanging from a spring scale whose spring constant is 240 N/m. (a) What is the potential energy of the spring? (b) If the fish were moved slowly upward to the equilibrium position of the spring, by how much would its gravitational potential energy change? (c) In case (b), by how much would the spring's potential energy change? Explain any apparent discrepancies.

19. The force exerted by a rubber band is given approximately by

$$F = F_0 \left[\frac{\ell + x}{\ell} - \frac{\ell^2}{(\ell + x)^2} \right],$$

where ℓ is the unstretched length and F_0 a constant. Find the potential energy of the rubber band as a function of the distance x it is stretched. Take the zero of potential energy in the unstretched position.

Section 8-3 Conservation of Mechanical Energy

20. A skier starts down a frictionless 30° slope. After a vertical drop of 22 m, the slope temporarily levels out, then drops at 20° an additional 31 m vertically before leveling out again. What is the skier's speed on the two level stretches?

21. A Navy jet of mass 10,000 kg lands on an aircraft carrier and snags a cable to slow it down. The cable is attached to a spring with spring constant 40,000 N/m. If the spring stretches 25 m to stop the plane, what was the landing speed of the plane?

22. A spring of constant k, compressed a distance x, is used to launch a mass m up a frictionless slope that makes an angle θ with the horizontal. Find an expression for the maximum distance along the slope that the mass moves after leaving the spring.

23. A 120-g arrow is shot vertically from a bow whose effective spring constant is 550 N/m. If the bow is drawn 64 cm before shooting the arrow, to what height does the arrow rise?

24. A child is on a swing whose 3.2-m-long chains make a maximum angle of 50° with the vertical. What is the child's maximum speed?

25. Derive Equation 2-14 using the conservation of energy principle.

26. In a switchyard, freight cars start from rest and roll down a 2.8-m incline and come to rest against a spring bumper at the end of the track (Fig. 8-24). If the spring constant is 4.3×10^6 N/m, how much is the spring compressed when hit by a 57,000-kg freight car?

FIGURE 8-24   Problem 26.

27. Rework the exercise following Example 8-6, now making the more realistic assumption that the ball swings through its lowest position to a 10° angle with the vertical when it hits the building.

28. A 200-g block slides back and forth on a frictionless surface between two springs, as shown in Fig. 8-25. The left-hand spring has $k = 130$ N/m and its maximum compression is 16 cm. The right-hand spring has $k = 280$ N/m. Find (a) the maximum compression of the right-hand spring and (b) the speed of the block as it moves between the springs.

FIGURE 8-25   Problem 28.

29. An initial speed of 2.4 km/s (the "escape speed") is required for an object launched from the moon to get arbitrarily far from the moon. At a mining operation on the moon, 1000-kg packets of ore are to be launched to a smelting plant in orbit around the Earth. If they are launched with a large spring whose maximum compression is 15 m, what should be the spring constant of the spring?

30. A runaway truck lane heads uphill at 30° to the horizontal. If a 16,000-kg truck goes out of control and enters the lane going 110 km/h, how far along the ramp does it go? Neglect friction.

31. A low-damage bumper on a 1500-kg car is mounted on springs whose total effective spring constant is 8.0×10^5 N/m. The springs can undergo a maximum compression of 18 cm without damage to the bumper, springs, or car. What is the maximum speed at which the car can collide with a stationary object without sustaining damage?

32. A block slides on the frictionless loop-the-loop track shown in Fig. 8-26. What is the minimum height h at which it can start from rest and still make it around the loop?

33. Show that the pendulum string in Example 8-6 can remain taut all the way to the top of its smaller loop only if $a \leq \frac{2}{5}\ell$. (Note that the maximum release angle is 90° for the string to be taut on the way down.)

34. The maximum speed of the pendulum bob in a grandfather clock is 0.55 m/s. If the pendulum makes a maximum angle of 8.0° with the vertical, what is the length of the pendulum?

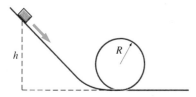

FIGURE 8-26 Problem 32.

35. A 2.0-kg mass rests on a frictionless table and is connected over a frictionless pulley to a 4.0-kg mass, as shown in Fig. 8-27. Use conservation of energy to calculate the speed of the masses after they have moved 50 cm.

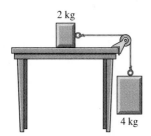

FIGURE 8-27 Problem 35.

36. The masses shown in Fig. 8-28 are connected by a massless string over a frictionless, massless pulley and are released from rest. Use energy conservation to find (a) the velocity of the 7.0-kg mass just before it hits the floor, (b) the maximum height reached by the 4.0-kg mass, and (c) the fraction of the system's initial mechanical energy lost when the 7.0-kg mass comes to rest on the floor.

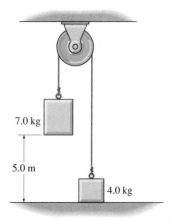

FIGURE 8-28 Problem 36.

37. A mass m is dropped from a height h above the top of a spring of constant k that is mounted vertically on the floor (Fig. 8-29). Show that the maximum compression of the spring is given by $(mg/k)(1 + \sqrt{1 + 2kh/mg})$. What is the significance of the other root of the quadratic equation?

FIGURE 8-29 Problem 37.

Section 8-4 Potential Energy Curves

38. A particle slides along the frictionless track shown in Fig. 8-30, starting at rest from point A. Find (a) its speed at B, (b) its speed at C, and (c) the approximate location of its right-hand turning point.

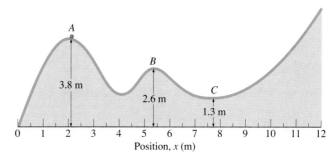

FIGURE 8-30 Problem 38.

39. A particle slides back and forth on a frictionless track whose height as a function of horizontal position x is given by $y = ax^2$, where $a = 0.56\ m^{-1}$. If the particle's maximum speed is 2.3 m/s, find the turning points of its motion.

40. A particle with total energy 3.0 J is trapped in a potential well described by $U = 6.0 - 6.0x + 1.5x^2$, where U is in joules and x in meters. Find its turning points.

41. The potential energy associated with a conservative force is shown in Fig. 8-31. Consider particles with total energies $E_1 = -1.5$ J, $E_2 = -0.5$ J, $E_3 = 0.5$ J, $E_4 = 1.5$ J, and $E_5 = 3.0$ J. Discuss the subsequent motion, including the approximate location of any turning points, if the particles are initially at point $x = 1$ m and moving in the $-x$ direction.

42. Make an accurate potential energy curve, covering the region $-8\ m < x < 8$ m, for potential energy $U = (ax^2 - b)e^{-x^2/c^2}$, where $a = 1.5$ J/m^2, $b = 5.0$ J, and $c = 3.0$ m. Discuss the subsequent motion of 1-kg particles starting from the origin and moving initially in the $+x$ direction with total energies of -3 J, 1 J, and 4 J. Include the location of any turning points. Determine also the speed of the highest energy particle when it is a great distance from the origin.

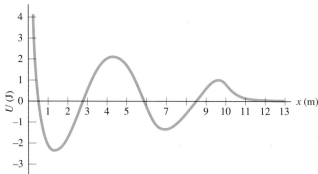

FIGURE 8-31 Problem 41.

43. (a) Derive an expression for the potential energy of an object subject to a force $F_x = ax - bx^3$, where $a = 5$ N/m and $b = 2$ N/m^3, taking $U = 0$ at $x = 0$. (b) Graph the potential energy curve for $x > 0$, and use it to find the turning points for an object whose total energy is -1 J.

Section 8-5 Force and Potential Energy

Note: In the following problems, motion is restricted to one dimension.

44. Figure 8-32 shows the potential energy curve for a certain particle. Find the force on the particle at each of the curve segments shown.

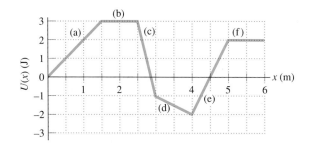

FIGURE 8-32 Problem 44.

45. A particle is trapped in a potential well described by $U(x) = 1.6x^2 - 4$, where U is in joules and x is in meters. Find the force on the particle when it's at (a) $x = 2.1$ m, (b) $x = 0$ m, and (c) $x = -1.4$ m.

46. Figure 8-33 shows the potential energy curve for a particle. At which of the labeled points (a) does the force have the greatest magnitude; (b) does the force point in the negative x direction; (c) is the force zero; (d) are the force and potential energy both zero?

47. The potential energy associated with a certain conservative force is given by $U = bx^2$, where b is a constant. Show that the force always tends to accelerate a particle toward the origin if b is positive and away from the origin if b is negative.

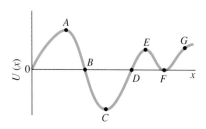

FIGURE 8-33 Problem 46.

48. A more accurate expression for the potential well in Fig. 8-17 than that provided in Example 8-7 is $U = 286(x - x_e)^2 - 6.22 \times 10^{12}(x - x_e)^3$ with U in joules and x in meters. Find the force on the hydrogen atoms when they are 0.10 nm apart.

49. The potential energy of a spring is given by $U = ax^2 - bx + c$, where $a = 5.20$ N/m, $b = 3.12$ N, and $c = 0.468$ J, and where x is the *overall* length of the spring (not the stretch). (a) What is the equilibrium length of the spring? (b) What is the spring constant?

Section 8-6 Nonconservative Forces

50. Repeat Problem 20 for the case when the coefficient of kinetic friction on both slopes is 0.11, while the level stretches remain frictionless.

51. A basketball dropped from a height of 2.00 m rebounds to a maximum height of 1.12 m. What fraction of the ball's initial energy is lost to nonconservative forces? Take the zero of potential energy at the floor.

52. A 1.5-kg block is launched up a 30° incline with an initial speed of 6.4 m/s. It comes to a halt after moving 3.4 m along the incline, as shown in Fig. 8-34. Find (a) the change in the block's kinetic energy; (b) the change in the block's potential energy; (c) the work done by friction; (d) the coefficient of kinetic friction.

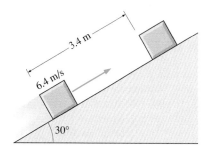

FIGURE 8-34 Problem 52.

53. A pumped-storage reservoir sits 140 m above its generating station and holds 8.5×10^9 kg of water. The power plant generates 330 MW of electric power while draining the reservoir over an 8.0-hour period. What fraction of the initial potential energy is lost to nonconservative forces (i.e., does not emerge as electricity)?

54. A spring of constant $k = 280$ N/m is used to launch a 2.5-kg block along a horizontal surface whose coefficient of sliding friction is 0.24. If the spring is compressed 15 cm, how far does the block slide?

55. A 2.5-kg block strikes a horizontal spring at a speed of 1.8 m/s, as shown in Fig. 8-35. The spring constant is 100 N/m. If the maximum compression of the spring is 21 cm, what is the coefficient of friction between the block and the surface on which it is sliding?

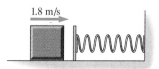

1.8 m/s

FIGURE 8-35 Problem 55.

56. A meteorite strikes Earth and embeds itself 1.7 m into the ground. Scientists dig up the meteorite and find that its mass is 400 g; they estimate that the ground exerted a retarding force of 10^6 N on the meteorite. Estimate the impact speed of the meteorite.

57. A surface is frictionless except for a region between $x = 1$ m and $x = 2$ m, where the coefficient of friction is given by $\mu = ax^2 + bx + c$, with $a = -2$ m^{-2}, $b = 6$ m^{-1}, and $c = -4$. A block is sliding in the $+x$ direction when it encounters this region. What is the minimum speed it must have to get all the way across the region?

58. A biologist uses a spring-loaded dart gun to shoot a 50-g tranquilizing dart into an elephant 21 m away. The gun's spring has spring constant $k = 690$ N/m and is pulled back 14 cm to launch the dart. The dart embeds itself 2.2 cm in the elephant. (a) What is the average stopping force exerted on the dart by the elephant's flesh? (b) How long does it take the dart to reach the elephant? Assume the dart's trajectory is nearly horizontal.

59. A skier starts from rest at the top of the left-hand peak in Fig. 8-36. What is the maximum coefficient of kinetic friction on the slopes that would allow the skier to coast to the second peak? (Your answer, of course, neglects air resistance.)

60. A bug slides back and forth in a hemispherical bowl of 11 cm radius, starting from rest at the top, as shown in Fig. 8-37. The bowl is frictionless except for a 1.5-cm-wide sticky patch at the bottom, where the coefficient of friction is 0.87. How many times does the bug cross the sticky region?

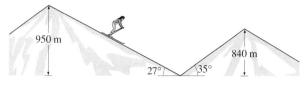

950 m

840 m

27°   35°

FIGURE 8-36 Problem 59.

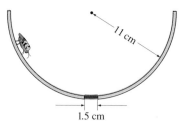

11 cm

1.5 cm

FIGURE 8-37 Problem 60.

61. A 190-g block is launched by compressing a spring of constant $k = 200$ N/m a distance of 15 cm. The spring is mounted horizontally, and the surface directly under it is frictionless. But beyond the equilibrium position of the spring end, the surface has coefficient of friction $\mu = 0.27$. This frictional surface extends 85 cm, followed by a frictionless curved rise, as shown in Fig. 8-38. After launch, where does the block finally come to rest? Measure from the left end of the frictional zone.

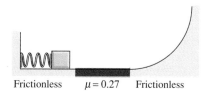

Frictionless $\mu = 0.27$ Frictionless

FIGURE 8-38 Problem 61.

Section 8-7 Conservation of Energy and Mass-Energy

62. Two deuterium nuclei fuse to form a helium nucleus. Each deuterium has mass 3.344×10^{-27} kg, and the helium has mass 6.645×10^{-27} kg. Find the energy released in this reaction.

63. A hypothetical power plant converts matter entirely into electrical energy. Each year, a worker at the plant buys a box of 1-g raisins, and each day drops one raisin into the plant's energy conversion unit. Estimate the average power output of the plant, and compare with a 500-MW coal-burning plant that consumes a 100-car trainload of coal every 3 days.

64. The Sun's total power output is 3.85×10^{-26} W. What is the associated rate at which the Sun loses mass?

Paired Problems

(Both problems in a pair involve the same principles and techniques. If you can get the first problem, you should be able to solve the second one.)

65. A block slides down a frictionless incline that terminates in a ramp pointing up at a 45° angle, as shown in Fig. 8-39. Find an expression for the horizontal range x shown in the figure, as a function of the heights h_1 and h_2 shown.

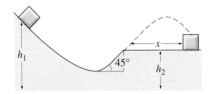

FIGURE 8-39   Problem 65.

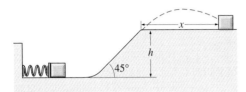

FIGURE 8-40   Problem 66.

66. A block of mass m is launched horizontally from a compressed spring on a frictionless track that turns upward at a 45° angle, as shown in Fig. 8-40. Find an expression for the horizontal range x shown in the figure, as a function of the distance d by which the spring is initially compressed and the height h of the ramp.

67. A ball of mass m is being whirled around on a string of length R in a vertical circle; the string does no work on the ball. (a) Show from force considerations that the speed at the top of the circle must be at least $\sqrt{Rg}$ if the string is to remain taut. (b) Show that, as long as the string remains taut, the speed at the bottom of the circle can be no more than $\sqrt{5}$ times the speed at the top.

68. An 840-kg roller-coaster car is launched from a giant spring of constant $k = 31$ kN/m into a frictionless loop-the-loop track of radius 6.2 m, as shown in Fig. 8-41. What is the minimum amount that the spring must be compressed if the car is to stay on the track?

FIGURE 8-41   Problem 68.

69. A pendulum consisting of a mass m on a string of length ℓ is pulled back so the string is horizontal, as shown in Fig. 8-42. The pendulum is then released. Find (a) the speed of the mass and (b) the magnitude of the string tension when the string makes a 45° angle with the horizontal.

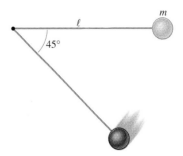

FIGURE 8-42   Problem 69.

70. A particle of mass m slides down a frictionless quarter-circular track of radius R as shown in Fig. 8-43. If it starts from rest at the top of the track, find (a) its speed and (b) the magnitude of the normal force exerted by the track when the radius vector from the center of the track to the particle makes a 45° angle, as shown.

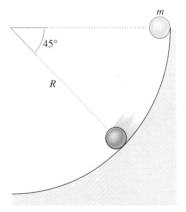

FIGURE 8-43   Problem 70.

71. A particle slides back and forth in a frictionless bowl whose height is given by $h(x) = 0.11x^2$, where x and h are both in meters. If the particle's maximum speed is 35 cm/s, find the x coordinates of its turning points.

72. A 1-kg particle slides on a frictionless track whose height is given by $y = ax^4 - bx^2$, where $a = 1$ m^{-3} and $b = 4$ m^{-1}. The particle's total energy is 20 J, where the zero of potential energy is at $y = 0$. Find the turning points of its motion.

73. A child sleds down a frictionless hill whose vertical drop is 7.2 m. At the bottom is a level but rough stretch where the coefficient of kinetic friction is 0.51. How far does she slide across the level stretch?

74. At the bottom of a frictionless ski slope is a 20-m wide stretch of rough snow where the coefficient of friction is 0.32. From what vertical height on the slope should a skier start at rest, in order to get through the rough patch and come out of it with a speed of 6.0 m/s?

Supplementary Problems

75. A uranium nucleus has a radius of 1.43×10^{-10} m. An alpha particle (mass 6.7×10^{-27} kg) leaves the surface of the nucleus with negligible speed, subject to a repulsive force whose magnitude is $F = A/x^2$, where $A = 4.1\times10^{-26}$ N·m² and where x is the distance from the alpha particle to the center of the nucleus. What is the speed of the alpha particle when it is (a) 4 nuclear radii from the nucleus; (b) 100 nuclear radii from the nucleus; (c) very far from the nucleus ($x \rightarrow \infty$)?

76. A mass m is attached to a spring of constant k that is hanging from the ceiling. (a) Taking the zero of potential energy with the spring in its normal unstretched position, derive an expression for the total potential energy (gravitational plus spring) as a function of distance y taken as positive downward. (b) Find the point where the potential energy is a minimum, and explain its significance. (c) Find a second point where the total potential energy is zero. Discuss its significance in terms of an experiment where you attach the mass to the unstretched spring and let go.

77. With the brick of Problem 9 standing on end, what is the minimum energy that can be given the brick to make it fall over?

78. A bug lands on top of the frictionless, spherical head of a bald man. It begins to slide down the head (Fig. 8-44). Show that the bug leaves the head when it has dropped a vertical distance one-third the radius of the head.

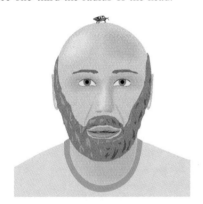

FIGURE 8-44 Problem 78.

79. Together, the springs in a 1200-kg car have an effective spring constant of 110,000 N/m and can compress a maximum distance of 40 cm. What is the maximum abrupt drop in road level (Fig. 8-45) that the car can tolerate without "bottoming out" —that is, without its springs reaching maximum compression? Assume the car is driving fast enough that it becomes temporarily airborne.

80. Show that the pendulum string in Example 8-6 will cease to be taut when the string has caught on the nail and makes an angle

FIGURE 8-45 Problem 79.

$$\theta = \cos^{-1}\left[\frac{2\ell}{3a}\left(\cos\theta_0 + \frac{a}{\ell} - 1\right)\right]$$

with the vertical. Show that your answer is consistent with that of Problem 33, in that when $\theta_0 = 90°$ and $a = \frac{2}{5}\ell$, the string remains taut all the way to the top of the small circle.

81. An electron with kinetic energy 0.85 fJ enters a region where its potential energy as a function of position is $U = ax^2 - bx$, where $a = 2.7$ fJ/cm² and $b = 4.2$ fJ/cm. (a) How far into the region does the electron penetrate? (b) At what position does the electron have its maximum speed? (c) What is this maximum speed?

82. A particle of mass m is subject to a force $\mathbf{F} = (a\sqrt{x})\mathbf{\hat{i}}$, where a is a constant. The particle is initially at rest at the origin, and is given a slight nudge in the positive x direction. Find an expression for the particle's speed as a function of position x.

83. (a) Repeat the previous problem for the case of a force $\mathbf{F} = (ax - bx^3)\mathbf{\hat{i}}$, where a and b are positive constants. (b) What is the significance of the negative square root that can occur for some values of x? (c) Find an expression for the particle's maximum speed.

84. A 17-m-long vine hangs vertically from a tree on one side of a 10-m-wide gorge, as shown in Fig. 8-46. Tarzan runs up, hoping to grab the vine, swing over the gorge, and drop vertically off the vine to land on the other side. At what minimum speed must he be running?

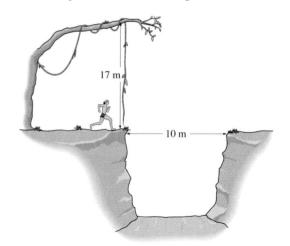

FIGURE 8-46 Problem 84.

85. A force points in the $-x$ direction with magnitude given by $F = ax^b$, where a and b are constants. Evaluate the potential energy as a function of position, taking $U = 0$ at some point $x_0 > 0$. Use your result to show that an object of mass m released at $x = \infty$ will reach x_0 with finite velocity provided $b < -1$. Find the velocity for this case.

86. A block slides on a horizontal surface with coefficient of sliding friction $\mu_k = 0.37$. It collides with a spring and stops at the point of maximum compression. If the block hit the spring at 1.79 m/s, and if the spring compressed 22 cm, and if these are the maximum speed and compression for which the block stops, show that the coefficient of static friction is twice the coefficient of sliding friction.

87. (a) Find the maximum speed of the particle in Problem 72. (b) At what position(s) does it occur?

88. The climbing rope described in Example 8-3 is securely fastened at its lower end. At the upper end is a 65-kg climber. At a height 2.4 m directly below the climber, the rope passes through a carabiner (essentially a frictionless metal loop), as shown in Fig. 8-47. If the climber falls, through what maximum total distance—including the

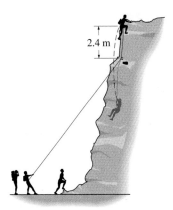

FIGURE 8-47 Before and after the fall. (Problem 88).

stretching of the rope—does he drop? *Hint:* Your statement of energy conservation should result in a cubic equation, which you can solve graphically, numerically, or analytically.

CHAPTER 9

GRAVITATION

Astronauts beginning repairs on the Hubble Space Telescope as it sits in the cargo bay of the space shuttle Endeavor in 1993. Precise application of Newton's laws of motion and gravity makes possible the space technology that is increasingly important in science, engineering, and communications.

Gravity is the most obvious of nature's fundamental forces. With your understanding of the laws of motion and mechanical energy, you are now in a position to explore some phenomena of gravitation. Historically, theories of gravity have led to clearer understandings of the nature and evolution of the universe beyond our planet Earth. With the advent of the space age, we've used our knowledge of gravity to explore the solar system and to engineer a host of new space-based technologies. In all but the most exotic astrophysical applications, we still use the theory of gravity developed by Isaac Newton in the late 1600s.

9-1 TOWARD A LAW OF GRAVITY

Newton's theory of gravity was the culmination of two centuries of scientific revolution that began in 1543 when the Polish astronomer Nicolaus Copernicus made his radical suggestion that the planets orbit not the Earth but the Sun. Before that time, views of the Greek philosopher-scientists Aristotle and Ptolemy dominated, holding that Earth was the center of the universe and the realm of imperfection (Fig. 9-1). Terrestrial objects naturally moved toward Earth. Celestial objects, in contrast, were perfect and moved only in perfect circles. Even Copernicus maintained this sharp distinction between celestial and terrestrial realms, holding that planetary motions involved only perfect circles.

FIGURE 9-1 In the Aristotelean scheme planets moved about Earth in perfect circles. To match the observed planetary behavior, Ptolemy posited additional circular motions about the main circular orbits.

Fifty years after Copernicus's work was published, the Danish noble Tycho Brahe began a program of accurate planetary observations. After Tycho's death in 1601, his assistant Johannes Kepler worked to make sense of the observations. Success came when Kepler took a radical step: He gave up the requirement that planetary motion involve only perfect circles. Kepler summarized his new insights in three laws of planetary motion:

Kepler's first law: **The planets orbit the Sun in ellipses, with the Sun at one focus (Fig. 9-2).**

Kepler's second law: **The line joining the Sun and a planet sweeps out equal areas in equal times (Fig. 9-3).**

Kepler's third law: **The square of a planet's orbital period is proportional to the cube of the semimajor axis of its orbit (Fig. 9-2).**

Kepler's laws were based solely on observation and had no theoretical basis. So Kepler knew *how* the planets moved but not *why* they did so.

Shortly after Kepler published his first two laws, Galileo trained his first telescopes on the heavens. Among his discoveries were four moons orbiting Jupiter,

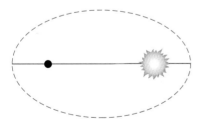

FIGURE 9-2 Kepler's first law states that planetary orbits are ellipses with the Sun at one focus; the other focus is a point in empty space. The ellipse shown here is highly exaggerated; all planetary orbits except Pluto's are very nearly circular. (A circle is an ellipse in which both foci coincide at the center.) Straight line is the major axis of the ellipse; half its length is the semimajor axis that figures in Kepler's third law.

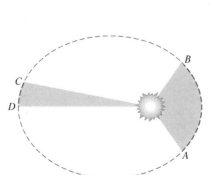

FIGURE 9-3 Kepler's equal-area law. If the planet takes the same time to go from A to B as from C to D, then the shaded areas are the same. This means the planet must be moving faster when it's closer to the Sun.

FIGURE 9-4 In an Earth-centered universe, Venus would not show the complete range of phases from crescent to full, and its apparent size would remain constant because of its constant distance from Earth. Galileo's observations of Venus provided strong evidence for Copernicus's heliocentric theory.

sunspots that blemished the supposedly perfect sphere of the Sun, and the phases of Venus (Fig. 9-4). His observations called into question the notion that all celestial objects were perfect and also lent credence to the Copernican view of the Sun as the center of planetary motion. Galileo's advocacy of the Copernican theory led to his 1633 conviction and house arrest on charges of heresy. Only in 1992 did the Vatican admit formally to having been in error and offer praise for Galileo.

By Newton's time the intellectual climate was ripe for the culmination of the revolution that had begun with Copernicus. Legend has it that Newton was sitting under an apple tree when an apple struck him on the head, causing him to discover gravity. That story is probably a myth, but if it were true the other half would be that Newton was staring at the moon when the apple struck (Fig. 9-5). Newton's stroke of genius was this:

> **Newton realized that the motion of the apple and the motion of the moon were the *same*; that both were "falling" toward Earth under the influence of the same force.**

Newton called this force **gravity,** after the Latin *gravitas,* meaning "heaviness." In one of the most sweeping syntheses in the history of human thought, Newton eliminated forever the distinction between terrestrial and celestial realms. Instead, he implied, the behavior of everything in the universe obeys the same set of physical laws.

FIGURE 9-5 Newton's stroke of genius was the realization that apple and moon move under the influence of the same force: gravity.

9-2 UNIVERSAL GRAVITATION

Newton generalized his new understanding of gravity to suggest that any two particles in the universe exert attractive forces on each other, with magnitude given by

$$F = \frac{Gm_1 m_2}{r^2}. \quad \text{(universal gravitation)} \qquad (9\text{-}1)$$

Here m_1 and m_2 are the particle masses, r the distance between them, and G the **constant of universal gravitation** whose value—determined after Newton's time—is 6.67×10^{-11} N·m²/kg². The constant G is truly a universal constant; observation and theory suggest that it has the same value throughout the universe.

The force of gravity acts *between* two particles. That is, m_1 exerts an attractive force on m_2, and m_2 exerts an equal but oppositely directed force on m_1. The two forces therefore obey Newton's third law and constitute an action-reaction pair.

Newton's law of universal gravitation applies strictly only to point particles that have no extent. But, as Newton showed using his newly developed calculus, it also holds for spherically symmetric objects of any size if the distance r is measured from their centers. It also applies approximately to arbitrarily shaped objects provided the distance between them is large compared with their sizes. For example, the gravitational force of Earth on a space shuttle is given accurately by Equation 9-1 because (1) Earth is essentially spherical and (2) the space shuttle, although irregular in shape, is much smaller than its distance from Earth's center.

● EXAMPLE 9-1 THE ACCELERATION OF GRAVITY

Use the law of universal gravitation to find the force of Earth's gravity on a mass m, and apply Newton's second law to determine the corresponding acceleration. Evaluate your results at Earth's surface and at the 250-km altitude of a space shuttle. Finally, adapt your calculations to find the acceleration due to the moon's gravity at the lunar surface.

Solution

Since Earth is essentially spherical, Equation 9-1 gives the gravitational force on a mass m:

$$F = \frac{GM_E m}{r^2},$$

where M_E is Earth's mass. Newton's second law, $F = ma$, shows that the mass m has acceleration given by

$$a = \frac{F}{m} = \frac{GM_E}{r^2}. \qquad (9\text{-}2)$$

The distance r is measured from the *center* of the gravitating object, so to find the acceleration at Earth's surface we use R_E, the radius of the Earth, for r. Taking R_E and M_E from inside the front cover (or from Appendix E), we have

$$a = \frac{GM_E}{R_E^2} = \frac{(6.67 \times 10^{-11}\ \text{N·m}^2/\text{kg}^2)(5.97 \times 10^{24}\ \text{kg})}{(6.37 \times 10^6\ \text{m})^2}$$

$$= 9.81\ \text{m/s}^2.$$

This, of course, is just the value of g—the acceleration of gravity at Earth's surface. Note the difference between the two numbers G and g commonly associated with gravity: G is a universal constant, while g describes the gravitational acceleration at a particular place—namely, Earth's surface—and happens to have the value it does because of the size and mass of the Earth.

At the space shuttle altitude, we have $r = R_E + 250$ km, so

$$a = \frac{GM_E}{r^2} = \frac{(6.67 \times 10^{-11}\ \text{N·m}^2/\text{kg}^2)(5.97 \times 10^{24}\ \text{kg})}{(6.37 \times 10^6\ \text{m} + 250 \times 10^3\ \text{m})^2}$$

$$= 9.09\ \text{m/s}^2,$$

or about 93% of its surface value. This calculation supports our contention in Chapter 5 that weightlessness doesn't mean the absence of gravitational force; rather, weightlessness arises because—as Equation 9-2 shows—the gravitational acceleration of an object is independent of its mass, so all objects "fall" together with a common acceleration.

Finally, using values for the moon from Appendix E, we have

$$g_{\text{moon}} = \frac{GM_m}{R_m^2} = \frac{(6.67 \times 10^{-11}\ \text{N·m}^2/\text{kg}^2)(7.35 \times 10^{22}\ \text{kg})}{(1.74 \times 10^6\ \text{m})^2}$$

$$= 1.62\ \text{m/s}^2.$$

Although the moon's mass is only about 1% of Earth's, it's also a lot smaller, and the two effects combine to give a gravitational acceleration about one-sixth that of Earth.

TIP Measure from the Center The law of universal gravitation applies to spherical objects when the distance r is measured *from the center*. If you're 100 m above Earth, the distance r in Equation 9-1 is $R_E + 100$ m, not just 100 m. "Distance between" in this context means between centers.

EXERCISE Calculate the gravitational acceleration at the surface of (a) Mars and (b) a neutron star with the Sun's mass crammed into a sphere 6.6 km in radius.

Answer: (a) $g_{\text{Mars}} = 3.75\ \text{m/s}^2$; (b) $g_{ns} = 3.0 \times 10^{12}\ \text{m/s}^2$

Some problems similar to Example 9-1: 1–4, 6 ●

● EXAMPLE 9-2 GRAVITY IS WEAK

Estimate the gravitational force between a 55-kg woman and a 75-kg man when they're 1.6 m apart. Find also the gravitational force between two electrons 1.0 cm apart, and compare with the 2.3×10^{-24} N repulsive electric force between the electrons.

Solution

The people are hardly spherical, and their separation is not large compared with their sizes, so Equation 9-1 provides at best an order-of-magnitude estimate of the force:

$$F = \frac{Gm_1m_2}{r^2} = \frac{(6.67\times10^{-11}\text{ N·m}^2/\text{kg}^2)(55\text{ kg})(75\text{ kg})}{(1.6\text{ m})^2}$$

$$= 10^{-7}\text{ N}.$$

This is far smaller than typical forces usually applied to people-size objects.

Electrons are essentially point particles, so Equation 9-1 applies exactly. Taking the electron mass from the table inside the front cover, we have

$$F = \frac{Gm_e^2}{r^2} = \frac{(6.67\times10^{-11}\text{ N·m}^2/\text{kg}^2)(9.11\times10^{-31}\text{ kg})^2}{(0.010\text{ m})^2}$$

$$= 5.5\times10^{-67}\text{ N},$$

about 10^{-43} times smaller than the electric force. Gravitational forces are completely negligible in the subatomic realm.

EXERCISE Two boulders, one of mass 3×10^5 kg and the other 9×10^4 kg, are roughly spherical and sit 21 m apart on a mountain ridge. Calculate the gravitational force between them, and compare with the weight of a 1-g raisin.

Answer: 4×10^{-3} N, about half the weight of the raisin

Some problems similar to Example 9-2: 5, 7, 8, 10, 11 ●

If gravity is so weak, why is it such an important force? Because, unlike the electric force, gravity is always attractive. No antigravity or negative mass has ever been found. So the gravitational force is always cumulative, and large concentrations of matter produce substantial gravitational forces. Electric forces, in contrast, can be attractive or repulsive because there are two kinds of electric charge. A large object like Earth contains nearly equal amounts of positive and negative charge and, therefore, exerts no significant electric force. We'll explore this fundamental distinction between electric and gravitational forces further in Chapter 23.

The Cavendish Experiment: Weighing the Earth

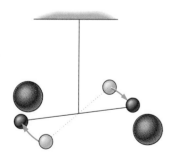

Given the mass and radius of the Earth and the measured value of g, we could use Equation 9-1 to determine the universal constant G. Unfortunately, the only way to determine Earth's mass accurately is to measure its gravitational effect and then use Equation 9-1. But that requires knowing G.

To determine G we need to measure the gravitational force of a *known* mass. Given the weak gravitational force of normal-size objects, this is a challenging task. It was accomplished in 1798 through an ingenious experiment by the British physicist Henry Cavendish. Cavendish mounted two 5-cm-diameter lead spheres on the ends of a rod suspended from a thin fiber. He then brought two 30-cm lead spheres nearby (Fig. 9-6). Their gravitational attraction caused a slight movement of the small spheres, twisting the fiber. Knowing the properties of the fiber, Cavendish could determine the force. With the known masses and their separation, he then used Equation 9-1 to calculate G. He used his result to determine the mass of the Earth; indeed, his published paper was entitled "On Weighing the Earth."

■ **APPLICATION** PROBING THE EARTH WITH GRAVITY

Variations in the thickness and density of Earth's crust are associated with geological features and give rise to small variations in the surface value of the gravitational acceleration g. Geologists use sensitive instruments called gravimeters to measure these variations and infer the underlying rock structures. Gravimeters themselves are based on simple physical principles; the most common types employ fixed masses on sensitive springs, while others use pendulums. Ingeniously designed, these devices detect variations in g as small as 10^{-7} m/s^2.

Petroleum geologists find gravity measurements particularly useful since they readily reveal the presence of low-density salt domes that often occur with oil and gas deposits (Fig. 9-7). Gravity measurements also show that mountains have lower density than their surroundings and that mountain roots extend deep into the crust. On a global scale, sensitive mapping of the height of the sea surface indicates the gravitational attraction of the underlying material and reveals a wealth of geological features (Fig. 9-8).

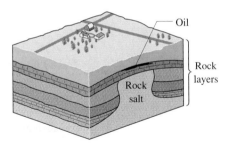

FIGURE 9-7 A "dome" of rock salt protrudes into the overlying rock layers. Since the salt has lower density, the value of g at the surface will be slightly lower than normal. Oil and gas deposits are often associated with such salt domes, so gravity measurements are widely used in petroleum exploration.

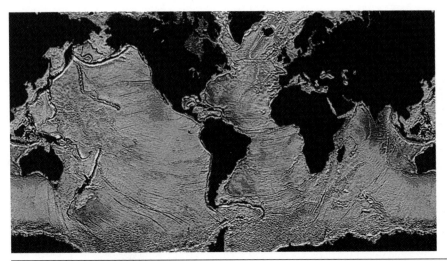

FIGURE 9-8 This map shows the gravitational acceleration at the ocean surface; lighter areas correspond to lower values of g and vice versa. The map was made using radar altimetry to determine the level of the ocean surface from the orbiting SEASAT satellite; accuracy of the surface-height measurement is about 10 cm. Regions of high ocean surface correspond to higher gravity, as the underlying mass (a mountain ridge, for example) attracts water, causing it to pile up. Ocean floor features like mid-ocean ridges and trenches produce variations to 10 m; variations up to 80 m in sea level arise from irregularities deep in Earth's mantle and core.

9-3 ORBITAL MOTION

Orbital motion occurs when gravity is the dominant force acting on a body. It's not just planets and spacecraft that are in orbit. An individual astronaut, floating outside a space shuttle, is in orbit about Earth. The Sun itself orbits the center of the galaxy, taking about 200 million years to complete one revolution. Neglecting air resistance, even a baseball is temporarily in orbit. Here we discuss quantitatively the special case of circular orbits, then describe qualitatively the general case.

FIGURE 9-9 Newton's "thought experiment" convinced him that projectile and orbital motion are essentially the same.

Newton's genius was to recognize that the moon is held in its circular orbit by the same force that pulls an apple to the ground. From there, it was a short step for Newton to realize that human-made objects could be put into orbit. Nearly 300 years before the first artificial satellites, he imagined a projectile launched horizontally from a high mountain (Fig. 9-9). The projectile moves in a curve, as gravity pulls it from the straight-line path it would follow if no force were acting. As its initial speed is increased, the projectile travels farther before striking Earth. Finally, there comes a speed for which the rate at which the projectile's path bends is exactly equal to the rate at which Earth's surface falls away beneath it. It is then in **circular orbit,** continuing forever unless a nongravitational force acts.

Why doesn't an orbiting object fall toward Earth? It does! Under the influence of gravity, it gets ever closer to Earth than it would be on a straight-line path (Fig. 9-10). It's behaving exactly as Newton's second law requires of an object under the influence of a force—by accelerating. For a *circular* orbit, that acceleration amounts to a change in the direction, but not the magnitude, of the orbiting object's velocity.

Remember that Newton's laws aren't so much about *motion* as they are about *changes* in motion. To ask why a satellite doesn't fall to Earth is to adopt the archaic Aristotelian view. The Newtonian question is this: Why doesn't the satellite move in a straight line? And the answer is simple: because a force is acting. That force—gravity—is exactly analogous to the tension force that keeps a ball on a string whirling on its circular path (Fig. 9-11).

We can analyze circular orbits quantitatively because we know that a force of magnitude mv^2/r is required to keep an object of mass m and speed v in a circular path of radius r. In the case of an orbit that force is gravity, so we have

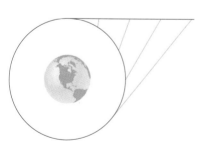

FIGURE 9-10 An object in circular orbit is falling away from the straight-line path it would take in the absence of gravity. Light lines connect points on the orbit with positions where the object would be in the absence of gravity.

FIGURE 9-11 A ball whirling on a string is like a satellite in circular orbit. The tension force pulls the ball toward the center of the circle, but it never gets any closer to the center. Similarly, the gravitational force pulls the satellite toward Earth's center, but the satellite never gets closer to Earth.

$$\frac{GMm}{r^2} = \frac{mv^2}{r},$$

where m is the mass of the orbiting object and M that of the object about which it's orbiting. We assume here that $M \gg m$, so the gravitating object can be considered essentially at rest—a reasonable approximation with Earth satellites or even for planets orbiting the much more massive Sun.

Solving for the orbital speed gives

$$v^2 = \frac{GM}{r}. \tag{9-3}$$

Often we're interested in the **orbital period,** or the time to complete one orbit. In one period T, the orbiting object moves the orbital circumference $2\pi r$, so its speed may be written $v = 2\pi r/T$. Then Equation (9-3) becomes

$$\left(\frac{2\pi r}{T}\right)^2 = \frac{GM}{r}$$

or

$$T^2 = \frac{4\pi^2 r^3}{GM}. \qquad \text{(orbital period, circular orbit)} \tag{9-4}$$

FIGURE 9-12 Astronaut Bruce McCandless floats freely outside a space shuttle. Because orbital parameters are independent of the orbiting mass, astronaut and shuttle remain in the same orbit, circling Earth at nearly 8 km/s.

In deriving Equation 9-4, we've proved Kepler's third law—that the square of the orbital period is proportional to the cube of the semimajor axis—for the special case of a circular orbit whose semimajor axis is just its radius.

Note that orbital speed and period are independent of the mass m of the orbiting object. This is another indication that all objects experience the same gravitational acceleration. An astronaut, for example, has the same orbital parameters as a space shuttle. That's why the astronaut is weightless inside the shuttle and why he doesn't float away if he steps outside (Fig. 9-12).

● **EXAMPLE 9-3** A SPACE SHUTTLE ORBIT

A space shuttle is in a circular orbit 250 km above Earth's surface. What are its orbital speed and period?

Solution
Remember that r appearing in Equations 9-3 and 9-4 is the distance from the center of the gravitating object. So here r is Earth's radius plus the 250-km altitude. Taking Earth's mass and radius from the table inside the front cover, Equation 9-3 gives

$$v = \sqrt{\frac{GM_E}{r}}$$

$$= \sqrt{\frac{(6.67\times10^{-11} \text{ N·m}^2/\text{kg}^2)(5.97\times10^{24} \text{ kg})}{(6.37\times10^6 \text{ m} + 250\times10^3 \text{ m})}}$$

$$= 7.8 \text{ km/s},$$

or about 17,500 mph. Astronauts have no choice; if they want a circular orbit 250 km up, they must have this speed.

We can get the orbital period from the speed and radius, or directly from Equation 9-4:

$$T = \sqrt{\frac{4\pi^2 r^3}{GM_E}}$$

$$= \sqrt{\frac{(4\pi^2)(6.37\times10^6 \text{ m} + 250\times10^3 \text{ m})^3}{(6.67\times10^{-11} \text{ N·m}^2/\text{kg}^2)(5.97\times10^{24} \text{ kg})}}$$

$$= 5.4\times10^3 \text{ s},$$

or 90 minutes. Again, there's no choice: As long as gravity is the only force acting, a circular orbit at this altitude must take this much time.

Since 250 km is small compared with Earth's radius, the orbital speed and 90-minute period found in this example are approximately correct for any near-Earth orbit. If it weren't for air resistance and mountains, a baseball could be put into orbit just above the ground—and its orbital period would be about 90 minutes.

EXERCISE Determine the orbital period of Saturn in its circular orbit of 1.43×10^{12}-m radius about the 1.99×10^{30}-kg Sun.

Answer: 9.33×10^8 s, or about 30 years

Example 9-3 shows that the orbital period in near-Earth orbit is about 90 minutes. The moon, on the other hand, takes 27 days to complete its nearly circular orbit. So there must be a distance where the orbital period is 24 hours—the same as Earth's rotation period. A satellite at this distance will remain fixed with respect to Earth's surface provided its orbit is parallel to the equator. Weather and communication satellites are often placed in this **geosynchronous orbit** (Fig. 9-13). Satellite TV, as well as some intercontinental telephone and data transmissions, use satellites in geosynchronous orbit (Fig. 9-14).

● EXAMPLE 9-4 GEOSYNCHRONOUS ORBIT

What altitude is required for geosynchronous orbit?

Solution
We want the period T to be 24 h or 8.64×10^4 s. Solving Equation 9-4 for r gives

$$r = \left(\frac{GM_E T^2}{4\pi^2} \right)^{1/3}$$

$$= \left(\frac{(6.67 \times 10^{-11} \text{ N·m}^2/\text{kg}^2)(5.97 \times 10^{24} \text{ kg})(8.64 \times 10^4 \text{ s})^2}{4\pi^2} \right)^{1/3}$$

$$= 4.22 \times 10^7 \text{ m},$$

or 42,200 km. This is the distance from Earth's center; the altitude above Earth's surface is then about 36,000 km or 22,000 miles.

A more careful calculation would use Earth's so-called sidereal rotation period, measured with respect to the distant stars rather than the Sun. Earth's sidereal period is 0.997 days. And, because Earth is not a perfect sphere, geosynchronous satellites drift from their intended positions. They fire small rocket thrusters every few weeks to correct for this effect.

EXERCISE Astronauts establish two separate bases on Mars, communicating via synchronous satellite. At what altitude should they place the satellite? (Consult Appendix E for Mars's mass, radius, and sidereal rotation period.)

Answer: 1.7×10^4 km.

Some problems similar to Examples 9-3 and 9-4:

12–15, 17, 18, 21–23

FIGURE 9-13 Weather over Africa and the South Atlantic as photographed by the European Space Agency's METEOSAT satellite from geosynchronous orbit.

FIGURE 9-14 This TV satellite dish is trained on a communications satellite in geosynchronous orbit 36,000 km above the equator.

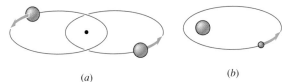

(a)   (b)

FIGURE 9-15 Two gravitating masses describe elliptical orbits with a focus at their common center of mass. In (a) the orbiting bodies have equal mass and the common focus is midway between them. In (b) one object is much more massive, and it remains essentially fixed at a focus of the other's elliptical orbit. This is the case for bodies orbiting the massive Sun, or artificial satellites orbiting Earth. More generally both bodies move, with the more massive one describing a smaller ellipse.

Elliptical Orbits

Using his laws of motion and gravity, Newton was able to prove Kepler's assertion that the planets move in elliptical paths with the Sun at one focus. More generally, Newton showed that two gravitating bodies describe elliptical orbits with one focus at their *common center of mass*—a concept we'll quantify in the next chapter (Fig. 9-15).

Circular orbits represent the special case where the two foci of the ellipse coincide, so the distance from the gravitating center remains constant. Most planetary orbits are nearly, but not quite, circular; Earth's distance from the Sun, for example, varies by about 3% throughout the year. But the orbits of comets and other smaller bodies are often highly elliptical (Fig. 9-16). Their orbital speeds also vary, as they gain speed "falling" toward the Sun, whip quickly around the Sun at the point of closest approach (the **perihelion**), and then "climb" ever more slowly to their most distant point (**aphelion**) before returning to the Sun's vicinity. Comet Halley, for example, visited the inner solar system in 1986 and will not do so again until 2061; most of its time will be spent moving slowly through the distant realms beyond the orbit of Uranus.

In Chapter 3 we showed that the trajectory of a projectile is a parabola. But our derivation neglected Earth's curvature and the associated variation in *g* with altitude. In fact, a projectile is just like any orbiting body. Neglecting air resistance, it too describes an elliptical orbit with Earth's center at one focus. Only for trajectories small compared with Earth's radius are the true elliptical path and the parabola of Chapter 3 essentially indistinguishable (Fig. 9-17).

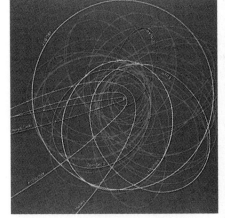

FIGURE 9-16 Orbits of most known comets are highly elliptical. Those with longer periods, like Comet Halley, spend much of their time moving slowly through the outer reaches of the solar system.

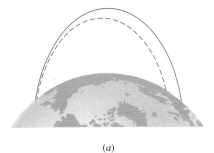   

(a)   (b)

FIGURE 9-17 (a) For projectiles of limited range, the true elliptical trajectory (solid line) differs only slightly from the parabolic trajectory calculated on the assumption of a flat Earth. (b) Long-range trajectories, like those of intercontinental missiles, are clearly elliptical. Were Earth's mass concentrated at the center, such a projectile would continue forever in its elliptical orbit.

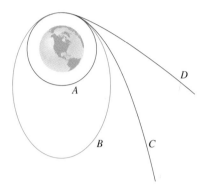

FIGURE 9-18 As projectile speed is increased above that needed for circular orbit (*A*), the trajectory becomes first an ellipse (*B*), then a parabola (*C*), and finally a hyperbola (*D*).

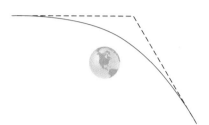

FIGURE 9-19 Hyperbolic orbit of a high-speed object passing a gravitating center like the Sun or Earth. Only near the gravitating center does the orbit deviate significantly from the straight-line asymptotes of the hyperbola.

Are missiles and baseballs really in orbit? Yes—but their orbits happen to intersect the Earth. At that point, nongravitational forces put an end to orbital motion. If Earth suddenly shrank to the size of a grapefruit, a baseball would continue happily in orbit, as Fig. 9-17*b* suggests. Newton's ingenious intuition was correct: Barring air resistance, there's truly no difference between the motion of everyday objects near Earth and the motion of celestial objects.

Open Orbits

With elliptical and circular orbits, the same motion is repeated indefinitely because the orbit is a closed path. But closed orbits aren't the only ones possible. Imagine again Newton's thought experiment, only now fire the projectile faster than necessary for a circular orbit (Fig. 9-18). It rises higher than before, describing an ellipse with its low point at the launch site. Faster, and the ellipse gets more elongated. But with great enough initial speed—which we'll derive in the next section—the projectile describes a hyperbolic trajectory that takes it ever farther from Earth and that approaches a straight line at great distances where the influence of Earth's gravity wanes (Fig. 9-19). A parabolic orbit marks the special case that divides elliptical from hyperbolic orbits. We'll see in the next section how energy determines the type of orbit.

9-4 GRAVITATIONAL ENERGY

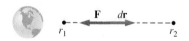

FIGURE 9-20 The change in gravitational potential energy between points r_1 and r_2 is found by integrating the gravitational force; along the path of integration the force **F** and path element $d\mathbf{r}$ are oppositely directed.

How much energy does it take to boost a satellite to geosynchronous altitude? Our simple answer *mgh* won't do here since *g* varies substantially over the distance involved. So, as we found in the preceding chapter, we have to integrate to determine the potential energy.

Figure 9-20 shows two points at distances r_1 and r_2 from the center of a gravitating mass *M*. Equation 8-2 gives the change in potential energy associated with moving a mass *m* from r_1 to r_2:

$$\Delta U_{1 \to 2} = -\int_{r_1}^{r_2} \mathbf{F} \cdot d\mathbf{r}.$$

Here the force points radially inward and has magnitude GMm/r^2, while the path element $d\mathbf{r}$ points radially outward. Then $\mathbf{F} \cdot d\mathbf{r} = -(GMm/r^2)dr$, where the minus sign comes from the factor $\cos 180°$ that appears in the dot product of oppositely directed vectors. Then the potential energy difference is

$$\Delta U_{1 \to 2} = \int_{r_1}^{r_2} \frac{GMm}{r^2} \, dr = GMm \int_{r_1}^{r_2} r^{-2} \, dr$$

$$= GMm \left. \frac{r^{-1}}{-1} \right|_{r_1}^{r_2} = GMm \left(\frac{1}{r_1} - \frac{1}{r_2} \right). \qquad (9\text{-}5)$$

Does this make sense? Yes: For $r_1 < r_2$, $\Delta U_{1 \to 2}$ is positive, showing that potential energy increases with height—consistent with our simpler result $\Delta U = mgh$ near Earth's surface. (In fact, as Problem 64 shows, $\Delta U = mgh$ is a special case of Equation 9-5 when $r_1 \simeq r_2 \simeq R_E$.) Although we derived Equation 9-5 for two points on a radial line, Fig. 9-21 shows that it holds for any two points at distances r_1 and r_2 from the gravitating center.

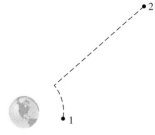

FIGURE 9-21 Since gravity is a conservative force, we can use any path to evaluate the potential energy change. For two points not on a radial line, we choose a path consisting of an arc and a radial line. The potential energy doesn't change along the arc since it's perpendicular to the gravitational force. All the change occurs along the line and is given by Equation 9-5. So Equation 9-5 holds for any two points at radii r_1 and r_2.

● EXAMPLE 9-5 STEPS TO THE MOON

Materials to construct an 11,000-kg lunar observatory are boosted from Earth to geosynchronous orbit. There they are assembled, then launched to the moon, 384,000 km from Earth. Compare the work that must be done against Earth's gravity on the two legs of the trip.

Solution

As we saw in the preceding chapter, the work done against a conservative force is equal to the change in potential energy; here that change is given by Equation 9-5. For the first step we have $r_1 = R_E$ and, from Example 9-4, $r_2 = 42{,}200$ km. Since the quantity $GM_E m$ that appears in Equation 9-5 will be used in both steps, we calculate it first: $GM_E m = 4.38 \times 10^{18}$ N·m². Then for the first step we have

$$W = \Delta U_{1 \to 2} = GM_E m \left(\frac{1}{r_1} - \frac{1}{r_2} \right)$$

$$= (4.38 \times 10^{18} \text{ N·m}^2) \left(\frac{1}{6.37 \times 10^6 \text{ m}} - \frac{1}{4.22 \times 10^7 \text{ m}} \right)$$

$$= 5.8 \times 10^{11} \text{ J}.$$

From geosynchronous orbit to the moon a similar calculation gives

$$W = (4.38 \times 10^{18} \text{ N·m}^2) \left(\frac{1}{4.22 \times 10^7 \text{ m}} - \frac{1}{3.84 \times 10^8 \text{ m}} \right)$$

$$= 9.2 \times 10^{10} \text{ J}.$$

Even though the second step is much longer, the rapid drop-off in the gravitational force means that less work is required than for the shorter boost to geosynchronous altitude.

Our calculations here include only the work done against Earth's gravity; additional energy would be required to attain a circular geosynchronous orbit. On the other hand, the moon's gravitational attraction would lower the required energy somewhat.

EXERCISE How much energy must a 790-kg space probe have to get from Earth's vicinity to Jupiter's? Consider only the effect of the Sun's gravity, not the planets'. Consult Appendix E.

Answer: 5.6×10^{11} J

Some problems similar to Example 9-5: 28, 29 ●

The Zero of Potential Energy

Equation 9-5 has an interesting feature—the potential energy difference remains finite even when the points are infinitely far apart, as you can see by setting either r_1 or r_2 to infinity. Although the gravitational force always acts, it weakens so

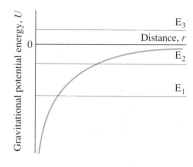

rapidly with distance that its cumulative effect is finite even over infinite distances. This property makes it convenient to set the zero of potential energy at infinity. Setting $r_1 = \infty$ and dropping the subscript on r_2, we then have an expression for the potential energy at an arbitrary distance r from the gravitating center:

$$U(r) = -\frac{GMm}{r}. \quad \text{(gravitational potential energy)} \quad (9\text{-}6)$$

The potential energy is negative because we chose $U = 0$ at $r = \infty$; any other point is closer to the gravitating center and, therefore, has lower potential energy.

Knowing the gravitational potential energy allows us to apply the powerful conservation of energy principle. Figure 9-22 shows the potential energy curve obtained by plotting Equation 9-6. Superposing several values of total energy shows immediately that an object with $E < 0$ is in a bound orbit, with a maximum distance beyond which it cannot go. In general, the orbit for $E < 0$ is an ellipse. For $E > 0$, the object is not bound since it's energetically possible for it to move all the way to infinity. In this case the orbit is hyperbolic. The special case $E = 0$ corresponds to a parabolic orbit.

EXAMPLE 9-6   BLAST-OFF

A rocket is launched vertically upward at 2.3 km/s. How high does it go?

$$= 6.603 \times 10^6 \text{ m}.$$

Again, this is the distance from Earth's center; subtracting Earth's radius then gives a peak altitude of 230 km.

Solution

This is a typical conservation of energy problem, like those of the preceding chapter. We write $K + U = K_0 + U_0$ to equate the total energies at two different points. At launch the rocket has both kinetic and potential energy, while at the peak of its trajectory its energy is all potential. As always, $K = \frac{1}{2}mv^2$, and here U is given by Equation 9-6. So we have

$$-\frac{GM_E m}{r} = \frac{1}{2}mv_0^2 - \frac{GM_E m}{R_E},$$

where m is the rocket's mass, r is the distance from Earth's center at the peak, and Earth's radius R_E is the distance at launch. Solving for r gives

$$r = \left(\frac{1}{R_E} - \frac{v_0^2}{2GM_E}\right)^{-1}$$

TIP   All Conservation of Energy Problems Are the Same   This problem is essentially the same as one from the preceding chapter where you throw a ball straight up and solve for its maximum height using $U = mgh$ for the potential energy. The only difference is the more complicated potential energy function $U = -GMm/r$ that arises because the variation in gravity becomes significant over the rocket's trajectory. Recognize what's common to all similar problems, and you'll begin to see physics for what it is—a powerful description based on a few simple principles, rather than a bewildering array of new and different things to learn.

EXAMPLE 9-7   BRINGING SATELLITE TV TO SIBERIA

Viewed from extreme northern and southern latitudes, geosynchronous satellites are too low in the sky for reliable communications. Service to these regions can be provided by satellites in highly elliptical orbits that move very slowly high over the intended recipients, then drop down to make a rapid, much lower pass around the other side of the Earth. Television satellites serving northern Russia, for example, are in 12-hour orbits with altitudes ranging from 500 km to 40,000 km. If such a satellite passes its lowest point moving at 9.94 km/s, how fast will it be moving at the peak of its trajectory?

Solution

Here's another conservation of energy problem, the only difference from Example 9-6 being that we have both kinetic and potential energy at both points, and we're asked for speed rather than distance. Writing $K + U = K_0 + U_0$, we have

$$\frac{1}{2}mv^2 - \frac{GM_Em}{r} = \frac{1}{2}mv_0^2 - \frac{GM_Em}{r_0},$$

where m is the satellite's mass. Solving for the speed v gives

$$v = \sqrt{v_0^2 + 2GM_E\left(\frac{1}{r} - \frac{1}{r_0}\right)}.$$

The radii appearing here are measured *from Earth's center,* so we add Earth's radius (6.37×10^6 m) to the altitudes given. Then $r - 6.37 \times 10^6$ m $+ 40,000 \times 10^3$ m $- 46.37 \times 10^6$ m, and similarly, $r_0 = 6.87 \times 10^6$ m. Using these figures along with $M_E = 5.97 \times 10^{24}$ kg then gives $v = 233$ m/s. This low figure makes sense since the satellite has slowed in "climbing" so far against Earth's gravity.

EXERCISE At its perihelion in February 1986, Comet Halley was 8.79×10^7 km from the Sun and was moving at 54.6 km/s. What was its speed a month later at its encounter with the Giotto spacecraft, 1.16×10^8 km from the Sun (Fig. 9-23)?

Answer: 47.4 km/s

Some problems similar to Examples 9-6 and 9-7: 30–32, 34, 43, 49 ●

Escape Speed

Throw a ball straight up, and eventually it comes down. Is this always true? Not if the ball has enough energy! Figure 9-22 shows that an object with total energy greater than zero can escape infinitely far from a gravitating body, never to return.

How fast should you throw the ball so it won't come back? If its total energy is exactly zero, it can just barely escape to infinity. Its speed is given by the conservation of energy principle, in this case with the total energy $K + U$ set to zero:

$$0 = K + U = \frac{1}{2}mv_{esc}^2 - \frac{GMm}{r},$$

where m is the mass of the ball or space probe or whatever, M the mass of the gravitating object, and r the distance from its center (i.e., $r = R_E$ for an object launched from Earth's surface). Solving for v_{esc} then gives

$$v_{esc} = \sqrt{\frac{2GM}{r}}. \qquad \text{(escape speed)} \qquad (9\text{-}7)$$

This is the **escape speed**—the speed that must be given to an object at distance r from the gravitating center if it's to escape all the way to infinity. At Earth's

FIGURE 9-24   In 1983 Pioneer 10 became the first artifact to leave the solar system. With its speed in excess of solar escape speed, Pioneer will travel forever through the galaxy.

surface, $v_{esc} = 11.2$ km/s. Earth-orbiting spacecraft have somewhat lower speeds. Moon-bound spacecraft have speeds just under v_{esc}, so if anything goes wrong (as happened to Apollo 13), they'll return to Earth. Planetary spacecraft have speeds greater than v_{esc}. Pioneer and Voyager missions to the outer solar system left Earth's vicinity at 14 km/s. They gained additional energy in their encounters with Jupiter (see "Application: Gravity's Slingshot" later in this section) and now have escape speed relative to the Sun (Fig. 9-24).

Energy in Circular Orbits

In the special case of a circular orbit, kinetic and potential energies are related in a simple way. In Section 9-3 we found that the speed in a circular orbit is given by

$$v^2 = \frac{GM}{r},$$

where r is the distance from a gravitating center of mass M. So the kinetic energy of the orbiting object is

$$K = \frac{1}{2}mv^2 = \frac{GMm}{2r},$$

while the potential energy is given by Equation 9-6:

$$U = -\frac{GMm}{r}.$$

Comparison of these two expressions shows that $U = -2K$ for a circular orbit. The total energy is therefore

$$E = U + K = -2K + K = -K, \qquad (9\text{-}8)$$

or, equivalently,

$$E = \frac{1}{2}U = -\frac{GMm}{2r}. \qquad (9\text{-}9)$$

The total energy given by these equations is negative, showing that circular orbits are—obviously—bound orbits. We stress that these results apply only to *circular* orbits; in elliptical orbits there's a continual interchange between kinetic and potential energy as the orbiting object moves relative to the gravitating center.

Equation 9-8 shows that *higher* kinetic energy corresponds to *lower* total energy. This surprising result occurs because *higher* orbital speed corresponds to a *lower* orbit, with lower potential energy. The implications for space flight are counterintuitive. To get into a faster orbit a spacecraft must *lose* energy—as if a car, to speed up, had to apply its brakes. Astronauts attempting orbital rendevous actually fire braking rockets in order to catch up with their target (Fig. 9-25).

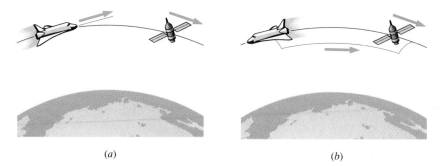

FIGURE 9-25 Orbital rendezvous. A space shuttle seeks to overtake a satellite further ahead in the same orbit; both spacecraft are moving from left to right. (*a*) Thrusting toward the target would take the shuttle into a higher-energy—and therefore higher but slower—orbit. (*b*) The correct maneuver is counterintuitive. The shuttle thrusts *away* from its target, decreasing its energy and changing its orbit to an ellipse. A second thrust circularizes the orbit, now lower and faster than the original circular orbit. After overtaking the target, the shuttle reverses these maneuvers.

● **EXAMPLE 9-8** LAUNCHING A COMMUNICATIONS SATELLITE

A 740-kg communications satellite is carried in a space shuttle to a 250-km-high circular orbit. How much energy is required to launch it from there to geosynchronous orbit?

Solution
In the shuttle orbit the satellite already has some orbital energy; the energy needed to get it from there to geosynchronous orbit is the difference between the orbital energies at the two altitudes. Those energies are given by Equation 9-9; their difference is then

$$\Delta E = -\frac{GM_E m}{2r_2} - \left(-\frac{GM_E m}{2r_1}\right) = \frac{GM_E m}{2}\left(\frac{1}{r_1} - \frac{1}{r_2}\right)$$

where r_2 and r_1 are the radii of the geosynchronous and shuttle orbits, respectively. In Example 9-4 we found that the geosynchronous orbit radius is $r_2 = 42{,}000$ km; putting $r_1 = R_E + $

250 km $= 6.62 \times 10^6$ m, the expression above then gives $\Delta E = 1.9 \times 10^{10}$ J. This is roughly the energy equivalent of 400 kg of chemical fuel.

Our answer in this problem includes everything—the energy needed to climb against Earth's gravity as well as the kinetic energy needed for the circular orbit. This is in contrast to Example 9-5, where we found only the work done against gravity.

EXERCISE A 73,000-kg space shuttle in a circular orbit at 250 km altitude is sent to repair the Hubble Space Telescope in its 610-km-altitude circular orbit. How much energy is required to get the shuttle into the new orbit?

Answer: 1.1×10^{11} J

Some problems similar to Example 9-8: 33, 36, 45
●

■ **APPLICATION** GRAVITY'S SLINGSHOT

The orbital kinetic energy of a massive body like a planet is substantial. It's possible for a spacecraft to tap that energy with negligible effect on the planet's motion. As a spacecraft passes closely by a planet, its trajectory is deflected and it may gain energy from the planet's orbital motion.

This effect is regularly used to increase the speed of spacecraft bound for the outer solar system and beyond. In their encounters with Jupiter, the Pioneer and Voyager spacecraft were

gravity-boosted to speeds sufficient to escape the Sun's gravity altogether. A more complex example of gravity boost is the Galileo probe to Jupiter. This spacecraft was originally designed for launch directly toward Jupiter from an orbiting shuttle, using a powerful liquid-fueled rocket. But after the 1986 explosion of the space shuttle Challenger, the carrying of liquid-fueled rockets within the shuttle was deemed unsafe, and Galileo was left without a sufficiently powerful launch vehicle. Instead, Galileo

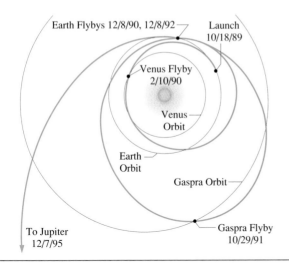

Earth Flybys 12/8/90, 12/8/92

Launch 10/18/89

Venus Flyby 2/10/90

Venus Orbit

Earth Orbit

Gaspra Orbit

To Jupiter 12/7/95

Gaspra Flyby 10/29/91

was launched in 1989 on a complicated trajectory taking it first toward Venus. A close encounter with Venus increased its orbital energy, and it returned in 1990 to Earth's vicinity, where it picked up additional energy. A second Earth encounter in 1992 finally gave Galileo the energy needed for its 1995 visit to Jupiter. During the 1992 encounter the spacecraft's speed increased by some 3.7 km/s. Galileo's gain came at the expense of Earth's orbital motion; as a result of the encounter, Earth's orbital speed dropped by a mere 10^{-18} km/s. Figure 9-26 shows Galileo's orbital gyrations.

FIGURE 9-26 A close encounter with Venus and two encounters with Earth gave the Galileo probe enough energy to reach Jupiter. On the way Galileo encountered the asteroid Gaspra.

9-5 THE GRAVITATIONAL FIELD

Our description of gravity so far suggests that a massive body like Earth somehow "reaches out" across empty space to pull on objects like falling apples, satellites, or the moon. This view—called **action-at-a-distance**—has bothered both physicists and philosophers for centuries. How can the moon, for example, "know" about the presence of the distant Earth?

An alternative view holds that Earth creates a **gravitational field** and that objects respond to the field in their immediate vicinity. The field is described by vectors that give the force per unit mass that would arise at each point if a mass were placed there. Near Earth's surface, for instance, the gravitational field vectors point vertically downward and have magnitude 9.8 N/kg. We could express this field vectorially by writing

$$\mathbf{g} = -g\hat{\mathbf{j}}. \quad \text{(gravitational field near Earth's surface)} \quad (9\text{-}10)$$

More generally, the field points toward a spherical gravitating center, and its strength drops inversely with the square of the distance:

$$\mathbf{g} = -\frac{GM}{r^2}\hat{\mathbf{r}}, \quad \text{(gravitational field of a spherical mass } M) \quad (9\text{-}11)$$

where $\hat{\mathbf{r}}$ is a unit vector that points radially outward. Figure 9-27 shows pictorial representations of Equations 9-10 and 9-11. Note that the units of gravitational field—N/kg—are equivalent to those of acceleration, so the field is really just a vectorial representation of g, the local acceleration of gravity.

What do we gain by this field description? As long as we deal with situations where nothing changes, the action-at-a-distance and field descriptions are equivalent. But what if, for example, Earth suddenly gains mass. How does the moon know to adjust its orbit? Under the field view its orbit doesn't change immediately; instead, it takes a small but nonzero time for the information about the more massive Earth to propagate out to the moon. The moon always responds to the

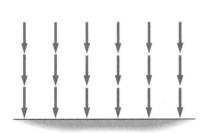

(a)

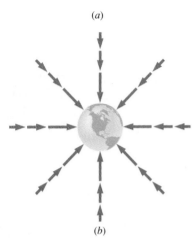

(b)

FIGURE 9-27 (a) Gravitational field vectors in a small region near Earth's surface have the same magnitude and direction. (b) On a larger scale the field vectors vary in both direction and magnitude.

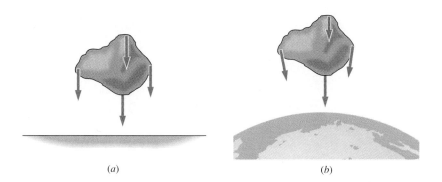

(a)   (b)

FIGURE 9-28 (*a*) In a uniform gravitational field, all parts of an object experience the same gravitational acceleration. (*b*) When the field is not uniform, gravity acts differently on different parts of the object, giving rise to a force that tends to stretch the object along the field direction and compress it at right angles to the field.

gravitational field *in its immediate vicinity,* and it takes a short time for the field itself to change. That description is consistent with Einstein's notion that instantaneous transmission of information is impossible; the action-at-a-distance view is not.

More generally, the field view provides a powerful way of describing the interactions of physics. We'll see fields again when we study electricity and magnetism, and you'll gain a sense that fields aren't just mathematical or philosophical conveniences but are every bit as real as matter itself.

9-6 TIDAL FORCES

When the gravitational field is uniform, all parts of a freely falling body experience exactly the same gravitational acceleration (Fig. 9-28*a*). But over large enough scales gravity is always nonuniform, and the acceleration of gravity varies from place to place. The result is a force that tends to stretch or compress an object (Fig. 9-28*b*).

Ocean tides are an important manifestation of this force, in which the oceans are stretched by the nonuniform gravitational forces of the moon and Sun (Fig. 9-29). For that reason the forces associated with the variation in gravitational field from place to place are called **tidal forces.**

Tidal forces arise not from gravity itself but from *differences* in gravity. As you can show in Problem 50, the tidal force therefore drops even more rapidly with distance than does gravity itself, being inversely proportional to the *cube* of the distance. That's why the moon is the dominant influence on ocean tides; even though the Sun's gravity is much stronger at Earth than is the moon's, the distance from the Sun is so great that the moon's *tidal* effect is greater.

Tidal forces exert stresses on any object in a nonuniform gravitational field, and the astronomical consequences are, therefore, numerous. The extensive volcanic activity of Jupiter's moon Io (Fig. 9-30) is thought to arise from heat generated by tidal stresses. Comets sometimes break apart under the influence of strong tidal forces as they pass near the Sun or even Jupiter. And stars in binary star systems are frequently distorted by tidal forces.

Since astronomical objects are generally held together by their own gravity, it makes sense to ask whether the tidal forces from a nearby body can ever be strong enough to overcome an object's own self-gravitation. The answer, explored in Problem 52, is yes: When a central body—for example, a planet—and its satellite have roughly the same densities, it turns out that tidal forces exceed self-gravita-

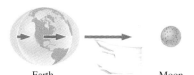

Earth   Moon

FIGURE 9-29 Ocean tides result primarily from the nonuniform gravitational field of the moon. The gravitational force on the moonward side of the ocean is stronger than on the Earth itself, which in turn is stronger than the force on the distant ocean. The ocean is therefore stretched into two "tidal bulges" that move as Earth rotates, giving rise to two high tides a day at most locations. (Not shown is the Sun, which produces a weaker tidal effect.)

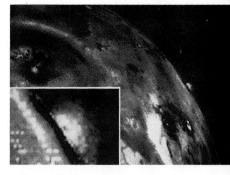

FIGURE 9-30 Volcanic activity on Jupiter's moon Io probably originates from internal heating caused by tidal stresses. Insert in this Voyager spacecraft photo shows an enlargement of the eruption just visible on Io's limb.

FIGURE 9-31 Planetary rings lie within the planet's Roche lobe, where tidal forces overcome the gravitational forces that would hold a moon together. Here, from left to right, are Voyager images of the rings of Saturn, Uranus, and Jupiter. An image of the Voyager spacecraft is superimposed on the Uranus picture.

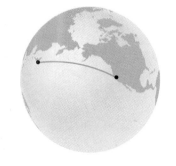

(a)

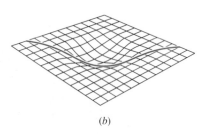

(b)

FIGURE 9-32 (*a*) The "straightest" path between San Francisco and Tokyo is a great circle skirting the Aleutian Islands. (*b*) Similarly, the "straightest" path in the curved space-time near a massive object is not a truly straight line but a curve. Depression in the surface represents the effect of a massive object on space and time.

tion when the satellite lies within about 2.5 planetary radii of the planet's center. This distance is called the **Roche limit,** after the French scientist Edouard Roche. In 1848 Roche suggested that satellites orbiting within the Roche limit would break up. And indeed, that's what we find: within the Roche limits of the planets Jupiter, Saturn, Uranus, and Neptune are rings rather than moons—showing that tidal forces prevent the formation of solid bodies held together by self-gravitation (Fig. 9-31). Beyond the Roche limit these planets have solid moons. We regularly place artificial satellites within the Roche limit of Earth, but these are held together by the strengths of materials, not by self-gravitation, so Roche's considerations don't apply.

9-7 GRAVITY AND THE GENERAL THEORY OF RELATIVITY

Newton's theory of gravity provided a brilliantly successful explanation of planetary motion and today guides space probes on billion-mile journeys throughout the solar system. But in very strong gravitational fields deviations from Newton's theory become evident. Those deviations are correctly described by the **general theory of relativity,** a radically different theory of gravity proposed by Einstein in 1916. Einstein's general theory is beyond the scope of this book; here we note only that Einstein describes gravity as a geometrical effect resulting from the curvature of space and time caused by the presence of matter. Objects moving under the influence of gravity follow the "straightest" possible paths in this curved space-time (Fig. 9-32).

For decades the only orbital evidence for Einstein's theory was a very slight shift in the orbit of Mercury (Fig. 9-33). Differences between Einsteinian and Newtonian gravity become substantial only in regions where the escape speed approaches the speed of light, and that doesn't happen in our solar system. But recently astrophysicists have discovered extraordinarily dense collapsed stars whose properties must be explained using general relativity. At the general relativistic extreme are **black holes**—objects of such density that the escape speed exceeds that of light, making it impossible for anything to escape their gravitational grip. Black holes with several times the Sun's mass have been tentatively identified in some binary star systems, and massive black holes—a million or more solar masses—are believed to lurk at the centers of many galaxies, including our own.

General relativity also predicts the existence of **gravitational waves.** Literally ripples in the fabric of space and time, these waves could be generated in violent stellar explosions and might be detected by the vibrations they induce in massive objects (Fig. 9-34). Indirect detection of gravitational waves has already occurred; by analyzing changes in the motion of a pulsar in close orbit around another object, Joseph Taylor and Russell Hulse showed that the pulsar is losing energy at just the rate that would result from the emission of gravitational waves. Their work won Taylor and Hulse the 1993 Nobel Prize in Physics.

Although general relativity is mathematically complex, it's based in a simple observation that, prior to Einstein, seemed a coincidence. Equation 9-1, the law of universal gravitation, states that the attractive force between two particles depends on their masses. Equation 5-3, Newton's second law, states that the force needed to produce a given acceleration is proportional to an object's mass. Why should the same property—mass—that determines the attraction to other objects also determine an object's response to applied forces? For years physicists puzzled over this seeming coincidence; they even coined the distinct terms "gravitational mass" and "inertial mass" to describe the two properties which experiments—like Galileo's dropping objects from the Tower of Pisa—showed to have the same value. In formulating general relativity, Einstein assumed that the equivalence of gravitational and inertial mass is no coincidence but reflects instead an underlying equivalence between the effects of gravity and of acceleration. General relativity is a theory not only about gravity but also about motion in arbitrary—including accelerated—reference frames, and it is this **principle of equivalence** that provides a coherent basis for the theory.

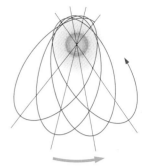

FIGURE 9-33 The axis (black lines) of Mercury's elliptical orbit precesses very slightly, an effect explained by general relativity but not by Newtonian gravitational theory. Both the shift and ellipticity are exaggerated; Mercury's orbit is nearly circular, and the shift amounts to only 43 seconds of arc (43/3600 degree) per century.

FIGURE 9-34 (*a*) This gravitational-wave detector at CERN, the European Center for Nuclear Research, consists of a 2300-kg aluminum bar using superconducting technology to detect minute vibrations that would be induced in the bar by gravitational waves. (*b*) Artist's conception of a black hole orbiting a large star. A disk of hot gas surrounds the hole.

CHAPTER SYNOPSIS

Summary

1. The concept of **universal gravitation** subsumes terrestrial and celestial motions under the same physical laws, providing a major philosophical shift in human understanding of the universe.

2. Newton's **law of universal gravitation** states that any two particles exert a mutually attractive force of magnitude

$$F = \frac{Gm_1m_2}{r^2},$$

where m_1 and m_2 are the particle masses and r the distance between them. Although the law applies strictly to point particles, Newton proved that it also holds exactly for the force between spherical masses. It is approximately true for arbitrarily shaped objects whose separation is large compared to their size. The **constant of universal gravitation** G, which was first determined by Cavendish's measurement of the attraction of lead spheres, has the approximate value $G = 6.67 \times 10^{-11}$ N·m²/kg².

3. An **orbit** is the path of a body whose motion is dominated by gravity. For gravitational interaction involving only two bodies, the possible orbits have simple shapes:

   a. Bodies whose total energy is negative travel in closed, **elliptical orbits.**

   b. A special case of an ellipse is the **circular orbit,** for which gravity provides the centripetal force mv^2/r needed to maintain circular motion. For an object in orbit about a much greater mass M, the orbital period and radius are related by

$$T^2 = \frac{4\pi^2 r^3}{GM}. \quad \text{(circular orbit)}$$

   An especially important orbit is the **geosynchronous orbit,** where a satellite's orbital period is equal to Earth's rotation period. The satellite then appears motionless above the equator. Geosynchronous orbit occurs about 36,000 km above Earth's surface.

   c. Objects whose total energy is greater than zero travel in open, **hyperbolic orbits** that take them infinitely far from the gravitating center. The intermediate case of zero energy corresponds to a parabolic orbit.

4. The gravitational potential energy of two masses M and m is given by

$$U = -\frac{GMm}{r}, \quad \text{(gravitational potential energy)}$$

   where r is the distance between their centers and where the zero of potential energy is taken at infinite separation.

   a. Because potential energy remains finite over infinite distances, it is possible to launch an object with enough energy that it will never return. The required **escape** speed a distance r from a gravitating center of mass M is given by

$$v_{esc} = \sqrt{\frac{2GM}{r}}. \quad \text{(escape speed)}$$

   b. For a circular orbit, the kinetic and potential energies are directly related:

$$U = -2K,$$

   so the total energy is

$$E = \frac{1}{2}U = -\frac{GMm}{2r} = -K.$$

5. The **gravitational field** describes the gravitational force per unit mass at each point in space. The field concept replaces the outmoded action-at-a-distance view, holding instead that a massive body creates a gravitational field and that other objects respond to the field in their immediate vicinities.

6. **Tidal forces** result from differences in gravitational force over an extended body. They give rise to stresses on astronomical objects, including ocean tides on Earth. Within the so-called **Roche limit,** tidal forces are strong enough to prevent self-gravitating satellites from forming.

7. The **general theory of relativity** is Einstein's description of gravity in terms of the geometry of space and time. It differs substantially from Newtonian theory in regions of very strong gravity—where the escape speed is comparable to the speed of light.

Terms You Should Understand

(Pairs are closely related terms whose distinction is important; number in parentheses is chapter section where term first appears.)

Copernicus, Brahe, Kepler, Galileo, Newton (9-1)	perihelion, aphelion (9-3)
Kepler's laws (9-1)	circular, elliptical, parabolic, hyperbolic orbits (9-3)
law of universal gravitation (9-2)	gravitational potential energy (9-4)
constant of universal gravitation (9-2)	escape speed (9-4)
G, g (9-2)	action-at-a-distance (9-5)
Cavendish experiment (9-2)	gravitational field (9-5)
orbital motion (9-3)	tidal forces (9-5)
orbital period (9-3)	Roche limit (9-5)
geosynchronous orbit (9-3)	general theory of relativity (9-6)

Symbols You Should Recognize

G (9-2)   **g** (9-5) $\hat{\mathbf{r}}$ (9-5)

Problems You Should Be Able to Solve

calculating gravitational forces using the law of universal gravitation (9-2)

solving for speeds, periods, distances in circular orbits (9-3)

evaluating changes in gravitational potential energy (9-4)

solving conservation of energy problems involving gravitational energy (9-4)

evaluating escape speed for different bodies (9-4)

Limitations to Keep in Mind

Newton's law of universal gravitation applies exactly only to point particles or to spherical objects; it's a good approximation for other bodies at separations large compared with their sizes.

The quantity r that appears in the law of universal gravitation and in other equations derived from it is always the distance *from the center* of the gravitating body, not from its surface.

Our description of orbits as circular, elliptical, parabolic, or hyperbolic applies exactly only when the interaction involves just two gravitating bodies.

Newton's law of universal gravitation provides an accurate description of gravity only in regions where the escape speed is much less than the speed of light—a condition valid throughout the solar system and in most of the universe.

QUESTIONS

1. Why is Newton's assertion that planetary orbits are elliptical more satisfying than Kepler's?
2. What do Newton's apple and the moon have in common?
3. Explain the difference between G and g.
4. Earth's orbital motion is fastest in January and slowest in June. When is Earth closest to the Sun? Explain.
5. When you stand on Earth, the distance between you and Earth is zero. So why isn't the gravitational force infinite?
6. The force of gravity on an object is proportional to the object's mass, yet all objects fall with the same gravitational acceleration. Why?
7. A friend who knows nothing about physics asks what keeps an orbiting space shuttle from falling to Earth. Give an answer that will satisfy your friend.
8. Could you put a satellite in an orbit that kept it stationary over the south pole? Explain.
9. Why are satellites generally launched eastward from low latitudes? *Hint:* Think about Earth's rotation.
10. It takes more energy to launch a satellite into the polar orbit shown in Fig. 9-35 than into the equatorial orbit shown. Why?
11. Can you launch a projectile that will travel three-fourths of the way around the Earth, as suggested in Fig. 9-36? Why or why not?

FIGURE 9-36 Question 11.

12. Why was the United States' first satellite launching facility (at Cape Canaveral, Florida) in the southern part of the country?
13. Given the mass of the Earth, the distance to the moon, the period of the moon's orbit, and the value of G, could you calculate the moon's mass? How or why not?
14. How should a satellite be launched so its orbit takes it over every point on the (rotating) Earth?
15. Does escape speed depend on the launch angle, as shown in Fig. 9-37?
16. Does the gravitational force of the Sun ever do work on a planet in a circular orbit? Explain.

FIGURE 9-35 Polar and equatorial orbits (Question 10).

FIGURE 9-37 Question 15.

17. Does the gravitational force of the Sun ever do work on a comet in an elliptical orbit? What is the net work done as the comet completes the entire orbit?

18. Could a satellite be placed in orbit that carried it around the arctic circle? Why or why not?

19. Why are the tides highest when Sun, moon, and Earth are in a line? At what phase(s) of the moon does this occur?

20. Why do we choose the zero of gravitational potential energy at infinity rather than, say, at Earth's surface?

21. Is it possible for a spacecraft to reach Jupiter if launched from Earth with less than escape speed relative to Earth? Explain.

22. Could you launch a projectile to the moon using a powerful slingshot mounted on Earth? Explain.

23. Could you launch a projectile into circular orbit using a powerful slingshot mounted on Earth? Explain.

24. Describe the maneuvers necessary for a spacecraft to catch up with another that is ahead of it in the same circular orbit.

25. If the Sun suddenly shrank to a white dwarf star with the same mass concentrated into the size of the Earth, what would happen to Earth's orbit?

26. Why is it significant whether the energy of a comet is more or less than zero?

27. Satellites in orbits lower than a few hundred kilometers experience frictional drag due to residual air at those altitudes. Assuming their orbits remain essentially circular, does this drag decrease or increase the satellites' speeds?

PROBLEMS

Section 9-2 The Law of Universal Gravitation

1. Space explorers land on a planet with the same mass as Earth, but they find they weigh twice as much as they would on Earth. What is the radius of the planet?

2. Jupiter's satellites Himalia and Io have masses of 9×10^{18} and 8.9×10^{22} kg, respectively. Their radii are 93 and 1820 km, respectively (see Fig. 9-38). Determine the acceleration of gravity at the surface of each.

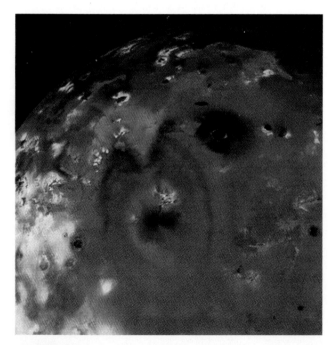

FIGURE 9-38 Jupiter's satellite Io (Problem 2).

3. To what fraction of its current radius would Earth have to be shrunk (with no change in mass) for the gravitational acceleration at its surface to triple?

4. Calculate the gravitational acceleration at the surface of (a) Mercury and (b) Saturn's moon Titan.

5. Two identical lead spheres are 14 cm apart and attract each other with a force of 0.25 μN. What is their mass?

6. What would be the initial acceleration of an object dropped from 1,000 km above Earth's surface?

7. What is the approximate value of the gravitational force between a 67-kg astronaut and a 73,000-kg space shuttle when they're 84 m apart?

8. Compare the gravitational attraction of the Earth for an astronaut on the surface of the moon with the gravitational attraction of the moon for the astronaut.

9. A sensitive gravimeter is carried to the top of Chicago's Sears Tower, where its reading for the acceleration of gravity is 0.00136 m/s^2 lower than at street level. Find the height of the building.

10. A roughly spherical volume of rock 1.3 km in radius is 30% denser than the surrounding rock, whose density is 2700 kg/m^3. The denser rock is centered 2.1 km below Earth's surface. By what percentage is the surface value of g directly above the denser rock increased due to its excess density? *Hint:* Treat the "extra" mass of the denser rock as a gravitating sphere, and calculate the gravitational acceleration it produces at Earth's surface.

11. If you're standing on the ground 15 m directly below the center of a spherical water tank containing 4×10^6 kg of water, by what fraction is your weight reduced due to the gravitational attraction of the water?

Section 9-3 Orbital Motion

12. At what altitude will a satellite complete a circular orbit of the Earth in 2.0 hours?

13. Find the speed of a satellite in geosynchronous orbit.
14. Mars's orbit has a diameter 1.52 times that of Earth's orbit. How long does it take Mars to orbit the Sun?
15. Calculate the orbital period for Jupiter's moon Io, which orbits 4.22×10^5 km from the center of the 1.9×10^{27} kg planet.
16. Given the orbital radius of 384,400 km and period of 27.3 days, calculate the acceleration of the moon in its circular orbit, and compare with the acceleration of gravity at Earth's surface. Show that the acceleration of the moon is smaller by the ratio of the square of Earth's radius to the square of the moon's orbital radius, thus confirming the inverse-square law for the gravitational force.
17. During the Apollo moon landings, one astronaut remained with the command module in lunar orbit, about 130 km above the surface. For half of each orbit, this astronaut was completely cut off from the rest of humanity, as the spacecraft rounded the far side of the moon (see Fig. 9-39). How long did this period last?

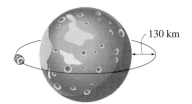

130 km

FIGURE 9-39 Problem 17.

18. A white dwarf is a collapsed star with roughly the mass of the Sun compressed into the size of the Earth. What would be (a) the orbital speed and (b) the orbital period for a spaceship in orbit just above the surface of such a white dwarf?
19. Given that our Sun orbits the galaxy with a period of 200 My at a distance of 2.6×10^{20} m from the galactic center, estimate the mass of the galaxy. Assume (incorrectly) that the galaxy is essentially spherical and that most of its mass lies interior to the Sun's orbit. To how many Sun-mass stars is your estimate equivalent?
20. Satellites A and B are in circular orbits, with A twice as far from Earth's center as B. How do their orbital periods compare?
21. Where should a satellite be placed to orbit the Sun in a circular orbit with a period of 100 days?
22. Determine the orbital period of the Hubble Space Telescope, which orbits Earth at an altitude of 610 km.
23. How far from the Sun's center would a satellite in circular orbit be heliosynchronous? *Hint:* Consult Appendix E, and remember that a synchronous orbit must parallel the equator. So which rotation period is appropriate?
24. Comets are thought to originate in a cloud of ice chunks in roughly circular orbits almost 1 light-year (the distance light travels in a year) from the Sun. Collisions among the ice

chunks and gravitational effects of other stars occasionally send a chunk into an elliptical orbit to become an observed comet. What is the orbital period of the cometary ice chunks in their remote orbits?
25. The asteroid Pasachoff orbits the Sun with a period of 1417 days. What is the semimajor axis of its orbit? Determine using Kepler's third law in comparison with Earth's orbital radius and period.
26. We derived Equation 9-4 on the assumption that the massive gravitating center remains fixed. Now consider the case of two objects of equal mass M orbiting each other as shown in Fig. 9-40. Show that the orbital period is given by

$$T^2 = \frac{16\pi^2 r^3}{GM},$$

where r is the orbital radius (half the distance between the objects).

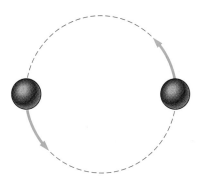

FIGURE 9-40 Problem 26.

27. Our galaxy belongs to a large group of galaxies known as the Virgo Cluster. The cluster contains about 1000 times the mass of our galaxy, which in turn contains about 10^{11} times the Sun's mass. We're roughly 50 million light-years from the center of the approximately spherical cluster. Could our galaxy have completed a full orbit of the cluster since the universe began, some 15 billion years ago?

Section 9-4 Gravitational Energy

28. Earth's distance from the Sun varies from 1.47×10^{11} m at perihelion to 1.52×10^{11} m at aphelion because its orbit is not quite circular. Find the change in potential energy as Earth goes from perihelion to aphelion.
29. How much energy does it take to launch a 230-kg instrument package on a vertical trajectory that peaks at an altitude of 1800 km?
30. One proposal for dealing with radioactive waste is to shoot it into the Sun. Suppose a waste canister were simply dropped, starting from rest in the vicinity of Earth's orbit. At what speed would it hit the Sun?
31. A rocket is launched vertically upward from Earth's surface at a speed of 3.1 km/s. What is its maximum altitude?

32. What vertical launch speed is necessary to get a rocket to 1100 km altitude?

33. Find the energy necessary to put 1 kg, initially at rest on Earth's surface, into geosynchronous orbit.

34. At its perihelion in February 1986, Comet Halley was 8.79×10^7 km from the Sun and was moving at 54.6 km/s. What was its speed when it crossed Saturn's orbit, in 1989?

35. Neglecting air resistance, to what height would you have to fire a rocket for the constant acceleration equations of Chapter 2 to give a height that is in error by 1%? Would those methods over- or underestimate the height?

36. What is the total mechanical energy associated with Earth's orbital motion?

37. Drag due to small amounts of residual air causes satellites in low Earth orbit to lose energy and eventually spiral to Earth. What fraction of its orbital energy is lost as a satellite drops from 300 to 100 km, assuming its orbit remains essentially circular?

38. Show that an object released from rest very far from Earth ($r \gg R_E$) reaches Earth's surface at essentially escape speed.

39. By what factor must the speed of an object in circular orbit be increased to reach escape speed from its orbital altitude?

40. Astronomers discover a new comet. As it crosses Earth's orbit the comet is moving at 45 km/s. Is the comet in an open or closed orbit?

41. The escape speed from a planet of mass 3.6×10^{24} kg is 9.1 km/s. What is the planet's radius?

42. Determine the escape speed from (a) Saturn's moon Iapetus, with mass 1.9×10^{21} kg and radius 7.3×10^5 m, and (b) a neutron star, with the Sun's mass crammed into a sphere 6.0 km in radius.

43. Two meteoroids are 250,000 km from Earth and moving at 2.1 km/s. One is headed straight for Earth, while the other is on a path that will come within 8500 km of Earth's center (Fig. 9-41). (a) What is the speed of the first meteoroid when it strikes Earth? (b) What is the speed of the second meteoroid at its closest approach to Earth? (c) Will the second meteoroid ever return to Earth's vicinity?

44. One component of the proposed "star wars" antimissile defense system calls for powerful laser beams aimed by mirrors in geosynchronous orbit. One simple countermeasure to this delicate technology is to put a truckload of rocks in the same orbit, going in the opposite direction. At what relative speed would the rocks hit the mirror?

45. Neglecting Earth's rotation, show that the energy needed to launch a satellite of mass m into circular orbit at altitude h is

$$\left(\frac{GM_E m}{R_E} \right) \left(\frac{R_E + 2h}{2(R_E + h)} \right).$$

46. Humanity currently consumes energy at the rate of about 10^{13} W. Suppose a method were developed to extract usable energy from the moon's orbital motion. If all our energy came from this source and if our energy use remained constant, by how much would (a) the moon's orbital radius and (b) its orbital period change in a century?

47. A projectile is launched vertically upward from a planet of mass M and radius R; its initial speed is twice the escape speed. Derive an expression for its speed as a function of the distance r from the center of the planet.

48. A spacecraft is in a circular Earth orbit at an altitude of 5500 km. By how much will its altitude decrease if it moves to a new circular orbit where (a) its orbital speed is 10% higher or (b) its orbital period is 10% shorter?

49. The Pioneer spacecraft left Earth's vicinity moving at about 38 km/s relative to the Sun (this figure combines the effect of rocket boost and Earth's orbital motion). How far out in the solar system could Pioneer get without additional rocket power or use of the "gravitational slingshot" effect?

Section 9-6 Tidal Forces

50. To get a feel for the behavior of tidal forces, consider an object consisting of two small masses m separated by a distance $2a$, a distance r from a body of mass M, as shown in Fig. 9-42. Use the law of universal gravitation to determine the gravitational force of the large mass on each of the small masses. Find an expression for the difference between the two forces, and put your expression over a common denominator. Show that, in the approximation where a^2 can be neglected compared with r^2, your expression for the force difference can be written

$$\Delta F = \frac{4GMma}{r^3}.$$

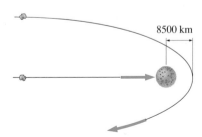

FIGURE 9-41 Partial trajectories of two meteoroids (Problem 43).

8500 km

r

$m \quad m$

M

$2a$

FIGURE 9-42 The tidal force on an object arises from the different gravitational forces experienced by different parts of the object (Problem 50).

This is the tidal force, and your expression shows that it drops as the *cube* of the distance.

51. Show that the force of the Sun's gravity on Earth is nearly 200 times that of the moon's gravity, but that the tidal force of the moon on Earth (see preceding problem) is about twice that of the Sun.

52. Suppose the composite object in Fig. 9-42 is held together by the gravitational attraction of the two small masses. Show that the tidal force of problem 50 is greater than the self-gravitational force of the object when $r^3 < 16Ma^3/m$. This defines the Roche limit for this simple example.

Paired Problems

(Both problems in a pair involve the same principles and techniques. If you get the first problem, you should be able to solve the second one.)

53. An astronaut hits a golf ball horizontally from the top of a lunar mountain so fast that it goes into circular orbit. What is its orbital period?

54. Find the period of a spacecraft in circular orbit 150 km above Mars's surface.

55. Two meteoroids are 160,000 km from Earth's center and heading straight toward Earth. One is moving at 10 km/s, the other at 20 km/s. At what speeds will they strike Earth?

56. Two rockets are launched from Earth's surface, one at 12 km/s and the other at 18 km/s. How fast is each moving when it crosses the moon's orbit?

57. A satellite is in an elliptical orbit at altitudes ranging from 230 to 890 km. At the high point it's moving at 7.23 km/s. How fast is it moving at the low point?

58. A missile's trajectory takes it to a maximum altitude of 1200 km. If its launch speed is 6.1 km/s, how fast is it moving at the peak of its trajectory?

59. To what radius would Earth have to be shrunk, with no loss of mass, for escape speed at its surface to be 30 km/s?

60. What is the mass of a planetary moon whose radius is 14 km and whose surface escape speed is 0.13 m/s?

61. A 720-kg spacecraft has total energy -5.3×10^{11} J and is in circular orbit about the Sun. Find (a) its orbital radius, (b) its kinetic energy, and (c) its speed.

62. A 2500-kg boulder is among the myriad particles making up Saturn's rings shown in Fig. 9-43. The boulder is in a circular orbit 95,000 km from the center of Saturn. Find (a) its total energy, (b) its orbital speed, and (c) its orbital period.

Supplementary Problems

63. Mercury's orbital speed varies from 38.8 km/s at aphelion to 59.0 km/s at perihelion. If the planet is 6.99×10^{10} m from the Sun's center at aphelion, how far is it at perihelion?

64. Show that the form $\Delta U = mg\Delta r$ follows from Equation 9-5 when $r_1 \simeq r_2$. *Hint:* Write $r_2 = r_1 + \Delta r$, and apply the binomial approximation (see Appendix A).

FIGURE 9-43 Saturn's rings (Problem 62).

65. A black hole is an object so dense that its escape speed exceeds the speed of light. Although a full description of black holes requires general relativity, the radius of a black hole can be calculated using Newtonian theory. (a) Show that the radius of a black hole of mass M is $2GM/c^2$, where c is the speed of light. What are the radii of black holes with (b) a the mass of the Earth and (c) the mass of the Sun?

66. When a star with twice the Sun's mass exhausts its nuclear fuel, it collapses into a neutron star with a radius of about 10 km. What are (a) the acceleration of gravity and (b) the escape speed at the surface of such a neutron star?

67. Two satellites are in geosynchronous orbit, but in diametrically opposite positions (Fig. 9-44). Into how much lower a circular orbit should one spacecraft descend if it is to catch up with the other after 10 complete orbits? Neglect rocket firing times and time spent moving between the two circular orbits.

68. A black hole (see Problem 65) has the same mass as the Sun. How close could you get to the black hole before the tidal force stretching you apart had a strength of 1 ton? (See Problem 50.)

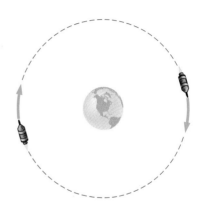

FIGURE 9-44 Problem 67.

CHAPTER 10

SYSTEMS OF PARTICLES

A gymnast's body is a system composed of many parts, and its motion is complex. Application of Newton's laws nevertheless allows us to understand this motion in simple terms.

So far we've treated most objects as point particles, ignoring the fact that they're actually composed of many smaller parts. Here we deal explicitly with systems composed of many particles. These include **rigid bodies**—objects like baseballs, cars, and planets whose constituent particles remain in fixed orientations relative to each other—as well as systems like human bodies, exploding fireworks, or flowing rivers, whose constituent parts may move relative to one another. Much contemporary research involves many-particle systems; for example, meteorologists are still struggling to understand and predict the motions of the air masses that determine our weather. In subsequent chapters we'll look at specific instances of many-particle systems, including the rotational motion of rigid bodies (Chapter 12) and the behavior of fluids (Chapter 18).

10-1 CENTER OF MASS

Figure 10-1 shows a hammer tossed into the air. The motion of most parts of the hammer is quite complex. But the curve superimposed on the photograph shows that one point seems to follow the parabola that we expect of a point projectile

(Section 4-3). This point is known as the **center of mass;** it represents a kind of "average position" of all the mass making up the hammer.

How do we find the center of mass? Figure 10-1 shows that the center of mass obeys Newton's second law just as would a point particle. If we can find a point whose acceleration **A** obeys Newton's second law $\mathbf{F} = M\mathbf{A}$, where **F** is now the net force on an entire system and M the mass of that system, then we will have found the center of mass. (Note that we use capital letters for quantities associated with the center of mass.)

Consider a system consisting of many particles. To find the center of mass, we want an equation that looks like Newton's second law but involves the total mass of the system and the net force on the entire system. If we apply Newton's second law to the ith particle in the system, we have

FIGURE 10-1 Most parts of the hammer move in complicated paths. But one point—the center of mass—follows the simple parabolic trajectory of a projectile.

$$\mathbf{F}_i = m_i\mathbf{a}_i = m_i\frac{d^2\,\mathbf{r}_i}{dt^2} = \frac{d^2 m_i\mathbf{r}_i}{dt^2},$$

where $\mathbf{F}_i$ is the net force on the particle, m_i its mass, and where we've written the acceleration $\mathbf{a}_i$ as the second derivative of the position $\mathbf{r}_i$. The total force on the entire system is the sum of the forces acting on all N particles. We write this sum compactly using the summation symbol Σ:

$$\mathbf{F}_{\text{total}} = \sum_{i=1}^{N} \mathbf{F}_i = \sum_{i=1}^{N} \frac{d^2 m_i r_i}{dt^2},$$

where the terms above and below the Σ indicate that the sum runs over all values of i from 1 to N—i.e., over all the particles comprising the system. But the sum of derivatives is the same as the derivative of the sum, so

$$\mathbf{F}_{\text{total}} = \frac{d^2\left(\sum m_i\mathbf{r}_i\right)}{dt^2}.$$

We can now put this equation in the form of Newton's second law. Multiplying and dividing the right-hand side by the total mass $M = \Sigma\, m_i$, and distributing this constant M through the differentiation, we have

$$\mathbf{F}_{\text{total}} = M\frac{d^2}{dt^2}\left(\frac{\sum m_i\mathbf{r}_i}{M}\right). \tag{10-1}$$

Equation 10-1 has the desired form $\mathbf{F} = M\mathbf{A} = M(d^2\mathbf{R}/dt^2)$ if we define

$$\mathbf{R} = \frac{\sum m_i\mathbf{r}_i}{M}. \quad \text{(center of mass)} \tag{10-2}$$

We've almost reached our goal of showing that the point **R** moves according to Newton's law $\mathbf{F} = Md^2\mathbf{R}/dt^2$. Why aren't we there now? Because the force $\mathbf{F}_{\text{total}}$ in Equation 10-1 is the sum of *all* the forces acting on all the particles of the system, including the **internal forces** that act between the particles. We would like the force **F** in Newton's law to be the net **external force**—the net force applied from outside the system. We can write the force $\mathbf{F}_{total}$ in Equation 10-1—the total force on all the particles of the system—as

$$\mathbf{F}_{total} = \Sigma\, \mathbf{F}_{ext} + \Sigma\, \mathbf{F}_{int},$$

where $\Sigma\, \mathbf{F}_{ext}$ is the sum of all the external forces and $\Sigma\, \mathbf{F}_{int}$ the sum of the internal forces. According to Newton's third law of motion, each of the internal forces has an equal but oppositely directed reaction force that itself acts on a particle of the system and is therefore included in the sum $\Sigma\, \mathbf{F}_{int}$. (Each external force also has a reaction force, but these reaction forces act *outside* the system and are therefore not included in the sum.) Added vectorially, the reaction forces therefore cancel in pairs, so $\Sigma\, \mathbf{F}_{int} = \mathbf{0},$ and the force $\mathbf{F}_{total}$ in Equation 10-1 is just the net *external* force applied to the system. So the point $\mathbf{R}$ defined in Equation 10-2 does obey Newton's law, written in the form

$$\mathbf{F}_{net\,ext} = M\,\frac{d^2\mathbf{R}}{dt^2}, \tag{10-3}$$

where $\mathbf{F}_{net\,ext}$ is the net external force applied to the system and M the total mass.

We have defined the center of mass $\mathbf{R}$ so we can apply Newton's second law to the entire system rather than to each individual particle. As far as its overall motion is concerned, a complex system acts as though all its mass were concentrated at the center of mass. The form of Equation 10-2 shows that the center of mass is indeed an average of the positions of the individual particles, each weighted according to its mass.

Finding the Center of Mass

Equation 10-2 expresses the center of mass in vector form; taking components gives three separate scalar equations for the xyz coordinates of the center of mass. If we consider only a one-dimensional situation, in which the constituent particles lie along a single line, then the center of mass must lie along that line. Similarly, for a two-dimensional object, the center of mass lies in the plane of the object. For the simple case of two point masses, we can take the x axis along the line joining the masses and write Equation 10-2 in the form

$$X = \frac{m_1 x_1 + m_2 x_2}{m_1 + m_2}, \tag{10-4}$$

● **EXAMPLE 10-1** WEIGHTLIFTER

You want to lift a barbell at its center of mass. (a) Suppose there are equal 60-kg masses at either end of the 1.5-m-long bar. (b) Suppose someone put unequal masses of 50 kg and 80 kg on the ends. Where is the center of mass in each case? Neglect the mass of the bar.

Solution

In (a) it's obvious that the center of mass must lie at the center of the barbell. If we choose the origin to lie at one of the masses, then $x_1 = 0$ m and $x_2 = 1.5$ m; applying Equation 10-4 then gives

$$X = \frac{m_1x_1 + m_2x_2}{m_1 + m_2} = \frac{(60 \text{ kg})(0 \text{ m}) + (60 \text{ kg})(1.5 \text{ m})}{60 \text{ kg} + 60 \text{ kg}}$$
$$= 0.75 \text{ m},$$

as expected. In case (b), with the origin at the 50-kg mass, we have

$$X = \frac{m_1x_1 + m_2x_2}{m_1 + m_2} = \frac{(50 \text{ kg})(0 \text{ m}) + (80 \text{ kg})(1.5 \text{ m})}{50 \text{ kg} + 80 \text{ kg}}$$
$$= 0.92 \text{ m}.$$

Again, the result makes sense: the center of mass is closer to the more massive constituent of the system (Fig. 10-2).

> **TIP Choosing Your Origin** In working this ex-
> ample we chose the origin at one of the masses. That
> choice was convenient because it made one of the
> terms in the sum $\Sigma\ m_ix_i$ zero. But, as always, the
> choice of origin is purely for convenience; it cannot
> influence the actual physical location of the center of
> mass.

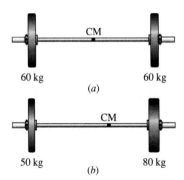

FIGURE 10-2   Center of mass of a barbell. (a) Two equal masses. (b) 50-kg and 80-kg masses.

EXERCISE   Where is the center of mass of a canoe if a 76-kg paddler is 0.85 m from the stern and a 52-kg paddler is 3.9 m from the stern? Neglect the mass of the canoe itself.

Answer: 2.1 m from the stern

Some problems similar to Example 10-1:   1, 4, 6, 8

● **EXAMPLE 10-2**   CENTER OF MASS IN A PLANE

Four equal masses m lie at the corners of a rectangle of width a and length b. Where is the center of mass?

Solution
A convenient coordinate system has two sides of the rectangle along the x and y axes, with one corner at the origin (Fig. 10-3). Then the x and y components of Equation 10-2 are

$$X = \frac{\Sigma\ m_ix_i}{M} = \frac{m(0 + b + b + 0)}{4m} = \frac{b}{2}$$

and

$$Y = \frac{\Sigma\ m_iy_i}{M} = \frac{m(0 + 0 + a + a)}{4m} = \frac{a}{2},$$

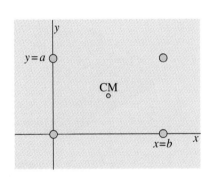

FIGURE 10-3   Example 10-2.

where in both cases we summed going counterclockwise around the rectangle, starting at the origin. Our result is hardly surprising: the center of mass lies at the center of the rectangle.

> **TIP Exploit Symmetries** It's no accident that the
> center of mass in this example lies at the center of the
> rectangle. Whenever an object has a symmetry axis,
> the center of mass must lie on that axis since individ-
> ual particles on either side cancel each other's effect in
> the center-of-mass averaging. The rectangle of Exam-
> ple 10-2 has *two* symmetry axes, namely vertical and
> horizontal lines through its center. The center of mass
> must lie on *both* axes and is therefore at their one
> point of intersection, the center. Attention to symme-
> try here and in many other areas of physics can save a
> great deal of calculation.

EXERCISE   Three equal masses m lie at the vertices of a triangle with coordinates $(-a, 0)$, $(a, 0)$, and $(0, a)$. Find the center of mass.

Answer: $X = 0$, $Y = \dfrac{a}{3}$

Some problems similar to Example 10-2:   2, 3, 5, 11

Continuous Distributions of Matter

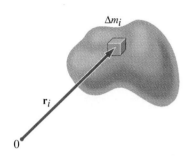

FIGURE 10-4 A chunk of continuous matter, showing one mass element Δm_i and its position vector $\mathbf{r}_i$ from an arbitrary origin.

We've expressed the center of mass in terms of sums over individual particles. Ultimately, matter *is* composed of individual particles. But it's often a convenient approximation to consider that matter is continuously distributed; after all, we don't want to deal with 10^{23} or so atoms to find the center of mass of a book or other macroscopic object. We can consider a chunk of continuous matter to be composed of individual pieces of mass Δm_i, with position vectors $\mathbf{r}_i$; we call these pieces **mass elements** (Fig. 10-4). The center of mass of the entire chunk is then given by Equation 10-2:

$$\mathbf{R} = \frac{\sum \Delta m_i \mathbf{r}_i}{M},$$

where $M = \sum \Delta m_i$ is the total mass. This expression is slightly vague, for we don't know exactly where within a mass element its position vector should terminate. But in the limit as the mass elements become arbitrarily small, like point particles, the expression becomes exact. As in Chapter 7, that limiting process defines an integral, giving

$$\mathbf{R} = \lim_{\Delta m_i \to 0} \frac{\sum \Delta m_i \mathbf{r}_i}{M} = \frac{1}{M} \int \mathbf{r}\, dm, \quad \left(\begin{array}{c}\text{center of mass,}\\ \text{continuous matter}\end{array}\right) \quad (10\text{-}5)$$

where the integration is over the entire volume of the object. Like the sum 10-2, the vector form 10-5 stands for three separate integrals for the three components of the center-of-mass position. Example 10-3 shows how to evaluate these integrals.

● **EXAMPLE 10-3**   AN AIRCRAFT WING

The wing of a supersonic aircraft is in the form of an isosceles triangle of length ℓ, width w, and negligible thickness (Fig. 10-5). It has total mass M, distributed uniformly over the wing. Where is its center of mass?

Solution
Figure 10-5 shows the wing with a suitable coordinate system. Since the wing is flat and of negligible thickness, its center of mass must lie in the x-y plane. The wing is symmetric about the x axis, so the center of mass must lie on the x axis; that is, $Y = 0$. So we need to integrate only to find the x coordinate. Figure 10-5 shows an appropriate mass element, here in the form of a strip of width dx located a distance x along the axis. To find the mass dm of the strip, we first find its area. Figure 10-5 shows that the equations describing the sloping sides of the wing are $y = \pm wx/2\ell$, so our strip extends from $y = -wx/2\ell$ to $+wx/2\ell$; its length is therefore wx/ℓ. Since its width is dx, its area is then $dA = (wx\, dx)/\ell$ (see Fig. 10-6).

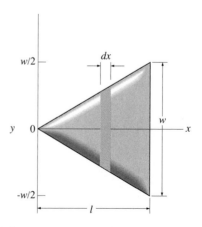

FIGURE 10-5 A supersonic aircraft wing (Example 10-3). Dark area is a mass element of width dx.

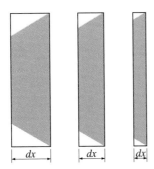

FIGURE 10-6 As the width dx of the strip becomes arbitrarily small, its area approaches that of a rectangle.

Since mass is distributed uniformly over the wing, the mass dm of the strip is the same fraction of the total mass as its area is of the total area:

$$\frac{dm}{M} = \frac{dA}{A} = \frac{(wx\,dx)/\ell}{\frac{1}{2}w\ell} = \frac{2x\,dx}{\ell^2},$$

so $dm = (2Mx\,dx)/\ell^2$. The x component of the center-of-mass position is given by the x component of Equation 10-5:

$$X = \frac{1}{M}\int x\,dm.$$

Using our expression for dm, and integrating over the entire triangle—that is, from $x = 0$ to $x = \ell$— we have

$$X = \frac{1}{M}\int_0^\ell \frac{2M}{\ell^2}x^2\,dx = \frac{2}{\ell^2}\frac{x^3}{3}\bigg|_0^\ell = \frac{2}{3}\ell.$$

Does this make sense? Yes. More of the wing's mass is concentrated near the wide end at $x = \ell$, so the center of mass should be located closer to the end.

> **TIP Setting Up an Integral** An integral like $\int x\,dm$ may seem confusing since you see an x and a dm under the integral sign, and the two don't seem re-

lated. But they must be, or the integral wouldn't make sense. Here we related them by expressing the mass dm of a mass element in terms of its x coordinate; note that the expression for dm increases with increasing x, as it should since the wing gets wider going to the right. We found that relation by forming the ratio of the mass element's mass dm to the total wing mass and equating that to the ratio of the mass element's area dA to the total wing area. You'll encounter similar situations involving not only mass but also electric charge and other quantities.

EXERCISE Find the center of mass of the triangular slab shown in Fig. 10-7, assuming its mass is distributed uniformly.

Answer: $X = \dfrac{a}{3}$, $Y = \dfrac{b}{3}$

Some problems similar to Example 10-3: 10, 18, 57

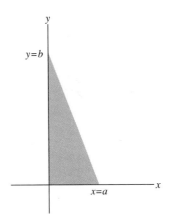

FIGURE 10-7 Find the center of mass of the triangular slab.

With still more complex objects, it's convenient to find the centers of mass of major subpieces, then treat those subpieces as point particles to find the center of mass of the entire object. If the wing of Example 10-3 is placed on a cylindrical airplane fuselage, for example, we can find the center of mass of the entire plane by treating the wing and fuselage as point particles located at their respective centers of mass (Fig. 10-8). And an irregular object like that of Fig. 10-9 is treated by locating the centers of mass of its three constituent rectangles (see Problem 10-17).

Figure 10-9 shows that the center of mass need not lie within the object. A high jumper (Fig. 10-10) takes advantage of this fact by arching her body into a shape qualitatively like that of Fig. 10-9. The jumper's center of mass never has to get as high as the bar, thereby reducing the amount of work she must do

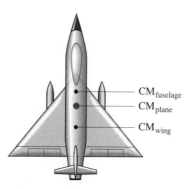

FIGURE 10-8 The center of mass of the airplane is found by treating the wing and fuselage as point particles located at their respective centers of mass. Given the center-of-mass location, is the wing or the fuselage more massive?

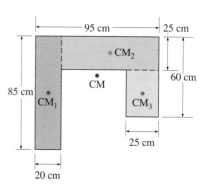

FIGURE 10-9 An irregular object, showing centers of mass of the constituent rectangles and of the object itself. Note that the center of mass lies outside the object.

FIGURE 10-10 Germany's Heike Henkel clears the high jump bar, but her center of mass doesn't. This is possible because her entire body is not above the bar at the same time.

against gravity. Similarly, a ballet dancer can appear to float through the air, her head and torso following a horizontal trajectory, by raising her legs and therefore the position of the center of mass relative to her torso. The center of mass, of course, follows a parabolic trajectory, but that's not what the audience notices (Fig. 10-11).

Motion of the Center of Mass

We defined the center of mass (Equation 10-2) so that its motion would obey Newton's law $\mathbf{F} = M\mathbf{A}$, where $\mathbf{F}$ is the net external force applied to a system and M the total mass. When, for example, gravity is the only external force acting on a system, the center of mass follows the trajectory appropriate to a point particle (as shown in Figs. 10-1 and 10-11). An example on a grander scale is the breakup of comet Biela as it passed the Sun in the mid-1800s. Today, the Andromedid swarm of meteors orbits in similar but not identical orbits. But the center of mass of the swarm remains in the orbit of the original comet, despite the fact that there is no longer any matter there!

When there is no net external force on a system, then the center of mass acceleration $\mathbf{A}$ is zero, and the center of mass moves with constant velocity. In the special case of a system at rest and subject to no net external force, the center of mass remains at rest despite any motions of the system's constituents.

FIGURE 10-11 Performing a grand jeté, Jennifer Davis of the Pittsburgh Ballet Theatre keeps her head and torso moving nearly horizontally while her center of mass describes a parabolic trajectory.

● EXAMPLE 10-4   LEAPING TO SHORE

You're standing 1.8 m from the shoreward end of a 4.2-m-long boat; from the end to shore is another 1.1 m (Fig. 10-12). Your mass is 70 kg and the boat's mass is 150 kg. The center of mass of the boat lies at its center. You walk to the shoreward end of the boat, preparing to leap to shore. How far do you have to leap? Where is the boat at the instant you reach shore? Neglect any frictional effects.

Solution

The net external force on the system consisting of you and the boat is zero, so the center of mass of the system remains fixed. Letting m_1 and x_1 be your mass and position, and m_2 and x_2 the mass and position of the boat (that is, of the boat's center of mass), and designating the initial and final values by the subscripts i and f, respectively, we equate the initial and final positions of the center of mass:

$$\frac{m_1 x_{1i} + m_2 x_{2i}}{m_1 + m_2} = \frac{m_1 x_{1f} + m_2 x_{2f}}{m_1 + m_2}.$$

The total mass $m_1 + m_2$ cancels, so

$$m_1 x_{1i} + m_2 x_{2i} = m_1 x_{1f} + m_2 x_{2f}. \qquad (10\text{-}6)$$

Since we're interested in the final position with respect to the shore, an appropriate choice of coordinate system has the origin at the shore. Then Fig. 10-12 shows that $x_{1i} = 1.1$ m + 1.8 m = 2.9 m and $x_{2i} = 1.1$ m + 2.1 m = 3.2 m. The points

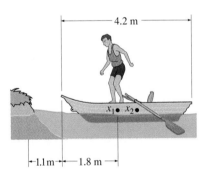

FIGURE 10-12   Example 10-4.

x_{1f} and x_{2f} are still unknown. But they're related: when you reach the end of the boat, you're half a boat length closer to shore than the center of the boat. Therefore, $x_{2f} = x_{1f} + \frac{1}{2}L$, where L is the 4.2-m length of the boat. Using this result in Equation 10-6 gives

$$m_1 x_{1i} + m_2 x_{2i} = m_1 x_{1f} + m_2(x_{1f} + \tfrac{1}{2}L).$$

Solving for x_{1f}, we then have

$$x_{1f} = \frac{m_1 x_{1i} + m_2 x_{2i} - \frac{1}{2}m_2 L}{m_1 + m_2}$$
$$= (70 \text{ kg})(2.9 \text{ m}) + (150 \text{ kg})(3.2 \text{ m})$$
$$- \tfrac{1}{2}(150 \text{ kg})(4.2 \text{ m})70 \text{ kg} + 150 \text{ kg}$$
$$= 1.7 \text{ m}.$$

This is the distance you have to leap from the end of the boat to shore. When you reach shore, you are at the origin: $x_{1f} = 0$. Solving Equation 10-6 for the boat's position then gives

$$x_{2f} = \frac{m_1 x_{1i} + m_2 x_{2i}}{m_2} = \frac{(70 \text{ kg})(2.9 \text{ m}) + (150 \text{ kg})(3.2 \text{ m})}{150 \text{ kg}}$$
$$= 4.6 \text{ m}.$$

After the instant you reach shore, we can no longer argue that the center of mass remains fixed. Why not? Because an external force—the stopping force of the ground—has acted on part of the system, so the center of mass is no longer fixed. Indeed, the boat continues to drift away from shore. Question 10 explores this situation further.

EXERCISE   A 940-kg horse trailer with essentially frictionless wheels is sitting in a level parking lot. The trailer is 5.1 m long, and its center of mass is at its center. Its passenger, a 480-kg horse, breaks free from its stall at one end of the trailer and walks to the other end. How far does the trailer move relative to the ground? Treat the horse as a point particle.

Answer: 1.7 m

Some problems similar to Example 10-4: 14–17 ●

10-2 MOMENTUM

In Chapter 5 we defined the linear momentum **p** of a particle as $\mathbf{p} = m\mathbf{v}$, and first wrote Newton's law in the form $\mathbf{F} = d\mathbf{p}/dt$. We subsequently applied the law to single particles, using the form $\mathbf{F} = m\mathbf{a}$, but suggested that the momentum concept would play an important role in many-particle systems. We're now ready to explore that role.

The momentum of a system of particles is just the vector sum of the individual momenta:

$$\mathbf{P} = \sum \mathbf{p}_i = \sum m_i \mathbf{v}_i,$$

where m_i and $\mathbf{v}_i$ are the masses and velocities of the individual particles. But we really don't want to keep track of all the particles in the system. Is there a simpler way to express the total momentum? There is, and it comes from writing the individual particle velocities as time derivatives of position: $\mathbf{v} = d\mathbf{r}/dt$. Then we have

$$\mathbf{P} = \sum m_i \frac{d\mathbf{r}_i}{dt} = \frac{d}{dt} \sum m_i \mathbf{r}_i,$$

where the last step follows because the individual particle masses are constant and because the sum of derivatives is the derivative of the sum. In Section 10-1 we defined the center-of-mass position $\mathbf{R}$ as $\sum m_i \mathbf{r}_i / M$, where M is the total mass. So the total momentum can be written

$$\mathbf{P} = \frac{d}{dt} M\mathbf{R},$$

or, assuming the system mass M remains constant,

$$\mathbf{P} = M\frac{d\mathbf{R}}{dt} = M\mathbf{V}, \tag{10-7}$$

where $\mathbf{V} = d\mathbf{R}/dt$ is the center-of-mass velocity. So the momentum of a system is given by an expression similar to that of a single particle; it is the product of the system mass with the system velocity—that is, with the velocity of the center of mass. If this seems so obvious as not to need deriving, watch out! We'll see in Section 10.3 that the same is *not* true for the system's total energy.

If we differentiate Equation 10-7 with respect to time, we have

$$\frac{d\mathbf{P}}{dt} = M\frac{d\mathbf{V}}{dt} = M\mathbf{A},$$

where $\mathbf{A}$ is the center-of-mass acceleration. But we defined the center of mass so that its motion obeyed Newton's second law, $\mathbf{F} = M\mathbf{A}$, with $\mathbf{F}$ the net external force on the system. So we can write simply

$$\mathbf{F} = \frac{d\mathbf{P}}{dt}, \tag{10-8}$$

showing that the momentum of a system of particles changes only if there is a net external force on the system. Remember the hidden role of Newton's third law in all this: only because the forces *internal* to the system cancel in pairs can we ignore them and consider just the external force.

Conservation of Momentum

In the special case when the net external force is zero, Equation 10-8 gives

$$\frac{d\mathbf{P}}{dt} = \mathbf{0}, \qquad \text{(no net external force)}$$

so $\qquad$ **P = constant.** $\qquad$ (conservation of linear momentum) $\qquad$ (10-9)

Equation 10-9 describes the **conservation of linear momentum,** one of the most fundamental laws of physics:

Conservation of linear momentum: In the absence of a net external force, the total momentum P of a system—the vector sum of the individual momenta mv of its constituent particles—remains constant.

Momentum conservation holds no matter how many particles are involved and no matter how they're moving. It applies to systems ranging in size from atomic nuclei to pool balls (Fig. 10-13), from colliding cars to galaxies. Although we derived Equation 10-9 from Newton's laws, momentum conservation is even more basic than those laws since it applies to subatomic and nuclear systems where the laws and even language of Newtonian physics are hopelessly inadequate. The following examples show the wide range and power of the momentum conservation principle.

FIGURE 10-13 Initially the momentum of a system of pool balls is that of the cue ball alone, shown coming from the left in this strobe photo. After that one ball has struck the others, the vector sum of all the balls' momenta remains what it was initially.

● EXAMPLE 10-5 CANOEING

Amy (mass 55 kg) and George (mass 70 kg) are sitting at opposite ends of a canoe (mass 22 kg) at rest on frictionless water. She tosses him a 14-kg pack, giving it a speed of 3.1 m/s relative to the water. What is the speed of the canoe (a) while the pack is in the air and (b) after George catches it?

Solution
We're interested here in the horizontal motion of the canoe. With no external forces acting in the horizontal direction, the horizontal component of the momentum is conserved. Originally, everything is at rest and the total momentum is zero. While the pack is in the air, it has horizontal momentum $m_p v_p$. The canoe and its passengers are all moving together; they have momentum $(m_a + m_g + m_c)v_c$, where m_a, m_g, and m_c are the masses of Amy, George, and the canoe, respectively, and where v_c is the canoe's horizontal velocity. Since horizontal momentum is conserved, it must still be zero when the pack is in the air, so

$$m_p v_p + (m_a + m_g + m_c)v_c = 0,$$

or

$$v_c = -\frac{m_p v_p}{m_a + m_g + m_c} = -\frac{(14 \text{ kg})(3.1 \text{ m/s})}{55 \text{ kg} + 70 \text{ kg} + 22 \text{ kg}}$$
$$= -0.30 \text{ m/s}.$$

The minus sign tells us that the canoe moves in the opposite direction from the pack.

There is no need to calculate the speed after George catches the pack: canoe, passengers, and pack all have the same velocity then. For the total momentum to remain zero, that velocity must be zero.

EXERCISE A 28-kg dog leaps horizontally out of a 72-kg rowboat at a speed of 1.6 m/s relative to the water. If the rowboat is initially at rest, how fast is it moving after the dog leaps?

Answer: 0.62 m/s

Some problems similar to Example 10-5: 22, 31, 51, 52

● EXAMPLE 10-6 THE SPEED OF A HOCKEY PUCK

A hockey coach enlists a physics student on the team to measure the speed of a hockey puck (Fig. 10-14). The student loads a small styrofoam chest with sand, giving it a total mass of 6.4 kg. He places the chest, at rest on frictionless ice, in the path of the hockey puck, whose mass is 160 g. The puck embeds itself in the styrofoam, and the chest moves off at 1.2 m/s. What was the speed of the puck?

Solution
Since there is no net external force, momentum is conserved. Originally, all the momentum of the puck + chest system is in the puck: $P = m_p v_p$, where we do not need vector notation because the motion is in one dimension. After the puck hits the chest, the two move off together at the same speed: $P = (m_p + m_c)v_c$. So the statement that momentum is conserved becomes

FIGURE 10-14 How fast is the puck moving? (Example 10-6)

$$m_p v_p = (m_p + m_c)v_c .$$

Then

$$v_p = \frac{(m_p + m_c)v_c}{m_p} = \frac{(0.16 \text{ kg} + 6.4 \text{ kg})(1.2 \text{ m/s})}{0.16 \text{ kg}} = 49 \text{ m/s} .$$

Variations on this technique are routinely used to measure the speeds of rapidly moving objects like bullets.

TIP Spare the Details   The power of conservation principles like conservation of linear momentum and conservation of energy lies in their ability to predict the outcomes of physical happenings without the need for lots of intermediate detail. In this example, for instance, we didn't need to know anything about the interaction between hockey puck and styrofoam chest—for example, how the puck slowed down, or the acceleration of the chest. All we knew was that the system had a certain momentum before the interaction, and the conservation of momentum principle told us it had to have that same momentum after the interaction. That alone was enough to solve the problem. When confronted with a conservation-law problem, go straight from the initial to the final situation without trying to analyze the intermediate details.

EXERCISE   A 42-kg child stands at rest on ice skates. She catches a 1.1-kg ball moving horizontally at 9.5 m/s. What is her speed after she's caught the ball?

Answer: 0.24 m/s

Some problems similar to Example 10-6:   23, 26 ●

● **EXAMPLE 10-7**   RADIOACTIVE DECAY

A lithium-5 nucleus (^{5}Li) is moving at 1.6×10^6 m/s when it decays into a proton (^{1}H) and an alpha particle (^{4}He). [The superscripted numbers are the total number of nucleons (neutrons and protons) in a nucleus, and give the approximate nuclear mass in unified atomic mass units (u).] The alpha particle is detected moving at 1.4×10^6 m/s, at 33° to the original velocity of the ^{5}Li nucleus. What are the magnitude and direction of the proton's velocity?

Solution
Again, momentum is conserved. Before the decay, we have only the lithium nucleus. After the decay, there are two particles whose individual momenta contribute to the total. So the statement that momentum is conserved becomes

$$m_{\text{Li}} \mathbf{v}_{\text{Li}} = m_p \mathbf{v}_p + m_\alpha \mathbf{v}_\alpha ,$$

where we keep the full vector notation because the motion involves two dimensions. If we choose the x axis along the direction of $\mathbf{v}_{\text{Li}}$, then we can write separately the two components of the momentum conservation equation:

x component: $\quad m_{\text{Li}} v_{\text{Li}} = m_p v_{px} + m_\alpha v_{\alpha x}$

y component: $\quad 0 = m_p v_{py} + m_\alpha v_{\alpha y}.$

The given information in this problem includes the speed v_{Li} of the lithium nucleus and the final speed v_α and direction ϕ of the alpha particle. From Fig. 10-15 we see that the components of the alpha particle velocity are $v_{\alpha x} = v_\alpha \cos \phi$ and $v_{\alpha y} = v_\alpha \sin \phi$. Solving our two equations for the unknown components of the proton velocity then gives

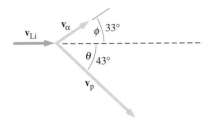

FIGURE 10-15 Velocity diagram for the decay of ^{5}Li nucleus (Example 10-7).

x component:

$$v_{px} = \frac{m_{Li}v_{Li} - m_\alpha v_{\alpha x}}{m_p} = \frac{m_{Li}v_{Li} - m_\alpha v_\alpha \cos\phi}{m_p}$$

$$= \frac{(5.0 \text{ u})(1.6 \text{ Mm/s}) - (4.0 \text{ u})(1.4 \text{ Mm/s})(\cos 33°)}{1.0 \text{ u}}$$

$$= 3.30 \text{ Mm/s}$$

y component: $v_{py} = -\dfrac{m_\alpha v_{\alpha y}}{m_p} = -\dfrac{m_\alpha v_\alpha \sin\phi}{m_p}$

$$= -\frac{(4.0 \text{ u})(1.4 \text{ Mm/s})(\sin 33°)}{1.0 \text{ u}}$$

$$= -3.05 \text{ Mm/s} .$$

Thus the proton's speed is

$$v_p = \sqrt{v_{px}^2 + v_{py}^2} = \sqrt{(3.30 \text{ Mm/s})^2 + (-3.05 \text{ Mm/s})^2}$$
$$= 4.5 \text{ Mm/s}$$

and its direction is given by

$$\theta = \tan^{-1}\left(\frac{v_{py}}{v_{px}}\right) = \tan^{-1}\left(\frac{-3.05 \text{ Mm/s}}{3.30 \text{ Mm/s}}\right) = -43°,$$

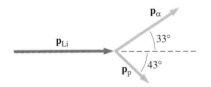

FIGURE 10-16 Momentum diagram for Example 10-7.

where the minus sign shows that the proton's velocity vector points below the x axis, as shown in Fig. 10-15.

We can see graphically that our results make sense by redrawing Fig. 10-15 as a momentum vector diagram—that is, by multiplying each velocity vector by the appropriate mass. The result, Fig. 10-16, shows that the y components of the alpha particle and proton momenta cancel, and that their x components sum to the momentum of the original nucleus.

With no external forces acting, momentum had to be conserved in this decay. But kinetic energy is not conserved, as you will find if you evaluate the kinetic energies before and after decay (see Problem 41). Internal potential energy of the ^{5}Li nucleus is converted to kinetic energy in the decay process.

EXERCISE A skater embeds a firecracker in a large snowball and pushes it across a frozen pond at 7.6 m/s. The firecracker explodes, breaking the snowball into two equal-mass chunks. One chunk moves off at 9.3 m/s at +19° to the original direction of motion. Find the speed and direction of the second chunk.

Answer: 7.1 m/s, −25° to the original direction

Some problems similar to Example 10-7: 24, 36, 49, 50, 70 ●

Defining the System: Internal and External Forces

How do we decide whether or not the forces acting on a system are external? That depends on our choice of what to include in the system. In Example 10-6, for instance, we chose to include both the hockey puck and the styrofoam chest in our system; the force exerted on the chest by the puck was then an internal force and we did not consider it explicitly. On the other hand, we could have chosen our system to include just the chest; then the force of the puck would be an external force, and the momentum of the system—the chest alone—would change. Of course, our definition of the system cannot affect what actually happens. Either description of the puck-chest interaction is equally valid; our choice was made so we could apply the conservation of momentum principle.

Sometimes we're interested in the force exerted on a particular object. Then it's convenient to redefine our system to include only that object, so the force of interest becomes an external force. The example below illustrates this situation.

● EXAMPLE 10-8 FIGHTING A FIRE

A firefighter directs a stream of water against the window of a burning building, hoping to break it so the water can get to the fire (Fig. 10-17). The hose delivers water at the rate of 45 kg/s; the water hits the window moving horizontally at 32 m/s. After hitting the window, the water drops vertically downward. What is the horizontal force exerted on the window?

Solution

The horizontal motion of the water stops abruptly at the window; since the water is moving initially at 32 m/s, each kg of water loses 32 kg·m/s of momentum. But water strikes the window at the rate of 45 kg/s, so that the rate of momentum loss is

$$\frac{dP}{dt} = (45 \text{ kg/s})(32 \text{ kg·m/s/kg}) = 1400 \text{ kg·m/s}^2.$$

If we regard the water alone as our system, then this change in the system's momentum means that an external force of magnitude 1400 N is acting. That force is provided by the window. By Newton's third law, the water must therefore exert an oppositely directed force of equal magnitude on the window. Because the window is attached to the building, and the building to the Earth, the force does not give rise to any significant acceleration. Once the window breaks, though, the glass fragments will be accelerated violently.

EXERCISE An airliner taxies directly away from the terminal building, its jet exhaust directed perpendicularly against the window of the terminal. The plane exhausts 94 kg of gas each second, and the gas strikes the window at 220 m/s and then moves parallel to the window surface. How much force must the window withstand?

Answer: 21 kN

Some problems similar to Example 10-8: 27, 32, 35, 53, 54

FIGURE 10-17 What force does the water exert? (Example 10-8).

Rockets

Rockets provide our means of escaping Earth's gravity. And in the vacuum of space, where there's nothing for a wheel or propeller to push against, rockets are, at present, our only practical means of propulsion. Here we analyze rocket propulsion using momentum conservation.

Consider a rocket whose total mass $M + \Delta m$ includes fuel, moving at speed v relative to some reference frame (Fig. 10-18a). Its total momentum is therefore $P = (M + \Delta m)v$. Now the rocket fires for a short time Δt, expelling a small quantity Δm of fuel moving at speed v_{ex} relative to the rocket and therefore at $v - v_{ex}$ relative to our reference frame (Fig. 10-18b). The momentum of the expelled fuel is then $\Delta m(v - v_{ex})$. The rocket's mass drops to M, and its speed increases by a small amount Δv, so its momentum is now $M(v + \Delta v)$. But in the absence of external forces the momentum of the (rocket + fuel) system is conserved, so the momenta of the rocket and the fuel must sum to the initial momentum $P = (M + \Delta m)v$:

$$(M + \Delta m)v = M(v + \Delta v) + \Delta m(v - v_{ex}),$$

or, multiplying out and simplifying,

$$M\Delta v = v_{ex}\Delta m.$$

We can recast this equation in terms of the change in rocket mass M by noting that the expelled fuel Δm represents a *decrease* in the rocket mass; that is, the change ΔM in rocket mass is given by $\Delta M = -\Delta m$, where the minus sign signifies a mass decrease. If we also divide by the time Δt over which the fuel is expelled, our equation becomes

$$M\frac{\Delta v}{\Delta t} = -v_{ex}\frac{\Delta M}{\Delta t}. \tag{10-10a}$$

Actually, the entire procedure leading to this equation is only approximate: The velocity of the rocket increases continuously as the fuel mass is expelled, and thus our assertion that the fuel velocity is $v - v_{ex}$ is itself only approximate. But in the limit of a vanishingly small fuel mass Δm expelled over a vanishingly small time Δt, the velocity change Δv becomes negligible, and the equation becomes exact. In that limit the ratios in Equation 10-10a become derivatives, and we have

$$M\frac{dv}{dt} = -v_{ex}\frac{dM}{dt}. \tag{10-10b}$$

Rocket engineers call the quantity $-v_{ex}\,dM/dt$ the **thrust** of the rocket. Thrust has the dimensions of force (check the units!) and is indeed a measure of the force accelerating the rocket. Equation 10-10b shows that thrust can be increased by expelling the gases at a higher speed or by ejecting a greater mass of gas per unit time. The minus sign shows that an increase in speed—positive dv/dt—corresponds to a *decrease* in rocket mass—negative dM/dt. In practice the exhaust gas does not point straight backward, but fans out somewhat, thereby reducing the forward thrust (Fig. 10-19).

What force actually accelerates the rocket? The high-pressure gases inside the rocket push on all sides of the rocket chamber—except the back, where the exhaust port is. So the force associated with gas pushing on the *front* of the rocket chamber is not canceled by a corresponding force on the rear, and that net force is what gives the rocket its forward acceleration (Fig. 10-20).

Our analysis so far considered only a short rocket firing. We can find the effect of a longer firing by first multiplying both sides of Equation 10-10b by dt/M:

$$dv = -v_{ex}\frac{dM}{M}.$$

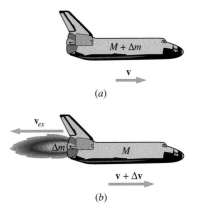

FIGURE 10-18 A rocket (*a*) before and (*b*) after it expels a small mass Δm of fuel. Rocket speed is relative to a nonaccelerating frame of reference; exhaust speed v_{ex} is relative to the rocket.

FIGURE 10-19 The exhaust plume of a rocket fans out slightly, reducing thrust from the theoretical value given by Equation 10-10. This photo shows the Space Shuttle Columbia as it lifts off.

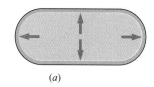

(a)

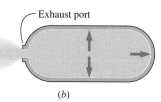

Exhaust port

(b)

FIGURE 10-20 (a) In a closed container of pressurized gas, the gas exerts forces on all the container walls, and the net force is zero. (b) With a hole at the left end of the container, gas rushes out and some of the pressure force on the right end is not canceled. The resulting net force is what accelerates the con: tainer to the right.

The overall change in velocity comes from summing—that is, integrating—all the small changes dv as the velocity goes from some initial value v_i to some final value v_f. During the same time the mass goes from M_i to M_f. So we have

$$\int_{v_i}^{v_f} dv = -v_{ex} \int_{M_i}^{M_f} \frac{dM}{M} .$$

The left-hand integral is just $v_f - v_i$, while the right-hand integral defines the natural logarithm. So

$$v_f - v_i = -v_{ex}[\ln(M_f) - \ln(M_i)] = v_{ex}[\ln(M_i) - \ln(M_f)] .$$

But $\ln(a) - \ln(b) = \ln(a/b)$, so

$$v_f = v_i + v_{ex} \ln\left(\frac{M_i}{M_f}\right) . \tag{10-11}$$

Since $M_i > M_f$, this equation gives the expected result that speed increases as the rocket ejects fuel.

A rocket carries only a finite amount of fuel, so the mass ratio from fully loaded to fully empty fuel tanks sets the maximum achievable speed—the **terminal speed** of the rocket. For a given mass ratio, the terminal speed can be increased by increasing the velocity of the exhaust gases.

● EXAMPLE 10-9 TERMINAL SPEED

A spacecraft is being designed to probe the interstellar medium. The craft is to be boosted from Earth and sent on an outbound trajectory by conventional rockets. At the orbit of Pluto, it is to be moving away from the Sun at 35 km/s; at that point an advanced rocket engine will fire to accelerate the craft to its terminal speed of 150 km/s. If the mass of the spacecraft alone is 750 kg, and if the advanced rocket exhausts fuel at 47 km/s, how much fuel must be carried to Pluto's orbit?

Solution
We want the final mass M_f to be that of the spacecraft alone, while M_i is the mass of the spacecraft plus fuel. To solve Equation 10-11 for the fuel mass, we first write

$$\ln\left(\frac{M_i}{M_f}\right) = \frac{v_f - v_i}{v_{ex}} = \frac{150 \text{ km/s} - 35 \text{ km/s}}{47 \text{ km/s}} = 2.45 ,$$

so

$$\frac{M_i}{M_f} = e^{2.45} = 11.6 ,$$

where we used the fact that natural logarithm and exponential are inverse functions: $e^{\ln(x)} = x$. So the total mass before the rocket firing is $11.6 M_f$, or 8700 kg. Subtracting the 750-kg spacecraft mass leaves 7950 kg of fuel. This example shows why the achievement of very high exhaust speeds is crucial to long-distance space travel; even in this case the useful payload is dwarfed by the fuel mass, and the situation would be far

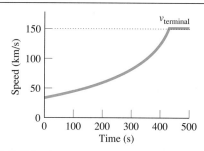

FIGURE 10-21 Speed versus time for the spacecraft of Example 10-9. The craft coasts at its terminal speed once all its fuel is exhausted. Note that the acceleration increases as the craft loses mass. Calculation assumes an exhaust rate of 19 kg/s.

worse with present-day rockets whose exhaust speeds are typically several km/s.

As Fig. 10-21 shows, most of the acceleration takes place in the later stages of the firing, when the mass of the rocket system is greatly reduced.

EXERCISE Find the terminal speed of a 750-kg rocket that starts from rest carrying 2600 kg of fuel and that expels its exhaust gases at 1.8 km/s.

Answer: 2.7 km/s

Some problems similar to Example 10-9: 38–40, 69

10-3 KINETIC ENERGY IN MANY-PARTICLE SYSTEMS

We've seen in this chapter how the momentum of a many-particle system is determined entirely by the motion of its center of mass; the detailed behavior of the individual particles doesn't matter. For example, a firecracker sliding on ice has the same total momentum before and after it explodes.

The same, however, is *not* true of a system's kinetic energy. Energetically, our firecracker is very different after it explodes; internal potential energy has been converted to kinetic energy of the fragments. Nevertheless, the center-of-mass concept remains useful in categorizing the kinetic energy associated with a system of particles.

The total kinetic energy of a system is simply the sum of the kinetic energies of the constituent particles:

$$K = \sum \frac{1}{2} m_i v_i^2 .$$

But the velocity $\mathbf{v}_i$ of a particle can be written as the vector sum of the center-of-mass velocity $\mathbf{V}$ and a velocity $\tilde{\mathbf{v}}_i$ of that particle relative to the center of mass:

$$\mathbf{v}_i = \mathbf{V} + \tilde{\mathbf{v}}_i .$$

Then the total kinetic energy of the system is

$$K = \sum \frac{1}{2} m_i (\mathbf{V} + \tilde{\mathbf{v}}_i) \cdot (\mathbf{V} + \tilde{\mathbf{v}}_i) \qquad (10\text{-}12)$$

$$= \sum \frac{1}{2} m_i V^2 + \sum m_i \mathbf{V} \cdot \tilde{\mathbf{v}}_i + \sum \frac{1}{2} m_i \tilde{v}_i^2 ,$$

where in the middle term the $\frac{1}{2}$ cancelled the 2 from the cross terms in multiplying out the dot product.

Let us examine the three sums making up the total kinetic energy. Since the center-of-mass speed V is common to all particles, it can be factored out of the first sum, so

$$\sum \frac{1}{2} m_i V^2 = \frac{1}{2} V^2 \sum m_i = \frac{1}{2} M V^2 ,$$

where M is the total mass. So what is this term? It is just the kinetic energy associated with a particle of mass M moving with speed V; we call this term K_{cm}, the **kinetic energy of the center of mass.**

The center-of-mass velocity can also be factored out of the second term in Equation 10-12, giving

$$\sum m_i \mathbf{V} \cdot \tilde{\mathbf{v}}_i = \mathbf{V} \cdot \sum m_i \tilde{\mathbf{v}}_i .$$

But what is $\sum m_i \tilde{\mathbf{v}}_i$? Because the $\tilde{\mathbf{v}}_i$'s are the particle velocities relative to the center of mass, that sum is just the total momentum measured in a frame of reference moving with the center of mass. But we found earlier that the total momentum is $\mathbf{P} = m\mathbf{V}$, with $\mathbf{V}$ the center of mass velocity. In a frame of reference where $\mathbf{V} = \mathbf{0}$— that is, a frame of reference moving with the center of mass—the total momentum must be zero. So the entire second term in Equation 10-12 is zero.

The third term in Equation 10-12, $\Sigma \frac{1}{2} m_i \tilde{v}_i^2$, is just the sum of the individual kinetic energies measured in a frame of reference moving with the center of mass. We call this term K_{int}, the **internal kinetic energy.**

With the middle term gone, Equation 10-12 shows that the kinetic energy of a system breaks down into two terms:

$$K = K_{cm} + K_{int}. \tag{10-13}$$

The first term, the kinetic energy of the center of mass, depends only on the center-of-mass motion. In our firecracker example, K_{cm} doesn't change when the firecracker explodes. The second term, the internal kinetic energy, depends only on the motions of the individual particles relative to the center of mass. Explosion of the firecracker dramatically increases the internal kinetic energy.

● **EXAMPLE 10-10** SEPARATION

A 325-kg booster rocket and a 732-kg satellite are coasting through space together at 5.22 km/s. Explosive bolts fire to separate the satellite from its booster, and after separation the satellite and booster are moving in their original direction at 6.69 km/s and 1.91 km/s, respectively. Find the center-of-mass and internal kinetic energies of the (booster + satellite) system before and after separation.

Solution

Before separation the two components are bolted together, so there is no internal kinetic energy. The center-of-mass kinetic energy is just the total kinetic energy:

$$K_{cm} = \frac{1}{2}MV^2 = \left(\frac{1}{2}\right)(325 \text{ kg} + 732 \text{ kg})(5.22 \times 10^3 \text{ m/s})^2$$
$$= 14.4 \text{ GJ}.$$

After separation, momentum conservation assures us that the center-of-mass velocity V remains 5.22 km/s. Since the total mass is unchanged, so, too, is the center-of-mass kinetic energy. To find the internal kinetic energy we need the velocities $\tilde{v}$ relative to the center of mass; we get these by subtracting the center-of-mass velocity from the individual velocities:

$$\tilde{v}_s = 6.69 \text{ km/s} - 5.22 \text{ km/s} = 1.47 \text{ km/s}$$
$$\tilde{v}_b = 1.91 \text{ km/s} - 5.22 \text{ km/s} = -2.31 \text{ km/s},$$

where the minus sign indicates that the booster is moving backwards relative to the center of mass. The internal kinetic energy is the sum of the energies associated with these internal motions:

$$K_{int} = \frac{1}{2}m_s\tilde{v}_s^2 + \frac{1}{2}m_b\tilde{v}_b^2$$
$$= \left(\frac{1}{2}\right)(682 \text{ kg})(1.47 \times 10^3 \text{ m/s})^2$$
$$+ \left(\frac{1}{2}\right)(325 \text{ kg})(-2.31 \times 10^3 \text{ m/s})^2 = 2.57 \text{ GJ}.$$

This extra energy comes from the conversion of chemical potential energy in the explosive bolt.

We can confirm that the center-of-mass energy remains constant by calculating the total energy after separation and subtracting the internal energy:

$$K_{total} - K_{internal} = \frac{1}{2}m_s v_s^2 + \frac{1}{2}m_b v_b^2 - K_{int}$$
$$= \left(\frac{1}{2}\right)(732 \text{ kg})(6.69 \times 10^3 \text{ m/s})^2$$
$$+ \left(\frac{1}{2}\right)(325 \text{ kg})(1.91 \times 10^3 \text{ m/s})^2 - 2.57 \text{ GJ}$$
$$= 14.4 \text{ GJ},$$

in agreement with our before-separation result for the center-of-mass kinetic energy.

TIP **Remember the SI Prefixes** What's a GJ? If you encounter an unfamiliar SI prefix, refer to Table 1-1 (also inside the front cover). There we find that G stands for "giga," signifying multiplication by 10^9. So the center-of-mass energy is 14.4×10^9 J.

EXERCISE   Find the center-of-mass and internal kinetic energies for the snowball in the exercise following Example 10-7. Take the mass of the snowball to be 14 kg. *Hint:* Here you'll need to do a vector subtraction to determine the velocities relative to the center of mass.

Answers: $K_{cm} = 404$ J; $K_{int} = 73$ J

Some problems similar to Example 10-10: 41, 42, 44

●

CHAPTER SYNOPSIS

Summary

1. A system of particles of total mass M has associated with it a point called the **center of mass** to which Newton's laws of motion apply as they do to a point particle:

$$\mathbf{F}_{net\,ext} = M\frac{d^2\mathbf{R}}{dt^2} = M\mathbf{A},$$

where $\mathbf{F}_{net\,ext}$ is the net external force on the system and $\mathbf{A}$ the acceleration of the center of mass. The position $\mathbf{R}$ of the center of mass is given by

$$\mathbf{R} = \frac{\sum m_i\mathbf{r}_i}{M}, \qquad \text{(center of mass)}$$

where m_i and $\mathbf{r}_i$ represent the masses and positions of the individual particles in the system. For continuously distributed matter, the center-of-mass position is given by an integral:

$$\mathbf{R} = \frac{1}{M}\int \mathbf{r}\,dm,$$

where the integration is taken over the entire system. That the center-of-mass concept is useful is a consequence of Newton's third law, which requires that internal forces cancel in pairs, leaving the overall system motion determined only by external forces.

2. In the absence of a net external force, the center-of-mass velocity $\mathbf{V} = \sum m_i\mathbf{v}_i/M$ remains constant. In particular, if the center of mass is at rest, then it remains at rest regardless of the motions of the constituent particles.

3. The total momentum $\mathbf{P}$ of a system is the vector sum of the momenta $\mathbf{p}_i = m_i\mathbf{v}_i$ of the constituent particles:

$$\mathbf{P} = \sum m_i\mathbf{v}_i.$$

Application of Newton's third law, as embodied in the center-of-mass concept, shows that the system obeys Newton's second law in the form

$$\mathbf{F} = \frac{d\mathbf{P}}{dt},$$

where $\mathbf{F}$ is the net external force applied to the system.

4. In the absence of a net external force, the momentum of a system remains constant. This statement is known as the **law of conservation of momentum.** It is a fundamental principle of physics, extending even into realms where Newtonian mechanics is not applicable.

5. The motion of a rocket is a manifestation of the momentum principle. The rate of change of a rocket's speed v is given by

$$M\frac{dv}{dt} = -v_{ex}\frac{dM}{dt},$$

where M is the rocket's mass, v_{ex} the speed of the exhaust gas relative to the rocket, and dM/dt the rate of change of the rocket's mass due to the ejection of gas.

6. The total kinetic energy of a system of particles includes the kinetic energy of the center of mass, given by

$$K_{cm} = \frac{1}{2}MV^2,$$

where M is the total mass and V the speed of the center of mass, and the **internal kinetic energy,** given by

$$K_{int} = \sum\frac{1}{2}m_i\tilde{v}_i^2,$$

where m_i is the mass of an individual particle and $\tilde{v}_i$ its speed relative to the center of mass. The total kinetic energy is the sum of these two:

$$K_{total} = K_{cm} + K_{int}.$$

Terms You Should Understand

(Pairs are closely related terms whose distinction is important; number in parentheses is chapter section where term first appears.)

rigid body (introduction)
center of mass (10-1)
internal force, external force (10-1)
mass element (10-1)
linear momentum (10-2)
conservation of linear momentum (10-2)
rocket (10-2)
thrust (10-2)
center-of-mass kinetic energy, internal kinetic energy (10-3)

Symbols You Should Recognize

$\mathbf{R}$, X, Y, Z (10-1)
M (10-1)
$\mathbf{F}_{ext}$, $\mathbf{F}_{int}$ (10-1)
$\int \mathbf{r}\,dm$ (10-1)
$\mathbf{P}$ (10-2)
$\mathbf{V}$, $\mathbf{v}_i$, $\tilde{\mathbf{v}}_i$ (10-3)
K_{cm}, K_{int} (10-3)

Problems You Should Be Able to Solve

determining the center of mass of a system consisting of point particles on a line or in a plane (10-1)
determining the center of mass of a composite object from the centers of mass of its constituent parts (10-1)
determining the center of mass of a continuous distribution of matter using integration (10-1)
using center of mass to determine positions of the constituent objects of a system as the system is reconfigured (10-1)
applying conservation of momentum to determine the behavior of individual parts of a system (10-2)
using conservation of momentum to determine forces on objects being bombarded by a stream of particles (10-2)
analyzing rocket motion (10-2)
evaluating center-of-mass and internal energies of systems of particles (10-3)

Limitations to Keep in Mind

Conservation of momentum applies to a system only when no external forces act on the system.

QUESTIONS

1. Roughly where is your center of mass when you are standing?

2. Can you form your body into a shape such that your center of mass is not within your body? If not, why not? If so, describe.

3. Where is the center of mass of a solid wooden wheel? Of a wagon wheel, with spokes?

4. Earth and moon are both in nearly circular orbits about a point 4700 km from the center of the Earth. Explain in terms of the center-of-mass concept as applied to the Earth-moon system.

5. Hunters drive a herd of bison over the edge of a cliff. As they fall, what happens to the center of mass of the Earth + bison system? What happens to the centers of mass of the herd and the Earth separately? Explain.

6. Explain why a high jumper's center of mass need not clear the bar.

7. Where would our derivation of the center-of-mass position $\mathbf{R}$ fail if the number of particles in the system were not constant?

8. The center of mass of a solid sphere is clearly at its center. If the sphere is cut in half, and the two halves are stacked as in Fig. 10-22, is the center of mass at the point where they touch? If not, roughly where is it? Explain.

9. How can a jet engine work when it takes air in the front as well as exhausting it out the back?

10. When you reach shore in Example 10-4, the center of mass of the (boat + you) system suddenly begins moving. What force accelerates it?

11. In Example 10-4, does the motion of the boat change when you reach shore?

12. Does the momentum of a basketball change as it rebounds off a backboard? What happens to the momentum of the backboard?

13. The momentum of a system of pool balls is the same before and after they are hit by the cue ball. Is it still the same after one of the balls strikes the edge of the table? Explain.

14. If all the cars in the world started driving eastward, what would happen to the length of the day?

15. When the chest of Example 10-6 hits the wall of the hockey rink, its momentum is no longer conserved. How would you redefine the system so momentum is still conserved?

16. A group of kids are pelting a wall with rocks. For a given rock speed and number of rocks thrown per unit time, will the force on the wall be greater (a) if the rocks stick to the wall; (b) if the rocks hit the wall and fall vertically downward; or (c) if the rocks bounce off the wall, heading back in the direction from which they came? Explain.

17. Consider the system consisting of a shot putter and his shot; the mass of the shot is about 10% that of the shot putter. Describe the motion of the system's center of mass (a) if the shot putter is standing on a concrete pad with high coefficient of friction and (b) if he is standing on a frictionless surface.

18. An asteroid 100 km in diameter is discovered to be on a collision course with Earth. In a show of international cooperation and mutual disarmament, nations launch all their nuclear missiles at the asteroid and succeed in blowing

FIGURE 10-22 Question 8.

it to smithereens. Will the center of mass of the asteroid still hit Earth?

19. Soon after the event of the previous question, a smaller asteroid is discovered heading for Earth. Having used all their nuclear missiles on the first asteroid, nations now cooperate in the manufacture of a huge rocket engine, which they transport to the asteroid. The rocket is strapped to the asteroid and fired long enough to move the asteroid off its collision course. Now does the center of mass of the asteroid hit Earth?

20. Can you make a sailboat go by blowing air onto the sails with a fan mounted on the boat? With air from a compressed-air cylinder mounted on the boat?*

21. When rocket pioneer Robert Goddard proposed using

*See *The Physics Teacher,* 1986, Vol. 24, pp. 38–39, 392–393.

rocket propulsion for space travel, the January 13, 1920 issue of *The New York Times* criticized Goddard with the scathing words "Professor Goddard . . . does not know the relation of action to reaction, and of the need to have something better than a vacuum against which to react . . . he . . . seems to lack the knowledge ladled out daily in high schools." How can a rocket work if it doesn't have anything to "push against?" Who was right, the *Times* or Goddard? Explain.

22. An hourglass is inverted and placed on a scale. Compare scale readings (a) before sand begins to hit the bottom; (b) while sand is hitting the bottom; (c) when all the sand is on the bottom.

23. The momentum of the center of mass of a system is the sum of the momenta of the individual particles, but the same is not true for kinetic energy. Why not?

PROBLEMS

Section 10-1 Center of Mass

1. A 28-kg child sits at one end of a 3.5-m long seesaw. Where should her 65-kg father sit so the center of mass will be at the center of the seesaw?

2. Two particles of equal mass m are at the vertices of the base of an equilateral triangle. The center of mass of the triangle is midway between the base and the third vertex. What is mass at the third vertex?

3. Four masses lie at the corners of a square 1.0 m on a side, as shown in Fig. 10-23. Where is the center of mass?

4. Rework Example 10-1 with the origin at the center of the barbell, showing that the physical location of the center of mass is independent of the choice of coordinate system.

5. Three equal masses lie at the corners of an equilateral triangle of side ℓ. Where is the center of mass?

6. How far from the center of the Earth is the center of mass of the Earth-moon system? *Hint:* Consult Appendix E.

7. Find the center of mass of the plane object shown in Fig. 10-9, assuming uniform density.

8. Find the center of mass of the solar system at a time when all the planets are lined up on the same side of the Sun. *Hint:* Consult Appendix E.

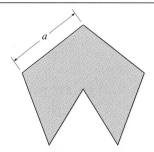

FIGURE 10-24 Problem 9.

9. Find the center of mass of a pentagon of side a with one triangle missing, as shown in Fig. 10-24. *Hint:* See Example 10-3, and treat the pentagon as a group of triangles.

10. A solid cube of side a has a density that varies linearly from zero at the bottom to ρ_0 at the top, as suggested in Fig. 10-25; that is, $\rho = \rho_0 \dfrac{y}{a}$. Find expressions for (a) the cube's mass and (b) the y coordinate of its center of mass. *Hint:* Use a square mass element of thickness dy, as shown in the figure.

FIGURE 10-23 Problem 3.

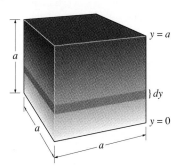

FIGURE 10-25 Cube for Problem 10. The dark band suggests a mass element in the form of a square slab with dimensions $a \times a \times dy$.

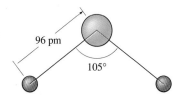

FIGURE 10-26 A water molecule (Problem 11).

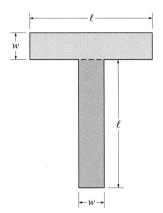

FIGURE 10-27 Problem 12.

cabinet directly overhead. The mouse climbs down the inside of the bowl to eat the crumbs at the bottom. If the bowl moves along the counter a distance equal to one-tenth of its diameter, how does the mouse's mass compare with the bowl's mass?

18. Find the center of mass of the uniform, solid cone of height h and constant density ρ shown in Fig. 10-28. *Hint:* Integrate over disk-shaped mass elements of thickness dy, as shown in the figure.

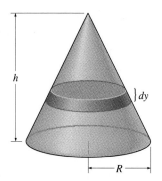

FIGURE 10-28 Cone for Problem 18 has base radius R and height h. Dark band shows a mass element for use in the integration. It is a disk with thickness dy and radius that depends on its position y.

11. A water molecule (H_2O) consists of two hydrogen atoms, each of mass 1.0 u, and one oxygen atom of mass 16 u. The hydrogen atoms are 96 pm from the oxygen and are separated by an angle of 105° (Fig. 10-26). Where is the center of mass of the molecule?

12. A flat, T-shaped structure is made from two identical rectangles of length ℓ and width w (Fig. 10-27). Find its center of mass.

13. Consider a system of three equal-mass particles moving in a plane; their positions are given by $a_i\hat{\imath} + b_i\hat{\jmath}$, where a_i and b_i are functions of time with the units of position. Particle 1 has $a_1 = 3t^2 + 5$ and $b_1 = 0$; particle 2 has $a_2 = 7t + 2$ and $b_2 = 2$; particle 3 has $a_3 = 3t$ and $b_3 = 2t + 6$. Find the position, velocity, and acceleration of the center of mass as functions of time.

14. A 50-kg woman on a 650-kg boat hooks a 150-kg blue marlin fish on the end of a nearly horizontal line 35 m long. If fish and boat are originally at rest on a frictionless sea, how far does the boat move as she reels in the fish?

15. Estimate how much the Earth would move if its entire human population jumped simultaneously into the 1-mile-deep Grand Canyon.

16. You are with 19 other people on a boat at rest in frictionless water. Your total mass is 1500 kg, and the mass of the boat itself is 12,000 kg. The entire party walks the 6.5-m distance from bow to stern. How far does the boat move?

17. A hemispherical bowl is at rest on a frictionless kitchen counter. A mouse drops onto the rim of the bowl from a

Section 10-2 Momentum

19. An object intially at rest at the origin bursts into two pieces with masses 560 g and 390 g. The 560-g piece moves off at 2.1 m/s in the positive x direction. Describe the motion of the other piece.

20. A 60-kg ice skater, at rest on frictionless ice, tosses a 12-kg snowball with velocity given by $\mathbf{v} = 3.0\hat{\imath} + 4.0\hat{\jmath}$ m/s, where the x and y axes are both in the horizontal plane. What is the subsequent velocity of the skater?

21. A firecracker, initially at rest, explodes into two fragments. The first, of mass 14 g, moves in the positive x direction at 48 m/s. The second moves at 32 m/s. Find its mass and the direction of its motion.

22. A rowboat with a mass of 94 kg carries a 61-kg rower and 34 kg of water that has accumulated in the bottom. It's drifting forward at 0.84 m/s when the rower scoops up 11 kg of water in a bucket of negligible mass and heaves it horizontally over the stern at a speed of 4.3 m/s relative to the boat. What is the subsequent speed of the boat?

23. A 780-g wood block is at rest on a frictionless table, when a 30-g bullet is fired into it. If the block with the embedded bullet moves off at 17 m/s, what was the original speed of the bullet?

24. A plutonium-239 nucleus at rest decays into a uranium-235 nucleus by emitting an alpha particle (^{4}He) with kinetic energy of 5.15 MeV. What is the speed of the uranium nucleus?

25. A runaway toboggan of mass 8.6 kg is moving horizontally at 23 km/h. As it passes under a tree, 15 kg of snow drop onto it. What is its subsequent speed?

26. A 950-kg airplane touches down on an icy runway, heading northward at 150 km/h. It collides with a 1200-kg plane taxiing eastward at 85 km/h. Find the speed and direction of the combined wreckage.

27. During a heavy storm, rain falls at the rate of 2.0 cm/hour; the speed of the individual raindrops is 25 m/s. (a) If the rain strikes a flat roof and then flows off the roof with negligible speed, what is the force exerted per square meter of roof area? (b) How much water would have to stand on the roof to exert the same force? The density of water is 1.0 g/cm³.

28. A circle of Ice Capades clowns contains six individuals with masses of 50, 55, 60, 65, 70, and 75 kg. If they are at rest and evenly spaced around a circle, then start skating radially outward each with the same momentum $p = 200$ kg·m/s, describe the subsequent motion of their center of mass.

29. An 11,000-kg freight car is resting against a spring bumper at the end of a railroad track. The spring has constant $k = 3.2 \times 10^5$ N/m. The car is hit by a second car of 9400 kg mass moving at 8.5 m/s, and the two cars couple together. (a) What is the maximum compression of the spring? (b) What is the speed of the two cars together when they rebound from the spring?

30. On an icy road, a 1200-kg car moving at 50 km/h strikes a 4400-kg truck moving in the same direction at 35 km/h. The combination is soon hit from behind by a 1500-kg car speeding at 65 km/h. If all three vehicles stick together, what is the speed of the wreckage?

31. A 1600-kg automobile is resting at one end of a 4500-kg railroad flatcar that is also at rest. The automobile then drives along the flatcar at 15 km/h relative to the flatcar. Unfortunately, the flatcar brakes are not set. How fast does it move?

32. A car of mass M is initially at rest on a frictionless surface. A jet of water carrying mass at the rate dm/dt and moving horizontally at speed v_0 strikes the rear window of the car, which makes a 45° angle with the horizontal; the water bounces off at the same relative speed with which it hit the window, as shown in Fig. 10-29. (a) Find an expression for

FIGURE 10-29 Problem 32.

the initial acceleration of the car. (b) What is the maximum speed reached by the car?

33. A 950-kg compact car is moving with velocity $\mathbf{v}_1 = 32\hat{\mathbf{i}} + 17\hat{\mathbf{j}}$ m/s. It skids on a frictionless icy patch, and collides with a 450-kg hay wagon moving with velocity $\mathbf{v}_2 = 12\hat{\mathbf{i}} + 14\hat{\mathbf{j}}$ m/s. If the two stay together, what is their velocity?

34. The 975-kg Giotto spacecraft, when it was close to Halley's Comet in 1986, passed through cometary dust of density 0.10 g/m³ at a speed of 200 m/s. Its frontal surface area was 10 m². Assuming that the dust stuck to the front of the spacecraft, at what rate was the spacecraft decelerated?

35. A biologist fires 20-g rubber bullets at a rhinoceros that is charging at 1.9 m/s. If the gun fires 5 bullets per second, with a speed of 1600 m/s, and if the biologist fires for 13 s to stop the rhino in its tracks, what is the mass of the rhino? Assume the bullets drop vertically after striking the rhino.

36. A ^{238}U nucleus is moving in the x direction at 5.0×10^5 m/s, when it decays into an alpha particle (^{4}He) and a ^{234}Th nucleus. If the alpha particle moves off at 22° above the x axis with a speed of 1.4×10^7 m/s, what is the recoil velocity of the thorium nucleus?

37. An Ariane rocket ejects 1.0×10^5 kg of fuel in the 90 s after launch. (a) How much thrust is developed if the fuel is ejected at 3.0 km/s with respect to the rocket? (b) What is the maximum total mass of the rocket if it is to get off the ground?

38. Ninety percent of a rocket's initial mass is fuel. If the fuel is exhausted at 3.5 km/s relative to the rocket, what is the terminal speed of the rocket?

39. If a rocket's exhaust speed is 200 m/s relative to the rocket, what fraction of its initial mass must be ejected to increase the rocket's speed by 50 m/s?

40. A space shuttle's main engines develop a thrust of 35×10^6 N as they eject gas at 2500 m/s relative to the shuttle. (a) How much fuel must the shuttle carry to permit a 5-minute engine firing? (b) At 200 kg/m³, how large a cubical fuel tank would be required?

Section 10-3 Kinetic Energy in Many-Particle Systems

41. Determine the center-of-mass and internal kinetic energies before and after decay of the lithium nucleus of Example 10-7. Treat the individual nuclei as point particles.

42. A 150-g trick baseball is thrown at 60 km/h. It explodes in flight into two pieces, with a 38-g piece continuing straight ahead at 85 km/h. How much energy do the pieces gain in the explosion?

43. A 1200-kg car moving at 88 km/h collides with a 7600-kg truck moving in the same direction at 65 km/h. The two stick together, continuing in their original direction at 68

km/h. Determine the center-of-mass and internal energies of the (car + truck) system before and after the collision.

44. A sealed can of air contains 10^{21} air molecules moving randomly, on average, at 340 m/s. If the can is thrown at 10 m/s, compare the center-of-mass and internal energies of the air.

Paired Problems

(Both problems in a pair involve the same principles and techniques. If you can get the first problem, you should be able to solve the second one.)

45. A drinking glass is in the shape of a cylinder whose inside dimensions are 9.0 cm high and 8.0 cm in diameter as shown in Fig. 10-30. Its base has a mass of 140 g, while the mass of the curved, cylindrical sides is 85 g. (a) Where is its center of mass? (b) If the glass is three-quarters filled with juice (density 1.0 g/cm³), where is the center of mass of the glass-juice system? Assume the thickness of the glass is negligible.

FIGURE 10-30   Problems 45 and 68.

46. A cylindrical concrete silo is 4.0 m in diameter and 30 m high. It consists of a 6,000-kg concrete base and 38,000-kg cylindrical concrete walls. Locate the center of mass of the silo (a) when it's empty and (b) when it's two-thirds full of silage whose density is 800 kg/m³. Neglect the thickness of the silo walls and base.

47. Olympic-champion ice dancers Marina Klimova (mass 50 kg) and Sergei Ponomarenko (mass 70 kg) start from rest and push off each other on essentially frictionless ice. If Ponomarenko's speed is 2.8 m/s, how far apart are they after 3.0 s?

48. Contestants in a "canoe joust" push at each other with boxing gloves mounted on long poles. Canoe A has a total mass of 150 kg including occupants. Canoe B has a total mass of 210 kg. Pushing as hard as possible, the contestant in canoe A gives canoe B a speed of 1.3 m/s relative to canoe A. What is canoe B's speed relative to shore, if both canoes were initially at rest?

49. A 42-g firecracker is at rest at the origin when it explodes into three pieces. The first, with mass 12 g, moves along the x axis at 35 m/s. The second, with mass 21 g, moves along the y axis at 29 m/s. Find the velocity of the third piece.

50. A 60-kg astronaut floating in space simultaneously tosses away a 14-kg oxygen tank and a 5.8-kg camera. The tank moves in the x direction at 1.6 m/s, and the astronaut recoils at 0.85 m/s in a direction 200° counterclockwise from the x axis. Find the velocity of the camera.

51. A 4100-kg bull elephant is at one end of a 16,000-kg circus train car initially at rest on a frictionless track. The ele: phant charges the 17-m length of the car at a speed of 6.5 m/s, then stops when it hits the far wall. (a) How far does the car move during the elephant's charge? (b) How fast is the car moving during the charge? (c) What is the car's speed after the elephant has stopped? Assume the center of mass of the car is at its center, and treat the elephant as a particle.

52. A 65-m long ferry boat has a mass of 32,000 kg and is at rest with its shoreward end 55 m from shore. The center of mass of the boat is at its center. A stunt driver starts from the far end of the ferry and heads toward the shoreward end, accelerating from rest to 23 m/s relative to the boat at the instant he leaves the boat. The mass of car and driver is 1100 kg. (a) How much farther from shore is the end of the boat when the car reaches it? (b) How fast is the boat moving when the car departs?

53. Firefighters spray water horizontally at the rate of 41 kg/s from a nozzle mounted on a 12,000-kg fire truck. The water speed is 28 m/s relative to the truck. (a) Neglecting friction, what is the initial acceleration of the truck? (b) If the 12,000-kg truck mass includes 2400 kg of water, how fast will the truck be moving when the water is exhausted? *Hint:* Think about rockets.

54. A 31-kg child stands on frictionless ice, holding a box of negligible mass containing twenty-four 250-g snowballs. If the child tosses the snowballs horizontally at the rate of one every two seconds and gives each ball a speed of 11 m/s relative to herself, how fast will she be moving once the snowballs are gone? Use the approximation that the child ejects snow continuously—i.e., the rocket equation.

Supplementary Problems

55. A 55-kg sprinter is standing at the left end of a 240-kg cart moving to the left at 7.6 m/s. She runs to the right end and continues horizontally off the cart. What should be her speed relative to the cart in order to leave the cart with no horizontal velocity component relative to the ground?

56. The coefficient of static friction between the boots of a 78-kg firefighter and the ground is 0.63. The firefighter is holding a hose that spews water at 19 m/s. At what rate, in kilograms per second, can the hose supply water without the firefighter sliding backward?

FIGURE 10-31 Problem 57.

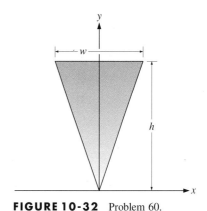

FIGURE 10-32 Problem 60.

57. Figure 10-31 shows a paraboloidal solid of constant dens: ity ρ. It extends a height h along the z axis, and is de: scribed by $z = ar^2$, where the units of a are m^{-1} and r is the radius in a plane perpendicular to the z axis. Find expressions for (a) the mass of the solid and (b) the z coordinate of its center of mass.

58. One plan for a space-based anti-ballistic-missile weapon involves shooting a beam of 100-MeV hydrogen atoms at an approaching missile. The power delivered by the beam is 100 MW. To destroy a missile, the beam should be turned on for 2 s and should cover an area of 1 m². Deter- mine (a) the momentum in the beam and (b) the average force exerted on the missile. Assume the beam is absorbed by the missile. (c) If the beam hits a 100-kg warhead mov: ing at 8 km/s, will the warhead's trajectory be affected significantly?

59. An alternative to the anti-missile weapon of the previous problem is to hit the missile with objects of about 1-kg mass traveling at 20 km/s. Assuming that such a "kinetic kill weapon" embeds itself in the missile, (a) how much momentum does it impart to the missile; (b) how much kinetic energy does it carry; (c) does it significantly affect the trajectory of a 100-kg warhead moving at 8 km/s? (d) Compare the energy and momentum in this problem with that of the preceding problem. Explain any similari: ties and differences.

60. The triangle of Fig. 10-32 has a density that varies in pro- portion to y^α, where α is a constant. Show that its center of mass is located at $y = (\alpha + 2)h/(\alpha + 3)$.

61. While standing on frictionless ice, you (mass 65.0 kg) toss a 4.50-kg rock with initial speed of 12.0 m/s. If the rock is 15.2 m from you when it lands, (a) at what angle did you toss it? (b) How fast are you moving?

62. A drunk driver in a 1600-kg car plows into a 1300-kg parked car with its brake set. Police measurements show that the two cars skid together a distance of 25 m before stopping. If the effective coefficient of friction is 0.77, how fast was the drunk going just before the collision?

63. A fireworks rocket is launched vertically upward at 40 m/s. At the peak of its trajectory, it explodes into two equal- mass fragments. One reaches the ground 2.87 s after the explosion. When does the second reach the ground?

64. A fire hose delivers 50 kg/s of water at 30 m/s. How many 75-kg firefighters are needed to hold the hose on muddy ground for which the coefficient of static friction is 0.35?

65. (a) Derive an expression for the thrust of a jet aircraft engine. Moving through the air with speed v, the engine takes in air at the rate dM_{in}/dt. It uses the air to burn fuel with a fuel/air ratio f (that is, f kg of fuel burned for each kg of air), and ejects the exhaust gases at speed v_{ex} with respect to the engine. (b) Use your result to find the thrust of a JT-8D engine on a Boeing 727 jetliner, for which $v_{ex} = 1034$ ft/s and $dM_{in}/dt = 323$ lb/s, and which con- sumes 3760 lb of fuel per hour while cruising at 605 mi/h.

66. Two blocks are sliding along a one-dimensional fric: tionless surface with speeds v_1 and v_2, when they collide. Use the conservation of momentum principle to show that the most kinetic energy is lost if they stick together.

67. An ideal spring of spring constant k rests on a frictionless surface. Blocks of mass m_1 and m_2 are pushed against the two ends of the spring until it is compressed a distance x from its equilibrium length. The blocks are then released. What are their speeds when they leave the spring?

68. (a) What depth of juice should be in the glass of Problem 45 to put the center of mass of the system at its lowest possible point? (b) Show that the center of mass is then at the surface of the juice.

CHAPTER 11

COLLISIONS

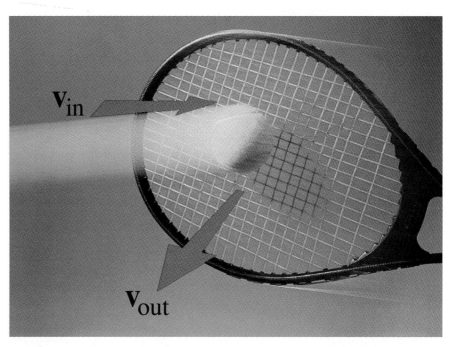

A collision is a brief interaction that involves large forces and abruptly changes the motion of one or both of the interacting bodies. Here a tennis ball and racket are about to collide; vectors represent incident and outgoing velocities.

On the highway, we try to avoid colliding with other cars. On the pool table, we tailor collisions to suit our purposes. In particle accelerators, we interpret collisions to reveal the innermost structure of matter. On a cosmic scale, we wonder if a collision with an asteroid led to the extinction of the dinosaurs (Fig. 11-1). Collisions are important features of our physical universe.

A **collision** is a relatively violent interaction of objects that lasts for a short time. By "violent interaction" we mean that forces associated with the collision are much greater than any other forces that may be acting on the objects. By "short time" we mean short compared with time scales appropriate in describing the overall motion of the objects. A collision between two airplanes, for example, involves forces much greater than the gravitational and aerodynamic forces; the time interval over which the collision occurs is short compared with the overall flight times. Similarly, a collision between two galaxies (Fig. 11-2) may take millions of years, but that period is still short compared with the ages of the galaxies.

11-1 IMPULSE AND COLLISIONS

In a collision of two objects—like the foot and football in Fig. 11-3—the velocity of one or both objects is abruptly changed. We know, from Newton's laws of motion, that such a change means a force has been applied. The strong but short-duration force associated with a collision is called an **impulsive force.** Although the gravitational force also acts on the ball, during the collision the impulsive force is overwhelmingly dominant.

We can relate the impulsive force to the change in an object's momentum through Newton's second law:

$$\mathbf{F} = \frac{d\mathbf{p}}{dt}.$$

Multiplying this equation by dt and integrating from some time t_1 before the collision to a time t_2 after the collision, we have

$$\int_{t_1}^{t_2} \mathbf{F}\, dt = \int_{t_1}^{t_2} d\mathbf{p}.$$

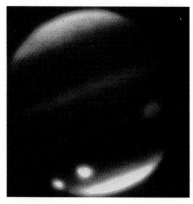

FIGURE 11-1 Celestial collisions are not uncommon in our solar system. This photo of Jupiter shows two bright spots (near bottom) that mark impacts of Comet Shoemaker-Levy 9 with the giant planet in 1994. The impact scars are approximately the size of Earth and were caused by comet fragments of roughly kilometer size, somewhat smaller than the object whose collision with Earth is believed responsible for the extinction of the dinosaurs. Jupiter's famous red spot shows above and to the right of the cometary impact sites, and the planet's polar caps appear brighter than their surroundings. Photo was taken in infrared light with the Keck telescope in Hawaii.

FIGURE 11-3 High-speed photo of a collision between a foot and a football. The strong impulsive force is evident in the deformation of the football. This force causes a rapid change in the ball's velocity. Vector **F** is the impulsive force on the ball; not shown is the reaction force of the ball on the foot.

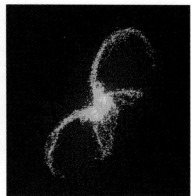

FIGURE 11-2 Computer simulation of colliding galaxies.

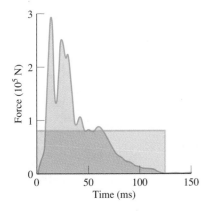

FIGURE 11-4 Impulsive force in the crash of a Suzuki Samurai into a rigid barrier. Horizontal line is the average force; area under the rectangle is the same as under the jagged curve.

The right-hand integral is just the change $\Delta\mathbf{p}$ that occurs during the collision. The integral on the left is known as the **impulse, I,** associated with the collision. Our result then shows that the change in an object's momentum is equal to the impulse:

$$\mathbf{I} \equiv \int_{t_1}^{t_2} \mathbf{F}\,dt = \Delta\mathbf{p}. \tag{11-1}$$

The units of impulse are the same as those of momentum: kg·m/s, or, equivalently, N·s. Since the definite integral corresponds to the area under a curve, Equation 11-1 shows that the impulse is given geometrically by the area under the force-versus-time curve. It doesn't much matter whether the force curve includes any nonimpulsive forces that may be present, for our definition of a collision ensures that the change in momentum associated with nonimpulsive forces is small during the time of the collision.

Although the impulsive force in a collision may be a very complicated function of time (Fig. 11-4), the overall effect of a collision on the motion of an object is summed up by the change in the object's momentum. We can work backward from that change to determine the **average impulsive force $\overline{\mathbf{F}}$**—the force that, if applied constantly during the collision time Δt, would give the same change in momentum. That is,

$$\overline{\mathbf{F}} = \frac{\mathbf{I}}{\Delta t} = \frac{\Delta\mathbf{p}}{\Delta t}. \tag{11-2}$$

Graphically, the average force is represented by the height of a rectangle whose area is the same as the actual impulse area $\int F\,dt$ (see Fig. 11-4).

● EXAMPLE 11-1 IMPULSE

A 150-g baseball is moving horizontally at 30 m/s just before reaching the bat. Bat and ball are then in contact for 2.0 ms (Fig. 11-5). Find the impulse and the average impulsive force (a) if the hit is a line drive to center field at 45 m/s and (b) if the hit is a pop foul that starts straight up at 22 m/s.

Solution

In a coordinate system with x axis along the line from home plate to center field, the incoming momentum is

$$\mathbf{p}_1 = m\mathbf{v}_1 = (0.15\text{ kg})(-30\hat{\imath}\text{ m/s}) = -4.5\hat{\imath}\text{ kg·m/s}.$$

For the line drive, the outgoing momentum is

$$\mathbf{p}_2 = m\mathbf{v}_2 = (0.15\text{ kg})(45\hat{\imath}\text{ m/s}) = 6.8\hat{\imath}\text{ kg·m/s},$$

giving an impulse

$$\mathbf{I} = \Delta\mathbf{p} = (6.8\hat{\imath}\text{ kg·m/s}) - (-4.5\hat{\imath}\text{ kg·m/s}) = 11\hat{\imath}\text{ kg·m/s}$$

and an average force

$$\overline{\mathbf{F}} = \frac{\Delta\mathbf{p}}{\Delta t} = \frac{11\hat{\imath}\text{ kg·m/s}}{2.0\times10^{-3}\text{ s}} = 5.5\times10^{3}\hat{\imath}\text{ N}.$$

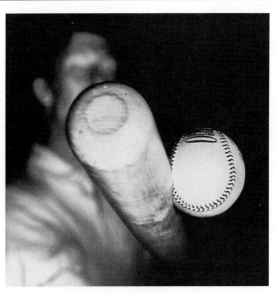

FIGURE 11-5 High-speed photo of ball and bat. The two are in contact for only about 2 ms.

This is a large force, nearly 4000 times the gravitational force on the ball.

For the pop foul, the outgoing momentum is in the *y* direction:

$$\mathbf{p}_2 = m\mathbf{v}_2 = (0.15 \text{ kg})(22\hat{\mathbf{j}} \text{ m/s}) = 3.3\hat{\mathbf{j}} \text{ kg·m/s},$$

so

$$\mathbf{I} = \Delta\mathbf{p} = (3.3\hat{\mathbf{j}} \text{ kg·m/s}) - (-4.5\hat{\mathbf{i}} \text{ kg·m/s})$$
$$= 4.5\hat{\mathbf{i}} + 3.3\hat{\mathbf{j}} \text{ kg·m/s}$$

and

$$\overline{\mathbf{F}} = \frac{\Delta\mathbf{p}}{\Delta t} = \frac{4.5\hat{\mathbf{i}} + 3.3\hat{\mathbf{j}} \text{ kg·m/s}}{2.0\times10^{-3} \text{ s}}$$
$$= 2.3\times10^3\hat{\mathbf{i}} + 1.7\times10^3\hat{\mathbf{j}} \text{ N},$$

or an average impulsive force of magnitude 2800 N. Large forces like this are typical of collisions, and explain why collisions often involve so much damage.

EXERCISE A 55-kg cyclist is riding a 10-kg bicycle at 7.5 m/s when she's struck by another cyclist. The collision lasts 0.11 s, after which the velocity of the first cyclist and her bicycle is $8.1\hat{\mathbf{i}} + 3.2\hat{\mathbf{j}}$ m/s, where the *x* direction is that of the initial motion. Find the impulse and the magnitude of the average impulsive force on the cyclist, and compare the latter with her weight.

Answers: $\mathbf{I} = 39\hat{\mathbf{i}} + 210\hat{\mathbf{j}}$ N·s; $F = 1900$ N, 3.5 times her weight

Some problems similar to Example 11-1: 2, 5, 6

11-2 COLLISIONS AND THE CONSERVATION LAWS

Often we know no more about the impulsive force of a collision than that it acts for a short time and produces abrupt changes in motion. Nevertheless, we can use the conservation laws developed in earlier chapters to relate the interacting objects' motions before and after collision. As always, conservation laws provide a "shortcut," giving us an overall picture lacking in some details. In the case of a collision involving an unknown impulsive force, those missing details are confined to the very short time of the collision.

Momentum

One thing we do know about impulsive forces between colliding objects is that they are *internal* to a system comprising those objects. We found in the preceding chapter that the momentum of a system can be changed only by external forces. Therefore, the impulsive forces of a collision leave the total momentum of the colliding objects unchanged. What about any external forces that may also act during a collision, like the gravitational force on a tennis ball as it collides with the racket? Our definition of a collision requires that the collision take place in a very short time, so that external forces do not have time to alter the system momentum significantly during the collision. To a very good approximation, therefore, the total momentum of a colliding system is conserved during a collision. This conclusion is valid regardless of the details of the impulsive force.

Energy

What about kinetic energy? It may or may not be conserved during a collision. If kinetic energy is conserved, the collision is termed **elastic.** Like a completely frictionless surface, an **elastic collision** is an impossible idealization in the macroscopic world. In a macroscopic collision, some kinetic energy is inevitably transformed to other forms like sound, random molecular motions, or the energy associated with permanently deforming materials (Fig. 11-6). But some interac-

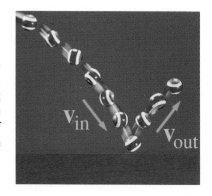

(*a*)

(*b*)

FIGURE 11-6 (*a*) A rubber ball undergoes a nearly elastic collision with the ground, rebounding with outgoing speed v_{out} nearly equal to its incident speed v_{in}. (*b*) An automobile collision is invariably inelastic, with much of the kinetic energy lost to permanent deformation of the colliding vehicles.

tions in the atomic, nuclear, and elementary particle realms are elastic. And macroscopic objects like billiard balls, trampolines, and hockey pucks undergo nearly elastic collisions. The important criterion is that in an elastic collision, the internal forces must be conservative. Then kinetic energy is stored temporarily as potential energy, all of which is again released as kinetic energy by the time the collision is over.

When mechanical energy is lost during a collision, the collision is **inelastic.** In a **totally inelastic collision,** the colliding objects stick together to form one composite object. Even then, mechanical energy is usually not all lost; complete loss of mechanical energy in most cases would violate conservation of momentum. But a totally inelastic collision involves the maximum loss of mechanical energy that is consistent with momentum conservation. In general, we can determine the motion after collision only for the case of completely elastic or totally inelastic collisions; in intermediate cases we must know just how much energy is lost before we can solve for the post-collision motion (see Problem 73).

11-3 INELASTIC COLLISIONS

The motion after a totally inelastic collision is determined entirely by the conservation of momentum principle. We analyzed some totally inelastic collisions in the previous chapter; for instance, measuring the speed of the hockey puck in Example 10-6 involved a totally inelastic collision.

Consider two objects with masses m_1 and m_2 and initial velocities $\mathbf{v}_{1i}$ and $\mathbf{v}_{2i}$ that undergo a totally inelastic collision. That is, after collision they stick together to form a single composite body of mass $m_1 + m_2$ and final velocity $\mathbf{v}_f$. Conservation of momentum states that the initial and final momenta of this two-particle system must be the same:

$$m_1\mathbf{v}_1 + m_2\mathbf{v}_2 = (m_1 + m_2)\mathbf{v}_f. \quad \text{(inelastic collision)} \quad (11\text{-}3)$$

Given four of the five quantities m_1, $\mathbf{v}_{1i}$, m_2, $\mathbf{v}_{2i}$, and $\mathbf{v}_f$, we can solve for the fifth. Often but not always, the unknown is the final velocity:

$$\mathbf{v}_f = \frac{m_1\mathbf{v}_{1i} + m_2\mathbf{v}_{2i}}{m_1 + m_2}. \quad (11\text{-}4)$$

Calculating the total kinetic energy before and after the collision will convince you that this collision is indeed inelastic (see Problem 21).

● EXAMPLE 11-2 THE BALLISTIC PENDULUM

The ballistic pendulum is a device used to measure the speeds of fast-moving objects like bullets. It consists of a wooden block of mass M suspended from vertical strings (Fig. 11-7). When a bullet strikes the block, it undergoes an inelastic collision and embeds itself in the block. If the bullet has mass m, and if the block rises a maximum height h after the collision, show that the speed of the bullet is $[(m + M)/m]\sqrt{2gh}$.

Solution

To solve this problem, we deal separately first with the collision and then with the subsequent rise of the block. Understanding the difference in the way we handle these two steps will strengthen your confidence in applying the conservation laws.

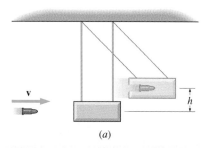

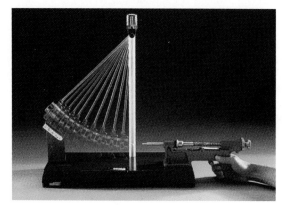

FIGURE 11-7 (*a*) A simplified ballistic pendulum (Example 11-2). (*b*) A ballistic pendulum used in physics labs.

1. The incoming bullet undergoes an inelastic collision with the block. Mechanical energy is not conserved. But because the collision takes such a short time, external forces—here gravity and the string tension—have little effect and the momentum of the bullet + block system is essentially conserved. So we can write:

$$mv = (m + M)V,$$

where v is the initial speed of the bullet and V the speed of the bullet/block combination after the collision. Solving for V gives $V = mv/(m + M)$.
2. As the block swings upward, the external forces of gravity and string tension act to alter the momentum of the bullet + block system. Momentum is not conserved. But

gravity is a conservative force, and the string tension acts at right angles to the motion and therefore does no work. So mechanical energy is conserved in the upward swing of the block. Taking the zero of potential energy in the block's initial position, we can then equate the kinetic energy just after the collision to the potential energy when the block is at its maximum height:

$$\frac{1}{2}(m + M)V^2 = (m + M)gh.$$

Now we relate the two parts of the problem. Using our expression for V in the statement of energy conservation, we have

$$\frac{1}{2}\left(\frac{mv}{m + M}\right)^2 = gh,$$

so

$$v = \left(\frac{m + M}{m}\right)\sqrt{2gh}.$$

TIP Know When To Apply Conservation Laws This example emphasizes the different conditions under which the conservation laws for momentum and energy apply. In the first part of the example we had an inelastic collision in which energy was explicitly *not* conserved. But momentum *was* conserved because the collision took place so rapidly that any impulse from external forces acting on the bullet + block system was negligible. In the second part the presence of a nonzero net force due to gravity and string tension meant that momentum was not conserved. But one of those forces was conservative and the other did no work, so mechanical energy was conserved. We needed both conservation laws—applied at different stages in the event being analyzed—to work this example.

EXERCISE A 950-kg car is stopped on an icy road at the bottom of a hill, facing uphill. It's hit from behind by a 7600-kg truck moving at 50 km/h. If the collision is totally inelastic, to what vertical distance up the hill does the combined wreckage slide?

Answer: 7.8 m

Some problems similar to Example 11-2: 17, 19, 20   ●

● **EXAMPLE 11-3** A FUSION REACTION

In a fusion reaction, two deuterium nuclei (^{2}H) combine to form a helium nucleus (^{4}He). One of the incident nuclei is initially moving in the $+x$ direction at 3.5 Mm/s, and the helium nucleus is subsequently detected moving at 0.23 Mm/s at an angle $\theta = 21°$ to the x axis (Fig. 11-8). Find the initial velocity of the second deuterium nucleus.

Solution
Momentum conservation, as stated in Equation 11-3, entirely determines the outcome of this totally inelastic collision. Solving that equation for the initial velocity $\mathbf{v}_2$ of the second deuterium nucleus, we have

$$\mathbf{v}_2 = \frac{(m_1 + m_2)\mathbf{v}_f - m_1\mathbf{v}_1}{m_2}.$$

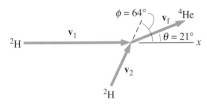

FIGURE 11-8 Velocity vectors for Example 11-3.

Breaking this equation into components gives

x component:

$$v_{2x} = \frac{(m_1 + m_2)v_{fx} - m_1 v_{1x}}{m_2}$$

$$= \frac{(2.0\,\text{u} + 2.0\,\text{u})(0.23\,\text{Mm/s})\cos 21° - (2.0\,\text{u})(3.5\,\text{Mm/s})}{2.0\,\text{u}}$$

$$= 0.79\,\text{Mm/s}$$

y component:

$$v_{2y} = \frac{(m_1 + m_2)v_{fy}}{m_2} = \frac{(2.0\,\text{u} + 2.0\,\text{u})(0.23\,\text{Mm/s})\sin 21°}{2.0\,\text{u}}$$

$$= 1.65\,\text{Mm/s},$$

where we've used the fact that $\mathbf{v}_1$ is entirely in the *x* direction. The magnitude of $\mathbf{v}_2$ is then $v_2 = \sqrt{v_{2x}^2 + v_{2y}^2} = 1.8\,\text{Mm/s}$,

while $\mathbf{v}_2$ makes an angle $\phi = \tan^{-1}(v_{2y}/v_{2x}) = 64°$ with the *x* axis. Figure 11-8 shows the velocity vectors for all three nuclei.

Does this problem remind you of the radioactive decay of Example 10-7? It should. Fusion and radioactive decay are essentially inverse processes. The same momentum conservation principle governs both, and the same mathematical techniques apply.

TIP Know When to Avoid Unit Conversions
Although we stress the importance of SI units, there are instances where unit conversions are unnecessary. The equations derived in this example have mass in both numerator and denominator, so mass units cancel, and therefore, it doesn't matter what units we use. The nuclear mass numbers (e.g., 2 in ^{2}H and 4 in ^{4}He) suffice, and there's no need to convert to kilograms.

EXERCISE A 9400-kg truck is moving eastward at 35 km/h on an icy road when it collides with a 2300-kg car. The two vehicles move off together at 14° north of east, at 41 km/h. Find the initial velocity of the car.

Answer: 78 km/h at 40° north of east.

Some problems similar to Example 11-3: 22, 25, 28 ●

11-4 ELASTIC COLLISIONS

We've seen that momentum is essentially conserved in any collision. In an elastic collision, kinetic energy is conserved as well. In the most general case of a two-body collision, we consider two objects of masses m_1 and m_2, moving initially with velocities $\mathbf{v}_{1i}$ and $\mathbf{v}_{2i}$, respectively. Their final velocities after collision are $\mathbf{v}_{1f}$ and $\mathbf{v}_{2f}$. Then the conservation statements for momentum and kinetic energy become

$$m_1\mathbf{v}_{1i} + m_2\mathbf{v}_{2i} = m_1\mathbf{v}_{1f} + m_2\mathbf{v}_{2f} \qquad (11\text{-}5)$$

and

$$\frac{1}{2}m_1 v_{1i}^2 + \frac{1}{2}m_2 v_{2i}^2 = \frac{1}{2}m_1 v_{1f}^2 + \frac{1}{2}m_2 v_{2f}^2. \qquad (11\text{-}6)$$

Given the initial velocities, we would like to be able to predict the outcome of a collision. In the totally inelastic two-dimensional collision of the preceding section, we had two equations (the two components of the vector momentum conservation equation) and two unknowns (the magnitude and direction of one velocity). Mathematically, we had enough information to solve the system. Here, in the two-dimensional elastic case, we have the two components of the momentum conservation equation 11-5 and the single scalar equation for energy conservation 11-6. But we have four unknowns—the magnitudes and directions of both final velocities. With three equations and four unknowns, we don't have enough information to solve the general two-dimensional elastic collision. Later in this section we will see how other information can help us solve such problems. First, though, we look at the special case of a one-dimensional elastic collision.

Elastic Collisions in One Dimension

When two objects collide head-on, the internal forces act along the same line as the incident motion, and the objects' subsequent motion must therefore be along that same line (Fig. 11-9). Although such one-dimensional collisions are a very special case, they do occur and they provide much insight into the more general case.

In the one-dimensional case, the momentum conservation equation 11-5 has only one nontrivial component:

$$m_1 v_{1i} + m_2 v_{2i} = m_1 v_{1f} + m_2 v_{2f}. \tag{11-5a}$$

where the v's stand for velocity components, rather than magnitudes, and can therefore be positive or negative. If we collect together the terms in Equations 11-5a and 11-6 that are associated with each mass, we have

$$m_1(v_{1i} - v_{1f}) = m_2(v_{2f} - v_{2i}) \tag{11-5b}$$

and

$$m_1(v_{1i}^2 - v_{1f}^2) = m_2(v_{2f}^2 - v_{2i}^2). \tag{11-6a}$$

But $a^2 - b^2 = (a + b)(a - b)$, so Equation 11-6a can be written

$$m_1(v_{1i} - v_{1f})(v_{1i} + v_{1f}) = m_2(v_{2f} - v_{2i})(v_{2f} + v_{2i}). \tag{11-6b}$$

Dividing the left and right sides of Equation 11-6b by the corresponding sides of Equation 11-5b then gives

$$v_{1i} + v_{1f} = v_{2f} + v_{2i}. \tag{11-7}$$

Rearranging Equation 11-7 shows that

$$v_{1i} - v_{2i} = v_{2f} - v_{1f}. \tag{11-8}$$

What does this equation tell us? Both sides describe the relative velocity between the two particles; the equation therefore shows that the relative speed remains unchanged after the collision, although the direction reverses. If one object is approaching another at a relative speed of 5 m/s, then after collision it will be receding at 5 m/s.

Continuing our search for the final velocities, we solve Equation 11-8 for v_{2f}.

$$v_{2f} = v_{1i} - v_{2i} + v_{1f},$$

and use this result in Equation 11-5a:

$$m_1 v_{1i} + m_2 v_{2i} = m_1 v_{1f} + m_2(v_{1i} - v_{2i} + v_{1f}).$$

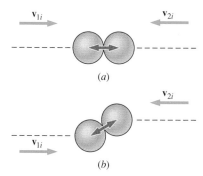

FIGURE 11-9 (a) When two objects collide head-on, internal forces act along the same line as the incident motion, and the subsequent motion is also along this line. The collision is one-dimensional. (b) If the collision is not head-on, internal forces act in new directions, and the motion is necessarily two-dimensional.

Case 1

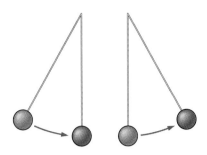

Case 2

Case 3

FIGURE 11-10 Three special cases of one-dimensional collisions. In case 1, a ping-pong ball strikes a bowling ball and rebounds with essentially its initial speed; the bowling ball remains at rest. In case 2 the colliding objects have equal masses. Here one pendulum bob strikes another of equal mass; the first stops, having transferred all its energy and momentum to the second. In case 3 a more massive object (here a baseball bat) strikes a much less massive one (here the ball). The bat's motion continues unchanged while the ball rebounds with greater speed. In a frame in which the ball was initially at rest, the ball would rebound with essentially twice the bat's speed.

Solving for v_{1f} then gives

$$v_{1f} = \frac{m_1 - m_2}{m_1 + m_2} v_{1i} + \frac{2m_2}{m_1 + m_2} v_{2i}. \qquad (11\text{-}9a)$$

Problem 46 asks you to show similarly that

$$v_{2f} = \frac{2m_1}{m_1 + m_2} v_{1i} + \frac{m_2 - m_1}{m_1 + m_2} v_{2i}. \qquad (11\text{-}9b)$$

Equations 11-9a and 11-9b are our desired result, expressing the final velocities in terms of the initial velocities alone. To see that these results make sense, we suppose that $v_{2i} = 0$. (This is really not a special case, since we can always work in a frame of reference in which m_2 is initially at rest.) Now consider several special cases illustrated in Figure 11-10.

Case 1: $m_1 \ll m_2$ For this case, picture a ping-pong ball colliding with a bowling ball, or any object colliding elastically with a perfectly rigid wall. Setting $v_{2i} = 0$ in Equations 11-9, and dropping m_1 as being negligible compared with m_2, Equations 11-9 become simply

$$v_{1f} = -v_{1i}$$

and

$$v_{2f} = 0.$$

That is, the lighter object rebounds with no change in speed, while the heavier object remains at rest. Does this make sense in light of the conservation laws that Equations 11-9 are supposed to reflect? Clearly energy is conserved: the kinetic energy of m_2 remains zero and the kinetic energy $\frac{1}{2}m_1 v_1^2$ is unchanged. But what about momentum? The momentum of the lighter object has changed, from $m_1 v_{1i}$ to $-m_1 v_{1i}$. But momentum *is* conserved; the momentum given up by the lighter object is absorbed by the heavier object. In the limit of an arbitrarily large m_2, though, the heavier object can absorb huge amounts of momentum without acquiring significant speed. If we "back off" from the extreme case that m_1 can be neglected altogether compared with m_2, we would find that the lighter object rebounds with slightly reduced speed and that the heavier object begins moving very slowly in the opposite direction (see Problem 67).

Case 2: $m_1 = m_2$ Again setting $v_{2i} = 0$, Equations 11-9 now give

$$v_{1f} = 0$$

and

$$v_{2f} = v_{1i}.$$

So the first object stops abruptly, transferring all its energy and momentum to the second object. Case 2 collisions have important implications for the design of nuclear reactors. The fission reactions that power a reactor are most easily initiated by slowly moving neutrons, while each uranium atom that fissions releases fast-moving neutrons. The reactor's fuel rods are therefore surrounded by a moderator—a substance that slows down the neutrons. Our Case 2 analysis shows that the most effective slowing-down is achieved in collisions of equal masses. Most nuclear reactors use water as a moderator; the protons comprising the nuclei of the hydrogen atoms have almost the same mass as a neutron and are therefore efficient absorbers of neutron energy.

For purposes of energy transfer, two equal-mass particles are perfectly "matched." We'll encounter analogous instances of energy transfer "matching" when we discuss wave motion and again in connection with electric circuits.

Case 3: $m_1 \gg m_2$ Now, with v_{2i} again set to zero, Equations 11-9 give

$$v_{1f} = v_{1i}$$

and

$$v_{2f} = 2v_{1i},$$

where we've neglected m_2 compared with m_1. So here the more massive object barrels right on with no change in motion, while the lighter one heads off in the same direction with twice the speed. How are momentum and energy conserved in this case? In the extreme limit where we neglect the mass m_2 its energy and momentum are negligible. Essentially all the energy and momentum remain with the more massive object, and both these quantities are unchanged in the collision. If we "back off" from the extreme case where we neglect m_2 altogether, we find that the more massive object's speed is reduced slightly, while the lighter object heads off with almost twice the initial speed of the heavier object (see Problem 68). The collision of a baseball bat and ball is a Case 3 collision; in a reference frame moving with the ball's initial velocity, the massive bat hits the ball, which then heads out at a higher speed than that of the bat. The bat's motion is hardly altered.

● **EXAMPLE 11-4** CAR-TRUCK COLLISIONS

A truck of mass $9M$ is moving at speed $v = 15$ km/h when it collides head-on with a parked car of mass M. Spring-mounted bumpers ensure that the collision is essentially elastic. Describe the subsequent motion of each vehicle. Repeat for the case of the car moving at speed $v = 15$ km/h and colliding with the stationary truck. What fraction of the incident vehicle's kinetic energy is transferred in each case?

Solution

For the first case, we can set $m_1 = 9M$, $v_{1i} = v = 15$ km/h, $m_2 = M$, and $v_{2i} = 0$. Then Equations 11-9 give

$$v_{1f} = \frac{9M - M}{9M + M} v, = \frac{4}{5} v = 12 \text{ km/h}$$

and

$$v_{2f} = \frac{(2)(9M)}{9M + M} v = \frac{9}{5} v = 27 \text{ km/h}.$$

So both vehicles move off in the same direction, with the truck's speed reduced by 20% and the car "kicked up" to 27 km/h. The

initial energy of the truck is $\frac{9}{2}Mv^2$ or $4.5Mv^2$; the final energy of the car is $\frac{1}{2}M(\frac{9}{5}v)^2$ or $1.6Mv^2$. So 1.6/4.5, or 36% of the truck's energy is transferred in this case.

For the second collision, we set $m_1 = M$ and $m_2 = 9M$, giving

$$v_{1f} = \frac{M - 9M}{M + 9M} v = -\frac{4}{5} v = -12 \text{ km/h}$$

and

$$v_{2f} = \frac{2M}{M + 9M} v = \frac{1}{5} v = 3.0 \text{ km/h},$$

so the car rebounds with 80% of its initial speed and the truck picks up a relatively small speed. Now the initial energy of the car is $\frac{1}{2}Mv^2$, while the final energy of the truck is $\frac{9}{2}M(\frac{1}{5}v)^2 = 0.18Mv^2$, so again 36% of the energy is transferred. Even though the mass ratio is quite large, neither collision approaches closely the Case 1 or Case 3 limit. Problem 40 explores the approach to those limits.

E X E R C I S E In a nuclear reactor, a neutron (mass 1 u) moving at 6.9×10^6 m/s collides elastically, head-on, with a carbon-12 nucleus (mass 12 u) initially at rest. Find the velocity of each particle after the collision, and the fraction of the neutron's kinetic energy transferred to the carbon nucleus.

Answers: $v_n = -5.8 \times 10^6$ m/s (minus indicates direction reversal); $v_c = 1.1 \times 10^6$ m/s; 28% of neutron energy transferred

Some problems similar to Example 11-4: 32, 34, 37, 40 ●

● **E X A M P L E 1 1 - 5** "WEIGHING" AN ELEMENTARY PARTICLE

In a nuclear experiment, a particle of unknown mass moving at 460 km/s undergoes a head-on elastic collision with a carbon nucleus (^{12}C; mass 12 u) moving at 220 km/s. After collision, the carbon continues in the same direction at 340 km/s. Find the mass and final velocity of the unknown particle.

Solution

If we designate the carbon as particle 2, we then know all quantities in Equation 11-9b except the unknown mass m_1. Multiplying that equation through by $(m_1 + m_2)$, we have

$$(m_1 + m_2)v_{2f} = 2m_1v_{1i} + (m_2 - m_1)v_{2i}.$$

Solving for m_1 then gives

$$m_1 = \frac{m_2(v_{2i} - v_{2f})}{v_{2f} - 2v_{1i} + v_{2i}}$$

$$= \frac{(12 \text{ u})(220 \text{ km/s} - 340 \text{ km/s})}{340 \text{ km/s} - (2)(460 \text{ km/s}) + 220 \text{ km/s}} = 4.0 \text{ u}.$$

The unknown particle is therefore probably an alpha particle (^{4}He, or helium nucleus). Knowing m_1, we use Equation 11-9a to find v_{1f}:

$$v_{1f} = \frac{4.0 \text{ u} - 12 \text{ u}}{16 \text{ u}}(460 \text{ km/s}) + \frac{(2)(12 \text{ u})}{16 \text{ u}}(220 \text{ km/s})$$

$$= 100 \text{ km/s}.$$

Are these results consistent with our earlier finding that a less massive particle should rebound off a more massive one? After all, both particles now seem to continue in the same direction. But there's no inconsistency; our earlier conclusion was based on one particle's being initially at rest, while here both have nonzero initial velocities. If we look at the situation in a reference frame where the carbon is initially at rest, we have $v_{1i} =$ 460 km/s − 220 km/s = 240 km/s, while $v_{1f} =$ 100 km/s − 220 km/s = −120 km/s. The less massive nucleus does rebound, as expected. Figure 11-11 summarizes the collision in both reference frames; the change of reference frames is explored further in Problem 41.

E X E R C I S E A 2.1×10^4 kg railroad freight car is moving at 32 km/h when it's struck from behind by a second freight car moving at 43 km/h. The collision is elastic, and afterwards the first car is moving at 38 km/h in its original direction. Find the mass and final velocity of the second car.

Answers: $m = 7.88 \times 10^3$ kg; $v = 27$ km/h (same direction)

Some problems similar to Example 11-5: 35, 36, 38

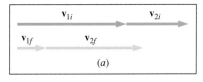

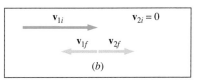

FIGURE 11-11 The collision of Example 11-5 (*a*) in the original reference frame and (*b*) in a reference frame moving with the initial 220 km/s velocity of the carbon nucleus. ●

Elastic Collisions in Two Dimensions

To analyze an elastic collision in two dimensions, we must use the full vector statement of momentum conservation (Equation 11-5), along with the energy conservation equation 11-6. But we found earlier that these equations alone don't provide enough information to solve the problem. In a collision between reasonably simple macroscopic objects, that information may be provided by the so-called **impact parameter,** a measure of how much the collision differs

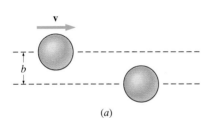

(a)

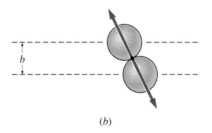

(b)

FIGURE 11-12 (a) The impact parameter is the perpendicular distance b between the centers of the colliding spheres. (b) The impact parameter determines where the spheres touch. Since the contact forces of the collision are perpendicular to the surfaces, the impact parameter carries information about the directions of the forces and therefore about the subsequent motion.

from being head-on (Fig. 11-12). More typically, though, and especially with atomic and nuclear interactions, the needed information must be supplied by measurements done after the collision. Knowing the direction of motion of one particle after collision, for example, provides enough information to analyze the collision if the masses and initial velocities are also known.

● **EXAMPLE 11-6** STARS IN A CLOSE ENCOUNTER

A star of mass $1.4M_\odot$ ($M_\odot$ is the Sun's mass, a convenient mass unit in astronomy) is a great distance from another star of equal mass and is approaching the second star at 680 km/s. The two stars undergo a close encounter, and much later the first star is moving at a 35° angle to its initial direction, as shown in Fig. 11-13. In the frame of reference in which the second star is initially at rest, find the final speeds of both stars and the direction of motion of the second star.

Solution
Since the force acting between the stars is the conservative gravitational force, the encounter is elastic. Designating the initially moving star by the subscript 1, and taking the x axis along the direction of its initial motion, the two components of the momentum conservation equation 11-5 become

x component: $m_1v_{1i} = m_1v_{1f}\cos\theta_1 + m_2v_{2f}\cos\theta_2$ (11-10)
y component: $0 = m_1v_{1f}\sin\theta_1 + m_2v_{2f}\sin\theta_2$, (11-11)

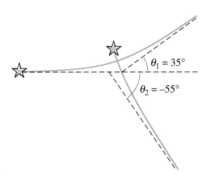

$\theta_1 = 35°$

$\theta_2 = -55°$

FIGURE 11-13 A close encounter between two stars, in a frame of reference in which one star is initially at rest. (Example 11-6).

where $\theta_{1,2}$ are the angles the final velocities make with the x axis. A third equation is provided by energy conservation:

$$\frac{1}{2}m_1v_{1i}^2 = \frac{1}{2}m_1v_{1f}^2 + \frac{1}{2}m_2v_{2f}^2, \quad (11\text{-}12)$$

so we have three equations in the three unknowns v_{1f}, v_{2f}, and θ_2. Since the masses are equal, they cancel from all three equations. Isolating the unknown θ_2 on one side of Equations 11-10 and 11-11 gives

$$v_{2f}\cos\theta_2 = v_{1i} - v_{1f}\cos\theta_1$$
$$v_{2f}\sin\theta_2 = -v_{1f}\sin\theta_1.$$

Squaring and adding these two equations, and using $\cos^2\theta + \sin^2\theta = 1$ on both sides of the resulting equation, we have

$$v_{2f}^2 = (v_{1i} - v_{1f}\cos\theta_1)^2 + v_{1f}^2\sin^2\theta_1$$
$$= v_{1i}^2 - 2v_{1i}v_{1f}\cos\theta_1 + v_{1f}^2.$$

Using this result in the energy conservation equation 11-12 gives

$$v_{1i}^2 = v_{1f}^2 + v_{2f}^2 = v_{1f}^2 + v_{1i}^2 - 2v_{1i}v_{1f}\cos\theta_1 + v_{1f}^2,$$

or $\quad v_{1f}^2 = v_{1i}v_{1f}\cos\theta_1,$

so $\quad v_{1f} = v_{1i}\cos\theta_1 = (680 \text{ km/s})(\cos 35°) = 557 \text{ km/s}.$

From Equation 11-12, we can then write

$$v_{2f} = \sqrt{v_{1i}^2 - v_{1f}^2} = \sqrt{(680 \text{ km/s})^2 - (557 \text{ km/s})^2}$$
$$= 390 \text{ km/s}.$$

Finally, the y momentum equation 11-11 gives

$$\theta_2 = \sin^{-1}\left(\frac{-v_{1f}\sin\theta_1}{v_{2f}}\right) = \sin^{-1}\left[\frac{-(557 \text{ km/s})(\sin 35°)}{390 \text{ km/s}}\right]$$
$$= -55°.$$

The minus sign indicates that this angle lies below the x axis—as it must since there's no net momentum in the y direction and therefore the y components of the two stars' momenta must cancel. Note that the angle between the outgoing velocity vectors is 90°. Problem 43 asks you to prove that this is always true in an elastic collision of equal masses, one initially at rest.

Is this event really a collision? After all, the two stars interact only through the gravitational force, which acts all the time. We certainly wouldn't consider an orbiting satellite to be "colliding" with Earth. But our stellar encounter really is a collision. The incoming star's speed greatly exceeds escape speed from the second star, so the two are not in closed orbits, and gravity plays a significant role only when they're really close. The interaction involves a strong force acting for a short time, and that satisfies our definition of a collision. In the astronomical and subatomic realms, most collisions actually involve such "close encounters" in which particles interact without touching.

EXERCISE A deuteron (^{2}H; mass 2 u) moving at 1.9×10^7 m/s collides elastically with a stationary proton (mass 1 u). A particle detector measurement shows that the deuteron moves off at 27° to the direction of its initial motion. Find the speeds of both particles and the direction of the proton's motion.

Answers: $v_d = 1.39 \times 10^7$ m/s, $v_p = 1.83 \times 10^7$ m/s, θ_p −43.9°

Some problems similar to Example 11-6: 44, 47, 48, 50, 55 ●

The Center-of-Mass Frame

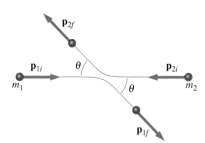

FIGURE 11-14 An elastic collision viewed in the center-of-mass frame. Since the system has zero momentum in this frame, the initial and final momentum vectors form pairs with equal magnitudes and opposite directions.

Two-dimensional collisions take a particularly simple form in a frame of reference moving with the center of mass of the colliding particles since the total momentum in such a frame must be zero. That remains true after a collision, which involves only *internal* forces which, as we found in the preceding chapter, cannot affect the motion of the center of mass. Therefore, both the initial and final momenta form pairs of oppositely directed vectors of equal magnitude, as shown in Fig. 11-14. In an elastic collision, energy conservation requires further that the incident and final momenta have the same values, so a single number—the angle θ in Fig. 11-14—completely describes the collision.

It is often easier to analyze a collision by transforming to the center-of-mass frame, doing the analysis, and then transforming the resulting momentum and velocity vectors back to the original or "lab" frame. High-energy physicists routinely make such transformations as they seek to understand the fundamental forces between elementary particles. Those forces are described most simply in the center-of-mass frame of colliding particles, but the physicists and their particle accelerators are usually not in that frame.

CHAPTER SYNOPSIS

Summary

1. A **collision** is a relatively violent interaction between two objects that takes place in a short time. An **impulsive force** acts between the colliding objects; this force is much stronger than any external force acting on either object.

2. The **impulse** imparted to an object is given by integrating the impulsive force over time; Newton's second law shows that the impulse is equal to the change in the object's momentum:

$$I = \int_{t_1}^{t_2} F \, dt = \Delta p.$$

The average value $\overline{F}$ of the impulsive force is given by the impulse divided by the time interval Δt over which the force acts:

$$\overline{F} = \frac{I}{\Delta t} = \frac{\Delta p}{\Delta t}.$$

3. Because impulsive forces acting during a collision are internal to the system of colliding objects, they cancel in pairs and therefore cannot change the momentum of the system. Since a collision takes place in a short time, any external forces have little effect on the system momentum, so to a good approximation **momentum is conserved during a collision.**

4. If kinetic energy is not conserved during a collision, the collision is **inelastic.** The maximum loss of kinetic energy occurs when the colliding objects stick together; then the collision is **totally inelastic.** In a totally inelastic collision, the final velocity $\mathbf{v}_f$ is determined entirely by the initial velocities through the momentum conservation equation:

$$m_1\mathbf{v}_{1i} + m_2\mathbf{v}_{2i} = (m_1 + m_2)\mathbf{v}_f,$$

where m_1 and m_2 are the masses of the colliding objects.
5. In an **elastic collision,** kinetic energy is conserved. The relation between the initial and final velocities is governed by conservation of momentum and conservation of energy:

$$m_1\mathbf{v}_{1i} + m_2\mathbf{v}_{2i} = m_1\mathbf{v}_{1f} + m_2\mathbf{v}_{2f}$$
$$\tfrac{1}{2}m_1v_{1i}^2 + \tfrac{1}{2}m_2v_{2i}^2 = \tfrac{1}{2}m_1v_{1f}^2 + \tfrac{1}{2}m_2v_{2f}^2.$$

In the general case, these equations are not sufficient to solve for the final velocities. Additional information, such as the direction of one of the final velocities or the details of the collision, must also be used. In the special case of a one-dimensional collision, the momentum and energy conservation equations combine to determine completely the final velocities:

$$v_{1f} = \frac{m_1 - m_2}{m_1 + m_2}v_{1i} + \frac{2m_2}{m_1 + m_2}v_{2i}$$
$$v_{2f} = \frac{2m_1}{m_1 + m_2}v_{1i} + \frac{m_2 - m_1}{m_1 + m_2}v_{2i}.$$

Terms You Should Understand

(Pairs are closely related terms whose distinction is important; number in parentheses is chapter section where term first appears.)
collision (introduction)
impulse (11-1)
impulsive force, average impulsive force (11-1)
elastic collision, inelastic collision (11-2)
totally inelastic collision (11-2)

Symbols You Should Recognize

$\mathbf{I}$ (11-1)

Problems You Should Be Able to Solve

calculating momentum changes given impulse (11-1)
calculating average impulsive force given momentum change and collision time (11-1)
solving for both components of the final velocity, or for any other pair of unknowns, in a totally inelastic collision (11-3)
solving for any two unknowns in a one-dimensional elastic collision (11-4)
solving for unknown quantities in a two-dimensional elastic collision, given sufficient information (11-4)

Limitations to Keep in Mind

Conservation of momentum holds during a collision only to the extent that the impulsive forces of the collision are far greater than any external forces acting on the system during the collision.
A perfectly elastic collision is an impossibility in the macroscopic realm.

QUESTIONS

1. Why do we require a short time interval and strong impulsive forces for an interaction to qualify as a collision? What is meant by "short" and "strong" in this context?
2. Comet Halley spends most of its 76-year orbital period moving slowly a long way from the Sun. Does it make sense to treat its roughly 6-month close encounter with the Sun as a collision? Discuss.
3. When a ball bounces off a wall, its momentum is reversed. How can this be, in view of the law of conservation of momentum?
4. A high jumper runs to the takeoff point with horizontal but no vertical momentum. When he jumps, where does his vertical momentum come from?
5. As you jump off a table, you bend your knees on landing. How does this lessen the shock to your body?
6. To what height does a bouncing ball return if the collision is elastic? Explain.
7. Must the analysis of a collision between two objects ever require consideration of all three dimensions? Explain.
8. Why are cars designed so that their front ends crush during an accident?
9. Discuss the relative advantages and disadvantages of air bags and seat belts for automotive safety during collisions.
10. Give three everyday examples of inelastic collisions.
11. Is it possible to have an inelastic collision in which *all* the kinetic energy of the colliding objects is lost? If not, why not? If so, give an example.

12. If you want to stop the neutrons in a reactor, why not use massive nuclei like lead?

13. Photons—particles of light—carry momentum and energy but have no intrinsic mass. When an electron and a positron (the electron's antiparticle) collide, they annihilate to form a pair of photons. If the electron and positron were initially moving straight toward each other at the same speed, what can you conclude about the directions of the two photons? Would it be possible for just one photon to be created in this annihilation? Why or why not? Is mass conserved in the annihilation? Could energy be conserved?

14. Discuss the conservation principles used in the analysis of the ballistic pendulum. When is each valid?

15. A truck collides with an identical-looking truck initially at rest. The two move off together with more than half the original speed of the moving truck. What can you conclude about their relative loads?

16. Is it possible to have a collision during which the kinetic energy of the colliding particles increases? Explain.

17. How could you generalize the concept of energy to reinstitute the conservation of energy principle for inelastic collisions?

18. Ty Cobb slides into second base. His kinetic energy has been dissipated as heat, but where did his momentum go?

19. Does a batted ball go farther if the pitch is thrown faster, all other factors being equal? Explain.

20. Why is it relatively safe (though not recommended, of course!) to drop a lead weight on your foot if your foot is not in direct contact with the ground?

21. A downward-moving gas molecule collides with a downward-moving piston in an automobile engine. Comment on the speeds of each after the collision, and on any transfer of energy that might take place. How are such collisions important to the operation of the engine?

22. Two identical satellites are going in opposite directions in the same circular orbit, when they collide head-on. Describe their subsequent motion if the collision is (a) elastic or (b) inelastic.

PROBLEMS

Section 11-1 Impulse and Collisions

1. What is the impulse associated with a 950-N force acting for 100 ms?

2. A 7.5-kg object is moving in the positive x direction at 34 m/s when it undergoes a collision that lasts 0.22 s and leaves it moving in the negative x direction at 51 m/s. Find (a) the impulse and (b) the average impulsive force associated with this collision.

3. A 62-kg parachutist hits the ground moving at 35 km/h and comes to a stop in 140 ms. Find the average impulsive force on the chutist, and compare with the chutist's weight.

4. (a) By how much does the momentum of a car change when it undergoes a collision during which an average impulsive force of 1.2×10^5 N acts for 0.25 s? (b) If the car's mass is 1800 kg, for what initial speed would this collision bring the car to a stop?

5. A 240-g ball is moving with velocity $\mathbf{v}_i = 6.7\hat{\mathbf{i}}$ m/s when it undergoes a collision lasting 52 ms. After the collision its velocity is $\mathbf{v}_f = -4.3\hat{\mathbf{i}} + 3.1\hat{\mathbf{j}}$ m/s. Find (a) the impulse and (b) the average impulsive force associated with this collision.

6. A proton moving in the positive x direction at 6.8×10^6 m/s collides with a nucleus. The collision lasts 1.3×10^{-17} s, and the average impulsive force is $-1.5\hat{\mathbf{i}} + 0.71\hat{\mathbf{j}}$ mN. (a) Find the velocity of the proton after the collision. (b) Through what angle has the proton's motion been deflected?

7. While doing aerobic dancing, you raise your center of mass 32 cm as you jump off the floor. On the way down, you bend your knees as your feet hit the ground, lowering your center of mass an additional 12 cm in the 50 ms it takes for your body to stop. If your mass is 60 kg, what are (a) the impulse and (b) the average impulsive force? Compare with your weight.

8. Explosive bolts separate a 950-kg communications satellite from its 640-kg booster rocket. If the impulse of the explosion is 350 N·s, at what relative speed do the satellite and booster separate?

9. A 727 jetliner in level flight with a total mass of 8.6×10^4 kg encounters a downdraft lasting 1.3 s. During this time, the plane acquires a downward velocity component of 85 m/s. Find (a) the impulse and (b) the average impulsive force on the plane.

10. Find the magnitude of the impulse imparted to the stars in Example 11-6.

11. (a) Estimate the impulse imparted by the force shown in Fig. 11-15. (b) What is the average impulsive force?

12. A 59-g tennis ball is thrown straight up, and at the peak of its trajectory is hit by a racket that exerts a horizontal force given by $F = at - bt^2$, where t is the time in milliseconds from the instant the racket first contacts the ball, and where $a = 1200$ N/ms and $b = 400$ N/ms². The ball separates from the racket after 3.0 ms. Find (a) the impulse, (b) the average impulsive force, and (c) the ball's speed just after it leaves the racket.

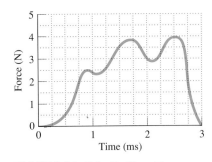

FIGURE 11-15 Problem 11.

Section 11-2 Collisions and the Conservation Laws

13. At the peak of its trajectory, a 1.0-kg projectile moving horizontally at 15 m/s collides with a 2.0-kg projectile at the peak of a vertical trajectory. If the collision takes 0.10 s, how good is the assumption that momentum is conserved during the collision? To find out, compare the change in momentum of the colliding system with the system's total momentum.

14. An object like a ball used in a sporting event can be characterized by its **coefficient of restitution,** defined as the ratio of outgoing to incident speed when the ball collides with a rigid surface. The coefficient of restitution of a typical tennis ball is about 0.7. What fraction of the ball's kinetic energy is lost at each bounce?

15. An 1800-kg car moving at 25 m/s collides with an identical car moving in the same direction at 15 m/s. If an external frictional force of 6.1 kN acts on both cars, what is the minimum collision time that will ensure the system momentum changes by less than 0.1% during the collision?

16. A 340-g ball moving at 7.40 m/s collides with a 230-g ball initially at rest. After the collision the first ball is moving at 1.60 m/s and the second at 8.57 m/s, both in the initial direction of the first ball. (a) Calculate the momenta before and after the collision and show that, to three significant figures, momentum is conserved. (b) Is the collision elastic, totally inelastic, or somewhere in between? Justify your answer.

Section 11-3 Inelastic Collisions

17. In a railroad switchyard, a 45-ton freight car is sent at 8.0 mi/h toward a 28-ton car that is moving in the same direction at 3.4 mi/h. (a) What is the speed of the pair after they couple together? (b) What fraction of the initial kinetic energy was lost in the collision?

18. In a totally inelastic collision between two equal masses, one of which is initially at rest, show that half the initial kinetic energy is lost.

19. A sled and child with a total mass of 33 kg are moving horizontally at 10 m/s when a second child leaps on with negligible speed. If the sled's speed drops to 6.4 m/s, what is the mass of the second child?

20. In an ice-show stunt, a 70-kg skater dressed as a baseball player catches a 150-g baseball moving at 23 m/s. (a) If the skater was initially at rest, what is his final speed? (b) If the catch takes 36 ms, what is the average impulsive force exerted by the ball?

21. A mass m collides totally inelastically with a mass M initially at rest. Show that a fraction $M/(m + M)$ of the initial kinetic energy is lost in the collision.

22. A 1200-kg Toyota and a 2200-kg Buick collide at right angles in an intersection. They lock together and skid 22 m; the coefficient of friction is 0.91. Show that at least one car must have exceeded the 25 km/h speed limit in effect at the intersection.

23. Astronomers warn that there is a nonzero chance of Earth colliding inelastically with a substantial asteroid (Fig. 11-1). Impact speed of the asteroid might be 10 km/s. Estimate the mass of an asteroid needed to alter Earth's orbital speed by 0.01%.

24. A 114-g Frisbee is caught in a tree. To dislodge it, you toss a 330-g lump of clay vertically upward. Clay and Frisbee stick together and rise to a maximum height of 4.1 m above the Frisbee's initial position. What was the speed of the clay as it hit the Frisbee?

25. A neutron (mass 1 u) strikes a deuteron (mass 2 u), and the two combine to form a tritium nucleus. If the neutron's initial velocity was $28\hat{\imath} + 17\hat{\jmath}$ Mm/s and if the tritium nucleus leaves the reaction with velocity $12\hat{\imath} + 20\hat{\jmath}$ Mm/s, what was the velocity of the deuteron?

26. Two identical trucks have mass 5500 kg when empty. One truck carries a 9500-kg load and is moving at 65 km/h. It collides inelastically with the second truck, which is initially at rest, and the pair moves off at 40 km/h. What is the load of the second truck?

27. Two identical pendulum bobs are suspended from strings of equal length, and one is released from a height h as shown in Fig. 11-16. When the first bob hits the second, the two stick together. Show that the maximum height to which the combination rises is $\frac{1}{4}h$.

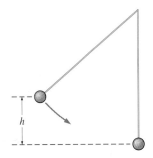

FIGURE 11-16 Problem 27.

28. A 5.4-kg hawk is diving at 18 m/s on a northward path at 65° to the horizontal when it grabs a 1.2-kg pigeon in its talons. If the pigeon was flying horizontally at 14 m/s in the same direction as the hawk's horizontal motion, what is the velocity of the pair just after the catch?

29. A 400-mg popcorn kernel is skittering across a nonstick frying pan at 8.2 cm/s when it pops and breaks into two equal-mass pieces. If one piece ends up at rest, how much energy was released in the popping?

30. A 1300-kg car moving at 10 km/h collides with a 1600-kg car moving in the same direction at 6.6 km/h. The first car is equipped with spring-loaded bumpers to prevent damage. If the spring constant is 28,000 N/m, find the maximum compression of the spring.

31. Two identical objects with the same initial speed collide and stick together. If the composite object moves with half the initial speed of either object, what was the angle between the initial velocities?

Section 11-4 Elastic Collisions

32. An alpha particle (^{4}He) strikes a stationary gold nucleus (^{197}Au) head-on. What fraction of the alpha particle's kinetic energy is transferred to the gold nucleus? Assume the collision is totally elastic.

33. While playing ball in the street, a child accidentally tosses a ball at 18 m/s toward the front of a car moving toward him at 14 m/s. What is the speed of the ball after it rebounds elastically from the car?

34. A block of mass m undergoes a one-dimensional elastic collision with a block of mass M initially at rest. If both blocks have the same speed after the collision, how are their masses related?

35. A proton moving at 6.9 Mm/s collides elastically and head-on with a second proton moving in the opposite direction at 11 Mm/s. Find their velocities after the collision.

36. A proton (mass 1 u) moving at 6.90 Mm/s collides elastically and head-on with a second particle moving in the opposite direction at 2.80 Mm/s. After the collision, the proton is moving opposite to its initial direction at 8.62 Mm/s. Find the mass and final velocity of the second particle.

37. Two objects, one initially at rest, undergo a one-dimensional elastic collision. If half the kinetic energy of the initially moving object is transferred to the other object, what is the ratio of their masses?

38. A 59.1-g tennis ball is moving at 14.5 m/s when it collides elastically and head-on with a basketball moving in the opposite direction at 9.63 m/s. If the tennis ball rebounds at twice its initial speed, find the mass and final velocity of the basketball.

39. Blocks B and C have masses $2m$ and m, respectively, and are at rest on a frictionless surface. Block A, also of mass m, is heading at speed v toward block B as shown in Fig.

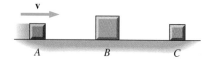

FIGURE 11-17 Problem 39.

11-17. If all subsequent collisions are elastic, determine the final velocity of each block.

40. A block of mass m_1 undergoes a one-dimensional elastic collision with an initially stationary block of mass m_2. Find an expression of the fraction of the initial kinetic energy transferred to the second block, and plot your result for mass ratios m_1/m_2 from 0 to 20. Show also that, for a given mass ratio, the energy transfer is the same no matter which mass is initially at rest.

41. Rework Example 11-5 using a frame of reference in which the carbon nucleus is initially at rest.

42. Two buses approach each other, each going 60 km/h relative to the road. The driver of one bus tosses a tennis ball straight ahead at 20 km/h relative to his bus. The ball bounces elastically off the vertical windshield of the second bus, and back to the driver who threw it. At what speed relative to the driver is it going when caught? Neglect gravity.

43. An object collides elastically with an equal-mass object initially at rest. If the collision is not head-on, show that the final velocity vectors are perpendicular.

44. On ice, a 3.2-kg rock moving at 1.0 m/s collides elastically with a 0.35-kg rock initially at rest. The smaller rock goes off at an angle of 45° to the larger rock's initial motion. What are the final speeds of the two rocks? What is the direction of the larger rock?

45. Two pendulums of equal length $\ell = 50$ cm are suspended from the same point. The pendulum bobs are steel spheres with masses of 140 and 390 g. The more massive bob is drawn back to make a 15° angle with the vertical (Fig. 11-18). When it is released the bobs collide elastically. What is the maximum angle made by the less massive pendulum?

FIGURE 11-18 Problem 45.

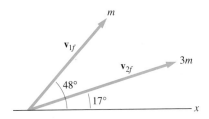

FIGURE 11-19   Problem 47.

46. Derive Equation 11-9b.
47. A particle of mass m is moving along the x axis when it collides elastically with a particle of mass $3m$. The more massive particle moves off at 1.2 m/s at 17° to the x axis, while the less massive particle moves off at 0.92 m/s at 48° to the x axis, as shown in Fig. 11-19. Find two sets of possible values for the initial speeds of both particles and the initial direction of the more massive particle.
48. A proton of mass 1.0 u is traveling through the bubble chamber of a particle accelerator at $2.4{\times}10^5$ m/s when it strikes a stationary deuteron (mass 2.0 u). The proton is deflected through an angle of 37°. Assuming the collision is elastic, what are the final velocities of the two particles?
49. A tennis ball moving at 18 m/s strikes the 45° hatchback of a car moving away at 12 m/s, as shown in Fig. 11-20. Both speeds are given with respect to the ground. What is the velocity of the ball with respect to the ground after it rebounds elastically from the car? *Hint:* Work in the frame of reference of the car; then transform to the ground frame.

FIGURE 11-20   Problem 49.

50. Rework Example 11-6 for the case when the star initially at rest is three times as massive as the other star.
51. Two identical billiard balls are initially at rest when they are struck symmetrically by a third identical ball moving with velocity $\mathbf{v}_0 = v_0\hat{\mathbf{i}}$, as shown in Fig 11-21. Find the velocities of all three balls after they undergo an elastic collision.
52. A particle of mass m, moving in the $+x$ direction at speed v_{1i}, undergoes an elastic collision with a particle of mass

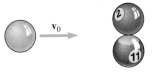

FIGURE 11-21   Problem 51.

$2m$, moving in the $+y$ direction. After the collision, the first particle is moving in the $+y$ direction and the second particle in the $+x$ direction. Find the final speeds of both particles and the initial speed of the more massive particle, all in terms of v_{1i}.

Paired Problems

(Both problems in a pair involve the same principles and techniques. If you can get the first problem, you should be able to solve the second one.)
53. A 590-g basketball is moving at 9.2 m/s when it hits a backboard at 45°. It bounces off at a 45° angle, still moving at 9.2 m/s. If the ball is in contact with the backboard for 22 ms, find the average impulsive force on the ball.
54. A 330-g snowball moving at 13 m/s hits you square in the forehead, after which it crumbles and falls straight to the ground. If the collision lasts 54 ms, what average impulsive force does the snowball exert on your forehead?
55. A 2100-kg car is moving east at 75 km/h when it undergoes a totally inelastic collision with a 7900-kg truck moving northeast at 50 km/h. Find the final velocity of the combined wreckage.
56. A 32-u oxygen molecule (O_2) moving in the $+x$ direction at 580 m/s collides with an oxygen atom (mass 16 u) moving at 870 m/s at 27° to the x axis. The particles stick together to form an ozone molecule. Find the velocity of the ozone.
57. In Fig. 11-22 a truck slams into a parked car, and the resulting one-dimensional collision is elastic and transfers 41% of the truck's kinetic energy to the car. Compare the masses of the two vehicles.

FIGURE 11-22   Problem 57.

58. A 4.0-u helium nucleus makes an elastic, head-on collision with another nucleus, initially at rest. If the helium's initial kinetic energy is 890 keV and if the collision transfers 240 keV to the second nucleus, what is the latter's mass?
59. A head-on, elastic collision between two particles with equal initial speed v leaves the more massive particle (mass m_1) at rest. Find (a) the ratio of the particle masses and (b) the final speed of the less massive particle.
60. A proton (mass 1.0 u) is moving at $9.7{\times}10^5$ m/s when it undergoes a head-on, elastic collision with an atomic nucleus moving in the same direction at one-fifth of the proton's speed. After the collision the two particles are moving in opposite directions, the proton at $3.6{\times}10^5$ m/s. What are (a) the mass and (b) the final speed of the nucleus?

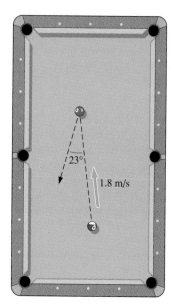

FIGURE 11-23 Problem 61.

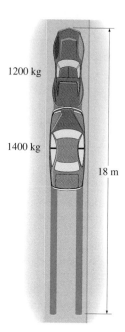

FIGURE 11-24 Problem 65.

61. A billiard ball moving at 1.8 m/s strikes an identical ball initially at rest as illustrated in Fig. 11-23. They undergo an elastic collision and the first ball moves off at 23° counterclockwise from its original direction. Find the final speeds of both balls and the direction of the second ball's motion.

62. A proton (mass 1.0 u) collides elastically with a deuteron (mass 2.0 u) initially at rest. After collision the proton is moving at 6.2×10^5 m/s at 31° clockwise from its initial direction of motion. Find the initial speed of the proton, the final speed of the deuteron, and the direction of the deuteron's motion.

Supplementary Problems

63. A 114-g Frisbee is lodged on a tree branch 7.65 m above the ground. To free it, you lob a 240-g wad of mud vertically upward. The mud leaves your hand at a point 1.23 m above the ground, moving at 17.7 m/s. It sticks to the Frisbee. Find (a) the maximum height reached by the Frisbee-mud combination and (b) the speed with which the combination hits the ground.

64. You set a small ball of mass m atop a large ball of mass $M \gg m$ and drop the pair from a height h. Assuming the balls are perfectly elastic, show that the smaller ball rebounds to a height $9h$.

65. A 1400-kg car moving at 75 km/h runs into a 1200-kg car moving in the same direction at 50 km/h (Fig. 11-24). The two cars lock together and both drivers immediately slam on their brakes. If the cars come to rest in a distance of 18 m, what is the coefficient of friction?

66. To show that elastic collisions among electrons and protons are not very effective at transferring energy, estimate the number of head-on collisions an electron must make with stationary protons ($m_p = 1800m_e$) before it has lost half its *initial* energy. *Hint:* Determine the fractional energy *remaining* with the electron. With each collision, the energy is further reduced by that factor. See Problem 40.

67. Consider a one-dimensional elastic collision with $m_1 \ll m_2$ and m_2 initially at rest. In discussing this extreme case, we neglected m_1 altogether and showed that m_1 then rebounds with its initial speed. Now use the binomial theorem (see Appendix A) to show that a better approximation gives a rebound speed that is less than the incident speed by an amount $2m_1v_1/m_2$ for the case $m_1 \ll m_2$. In applying the binomial theorem, keep terms of order m_1/m_2, but neglect terms of order m_1^2/m_2^2.

68. Repeat the preceding problem for the case $m_1 \gg m_2$, and show that the less massive object moves off with speed given approximately by $2v_1(1 - m_2/m_1)$, where v_1 is the initial speed of the more massive object.

69. How many head-on collisions must a neutron (mass 1.0087 u) make with stationary ^{12}C nuclei (mass 11.9934 u) in a graphite-moderated nuclear reactor in order to lose as much energy as it would in a water-moderated reactor where it collides with a single proton (mass 1.0073 u)? *Hint:* See hint in Problem 66.

70. A 200-g block is released from rest 25 cm high on a frictionless 30° incline. It slides down the incline, then along a frictionless surface until it collides elastically with an

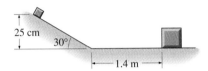

FIGURE 11-25 Problem 70.

800-g block at rest 1.4 m from the bottom of the incline (Fig. 11-25). How much later do the two blocks collide again?

71. A 1.0-kg particle is moving in the $+x$ direction at 4.0 m/s when it collides elastically with a 4.0-kg particle moving in the $-x$ direction at 1.0 m/s. After colliding, the 1-kg particle moves off at 130° counterclockwise from the positive x axis. Find the final speeds of both particles and the direction of the more massive one.

72. When two particles of equal mass undergo an elastic collision in two dimensions, the angle between the outgoing velocity vectors is always 90° in a frame of reference where one particle is initially at rest. (Problem 43 asks you to prove this.) Is the angle between the outgoing velocity vectors constant in a similar collision where the masses are not equal? To find out, determine the angle between the outgoing velocities when a mass m collides elastically with a mass $1.1m$, for the cases where the mass m moves off at angles of (a) 30°, (b) 60°, and (c) 90° to its original direction of motion.

73. A 1200-kg car moving at 25 km/h undergoes a one-dimensional collision with an 1800-kg car initially at rest. The collision is neither elastic nor totally inelastic; the kinetic energy lost is 5800 J. Find the speeds of both cars after the collision.

74. A 14-kg projectile is launched at 380 m/s at a 55° angle to the horizontal. At the peak of its trajectory it collides with a second projectile moving horizontally, in the opposite direction, at 140 m/s. The two stick together and land 9.6 km horizontally downrange from the first projectile's launch point. Find the mass of the second projectile.

75. A block of mass M is moving at speed v_0 on a frictionless surface that ends in a rigid wall. Farther from the wall is a more massive block of mass αM, initially at rest (Fig. 11-26). The less massive block undergoes elastic collisions with the other block and with the wall, and the motion of both blocks is confined to one dimension. (a) Show that the two blocks will undergo only one collision if $\alpha \le 3$. (b) Show that the two blocks will undergo two collisions if $\alpha = 4$, and determine their final speeds. (c) Find out how many collisions the two blocks will undergo if $\alpha = 10$, and determine their final speeds.

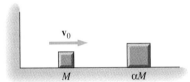

FIGURE 11-26 Problem 75.

CHAPTER 12

ROTATIONAL MOTION

Amusement park rides involve a variety of rotational motions.

You're sitting on a rotating planet. The wheels of your car rotate. Your favorite music comes from a rotating compact disc. A circular saw rotates at high speed to rip its way through a board. A dancer pirouettes, and a satellite spins about its axis. Even molecules rotate. Rotational motion is commonplace throughout the physical universe.

In principle, we could treat rotational motion by analyzing the motion of each particle making up a composite, rotating object. But that would be a hopeless task for any but the simplest objects. Instead, we'll develop a description of rotational motion that closely parallels our understanding of motion as described by Newton's laws.

12-1 ANGULAR SPEED AND ACCELERATION

You slip a compact disc into its player, and it starts spinning. You could describe its motion by giving the speed and direction of each point on the disc. But it's much easier just to say that the disc is rotating at 200 revolutions per minute (rpm). As long as the disc is a **rigid body**—one whose parts remain in fixed positions relative to one another—then that single statement suffices to describe the motion of the entire disc.

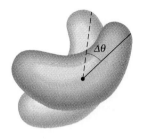

FIGURE 12-1 A rigid body rotates through the angle $\Delta\theta$ in time Δt. Its average angular speed is $\Delta\theta/\Delta t$.

Angular Speed

The rate at which a body rotates is its **angular speed**—a quantity that describes how rapidly the angular position of any point on the body changes. With our 200-rpm CD, our unit of angle was one full revolution (360°, or 2π radians), and our unit of time was the minute. But we could equally well express angular speed in revolutions per second (rev/s), degrees per second (°/s), or radians per second (rad/s or simply s^{-1} since radians are dimensionless). Because of the mathematically simple status of radian measure, we'll often use radians in calculations involving rotational motion.

We use the Greek symbol ω (omega) for angular speed and define **average angular speed** $\overline{\omega}$ as:

$$\overline{\omega} = \frac{\Delta\theta}{\Delta t}, \tag{12-1}$$

where $\Delta\theta$ is the change in angle occurring in the time Δt (Fig. 12-1). When the angular speed is constant, its value at any instant is the same as its average value. When angular speed is changing, we define **instantaneous angular speed** as the limit of the average angular speed taken over arbitrarily short time intervals:

$$\omega = \lim_{\Delta t \to 0} \frac{\Delta\theta}{\Delta t} = \frac{d\theta}{dt}. \tag{12-2}$$

These definitions are analogous to our definitions of average and instantaneous linear speed introduced in Chapter 2. The only difference is the use of angular displacement $\Delta\theta$, rather than linear displacement Δx.

Knowing the angular speed of a rotating object, we can easily find the linear speed of any point on the object. Recall that an angle in radian measure is defined as the ratio of subtended arc length to radius (Fig. 12-2):

$$\theta = \frac{s}{r}. \tag{12-3}$$

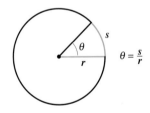

FIGURE 12-2 The angle θ in radians is the ratio of the arc length s to the radius r.

Differentiating this expression with respect to time gives

$$\frac{d\theta}{dt} = \frac{1}{r}\frac{ds}{dt},$$

because the radius r is constant. But $d\theta/dt$ is the angular speed, and ds/dt is the linear speed, v, so $\omega = v/r$, or

$$v = \omega r. \tag{12-4}$$

FIGURE 12-3 The linear speed of a point on a rotating object is proportional to the angular speed and the distance from the rotation axis.

Thus the linear speed of any point on a rotating object is proportional both to the angular speed of the object and to the distance from that point to the axis of rotation (Fig. 12-3).

> **TIP** **Know When To Use Radian Measure** Equation 12-4 was derived using the definition of angle *in radians* and therefore holds only for angular speed measured in radians per unit time. For other measures of angle, a proportionality constant must be included in the relation among v, ω, and r (see Problem 3). Furthermore, Equation 12-4 is valid only for rotation about a fixed axis. In Section 12-5 we'll consider the more general case of an object—like a rolling wheel—whose rotation axis is itself in motion.

● EXAMPLE 12-1 ANGULAR SPEED

A turbine in an electric power plant rotates at 3600 rpm. Find its angular speed in rad/s, and determine the linear speed of a point 72 cm from the rotation axis.

Solution
One revolution is 2π radians, and 1 minute is 60 s, so we have

$$\omega = 3600 \text{ rpm} = \frac{(3600 \text{ rev/min})(2\pi \text{ rad/rev})}{60 \text{ s/min}} = 377 \text{ rad/s}.$$

Then the speed of a point 72 cm from the axis is given by Equation 12-4:

$$v = \omega r = (377 \text{ rad/s})(0.72 \text{ m}) = 270 \text{ m/s},$$

where the units work out because radians are dimensionless.

EXERCISE An engineer is designing a helicopter blade to rotate at 260 rpm. (a) Find the rotation rate in rad/s, and (b) determine the maximum blade length (measured from the rotation axis) that will keep the blade speed everywhere subsonic (less than the speed of sound in air, about 340 m/s).

Answers: (a) 27.2 rad/s (b) 12.5 m

Some problems similar to Example 12-1: 2, 4, 7, 8 ●

Angular Acceleration

If the angular speed of a rotating object changes, then the object undergoes an **angular acceleration,** α, defined analogously to linear acceleration:

$$\alpha = \lim_{\Delta t \to 0} \frac{\Delta \omega}{\Delta t} = \frac{d\omega}{dt}. \tag{12-5}$$

Taking the limit in Equation 12-5 gives the instantaneous angular acceleration; if we don't take the limit, then we have an average value over the time interval Δt. The SI units for angular acceleration are rad/s², although we will sometimes use other units such as rpm/s or rev/s². Having defined angular acceleration, we can differentiate Equation 12-4 to get a relation between angular acceleration and tangential acceleration a_t of a point a distance r from the axis of rotation:

$$a_t = \frac{dv}{dt} = r\frac{d\omega}{dt} = r\alpha. \tag{12-6}$$

▲ **TABLE 12-1** ANGULAR AND LINEAR POSITION, SPEED, AND ACCELERATION

LINEAR QUANTITY OR EQUATION	ANGULAR QUANTITY OR EQUATION
Position x	Angular position θ
Speed $v = \dfrac{dx}{dt}$	Angular speed $\omega = \dfrac{d\theta}{dt}$
Acceleration $a = \dfrac{dv}{dt} = \dfrac{d^2x}{dt^2}$	Angular acceleration $\alpha = \dfrac{d\omega}{dt} = \dfrac{d^2\theta}{dt^2}$

EQUATIONS FOR CONSTANT ACCELERATION			
$\bar{v} = \frac{1}{2}(v_0 + v)$	(2-8)	$\bar{\omega} = \frac{1}{2}(\omega_0 + \omega)$	(12-8)
$v = v_0 + at$	(2-7)	$\omega = \omega_0 + \alpha t$	(12-9)
$x = x_0 + v_0 t + \frac{1}{2}at^2$	(2-10)	$\theta = \theta_0 + \omega_0 t + \frac{1}{2}\alpha t^2$	(12-10)
$v^2 = v_0^2 + 2a(x - x_0)$	(2-11)	$\omega^2 = \omega_0^2 + 2\alpha(\theta - \theta_0)$	(12-11)

This equation gives only the tangential acceleration—the component tangent to the circular path of a rotating point. There is also a radial acceleration a_r given by

$$a_r = \frac{v^2}{r} = \omega^2 r, \qquad (12\text{-}7)$$

where we've used Equation 12-4 to express v in terms of ω. Since Equations 12-6 and 12-7 were derived using Equation 12-4, they, too, require that the angular speed ω be expressed in rad/s.

Because angular speed and acceleration are defined in a way that is mathematically analogous to linear speed and acceleration, all the relations among linear position, speed, and acceleration automatically apply among angular position, angular speed, and angular acceleration. For example, angular acceleration is the second time derivative of angular position; angular speed is the integral of angular acceleration. And if angular acceleration is constant, then all our constant-acceleration formulas of Chapter 2 apply when we make the substitutions θ for x, ω for v, and α for a. Table 12-1 summarizes this direct analogy between linear and rotational quantities. With the analogies of Table 12-1, problems involving rotational motion are just like the one-dimensional linear problems you solved in Chapter 2.

● **EXAMPLE 12-2** A POWER PLANT SHUTDOWN

During a power plant shutdown, the turbine of Example 12-1 coasts to a halt in 590 s. If the angular acceleration is constant, how many revolutions does the turbine make during this time?

Solution

With constant angular acceleration, the average angular speed during the 590-s stopping time is just the average of the initial 3600 rpm and the final 0 rpm (Equation 12-8). So the number of revolutions is

$$\theta = \bar{\omega}t = \frac{1}{2}(\omega_0 + \omega)t = \frac{1}{2}(3600 \text{ rpm} + 0 \text{ rpm})\left(\frac{590 \text{ s}}{60 \text{ s/min}}\right)$$
$$= 1.8 \times 10^4 \text{ rev}.$$

Alternatively, we could have solved for the angular acceleration using $\alpha = \Delta\omega/\Delta t$. Then Equation 12-10 would give the angular displacement $\theta - \theta_0$ during the stopping time. Or we could have used Equation 12-11, setting the final angular speed ω to zero and solving for $\theta - \theta_0$.

TIP   You Don't Always Need To Work in Radians   Here, with speed given in rpm and the desired answer in revolutions, there was no need to convert angular measure to radians.

EXERCISE   The blade in a food processor spins at 1700 rpm. After the machine is turned off, the blade comes to a stop in 3.7 s. How many revolutions does it make during this time, assuming constant angular deceleration?

Answer: 52

Some problems similar to Example 12-2:   6, 10, 11, 61, 62

● EXAMPLE 12-3   A QUICK STOP

A rotating wheel 80 cm in diameter is decelerating at 0.21 rad/s^2. What should be the initial angular speed if the wheel is to stop after exactly one revolution? What is the tangential linear deceleration of a point on the wheel's rim?

Solution
Can you see that this problem is exactly like one in which you're asked to find the initial vertical speed needed to throw a ball to a given height? We apply Equation 12-11, setting the final angular speed ω to zero and taking $\theta_0 = 0$ for convenience:

$$0 = \omega_0^2 + 2\alpha\theta.$$

Setting θ to 2π because we want exactly one revolution, and noting that α is negative because it's a deceleration, we solve for ω_0 to get

$$\omega_0 = \sqrt{-2\alpha\theta} = \sqrt{(-2)(-0.21 \text{ rad/s}^2)(2\pi)} = 1.6 \text{ rad/s}.$$

The tangential acceleration of a point on the rim is given by Equation 12-6:

$$a_t = r\alpha = (40 \text{ cm})(-0.21 \text{ rad/s}^2) = -8.4 \text{ cm/s}^2,$$

where the minus sign tells us that the tangential acceleration is directed oppositely to the linear velocity of the rim.

EXERCISE   A circular saw blade is 25 cm in diameter and rotates at 3600 rpm. A magnetic brake brings the blade to a halt in just 10 revolutions. What are the magnitudes of (a) the angular deceleration and (b) the tangential deceleration of the blade teeth?

Answers: (a) 1.13 krad/s^2; (b) 141 m/s^2

Some problems similar to Example 12-3:   5, 9, 14

12-2 TORQUE

Newton's second law, $\mathbf{F} = m\mathbf{a}$, proved very powerful in our study of motion. Ultimately Newton's law governs all motion, but its application to every particle in a rotating object would be terribly cumbersome. Can we instead formulate an analogous law that deals with rotational quantities?

To develop such a law, we need rotational analogs of force, mass, and acceleration. Angular acceleration α is the analog of linear acceleration; in the next two sections we develop analogs for force and mass. In this chapter we consider only the one-dimensional form of Newton's law and its rotational analog; in the next chapter we'll develop the full vector description of rotational motion.

Figure 12-4 shows a child balancing her big sister on a seesaw. How can she do this? By sitting far from the seesaw's rotation axis, she's increased the effectiveness of the force she applies. Quite generally, the effectiveness of a force in bringing about changes in rotational motion depends not only on the magnitude of the force, but also on how far from the rotation axis it's applied (Fig. 12-5). Effectiveness of the force also depends on the *direction* in which it's applied; pushing at right angles to the line from the rotation axis to the force application point is most effective (Fig. 12-6). In fact, as Fig. 12-6 suggests, only the component of force at right angles to that line has any rotational effect.

FIGURE 12-4   The effectiveness of a force in effecting rotational motion depends on how far from the axis it's applied. That's why the small child can balance her big sister on the seesaw.

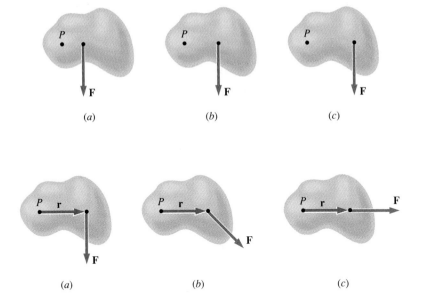

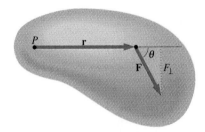

FIGURE 12-5 Torque—a measure of a force's effectiveness in altering rotational motion—increases with the distance from the rotation axis *P* to the force application point. The same force produces the greatest torque in (*c*).

FIGURE 12-6 The torque is greatest with **F** and **r** at right angles, and diminishes to zero as they become colinear. You can see that pulling with the force shown in (*c*) produces no tendency for the object to rotate about the axis *P*. In general, only the component of force perpendicular to **r** contributes to the torque.

Based on these considerations, we measure the effectiveness of a given force in producing rotational motion by the product of the distance *r* from the rotation axis with the component of force perpendicular to that axis. This quantity is called **torque** and is given the symbol τ (Greek tau, pronounced to rhyme with "how"). Then we can write

$$\tau = rF \sin\theta, \tag{12-12}$$

where θ is the angle between the force vector and the vector **r** from the rotation axis to the force application point, as shown in Fig. 12-7. Torque—you can think of it as a "twisting force"—is the rotational analog of force. Equation 12-12 shows that torque is measured in N·m.

FIGURE 12-7 Torque is given by the product of the distance *r* and the component of force at right angles to the vector **r**. Here $\tau = rF \sin\theta$, where θ is the angle between the vectors **r** and **F**.

● **EXAMPLE 12-4** CHANGING A TIRE

You're tightening your car's wheel nuts after changing a flat tire. The instructions specify a tightening torque of 95 N·m to ensure the nuts won't come loose. If your 45-cm-long wrench makes a 67° angle with the horizontal, as shown in Fig. 12-8, with what force must you pull horizontally to produce the required torque?

Solution
With the force applied horizontally, the angle θ in Equation 12-12 is $180° - 67° = 113°$ (see Fig. 12-8), so we can solve Equation 12-12 for the force *F*:

$$F = \frac{\tau}{r \sin\theta} = \frac{95 \text{ N·m}}{(0.45 \text{ m})(\sin 113°)} = 230 \text{ N}.$$

Tightening torques, as in this example, are often specified for nuts and bolts in critical applications. Mechanics use specially

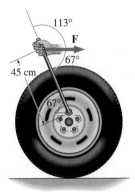

FIGURE 12-8 Tightening the wheel nut (Example 12-4).

FIGURE 12-9 A bicycle wheel. What torque arises from the weight of the valve stem?

designed "torque wrenches" that provide direct indication of the applied torque.

EXERCISE You have your bicycle upside down for repairs. The front wheel is free to rotate and is perfectly balanced

except for the 25-g valve stem. If the valve stem is 32 cm from the rotation axis and is located at 24° below the horizontal, as shown in Fig. 12-9, what is the resulting torque about the wheel's axis?

Answer: 0.072 N·m

Some problems similar to Example 12-4: 16, 18–20

TIP Specify the Axis Torque depends not only on force but also on where the force is applied *relative to some rotation axis*. The same physical force gives rise to different torques about different axes. Any calculation involving torque implicitly or explicitly assumes a particular rotation axis. If the rotation axis isn't obvious, be sure to specify it before working problems involving torque.

Does torque have direction? In Fig. 12-8 the torque tends to rotate the nut clockwise, while in Fig. 12-9 the weight of the valve stem will cause a counterclockwise rotation. For now we'll specify the direction of torque as being either clockwise or counterclockwise; these are analogous to the two directions $+x$ and $-x$ in one-dimensional motion. In the next chapter we'll expand our notion of torque to provide a full vector description of its direction.

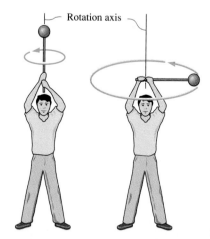

FIGURE 12-10 A composite object consisting of a mass on the end of a rigid rod. Its resistance to changes in rotational motion depends on how the mass is distributed relative to the rotation axis. It's easier to set it rotating in the left-hand case, where the mass is concentrated near the axis.

12-3 ROTATIONAL INERTIA AND THE ANALOG OF NEWTON'S LAW

Torque and angular acceleration are the rotational analogs of force and linear acceleration; to develop a rotational analog of Newton's law, we still need the rotational analog of mass.

The mass m in Newton's law is a measure of a body's inertia—of its resistance to changes in motion. We want a quantity that, analogously, describes resistance to changes in rotational motion. Figure 12-10 shows that it's easier to set an object rotating when its mass is concentrated near the rotation axis. So our rotational analog of inertia must depend not only on mass itself but also on the distribution of mass in relation to the rotation axis.

Suppose the composite object shown in Fig. 12-10 consists of an essentially massless rod of length R with a ball of mass m on the end. We allow the object to rotate about an axis through the free end of the rod and apply a force **F** to the ball, always at right angles to the rod (Fig. 12-11). The ball undergoes a tangential acceleration given by Newton's law: $F = ma_t$. (There's also a centripetal force—the tension in the rod—but because it acts along the rod, it doesn't contribute to the torque or angular acceleration.) We can use Equation 12-6 to express the tangential acceleration in terms of the angular acceleration α and the distance R from the rotation axis:

$$F = ma_t = m\alpha R. \qquad (12\text{-}13)$$

We can also express the force F in terms of its associated torque. Since the force is perpendicular to the rod, Equation 12-12 gives

$$\tau = RF.$$

Using F from Equation 12-13, we have

$$\tau = (mR^2)\alpha.$$

Here we have Newton's law, $F = ma$, written in terms of rotational quantities. The torque—analogous to force—is the product of the angular acceleration and the quantity mR^2, which must therefore be the rotational analog of mass. We call this quantity the **rotational inertia** or **moment of inertia** and give it the symbol I. Rotational inertia is measured in kg·m² and accounts for both an object's mass and the distribution of that mass. Like torque, the value of the rotational inertia depends on the location of the rotation axis. Given the rotational inertia I, our rotational analog of Newton's law becomes

$$\tau = I\alpha. \qquad (12\text{-}14)$$

Although we derived Equation 12-14 for a single, localized mass, we can apply it to extended objects if we interpret τ as the net torque on the object and I as the sum of the rotational inertias of the individual mass elements making up the object. A rigorous proof that Equation 12-14 applies to extended objects invokes Newton's third law to show that internal forces—and therefore torques—cancel in pairs, making the net torque indeed the sum of the external torques acting. That proof requires the additional condition—nearly always satisfied—that the internal forces act along the line joining two mass elements (Fig. 12-12).

In the next two sections we show first how to calculate the rotational inertia of extended objects, and then how to apply Equation 12-14 to the dynamics of rotating objects.

Calculating the Rotational Inertia

When an object consists of a number of discrete mass points, its rotational inertia about an axis is just the sum of the rotational inertias of the individual mass points about that axis:

$$I = \sum m_i r_i^2. \qquad (12\text{-}15)$$

Here m_i is the mass of the ith mass point, and r_i its distance from the rotation axis.

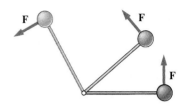

FIGURE 12-11 A force applied perpendicular to the rod results in angular acceleration.

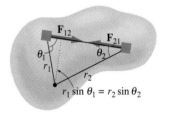

FIGURE 12-12 Two mass elements exert equal but opposite forces on each other. If the forces act along the line joining the mass elements, then $r_1 \sin \theta_1 = r_2 \sin \theta_2$. Because the forces are equal but opposite, so are the torques $Fr \sin \theta$. Thus internal torques cancel in pairs, and the net torque on the object is the sum of the external torques only. (The angles shown are the supplements of the angles used in Equation 12-12, but that's ok since an angle and its supplement have the same sine.)

● **EXAMPLE 12-5**   ROTATIONAL INERTIA

A dumbbell-shaped object consists of two equal masses $m =$ 0.64 kg on the ends of a massless rod of length $\ell = 85$ cm. Calculate the rotational inertia of this object about an axis one-fourth of the way from one end of the rod and perpendicular to it. If the object is accelerated from rest by a torque of 1.2 N·m about the same axis, how many revolutions will it turn in 2.5 s?

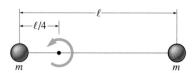

FIGURE 12-13   Example 12-5.

Solution
The situation is shown in Fig. 12-13. Summing the individual rotational inertias, we have

$$I = \sum m_i r_i^2 = m\left(\frac{1}{4}\ell\right)^2 + m\left(\frac{3}{4}\ell\right)^2 = \frac{5}{8}m\ell^2$$

$$= \frac{5}{8}(0.64 \text{ kg})(0.85 \text{ m})^2 = 0.29 \text{ kg·m}^2.$$

From Equation 12-14, the angular acceleration is

$$\alpha = \frac{\tau}{I}.$$

Using this result in Equation 12-10 then gives the angular displacement:

$$\theta = \frac{1}{2}\alpha t^2 = \frac{1}{2}\left(\frac{\tau}{I}\right)t^2 = \frac{1}{2}\left(\frac{1.2 \text{ N·m}}{0.29 \text{ kg·m}^2}\right)(2.5 \text{ s})^2$$

$$= 13 \text{ rad} = 2.1 \text{ rev}.$$

EXERCISE   Three 1.2-kg masses are located at the vertices of an equilateral triangle 88 cm on a side, connected by rods of negligible mass. Find the rotational inertia of this object (a) about an axis through the center of the triangle and perpendicular to its plane and (b) about an axis that passes through one vertex and the midpoint of the opposite side.

Answers: (a) 0.93 kg·m²; (b) 0.46 kg·m²

Some problems similar to Example 12-5:   24, 27
                                                                                 ●

When an object consists of a continuous distribution of matter, then we consider a large number of very small mass elements dm throughout the object, and sum the individual rotational inertias $r^2 dm$ over the entire object (Fig. 12-14). In the limit of an arbitrarily large number of very small mass elements, the sum becomes an integral, and we have

$$I = \int r^2 dm, \tag{12-16}$$

where the limits of integration are chosen to include the entire object.

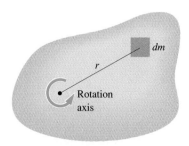

FIGURE 12-14   The rotational inertia of a solid object can be found by integrating the rotational inertias $r^2 dm$ of all the infinitesimal mass elements making up the object.

● **EXAMPLE 12-6** ROTATIONAL INERTIA OF A ROD

Find the rotational inertia of a uniform, narrow rod of mass M and length ℓ about an axis through its center and perpendicular to the rod.

Solution

Let the rod coincide with the x axis, with origin at the center of the rod, and consider mass elements of mass dm and length dx, as shown in Fig. 12-15. The distance x from the center of the rod to the mass element dm plays the role of r in Equation 12-16, so

$$I = \int_{x=-\ell/2}^{x=\ell/2} x^2 dm , \qquad (12\text{-}17)$$

where we've chosen the limits to include the entire rod. To do the integration, we must relate x and dm so we have a single variable under the integral sign. The length dx of the mass element is some tiny fraction of the total length ℓ. Since the rod is uniform, the mass dm is the same fraction of the total mass M. Therefore

$$\frac{dm}{M} = \frac{dx}{\ell}.$$

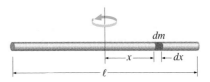

FIGURE 12-15 A uniform rod, showing a mass element of mass dm and length dx.

Solving for dm and using the result in Equation 12-17, we have

$$I = \int_{-\ell/2}^{\ell/2} \frac{M}{\ell} x^2 \, dx = \frac{M}{\ell} \int_{-\ell/2}^{\ell/2} x^2 \, dx$$

$$= \frac{M}{\ell} \frac{x^3}{3} \bigg|_{-\ell/2}^{\ell/2} = \frac{1}{12} M\ell^2. \quad \text{(rod about bisector)} \qquad (12\text{-}18)$$

Does this make sense? Yes. Much of the rod's mass is near the axis of rotation, and therefore contributes little to the rotational inertia. The result is a much lower rotational inertia than if all the mass were at the ends of the rod. ●

● **EXAMPLE 12-7** ROTATIONAL INERTIA OF A RING

Find the rotational inertia of a thin ring of radius R and mass M about the ring's axis (Fig. 12-16).

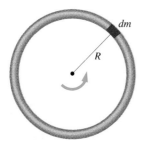

FIGURE 12-16 A thin ring, showing one mass element dm.

Solution

We divide the ring into mass elements dm, one of which is shown in Fig. 12-16. All the mass elements in the ring are the same distance R from the rotation axis, so r in Equation 12-16 is the constant R, and the equation becomes

$$I = \int R^2 dm = R^2 \int dm ,$$

where the integration is over the ring. But the sum of mass elements over the ring is just the total mass, M, so

$$I = MR^2. \quad \text{(thin ring)} \qquad (12\text{-}19)$$

The rotational inertia of the ring is the same as if all the mass were concentrated in one place a distance R from the rotation axis; the angular distribution of the mass about the axis doesn't matter. Notice, too, that it doesn't matter whether the ring is narrow like a loop of wire or long like a section of hollow pipe, as long as it's thin enough that all of it is essentially equidistant from the rotation axis (Fig 12-17).

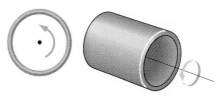

FIGURE 12-17 The rotational inertia is MR^2 for any thin ring, whether it's narrow like a wire loop (left) or long like a section of pipe (right). ●

● **EXAMPLE 12-8** ROTATIONAL INERTIA OF A DISK

A disk of radius R and mass M has uniform density. Find the rotational inertia of the disk about an axis through its center and perpendicular to the disk.

Solution
Not all parts of the disk are the same distance from the axis, so r is not constant in Equation 12-16. Since we already know the rotational inertia of a ring, we can divide the disk into mass elements that are themselves rings, as shown in Fig. 12-18. A given mass element is a ring of radius r, width dr, and mass dm. Its rotational inertia is $r^2\,dm$, so the total rotational inertia of the disk is

$$I = \int_{r=0}^{r=R} r^2\,dm, \qquad (12\text{-}20)$$

where the limits are chosen to include all the ring-shaped mass elements in the entire disk. To evaluate this integral, we must relate the variables m and r. The mass element dm is a tiny fraction of the total mass M—the *same* fraction of the total mass as its area is of the total disk area, πR^2. Since the ring dm is very thin, we can imagine "unwinding" it to get a thin rectangle of length $2\pi r$ and width dr (Fig. 12-19). Then the area of the ring is $2\pi r\,dr$, and we have

$$\frac{dm}{M} = \frac{2\pi r\,dr}{\pi R^2}.$$

Solving for dm and using the result in Equation 12-20 leaves only the variable r under the integral, allowing us to do the integration:

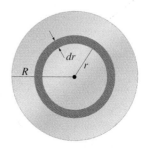

FIGURE 12-18 A disk may be divided into ring-shaped mass elements of mass dm, radius r, and width dr.

$$I = \int_{r=0}^{r=R} r^2\,dm = \int_0^R r^2 \frac{2M}{R^2} r\,dr = \frac{2M}{R^2}\int_0^R r^3\,dr$$

$$(12\text{-}21)$$

$$= \frac{2M}{R^2}\frac{r^4}{4}\Big|_0^R = \frac{1}{2}MR^2. \qquad \text{(disk)}$$

Does this make sense? In the disk, mass is distributed from the axis out to the edge. Mass nearer the axis contributes less to the rotational inertia, so we expect a lower rotational inertia for the disk than for a ring of the same mass and radius.

EXERCISE Find the rotational inertia of a thin rod of length ℓ and mass M when it's rotated about an axis through one end (Fig. 12-20). Why isn't your answer twice the answer for Example 12-5?

Answer: $I = \frac{1}{3}M\ell^2$

Some problems similar to Examples 12-6 through 12-8: 30, 36, 72, 73

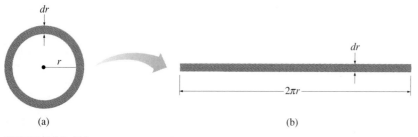

(a)   (b)

FIGURE 12-19 The mass element dm may be "unwound" into a thin rectangle of length $2\pi r$ and width dr.

FIGURE 12-20 The rotational inertia of a rod about an axis through one end is *not* twice that about an axis through the center. Why not?

●

The rotational inertias of other shapes about various axes may be found by integration as in these examples. Table 12-2 lists results for some common shapes. Note that more than one rotational inertia is listed for some shapes, since the rotational inertia depends on the rotation axis.

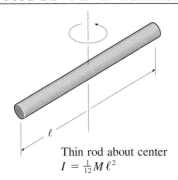

Thin rod about center
$I = \frac{1}{12}M\ell^2$

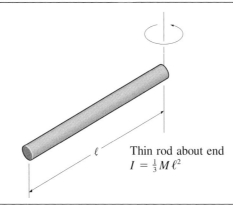

Thin rod about end
$I = \frac{1}{3}M\ell^2$

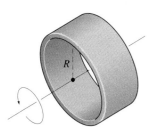

Thin ring or hollow cylinder
about its axis
$I = MR^2$

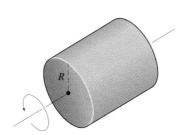

Disk or solid cylinder
about its axis
$I = \frac{1}{2}MR^2$

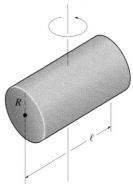

Solid cylinder about its perpendicular bisector
$I = \frac{1}{4}MR^2 + \frac{1}{12}M\ell^2$

Solid sphere about diameter
$I = \frac{2}{5}MR^2$

Hollow spherical shell about diameter
$I = \frac{2}{3}MR^2$

Solid sphere about tangent line
$I = \frac{7}{5}MR^2$

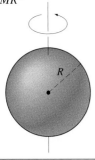

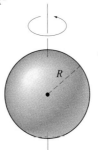

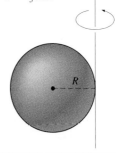

Flat plate about perpendicular axis
$I = \frac{1}{12}M(a^2 + b^2)$

Flat plate about central axis
$I = \frac{1}{12}Ma^2$

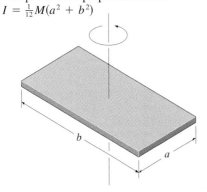

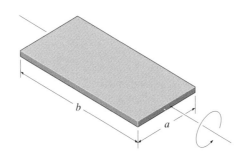

If we know the rotational inertia, I_{cm}, about an axis through the center of mass of a body, a useful relation called the **parallel-axis theorem** allows us to calculate the rotational inertia, I, through any parallel axis. The parallel-axis theorem states that

$$I = I_{cm} + Mh^2, \tag{12-22}$$

where h is the distance from the center of mass axis to the parallel axis and M the total mass of the object. Problem 37 explores the proof of this theorem.

● **EXAMPLE 12-9**   THE PARALLEL-AXIS THEOREM

Use the parallel-axis theorem, along with Table 12-2's entry for the rotational inertia of a solid sphere about a diameter, to verify the table's entry for the sphere about a tangent line.

Solution
Table 12-2 gives $I_{cm} = \frac{2}{5}MR^2$ for the sphere about a diameter; we write I_{cm} since any diameter passes through the center of mass. The tangent line is one radius R from the center, and parallel to a diameter, so we can apply the parallel-axis theorem:

$$I = I_{cm} + MR^2 = \frac{2}{5}MR^2 + MR^2 = \frac{7}{5}MR^2.$$

TIP   Save Work   Integrating to get the result of this example would have been exceedingly tedious. The parallel-axis theorem can save a lot of work! Similarly, you can often find the rotational inertia of a composite object by adding contributions from different parts if they have the shapes shown in Table 12-2.

EXERCISE   Use the parallel-axis theorem and the result of Example 12-6 to verify the answer to the exercise following Example 12-8.

Some problems similar to Example 12-9:   31, 63, 64   ●

Rotational Dynamics

Knowing a body's rotational inertia, we can use the rotational analog of Newton's second law (Equation 12-14) to determine its behavior, just as we used Newton's law itself to analyze linear motion. As with the force in Newton's law, the torque in Equation 12-14 is the *net* external torque—the sum of all external torques acting on the body.

● **EXAMPLE 12-10**   DE-SPINNING A SATELLITE

A cylindrical satellite is 1.4 m in diameter, with its mass of 940 kg distributed approximately uniformly throughout its volume. The satellite is spinning at 10 rpm about its axis but must be stopped so that a space shuttle crew can make repairs. Two small gas jets, each with a thrust of 20 N, are mounted on opposite sides of the satellite and are aimed tangent to the satellite's surface (Fig 12-21). How long must the jets be fired in order to stop the satellite's rotation?

Solution
The satellite's angular speed must change by $\Delta\omega = 10$ rpm. At constant angular acceleration α, this takes time

$$\Delta t = \frac{\Delta\omega}{\alpha} = \frac{\Delta\omega I}{\tau}, \tag{12-23}$$

where we've used Equation 12-14 to write $\alpha = \tau/I$. From Table 12-2 the rotational inertia of the cylindrical satellite is $I = \frac{1}{2}MR^2$. The torque exerted by both jets, each a distance R from the rotation axis and directed perpendicular to the radius, is $\tau = 2RF$, with F the thrust of one jet. Then Equation 12-23 becomes

$$\Delta t = \frac{(\Delta\omega)(\frac{1}{2}MR^2)}{2RF} = \frac{\Delta\omega MR}{4F}$$

$$= \frac{(10 \text{ rpm})(2\pi \text{ rad/rev})}{60 \text{ s/min}} \frac{(940 \text{ kg})(0.70 \text{ m})}{(4)(20 \text{ N})} = 8.6 \text{ s},$$

where we have converted the change in angular speed from rpm to rad/s.

EXERCISE Essentially all the mass of a 740-g bicycle wheel is concentrated in its rim, 34 cm from the rotation axis. You have the bicycle upside down, with the wheel spinning freely at 210 rpm, in order to adjust the brakes. What frictional force must the brakes apply to the wheel rim if they're to bring the wheel to a stop in 0.92 s?

Answer: 6.0 N

Some problems similar to Example 12-10: 39, 44

FIGURE 12-21   Satellite with gas jets (Example 12-10).   ●

Often, rotational and linear motion are coupled in a single problem. Then we use the relations between linear and rotational motion to formulate equations containing all the unknowns of the problem. The following example illustrates this procedure:

● **EXAMPLE 12-11**   ROTATIONAL AND LINEAR DYNAMICS

A solid cylinder of mass M and radius R is mounted on a frictionless horizontal axle over a well, as shown in Fig. 12-22. A rope of negligible mass is wrapped around the cylinder, and a bucket of mass m is suspended from the rope. Find an expression for the acceleration of the bucket as it falls down the empty well shaft.

Solution
Were it not connected to the cylinder, the bucket would, of course, accelerate downward with linear acceleration g. But now the rope exerts an upward tension force T on the bucket, reducing the net downward force and at the same time exerting a torque on the cylinder. The downward force on the bucket is $F = mg - T$ (Fig. 12-23), so Newton's second law gives

$$ma = mg - T, \tag{12-24}$$

where a is the linear acceleration of the bucket. Looking at the cylinder end-on (Fig. 12-24), we see that the rope exerts a torque $\tau = RT$ on the cylinder, giving it an angular acceleration

FIGURE 12-22   Example 12-11.

FIGURE 12-23 Force diagram for the bucket. The net downward force is $mg - T$.

$$\alpha = \frac{\tau}{I} = \frac{RT}{I}, \tag{12-25}$$

where I is the rotational inertia.

As the rope unwinds, the tangential acceleration of the cylinder edge must be equal to the bucket's linear acceleration. Therefore $a = \alpha R$, or, using Equation 12-25,

$$a = \alpha R = \frac{R^2 T}{I}. \tag{12-26}$$

The tension T is the same throughout the massless rope. Solving Equation 12-26 for this tension and using the result in Equation 12-24 gives

$$ma = mg - \frac{I}{R^2}a,$$

so

$$\left(m + \frac{I}{R^2}\right)a = mg.$$

Solving for the acceleration a, and noting that the rotational inertia of a solid cylinder about its axis is $I = \frac{1}{2}MR^2$, we have

$$a = \frac{mg}{m + \frac{1}{2}M}.$$

Is this result reasonable? If the cylinder mass M is small compared with the bucket mass m, the rotation of the cylinder is unimportant and the acceleration is simply g. But more generally, the gravitational force on the bucket provides not only the bucket's linear acceleration, but also the angular acceleration of the cylinder. As a result, the linear acceleration is decreased.

EXERCISE Two masses m_1 and m_2 are connected by a rope that passes without slipping over a massive pulley of radius R and rotational inertia I mounted on frictionless bearings, as shown in Fig. 12-25. If $m_1 > m_2$, find an expression for the downward acceleration of m_1, and show that your result makes sense as $I \to 0$. *Hint:* The rope tensions on m_1 and m_2 aren't the same since it's the difference in rope tensions that provides a net torque on the pulley.

Answer: $a = \dfrac{g(m_1 - m_2)}{m_1 + m_2 + (I/R^2)}$

Some problems similar to Example 12-11: 40–42, 45

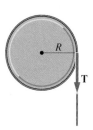

FIGURE 12-24 End view of the cylinder, showing that the rope tension T exerts a torque RT about the cylinder axis.

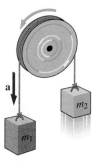

FIGURE 12-25 As m_1 falls, it not only causes m_2 to accelerate upward but also imparts an angular acceleration to the pulley.

12-4 ROTATIONAL ENERGY

A rotating object clearly has kinetic energy. We define an object's **rotational kinetic energy** as the sum of the kinetic energies of all its mass elements, taken with respect to the rotation axis. Figure 12-26 shows that an individual mass element dm a distance r from the rotation axis has kinetic energy given by

$$dK = \frac{1}{2}(dm)(v^2) = \frac{1}{2}(dm)(\omega r)^2.$$

The rotational kinetic energy is given by summing—that is, integrating—over the entire object:

$$K_{\text{rot}} = \int dK = \int \frac{1}{2}(dm)(\omega r)^2 = \frac{1}{2}\omega^2 \int r^2 \, dm,$$

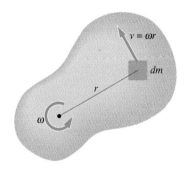

FIGURE 12-26 A mass element dm a distance r from the rotation axis has linear speed $v = \omega r$, giving it kinetic energy $dK = \frac{1}{2}(dm)(\omega r)^2$.

where we've taken ω^2 outside the integral because it's a constant characterizing the rotation of the entire object. But the remaining integral is just the rotational inertia I, so we have

$$K_{\text{rot}} = \frac{1}{2}I\omega^2. \tag{12-27}$$

This formula should come as no surprise: since I and ω are the rotational analogs of mass and speed, Equation 12-27 is the rotational equivalent of $K = \frac{1}{2}mv^2$.

● **EXAMPLE 12-12** EARTH'S ROTATIONAL ENERGY

Estimate the rotational kinetic energy of the Earth, assuming it to be a solid sphere of uniform density.

Solution
To use Equation 12-27 for rotational kinetic energy, we need the rotational inertia and angular velocity of the Earth. From Table 12-1, we find that $I = \frac{2}{5}MR^2$ for a uniform solid sphere, so

$$K_{\text{rot}} = \frac{1}{2}I\omega^2 = \frac{1}{5}MR^2\omega^2.$$

Obtaining M and R from Appendix E, and converting Earth's angular speed of 1 rev/day to rad/s, we have

$$K_{\text{rot}} = \frac{(5.97 \times 10^{24} \text{ kg})(6.37 \times 10^6 \text{ m})^2}{5}$$

$$\times \left(\frac{(1.0 \text{ rev/day})(2\pi \text{ rad/rev})}{(3600 \text{ s/h})(24 \text{ h/day})} \right)^2 = 2.56 \times 10^{29} \text{ J}.$$

Actually, more of Earth's mass is concentrated toward the center, reducing the rotational inertia and making this value an overestimate.

TIP Know When To Use Radians We used Equation 12-5, $v = \omega r$, in deriving Equation 12-27. Since Equation 12-5 works only with radian measure, the same is true for Equation 12-27.

EXERCISE In the magnet laboratory at the Massachusetts Institute of Technology (MIT), huge flywheels are used to store energy that eventually powers strong electromagnets. Each flywheel is a solid disk 2.4 m in radius, with a mass of 7.7×10^4 kg. How much energy does each wheel store when it's rotating at 360 rpm?

Answer: 158 MJ

Some problems similar to Example 12-12: 47, 49

●

Energy and Work in Rotational Motion

A force acting in one dimension does work on an object and changes the kinetic energy of its straight-line motion. Similarly, a torque does work on an object, changing its rotational kinetic energy. Suppose a torque τ acts to rotate a rigid body through a small angle $d\theta$, as suggested in Fig. 12-27. We can think of this torque as arising from the force $\mathbf{F}$ shown in the figure. If the force is applied a distance r from the rotation axis, then the force application point moves through a distance $ds = r\,d\theta$. Acting over this distance, the force does work $dW = F\,ds$. But $\tau = rF$, so $F = \tau/r$, and $dW = F\,ds$ becomes $dW = \tau\,d\theta$. If the torque is constant, the work done as the object rotates through any angle $\Delta\theta$ is then

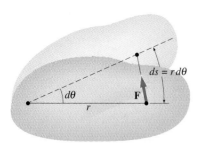

FIGURE 12-27 The force $\mathbf{F}$ acts over a distance $ds = r\,d\theta$ while rotating the rigid body through a very small angle $d\theta$. The work done is $F\,ds$, and the torque is rF.

$$W = \tau\Delta\theta. \quad \text{(constant torque)} \quad (12\text{-}28a)$$

If the torque varies with angular position then we need to integrate:

$$W = \int dW = \int \tau\,d\theta. \quad (12\text{-}28b)$$

Equations 12-28 are equivalent, respectively, to $W = F\,\Delta x$ and $W = \int F\,dx$ for straight-line motion. Problem 78 shows how Equation 12-28b leads to the rotational version of the work-energy theorem, which states that the work done by torques on a body rotating about a fixed axis is equal to the change in rotational kinetic energy:

$$W = \int_{\theta_0}^{\theta_f} \tau\,d\theta = \Delta K_{\text{rot}} = \frac{1}{2}I\omega_f^2 - \frac{1}{2}I\omega_0^2, \quad (12\text{-}29)$$

where the subscripts refer to the initial and final values. Equation 12-29 is the rotational analog of the work-energy theorem we derived in Chapter 7.

● **EXAMPLE 12-13** SPINNING A GYROSCOPE

A gyroscope consists of a ring of mass $M = 85$ g and radius $R = 2.6$ cm, connected by essentially massless spokes to an axle of negligible mass (Fig. 12-28). A massless string of length $\ell = 45$ cm is wrapped around the axle and pulled with a steady force $F = 3.7$ N. If the string unwinds without slipping, find the final speed of the gyroscope. If the torque due to the string is 7.4×10^{-3} N, how many revolutions does the gyroscope make while the string is being pulled?

Solution

We don't know the angular displacement $\Delta\theta$, so we can't immediately use $W = \tau\Delta\theta$ to determine the work. But we do know the work done by the agent pulling the string—with constant force, that's just $W = F\ell$. And since the string is massless, energy conservation ensures that all that work ends up as rotational kinetic energy. So we have

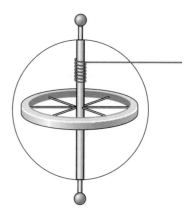

FIGURE 12-28 A gyroscope (Example 12-13).

$$\frac{1}{2}I\omega^2 = F\ell.$$

Solving for ω and using $I = MR^2$ for the ring gives

$$\omega = \sqrt{\frac{2F\ell}{I}} = \sqrt{\frac{2F\ell}{MR^2}} = \sqrt{\frac{(2)(3.7 \text{ N})(0.45 \text{ m})}{(0.085 \text{ kg})(0.026 \text{ m})^2}}$$

$$= 240 \text{ rad/s}.$$

As with linear motion, conservation of energy provides a quick route to the final answer.

Since the work $F\ell$ done in pulling the string is equal to the work $\tau\Delta\theta$ done on the gyroscope, we must have

$$\Delta\theta = \frac{F\ell}{\tau} = \frac{(3.7 \text{ N})(0.45 \text{ m})}{7.4 \times 10^{-3} \text{ N·m}} = 225 \text{ rad} = 36 \text{ rev}.$$

EXERCISE Mounted on its rim, an automobile tire has a radius of 38 cm and a rotational inertia of 2.7 kg·m². A tire testing machine spins up the tire by applying a force at the outer edge of the tire. If the force is at right angles to a radius, what should its magnitude be if the tire is to reach 690 rpm in 30 revolutions?

Answer: 98 N

Some problems similar to Example 12-13: 51, 65, 66   ●

12-5 ROLLING MOTION

A rolling object has both rotational and translation motion—motion of the whole object from place to place (Fig. 12-29). How much kinetic energy is associated with each of these motions?

Figure 12-29 shows a rolling wheel. If it's truly rolling, without sliding or skidding, then the point in contact with the ground is instantaneously at rest. The whole wheel is rotating about that point, as the velocity vectors in Fig. 12-29 suggest. The *total* kinetic energy is the rotational kinetic energy *about this contact point:*

$$K_{\text{total}} = \frac{1}{2}I_c\omega_c^2,$$

FIGURE 12-29 A rolling object has both translational and rotational motion.

where the rotational inertia I_c and angular speed ω_c are measured with respect to the contact point.

We can write this energy in terms of the translational energy of the center of mass and the rotational energy about the center of mass. Using the parallel axis theorem (Equation 12-22), we write the rotational inertia I_c about the contact point in terms of the rotational inertial I_{cm} about the center of mass:

$$I_c = I_{\text{cm}} + MR^2,$$

where M and R are the mass and radius of the rolling object.

What about the angular speed about the center of mass? Figure 12-30 shows that the center of mass is moving at speed $v = \omega_c R$ with respect to the ground; in the center-of-mass frame of reference, that means the ground and the contact point are moving backwards at the *same* speed v. So v is also given by $v = \omega_{\text{cm}}R$, and thus $\omega_c = \omega_{\text{cm}}$.

Using our expressions for I_c and ω_c in the total energy gives

$$K_{\text{total}} = \frac{1}{2}(I_{\text{cm}} + MR^2)\omega^2 = \frac{1}{2}I_{\text{cm}}\omega^2 + \frac{1}{2}M(\omega R)^2,$$

where we've dropped the subscript on ω since $\omega_c = \omega_{\text{cm}}$. But $\omega R = v$, the translational speed of the center of mass, so we have

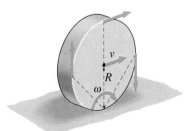

FIGURE 12-30 The point of contact with the ground is instantaneously at rest. The motion may be thought of as a pure rotation about that point. Arrows represent the linear velocities of several points on the object.

$$K_{\text{total}} = \frac{1}{2}I_{\text{cm}}\omega^2 + \frac{1}{2}Mv^2 = K_{\text{rot}} + K_{\text{trans}}. \qquad (12\text{-}30)$$

So the total energy of a rolling object is simply the sum of the translational energy of its center of mass and the rotational energy about the center of mass.

Rolling motion requires friction, as the frictional force acts to keep the contact point of the rotating object at rest; a wheel sent down a perfectly frictionless incline, for example, would not roll. But precisely because the contact point is at rest, the frictional force does no work and therefore mechanical energy is conserved. Equation 12-30 is therefore useful in conservation-of-energy problems involving rolling motion, as Example 12-14 illustrates.

● **EXAMPLE 12-14**   ROLLING DOWNHILL

A solid ball of M and radius R starts from rest and rolls without slipping down a hill; its center of mass drops a total distance h (Fig. 12-31). Find the ball's speed at the bottom of the hill.

Solution
Mechanical energy is conserved, so as usual we can equate the initial potential energy Mgh to the final kinetic energy. Now, however, the final kinetic energy consists of both rotational and translational energy, as given by Equation 12-30. So we have

$$Mgh = \frac{1}{2}I_{\text{cm}}\omega^2 + \frac{1}{2}Mv^2.$$

But $\omega = v/R$ for a rolling object, and Table 12-2 gives $I_{\text{cm}} = \frac{2}{5}MR^2$ for a solid sphere, so

$$Mgh = \frac{1}{2}MR^2\left(\frac{v}{R}\right)^2 + \frac{1}{2}Mv^2 = \frac{7}{10}Mv^2.$$

Solving for v then gives

$$v = \sqrt{\frac{10}{7}gh} \ .$$

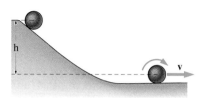

FIGURE 12-31   How fast is the ball moving at the bottom of the hill (Example 12-14)?

This is less than the speed $v = \sqrt{2gh}$ for an object that slides without rolling down frictionless incline. Why? Because some of the energy the rolling object gains goes into rotation, leaving less for translational motion.

EXERCISE   You're about to change a flat tire, but when you get out the spare tire it rolls away down a hill. The 28-kg spare is 70 cm in diameter and has rotational inertia 1.7 kg·m². How fast is it rolling after it goes a vertical distance of 15 m?

Answer: 14 m/s

Some problems similar to Example 12-14:   53–55, 67, 68   ●

CHAPTER SYNOPSIS

Summary

Rotational motion is described using quantities analogous to those of linear motion. These rotational quantities satisfy the same relations as their linear counterparts, making problems involving rotational motion similar to the linear motion problems of previous chapters.

1. The rotational analog of displacement is **angular displacement,** defined as the angle through which an object has rotated. The SI unit of angular displacement is the radian. The first and second time derivatives of angular displacement are **angular speed,** ω, and **angular acceleration,** α.

2. The rotational analog of mass is **rotational inertia,** a quantity that measures the resistance of a body to changes in rotational motion. Rotational inertia depends on the mass of a body and on the distribution of mass about the rotation axis, and is given by

$$I = \sum m_i r_i^2$$

for a body consisting of discrete masses, and by

▲ **TABLE 12-3** LINEAR AND ANGULAR QUANTITIES

LINEAR QUANTITY OR EQUATION	ANGULAR QUANTITY OR EQUATION	RELATION BETWEEN LINEAR AND ANGULAR QUANTITIES
Position x	Angular position θ	
Speed $v = dx/dt$	Angular speed $\omega = d\theta/dt$	$v_t = \omega r$
Acceleration a	Angular acceleration α	$a_t = \alpha r$
Mass m	Rotational inertia I	$I = \int r^2 dm$
Force F	Torque τ	$\tau = r F \sin\theta$

Newton's second law (constant mass or rotational inertia):
$$F = ma \qquad \tau = I\alpha$$

Kinetic energy:
$$K_{\text{trans}} = \tfrac{1}{2}mv^2 \qquad K_{\text{rot}} = \tfrac{1}{2}I\omega^2$$

$$I = \int r^2 \, dm$$

for a continuous distribution of matter.

3. **Torque** is the rotational analog of force; torque, rotational inertia, and angular acceleration are related by the rotational analog of Newton's second law:

$$\tau = I\alpha.$$

4. **Rotational kinetic energy** is calculated from rotational inertia and angular speed in the same way that translational kinetic energy is calculated from mass and linear speed.

$$K_{\text{rot}} = \frac{1}{2}I\omega^2.$$

The total kinetic energy of a rigid, rotating object may be written as the sum of the translational kinetic energy of its center of mass and the rotational kinetic energy about its center of mass.

Table 12-3 summarizes the analogy between linear and rotational quantities.

Terms You Should Understand

(Pairs are closely related terms whose distinction is important; number in parentheses is chapter section where term first appears.)

angular speed (12-1)
average angular speed, instantaneous angular speed (12-1)
angular acceleration (12-1)
torque (12-2)
rotational inertia (12-3)
parallel-axis theorem (12-3)
rotational kinetic energy (12-4)
rolling motion (12-5)

Symbols You Should Recognize

ω (12-1)
α (12-1)
τ (12-2)
I (12-3)
K_{rot} (12-4)

Problems You Should Be Able to Solve

calculating angular speed and acceleration (12-1)
relating angular and linear speeds and accelerations (12-1)
solving rotational analogs of problems in one-dimensional motion with constant acceleration (12-1)
calculating torques (12-2)
calculating rotational inertias by summing over discrete masses (12-3)
calculating rotational inertias by integrating over solid objects (12-3)
solving for rotational motion given torque and rotational inertia (12-3)
solving problems that mix linear and rotational motion (12-3)
calculating rotational energies (12-4)
applying the work-energy theorem for rotational motion (12-3)
applying conservation of energy to the motion of rolling objects (12-4)

Limitations to Keep in Mind

Some problems, namely, all those derived using the relation $v = \omega r$ are valid only when angular quantities are given using radian measure.

The results given in this chapter apply only when the orientation of the rotation axis remains fixed.

QUESTIONS

1. Do all points on a rigid, rotating object have the same angular speed? linear speed? radial acceleration?

2. Suppose you mount tires on your car that are slightly larger in diameter than factory-supplied tires. If you are in a 55 mi/h zone and your speedometer says 55 mi/h, are you speeding? Explain.

3. A wheel undergoes a constant angular acceleration. How do each of the following depend on time: (a) tangential acceleration of a point on the rim? (b) radial acceleration of a point on the rim? (c) angular speed of the wheel?

4. Part of a train wheel extends below the point of contact with the rails, as shown in Fig. 12-32. It is often said that this part of the train moves backward. Explain.

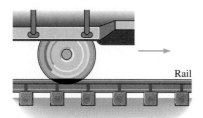

FIGURE 12-32   Question 4.

5. A point on the rim of a rotating wheel has nonzero centripetal acceleration. Does it necessarily follow that the wheel is undergoing angular acceleration?

6. Why doesn't it make sense to talk about a body's rotational inertia unless a rotation axis is specified?

7. Two different forces act on an object, but the net force is zero. Must the net torque be zero? If so, why? If not, give a counterexample.

8. Is it possible to apply a counterclockwise torque to an object that's rotating clockwise? If not, why not? If so, how will the object's motion change?

9. A solid sphere and a hollow sphere, of the same mass and radius, are rolling along level ground. If they have the same kinetic energy, which is moving faster?

10. A solid cylinder and a hollow cylinder, of the same mass and radius, are rolling along level ground at the same speed. Which has more kinetic energy?

11. As the wheel in Fig. 12-9 rotates, does the torque due to the weight of the valve stem remain constant? Explain.

12. What is meant by the statement that the point where a rotating wheel contacts the ground is instantaneously at rest? If it's true, how can the wheel be moving?

13. An electric circular saw takes a long time to stop rotating after the power is turned off. Without the saw blade mounted, the motor stops much more quickly when turned off. Why?

14. A solid sphere and a solid cube have the same mass, and the side of the cube is equal to the diameter of the sphere.

Which has the greatest rotational inertia about an axis through the center of mass? The rotational axis for the cube is perpendicular to two of the cube faces.

15. A badly unbalanced wheel has its center of mass located closer to its rim than to its geometrical center. The wheel is released from rest on a slope where it is free to roll without slipping. Describe its initial motion when released from each of the orientations shown in Fig. 12-33.

FIGURE 12-33   How will the wheel move? Red dot indicates its center of mass (Question 15).

16. The lower leg of a horse contains essentially no muscle. How does this help the horse to run fast? Explain in terms of rotational inertia.

17. You wish to store energy in a rotating flywheel. Given a fixed amount of a material, what shape should you make the flywheel so it will store the most energy at a given angular speed?

18. A ball starts from rest and rolls without slipping down a slope, then starts up a *frictionless* slope (Fig. 12-34). Compare its maximum height on the frictionless slope with its starting height on the first slope.

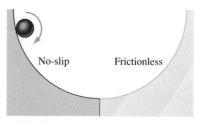

FIGURE 12-34   Question 18.

19. Two solid cylinders of the same mass but different radii are rolling with the same linear speed. How do their translational and rotational kinetic energies compare?

20. A raw egg and a hard-boiled egg, otherwise identical, are rolled down an incline. Which hits the bottom first? Explain.

21. What happens to the rotational inertia of a washing machine tub and its contents when the spin portion of the cycle starts?

PROBLEMS

Section 12-1   Angular Speed and Acceleration

1. Determine the angular speed, in rad/s, of (a) Earth about its axis; (b) the minute hand of a clock; (c) the hour hand of a clock; (d) an egg beater turning at 300 rpm.
2. What is the linear speed (a) of a point on Earth's equator? (b) At your latitude?
3. Express Equation 12-4 with constants appropriate to angular speed in (a) rpm, (b) rev/s, (c) °/s.
4. A 25-cm-diameter circular saw blade spins at 3500 rpm. How fast would you have to push a straight hand saw to have the teeth move through the wood at the same rate as the circular saw teeth?
5. A wheel is turned through 2.0 revolutions while being accelerated from rest at 18 rpm/s. (a) What is the final angular speed? (b) How long does it take to turn the 2.0 revolutions?
6. You switch a food blender from its high to its low setting; the blade speed drops from 3600 rpm to 1800 rpm in 1.4 s. How many revolutions does it make during this time?
7. A compact disc (CD) player varies the rotation rate of the disc in order to keep the part of the disk from which information is being read moving at a constant linear speed of 1.30 m/s (Fig. 12-35). Compare the rotation rates of a 12.0-cm-diameter CD when information is being read from (a) its outer edge and (b) a point 3.75 cm from the center. Give your answers in rad/s and rpm.

FIGURE 12-35   Problem 7.

8. The maximum rotation rate of a compact disc is 500 rpm. What is the minimum distance from the center at which information is read from the disc? Refer to the preceding problem for necessary information.
9. The rotation rate of a compact disc varies from about 200 rpm to 500 rpm (see Problem 7). If the disc plays for 74 min, what is its average angular acceleration in (a) rpm/s and (b) rad/s²?
10. You rev your car's engine and watch the tachometer climb steadily from 1200 rpm to 5500 rpm in 2.7 s. (a) What is the angular acceleration of the engine? (b) What is the tangential acceleration of a point on edge of the engine's 3.5-cm-diameter crankshaft? (c) How many revolutions does the engine make during this time?
11. During startup of a power plant, a turbine accelerates from rest at 0.52 rad/s². (a) How long does it take to reach its 3600-rpm operating speed? (b) How many revolutions does it make during this time?
12. A wheel is spinning at 47 rad/s when an acceleration of 0.72 rad/s² is applied. If the acceleration lasts for 140 revolutions, what is the final angular speed?
13. A piece of machinery is spinning at 680 rpm. When a brake is applied, its rotation rate drops to 440 rpm while it turns through 180 revolutions. What is the magnitude of the angular deceleration?
14. A circular saw blade completes 1200 revolutions in 40 s while coasting to a stop after being turned off. Assuming constant deceleration, what are (a) the angular deceleration and (b) the initial angular speed?
15. The angular acceleration of a wheel in rad/s² is given by $24t^2 - 16t^3$, where t is the time in seconds. The wheel starts from rest at $t = 0$. (a) When is it again at rest? (b) How many revolutions has it turned between $t = 0$ and when it is again at rest?

Section 12-2 Torque

16. A frictional force of 320 N acts on the rim of a 1.0-m-diameter wheel to oppose its rotational motion. What is the torque about the wheel's central axis?
17. A torque of 110 N·m is required to start a revolving door rotating. If a child can push with a maximum force of 90 N, how far from the door's rotation axis must she apply this force?
18. A car tune-up manual calls for tightening the spark plugs to a torque of 35.0 N·m. To achieve this torque, with what force must you pull on the end of a 24.0-cm-long wrench if you pull (a) at right angles to the wrench shaft and (b) at 110° to the wrench shaft?
19. A 55-g mouse runs out to the end of the 17-cm-long minute hand of a grandfather clock when the clock reads 10 minutes past the hour. What torque does the mouse's weight exert about the rotation axis of the clock hand?
20. In a large steam engine, a horizontal rod is attached to a flywheel by means of a frictionless pivot 95 cm from the wheel's axis. If the rod pushes with 13 kN force, what is the torque about the wheel's axis in the configuration shown in Fig. 12-36?

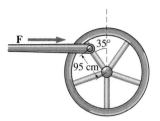

FIGURE 12-36 Problems 20, 29.

21. A pulley 12 cm in diameter is free to rotate about a horizontal axle. A 220-g mass and a 470-g mass are tied to either end of a massless string, and the string is hung over the pulley. If the string does not slip, what torque must be applied to keep the pulley from rotating?

22. Two arm wrestlers are deadlocked in the position shown in Fig. 12-37. Each arm is 34 cm long and makes a 60° angle with the horizontal; the contact force at the hands is 180 N. What is the torque each exerts about the other's elbow? Neglect the weights of the wrestlers' arms.

23. A 1.5-m-diameter wheel is mounted on an axle through its center. (a) Find the net torque about the axle due to the forces shown in Fig. 12-38. (b) Are there any forces that don't contribute to the torque?

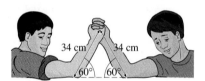

FIGURE 12-37 Problem 22.

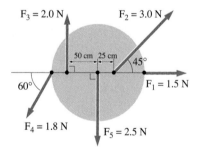

FIGURE 12-38 Problem 23.

Section 12-3 Rotational Inertia and the Analog of Newton's Law

24. Four equal masses m are located at the corners of a square of side ℓ, connected by essentially massless rods. Find the rotational inertia of this system about an axis (a) that coincides with one side and (b) that bisects two opposite sides.

25. What is the radius of a solid cylinder of mass 6.2 kg and rotational inertia 0.15 kg·m² about its axis?

26. The chamber of a rock-tumbling machine is a hollow cylinder with mass 65 g and radius 7.1 cm. The chamber is closed off by end caps in the form of uniform circular disks each of mass 22 g. (a) What is the rotational inertia of the chamber about its central axis? (b) What torque is necessary to give the chamber an angular acceleration of 3.4 rad/s²?

27. A square frame is made from four thin rods, each of length ℓ and mass m. Calculate its rotational inertia about the three axes shown in Fig. 12-39.

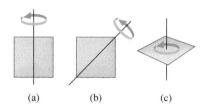

(a) (b) (c)

FIGURE 12-39 Problem 27.

28. The full diameter of a wheel is 92 cm, and its rotational inertia is 7.8 kg·m². (a) What is the minimum mass it could have? (b) How could it have more mass?

29. The wheel shown in Fig. 12-36 consists of a 120-kg outer rim 55 cm in radius, connected to the center by five 18-kg spokes. Treating the rim as a thin ring and the spokes as thin rods, determine the rotational inertia of the wheel.

30. Use integration to show that the rotational inertia of a thick ring of mass M and inner and outer radii R_1 and R_2 is given by $\frac{1}{2}M(R_1^2 + R_2^2)$. *Hint:* See Example 12-8.

31. A uniform rectangular flat plate has mass M and dimensions a by b. Use the parallel-axis theorem in conjunction with Table 12-2 to show that its rotational inertia about the side of length b is $\frac{1}{3}Ma^2$.

32. A rotating glass door is made from four rectangular panes, each 1.2 m wide by 2.4 m high, joined together along one long edge and at right angles to each other. The mass of each pane is 63 kg. Use the result of the previous problem to find the rotational inertia of the door about its central axis.

33. (a) Estimate the rotational inertia of the Earth, assuming it to be a uniform solid sphere. (b) What torque would have to be applied to Earth to cause the length of the day to change by one second every century?

34. Each propeller on a King Air twin engine airplane consists of three blades, each of mass 10 kg and length 125 cm. The blades may be treated approximately as uniform, thin rods. (a) What is the rotational inertia of the propeller? (b) If the propeller is driven by an engine that develops a torque of 2700 N·m, how long will it take to change the propeller's angular speed from 1400 rpm to 1900 rpm? Neglect the rotational inertia of the engine and any aerodynamic forces on the propeller.

35. A neutron star is an extremely dense, rapidly spinning object that results from the collapse of a star at the end of

its life. A neutron star of 1.8 times the Sun's mass has an approximately uniform density of 1×10^{18} kg/m³. (a) What is its rotational inertia? (b) The neutron star's spin rate slowly decreases as a result of torque associated with magnetic forces. If the spin-down rate is 5×10^{-5} rad/s², what is the magnetic torque?

36. Verify by direct integration the formula given in Table 12-2 for the rotational inertia of a flat plate about a central axis. *Hint:* Divide the plate into strips parallel to the axis.

37. Proof of the parallel-axis theorem: Fig 12-40 shows an object of mass M with axes through the center of mass and through an arbitrary point A. Both axes are perpendicular to the page. Let **h** be a vector from the axis through the center of mass to the axis through point A, $\mathbf{r}_{cm}$ a vector from the axis through the CM to an arbitrary mass element dm, and **r** a vector from the axis through point A to the mass element dm, as shown. (a) Use the law of cosines to show that

$$r^2 = r_{cm}^2 + h^2 - 2\mathbf{h} \cdot \mathbf{r}_{cm}.$$

(b) Use this result in the expression $I = \int r^2 \, dm$ to calculate the rotational inertia of the object about the axis through A. Each of the three terms in your expression for r^2 leads to a separate integral. Identify one as the rotational inertia about the CM, another as the quantity Mh^2, and show that the third is zero because it involves the position of the center of mass relative to itself. Your result is then a statement of the parallel-axis theorem.

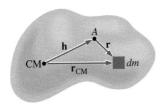

FIGURE 12-40 Problem 37.

38. A 108-g Frisbee is 24 cm in diameter and has about half its mass spread uniformly in a disk, and the other half concentrated in the rim. With a quarter-turn flick of the wrist, a student sets the Frisbee rotating at 550 rpm. (a) What is the rotational inertia of the Frisbee? (b) What is the magnitude of the torque, assumed constant, that the student applies?

39. A space station is constructed in the shape of a wheel 22 m in diameter, with essentially all of its 5.0×10^5-kg mass at the rim (Fig. 12-41). Once the station is completed, it is set rotating at a rate that requires an object at the rim to have radial acceleration g, thereby simulating Earth's surface gravity. This is accomplished using two small rockets, each with 100 N thrust, that are mounted on the rim of the station as shown. (a) How long will it take to reach the desired spin rate? (b) How many revolutions will the station make in this time?

FIGURE 12-41 Problem 39.

40. A motor is connected to a solid cylindrical drum with a diameter of 1.2 m and a mass of 51 kg. A massless rope is attached to the drum and tied at the other end to a 38-kg weight, so the rope will wind onto the drum as it turns. What torque must the motor apply if the weight is to be lifted with an acceleration of 1.1 m/s²?

41. The crane shown in Fig. 12-42 contains a hollow drum of mass 150 kg and radius 0.80 m that is driven by an engine to wind up a cable. The cable passes over a solid cylindrical 30-kg pulley 0.30 m in radius to lift a 2000-N weight. How much torque must the engine apply to the drum to lift the weight with an acceleration of 1.0 m/s²? Neglect the rotational inertia of the engine and the mass of the cable.

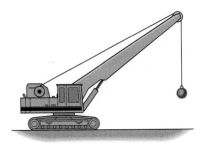

FIGURE 12-42 Problems 41, 66.

42. A 2.4-kg block rests on a 30° slope and is attached by a string of negligible mass to a solid drum of mass 0.85 kg and radius 5.0 cm, as shown in Fig. 12-43. When released, the block accelerates down the slope at 1.6 m/s². What is the coefficient of friction between block and slope?

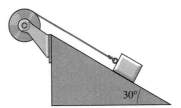

FIGURE 12-43 Problem 42.

43. A bicycle is under repair and is upside-down with its 66-cm-diameter wheel spinning freely at 230 rpm. The mass of the wheel is 1.9 kg and is concentrated mostly at

the rim. The cyclist holds a wrench against the tire for 3.1 s, with a normal force of 2.7 N. If the coefficient of friction between the wrench and the tire is 0.46, what is the final angular speed of the wheel?

44. At the MIT Magnet Laboratory, energy is stored in huge solid flywheels of mass 7.7×10^4 kg and radius 2.4 m. The flywheels ride on shafts 41 cm in diameter. If a frictional force of 34 kN acts tangentially on the shaft, how long will it take the flywheel to coast to a stop from its normal rotation rate of 360 rpm?

45. Two blocks of mass m_1 and m_2 are connected by a massless string that passes over a solid cylindrical pulley, as shown in Fig. 12-44. The surface under the block m_2 is frictionless, and the pulley rides on frictionless bearings. The string passes over the pulley without slipping. When released, the masses accelerate at $\frac{1}{3}g$. The tension in the lower half of the string is 2.7 N and that in the upper half, 1.9 N. What are the masses of the pulley and of the two blocks?

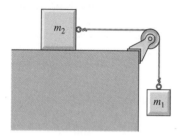

FIGURE 12-44 Problem 45.

Section 12-4 Rotational Energy

46. The flywheel of Problem 44 is used to generate a large amount of electric power for a short time. If the wheel's rotation rate is dropped from 390 to 300 rpm in 5.0 s, (a) what is the average power output and (b) what fraction of the rotational kinetic energy has been removed?

47. A 25-cm-diameter circular saw blade has a mass of 0.85 kg, distributed uniformly as in a disk. (a) What is its rotational kinetic energy at 3500 rpm? (b) What average power must be applied to bring the blade from rest to 3500 rpm in 3.2 s?

48. Humanity uses energy at the rate of about 10^{13} W. If we found a way to extract this energy from Earth's rotation, how long would it take before the length of the day increased by 1 minute?

49. A 150-g baseball is pitched at 33 m/s, spinning at 42 rad/s. What fraction of its kinetic energy is rotational? Treat the baseball as a uniform solid sphere of radius 3.7 cm.

50. A potter's wheel is a stone disk 90 cm in diameter with a mass of 120 kg. If the potter's foot pushes at the outer edge of the initially stationary wheel with a 75-N force for one-eighth of a revolution, what will be the final speed? Use the work-energy theorem.

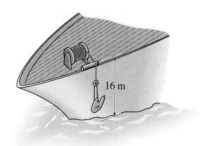

FIGURE 12-45 Problem 51.

51. A ship's anchor weighs 5000 N. Its cable passes over a roller of negligible mass and is wound around a hollow cylindrical drum of mass 380 kg and radius 1.1 m, as shown in Fig. 12-45. The drum is mounted on a frictionless axle. The anchor is released and drops 16 m to the water. Use energy considerations to determine the drum's rotation rate when the anchor hits the water. Neglect the mass of the cable.

52. A jetliner lands, and one of its 1.3-m-diameter wheels spins up to an angular speed of 110 rad/s while the wheel undergoes 12 revolutions. If the wheel's rotational inertia is 30 kg·m², find (a) the rotational kinetic energy gained, and (b) the applied torque.

Section 12-5 Rolling Motion

53. A basketball rolls down a 30° incline. If it starts from rest, what is its speed after it's gone 8.4 m along the incline? (The basketball is hollow.)

54. Repeat the preceding problem for a solid baseball, assuming uniform density.

55. As long as its mass is distributed symmetrically about a central axis, a round object like a sphere or wheel may be characterized by a rotational inertia of the form $I = \alpha MR^2$, where α is a constant. Derive a formula for the final speed of such an object if it starts from rest and rolls without slipping down an incline of vertical height h.

56. A hollow ball is rolling along a horizontal surface at 3.7 m/s when it encounters an upward incline. If it rolls without slipping up the incline, what maximum height will it reach?

57. The rotational kinetic energy of a rolling automobile wheel is 40% of its translational kinetic energy. The wheel is then redesigned to have 10% lower rotational inertia and 20% less mass, while keeping its radius the same. By what percentage does its total kinetic energy at a given speed decrease?

58. Two solid cylinders of the same mass roll without slipping down an incline. If they start from the same height h, show that they reach the bottom with the same linear speed, even if their radii are not the same. What is the linear speed at the bottom?

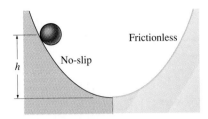

FIGURE 12-46 Problem 60.

FIGURE 12-48 Problem 64.

59. A ball rolls without slipping down a slope of vertical height 34 cm, and reaches the bottom moving at 2.0 m/s. Is the ball hollow or solid?

60. A solid ball of mass M and radius R starts at rest at height h above the bottom of the path shown in Fig. 12-46. It rolls without slipping down the left side of the path. The right side of the path, starting at the bottom, is frictionless. To what height does the ball rise on the right?

Paired Problems

(Both problems in a pair involve the same principles and techniques. If you can get the first problem, you should be able to solve the second one.)

61. A merry-go-round starts from rest and accelerates with angular acceleration 0.010 rad/s² for 14 s. (a) How many revolutions does it make during this time? (b) What is its average angular speed during the spin-up time?

62. A wheel spinning initially at 440 rpm is slowed with a constant angular deceleration of 2.8 rad/s² applied for 6.1 s. (a) How many revolutions does it turn during this time? (b) What is its average angular speed during the slowdown?

63. A disk of radius R has an initial mass M. Then a hole of radius $\frac{1}{4}R$ is drilled, with its edge at the disk center (Fig. 12-47). Find the new rotational inertia about the central axis. *Hint:* Find the rotational inertia of the missing piece, and subtract from that of the whole disk. You'll need to determine what fraction the missing mass is of the total M, and you'll need to use the parallel-axis theorem.

64. To reduce its rotational inertia, an engineer proposes drilling three holes in a large wheel of radius R. The holes are to be centered midway between the center and edge of the wheel, as shown in Fig. 12-48. What should be their radii if drilling the holes is to reduce the rotational inertia by 15%?

65. A 50-kg mass is tied to a massless rope wrapped around a solid cylindrical drum. The drum is mounted on a frictionless horizontal axle. When the mass is released, it falls with acceleration $a = 3.7$ m/s². Find (a) the tension in the rope and (b) the mass of the drum.

66. A crane like that shown in Fig. 12-42 has a hollow drum of mass 110 kg and radius 0.35 m. The pulley at the top of the crane has negligible mass. If a 520 N·m torque is applied to the drum, it undergoes an angular acceleration $\alpha = 1.2$ rad/s². Find the mass being lifted by the crane.

67. Your little sister is building a toy car. She attaches four spools to a 48-g milk carton. The spools are essentially solid cylinders, each with mass 25 g and radius 1.8 cm and are mounted on frictionless axles of negligible mass. Starting from rest, the contraption rolls without slipping down a slope. (a) How fast is it going after it's dropped a vertical distance of 85 cm? (b) What per cent of its total kinetic energy is in the translational motion of the milk carton alone?

68. A 320-kg motorcycle includes two wheels each of which is 52 cm in diameter and has rotational inertia 2.1 kg·m². The cycle and its 75-kg rider are coasting at 85 km/h on a flat road when they encounter a hill. If the cycle rolls up the hill with no applied power and no significant internal friction, what vertical height will it reach?

Supplementary Problems

69. A solid marble starts from rest and rolls without slipping on the loop-the-loop track shown in Fig. 12-49. Find the minimum starting height of the marble from which it will remain on the track through the loop. Assume the marble radius is small compared with R.

70. A 100-g yo-yo consists of two disks joined by a narrower shaft. The rotational inertia of the yo-yo is 600 g·cm². A string 80 cm long is wound around the narrow shaft of the yo-yo. With the yo-yo initially 50 cm off the ground, the

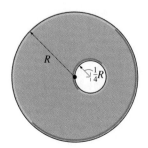

FIGURE 12-47 Problems 63, 76.

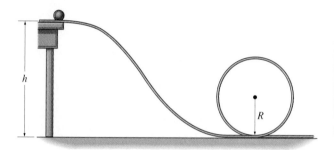

FIGURE 12-49 Problem 69.

string is pulled vertically upward with a force of 0.98 N until the string comes entirely free of the yo-yo. The yo-yo then falls freely. What are its rotational and translational kinetic energies just before it hits the ground?

71. An object of rotational inertia I is initially at rest. A torque is then applied to the object, causing it to begin rotating. The torque is applied for only one-quarter of a revolution, during which time its magnitude is given by $\tau = A \cos\theta$, where A is a constant and θ is the angle through which the object has rotated. What is the final angular speed of the object?

72. A disk of radius R and thickness w has a mass density that increases from the center outward, being given by $\rho = \rho_0 r/R$, where r is the distance from the axis of the disk. (a) Calculate the total mass M of the disk. (b) Calculate the rotational inertia about the disk axis, in terms of M and R. Compare with the results for a solid disk of uniform density and for a ring.

73. A thin, uniform vane of mass M is in the shape of a right triangle, as shown in Fig. 12-50. Find the rotational inertia about a vertical axis through its apex, as shown in the figure. Express your answer in terms of the triangle's base width b and its mass M.

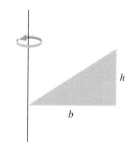

FIGURE 12-50 Problem 73.

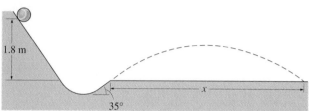

FIGURE 12-51 Problem 74.

74. A marble (solid) and a tennis ball (hollow) roll without slipping on the track shown in Fig. 12-51. Both start at a height of 1.8 m above the level stretch at the right. Find the horizontal range x of each ball.

75. A uniform disk of radius R is free to rotate about a horizontal axis at its edge. If it's at its lowest position, as shown in Fig. 12-52, what is the minimum angular speed ω necessary to ensure that its motion describes a complete circle?

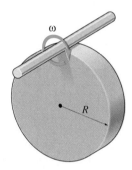

FIGURE 12-52 Problem 75.

76. The wheel shown in Fig. 12-47 is rotating freely about a frictionless horizontal axle. Since the wheel is unbalanced, its angular speed varies as it rotates. If the maximum angular speed is ω_{max}, find an expression for the minimum speed. *Hint:* How does the potential energy change as the wheel rotates?

77. A thin solid rod of length ℓ and mass M is free to pivot about one end. If it makes an angle ϕ with the horizontal, find the torque due to gravity about the pivot point. You'll need to integrate the torques on the individual mass elements comprising the rod.

78. Proof of the work-energy theorem for rotational motion: Start with the expression $\tau = I\alpha$ and substitute $\alpha = d\omega/dt$. Then apply the chain rule for derivatives to write $d\omega/dt = (d\omega/d\theta)(d\theta/dt)$. Identify the term $d\theta/dt$, and substitute the quantity it's equal to. Then multiply the whole equation through by $d\theta$. You should then be able to integrate to get Equation 12-29.

CHAPTER 13

ROTATIONAL VECTORS AND ANGULAR MOMENTUM

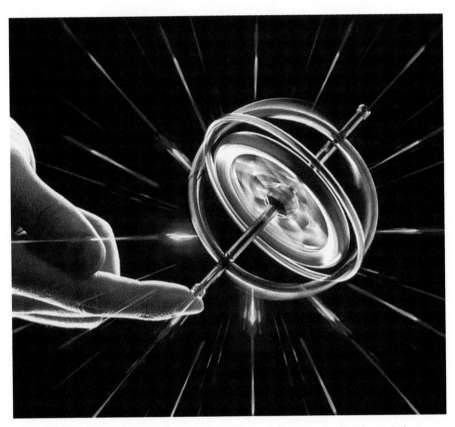

The spinning gyroscope seems to defy gravity, but actually it's just obeying Newton's laws.

The rotational motion we've dealt with so far is analogous to one-dimensional linear motion, in that torques change only the rotational speed but not the direction of rotation. Here we treat more general cases involving changes in direction as well. To do so we introduce a vector description of rotational motion that parallels the vector description of linear motion first used in Chapter 3.

FIGURE 13-1 The direction of the angular velocity vector is given by the right-hand rule.

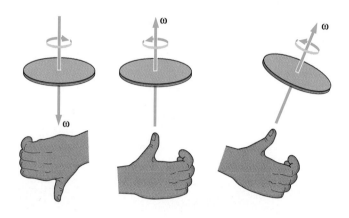

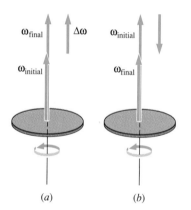

FIGURE 13-2 (a) An increase in angular speed alone corresponds to a change $\Delta\boldsymbol{\omega}$ parallel to the angular velocity $\boldsymbol{\omega}$ itself; therefore the angular acceleration vector $\boldsymbol{\alpha}$ is also parallel to $\boldsymbol{\omega}.$ (b) With a decrease in ω, $\Delta\boldsymbol{\omega}$ and $\boldsymbol{\alpha}$ are antiparallel to $\boldsymbol{\omega}.$

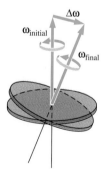

FIGURE 13-3 When the angular velocity $\boldsymbol{\omega}$ changes only in direction, the small change $\Delta\boldsymbol{\omega}$ and, therefore, the angular acceleration are perpendicular to $\boldsymbol{\omega}.$

13-1 ANGULAR VELOCITY AND ACCELERATION VECTORS

So far we've ascribed direction to rotational motion using the terms "clockwise" and "counterclockwise." What more could we mean by "direction?" Rotation of a wheel about a horizontal axis is not quite the same motion as rotation about a vertical axis. So we need to specify the direction of the rotation axis to describe fully the rotational motion. We therefore define **angular velocity $\boldsymbol{\omega}$** as a vector whose magnitude is the angular speed ω and whose direction is parallel to the rotation axis.

There's an ambiguity in this definition since there are two possible directions parallel to the axis. We resolve the ambiguity with the **right-hand rule:** If you curl the fingers of your right hand to follow the rotation, then your right thumb points in the direction of the angular velocity (Fig. 13-1). This refinement means that $\boldsymbol{\omega}$ gives not only the angular speed and the direction of the rotation axis but distinguishes what we would have described previously as clockwise or counterclockwise rotation.

By analogy with the linear acceleration vector, we define angular acceleration as the rate of change of the angular velocity vector:

$$\boldsymbol{\alpha} = \frac{d\boldsymbol{\omega}}{dt}. \tag{13-1}$$

This says that angular acceleration points in the direction of the *change* in angular velocity. If that change is only in magnitude, then $\boldsymbol{\omega}$ simply grows or shrinks, and $\boldsymbol{\alpha}$ is parallel or antiparallel to the rotation axis (Fig. 13-2). But a change in *direction* is also a change in angular velocity. When the angular velocity $\boldsymbol{\omega}$ changes only in direction, then the angular acceleration vector is perpendicular to $\boldsymbol{\omega}$ (Fig. 13-3). More generally, both the magnitude and direction of $\boldsymbol{\omega}$ may change; then $\boldsymbol{\alpha}$ is neither parallel nor perpendicular to $\boldsymbol{\omega}.$ These cases are exactly analogous to the situations we treated in Chapter 4, where acceleration parallel to velocity changes only the speed, while acceleration perpendicular to velocity changes only the direction of motion.

13-2 TORQUE AND THE VECTOR CROSS PRODUCT

Figure 13-4 shows a wheel, initially stationary, with a force applied at its rim. The torque associated with this force sets the wheel rotating in the direction shown; applying the right-hand rule we see that angular velocity vector **ω** points upward. Since the angular speed is increasing, the angular acceleration **α** also points upward. We'd like the torque to have an upward direction, too, so our rotational analog of Newton's law—angular acceleration proportional to torque—will hold for directions as well as magnitudes.

We already know the magnitude of the torque: from Equation 12-12, it is $\tau = rF\sin\theta$, where r is the distance from the rotation axis to the force application point and θ is the angle between the corresponding vector **r** and the force vector **F**. To get the upward direction from these same two vectors, we start by putting them tail to tail. We then define the direction of the torque as being perpendicular to both **r** and **F**; specifically, the direction is given by the right-hand-rule as shown in Figs. 13-5 and 13-6.

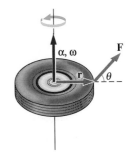

FIGURE 13-4 A force applied at the rim of the wheel will set it rotating with angular velocity **ω** and angular acceleration **α** both upward. For consistency with Newton's law, the associated torque vector should be in the same direction as the angular acceleration.

> **TIP** **Choose a Right-Hand Rule** The right-hand rules in Fig. 13-5 are fully equivalent. Choose the one you're most comfortable with and get familiar with it. You'll encounter the same rule in other areas of physics.

The Cross Product

The magnitude of the torque, $\tau = rF\sin\theta$, is determined by the magnitudes of the vectors **r** and **F** and the angle between them; we've now seen that the direction of the torque is also determined by the vectors **r** and **F** through the

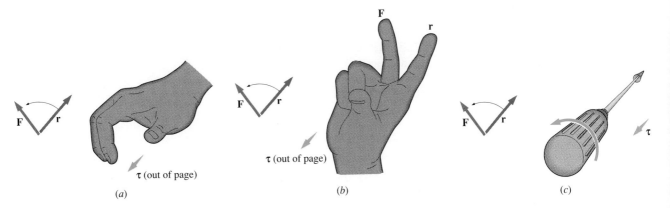

(a) (b) (c)

FIGURE 13-5 Three equivalent right-hand rules for the direction of torque and, more generally, for the cross product of two vectors. (a) Curl your right fingers in a direction that would rotate the first vector (**r**) onto the second (**F**); then your right thumb points in the direction of **τ** = **r** × **F**. (b) Point the first two fingers of your right hand in the directions of the first and second vectors (**r** and **F**, respectively); again, your right thumb points in the direction of the torque. (c) Imagine turning a screwdriver in a way that would rotate the first vector onto the second; the direction the screw would move is the direction of the torque.

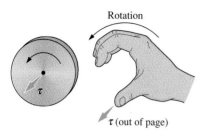

FIGURE 13-6 It's also appropriate to get the direction of the torque by curling your right fingers in the direction that the torque would rotate an initially stationary body; now your right thumb points along the torque vector.

right-hand rule. So the torque vector is completely determined by the vectors **r** and **F**. This operation—forming from two vectors **A** and **B** a third vector **C** of magnitude $C = AB \sin \theta$ and direction given by the right-hand rule—occurs frequently in physics and is called the **cross product:**

> **The cross product C of two vectors A and B is written**
>
> $$\mathbf{C} = \mathbf{A} \times \mathbf{B},$$
>
> **and is a vector with magnitude $AB \sin \theta$, where θ is the angle between A and B, and where the direction of C is given by the right-hand rule of Fig. 13-5.**

Torque is an instance of the cross product, and we can write the torque vector simply as

$$\boldsymbol{\tau} = \mathbf{r} \times \mathbf{F}. \tag{13-2}$$

Both direction and magnitude are described succinctly in this equation.

BOX 13-1 VECTOR PRODUCTS

The cross product **A** × **B** is the second type of vector product we've encountered, after the vector dot product introduced in Chapter 7. Both depend on the product of the vector magnitudes and on the angle between them. But where the dot product depends on the *cosine* of the angle and is therefore maximum when the two vectors are parallel, the cross product depends on the *sine* and is therefore maximum for perpendicular vectors. And most significantly, the dot product is a *scalar*—a single number, with no direction, while the cross product is a *vector*.

The cross product obeys the usual distributive rule:

$$\mathbf{A} \times (\mathbf{B} + \mathbf{C}) = \mathbf{A} \times \mathbf{B} + \mathbf{A} \times \mathbf{C},$$

but it's *not* commutative; in fact, as you can see by rotating **F** onto **r** instead of **r** onto **F** in Fig. 13-5:

$$\mathbf{B} \times \mathbf{A} = -\mathbf{A} \times \mathbf{B}.$$

● **EXAMPLE 13-1** TORQUE AS A CROSS PRODUCT

A force $\mathbf{F} = \sqrt{3}\hat{\imath} + \hat{\jmath}$ N acts on a wheel whose rotation axis coincides with the z axis; the force is applied at the point $x = 3$ m, $y = 0$ m. Find the torque due to this force.

Solution

The torque is given by Equation 13-2: $\boldsymbol{\tau} = \mathbf{r} \times \mathbf{F}$. With the force applied at the point (3, 0), we have $\mathbf{r} = 3\hat{\imath}$ m and the magnitude $r = 3$ m. The magnitude of **F** is given by the pythagorean theorem:

$$F = \sqrt{F_x^2 + F_y^2} = \sqrt{3 \text{ N}^2 + 1 \text{ N}^2} = 2 \text{ N},$$

and the angle **F** makes with the x axis (and thus with **r**) is

$$\theta = \tan^{-1}\left(\frac{1}{\sqrt{3}}\right) = 30°.$$

Then the magnitude of the cross product is

$$\tau = |\mathbf{r} \times \mathbf{F}| = rF \sin \theta = (3 \text{ m})(2 \text{ N}) \sin 30° = 3 \text{ N·m}.$$

Rotating **r** into **F** shows that $\boldsymbol{\tau}$ is in the z direction (see Fig. 13-7); therefore, we can write

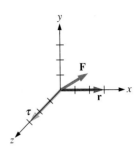

FIGURE 13-7 With **r** along the x axis, and **F** in the x-y plane, the torque $\boldsymbol{\tau} = \mathbf{r} \times \mathbf{F}$ is in the z direction.

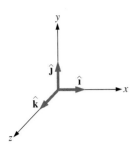

FIGURE 13-8 Rotate one unit vector into another, and the cross product is plus or minus the third unit vector, with the plus sign taken when the vectors are in cyclic order $\hat{\mathbf{i}}\,\hat{\mathbf{j}}\,\hat{\mathbf{k}}\,\hat{\mathbf{i}}\dots$. The signs are correct only in a right-handed coordinate system; alternately, you can define a right-handed coordinate system as being one in which it's true, for example, that $\hat{\mathbf{i}} \times \hat{\mathbf{j}} = \hat{\mathbf{k}}$.

$$\boldsymbol{\tau} = \mathbf{r} \times \mathbf{F} = 3\hat{\mathbf{k}} \text{ N·m.}$$

We could also obtain this result using unit vectors:

$$\boldsymbol{\tau} = \mathbf{r} \times \mathbf{F} = (3\hat{\mathbf{i}} \text{ m}) \times (\sqrt{3}\hat{\mathbf{i}} + \hat{\mathbf{j}} \text{ N})$$
$$= 3\sqrt{3}\hat{\mathbf{i}} \times \hat{\mathbf{i}} + 3\hat{\mathbf{i}} \times \hat{\mathbf{j}} \text{ N·m.}$$

But the unit vector $\hat{\mathbf{i}}$ is parallel to itself, so the cross product $\hat{\mathbf{i}} \times \hat{\mathbf{i}}$ is zero. On the other hand, $\hat{\mathbf{i}} \times \hat{\mathbf{j}} = \hat{\mathbf{k}}$, as Fig. 13-8 will convince you. So we again have $\boldsymbol{\tau} = 3\hat{\mathbf{k}}$ N·m.

TIP   Stick with Right-Handed Coordinate Systems   If you want to add a z axis to an x-y coordinate system, there are two possible directions for the positive z axis to point. You need to choose the one that makes $\hat{\mathbf{i}} \times \hat{\mathbf{j}} = \hat{\mathbf{k}}$ according to the right-hand rule.

The result is a right-handed coordinate system, consistent with all the right-hand rules used in physics.

If you're left-handed don't feel snubbed. Physics would have been equally consistent had we defined rotational vectors and cross products in terms of a left-hand rule and then used only left-handed coordinate systems.

EXERCISE   A force $\mathbf{F} = 4.6\hat{\mathbf{k}}$ N is applied at the point $\mathbf{r} = 1.3\hat{\mathbf{j}} + 2.4\hat{\mathbf{k}}$ m. Find the torque exerted about the origin.

Answer: $6.0\hat{\mathbf{i}}$ N·m

Some problems similar to Example 13-1: 7, 9, 11, 12

13-3 ANGULAR MOMENTUM

We first introduced Newton's law in the form $\mathbf{F} = m\mathbf{a}$, but later found the form $\mathbf{F} = d\mathbf{P}/dt$, with **P** the linear momentum of a system of particles, especially powerful. The same is true in rotational motion: To explore fully some surprising aspects of rotational dynamics, we need to define **angular momentum** and develop a relation between its rate of change and the applied torque.

Like other rotational quantities, angular momentum is specified with respect to a given point or axis. We begin with the angular momentum of a single particle:

If a particle with linear momentum p is at position r with respect to some point, then its angular momentum L about that point is

$$\mathbf{L} = \mathbf{r} \times \mathbf{p}. \tag{13-3}$$

● EXAMPLE 13-2   ANGULAR MOMENTUM OF A PARTICLE

A particle of mass m moves counterclockwise at speed v around a circle of radius r in the x-y plane. Find its angular momentum about the center of the circle, and express in terms of its angular velocity.

Solution

From Fig. 13-9 we see that the particle's linear momentum $m\mathbf{v}$ and the radius $\mathbf{r}$ are always at right angles, so the magnitude of the angular momentum $\mathbf{L}$ is just rmv. Applying the right-hand rule shows that $\mathbf{L}$ points out of the page, or in the z direction. So we have $\mathbf{L} = rmv\hat{\mathbf{k}}$. We can write this in terms of the angular velocity $\boldsymbol{\omega}$ by noting that $v = \omega r$ and by applying the right-hand rule to see that $\boldsymbol{\omega}$, too, points in the z direction; that is, $\boldsymbol{\omega} = \omega\hat{\mathbf{k}}$. Then we have

$$\mathbf{L} = rmv\hat{\mathbf{k}} = mr^2\omega\hat{\mathbf{k}} = mr^2\boldsymbol{\omega}.$$

EXERCISE Determine the angular momentum about Earth's center of a 540-kg communications satellite in geosynchronous orbit. *Hint:* Consult Example 9-4 and the answer to Problem 9-13.

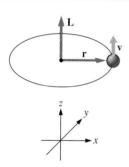

FIGURE 13-9   The angular momentum $\mathbf{L}$ of a particle moving at speed v around a circle of radius r has magnitude rmv and is perpendicular to the plane of the circle. Here $\mathbf{L}$ is calculated about the center of the circle.

Answer: 7.1×10^{13} kg·m²/s, northward

Some problems similar to Example 13-2:   19, 24, 29   ●

You might expect that we could write

$$\mathbf{L} = I\boldsymbol{\omega} \tag{13-4}$$

in analogy with the definition $\mathbf{p} = m\mathbf{v}$. Since the rotational inertia of a single particle is mr^2, the result of Example 13-2 can indeed be written $\mathbf{L} = I\boldsymbol{\omega}$. By defining the angular momentum of a system as the sum of the angular momenta of its constituent particles, Equation 13-4 can be shown to hold in the case of a symmetric object like a wheel or sphere rotating about a symmetry axis that is itself stationary. But in the more general case, Equation 13-4 does not necessarily hold; surprisingly, $\mathbf{L}$ and $\boldsymbol{\omega}$ can actually have different directions. Even vector mathematics fails to describe such cases adequately, and we need new mathematical tools that are best left for advanced courses in mechanics.

We'll deal mostly with situations for which Equation 13-4 does apply. In those cases it's usually easiest to calculate the angular momentum from Equation 13-4. But Equation 13-3—summed over all particles of a system—remains the fundamental definition of angular momentum.

Torque and Angular Momentum

We're now ready to develop the full vector analog of Newton's law in the form $\mathbf{F} = d\mathbf{P}/dt$. Recall that $\mathbf{F}$ here is the *net* external force on a system, and $\mathbf{P}$ is the system's momentum—the vector sum of the momenta of its constituent particles. Can we write, by analogy,

$$\boldsymbol{\tau} = \frac{d\mathbf{L}}{dt} ?$$

To see that we can, we write the angular momentum of a system as the sum of the angular momenta of its constituent particles:

$$\mathbf{L} = \sum \mathbf{L}_i = \sum (\mathbf{r}_i \times \mathbf{p}_i),$$

where the subscript i refers to the ith particle of the system. Differentiating this equation with respect to time gives

$$\frac{d\mathbf{L}}{dt} = \sum \left(\mathbf{r}_i \times \frac{d\mathbf{p}_i}{dt} + \frac{d\mathbf{r}_i}{dt} \times \mathbf{p}_i \right),$$

where we've applied the product rule for differentiation, being careful to preserve the order of the cross product since it is not commutative. But $d\mathbf{r}_i/dt$ is the velocity of the ith particle, so the second term in the sum is the cross product of velocity $\mathbf{v}$ and momentum $\mathbf{p} = m\mathbf{v}$. Since these two vectors are parallel, their cross product is zero, and we're left with only the first term in the sum:

$$\frac{d\mathbf{L}}{dt} = \sum \left(\mathbf{r}_i \times \frac{d\mathbf{p}_i}{dt} \right) = \sum \left(\mathbf{r}_i \times \mathbf{F}_i \right),$$

where we've used Newton's law to write $d\mathbf{p}_i/dt = \mathbf{F}_i$. But $\mathbf{r}_i \times \mathbf{F}_i$ is the torque, $\boldsymbol{\tau}_i$, on the ith particle, so

$$\frac{d\mathbf{L}}{dt} = \sum \boldsymbol{\tau}_i \, .$$

As we discussed in Section 12-3, the internal torques cancel in pairs as long as the internal forces in a system of particles act along the lines between the particles. Under these conditions, the sum of all the torques on all the particles reduces to the net *external* torque on the system, and we have

$$\frac{d\mathbf{L}}{dt} = \boldsymbol{\tau} \, . \tag{13-5}$$

where $\boldsymbol{\tau}$ is the net external torque. Thus our analogy between linear and rotational motion holds for momentum as well as for the other quantities we've discussed.

13-4 CONSERVATION OF ANGULAR MOMENTUM

When there is no external torque on a system, Equation 13-5 tells us that angular momentum is constant. This statement—that the angular momentum of an isolated system cannot change—is of fundamental importance in physics, and applies to systems ranging from subatomic particles to galaxies. In deriving Equation 13-5, we did not require that the system in question be a rigid object, so conservation of angular momentum applies even to systems that undergo changes in configuration and therefore in rotational inertia. The classic example is a figure skater, who starts spinning relatively slowly with arms extended, then pulls her arms in to spin much more rapidly (Fig. 13-10). Why? As the skater's arms move in, her mass is concentrated more toward the rotation axis, lowering her rotational inertia, I. But her angular momentum $I\omega$ is conserved, so her

FIGURE 13-10 As skater Kristi Yamaguchi pulls in her arms, her rotational inertia I decreases. To conserve her angular momentum $I\omega$, her angular speed ω must increase.

angular speed ω must increase. A much more dramatic example is illustrated in Example 13-3: the collapse of a star at the end of its lifetime. Again, the rotational inertia decreases, requiring an increase in the star's rotational speed.

● **EXAMPLE 13-3** PULSARS

A certain star rotates about its axis once every 45 days. At the end of its life it undergoes a colossal outburst called a supernova explosion. Much of the star's mass is flung outward into the interstellar medium, where it will eventually condense into new stars and planets and perhaps find its way into living creatures (Fig. 13-11). But the inner core of the star, whose initial radius was 2×10^7 m, collapses into a neutron star, only 6 km in radius. Calculate the rotation rate of the neutron star, assuming that no torque acts on the stellar core during the collapse. Consider the core to be a uniform sphere both before and after collapse.

Solution

This situation is exactly like that of the figure skater: a decrease in rotational inertia requires an increase in angular speed. Conservation of angular momentum states that

$$I_1\omega_1 = I_2\omega_2,$$

where the subscripts 1 and 2 refer to conditions before and after the collapse, respectively. For a uniform sphere, $I = \frac{2}{5}MR^2$, so

$$\frac{2}{5}MR_1^2\omega_1 = \frac{2}{5}MR_2^2\omega_2.$$

Solving for ω_2 and using the known values of ω_1, R_1, and R_2 gives

$$\omega_2 = \omega_1\left(\frac{R_1}{R_2}\right)^2 = \left(\frac{1\text{ rev}}{45\text{ day}}\right)\left(\frac{2\times10^7\text{ m}}{6\times10^3\text{ m}}\right)^2$$
$$= 2.5\times10^5\text{ rev/day} = 2.9\text{ rev/s}.$$

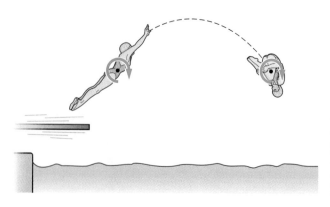

FIGURE 13-12 How fast is the diver rotating when she goes into the tuck position?

FIGURE 13-11 The Crab Nebula is the remains of a supernova explosion as it looks 1000 years after the explosion. At its center is a rapidly rotating neutron star.

This neutron star is a fantastic thing—an object with more mass than the entire Sun, rotating three times a second. Often intense radio waves, light, or x rays radiate in a narrow beam from the magnetic axis of the neutron star. As the star rotates, this beam sweeps through space like a searchlight, and we at Earth detect regular pulses of radiation that give the neutron star the name "pulsar."

TIP You Don't Always Need Vectors
Problems involving angular momentum where the direction of rotation doesn't change can be treated like other one-dimensional problems, without the need for vector notation. This example and the next one are both instances of one-dimensional angular momentum problems.

EXERCISE A 55-kg diver leaves the diving board in an essentially straight configuration, with an angular speed of 0.25 rev/s about her center of mass. She then goes into a tuck position with a radius of approximately 31 cm (Fig. 13-12). Approximating the diver with her outstretched arms first as a uniform thin rod 2.0 m long, then as a uniform sphere, estimate her rotation rate while in the tuck position.

Answer: 2.2 rev/s

Some problems similar to Example 13-3: 35, 36, 49, 50 ●

● **EXAMPLE 13-4** A MERRY-GO-ROUND

A merry-go-round of radius $R = 1.3$ m has a rotational inertia $I = 240$ kg·m^2 and is rotating freely at $\omega_1 = 11$ rpm. A boy of mass $m_1 = 28$ kg leaps onto the edge of the merry-go-round, heading directly toward the center (Fig. 13-13). What is the new angular speed of the merry-go-round? Later a girl, of mass $m_2 = 32$ kg, leaps onto the merry-go-round by running tangentially at speed $v = 3.7$ m/s. If the direction of her motion is the same as that of the merry-go-round (see Fig. 13-13), what is the new angular speed?

Solution

Moving directly toward the rotation axis, the boy carries no angular momentum relative to the axis, so the total angular momentum is $I\omega_1$. But sitting at the edge of the merry-go-round, he adds rotational inertia m_1R^2 to the system. Then conservation of angular momentum may be written

$$I\omega_1 = (I + m_1R^2)\omega_2.$$

where ω_2 is the angular speed after the boy is on the merry-go-round. Solving for ω_2 gives

$$\omega_2 = \frac{I\omega_1}{I + m_1R^2} = \frac{(240 \text{ kg·m}^2)(11 \text{ rpm})}{240 \text{ kg·m}^2 + (28 \text{ kg})(1.3 \text{ m})^2}$$
$$= 9.2 \text{ rpm}.$$

In this calculation, where we dealt with a ratio of angular speeds, we didn't need to convert from rpm to rad/s.

The girl, running tangentially to the merry-go-round, carries angular momentum m_2vR, in the same direction as the merry-go-round's angular momentum, so the total angular momentum of the system is now $I\omega_1 + m_2vR$. After she leaps on, the total rotational inertia includes that of the merry-go-round along with a contribution mR^2 from each child. Calling the final angular speed ω_3, conservation of angular momentum may then be written

$$I\omega_1 + m_2vR = (I + m_1R^2 + m_2R^2)\omega_3,$$

so
$$\omega_3 = \frac{I\omega_1 + m_2vR}{I + (m_1 + m_2)R^2}.$$

With m_2 in kilograms, v in meters per second, and R in meters, the angular momentum m_2vR of the second child has the units of kg·m^2·rad/s, while all our other angular momenta have been expressed implicitly in kg·m^2·rpm. To solve for ω_3 in rpm, we

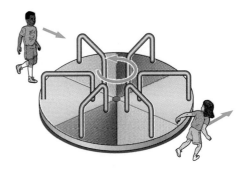

FIGURE 13-13 Running straight toward its axis, the boy brings no angular momentum to the merry-go-round. But the girl, running tangentially at speed v, brings angular momentum mvR.

convert the girl's angular momentum into the latter unit, multiplying by the conversion factor $\omega = (1 \text{ rev}/2\pi \text{ rad})(60 \text{ s/min})$. The equation above then gives $\omega_3 = 12$ rpm.

Is mechanical energy conserved as the children leap onto the merry-go-round? No! In each case frictional forces are involved in bringing children and merry-go-round to rest with respect to each other, so that the situation is like an inelastic collision.

TIP You Don't Have To Be Rotating To Have Angular Momentum   The girl in this example, was running in a straight line, yet she had nonzero angular momentum of magnitude mvR with respect to the merry-go-round axis. This is generally true: As Problem 26 shows, a particle moving in a straight line at constant speed has constant, nonzero angular momentum with respect to a point not on that line.

EXERCISE   A circular bird feeder 19 cm in radius has rotational inertia 0.12 kg·m^2. The feeder is suspended by a thin wire and is spinning slowly at 5.6 rpm. A 140-g bird lands on the rim of the feeder, coming in tangent to the rim at 1.1 m/s in a direction opposite the feeder's rotation. What is the rotation rate after the bird lands?

Answer: 5.28 rpm

Some problems similar to Example 13-4: 30, 31, 51, 52 ●

13-5 ROTATIONAL DYNAMICS IN THREE DIMENSIONS

So far we've explored rotational dynamics only in cases where the direction of rotation doesn't change. But when torque and angular momentum are in different directions, new and often startling behavior results. However strange this

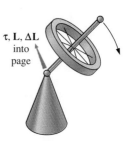

$\tau, \mathbf{L}, \Delta\mathbf{L}$
into
page

FIGURE 13-14 Why doesn't the spinning gyroscope fall over?

FIGURE 13-15 If it weren't spinning, the torque due to the gravitational force *would* cause it to fall. It would gain angular momentum about its pivot, and both the torque and change in angular momentum would point into the page, obeying the rotational analog of Newton's law.

behavior may seem, you can understand it if you really accept the rotational analog of Newton's law—that the rate of change of the angular momentum *vector* is equal to the net torque *vector*.

Precession

A gyroscope seems to defy gravity, balanced on a single point with most of its mass unsupported (Fig. 13-14). Surprising though it is, this behavior follows directly from the rotational analog of Newton's law. The downward force of gravity produces a torque whose direction in Fig. 13-14 is into the page. Were the gyroscope not rotating, it would simply fall over, gaining angular momentum into the page about its bottom pivot (Fig. 13-15). The change in angular momentum—here simply an increase in magnitude—is indeed in the same direction as the torque, as the rotational analog of Newton's law requires.

When the gyroscope is rotating, the torque is still into the page, so that must still be the direction of the *change* in angular momentum as well. But now the gyroscope already has angular momentum that points along its rotation axis, so now the change in angular momentum is at right angles to the angular momentum itself. That means that only the direction, not the magnitude, of the angular momentum vector changes. As a result the gyroscope moves so its rotation axis sweeps out a circle—a phenomenon called **precession** (Fig. 13-16).

So why doesn't the gyroscope fall over? That's like asking why a satellite in circular orbit doesn't fall to Earth. There's a gravitational force pulling the satellite toward Earth, but it's just the right strength to keep the satellite in a circular path. Similarly, there's a torque on the gyroscope, but it's just sufficient to change the angular momentum at the rate that maintains precession.

We can analyze the gyroscope's behavior quantitatively through Fig. 13-17, which shows a gyroscope spinning with its axis in a horizontal plane. The downward force of gravity on the unbalanced gyroscope produces a torque into the page in Fig. 13-17a—that is, at right angles to the angular momentum vector. In a short time dt, the angular momentum vector changes by an amount $d\mathbf{L}$ given by Equation 13-5:

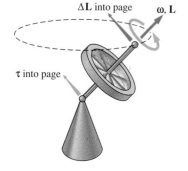

$\Delta\mathbf{L}$ into page $\omega, \mathbf{L}$

τ into page

FIGURE 13-16 With $\mathbf{L}$ and τ at right angles, the gyroscope precesses, the end of its rotation axis describing a circle. The *change* in angular momentum is still in the direction of the torque.

$$dL = \tau \, dt. \tag{13-6}$$

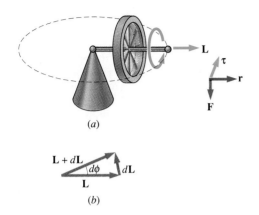

FIGURE 13-17 (*a*) A gyroscope spinning with its axis horizontal. (*b*) Top view, showing change in the angular momentum vector.

This change is shown in Fig. 13-17*b*, and represents a rotation of the angular momentum vector through a small angle $d\phi$. For an arbitrarily small time interval dt, and correspondingly small angle $d\phi$, the magnitude dL of the change in angular momentum is essentially equal to the arc length through which the tip of the angular momentum vector swings. From the definition of angle (Equation 12-3), this arc length is just $L\,d\phi$, so

$$dL = L\,d\phi.$$

Combining this result with Equation 13-6 gives

$$\tau\,dt = L\,d\phi.$$

The angular speed of precession, called Ω, is the rate at which the angular momentum vector rotates, or $d\phi/dt$. Solving for this quantity gives

$$\Omega = \frac{d\phi}{dt} = \frac{\tau}{L} = \frac{mgd}{L}, \qquad (13\text{-}7)$$

where we wrote $\tau = mgd$ with d the distance from the pivot to the gyroscope's center of gravity. Although we derived Equation 13-7 for a gyroscope with its rotation axis horizontal, Problem 44 shows that the form $\Omega = mgd/L$ holds for any angle.

Our assumption that the gyroscope's angular momentum is exactly horizontal isn't quite right. Because it's precessing, it also has a small vertical component of angular momentum. Equation 13-7 is therefore an approximation that holds when the precessional angular momentum is much smaller than that associated with rotation—an approximation that is usually valid.

Earth itself provides an important example of precession. Our planet bulges slightly at the equator, and the gravitational force of Sun and moon on this bulge gives rise to a torque on the tilted, spinning Earth. In response, Earth's rotation axis precesses, taking about 26,000 years to describe a complete circle (Fig. 13-18). The axis now points toward the star Polaris, which we consequently call the North Star, but Polaris won't always be the North Star!

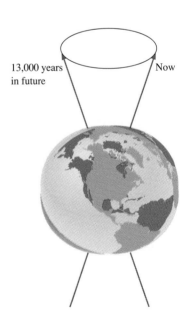

13,000 years in future

Now

FIGURE 13-18 Torque due to the Sun's gravity on Earth's equatorial bulge (shown highly exaggerated) results in precession of the planet's rotation axis with a period of about 26,000 years.

● **EXAMPLE 13-5** PRECESSION

A gyroscope consists of a solid disk of radius $R = 10$ cm, mounted in the center of a shaft whose length $\ell = 32$ cm. One end of the shaft is mounted in a frictionless pivot, and the other end is free; the shaft has negligible mass. The disk mass $M = 840$ g. The disk is precessing horizontally at $\Omega = 1.2$ rad/s. Find the angular speed ω of rotation about the shaft.

Solution

We want the angular speed of the disk, but Equation 13-7 contains only the angular momentum, torque, and precession rate. But we can relate the angular speed and angular momentum through $L = I\omega$. (Although this relation is strictly correct only when the rotation axis is stationary, Equation 13-7's assumption that the precessional angular momentum is small allows us to use $L = I\omega$ here.) Solving Equation 13-7 for L then gives

$$L = I\omega = \frac{\tau}{\Omega}.$$

But $I = \frac{1}{2}MR^2$ for the solid disk, and $\tau = \frac{1}{2}\ell Mg$ because the disk (weight Mg) is mounted $\frac{1}{2}\ell$ from the pivot. Solving our equation for ω and using these expressions for I and τ, we then have

$$\omega = \frac{\tau}{I\Omega} = \frac{\frac{1}{2}\ell Mg}{\frac{1}{2}MR^2\Omega} = \frac{\ell g}{R^2\Omega} = \frac{(0.32 \text{ m})(9.8 \text{ m/s}^2)}{(0.10 \text{ m})^2(1.2 \text{ rad/s})}$$
$$= 260 \text{ rad/s}.$$

That $\omega \gg \Omega$ suggests that our assumption that the precessional angular momentum is small is a good one.

TIP You Don't Always Need All the Information Sometimes a problem contains superfluous information, in this case the disk mass M. It may not be obvious at first that the information is unnecessary, but if it cancels from your final equations then it indeed was. You can sometimes gain valuable physical insight by considering why the given information doesn't matter. In this case it's because an increased mass gives rise to a greater angular momentum, but also proportionately to a greater torque, so the relationship expressed in Equation 13-7 remains unchanged.

EXERCISE A solid sphere is mounted on a shaft of negligible mass and length 15 cm; the shaft end rests on a frictionless pivot, as shown in Fig. 13-19. The sphere is spinning at 280 rad/s and precessing in a horizontal circle at 1.9 rad/s. Find its radius.

Answer: 11 cm

Some problems similar to Example 13-5: 42, 43, 45

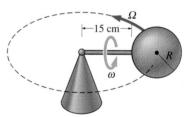

FIGURE 13-19 How big is the sphere? ●

Conservation of Angular Momentum in Three Dimensions

When no external torques act on a system, both the direction and magnitude of its angular momentum remain constant. A spinning gyroscope, suspended so as to be free of torques, keeps its rotation axis in the same direction even as it moves about the Earth (Fig. 13-20). Gyroscopes are therefore used extensively

FIGURE 13-20 A gyrocompass. When no external torques act, the rotation axis of the shipboard gyroscope always points in the same direction.

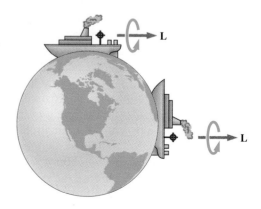

in navigation systems. Changes in the orientation of the rotation axis relative to a submarine, ship, aircraft, or missile are "remembered" by a computer; combining information from on-board accelerometers allows the computer to calculate the vector displacement from the starting point.

■ **APPLICATION** POINTING A SATELLITE

Some satellites use angular momentum conservation to orient themselves in space (Fig. 13-21). Such a satellite contains three wheels whose axes are mutually perpendicular. Each wheel has a motor and brakes to control its rotation. When a wheel starts rotating, the satellite rotates in the opposite direc-tion to conserve angular momentum. After the satellite has ro-tated through the desired angle, the wheel is stopped and the satellite also stops. Using three wheels allows the satellite to point in any direction. Electricity generated by solar panels pow-ers the motors and brakes, so there's no fuel to run out.

FIGURE 13-21 The Solar Maxi-mum Mission satellite became the first to be repaired in space after an electrical failure crippled the reaction-wheel sys-tem that used angular momentum con-servation to point the satellite at the Sun. Here a shuttle astronaut attempts to link with the slowly spinning satellite.

13-6 ANGULAR MOMENTUM AT THE SUBATOMIC LEVEL

Angular momentum is among the most fundamental quantities in physics—more fundamental, in a sense, than linear momentum. At the subatomic level, individual elementary particles possess intrinsic angular momentum that is an essential part of their basic nature. Complex particles like nuclei and atoms have angular momenta determined by summing—in a fashion dictated by quantum mechanics—both the intrinsic angular momentum vectors of their individual particles as well as angular momenta arising from particle motion. In all such cases the angular momenta are ultimately written in terms of a fundamental unit of angular momentum designated $\hbar$; in SI units $\hbar$ has the approximate value 1.05×10^{-34} kg·m²/s. Quantum mechanics shows that the component of any angular momentum vector on any axis must be a multiple of half the fundamental unit $\hbar$. Readers of the extended version of this text will explore this curious situation further in Chapter 41.

CHAPTER SYNOPSIS

Summary

1. Full description of rotational motion requires that rotational quantities be treated as vectors. **Angular velocity** is a vector whose magnitude is the angular speed ω and whose direction is that of the rotation axis, as established by the right-hand rule. The **angular acceleration** vector is the rate of change of the angular velocity.

2. The **torque** vector $\boldsymbol{\tau}$ has magnitude $rF\sin\theta$ as in Chapter 12, with its direction determined by applying the right-hand rule to the vectors $\mathbf{r}$ and $\mathbf{F}$ as shown in Fig. 13-5. The operation of forming from two vectors $\mathbf{r}$ and $\mathbf{F}$ a third vector with magnitude $rF\sin\theta$ and direction given by the right-hand rule is called the **cross product.** Torque as the cross product of $\mathbf{r}$ and $\mathbf{F}$ is written

$$\boldsymbol{\tau} = \mathbf{r} \times \mathbf{F}.$$

3. The **angular momentum, L,** of a particle about some point is the cross product of the radius vector from that point to the particle's location with the particle's linear momentum $\mathbf{p}$:

$$\mathbf{L} = \mathbf{r} \times \mathbf{p}.$$

The angular momentum of a system is the sum of the angular momenta of its constituent particles. For symmetrical systems with stationary rotation axes, the angular momentum can also be written $\mathbf{L} = I\boldsymbol{\omega}$.

4. The rate at which a system's angular momentum changes is equal to the net external torque applied to the system:

$$\frac{d\mathbf{L}}{dt} = \boldsymbol{\tau}.$$

This is Newton's law for rotational motion, in its full vector form.

5. In the absence of external torques, $d\mathbf{L}/dt = \mathbf{0}$ and a system's angular momentum remains constant in direction and magnitude.

6. **Precession**—circular motion of a spinning object's rotation axis—occurs when the applied torque has a component perpendicular to the rotation axis.

Terms You Should Understand

(Pairs are closely related terms whose distinction is important; number in parentheses is chapter section where term first appears.)

angular velocity, angular acceleration vectors (13-1)
vector cross product (13-2)
torque vector (13-2)
angular momentum (13-3)
precession (13-5)

Symbols You Should Recognize

$\boldsymbol{\omega}$ (13-1)
$\boldsymbol{\alpha}$ (13-1)
$\mathbf{A} \times \mathbf{B}$ (13-2)
$\mathbf{L}$ (13-3)
Ω (13-5)

Problems You Should Be Able to Solve

applying the right-hand rule to determine the direction of angular velocity and acceleration vectors (13-1)
evaluating vector cross products (13-2)
calculating torque vectors (13-2)
calculating angular momentum from linear velocity and position (13-3)
calculating angular momentum from rotational inertia and angular velocity (13-3)
solving problems involving the conservation of angular momentum in one dimension (13-4)
solving problems involving precession (13-5)

Limitations to Keep in Mind

The expression $\mathbf{L} = I\boldsymbol{\omega}$ applies exactly only when the rotation axis is stationary and is a symmetry axis.

The expression $\Omega = \tau/L$ applies only when the angular momentum associated with precession is much less than the rotational angular momentum.

QUESTIONS

1. Does Earth's angular velocity vector point north or south?
2. Must a torque necessarily change the angular speed of a rotating object?
3. Figure 13-22 shows four forces acting on a body. What are the directions of each of the associated torques about the point O? About the point P?

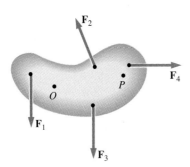

FIGURE 13-22 Question 3.

4. A satellite is small enough to be treated as a point particle. Can Earth's gravity exert a torque on the satellite about Earth's center?
5. From the result of the previous question, what can you conclude about the angular momentum associated with a satellite's orbital motion?
6. You're sitting on a merry-go-round, holding a spinning top with its rotation axis vertical. Is it possible for the top to have zero angular momentum with respect to the axis of the merry-go-round?
7. You tie a ball to the end of a string and whirl it in a vertical circle. Is its angular momentum constant? If not, why not?
8. Why isn't the cross product commutative?
9. Although it contains no parentheses, the expression **A** × **B** · **C** is unambiguous. Why?
10. Two vectors are parallel. What can you conclude about their cross product?
11. The dot product of two vectors is equal to the magnitude of their cross product. What can you conclude about the angle between them?
12. Why does a tether ball move faster as it winds up its pole?
13. A cloud of interstellar matter collapses to form a star. Under what conditions will the star *not* be rotating?
14. Why do helicopters have two rotors?
15. A bug, initially at rest on a stationary turntable, walks halfway around the circumference. Describe the motion of the turntable while the bug is walking and after it stops.
16. A group of polar bears are standing around the edge of a circular ice floe that's slowly rotating. If the bears all walk to the center, what will happen to the rotation rate?

17. Tornadoes in the northern hemisphere rotate counterclockwise as viewed from above. A far-fetched idea suggests that driving on the right side of the road may increase the frequency of tornadoes. Does this idea have *any* merit? Explain in terms of the angular momentum imparted to the air by passing cars.
18. How does angular momentum conservation help explain Earth's seasons?
19. A student is standing on a nonrotating turntable that's free to rotate and is holding a spinning wheel (Fig. 13-23). What happens if she turns the wheel upside down?

FIGURE 13-23 Question 19.

20. Does a particle moving at constant speed in a straight line have angular momentum about a point on the line? About a point not on the line? In either case, is its angular momentum constant?
21. Spacecraft are often set rotating. How does this help stabilize their orientations?
22. When you turn on a high-speed power tool like a router, you feel the tool tending to twist in your hands. Why?
23. Does the center of mass of a car have angular momentum about the front axle? What does this have to do with the car's tendency to nose downward when braking hard? (*Note:* The front brakes do most of the stopping.)
24. A huge turbine in the engine room of a ship has its axis aligned with the ship and is spinning rapidly with angular momentum vector toward the ship's bow. If its lubrication fails and the turbine suddenly "seizes" and stops abruptly, what will happen to the ship?

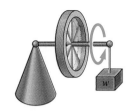

FIGURE 13-24 Question 25.

FIGURE 13-25 Question 26.

25. A gyroscope is spinning with its rotation axis horizontal and is precessing about a vertical axis. If a weight is hung from the end of the gyroscope (Fig. 13-24), what will happen to the precession rate?

26. What happens if you push horizontally at right angles to the axis of the spinning gyroscope shown in Fig. 13-25?

PROBLEMS

Section 13-1 Angular Velocity and Acceleration Vectors

1. A car is headed north at 70 km/h. Give the magnitude and direction of the angular velocity of its 62-cm-diameter wheels.

2. If the car of Problem 1 makes a 90° left turn lasting 25 s, determine the angular acceleration of the wheels.

3. A wheel is spinning at 45 rpm with its spin axis vertical. After 15 s, it's spinning at 60 rpm with its axis horizontal. Find (a) the magnitude of its average angular acceleration and (b) the angle the average angular acceleration vector makes with the horizontal.

4. A wheel is spinning about a horizontal axis, with angular speed 140 rad/s and with its angular velocity pointing east. Find the magnitude and direction of its angular velocity after an angular acceleration of 35 rad/s², pointing 68° west of north, is applied for 5.0 s.

5. A wheel is spinning with angular speed $\omega = 5.0$ rad/s, when a constant angular acceleration $\alpha = 0.85$ rad/s² is applied at right angles to the initial angular velocity. How long does it take for the angular speed to increase by 10 rad/s?

Section 13-2 Torque and the Vector Cross Product

6. A rod is free to pivot about one end, and a force **F** is applied at the other end, as shown in Fig. 13-26. What are the magnitude and direction of the torque about the pivot point?

7. A coordinate system lies with its x-y plane in the plane of this page, and its z axis coming out of the page. A force **F** is applied at the point $x = 1$, $y = 1$. What is the direction of the torque about the origin if **F** points (a) in the x direction, (b) in the y direction, and (c) in the z direction?

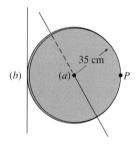

FIGURE 13-26 Problem 6.

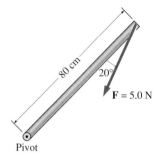

FIGURE 13-27 Problem 8.

8. The disk in Fig. 13-27 has radius 35 cm. What force (magnitude and direction) would you apply at point P in order to produce a torque of 1.2 N·m (a) about an axis through the center of the disk and perpendicular to the page and (b) about a vertical axis tangent to the left edge of the disk?

9. A 12-N force is applied at the point $x = 3$ m, $y = 1$ m. Find the torque about the origin if the force points in (a) the x direction, (b) the y direction, and (c) the z direction.

10. A sphere 1.0 m in radius is centered at the origin. At what x, y, and z coordinates on the surface of the sphere could you apply a 5.0-N force pointing in the y direction in order to produce, about the origin, (a) a 5.0 N·m torque pointing in the $-z$ direction, (b) a 3.4 N·m torque pointing in the $+z$ direction, and (c) a 3.4 N·m torque pointing in the $-x$ direction?

11. A force $\mathbf{F} = 2.0\hat{\imath} - 8.0\hat{\jmath}$ N is applied at the point $x = 1.2$ m, $y = 3.6$ m. Find the resulting torque about the origin.

12. A force $\mathbf{F} = 1.3\hat{\imath} + 2.7\hat{\jmath}$ N is applied at the point $x = 3.0$ m, $y = 0$ m. Find the torque about (a) the origin and (b) the point $x = -1.3$ m, $y = 2.4$ m.

13. Vector $\mathbf{A}$ points 30° counterclockwise from the x axis. Vector $\mathbf{B}$ is twice as long as $\mathbf{A}$. Their product $\mathbf{A} \times \mathbf{B}$ has length A^2, and points in the negative z direction. What is the direction of vector $\mathbf{B}$?

14. Show that the cross product of two arbitrary vectors $\mathbf{A} = A_x\hat{\imath} + A_y\hat{\jmath} + A_z\hat{\mathbf{k}}$ and $\mathbf{B} = B_x\hat{\imath} + B_y\hat{\jmath} + B_z\hat{\mathbf{k}}$ can be written $\mathbf{A} \times \mathbf{B} = (A_yB_z - A_zB_y)\hat{\imath} + (A_zB_x - A_xB_z)\hat{\jmath} + (A_xB_y - A_yB_x)\hat{\mathbf{k}}$.

15. Show that $\mathbf{A} \cdot (\mathbf{A} \times \mathbf{B}) = 0$ for any vectors $\mathbf{A}$ and $\mathbf{B}$.

16. Find the cross product of the vectors $\mathbf{A} = \hat{\imath} - 2\hat{\jmath} + 3\hat{\mathbf{k}}$ and $\mathbf{B} = 2\hat{\imath} + 4\hat{\jmath} + 2\hat{\mathbf{k}}$.

17. Find a force vector that, when applied at the point $x = 2.0$ m, $y = 1.5$ m, will produce a torque $\boldsymbol{\tau} = 4.7\hat{\mathbf{k}}$ N·m about the origin. (There are many answers.)

Section 13-3 Angular Momentum

18. Express the units of angular momentum (a) using only the fundamental units kilogram, meter, and second; (b) in a form involving newtons; (c) in a form involving joules.

19. In the Olympic hammer throw (Fig. 13-28), a contestant whirls a 7.3-kg steel ball on the end of a 1.2-m cable. If the contestant's arms reach an additional 90 cm from his axis of rotation and if the speed of the ball just prior to release

FIGURE 13-28 The hammer throw (Problem 19).

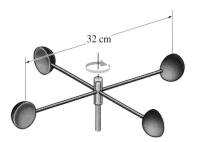

FIGURE 13-29 Problem 20.

is 27 m/s, what is the magnitude of its angular momentum?

20. An anemometer for measuring wind speed consists of four small metal cups, each of mass 120 g, mounted on the ends of essentially massless rods 32 cm long (Fig. 13-29). Find (a) the anemometer's rotational inertia about its central axis and (b) its angular momentum when it's spinning at 12 rev/s in the direction indicated. Treat the cups as point masses.

21. A gymnast of rotational inertia 62 kg·m² is tumbling head over heels. If her angular momentum is 470 kg·m²/s, what is her angular speed?

22. A 640-g hoop 90 cm in diameter is rotating at 170 rpm about its central axis. What is its angular momentum?

23. A 7.4-cm-diameter baseball has a mass of 150 g and is spinning at 210 rad/s. What is the magnitude of its angular momentum? Treat the baseball as a solid sphere.

24. What is the magnitude of the angular momentum associated (a) with Earth's rotation about its own axis and (b) with its revolution about the Sun? See Appendix E, and consider Earth to be a uniform sphere.

25. A weightlifter's barbell consists of two 25-kg masses on the ends of a 15-kg rod 1.6 m long. The weightlifter holds the rod at its center and spins it at 10 rpm about an axis perpendicular to the rod. What is the magnitude of the barbell's angular momentum?

26. A particle of mass m moves in a straight line at constant speed v. Show that its angular momentum about a point located a perpendicular distance b from its line of motion is mvb regardless of where the particle is on the line.

27. Two identical 1800-kg cars are traveling in opposite directions at 90 km/h. Each car's center of mass is 3.0 m from the center of the highway (Fig. 13-30). What are the mag-

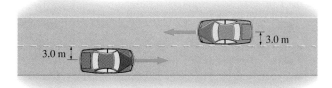

FIGURE 13-30 Problem 27.

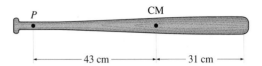

FIGURE 13-31 Problem 28.

nitude and direction of the angular momentum of the system consisting of the two cars, about a point on the centerline of the highway?

28. Figure 13-31 shows dimensions of a 880-g wooden baseball bat whose rotational inertia about its center of mass is 0.048 kg·m². If the bat is swung so its far end moves at 50 m/s, find (a) its angular momentum about the pivot point P and (b) the constant torque applied about P to achieve this angular momentum in 0.25 s. *Hint:* Remember the parallel axis theorem.

29. A 1.0-kg particle is moving at a constant 3.5 m/s along the line $y = 0.62x + 1.4$, where x and y are in meters and where the motion is toward the positive x and y directions. Find its angular momentum (a) about the origin (b) about the point $x = 0$, $y = 1.4$, and (c) about the point $x = 2.0$, $y = 4.8$.

Section 13-4 Conservation of Angular Momentum

30. A potter's wheel, with rotational inertia 6.40 kg·m², is spinning freely at 19.0 rpm. The potter drops a 2.70-kg lump of clay onto the wheel, where it sticks a distance of 46.0 cm from the rotation axis. What is the subsequent angular speed of the wheel?

31. A 3.0-m-diameter merry-go-round with rotational inertia 120 kg·m² is spinning freely at 0.50 rev/s. Four 25-kg children sit suddenly on the edge of the merry-go-round. (a) Find the new angular speed, and (b) determine the total energy lost to friction between the children and the merry-go-round.

32. Two ice skaters, both of mass 60 kg, approach on parallel paths 1.4 m apart. Both are moving at 3.2 m/s with their arms outstretched (Fig. 13-32). They join hands as they pass, still maintaining their 1.4-m separation, and begin rotating about one another. What is their angular speed?

33. In Fig. 13-33 the lower disk, of mass 440 g and radius 3.5 cm, is rotating at 180 rpm on a frictionless shaft of negligible radius. The upper disk, of mass 270 g and radius 2.3 cm, is initially not rotating. It drops freely down the shaft onto the lower disk, and frictional forces act to bring the two disks to a common rotational speed. (a) What is that speed? (b) What fraction of the initial kinetic energy is lost to friction?

34. In the apparatus shown in Fig. 13-33, the lower disk is initially spinning about a frictionless shaft while the upper disk is stationary. The upper disk drops onto the lower one, and they come to a common angular speed. If one-third of the initial energy is lost in the process, how do the rotational inertias of the two disks compare?

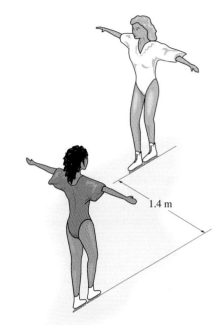

FIGURE 13-32 Problem 32.

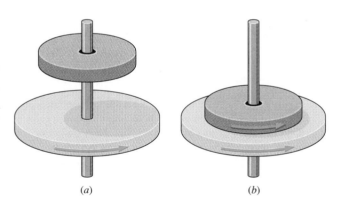

FIGURE 13-33 Problems 33, 34.

35. A uniform, spherical cloud of interstellar gas has mass 2.0×10^{30} kg and radius 1.0×10^{13} m, and it is rotating with period 1.4×10^6 years. If the cloud collapses to form a star 7.0×10^8 m in radius, what will be the star's rotation period?

36. A skater's body has rotational inertia 4.2 kg·m² with his fists held to his chest and 5.7 kg·m² with arms outstretched. The skater is twirling at 3.0 rev/s while holding 2.5-kg weights in each outstretched hand; the weights are 76 cm from the his rotation axis. If he pulls his hands in to his chest, how fast will he be twirling?

37. A turntable of radius 25 cm and rotational inertia 0.0154 kg·m² is spinning freely at 22.0 rpm about its central axis, with a 19.5-g mouse on its outer edge. The mouse walks from the edge to the center. Find (a) the new rotation speed and (b) the work done by the mouse.

38. A 17-kg dog is standing on the edge of a stationary, frictionless turntable of rotational inertia 95 kg·m² and radius 1.81 m. The dog walks once around the turntable. What fraction of a full circle does the dog's motion make with respect to the ground?

39. Two small beads of mass m are free to slide on a frictionless rod of mass M and length ℓ, as shown in Fig. 13-34. Initially the beads are held together at the rod center, and the rod is set spinning freely with initial angular speed ω_0 about a vertical axis coming out of the page in Fig. 13-34. The beads are released, and they slide to the ends of the rod and then off. Find the expressions for the angular speed of the rod (a) when the beads are halfway to the ends of the rod, (b) when they're at the ends, and (c) after the beads are gone. *Hint:* Two of the answers are the same. Why?

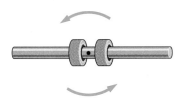

FIGURE 13-34 Problem 39.

40. A physics student is standing on an initially motionless, frictionless turntable with rotational inertia 0.31 kg·m². He's holding a wheel of rotational inertia 0.22 kg·m² spinning at 130 rpm about a vertical axis, as shown in Fig. 13-35. When he turns the wheel upside down, student and turntable begin rotating at 70 rpm. (a) What is the student's mass, considering him to be a cylinder 30 cm in diameter? (b) How much work did he do in turning the wheel upside down? Neglect the distance between the axes of the turntable and wheel.

FIGURE 13-35 Problem 40.

41. Eight 60-kg skaters join hands and skate down an ice rink at 4.6 m/s. Side by side, they form a line 12 m long. The skater at one end stops abruptly, and the line proceeds to rotate rigidly about that skater. Find (a) the angular speed, (b) the linear speed of the outermost skater, and (c) the force that must be exerted on the outermost skater.

Section 13-5 Rotational Dynamics in Three Dimensions

42. A gyroscope with rotational inertia 6.8×10^{-3} kg·m² is precessing in a horizontal plane at 0.42 rad/s under the influence of a 0.31 N·m torque caused by a force pointing vertically downward. How fast is the gyroscope spinning on its axis?

43. A gyroscope consists of a solid disk of radius 7.5 cm, mounted on one end of a shaft of negligible mass. The far end of the shaft is placed on a frictionless pivot and the disk set spinning at 950 rpm. The gyroscope then precesses at 2.1 rad/s. How long is the shaft?

44. Show that the final equality of Equation 13-7 holds for a gyroscope whose angular momentum vector is not necessarily in a horizontal plane.

45. A gyroscope consists of a disk and shaft mounted on frictionless bearings in a frame of diameter d, as shown in Fig. 13-36. Initially the gyroscope is spinning with angular speed ω and is perfectly balanced so it's not precessing. When a mass m is hung from the frame at one of the shaft bearings, the gyroscope precesses about a vertical axis with angular speed Ω. Find an expression for the rotational inertia of the gyroscope.

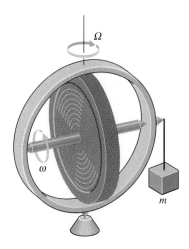

FIGURE 13-36 Problem 45.

46. Because it bulges slightly at the equator, Earth experiences a gravitational torque that causes its axis to precess with a period of about 26,000 years. Use this fact, and the result of Problem 24, to estimate the magnitude of the torque.

Section 13-6 Angular Momentum at the Subatomic Level

47. A crude model for the hydrogen atom has the electron in a circular orbit 52.9 pm from the proton, moving at 2.18 Mm/s. What would be the angular momentum associated with this electron orbit? How does your answer compare with the fundamental unit of angular momentum $\hbar$?

48. An oxygen molecule can be modeled crudely as a pair of point masses (the two oxygen atoms) each with mass 2.7×10^{-26} kg, separated by 120 pm. What should be its angular speed about its center of mass if the molecule is to possess exactly two fundamental units ($2\hbar$) of angular momentum?

Paired Problems

(Both problems in a pair involve the same principles and techniques. If you can get the first problem, you should be able to solve the second one.)

49. The dot product of a pair of vectors is twice the magnitude of their cross product. What is the angle between the vectors?

50. A force **F** applied at the point $x = 2.0$ m, $y = 0$ produces a torque $\boldsymbol{\tau} = 4.6\hat{\mathbf{k}}$ N·m about the origin. If the x component of **F** is 3.1 N, what angle does it make with the x axis?

51. A student's rotational inertia about a vertical axis through his center is 4.5 kg·m^2 with his arms held to his chest and 5.6 kg·m^2 with his arms outstretched. In a physics demonstration, the student stands on a turntable rotating at 1.0 rev/s, clutching two 7.5-kg weights to his chest. The turntable's rotational inertia is 4.0 kg·m^2. If the student extends his arms fully so the weights are each 95 cm from his rotation axis, what will be his new angular speed?

52. Five ice skaters of equal mass join hands. With their elbows bent they form a tight circle 1.0 m in diameter, rotating once every 1.3 s about its center. If the skaters now extend their arms so the circle grows to 2.1 m in diameter, what is the new rotation period?

53. The turntable in Fig. 13-37 has rotational inertia 0.021 kg·m^2, and is rotating at 0.29 rad/s about a frictionless vertical axis. A wad of clay is tossed onto the turntable and sticks 15 cm from the rotation axis. The clay impacts with horizontal velocity component 1.3 m/s, at right angles to the turntable's radius, and in a direction that opposes the rotation, as suggested in Fig. 13-37. After the clay lands the turntable has slowed to 0.085 rad/s. Find the mass of the clay.

54. A uniform, solid, spherical asteroid with mass 1.2×10^{13} kg and radius 1.0 km is rotating with a period of 4.3 hours. A meteoroid moving in the asteroid's equatorial plane crashes into the equator at 8.4 km/s. It impacts at a 58° angle to the vertical and embeds itself at the surface. After the impact the asteroid's rotation period is 3.9 hours. Find the meteoroid's mass.

55. A dog of mass m is standing on the edge of a stationary, frictionless turntable of rotational inertia I and radius R. The dog walks once around the turntable. What fraction of a full circle does the dog's motion make with respect to the ground?

56. A dog of mass m is standing on the edge of a stationary, frictionless turntable of rotational inertia I and radius R. On the ground next to the dog is a tree. The dog walks around the turntable until it's again next to the tree. How many times around the turntable has the dog walked?

Supplementary Problems

57. An advanced civilization lives on a solid spherical planet of uniform density. Running out of room for their expanding population, the civilization's government calls an engineering firm specializing in planetary reconfiguration. Without adding any material or angular momentum, the engineers reshape the planet into a hollow shell whose thickness is one-fifth of its outer radius. Find the ratio of the new to the old (a) surface area and (b) length of day.

58. A massless spring of spring constant k is mounted on a turntable of rotational inertia I, as shown in Fig. 13-38. The turntable is on a frictionless vertical axle, though initially it's not rotating. The spring is compressed a distance Δx from its equilibrium, with a mass m placed against it. When the spring is released, the mass leaves the spring moving at right angles to a line through the center of the turntable, at a distance b from the center, and slides without friction across the table and off. Find expressions for (a) the linear speed of the mass and (b) the rotational speed of the turntable. *Hint:* What's conserved?

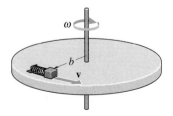

FIGURE 13-38 Problem 58.

59. If you're familiar with determinants, show that the cross product **A** × **B** can be written as a determinant:

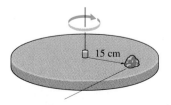

FIGURE 13-37 Problem 53.

$$\mathbf{A} \times \mathbf{B} = \begin{vmatrix} \hat{\mathbf{i}} & \hat{\mathbf{j}} & \hat{\mathbf{k}} \\ A_x & A_y & A_z \\ B_x & B_y & B_z \end{vmatrix}.$$

60. A 150-g baseball has angular momentum 7.7 kg·m²/s about a vertical axis through the batter. If it's moving at 43 m/s without spinning, by how much will it miss the batter's axis?

61. About 99.9% of the solar system's total mass lies in the Sun. Using data from Appendix E, estimate what fraction of the solar system's angular momentum about its center is associated with the Sun. Where is most of the rest of the angular momentum?

62. A 55-kg diver is spinning forward in a tuck position at 2.1 rev/s, as shown in Fig. 13-39a. In this position, the diver's rotational inertia is 3.9 kg·m². At the peak of her trajectory, the diver suddenly straightens out in a horizontal position as shown in Fig. 13-39b. In this new position, the diver has an overall length of 2.2 m with center of mass midway along this length, and a rotational inertia of 21 kg·m² about the rotation axis. What is the minimum height of the diver above the water if she is to enter the water hands first? Neglect the diver's horizontal translational motion.

63. A rod of length ℓ and mass M is suspended from a pivot, as shown in Fig. 13-40. The rod is struck midway along its length by a wad of putty of mass m moving horizontally at

FIGURE 13-40 Problem 63.

speed v. The putty sticks to the rod. Find an expression for the minimum speed v that will result in the rod's making a complete circle rather than swinging like a pendulum.

64. A 30-cm-diameter phonograph record is dropped onto a turntable being driven at $33\frac{1}{3}$ rpm. If the coefficient of friction between record and turntable is 0.19, how far will the turntable rotate between the time when the record first contacts it and when the record is rotating at the full $33\frac{1}{3}$ rpm? Assume that the record is a homogeneous disk. *Hint:* You'll need to do an integral to calculate the torque.

65. The contraption shown in Fig. 13-41 consists of two solid rubber wheels each of mass M and radius R mounted on an axle of negligible mass in a rigid square frame made of thin rods of mass m and length ℓ. The axle bearings are frictionless, and the wheels are mounted symmetrically about the center line of the frame, just far enough apart that they don't touch. The whole contraption is floating freely in space, and the frame is not rotating. The wheels are rotating with angular speed ω in the same direction, as shown. A mechanism built into the frame moves the axles very slightly so the wheels touch and frictional forces act between them. Describe quantitatively the motion of the system after the wheels have stopped.

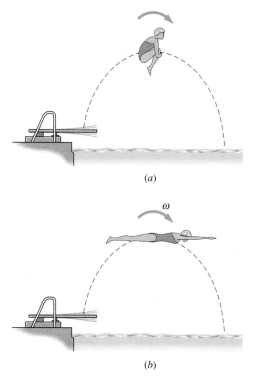

(a)

(b)

FIGURE 13-39 Problem 62.

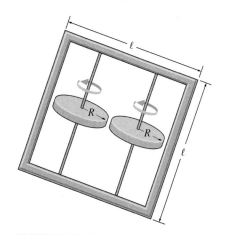

FIGURE 13-41 Problem 65.

CHAPTER 14

STATIC EQUILIBRIUM

A delicate balance of forces and torques keeps this tightrope artist in equilibrium.

W hat keeps the steelworker of Fig. 14-1 alive? Two things: First, the cable is strong enough that its tension force can balance the weight of the beam and steelworker, giving zero net force. Second, there's no tendency of the beam to rotate—that is, there is zero net torque on the system. Thus the steelworker remains safely in equilibrium. Understanding equilibrium is important in many fields of science and engineering; for example, it is crucial in designing bridges, buildings, and other structures so they won't fall down. This chapter explores conditions necesary for equilibrium and their application in practical situations.

14-1 CONDITIONS FOR EQUILIBRIUM

A body is in **equilibrium** when the net external force and torque on the body are both zero. According to Newton's law, it may still be moving or rotating uniformly. In the special case where the body is also at rest and not rotating, then it is in **static equilibrium:**

A body is in static equilibrium when it is stationary and when the net external force and torque on the body are both zero.

We can write the conditions for static equilibrium mathematically by setting the sums of all the external forces and torques both to zero:

$$\sum \mathbf{F}_i = \mathbf{0} \qquad (14\text{-}1)$$

and

$$\sum \boldsymbol{\tau}_i = \sum (\mathbf{r}_i \times \mathbf{F}_i) = \mathbf{0}. \qquad (14\text{-}2)$$

Here the subscripts i label the different forces that act on a body, the radius vectors from a pivot point to the force application points, and the associated torques.

There seems to be an ambiguity in Equation 14-2 since we haven't specified the point to be used in evaluating the torques. But a body in equilibrium can't rotate about *any* point, so Equation 14-2 must hold no matter what point we choose. Must we then apply the equation at every possible point? Fortunately not. If the first equilibrium condition ($\sum \mathbf{F}_i = \mathbf{0}$) holds, then it suffices to show that the sum of torques about any one point is zero. We can prove this using Fig. 14-2, which shows a body subject to several forces whose vector sum is zero. Suppose also that the net torque about point O is zero. What's the torque about point P? We can write this torque

$$\boldsymbol{\tau}_P = \sum (\mathbf{r}_{Pi} \times \mathbf{F}_i), \qquad (14\text{-}3)$$

where $\mathbf{r}_{Pi}$ is the vector from P to the application point of $\mathbf{F}_i$. Let $\mathbf{R}$ be the vector from P to O. Then, as suggested in Fig. 14-2 for the case $i = 2$, we can write

$$\mathbf{r}_{Pi} = \mathbf{R} + \mathbf{r}_{Oi},$$

where $\mathbf{r}_{Oi}$ is the vector from O to the application point of $\mathbf{F}_i$. Using this expression in Equation 14-3, we have

$$\boldsymbol{\tau}_P = \sum (\mathbf{r}_{Oi} + \mathbf{R}) \times \mathbf{F}_i$$
$$= \sum (\mathbf{r}_{Oi} \times \mathbf{F}_i) + \sum (\mathbf{R} \times \mathbf{F}_i) = \sum (\mathbf{r}_{Oi} \times \mathbf{F}_i) + \mathbf{R} \times \sum \mathbf{F}_i, \qquad (14\text{-}4)$$

where, in the last step, we have taken $\mathbf{R}$ outside the summation since it's the same vector for each term in the sum. The first term on the rightmost side of Equation 14-4 is just the net torque about O, which we've assumed to be zero. The second term is the cross product of the vector $\mathbf{R}$ with the net force on the body. But we've assumed the net force to be zero. Thus the net torque about P is also zero. Since the location of P is quite arbitrary, our result shows that if the net force on an object is zero, and the net torque about *some* point is zero, then the net torque about *any* point is also zero.

This result is useful in solving equilibrium problems because it allows us to choose any convenient point about which to evaluate the torques. An appropriate point is usually the application point of one force, because then the torque due to that force is zero. This leaves Equation 14-2 with one fewer term than it would otherwise have.

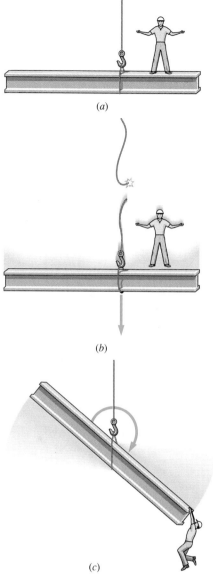

FIGURE 14-1 (*a*) With no net force or torque, steelworker and beam are in static equilibrium. A nonzero net force (*b*) or nonzero net torque (*c*) results in disaster.

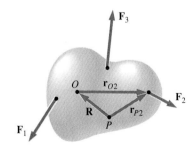

FIGURE 14-2 The forces sum to zero, and the torques about O sum to zero. Is the net torque about P also zero?

● EXAMPLE 14-1 CHOOSING THE PIVOT POINT

A 5.0-m-long canoe is tied at bow and stern to a pair of posts so it rides parallel to the shore, as shown in Fig. 14-3. The tie ropes are perpendicular to the shore. A paddler boards the canoe 1.0 m from the stern, in the process exerting an outward force of 200 N at right angles to the shore. Find the tension in the bow ropes if the canoe remains in static equilibrium in the horizontal plane.

Solution

All three horizontal forces are perpendicular to the shore, so the condition that there be no net force (Equation 14-1) becomes

$$T_b + T_s - 200 \text{ N} = 0,$$

where we take the shoreward direction as positive. Choosing the application point of one of the unknown tensions as our pivot will eliminate that unknown from the torque equation. If we choose the stern, the condition that there be no net torque (Equation 14-2) becomes

$$(5.0 \text{ m})(T_b) - (200 \text{ N})(1.0 \text{ m}) = 0,$$

where the first term is the torque due to the bow-rope tension and the second is due to the force of the paddler. Here we take a positive torque as one tending to produce counterclockwise rotation (torque vector out of the page in Fig. 14-3), so the paddler torque is negative. Solving for T_b gives

$$T_b = \frac{(200 \text{ N})(1.0 \text{ m})}{5.0 \text{ m}} = 40 \text{ N}.$$

Suppose we had chosen the center of the canoe as our pivot point. The force-balance equation would be unchanged, but the torque equation would now contain three terms:

$$(2.5 \text{ m})(T_b) + (1.5 \text{ m})(200 \text{ N}) - (2.5 \text{ m})(T_s) = 0.$$

The three terms are due to the bow rope, the paddler, and the stern rope, respectively, and the distances are measured from the canoe center. Note that now the bow-rope and paddler torques have the same sign, since both tend to produce counterclockwise rotation about the canoe center. We can solve the force equation for T_s to get

$$T_s = 200 \text{ N} - T_b.$$

Using this result in the torque equation gives

$$(2.5 \text{ m})(T_b) + (1.5 \text{ m})(200 \text{ N}) - (2.5 \text{ m})(200 \text{ N} - T_b) = 0,$$

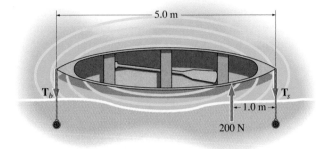

FIGURE 14-3 Top view of the canoe in Example 14-1.

so

$$T_b = \frac{(2.5 \text{ m})(200 \text{ N}) - (1.5 \text{ m})(200 \text{ N})}{5.0 \text{ m}} = 40 \text{ N},$$

as before. Although we get the same answer, the algebra is considerably more complicated with this choice of pivot point.

EXERCISE Inside a piece of machinery, a metal rod is secured in the horizontal plane by three horizontal springs, as shown in Fig. 14-4. Find the tension force $\mathbf{F}_3$ and point of attachment x for the upper spring that will maintain the rod in static equilibrium.

Answers: $F_3 = 14$ N, upward in Fig. 14-4; $x = 2.9$ cm

Some problems similar to Example 14-1: 2, 8, 13

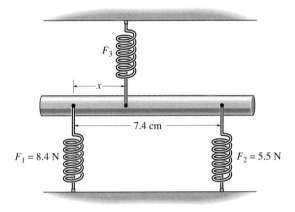

FIGURE 14-4 Where to attach $\mathbf{F}_3$? How big should it be? (The entire structure lies in the horizontal plane.)

14-2 CENTER OF GRAVITY

One force often present is gravity, which acts on all parts of a body (Fig. 14-5). The vector sum of these gravitational forces is just $M\mathbf{g}$. But what about the torques due to the gravitational forces acting on different parts of the body? Summing these torques, we have

$$\boldsymbol{\tau} = \sum \mathbf{r}_i \times \mathbf{F}_i = \sum \mathbf{r}_i \times m_i\mathbf{g} = \left(\sum m_i\mathbf{r}_i\right) \times \mathbf{g},$$

where $\mathbf{F}_i$ is the gravitational force on the ith mass element, and $\mathbf{r}_i$ is the vector from some pivot point to that mass element (Fig. 14-6). We can rewrite this equation by multiplying the right-hand side by M/M, where M is the total mass of the body:

$$\boldsymbol{\tau} = \left(\frac{\sum m_i\mathbf{r}_i}{M}\right) \times M\mathbf{g}.$$

FIGURE 14-5 The gravitational force acts everywhere on an extended body.

The term in parentheses is just the center of mass (see Section 10-1), while the right-hand term is the total weight of the body. Therefore, the net torque on the body due to gravity is just that of the gravitational force $M\mathbf{g}$ acting at the center of mass. In general, the point at which the gravitational force seems to act is called the **center of gravity.** We have just proven that the center of gravity coincides with the center of mass when the gravitational field is uniform—that is, when $\mathbf{g}$ is the same for all mass elements in the body. If a body is large enough that the variation in magnitude or direction of $\mathbf{g}$ is significant, then the two points may not coincide (see Problem 59), although this is not a situation we usually encounter.

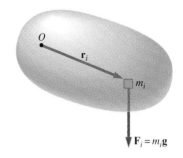

FIGURE 14-6 The gravitational force on the mass element m_i produces a torque about the point O.

We can easily locate the center of gravity by suspending an object from a string attached to its edge. The torque about the suspension point is due entirely to gravity and is zero only if the center of gravity lies directly beneath that point (Fig. 14-7). In equilibrium, a vertical line from the suspension point must therefore pass through the center of gravity. If we now change the suspension

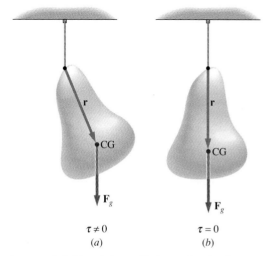

FIGURE 14-7 A suspended object is in equilibrium only when its center of gravity lies directly below the suspension point.

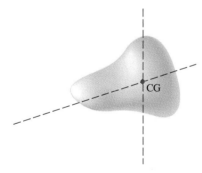

FIGURE 14-8 The intersection of vertical lines from two suspension points marks the center of gravity.

point, we can determine another line that passes through the center of gravity. These two lines intersect in a single point, which must be the center of gravity (Fig. 14-8).

14-3 EXAMPLES OF STATIC EQUILIBRIUM

Example 14-1 is particularly simple because all the forces have a single common nonzero component (perpendicular to shore), and so do the torques (vertical), reducing the equilibrium conditions to two equations in two unknowns. More generally, other components of Equations 14-1 and 14-2 give rise to additional equations. A common situation involves forces that lie in a plane; then the force equation has two components while the torque equation still has only one, so we end up with three equations in three unknowns.

Whatever the geometry, solving a static equilibrium problem is much like solving a Newton's law problem; after all, the equations for static equilibrium are Newton's law and its rotational analog, with the linear and angular acceleration set to zero. Here we adapt the strategy for Newton's law problems given in Chapter 6 to problems of static equilibrium:

> ### **STRATEGY** Static Equilibrium Problems
>
> 1. Identify the object in equilibrium.
> 2. Identify all the forces acting on the object, and draw a free-body diagram showing the force vectors. If the direction of any force is an unknown, draw a force vector you think looks appropriate and let the algebra take care of signs and angles.
> 3. Choose an appropriate coordinate system, and resolve all vectors into components.
> 4. Write Equation 14-1 (the condition that the forces sum to zero) as a set of equations, one for each component.
> 5. Choose a pivot point for calculating torques. The choice is yours, but a smart choice—usually at the application point of one of the forces—will simplify the algebra.
> 6. Calculate the torque associated with each force, and formulate Equation 14-2 (the condition that the torques sum to zero).
> 7. Solve your system of force and torque equations for the unknowns of the problem.

● **EXAMPLE 14-2** A DROP-LEAF TABLE

A dining table extends with a leaf of length ℓ supported by a diagonal brace at angle $\theta = 40°$, as shown in Fig. 14-9a. If the maximum load is 120 N, centered on the center of the leaf, what is the force exerted by the hinge that connects the leaf to the table?

Solution

In Fig. 14-9b we have drawn the forces acting on the leaf; these include the force $\mathbf{F}_1$ of the hinge, the load weight $\mathbf{W}$, and the compression force $\mathbf{F}_2$ of the diagonal brace. Equilibrium requires that the forces sum to zero, giving two component equations:

horizontal: $\qquad F_{1x} + F_2\cos\theta = 0$

vertical: $\qquad F_{1y} - W + F_2\sin\theta = 0$

on the coordinate system shown in Fig. 14-9b. It's convenient to take the pivot at the left end of the overhang; then $\mathbf{F}_1$ doesn't enter the torque equation. All the torques are perpendicular to the plane of the page, so the torque equation has a single component:

$$-\frac{1}{2}\ell W + F_2\ell\sin\theta = 0,$$

where the first term is the torque due to the load weight, applied at $x = \frac{1}{2}\ell$ and perpendicular to the radius vector from the pivot, and the second is due to the force $\mathbf{F}_2$ of the diagonal brace. The load torque is negative since it points into the page in Fig. 14-9, or in the $-z$ direction.

Solving the torque equation for F_2 gives $F_2 = W/2\sin\theta$; using this result in the horizontal force equation gives

$$F_{1x} = -F_2\cos\theta = -\frac{W\cos\theta}{2\sin\theta} = -\frac{1}{2}W\cot\theta$$

$$= -\left(\frac{1}{2}\right)(120\ \text{N})\cot 40° = -71.5\ \text{N}.$$

Does this make sense? With the positive x direction to the right, F_{1x} had better be negative to balance the rightward component from the diagonal brace. Similarly, the vertical equation gives

$$F_{1y} = W - F_2\sin\theta = W - \frac{1}{2}W = \frac{1}{2}W = 60\ \text{N}.$$

This, too, makes sense. With the load weight in the center of the leaf, the vertical force components at the two ends should each equal half the load. Finally, we can find the magnitude of the hinge force $\mathbf{F}_1$:

$$F_1 = \sqrt{F_{1x}^2 + F_{1y}^2} = \sqrt{(-71.5\ \text{N})^2 + (60\ \text{N})^2} = 93\ \text{N}.$$

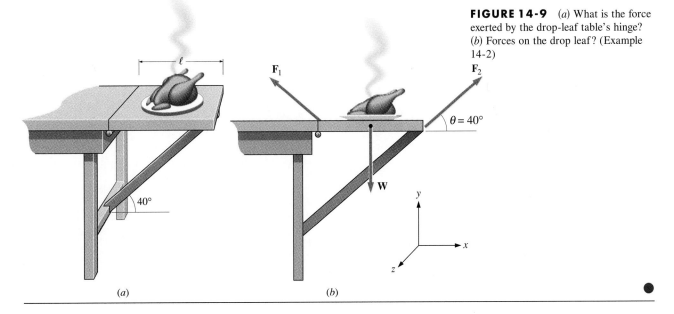

FIGURE 14-9 (a) What is the force exerted by the drop-leaf table's hinge? (b) Forces on the drop leaf? (Example 14-2)

● **EXAMPLE 14-3** A DRAWBRIDGE

The raised span of the drawbridge shown in Fig. 14-10a has its mass of 11,000 kg distributed uniformly over its 14-m length. Find the magnitude of the tension in the supporting cable and the magnitude and direction of the force exerted by the hinge.

Solution

Figure 14-10b shows the forces on the raised span. Resolving the forces into components and summing them to zero gives

horizontal: $\qquad F_h\cos\phi_1 - T\cos\phi_2 = 0$

vertical: $\qquad F_h\sin\phi_1 - mg - T\sin\phi_2 = 0.$

Choosing the pivot at the hinge eliminates the hinge force from the torque equation. Figure 14-10b shows that the angles θ_1 and θ_2 for calculating the torques are 120° and 165°, respectively. The two individual torques are both perpendicular to the plane of Fig. 14-10b, so the torque equation has only a single component:

$$-\frac{L}{2}mg\sin\theta_1 + LT\sin\theta_2 = 0,$$

where L is the length of the span and where $L/2$ appears in the first term since the gravitational force acts at the center of mass, halfway along the span. The minus sign appears because the gravitational torque is into the page in Fig. 14-10b.

We're now ready to solve for the unknowns F_h, T, and ϕ_1. Our choice of pivot leaves the single unknown T in the torque equation, which we can solve to get

$$T = \frac{mg\sin\theta_1}{2\sin\theta_2} = \frac{(11{,}000 \text{ kg})(9.8 \text{ m/s}^2)(\sin 120°)}{(2)(\sin 165°)}$$
$$= 1.80 \times 10^5 \text{ N}.$$

If we isolate the terms in the unknowns F_h and ϕ_1 on the left-hand sides of the two force equations and divide the resulting vertical equation by the horizontal equation, we have

$$\frac{\sin\theta_1}{\cos\theta_1} \equiv \tan\theta_1 = \frac{mg + T\sin\phi_2}{T\cos\phi_2}$$
$$= \frac{(11{,}000 \text{ kg})(9.8 \text{ m/s}) + (1.80\times10^5 \text{ N})(\sin 15°)}{(1.80\times10^5 \text{ N})(\cos 15°)} = 0.887,$$

giving $\phi_1 = \tan^{-1}(0.887) = 41.6°$. Finally, we can solve the x component force equation for F_h:

$$F_h = \frac{T\cos\phi_2}{\cos\phi_1} = \frac{(1.80\times10^5 \text{ N})(\cos 15°)}{\cos 41.6°} = 2.3\times10^5 \text{ N}.$$

Note that both the hinge force and the cable tension are considerably larger than the 1.1×10^5-N weight of the bridge span. Engineers are responsible for ensuring that the bridge components can handle these loads.

Some problems similar to Examples 14-2 and 14-3: 29, 41, 58

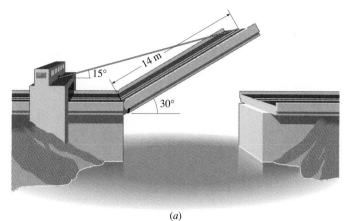

(a)

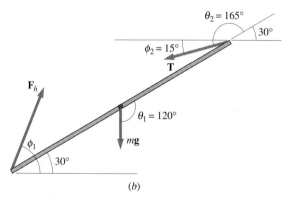

(b)

FIGURE 14-10 (a) How should the drawbridge cable and hinge be engineered to ensure adequate strength? (b) Side view of the bridge span, showing the force $\mathbf{F}_h$ exerted by the hinge, the gravitational force $m\mathbf{g}$ acting at the span's center of gravity, and the cable tension $\mathbf{T}$. Also shown are the unknown angle ϕ_1 and the angles θ_1 and θ_2 used in calculating the torques; these follow from the 30° inclination angle and the 15° cable angle.

● **EXAMPLE 14-4** LEANING A BOARD

A board of mass m and length L is leaning against a wall, as shown in Fig. 14-11a. The wall is frictionless, and the coefficient of static friction between board and floor is μ. Find an expression for the minimum angle ϕ at which the board can be leaned without slipping.

Solution

Figure 14-11b shows the force diagram for the board. In addition to the board's weight $m\mathbf{g}$, there are normal forces $\mathbf{F}_1$ and $\mathbf{F}_2$ at the floor and wall, and a frictional force at the floor whose maximum possible magnitude μF_1 corresponds to the minimum board angle. Then the components of the force equation are

horizontal: $\mu F_1 - F_2 = 0$
vertical: $F_1 - mg = 0$.

A smart choice for the pivot point is the bottom of the board since then we eliminate two forces from the torque equation. The remaining torques are due to $\mathbf{F}_2$ and the gravitational force $m\mathbf{g}$. Taking a torque out of the page as positive, we have

$$LF_2\sin\phi - \frac{L}{2}mg\cos\phi = 0,$$

where Fig. 14-11b shows how the angles used to calculate the torques are related to the board angle ϕ. The actual angle

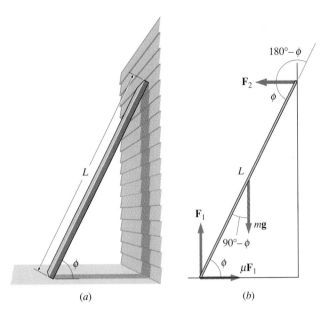

(a)   (b)

FIGURE 14-11 (*a*) At what angle will the board slip? (*b*) Forces on the board include its weight, the normal forces $\mathbf{F}_1$ and $\mathbf{F}_2$ at the floor and wall, and the frictional force of magnitude μF_1 at the floor. (Absence of friction at the wall means the contact force with the wall is entirely normal.) The angle used in calculating the gravitational torque is actually the supplement of $90 - \phi$, but the two have the same sine, namely $\sin(90° - \phi) = \cos\phi$ (Example 14-4).

for the first term is $180° - \phi$, but we wrote $\sin\phi$ since $\sin(180° - \phi) = \sin\phi$.

Having written the force and torque equations, we've completed the essential physics. The rest is algebra. The vertical force equation gives $F_1 = mg$; substituting this in the horizontal force equation then gives $F_2 = \mu mg$. Substituting this result in the torque equation then allows us to solve for the tangent of the unknown angle:

$$\frac{\sin\phi}{\cos\phi} \equiv \tan\phi = \frac{mg}{2F_2} = \frac{mg}{2\mu mg} = \frac{1}{2\mu},$$

giving $\phi = \tan^{-1}(1/2\mu)$.

This is a remarkably simple result—it doesn't depend on the mass or length of the board, nor on g. Does it make sense? As the coefficient of friction goes to zero, $\tan\phi$ grows without bound, and ϕ approaches 90°. With no friction, that says, we can lean the board against the wall only if it's vertical. Increasing the friction, on the other hand, allows us to lean the board at ever lower angles.

EXERCISE Climbers attempting to cross a stream place a 340-kg log against a vertical, frictionless ice cliff on the opposite side (Fig. 14-12). The log is inclined at 27°, and its center of gravity is one-third of the way along its 6.3-m length. If the coefficient of friction between the left end of the log and the ground is 0.92, what is the maximum mass for a climber and pack to cross without the log slipping?

Answer: 87 kg

FIGURE 14-12 How heavy a climber can cross the log? Dot marks the log's center of gravity.

Some problems similar to Example 14-4: 24, 28, 43, 44, 50

● **EXAMPLE 14-5** EQUILIBRIUM IN THE HUMAN BODY

Figure 14-13 depicts a human arm. The forearm measures 32 cm from elbow joint to center of palm and has a mass $m = 2.7$ kg, with center of mass located 14 cm from the elbow joint. The biceps muscle attaches to the radius (forearm bone) 3.6 cm from the elbow joint. With the humerus (upper arm bone) vertical and the forearm horizontal, the biceps muscle makes an 80° angle with the horizontal. If a pumpkin of mass

$M = 4.5$ kg rests on the palm, what are the tension in the biceps and the force exerted on the elbow joint?

Solution
Figure 14-14 shows the forces acting on the forearm, including the weights $m\mathbf{g}$ and $M\mathbf{g}$ of the arm and pumpkin, the muscle

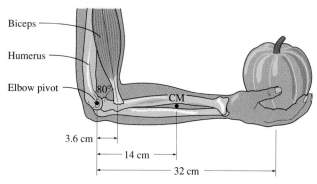

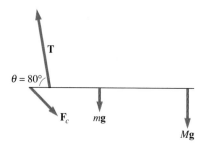

FIGURE 14-13 A human arm.

FIGURE 14-14 Forces on the forearm include its own weight $m\mathbf{g}$, the pumpkin weight $M\mathbf{g}$, the muscle tension $\mathbf{T}$, and the elbow contact force $\mathbf{F}_c$.

tension $\mathbf{T}$, and the contact force $\mathbf{F}_c$. We can read the horizontal and vertical components of force balance from the diagram:

horizontal: $\quad\quad\quad F_{cx} - T\cos\theta = 0$
vertical: $\quad T\sin\theta - F_{cy} - mg - Mg = 0.$

TIP You Can Avoid Solving for Trig Functions
Here we've chosen to treat the two components of the contact force as unknowns rather than its magnitude and direction as we did with the hinge force in Example 14-2. That way we avoid solving for trig functions, but in the end we'll have to calculate the magnitude of the force from its components. It doesn't make much difference in these examples, but sometimes it's more straightforward if you can avoid solving for an unknown angle.

Choosing the pivot at the elbow gives a torque equation in which the contact force doesn't appear:

$$d_1 T\sin\theta - d_2 mg - d_3 Mg = 0,$$

where the d's are the distances from the elbow to the force application points. Solving for the biceps tension T gives

$$T = \frac{(d_2 m + d_3 M)g}{d_1 \sin\theta}$$
$$= \frac{[(0.14\text{ m})(2.7\text{ kg}) + (0.32\text{ m})(4.5\text{ kg})](9.8\text{ N/kg})}{(0.036\text{ m})(\sin 80°)}$$
$$= 500\text{ N}.$$

The horizontal and vertical force equations then give the components of the elbow contact force:

$$F_{cx} = T\cos\theta = (500\text{ N})(\cos 80°) = 87\text{ N},$$

and

$$F_{cy} = T\sin\theta - (m + M)g$$
$$= (500\text{ N})(\sin 80°) - (2.7\text{ kg} + 4.5\text{ kg})(9.8\text{ N/kg})$$
$$= 420\text{ N},$$

making the total force on the elbow $F_c = \sqrt{(87\text{ N})^2 + (420\text{ N})^2} = 430\text{ N}$. This example shows how the human body often sustains forces far larger than the weights of objects it may be lifting.

EXERCISE A 95-kg horizontal tree branch has its center of mass 2.2 m from the tree trunk. A rope helps support the branch, attached as shown in Fig. 14-15. A 4.5-kg swing hangs from a point 7.3 m out along the branch. How massive a child can sit stationary on the swing if the rope tension is not to exceed 1750 N? Neglect any supporting torque where the branch joins the trunk.

Answer: 33.4 kg.

Some problems similar to Example 14-5: 19, 39, 40

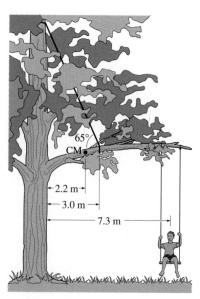

FIGURE 14-15 How heavy a child can sit on the swing?

■ **APPLICATION** RESTORING THE STATUE OF LIBERTY

In 1986, workers completed major renovation of the Statue of Liberty, France's famous gift to the United States that was first dedicated in 1886. Although its designer, French sculptor Frédéric-Auguste Bartholdi, suggested that his creation should last as long as Egypt's pyramids, many factors conspired to make major renovation necessary after only one century. These include corrosion from air pollution and from a chemical reaction between the iron framework and the copper skin, as well as an assembly change that resulted in excess torques on the statue's structural members.

Sculptor Bartholdi was no engineer, and without the work of the French engineer Eiffel (of tower fame) the statue could not have maintained itself in static equilibrium. Eiffel designed an inner skeleton of iron to provide the forces necessary to counteract the weights and torques associated with components of the statue, as well as with the force exerted by wind (Fig. 14-16). The statue was constructed in France in 300 separate pieces, then shipped to New York. During assembly, probably as a conscious aesthetic decision, Liberty's head and upraised arm were mounted 2 feet from their locations on Eiffel's plan,

with the arm making a greater angle to the vertical than planned (Fig 14-17). This made the arm exert a much greater torque about the shoulder than planned, resulting in greater forces on structural components. For historical integrity, renovators did not return to Eiffel's original design; instead, they reinforced the support structure to withstand better the excess forces and torques.

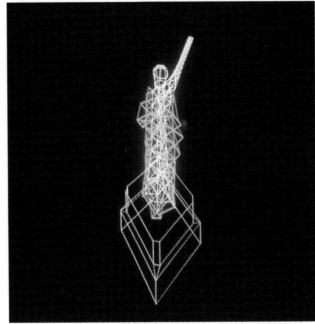

FIGURE 14-16 The Statue of Liberty's interior skeleton counteracts forces and torques acting on the statue. Computer-drawn images of Liberty's skeleton helped engineers and architects plan its renovation.

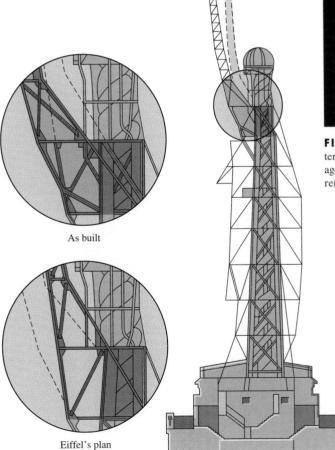

As built

Eiffel's plan

FIGURE 14-17 The statue's head and arm are offset from their planned positions, resulting in greater than anticipated forces and torques on structural members.

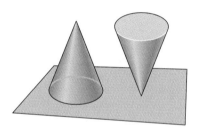

FIGURE 14-18 Stable (left) and unstable (right) equilibria.

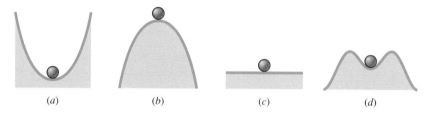

FIGURE 14-19 (*a*) Stable, (*b*) unstable, (*c*) neutrally stable, and (*d*) metastable equilibria.

14-4 STABILITY OF EQUILIBRIA

What if a body is disturbed from static equilibrium? Then, in general, it experiences nonzero net forces or torques, and it accelerates or begins to rotate. Figure 14-18 shows two very different possibilities for the subsequent behavior of a body displaced slightly from equilibrium. On the left, a slight displacement results in a torque that moves the body back toward its equilibrium position. Although the body may rock back and forth for a while, any friction or other dissipation will soon sap the body's energy and it will settle back into its original equilibrium. On the right in Fig.14-18, the torque arising from a slight displacement swings the body away from its original equilibrium. The former situation is an example of **stable equilibrium,** the latter of **unstable equilibrium.** Nearly all the equilibria we encounter in nature are stable, for a body in unstable equilibrium will not remain so for any significant time. The slightest disturbance will set it in motion, and it will eventually stop in a very different equilibrium state.

The distinction between stable and unstable equilibria is often subtler than Fig. 14-18 would imply. Figure 14-19 shows a ball in four different equilibrium situations. Clearly situation (*a*) is stable, while (*b*) is unstable. Situation (*c*) is neither stable nor unstable; it is termed **neutrally stable.** But what about situation (*d*)? For very small disturbances, the ball will return to its original state, so the equilibrium is stable. But for larger disturbances—large enough to push the ball just over the highest points on the hill—it's unstable. Such an equilibrium is called **conditionally stable** or **metastable.** Many common equilibria are actually metastable, as Fig. 14-20 suggests.

Stability is closely associated with the potential energy of the system in equilibrium. In Fig. 14-19, for example, the shapes of the hills and valleys reflect the gravitational potential energy of the ball as a function of position. In all cases of equilibrium, the ball is at a minimum or maximum of the potential energy curve—at a place where the force (that is, the derivative of potential energy with respect to position) is zero. For the stable and metastable equilibria, the potential energy at equilibrium is a minimum, at least with respect to positions immediately adjacent to equilibrium. A deviation from equilibrium requires that work be done against a net force that tends to restore the ball to its equilibrium position. The unstable equilibrium, in contrast, occurs at a maximum in potential energy. Here, a deviation from equilibrium results in lower potential energy and in a net force that tends to accelerate the ball farther from

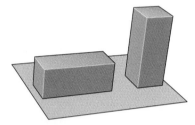

FIGURE 14-20 Both equilibria are stable against very small disturbances, but a larger disturbance will knock the vertical block over. The vertical equilibrium is therefore metastable.

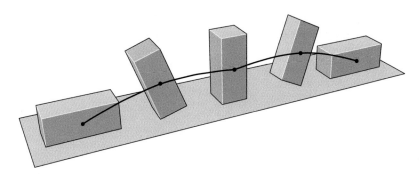

FIGURE 14-21 As the block is tilted, its center of mass traces out a potential energy curve. Shown are two stable equilibria (ends), a metastable equilibrium (center), and two unstable equilibria (tilted blocks).

equilibrium. For the neutrally stable equilibrium, there's no change whatever in potential energy as we move away from equilibrium; consequently the ball experiences no force.

The potential energy curve for the rectangular block of Fig. 14-20 is less obvious. In both cases, tipping the block requires that the center of mass be raised, thereby raising the potential energy. Figure 14-21 shows the center of mass as we tilt the block. The path traced out by the center of mass is the potential energy curve for the block. From this curve we see immediately that there are two stable equilibria of lowest potential energy corresponding to the block lying on either side. Similarly, there are two metastable states corresponding to the block standing vertically on either end. Between are unstable equilibria that occur when the block is balanced on a corner with its center of mass just over the balance point. Disturbed from these unstable states, the block experiences a net torque that rotates it toward one of the stable or metastable states.

We can sum up our understanding of equilibrium and potential energy in two simple mathematical statements. First, the force must be zero; that requires a local maximum or minimum in the potential energy:

$$\frac{dU}{dx} = 0, \qquad \text{(equilibrium condition)} \qquad (14\text{-}5)$$

where U is the potential energy of a system and x is a variable describing the system's configuration. For the simple systems we've been considering, x measures the position or orientation of an object, but with more complicated systems it may represent other quantities such as the system's volume or even its composition. For a stable equilibrium, we require a local minimum, so the potential energy curve is concave upward. Mathematically,

$$\frac{d^2U}{dx^2} > 0. \qquad \text{(stable equilibrium)} \qquad (14\text{-}6)$$

This condition applies to metastable equilibria as well, for they are *locally stable*. In contrast, unstable equilibrium occurs where the potential energy has a local maximum, or

$$\frac{d^2U}{dx^2} < 0. \quad \text{(unstable equilibrium)} \quad (14\text{-}7)$$

The intermediate case $d^2U/dx^2 = 0$ corresponds to neutral stability.

● EXAMPLE 14-6   SEMICONDUCTOR ENGINEERING

Solid-state physicists developing a new semiconductor device theorize that the potential energy of an electron in a region of the device should be given by

$$U(x) = ax^2 - bx^4,$$

where x is the electron's position measured in nm, U is its energy in aJ (1 aJ $= 10^{-18}$ J), and constants a and b are given by $a = 8$ aJ/nm^3 and $b = 1$ aJ/nm^4. Locate the equilibrium positions for the electron, and describe their stability.

Solution
Equilibrium occurs where the potential energy has a minimum or maximum, so $dU/dx = 0$. Taking the derivative of the potential energy and setting it to zero gives

$$0 = \frac{dU}{dx} = 2ax - 4bx^3 = 2x(a - 2bx^2) = 2x(8 - 2x^2)$$

$$= 4x(4 - x^2) \text{ aJ/nm},$$

where for mathematical clarity we left off the units except in the final answer. This equation is satisfied if $x = 0$ or $x = \pm 2$. Therefore, the electron is in equilibrium at $x = -2$ nm, at $x = 0$ nm, or at $x = 2$ nm. To determine the stability of these equilibria, we evaluate the second derivative at each equilibrium position:

$$\frac{d^2U}{dx^2} = 2a - 12bx^2 = (16 - 12x^2) \text{ aJ/nm}^2,$$

so

$$\left(\frac{d^2U}{dx^2}\right)_{x=\pm 2} = 16 - (12)(4) = -32 \text{ aJ/nm}^2$$

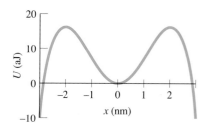

FIGURE 14-22   Potential energy curve for Example 14-6, showing that unstable equilibria occur at $x = \pm 2$ nm and stable equilibrium at $x = 0$.

and

$$\left(\frac{d^2U}{dx^2}\right)_{x=0} = 16 - 0 = 16 \text{ aJ/nm}^2.$$

From conditions 14-6 and 14-7, we see that the equilibria at $x = \pm 2$ nm are unstable, while the equilibrium at $x = 0$ nm is locally stable. Figure 14-22 is a plot of the potential energy curve, which makes clear the location and stability of the equilibria.

EXERCISE   A particle's potential energy as a function of position is given by $U = 2x^3 - 2x^2 - 7x + 10$, with x in meters and U in joules. Find the positions of any stable and unstable equilibria, and give the energy of each.

Answers: Stable at $x = 1.46$ m, $U = 1.74$ J; unstable at $x = -0.797$ m, $U = 13.3$ J

Some problems similar to Example 14-6:   34, 36, 37   ●

Stability considerations apply to the overall arrangement of the particles that comprise matter. A mixture of hydrogen and oxygen, for example, is in metastable equilibrium at room temperature. Lighting a match puts some atoms over the maxima in their potential energy curves, at which point they rearrange into a state of lower potential energy—the state we call H_2O. Similarly, a uranium nucleus is at a local minimum of its potential energy curve, and a little excess energy can result in its splitting into two smaller nuclei whose total potential energy is much lower. That transition from a less stable to a more stable equilibrium describes the basic physics of nuclear fission.

Potential energy curves for complex structures like molecules or skyscrapers can't be described fully with one-dimensional graphs. If potential energy varies in different ways when the structure is altered in different directions, then we need to consider all possible variations to determine stability. For example, a snowball sitting on a saddle-shaped mountain pass—or any system with a saddle-shaped potential energy curve—is stable against displacements in one direction but not another (Fig. 14-23). Stability analysis of complex physical systems, ranging from nuclei and molecules to bridges and buildings and machinery, and on to stars and galaxies, is an important part of contemporary work in engineering and science.

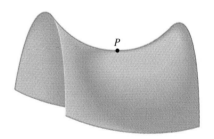

FIGURE 14-23 A saddle-shaped potential energy curve. Point P is an equilibrium position that's stable in one direction and unstable in another.

CHAPTER SYNOPSIS

Summary

1. A body at rest is in **static equilibrium** when the net external force and torque on the body are both zero:

$$\sum \mathbf{F} = 0$$

and

$$\sum \tau = 0.$$

If the first of these conditions is met, and the second holds for some point about which the torques are calculated, then it holds for all such points.

2. When the gravitational force acts on an extended body, the torque it produces is that of the body's weight acting at the **center of gravity.** When gravity is uniform over the entire body, then the center of gravity coincides with the center of mass.

3. Solving static equilibrium problems requires writing equations that set both the net force and net torque equal to zero. The result is a set of simultaneous equations that set conditions on the unknown quantities.

4. An equilibrium is **stable** if a small displacement from equilibrium results in forces and/or torques that tend to restore the equilibrium. An equilibrium is **unstable** if small displacements result in forces and/or torques that drive the system further from equilibrium. A **metastable** equilibrium is stable against very small displacements but unstable if larger displacements occur. Stable and metastable equilibria occur at local minima of a system's potential energy curve, while unstable equilibria occur at maxima. In one dimension, that is:

$$\frac{dU}{dx} = 0, \quad \text{(equilibrium condition)}$$

$$\frac{d^2U}{dx^2} > 0, \quad \text{(stable or metastable equilibrium)}$$

$$\frac{d^2U}{dx^2} < 0. \quad \text{(unstable equilibrium)}$$

The intermediate case $d^2U/dx^2 = 0$ is termed **neutral stability.**

Terms You Should Understand

(Pairs are closely related terms whose distinction is important; number in parentheses is chapter section where term first appears.)

static equilibrium (14-1)
center of gravity (14-2)
stable, unstable (14-4)
metastable (14-4)
neutrally stable (14-4)

Symbols You Should Recognize

$\dfrac{dU}{dx}, \dfrac{d^2U}{dx^2}$ (14-4)

Problems You Should Be Able to Solve

setting up and solving static equilibrium problems for any appropriate unknowns, including forces, torques, angles, and distances to force application points (14-2 and 14-3) determining equilibrium points and their stability from potential energy curves (14-4)

Limitations to Keep in Mind

A body's center of gravity coincides with its center of mass only when gravity doesn't change significantly over the extent of the body.

The conditions for equilibrium give at most six equations, which may not be enough to solve for all unknowns in situations involving many forces.

QUESTIONS

1. Give an example of an object on which the net force is zero but which is not in static equilibrium.
2. Give an example of an object on which the net torque about the center of gravity is zero, but which is not in static equilibrium.
3. The net torque about a body's center of gravity is zero, but the body is undergoing linear acceleration. Is the net torque about a point other than the center of gravity necessarily zero? Explain.
4. The best way to lift a heavy weight is to squat with your back vertical, rather than to lean over. Why?
5. Pregnant women often assume a posture in which the shoulders are held well back of their normal position. Explain in terms of torque and center of gravity.
6. A large, irregularly shaped asteroid of uniform density is falling toward Earth in the orientation shown in Fig. 14-24. Which is closer to Earth, the asteroid's center of gravity or its center of mass? Explain.

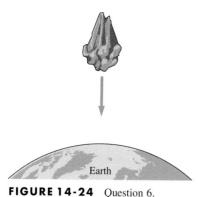

FIGURE 14-24   Question 6.

7. When you carry a bucket of water with one hand, you often find it easier if you extend your opposite arm. Why?
8. Is a ladder more likely to slip when you stand near the top or the bottom? Explain.
9. How does a heavy keel on a boat help keep the boat from tipping over?
10. In addition to the wings, most airplanes have a smaller set of horizontal surfaces near the tail. Why? What does this suggest about the center of gravity of the airplane?

11. Does choosing a pivot point in an equilibrium problem mean that something is necessarily going to rotate about that point?
12. If you take the pivot at the application point of a force in a static equilibrium problem, that force doesn't enter the torque equation. Does that make the force irrelevant to the problem? Explain.
13. You're hanging a heavy picture on a wall, using wire attached to the top corners of the picture. Is the wire more likely to break if you run it tightly between the corners or if you give it some slack? Explain.
14. A projectile is shot vertically upward. Is it in equilibrium at the top of its trajectory, when it's instantaneously at rest? Why or why not?
15. Suppose that the floor in Example 4-3 were frictionless but the wall were not. Would this change the minimum angle at which the board could lean? Explain.
16. In Fig. 14-25, why is the dancer's body leaning slightly to the right?

FIGURE 14-25   Why is the dancer leaning to the right? (Question 16)

17. A short dog and a tall person are standing on a slope. If the angle of the slope is increased, which will fall over first? Why?
18. A stiltwalker is standing motionless on one stilt. What can you say about the location of the stiltwalker's center of mass?
19. Why is a bottle less stable if stood on its top rather than on its bottom?

20. A metastable equilibrium represents a local minimum in the potential energy curve. What is meant here by the term *local?*
21. Given that hydrogen nuclei can fuse together with enormous energy release, is water really a fully stable configuration of matter?

PROBLEMS

Section 14-1 Conditions for Equilibrium

1. Five forces act on a rod, as shown in Fig. 14-26. Write the torque equations that must be satisfied for the rod to be in static equilibrium taking the torques (a) about the top of the rod and (b) about the center of the rod.

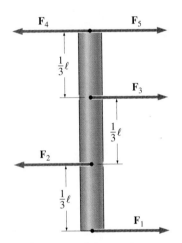

FIGURE 14-26 Problem 1.

2. A body is subject to three forces: $F_1 = 2\hat{i} + 2\hat{j}$ N, applied at the point $x = 2$ m, $y = 0$ m; $F_2 = -2\hat{i} - 3\hat{j}$ N, applied at $x = -1$ m, $y = 0$; and $F_3 = 1\hat{j}$ N, applied at $x = -7$ m, $y = 1$ m. (a) Show explicitly that the net force on the body is zero. (b) Show explicitly that the net torque about the origin is zero. (c) To confirm the assertion following Equation 14-4 that the net torque must be zero about any other point, evaluate the net torque about the point (3 m, 2 m), the point (−7 m, 1 m), and about any other point of your choosing.
3. Suppose the force F_3 in the preceding problem is doubled so the forces no longer balance and the body is therefore accelerating. Show that (a) the torque about the point (−7 m, 1 m) is still zero, but that (b) the torque about the origin is no longer zero. What is the torque about the origin?
4. A rod of mass m and length ℓ is falling freely in a horizontal orientation, with no torque about its center of mass.

Find the magnitude of the torque about either end. Why does your answer not violate the discussion following Equation 14-4?
5. In Fig. 14-27 the forces shown all have the same magnitude F. For each of the cases shown, is it possible to place a third force so the three forces meet both conditions for static equilibrium? If so, specify the force and a suitable application point; if not, why not?

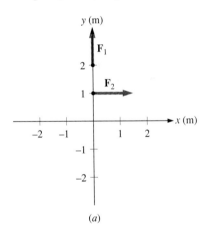

(a)

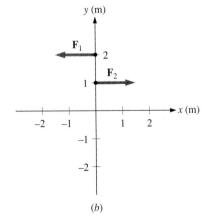

(b)

FIGURE 14-27 Problem 5.

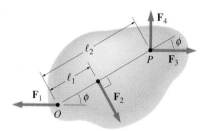

FIGURE 14-28 Problem 7.

6. Are there any other application points for the force $\mathbf{F}_3$ in Problem 2 that will ensure that both static equilibrium conditions are met?

7. Four forces act on a body, as shown in Fig. 14-28. Write the set of scalar equations that must hold for the body to be in equilibrium, evaluating the torques (a) about point O and (b) about point P.

8. The center of mass of an unoccupied seesaw is directly over its pivot. A 50-kg child sits on the left end, 2.1 m from the pivot. A 35-kg child sits on the right, 1.4 m from the pivot. How heavy a child should sit on the right end, 2.1 km from the pivot, in order to balance the seesaw? Solve (a) using the central pivot as the point for calculating the torques and (b) using the left end as the point for calculating the torques.

Section 14-2 Center of Gravity

9. Figure 14-29a shows a thin, uniform square plate of mass m and side ℓ. The plate is in a vertical plane. Find the magnitude of the gravitational torque on the plate about each of the three points shown.

10. Figure 14-29b shows a thin, uniform plate of mass m in the shape of an equilateral triangle of side ℓ. The plate is in a vertical plane. Find the magnitude of the gravitational torque on the plate about each of the three points shown.

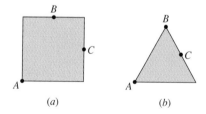

FIGURE 14-29 Problems 9, 10.

11. Fig. 14-30a shows an antique toy horse that balances on its hind legs. Fig. 14-30b shows a crude model for the horse, taken as a uniform rectangle with dimensions shown. A wire of negligible mass connects the rectangle's center of gravity to a balance mass. For the rectangle to be in equilibrium in the orientation shown, how should the balance mass compare with the rectangle's mass?

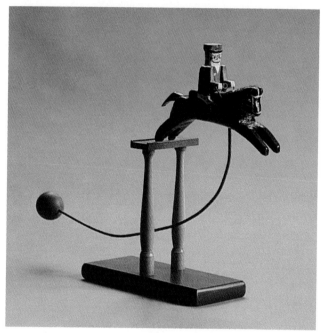

(a)

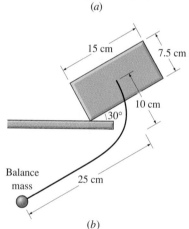

(b)

FIGURE 14-30 (a) An antique balancing horse. (b) A crude model for estimating the relative masses (Problem 11).

12. A 60-kg uniform tabletop 2.4 m long is supported by a pivot 80 cm from the left end, and by a scale at the right end (Fig. 14-31). How far from the left end should a 40-kg child sit if the scale is to read zero?

Section 14-3 Examples of Static Equilibrium

13. Where should the child in Fig. 14-31 sit if the scale is to read (a) 100 N and (b) 300 N?

14. A 4.2-m-long beam is supported by a cable at its center. A 65-kg steelworker stands at one end of the beam. Where

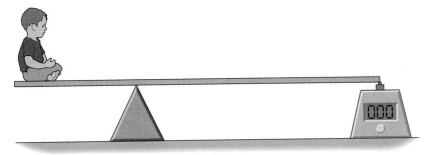

FIGURE 14-31 Problems 12, 13.

should a 190-kg bucket of concrete be suspended if the beam is to be in static equilibrium?

15. Two pulleys are mounted on a horizontal axle, as shown in Fig. 14-32. The inner pulley has a diameter of 6.0 cm, the outer a diameter of 20 cm. Cords are wrapped around both pulleys so they don't slip. In the configuration shown, with what force must you pull on the outer rope in order to support the 40-kg mass?

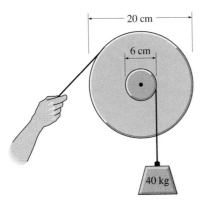

FIGURE 14-32 Problem 15.

16. A 23-m-long log of irregular cross section is lying horizontally, supported by a wall at one end and a cable attached 4.0 m from the other end, as shown in Fig. 14-33. The log weighs 7.5×10^3 N, and the tension in the cable is 6.2×10^3 N. Where is the log's center of gravity?

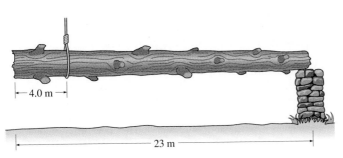

FIGURE 14-33 Problem 16.

17. Figure 14-34 shows a traffic signal, with masses and positions of its various members indicated. The structure is mounted with two bolts, located symmetrically about the vertical member's centerline, as indicated. What tension force must the left-hand bolt be capable of withstanding?

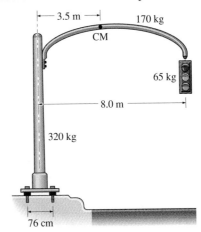

FIGURE 14-34 Problem 17.

18. Figure 14-35 shows how a scale with a capacity of only 250 N can be used to weigh a heavier person. The board is 3.0 m long, has a mass of 3.4 kg, and is of uniform density. It is free to pivot about the end farthest from the scale. What is the weight of a person standing 1.2 m from the pivot end, if the scale reads 210 N? Assume that the beam remains nearly horizontal.

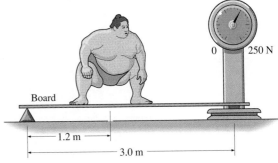

FIGURE 14-35 Problem 18.

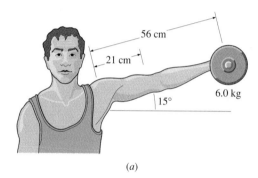

(a)

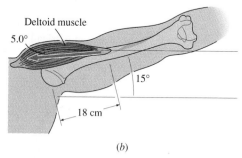

(b)

FIGURE 14-36 Problem 19.

19. Figure 14-36a shows an outstretched arm with a mass of 4.2 kg. The arm is 56 cm long, and its center of gravity is 21 cm from the shoulder. The hand at the end of the arm holds a 6.0-kg mass. (a) What is the torque about the shoulder due to the weights of the arm and the 6.0-kg mass? (b) If the arm is held in equilibrium by the deltoid muscle, whose force on the arm acts 5.0° below the horizontal at a point 18 cm from the shoulder joint (Fig. 14-36b), what is the force exerted by the muscle?

20. Figure 14-37 shows a portable infant seat that is supported by the edge of a table. The mass of the seat is 1.5 kg, and its center of mass is located 16 cm from the table edge. A 12-kg baby is sitting in the seat with her center of mass

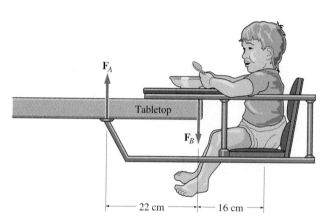

FIGURE 14-37 Problem 20.

over the seat's center of mass. Find the forces F_A and F_B that the seat exerts on the table.

21. A 15.0-kg door measures 2.00 m high by 75.0 cm wide. It hangs from hinges mounted 18.0 cm from top and bottom. Assuming that each hinge carries half the door's weight, determine the horizontal and vertical forces that the door exerts on each hinge.

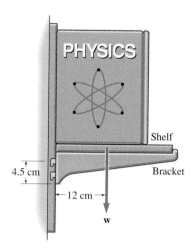

FIGURE 14-38 Problem 22.

22. Figure 14-38 shows a popular system for mounting bookshelves. An aluminum bracket is mounted on a vertical aluminum support by small tabs inserted into vertical slots. If each bracket in a shelf system supports 32 kg of books, with the center of gravity 12 cm out from the vertical support, what is the horizontal component of the force exerted on the upper of the two bracket tabs? Assume contact between the bracket and support occurs only at the upper tab and at the bottom of the bracket, 4.5 cm below the upper tab.

23. Figure 14-39 shows a house designed to have high "cathedral" ceilings. Following a heavy snow, the total mass supported by each diagonal roof rafter is 170 kg, including building materials as well as snow. Under these conditions, what is the force in the horizontal tie beam near the roof peak? Is this force a compression or a tension? Neglect any horizontal component of force due to the vertical walls below the roof. Ignore the widths of the various structural components, treating contact forces as though they were concentrated at the roof peak and at the outside edge of the rafter/wall junction.

24. Repeat Example 14-4, now assuming that the coefficient of friction at the floor is μ_1 and that at the wall μ_2. Show that the minimum angle at which the board will not slip is now given by

$$\phi = \tan^{-1}\left(\frac{1 - \mu_1 \mu_2}{2\mu_1}\right).$$

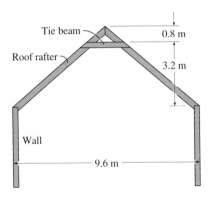

FIGURE 14-39 Problem 23.

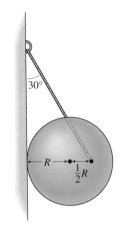

FIGURE 14-40 Problem 25.

25. A uniform sphere of radius R is supported by a rope attached to a vertical wall, as shown in Fig 14-40. The point where the rope is attached to the sphere is located so a continuation of the rope would intersect a horizontal line through the sphere's center a distance $R/2$ beyond the center, as shown. What is the smallest possible value for the coefficient of friction between wall and sphere?

26. Show that if the wall in the previous problem is frictionless, then a continuation of the rope line must pass through the center of the sphere.

27. A garden cart loaded with firewood is being pushed horizontally when it encounters a step 8.0 cm high, as shown in Fig. 14-41. The mass of the cart and its load is 55 kg, and the cart is balanced so that its center of mass is directly over the axle. The wheel diameter is 60 cm. What is the minimum horizontal force that will get the cart up the step?

28. A uniform 5.0-kg ladder is leaning against a frictionless vertical wall, with which it makes a 15° angle. The coefficient of friction between ladder and ground is 0.26. Can a 65-kg person climb to the top of the ladder without it slipping? If not, how high can the person climb? If so, how massive a person would make the ladder slip?

FIGURE 14-41 Problem 27.

29. The boom in the crane of Fig. 14-42 is free to pivot about point P and is supported by the cable that joins halfway along its 18-m total length. The cable passes over a pulley and is anchored at the back of the crane. The boom has mass 1700 kg, distributed uniformly along its length, and the mass hanging from the end of the boom is 2200 kg. The boom makes a 50° angle with the horizontal. What is the tension in the cable that supports the boom?

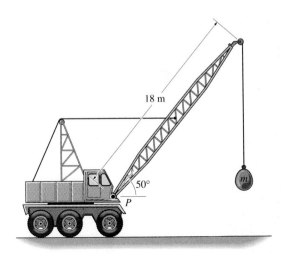

FIGURE 14-42 Problem 29.

30. A uniform board of length ℓ and weight W is suspended between two vertical walls by ropes of length $\ell/2$ each. When a weight w is placed on the left end of the board, it assumes the configuration shown in Fig. 14-43. Find the weight w in terms of the board weight W.

31. Figure 14-44 shows a 1250-kg car that has slipped over the edge of an embankment. A group of people are trying to hold the car in place by pulling on a horizontal rope, as shown. The car's bottom is pivoted on the edge of the embankment, and its center of mass lies further back, as shown. If the car makes a 34° angle with the horizontal, what force must the group apply to hold it in place?

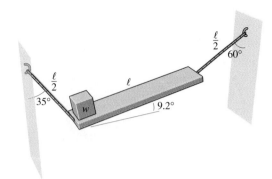

FIGURE 14-43 Problem 30.

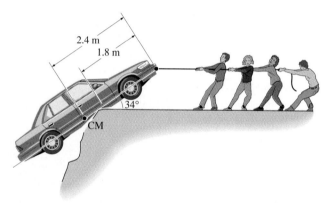

FIGURE 14-44 Problem 31.

32. A uniform board of length ℓ is dangling over a *frictionless* edge, secured by a *horizontal* rope, as shown in Fig. 14-45. Show that the angle it makes with the horizontal must be

$$\theta = \sin^{-1} \sqrt{\frac{2d}{\ell}},$$ where d is the distance from the edge to the center of the board.

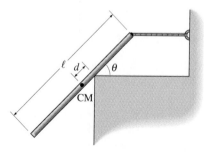

FIGURE 14-45 Problems 32, 33.

33. Figure 14-45 shows a uniform board dangling over a *frictionless* edge, secured by a *horizontal* rope. If the angle θ in Fig. 14-45 were 30°, what fraction would the distance d shown in the figure be of the board length ℓ?

Section 14-4 Stability of Static Equilibria

34. A portion of a roller coaster track is described by $h = 0.94x - 0.010x^2$, where h and x are the height and horizontal position in meters. (a) Find a point where the roller coaster car could be in static equilibrium on this track. (b) Is the equilibrium stable or unstable? (c) What is the height of the track at the equilibrium point?

35. A roly-poly toy clown is made from part of a sphere topped by a cone. The sphere is truncated at just the right point so that there is no discontinuity in angle as the surface changes from sphere to cone (Fig. 14-46a). If the clown always returns to an upright position, what is the maximum possible height for its center of mass? Would your answer change if the continuity-of-angle condition were not met, as in Fig. 14-46b?

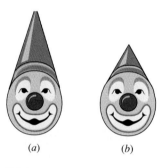

(a)   (b)

FIGURE 14-46 Problem 35.

36. A uniform rectangular block is twice as long as it is wide. Letting θ be the angle that the long dimension makes with the horizontal, determine the angular positions of any static equilibria, and comment on their stability.

37. The potential energy as a function of position for a certain particle is given by

$$U(x) = U_0\left(\frac{x^3}{x_0^3} + a\frac{x^2}{x_0^2} + 4\frac{x}{x_0}\right),$$

where U_0, x_0, and a are constants. For what values of a will there be two static equilibria? Comment on the stability of these equilibria.

38. A cubical block rests on an inclined board with two sides parallel to the direction of the incline. The coefficient of static friction between block and board is 0.95. If the inclination angle of the board is increased, will the block first slide or first tip over?

Paired Problems

(Both problems in a pair involve the same principles and techniques. If you can get the first problem, you should be able to solve the second one.)

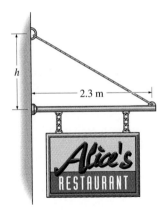

FIGURE 14-47 Problem 39 (the sign is centered beneath the rod).

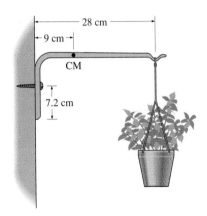

FIGURE 14-49 Problem 41.

39. Fig. 14-47 shows a 66-kg sign hung centered from a uniform rod of mass 8.2 kg and length 2.3 m. At one end the rod is attached to the wall by a pivot; at the other end it's supported by a cable that can withstand a maximum tension of 800 N. What is the minimum height *h* above the pivot for anchoring the cable to the wall?

40. A crane in a marble quarry is mounted on the rock walls of the quarry and is supporting a 2500-kg slab of marble as shown in Fig. 14-48. The center of mass of the 830-kg boom is located one-third of the way from the pivot end of its 15 m length, as shown. Find the tension in the horizontal cable that supports the boom.

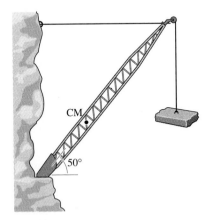

FIGURE 14-48 Problem 40.

41. A 4.2-kg plant hangs from the bracket shown in Fig. 14-49. The bracket has a mass of 0.85 kg, and its center of mass lies 9.0 cm from the wall. A single screw holds the bracket to the wall, as shown. Find the horizontal tension

force in the screw. *Hint:* Imagine that the bracket is slightly loose and pivoting about its bottom end. Assume the wall is frictionless.

42. A 160-kg highway sign of uniform density is 2.3 m wide and 1.4 m high. At one side it is secured to a pole with a single bolt, mounted a distance *d* from the top of the sign. The only other place where the sign contacts the pole is at its bottom corner. If the bolt can sustain a horizontal tension of 2100 N, what is the maximum permissible value for the distance *d*?

43. A 5.0-m-long ladder has mass 9.5 kg and is leaning against a frictionless wall, making a 66° angle with the horizontal. If the coefficient of friction between the ladder and ground is 0.42, what is the mass of the heaviest person who can safely ascend to the top of the ladder? (The center of mass of the ladder is at its center.)

44. To what vertical height on the ladder in the preceding problem could a 95-kg person reach before the ladder starts to slip?

45. A uniform, solid cube of mass *m* and side *s* is in stable equilibrium when sitting on a level tabletop. How much energy is required to bring it to an unstable equilibrium where it's resting on its corner?

46. An isosceles triangular block of mass *m* and height *h* is in stable equilibrium, resting on its base on a horizontal surface. How much energy does it take to bring it to unstable equilibrium, resting on its apex? *Hint:* Consult Example 10-3.

Supplementary Problems

47. A uniform pole of mass *M* is at rest on an incline of angle *θ*, secured by a horizontal rope as shown in Fig. 14-50. What is the minimum coefficient of friction that will keep the pole from slipping?

48. For what angle does the situation of Problem 47 require the greatest coefficient of friction?

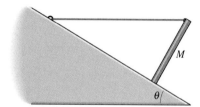

FIGURE 14-50 Problem 47.

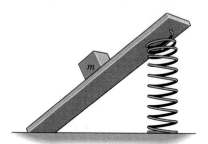

FIGURE 14-51 Problem 49.

49. One end of a board of negligible mass is attached to a spring of spring constant k, while its other end rests on a frictionless surface, as shown in Fig. 14-51. If a mass m is placed on the middle of the board, by how much will the spring compress?
50. A uniform ladder of mass m is leaning against a frictionless vertical wall with which it makes an angle θ. The coefficient of static friction at the floor is μ. Find an expression for the maximum mass for a person who is able to climb to the top of the ladder without its slipping. Use your result to show that *anyone* can climb to the top if $\mu \geq \tan\theta$, but that *no one* can if $\mu < \frac{1}{2}\tan\theta$.
51. Figure 14-52 shows a wheel on a slope with inclination angle $\theta = 20°$, where the coefficient of friction is adequate to prevent the wheel from slipping; however, it might still roll. The wheel is a uniform disk of mass 1.5 kg, and it is weighted at one point on the rim with an additional 0.95-kg mass. Find the angle ϕ shown in the figure such that the wheel will be in static equilibrium.
52. The wheel in Fig. 14-52 has mass M and is weighted with an additional mass m spread uniformly around the rim.

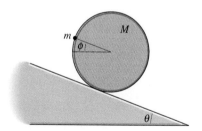

FIGURE 14-52 Problems 51, 52.

(The extra point mass of the preceding problem has been removed.) The slope angle is θ. Show that static equilibrium is possible only if $m > \dfrac{M\sin\theta}{1 - \sin\theta}$.
53. A 2.0-m-long rod has a density described by $\lambda = a + bx$, where λ is the density in kilograms per meter of length, $a = 1.0$ kg/m, $b = 1.0$ kg/m², and x is the distance in meters from the left end of the rod. The rod rests horizontally with its ends each supported by a scale. What do the two scales read?
54. What horizontal force applied at its highest point is necessary to keep a wheel of mass M from rolling down a slope inclined at angle θ to the horizontal?
55. A rectangular block twice as high as it is wide is resting on a board. The coefficient of static friction between board and incline is 0.63. If the board is tilted as shown in Fig. 14-53, will the block first tip over or first begin sliding?

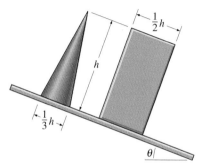

FIGURE 14-53 Problems 55 (block), 56 (block), 57 (cone).

56. What condition on the coefficient of friction in the preceding problem will cause the block to slide before it tips?
57. A uniform solid cone of height h and base diameter $\frac{1}{3}h$ is placed on the board of Fig. 14-53. The coefficient of static friction between the cone and incline is 0.63. As the slope of the board is increased, will the cone first tip over or begin sliding? *Hint:* Begin with an integration to find the center of mass.
58. In Fig. 14-54 a uniform boom of mass 350 kg is attached to a vertical wall by a pivot, and its far end is supported by a cable as shown. If the cable can withstand a maximum tension of 4.0 kN, what is the maximum value for the angle ϕ?
59. An interstellar spacecraft from an advanced civilization is hovering above Earth, as shown in Fig. 14-55. The ship consists of two pods of mass m separated by a rigid shaft of negligible mass that is one Earth radius (R_E) long. Find (a) the magnitude and direction of the net gravitational force on the ship and (b) the net torque about the center of mass. (c) Show that the ship's center of gravity is displaced approximately $0.083R_E$ from its center of mass.

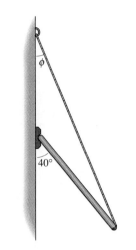

FIGURE 14-54   Problem 58.

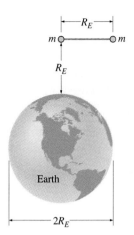

FIGURE 14-55   Problem 59.

P A R T 1 CUMULATIVE PROBLEMS

These problems combine material from chapters throughout the entire part or, in addition, from chapters in earlier parts, or they present special challenges.

1. A 170-g apple sits atop a 2.8-m-high post. A 45-g arrow moving horizontally at 130 m/s passes horizontally through the apple and strikes the ground 36 m from the base of the post, as shown in Fig. 1. Where does the apple hit the ground? Neglect the effect of air resistance on either object as well as any friction between apple and post.

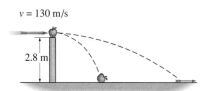

FIGURE 1 Cumulative Problem 1.

2. A fire department's tanker truck has a total mass of 21×10^3 kg, including 15×10^3 kg of water. Its brakes fail at the top of a long 3° slope and it begins to roll downward, starting from rest. In an attempt to stop the truck, firefighters direct a stream of water parallel to the slope, as shown in Fig. 2, beginning as soon as the truck starts to roll. The water leaves the 6.0-cm-diameter hose nozzle at 50 m/s. Will the truck stop before it runs out of water? If so, when? If not, what is the minimum speed reached?

FIGURE 2 Cumulative Problem 2.

3. A block of mass m_1 is attached to the axle of a uniform solid cylinder of mass m_2 and radius R by massless strings. The two accelerate down a slope that makes an angle θ with the horizontal, as shown in Fig. 3. The cylinder rolls without slipping and the block slides with coefficient of kinetic friction μ between block and slope. The strings are attached to the cylinder's axle with frictionless loops so that the cylinder can roll freely without any torque from the string. Find an expression for the acceleration of the pair, assuming that the string remains taut.

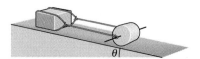

FIGURE 3 Cumulative Problem 3.

4. A missile is launched from point A in Fig. 4, heading for target C. The launch angle is 45.0°, and the launch speed is 1.90 km/s. An antimissile defense system is located at point B, 310 km downrange from the launch site. It fires an interceptor rocket at a 65.0° launch angle, with the intention of hitting the attacking missile when the latter is 270 km downrange of its launch site. (a) What should be the interceptor's launch speed? (b) How long after the launch

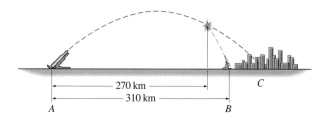

FIGURE 4 Cumulative Problem 4.

of the attacking missile should the interceptor be launched? (c) At what altitude does the interception take place?

5. A solid ball of radius R is set spinning with angular speed ω about a horizontal axis. The ball is then lowered vertically with negligible speed until it just touches a horizontal surface and is released (Fig. 5). If the coefficient of kinetic friction between the ball and the surface is μ, find (a) the linear speed of the ball once it achieves pure rolling motion and (b) the distance it travels before its motion is pure rolling.

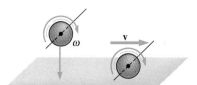

FIGURE 5 Cumulative Problem 5.

PART 2

OSCILLATIONS, WAVES, AND FLUIDS

High-speed photo taken just after a drop has struck calm water shows complex fluid behavior and circular waves propagating outward.

CHAPTER 15

OSCILLATORY MOTION

A ball attached to a spring undergoes oscillatory motion. Photo was taken with a strobe light flashing to image the ball at different points in its oscillation cycle. At the same time the camera panned sideways, displacing successive images horizontally and thus producing a plot of the ball's vertical position as a function of time. The position is a sinusoidal function of time. Interval between strobe flashes is 90 ms.

(a) *(b)*

FIGURE 15-1 (*a*) A ball in stable equilibrium at the bottom of a valley. (*b*) Disturbed from equilibrium, the ball oscillates about its equilibrium position.

When a system is disturbed from stable equilibrium, forces within the system tend to restore the equilibrium. But, like the ball in Fig. 15-1, the system may overshoot its equilibrium position. Again a restoring force develops and eventually the system turns back toward equilibrium. Again it may overshoot. The result is **oscillatory motion**—back and forth motion about equilibrium. In the absence of friction or other energy losses oscillatory motion would continue forever; in reality the system eventually settles back into equilibrium.

A wide variety of physical systems exhibit oscillatory motion. A uranium nucleus oscillates before it fissions. Oscillatory motion of water molecules is what cooks the food in a microwave oven. The horrible squeaking noise of chalk on a blackboard results from oscillatory motion. The timekeeping mechanism of a watch—whether an old-fashioned spring and balance wheel or a modern quartz crystal—is a carefully engineered system in oscillatory motion. Even large-scale structures like buildings and bridges undergo oscillatory motion, sometimes with disastrous results.

Oscillatory motion is a universal phenomenon. In this chapter we develop a detailed description of oscillatory motion that applies to a great many physical systems.

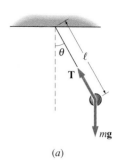

(a)

15-1 PROPERTIES OF OSCILLATORY MOTION

Two quantities characterize oscillatory motion: **amplitude** is the maximum displacement from equilibrium, and **period** is the time it takes for the motion to repeat itself. Figure 15-2 shows both these quantities on a plot of position versus time. Closely related to the period is the **frequency,** or number of oscillation cycles per unit time. Frequency and period are simply inverses:

FIGURE 15-2   Position-versus-time graphs for two oscillatory motions. Both have the same amplitude (A) and frequency (and therefore period). Their differences reflect the different restoring forces acting on each system.

$$f = \frac{1}{T}, \tag{15-1}$$

where f is the frequency and T the period. The unit of frequency is the **hertz** (Hz), named after the German physicist, engineer, and mathematician Heinrich Hertz (1857–1894), who was the first to produce and detect radio waves. One hertz is equal to one oscillation cycle per second.

● EXAMPLE 15-1   A BORING LECTURE

Bored by a physics lecture, a student holds one end of a flexible plastic ruler against a table and idly strikes the other end, setting it into oscillation (Fig. 15-3). The student notes that 28 complete cycles occur in 10 s, and that the end of the ruler moves a total distance of 8.0 cm. What are the amplitude, period, and frequency of this oscillatory motion?

Solution
We've defined amplitude as the maximum displacement from

equilibrium. Because the end of the ruler moves both sides of equilibrium, the amplitude of the motion is 4.0 cm. (The full 8.0-cm motion is called the **peak-to-peak** amplitude.) With 28 cycles in 10 s, the time per cycle, or period, is

$$T = \frac{10 \text{ s}}{28} = 0.36 \text{ s}.$$

The frequency in Hz is the inverse of the period:

$$f = \frac{1}{T} = \frac{1}{0.36 \text{ s}} = 2.8 \text{ Hz}.$$

EXERCISE   The top of a skyscraper sways back and forth, completing 9 oscillation cycles in one minute. Find the period and frequency of the motion.

Answers: $T = 6.7$ s; $f = 0.15$ Hz

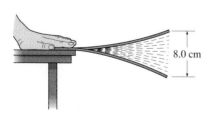

8.0 cm

FIGURE 15-3   Side view of a ruler undergoing oscillatory motion (Example 15-1).

Some problems similar to Example 15-1:   1–3   ●

Amplitude and frequency don't provide all the details of oscillatory motion, since two quite different motions can have the same frequency and amplitude (Fig. 15-2). The differences reflect details of the restoring forces that tend to return systems to equilibrium. Remarkably, though, many physical systems exhibit simple restoring forces that result in the same type of oscillatory motion. The rest of the chapter concentrates on this important case.

15-2 SIMPLE HARMONIC MOTION

In a wide variety of physical systems, the force that develops when a body is displaced from stable equilibrium is directly proportional to the displacement. The type of motion that results is called **simple harmonic motion:**

> *Simple harmonic motion is the motion that results when an object is subject to a restoring force proportional to its displacement from equilibrium.*

Mathematically, we describe such a force by writing

$$F = -kx, \tag{15-2}$$

where F is the force, x the displacement, and k a constant giving the proportionality between them. The minus sign in Equation 15-2 indicates a *restoring* force: If the object is displaced in one direction, the force is in the *opposite* direction, so it tends to restore the equilibrium.

We've seen Equation 15-2 before: It's the force exerted by an ideal spring of spring constant k. So a system consisting of a mass attached to a spring undergoes simple harmonic motion (Fig. 15-4). Many other systems—including atoms and molecules—can often be understood by modeling them as miniature mass-spring systems.

We've defined simple harmonic motion, but we haven't described it. How does a body in simple harmonic motion actually move? We can find out by applying Newton's second law, $\mathbf{F} = m\mathbf{a}$, to the mass-spring system of Fig. 15-4. Here the force on the mass m is $-kx$, so Newton's law becomes

$$-kx = ma, \tag{15-3}$$

where we take the x axis along the direction of motion, with $x = 0$ at the equilibrium position. We want to know how the position x depends on time. Although time doesn't appear explicitly in Equation 15-3, it's implicit because the acceleration is the second time derivative of position. Thus Equation 15-3 may be written

$$m\frac{d^2x}{dt^2} = -kx. \tag{15-4}$$

The solution to this equation is the position x as a function of time. What sort of function might it be? Physical insight tells us to expect periodic motion, so we could try periodic functions—like sine and cosine. Suppose we pull the mass in

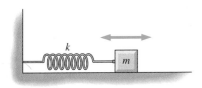

FIGURE 15-4 A mass attached to an ideal spring undergoes simple harmonic motion. The mass rides on a frictionless surface, and the spring exerts a force if it's either stretched or compressed from equilibrium.

Fig. 15-4 to the right and, at time $t = 0$, release it. Since it starts with a nonzero displacement, we might expect cosine to be the appropriate function (recall that $\cos(0) = 1$, while $\sin(0) = 0$). We still don't know the amplitude or frequency, so we'll try a form that has two unknown constants:

$$x = A \cos\omega t. \tag{15-5}$$

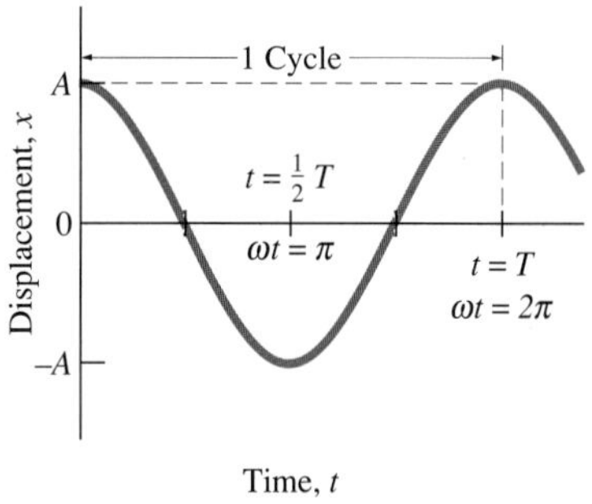

What's the physical significance of the constants in Equation 15-5? Because the cosine function itself varies between $+1$ and -1, the quantity A is just the amplitude—the greatest displacement from equilibrium (Fig. 15-5).

What about the constant ω? The cosine function undergoes a full cycle as its argument increases by 2π radians, or $360°$, as shown in Fig. 15-5. In Equation 15-5, the argument of the cosine is ωt. Since the time for a full cycle is the period T, the argument ωt must go from 0 to 2π as the time t goes from 0 to T. So we have

$$\omega T = 2\pi,$$

or

$$T = \frac{2\pi}{\omega}. \tag{15-6}$$

The frequency of the motion is then

$$f = \frac{1}{T} = \frac{\omega}{2\pi}. \tag{15-7}$$

Equation 15-7 shows that ω is a measure of the frequency, although it differs from the frequency f by a factor of 2π. The quantity ω is called the **angular frequency**, and its units are radians per second or, since radians are dimensionless, simply inverse seconds (s^{-1}).

> **TIP** **Why Radians?** Here, as in Chapter 12's description of rotational motion, we use the angular quantity ω because it provides the simplest mathematical description of the motion. In fact, the relation between angular frequency and frequency in hertz is the same as Chapter 12's relation between angular speed in radians per second and in revolutions per second. Although there's no physical angle involved in our one-dimensional simple harmonic motion, the similarity between the descriptions of simple harmonic motion and of rotational motion is no coincidence. We'll explore that similarity further in Section 15-4.

Writing the displacement x in the form 15-5 doesn't guarantee that we have a solution; we must now find out if the assumed solution 15-5 satisfies Equation 15-4. If x is given by Equation 15-5, then its first derivative is

$$\frac{dx}{dt} = \frac{d}{dt}(A \cos\omega t) = -A\omega \sin \omega t, \tag{15-8}$$

FIGURE 15-5 The function $A \cos\omega t$. This function varies between $\pm A$, and undergoes one full cycle as ωt increases by 2π.

where we've used the chain rule for differentiation (see Appendix A). Then the second time derivative is

$$\frac{d^2x}{dt^2} = \frac{d}{dt}\left(\frac{dx}{dt}\right) = \frac{d}{dt}(-A\omega \sin\omega t) = -A\omega^2 \cos\omega t. \qquad (15\text{-}9)$$

We can now try out the assumed solution for x (Equation 15-5), and its second time-derivative (Equation 15-9), in Equation 15-4: Substituting x and d^2x/dt^2 in the appropriate places gives

$$m(-A\omega^2 \cos\omega t) \overset{?}{=} -k(A \cos\omega t), \qquad (15\text{-}10)$$

where the ? indicates that we're still trying to find out if this is indeed an equality. Now Equation 15-10 must be true *for all values of time t.* Why? Because Newton's law holds at all times, and Equation 15-10 is derived from Newton's law. Fortunately, the time-dependent term $\cos\omega t$ appears on both sides of the equation, so we can cancel it. Also the amplitude A and the minus sign cancel from the equation, leaving only $m\omega^2 = k$, or

$$\omega = \sqrt{\frac{k}{m}}. \qquad (15\text{-}11)$$

Thus, Equation 15-5 *is* a solution of Equation 15-4, *provided* the angular frequency ω is given by Equation 15-11.

Frequency and Period in Simple Harmonic Motion

We can recast Equation 15-11 in terms of the more familiar frequency f and period T using Equation 15-7, $f = \omega/2\pi$. This gives

$$f = \frac{\omega}{2\pi} = \frac{1}{2\pi}\sqrt{\frac{k}{m}}, \qquad (15\text{-}12)$$

for the frequency in hertz, and

$$T = \frac{1}{f} = 2\pi\sqrt{\frac{m}{k}} \qquad (15\text{-}13)$$

for the period.

Do these relations make sense? If we increase the mass m, it becomes harder to accelerate and we expect slower oscillations. This is indeed reflected in Equations 15-11 and 15-12, where m appears in the denominator. Increasing k, on the other hand, makes the spring stiffer and increases the oscillation frequency—as shown by the presence of k in the numerators of Equations 15-11 and 15-12.

Physical systems display a wide range of m and k values and a correspondingly large range of oscillation frequencies. An atom, with its small mass and "springiness" provided by electric forces, may oscillate at 10^{15} Hz or more. A massive skyscraper, in contrast, typically oscillates at about 0.1 Hz.

Amplitude in Simple Harmonic Motion

The amplitude A canceled from our equations, showing that our analysis works for *any* value of A. This means that the oscillation frequency does not depend on the oscillation amplitude. Independence of frequency and amplitude is a particularly simple feature of simple harmonic motion, and arises because the restoring force is *directly proportional* to displacement. When the restoring force does not have the simple form $F = -kx$, then frequency *does* depend on amplitude and the analysis of oscillatory motion becomes much more complicated (see Problem 72). In many systems the relation $F = -kx$ breaks down if the displacement x gets too big; for this reason, simple harmonic motion is often an approximation valid for relatively small oscillation amplitudes.

Phase

Equation 15-5 is not the only possible solution to Equation 15-4; you can readily show, for example, that $x = A \sin\omega t$ works just as well. We chose the cosine because we took time $t = 0$ at the point of maximum displacement. Had we set $t = 0$ as the mass passed through its equilibrium point, sine would have been the appropriate function. More generally we could take the zero of time at some arbitrary point in the oscillation cycle. Then, as Fig. 15-6 shows, we can represent the motion by the form

$$x = A\cos(\omega t + \phi), \tag{15-14}$$

where the **phase constant** ϕ has the effect of shifting the cosine curve to the left or right but does not affect the frequency or amplitude.

Velocity and Acceleration in Simple Harmonic Motion

Since Equation 15-5 gives the position of the oscillating mass as a function of time, its first and second derivatives—Equations 15-8 and 15-9—must be the velocity and acceleration, respectively. Equation 15-8 shows that the velocity is a sine when the displacement is a cosine. Velocity is therefore a maximum when displacement is zero, and vice versa (Fig. 15-7). Thus the mass has its maximum speed as it moves through equilibrium, while at its extreme displacements it is instantaneously stopped. Equation 15-8 also shows that the maximum velocity is $A\omega$. This makes sense because a higher frequency oscillation requires a higher speed in order to traverse the distance A in a shorter time.

Similarly, Equations 15-5 and Fig. 15-8 show that displacement and acceleration reach their extremes at the same time, but with opposite signs. This is a reflection of the restoring force $F = -kx$: At the extremes of displacement the force has its greatest magnitude, and it points opposite the displacement because it accelerates the system back toward equilibrium. Equation 15-9 shows that the maximum acceleration is $A\omega^2$.

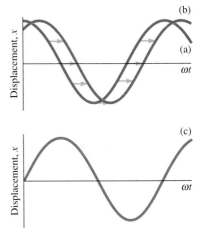

FIGURE 15-6 The phase constant ϕ determines the displacement x at time $t = 0$. (a) $\phi = 0$ corresponds to oscillation with maximum displacement at $t = 0$, while (b) $\phi = -\pi/4$ shifts the curve one-eighth cycle to the right. (c) $\phi = -\pi/2$ shifts the curve by one-fourth cycle and corresponds to motion that passes through equilibrium at $t = 0$. Since $\cos\left(\alpha + \dfrac{\pi}{2}\right) = \sin\alpha$, case (c) is equally well described by the form $x = A\sin\omega t$.

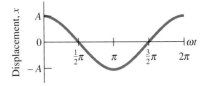
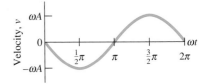
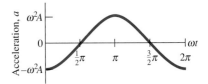

FIGURE 15-7 Displacement, velocity, and acceleration in simple harmonic motion. Note that the speed is greatest when the displacement is zero, and vice versa.

● **EXAMPLE 15-2**   A SWAYING SKYSCRAPER

The huge mass-spring system shown in Fig. 15-8 is on the top floor of New York's 300-m high Citicorp Tower. It is engineered to oscillate at the same frequency as the building itself, but out of phase, thereby reducing the overall oscillation amplitude of the building. The 3.73×10^5-kg concrete block completes one oscillation in 6.80 s, and the oscillation amplitude in a high wind is 110 cm. Determine the spring constant and the maximum speed and acceleration experienced by the concrete block.

Solution

Since we know the mass and period, we can solve Equation 15-13 for the spring constant k:

$$k = \frac{4\pi^2 m}{T^2} = \frac{(4\pi^2)(3.73 \times 10^5 \text{ kg})}{(6.80 \text{ s})^2} = 3.18 \times 10^5 \text{ N/m}.$$

Equations 15-8 and 15-9 show that the maximum speed and acceleration are $v_{max} = \omega A$ and $a_{max} = \omega^2 A$. In terms of the period, the angular frequency may be obtained from Equation 15-6:

$$\omega = \frac{2\pi}{T} = \frac{2\pi}{6.80 \text{ s}} = 0.924 \text{ s}^{-1},$$

so $v_{max} = A\omega = (1.10 \text{ m})(0.924 \text{ s}^{-1}) = 1.02 \text{ m/s}$

and $a_{max} = A\omega^2 = (1.10 \text{ m})(0.924 \text{ s}^{-1})^2 = 0.939 \text{ m/s}^2$.

EXERCISE   A hydrogen chloride molecule (HCl) can be modeled as a hydrogen atom attached by a spring to a chlorine atom; the chlorine atom is so massive that it remains essentially fixed as the hydrogen oscillates at a frequency of 8.66×10^{13} Hz. (a) Find the effective spring constant. (b) If the maximum speed of the hydrogen atom is 2.50×10^5 m/s, what is its maximum displacement from equilibrium?

Answers: (a) 494 N/m; (b) 0.459 nm

Some problems similar to Example 15-2:   6–9

(a)

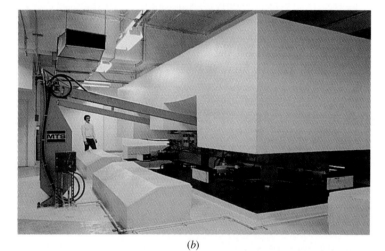

(b)

(c)

FIGURE 15-8   (a) New York's 59-floor Citicorp Tower. (b) To reduce the building's tendency to oscillate, engineers designed this huge mass-spring system, mounted on the top floor. The 410-ton mass and springs have the same oscillation frequency as the building. The mass rides on a film of oil and is set oscillating 180° out of phase with the building, thereby reducing the amplitude of the building's oscillation. (c) A similar "tuned mass damper" in action in Boston's John Hancock Building.

●

15-3 APPLICATIONS OF SIMPLE HARMONIC MOTION

Simple harmonic motion occurs in any system where a restoring force or analogous quantity is directly proportional to the displacement from equilibrium. Analysis of such systems follows that of the mass-spring system we just considered, but often involves different physical quantities. Here we explore a variety of systems that exhibit simple harmonic motion.

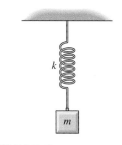

FIGURE 15-9 A vertical mass-spring system.

The Vertical Mass-Spring System

A mass hanging vertically from a spring is subject to gravity as well as the spring force (Fig 15-9). In equilibrium the spring must be stretched enough for its force to balance gravity:

$$mg - kx_1 = 0,$$

where x_1 is the new equilibrium position and where we take the x axis vertically downward. Suppose we stretch the spring an additional amount Δx from its new equilibrium. Then the spring force is $F_{\text{spring}} = -k(x_1 + \Delta x)$ and the net force, including gravity, is

$$F = mg - k(x_1 + \Delta x) = mg - kx_1 - k\Delta x = -k\Delta x,$$

where the last step follows because the equilibrium position x_1 satisfies the condition $mg - kx_1 = 0$. Thus the system is subject to a restoring force $-k\Delta x$ that is directly proportional to the displacement Δx from the new equilibrium. We can apply the analysis of Section 15-2, now with Δx playing the role of x, and the result is again simple harmonic motion with the same frequency $\omega = \sqrt{k/m}$. Thus gravity changes only the equilibrium position and doesn't affect the frequency at all.

FIGURE 15-10 The balance wheel in a mechanical watch is a torsional oscillator whose period is the basic time interval of the watch. The coiled hairspring provides the restoring force.

The Torsional Oscillator

A torsional oscillator is a system that is subject to a restoring torque when it is given an angular displacement from equilibrium. A common example is the balance wheel in a mechanical watch (Fig. 15-10). In this and all other torsional oscillators, the restoring torque sets up a back-and-forth rotational motion about the torque-free equilibrium state. When the restoring torque is directly proportional to the angular displacement from equilibrium, then the result is simple harmonic motion.

A simple example of a torsional oscillator is an object of rotational inertia I suspended from a fiber that develops a torque when twisted (Fig. 15-11). When the restoring torque is directly proportional to the angular displacement, we can write

$$\tau = -\kappa\theta, \tag{15-15}$$

where τ is the torque and θ the angular displacement. The constant κ is the **torsional constant,** and is the rotational analog of the spring constant.

FIGURE 15-11 A torsional oscillator consisting of a disk suspended from a wire.

The behavior of the torsional oscillator is described by the rotational analog of Newton's law (Equation 12-14):

$$\tau = I\alpha,$$

where α is the angular acceleration. Using Equation 15-15 for τ and writing α as the second time derivative of angular displacement, the analog of Newton's law becomes

$$I\frac{d^2\theta}{dt^2} = -\kappa\theta. \tag{15-16}$$

This equation is identical to Equation 15-4 for the linear oscillator, except that I replaces m, θ replaces x, and κ replaces k. In direct analogy with the solution of Equation 15-4, we can then write

$$\theta = A\cos\omega t,$$

where A of Equation 15-4 is the amplitude and where ω is given by analogy with Equation 15-11:

$$\omega = \sqrt{\frac{\kappa}{I}}. \tag{15-17}$$

Expressions for the period and frequency in Hz follow directly from Equations 15-6 and 15-7.

In addition to its timekeeping functions, the torsional oscillator provides an accurate way of measuring rotational inertia, as the example below illustrates.

● **EXAMPLE 15-3** SWINGING

A child mounts a swing consisting of a tire suspended from a rope of torsional constant 6.1 N·m/rad, and sets herself into torsional oscillations, rotating back and forth about the rope axis. If 35 complete oscillation cycles occur in 2 minutes, what is the rotational inertia of child and swing about the rope?

Solution

Solving Equation 15-17 for I gives

$$I = \frac{\kappa}{\omega^2} = \frac{\kappa}{(2\pi f)^2} = \frac{6.1\ \text{N·m/rad}}{[(2\pi)(35\ \text{cycles}/120\ \text{s})]^2} = 1.8\ \text{kg·m}^2.$$

Torsional oscillators are often used to measure the rotational inertias of irregular objects, as in this example.

EXERCISE The balance wheel in a watch has rotational inertia 1.24×10^{-7} kg·m². What should be the torsional constant of the hairspring if the period of the wheel's torsional oscillations is to be 1.00 s?

Answer: 4.9×10^{-6} N·m/rad

Some problems similar to Example 15-3: 19, 25, 59, 60 ●

The Pendulum

A **simple pendulum** consists of a point mass suspended from a massless cord. This idealization is approximated by many real systems in which a mass of small extent is suspended by a long structure of much lower mass.

Figure 15-12 shows a pendulum of mass m and length ℓ displaced slightly from equilibrium. The tension force **T** exerts no torque about the suspension

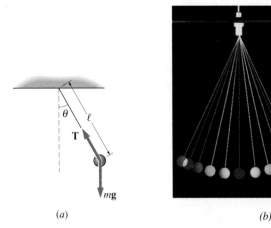

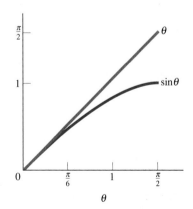

(a)   *(b)*

FIGURE 15-12   (*a*) Forces acting on a simple pendulum. The gravitational force produces a torque of magnitude $mg\ell \sin\theta$ about the suspension point. (*b*) The torque tends to rotate the displaced pendulum back toward its equilibrium, resulting in oscillatory motion. Here a multi-colored strobe light was used to image a pendulum over one-half cycle of its motion. Distance between successive images shows that the speed is greatest as the pendulum swings through its equilibrium position.

point because it lies on the line between the mass and suspension point. But the gravitational force $m\mathbf{g}$ does exert a torque, given by

$$\tau = -mg\ell \sin\theta,$$

where the minus sign indicates that the torque tends to rotate the pendulum back toward its equilibrium. The rotational analog of Newton's law, $\tau = I\alpha$, then becomes

$$I\frac{d^2\theta}{dt^2} = -mg\ell \sin\theta, \tag{15-18}$$

where we've written the angular acceleration as the second derivative of the angular displacement.

Equation 15-18 looks somewhat like Equation 15-16 for the torsional oscillator—but not quite, since the torque involves $\sin\theta$ rather than θ itself. Thus the restoring torque is not *directly* proportional to the angular displacement, and the motion is therefore *not* simple harmonic. Problem 72 explores this more complex situation.

If, however, the amplitude of the motion is small, then it *approximates* simple harmonic motion. Figure 15-13 shows that for small angles—much less than 1 radian—$\sin\theta$ and θ are essentially equal. For a small-amplitude pendulum we can therefore replace $\sin\theta$ in Equation 15-18 with θ to get

$$I\frac{d^2\theta}{dt^2} = -mg\ell\theta.$$

This is essentially Equation 15-16, with $mg\ell$ playing the role of κ. So the small-amplitude pendulum undergoes simple harmonic motion, its angular frequency given by Equation 15-17 with $\kappa = mg\ell$:

FIGURE 15-13   For small values of θ measured in radians, $\sin\theta$ and θ are nearly equal. Even at 30° ($\pi/6$, or 0.52 rad) the value of $\sin\theta$ (= 0.50) is within 4% of θ itself.

$$\omega = \sqrt{\frac{mg\ell}{I}}. \tag{15-19}$$

For a *simple* pendulum, the rotational inertia I is that of a point mass m a distance ℓ from the rotational axis, or $I = m\ell^2$. Then we have

$$\omega = \sqrt{\frac{mg\ell}{m\ell^2}} = \sqrt{\frac{g}{\ell}}, \qquad \text{(simple pendulum)} \tag{15-20}$$

or, from Equation 15-6,

$$T = \frac{2\pi}{\omega} = 2\pi\sqrt{\frac{\ell}{g}}. \qquad \text{(simple pendulum)} \tag{15-21}$$

These equations show that the frequency and period of a simple pendulum are independent of its mass, depending only on length and gravitational acceleration.

● EXAMPLE 15-4   RESCUING TARZAN

Tarzan stands on a branch, petrified with fear as a leopard approaches. Fortunately, Jane is on a branch of the same height in a nearby tree, 8.0 m away, holding a 25-m-long vine of negligible mass attached directly above the point midway between her and Tarzan (Fig. 15-14). Jane grasps the vine and steps off her branch with negligible initial velocity. How soon does she reach Tarzan?

Solution
Jane and the vine constitute a pendulum that takes half a period to reach Tarzan. The period is given by Equation 15-21:

$$T = 2\pi\sqrt{\frac{\ell}{g}} = 2\pi\sqrt{\frac{25 \text{ m}}{9.8 \text{ m/s}^2}} = 10 \text{ s},$$

so Jane reaches Tarzan in 5.0 s. Since the period of a simple pendulum is independent of its mass, the return trip with Tarzan takes the same time.

How good is the approximation $\sin\theta \approx \theta$? From Fig. 15-14, we see that $\sin\theta_0 = \frac{4.0 \text{ m}}{25 \text{ m}} = 0.16$. Then $\theta_0 = \sin^{-1}(0.16) = 0.1607$ rad, for a difference of only 0.43 per cent at the greatest angular displacement.

> **TIP   Set Your Calculator for Radians**   Most calculators assume that the arguments of trig functions are given in degrees. It's usually possible to set the calculator for radians instead. Make sure you do this before working with radian measure.

FIGURE 15-14   Tarzan in trouble (Example 15-4). Lengths are not to scale.

EXERCISE   A 650-kg wrecking ball is suspended from a 20-m chain of negligible mass. The ball is hanging vertically, resting against the wall of a building. It is then withdrawn a distance much smaller than the chain length and released. How soon does it strike the building?

Answer: 2.2 s

Some problems similar to Example 15-4:   17, 18, 21, 63

A **physical pendulum** is an object of arbitrary shape that is free to swing (Fig. 15-15). It differs from a simple pendulum in that its mass is distributed over its entire length, rather than being concentrated at the bottom. In our analysis, we used this distinction only at the very end, when we wrote $m\ell^2$ for the rotational inertia. Our analysis before that point therefore applies to the physical pendulum as well. In particular, Equation 15-19 gives the frequency of a physical pendulum executing small-amplitude oscillations, provided we interpret ℓ as the distance from the pivot to the center of gravity.

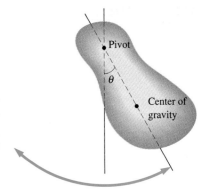

FIGURE 15-15 A physical pendulum.

● EXAMPLE 15-5   WALKING

When walking, the leg not in contact with the ground swings forward, acting essentially like a physical pendulum. Making the crude approximation that the leg acts like a uniform rod, find the period of this pendulum motion for leg length $L = 90$ cm.

Solution

With the uniform-rod assumption the center of gravity lies halfway along the leg, so ℓ in Equation 15-19 is $L/2$. Combining Equations 15-19 and 15-6 then gives the period:

$$T = \frac{2\pi}{\omega} = 2\pi\sqrt{\frac{I}{mgL/2}} = 2\pi\sqrt{\frac{mL^2}{3mgL/2}} = 2\pi\sqrt{\frac{2L}{3g}},$$

where we used Table 12-2's result $I = \frac{1}{3}mL^2$ for the rotational inertia. Using the 90-cm leg length gives $T = 1.55$ s, which is indeed about the time for one complete stride with both legs.

E X E R C I S E   A uniform solid ball of radius R hangs from a pivot attached to its surface (Fig. 15-16). What will be the period of small-amplitude oscillations about its equilibrium position?

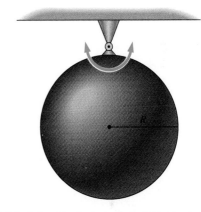

FIGURE 15-16 What's the period of this physical pendulum?

Answer: $2\pi\sqrt{\dfrac{7R}{5g}}$

Some problems similar to Example 15-5: 20, 28, 29, 57, 58

15-4 SIMPLE HARMONIC MOTION IN TWO DIMENSIONS

For many systems, simple harmonic motion is possible in several directions at once. Figure 15-17, for example, shows a pendulum free to swing in any direction. Suppose we displace this pendulum in the x direction, and as we release it give it an impulse in the y direction. The result is a composite motion whose maximum x displacement occurs when the y displacement is zero and the y velocity a maximum. As a result the pendulum traces out a curved path.

If the amplitudes of the pendulum's x and y motions are equal, then that path is a circle (Fig. 15-18). We can prove this as follows: With the maximum x displacement at time $t = 0$, we have $x = A \cos \omega t$, where A is the amplitude for both x and y motions. With maximum speed but zero displacement at $t = 0$, the y motion is given by $y = A \sin \omega t$. So the magnitude of the pendulum's displace-

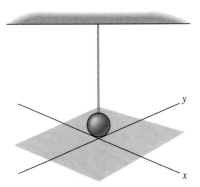

FIGURE 15-17 The pendulum is free to swing in any direction. We can analyze its motion into components along the x and y axes.

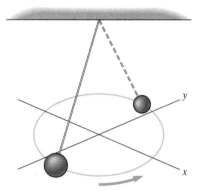

FIGURE 15-18 When the motions have equal amplitude but are $\pi/2$ radians out of phase, then the pendulum traces out a circle.

FIGURE 15-19 If we look in the plane of the circular motion, we see only one component of the motion— and that component is simple harmonic motion.

ment from the origin is given by $x^2 + y^2 = A^2 \sin^2 \omega t + A^2 \cos^2 \omega t = A^2$, where the last step follows because $\cos^2 \alpha + \sin^2 \alpha = 1$ for any α. Thus the pendulum remains a fixed distance from the origin, and its curved path is therefore circular.

We could have analyzed this circular motion without any reference to simple harmonic motion. In fact, we did so in Example 6-6, to illustrate purely circular motion. The two approaches show the link between simple harmonic motion and circular motion, and why we speak of angular frequency in simple harmonic motion even though no angles are involved. Our analysis of the two-dimensional pendulum shows that its motion may be resolved into two simple harmonic motions, at right angles in space and $\pi/2$ radians out of phase. If we look at the pendulum from above, we see it trace out a circle. But if we peer in a horizontal plane, we see only the projection of that circular motion in one direction (Fig. 15-19). That projection is itself simple harmonic motion. The angular frequency of the circular motion and of its two simple harmonic components is the same, showing that the "angle" ωt in simple harmonic motion corresponds to the actual angular displacement in circular motion, of which the simple harmonic motion could be a component. Figure 15-20 summarizes these relations between circular and simple harmonic motion.

Many physical systems exhibit different oscillation frequencies in different directions. The resulting motions are then interestingly complex, as Fig. 15-21 suggests.

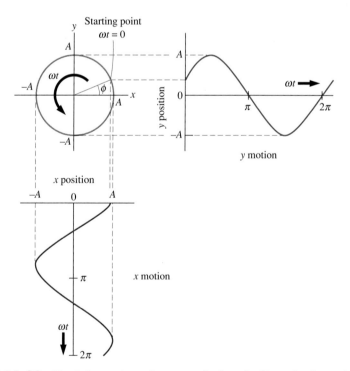

FIGURE 15-20 Simple harmonic motion as a projection of uniform circular motion. The quantities ωt and ϕ in simple harmonic motion correspond to physical angles in the associated uniform circular motion.

15-5 ENERGY IN SIMPLE HARMONIC MOTION

If we displace a mass-spring system from equilibrium, we do work as we build up potential energy in the spring. When we release the mass it accelerates toward equilibrium, gaining kinetic energy at the expense of potential energy. The mass passes through its equilibrium position with maximum kinetic energy and no potential energy, then slows and builds potential energy as it compresses the spring. If there is no energy loss, this process continues indefinitely. In oscillatory motion, energy is continually transferred back and forth between its kinetic and potential forms (Fig. 15-22).

For a mass-spring system, the potential energy is given by Equation 8-4:

$$U = \tfrac{1}{2}kx^2,$$

where x is the displacement from equilibrium. Meanwhile, the kinetic energy is $K = \tfrac{1}{2}mv^2$. We can illustrate explicitly the interchange of kinetic and potential energy in simple harmonic motion by using x from Equation 15-5 and v (that is, dx/dt) from Equation 15-8 in the expressions for potential and kinetic energy. Then we have

$$U = \tfrac{1}{2}kx^2 = \tfrac{1}{2}k(A \cos\omega t)^2 = \tfrac{1}{2}kA^2 \cos^2\omega t$$

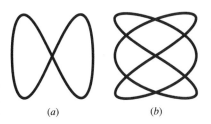

FIGURE 15-21 When oscillation frequencies differ in different directions, the oscillating object traces out a complex path. Here we see paths for ratios of (*a*) 2:1 and (*b*) 2:3 for the vertical to horizontal frequency.

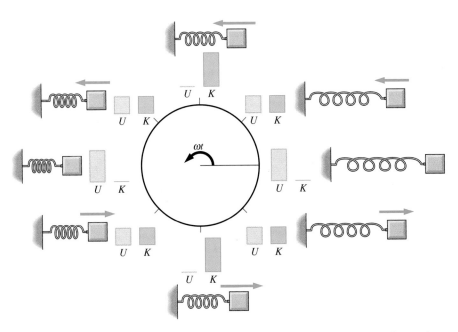

FIGURE 15-22 Kinetic and potential energy in a simple harmonic oscillator. Figures show the extension or compression of the spring, the velocity of the mass, and the balance of kinetic and potential energy at eight points in the cycle. Top and bottom configurations represent the equilibrium position; ωt is measured from the point of maximum displacement.

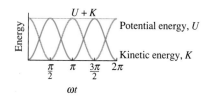

FIGURE 15-23 The potential and kinetic energy of a simple harmonic oscillator vary with time, but their sum remains constant.

and

$$K = \tfrac{1}{2}mv^2 = \tfrac{1}{2}m(-A\omega \sin\omega t)^2 = \tfrac{1}{2}m\omega^2 A^2 \sin^2\omega t = \tfrac{1}{2}kA^2 \sin^2\omega t,$$

where we have used the fact that $\omega^2 = k/m$. Comparing our expressions for potential and kinetic energy, we see that both have the same maximum value—$\tfrac{1}{2}kA^2$—equal to the initial potential energy of the stretched spring. The only difference lies in their relative phase—the potential energy is a maximum when the kinetic energy is zero, and vice versa. What about the total energy? It is

$$E = U + K = \tfrac{1}{2}kA^2 \cos^2\omega t + \tfrac{1}{2}kA^2 \sin^2\omega t = \tfrac{1}{2}kA^2,$$

where we have used $\sin^2 \omega t + \cos^2 \omega t = 1$. Although both kinetic and potential energy vary with time, their sum remains constant (Fig. 15-23).

● **EXAMPLE 15-6** ENERGY IN SIMPLE HARMONIC MOTION

A mass-spring system is undergoing simple harmonic motion with angular frequency ω and amplitude A. Find its speed at the point where the kinetic and potential energies are equal.

Solution

When the energies are equal, each must be half the total energy. Equation 15-8 shows that the maximum speed is $v_{max} = A\omega$. Since the maximum speed occurs when the energy is all kinetic, the total energy must be $E = \tfrac{1}{2}mv_{max}^2 = \tfrac{1}{2}mA^2\omega^2$. When the kinetic energy is half the total energy we then have

$$K = \tfrac{1}{2}mv^2 = \tfrac{1}{2}E = \tfrac{1}{2}\left(\tfrac{1}{2}mA^2\omega^2\right) = \tfrac{1}{4}mA^2\omega^2.$$

Solving for v gives the desired answer:

$$v = \frac{A\omega}{\sqrt{2}}.$$

EXERCISE Find the position x of the mass in Example 15-6 at the point when the kinetic and potential energies are equal.

Answer: $A/\sqrt{2}$

Some problems similar to Example 15-6: 41, 61, 62

FIGURE 15-24 Near their minima, potential energy curves often approximate parabolas. Simple harmonic motion can therefore occur about these minima, which are points of stable equilibrium.

Potential Energy Curves and Simple Harmonic Motion

We arrived at the expression $U = \tfrac{1}{2}kx^2$ for the potential energy of a spring by integrating the spring force, $-kx$, over distance. Since every simple harmonic oscillator has a restoring force or torque proportional to displacement, integration always results in a potential energy proportional to the *square* of the displacement—that is, in a parabolic potential energy curve. Conversely, any system with a parabolic potential energy curve exhibits simple harmonic motion. Potential energy functions for complex systems are often approximately quadratic near their stable equilibrium points (Fig. 15-24), and small disturbances from these equilibria therefore result in simple harmonic motion.

15-6 DAMPED HARMONIC MOTION

In real oscillating systems, forces like friction normally dissipate the oscillation energy. The motion in this case is said to be **damped.** If the dissipation is sufficiently weak that only a small fraction of the system's energy is removed in each oscillation cycle, then we expect that the system should behave essentially as in the undamped case, except for a gradual decrease in the oscillation amplitude (Fig. 15-25).

In many systems the damping force is approximately proportional to the velocity, and in the opposite direction:

$$F_d = -bv = -b\frac{dx}{dt},$$

where b is a constant giving the strength of the damping. We can write Newton's law as before, now including the damping force along with the restoring force. For a mass-spring system, we have

$$m\frac{d^2x}{dt^2} = -kx - b\frac{dx}{dt},$$

or

$$m\frac{d^2x}{dt^2} + b\frac{dx}{dt} + kx = 0. \qquad (15\text{-}22)$$

We will not solve this equation, but simply state that its solution, provided the damping is not too large, is

$$x = Ae^{-bt/2m}\cos(\omega t + \phi). \qquad (15\text{-}23)$$

This solution describes sinusoidal motion whose amplitude decreases exponentially with time. How fast the amplitude drops depends on the damping constant b and mass m: when $t = 2m/b$, the amplitude has dropped to $1/e$ of its original value. When the damping is so weak that only a small fraction of the total energy is lost in each cycle, then the frequency ω in Equation 15-23 is essentially equal to the undamped frequency $\sqrt{k/m}$. But with stronger damping, the damping force slows the motion, and the frequency becomes lower. Also, the oscillation amplitude decreases more rapidly with time (Fig. 15-26, curve a). As long as oscillation occurs, the motion is said to be **underdamped.** For sufficiently strong damping, though, the effect of the damping force is as great as that of the spring force. Under this condition, called **critical damping,** the system returns to its equilibrium state without undergoing any oscillations (Fig. 15-26, curve b). If the damping is made still stronger, the system becomes **overdamped.** The damping force now dominates, so the system returns more slowly to equilibrium (Fig. 15-26, curve c).

Many physical systems, from atoms to the human leg, can be modeled as damped oscillators. Engineers often design systems with specific amounts of damping. Automobile shock absorbers, for example, are designed with the springs to give critical damping. This results in rapid return to equilibrium while absorbing energy imparted by road bumps (Fig. 15-27). Similarly, the strings of a piano are damped—although not critically—so notes die out relatively quickly.

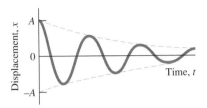

FIGURE 15-25 Weakly damped motion, showing sinusoidal oscillations confined within the "envelope" of a decaying exponential.

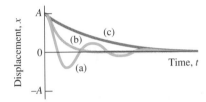

FIGURE 15-26 As the damping increases the amplitude falls more rapidly and the period increases slightly, as shown in curve a. With critical damping the oscillations cease altogether, as shown in curve b. Overdamping is shown in curve c.

FIGURE 15-27 This motorcycle spring cushions the rider from road bumps. Cylinder inside the spring is a shock absorber that provides critical damping, thus preventing continuing oscillation.

● EXAMPLE 15-7 BAD SHOCKS

A car and its suspension system act like a mass-spring system with $m = 1200$ kg and $k = 58$ kN/m. Its worn-out shock absorbers provide a damping constant $b = 230$ kg/s. The car hits a pothole that sets it oscillating. How many oscillations will it make before the oscillation amplitude drops to half its initial value?

Solution

The amplitude will be half its original value when the factor $e^{-bt/2m}$ in Equation 15-23 has the value one-half:

$$e^{-bt/2m} = \frac{1}{2}.$$

Taking the natural logarithm of both sides gives

$$\frac{-bt}{2m} = -\ln 2,$$

where we have used the facts that $\ln(x)$ and e^x are inverse functions and that $\ln(1/x) = -\ln(x)$. Then

$$t = \frac{2m}{b}\ln 2 = \frac{(2)(1200 \text{ kg})}{230 \text{ kg/s}}\ln 2 = 7.23 \text{ s}$$

is the time for the amplitude to drop to half its original value. For weak damping, the period is very close to the undamped period, which is

$$T = 2\pi\sqrt{\frac{m}{k}} = 2\pi\sqrt{\frac{1200 \text{ kg}}{58\times10^3 \text{ N/m}}} = 0.904 \text{ s}.$$

Then the number of cycles during the 7.23 s it takes the amplitude to drop in half is

$$\frac{7.23 \text{ s}}{0.904 \text{ s}} = 8.$$

That the number of oscillations is much greater than 1 tells us that the damping is weak, justifying our use of the undamped period.

EXERCISE Find the time it takes the system of Example 15-7 to drop to one-tenth of its initial amplitude.

Answer: 24 s

Some problems similar to Example 15-7: 47, 48

●

15-7 DRIVEN OSCILLATIONS AND RESONANCE

Pushing a child on a swing, you can build up a large amplitude by giving a relatively small push once each oscillation cycle. If your pushing were not in step with the swing's natural oscillatory motion, then the same force would have little effect.

When an external force acts on an oscillatory system, we say that the system is **driven.** Consider a mass-spring system, and suppose the driving force is given by $F_0 \cos \omega_d t$, where ω_d is called the **driving frequency.** Then Newton's law is

$$m\frac{d^2x}{dt^2} = -kx - b\frac{dx}{dt} + F_0 \cos \omega_d t, \tag{15-24}$$

where the first term on the right-hand side is the restoring force, the second the damping force, and the third the driving force. Since the system is being pushed at the driving frequency ω_d, we expect it to undergo oscillatory motion at this frequency. So we guess that the solution to Equation 15-24 might have the form

$$x = A \cos(\omega_d t + \phi). \tag{15-25}$$

Substituting this expression and its derivatives into Equation 15-24 (see Problem 51) shows that the equation is satisfied if

$$A = \frac{F_0}{m \sqrt{(\omega_d^2 - \omega_0^2)^2 + b^2 \, \omega_d^2/m^2}}, \tag{15-26}$$

where ω_0 is the undamped **natural frequency** $\sqrt{k/m}$, as distinguished from the driving frequency ω_d.

Equation 15-26 shows that the amplitude of the motion varies with driving frequency for a fixed amplitude F_0 of the driving force. Figure 15-28 shows a **resonance curve,** which is just a plot of Equation 15-26 as a function of driving frequency, for several values of the damping constant. As long as the system is underdamped, the curve has a maximum at some nonzero frequency (see Problem 52), and for weak damping that maximum occurs at very nearly the natural frequency. The weaker the damping, the more sharply peaked is the resonance curve. Thus it is possible to build up large-amplitude oscillations with relatively small driving forces. Such **resonance** can cause serious problems in physical systems. For example, we would not want the wind to build up large amplitude oscillations in a skyscraper; in fact, the device described in Example 15-2 is designed to minimize just such oscillations. A famous disaster involving resonance is the 1940 collapse of the Tacoma Narrows Bridge in Washington state (Fig. 15-29). Figure 15-30 shows the rebuilt bridge. Resonance is also important at the microscopic level. Resonant absorption of radio-wave energy heats ionized gases in some experiments designed to harness fusion energy. At the nuclear level, the process called nuclear magnetic resonance (NMR) uses resonant behavior of protons to probe the structure of matter and to produce images used in medical diagnosis (Fig. 15-31).

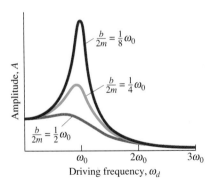

FIGURE 15-28   Resonance curves for several damping strengths. The quantity ω_0 is the undamped natural frequency $\sqrt{k/m}$.

FIGURE 15-30   The rebuilt bridge includes a much stiffer deck.

FIGURE 15-29   Collapse of the Tacoma Narrows Bridge—only four months after its opening in 1940—followed the growth of large-amplitude oscillations. That growth resulted from complex interactions between aerodynamic forces and the bridge itself.

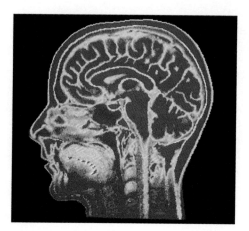

FIGURE 15-31 Magnetic resonance imaging of a human brain involves the resonant absorption of energy by protons in the brain tissue.

CHAPTER SYNOPSIS

Summary

1. **Oscillatory motion** is periodic motion back and forth about a system's equilibrium position. It occurs when a system subject to a restoring force is displaced from stable equilibrium.

2. **Simple harmonic motion** is the special case that occurs when the restoring force is proportional to displacement from equilibrium. Simple harmonic motion is described by

$$x = A \cos(\omega t + \phi),$$

where A is the maximum displacement, or **amplitude,** ϕ the **phase constant,** and ω the **angular frequency** of oscillation. For a mass-spring system, $\omega = \sqrt{k/m}$. Analogous expressions hold for other systems, including torsional oscillators and pendulums. The frequency in hertz (or cycles per second) and the period are related to the angular frequency by the equations

$$f = \frac{\omega}{2\pi} \quad \text{and} \quad T = \frac{1}{f} = \frac{2\pi}{\omega}.$$

3. Complex motion results when a system undergoes simple harmonic motion in two perpendicular directions. But when perpendicular motions with the same frequency differ by $\pi/2$ in phase, then the resultant motion is circular—showing that simple harmonic motion may be considered a component of circular motion.

4. In oscillatory motion, energy is transferred back and forth between kinetic and potential forms; in the absence of loss mechanisms, the total energy remains constant. The special case of simple harmonic motion results when the potential energy is a quadratic function of displacement—a situation that is approximated for many systems near stable equilibrium.

5. Energy loss results in **damped harmonic motion.** In an **underdamped** system, the damping is relatively weak and the motion is described by

$$x = A \, e^{-bt/2m} \cos(\omega t + \phi),$$

where b is a constant describing the strength of the damping. For very weak damping, ω is essentially the natural oscillation frequency of the system. As the damping increases, the frequency drops and eventually the system becomes **critically damped,** returning to equilibrium without oscillating.

6. A driven oscillatory system responds with motion at the driving frequency. The amplitude peaks for driving frequencies near the system's natural frequency—the phenomenon of **resonance.** The amplitude of resonant oscillations is given by

$$A = \frac{F_0}{m \sqrt{(\omega_d^2 - \omega_0^2)^2 + b^2 \, \omega_d^2/m^2}},$$

where F_0 is the amplitude of the driving force, ω_d its frequency, m the mass, b the damping constant, and $\omega_0 = \sqrt{k/m}$ the undamped natural frequency of the system.

Terms You Should Understand

(Pairs are closely related terms whose distinction is important; number in parentheses is chapter section where term first appears.)

oscillatory motion (introduction)
amplitude (15-1)
period, frequency (15-1)
hertz (15-1)
simple harmonic motion (15-2)

phase (15-2)
torsional oscillator (15-3)
simple pendulum, physical pendulum (15-3)
damping (15-6)
underdamping, critical damping, overdamping (15-6)
driven oscillations (15-7)
driving frequency, natural frequency (15-7)
resonance (15-7)

Symbols You Should Recognize

ω, f, T (15-1)
ϕ (15-2)
κ (15-3)
ω_d, ω_0 (15-7)

Problems You Should Be Able to Solve

calculating frequency from period, and vice versa (15-1)
calculating frequency and period of mass-spring systems (15-2)
determining velocity and acceleration in simple harmonic motion (15-2)

solving problems involving torsional oscillators, simple pendulums, and physical pendulums (15-3)
finding kinetic and potential energies in simple harmonic motion (15-5)
analyzing damped harmonic motion (15-6)
determining amplitudes of resonant motion (15-7)

Limitations to Keep in Mind

Simple harmonic motion occurs only when the restoring force is strictly proportional to displacement from equilibrium. For many systems this is a reasonable approximation only for small amplitudes.

The expression $T = 2\pi\sqrt{g/\ell}$ for the period of a pendulum holds only for small angular displacements from vertical.

The total energy of most oscillating systems is only approximately constant, as energy loss mechanisms damp the motion.

Equation 15-23 for damped harmonic motion applies only to underdamped systems, and the frequency ω is only approximately the natural frequency of the system.

QUESTIONS

1. Is a vertically bouncing ball an example of oscillatory motion? Of simple harmonic motion? Explain.
2. The vibration frequencies of atomic-sized systems are much higher than those of macroscopic mechanical systems. Why should this be?
3. What happens to the frequency of a simple harmonic oscillator when the spring constant is doubled? When the mass is doubled?
4. If the spring of a simple harmonic oscillator is cut in half, what happens to the frequency?
5. How does the frequency of a simple harmonic oscillator depend on its amplitude?
6. How would the frequency of a horizontal mass-spring system change if it were taken to the moon? Of a vertical mass-spring system? Of a simple pendulum?
7. In what ways is the motion of a damped pendulum not exactly periodic?
8. When is the acceleration of an undamped simple harmonic oscillator zero? When is the velocity zero?
9. Is the acceleration of a damped harmonic oscillator ever zero? Explain.
10. If the spring in a mass-spring system is not massless, how will this affect the frequency?
11. What will happen to the period of a mass-spring system if it is placed in a jetliner accelerating down the runway? What will happen to the period of a pendulum in the same situation?
12. Explain how simple harmonic motion might be used to determine the mass of objects in an orbiting spacecraft.
13. One pendulum consists of a solid rod of mass m and length ℓ, another of a compact ball of the same mass m on the end of a massless string of the same length ℓ. Which has the greater period? Why?
14. Why doesn't the period of a simple pendulum depend on its mass?
15. When the amplitude of a pendulum's motion becomes large enough that the approximation $\sin\theta \simeq \theta$ is no longer valid, the period lengthens with increasing amplitude. Why should this be?
16. Could a mass suspended from a string ever undergo non-oscillatory motion?
17. The x and y components of motion of a body are both simple harmonic with the same frequency and amplitude. What shape is the path of the body if the component motions are (a) in phase; (b) $\pi/2$ out of phase; (c) $\pi/4$ out of phase?
18. List five oscillatory systems and identify the damping forces that act on each.
19. Why is critical damping desirable in many mechanical systems?
20. Explain why the frequency of a damped system is lower than that of the equivalent undamped system.

FIGURE 15-32   A sewing machine needle executes simple harmonic motion (Question 21).

21. The needle of a sewing machine moves up and down, yet its driving force comes from an electric motor executing rotary motion (Fig 15-32). What sort of mechanism might convert rotary to up-and-down motion? *Hint:* What relevance would this conversion mechanism have for the present chapter?
22. The quantity $2m/b$ in Equation 15-23 has the units of time. How should this time compare with the period of the motion if damping is to be considered small?
23. Opera singers have been known to break glasses with their voices. How?
24. Depressing the rightmost pedal on a piano causes notes to sound much longer than usual. What quantity changes in Equation 15-23 when the pedal is pressed?
25. Real physical systems often have more than one resonant frequency. How can this come about?

PROBLEMS

Sections 15-1 and 15-2 Oscillations and Simple Harmonic Motion

1. A doctor counts 77 heartbeats in one minute. What are the period and frequency of the heart's oscillations?
2. A violin string playing the note "A" oscillates at 440 Hz. What is the period of its oscillation?
3. The vibration frequency of a hydrogen chloride molecule is 8.66×10^{13} Hz. How long does it take the molecule to complete one oscillation?
4. Write expressions for simple harmonic motion (a) with amplitude 10 cm, frequency 5.0 Hz, and with maximum displacement at $t = 0$ and (b) with amplitude 2.5 cm, angular frequency 5.0 s^{-1}, and with maximum velocity at $t = 0$.
5. Determine the amplitude, angular frequency, and phase constant for each of the simple harmonic motions shown in Fig. 15-33.

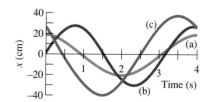

FIGURE 15-33   Problem 5.

6. A 200-g mass is attached to a spring of constant $k = 5.6$ N/m and set into oscillation with amplitude $A = 25$ cm. Determine (a) the frequency in hertz, (b) the period, (c) the maximum velocity, and (d) the maximum force in the spring.

7. An astronaut in an orbiting spacecraft is "weighed" by being strapped to a spring of constant $k = 400$ N/m, and set into simple harmonic motion. If the oscillation period is 2.5 s, what is the astronaut's mass?
8. A simple model for an automobile suspension consists of a mass attached to a spring. If the mass is 1900 kg and the spring constant is 26 kN/m, with what frequency and period will the car undergo simple harmonic motion?
9. A simple model of a carbon dioxide (CO_2) molecule consists of three mass points (the atoms) connected by two springs (electrical forces), as suggested in Fig. 15-34. One way this system can oscillate is if the carbon atom stays fixed and the two oxygens move symmetrically on either side of it. If the frequency of this oscillation is 4.0×10^{13} Hz, what is the effective spring constant? The mass of an oxygen atom is 16 u.

FIGURE 15-34   Problem 9.

10. Sketch the following simple harmonic motions on the same graph: (a) $x = (15$ cm$)[\cos(2.5t + \pi/2)]$, (b) motion with amplitude 30 cm, period 5.0 s, phase constant 0; (c) motion with amplitude 15 cm, frequency 0.40 Hz, phase constant 0.
11. Two identical mass-spring systems consist of 430-g masses on springs of constant $k = 2.2$ N/m. Both are displaced from equilibrium and the first released at time $t = 0$. How much later should the second be released so the two oscillations differ in phase by $\pi/2$?
12. The quartz crystal in a digital quartz watch executes simple harmonic motion at 32,768 Hz. (This is 2^{15} Hz, chosen so

FIGURE 15-35   Problem 13.

15 divisions by 2 give a signal at 1.00000 Hz.) If each face of the crystal undergoes a maximum displacement of 100 nm, find the maximum velocity and acceleration of the crystal faces.

13. A mass m slides along a frictionless horizontal surface at speed v_0. It strikes a spring of constant k attached to a rigid wall, as shown in Fig. 15-35. After a completely elastic encounter with the spring, the mass heads back in the direction it came from. In terms of k, m, and v_0, determine (a) how long the mass is in contact with the spring and (b) the maximum compression of the spring.

14. A 50-g mass is attached to a spring and undergoes simple harmonic motion. Its maximum acceleration is 15 m/s^2 and its maximum speed is 3.5 m/s. Determine (a) the angular frequency, (b) the spring constant, and (c) the amplitude of the motion.

15. Show by substitution that $x = A \sin \omega t$ is a solution to Equation 15-4.

16. Show by substitution that $x = a \cos \omega t - b \sin \omega t$ is a solution to Equation 15-4, and that this form is equivalent to Equation 15-14 with $A = \sqrt{a^2 + b^2}$ and $\phi = \tan^{-1}(b/a)$.

Section 15-3 Applications of Simple Harmonic Motion

17. How long should you make a simple pendulum so its period is (a) 200 ms; (b) 5.0 s; (c) 2.0 min?

18. At the heart of a grandfather clock is a simple pendulum 1.45 m long; the clock ticks each time the pendulum reaches its maximum displacement in either direction. What is the time interval between ticks?

19. A 640-g hollow ball 21 cm in diameter is suspended by a wire and is undergoing torsional oscillations at a frequency of 0.78 Hz. What is the torsional constant of the wire?

20. A physics student, bored by a lecture on simple harmonic motion, idly picks up his pencil (mass 9.2 g, length 17 cm) by the tip with his frictionless fingers, and allows it to swing back and forth with small amplitude. If it completes 6279 full cycles during the lecture, how long does the lecture last? (*Hint:* See Example 15-5.)

21. A pendulum of length ℓ is mounted in a rocket. What is its period if the rocket is (a) at rest on its launch pad; (b) accelerating upward with acceleration $a = \frac{1}{2}g$; (c) accelerating downward with acceleration $a = \frac{1}{2}g$; (d) in free fall?

22. A 340-g mass is attached to a vertical spring and lowered slowly until it rests at a new equilibrium position, which is 30 cm below the spring's original equilibrium. The system

is then set into simple harmonic motion. What is the period of the motion?

23. A mass is attached to a vertical spring, which then goes into oscillation. At the high point of the oscillation, the spring is in the original unstretched equilibrium position it had before the mass was attached; the low point is 5.8 cm below this. What is the period of oscillation?

24. Derive the period of a simple pendulum by considering the horizontal displacement x and the force acting on the bob, rather than the angular displacement and torque.

25. A solid disk of radius R is suspended from a spring of linear spring constant k and torsional constant κ, as shown in Fig. 15-36. In terms of k and κ, what value of R will give the same period for the vertical and torsional oscillations of this system?

FIGURE 15-36   Problem 25.

26. A thin steel beam 8.0 m long is suspended from a crane and is undergoing torsional oscillations. Two 75-kg steelworkers leap onto opposite ends of the beam, as shown in Fig. 15-37. If the frequency of torsional oscillations diminishes by 20%, what is the mass of the beam?

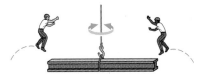

FIGURE 15-37   Problem 26.

27. Geologists use an instrument called a **gravimeter** to measure the local acceleration of gravity. A particular gravimeter uses the period of a 1-m-long pendulum to determine g. If g is to be measured to within 1 mgal (1 gal = 1 cm/s^2) and if the period can be measured with arbitrary accuracy, how accurately must the length of the pendulum be known?

28. A pendulum consists of a 320-g solid ball 15.0 cm in diameter, suspended by an essentially massless string 80.0 cm long. Calculate the period of this pendulum, treating it first as a simple pendulum and then as a physical pendulum. How much error is introduced by the simple pendu-

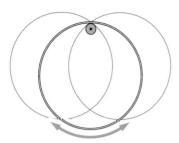

FIGURE 15-38 Problem 29.

lum approximation? *Hint:* Remember the parallel-axis theorem.

29. A thin, uniform hoop of mass M and radius R is suspended from a thin horizontal rod and set oscillating with small amplitude, as shown in Fig. 15-38. Show that the period of the oscillations is $2\pi\sqrt{2R/g}$. *Hint:* You may find the parallel-axis theorem useful.

30. A solid disk of mass M and radius R is mounted on a horizontal axle, as shown in Fig. 15-39. A spring of spring constant k is connected to the disk at a point $\frac{1}{2}R$ above the axle, and in equilibrium runs horizontally to a wall. If the disk is rotated slightly away from equilibrium, what is the angular frequency of the resulting oscillations? *Hint:* For small θ, $\sin\theta \simeq \theta$ and $\cos\theta \simeq 1$.

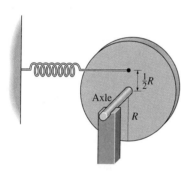

FIGURE 15-39 Problem 30.

31. A point mass m is attached to the rim of an otherwise uniform solid disk of mass M and radius R (Fig. 15-40). The disk is rolled slightly away from its equilibrium position and released. It rolls back and forth without slipping. Show that the period of this motion is given by

$$T = 2\pi\sqrt{\frac{3MR}{2\,mg}}.$$

32. Repeat the previous problem for the case when the disk does not contact the ground but is mounted on a frictionless horizontal axle through its center. Why is your answer different?

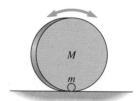

FIGURE 15-40 Problem 31.

33. A cyclist turns her bicycle upside down to tinker with it. After she gets it upside down, she notices the front wheel executing a slow, small-amplitude back-and-forth rotational motion with a period of 12 s. Considering the wheel to be a thin ring of mass 600 g and radius 30 cm, whose only irregularity is the presence of the tire valve stem, determine the mass of the valve stem.

34. A mass m is mounted between two springs of constants k_1 and k_2, as shown in Fig. 15-41. Show that the angular frequency of oscillation is given by

$$\omega = \sqrt{\frac{k_1 + k_2}{m}}.$$

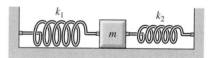

FIGURE 15-41 Problem 34.

35. Repeat the previous problem for the case when the springs are connected as in Fig. 15-42.

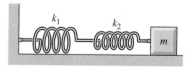

FIGURE 15-42 Problem 35.

Section 15-4 Simple Harmonic Motion in Two Dimensions

36. The position of a simple harmonic oscillator is described by $\mathbf{r} = B\sin\omega t\,\hat{\mathbf{i}} + B\cos\omega t\,\hat{\mathbf{j}}.$ (a) What is the amplitude of the motion? (b) Find an expression for the velocity. (c) Show that the speed remains constant, and find its value.

37. The equation for an ellipse is

$$\frac{x^2}{a^2} + \frac{y^2}{b^2} = 1.$$

Show that two-dimensional simple harmonic motion whose two components have different amplitudes and are $\pi/2$ out of phase gives rise to elliptical motion. How are a and b related to the amplitudes?

38. The x and y components of motion of a body are harmonic with frequency ratio 1.75 : 1. How many oscillations must each component undergo before the body returns to its initial position?

Section 15-5 Energy in Simple Harmonic Motion

39. A 1400-kg car with poor shock absorbers is bouncing down the highway at 20 m/s, executing vertical harmonic motion at 0.67 Hz. If the amplitude of the oscillations is 18 cm, what is the total energy in the oscillations? What fraction of the car's kinetic energy is this? Neglect rotational energy of the wheels and the fact that not all the car's mass participates in the oscillation.

40. A 450-g mass on a spring is oscillating at 1.2 Hz. The total energy of the oscillation is 0.51 J. What is the amplitude of oscillation?

41. The motion of a particle is described by

$$x = (45 \text{ cm})[\sin(\pi t + \pi/6)],$$

with x in cm and t in seconds. At what time is the potential energy twice the kinetic energy? What is the position of the particle at this time?

42. A torsional oscillator of rotational inertia 1.6 kg·m² and torsional constant 3.4 N·m/rad has a total energy of 4.7 J. What are its maximum angular displacement and maximum angular speed?

43. Show that the potential energy of a simple pendulum is proportional to the square of the angular displacement in the small-amplitude limit.

44. The total energy of a mass-spring system is the sum of its kinetic and potential energy: $E = \frac{1}{2}mv^2 + \frac{1}{2}kx^2$. Assuming E remains constant, differentiate both sides of this expression with respect to time and show that Equation 15-4 results. *Hint:* Remember that $v = dx/dt$.

45. A solid cylinder of mass M and radius R is mounted on an axle through its center. The axle is attached to a horizontal spring of constant k, and the cylinder rolls back and forth without slipping (Fig. 15-43). Write the statement of energy conservation for this system, and differentiate it to

obtain an equation analogous to Equation 15-4 (see previous problem). Comparing your result with Equation 15-4, determine the angular frequency of the motion.

46. A mass m is free to slide on a frictionless track whose height y as a function of horizontal position x is given by $y = ax^2$, where a is a constant with the units of inverse length. The mass is given an initial displacement from the bottom of the track and then released. Find an expression for the period of the resulting motion.

Section 15-6 Damped Harmonic Motion

47. A 250-g mass is mounted on a spring of constant $k = 3.3$ N/m. The damping constant for this system is $b = 8.4 \times 10^{-3}$ kg/s. How many oscillations will the system undergo during the time the amplitude decays to $1/e$ of its original value?

48. The vibration of a piano string can be described by an equation analogous to Equation 15-23. If the quantity analogous to $b/2m$ in that equation has the value 2.8 s⁻¹, how long will it take the vibration amplitude to drop to half its original value?

Section 15-7 Driven Oscillations and Resonance

49. A mass-spring system has $b/m = \omega_0/5$, where b is the damping constant and ω_0 the natural frequency. How does its amplitude when driven at frequencies 10% above and below ω_0 compare with its amplitude at ω_0?

50. A car's front suspension has a natural frequency of 0.45 Hz. The car's front shock absorbers are worn out, so they no longer provide critical damping. The car is driving on a bumpy road with bumps 40 m apart. At a certain speed, the driver notices that the car begins to shake violently. What speed?

51. Show by direct substitution that Equation 15-25 satisfies Equation 15-24 with A given by Equation 15-26.

52. A harmonic oscillator is underdamped provided that the damping constant b is less than $\sqrt{2}m\omega_0$, where ω_0 is the natural frequency of undamped motion. Show that for an underdamped oscillator, Equation 15-26 has a maximum for a driving frequency less than ω_0.

Paired Problems

(Both problems in a pair involve the same principles and techniques. If you can get the first problem, you should be able to solve the second one.)

53. A particle undergoes simple harmonic motion with amplitude 25 cm and maximum speed 4.8 m/s. Find (a) the angular frequency, (b) the period, and (c) the maximum acceleration.

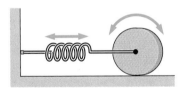

FIGURE 15-43 Problem 45.

54. A particle undergoes simple harmonic motion with maximum speed 1.4 m/s and maximum acceleration 3.1 m/s². Find (a) the angular frequency, (b) the period, and (c) the amplitude of the motion.

55. A massless spring of spring constant $k = 74$ N/m is hanging from the ceiling. A 490-g mass is hooked onto the unstretched spring and allowed to drop. Find (a) the amplitude and (b) the period of the resulting motion.

56. A massless spring is hanging from the ceiling. A mass is hooked onto the end of the spring and allowed to drop. If the amplitude of the resulting motion is 20 cm, what is its frequency?

57. A meter stick is suspended from one end and set swinging. What is the period of the resulting oscillations, assuming they have small amplitude?

58. A meter stick is suspended from a frictionless rod inserted through a small hole at the 25 cm mark. What is the period of small-amplitude oscillations about the stick's equilibrium position?

59. Two balls each of unknown mass m are mounted on opposite ends of a 1.5-m-long rod of mass 850 g. The system is suspended from a wire attached to the center of the rod and set into torsional oscillations. If the torsional constant of the wire is 0.63 N·m/rad and the period of the oscillations is 5.6 s, what is the unknown mass m?

60. Figure 15-44 shows a bird feeder that consists of a 340-g solid circular disk 50 cm in diameter suspended by a wire attached at the center. Two 65-g birds land at opposite ends of a diameter, and the system goes into torsional oscillation at 2.6 Hz. What is the torsional constant of the wire?

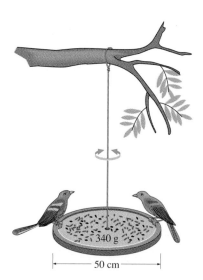

FIGURE 15-44   Problem 60.

61. Two mass-spring systems with the same mass are undergoing oscillatory motion with the same amplitudes. System 1 has twice the frequency of system 2. How do (a) their energies and (b) their maximum accelerations compare?

62. Two mass-spring systems have the same mass and the same total energy. The amplitude of system 1 is twice that of system 2. How do (a) their frequencies and (b) their maximum accelerations compare?

Supplementary Problems

63. While waiting for your plane to take off, you suspend your keys from a thread and set the resulting pendulum oscillating. It completes exactly 90 cycles in 1 minute. You repeat the experiment as the plane accelerates down the runway, and now find the pendulum completes exactly 91 cycles in 1 minute. Find the plane's acceleration.

64. Integrating the nonconstant acceleration of a harmonic oscillator over time from the time of maximum displacement to the time of zero displacement should give the velocity at zero displacement. Carry out this integration, using Equation 15-9 for the acceleration, and show that your answer is just the maximum velocity.

65. A 500-g block on a frictionless surface is connected to a rather limp spring of constant $k = 8.7$ N/m. A second block rests on the first, and the whole system executes simple harmonic motion with a period of 1.8 s. When the amplitude of the motion is increased to 35 cm, the upper block just begins to slip. What is the coefficient of static friction between the blocks?

66. The potential energy of a 75-g particle is given by $U = ax^2 - bx^4$, where $a = 3.5$ J/m², $b = 1.2$ J/m⁴, and x is in meters. (a) Show that there is a metastable equilibrium at $x = 0$. (b) Find the frequency of small amplitude oscillations about this equilibrium. *Hint:* For small x, $x^4 \ll x^2$. (c) What is the maximum amplitude for oscillatory (but not necessarily simple harmonic) motion to occur?

67. Repeat Problem 46 for a small solid ball of mass M and radius R that rolls without slipping on the parabolic track.

68. A child twirls around on a swing, twisting the swing ropes, as shown in Fig. 15-45. As a result, the child and swing rise slightly, with the rise, h, in cm equal to the square of the number of full turns of the swing. When the child stops twisting up the swing, it goes into torsional oscillation. What is the period of this oscillation, assuming that all the potential energy of the system is gravitational? The combined mass of the child and swing is 20 kg, and the rotational inertia of the pair about the appropriate vertical axis is 0.12 kg·m².

69. A 1.2-kg block rests on a frictionless surface and is attached to a horizontal spring of constant $k = 23$ N/m (Fig. 15-46). The block is oscillating with amplitude 10 cm and with phase constant $\phi = -\pi/2$. A block of mass 0.80 kg is moving from the right at 1.7 m/s. It strikes the first block

FIGURE 15-45 Problem 68.

FIGURE 15-46 Problem 69.

when the latter is at the rightmost point in its oscillation. The collision is completely inelastic, and the two blocks stick together. Determine the frequency, amplitude, and phase constant (relative to the *original* $t = 0$) of the resulting motion.

70. The motion of a two-dimensional simple harmonic oscillator is described by

$$\mathbf{r} = A \sin \omega t \, \hat{\mathbf{i}} + A \sin(\omega t + \pi/4) \, \hat{\mathbf{j}} \, .$$

The path is an ellipse; find the orientation of its major axis. *Hint:* At what time is the magnitude of the displacement a maximum? What are the components of the displacement at this time?

71. A small object of mass m slides without friction in a circular bowl of radius R. Derive an expression for small-amplitude oscillations about equilibrium, and compare with that of a simple pendulum.

72. A more exact expression than Equation 15-22 for the period of a simple pendulum is

$$T = T_0[1 + \tfrac{1}{4}\sin^2(\tfrac{1}{2}\,\theta_0) + \tfrac{9}{64}\sin^4(\tfrac{1}{2}\,\theta_0) + \cdots],$$

where $T_0 = 2\pi \sqrt{\ell/g}$ is the period in the limit of arbitrarily small amplitude, and θ_0 is the amplitude. The $\cdots$ indicates that additional terms (in fact, infinitely many more) are needed for an exact expression. For a pendulum with $T_0 = 1.00$ s, plot the period given above versus amplitude for amplitudes from 0 to 45°. By what

percentage does the plotted period differ from T_0 for θ_0 of 30° and 45°?

73. A mass m is connected between two springs of length L, as shown in Fig. 15-47. At equilibrium, the tension force in each spring is F_0. Find the period of oscillations *perpendicular* to the springs, assuming sufficiently small amplitude that the magnitude of the spring tension is essentially unchanged.

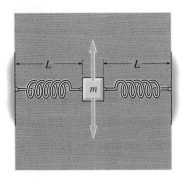

FIGURE 15-47 Problem 73. This is a top view; the mass is moving on a frictionless horizontal surface.

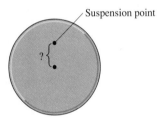

FIGURE 15-48 Problem 74.

74. A disk of radius R is suspended from a pivot somewhere between its center and edge (Fig. 15-48). For what pivot point will the period of this physical pendulum be a minimum?

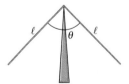

FIGURE 15-49 Problem 75.

75. A uniform piece of wire is bent into a V-shape with angle θ between two legs of length ℓ. The wire is placed over a pivot, as shown in Fig. 15-49. Show that the angular frequency of small-amplitude oscillations about this equilibrium is given by

$$\omega = \sqrt{\frac{3g \cos (\theta/2)}{2\ell}} \, .$$

WAVE MOTION

An ocean wave carries energy across thousands of kilometers of open water before breaking.

Disturbing the simple mass-spring system in Fig. 16-1a results in oscilla-tory motion. In the more complicated system of Fig. 16-1b, the motion of one mass affects the others. The result is a **wave**—a disturbance that moves or **propagates** through the system. Although the individual masses oscil-late about their equilibrium positions, they do not move with the wave. What moves is energy, as evidenced by the temporary motions of the masses and stretching of the springs as the wave passes. We clarify this point in defining a wave:

A wave is a traveling disturbance that transports energy but not matter.

You're familiar with many kinds of waves. Sound waves carry energy—and information—through the air, but the air itself doesn't move. Similarly, ocean

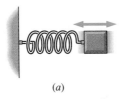

(a)

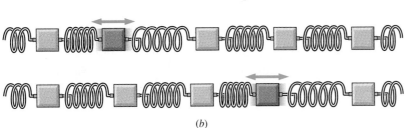

(b)

FIGURE 16-1 (*a*) Disturbance of the simple mass-spring system results in oscillatory motion. (*b*) A disturbance in this coupled mass-spring system is communicated to adjacent masses, resulting in a propagating wave.

waves bring to shore the energy imparted by distant winds, but the water itself doesn't travel with the wave (Fig. 16-2). Earthquakes set up wave motion in Earth itself, while ultrasound waves probe the human body (Fig. 16-3).

These examples are all of **mechanical waves,** involving the disturbance of some mechanical medium like air, water, or the solid Earth. This chapter and the next explore the behavior of mechanical waves. Many of their properties are also shared by electromagnetic waves, which include visible light, radio waves, microwaves, x rays, and others. We will explore electromagnetic waves in Chapters 34 through 38.

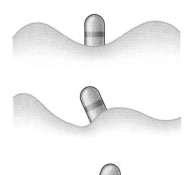

FIGURE 16-2 Waves disturb the buoy, but the buoy does not participate in their shoreward motion.

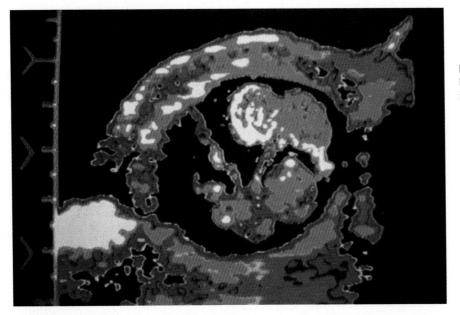

FIGURE 16-3 Twin fetuses are evident in this image, produced by ultrasound waves reflecting in the human body.

FIGURE 16-4 (*a*) Longitudinal and (*b*) transverse waves.

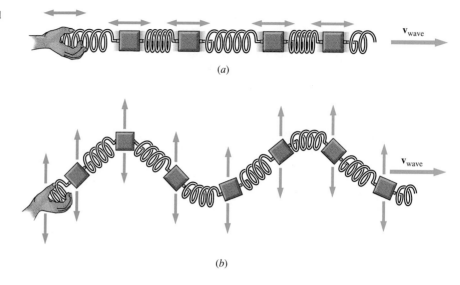

(*a*)

(*b*)

16-1 WAVE PROPERTIES

Amplitude

A wave disturbs a medium from its equilibrium state. The maximum value of the disturbance is the wave **amplitude.** Amplitude measures whatever physical quantity is affected by the wave—the height of a water wave, for example, or the pressure of a sound wave.

Longitudinal and Transverse Waves

We could start a wave on the coupled mass-spring system of Fig. 16-1*b* by displacing a mass either along the direction of the spring or at right angles to it. In the first case we get a **longitudinal wave,** in which the masses move back and forth in the direction in which the wave itself travels (Fig. 16-4*a*). Figure 16-4*b* shows a **transverse wave,** with masses moving perpendicularly to the direction of wave propagation. Some waves are neither fully longitudinal nor transverse but include components of each (Fig. 16-5).

Waveforms

Wave disturbances come in many shapes, called **waveforms** (Fig. 16-6). An isolated disturbance, traveling through an otherwise undisturbed medium, is a **pulse.** A pulse occurs when a medium is disturbed only briefly. At the opposite

FIGURE 16-5 A water wave has both longitudinal and transverse components, as parcels of water describe nearly circular paths around their equilibrium positions.

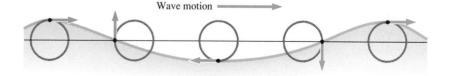

Wave motion

extreme is a **continuous wave,** produced when a medium is disturbed in a regular, periodic way. A **wave train** results in the more realistic case when a periodic disturbance lasts for a finite time.

Wavelength and Period

A continuous wave repeats itself in both space and time. The **wavelength** λ is the *distance* over which the wave pattern repeats, as measured at a *fixed time* (Fig. 16-7). The wave **period** T is the *time* for one complete wave cycle to pass a *fixed position* (Fig. 16-8). The **frequency** f, or number of wave cycles passing a given point per unit time, is the inverse of the period.

Wave Speed

A wave travels at a characteristic speed through its medium. Under typical conditions, the speed of sound in air is 340 m/s. Small ripples on the surface of a pond move at about 20 cm/s, while earthquake waves move through Earth's outer crust at about 6 km/s. Wave speed, wavelength, and period are related. In one wave period, a fixed observer sees one complete wavelength go by (Fig 16-9). Thus, the wave moves one wavelength in one period, so its speed is

$$v = \frac{\lambda}{T} = \lambda f, \tag{16-1}$$

where the second equality follows because the period and frequency are inverses.

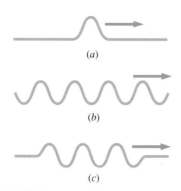

FIGURE 16-6 (*a*) A pulse, (*b*) a continuous wave, and (*c*) a wave train.

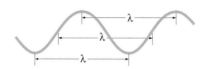

FIGURE 16-7 The wavelength λ is the distance over which the wave pattern repeats.

FIGURE 16-8 (*a*) A continuous wave, seen at successive instants of time. At a fixed position, the medium undergoes oscillatory motion. (*b*) The period of this motion is the wave period T.

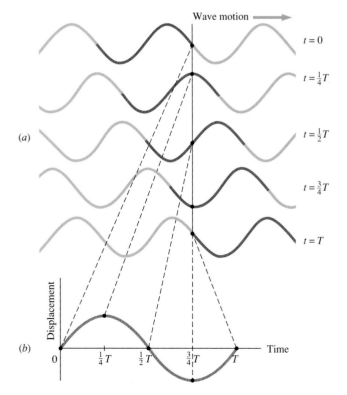

FIGURE 16-9 One full cycle—occupying a distance of one wavelength—
passes a given point in one wave period.
The wave speed is therefore $v = \lambda/T$.

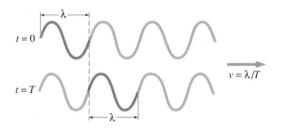

16-2 MATHEMATICAL DESCRIPTION OF WAVE MOTION

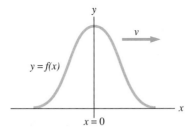

FIGURE 16-10 A "snapshot" of a
wave pulse at time $t = 0$. The pulse is
moving to the right with speed v.

Figure 16-10 shows a "snapshot" of a wave pulse at time $t = 0$. The pulse is
moving to the right with speed v, and at $t = 0$ its peak coincides with $x = 0$. We
can describe the pulse by giving the displacement y as a function of position x:
$y = f(x)$, where the peak of the function f occurs where its argument is zero.

Figure 16-11 shows the pulse some time t later. Moving at speed v, its peak
is now located at $x = vt$. We can describe this displaced pulse by adjusting the
function $y = f(x)$ so it peaks at $x = vt$ instead of at $x = 0$. The function f itself
peaks where its argument is zero, so if we change the argument to $x - vt$ we will
have just what we want: a function whose peak occurs when $x - vt = 0$, or $x =
vt$. As time increases, so does vt and, therefore, the value of x that gives the
location of the peak. Thus, the function $f(x - vt)$ correctly describes the moving pulse.

Although we considered a single pulse, the same argument applies to *any*
function $f(x)$, including continuous waves. Quite generally, a wave moving in the
positive x direction is described by some function of the form $f(x - vt)$, with v
the wave speed. You can convince yourself that a wave moving in the negative
x direction is described by $f(x + vt)$.

A particularly important case is the **simple harmonic wave,** so called because the wave medium executes simple harmonic motion at the wave period T.
A simple harmonic wave is sinusoidal in shape, with one full cycle occupying
one wavelength λ. At time $t = 0$ we can therefore describe such a wave by

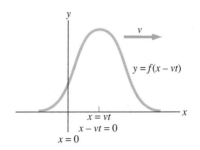

FIGURE 16-11 At a later time t
the wave pulse has moved to the right a
distance vt. It is described by the same
function f, but now with argument
$x - vt$.

$$y(x, 0) = A\cos\left(\frac{2\pi x}{\lambda}\right),$$

where y is the displacement from equilibrium and A the amplitude (Fig. 16-12a).
We chose the argument of the cosine so that it becomes 2π when $x = \lambda$ to ensure
that the function repeats itself in one wavelength. This expression holds at
$t = 0$; to get an expression valid for all times, we replace x by $x \pm vt$:

$$y(x, t) = A\cos\left[\frac{2\pi(x \pm vt)}{\lambda}\right] = A\cos\left(\frac{2\pi x}{\lambda} \pm 2\pi ft\right), \qquad (16\text{-}2)$$

where the last step follows using $v/\lambda = f$ from Equation 16-1. Equation 16-2
shows that if we sit at a fixed position—say, $x = 0$—then the displacement

varies in time as $y(0, t) = A\cos(2\pi ft)$, which we recognize as simple harmonic motion with **angular frequency** given by

$$\omega = 2\pi f = \frac{2\pi}{T}, \qquad (16\text{-}3)$$

as shown in Fig. 16-12b. Similarly, it is convenient to define the **wave number,** k:

$$k = \frac{2\pi}{\lambda}. \qquad (16\text{-}4)$$

The wave number is the spatial analog of frequency; the latter describes the number of radians of wave cycle per unit *time,* the former the number of radians per unit *distance.* The definition of k shows that its units are m^{-1}. Using ω and k, we can rewrite Equation 16-2:

$$y(x, t) = A\cos(kx \pm \omega t). \qquad (16\text{-}5)$$

The simple way in which k and ω enter this description of the wave is the reason these quantities are so often used. We can use Equation 16-1 with Equations 16-3 and 16-4 to write the wave speed in terms of ω and k:

$$v = \frac{\lambda}{T} = \frac{2\pi/k}{2\pi/\omega} = \frac{\omega}{k}. \qquad (16\text{-}6)$$

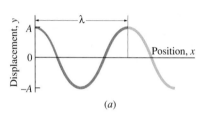

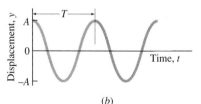

FIGURE 16-12 (a) "Snapshot," at time $t = 0$, of the simple harmonic wave described by Equation 16-2. The graph is the wave displacement as a function of position x. (b) The wave displacement at a fixed position, plotted as a function of *time.* The medium oscillates with angular frequency $\omega = 2\pi f = 2\pi/T$. Note the difference between (a) and (b); (a) shows the *spatial* variation at a fixed time, while (b) shows the *temporal* variation at a *fixed position.*

● **EXAMPLE 16-1** SURF'S UP

A surfing physicist paddles out beyond the breaking surf to a deep-water region where the ocean waves are sinusoidal in shape, with crests 14 m apart. The surfer rises a vertical distance 3.6 m from wave trough to crest, a process that takes 1.5 s. Find the wave speed, and describe it with an equation of the same form as Equation 16-5. Take a wave crest to be at $x = 0$ at time $t = 0$, with the positive x direction toward the open ocean.

Solution
Figure 16-12b shows that the time from trough to crest is half a period, so $T = 3.0$ s. Then Equation 16-1 gives

$$v = \frac{\lambda}{T} = \frac{14 \text{ m}}{3.0 \text{ s}} = 4.67 \text{ m/s}.$$

To use the form of Equation 16-5 we need the amplitude A, wavenumber k, and angular frequency ω. Figure 16-12 shows that the amplitude is half the trough-to-crest displacement, or 1.8 m, while Equations 16-3 and 16-4 give ω and k:

$$\omega = \frac{2\pi}{T} = \frac{2\pi}{3.0 \text{ s}} = 2.09 \text{ s}^{-1}, \, k = \frac{2\pi}{\lambda} = \frac{2\pi}{14 \text{ m}} = 0.449 \text{ m}^{-1}.$$

With the wave propagating in the $-x$ direction—toward shore—we take the plus sign in Equation 16-5, so the wave description is

$$y(x, t) = 1.8\cos(0.449x + 2.09t) \text{ m}.$$

EXERCISE The temporal and spatial variation of the waves in a microwave oven are described by $\cos(50.3x - 15.1t)$, with t in nanoseconds and x in meters. Find (a) the wave frequency in hertz, (b) the wavelength, and (c) the wave speed.

Answers: (a) 2.4 GHz; (b) 12.5 cm; (c) 3.00×10^8 m/s

Some problems similar to Example 16-1: 10, 13, 14
●

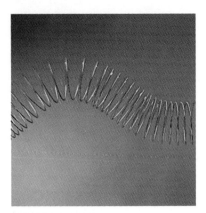

FIGURE 16-13 A pulse on a Slinky is an example of a transverse wave on an elongated structure.

16-3 WAVES ON A STRING

A Stretched String

What determines how fast waves propagate in a mechanical medium? The answer lies in the physical properties of the medium. To find that answer we apply Newton's laws, as we now show for transverse waves on a stretched string. Our results are directly applicable to strings and wires on musical instruments, as well as to other elongated structures (Fig. 16-13).

Our string has mass per unit length μ in kilograms per meter, and is stretched to a tension force F in newtons. In equilibrium, the string lies along the x axis. Suppose we distort a section of it slightly by displacing it in the y direction. We want to show that wave motion results when the string is released and to determine the wave speed.

To find the speed, we need to apply Newton's law to the motion of the string. Figure 16-14a shows the wave pulse we've created by disturbing the string; we assume the pulse is moving to the right with some speed v. It is easiest to apply Newton's law in a frame of reference moving with the uniform velocity of the pulse; in that frame the entire string is moving to the *left* at speed v. As a section of the string encounters the pulse, which is stationary in this frame of reference, its motion deviates from a straight line as it rides up over the pulse (Fig. 16-14b).

Whatever the pulse shape, we can describe a small enough section at the top as a circular arc of some radius R, as shown in Fig. 16-14c. Then a small section of string right at the top of the pulse undergoes circular motion with speed v and radius R; if its mass is m, then Newton's law requires that a force of magnitude mv^2/R act toward the center of curvature in order to keep the string section on its circular path. This force is provided by the difference in string tension between the two ends of the section; as Fig. 16-14c shows, the section's curva-

FIGURE 16-14 (a) A pulse moving to the right with speed v on a stretched string. (b) In the reference frame of the moving pulse, the string moves to the left. Individual segments of the string follow curved paths as they move through the pulse. (c) A blow-up of a small string segment at the top of the pulse. The net force on the segment is the vector sum of the tension forces **F** at the two ends; this force keeps the string segment in its path through the curved pulse. The angle θ is exaggerated; it is actually so small that $\sin\theta \simeq \theta$, making $F_{\text{net}} \simeq 2F\theta$. The quantity R is the curvature radius at the top of the pulse.

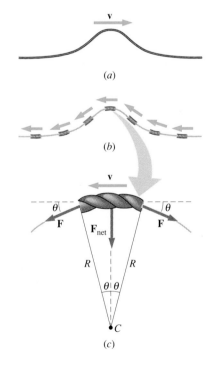

ture means the tension vectors at the two ends point in different directions. The tension at each end contributes a component $F \sin\theta$ toward the center of curvature, where θ is shown in Fig. 16-14c. Then the net force on the segment is $2F \sin\theta$.

Now we make an additional assumption: that the disturbance of the string is small, in the sense that the string remains almost horizontal even at the pulse. Then the angle θ is small, and we can apply the approximation $\sin\theta \simeq \theta$. Therefore the net force on the string section becomes approximately $2F\theta$. Furthermore, the small-disturbance approximation means that the tension doesn't vary significantly from its undisturbed value, so the F in this expression is essentially the same F we're using to characterize the tension throughout the string. Finally, our curved string section forms a circular arc whose length, from Fig. 16-14c, is $2\theta R$. Multiplying by the mass per unit length μ gives its mass: $m = 2\theta R\mu$. Now we can apply Newton's law, equating the net force $2F\theta$ to the mass times acceleration:

$$2F\theta = \frac{mv^2}{R} = \frac{2\theta R\mu v^2}{R} = 2\theta\mu v^2.$$

Solving for the wave speed v then gives

$$v = \sqrt{\frac{F}{\mu}}. \tag{16-7}$$

Does this result make sense? The string tension provides the restoring force that drives the disturbed part of the string toward equilibrium. The greater the tension F, the greater the acceleration of the disturbed string, and thus the more rapidly the wave should propagate. The string's inertia, on the other hand, limits the acceleration with which the string responds to the force, and therefore a greater mass per unit length should slow the wave propagation. Equation 16-7, with F in the numerator and μ in the denominator, reflects both these trends.

We have made no assumption about wave shape other than to assume that the disturbance is small; therefore Equation 16-7 applies to small-amplitude pulses, continuous waves, and wave trains of any shape. In Section 16-6 we will derive Equation 16-7 using a more advanced analysis that gives a general equation shared by many systems that support propagating waves and will show explicitly that simple harmonic waves are among its solutions.

● **EXAMPLE 16-2** ROCK CLIMBING

Two climbers are joined by a 35-m-long rope of total mass 5.0 kg. One climber strikes the rope with a fist, and 1.8 s later the second climber feels the effect. What is the rope tension?

Solution
Solving Equation 16-7 for the tension gives $F = \mu v^2$. Here $\mu = m/L$ and $v = L/t$, with L and m the rope length and mass and t the time interval. So we have

$$F = \mu v^2 = \left(\frac{m}{L}\right)\left(\frac{L}{t}\right)^2 = \frac{mL}{t^2} = \frac{(5.0 \text{ kg})(35 \text{ m})}{(1.8 \text{ s})^2} = 54 \text{ N}.$$

Is this number reasonable? A typical adult weighs between 500 and 1000 N, so the rope is supporting only a small fraction of the lower climber's weight—a reasonable situation.

EXERCISE A 4.5-g piano wire is under 680 N tension. If it supports waves propagating at 320 m/s, what is its length?

Answer: 68 cm

Some problems similar to Example 16-2: 21, 24, 27, 29 ●

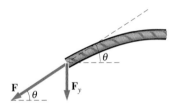

FIGURE 16-15 The force acting on the left end of the string segment is the vertical component of the tension force **F**, with magnitude $F\sin\theta$, or approximately $F\theta$ for small θ. The tangent of θ, which is also essentially equal to θ, is the slope—that is, the derivative of the displacement y with respect to x—at the end of the segment.

16-4 WAVE POWER AND INTENSITY

A wave propagates because part of a medium communicates its motion to adjacent parts. In the process, energy passes through the medium. As always in mechanics, that energy is transferred by forces that do work.

In the case of a stretched string, it is the tension force that does work on the string, and this work results in energy transfer along the string in the direction of the wave propagation. As we showed in Section 7-7, power—the rate of doing work—is the dot product of the force with the velocity. Figure 16-15 shows the section of string we considered earlier, now in the reference frame of the string. In that frame the left end of the displaced segment is moving downward as the pulse passes to the right, and, as we found earlier, the vertical component of the tension force on the left end alone is very nearly $-F\theta$, where the minus sign designates downward. Then the rate at which work is done by this force is

$$P = F_y u = -F\theta u,$$

where u is the vertical speed of the string segment.

> **TIP** **The Medium Is Not the Wave** The speed u is the speed at which the disturbed medium moves and is *not* the same as the wave speed v. For this transverse wave, the string motion is in fact at right angles to the wave propagation.

This expression for power holds for any wave shape; we now specialize to the case of a simple harmonic wave, described by Equation 16-5. The speed u at which the string moves vertically is simply the derivative of the string displacement $y(x, t) = A\cos(kx - \omega t)$ with respect to time:

$$u = \frac{\partial y}{\partial t} = \omega A \sin(kx - \omega t),$$

where we used the chain rule for differentiation, differentiating cosine to get $-$sine, then multiplying by the derivative, $-\omega$, of the cosine's argument $kx - \omega t$. As Fig. 16-15 shows, the tangent of the angle θ is the slope, $\partial y/\partial x$, at the left end of the segment. In the small angle approximation, $\theta \simeq \tan\theta$. Then we have

$$\theta \simeq \frac{\partial y}{\partial x} = -kA \sin(kx - \omega t).$$

The minus sign shows that θ is *positive*—as in Fig. 16-15—when $\sin(kx - \omega t)$ is *negative*. Thus, the force is downward when the velocity **u** is downward, indicating that the tension force does work on the string segment.

TIP **What's That Symbol ∂?** Equation 16-5 gives the displacement in a simple harmonic wave as a function of *both position and time*. Figure 16-15 is a "snapshot" of the string *at an instant in time*. Therefore, the derivative is to be taken *with time fixed*. This qualification is important because the string displacement is a function of both x and t and changes if *either* of these variables changes. With functions of more than one variable, we use the symbol ∂ instead of the usual d to indicate a rate of change with respect to one variable while other variables are held fixed. Such derivatives, like $\partial y/\partial t$ and $\partial y/\partial x$, are called **partial derivatives.**

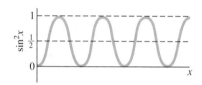

FIGURE 16-16 The function $\sin^2 x$ swings symmetrically between 0 and 1, so its average value is $\frac{1}{2}$.

Using our expressions for u and θ, we can then write the power transmitted along the wave as

$$P = -F\theta u = F\omega k A^2 \sin^2(kx - \omega t).$$

The term $\sin^2(kx - \omega t)$ shows that the power fluctuates throughout the wave cycle. We're usually more interested in the *average* power, which we get by noting that the average value of $\sin^2$ is $\frac{1}{2}$ (see Fig. 16-16). Then the average power is simply

$$\overline{P} = \frac{1}{2} F\omega k A^2.$$

We can give this expression more physical meaning if we use Equations 16-6 and 16-7 to write $k = \omega/v$ and $F = \mu v^2$, where v is the wave speed. Then we have

$$\overline{P} = \frac{1}{2} \mu \omega^2 A^2 v. \qquad (16\text{-}8)$$

This equation makes sense since it shows that the power—the rate at which energy moves along the wave—is directly proportional to the wave speed v.

Although we derived Equation 16-8 for waves on a string, some aspects of the equation are common to all waves. In particular, wave power is always proportional to the wave speed and to the *square* of the wave amplitude.

● **EXAMPLE 16-3** WAVE POWER

A garden hose with 440 g of mass per meter of length is lying on frictionless ground. You pull on one end of the hose until the tension is 12 N. You then shake it from side to side, sending waves along the hose (Fig. 16-17). At what average rate must you do work if you displace the hose 25 cm either side of its equilibrium position, completing two full back-and-forth cycles every second? What will be the distance between wave crests?

Solution

The work you do is transmitted along the hose as the energy of the wave, so you do work at the rate given by Equation 16-8. To use this equation we need the wave speed v and angular frequency ω. The former is given by Equation 16-7:

$$v = \sqrt{\frac{F}{\mu}} = \sqrt{\frac{12 \text{ N}}{0.44 \text{ kg/m}}} = 5.22 \text{ m/s},$$

FIGURE 16-17   *How much power does it take to send waves along the hose?*

while the angular frequency is $\omega = 2\pi f = (2\pi)(2.0 \text{ Hz}) = 12.6 \text{ s}^{-1}$. Then the average power is

$$\overline{P} = \frac{1}{2}\mu\omega^2 A^2 v$$

$$= \frac{1}{2}(0.44 \text{ kg/m})(12.6 \text{ s}^{-1})^2(0.25 \text{ m})^2(5.22 \text{ m/s})$$

$$= 11 \text{ W}.$$

Finally, Equation 16-1 gives the wavelength—the distance between wave crests:

$$\lambda = \frac{v}{f} = \frac{5.22 \text{ m/s}}{2.0 \text{ Hz}} = 2.6 \text{ m}.$$

E X E R C I S E   A suspension bridge cable has 3800 kg/m and is under 230 MN tension. What power is required to send a 15-Hz wave with amplitude 5.0 mm along this cable?

Answer: 100 kW

Some problems similar to Example 16-3:   33–35, 38   ●

Wave Intensity

The total power is useful in describing a wave confined to a narrow structure like a string for mechanical waves or a cable or optical fiber for electromagnetic waves. But when waves travel throughout a three-dimensional medium, it makes more sense to talk about the rate at which the wave carries energy across a unit area. This quantity defines the wave **intensity:**

> *The intensity of a wave is the average rate at which the wave carries energy per unit area across a surface perpendicular to the wave propagation.*

Since the rate of energy flow is power, the intensity is just the power per unit area carried by the wave; its units are W/m².

A **wavefront** is a surface over which the argument of the wave function ($x \pm vt$ in the simple cases we've been discussing) has the same value. For example, a surface where the wave displacement is at its peak value is a wavefront. A **plane wave** is one whose wavefronts are planes. Since the wave energy doesn't spread out as the wave propagates, the intensity remains constant (Fig. 16-18a). But with waves from a localized source, the wave energy spreads over ever greater areas, and therefore, the intensity decreases. **Spherical waves** originate from point-like sources and their spherical wavefronts spread in all directions; since the area increases as the square of the distance from the source, the intensity decreases as the inverse square of the distance (Fig. 16-18b). If P is the total power emitted by the source, then the intensity at a distance r is given by

$$I = \frac{P}{A} = \frac{P}{4\pi r^2}, \quad \text{(spherical wave)} \qquad (16\text{-}9)$$

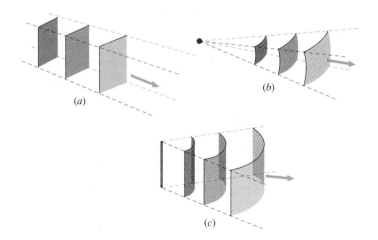

FIGURE 16-18 (*a*) A plane wave does not spread, so its intensity remains constant. (*b*) As a spherical wave spreads from a point source, its intensity decreases as the inverse square of the distance from the source. (*c*) Intensity decreases as the inverse of the distance from a line source.

since the area of a sphere is $4\pi r^2$. Similarly, for a line-like source, intensity decreases as the inverse of the distance (Fig. 16-18*c*). It isn't that wave energy is being lost; rather, the same energy is spread over ever greater areas as the waves propagate outward.

Table 16-1 lists the intensities of various waves.

● **EXAMPLE 16-4** READING LIGHT

The page of your book is 1.9 m from a 75-W light bulb, and the light is just barely adequate for reading. How far from a 40-W bulb would the page have to be to get the same light intensity? Treat the bulbs as sources of spherical light waves.

Solution
Equation 16-9 gives the intensity of a spherical wave in terms of the source power and distance. We want the same intensity from both bulbs; equating the intensities as given by Equation 16-9, we have

$$\frac{P_{75}}{4\pi r_{75}^2} = \frac{P_{40}}{4\pi r_{40}^2}$$

Solving for the unknown distance r_{40} then gives

$$r_{40} = r_{75}\sqrt{\frac{P_{40}}{P_{75}}} = (1.9 \text{ m})\sqrt{\frac{40 \text{ W}}{75 \text{ W}}} = 1.4 \text{ m}.$$

Although the 40-W bulb has only about half the power output, the decrease in distance is not as great because the intensity depends on the inverse *square* of the distance.

EXERCISE Use the entry in Table 16-1 for sunlight at Earth's orbit, along with Earth's distance from the Sun, to calculate the total power emitted by the Sun.

Answer: 3.9×10^{26} W

Some problems similar to Example 16-4: 39–41, 59

▲ **TABLE 16-1** WAVE INTENSITIES

WAVE	INTENSITY, W/m²
Sound, 4 m from loud rock band	1
Sound, jet aircraft at 50 m	10
Sound, whisper at 1 m	10^{-10}
Light, sunlight at Earth's orbit	1368
Light, sunlight at Jupiter's orbit	50
Light, 1 m from typical camera flash	4000
Light, at target of laser fusion experiment	10^{18}
TV signal, 5 km from 50 kW transmitter	1.6×10^{-4}
Microwaves, inside microwave oven	6000
Earthquake wave, 5 km from Richter 7.0 quake	4×10^4

FIGURE 16-19 (*a*) Sinusoidal wave trains propagating in opposite directions. (b) As their crests overlap, they interfere constructively, producing the darker waveform. (*c*) A quarter cycle later, they interfere destructively; the result is zero displacement (dark line). (*d*) The wave trains continue on their way, unaffected by their encounter.

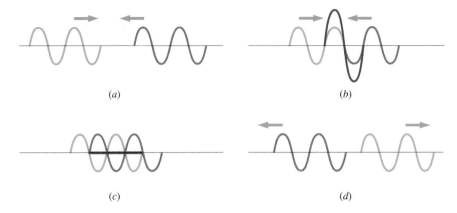

(*a*)   (*b*)

(*c*)   (*d*)

16-5 THE SUPERPOSITION PRINCIPLE AND WAVE INTERFERENCE

What happens if we launch two wave trains from opposite ends of our stretched string? Where they meet, experiment shows that the displacement of the string is just the sum of the displacements of the two wave trains (Fig. 16-19). It's also possible to prove this mathematically, under the small-amplitude approximation that led to Equation 16-7; we'll see this in the next section. The combining of two waves to form a composite wave is called **interference.** The interference is **constructive** if the waves reinforce each other (Fig. 16-19*b*) and **destructive** if they tend to cancel (Fig. 16-19*c*). When the displacements of the two or more interfering waves add algebraically to produce the composite, the waves are said to obey the **superposition principle.** Waves on a stretched string, light waves, sound waves, and many others obey the superposition principle provided the wave amplitude is not too great; such waves are called linear waves. Large amplitude waves, or those propagating in complicated media, often do not obey the superposition principle; mathematical analysis of these so-called nonlinear waves is quite difficult.

Wave interference and superposition occur throughout physics, from mechanical waves to light and sound, and even in the quantum-mechanical waves that govern behavior in the atomic and subatomic realms. Here and in the next chapter we present some interference phenomena, and we will consider the interference of light waves in much greater detail in Chapter 37.

Analysis of Complex Waves

The superposition principle allows us to build complex wave shapes by superposing simpler ones—in particular, simple harmonic waves (Fig. 16-20). Analysis of a complex wave in terms of its harmonic components is called **Fourier analysis,** after the French mathematician Jean Baptiste Joseph Fourier (1768–1830). Fourier showed that *any* periodic wave can be decomposed into a sum of simple harmonic terms. Fourier analysis has many applications, ranging from music to structural engineering and communications because it helps us understand how a complex wave behaves if we know how its simple harmonic components behave (Fig. 16-21).

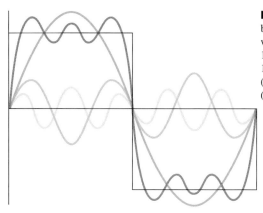

FIGURE 16-20 A square wave built up as a sum of simple harmonic waves, with frequencies in the ratio $1:3:5:\ldots$ and amplitudes in the ratio $1:\frac{1}{3}:\frac{1}{5}:\ldots$ Shown are the square wave (black), the first three harmonic waves (green), and their sum (purple).

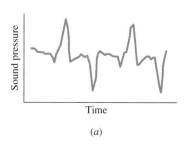

(a)

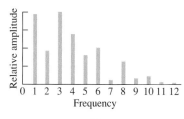

(b)

FIGURE 16-21 An electric guitar plays the note E, producing a complex waveform of which two cycles are shown in (a) as a plot of sound wave pressure variation versus time. (b) Fourier analysis shows the relative strengths of individual sine waves whose sum forms the complex waveform. Numbers on the horizontal axis are multiples of the fundamental frequency, and the height of each bar measures the amplitude of that frequency component. To reproduce the waveform faithfully, an audio system must be capable of reproducing not only the fundamental frequency but also the higher harmonics.

Dispersion

When wave speed is independent of wavelength, then the simple harmonic components making up a complex waveform travel at the same speed. As a result, the waveform maintains its shape. But for some waves, speed depends on wavelength. Then, individual harmonic waves travel at different speeds, and a complex waveform changes shape as it moves. This phenomenon is called **dispersion,** and is illustrated in Fig. 16-22. Waves on the surface of deep water, for example, have speed given by

$$v = \sqrt{\frac{\lambda g}{2\pi}}, \qquad (16\text{-}10)$$

where λ is the wavelength and g the acceleration of gravity. Because v depends on λ, the waves are dispersive. Long-wavelength waves from a storm at sea have the highest speeds and therefore reach shore well in advance of both the storm and the shorter wavelength waves. Dispersion is also important in communications systems; for example, dispersion of the square pulses carrying digital data sets the maximum lengths for wires and optical fibers used in computer networks.

Beats

When two waves of slightly different frequencies are superposed, their interference is constructive at some points and destructive at others, resulting in a waveform whose amplitude itself varies (Fig. 16-23). We can analyze this phenomenon by writing the composite waveform as the sum of two equal-amplitude waves of slightly different frequencies. If we consider just the time variation at the fixed position $x = 0$, then this sum is

$$y = A \cos\omega_1 t + A \cos\omega_2 t.$$

We can express this in a more enlightening form using the trigonometric identity $\cos\alpha + \cos\beta = 2\cos\left(\dfrac{\alpha - \beta}{2}\right)\cos\left(\dfrac{\alpha + \beta}{2}\right)$ given in Appendix A. Then we have

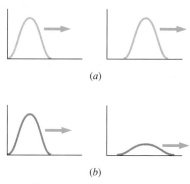

(a)

(b)

FIGURE 16-22 (a) A wave pulse in a nondispersive medium holds its shape as it propagates. (b) In a dispersive medium, the pulse changes shape as it propagates.

FIGURE 16-23 (*a*) Two waves with slightly different frequencies. Where the two are in phase, they interfere constructively. Where they are out of phase, the interference is destructive. The amplitude of the resulting waveform (*b*) varies with a frequency equal to the difference of the frequencies of the two interfering waves. The equation that results in the gray curve is the term $2A\cos[\frac{1}{2}(\omega_1 - \omega_2)t]$.

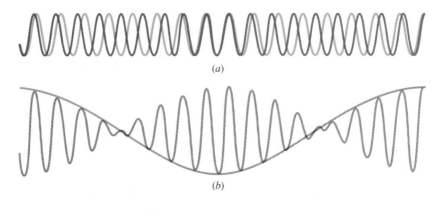

(*a*)

(*b*)

$$y = 2A \cos\left(\frac{1}{2}(\omega_1 - \omega_2)t\right)\cos\left(\frac{1}{2}(\omega_1 + \omega_2)t\right).$$

The second term in this equation represents a sinusoidal oscillation at the average of the two individual frequencies. The first term oscillates at a lower frequency—half the difference of the individual frequencies. If we think of the entire term $2A\cos(\frac{1}{2}(\omega_1 - \omega_2)t)$ as the "amplitude" of the higher frequency oscillation, then this amplitude itself varies with time, as Fig. 16-23 shows. Note that there are *two* amplitude peaks for each cycle of the slow oscillation, so the frequency with which the amplitude varies is simply $\omega_1 - \omega_2$.

For sound waves, interference of two nearly equal frequencies results in variations in sound intensity called **beats;** the closer the two frequencies, the longer the period between beats. In twin-engine aircraft, for example, pilots synchronize engine speeds by reducing the beat frequency toward zero. Beating of electromagnetic waves forms the basis for some very sensitive measurements; modern optical techniques, for example, allow us to detect beat frequencies in

FIGURE 16-24 Interference of circular water waves from two point sources (at rear) shows clearly the pattern of nodal lines (straight lines radiating from back toward front). These lines are regions of low wave amplitude resulting from destructive interference.

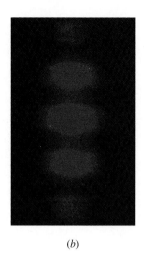

(a)   (b)

FIGURE 16-25 (*a*) Interference pattern produced by passing plane waves through two apertures. Dark curves (purple) are nodal lines, with low wave amplitude; light curves (pink) mark constructive interference giving rise to high amplitude. (*b*) Interference pattern produced by shining laser light through two narrow slits. Photographic film was placed at the right end of an optical setup analogous to (*a*), thus capturing a pattern of alternating light and dark regions.

the millihertz range arising from the interference of light waves whose frequencies are around 10^{14} Hz.

Interference in Two Dimensions

Waves propagating in two and three dimensions exhibit a rich variety of interference patterns. Figure 16-24 shows one of the simplest and most important examples—the interference of waves from two point sources oscillating at the same frequency. Points on a line passing perpendicular midway between the sources are equidistant from both sources, and therefore waves arrive at this line in phase. Thus, they interfere constructively, producing a high amplitude on this center line. Some distance away, though, is a set of points on which waves arrive exactly half a period out of phase. They therefore interfere destructively, giving rise to a **nodal line** on which the wave amplitude is very small. Since waves travel half a wavelength in half a period, the nodal line occurs where the distances to the two sources differ by half a wavelength. Additional nodal lines occur where those distances differ by $1\frac{1}{2}$ wavelengths, $2\frac{1}{2}$ wavelengths, and so forth.

Two-source interference also arises when plane waves pass through two small apertures and then spread as circular or spherical waves. (Fig. 16-25*a*). Such two-slit interference experiments are of considerable importance in optics and in modern physics and are of historical interest because they were first used to demonstrate the wave nature of light (Fig. 16-25*b*). We will explore optical interference further in Chapter 37.

● EXAMPLE 16-5   CALM WATER

Ocean waves pass through two small openings in a breakwater. The openings are $d = 20$ m apart, as shown in Fig. 16-26. You're in a small boat on a perpendicular line midway between the two openings, 75 m from the breakwater. You row 33 m parallel to the breakwater and, for the first time, find yourself in relatively calm water. What is the wavelength of the waves?

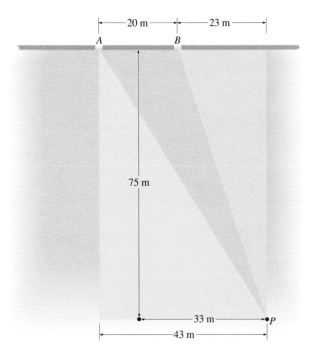

FIGURE 16-26 Calm water at P implies that paths AP and BP differ by half a wavelength.

Solution
Finding calm water means finding a point P where the waves interfere destructively. That means waves from openings A and B arrive at P out of phase by half a cycle—that is, waves from A arrive half a period earlier than those from B. The first time that happens is when AP is longer than BP by the distance the waves go in half a period—that is, by half a wavelength. Mathematically, $AP - BP = \frac{1}{2}\lambda$. In Fig. 16-26 we've lightly shaded two right triangles whose hypotenuses are the lengths AP and BP. Applying the Pythagorean theorem gives

$$AP = \sqrt{(75 \text{ m})^2 + (43 \text{ m})^2} = 86.5 \text{ m}$$
$$BP = \sqrt{(75 \text{ m})^2 + (23 \text{ m})^2} = 78.4 \text{ m}.$$

Then

$$\lambda = 2(BP - AP) = 2(86.5 \text{ m} - 78.4 \text{ m}) = 16 \text{ m}.$$

EXERCISE On another day waves of a different wavelength impinge on the breakwater in Fig. 16-26. Still 75 m from the breakwater, you now row 38 m from the center line and find yourself in the first region of *maximum* wave amplitude. What is the wavelength on this day?

Answer: 9.0 m

Some problems similar to Example 16-5: 51, 52, 69, 70 ●

16-6 THE WAVE EQUATION

Our only quantitative analysis of wave propagation in this chapter involved waves on a stretched string, and in that analysis we found only the speed of the waves. We did not explicitly show that wave functions of the form $f(x \pm vt)$ are compatible with Newton's law applied to the string. Here we derive the so-called linear **wave equation** that is satisfied not only by waves on a stretched string but by other linear waves as well. We will see the wave equation again in Chapter 34 when we consider electromagnetic waves.

Figure 16-27 again shows our stretched string and the forces at opposite ends of a small segment of the string with mass dm and extending from x to $x + dx$. Under the small-angle approximation, the net force is nearly vertical and Fig. 16-27 shows that its magnitude is

$$dF_y \simeq F\sin\theta_2 - F\sin\theta_1 \simeq F(\tan\theta_1 - \tan\theta_2),$$

where the last step follows because for small angles $\sin\theta \simeq \theta \simeq \tan\theta$. But the tangents are the slopes dy/dx at opposite ends of the string segment, so

$$dF_y = F(\tan\theta_1 - \tan\theta_2) = F\left(\frac{\partial y}{\partial x}\bigg|_{x+\Delta x} - \frac{\partial y}{\partial x}\bigg|_x\right).$$

(See the tip on page 383 for the meaning of the symbol ∂.) The term in parentheses is the change in the first derivative over the interval dx; dividing through by

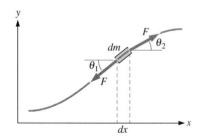

FIGURE 16-27 A portion of the stretched string, showing the forces on a short segment of mass dm. The forces at opposite ends point in slightly different directions, producing a net downward force. The vertical distortion is exaggerated for clarity; actually, the string remains very nearly horizontal.

dx would give the rate of change of the first derivative—that is, the second derivative $\partial^2 y/\partial x^2$. So we can write

$$dF_y = F\frac{\partial^2 y}{\partial x^2}dx$$

for the net force on the segment. Since the string has mass per unit length μ and is nearly horizontal, the segment's length is essentially dx and therefore its mass is $dm = \mu dx$. Knowing the force, we can then apply Newton's law $F_{net} = ma$ to get

$$F\frac{\partial^2 y}{\partial x^2}dx = \mu\, dx\frac{\partial^2 y}{\partial t^2},$$

where we have written the acceleration as the second time derivative of the displacement. Canceling dx, we can write

$$\frac{\partial^2 y}{\partial x^2} = \frac{\mu}{F}\frac{\partial^2 y}{\partial t^2}.$$

But Equation 16-7 shows that $\mu/F = 1/v^2$, with v the wave speed. So we have

$$\frac{\partial^2 y}{\partial x^2} = \frac{1}{v^2}\frac{\partial^2 y}{\partial t^2}. \qquad (16\text{-}11)$$

This is the linear **wave equation.** Any time the analysis of a physical system leads to an equation of this form, then that system supports waves propagating with speed v—that is, with speed given by the inverse of the square root of whatever quantity multiplies $\partial^2 y/\partial t^2$.

It's a simple matter to show that any harmonic wave of the form $y = A\cos(kx - \omega t)$ satisfies the wave equation. To do so, we twice differentiate this function with respect to x and t:

$$\frac{\partial^2}{\partial x^2}[A\cos(kx - \omega t)] = \frac{\partial}{\partial x}[-kA\sin(kx - \omega t)] = -k^2 A\cos(kx - \omega t)$$

$$\frac{\partial^2}{\partial t^2}[A\cos(kx - \omega t)] = \frac{\partial}{\partial t}[\omega A\sin(kx - \omega t)] = -\omega^2 A\cos(kx - \omega t).$$

Substituting these results in Equation 16-11 gives

$$-k^2 A\cos(kx - \omega t) = \frac{1}{v^2}[-\omega^2 A\cos(kx - \omega t)]$$

or, canceling the cosine term and the minus sign,

$$v^2 = \frac{\omega^2}{k^2}.$$

But we found earlier that $v = \omega/k$ for a harmonic wave, so this equation is an identity. Thus, we've shown by direct substitution that the harmonic wave $A\cos(kx - \omega t)$ satisfies the wave equation 16-11. In Problem 54 you can show that *any* function of the form $f(x \pm vt)$ also satisfies the wave equation. You can also show that the sum of two different solutions to the wave equation is also a solution, thus verifying that waves that satisfy this equation obey the superposition principle.

CHAPTER SYNOPSIS

Summary

1. A **wave** is a propagating disturbance that carries energy but not matter. A wave is **longitudinal** if the disturbance displaces the medium in the direction of propagation, and **transverse** if the displacement is perpendicular to the wave propagation. A wave may have any shape, from a single **pulse** to a **continuous wave.** The peak value of the wave disturbance is the wave **amplitude.**

2. A continuous wave is a periodic disturbance characterized by its **wavelength** λ—the distance between wave crests— and its **period** T—the time between crests measured at a fixed point. A wave travels one wavelength in one period, so its speed is

$$v = \frac{\lambda}{T} = f\lambda,$$

where $f = 1/T$ is the wave **frequency.**

3. A wave propagating in the x direction is described by a function of the quantity $x \pm vt$. In the special case of a **simple harmonic wave,** that function is sinusoidal and may be written

$$y(x, t) = A\cos(kx \pm \omega t),$$

where y is the displacement of the medium, A the amplitude, $k = 2\pi/\lambda$ the **wave number,** and $\omega = 2\pi f = 2\pi/T$ the **angular frequency.** The wave speed v is then $v = \lambda/T = \omega/k$.

4. Applying Newton's law for small-amplitude displacements of a stretched string shows that the waves propagate with speed $v = \sqrt{F/\mu}$, where F is the tension and μ the mass per unit length.

5. Waves carry energy at a rate proportional to the wave speed and to the square of the wave amplitude. For waves propagating in three dimensions, the rate of energy flow is characterized by the **intensity,** or power per unit area through an area at right angles to the wave propagation. For **plane waves** the intensity remains constant as the wave propagates without spreading, but for the more realistic case of spherical waves from a localized source, the waves spread and their intensity drops as the inverse square of the distance from the source.

6. Many waves obey the **superposition principle,** meaning that their wave displacements simply add when the waves overlap or **interfere.** This allows complex waves to be analyzed as sums of simple harmonic waves of different wavelengths. When wave speed is independent of wavelength, a complex wave retains its shape. But if wave speed varies with wavelength, a complex waveform exhibits **dispersion,** changing shape as it propagates. Wave interference is **constructive** if the overall amplitude is enhanced and **destructive** when it is diminished. In two dimensions, wave interference results in patterns including **nodal lines** where the combined wave amplitude is a minimum.

7. Many waves satisfy the linear **wave equation,**

$$\frac{\partial^2 y}{\partial x^2} = \frac{1}{v^2}\frac{\partial^2 y}{\partial t^2}$$

for sufficiently small wave amplitudes and those that do then obey the superposition principle.

Terms You Should Understand

(Pairs are closely related terms whose distinction is important; number in parentheses is chapter section where term first appears.)

wave, mechanical wave (introduction)
amplitude (16-1)
longitudinal wave, transverse wave (16-1)
pulse, continuous wave, wave train (16-1)
wavelength, period, frequency (16-1)
simple harmonic wave (16-2)
wave number, angular frequency (16-2)
partial derivative (16-3)
wave equation (16-3)
intensity (16-4)
plane wave, spherical wave (16-4)
superposition principle (16-5)
Fourier analysis (16-5)
dispersion (16-5)
interference, constructive and destructive (16-6)
beats (16-6)
nodal line (16-6)

Symbols You Should Recognize

k, ω (16-2)
$kx \pm \omega t$ (16-2)
μ (16-3)
∂ (16-4)

Problems You Should Be Able to Solve

solving for wavelength, period, or speed from the other two quantities (16-1)

solving for frequency, wavelength, or speed from the other two quantities (16-1)

converting among period, frequency, and angular frequency (16-2)

converting between wavelength and wave number (16-2)

solving for angular frequency, wave number, or speed from the other two quantities (16-2)

writing mathematical expressions for simple harmonic waves (16-2)

solving for speed, tension, or mass per unit length involved with waves on a stretched string, given two of those quantities (16-3)

calculating the wave power propagating on a stretched string (16-4)

calculating the intensity of spherical waves (16-4)

describing quantitatively simple interference situations, including beats and nodal lines in two dimensions (16-5)

Limitations to Keep in Mind

Many of the results developed in this chapter—including the wave speed on stretched strings and the superposition principle—apply only for small-amplitude waves.

QUESTIONS

1. What distinguishes a wave from an oscillation?

2. Red light has a longer wavelength than blue light. Compare their frequencies.

3. Consider a light wave and a sound wave with the same wavelength. Which has the higher frequency?

4. In what sense is "the wave" passing through the crowd at a football game really a wave?

5. A car stops suddenly. The subsequent stopping of cars behind it can be described as a "wave" that propagates back into the traffic. In what sense is this a wave? What are some factors that determine its speed?

6. Must a wave be either transverse or longitudinal? Explain.

7. Does a wave pulse have a period? A frequency? A speed? An amplitude? Explain.

8. Explain in words why the speed of a wave is given by λ/T.

9. As a wave propagates on a stretched string, the string moves. Is the string speed related to the wave speed? Explain.

10. If you doubled the tension in a string, what would happen to the speed of waves on the string?

11. The wave number k is sometimes called the *spatial frequency*. Why is this name appropriate?

12. A heavy cable is hanging vertically, its bottom end free. How will the speed of transverse waves near the top and bottom of the cable compare? Why?

13. Why won't waves propagate on a string that's not under tension?

14. If you halve the amplitude of your side-to-side shaking of the garden hose in Example 16-3, what would happen to the power you need to supply?

15. As a wave propagates down an ideal, uniform string, does the rate at which it carries energy change?

16. The intensity of waves from a spherical source decreases as the inverse square of the distance from the source. How does the wave amplitude depend on distance?

17. The intensity of light from a localized source decreases as the inverse square of the distance from the source. Does this mean that the light loses energy as it propagates?

18. The speed of small-amplitude waves on a stretched string is independent of wavelength. Do you expect that the speed of large-amplitude waves should increase or decrease with increasing wavelength? Give a physical reason for your answer.

19. An upward and a downward pulse, otherwise identical in shape, are propagating in opposite directions along a stretched string. At the instant they overlap completely, the displacement of the string is exactly zero everywhere. How does this situation differ from true equilibrium? *Hint:* Where is the wave energy?

20. A lightning flash in Earth's southern hemisphere produces radio waves that give a burst of static in nearby radios. But a radio receiver appropriately located in the northern hemisphere detects a long whistle whose pitch decreases with time. What does this say about the speed of these radio waves as a function of frequency? (Electrical properties of the upper atmosphere cause this phenomenon; in vacuum all radio and other electromagnetic waves travel at the same speed.)

21. Do the nodal lines in Fig. 16-24 move as the waves propagate?

22. Is the wave amplitude on a nodal line exactly zero? Why?

23. The maximum frequency the human ear can detect is about 20 kHz. If you walk into a room in which two sources are emitting sound waves at 100 kHz and 102 kHz, will you hear anything? Explain.

PROBLEMS

Section 16-1 Wave Properties

1. Ocean waves with 18-m wavelength travel at 5.3 m/s. What is the time interval between wave crests passing under a boat moored at a fixed location?

2. Ripples in a shallow puddle are propagating at 34 cm/s. If the wave frequency is 5.2 Hz, what are (a) the period and (b) the wavelength?

3. An 88.7-MHz FM radio wave propagates at the speed of light. What is its wavelength?

4. One end of rope is tied to a wall. You shake the other end with a frequency of 2.2 Hz, producing waves whose wavelength is 1.6 m. What is their propagation speed?

5. A 145-MHz radio signal propagates along a cable. Measurement shows that the wave crests are spaced 1.25 m apart. What is the speed of the waves on the cable? Compare with the speed of light in vacuum.

6. Calculate the wavelengths of (a) a 1.0-MHz AM radio wave, (b) a channel 9 TV signal (190 MHz), (c) a police radar (10 GHz), (d) infrared radiation from a hot stove (4.0×10^{13} Hz), (e) green light (6.0×10^{14} Hz), and (f) 1.0×10^{18} Hz x rays. All are electromagnetic waves that propagate at 3.0×10^8 m/s.

7. Detecting objects by reflecting waves off them is effective only for objects larger than about one wavelength. (a) What is the smallest object that can be seen with visible light (maximum frequency 7.5×10^{14} Hz)? (b) What is the smallest object that can be detected with a medical ultrasound unit operating at 5 MHz? The speed of ultrasound waves in body tissue is about 1500 m/s.

8. A seismograph located 1200 km from an earthquake detects waves from the quake 5.0 min after the quake occurs. The seismograph oscillates in step with the waves, at a frequency of 3.1 Hz. Find the wavelength of the waves.

9. In Fig. 16-28 two boats are anchored offshore and are bobbing up and down on the waves at the rate of six complete cycles each minute. When one boat is up the other is down. If the waves propagate at 2.2 m/s, what is the minimum distance between the boats?

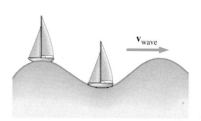

FIGURE 16-28 Problem 9.

Section 16-2 Mathematical Description of Wave Motion

10. Find the wave number and angular frequency for a sinusoidal wave with frequency 200 Hz and wavelength 1.7 m.

11. An ocean wave has period 4.1 s and wavelength 10.8 m. Find (a) its wave number and (b) its angular frequency.

12. Find (a) the amplitude, (b) the wavelength, (c) the period, and (d) the speed of a wave whose displacement is given by $y = 1.3\cos(0.69x + 31t)$, where x and y are in cm and t is in seconds. (e) In which direction is the wave propagating?

13. A simple harmonic wave of wavelength 16 cm and amplitude 2.5 cm is propagating along a string in the negative x direction at 35 cm/s. Find (a) the angular frequency and (b) the wave number. (c) Write a mathematical expression describing the displacement y of this wave (in centimeters) as a function of position and time. Assume the displacement at $x = 0$ is a maximum when $t = 0$.

14. Figure 16-29 shows a simple harmonic wave at time $t = 0$ and later at $t = 2.6$ s. Write a mathematical description of this wave.

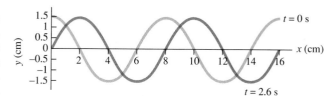

FIGURE 16-29 Problem 14.

15. What are (a) the amplitude, (b) the frequency in hertz, (c) the wavelength, and (d) the speed of a water wave whose displacement is $y = 0.25\sin(0.52x - 2.3t)$, where x and y are in meters and t in seconds?

16. For the wave given in the previous problem, find two points where the wave displacement is a maximum (a) when $t = 0$ and (b) when $t = 1.12$ s.

17. At time $t = 0$, the displacement in a transverse wave pulse is described by $y = \dfrac{2}{x^4 + 1}$, with both x and y in cm. Write an expression for the pulse as a function of position x and time t if it is propagating in the positive x direction at 3 cm/s.

18. Plot the answer to the previous problem as a function of position x for the two cases $t = 0$ and $t = 4$ s.

19. Figure 16-30a shows a wave plotted as a function of position at time $t = 0$, while Fig. 16-30b shows the same wave plotted as a function of time at position $x = 0$. Find (a) the

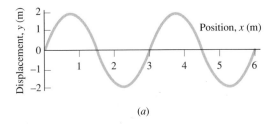

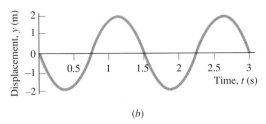

FIGURE 16-30 Problems 19, 20.

wavelength, (b) the period, (c) the wave speed, and (d) the direction of propagation.

20. Write a mathematical description of the wave in the preceding problem.

Section 16-3 Waves on a String

21. The main cables supporting New York's George Washington Bridge have a mass per unit length of 4100 kg/m and are under tension of 250 MN. At what speed would a transverse wave propagate on these cables?

22. A transverse wave 1.2 cm in amplitude is propagating on a string; the wave frequency is 44 Hz. The string is under 21 N tension and has mass per unit length of 15 g/m. Determine (a) the wave speed and (b) the maximum speed of a point on the string.

23. A transverse wave with 3.0-cm amplitude and 75-cm wavelength is propagating on a stretched spring whose mass per unit length is 170 g/m. If the wave speed is 6.7 m/s, find (a) the spring tension and (b) the maximum speed of any point on the spring.

24. A rope is stretched between supports 12 m apart; its tension is 35 N. If one end of the rope is tweaked, the resulting disturbance reaches the other end 0.45 s later. What is the total mass of the rope?

25. A 3.1-kg mass hangs from a 2.7-m-long string whose total mass is 0.62 g. What is the speed of transverse waves on the string? *Hint:* You can ignore the string mass in calculating the tension but not in calculating the wave speed. Why?

26. Transverse waves propagate at 18 m/s on a string whose tension is 14 N. What will be the wave speed if the tension is increased to 40 N?

27. The density of copper is 8.29 g/cm^3. What is the tension in a 1.0-mm-diameter copper wire that propagates transverse waves at 120 m/s?

28. A 100-m-long wire has a mass of 130 g. A sample of the wire is tested and found to break at a tension of 150 N. What is the maximum propagation speed for transverse waves on this wire?

29. A 25-m-long piece of 1.0-mm-diameter wire is put under 85 N tension. If a transverse wave takes 0.21 s to travel the length of the wire, what is the density of the material comprising the wire?

30. A mass m_1 is attached to a wire of linear density 5.6 g/m, and the other end of the wire run over a pulley and tied to a wall as shown in Fig. 16-31. The speed of transverse waves on the horizontal section of wire is observed to be 20 m/s. If a second mass m_2 is added to the first, the wave speed increases to 45 m/s. Find the second mass. Assume the string does not stretch appreciably.

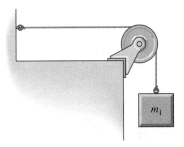

FIGURE 16-31 Problem 30.

31. A steel wire can tolerate a maximum tension per unit cross-sectional area of 2.7 GN/m^2 before it undergoes permanent distortion. What is the maximum possible speed for transverse waves in a steel wire if it is to remain undistorted? Steel has a density of 7.9 g/cm^3.

32. A uniform cable hangs vertically under its own weight. Show that the speed of waves on the cable is given by $v = \sqrt{yg}$, where y is the distance from the bottom of the cable.

Section 16-4 Wave Power and Intensity

33. A rope with 280 g of mass per meter is under 550 N tension. A wave with frequency 3.3 Hz and amplitude 6.1 cm is propagating on the rope. What is the average power carried by the wave?

34. A motor drives a mechanism that produces simple harmonic motion at one end of a stretched cable. The frequency of the motion is 30 Hz, and the motor can supply energy at an average rate of 350 W. If the cable has linear density 450 g/m and is under 1.7 kN tension, (a) what is the maximum wave amplitude that can be driven down the cable? (b) If the motor were replaced by a larger one

capable of supplying 700 W, how would the maximum amplitude change?

35. A 600-g Slinky is stretched to a length of 10 m. You shake one end at the frequency of 1.8 Hz, applying a time-average power of 1.1 W. The resulting waves propagate along the Slinky at 2.3 m/s. What is the wave amplitude?

36. A simple harmonic wave of amplitude 5.0 cm, wavelength 70 cm, and frequency 14 Hz is propagating on a wire with linear density 40 g/m. Find the wave energy per unit length of the wire.

37. Figure 16-32 shows a wave train consisting of two cycles of a sine wave propagating along a string. Obtain an expression for the total energy in this wave train, in terms of the string tension F, the wave amplitude A, and the wavelength λ.

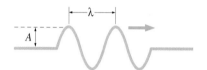

FIGURE 16-32 Problem 37.

38. A steel wire with linear density 5.0 g/m is under 450 N tension. What is the maximum power that can be carried by transverse waves on this wire if the wave amplitude is not to exceed 10% of the wavelength?

39. A loudspeaker emits energy at the rate of 50 W, spread in all directions. What is the intensity of sound 18 m from the speaker?

40. The light intensity 3.3 m from a light bulb is 0.73 W/m². What is the power output of the bulb, assuming it radiates equally in all directions?

41. Use data from Appendix E to determine the intensity of sunlight at (a) Mercury and (b) Pluto.

42. A 9-W laser produces a beam 2 mm in diameter. Compare its light intensity with that of sunlight at noon, about 1 kW/m².

43. Light emerges from a 5.0-mW laser in a beam 1.0 mm in diameter. The beam shines on a wall, producing a spot 3.6 cm in diameter. What are the beam intensities (a) at the laser and (b) at the wall?

44. A large boulder drops from a cliff into the ocean, producing circular waves. A small boat 18 m from the impact point measures the wave amplitude at 130 cm. At what distance will the amplitude be 50 cm?

45. Use Table 16-1 to determine how close to a rock band you should stand for it to sound as loud as a jet plane at 200 m. Treat the band and the plane as point sources. Is this assumption reasonable?

Section 16-5 The Superposition Principle and Wave Interference

46. Consider two functions $f(x \pm vt)$ and $g(x \pm vt)$ that both satisfy the wave equation (Equation 16-11). Show that their sum also satisfies the wave equation.

47. Two wave pulses are described by $y_1(x, t) = \dfrac{2}{(x - t)^2 + 1}$, $y_2(x, t) = \dfrac{-2}{(x - 5 + t)^2 + 1}$, where x and y are in cm and t in seconds. (a) What is the amplitude of each pulse? (b) At $t = 0$, where is the peak of each pulse, and in what direction is it moving? (c) At what time will the two pulses exactly cancel?

48. The triangular wave of Fig. 16-33 can be described by the following sum of simple harmonic terms:

$$y = \frac{8}{\pi^2}\left(\frac{\sin x}{1^2} - \frac{\sin 3x}{3^2} + \frac{\sin 5x}{5^2} - \cdots\right).$$

Plot the sum of the first three terms in this series for x ranging from 0 to 2π, and compare with the first cycle shown in Fig. 16-33.

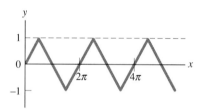

FIGURE 16-33 A triangular wave (Problem 48).

49. You're in an airplane whose two engines are running at 560 rpm and 570 rpm. How often do you hear the sound intensity increase as a result of wave interference?

50. Two waves have the same angular frequency ω, wave number k, and amplitude A, but they differ in phase: $y_1 = A\cos(kx - \omega t)$ and $y_2 = A\cos(kx - \omega t + \phi)$. Show that their superposition is also a simple harmonic wave, and determine its amplitude A_s as a function of the phase difference ϕ.

51. What is the wavelength of the ocean waves in Example 16-5 if the calm water you encounter at 33 m is the *second* calm region on your voyage from the center line?

52. The two loudspeakers shown in Fig. 16-34 emit identical 500-Hz sound waves. Point P is on the first nodal line of the interference pattern. Use the numbers shown to calculate the speed of the sound waves.

Section 16-6 The Wave Equation

53. The following equation arises in analyzing the behavior of shallow water:

$$\frac{\partial^2 y}{dx^2} - \frac{1}{gh}\frac{\partial^2 y}{dt^2} = 0,$$

where h is the equilibrium depth and y the displacement from equilibrium. Give an expression for the speed of waves in shallow water. (Here shallow means the water depth is much less than the wavelength.)

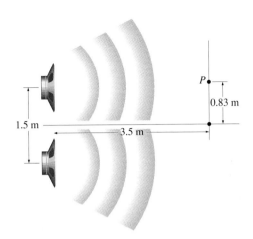

FIGURE 16-34 Problem 52.

FIGURE 16-35 Problem 60.

running at 3600 rpm and the other at 3602 rpm, how often will workers hear a peak in the sound intensity?

62. Two radio waves with frequencies of approximately 50 MHz interfere. The composite wave is detected and fed to a loudspeaker, which emits audible sound at 500 Hz. What is the percentage difference between the frequencies of the two radio waves?

54. Use the chain rule for differentiation to show explicitly that any function of the form $f(x \pm vt)$ satisfies the wave equation (Equation 16-11).

Paired Problems

(Both problems in a pair involve the same principles and techniques. If you can get the first problem, you should be able to solve the second one.)

55. A wave on a taut wire is described by the equation $y = 1.5\sin(0.10x - 560t)$, where x and y are in cm and t is in seconds. If the wire tension is 28 N, what are (a) the amplitude, (b) the wavelength, (c) the period, (d) the wave speed, and (e) the power carried by the wave?

56. A wave give by $y = 23\cos(0.025x - 350t)$, with x and y in mm and t in seconds, is propagating on a cable with mass per unit length 410 g/m. Find (a) the amplitude, (b) the wavelength, (c) the frequency in Hz, (d) the wave speed, and (e) the power carried by the wave.

57. A spring of mass m and spring constant k has an unstretched length ℓ_0. Find an expression for the speed of transverse waves on this spring when it has been stretched to a length ℓ.

58. When a 340-g spring is stretched to a total length of 40 cm, it supports transverse waves propagating at 4.5 m/s. When it's stretched to 60 cm, the waves propagate at 12 m/s. Find (a) the unstretched length of the spring and (b) its spring constant.

59. At a point 15 m from a source of spherical sound waves, you measure a sound intensity of 750 mW/m². How far do you need to walk, directly away from the source, until the intensity is 270 mW/m²?

60. Figure 16-35 shows two observers 20 m apart, on a line that connects them and a spherical light source. If the observer nearest the source measures a light intensity 50% greater than the other observer, how far is the nearest observer from the source?

61. Two motors in a factory produce sound waves with the same frequency as their rotation rates. If one motor is

Supplementary Problems

63. For a transverse wave on a stretched string, the requirement that the string be nearly horizontal is met if the amplitude is much less than the wavelength. (a) Show this by drawing an appropriate sketch. (b) Show that, under this approximation that $A \ll \lambda$, the maximum speed u of the string must be considerably less than the wave speed v. (c) If the amplitude is not to exceed 1% of the wavelength, how large can the string speed u be in relation to the wave speed v?

64. A 64-g spring has unstretched length 25 cm. With a 940-g mass attached, the spring undergoes simple harmonic motion with angular frequency 6.1 s⁻¹. What will be the speed of transverse waves on this spring when it's stretched to a total length of 40 cm?

65. An ideal spring is compressed until its total length is ℓ_1, and the speed of transverse waves on the spring is measured. When it's compressed further to a total length ℓ_2, waves propagate at the *same* speed. Show that the uncompressed spring length is just $\ell_1 + \ell_2$.

66. An ideal spring is stretched to a total length ℓ_1. When that length is doubled, the speed of transverse waves on the spring triples. Find an expression for the unstretched length of the spring.

67. A 1-megaton nuclear explosion produces a shock wave whose amplitude, measured as excess air pressure above normal atmospheric pressure, is 1.4×10^5 Pa (1 Pa = 1 N/m²) at a distance of 1.3 km from the explosion. An excess pressure of 3.5×10^4 Pa will destroy a typical wood-frame house. At what distance from the explosion will such houses be destroyed? Assume the wavefront is spherical.

68. Show that the time it takes a wave to propagate up the cable in Problem 32 is $t = 2\sqrt{\ell/g}$, where ℓ is the cable length.

69. In Example 16-5, how much farther would you have to row to reach a region of maximum wave amplitude?

70. Suppose the wavelength of the ocean waves in Example 16-5 were 8.4 m. How far would you have to row from the center line, staying 75 m from the breakwater, in order to find (a) the first and (b) the second region of relative calm?

CHAPTER 17

SOUND AND OTHER WAVE PHENOMENA

High-intensity sound waves from a trumpet set up resonant oscillations that shatter a wine glass.

Sound waves are longitudinal mechanical waves that propagate through gases, liquids, and solids. We think of sound waves primarily in connection with the sounds we hear, but **audible** sound comprises only a small part of the range of sound waves important both in the natural world and in technology. The human ear is sensitive to sounds in the range from approximately 20 Hz to 20 kHz; below this range is **infrasound** arising, for example, from the vibration of massive structures like buildings. Above the audible range lies **ultrasound,** used in medical diagnostics and treatment, location of underwater objects, analysis of materials, and even microscopy. Many animals can detect sounds above the range audible to humans; bats, in particular, use ultrasound for navigation (Fig. 17-1).

The generation and propagation of sound waves involves a number of wave phenomena beyond those introduced in Chapter 16. Here we describe the properties of sound waves, then explore these additional wave phenomena.

17-1 SOUND WAVES IN GASES

All materials are, to some degree, compressible. That means a force applied to the material results in a change in volume. According to Newton's third law, the material in turn exerts a reaction force. In gases it's simplest to characterize that force in terms of **pressure,** or force per unit area.

Applying a force to one part of a gas immediately compresses only that part. But the increased pressure is communicated to adjacent regions, and the result is a propagating longitudinal wave—a sound wave. Figure 17-2 shows a hollow pipe closed at one end by a movable piston. Moving this piston back and forth alternately rarefies and compresses the air, with the pressure varying slightly about its equilibrium value. That equilibrium pressure itself arises from random motion of the air molecules, which we'll examine in detail in Chapter 20.

FIGURE 17-1 A bat uses sound waves to home in on a moth.

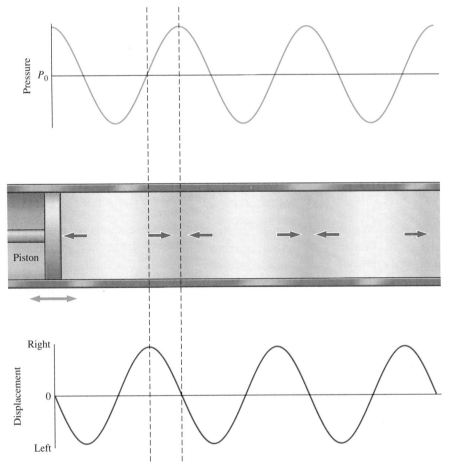

FIGURE 17-2 A sound wave in a hollow pipe, showing alternating regions of compression and rarefaction propagating through air. The air undergoes oscillatory motion as the wave passes. Arrows show direction of the instantaneous displacements from equilibrium and indicate that points of zero displacement correspond to pressure maxima and minima. The displacement is highly exaggerated; normally sound wave amplitude is well under 1% of the wavelength.

Collisions among the rapidly moving molecules communicate changes in pressure to adjacent volumes of air, with the result that the compression and rarefaction induced by the piston propagate down the pipe as a sound wave.

The piston's movement communicates not only pressure changes but also motion to the air. As Fig. 17-2 shows, adjacent volumes of air have displacements in opposite directions; the region where those air volumes meet is either a pressure maximum or minimum, depending on whether air has moved into or out of the region. Since the displacement must be zero where oppositely displaced air masses meet, the displacement and pressure variation are therefore out of phase. So the sound wave consists of alternating regions of compression and rarefaction, and correspondingly of back-and-forth air motion, propagating down the pipe.

> **TIP** **The Air Motion Is Not the Wave Motion Is Not the Molecular Motion** As always, a wave transports energy but not matter. The back-and-forth motion of the air is a manifestation of the wave's passage, just as is the sideways motion of a string as a transverse wave passes. Although the air moves back and forth, it undergoes no overall change in position. And its speed is *not* the wave speed.
>
> There's another important distinction to be made. The back-and-forth air motion refers to the *average* motion of a group of air molecules. Individually, those molecules are whizzing around with random motion at much higher speeds. The back-and-forth motion is just a small effect superposed on the individual random molecular motions.

The variations suggested in Fig. 17-2 are highly exaggerated. The pressure changes even in the loudest audible sound are about one ten-thousandth of the equilibrium air pressure, and the maximum displacement is a tiny fraction of a wavelength.

17-2 THE SPEED OF SOUND IN GASES

In the preceding chapter we found the speed of transverse waves on a stretched string to be given by $v = \sqrt{F/\mu}$. Here the string tension F is the force that tends to restore the string to equilibrium, while μ—the mass per unit length—measures the inertia associated with the string's mass. We might expect the speed of sound to depend on analogous factors—a restoring force and inertia. In a gas that restoring force is provided by the gas pressure P, while the density ρ—mass per unit volume—measures the inertia. So is the sound speed given by $v = \sqrt{P/\rho}$? Almost: dimensional analysis (see Problem 1) shows that this quantity does indeed have the units of velocity, but dimensional analysis cannot tell whether we have exactly the right constant factors. In fact, for reasons we'll see in Chapter 21, the exact expression is

$$v = \sqrt{\frac{\gamma P}{\rho}}, \quad \text{(sound speed in a gas)} \tag{17-1}$$

where γ is a constant that depends on the nature of the gas. For a gas consisting of simple particles—helium atoms, for example, or the electron-proton mixtures that characterize matter at very high temperatures—γ has the value $\frac{5}{3}$. For diatomic molecules like N_2 and O_2 that comprise air, γ takes the lower value $\frac{7}{5}$. And for more complicated molecules γ is lower still; in triatomic CO_2, for example, γ is close to $\frac{4}{3}$.

● EXAMPLE 17-1 "DONALD DUCK" TALK

Under normal atmospheric conditions, air pressure is 1.01×10^5 N/m², and there are 2.51×10^{25} air molecules per cubic meter. Air is essentially 21% oxygen (O_2) and 79% nitrogen (N_2). (a) Find the speed of sound in air and in helium (He) with the same pressure and number of molecules per unit volume. (b) For each gas, find the frequency of a sound wave whose wavelength is 64 cm.

Solution
The masses of oxygen and nitrogen molecules are 32 u and 28 u respectively, so the density of air is

$$\rho = [(0.21)(32 \text{ u}) + (0.79)(28 \text{ u})]$$
$$\times (1.66\times10^{-27} \text{ kg/u})(2.51\times10^{25} \text{ m}^{-3}) = 1.20 \text{ kg/m}^3,$$

where we've weighted the masses of the individual molecules according to their abundances. Then we have

$$v = \sqrt{\frac{\gamma P}{\rho}} = \sqrt{\frac{(7/5)(1.01\times10^5 \text{ N/m}^2)}{1.20 \text{ kg/m}^3}} = 343 \text{ m/s}.$$

The atomic weight of helium is 4 u, giving $\rho = (4 \text{ u})(1.66\times10^{-27} \text{ kg/u})(2.51\times10^{25} \text{ m}^{-3}) = 0.167 \text{ kg/m}^3$. Then, using $\gamma = \frac{5}{3}$ for a monatomic gas, we have

$$v = \sqrt{\frac{\gamma P}{\rho}} = \sqrt{\frac{(5/3)(1.01\times10^5 \text{ N/m}^2)}{0.167 \text{ kg/m}^3}} = 1000 \text{ m/s}.$$

Both these speeds are, incidentally, close to the mean speeds of the random molecular motion in the two gases—not surprising, since it's ultimately collisions between gas molecules that transmit the wave energy.

Writing Equation 16-1 in terms of the frequency $f = 1/T$ then gives

$$f_{air} = \frac{v}{\lambda} = \frac{343 \text{ m/s}}{0.64 \text{ m}} = 536 \text{ Hz}$$

$$f_{He} = \frac{v}{\lambda} = \frac{1000 \text{ m/s}}{0.64 \text{ m}} = 1560 \text{ Hz}.$$

If you've ever heard someone speak after inhaling helium, you've experienced directly the effect of the different sound speeds calculated in this example. Wavelengths in human speech are set by the geometry of the vocal cords and throat, so a rise in sound speed increases the frequency—resulting in the high-pitched, "Donald Duck" speech of the helium inhaler.

EXERCISE Find the sound speed in hydrogen (H_2) under standard conditions. The number of particles per cubic meter in this case is the same as for air. Take $\gamma = \frac{7}{5}$ since hydrogen is diatomic.

Answer: 1.3 km/s

Some problems similar to Example 17-1: 6–8, 12, 14

17-3 SOUND INTENSITY

In Section 16-4 we calculated the power carried in a wave on a stretched string by finding the rate at which one section of string did work on the next section. That rate was $F_y u$, with F_y the y component of the string tension and u the speed *of the string.* Here we do the analogous calculation for sound waves.

Let s be the air's displacement from equilibrium, so $u = \partial s/\partial t$ is its speed. The air motion in a sound wave is driven by the force associated with the pressure *change* ΔP from the equilibrium value. Since pressure is force per unit area, the force exerted *by* one region of the air *on* an adjacent region is ΔPA, and the power is therefore $\Delta PAu = \Delta PA(\partial s/\partial t)$. Dividing by A gives the power per unit area, or intensity:

$$I = \Delta P \frac{\partial s}{\partial t}.$$

In a simple harmonic wave both pressure and displacement vary sinusoidally in space and time with amplitudes ΔP_0 and s_0:

$$\Delta P = \Delta P_0 \cos(kx - \omega t) \tag{17-2a}$$

$$s = -s_0 \sin(kx - \omega t), \tag{17-2b}$$

where we've chosen cosine and sine, and the minus sign on s, for consistency with Fig. 17-2. Using these expressions in our equation for intensity gives

$$I = \Delta P \frac{\partial s}{\partial t} = \Delta P_0 \cos(kx - \omega t) \frac{\partial}{\partial t}[-s_0 \sin(kx - \omega t)]$$

$$= s_0 \omega \Delta P_0 \cos^2(kx - \omega t).$$

As in Chapter 16, we're more interested in the intensity averaged over one cycle; since the average of $\cos^2$ is $\frac{1}{2}$, we have

$$\bar{I} = \frac{1}{2} s_0 \omega \Delta P_0. \tag{17-3a}$$

Finally, we can relate the pressure amplitude ΔP_0 and displacement amplitude s_0. Figure 17-3 shows a thin slab of air in a sound wave. On one side of the slab, whose thickness is dx, the pressure is P; on the other side, it is $P + dP$. The slab has area A, so the net force on it is $dF = [P - (P + dP)]A = -A\,dP$, where the minus sign indicates a leftward force in Fig. 17-3 when dP is positive. The volume of the slab is $A\,dx$, so its mass is $dm = A\rho\,dx$, with ρ the density. Solving Newton's law for acceleration gives $a = F/m$; here the force is dF and the mass is dm, so

$$a = \frac{dF}{dm} = \frac{-A\,dP}{A\rho\,dx} = -\frac{1}{\rho}\frac{\partial P}{\partial x}$$

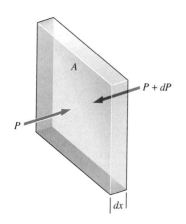

FIGURE 17-3 A thin slab of air in a sound wave has area A and thickness dx. The pressure on one side is P and on the other side $P + dP$, giving a net force of magnitude $A\,dP$ on the slab.

is the acceleration of the slab. We've used the partial derivative because the pressure is a function of both position and time, and here we want its variation with position only. Using Equations 17-2, we can write $a = \dfrac{\partial^2 s}{\partial t^2} = -s_0 \omega^2 \cos(kx - \omega t)$ and $\dfrac{\partial P}{\partial x} = -k\Delta P_0 \sin(kx - \omega t)$. So the acceleration amplitude is $s_0 \omega^2$ and the amplitude of $\partial P/\partial x$ is $k\Delta P_0$. Using our Newton's law equation to relate these amplitudes gives

$$s_0 \omega^2 = \frac{k\Delta P_0}{\rho}.$$

Solving this equation first for s_0 and then for ΔP_0 and using the results in Equation (17-3a) gives two alternative forms for the sound wave intensity:

$$\bar{I} = \frac{1}{2} s_0 \omega \Delta P_0 = \frac{\Delta P_0^2}{2\rho v} \qquad (17\text{-}3b)$$

$$\bar{I} = \frac{1}{2} s_0 \omega \Delta P_0 = \frac{1}{2} \rho \omega^2 s_0^2 v, \qquad (17\text{-}3c)$$

where we've used Equation 16-6 to write the wave speed $v = \omega/k$. Note that both these expressions depend on the *square* of the disturbed quantity, as advertised in Section 16-4.

Sound and the Human Ear

The intensity of normal sound waves ranges widely from small to truly minuscule values, and the ear is a remarkable instrument both in its ability to respond to these weak waves and in the wide range of amplitudes it can handle. Figure 17-4 shows the minimum intensity and the threshold of pain for the human ear, as functions of frequency.

■ APPLICATION STEREO CONTROLS

Your stereo system probably has a "loudness" switch that you're supposed to turn on at low volumes. The shape of the curves in Fig. 17-4 is the reason for this switch. At high volumes (i.e., high intensities), the curves are not far from being flat, meaning that sounds of the same power but different frequencies are perceived at essentially the same volume. But at low volumes the curves turn upward at both ends of the frequency scale, showing that it takes greater intensity at high and low frequencies to give the same perceived volume. The "loudness" control switches in a circuit that boosts the high and low frequencies to compensate for this effect.

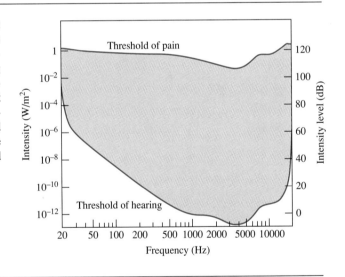

FIGURE 17-4 Minimum and maximum sound intensities for a typical human ear, as functions of frequency. The ear responds to sounds in the entire shaded region. Note the logarithmic scales. Intensities are shown in both watts per square meter and in decibels. Sounds above the threshold of pain are also heard, but prolonged exposure can damage the ears.

● EXAMPLE 17-2   RESPONSE OF THE HUMAN EAR

(a) Find the displacement and pressure amplitudes of the weakest sound to which the typical human ear can respond. (b) In the frequency range around 1 kHz, the threshold of pain is about 1 W/m². Find the displacement and pressure amplitudes for this sound.

Solution

(a) Figure 17-4 shows that the weakest sound for human perception is in the frequency range from 3 kHz to 4 kHz, where the minimum intensity is about 10^{-13} W/m². Equation 17-3b shows that the pressure amplitude is independent of frequency; solving for ΔP_0 then gives

$$\Delta P_0 = \sqrt{2\rho v \bar{I}} = \sqrt{(2)(1.2 \text{ kg/m}^3)(343 \text{ m/s})(10^{-13} \text{ W/m}^2)}$$

$$= 10^{-5} \text{ N/m}^2,$$

where we used the density and sound speed from Example 17-1. Since normal air pressure is 10^5 N/m², this amplitude represents a variation of only one part in ten billion! The corresponding displacement is also impressively small. Taking $f = 3.5$ kHz as typical of the frequency range for minimum intensity and solving Equation 17-3c for s_0 gives

$$s_0 = \sqrt{\frac{2\bar{I}}{\rho \omega^2 v}} = \sqrt{\frac{(2)(10^{-13} \text{ W/m}^2)}{(1.2 \text{ kg/m}^3)[2\pi(3500 \text{ Hz})]^2(343 \text{ m/s})}}$$

$$= 10^{-12} \text{ m},$$

where we used $\omega = 2\pi f$. This number is one hundredth of an atomic diameter! That your eardrum can detect and transmit to your brain information about such small amplitude motions is truly astounding. (b) With 1-kHz waves at the 1 W/m² threshold of pain, similar calculations give $\Delta P_0 = 29$ N/m² and $s_0 = 1\times10^{-5}$ m, or one hundredth of a millimeter.

EXERCISE   In a whispered conversation, the sound intensity 1 m from the source is about 100 pW/m². Under these conditions, what are (a) the total power falling on a typical adult eardrum of area 60 mm² and (b) the displacement amplitude, assuming a frequency of 850 Hz?

Answers: (a) 6×10^{-15} W; (b) 0.13 nm

Some problems similar to Example 17-2:   15, 16, 18, 20   ●

Decibels

The human ear responds to sound intensities and frequencies covering many orders of magnitude; that's why Fig. 17-4 is plotted with logarithmic scales. Because of this wide range, it makes sense to quantify sound levels with a unit proportional to the logarithm of the actual intensity. Such a unit is the **decibel** (dB), defined by

$$\beta = 10 \log\left(\frac{I}{I_0}\right), \tag{17-4}$$

where β is the sound **intensity level** in decibels and where $I_0 = 10^{-12}$ W/m² is a reference level chosen as the threshold of hearing at 1 kHz. The decibel is one-tenth of a bel, a less frequently used unit named for telephone inventor Alexander Graham Bell. Since the logarithm of 10 is 1, each factor-of-10 increase in sound intensity over the reference level of 10^{-12} W/m² represents an increase of 10 dB. Your actual perception of loudness is somewhat subjective, depending not only on the intensity level but also on frequency and duration; in general, for intensity levels above about 40 dB, the perceived loudness approximately doubles for every 10 dB increase in intensity level. One consequence of this relation is the enormous power increase needed to produce really loud sounds (Fig. 17-5).

FIGURE 17-5   Is this scene realistic? It takes enormous power to deliver really loud sound.

▲ **TABLE 17-1** INTENSITY LEVELS

SOURCE	INTENSITY LEVEL, dB	
Space shuttle, at 100 m	165	
Eardrum ruptures	160	
Jet aircraft, at 50 m	130	
Rock band, at 4 m	120	← Threshold of pain
Subway	100	
City traffic	80	
Normal conversation, at 1m	60	
Mosquito	40	
Whisper, at 1 m	20	
Normal breathing	10	
Threshold of hearing	0	

Sound levels over 120 dB are generally painful, while continuous exposure to levels in excess of 90 dB can lead to hearing loss. Table 17-1 lists some common sound intensity levels in dB; we've also included a decibel scale on Fig. 17-4.

● **EXAMPLE 17-3** TURN DOWN THE TV

Your little sister is watching TV, and the sound is blasting your ears at 75 dB. You yell to her to turn it down, and she drops the intensity level to 60 dB. By what factor has the power dropped?

Solution

The intensity level has dropped by 15 dB, corresponding to 1.5 orders of magnitude in actual intensity. So the intensity—and therefore the power emitted by the speakers—drops by a factor $10^{-1.5}$, or to about one thirtieth of its original value. We can get this result more formally by writing Equation 17-4 for each value and subtracting. Designating the initial and final values by the subscripts 1 and 2, we have:

$$\beta_2 - \beta_1 = 10\log\left(\frac{I_2}{I_0}\right) - 10\log\left(\frac{I_1}{I_0}\right) = 10\log\left(\frac{I_2}{I_1}\right).$$

Solving for the logarithm gives $\log(I_2/I_1) = (\beta_2 - \beta_1)/10$; exponentiating in base 10 then gives

$$10^{\log(I_2/I_1)} = \frac{I_2}{I_1} = 10^{(\beta_2-\beta_1)/10} = 10^{(60\,dB-75\,dB)/10}$$

$$= 10^{-1.5} = 0.032.$$

This substantial power reduction results in only a modest reduction in perceived loudness; since the loudness changes by roughly a factor of 2 for each 10 dB change in intensity level, the reduced volume will sound between one-fourth and one-half as loud as before.

TIP Working with Decibels Decibels are designed for ease in calculating sound intensity levels. That's why we could do this example in a few sentences and get the same result as a paragraph of equations. Note also that decibels are dimensionless because the decibel measure always involves intensity ratios.

EXERCISE At the softest passage in a piece of music, your loudspeakers put out a sound level of 48 dB; the actual power delivered by the speakers at this time is 7.8 mW. If you perceive the loudest music to be eight times as loud as the softest, what will be the greatest power the speakers must deliver during this piece?

Answer: 7.8 W

Some problems similar to Example 17-3: 22, 24, 25, 28

17-4 SOUND WAVES IN LIQUIDS AND SOLIDS

Although less compressible than gases, liquids and solids also support longitudinal sound waves. In general, a material's compressibility is described by its **bulk**

▲ TABLE 17-2
SOUND SPEEDS IN
SELECTED MATERIALS*

MATERIAL	SOUND SPEED (m/s)
Gases	
Air (0°C)	331
Air (20°C)	343
Carbon dioxide	259
Hydrogen	1284
Liquids (at 25°)	
Benzene	1295
Mercury	1450
Water	1497
Solids	
Aluminum	6420
Copper	5010
Glass (Pyrex)	5640
Lead	1960
Neoprene	1600
Steel	5940

*Assuming normal atmospheric pressure. Sound speeds for metals depend on how the metals are produced.

modulus of elasticity, or the ratio of the change in pressure to the fractional change in volume:

$$B = -\frac{\Delta P}{\Delta V/V}, \tag{17-5}$$

where the minus sign arises because a positive pressure increase corresponds to a decrease in volume. Because the denominator in this expression is dimensionless, B has the units of pressure. The bulk modulus is essentially a measure of a material's elasticity, with a large value of B describing a "stiff" material, one that is hard to compress. The stiffer the material, the stronger the restoring forces that develop in response to compression; at the same time, response to compressional forces depends on inertia as measured by the material density. Therefore, it's not surprising that the speed of sound in a material is given by

$$v = \sqrt{\frac{B}{\rho}}. \tag{17-6}$$

This is the analog of Equation 17-1 for solids and liquids. Table 17-2 lists sound speeds in some common materials. The stiffness of solids and liquids results in greater sound speeds than in gases, although greater density makes the increase less dramatic than it would otherwise be.

Wave propagation in solids is a complex subject. A simple model for a solid consists of individual atoms linked by springs representing the interatomic electrical forces (Fig. 17-6). Such a structure supports not only longitudinal but also transverse waves. Furthermore, unless the structure is perfectly symmetric, wave properties differ depending on propagation direction. The analysis of possible wave modes is the focus of much contemporary work by solid-state physicists.

17-5 WAVE REFLECTION

You shout in a mountain valley and hear echoes reverberating off the valley walls. You look in a mirror and see light from your face reflected off the silvered surface. A metal screen in your microwave oven's door turns back potentially

FIGURE 17-6 A simple model for a solid, showing atoms connected by springs. This structure supports both longitudinal and transverse waves.

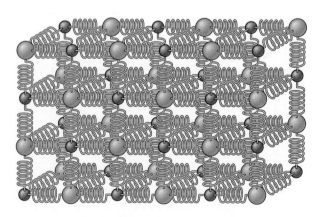

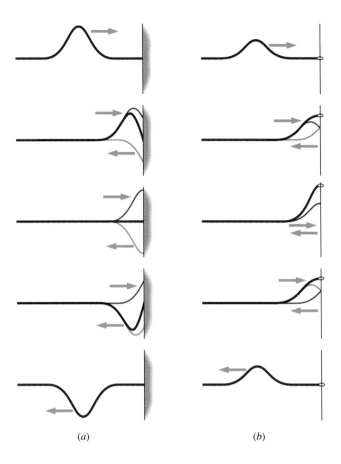

(a)   (b)

FIGURE 17-7 Reflection of a pulse at the end of a string, showing incident and reflected pulses and their superposition. Conditions at the end determine the details of the reflection. In (a) the string is rigidly clamped, so the wave amplitude must be zero at the end and therefore the incident and reflected pulses (pink and blue, respectively) interfere destructively. In (b) the string is free to move up and down on a frictionless rod, and the wave displacement at the end is a maximum; therefore the pulses interfere constructively.

harmful waves and keeps them where they'll heat your food. A doctor's ultrasound probes your body, bouncing back to reveal internal structures. A bat navigates flawlessly in response to echoes from a cave wall. All these are examples of wave **reflection.**

You can see that wave reflection *must* occur when a wave hits a medium in which it can't propagate; otherwise, where would the wave energy go? Figure 17-7 details the reflection process for waves on a stretched string, in the two cases where the string end is (a) clamped at a rigid wall and (b) free to move up and down. In the first case the wave amplitude must remain zero at the end, so the incident and reflected pulses interfere destructively and the reflected wave is therefore inverted. In the second case the displacement is a maximum at the free end, and the waves interfere constructively.

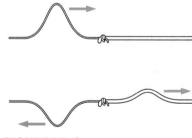

FIGURE 17-8 Partial reflection at a junction between two different strings. Here the right-hand string is heavier, so the reflected pulse is inverted.

Between the extremes of a rigid wall and a perfectly free end lies the case of one string connected to another with a different mass per unit length. In this case some wave energy is transmitted to the second string and some reflected back along the first (Fig. 17-8). Problem 82 shows that the amplitudes of the transmitted and reflected waves are given by

$$A_R = \frac{\sqrt{\mu_1} - \sqrt{\mu_2}}{\sqrt{\mu_1} + \sqrt{\mu_2}} A_I \qquad (17\text{-}7)$$

FIGURE 17-9 Even though glass is transparent, partial reflection of light occurs at the interface between air and glass. Here the Dallas skyline is reflected in a glass-walled building.

and

$$A_T = \frac{2\sqrt{\mu_1}}{\sqrt{\mu_1} + \sqrt{\mu_2}} A_I, \qquad (17\text{-}8)$$

where A_I, A_R, and A_T are the incident, reflected, and transmitted wave amplitudes, and where the wave is incident from a string of mass per unit length μ_1 onto a string with mass per unit length μ_2. Letting μ_2 become indefinitely large gives $A_R = -A_I$; this corresponds to total reflection at a rigid wall, with the minus sign signifying the phase change described in Fig. 17-7a. Letting $\mu_2 = 0$ gives $A_R = A_I$, corresponding to reflection with the end of the string free; as in Fig. 17-7b, there is no phase change. Finally, putting $\mu_1 = \mu_2$ gives $A_R = 0$ and $A_T = A_I$, showing that the wave simply continues on if the strings have the same mass per unit length. More generally, when $\mu_2 > \mu_1$, we get reflection with a phase change; otherwise there is no phase change.

The phenomenon of partial reflection and transmission at a junction of strings has its analog in the behavior of all sorts of waves at interfaces between different media. For example, shallow water waves are partially reflected if the water depth changes abruptly. Light incident on even the clearest glass undergoes partial reflection because of the difference in light-transmitting capabilities of air and glass (Fig. 17-9). Partial reflection of ultrasound waves at interfaces of body tissues with different densities makes ultrasound a valuable medical diagnostic.

When waves strike an interface at an oblique angle, the phenomenon of **refraction**—changing the direction of wave propagation—also occurs. We discuss refraction in Chapter 35.

■ APPLICATION PROBING THE EARTH

Waves propagating and reflecting inside the Earth allow geologists to deduce the structure of the planet's interior, from its largest features to the locations of oil and gas deposits. The waves used include those generated naturally by earthquakes, as well as waves produced intentionally with explosives.

Recall that solids support both longitudinal and transverse waves, while only longitudinal waves propagate in liquids. This fact provides our most direct evidence that Earth has a liquid core. Transverse waves—geologists call them S-waves—do not propagate in the liquid core. S-waves from an earthquake therefore leave a "shadow" zone as shown in Fig. 17-10. Measuring the extent of the shadow zone with a network of seismic stations lets geologists infer the size of the liquid core.

Longitudinal waves—or P-waves—do propagate through the liquid core. Detection of P-waves partially reflected from within the core led geologists to propose an inner core structure. Propagation times for waves passing through this inner core proved shorter than for those passing through the outer core alone, giving evidence that the inner core is solid, with a higher sound speed. Detailed analysis of P-waves from nuclear weapons tests led to geologists' best measurements of the core

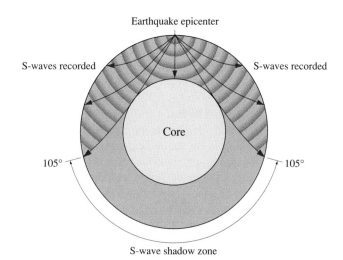

FIGURE 17-10 Transverse waves (S-waves) cannot propagate in liquids, so the presence of a shadow zone for earthquake-generated S-waves is evidence for Earth's liquid core. Bending of the wave paths occurs because of changes in wave speed with depth.

sizes; the solid inner core is about 1200 km in radius and is surrounded by the 2300-km-thick liquid outer core (Fig. 17-11).

Probing with seismic waves on a smaller scale helps in the search for fuel and mineral deposits. Using small explosive charges or special machinery that vigorously "thumps" the ground, geologists generate seismic waves that are detected a short distance away (Fig. 17-12). Measurement of wave propagation times and reflected wave amplitudes provides information on the location and nature of changes in rock density.

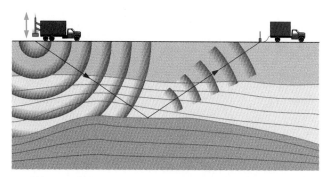

FIGURE 17-12 Prospecting with seismic waves. Truck at left "thumps" the ground, producing seismic waves. Seismograph at right detects reflected waves, and electronic equipment in the van times the wave propagation to give information about the location of different rock layers.

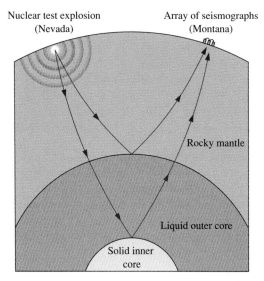

FIGURE 17-11 Longitudinal waves (P-waves) do propagate in the liquid core. The P-waves from nuclear explosions are partially reflected at the boundaries of the inner and outer cores; precise timing of these "echoes" gives the sizes of the cores.

17-6 STANDING WAVES

When a string is clamped at both ends, waves can propagate back and forth by reflecting at the ends. But now, the only waves possible are those whose displacement at the ends is always zero. To see what this condition implies, consider a string of length L lying along the x axis with one end at $x = 0$. If a simple harmonic wave is propagating in the $+x$ direction, the string displacement is

$$y_1(x, t) = A\cos(kx - \omega t).$$

When this wave reflects from the rigidly clamped end at $x = L$, the result is a wave propagating in the $-x$ direction:

$$y_2(x, t) = -A\cos(kx + \omega t),$$

where the minus sign accounts for the phase change that, as we saw in the preceding section, occurs on reflection at a rigid boundary. At the other end, this wave also reflects. So both waves exist simultaneously on the string, whose displacement is therefore their superposition:

$$y(x, t) = y_1 + y_2 = A[\cos(kx - \omega t) - \cos(kx + \omega t)].$$

Appendix A lists a trig identity for the difference of two cosines:

$$\cos\alpha - \cos\beta = -2\sin[\tfrac{1}{2}(\alpha + \beta)]\sin[\tfrac{1}{2}(\alpha - \beta)];$$

applying this identity with $\alpha = kx - \omega t$ and $\beta = kx + \omega t$ gives

$$y(x, t) = 2A\sin kx \sin\omega t, \tag{17-9}$$

where we used $\sin(-\theta) = -\sin\theta$ to eliminate the minus sign.

What sort of wave does this equation describe? At each position x, the string oscillates up and down with frequency ω and amplitude $2A\sin kx$. The maximum amplitude always occurs at the same position x, so the disturbance does not move along the string. This oscillation is a **standing wave**—a nonpropagating structure that results from the superposition of two waves propagating in opposite directions.

Because the string ends are clamped rigidly, they must correspond to points of zero displacement; such points are called **nodes.** In general, nodes lie half a wavelength apart. Therefore, a wave whose wavelength was twice the string length L could form a standing wave. So could a wave with $\lambda = L$; in addition to nodes at the clamped ends, this standing wave would have a node at the center. More generally, if the string length L is any multiple of half a wavelength, then a standing wave can fit on the string (Fig. 17-13):

$$L = \frac{m\lambda}{2}, \quad m = 1, 2, 3, 4, \ldots$$

or

$$\lambda = \frac{2L}{m}. \tag{17-10}$$

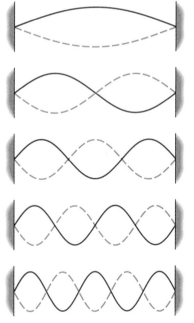

FIGURE 17-13 Standing waves on a string. Clamped rigidly at both ends, the string can accommodate only an integer number of half wavelengths. Shown are the fundamental and four harmonics. Solid line is the string at one instant, dashed line one-half cycle later. Nodes—points where the displacement is always zero—occur at the ends in all cases, and in between for the harmonics.

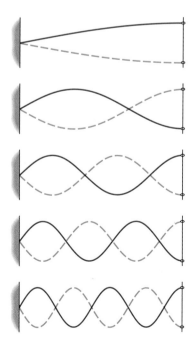

FIGURE 17-14 When one end of the string is fixed and the other free, the string can accommodate only an odd number of quarter wavelengths. (Here the string is clamped at the left end, but at the right end is attached to a ring that slides without friction along a vertical pole.)

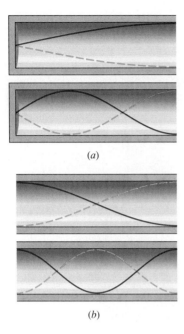

(a)

(b)

FIGURE 17-15 Graphs showing displacement of standing sound waves in wind instruments. Solid and dashed lines are one-half cycle apart. Shown are the fundamental and the lowest harmonic that the instrument can support. (*a*) An instrument with one end open must have a node at the closed end and an antinode at the open end, so it supports odd integer multiples of a quarter wavelength. (*b*) With both ends open antinodes occur at both ends, so the instrument supports integer multiples of a half wavelength.

The integer m in Equation 17-10 is called the **mode number.** The $m = 1$ mode is the **fundamental** and represents the longest wavelength and lowest frequency wave that can exist on the string. The higher modes are called **harmonics.**

When a string is fixed rigidly at one end and is free at the other, its free end must be a point of maximum amplitude—an **antinode.** Figure 17-14 shows that now the string length is an odd multiple of a quarter wavelength. The corresponding frequencies are odd harmonics of the fundamental frequency, which is itself half the fundamental frequency for a string clamped at both ends (see Problem 45).

Musical Instruments

Our analysis of standing waves on strings applies directly to stringed musical instruments like violins, guitars, and pianos. Standing-wave vibrations in the instrument strings are communicated to the air as sound waves, usually through the intermediary of a sounding box or electronic amplifiers. Wind instruments, in contrast, generate standing sound waves in air columns. These must be open at one end to allow sound to escape; in many instruments the column is effectively open at both ends. An open end has its pressure fixed at atmospheric pressure; it is therefore a pressure node and thus, from Fig. 17-2, a displacement antinode. As a result, an instrument open at one end supports odd integer multiples of a quarter wavelength (Fig. 17-15*a*), in analogy with Fig. 17-14. An instrument open at both ends, on the other hand, supports integer multiples of a half wavelength (Fig. 17-15*b*).

● **EXAMPLE 17-4**   THE DOUBLE BASSOON

The double bassoon (Fig. 17-16) is the lowest pitched instrument in a normal orchestra. The instrument is "folded" to achieve an effective air column 5.5 m long, and it acts like a pipe open at both ends. What is the frequency of the double bassoon's fundamental note?

Solution

Because the double bassoon acts like a pipe open at both ends, Fig. 17-15b shows that its fundamental mode has a wavelength twice that of the air column, or 11 m. Equation 16-1, $v = \lambda f$, then gives

$$f = \frac{v}{\lambda} = \frac{343 \text{ m/s}}{11 \text{ m}} = 31 \text{ Hz},$$

where we found the sound speed in Table 17-2. This frequency is the note B_0, and lies near the low-frequency limit of the human ear.

The actual length of the double bassoon's air column is 4.8 m, but the instrument's detailed shape alters the standing wave modes, giving a greater effective length. Like most wind instruments, the bassoon has a number of holes that, when uncovered, alter the positions of the antinodes and therefore change the pitch.

EXERCISE   The C clarinet has an effective length of 32.8 cm, and for its lowest notes it functions as a pipe closed at one end. What is the fundamental frequency of this instrument?

Answer: 261 Hz

Some problems similar to Example 17-4:   50, 52–55

FIGURE 17-16   The double bassoon, whose air column is about 5 m long, is the lowest pitched instrument in a standard orchestra. The instrument is "folded" to achieve its long air column in a more compact space.

A vibrating string or air column can support many modes, each corresponding to a different frequency and wavelength. In general, a mixture of modes exists simultaneously, a superposition of many simple harmonic waves. The method of Fourier analysis, introduced in Section 16-5, is useful in describing the distribution of individual modes. Different musical instruments favor different modes, which is what gives each its distinctive sound (Fig. 17-17).

Other Standing Waves

Standing waves are common phenomena. Water waves in confined spaces exhibit standing waves, and entire lakes can develop very slow oscillations corresponding to low-mode-number standing waves. Standing electromagnetic waves occur inside closed metal cavities; in microwave ovens the nodes of the standing-wave pattern would result in "cold" spots were not either the food or the source of microwaves kept in motion. Standing waves are common in mechanical systems. And even atomic structure can be understood in terms of standing waves associated with electrons.

FIGURE 17-17 Waveforms of a clarinet (above) and oboe (below) playing the same note. Although the fundamental frequencies are the same, the waveforms and the resulting sounds are quite different because of the different mixtures of standing wave modes in the two instruments.

In a novel variant on musical wind instruments, engineers seeking to eliminate ozone-destroying chemicals now used in refrigerators have developed so-called thermoacoustic refrigerators, in which high-amplitude standing sound waves provide the compression and expansion necessary to extract heat (more on this in Chapter 22).

Standing-wave patterns in two- and three-dimensional systems can be quite complex, as Fig. 17-18 suggests.

Standing Wave Resonance

Initiation and buildup of standing waves occurs whenever a system is supplied with energy at just the right frequency—namely, the frequency of *any* allowed standing-wave mode. In a wind instrument, for example, the motion of the lips or of a vibrating reed supplies energy to the air column, allowing the buildup of standing waves. This phenomenon is like resonance in a simple harmonic oscillator, except that the system can be driven to resonance at any of its many allowed frequencies. The oscillations of the Tacoma Narrows bridge shown in Figs. 15-29 and 15-30 are actually torsional standing waves.

FIGURE 17-18 Complex standing-wave patterns arise in two- and three-dimensional systems. (*a*) Standing waves on a drum. (*b*) Standing-wave pattern on a guitar, imaged using holographic interference of laser light. (*c*) Standing-wave oscillations provide a direct probe of conditions deep inside the Sun's interior. Shown is a computer simulation of solar oscillations, in which red and blue represent motion in opposite directions.

(*a*)

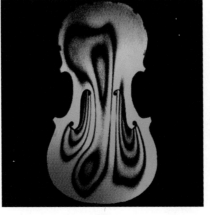

(*b*)

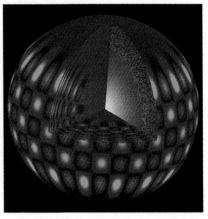

(*c*)

● EXAMPLE 17-5   MEASURING THE SOUND SPEED

Resonance in hollow pipes provides an accurate means of measuring the speed of sound. In a particular experiment, a pipe 1.300 m long is closed at one end with a diaphragm that can be made to oscillate at an arbitrary frequency, while the other end is open to the atmosphere. A microphone monitors the sound intensity at the open end. Experimentally, the lowest diaphragm frequency that causes the intensity to peak is found to be 66.29 Hz. (a) What is the sound speed in the pipe? (b) At what frequency will the next resonance occur?

Solution

Figure 17-15a shows that a pipe open at one end supports standing waves with the pipe length L an odd quarter multiple of a wavelength. Thus, for the 66.29-Hz fundamental, $L = \frac{1}{4}\lambda$, so $\lambda = 4L$, and the sound speed is

$$v = f\lambda = (66.29 \text{ Hz})(4)(1.300 \text{ m}) = 344.7 \text{ m/s}.$$

The next resonance occurs when $L = \frac{3}{4}\lambda$ or $\lambda = \frac{4}{3}L$; the corresponding frequency is

$$f = \frac{v}{\lambda} = \frac{344.7 \text{ m/s}}{(\frac{4}{3})(1.300)} = 198.9 \text{ Hz}.$$

EXERCISE   In air at 20°C and 30% humidity, the speed of sound is 343.8 m/s. Under these conditions the fundamental resonant frequency in a hollow pipe is 1148 Hz. When the humidity increases to 80%, the resonant frequency rises to 1150 Hz. What is the sound speed at 80% humidity?

Answer: 344.4 m/s

Some problems similar to Example 17-5:   43, 55

●

17-7 THE DOPPLER EFFECT

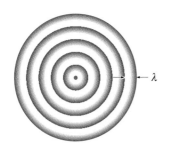

FIGURE 17-19   Circular waves from a source at rest with respect to the medium.

The speed v of a wave is its speed relative to the medium through which it propagates. A point source at rest in the medium radiates waves uniformly in all directions (Fig. 17-19). But when the source moves, wave crests bunch up in the direction toward which the source is moving, resulting in a decreased wavelength (Fig. 17-20). In the opposite direction, wave crests spread out and the wavelength increases.

The wave speed is determined by the properties of the medium, so it doesn't change with source motion. Thus the relation $v = \lambda f$ still holds. This means that an observer in front of the moving source, where λ is smaller, experiences a higher wave frequency as more wave crests pass per unit time. Similarly, an

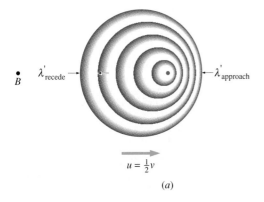

(a)

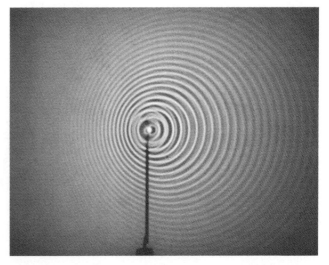

(b)

FIGURE 17-20   (a) When the source moves through the medium, the wavelength is shortened in the direction of motion and increased in the opposite direction. Observers at A and B therefore detect higher and lower frequencies, respectively. In the case shown, the source speed u is half the wave speed v. (b) Doppler effect in water waves produced by a moving source tapping the water surface at regular intervals. Which way is the source moving?

observer behind the source experiences a lower frequency. This change in wavelength and frequency from a moving source is called the **Doppler effect.**

To analyze the Doppler effect, let λ be the wavelength measured when the source is stationary, and λ' the wavelength when the source is moving at speed u through a medium where the wave speed is v. At the source, the time between wave crests is the wave period T, and a wave crest moves one wavelength λ in this time. But during the same time T, the moving source covers a distance uT, after which it emits the next wave crest. So the distance between wave crests, as seen by an observer in front of the moving source, is

$$\lambda' = \lambda - uT.$$

Writing $T = \lambda/v$, this becomes

$$\lambda' = \lambda - u\frac{\lambda}{v} = \lambda\left(1 - \frac{u}{v}\right). \quad \text{(source approaching)} \qquad (17\text{-}11a)$$

The situation is similar in the direction opposite the source motion, except that now the wavelength increases by the amount $\lambda u/v$, giving

$$\lambda' = \lambda\left(1 + \frac{u}{v}\right). \quad \text{(source receding)} \qquad (17\text{-}11b)$$

We can recast these expressions in terms of frequency using the relations $\lambda = v/f$ and $\lambda' = v/f'$, where f' is the frequency of waves from the moving source as measured by an observer at rest in the medium. Substituting these relations in our expressions for λ' and solving for f' gives

$$f' = \frac{f}{1 \pm u/v}, \qquad (17\text{-}12)$$

for the Doppler-shifted frequency, where the $+$ and $-$ signs correspond to receding and approaching sources, respectively.

You've probably experienced the Doppler effect for sound when standing near a highway. A loud truck approaches with a high-pitched sound "aaaaaaaaaaa." As it passes, the pitch drops abruptly: "aaaaaaaaeiooooooooooo," and stays low as the truck recedes. Practical uses of the Doppler effect are numerous. Measuring the Doppler shift in reflected ultrasound allows measurement of blood flow and fetal heartbeat. Police radar works by measuring the Doppler shift of high-frequency radio waves reflected from moving cars (Fig. 17-21). The Doppler shift of starlight reveals stellar motions, while Doppler-shifted light from distant galaxies is evidence that our entire universe is expanding.

FIGURE 17-21 Speed trap! Police radar works by measuring the Doppler shift of radio waves reflected off moving cars.

● **EXAMPLE 17-6** THE WRONG NOTE

A car speeds down the highway with its stereo blasting. An observer with perfect pitch is standing by the roadside and, as the car approaches, notices that a musical note that should be G ($f = 392$ Hz) sounds like A (440 Hz). How fast is the car moving?

Solution
Solving Equation 17-12 for the source speed u, we have

$$u = v\left(1 - \frac{f}{f'}\right) = (343 \text{ m/s})\left(1 - \frac{392 \text{ Hz}}{440 \text{ Hz}}\right) = 37.4 \text{ m/s}$$
$$= 135 \text{ km/h}.$$

EXERCISE What is the shift in frequency of a 1.2-kHz m ambulance siren when the ambulance is approaching at 130 km/h?

Answer: $\Delta f = 140$ Hz

Some problems similar to Example 17-6: 56, 57, 59

●

A shift in frequency, but not wavelength, also occurs when a moving observer approaches a stationary source. Waves from a stationary source reflected off a moving object are therefore Doppler shifted *twice* —once because they arrive at the object with a shifted frequency and again because the reflecting object acts like a moving source. Problems 76 and 77 explore this situation.

Although light and other electromagnetic waves do not require a material medium, they, too, are subject to the Doppler shift. However, the Doppler formulas we've derived here apply to electromagnetic waves only in the limit where the relative speed between source and observer is much less than the 3×10^8 m/s speed of light.

17-8 SHOCK WAVES

Equation 17-11a suggests that wavelength goes to zero if a source approaches at exactly the wave speed. This happens because wave crests can't get away from the source, so they pile up just ahead of it to form a large-amplitude wave called a **shock wave** (Fig. 17-22a). When the source moves faster than the wave speed, waves pile up on a cone whose half angle is given by $\sin\theta = v/u$, as shown in Fig. 17-22b. The ratio u/v is called the **Mach number,** and the cone angle is the **Mach angle.**

FIGURE 17-22 (a) When the source moves at the wave speed, wave crests pile up to create a shock wave. (b) Shock wave formed when source speed u exceeds wave speed v. In a time Δt from the generation of the largest circular wavefront shown, that wavefront moves a distance $v\Delta t$. At the same time, the source moves a greater distance $u\Delta t$. As a result, a shock wave forms on a cone with half angle $\theta = \sin^{-1}(v/u)$.

$u = v$

(a)

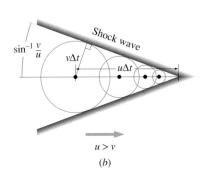

$u > v$

(b)

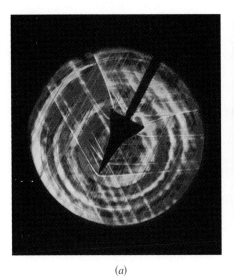

(a)

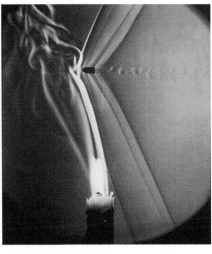

(b)

(c)

FIGURE 17-23 Shock waves. (a) Shock waves (red lines) trail from a model of a delta-winged orbiter in a wind tunnel. Shocks from real super-sonic aircraft are the sonic booms that can damage buildings and marine life. (b) A bullet passes through a candle flame, trailing shock waves and disrupting the air flow above the flame. Photos (a) and (b) were both made with a technique that images regions of enhanced density, such as the air that "piles up" behind a shock wave. (c) A powerboat's wake is a shock wave.

Shock waves occur in a wide variety of physical situations (Fig. 17-23). Sonic booms are shock waves from supersonic aircraft. The bow wave of a boat is a shock wave on the water surface. On a much larger scale, a huge shock wave forms in space as the solar wind—a high-speed flow of particles from the Sun—encounters Earth's magnetic field.

CHAPTER SYNOPSIS

Summary

1. **Sound waves** are longitudinal mechanical waves propagating in gases, liquids, or solids. The speed of sound in gases is given by

$$v = \sqrt{\frac{\gamma P}{\rho}},$$

with P the pressure, ρ the density, and γ a constant that depends on the nature of the gas molecules. More generally, the speed of sound is given by $v = \sqrt{B/\rho}$, where the **bulk modulus** B measures the elastic properties of the medium.

2. The intensity of sound, like that of any wave, depends on the square of the wave disturbance. In sound the medium undergoes both displacement and pressure changes, and the intensity can be written as

$$\bar{I} = \frac{\Delta P_0^2}{2\rho v} = \frac{1}{2}\rho\omega^2 s_0^2 v,$$

where ΔP_0 and s_0 are the pressure and displacement amplitudes, ω the angular frequency, and v the wave speed.

3. Normal sound covers many orders of magnitude in intensity, and the human ear responds logarithmically to sound intensity. A convenient measure of intensity is the **decibel** level, defined by

$$\beta = 10 \log\left(\frac{I}{I_0}\right),$$

where $I_0 = 10^{-12} \text{ W/m}^2$ is approximately the threshold of hearing at 1 kHz.

4. Wave **reflection** occurs whenever the medium changes abruptly. Complete reflection occurs if a wave cannot propagate in the new medium. More generally, a wave is partially reflected and partially transmitted at an interface between two media.

5. **Standing waves** arise when a vibrating medium is bounded. Each point in a standing wave undergoes simple harmonic motion. When the system is constrained so that there can be no displacement at either end, then its length must be an integer number of half wavelengths:

$$L = \frac{n\lambda}{2}.$$

If one end is fixed and the other free, then an odd integer number of quarter wavelengths must fit between the ends.

The longest wavelength—and lowest frequency—standing wave is the **fundamental;** higher frequencies are the **harmonics.**

6. The **Doppler effect** is the change in frequency due to the motion of a wave source or observer. For mechanical waves, the Doppler shifted frequency from a moving source is

$$f' = \frac{f}{1 \pm u/v},$$

where f is the frequency emitted when the source is at rest, v the wave speed, and u the source speed with respect to the medium. The positive sign applies to a source receding from the observer, the minus sign to an approaching source.

7. **Shock waves** are formed by sources moving through a medium at speeds greater than the wave speed. Shock waves are large-amplitude disturbances that form a cone of half angle $\theta = \sin^{-1}(v/u)$ with apex at the moving source.

Terms You Should Understand

(Pairs are closely related terms whose distinction is important; number in parentheses is chapter section where term first appears.)

audible sound, infrasound, ultrasound (introduction)
pressure (17-1)
decibel (17-3)
bulk modulus (17-4)
reflection, partial reflection (17-5)
standing wave (17-6)
node, antinode (17-6)
mode (17-6)

Doppler effect (17-7)
shock wave (17-8)
Mach number, Mach angle (17-8)

Symbols You Should Recognize

γ (17-2)	s (17-3)
P (17-2)	β (17-3)
ρ (17-2)	f', λ' (17-7)

Problems You Should Be Able to Solve

calculating sound speed from pressure, density, and constant γ (17-2)
relating frequency and wavelength of waves in different gases (17-2)
solving for sound intensity, displacement amplitude, or pressure amplitude (17-3)
working with intensity level in decibels (17-3)
solving for reflected and transmitted amplitude with waves on strings (17-5)
finding frequency and wavelength of allowed standing wave modes (17-6)
determining Doppler-shifted frequency and wavelength (17-7)
evaluating shock-wave angles (17-8)

Limitations to Keep in Mind

Expressions for sound speed apply only for small-amplitude waves.

When applied to light and other electromagnetic waves, the Doppler shift formulas developed in this chapter are approximations good when the source speed is small compared with the speed of light.

QUESTIONS

1. Consider a light wave and a sound wave with the same wavelength. Which has the higher frequency?

2. If you double the pressure of a gas while keeping its density the same, what happens to the sound speed?

3. Water is about 1000 times more dense than air, yet the speed of sound in water is greater than in air. How is this possible?

4. Why can't astronauts on the moon communicate with sound?

5. In a sound wave, do the displacement and pressure peak at the same point or at different points? Why?

6. If the sound intensity level increases from 20 dB to 30 dB, you perceive a certain change in loudness. How does that compare with the change you perceive when the level increases from 30 dB to 40 dB?

7. Does an increase of 1 dB in sound level correspond to a definite change in the intensity as measured in W/m²? Explain.

8. If the decibel level drops by half, does the sound intensity in W/m² also drop by half? Explain.

9. What property of the human ear makes the decibel scale particularly useful?

10. A loudspeaker is blaring at only 10 dB below the threshold of pain. By what factor must its power be increased to reach that threshold?

11. Figure 17-4 is drawn for a typical human ear. How might the figure change as a person ages?

12. Is is meaningful to talk about sound with a negative decibel level? Explain.

13. Why do you see a reflection of yourself in a perfectly clear glass window?

14. A light string is attached to a heavier one, and a wave sent down the light string toward the heavy one. Will its reflection be inverted or upright?

15. A light string is attached to a heavier one, and a wave sent down the heavy string toward the light one. Will its reflection be inverted or upright?

16. Which string in Fig. 17-8 has the greater mass per unit length?

17. If you place a perfectly clear piece of glass in perfectly clear water, you can still see the glass. Why?

18. Standing waves don't propagate. In what sense, then, are they waves?

19. Do standing waves satisfy the wave equation? You can answer without doing any math; just think about the superposition principle.

20. Light from distant galaxies is red shifted relative to nearby galaxies. Assuming this red shift is due to the Doppler shift, what does it say about the distant galaxies' motion relative to Earth?

21. Why does a boat easily produce a shock wave on the water surface, while it takes very high speed aircraft to produce sonic booms?

PROBLEMS

Sections 17-1 and 17-2 Sound Waves and the Speed of Sound in Gases

1. Show that the quantity $\sqrt{P/\rho}$ has the units of speed.

2. Dimensional analysis alone suggests that the sound speed in a gas should be given roughly by $\sqrt{P/\rho}$. By how much would an estimate based on this simple analysis be in error for a gas with $\gamma = \frac{7}{5}$?

3. Find the wavelength, period, angular frequency, and wave number of a 1.0-kHz sound wave in air under the conditions of Example 17-1.

4. (a) Determine an approximate value for the speed of sound in miles per second. (b) Suppose you see a lightning flash and, 10 s later, hear the thunder. How many miles away did the flash occur? Neglect the travel time for the light (why?)

5. Timers in sprint races start their watches when they see smoke from the starting gun, not when they hear the sound (Fig. 17-24). Why? How much error would be introduced by timing a 100-m race from the sound of the shot?

6. The factor γ for nitrogen dioxide (NO_2) is 1.29. Find the sound speed in NO_2 at a pressure of 4.8×10^4 N/m^2 and density 0.35 kg/m^3.

7. At standard atmospheric pressure (1.0×10^5 N/m^2), what density of air would make the sound speed 1.0 km/s?

8. The Sun's outer atmosphere, or corona, has about 10^8 electrons and an equal number of protons per cubic centimeter. The pressure of this electron-proton gas is about 3×10^{-3} N/m^2, and it behaves like a monatomic gas with $\gamma = \frac{5}{3}$. To one significant figure, what is the sound speed in the corona?

9. A gas with density 1.0 kg/m^3 and pressure 8.0×10^4 N/m^2 has sound speed 365 m/s. Are the gas molecules monatomic or diatomic?

10. By what percentage would the sound speed in pure oxygen (O_2) differ from that in air with the same pressure and number of particles per unit volume? Consult Example 17-1.

11. Saturn's moon Titan has one of the solar system's thickest atmospheres. Near Titan's surface, atmospheric pressure is 50% greater than standard atmospheric pressure on Earth, while the density in molecules per unit volume is one-third that of Earth's atmosphere. If Titan's atmosphere is essentially all nitrogen (N_2), what is the sound speed?

12. Divers in an underwater habitat breathe a special mixture of oxygen and neon, with pressure 6.2×10^5 N/m^2 and density 4.5 kg/m^3. The effective γ value for the mixture is 1.61. Find the frequency in this mixture for a 50-cm-wavelength sound wave, and compare with its frequency in air under normal conditions.

13. You see an airplane straight overhead at an altitude of 5.2 km. Sound from the plane, however, seems to be coming from a point back along the plane's path at a 35° angle to the vertical (Fig. 17-25). What is the plane's speed, assuming an average 330 m/s sound speed?

14. A mixture of oxygen and nitrogen with the pressure and molecules per cubic meter given in Example 17-1 has

FIGURE 17-24 Problem 5.

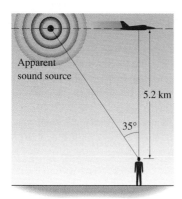

FIGURE 17-25 Problem 13.

sound speed 339 m/s. What fraction of the molecules are oxygen?

Section 17-3 Sound Intensity

15. Sound intensity in normal conversation is about 1 μW/m². What is the displacement amplitude of air in a 2.5-kHz sound wave with this intensity?

16. The eardrum ruptures at a sound intensity of 10 kW/m². What is the amplitude of the sound-wave pressure variation at this intensity, in air under normal conditions? Compare with standard atmospheric pressure, about 10⁵ N/m².

17. A speaker produces 440-Hz sound with total power 1.2 W, radiating equally in all directions. At a distance of 5.0 m, what are (a) the average intensity, (b) the decibel level, (c) the pressure amplitude, and (d) the displacement amplitude?

18. A "tweeter" loudspeaker emitting 5.0 kHz sound has an oscillation amplitude of 1 μm. What must be the oscillation amplitude of a "woofer" speaker producing the same sound intensity at 30 Hz?

19. What is the approximate frequency range over which sound with intensity 10⁻¹² W/m² can be heard? Consult Fig. 17-4.

20. The radius of the hydrogen atom is 0.0529 nm. (a) What is the intensity of a 100-Hz sound wave whose displacement amplitude is this size? (b) Repeat for a 2.0 kHz wave. Is either sound audible? Consult Fig. 17-4.

21. What are the intensity and pressure amplitudes in sound waves with intensity levels of (a) 65 dB and (b) −5 dB?

22. A 1-dB increase in sound level is about the minimum change the human ear can perceive. By what factor is the sound intensity increased if the level changes by 1 dB?

23. (a) What is the decibel level of a sound wave whose pressure amplitude is 2.9×10^{-4} N/m²? (b) Consult Fig. 17-4 to determine the approximate lowest frequency at which this sound would be audible.

24. If the pressure amplitude of a sound wave doubles, what happens to (a) the intensity in W/m² and (b) the decibel level?

25. Show that a doubling of sound intensity corresponds to very nearly a 3 dB increase in the decibel level.

26. Sound intensity from a localized source decreases as the inverse square of the distance, according to Equation 16-9. If the distance from the source doubles, what happens to (a) the intensity and (b) the decibel level?

27. At a distance 2.0 m from a localized sound source you measure the intensity level as 75 dB. How far away must you be for the perceived loudness to drop in half (i.e., to an intensity level of 65 dB)?

28. An amplifier is supplying a loudspeaker with 50 mW of power, and the resulting sound level where you are sitting is 46 dB. How much power must the amplifier supply for you to perceive a fourfold increase in loudness?

29. Sound intensity from a certain extended source drops as $1/r^n$, where r is the distance from the source. If the intensity level drops by 3 dB every time the distance is doubled, what is n?

30. (a) Show that the decibel level may be written

$$\beta = 20 \log\left(\frac{\Delta P}{P_0}\right),$$

where ΔP is the sound-wave pressure amplitude and P_0 is a reference pressure. (b) Find the value of P_0.

Section 17-4 Sound Waves in Liquids and Solids

31. The bulk modulus for tungsten is 2.0×10^{11} N/m², and its density is 1.94×10^4 kg/m³. Find the sound speed in tungsten.

32. The density of aluminum is 2700 kg/m³. Use the sound speed in Table 17-2 to find aluminum's bulk modulus.

33. The speed of sound in body tissues is essentially the same as in water. Find the wavelength of 2.0 MHz ultrasound used in medical diagnostics.

34. A 1-m-long lead bar and a 1-m-long steel bar are side by side. If they are simultaneously struck a blow on one end, how much sooner will the resulting pressure pulse arrive at the far end of the steel bar?

35. Mechanical vibration induces a sound wave in a mechanism consisting of a 12-cm-long steel rod attached to a 3.0-cm-long neoprene block. How long does it take the wave to propagate through this structure?

Section 17-5 Wave Reflection

36. A 5.0-cm-amplitude wave propagates along a string with mass per unit length 6.8 g/m. The string is joined to a second string with 9.2 g/m. Find the amplitudes of (a) the transmitted and (b) the reflected waves.

37. A string with mass per unit length μ is joined to another with 4μ. If a wave of amplitude A propagates from the lighter toward the heavier string, find the amplitudes of (a) the transmitted and (b) the reflected waves.

38. When a wave reaches the junction of two different strings, the transmitted and reflected waves have equal amplitudes, with the reflected wave inverted. How do the masses per unit length of the two strings compare?

39. A 3.3-m long string of total mass 56 g is spliced to a 2.6-m cord of total mass 280 g. The two are under a uniform 110-N tension. A 2.0-cm-high pulse is launched from the end of the lighter string at time $t = 0$. (a) At what time will the pulse reach the far end of the heavier cord? (b) What will be the amplitude in the heavier cord? (c) What will be the amplitude of the pulse reflected back into the lighter string?

40. A wave pulse is propagating down a string with $\mu = 14$ g/m, when it encounters a splice to a heavier string with $\mu = 32$ g/m. What per cent of the initial pulse energy is carried down the heavier string?

41. A rope is made from a number of identical strands twisted together. The rope is frayed, with only a single strand continuing, as shown in Fig. 17-26. The rope is held under tension, and a 2.0-cm-high pulse is sent from the single strand. The first pulse reflected back along the string has 0.90 cm amplitude. How many strands are in the rope?

FIGURE 17-26   Problem 41.

Section 17-6 Standing Waves

42. A 2.0-m-long string is clamped at both ends. (a) What is the longest wavelength standing wave that can exist on this string? (b) If the wave speed is 56 m/s, what is the lowest standing-wave frequency?

43. When a stretched string is clamped at both ends, its fundamental standing-wave frequency is 140 Hz. (a) What is the next higher frequency? (b) If the same string, with the same tension, is now clamped at one end and free at the other, what is the fundamental frequency? (c) What is the next higher frequency in case (b)?

44. The A-string (440 Hz) on a piano is 38.9 cm long and is clamped tightly at both ends. If the string is under 667 N tension, what is its mass?

45. Show that only odd harmonics are allowed on a taut string with one end tight and the other free.

46. A string is clamped at both ends and tensioned until its fundamental vibration frequency is 85 Hz. If the string is then held rigidly at its midpoint, what is the lowest frequency at which it will vibrate?

47. Show that the standing-wave condition of Equation 17-10 is equivalent to the requirement that the time it takes a wave to make a round trip from one end of the medium to the other and back be an integer multiple of the wave period.

48. The wave speed on a taut wire is 320 m/s. If the wire is 80 cm long, and is vibrating at 800 Hz, how far apart are the nodes?

49. "Vibrato" in a violin is produced by sliding the finger back and forth along the vibrating string. The G-string on a particular violin measures 30 cm between the bridge and its far end and is clamped rigidly at both points. Its fundamental frequency is 197 Hz. (a) How far from the end should the violinist place a finger so that the G-string plays the note A (440 Hz)? (b) If the violinist executes vibrato by moving the finger 0.50 cm to either side of the position in part (a), what range of frequencies results?

50. Estimate the fundamental frequency of the human vocal tract by assuming it to be a cylinder 15 cm long that is closed at one end.

51. A bathtub 1.7 m long contains 13 cm of water. By sloshing water back and forth with your hand, you can build up large-amplitude oscillations. Determine the lowest frequency possible for such a resonant oscillation, using the fact that the speed of waves in shallow water of depth h is $v = \sqrt{gh}$. *Hint:* At resonance in this case, the wave has a crest at one end and a trough at the other.

52. What length is necessary for an organ pipe (Fig. 17-27) to produce a 22-Hz tone (a) if the pipe is closed at one end and (b) if it's open at both ends?

FIGURE 17-27   What length pipe organ will produce a 22-Hz tone? (Problem 52)

53. What would be the fundamental frequency of the double bassoon of Example 17-4, if it were played in helium under conditions of Example 17-1?

54. An organ pipe is closed at one end and has a fundamental frequency of 50 Hz. What are the mode numbers and frequencies of the next two higher frequencies it can play?

55. An astronaut smuggles a double bassoon (Example 17-4) to Mars and plays the instrument's fundamental note. If it sounds at 23 Hz, what is the sound speed on Mars?

Section 17-7 The Doppler Effect

56. A car horn emits 380-Hz sound. If the car moves at 17 m/s with its horn blasting, what frequency will a person standing in front of the car hear?

57. A fire truck's siren at rest wails at 1400 Hz; standing by the roadside as the truck approaches, you hear it at 1600 Hz. How fast is the truck going?

58. Red light emitted by hydrogen atoms at rest in the laboratory has wavelength 656 nm. Light emitted in the same process on a distant galaxy is received at Earth with wavelength 708 nm. Describe the galaxy's motion relative to Earth.

59. The dominant frequency emitted by an airplane's engines is 1400 Hz. (a) What frequency will you measure if the plane approaches you at half the sound speed? (b) What frequency will you measure if the plane recedes at half the sound speed?

60. A wave source approaches you at constant speed, and you measure a wave frequency f_1. As the source passes and then recedes, you measure frequency f_2. Find expressions for (a) the source speed and (b) the frequency emitted if the source were stationary, both in terms of f_1, f_2 and the wave speed v.

61. You're standing by the roadside as a truck approaches, and you measure the dominant frequency in the truck noise at 1100 Hz. As the truck passes the frequency drops to 950 Hz. What is the truck's speed?

Section 17-8 Shock Waves

62. What will be the cone angle for a supersonic aircraft traveling at Mach 2.5 (2.5 times the sound speed)?

63. Figure 17-28 shows a projectile in supersonic flight, with shock waves clearly visible. By making appropriate measurements, determine the projectile's speed as compared with the sound speed.

64. A supersonic plane flies directly over you at 2.2 times the sound speed. You hear its sonic boom 19 s later. What is the plane's altitude, assuming a constant 340 m/s sound speed?

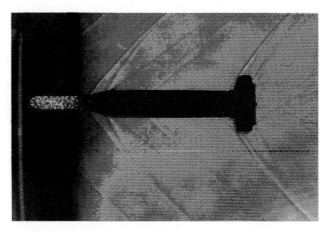

FIGURE 17-28 Problem 63.

Paired Problems

(Both problems in a pair involve the same principles and techniques. If you can get the first problem, you should be able to solve the second one.)

65. A 1.0-W sound source emits uniformly in all directions. Find (a) the intensity and (b) the decibel level 12 m from the source.

66. At a distance 3.5 m from a localized sound source the intensity level is 84 dB. (a) What will it be 8.0 m from the source? (b) What is the total power emitted by the source?

67. A pipe 80 cm long is open at both ends. When the pipe is immersed in a gas mixture, the frequency of a certain harmonic is 280 Hz and the next higher harmonic is 350 Hz. Determine (a) the sound speed and (b) the mode numbers of the two harmonics.

68. A 1.5-m-long pipe has one end open. Among its possible standing-wave frequencies is 225 Hz; the next higher frequency is 375 Hz. Find (a) the fundamental frequency and (b) the sound speed.

69. Find the wave speed in a medium where a 28 m/s source speed causes a 3% increase in frequency measured by a stationary observer.

70. A wave source recedes from you at 8.2 m/s, and the wavelength you measure is 20% greater than what you would measure were the source at rest. What is the wave speed?

Supplementary Problems

71. The sound speed in air at 0°C is 331 m/s, and for temperatures within a few tens of degrees of 0°C it increases at the rate 0.590 m/s for every °C increase in temperature. How long would it take a sound wave to travel 150 m over a path where the temperature rises linearly from 5°C at one end to 15°C at the other end?

72. Rods of lead and steel are joined end-to-end; the total length is 6.0 m. If a sound wave takes 1.8 ms to traverse this structure, how long is the lead rod?

73. A rectangular trough is 2.5 m long and is much deeper than its length, so Equation 16-10 applies. Determine the wavelength and frequency of (a) the longest and (b) the next longest standing waves possible in this trough. Why isn't the higher frequency twice the lower?

74. Show by direct substitution (with appropriate differentiation) that Equation 17-9 satisfies the wave equation 16-11.

75. A supersonic airplane flies directly over you at 6.5 km altitude. You hear its sonic boom 13 s later. What is the plane's Mach number?

76. Show that an observer moving toward a source of sound waves that is stationary with respect to the medium will measure a higher frequency given by $f' = f(1 + u/v)$.

77. Consider an object moving at speed u through a medium, and reflecting sound waves from a stationary source back toward the source. The object receives the waves at the shifted frequency given in the preceding problem, and when it re-emits them they are shifted once again, this time according to Equation 17-12. Find an expression for the overall frequency shift that results, and show that, for $u \ll v$, this shift is approximately $2fu/v$.

78. Obstetricians use ultrasound to monitor fetal heartbeat. If 5.0-MHz ultrasound reflects off the moving heart wall with a 100-Hz frequency shift, what is the speed of the heart wall? *Hint:* The heart *reflects* the waves, so consider the result of the preceding problem.

79. What is the frequency shift of a 70-GHz police radar signal when it reflects off a car moving at 120 km/h? (Radar waves travel at the speed of light.) *Hint:* See Problem 77.

80. In an extraordinarily sensitive experiment, physicists at Harvard University recently measured the growth rate of crystals by reflecting a laser beam off the crystal surface. The reflected and transmitted beams were combined, and the resulting beat frequency—arising from the very small

Doppler shift—was measured. (a) For a laser frequency of 5×10^{14} Hz, what beat frequency results from a crystal growth rate of 0.2 nm/s? (b) What fraction is the frequency shift of the original laser frequency?

81. A wave pulse of total energy E is propagating on a string with mass per unit length μ_1 toward the junction with a second string whose mass per unit length is μ_2. Find an expression for the fraction of the pulse energy transmitted into the second string.

82. A string with mass per unit length μ_1 is joined at $x = 0$ to a second string with mass per unit length μ_2 (Fig. 17-29). An incident wave whose displacement is $y_I(x, t) = A_I \cos(k_1 x - \omega t)$ is incident on the junction, resulting in reflected and transmitted waves described by $y_R(x, t) = A_R \cos(k_1 x + \omega t)$ and $y_T(x, t) = A_T \cos(k_2 x - \omega t)$, respectively. The wave numbers k_1 and k_2 are not the same because of the different wave speeds in the two media. If the spring is not to break, the total displacement approaching the junction from the left must equal that approaching it from the right, so $y_I(0, t) + y_R(0, t) = y_T(0, t)$. Furthermore, the derivatives $\partial y/\partial x$ must be continuous at the junction, or the string's acceleration would be infinite. Use these continuity conditions to derive Equations 17-7 and 17-8.

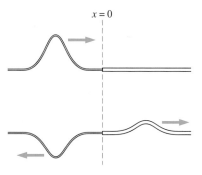

FIGURE 17-29 Problem 82.

CHAPTER 18

FLUID MOTION

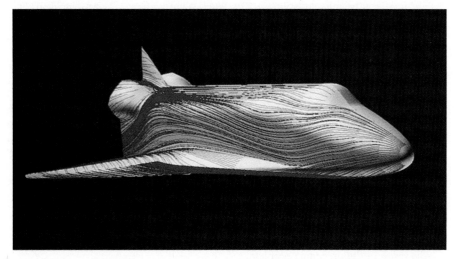

Computer simulation showing airflow over a space shuttle. The behavior of fluids, like the air in this simulation, is ultimately governed by the laws of mechanics.

A tornado whirls across a darkened sky. Earth's continents drift about the planet, riding a layer of liquid rock. A jetliner takes off, propelled by hot gases and supported by air pressure on its wings. A stream of gas leaves the surface of a giant star and forms a cosmic whirlpool before it plunges into a black hole. Your car brakes to a stop, the force of your foot on the brake pedal amplified by the effect of fluid in the brake cylinders. Your own body is sustained by air moving in and out of your lungs, and by the flow of blood throughout your tissues. All these examples involve fluid motion.

Fluid is matter that flows under the influence of external forces. The intermolecular forces are weaker in fluids than in solids, and as a result the molecules move around readily rather than being locked into the rigid structure that characterizes a solid. In a liquid, those forces are still significant enough to keep the molecules in close contact, while in a gas they are almost negligible and the molecules are usually widely spaced. We will see in Chapter 20, though, that the distinction between liquids and gases is not quite so clear-cut.

18-1 DESCRIBING FLUIDS: DENSITY AND PRESSURE

If we could observe a fluid on the molecular scale, we would find large numbers of molecules in continuous motion, colliding frequently with each other and with the walls of their containers. This molecular behavior is governed by the laws of mechanics, and in principle we could study fluids by applying those laws to all the individual molecules. But even a drop of water contains about 10^{21} molecules; to calculate the motions of all those molecules would take the fastest computers many times the age of the universe!

Because the number of molecules is so large, we approximate a fluid by considering it to be continuous rather than composed of discrete particles. In this approximation, valid for fluid samples large compared with the distance between molecules, we describe the fluid by specifying macroscopic properties like density and pressure.

Density

Density (symbol ρ, the Greek rho) measures the mass per unit volume; its SI units are kg/m³. The density of water under normal conditions is about 1000 kg/m³; that of air is about a factor of 1000 smaller. Because their molecules are essentially in contact, liquids are **incompressible,** meaning that their densities remain nearly constant. Gases, with relatively large distances between their molecules, are **compressible;** their densities change readily.

Pressure

Pressure measures the normal force per unit area exerted by a fluid (Fig. 18-1a). If the pressure is uniform over an area, then we have

$$P = \frac{F}{A};$$
(18-1a)

if pressure varies, we consider the force dF on an infinitesimal area dA and write

$$P = \frac{dF}{dA}.$$
(18-1b)

The SI pressure unit is N/m², given the name **pascal** (Pa) after the French mathematician, scientist, and philosopher Blaise Pascal (1623–1662). Another commonly used pressure unit is the **atmosphere** (atm), defined as Earth's normal atmospheric pressure at sea level and equal to 1.013×10^5 Pa (14.7 pounds per square inch). Problems 6 and 7 introduce several other pressure units.

Pressure is a scalar quantity; at a given point in a fluid, pressure is exerted equally in all directions, so it makes no sense to associate a direction with it (Fig. 18-1b).

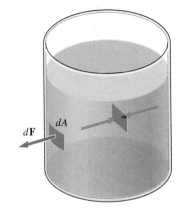

(a)

(b)

FIGURE 18-1 (a) Pressure is the force per unit area exerted by a fluid, either on its container walls or an adjacent volume of fluid. (b) At a given point, a fluid exerts the same pressure in all directions.

FIGURE 18-2 (*a*) If pressure varies with position, then there is a net pressure force on a volume of fluid. In the absence of external forces, the fluid then cannot be in hydrostatic equilibrium. (*b*) When the pressure is constant, the net pressure force is zero.

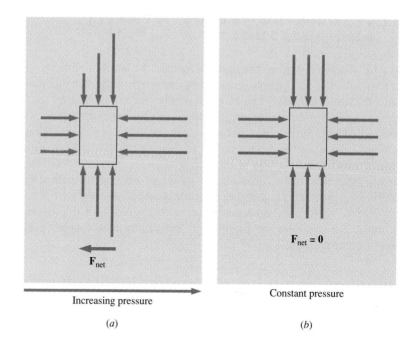

(*a*)

Increasing pressure

$\mathbf{F}_{net} = 0$

Constant pressure

(*b*)

18-2 FLUIDS AT REST: HYDROSTATIC EQUILIBRIUM

For a fluid to remain at rest, the net force everywhere in the fluid must be zero. When this condition is met, the fluid is in **hydrostatic equilibrium.** In the absence of any external forces, hydrostatic equilibrium requires that the pressure be constant throughout the fluid; otherwise pressure differences would result in forces acting on the fluid. As Fig. 18-2 suggests, it is pressure *difference,* rather than pressure itself, that gives rise to net forces within fluids.

Hydrostatic Equilibrium with Gravity

For a fluid to be in hydrostatic equilibrium in the presence of gravity, there must be a pressure force to counteract the gravitational force. Since pressure forces arise only from pressure differences, the fluid pressure must therefore vary with depth.

Figure 18-3 shows the forces on a fluid element of area A, thickness dh, and mass dm. A gravitational force acts downward on this fluid element; for it to be in equilibrium there must therefore be an upward pressure force—and that requires a greater pressure on the lower side of the fluid element. Suppose the pressures at the top and bottom are P and $P + dP$, respectively. Since pressure is force per unit area, the net pressure force on the fluid element is therefore

$$dF_{press} = (P + dP)A - PA = A\,dP.$$

The gravitational force is $dF_{grav} = -g\,dm$, where the minus sign designates the downward direction. But the mass dm is the density times the volume, so

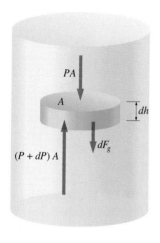

FIGURE 18-3 Forces on a fluid element in hydrostatic equilibrium include gravity and the pressure forces on top and bottom.

$$dF_{grav} = -g \, dm = -g\rho A \, dh.$$

Hydrostatic equilibrium requires that these forces sum to zero:

$$A \, dP - g\rho A \, dh = 0,$$

or

$$\frac{dP}{dh} = \rho g. \qquad (18\text{-}2)$$

This equation shows that dP/dh—the variation in pressure with depth h—is positive, confirming our expectation that pressure increases with depth. The exact form of this variation depends on the nature of the fluid. For a liquid, which is essentially incompressible, ρ is constant and Equation 18-2 shows that pressure increases linearly with depth:

$$P = P_0 + \rho gh, \qquad (18\text{-}3)$$

where P_0 is the pressure at the liquid surface.

● EXAMPLE 18-1 OCEAN DEPTHS

(a) At what water depth is the pressure twice atmospheric pressure? (b) What is the water pressure at the bottom of the 11.3-km-deep Marianas trench in the North Pacific Ocean? Atmospheric pressure is 1.0×10^5 Pa, and the density of water is 1000 kg/m³.

Solution
At twice atmospheric pressure, $P = 2P_0$ in Equation 18-2. Solving for the depth h then gives

$$h = \frac{P - P_0}{\rho g} = \frac{2.0 \times 10^5 \text{ Pa} - 1.0 \times 10^5 \text{ Pa}}{(1000 \text{ kg/m}^3)(9.8 \text{ m/s}^2)} = 10 \text{ m}.$$

Since pressure increases linearly with depth, the pressure continues to increase by 1.0×10^5 Pa for every 10 m of depth. In the Marianas trench, 11.3×10^3 m deep, the pressure increase is then

$$P - P_0 = (11.3 \times 10^3 \text{ m})(1.0 \times 10^5 \text{ Pa}/10 \text{ m}) = 1.1 \times 10^9 \text{ Pa}.$$

This is over 1000 times atmospheric pressure, or more than 8 tons per square inch! Creatures living at these depths are in pressure equilibrium with their surroundings; to bring them to the surface for study, scientists must maintain their natural pressure or they will explode (Fig. 18-4). A similar plight awaits scuba divers who hold their breath while ascending; air in the lungs expands, bursting the alveoli.

FIGURE 18-4 Oceanographers bring deep ocean water to the surface in a container that maintains deep-sea creatures at their natural pressure.

EXERCISE A submarine is designed for depths to 1500 m. How much pressure must it withstand if its interior is maintained at normal atmospheric pressure?

Answer: 1.5×10^7 Pa
Some problems similar to Example 18-1: 17–19
●

● EXAMPLE 18-2 A SWIMMING POOL

A swimming pool is 15 m wide and, at the deep end, 3.0 m deep. Find the force of the water on the pool wall at the deep end.

Solution

The pressure in the pool varies from surface to bottom, so we can't simply multiply the pressure by the area to find the force. Instead, we consider the force on a small strip of height dh and pool width W, as shown in Fig. 18-5. This strip has area $dA = W dh$, so the force on it is $dF = P dA = PW dh$. Using Equation 18-3 for the pressure, and integrating to get the total force, we have

$$F = \int dF = \int_{h=0}^{h=H} (P_0 + \rho g h)W dh = W(P_0 h + \tfrac{1}{2}\rho g h^2)\big|_0^H$$
$$= P_0 WH + \tfrac{1}{2}\rho g WH^2,$$

where H is the pool depth. Using $P_0 = 1.0 \times 10^5$ Pa for atmospheric pressure, $\rho = 1000$ kg/m^3 for water, and the pool dimensions gives $F = 5.2$ MN.

EXERCISE The spent fuel storage pool at a nuclear reactor measures 20 m long by 10 m wide by 6.5 m deep and is filled with water. What force must its longer sides be capable of withstanding?

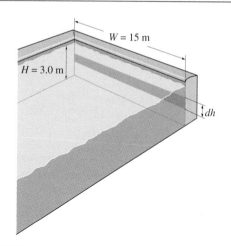

FIGURE 18-5 Since pressure varies with depth, calculating the force on the pool wall involves integrating over the vertical position. A suitable area element is a strip of width w and height dh.

Answer: 17 MN

Some problems similar to Example 18-2: 23, 24

●

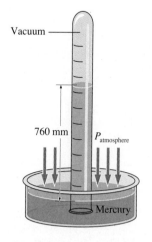

FIGURE 18-6 A mercury barometer. Normal air pressure supports a mercury column 760 mm high.

Measuring Pressure

The variation in liquid pressure with depth is the basis of several common pressure-measuring instruments. Figure 18-6 shows a simple **barometer.** Atmospheric pressure acts on the open pool of liquid mercury, pushing the liquid into the evacuated tube. Since $P_0 = 0$ in the vacuum at the top of the tube, Equation 18-3 becomes simply $P = \rho g h$, showing that the height h of the mercury column is directly proportional to the atmospheric pressure P at the bottom of the column. In mercury, standard atmospheric pressure of 1.013×10^5 Pa supports a mercury column 760 mm or 29.92 inches high (see Problem 6). Weather forecasters often report atmospheric pressure in mm or inches of mercury.

Barometers use mercury because its high density makes for a relatively short instrument; Example 18-1 shows that a water-filled barometer would need to be 10 m long to measure atmospheric pressure. The same underlying fact—that atmospheric pressure can support a water column 10 m high—puts a limitation on water pumps that operate by suction. No matter how good a vacuum a suction pump can produce, atmospheric pressure can't push water up more than 10 m. Pumping from greater depths requires a submerged pump or one that somehow increases the pressure at the pumping depth.

Measurement of pressure differences is accomplished with a **manometer,** a U-shaped tube containing liquid and open at both ends (Fig. 18-7). A pressure

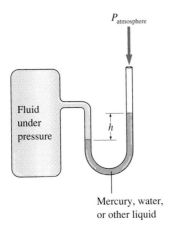

$P_{atmosphere}$

Fluid under pressure

h

Mercury, water, or other liquid

FIGURE 18-7 A manometer used to measure pressure difference between a closed container and the ambient atmosphere. Height difference h is proportional to the pressure difference.

difference between the ends of the tube results in a difference in height of the two liquid surfaces; Equation 18-3 shows that this height difference is directly proportional to the pressure difference.

When one end of a manometer, or other pressure-difference instrument, senses atmospheric pressure, the pressure it reads is called **gauge pressure,** meaning the pressure in excess of atmospheric pressure. Inflation pressures for tires and sports equipment are given in gauge pressure; a tire inflated to 200 kPa (about 30 pounds/in²) actually has an absolute pressure of about 300 kPa because the 100-kPa atmospheric pressure is not included in the specification. Tire gauges read gauge, not absolute, pressure.

Pascal's Law

Equation 18-3 shows that an increase in surface pressure P_0 results in the same pressure increase throughout the fluid. More generally, a pressure increase anywhere in a fluid is felt throughout the fluid—a fact first recognized by Pascal and now known as **Pascal's law.** Pascal applied this principle in his invention of the hydraulic press; today hydraulic systems, based on Pascal's law, are used to control machinery ranging from automobile brake systems, to aircraft wings, to bulldozers, cranes, and robots (Fig. 18-8).

FIGURE 18-8 Arrows indicate the hydraulic cylinders that actuate this power shovel.

● EXAMPLE 18-3 A HYDRAULIC LIFT

In the hydraulic lift shown in Fig. 18-9, a 60-cm-diameter piston supports a car. If the total mass of the piston and car is 3200 kg, what should be the diameter of the smaller piston if an applied force of 450 N is to maintain the system in equilibrium? Neglect the mass of the smaller piston and any pressure variation with height.

Solution

The small piston exerts a pressure $P = F_1/A_1 = F/\pi r_1^2$, with r_1 its radius and F_1 the 450-N applied force. Pascal's law says that this pressure is transmitted equally throughout the system; at

the large piston it produces a force $F_2 = PA_2$. This force supports the weight mg of piston and car; therefore we have

$$mg = PA_2 = P\pi r_2^2 = \frac{F_1}{\pi r_1^2}\pi r_2^2 = F_1\frac{r_2^2}{r_1^2}.$$

Solving for r_1 gives

$$r_1 = r_2\sqrt{\frac{F_1}{mg}} = (30\ \text{cm})\sqrt{\frac{450\ \text{N}}{(3200\ \text{kg})(9.8\ \text{N/kg})}} = 3.6\ \text{cm},$$

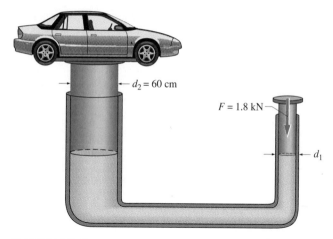

FIGURE 18-9 A hydraulic lift (Example 18-3).

has an interior diameter of 8.0 mm, and it operates two slave cylinders each with diameter 3.2 cm. (Two separate systems each operating two wheels are provided on modern cars for safety in the event of a hydraulic fluid leak.) If a 750-N force is applied to the master cylinder, what force does each of the two slave cylinders exert on its brake pads?

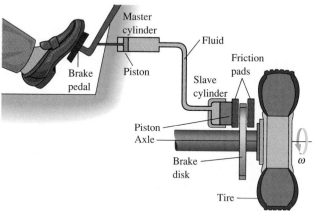

FIGURE 18-10 Simplified diagram of an automobile braking system, showing master cylinder and one of the slave cylinders. The slave cylinder pushes the brake pads against the rotating brake disk, stopping it and the attached wheel.

or a diameter of 7.2 cm. Note that the 450-N force is effectively multiplied by the ratio of the piston areas, ultimately supporting a weight of nearly 32,000 N.

Are we getting something for nothing here? No. Work done in moving the small piston—the product of force with distance moved—is equal to the work done on the large piston. Since the force on the large piston is much greater, the distance moved is much smaller, and energy is conserved.

EXERCISE Figure 18-10 shows an automobile braking system. The brake pedal applies a force to a piston in the "master cylinder," and the resulting pressure increase is transmitted to "slave cylinders" that operate the brakes. The master cylinder

Answer: 12 kN

Some problems similar to Example 18-3: 28–30

●

18-3 ARCHIMEDES' PRINCIPLE AND BUOYANCY

Drop a rock into water and it sinks. Drop a block of wood and it floats. Why the difference?

Figure 18-11*a* shows the upward pressure force on an arbitrary fluid volume balancing the downward gravitational force. Now imagine replacing the fluid volume by a solid object of identical shape (Fig. 18-11*b*). The remaining fluid hasn't changed, so it continues to exert an upward force on the object—a force whose magnitude equals the weight of the *original fluid volume.* This force is called the **buoyancy force,** and in giving its magnitude we've stated **Archimedes' principle:**

> **Archimedes' principle: The buoyant force on an object is equal to the weight of the fluid displaced by the object.**

If the submerged object weighs more than the displaced fluid, then the gravitational force exceeds the buoyancy force and it sinks. If the object weighs less

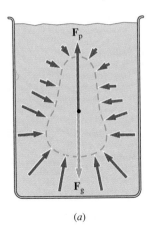

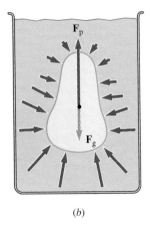

FIGURE 18-11 (*a*) In hydrostatic equilibrium, the upward pressure force $\mathbf{F}_p$ is equal to the weight $\mathbf{F}_g$ of any fluid volume. (*b*) If the fluid is replaced by a solid object, the pressure force remains the same but the gravitational force generally changes. The direction of the net force on the object depends on whether it is more or less dense than the fluid.

(*a*)   (*b*)

FIGURE 18-12 (*a*) A hot-air balloon at constant altitude is in neutral buoyancy. (*b*) So is a fish, whose gas-filled swim bladder compresses or expands to maintain the same average density as water.

(*a*)   (*b*)

than the displaced fluid, the buoyancy force is greater and it rises. Therefore, an object floats or sinks depending on whether its average density is greater or less than that of water. In between is the case of **neutral buoyancy,** when an object's average density is the same as that of water. Fish, submarines, and balloons are often in neutral buoyancy (Fig. 18-12).

● **EXAMPLE 18-4** AEROGELS, ARCHIMEDES, AND THE KING'S CROWN

Physicists have developed new materials called *aerogels* that are at once amazingly light (Fig. 18-13*a*) and surprisingly strong (Fig. 18-13*b*). A particular piece of aerogel is made from 1.00 g of silicon dioxide. Yet its apparent weight on a scale is 7.35 mN. How does the aerogel's density compare with that of air?

Solution

The aerogel's apparent weight is less than its actual weight $m_{gel}g = 9.8$ mN because of the upward buoyant force of the air. That force is equal to the weight of displaced air, or $F_b = m_{air}g$. The *volume* of displaced air is that of the aerogel, so $m_{air} = \rho_{air}V_{gel}$. But the aerogel's volume is $V_{gel} = m_{gel}/\rho_{gel}$. Putting this all together gives the buoyant force:

$$F_b = m_{air}g = \rho_{air}V_{gel}g = \rho_{air}\frac{m_{gel}}{\rho_{gel}}g = W_{gel}\frac{\rho_{air}}{\rho_{gel}},$$

where $W_{gel} = m_{gel}g$ is the gel's actual weight. Its apparent weight is then $W_{apparent} = W_{gel} - F_b$. Using our expression for F_b and solving for the density ratio ρ_{gel}/ρ_{air} then gives

$$\frac{\rho_{gel}}{\rho_{air}} = \frac{W_{gel}}{W_{gel} - W_{apparent}} = \frac{9.8 \text{ mN}}{9.8 \text{ mN} - 7.35 \text{ mN}} = 4.0.$$

The density of air is about 1.2 kg/m^3, so the aerogel's density is only 4.8 kg/m^3—compared with 1000 kg/m^3 for water.

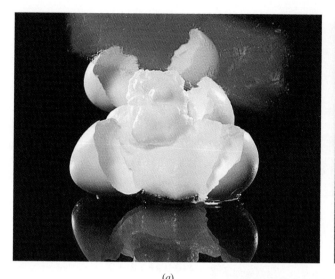

(a)

(b)

FIGURE 18-13 (a) Aerogels, made from silicon dioxide, are so tenuous that the buoyant force of air significantly reduces the apparent weight. Here a ghostly blue slab of aerogel rests on freshly beaten egg whites. (b) Aerogels are surprisingly strong. Here an aerogel slab with mass under 1 g supports a 100-g mass.

The buoyant force of air significantly reduces the aerogel's apparent weight because its density is only a few times that of air. Buoyancy actually lowers the apparent weights of all objects when weighed in air, but usually the effect is negligible because typical densities are on the order of 1000 times that of air.

Archimedes, one of the greatest scientists of ancient Greece, purportedly used his principle to determine the density of the king's crown, and thus verify that it was indeed made of gold. Archimedes achieved a nonnegligible buoyancy force by suspending the crown in water (see Fig. 18-14 and Problem 34).

EXERCISE Salvage experts recover a shipment of steel beams from a sunken ship. What will be the tension in a cable lifting a 750-kg beam through water with density 1.0 g/cm³? The density of the steel is 7.9 g/cm³ and the lifting is at constant speed.

Answer: 6.4 kN

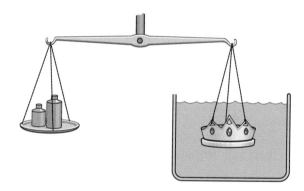

FIGURE 18-14 Archimedes purportedly verified that the king's crown was pure gold by finding its apparent weight when submerged in water.

Some problems similar to Example 18-4: 31–34

Floating Objects

Archimedes' principle still holds for an object floating on a fluid surface. But now the buoyant force must balance the object's weight—and this can happen only if the displaced fluid has a weight equal to that of the object. This condition determines how high in the water the object floats, as Fig. 18-15 and the example below illustrate.

(a)   (b)

FIGURE 18-15 Supertankers, (a) empty and (b) fully loaded. In both cases the buoyant force supports the full weight of the ship and cargo. The buoyant force is equal to the weight of the displaced water, so the fully loaded ship must displace more water and therefore floats lower.

● EXAMPLE 18-5   THE TIP OF THE ICEBERG

The average density of a typical arctic iceberg is 0.86 that of sea water. What fraction of an iceberg's volume is submerged?

Solution
The buoyant force supports the entire weight of the iceberg, so the weight of the displaced water is equal to the iceberg weight. If V_{sub} is the volume of ice that's submerged, then the weight of an equal volume of water—and therefore the magnitude of the buoyant force—is

$$F_b = W_{water} = m_{water}g = \rho_{water}V_{sub}g.$$

The weight of the iceberg is

$$W_{ice} = m_{ice}g = \rho_{ice}V_{ice}g,$$

where V_{ice} is the volume of the entire iceberg. Equating the iceberg weight to the magnitude of the buoyant force, we have

$$\rho_{ice}V_{ice}g = \rho_{water}V_{sub}g,$$

so

$$\frac{V_{sub}}{V_{ice}} = \frac{\rho_{ice}}{\rho_{water}} = 0.86.$$

FIGURE 18-16 Most of an iceberg's volume is submerged because its density is close to that of water.

Roughly six-sevenths of an arctic iceberg is therefore submerged (Fig. 18-16).

● EXAMPLE 18-6   AN OVERLOADED BOAT

A flat-bottomed rowboat has length $\ell = 3.2$ m, width $w = 1.1$ m, and is 28 cm deep. Empty, it has a mass of 130 kg. (a) How much of the boat is submerged when it is empty? (b) What is the total mass it could carry without taking on water?

Solution
In equilibrium, the total weight of the boat and its load must be supported by the buoyant force of the water. By Archimedes' principle, this force is equal to the weight of water displaced by

the boat. Letting y be the depth to which the boat is submerged, we have

$$\ell wy\rho_{water}g = Mg.$$

Here ρ_{water} is the density of water, and ℓwy the submerged volume, so the term on the left is the weight of displaced water and is therefore equal to the magnitude of the buoyant force. The quantity M is the total mass of the boat plus its load, so the term on the right is the weight supported by the buoyant force. Solving for y, we have

$$y = \frac{M}{\ell w\rho_{water}} = \frac{(130 \text{ kg})}{(3.2 \text{ m})(1.1 \text{ m})(1000 \text{ kg/m}^3)} = 3.7 \text{ cm}$$

for the empty boat.

To find the absolute maximum load, we set y equal to the boat's 28-cm depth, and solve for M:

$$M = \ell wy\rho_{water} = (3.2 \text{ m})(1.1 \text{ m})(0.28 \text{ m})(1000 \text{ kg/m}^3)$$
$$= 986 \text{ kg}.$$

This is the total mass; subtracting the 130-kg boat mass gives 856 kg for the maximum load.

EXERCISE A rectangular bar of soap floats with 3.5 cm extending below the water surface and 1.5 cm above. What is its density?

Answer: 700 kg/m^3

Some problems similar to Examples 18-5 and 18-6: 35–37, 40 ●

18-4 FLUID DYNAMICS

We now turn our attention to moving fluids, which are described by the flow velocity at each point in the fluid and at each instant of time. Figure 18-17 shows some flow velocity vectors in a river. We can describe flow velocity either with individual vectors, as in Fig. 18-17*a*, or by drawing continuous lines called **streamlines** that are everywhere tangent to the local flow direction (Fig. 18-17*b*). Their spacing is a measure of flow speed, with closely spaced streamlines indicating high speed. Small particles, like smoke or dyes, are often introduced into moving fluids because they follow streamlines and therefore give a visual indication of the flow velocity pattern (Fig. 18-18).

We distinguish two types of fluid motion. In **steady flow,** the pattern of fluid motion remains the same at each point, even though individual fluid elements are in continual motion. When you look at a river in steady flow, for example, it always looks the same, even though you're not seeing the same water each time you look. At a given point in the river, the water velocity, density, and pressure are always the same. **Unsteady flow,** in contrast, involves fluid motion that changes with time even at a fixed point. The blood in your arteries is in unsteady flow; with each contraction of the heart ventricles, the pressure rises, and the flow velocity increases. Figure 18-19 shows other examples of steady and unsteady flow. We will restrict our quantitative description of fluid motion to steady flow.

FIGURE 18-17 (*a*) Flow vectors in a river. Note the higher speed where the river narrows. (*b*) Streamlines are an alternative way of representing the flow. Note that lines are closer where the flow speed is higher.

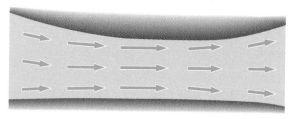

(a)

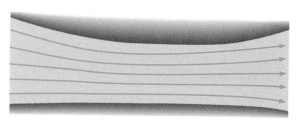

(b)

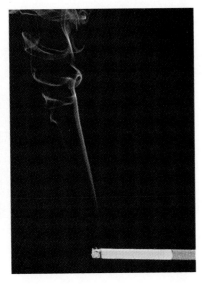

(a)

FIGURE 18-18 In this aerodynamic test, smoke particles trace streamlines in the airflow over a car. Can you tell where the flow is fastest?

Like all other motion in classical physics, fluid motion is governed by Newton's laws. Indeed, it is possible to write Newton's second law in a form that involves explicitly the fluid velocity as a function of position and time. But the resulting equation is difficult to solve in any but the simplest cases. Instead of applying Newton's law directly, we will approach fluid dynamics from the point of view of energy conservation. As in mechanics, energy conservation provides a shortcut to problem solving; the price of that shortcut is the lack of some details.

Conservation of Mass: The Continuity Equation

In mechanics we had no trouble keeping track of the individual objects. But a fluid is continuous and deformable, so it's not easy to follow an individual fluid element as it moves. Yet fluid is conserved; as it moves and deforms, new fluid is neither created nor destroyed. The **continuity equation** is a mathematical statement of this fact.

To develop the continuity equation, consider a steady fluid flow represented by streamlines, as shown in Fig. 18-20a. We have shaded a **flow tube**—a small tube-like region bounded on its sides by a set of streamlines and on its ends by areas at right angles to the streamlines. We choose the flow tube to have sufficiently small cross-sectional area that fluid velocity and other fluid properties do not vary significantly over any cross section; however, fluid properties may vary along the flow tube. Although our flow tube has no physical boundaries, it nevertheless acts like a pipe of the same shape because fluid flows *along,* not across the streamlines. In steady flow, the rate at which fluid enters the tube at its left end equals the rate at which it exits at the right; otherwise, the amount of fluid in the pipe would change and the fluid properties would not be independent of time.

Figure 18-20b shows a small fluid element just about to enter the flow tube, a process that will take some time Δt. Suppose the fluid is moving at speed v_1;

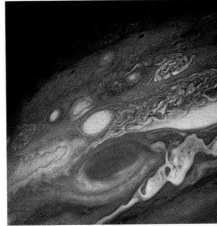

(b)

FIGURE 18-19 (a) Smoke from a burning cigarette first rises in a steady flow, but soon the flow becomes unsteady. (b) Unsteady flows in Jupiter's atmosphere, photographed by a Voyager spacecraft.

FIGURE 18-20 (a) A flow tube. In steady flow, the rate at which fluid enters the tube equals the rate at which it leaves the tube. (b) Fluid elements of equal mass entering and leaving the tube.

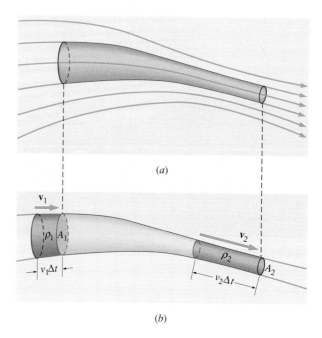

since it takes time Δt to cross the tube end, its length is $v_1\Delta t$. With cross-sectional area A_1, length $v_1\Delta t$, and density ρ_1, the mass of the entering fluid element is

$$m = \rho_1 A_1 v_1 \Delta t.$$

Another fluid element is shown just about to leave the tube. Suppose it has the *same* mass m as the entering fluid element. Then it must exit the tube in the *same* time Δt, in order to keep the total mass in the tube constant. But its mass can be written

$$m = \rho_2 A_2 v_2 \Delta t.$$

Equating our two expressions for m shows that

$$\rho_1 v_1 A_1 = \rho_2 v_2 A_2. \tag{18-4a}$$

Since the endpoints of the tube are arbitrary, we conclude that the quantity $\rho v A$ must have the same value anywhere along the flow tube:

$$\rho v A = \text{constant along a flow tube.} \quad \text{(any fluid)} \tag{18-4b}$$

What is this quantity $\rho v A$ that is constant along a flow tube? Its SI units are $(\text{kg/m}^{-3})(\text{m/s})(\text{m}^2)$, or simply kg/s. The quantity $\rho v A$ is therefore the **mass flow rate** or mass of fluid per unit time passing through the flow tube. Equations 18-4a and 18-4b are equivalent expressions of the continuity equation, both stating that the mass flow rate is constant along a flow tube in steady flow.

For a liquid, the density ρ is essentially constant, and the continuity equation 18-5 becomes simply

$$vA = \text{constant along a flow tube.}\quad\text{(liquid)}\qquad(18\text{-}5)$$

Now the constant quantity is just vA, with units of $(m/s)(m^2)$, or m^3/s. The quantity vA is the **volume flow rate;** in a fluid of constant density, constancy of mass flow rate also implies constancy of volume flow rate because a given mass of fluid does not change volume. In the form 18-5, the continuity equation makes obvious physical sense. Where the liquid has a large cross-sectional area, it can flow relatively slowly to transport a given mass of fluid per unit time. But where the area is more constricted, the flow must be faster to carry the same mass per unit time. With a gas, obeying Equation 18-4 but not necessarily 18-5, the situation is slightly more ambiguous, as density variations also play a role in the continuity equation. For flow speeds below the speed of sound in a gas, it turns out that lower area implies a higher flow speed just as for a liquid. But when gas flow speed exceeds the sound speed, density changes become so great that flow speed actually decreases with lower area.

● **EXAMPLE 18-7** AUSABLE CHASM

In the lower part of its valley, the Ausable River in upstate New York is about 40 m wide. Under typical early summer conditions, it is 2.2 m deep and flows at 4.5 m/s. Just before it reaches Lake Champlain, the river enters Ausable Chasm, a deep gorge cut through rock (Fig. 18-21). At its narrowest, the gorge is only 3.7 m wide at the river surface. If the flow rate in the gorge is 6.0 m/s, how deep is the river at this point? Assume that the river has a rectangular cross section, with uniform flow speed over a cross section.

Solution
Writing the cross-sectional area as the product of width w and depth d, Equation 18-5 becomes

$$v_1 w_1 d_1 = v_2 w_2 d_2.$$

Solving for the gorge depth d_2 gives

$$d_2 = \frac{v_1 w_1 d_1}{v_2 w_2} = \frac{(4.5\ \text{m/s})(40\ \text{m})(2.2\ \text{m})}{(6.0\ \text{m/s})(3.7\ \text{m})} = 18\ \text{m},$$

or about 60 feet!

EXERCISE A 2.5-cm-diameter pipe is full of water flowing at 1.8 m/s. If the pipe narrows to 2.0-cm diameter, what is the flow speed in the narrow section?

FIGURE 18-21 The Ausable River in upstate New York cuts through a narrow chasm. To accommodate the flow, water depth in the chasm is much greater than elsewhere.

Answer: 2.8 m/s

Some problems similar to Example 18-7: 43–45

FIGURE 18-22 A flow tube showing the same fluid element entering and leaving. The work done by pressure and gravitational forces equals the change in kinetic energy of the fluid element.

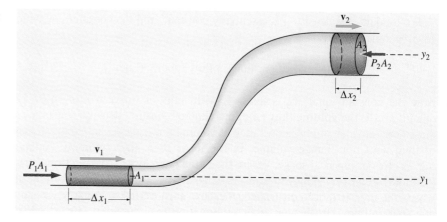

Conservation of Energy: Bernoulli's Equation

We now turn to conservation of fluid energy. Figure 18-22 shows the same fluid element of mass m as it enters and again as it leaves a flow tube. If it enters with speed v_1 and leaves with speed v_2, the change in its kinetic energy is

$$\Delta K = \frac{1}{2}m(v_2^2 - v_1^2).$$

The work-energy theorem (Equation 7-16) equates this change to the net work done on the fluid element. As the element enters the tube, it's subject to a pressure force P_1A_1 from the fluid to its left. This external force acts over the length Δx_1 of the fluid element as it enters, so it does work $W_1 = P_1A_1\Delta x_1$. Similarly, as it leaves the tube the fluid element experiences a force P_2A_2 from the fluid to its right. Because this force is opposite to the flow direction, it does negative work $W_2 = -P_2A_2\Delta x_2$. There are also forces on the fluid within the flow tube, but these internal forces cancel in pairs and therefore do no net work. Forces from adjacent flux tubes act at right angles to the flow, so they, too, do no net work. Finally, the fluid element rises a distance $y_2 - y_1$ as it traverses the tube; therefore gravity does negative work $W_g = -mg(y_2 - y_1)$. (Here we assume the flow tube is narrow enough that height variations across either end of the tube are insignificant.) Summing the three contributions to the work and applying the work-energy theorem, we have

$$W_1 + W_2 + W_3 = \Delta K,$$

or

$$P_1A_1\Delta x_1 - P_2A_2\Delta x_2 - mg(y_2 - y_1) = \frac{1}{2}m(v_2^2 - v_1^2).$$

The quantities $A_1\Delta x_1$ and $A_2\Delta x_2$ are the volumes of the fluid element as it enters and leaves the flow, respectively. If we restrict ourselves to incompressible fluids, then those two volumes are equal. Dividing through by this common volume $V = A\Delta x$ and noting that $m/V = \rho$, our equation becomes

$$P_1 + \frac{1}{2}\rho v_1^2 + \rho g y_1 = P_2 + \frac{1}{2}\rho v_2^2 + \rho g y_2, \qquad (18\text{-}6a)$$

or

$$P + \frac{1}{2}\rho v^2 + \rho g y = \text{constant along a flow tube.} \qquad (18\text{-}6b)$$

This is called **Bernoulli's equation,** after the Swiss mathematician Daniel Bernoulli (1700–1782).

What do the terms in Bernoulli's equation mean? The quantity $\frac{1}{2}\rho v^2$ looks like the kinetic energy $\frac{1}{2}mv^2$, except it has mass per unit volume ρ instead of mass m. It's therefore the kinetic energy per unit volume, or kinetic energy density. Similarly, $\rho g y$ is the gravitational potential energy per unit volume. What about the pressure P? It, too, has the units of energy density, and represents the internal energy of the fluid. Bernoulli's equation therefore says that the total energy per unit volume of fluid is conserved as the fluid moves.

Bernoulli's equation applies to incompressible fluids in the absence of energy loss mechanisms or external sources of energy. If the flow is irrotational— meaning that the fluid element at a point has no angular momentum about that point—then it turns out that the total energy per unit volume has the same value not only along each flow tube but throughout the entire fluid.

18-5 APPLICATIONS OF FLUID DYNAMICS

The laws of mass and energy conservation we have just written for fluids can be used to analyze a wide variety of natural and technological phenomena involving fluid motion. As we give examples of their applications, remember that those laws are based in the same physical principles—particularly Newton's laws— that we used to describe mechanical systems.

● EXAMPLE 18-8 DRAINING A TANK

A large, open tank is filled to a height h with liquid of density ρ. What is the speed of the liquid emerging from a small hole at the base of the tank (Fig. 18-23)?

Solution

We apply Bernoulli's equation to the emerging fluid, calculating the quantity $P + \frac{1}{2}\rho v^2 + \rho g y$ at the hole ($y = 0$) and at the tank top ($y = h$). At both these points the fluid is open to the atmosphere, so the pressure at both points is atmospheric pressure, P_a. If the tank area is large compared with the hole area, fluid at the tank top will be moving very slowly and we can make the approximation $v_{\text{top}} \approx 0$. Then Bernoulli's equation becomes

$$P_a + \frac{1}{2}\rho v_{\text{hole}}^2 = P_a + \rho g h.$$

Solving for the outflow speed at the hole gives

$$v_{\text{hole}} = \sqrt{2gh}.$$

This is the same result we would get by dropping an object through the distance h—and for the same reason: conservation

FIGURE 18-23 Example 18-8. How fast does the liquid emerge from the tank?

of energy. Draining a gram of water from the hole is energetically equivalent to removing a gram of water from the top and dropping it through the distance h. Just as the speed of a falling object is independent of its mass, so is the speed of the liquid independent of its density.

As the liquid drains the height h decreases, and so does the flow rate. Problem 67 explores this situation further.

EXERCISE   Suppose the top of the tank in Example 18-8 is sealed and pressurized to twice atmospheric pressure. What now is the outflow speed at the hole?

Answer: $\sqrt{2(P_a/\rho + gh)}$

Some problems similar to Example 18-8:   48, 49, 59, 67

Venturi Flows

A constriction in a pipe carrying incompressible fluid—a liquid or a gas moving at well below its sound speed—requires that flow speed increase in order to maintain constant mass flow. Such a constriction is called a **venturi.** Bernoulli's equation requires that the pressure be lower in the venturi. Example 18-9 shows how this effect is used to measure fluid flow.

● EXAMPLE 18-9   A VENTURI FLOWMETER

Figure 18-24 shows a pipe of cross sectional area A_1 with a venturi constriction of area A_2. A pressure gauge senses pressure differences between the venturi and the unconstricted pipe. The pipe carries an incompressible fluid of density ρ. Find an expression for the flow speed in the unconstricted pipe as a function of the pressure difference ΔP.

Solution
The gravitational term in Bernoulli's equation is the same in the pipe and venturi, so the equation becomes simply

$$P_1 + \frac{1}{2}\rho v_1^2 = P_2 + \frac{1}{2}\rho v_2^2,$$

where the subscripts 1 and 2 refer to the unconstricted pipe and venturi, respectively. We can relate the two speeds using the continuity equation for an incompressible fluid, Equation 18-5:

$$v_1 A_1 = v_2 A_2.$$

Solving for v_2 and using the result in Bernoulli's equation gives

$$P_1 - P_2 = \frac{1}{2}\rho v_2^2 - \frac{1}{2}\rho v_1^2 = \frac{1}{2}\rho\left(\frac{v_1 A_1}{A_2}\right)^2 - \frac{1}{2}\rho v_1^2$$
$$= \frac{1}{2}\rho v_1^2\left[\left(\frac{A_1}{A_2}\right)^2 - 1\right].$$

Solving for v_1 in terms of the pressure difference $\Delta P = P_1 - P_2$ then gives

$$v_1 = \sqrt{\frac{2\Delta P}{\rho[(A_1/A_2)^2 - 1]}}.$$

This is the flow speed; multiplying by the area A_1 would give the volume flow rate.

> **TIP   Combine Information**   Bernoulli's equation alone could not solve this example; instead we needed

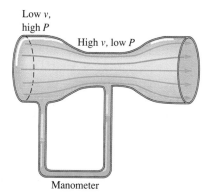

FIGURE 18-24   A venturi flowmeter, with a manometer used to measure the pressure difference between unconstricted pipe and venturi. Bernoulli's equation shows that the pressure difference depends on flow speed.

both Bernoulli's equation (conservation of energy) and the continuity equation (conservation of mass). You've frequently encountered problems like this where you need to invoke several distinct physical laws. As you attempt new problems, be sure to consider all the laws that might be applicable.

EXERCISE   A 1.0-cm-diameter venturi flowmeter is inserted in a 2.0-cm-diameter pipe carrying water (density 1000 kg/m³). What are (a) the flow speed in the pipe and (b) the volume flow rate if the pressure difference between venturi and unconstricted pipe is 17 kPa?

Answers: (a) 1.5 m/s; (b) 0.473 L/s

Some problems similar to Example 18-9:   46, 47, 51, 57, 58

The venturi phenomenon is put to good use in the carburetor, a device that mixes fuel with air to provide a combustible mixture to a gasoline engine. Liquid fuel is introduced at a venturi in the air stream heading to the engine. Lower pressure at the venturi causes fuel to be forced through a small nozzle, where it mixes thoroughly with the air (Fig. 18-25; see also Problem 65). Small gasoline engines and older cars use carburetors, but fuel injection systems have largely supplanted carburetors in modern cars.

The occurrence of lower pressure with higher flow speeds, and vice versa—often called the **Bernoulli effect**—has numerous manifestations. The dirt around a prairie dog's hole is mounded up in a way that forces wind to accelerate over the hole, resulting in lower pressure above the hole (Fig. 18-26). Biologists speculate that prairie dogs have evolved this design to provide natural ventilation. Sometimes the Bernoulli effect can be strikingly counterintuitive. Figure 18-27 shows a ping-pong ball suspended by a *downward* blowing airflow in an inverted funnel. Rapid divergence of the flow results in lower speed and therefore higher pressure below the ball.

Flight and Lift

Airplanes, helicopters, and birds are supported in flight by aerodynamic forces resulting from their dynamic interaction with the air. Hydrofoil boats, water skis, and high-performance sailboards have analogous interactions with water. Projectiles like baseballs and missiles, although not supported by the air, have their trajectories substantially modified by aerodynamic forces.

One of the simplest examples of aerodynamic **lift** is the helicopter. Its whirling blades are tilted so they force air downward as they move, just like a giant fan (Fig. 18-28). By Newton's third law, the air exerts an upward force on the blades, ultimately supporting the helicopter. An airplane wing works in the same way, except that it moves forward in a straight line instead of describing a circle. Wings are shaped to maximize the downward deflection of the air even with the wing horizontal, but in principle even a flat board would function as a

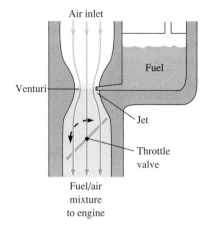

FIGURE 18-25 Simplified view of a carburetor. Pressure difference between the top of the fuel, at atmospheric pressure, and lower pressure in the venturi forces fuel into the air stream. Throttle plate controls the rate of air flow and therefore the engine speed; it's what's connected to the gas pedal in automobile engines using carburetors.

FIGURE 18-26 Do prairie dogs shape the entrances to their holes to provide ventilation by lowering air pressure above the hole?

FIGURE 18-27 A ping-pong ball supported by a downward-flowing air stream.

FIGURE 18-28 Helicopter blades are tilted to deflect air downward. By Newton's third law, the deflected air exerts an upward force on the blade, ultimately supporting the copter. Airflow is depicted in the frame of reference of the moving blade.

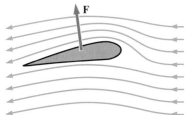

FIGURE 18-29 Flow past a wing. The wing deflects the initially horizontal flow downward; in reaction, the air exerts an upward lift force on the wing. Curved shape enhances lift, but is not essential. Note closer-spaced streamlines above the wing, indicating higher speed and therefore lower pressure.

FIGURE 18-30 Vertical component (lift) of the aerodynamic force $\mathbf{F}_a$ supports the entire weight $\mathbf{F}_g$ (140 MN or 150 tons) of this A300 Airbus. Engine force $\mathbf{F}_e$ balances the horizontal component (drag) to keep the plane moving at constant speed.

wing if it were tilted to the oncoming air. Figure 18-29 shows the airflow around a wing. Note how the flow, initially horizontal, leaves the wing moving downward—a clear indication that the wing has exerted a downward force on the air. The upward reaction to this downward force is what supports the airplane (Fig. 18-30).

A spinning ball provides another example of aerodynamic lift. Figure 18-31a shows the airflow around a ball that's not spinning; the flow is symmetric on top and bottom of the ball. But when the ball spins in the direction shown in Fig. 18-31b, air dragged around the top is deflected downward. This does not

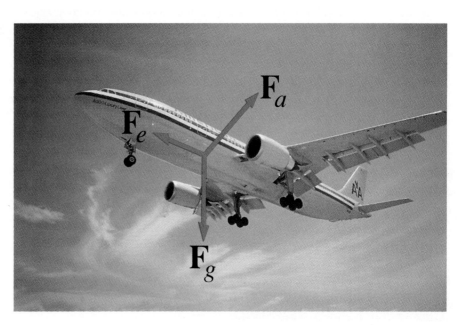

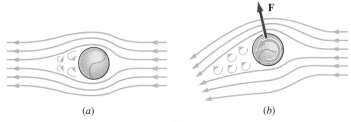

(a)   *(b)*

FIGURE 18-31   (*a*) Flow is symmetric around a non-spinning ball, passing the ball with no net deflection. Therefore there is no lift force on the ball. (*b*) When the ball spins it drags air with it, causing a downward air deflection and therefore giving rise to a lift force. Note also the higher flow speed and therefore lower pressure above the ball. Small curves represent turbulent eddies that form behind the ball.

happen symmetrically on the bottom because of differences in the flow pattern between the forward and backward directions. The result is a net downward air deflection and, therefore, an upward force on the ball. If the spin is reversed, the "lift" force will be downward; if the spin axis is turned vertical, the ball will veer left or right (Fig. 18-32).

Bernoulli's equation is frequently invoked to explain lift forces. It is true, as Figs. 18-29 and 18-31b suggest, that flow speeds are higher, and therefore—according to Bernoulli's equation—pressures are lower on top of a wing or a spinning ball. Forces associated with that pressure difference provide the lift, so Bernoulli's equation does help explain what's going on. But those pressure differences and the flow speeds associated with them are manifestations of a simpler underlying phenomenon—the action-reaction forces of Newton's third law.

FIGURE 18-32   Aerodynamic forces (**F**$_a$) give the spinning curve ball its curved trajectory.

■ **APPLICATION** WIND ENERGY

Bernoulli's equation states that fluid energy is conserved. If, however, we put moving machinery in a fluid flow, that machinery may add energy, as do pumps, or it may extract energy, as do windmills and other turbines (Fig. 18-33).

The air motion we call wind arises ultimately from the heating of the Earth by sunlight. Where the wind is blowing with speed v, Bernoulli's equation shows that kinetic energy of fluid motion is present with density $\frac{1}{2}\rho v^2$. A wind turbine can extract fluid kinetic energy from the flow, converting it to the mechanical energy of moving machinery and ultimately to electricity.

What is the rate at which a turbine can extract energy from the wind? If the air approaching the turbine in Fig. 18-33 has speed v, then in time Δt a column of air with length $v\Delta t$ flows past the turbine. If the turbine sweeps out an area A, then the volume passing in time Δt is $vA\Delta t$. Since the kinetic energy density is $\frac{1}{2}\rho v^2$, the kinetic energy ΔK passing the turbine is

$$\Delta K = \left(\frac{1}{2}\rho v^2\right)(vA\Delta t) = \frac{1}{2}\rho v^3 A\Delta t.$$

Dividing through by $A\Delta t$ gives the energy per time per unit area—that is, the power per unit area available from the wind:

$$\text{Wind power per unit area} = \frac{1}{2}\rho v^3.$$

Unfortunately the wind turbine can't extract *all* the kinetic energy, for if it did the air would come to a complete stop behind the turbine and the pressure would build up, halting the flow. A thorough analysis shows that the maximum rate for wind energy extraction is $\frac{8}{27}\rho v^3$, amounting to about 59% of the total kinetic energy. For air (density 1.2 kg/m³) with a wind speed of 10 m/s, this gives 356 watts for every square meter swept out by the turbine. Practical wind turbines achieve only about 20% of this theoretical maximum.

Because wind power scales as the cube of the wind speed, regions where winds are strong and steady have considerable potential for producing wind energy. In the United States those regions are concentrated near the coasts and in the Great Plains; estimates suggest that large-scale wind development could supply 20% or more of the nation's electrical energy. Thanks to intensive wind development in the 1980s, California already gets 1% of its electricity from the wind (Fig. 18-34).

Wind is rarely constant, so wind-generated energy must

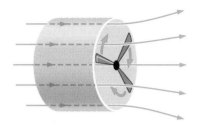

FIGURE 18-33 A set of rotating blades—a turbine—in a fluid flow. If external forces turn the turbine, then the turbine acts as a pump, doing work on the fluid and increasing fluid energy. If the fluid turns the turbine, then the turbine extracts energy from the fluid. Can you tell from the streamlines which this turbine is doing?

either be stored somehow or used to supplant other energy sources at those times when wind is available. The effect of wind fluctuations can be minimized by connecting generators at diverse locations to a common electrical grid.

Wind generators have fewer adverse environmental effects than most other means of generating electricity. They may present aesthetic problems, especially on mountain ridges, and their mounting towers do take land area. On the other hand, activities like farming and livestock grazing can continue even in dense "wind farms" like the one shown in Fig. 18-34.

FIGURE 18-34 "Wind farm" at Altamont Pass, California, has more than 6,000 wind turbines and produces an average power of more than 100 MW.

18-6 VISCOSITY AND TURBULENCE

Our discussion of the spinning baseball suggests that air is "dragged" by objects moving through it. This indeed happens, not only when a fluid is in motion relative to a solid object but even when it moves with different speeds on different streamlines. This phenomenon—called fluid friction or **viscosity**—ultimately arises from momentum transfers among individual molecules with components of motion perpendicular to the flow.

We can often neglect viscosity, especially in large-volume flows where not much of the fluid is near a solid boundary. But in fluid confined to a narrow channel, viscosity plays a substantial role. The importance of viscosity depends not only on the dimensions of the flow but also on the fluid itself; water is relatively inviscid, while molasses is extremely viscous. At a solid surface, viscosity results in a drag force that slows the fluid, bringing it to a complete stop right at the surface. The result is a nonuniform flow profile in enclosed structures like pipes (Fig. 18-35). Viscosity is the dominant force in fluids confined to very small spaces, like the lubricating oils separating metal surfaces in machinery.

Viscosity also plays an important role in stabilizing fluid flows that would otherwise become **turbulent,** or chaotically unsteady (Fig. 18-36). Turbulence ultimately results from the growth of waves that gain energy at the expense of the flow. When many wave modes are present at once the flow becomes completely chaotic, changing continuously in unpredictable ways. There is not yet a fully satisfactory theory of turbulence, so this area of fluid dynamics is the focus of much contemporary research. Researchers continue to find remarkable connections between the behavior of turbulent fluids and of seemingly unrelated phenomena like the growth of crystals, the shapes of mountain ranges, and population fluctuations in ecological systems.

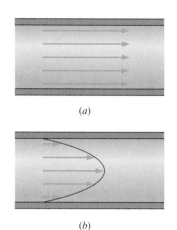

(a)

(b)

FIGURE 18-35 (*a*) Inviscid flow in a pipe would be uniform. (*b*) Viscosity reduces the flow velocity, especially near the walls, resulting in a nonuniform flow profile.

FIGURE 18-36 (*a*) An initially uniform flow of liquid (red area) soon becomes turbulent. (*b*) This turbulent flow shows a striking pattern of whirlpool-like eddies.

CHAPTER SYNOPSIS

Summary

1. **Fluid** is matter that readily deforms and flows under the influence of forces. Fluids are characterized by **density,** or mass per unit volume, and **pressure,** or force per unit area. Liquids are nearly **incompressible,** meaning liquid density hardly changes. Gases are **compressible,** capable of large density changes, but such changes generally occur only when flow speeds approach or exceed the sound speed.

2. In **hydrostatic equilibrium** there is no net force on any element of a fluid. In the absence of external forces, hydrostatic equilibrium implies uniform pressure throughout the fluid. In the presence of gravity, fluid pressure increases with depth so pressure forces balance the gravitational force; in a liquid, that increase is described by

$$P = P_0 + \rho gh,$$

where P_0 is the surface pressure and h the depth.

3. In hydrostatic equilibrium, a pressure increase at any point is transmitted throughout the fluid, a fact known as **Pascal's law.**

4. **Buoyancy** is the upward pressure force on an object wholly or partly immersed in a fluid. The buoyancy force is equal to the weight of fluid displaced by the object—a fact known as **Archimedes' principle.** If an object is less dense than the fluid, the buoyancy force exceeds the gravitational force, and the object rises.

5. A moving fluid is characterized by its flow velocity at each point in space and time. In **steady flow** the velocity is always the same at a given point; such flow is represented by **streamlines** that mark the paths of the fluid elements. In **unsteady flow** the flow velocity varies with time as well as position.

6. The laws of conservation of mass and conservation of energy provide a simplified description of a fluid in steady flow. Both laws are applied to a narrow **flow tube**—a volume bounded by nearby streamlines. Conservation of mass results in the **continuity equation:**

$$\rho vA = \text{constant along a flow tube,}$$

where A is the tube area and ρvA the **mass flow rate.** In a liquid, or a gas with flow speed well below the sound speed, the density ρ is constant and therefore the **volume flow rate** vA remains constant along a flow tube.

In steady, incompressible flow in the absence of viscosity or other forms of energy loss or addition, conservation of energy yields **Bernoulli's equation:**

$$P + \frac{1}{2}\rho v^2 + \rho gh = \text{constant along a flow tube}.$$

The continuity equation and Bernoulli's equation together help explain a great many fluid phenomena; other phenom-ena, like airplane flight, are more simply explained through the action-reaction forces of Newton's third law.

7. Fluid friction, or **viscosity,** is especially important near fluid boundaries, particularly in narrowly confined flows. Viscosity exerts a stabilizing influence on flows that would otherwise become **turbulent,** or chaotically unstable.

Terms You Should Understand

(Pairs are closely related terms whose distinction is important; number in parentheses is chapter section where term first appears.)

fluid (introduction)
density (18-1)
pressure (18-1)
pascal (18-1)
hydrostatic equilibrium (18-2)
barometer, manometer (18-2)
gauge pressure (18-2)
Pascal's law (18-2)
buoyancy force (18-3)
Archimedes' principle (18-3)
neutral buoyancy (18-3)
streamlines (18-4)
steady flow, unsteady flow (18-4)
flow tube (18-4)
continuity equation (18-4)
mass flow rate, volume flow rate (18-4)
Bernoulli's equation (18-4)
venturi (18-5)
Bernoulli effect (18-5)
lift (18-5)
viscosity (18-6)
turbulence (18-6)

Symbols You Should Recognize

P (18-1)
ρ (18-1)
Pa (18-1)

Problems You Should Be Able to Solve

calculating pressure from force and vice versa (18-1)
calculating pressure as a function of depth in liquids (18-2)
analyzing simple hydraulic systems (18-2)
determining density given apparent weight of submerged objects (18-3)
determining floating position of buoyant objects (18-3)
using the continuity equation to determine flow speeds (18-4)
using the continuity equation and Bernoulli's equation together to solve fluid dynamics problems (18-5)

Limitations to Keep in Mind

Treating matter as a continuous fluid is an approximation valid only when the spacing between molecules is much smaller than any length of interest—including the wavelength of any significant wave motion.

Bernoulli's equation applies only to incompressible flows—that is, to liquids or to gases moving at much less than the sound speed.

QUESTIONS

1. Explain the difference between hydrostatic equilibrium, steady flow, and unsteady flow.
2. Why do your ears "pop" when you drive up a mountain?
3. The cabins of commercial jet aircraft are usually pressurized to the pressure of the atmosphere at about 2 km above sea level. Why don't you feel the lower pressure on your entire body?
4. Water pressure at the bottom of the ocean arises from the weight of the overlying water. Does this mean that the water exerts pressure only in the downward direction? Explain.
5. The three containers in Fig. 18-37 are filled to the same level and are open to the atmosphere. How do the pressures at the bottoms of the three containers compare?

FIGURE 18-37 Question 5.

6. Municipal water systems often include tanks or reservoirs mounted on hills or towers. Besides water storage, what function might these reservoirs have?
7. Why is it easier to float in the ocean than in fresh water?
8. Figure 18-38 shows a cork suspended from the bottom of a sealed container of water. The container is on a turntable rotating about a vertical axis, as shown. Explain the position of the cork.

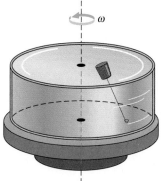

FIGURE 18-38 Question 8.

FIGURE 18-39 Why don't snorkels work at more than a meter or so of depth (Question 13)?

9. An ice cube is floating in a cup of water. Will the water level rise, fall, or remain the same when the cube melts?
10. Meteorologists in the United States usually report barometer readings in "inches." What are they talking about?
11. A mountain stream, frothy with entrained air bubbles, presents a serious hazard to hikers who fall into it, for they may sink in the stream where they would float in calm water. Why?
12. Why are dams thicker at the bottom than at the top?
13. It's not possible to breathe through a snorkel from a depth greater than a meter or so (Fig. 18-39). Why not?
14. Most humans float naturally in fresh water. Yet the body of a drowning victim generally sinks, often rising several days later after bodily decomposition has set in. What might explain this sequence of floating, sinking, and floating again?
15. A helium-filled balloon stops rising long before it reaches the "top" of the atmosphere, while a cork released from the bottom of a lake rises all the way to the surface of the water. Explain the difference between these two behaviors.
16. A barge filled with steel beams overturns in a lake, spilling its cargo. Does the water level in the lake rise, fall, or remain the same?
17. Imagine a vertical cylinder filled with water and set rotating about its axis. If pieces of wood and stone are introduced into the cylinder, where will each end up?
18. When gas in steady, subsonic flow through a tube encounters a constriction, its flow speed increases. When it flows supersonically in the same situation, flow speed decreases in the constriction. What must be happening to the gas density at the constriction in the supersonic case?
19. A ball moves horizontally through the air without spinning. Where on the ball's surface is the air pressure greatest?

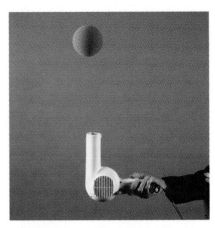

FIGURE 18-40 A ball supported by air from a hair dryer. Why is it in stable equilibrium? (Question 20)

20. A ball supported by an upward-flowing air column (Fig. 18-40) is essentially in stable equilibrium; if the ball is displaced slightly in any direction, it returns to its origi-nal position. Explain. *Hint:* How do you expect the flow speed to vary with horizontal position from the center of the air column?

21. Under what conditions can a gas be treated as incompress-ible?

22. As you drive along a highway and are passed by a large truck, your car may experience a force toward the truck. Explain the origin of this force in terms of the Bernoulli effect.

23. A pump is submerged at the bottom of a 200-ft-deep well. Does it take more power to pump water to the surface when the well is full of water or nearly empty? Or doesn't it matter?

24. Why do airplanes take off into the wind?

25. How does a kite fly?

26. Is the turbine in Fig. 18-33 adding or removing energy from the flow? How can you tell?

27. Is the flow speed behind a wind turbine greater or less than the flow speed in front? Is the pressure behind the turbine greater or less than in front? Is there a violation of Bernoulli's equation here? Explain.

PROBLEMS

Section 18-1 Describing Fluids: Density and Pressure

1. The density of molasses is 1600 kg/m^3. Find the mass of the molasses in a 0.75-liter jar.

2. Salad dressing is made from one part (by volume) of vine-gar (density 1.0 g/cm^3) to three parts olive oil (density 0.92 g/cm^3). What is the average density of the dressing?

3. The density of atomic nuclei is about 10^{17} kg/m^3, while the density of water is 10^3 kg/m^3. Roughly what fraction of the volume of water is *not* empty space?

4. Compressed air with mass 8.8 kg is stored in a gas cylinder with a volume of 0.050 m^3. (a) What is the density of the compressed air? (b) How large a volume would the same gas occupy at typical atmospheric density of 1.2 kg/m^3?

5. A plant hangs from a 3.2-cm-diameter suction cup affixed to a smooth horizontal surface (Fig. 18-41). What is the maximum weight that can be suspended (a) at sea level and (b) in Denver, where atmospheric pressure is about 0.80 atm?

6. The pressure unit **torr** is defined as the pressure that will support a column of mercury 1 mm high. Meteorologists often give barometric pressure in **inches of mercury,** defined analogously. Express each of these units in SI. The density of mercury is 1.36×10^4 kg/m^3.

7. Measurement of small pressure differences, for example, between the interior of a chimney and the ambient atmo-sphere, is often given in **inches of water,** where one inch of water is the pressure that will support a 1-in.-high water column. Express this unit in SI.

8. (a) What is the weight of a column of air with cross-sectional area 1 m^2 extending from Earth's surface to the top of the atmosphere? (b) What is the weight of the entire atmosphere?

9. The fuselage of a 747 jumbo jet is roughly a cylinder 60 m long and 6 m in diameter. If the interior of the plane is pressurized to 0.75 atm, what is the net pressure force tending to separate half the cylinder from the other half when the plane is flying at 10 km, where air pressure is about 0.25 atm? (The earliest commercial jets suffered structural failure from just such forces; modern planes are better engineered.)

10. A 4300-kg circus elephant balances on one foot (Fig. 18-42). If the foot is a circle 30 cm in diameter, what pressure does it exert on the ground?

Suction cup

FIGURE 18-41 Problem 5.

FIGURE 18-42 Problem 10.

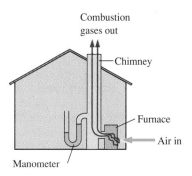

FIGURE 18-43 Water manometer used on a chimney (Problem 20).

11. A paper clip is made from wire 1.5 mm in diameter. You unbend a paper clip and push the end against the wall. What force must you exert to give a pressure of 120 atm?

12. Continuously running exhaust hoods in a university chemistry building keep the inside air pressure slightly lower than outside. A student notices that a 150-N force is necessary to start opening a 1.2 m by 2.3 m door. The door is hinged on one of its 2.3-m vertical sides, and the handle is mounted on the other side. What is the pressure difference between inside and outside? Give in pascals and as a fraction of atmospheric pressure. *Hint:* Think about torque.

13. When a couple with a total mass of 120 kg lies on a waterbed, the pressure in the bed increases by 4700 Pa. What surface area of the two bodies is in contact with the bed?

14. A fully loaded Volvo station wagon has a mass of 1950 kg. If each of its four tires is inflated to a gauge pressure of 230 kPa, what is the total tire area in contact with the road?

15. The emergency escape window of a DC-9 jetliner measures 50 cm by 90 cm. The interior pressure is 0.75 atm, and the plane is at an altitude where atmospheric pressure is 0.25 atm. Is there any danger that a passenger could open the window? Answer by calculating the force needed to pull the window straight inward.

Section 18-2 Fluids at Rest: Hydrostatic Equilibrium

16. Calculate the height of a mercury column that can be supported by air at 1.0 atm pressure. Mercury's density is 1.36×10^4 kg/m^3.

17. What is the density of a fluid whose pressure increases at the rate of 100 kPa for every 6.0 m of depth?

18. A research submarine can withstand external pressures of 50 MPa when its internal pressure is 100 kPa. How deep can it dive?

19. Scuba equipment provides the diver with air at the same pressure as the surrounding water. But at pressures greater than about 1 MPa, the nitrogen in air becomes dangerously narcotic. At what depth does nitrogen narcosis become a hazard?

20. Hot gases rising in a chimney produce a slightly lower air pressure in the chimney, which in turn ensures that these gases can't escape into the surrounding building. The pressure difference, called draft, is often measured with a manometer as shown in Fig. 18-43. If the difference in water levels in the manometer tube is 0.04 in. (typical of an oil furnace), by how much does the chimney pressure differ from atmospheric pressure?

21. A vertical tube open at the top contains 5.0 cm of oil (density 0.82 g/cm^3) floating on 5.0 cm of water. Find the *gauge* pressure at the bottom of the tube.

22. A vertical tube 1.0 cm in diameter and open at the top contains 5.0 g of oil (density 0.82 g/cm^3) floating on 5.0 g of water. Find the *gauge* pressure (a) at the oil-water interface and (b) at the bottom.

23. A 1500-m-wide dam holds back a lake 95 m deep. What force does the water exert on the dam?

24. Show that the force on the wall in Example 18-2 can be found by multiplying the wall area by the average of the pressures at top and bottom of the wall. Why does this work?

25. A U-shaped tube open at both ends contains water and a quantity of oil occupying a 2.0-cm length of the tube, as shown in Fig. 18-44. If the oil's density is 0.82 times that of water, what is the height difference h?

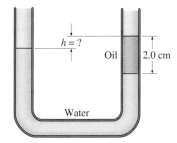

FIGURE 18-44 Problem 25.

26. A child attempts to drink water through a 100-cm-long straw, but finds that the water rises only 75 cm. By how much has the child reduced the pressure in her mouth below atmospheric pressure?

27. Barometric pressure in the eye of a hurricane is 0.91 atm (27.2 inches of mercury). How does the level of the ocean surface under the eye compare with that under a distant fair-weather region where the pressure is 1.0 atm?

28. A hydraulic cylinder moving a robotic arm has a diameter of 5.0 cm and can exert a maximum force of 5.6 kN. (a) What pressure must the hydraulic lines be capable of withstanding? (b) The cylinder at the other end of the hydraulic system has a 1.0-cm diameter. What force must be applied to it to get the maximum force out of the cylinder driving the arm?

29. A garage lift has a 45-cm-diameter piston supporting the load. Compressed air with a maximum pressure of 500 kPa is applied to a small piston at the other end of the hydraulic system. What is the maximum mass the lift can support?

30. Fig. 18-45 shows a hydraulic lift operated by pumping fluid into the hydraulic system. The large cylinder is 40 cm in diameter, while the small tube leaving the pump is 1.7 cm in diameter. A total load of 2800 kg is raised 2.3 m at a constant rate. (a) What volume of fluid passes through the pump? (b) What is the pressure at the pump outlet? (c) How much work does the pump do? (d) If the lifting takes 40 s, what is the pump power? Neglect pressure variations with height, and neglect also the weight of the fluid raised.

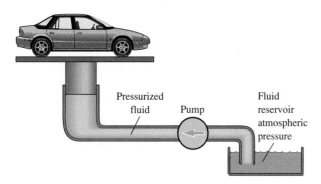

FIGURE 18-45 Problem 30.

Section 18-3 Archimedes' Principle and Buoyancy

31. On land, the most massive concrete block you can carry is 25 kg. How massive a block could you carry underwater, if the density of concrete is 2300 kg/m^3?

32. A 5.4-g jewel has an apparent weight of 32 mN when submerged in water. Could the jewel be diamond (density 3.51 g/cm^3)?

33. The density of styrofoam is 160 kg/m^3. What per cent error is introduced by weighing a styrofoam block in air, which exerts an upward buoyancy force, rather than in vacuum? The density of air is 1.2 kg/m^3.

34. Archimedes purportedly used his principle to verify that the king's crown was pure gold, by weighing the crown while it was submerged in water (see Fig. 18-14). If the crown's actual weight was 25.0 N, what should its apparent weight have been? The density of gold is 19.3 g/cm^3, and that of water is 1.00 g/cm^3.

35. A partially full beer bottle with interior diameter 52 mm is floating upright in water, as shown in Fig. 18-46. A drinker takes a swig and replaces the bottle in the water, where it now floats 28 mm higher than before. How much beer did the drinker drink?

FIGURE 18-46 Problem 35.

36. A glass beaker measures 10 cm high by 4.0 cm in diameter. Empty, it floats in water with one-third of its height submerged. How many 15-g rocks can be placed in the beaker before it sinks?

37. A typical supertanker has mass 2.0×10^6 kg and carries twice that much oil. If 9.0 m of the ship is submerged when it's empty, what is the minimum water depth needed for it to navigate when full? Assume the sides of the ship are vertical.

38. A balloon contains gas of density ρ_g and is to lift a mass M, including the balloon but not the gas. Show that the minimum mass of gas required is

$$m = \frac{M\rho_g}{\rho_a - \rho_g},$$

where ρ_a is the atmospheric density.

39. (a) How much helium (density 0.18 kg/m^3) is needed to lift a balloon carrying two people in a basket, if the total mass of people, basket, and balloon (but not gas) is 280 kg? (b) Repeat for a hot air balloon, whose air density is 10% less than that of the surrounding atmosphere.

40. A 55-kg swimmer climbs onto a styrofoam block whose density is 160 kg/m^3. If the water level comes right to the top of the styrofoam, what is its volume?

Sections 18-4 and 18-5 Fluid Dynamics and Applications

41. A fluid is flowing steadily, roughly from left to right. At left it is flowing rapidly; it then slows down, and finally speeds

up again. Its final speed at right is not as great as its initial speed at left. Sketch a streamline pattern that could represent this flow.

42. Show that pressure has the units of energy density.

43. A typical mass flow rate for the Mississippi River is 1.8×10^7 kg/s. Find (a) the volume flow rate and (b) the flow speed in a region where the river is 2.0 km wide and an average of 6.1 m deep.

44. A fire hose 10 cm in diameter delivers water at the rate of 15 kg/s. The hose terminates in a nozzle 2.5 cm in diameter. What are the flow speeds (a) in the hose and (b) in the nozzle?

45. A typical human aorta, or main artery from the heart, is 1.8 cm in diameter and carries blood at a speed of 35 cm/s. What will be the flow speed around a clot that reduces the flow area by 80%?

46. If the blood pressure in the unobstructed artery of the previous problem is 16 kPa gauge (about 120 mm of mercury, in the unit commonly reported by doctors), what will it be at the clot? The density of blood is 1060 kg/m^3. (Too low a pressure can actually collapse the artery, momentarily cutting off blood flow.)

47. In Fig. 18-47 a horizontal pipe of cross-sectional area A is joined to a lower pipe of cross-sectional area $\frac{1}{2}A$. The entire pipe is full of liquid with density ρ, and the left end is at atmospheric pressure P_a. A small open tube extends upward from the lower pipe. Find the height h_2 of liquid in the small tube (a) when the right end of the lower pipe is closed, so the liquid is in hydrostatic equilibrium, and (b) when the liquid flows with speed v in the upper pipe.

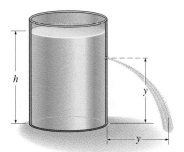

FIGURE 18-48 Problem 48.

FIGURE 18-49 The diameter of the water column varies as the fourth root of the distance from the faucet (Problem 50).

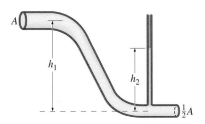

FIGURE 18-47 Problem 47.

48. A can of height h is full of water. At what height y should a small hole be cut so the water initially goes as far horizontally as it does vertically, as shown in Fig. 18-48?

49. The water in a garden hose is at a gauge pressure of 140 kPa and is moving at negligible speed. The hose terminates in a sprinkler consisting of many small holes. What is the maximum height reached by the water emerging from the holes?

50. Water emerges from a faucet of diameter d_0 in steady, near vertical flow with speed v_0. Show that the diameter of the falling water column is given by $d = d_0[v_0^2/(v_0^2 + 2gh)]^{1/4}$, where h is the distance below the faucet (Fig. 18-49).

51. The venturi flowmeter shown in Fig. 18-50 is used to measure the flow rate of water in a solar collector system. The flowmeter is inserted in a pipe with diameter 1.9 cm; at the venturi of the flowmeter the diameter is reduced to 0.64 cm. The manometer tube contains oil with density 0.82 times that of water. If the difference in oil levels on the two sides of the manometer tube is 1.4 cm, what is the volume flow rate?

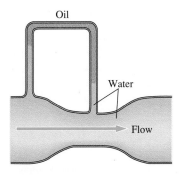

FIGURE 18-50 Problem 51.

52. A drinking straw 20 cm long and 3.0 mm in diameter stands vertically in a cup of juice 8.0 cm in diameter. A section of straw 6.5 cm long extends above the juice. A child sucks on the straw, and the level of juice in the glass begins dropping at 0.20 cm/s. (a) By how much does the pressure in the child's mouth differ from atmospheric pressure? (b) What is the greatest height from which the child could drink, assuming this same mouth pressure?

Paired Problems

(Both problems in a pair involve the same principles and techniques. If you can get the first problem, you should be able to solve the second one.)

53. A steel drum has volume 0.23 m³ and mass 16 kg. Will it float in water when filled with (a) water or (b) gasoline (density 860 kg/m³)? Neglect the thickness of the steel.

54. A 260-g circular pan 20 cm in diameter has straight sides 6.0 cm high and is made from metal of negligible thickness. To what maximum depth can the pan be filled with water and still float on water?

55. A spherical rubber balloon with mass 0.85 g and diameter 30 cm is filled with helium (density 0.18 kg/m³). How many 1.0-g paper clips can you hang from the balloon before it loses its buoyancy?

56. A string of negligible diameter has mass per unit length 1.4 g/m. You tie a 3-m-long piece of the string to a spherical helium balloon 23 cm in diameter and find that the balloon floats with 1.8 m of string off the floor. Find the combined mass of the balloon and helium.

57. Water at a pressure of 230 kPa is flowing at 1.5 m/s through a pipe, when it encounters an obstruction where the pressure drops by 5%. What fraction of the pipe's area is obstructed?

58. A venturi flowmeter in an oil pipeline has a radius half that of the pipe. The flow speed in the unconstricted flow is 1.9 m/s. If the pressure difference between the unconstricted flow and the venturi is 16 kPa, what is the density of the oil?

59. Find an expression for the volume flow rate from the siphon shown in Fig. 18-51, assuming the siphon area A is much less than the tank area.

60. (a) Find the initial siphon flow speed in Fig. 18-51 if the tank is sealed, with its top at only one-fourth of atmospheric pressure. Answer in terms of atmospheric pressure P_a, liquid density ρ, height h, and g. (b) What is the maximum distance between the bend at the top of the siphon and the liquid level in the tank for which the siphon will work under these conditions, assuming the liquid is water? (Give a numerical value here.)

Supplementary Problems

61. A 1.0-m-diameter tank is filled with water to a depth of 2.0 m and is open to the atmosphere at the top. The water

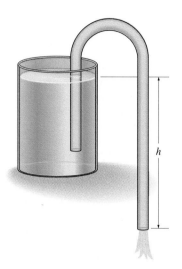

FIGURE 18-51 Problems 59, 60.

drains through a 1.0-cm-diameter pipe at the bottom; that pipe then joins a 1.5-cm-diameter pipe open to the atmosphere, as shown in Fig. 18-52. Find (a) the flow speed in the narrow section and (b) the water height in the *sealed* vertical tube shown.

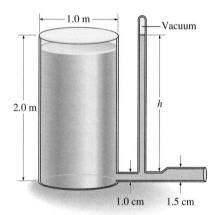

FIGURE 18-52 Problem 61.

62. How massive an object can be supported by a 5.0-cm-diameter suction cup mounted on a vertical wall, if the coefficient of friction between cup and wall is 0.72? Assume normal atmospheric pressure.

63. Figure 18-53 shows a simplified diagram of a Pitot tube, used for measuring aircraft speeds. The tube is mounted on the underside of the aircraft wing with opening A at right angles to the flow and opening B pointing into the flow. The gauge prevents airflow through the tube. Use Bernoulli's equation to show that the air speed relative to the wing is given by $v = \sqrt{2\Delta P/\rho}$, where ΔP is the pressure difference between the tubes and ρ is the density of air. *Hint:* The flow must be stopped at B, but continues past A with its normal speed.

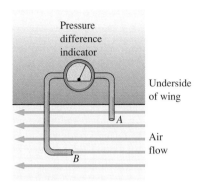

FIGURE 18-53   Problem 63.

64. A ship's hull has a V-shaped cross section, as shown in Fig. 18-54. The ship has vertical height h_0 from keel to deck and a total length ℓ perpendicular to the plane of Fig. 18-53. Empty, the hull extends underwater to a depth h_1, as shown. Find an expression for the maximum load the ship can carry, in terms of the water density ρ and the quantities h_0, h_1, ℓ, and θ shown in the figure.

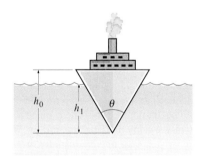

FIGURE 18-54   Problem 64.

65. With its throttle valve wide open, an automobile carburetor has a throat diameter of 2.4 cm. With each revolution, the engine draws 0.50 L of air through the carburetor. At an engine speed of 3000 rpm, what are (a) the volume flow rate, (b) the airflow speed, and (c) the difference between atmospheric pressure and air pressure in the carburetor throat? The density of air is 1.2 kg/m³.

66. A wind turbine has a blade diameter of 65 m. What is the theoretical maximum power output of this turbine in winds of (a) 10 m/s and (b) 15 m/s? (c) If the machine actually achieves only 14% of its theoretical maximum output, how many such turbines would be needed to displace a 1-GW nuclear power plant, assuming an average wind speed of 12 m/s?

67. A can of height h and cross-sectional area A_0 is initially full of water. A small hole of area $A_1 \ll A_0$ is cut in the bottom of the can. Find an expression for the time it takes all the water to drain from the can. *Hint:* Call the water depth y, use the continuity equation to relate dy/dt to the outflow speed at the hole, then integrate.

68. A pencil is weighted so it floats vertically with length ℓ submerged. It's pushed vertically downward without being totally submerged, then released. Show that it undergoes simple harmonic motion with period $T = 2\pi\sqrt{\ell/g}$.

69. A circular pan of liquid (density ρ) is centered on a horizontal turntable rotating with angular speed ω. Its axis coincides with the rotation axis, as shown in Fig. 18-55. Atmospheric pressure is P_a. Find expressions for (a) the pressure at the bottom of the pan and (b) the height of the liquid surface as functions of the distance r from the axis, given that the height at the center is h_0.

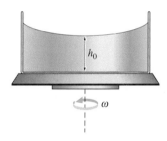

FIGURE 18-55   Problem 69.

70. Density and pressure in Earth's atmosphere are proportional: $\rho = P/h_0 g$, where h_0 is a constant with the approximate value 8200 m and g is the acceleration of gravity. (a) Integrate Equation 18-2 for this case to show that atmospheric pressure as a function of height h above the surface is given by $P = P_0 e^{-h/h_0}$, where P_0 is the surface pressure. (b) At what height will the pressure have dropped to half its surface value?

71. (a) Use the result of the preceding problem to express Earth's atmospheric density as a function of height (this is simple). (b) Use the result of (a) to find the height below which half of Earth's atmospheric mass lies (this will require integration).

P A R T 2 CUMULATIVE PROBLEMS

1. A cylindrical log of total mass M and uniform diameter d has an uneven mass distribution that causes it to float in a vertical position, as shown in Figure 1. (a) Find an expression for the length ℓ of the submerged portion of the log when it is floating in equilibrium, in terms of M, d, and the water density ρ. (b) If the log is displaced vertically from its equilibrium position and released, it will undergo simple harmonic motion. Find an expression for the period of this motion, neglecting viscosity and other frictional effects.

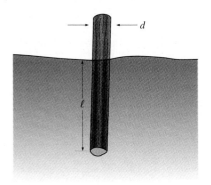

FIGURE 1 Cumulative Problem 1.

2. A cable of total mass m and length ℓ hangs vertically, with a mass M attached to its bottom end, as shown in Fig 2. The mass is given a sudden sideways blow that starts a low-amplitude pulse propagating up the cable. Show that the time it takes the pulse to reach the top of the cable is

$$t = 2\left(\sqrt{\frac{(m + M)\ell}{mg}} - \sqrt{\frac{M\ell}{mg}} \right).$$

FIGURE 2 Cumulative Problem 2.

3. Let P_0 and ρ_0 be the atmospheric pressure and density at Earth's surface. Assume that the ratio P/ρ is the same throughout the atmosphere (this implies that the tempera-ture is uniform). Show that the pressure a vertical height z above the surface is given by $P(z) = P_0 e^{-\rho_0 g z/P_0}$, for z much less than Earth's radius (this amounts to neglecting Earth's curvature, and thus taking g to be constant).

4. A piece of rope of length ℓ and mass m has its two ends spliced together to form a continuous loop. The loop is set spinning at so high a rate that it forms a circle with essentially uniform tension. It is then placed in contact with the ground, where it rolls, without slipping, like a rigid hoop. The loop is rolling on level ground when it rolls over a stick that produces a small distortion (see Fig. 3). As a result, two pulses, initially coinciding, propagate along the loop in opposite directions. (a) Where will they again coincide? (b) Through what angle will the loop have rotated while the pulses are separated?

FIGURE 3 Cumulative Problem 4.

5. A U-shaped tube containing liquid is mounted on a table that tilts back and forth through a slight angle, as shown in Fig. 4. The diameter of the tube is much less than either the height of its arms or their separation. When the table is rocked very slowly or very rapidly, nothing particularly dramatic happens. But when the rocking takes place at a few times per second, the liquid level in the tube oscillates violently, with maximum amplitude at a rocking frequency of 1.7 Hz. Explain what is going on, and find the total length of the liquid including both vertical and horizontal portions.

FIGURE 4 Cumulative Problem 5.

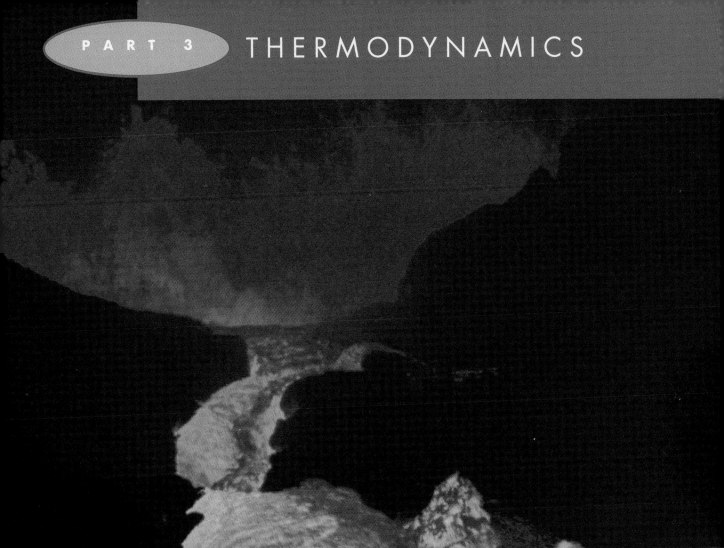

PART 3 THERMODYNAMICS

Thermodynamics is the study of heat. Here, thermodynamic activity in Earth's interior results in spectacular lava fountains and a lava river during an eruption of Hawaii's Kilauea volcano.

CHAPTER 19

TEMPERATURE AND HEAT

Thermogram of a house, taken with an infrared scanning device, shows regions of greatest heat loss in red and yellow, lowest in blue and black. Understanding heat flow allows engineers to design systems that keep buildings comfortable and helps scientists predict the effect of human activities on Earth's climate.

The next four chapters deal with the physics of heat, temperature, and related phenomena. Our own bodies provide a qualitative sense of what it means for things to be "hot" or "cold." Here we'll quantify this sense and explore the behavior of matter under different thermal conditions. And we will learn something our sense of temperature could never tell us—namely, how thermal phenomena reflect the laws of mechanics operating on a microscopic scale.

19-1 MACROSCOPIC AND MICROSCOPIC DESCRIPTIONS

Some physical quantities—for example, mass—apply equally to microscopic objects like atoms and molecules and to macroscopic objects like cars, gas cylinders, and planets. But others—for example, temperature and pressure—have no meaning on a microscopic scale. We speak of the temperature and pressure of air, but it makes no sense to ask about the temperature or pressure of an individual air molecule. The study of temperature, heat, and related macroscopic quantities comprises the branch of physics called **thermodynamics.**

The thermodynamic behavior of matter is determined by the behavior of its constituent atoms and molecules in response to the laws of mechanics. **Statistical mechanics** is the branch of physics that relates the macroscopic description of matter to the underlying microscopic processes. Historically, thermodynamics was developed before the atomic theory of matter was well established. The subsequent explanation of thermodynamics in terms of statistical mechanics—the mechanics of atoms and molecules—was a triumph for classical physics. At the same time, the few thermodynamic phenomena that could not be explained successfully with classical physics helped point the way to the development of the quantum theory. In our study of thermal phenomena, we will interweave the macroscopic and microscopic descriptions to provide the fullest understanding of both viewpoints.

19-2 TEMPERATURE AND THERMODYNAMIC EQUILIBRIUM

Take a bottle of soda from the refrigerator, and leave it on the kitchen counter. Eventually it reaches room temperature and then stays there. We don't need a scientific definition of temperature to recognize this; our sense of touch suffices to tell that the soda's temperature eventually stops changing. We generalize this simple experiment to say that when two systems—in this case the soda and the surrounding air—remain in contact until no further change occurs in any macroscopic property, then the two are in **thermodynamic equilibrium.**

The concept of thermodynamic equilibrium does not require a prior notion of temperature. To determine whether two systems are in equilibrium, we can consider *any* macroscopic properties. Typical properties include length, volume, pressure, and electrical resistance, as well as temperature. If any macroscopic property changes when two systems are placed in contact, then they were not originally in equilibrium. Once changes cease, then the systems are in equilibrium.

When we speak of systems in contact, we mean specifically **thermal contact.** Two systems are in thermal contact if heating one of them results in macroscopic changes in the other. For example, if we place two metal cups of water in physical contact, and heat one cup with a flame, the water in both cups gets hotter. In this case, the cups are in thermal contact. If we were to separate our

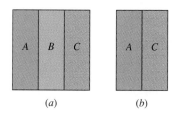

FIGURE 19-1 (*a*) Systems *A* and *C* are in thermal contact with *B*, but not with each other. (*b*) When they're brought together, *A* and *C* are already in thermodynamic equilibrium.

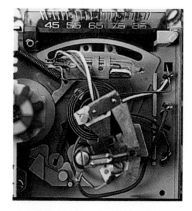

FIGURE 19-2 The coil at the center of this thermostat is a bimetallic strip that bends with increasing temperature, actuating a switch that controls a furnace.

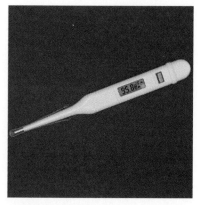

FIGURE 19-3 An electronic digital fever thermometer.

cups by a layer of styrofoam, then heating one would have little effect on the other. In this case, the two systems are **thermally insulated.**

Using the concept of thermodynamic equilibrium, we can now define quantitatively what we mean by temperature. We first state what it means for two systems to have the same temperature:

> **Two systems have the same temperature if they are in thermodynamic equilibrium.**

Conversely, two systems that are not in thermodynamic equilibrium do not have the same temperature.

A simple experimental fact makes it possible to compare systems that aren't in thermal contact. Consider two systems *A* and *C* in thermal contact with a third system *B*, but not with each other (Fig. 19-1*a*). If we wait until *A* and *C* are both in thermodynamic equilibrium with *B*, then bring *A* and *C* into contact (Fig. 19-1*b*), we find that no further changes occur—indicating that *A* and *C* are already in equilibrium. This fact is so fundamental to the rest of thermodynamics that it is called the **zeroth law of thermodynamics:**

> ***Zeroth law of thermodynamics: If two systems A and C each are in thermodynamic equilibrium with system B, then A and C are in thermodynamic equilibrium with each other.***

19-3 MEASURING TEMPERATURE

Rephrased in terms of temperature, the zeroth law says that if system *A* has the same temperature as system *B*, and system *B* has the same temperature as system *C*, then *A* and *C* have the same temperature. This property allows us to make quantitative measurements of temperature.

A **thermometer** is a system with some conveniently observed macroscopic property that changes with temperature. The length of the mercury column indicates temperature in a mercury thermometer. A strip made of two different metals bends when heated, turning the indicating needle on a dial-type thermometer or closing a switch that starts a furnace (Fig. 19-2). Electrical resistance has long been used in science and engineering to measure temperature and is rapidly replacing mercury in medical and other applications (Fig. 19-3).

To use a thermometer, we simply put it in contact with the system whose temperature we're trying to measure and let the two come to thermodynamic equilibrium. The value of whatever macroscopic property has changed then gives a quantitative indication of temperature. The zeroth law assures us that this process is consistent, in that two systems for which the thermometer gives the same reading must have the same temperature.

Gas Thermometers and the Kelvin Scale

We need a standard against which to calibrate different types of thermometers. Gas thermometers—using either the pressure or the volume of a gas to indicate temperature—operate over an extremely wide temperature range and are therefore used to define the temperature scale.

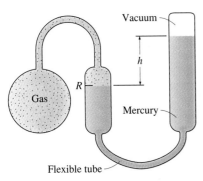

FIGURE 19-4 A constant-volume gas thermometer. As the gas pressure increases with temperature, the closed tube at right is raised to keep the mercury at the reference level R, thus maintaining constant gas volume. The height difference h is then a measure of gas pressure and therefore of temperature.

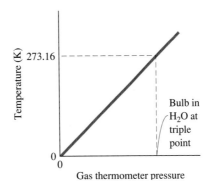

FIGURE 19-5 Kelvin temperature scale defined using a constant-volume gas thermometer.

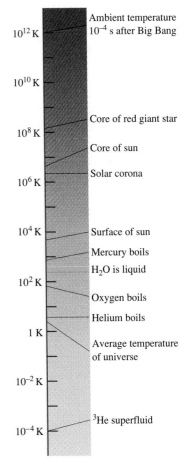

Figure 19-4 shows a **constant-volume gas thermometer,** whose temperature indication is the pressure of gas held at constant volume. Temperature is taken to be a linear function of the gas pressure. We need two points to define a linear function. The zero of temperature corresponds to zero gas pressure, and the second point is established by the **triple point** of water—the unique temperature at which water can exist wth solid, liquid, and gas simultaneously in equilibrium. The SI temperature unit, the **kelvin,** is defined by setting the triple point at exactly 273.16 K. The temperature of a constant-volume gas thermometer is then

$$T = 273.16\frac{P}{P_3}, \tag{19-1}$$

where P is the thermometer pressure and P_3 its pressure at the triple point (Fig. 19-5).

Since a gas cannot have negative pressure, the zero of the kelvin scale represents an absolute lower limit on temperature, and is called **absolute zero.** We'll explore the meaning of absolute zero further in Chapter 22. Figure 19-6 shows some important physical situations on the kelvin scale.

Other Temperature Scales

Other scales in common use include the Celsius (°C), Fahrenheit (°F), and Rankine (°R) scales, although the latter two are largely limited to the United States. One Celsius degree represents the same temperature difference as one kelvin (K), but the zero of the Celsius scale occurs at 273.15 K, so

$$T_C = T - 273.15, \tag{19-2}$$

where T_C is the Celsius temperature and T the absolute temperature in kelvins. The Celsius scale was chosen so the melting point of ice at standard atmospheric

FIGURE 19-6 Some physical processes on the kelvin scale. Chart is logarithmic to display a wide temperature range.

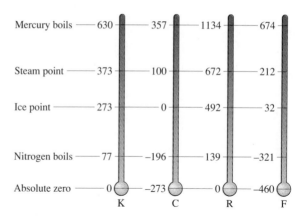

	K	C	R	F
Mercury boils	630	357	1134	674
Steam point	373	100	672	212
Ice point	273	0	492	32
Nitrogen boils	77	−196	139	−321
Absolute zero	0	−273	0	−460

FIGURE 19-7 Relations among the four temperature scales.

pressure—the ice point—is at exactly 0°C, while the boiling point of water at standard atmospheric pressure—the steam point—is at exactly 100°C. The triple point of water occurs at 0.01°C, which accounts for the 273.15 difference between the kelvin and Celsius scales as well as for the difference of 0.01 in the constants of Equations 19-1 and 19-2.

The Fahrenheit scale, which is part of the British system of units, is defined with the ice point at 32°F and the steam point at 212°F. As a result, the relation between the Fahrenheit and Celsius scales is

$$T_F = \tfrac{9}{5}T_C + 32. \tag{19-3}$$

A fourth scale, often used in engineering work in the United States, is the Rankine scale. A Rankine degree is the same size as a Fahrenheit degree, but the zero of the Rankine scale coincides with the zero of the kelvin scale.

Figure 19-7 summarizes the four temperature scales in relation to some important physical processes.

19-4 TEMPERATURE AND HEAT

A lighted match will burn your finger, yet it wouldn't provide much heat in a cold room. A large vat of water, although much cooler than the match, would do a better job heating the room. This example shows our intuitive sense of the difference between temperature and heat: Heat measures an *amount* of "something," while temperature measures the *intensity* of that "something."

What is that "something"? Before the early 1800s, heat was considered a material fluid, called **caloric,** that flowed from hot bodies to colder ones. But observations made by the American-born scientist Benjamin Thompson in the late 1700s began to shed doubt on the caloric theory. Thompson was appointed director of the Bavarian arsenal, where he supervised the boring of cannon. Thompson recognized that essentially limitless amounts of heat could be produced in the boring process, and he concluded that heat could not be a conserved fluid. Instead, he suggested, heating was associated with mechanical work done by the boring tool.

● **EXAMPLE 19-1** TEMPERATURE SCALES

What is normal body temperature (98.6°F) on the Celsius and kelvin scales? If you have a fever of 101.6°F, by how much has your temperature risen on each of these scales?

Solution
Solving Equation 19-3 for T_C gives

$$T_C = \tfrac{5}{9}(T_F - 32) = \tfrac{5}{9}(98.6 - 32.0) = 37.0°C.$$

Then Equation 19-2 gives the kelvin temperature:

$$T = T_C + 273.2 = 310.2 \text{ K}.$$

A kelvin and a degree Celsius are the same size, and are both larger than a degree Fahrenheit by the factor $\tfrac{9}{5}$, so a rise of 3.0°F is equivalent to a rise of

$$\frac{3.0}{9/5} = 1.7°C = 1.7 \text{ K}.$$

EXERCISE A meteorologist on a Canadian radio station reports a high temperature of 24°C. What is the corresponding Fahrenheit temperature?

Answer: 75°F

Some problems similar to Example 19-1: 2–8 ●

In the half-century following Thompson's observations, the caloric theory gradually faded in popularity as a series of experiments confirmed the association between heating and mechanical energy. These experiments culminated in the work of the British physicist and brewer James Joule (1818–1889), who explored the relation between heat and mechanical, electrical, and chemical energy. In 1843, Joule quantified the relation between heat and energy, bringing thermal phenomena under the powerful conservation-of-energy law. In recognition of this major synthesis in physics, the SI unit of energy is named after Joule (Fig. 19-8). In Chapter 21 we will explore in detail the relation between heat and energy.

Our everyday experience of heat is nearly always associated with the movement of energy from one body to another. We rarely make statements about the actual amount of "heat" in an object—we're concerned instead that the temperature be appropriate. We want the furnace to transfer energy to our house in the winter and are satisfied when the house reaches a certain temperature, not when it contains a certain amount of "heat." The scientific definition of heat reflects this natural inclination to think of heat as energy in transit:

> **Heat is energy being transferred from one object to another because of temperature differences alone.**

Strictly speaking, the word **heat** refers only to energy in transit. Once heat has been transferred to an object, we say that the **internal energy** of the object has increased, but not that it contains more heat. This distinction reflects the fact that other processes than heating—such as the transfer of mechanical or electrical energy—can also change an object's temperature. The next chapter explores this relation between heat and internal energy.

FIGURE 19-8 In SI-based Australia, energy content of foods is given in joules rather than calories.

19-5 HEAT CAPACITY AND SPECIFIC HEAT

Experimentally, we find the heat ΔQ transferred to an object and the resulting change ΔT in the object's temperature are directly proportional. We write

$$\Delta Q = C\Delta T, \tag{19-4}$$

▲ **TABLE 19-1** SPECIFIC HEATS (TEMPERATURE RANGE 0°C TO 100°C EXCEPT AS NOTED)

SUBSTANCE	SPECIFIC HEAT (J/kg·K)	SPECIFIC HEAT (cal/g·°C, kcal/kg·°C, Btu/lb·°F)
Aluminum	900	0.215
Copper	386	0.0923
Iron	447	0.107
Glass	753	0.18
Mercury	140	0.033
Steel	502	0.12
Stone (granite)	840	0.20
Water:		
Liquid	4184	1.00
Ice, −10°C	2050	0.49
Wood	1400	0.33

where C is called the **heat capacity** of the object. Since heat is a measure of energy transfer, its unit is the joule, and therefore the units of heat capacity are J/K. The heat capacity C applies to a specific object and depends on its mass and on the substance from which it's made. We characterize different substances in terms of **specific heat** c, or heat capacity per unit mass. The heat capacity of an object is then the product of its mass and specific heat, so we can write

$$\Delta Q = mc\Delta T. \tag{19-5}$$

The SI units of heat capacity are J/kg·K.

Historically, thermodynamic phenomena were first studied before scientists knew the relation between heat and energy. Consequently other units were defined for heat. The **calorie** (cal) was defined as the heat needed to raise the temperature of one gram of water from 14.5°C to 15.5°; consequently, the specific heat of water in this temperature range is 1 cal/g·°C. Several different definitions of the calorie exist today, based on different methods for establishing the heat-energy equivalence. In this book we use the so-called thermochemical calorie, defined as exactly 4.184 J. The "calorie" used in describing the energy content of foods is actually a kilocalorie. In the English system, still widely used in engineering in the United States, the unit of heat is the **British thermal unit** (Btu). One Btu is the amount of heat needed to raise the temperature of one pound of water from 63°F to 64°F, and is equal to 1055 J.

Table 19-1 lists specific heats of some common materials.

● **EXAMPLE 19-2** WAITING TO SHOWER

Your whole family has taken showers before you, dropping the temperature in the water heater to 18°C. If the heater holds 150 kg of water, how much energy will it take to bring it to 50°C? If the energy is supplied by a 5.0-kW electric heating element, how long will that take?

Solution

The heat required is given by Equation 19-5:

$$\Delta Q = mc\Delta T = (150 \text{ kg})(4184 \text{ J/kg·K})(50°C − 18°C)$$
$$= 2.0 \times 10^7 \text{ J},$$

where we found the specific heat of water in Table 19-1. The heating element supplies energy at the rate of 5.0 kW, or 5.0×10^3 J/s; at that rate the time needed to supply 2.0×10^7 J is

$$\Delta t = \frac{2.0 \times 10^7 \text{ J}}{5.0 \times 10^3 \text{ J/s}} = 4000 \text{ s}, \quad \text{or a little over an hour.}$$

by a stove burner at the rate of 2.0 kW, how long will it take to heat the pan?

Answers: (a) 166 kJ; (b) 84 s

EXERCISE (a) How much heat does it take to bring a 3.4-kg iron skillet from 20°C to 130°C? (b) If the heat is supplied

Some problems similar to Example 19-2: 19, 23–25, 28 •

For many substances the specific heat is nearly independent of temperature over wide temperature ranges. When the specific heat does vary significantly with temperature, we write Equation 19-5 in the limit of very small temperature change dT and heat dQ, and integrate to find the heat required for larger changes. Problem 73 explores this situation.

The Equilibrium Temperature

When two objects at different temperatures are in thermal contact, heat flows from the hotter object to the cooler one until they reach thermodynamic equilibrium. If the two objects are thermally insulated from their surroundings, then all the energy leaving the hotter object ends up in the cooler one. Mathematically, this statement may be written

$$m_1 c_1 \, \Delta T_1 + m_2 c_2 \, \Delta T_2 = 0, \qquad (19\text{-}6)$$

where mc is the heat capacity of an object of mass m and specific heat c, and ΔT is the temperature change of that object. For the hotter object, ΔT is negative, so the two terms in Equation 19-6 have opposite signs. One term represents the outflow of heat from the hotter object, the other inflow into the cooler object.

● **EXAMPLE 19-3** COOLING DOWN

An aluminum frying pan with mass 1.5 kg is heated on a stove to 180°C, then plunged into a sink containing 8.0 kg of water at room temperature (20°C). Assuming that none of the water boils, and that no heat is lost to the surroundings, what is the equilibrium temperature of the water and pan?

Solution
We write Equation 19-6 in the form

$$m_p c_p (T - T_p) + m_w c_w (T - T_w) = 0,$$

where the subscripts p and w refer to the pan and the water, and where the equilibrium temperature T is the same for both. Solving for T then gives

$$T = \frac{m_p c_p T_p + m_w c_w T_w}{m_p c_p + m_w c_w}.$$

Using the given values of m_p, T_p, m_w, m_w, and T_w, and taking c_p and c_w from Table 19-1, we find that $T = 26°C$.

EXERCISE You take a 1.0-L bottle of soda from a 4.0°C refrigerator and put it in an insulated cooler chest whose heat capacity is negligible. Also in the chest is a wooden cutting board of mass 2.1 kg and temperature 24°C. When you open the chest much later, what will be the temperature of the soda? Soda has essentially the same density (1.0 kg/L) and specific heat as water. Neglect the heat capacity of the soda bottle itself.

Answer: 12°C

Some problems similar to Example 19-3: 34–36, 59, 60 •

19-6 HEAT TRANSFER

How is heat transferred between two objects? Engineers need to know so they can design systems to control heat transfer, as in heating or cooling a building. Scientists need to know so they can anticipate temperature changes, as in the global warming that may now be arising from human activity.

Three heat-transfer mechanisms commonly occur. These are conduction, a process involving direct physical contact; convection, involving energy transfer by the bulk flow of a fluid; and radiation, or energy transfer by electromagnetic waves. In a given situation, one of the three may dominate, or we may have to take all three into account.

Conduction

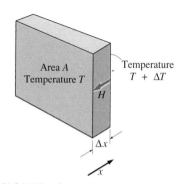

FIGURE 19-9 Heat flows from the hotter to the cooler face of the slab, at a rate H that depends on the slab area, the temperature difference ΔT, and the thermal conductivity k.

Conduction is the transfer of heat through direct physical contact. Microscopically, conduction occurs because molecules in a hotter region transfer energy to those of an adjacent cooler region through collisions. The effect of these collisions in a given material is quantified in the material's **thermal conductivity, k**, whose SI units are W/m·K. Common materials exhibit a broad range of thermal conductivities, from about 400 W/m·K for copper—a good conductor of heat—to 0.029 W/m·K for styrofoam, a good thermal insulator.

Figure 19-9 shows a slab of material of thickness Δx and cross-sectional area A. Suppose one face of the material is held at temperature T and the other at $T + \Delta T$. Intuitively, we might expect the rate of heat flow through the slab to increase with increasing area A and to decrease with increasing thickness Δx. Of course, the heat-flow rate must also depend on the thermal conductivity of the material. We expect further that heat will flow from the hotter to the cooler face of the slab, with the heat-flow rate dependent also on the temperature difference between the two faces. Our intuition is borne out experimentally: for a rectangular slab of material, the heat-flow rate H is given by

$$H = -kA\frac{\Delta T}{\Delta x}. \tag{19-7}$$

Since H is a rate of energy flow, its units are joules/second, or watts. The minus sign in Equation 19-7 shows that heat transfer is opposite to the direction of increasing temperature, that is, from hotter to cooler.

Table 19-2 lists thermal conductivities for some common materials. Both SI and British units are listed since the latter are commonly used in calculations involving heat loss in buildings. Note the wide range of thermal conductivities in Table 19-2. Metals are exceptionally good conductors because they contain free electrons that move quickly, transferring energy with them. Gases, on the other hand, are poor conductors because the wide spacing of their molecules makes for infrequent energy transfers. Good heat insulators, like fiberglass and styrofoam, owe their insulating properties to a physical structure that traps small volumes of air or other gas.

▲ **TABLE 19-2** THERMAL CONDUCTIVITIES

MATERIAL	k(W/m·K)	k(Btu·in./h·ft²·°F)
Air	0.026	0.18
Aluminum	237	1644
Concrete (varies with mix)	1	7
Copper	401	2780
Fiberglass	0.042	0.29
Glass	0.7–0.9	5–6
Goose down	0.043	0.30
Helium	0.14	0.97
Iron	80.4	558
Steel	46	319
Styrofoam	0.029	0.20
Water	0.61	4.2
Wood (pine)	0.11	0.78

● **EXAMPLE 19-4** WARMING A LAKE

A lake with a flat bottom and steep sides has surface area 1.5 km² and is 8.0 m deep. On a summer day, the surface water is at a temperature of 30°C, while the bottom water is at 4.0°C. What is the rate of heat conduction through the lake? Assume that the temperature declines uniformly from surface to bottom.

Solution
The lake resembles the slab of Fig. 19-9. Taking the thermal conductivity of water from Table 19-2, Equation 19-7 gives

$$H = -kA\frac{\Delta T}{\Delta x}$$

$$= -(0.61 \text{ W/m·K})(1.5\times10^6 \text{ m}^2)\frac{30°C - 4.0°C}{8.0 \text{ m}}$$

$$= -3.0\times10^6 \text{ W}.$$

The minus sign indicates that heat flows in the direction of decreasing temperature, or downward into the lake. Were the surface and bottom temperatures not maintained, heat flow would eventually bring the entire lake to a uniform temperature.

EXERCISE A concrete wall 20 cm thick measures 2.2 m high by 11 m wide. On the inside, the temperature is 20°C; outside, it's −10°C. What is the heat-flow rate through the wall?

Answer: 3.6 kW

Some problems similar to Example 19-4: 38, 39, 42, 46 ●

Equation 19-7 is strictly correct only when the temperature varies uniformly from one surface to the other. This is the case in practical problems where two surfaces at different temperatures have the same area. With other geometries— as in evaluating heat loss through the insulation surrounding a cylindrical pipe— we need to write $\Delta T/\Delta x$ as the derivative dT/dx and integrate to relate the heat loss and the temperature difference. Problems 74 and 75 explore this situation.

Often heat flows through several different materials. A building wall, for example, may contain wood, plaster, fiberglass insulation, and other materials. Figure 19-10 shows such a composite structure, with temperature T_1 on one side and T_3 on the other. The heat-flow rate H must be the same through both slabs since energy doesn't accumulate or disappear at the interface between the two. Then Equation 19-7 gives

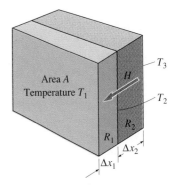

FIGURE 19-10 A composite slab. The heat-flow rate H is the same through both sections.

$$H = -k_1 A \frac{T_2 - T_1}{\Delta x_1} = -k_2 A \frac{T_3 - T_2}{\Delta x_2}, \tag{19-8}$$

where k_1 and k_2 are the thermal conductivities of the two materials, and T_2 is the temperature at the interface. We would like to express the heat-flow rate in terms of the surface temperatures T_1 and T_3 alone, without having to worry about the intermediate temperature T_2. To do so, it's convenient to define the **thermal resistance,** R, of each slab:

$$R = \frac{\Delta x}{kA}. \tag{19-9}$$

The SI units of R are K/W. Unlike the thermal conductivity, k, which is a property of a *material*, R is a property of a *particular piece* of material, reflecting both its conductivity and its geometry. In terms of thermal resistance, Equation 19-8 becomes

$$H = -\frac{T_2 - T_1}{R_1} = -\frac{T_3 - T_2}{R_2},$$

so

$$R_1 H = T_1 - T_2$$

and

$$R_2 H = T_2 - T_3.$$

Adding these two equations gives

$$(R_1 + R_2)H = T_1 - T_2 + T_2 - T_3 = T_1 - T_3,$$

or

$$H = \frac{T_1 - T_3}{R_1 + R_2}. \tag{19-10}$$

Equation 19-10 shows that the composite slab acts like a single slab whose thermal resistance is the sum of the resistances of the two slabs that compose it. We could easily extend this treatment to show that the thermal resistances of three or more slabs add when the slabs are arranged so the same heat flows through all of them.

In the United States, the insulating properties of building materials are usually described in terms of the **R-factor, $\mathcal{R}$,** which is the thermal resistance for a slab of unit area:

$$\mathcal{R} = RA = \frac{\Delta x}{k}. \tag{19-11}$$

The units of $\mathcal{R}$, although rarely stated, are ft²·°F·h/Btu. This means that $\mathcal{R}$-19 fiberglass insulation, now in common use as a wall insulation in the northern part of the United States, has a heat loss of $\frac{1}{19}$ Btu per hour for each square foot of insulation for each degree Fahrenheit temperature difference across the insulation.

● EXAMPLE 19-5 INSULATION, HEAT LOSS, AND THE PRICE OF OIL

Figure 19-11 shows a house whose walls consist of plaster ($\mathfrak{R} = 0.17$), $\mathfrak{R} = 11$ fiberglass insulation, plywood ($\mathfrak{R} = 0.65$), and cedar shingles ($\mathfrak{R} = 0.55$). The roof is the same construction except it uses $\mathfrak{R} = 30$ fiberglass insulation. The average outdoor temperature in winter is 20°F, and the house is maintained at 70°F. The house's oil furnace produces 100,000 Btu for every gallon of oil, and oil costs 94¢ per gallon. How much does it cost to heat the house for a month?

Solution
The R-factors for the walls and roof are

$$\mathfrak{R}_{wall} = \mathfrak{R}_{plaster} + \mathfrak{R}_{fiberglass} + \mathfrak{R}_{plywood} + \mathfrak{R}_{shingles}$$
$$= 0.17 + 11 + 0.65 + 0.55 = 12.4,$$

and

$$\mathfrak{R}_{roof} = 0.17 + 30 + 0.65 + 0.55 = 31.4.$$

The total wall area is

$$A_{wall} = (28\ ft + 36\ ft + 28\ ft + 36\ ft)(10\ ft)$$
$$+ (28\ ft)(8.08\ ft) = 1506\ ft^2,$$

where the second term in the sum is the area of the two triangular portions of the side walls, whose height is $(14\ ft)(\tan 30°)$, or 8.08 ft. These $\mathfrak{R}$-12.4 walls lose $1/12.4$ Btu/h/ft²/°F, and the temperature difference across them is 50°F, so the total heat-loss rate through the walls is

$$H_{walls} = (\tfrac{1}{12.4}\ Btu/h/ft^2/°F)(1506\ ft^2)(50°F) = 6073\ Btu/h.$$

The area of the pitched roof is increased over that of a flat roof by the factor $1/\cos 30°$, so the heat-loss rate through the roof is

$$H_{roof} = (\tfrac{1}{31.4}\ Btu/h/ft^2/°F)\frac{(36\ ft)(28\ ft)}{\cos 30°}(50°F)$$
$$= 1853\ Btu/h.$$

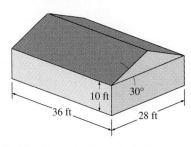

FIGURE 19-11 House for Example 19-5.

The total heat-loss rate is then

$$H = 6073\ Btu/h + 1853\ Btu/h = 7926\ Btu/h.$$

In a month, this results in a total heat loss of

$$Q = (7926\ Btu/h)(30\ days/month)(24\ h/day)$$
$$= 5.71 \times 10^6\ Btu.$$

With 10^5 Btu per gallon of oil burned, this requires 57.1 gallons of oil, at a cost of (57.1 gal) ($0.94/gal) = $53.58.

   This estimate is low; losses through windows and doors are substantial, and cold air infiltration results in additional heat loss. A more accurate analysis would also consider heat lost to the ground and solar energy gained through the windows. Problem 67 provides a more realistic look at heat loss in this house.

EXERCISE In the house of Example 19-5, how much money would be saved each month if the wall insulation were increased to the same $\mathfrak{R} = 30$ as in the roof?

Answer: $24.80

Some problems similar to Example 19-5: 43, 45, 47, 48, 67 ●

Convection

Convection is heat transfer by the bulk motion of a fluid. Convection occurs because a fluid becomes less dense when heated and therefore rises. Figure 19-12 shows two plates held at different temperatures, with fluid between them. Fluid heated by the lower plate rises and transfers heat to the upper plate. The cooled fluid then sinks, and the process repeats. The pattern of rising and sinking fluid often acquires a striking regularity, as shown in Fig. 19-13.

   Convection is important in many technological and natural environments. When you heat water on a stove, convection carries heat from the bottom of the pan to the top. Houses usually rely on convection from heat sources near floor level to circulate warm air throughout a room. Insulating materials like fiberglass and goose down trap air and thereby inhibit convection that would

FIGURE 19-12 Convection between two plates at different temperatures.

FIGURE 19-13 Top view of convection cells in a laboratory experiment.

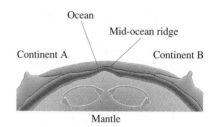

FIGURE 19-14 Slow convection in Earth's mantle brings up material at the mid-ocean ridges. Here continents *A* and *B* are drifting apart as the ocean widens.

otherwise cause excessive heat loss from our houses and our bodies. Convection associated with solar heating of Earth's surface drives the vast air movements that establish our overall climate. Violent convective movements, such as those in thunderstorms, are associated with localized temperature differences. On a much longer time scale, convection in Earth's mantle is involved in continental drift (Fig. 19-14). Convection plays a crucial role in many astrophysical processes, including the generation of magnetic fields in stars and planets.

As with conduction, the convective heat-loss rate often is approximately proportional to the temperature difference. But the calculation of convective heat loss is complicated because of the need to understand the details of the associated fluid motion. The study of convection processes is an important research area in many fields of contemporary science and engineering.

Radiation

Turn an electric stove burner to "high," and it glows a brilliant red-orange. Set it on "low," and it does not appear hot, yet you can still tell that it is by holding your hand near it. In both cases the burner is losing energy by emitting **radiation.** In later chapters, we will investigate the nature of radiation in terms of electromagnetic phenomena and atomic theory. Our stove-burner example suggests that radiation increases rapidly with temperature. Experiment confirms this: the rate of energy loss by radiation is given by the **Stefan-Boltzmann law:**

$$P = e\sigma AT^4, \tag{19-12}$$

where *P* is in watts, *A* is the surface area of the emitting surface, *T* the temperature in kelvins, and σ a constant called the **Stefan-Boltzmann** constant, approximately 5.67×10^{-8} W/m²·K.

The quantity e is called the **emissivity** of the material; it ranges from 0 to 1 and measures the material's effectiveness in emitting radiation. Materials not only emit but also absorb radiation, and it turns out that the same quantity e describes a material's effectiveness as an absorber. Thus a material with $e = 1$ absorbs all radiation incident on it. At normal temperatures it would therefore appear black, and is therefore called a **blackbody.** When heated sufficiently, though, a blackbody glows brightly.

The rate of energy loss by radiation depends on the fourth power of the temperature. Therefore radiation is generally less important at low temperatures but dominates at high temperatures. A house loses only modest heat through radiation, while very hot things like the Sun, light bulb filaments, or molten metal lose most of their energy by radiation (Fig. 19-15). In vacuum, where conduction and convection cannot occur, all energy loss is by radiation. That's why Thermos bottles and Dewar flasks—whose insulation is the vacuum between layers of glass—are coated with shiny material that reflects radiation.

FIGURE 19-15 Foundry worker pouring molten gold concentrate. Radiation is the dominant heat loss from the hot, liquid metal, as evidenced by its bright glow.

● **EXAMPLE 19-6** THE SUN'S TEMPERATURE

The Sun radiates energy at the rate $P = 3.9 \times 10^{26}$ W, and its radius is 7.0×10^8 m (Fig. 19-16). Assuming the Sun to be a blackbody (emissivity = 1), what is its surface temperature?

Solution
The radiated power is given by the Stefan-Boltzmann law, Equation 19-12. Solving this equation for T and using $4\pi R^2$ for the surface area of the Sun gives

$$T = \left(\frac{P}{4\pi R^2 \sigma} \right)^{1/4}$$
$$= \left[\frac{3.9 \times 10^{26} \text{ W}}{4\pi (7.0 \times 10^8 \text{ m})^2 (5.7 \times 10^{-8} \text{ W/m}^2\cdot\text{K}^4)} \right]^{1/4}$$
$$= 5.8 \times 10^3 \text{ K},$$

in agreement with observational measurements.

EXERCISE What is the surface area of a 100-W light bulb filament at a temperature of 3100 K? Assume the filament has emissivity $e = 1$.

Answer: 1.9×10^{-5} m^2

FIGURE 19-16 What is the Sun's surface temperature?

Some problems similar to Example 19-6: 52–54, 63, 64
●

Materials not only emit radiation but also absorb it. In thermodynamic equilibrium an object is at the same temperature as its surroundings, so it must absorb and emit radiation at the same rate—both given by Equation 19-12. When an object is not in equilibrium with its surroundings, then there is a net energy transfer by radiation, given by the difference between its emission and absorption. The Sun and the filament in Example 19-6 and its associated exercise are decidedly not in equilibrium with their surroundings, and we neglected the minuscule amounts of radiation absorbed by these hot objects from their much cooler surroundings.

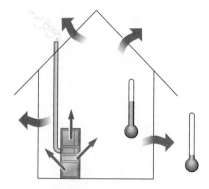

FIGURE 19-17 A house in thermal energy balance. Energy from the furnace balances heat loss, maintaining the house at a constant temperature higher than that of its surroundings.

19-7 THERMAL ENERGY BALANCE

When an object is in thermodynamic equilibrium with its surroundings, its temperature remains constant. But it could also maintain a constant temperature, different from its surroundings, by balancing energy gain and loss. A house in winter provides a good example of this **thermal energy balance.** Unheated, the house would come to equilibrium at the outdoor temperature. Heated, it loses heat because of the temperature difference between itself and its environment, but this loss is balanced by energy input from a furnace, solar collector, electricity, or other source (Fig. 19-17).

The principle of thermal energy balance is used throughout science and engineering to predict temperatures under different conditions. It's also used to determine the sizes of heat sources needed to achieve desired temperatures. When heat is supplied to an object, its temperature initially rises. But so does the heat-loss rate, which increases with the temperature difference between the object and its surroundings. Eventually the loss rate equals the rate of energy input. The object is then in thermal energy balance at a constant temperature.

● **EXAMPLE 19-7** A WATER HEATER

A poorly insulated electric water heater loses heat at the rate of 40 W for each Celsius degree difference between the water and its surroundings. The tank is heated by a 2.5-kW electric heating element. If the heater is located in a 15°C basement, what will be its temperature if the heating element operates continuously?

Solution

The heating element supplies energy at the rate $H_{in} = 2.5$ kW, while the heater loses energy at the rate $H_{out} = (40 \text{ W/°C})(\Delta T)$, where ΔT is the temperature difference between the water and its surroundings. In thermal energy balance the heat input equals the heat loss, so we have

$$(40 \text{ W/°C})(\Delta T) = 2.5 \text{ kW},$$

or

$$\Delta T = \frac{2.5 \text{ kW}}{40 \text{ W/°C}} = 63°C.$$

With the basement at 15°C, the water temperature is then 78°C.

EXERCISE A home heating system supplies heat at the maximum rate of 40 kW. If the house loses 1.1 kW for each °C between inside and outside, what is the minimum outdoor temperature for which the heating system can maintain 20°C inside?

Answer: −16°C

Some problems similar to Example 19-7: 51, 54, 58, 65, 66 ●

● **EXAMPLE 19-8** A SOLAR GREENHOUSE

Figure 19-18 shows a solar-heated greenhouse. The south-facing diagonal wall is made of double-pane glass ($\Re = 1.8$), while the other walls are opaque and have R-factors of 30. Heat loss through the floor is negligible. The intensity of sunlight on the glass, averaged over a typical day, is 120 W/m². If the outdoor temperature is 15°F, what will be the temperature in the greenhouse?

Solution

We first calculate the heat-loss rate for the greenhouse. The total area of insulated wall is

$$A_w = (10 \text{ ft})(20 \text{ ft}) + 2[\tfrac{1}{2}(10 \text{ ft})^2] = 300 \text{ ft}^2,$$

so the heat-loss rate through these $\Re$-30 walls is

$$H_w = \frac{A_w \Delta T}{\Re_w} = \frac{300\Delta T}{30} = 10\Delta T \text{ Btu/h·°F}.$$

Similarly, the glass area is

$$A_g = (20 \text{ ft})(10\sqrt{2} \text{ ft}) = 283 \text{ ft}^2,$$

so the heat-loss rate through the glass is

$$H_g = \frac{A_g \Delta T}{\Re_g} = \frac{283\Delta T}{1.8} = 157\Delta T \text{ Btu/h·°F}.$$

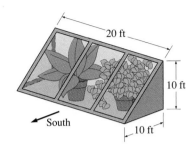

FIGURE 19-18 Solar greenhouse for Example 19-8.

If the greenhouse is to be in thermal energy balance, the total heat-loss rate of $167\Delta T$ Btu/h·°F must be balanced by the energy gained from the Sun. The average solar input, which we denote S, is 120 W/m², or

$$S = (120 \text{ W/m}^2)[3.413(\text{Btu/h})/\text{W}](0.0929 \text{ m}^2/\text{ft}^2)$$
$$= 38.0 \text{ Btu/h·ft}^2,$$

where we obtained the conversion factors from Appendix C. Then the total solar gain on the 283-ft² glass is

$$(283 \text{ ft}^2)(38.0 \text{ Btu/h·ft}^2) = 1.075 \times 10^4 \text{ Btu/h}.$$

Equating the heat loss to the solar gain gives

$$167\Delta T \text{ Btu/h·°F} = 1.075 \times 10^4 \text{ Btu/h},$$

so
$$\Delta T = \frac{1.075 \times 10^4 \text{ Btu/h}}{167 \text{ Btu/h·°F}} = 64°\text{F}$$

With the outdoors at 15°F, the greenhouse temperature is then 79°F.

EXERCISE A cubical doghouse measures 4.0 ft on a side, and its four walls and roof have $\Re = 8.5$, including an insulated door. Heat loss to the ground is negligible. If the dog's metabolism produces energy at the average rate of 45 W, what will be the doghouse temperature when the outside temperature is −5°F?

Answer: 11°F

Some problems similar to Example 19-8: 55, 67 ●

■ APPLICATION THE GREENHOUSE EFFECT AND GLOBAL WARMING

Among the most serious threats facing civilization are the consequences of global climate change, brought about as human activity alters the composition of Earth's atmosphere. The most likely result of these alterations appears to be an increase in global temperature. That, in turn, could bring a rise in sea level, along with shifts in rainfall patterns that could jeopardize our ability to feed the burgeoning human population (Fig. 19-19). The human species would not be the only one to suffer; with warming rates far in excess of naturally occurring climate change, many plant and animal species could perish from inability to adapt or migrate fast enough to escape inhospitable conditions.

Scientific analysis of global warming is complex and not without controversy, although most scientists now agree that warming is almost certain to occur through the twenty-first century. But if the details are complex, the basic physics is straightforward. Earth itself is essentially in thermal energy balance. Sunlight heats the planet, which in turn radiates energy as infrared radiation. Earth's temperature is established by the balance between solar input and infrared radiation into space. Figure 19-20 shows that our planet presents an effective area πR_E^2 to the incident sunlight. The rate of solar energy input at the top of the atmosphere is then $\pi R_\text{E}^2 S$, where S is the power per unit area in sunlight. Above the atmosphere, $S = 1.37$ kW/m², but about 30% of the sunlight is reflected back into space, giving an effective solar intensity $S = 960$ W/m². Figure 19-20 also shows that Earth radiates from its entire surface

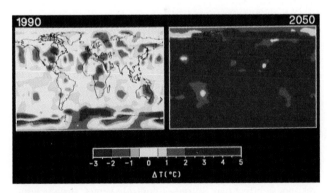

FIGURE 19-19 Global temperature changes from 1900 to 1990 (left) and, as predicted by a computer model, to 2050 (right). The scale below indicates the temperature *change* from the values in 1900.

area $4\pi R_\text{E}^2$. If Earth were a blackbody ($e = 1$), Equation 19-12 would give a total energy loss rate $P = \sigma 4\pi R_\text{E}^2 T^4$. Equating energy gain and loss rates would give

$$\pi R_\text{E}^2 S = \sigma 4\pi R_\text{E}^2 T^4,$$

or

$$T = \left(\frac{S}{4\sigma}\right)^{1/4} = \left(\frac{960 \text{ W/m}^2}{(4)(5.67 \times 10^{-8} \text{ W/m}^2\cdot\text{K})}\right)^{1/4}$$
$$= 255 \text{ K} = -18°\text{C}.$$

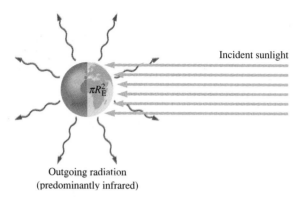

FIGURE 19-20 Earth's temperature is determined by the balance of incoming sunlight and outgoing infrared radiation. The planet presents an effective area πR_E^2 to sunlight, but radiates from its entire surface.

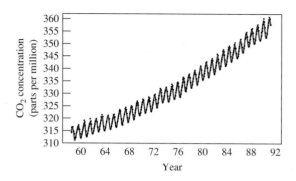

FIGURE 19-21 Atmospheric CO_2 concentration as measured at Mauna Loa, Hawaii, over roughly the past three decades. Annual variations reflect seasonal changes in vegetation. The overall trend shows an increase that has been occurring since the beginning of the industrial age.

This number is certainly the right order of magnitude, but is considerably lower than the measured global average temperature of about 14°C.

What's missing in our simple calculation that would account for the 32-°C discrepancy? The answer lies in the composition of Earth's atmosphere. A number of atmospheric molecules absorb infrared radiation, reradiating some to space but some back to Earth. As a result, Earth's surface temperature must be somewhat higher than it would be in the absence of those gases in order to get the same rate of infrared radiation escaping to outer space. This phenomenon is called the **greenhouse effect** because the infrared-absorbing gases play the same role as greenhouse glass: they let incident sunlight through but impede escaping infrared, helping to keep the greenhouse temperature higher than it otherwise would be.

It's a good thing we have the greenhouse effect; otherwise Earth would be intolerably cold, as our calculation suggests. The planet Mars has insignificant quantities of greenhouse gases, and it's much colder relative to Earth than its greater distance from the Sun would imply. On the other hand the greenhouse effect can get out of control; Venus' surface temperature is hotter than an oven, thanks to an overabundance of greenhouse gases. Earth's temperature is in a delicate balance between too little and too much greenhouse effect.

Unfortunately that balance is changing. The predominant greenhouse gas is carbon dioxide, which occurs naturally in the atmosphere but which also arises from burning fossil fuels. The CO_2 content of Earth's atmosphere has risen steadily throughout the industrial age (Fig. 19-21). And technology has unleashed far more potent greenhouse gases, particularly the chloro-fluorocarbons (CFCs) used in refrigeration. Figure 19-22 shows the mix of greenhouse gases important in today's atmosphere.

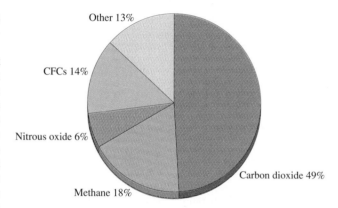

FIGURE 19-22 Contribution of different gases to the greenhouse effect. Historically, CO_2 accounted for two-thirds of the effect, but others—especially the chlorofluorocarbons (CFCs) used in refrigeration systems—are now increasingly important.

It's probably too late to prevent global warming altogether. With a concerted effort, though, the average temperature increase might be held to about 2°C—a rise that, although small, still implies significant and rapid climatic change. This effort would require international cooperation on several fronts:

- An almost immediate halt in the production of chlorofluorocarbons.
- A reduction in global CO_2 emissions by 30% over several decades.
- A halt to deforestation within a decade, followed by a decade of reforestation, since tree growth absorbs atmospheric CO_2.

In the absence of the political will and technological means to achieve these goals, a world going about its "business as usual" will probably see a 3–5°C temperature increase by the year 2050. These numbers may seem modest, but they represent a rate of change tens to hundreds of times what the planet has heretofore experienced (Fig. 19-23). Serious climatic effects would almost certainly follow.

There remain great uncertainties in predicting both the extent of global warming and its effects. Physical, chemical, and biological mechanisms, many poorly understood, may help stabilize global warming, or they may exacerbate it. New technologies may help alleviate the problem or they may contribute to it. Global warming presents an enormous challenge to scientists and engineers—and to all the world's citizens.

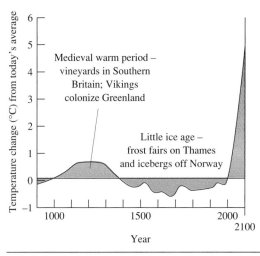

FIGURE 19-23   Global average temperature change over 1200 years, including predictions for the next century by the Intergovernmental Panel on Climate Change. Curve shown is an upper estimate in the absence of feedback mechanisms, which IPCC scientists think are likely to increase the temperature change even further. The more modest natural changes over the past thousand years had obvious effects on climate.

CHAPTER SYNOPSIS

Summary

1. Heat and related phenomena may be viewed from either the macroscopic viewpoint of **thermodynamics** or the microscopic viewpoint of **statistical mechanics.**
2. Two systems are in **thermodynamic equilibrium** if none of their macroscopic properties change when they are brought into thermal contact. Systems in thermodynamic equilibrium have the same **temperature.**
3. The **zeroth law of thermodynamics** states that two systems in thermodynamic equilibrium with a third system are in thermodynamic equilibrium with each other. This law allows the establishment of temperature scales.
4. A **thermometer** is a system with a convenient macroscopic property that serves to indicate temperature. **Gas thermometers** are used to establish the **kelvin** temperature scale of the SI system.
5. **Heat** is energy being transferred between objects as a result of a temperature difference.
6. **Specific heat** measures the ratio of heat to temperature change for a unit mass of a given material:

$$c = \frac{1}{m}\frac{\Delta Q}{\Delta T}.$$

Heat capacity is the ratio of heat to temperature change for a given object:

$$C = mc = \frac{\Delta Q}{\Delta T}.$$

7. Heat transfer occurs by three distinct mechanisms:
   a. **Conduction** is direct transfer of heat through physical contact, and involves energy exchange through molecular collisions. The conduction rate H through a slab of material depends on the **thermal conductivity** k of the material, the slab thickness Δx and slab area A, and on the temperature difference across the slab:

$$H = -kA\frac{\Delta T}{\Delta x}.$$

   Thermal resistance, $R = \Delta x/kA$ and **R-factor,** $\Re = \Delta x/k$, also characterize heat conduction through materials.
   b. **Convection** is heat transfer through bulk motion of a fluid, as when heated air rises. Convection is difficult to describe quantitatively but is an important process in many natural and technological systems.

c. **Radiation** is energy in the form of electromagnetic waves that can travel through empty space. The rate at which a hot surface of temperature T and area A loses energy by radiation is

$$P = e\sigma AT^4,$$

where e is a number between 0 and 1 that describes the effectiveness of the surface as a radiation emitter and σ is the universal **Stefan-Boltzmann constant,** $\sigma = 5.67 \times 10^{-8}$ W/m²·K.

8. **Thermal energy balance** exists when an object gains and loses energy at the same rate, thereby maintaining a constant temperature. If either energy gain or loss changes, the temperature rises or falls until heat loss again balances gain.

Terms You Should Understand

(Pairs are closely related terms whose distinction is important; number in parentheses is chapter section where term first appears.)

thermodynamics, statistical mechanics (19-1)
thermodynamic equilibrium (19-2)
thermal contact, thermal insulation (19-2)
thermometer, gas thermometer (19-3)
kelvin (19-3)
calorie, British thermal unit (19-4)
specific heat, heat capacity (19-5)
conduction, convection, radiation (19-6)
thermal conductivity (19-6)
thermal resistance, R-factor (19-6)
Stefan-Boltzmann law (19-6)
emissivity (19-6)
blackbody (19-6)
thermal energy balance (19-7)
greenhouse effect (19-7)

Symbols You Should Recognize

T (19-3)
K, °C, °F (19-3)
cal, Btu (19-4)
c, C (19-5)
H (19-6)
k (19-6)
R, $\Re$ (19-6)
e (19-6)
σ (19-6)

Problems You Should Be Able to Solve

converting among different temperature scales (19-3)
converting among different energy units (19-4)
calculating the equilibrium temperature of a mixture (19-5)
evaluating heat loss given thermal conductivity, resistance, or R-factor (19-6)
evaluating radiant heat loss (19-6)
determining temperature in thermal energy balance (19-7)

Limitations to Keep in Mind

When specific heat varies significantly over a temperature range, Equation 19-4 must be written in terms of the derivative dQ/dT and then integrated to find the heat.

Equation 19-7 applies exactly only when the thermal conductivity is independent of temperature and the geometry is that of a rectangular slab. Otherwise it must be written in terms of the derivative dT/dx and integrated to relate heat loss and temperature difference.

Equation 19-12 gives only the radiation loss; there is also an energy gain given by the same equation with T the temperature of the surroundings. The gain is negligible if an object is significantly hotter than its surroundings.

QUESTIONS

1. Two identical-looking physical systems are in the same macroscopic state. Must they be in the same microscopic state? Explain.
2. Two identical-looking physical systems are in the same microscopic state. Must they be in the same macroscopic state? Explain.
3. If system A is not in thermodynamic equilibrium with system B, and B is not in equilibrium with C, can you draw any conclusions about the temperatures of the three systems?
4. Given that there are three mechanisms of heat transfer, how would you construct a good insulator?
5. Does a thermometer measure its own temperature, or the temperature of its surroundings? Explain.
6. To get an accurate body temperature measurement, you must hold a glass-and-mercury fever thermometer under your tongue for about 3 minutes. Some electronic fever thermometers require less than a minute. What might account for the difference?
7. Why is it better to define temperature scales in terms of a physical state, like the triple point of water, rather than by having an official, standard thermometer stored at the International Bureau of Weights and Measures?
8. Compare the relative sizes of the kelvin, the degree Celsius, the degree Fahrenheit, and the degree Rankine.

9. Does a vacuum have temperature? Explain.

10. If you put a thermometer in direct sunlight, what do you measure? The air temperature? The temperature of the Sun? Some other temperature?

11. Why does the temperature in a stone building usually vary less than in a wooden building?

12. Why do large bodies of water exert a temperature-moderating effect on their surroundings?

13. A Thermos bottle consists of an evacuated, double-wall glass liner. The glass is coated with a thin layer of aluminum. How does a Thermos bottle work?

14. Stainless-steel cookware often has a layer of aluminum or copper embedded in the bottom. Why?

15. What method of energy transfer is involved in baking? In broiling?

16. After a calm, cold night the temperature a few feet above ground often drops just as the Sun comes up. Explain in terms of convection.

17. Solar collectors often have a copper absorber surface coated with a black paint, whose emissivity is nearly 1 for visible and infrared radiation. Better results can be achieved, though, with a "selective surface," whose emissivity is high for visible light but low for the infrared radiation associated with lower temperatures than the Sun's. Why is this?

18. Glass and fiberglass are made from the same material, yet have dramatically different thermal conductivities. Why?

19. A thin layer of glass does not offer much resistance to heat conduction. Yet double-glazed windows provide substantial energy savings. Why?

20. The insulating value of a double-glazed window actually decreases if the spacing is made too great. Why might this be?

21. To keep your hands warm while skiing, you should wear mittens instead of gloves. Why?

22. Since Earth is exposed to solar radiation, why doesn't Earth have the same temperature as the Sun?

23. On a clear night, a solar collector will often cool well below the ambient air temperature. Why? Why doesn't this happen on a cloudy night?

24. Is Earth in perfect energy balance? If not, why not?

PROBLEMS

Section 19-1 Macroscopic and Microscopic Descriptions

1. The macroscopic state of a carton capable of holding a half-dozen eggs is specified by giving the number of eggs in the carton. The microscopic state is specified by telling where each egg is in the carton. How many microscopic states correspond to the macroscopic state of a full carton?

Section 19-3 Measuring Temperature

2. A Canadian meteorologist predicts an overnight low of $-15°C$. What would a U.S. meteorologist predict for the same location?

3. Normal room temperature is 68°F. What is this in Celsius?

4. The outdoor temperature rises by 10°C. What is the rise in Fahrenheit?

5. At what temperature do the Fahrenheit and Celsius scales coincide?

6. Give equations for Rankine temperature in terms of (a) kelvins and (b) °F.

7. The normal boiling point of nitrogen is 77.3 K. Express this in Celsius and Fahrenheit.

8. The temperature at the Sun's center is about 16 MK. Express this in Celsius and Fahrenheit.

9. A constant-volume gas thermometer is filled with air whose pressure is 101 kPa at the normal melting point of ice. What would its pressure be at (a) the normal boiling point of water, (b) the normal boiling point of oxygen (90.2 K), and (c) the normal boiling point of mercury (630 K)?

10. A constant-volume gas thermometer is at 55 kPa pressure at the triple point of water. By how much does its pressure change for each kelvin temperature change?

11. The temperature of a constant-pressure gas thermometer is directly proportional to the gas volume. If the volume is 1.00 L at the triple point of water, what is it at water's normal boiling point?

12. In the gas thermometer of Fig. 19-24, the height h is 60.0 mm at the triple point of water. When the thermometer is immersed in boiling sulfur dioxide the height drops to 57.8 mm. What is the boiling point of SO_2 in kelvins and in degrees Celsius?

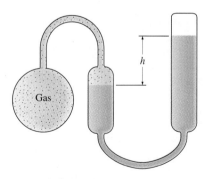

FIGURE 19-24   Problem 12.

13. A constant-volume gas thermometer supports a 72.5-mm-high mercury column when it's immersed in liquid nitro-

gen at $-196°C$. What will be the column height when the thermometer is in molten lead at $350°C$?

Sections 19-4 and 19-5 Temperature and Heat, Heat Capacity and Specific Heat

14. The average human diet contains about 2000 kcal per day. If all this food energy is released rather than being stored as fat, what is the approximate average power output of the human body?

15. If your mass is 60 kg, what is the minimum number of calories you would "burn off" climbing a 1700-m-high mountain? (The actual metabolic energy used would be much greater.)

16. Walking at 3 km/h requires an energy expenditure rate of about 200 W. How far would you have to walk to "burn off" a 300-kcal hamburger?

17. Typical fats contain about 9 kcal per gram. If the energy in body fat could be utilized with 100% efficiency, how much mass could a 78-kg person lose running a 26.2-mile marathon? The energy expenditure rate for that mass is 125 kcal/mile.

18. You expend 150 kcal while exercising for 5 minutes. If your mass is 60 kg and you're essentially water, by how much would your temperature increase if you had no heat loss?

19. A circular lake 1.0 km in diameter is 10 m deep (Fig. 19-25). Solar energy is incident on the lake at an average rate of 200 W/m². If the lake absorbs all this energy and does not exchange heat with its surroundings, how long will it take to warm from $10°C$ to $20°C$?

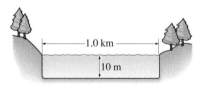

FIGURE 19-25   Problem 19.

20. To raise the temperature of 180 g of a substance by $15°C$, 1700 J of heat are required. (a) What is the heat capacity of this piece of the substance? (b) What is the specific heat of the substance?

21. How much heat is required to raise an 800-g copper pan from $15°C$ to $90°C$ if (a) the pan is empty; (b) the pan contains 1.0 kg of water; (c) the pan contains 4.0 kg of mercury?

22. Initially, 100 g of water and 100 g of another substance listed in Table 19-1 are at $20°C$. Heat is then transferred to each substance at the same rate for 1.0 min. At the end of that time the water is at $32°C$ and the other substance at $76°C$. (a) What is the other substance? (b) What is the heating rate?

23. How much power does it take to raise the temperature of a 1.3-kg copper pipe by $15°C/s$?

24. How long will it take a 625-W microwave oven to bring 250 mL of water from $10°C$ to the boiling point?

25. You insert your microwave oven's temperature probe in a roast and start it cooking. You notice that the temperature goes up $1°C$ every 20 s. If the roast has the same specific heat as water, and if the oven power is 500 W, what is the mass of the roast? Neglect heat loss.

26. Two neighbors return from Florida to find their houses at a frigid $15°F$. Each house has a furnace with heat output of 100,000 Btu/h. One house is made of stone and weighs 75 tons. The other is made of wood and weighs 15 tons. How long does it take each house to reach $65°F$? Neglect heat loss, and assume the entire house mass reaches the same $65°F$ temperature.

27. A stove burner supplies heat at the rate of 1.0 kW, a microwave oven at 625 W. You can heat water in the microwave in a paper cup of negligible heat capacity, but the stove requires a pan whose heat capacity is 1.4 kJ/K. (a) How much water do you need before it becomes quicker to heat on the stovetop? (b) What will be the rate at which the temperature of this much water rises?

28. When a nuclear power plant's reactor is shut down, radioactive decay continues to produce heat at about 10% of the reactor's normal power level of 3.0 GW. In a major accident, a pipe breaks and all the reactor cooling water is lost. The reactor is immediately shut down, the break sealed, and 420 m³ of $20°C$ water injected into the reactor. If the water were not actively cooled, how long would it take to reach its normal boiling point?

29. A 1.2-kg iron tea kettle sits on a 2.0-kW stove burner. If it takes 5.4 min to bring the kettle and the water in it from $20°C$ to the boiling point, how much water is in the kettle?

30. The volume of Lake Erie is 480 km³. If a 20-megaton nuclear bomb were exploded in the lake, with all its energy going into heat, what would be the average rise in the lake temperature? (Consult Appendix C for the conversion factor for megatons.)

31. Two cars collide head-on at 90 km/h. If all their kinetic energy ended up as heat, what would be the temperature increase of the wrecks? The specific heat of the cars is essentially that of iron.

32. A 1500-kg car moving at 40 km/h is brought to a sudden stop. If all the car's energy is dissipated in heating its four 5.0-kg steel brake disks, by how much do the disk temperatures increase?

33. A leaf absorbs sunlight with intensity 600 W/m². The leaf has a mass per unit area of 100 g/m², and its specific heat is 3800 J/kg·K. In the absence of any heat loss, at what rate would the leaf's temperature rise?

34. A child complains that her cocoa is too hot. The cocoa is at $90°C$. Her father pours 2 oz of milk at $3°C$ into the 6 oz

of cocoa. Assuming milk and cocoa have the same specific heat as water, what is the new temperature of the cocoa?

35. A piece of copper at 300°C is dropped into 1.0 kg of water at 20°C. If the equilibrium temperature is 25°C, what is the mass of the copper?

36. Vegetables with mass 2.2 kg, specific heat 0.92 cal/g·°C, and temperature 4.0°C are added to 4.5 kg of soup stock (essentially water) at 80°C. What is the equilibrium temperature?

37. A thermometer of mass 83.0 g is used to measure the temperature of a 150-g water sample. The thermometer's specific heat is 0.190 cal/g·°C, and it reads 20.0°C before immersion in the water. The water temperature is initially 60.0°C. What does the thermometer read after it comes to equilibrium with the water?

Sections 19-6 and 19-7 Heat Transfer and Thermal Energy Balance

38. Find the heat-loss rate through a 1.0-m² slab of (a) wood and (b) styrofoam, each 2.0 cm thick, if one surface is at 20°C and the other at 0°C.

39. A steel plate measures 20 cm by 20 cm by 5.0 mm thick. One face is maintained at 140°C. How much power would be needed to maintain the other face at 150°C?

40. How thick a concrete wall would be needed to give the same insulating value as $3\frac{1}{2}$ inches of fiberglass?

41. Building heat loss in the United States is usually expressed in Btu/h. What is 1 Btu/h in SI?

42. An 8.0 m × 12 m house is built on a concrete slab 23 cm thick. What is the heat-loss rate through the floor if the interior is at 20°C while the ground is at 10°C?

43. What is the R-factor of a wall that loses 0.040 Btu each hour through each square foot for each °F temperature difference?

44. Compute the R-factors for 1-inch thicknesses of air, concrete, fiberglass, glass, styrofoam, and wood.

45. A biology lab's walk-in cooler measures 3.0 m × 2.0 m × 2.3 m and is insulated with 8.0-cm-thick styrofoam. If the surrounding building is at 20°C, at what average rate must the cooler's refrigeration unit remove heat in order to maintain 4.0°C in the cooler?

46. One end of an iron rod 40 cm long and 3.0 cm in diameter is in ice water, the other in boiling water. (See Fig. 19-26.) The pipe is well insulated so there's no heat lost out the sides. What is the heat-flow rate along the rod?

47. (a) What is the R-factor for a wall consisting of $\frac{1}{4}$-in. pine paneling, $\Re$-11 fiberglass insulation, $\frac{3}{4}$-in. pine sheathing, and 2.0-mm aluminum siding? (b) What is the heat-loss rate through a 20 ft × 8 ft section of wall when the temperature difference across the wall is 55°F?

48. You're considering installing a 5 ft × 8 ft picture window in a north-facing wall that now has $\Re$-19 insulation. The

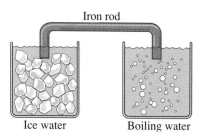

FIGURE 19-26 Problem 46.

window has $\Re$ = 2.1 ft²·°F·h/Btu. If you install the window, how much more oil, at 100,000 Btu per gallon, will you have to burn in a winter month when the outdoor temperature averages 15°F and the indoor temperature is 68°F?

49. Repeat the preceding problem for a south-facing window where the average sunlight intensity is 180 W/m².

50. A flat-roofed house measures 28 ft × 28 ft × 9 ft high and has $\Re$ = 30 for its roof and $\Re$ = 20 for its walls. The house next door has the same height, the same wall and roof construction, the same square shape, but twice the floor area. If the houses are kept at the same temperature, compare their heating bills. Neglect losses through the floors.

51. A house is insulated so its total heat loss is 370 W/°C. On a night when the outdoor temperature is 12°C the owner throws a party, and 40 people come. The average power output of the human body is 100 W. If there are no other heat sources in the house, what will the house temperature during the party?

52. A 1000-W electric clothes iron has surface area 300 cm² and emissivity e = 0.97. If the surface temperature of the iron is 500 K and its surroundings are at 300 K, (a) what is the *net* rate of energy transfer by radiation from the iron? (b) What must be the rate of energy transfer by conduction and convection?

53. An electric stove burner has surface area 325 cm² and emissivity e = 1.0. The burner is at 900 K and the electric power input to the burner is 1500 W. If room temperature is 300 K, what fraction of the burner's heat loss is by radiation?

54. An electric current passes through a metal strip 0.50 cm × 5.0 cm × 0.10 mm, heating it at the rate of 50 W. The strip has emissivity e = 1.0 and its surroundings are at 300 K. What will be the temperature of the strip if it's enclosed in (a) a vacuum bottle transparent to all radiation and (b) an insulating box with thermal resistance R = 8.0 K/W that blocks all radiation?

55. The average human body produces heat at the rate of 100 W and has total surface area of about 1.5 m². What is the coldest outdoor temperature in which a down sleeping bag with 4.0-cm loft (thickness) can be used without the body

temperature dropping below 37°C? Consider only conductive heat loss.

56. A crude way of characterizing the effect of atmospheric carbon dioxide is by reducing the emissivity e in Equation 19-12. If e dropped by 5%, what would happen to Earth's average temperature, now 287 K?

57. Scientists worry that a nuclear war could inject enough dust into the upper atmosphere to reduce significantly the amount of solar energy reaching Earth's surface. If an 8% reduction in solar input occurred, what would happen to Earth's 287-K average temperature?

58. Sunlight intensity varies with the inverse square of the distance from the Sun; at Earth's orbit the intensity is 1.37 kW/m². Use data from Appendix E to calculate the average temperatures of Mars and Venus if those planets' atmospheres contained no greenhouse gases, and treating each as a blackbody ($e = 1$). Compare with the measured surface temperatures of about 220 K and 730 K, respectively. Does either planet have a significant greenhouse effect?

Paired Problems

(Both problems in a pair involve the same principles and techniques. If you can get the first problem, you should be able to solve the second one.)

59. A blacksmith heats a 1.1-kg iron horseshoe to 550°C, then plunges it into a bucket containing 15 kg of water at 20°C. What is the final temperature?

60. A 2.3-kg cast-iron frying pan is plunged into 7.5 kg of water at 22°C. If the equilibrium temperature is 30°C, what was the pan's initial temperature?

61. What is the power output of a microwave oven that can heat 430 g of water from 20°C to the boiling point in 5.0 minutes? Neglect the heat capacity of the container.

62. A nuclear power plant is being started up after refueling. If it takes 1.5 hours to bring the reactor's 5.4×10^6 kg of cooling water from 10°C to its 350°C operating temperature, what is the reactor's thermal power output? Neglect the heat capacity of the reactor vessel and plumbing.

63. A cylindrical log 15 cm in diameter and 65 cm long is glowing red hot in a fireplace. If it's emitting radiation at the rate of 34 kW, what is its temperature? The log's emissivity is essentially 1.

64. A star whose surface temperature is 50 kK radiates 4.0×10^{27} W. If the star behaves as a blackbody, what is its radius?

65. An enclosed rabbit hutch has a thermal resistance of 0.25 K/W. If you put a 50-W heat lamp in the hutch on a day when the outside temperature is -15°C, what will be the hutch temperature? Neglect the rabbit's metabolism.

66. A home refrigerator has a thermal resistance of 0.090 K/W. What is the average rate at which it must transfer heat in order to maintain the interior at 4°C in a room where the temperature is 21°C?

Supplementary Problems

67. Rework Example 19-5, now assuming that the house has 10 single-glazed windows, each measuring 2.5 ft × 5.0 ft. Four of the windows are on the south, and admit solar energy at the average rate of 30 Btu/h·ft². *All* the windows lose heat; their R-factor is 0.90. (a) What is the total heating cost for the month? (b) How much is the solar gain worth?

68. What are the SI units of R-factor? Express the R-factor for 6-in fiberglass insulation ($\mathfrak{R} = 19$ ft²·°F·h/Btu) in SI.

69. My house currently burns 160 gallons of oil in a typical winter month when the outdoor temperature averages 15°F and the indoor temperature averages 66°F. Roof insulation consists of $\mathfrak{R} = 19$ fiberglass, and the roof area is 770 ft². If I double the thickness of the roof insulation, by what percentage will my heating bills drop? A gallon of oil yields about 100,000 Btu of heat.

70. A black woodstove with surface area 4.6 m² is made from cast iron 4.0 mm thick. The interior wall of the stove is at 650°C while the exterior is at 647°C. (a) What is the rate of heat conduction through the stove wall? (b) What is the rate of heat loss by radiation from the stove? (c) Use the results of (a) and (b) to find how much heat the stove loses by a combination of conduction and convection in the surrounding air.

71. A copper pan 1.5 mm thick and a cast iron pan 4.0 mm thick are sitting on electric stove burners; the bottom area of each pan is 300 cm². Each contains 2.0 kg of water whose temperature is rising at the rate of 0.15 K/s. Find the temperature difference between the inside and outside bottom of each pan.

72. What should be the average temperature on Pluto, at its 5.9×10^{12}-m distance from the Sun? Treat Pluto as a blackbody.

73. At low temperatures the specific heat of a solid is approximately proportional to the cube of the temperature; for copper the specific heat is given by $c = 31(T/343 \text{ K})^3$ J/g·K. When heat capacity is not constant, Equations 19-4 and 19-5 must be written in terms of the derivative dQ/dT and integrated to get the total heat involved in a temperature change. Find the heat required to bring a 40-g sample of copper from 10 K to 25 K.

74. When area or heat conductivity are not constant, Equation 19-7 must be written in terms of the derivative dT/dx:

$$H = -kA \frac{dT}{dx} \qquad (19\text{-}13)$$

and must be integrated to get the relation between the temperature difference and heat loss. Consider a truncated cone with faces of radii R_1 and R_2 and length ℓ, as shown in Fig. 19-27. If the temperatures on the faces are T_1 and T_2, respectively, and if insulation prevents heat loss out the sides, show that the heat flow rate is given by $H = \pi k R_1 R_2 (T_1 - T_2)/\ell$. *Hint:* Divide the cone into thin slabs and write the area of each in terms of its distance x from

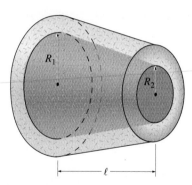

FIGURE 19-27 What is the heat flow rate through the cone? The insulation prevents heat loss out the sides (Problem 74).

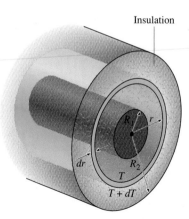

FIGURE 19-28 An insulated pipe. Bright cylinder is a layer of thickness dr for use in calculating the heat loss (Problem 75).

the left face. Use this expression for the area in Equation 19-13, multiply through by dx/A, and integrate, noting that H must be the same throughout the cone.

75. A pipe of length ℓ and radius R_1 is surrounded by insulation of outer radius R_2 and thermal conductivity k. Use the methods of the preceding problem to show that the heat loss rate through the insulation is

$$H = \frac{2\pi k \ell (T_1 - T_2)}{\ln(R_2/R_1)}.$$

Hint: Consider the heat flow through a thin layer of thickness dr and temperature difference dT as shown in Fig. 19-28.

76. An object of heat capacity C J/K is surrounded by insulation with thermal resistance R K/W. The object is initially at temperature T_1 and its surroundings are at the constant temperature T_0. (a) If there is no source of heat in the object, show that its temperature T as a function of time t is described by the equation

$$C\frac{dT}{dt} = -\frac{1}{R}(T - T_0).$$

(b) Show by substitution that the solution to this equation is

$$T = T_0 + (T_1 - T_0)e^{-t/RC}.$$

Show that this solution gives reasonable results for $t = 0$ and as $t \to \infty$.

77. A house is at 20°C on a winter day when the outdoor temperature is -15°C. Suddenly the furnace fails. Use the result of the previous problem to determine how long it will take the house temperature to reach the freezing point. The heat capacity of the house is 6.5 MJ/K, and its thermal resistance is 6.67 mK/W.

CHAPTER 20

THE THERMAL BEHAVIOR OF MATTER

Melting is one of the changes matter undergoes when its temperature increases. Here molten steel pours from a red-hot crucible.

Matter responds to heating in a variety of ways. It may get hotter or it may melt. It may experience changes in size or shape, or in pressure. In this chapter, we seek to understand the thermal behavior of matter. We start with a particularly simple state—the gaseous state—whose behavior can be explained by applying Newtonian mechanics at the molecular level. We then move to more complicated situations whose explanation is still grounded in the molecular properties of matter, but whose description is necessarily more empirical.

20-1 GASES

Gas is matter in a rarefied state. The rarefied nature of a gas results in weak interactions among its constituent particles, making its thermal behavior particularly simple. In studying this behavior, we will develop a clear understanding of the relation between macroscopic properties—like temperature and pressure—and the underlying microscopic properties of the constituent particles.

The Ideal Gas Law

In equilibrium, the macroscopic state of a gas is determined completely by its temperature, pressure, and volume. A simple system for studying gas behavior consists of a cylinder sealed by a movable piston (Fig. 20-1). If we maintain the gas at a constant temperature—by immersing the cylinder in a large reservoir of water, for example—and push slowly on the piston to decrease the gas volume, we find that the pressure rises in inverse proportion to the volume:

$$P \propto \frac{1}{V}. \qquad \text{(at fixed } T \text{)}$$

FIGURE 20-1 A piston-cylinder system.

This relation is **Boyle's law**, first described by the English scientist Robert Boyle in 1660. If we hold the volume constant and heat the gas, we find that the pressure is proportional to the absolute temperature:

$$P \propto T. \qquad \text{(at fixed } V \text{)}$$

Combining our two proportionalities, we have

$$P \propto \frac{T}{V}.$$

If we now hold T and V constant, but introduce more gas into the cylinder, we find that the pressure increases in proportion to the amount of gas. If N is the actual number of gas molecules, we can then write

$$P = \frac{NkT}{V},$$

or
$$PV = NkT, \qquad (20\text{-}1)$$

where k is a constant.

Equation 20-1 is the **ideal gas law.** The constant k is found experimentally to have essentially the same value for all gases:

$$k = 1.38 \times 10^{-23} \text{ J/K},$$

and is called **Boltzmann's constant,** after the Austrian physicist Ludwig Boltzmann (1844–1906), who was instrumental in developing the microscopic description of thermal phenomena.

Because the number of molecules, N, in a typical sample of gas is so large, we often express the ideal gas law in terms of the number of moles (mol) of gas. A mole of anything consists of Avogadro's number of that thing, where Avogadro's number is $N_A = 6.022 \times 10^{23}$. If there are n moles in a gas, then $N = nN_A$ is the number of molecules, so the ideal gas law becomes

$$PV = nN_A kT = nRT, \qquad (20\text{-}2)$$

where the constant $R = N_A k = 8.314$ J/K·mol is called the **universal gas constant.**

● **EXAMPLE 20-1**   STP

What is the volume occupied by 1.00 mol of an ideal gas at standard temperature and pressure (STP), where $T = 0°C$ and $P = 101.3$ kPa?

Solution

Solving the ideal gas law for the volume V, and expressing temperature in kelvins, we have

$$V = \frac{nRT}{P} = \frac{(1.00 \text{ mol})(8.314 \text{ J/K·mol})(273 \text{ K})}{1.013 \times 10^5 \text{ Pa}}$$

$$= 22.4 \times 10^{-3} \text{ m}^3 = 22.4 \text{ L}.$$

EXERCISE If the gas of Example 20-1 is cooled to $-120°C$ and compressed into a 2.0-L container, what will be its pressure?

Answer: 636 kPa

Some problems similar to Example 20-1: 3, 6, 8, 11

●

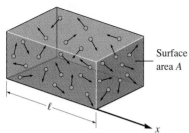

FIGURE 20-2 Gas molecules confined to a rectangular box.

The ideal gas law is remarkably simple. Neither the form of the law nor the constants k and R depend on the substance making up the gas or on the mass of the individual gas molecules. Yet most real gases follow the ideal gas law very closely over a wide range of pressures. This nearly ideal behavior is what gives gas thermometers their high precision over a wide temperature range.

Kinetic Theory of the Ideal Gas

Why do gases obey such a simple relation among temperature, pressure, and volume? Here we answer that question with a microscopic analysis of the ideal gas that is based ultimately in the laws of Newtonian mechanics.

To understand how matter in the gaseous state should behave, we make a number of simplifying assumptions:

1. The gas consists of a very large number of identical particles, each with mass m but having no size or internal structure. This assumption is approximately true for real gases provided the distance between molecules is large compared with their size.
2. The particles don't interact with each other. This assumption is approximately valid when potential-energy changes associated with intermolecular forces are small compared with the molecules' kinetic energies—as happens when the molecules are far apart.
3. The particles are moving in random directions with a range of speeds.
4. Collisions with the container walls are elastic, conserving the particles' momentum and energy. Here is where our gas model is tied to the laws of Newtonian mechanics.

Consider a gas confined inside a rectangular container (Fig. 20-2). Each time a gas molecule collides with a container wall, it exerts a force on the wall. There are so many molecules that individual collisions aren't evident on the macroscopic scale; instead the wall experiences a force that is essentially constant. The gas pressure is simply a measure of this force on a unit area.

To calculate the pressure, consider one molecule colliding with a wall (Fig. 20-3). Since the collision is elastic, the y component of the molecule's velocity remains unchanged, but the x component changes sign. Thus the molecule undergoes a change in momentum of magnitude $2mv_{xi}$, where v_{xi} is the magnitude of the x velocity of this particular molecule. Now suppose there are N_i molecules in the container whose x components of velocity have the same

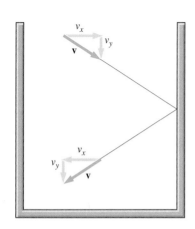

FIGURE 20-3 A molecule undergoes an elastic collision with the container wall, reversing its x component of velocity and transferring momentum $2mv_x$ to the wall.

magnitude v_{xi}. Because the molecules are moving in random directions, on the average half of them have x components of velocity in the $+x$ direction and half in the $-x$ direction. Then in the time it takes one of these N_i molecules to cross the entire container, all $N_i/2$ of them that are moving toward the wall will hit it. The time in which these $N_i/2$ collisions take place is

$$\Delta t = \frac{\ell}{v_{xi}},$$

where ℓ is the length of the container in the x direction. The change in momentum of all these molecules during this time is then

$$\Delta p_i = \frac{N_i}{2} 2mv_{xi} = N_i m v_{xi}.$$

By Newton's second law, the wall exerts a force whose average magnitude is equal to the average rate of change of the molecules' momentum:

$$F_i = \frac{\Delta p_i}{\Delta t} = \frac{N_i m v_{xi}}{\ell/v_{xi}} = \frac{N_i m v_{xi}^2}{\ell}.$$

By Newton's third law, this is also the magnitude of the average force exerted on the wall by the molecules whose x component of velocity is v_{xi}. The total force on the wall is the sum over all values of v_{xi}:

$$F = \sum F_i = \frac{\sum N_i m v_{xi}^2}{\ell}.$$

If the wall has area A, then the pressure, or force per unit area, is

$$P = \frac{F}{A} = \frac{\sum N_i m v_{xi}^2}{\ell A} = \frac{\sum N_i m v_{xi}^2}{V},$$

where $V = \ell A$ is the volume of the container. We can rewrite the pressure in a more meaningful way if we consider the average value of v_x^2 over all the molecules. This average, designated by $\overline{v_x^2}$, is obtained by summing the individual values v_{xi}^2, weighted by the number of molecules N_i with each value, and then dividing by N, the total number of molecules:

$$\overline{v_x^2} = \frac{\sum N_i v_{xi}^2}{N}.$$

Comparing this average with our expression for pressure shows that the pressure can be written

$$P = \frac{Nm\overline{v_x^2}}{V},$$

so

$$PV = Nm\overline{v_x^2} = 2N\left(\frac{1}{2}m\overline{v_x^2}\right).$$

We recognize the quantity $\frac{1}{2}m\overline{v_x^2}$ as the average kinetic energy per molecule associated with the motion of molecules in the x direction. In this equation we have the ideal gas law expressed in microscopic terms! Comparing with the macroscopic expression, $PV = NkT$, shows that

$$kT = 2(\tfrac{1}{2}m\overline{v_x^2}),$$

or

$$\tfrac{1}{2}m\overline{v_x^2} = \tfrac{1}{2}kT.$$

Because our molecules are moving in random directions, the average kinetic energy associated with motion in the x direction is the same as in the y and z directions. The average total kinetic energy of a molecule is thus three times the average kinetic energy associated with its motion in any one direction, or

$$\tfrac{1}{2}m\overline{v^2} = \tfrac{3}{2}kT. \tag{20-3}$$

Our derivation of Equation 20-3 shows us why, in terms of simple Newtonian mechanics, we should expect a gas to obey the ideal gas law. In Equation 20-3 we get an added bonus: a microscopic understanding of the meaning of temperature:

> **Temperature measures the average kinetic energy associated with random translational motion of the molecules.**

● EXAMPLE 20-2 MOLECULAR ENERGY

What is the average kinetic energy of a molecule in air at room temperature? What would be the speed of a nitrogen molecule with this energy?

Solution
The average kinetic energy is given by Equation 20-3, where room temperature is about 20°C (293 K):

$$\frac{1}{2}m\overline{v^2} = \frac{3}{2}kT = \frac{3}{2}(1.38\times10^{-23}\ \text{J/K})(293\ \text{K})$$
$$= 6.07\times10^{-21}\ \text{J}.$$

A nitrogen molecule consists of two nitrogen atoms (atomic mass 14 u; see Appendix D), so its mass is

$$m = 2(14\ \text{u})(1.66\times10^{-27}\ \text{kg/u}) = 4.65\times10^{-26}\ \text{kg}.$$

The kinetic energy is $K = \frac{1}{2}mv^2$, so

$$v = \sqrt{\frac{2K}{m}} = \sqrt{\frac{2(6.07\times10^{-21}\ \text{J})}{4.65\times10^{-26}\ \text{kg}}} = 511\ \text{m/s}.$$

Not surprisingly, this **thermal speed** is of the same order of magnitude as the sound speed ($\approx$340 m/s) in air at room temperature. At the microscopic level, the speed of the individual molecules sets an approximate upper limit on the maximum rate at which information can be transmitted by disturbances— that is, sound waves—propagating through the gas.

EXERCISE Find the thermal speed of hydrogen (H_2) at 300 K.

Answer: 1.9 km/s

Some problems similar to Example 20-2: 14–18 ●

The Distribution of Molecular Speeds

Our derivation of Equation 20-3 is fundamentally statistical. The macroscopic pressure is not associated with any one molecular collision, but only with huge numbers of collisions occurring so frequently, and each transferring so little momentum, that we sense only the average force and not the many individual impacts. Similarly, temperature doesn't tell us much about the energy of any individual molecule, but only about the average over all the molecules.

Sometimes, though, we want to know more than just these average quantities. For example, is the range of molecular speeds limited to a narrow band about the thermal speed $v_{th} = \sqrt{3kT/m}$ obtained from Equation 20-3? Or are there many molecules moving much faster than the thermal speed? Such fast molecules would be important in determining, for example, the chemical reaction rates in a gas mixture, or the tendency of an atmosphere to escape the gravitational field of its planet.

In the 1860s, the Scottish physicist James Clerk Maxwell considered the sharing of energy that must result from collisions among the molecules of a gas. Although we've neglected such collisions, their presence changes none of our earlier conclusions as long as the collisions are elastic; in fact, collisions are responsible for bringing a gas into thermodynamic equilibrium. Maxwell showed that molecular collisions result in a speed distribution where the number of molecules in a small speed range Δv about some speed v is given by

$$N(v)\,\Delta v = 4\pi N\left(\frac{m}{2\pi kT}\right)^{3/2} v^2 e^{-mv^2/2kT}\,\Delta v, \qquad (20\text{-}4)$$

where N is the total number of molecules in the gas, m the molecular mass, k Boltzmann's constant, and T the absolute temperature. The quantity $N(v)$ is the number of molecules per unit speed range; multiplying by a small speed range Δv gives $N(v)\Delta v$, the actual number of molecules in that range. Equation 20-4 is known as the **Maxwell-Boltzmann distribution.** Figure 20-4 shows plots of this distribution for the same gas sample at two different temperatures. Each curve exhibits a single peak; this is the most probable value for a molecule's speed. Because the curve is not symmetric about the peak, the most probable speed lies below the mean thermal speed (see Problem 74). Figure 20-4 shows what Equation 20-3 already told us: that an increase in temperature corresponds

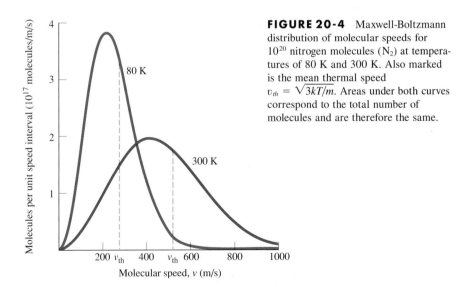

FIGURE 20-4 Maxwell-Boltzmann distribution of molecular speeds for 10^{20} nitrogen molecules (N_2) at temperatures of 80 K and 300 K. Also marked is the mean thermal speed $v_{th} = \sqrt{3kT/m}$. Areas under both curves correspond to the total number of molecules and are therefore the same.

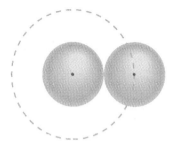

FIGURE 20-5 Two spherical gas molecules cannot get closer than a center-to-center separation of twice their radius. This effect reduces the total volume available to the molecules.

to an increase in mean thermal speed. But it also shows something averages alone could not reveal: that an increase in temperature results in a broader speed distribution, with relatively fewer molecules near the most probable speed.

The Maxwell-Boltzmann distribution shows why chemical reaction rates increase dramatically with temperature; it is the most energetic molecules, of which there are more at high temperatures, that are largely responsible for sustaining chemical reactions. A related phenomenon explains evaporative cooling in liquids; molecules in the high-energy "tail" of the molecular speed distribution can escape the liquid, leaving lower-speed molecules—and therefore cooler liquid—behind.

Real Gases

Though the ideal gas law is a good approximation to the behavior of real gases, the assumptions leading to its derivation are not entirely realistic. Real gases therefore exhibit deviations from ideal behavior.

Our derivation of the ideal gas law assumed gas molecules are noninteracting point particles. But real molecules take up space, and they do collide. Figure 20-5 shows that each molecule prevents others from approaching closer than one molecular radius, thereby reducing the total volume available to the molecules. When molecules are close, furthermore, electrical effects that we'll analyze in Chapter 25 give rise to a weak attractive force called the **van der Waals force.** As molecules move apart they do work to overcome the van der Waals force, and as a result the molecular kinetic energy drops. Correcting for both the nonzero molecular size and the effect of the van der Waals force gives the **van der Waals equation,** which we state without further derivation:

$$\left(P + \frac{n^2 a}{V^2}\right)(V - nb) = nRT, \tag{20-5}$$

where a and b are constants that depend on the particular gas. For low particle densities n/V, both correction terms in the van der Waals equation become negligible, showing that a rarefied gas closely follows the ideal gas law.

● EXAMPLE 20-3 VAN DER WAALS EFFECTS

The constants a and b in the van der Waals equation for nitrogen have the values $a = 0.14$ Pa·m^6/mol^2, $b = 3.91 \times 10^{-5}$ m^3/mol. If 1.000 mol of nitrogen is confined to a volume of 2.000 L, and is at pressure of 10.00 atm, by how much does its temperature as predicted by the van der Waals equation differ from the ideal gas prediction?

Solution
Converting to SI units, we have $P = 1.013 \times 10^6$ Pa and $V = 2.000 \times 10^{-3}$ m^3. Then the ideal gas law predicts a temperature

$$T_{\text{ideal}} = \frac{PV}{nR} = \frac{(1.013 \times 10^6 \text{ Pa})(2.000 \times 10^{-3} \text{ m}^3)}{(1.000 \text{ mol})(8.314 \text{ J/K·mol})} = 244 \text{ K}.$$

The van der Waals equation predicts a temperature

$$T_{\text{van}} = \frac{(P + n^2 a/V^2)(V - nb)}{nR} = 247 \text{ K},$$

Using the given values of a, b, n, P, and V. Under these conditions, nitrogen deviates from ideal gas behavior by only about 1%.

EXERCISE If the gas volume in Example 20-3 is halved while the pressure is held constant, by what percentage will the ideal and van der Waals temperature calculations differ?

Answer: 9.4%

Some problems similar to Example 20-3: 19–21 ●

20-2 PHASE CHANGES

At high density, the van der Waals equation provides a much better description of gas behavior than the ideal gas law. But at high enough density or low enough temperature, intermolecular forces become dominant and lock the molecules so tightly together that the gas condenses into a liquid or even a solid. These different states of matter are called **phases.** A full description of **phase changes** that substances undergo is well beyond the scope of this book, but we can gain some understanding by examining the relation of pressure, volume, and temperature over a wide range of values.

PVT Diagrams

A graph of pressure versus volume at fixed temperature is called an **isotherm.** In an ideal gas at fixed temperature the pressure is inversely proportional to the volume, and the isotherms are therefore hyperbolas. Figure 20-6 is a *PV* **diagram** showing some isotherms. As the temperature is lowered the gas volume drops and van der Waals effects become important, altering the shape of the isotherms. Eventually we reach a **critical temperature** T_C at which even the van der Waals description fails. Here the *PV* diagram exhibits a most unusual behavior: the pressure versus volume curves become flat over a range of volumes.

What's going on here? One clue comes from the left side of the *PV* curves for temperatures below T_C. These curves are almost vertical, indicating that volume hardly changes with changing pressure. The substance is virtually incompressible—it's become a liquid. As we drop down one of these steep isotherms the volume is essentially constant until we reach the point marked *A* where the isotherm becomes horizontal. At this point the liquid begins turning to gas and the volume increases while the pressure remains constant. Finally, at point *B*, the liquid is gone, and the isotherm shows the compressibility characteristic of the gas phase. Figure 20-7 is a redrawn version of Fig. 20-6 that emphasizes the division between liquid and gas phases.

The horizontal portion of an isotherm in Fig. 20-7 is a region where liquid and gas can coexist in equilibrium. The available volume determines the mix of liquid and gas, from all liquid at point *A* to all gas at point *B*. Given an arbitrarily large volume in which to expand, a liquid will turn rapidly to gas in the process we call boiling. Under normal atmospheric pressure the boiling point for water is 100°C, but water boils at different temperatures if the pressure is different (Fig. 20-8).

That we see a clear distinction between liquids and gases is partly a matter of coincidence arising from the particular value of Earth's atmospheric pressure. As Fig. 20-7 shows, only isotherms below the critical temperature exhibit a region of liquid-gas equilibrium. Above the critical temperature the volume drops and density increases smoothly as we move up an isotherm. In this region gas and liquid are essentially indistinguishable. At the top of the liquid-gas equilibrium region is the **critical point;** if we raise the pressure enough to bring a liquid-gas mixture to this point the distinction between liquid and gas simply disappears (Fig. 20-9).

Including the solid phase in our *PV* diagram would make it extremely complicated. Instead, we resort to a three-dimensional *PVT* surface with pressure, volume, and temperature along the three axes (Fig. 20-10). The shape of the *PVT* surface characterizes all possible relations among these three quantities.

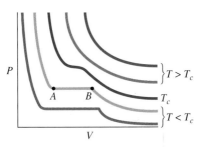

FIGURE 20-6 Isotherms in a *PV* diagram. Curve at upper right corresponds to the highest temperature; here the gas behaves ideally, with pressure and volume inversely proportional. Deviations from ideal behavior occur at lower temperatures. Below the critical-temperature isotherm the substance exhibits both liquid and vapor phases; along the horizontal portions of the lower isotherms the pressure and temperature remain constant as the liquid turns to vapor.

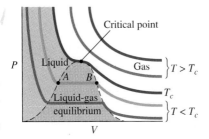

FIGURE 20-7 *PV* diagram showing boundaries between liquid and gas phases.

FIGURE 20-8 Water boils below 100°C under the reduced pressure atop Hawaii's Mauna Kea.

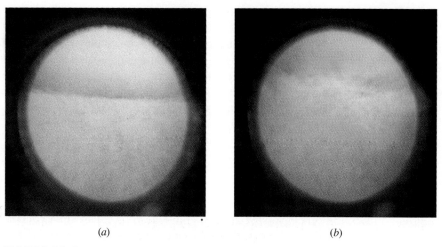

(a)

(b)

FIGURE 20-9 (a) A fluid below its critical point shows a clear demarcation between liquid and gas. (b) At the critical point, the distinction disappears.

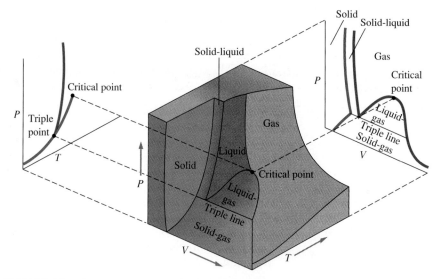

FIGURE 20-10 A *PVT* surface for a substance that expands as it melts. Also shown are projections onto the *PV* plane—giving a *PV* diagram analogous to Fig. 20-7—and onto the *PT* plane, giving a *PT*, or phase, diagram.

FIGURE 20-11 Carbon dioxide subliming from solid to gas. The gray mist is actually a cloud of water droplets that condensed on contact with the cool CO_2 gas.

Analysis of the *PVT* surface shows not only the region of liquid-gas equilibrium we found in Fig. 20-7, but also regions of solid-gas and solid-liquid equilibrium. At a particular temperature and pressure—comprising the **triple point**—all three phases can coexist.

Our everyday experience suggests that heating brings a solid first to its melting point and then eventually to boiling, carrying the substance through all three phases. The *PVT* diagram shows that this sequence doesn't always occur; below the triple-point pressure the solid turns directly into gas. At atmospheric pressure, for example, carbon dioxide never enters the liquid phase (Fig. 20-11). Only under higher pressure, as in a fire extinguisher, does CO_2 liquify.

Relations among the various phases are summarized in a **phase diagram**, which is projection of the *PVT* surface onto the *PT* plane (Fig. 20-12). In this two-dimensional picture we lose information about volume changes but show more clearly the temperature and pressure regimes where the different phases exist.

Our *PVT* surface and phase diagram are for a particularly simple substance. Many substances exhibit more complicated diagrams associated with subphases of liquid and solid that reflect changes in crystalline structure and molecular bonding. Materials scientists and engineers exploit these phase changes in a variety of applications.

Heat and Phase Changes

Once we have heated a solid to its melting point, we can continue to add heat without any corresponding change in temperature until the solid has entirely liquefied. The same thing happens at the boiling point: At a pressure of 1 atm, a pot of boiling water stays at 100°C until all the water is gone.

What's happening to the heat we add during these phase changes? On a molecular level, the energy goes into breaking the bonds that hold molecules in the tight configurations of solid or liquid. During a phase change, the potential energy of molecular separation changes, rather than the kinetic energy that's associated with temperature. Macroscopically, we characterize phase changes by the energy required to melt or vaporize a given amount of material. We call this quantity the **heat of fusion, L_f,** or **heat of vaporization, L_v.** Both are **heats of transformation.** If we determine that a mass m of a given substance requires heat Q to change completely from one phase to another with no temperature change, then the heat of transformation is given by:

$$L = \frac{Q}{m}. \tag{20-6}$$

To reverse the phase change—for example, to freeze water—we must remove the energy implied by Equation 20-6. In addition to the heats of fusion and vaporization, we specify a heat of sublimation when a substance goes directly from the solid to the vapor state. Table 20-1 lists heats of transformation for some common materials at atmospheric pressure.

Heat transfers associated with phase changes are typically quite large. The heat of fusion of water, for example, is 334 kJ/kg or 80 cal/g—meaning that it takes as much energy to melt one gram of ice as it does to raise the resulting water from 0°C to 80°C.

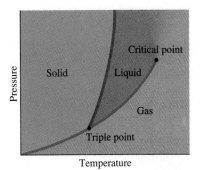

FIGURE 20-12 A phase diagram is the projection of a *PVT* surface onto the *PT* plane. Solid lines mark phase changes; note that the liquid-gas distinction ends at the critical point.

▲ **TABLE 20-1** HEATS OF TRANSFORMATION (AT ATMOSPHERIC PRESSURE)

SUBSTANCE	MELTING POINT, K	L_f, kJ/kg	BOILING POINT, K	L_v, kJ/kg
Alcohol, ethyl	159	109	351	879
Copper	1357	205	2840	4726
Lead	601	24.7	2013	858
Mercury	234	11.3	630	296
Oxygen	54.8	13.8	90.2	213
Sulfur	388	38.5	718	287
Water	273	334	373	2257
Uranium	1406	82.8	4091	1875

● EXAMPLE 20-4 MELTDOWN

In a serious accident, a nuclear power plant's reactor vessel cracks, and all the cooling water drains out (Fig. 20-13). Although nuclear fission stops, radioactive decay continues to heat the reactor's 2.5×10^5-kg uranium core at the rate of 120 MW. Once the core reaches its melting point, how much energy will it take to melt the core? How long will the melting take?

Solution

Using the heat of fusion from Table 20-1 in Equation 20-6 gives

$$Q = mL_f = (2.5 \times 10^5 \text{ kg})(82.8 \text{ kJ/kg}) = 2.07 \times 10^{10} \text{ J}.$$

At a heating rate of 120 MW ($= 1.2 \times 10^8$ J/s), the melting will take

$$\frac{2.07 \times 10^{10} \text{ J}}{1.2 \times 10^8 \text{ J/s}} = 172 \text{ s},$$

or less than 3 minutes! Failsafe emergency cooling systems are essential in preventing nuclear meltdowns.

EXERCISE (a) Find the energy needed to melt a 500-g block of lead at its melting point. (b) If the energy is supplied with a 1.1-kW torch, how long will the melting take?

Answers: (a) 12.4 kJ; (b) 11 s

Some problems similar to Example 20-4: 30, 32, 34–37

FIGURE 20-13 Workers practice picking up radioactive debris in a full-scale model of the damaged Three Mile Island nuclear reactor. The TMI reactor experienced a partial meltdown, although its pressure vessel did not crack. ●

The heats of transformation measure only the energy associated with phase changes, during which temperature doesn't change. To calculate the heat needed to change first the temperature and then the phase, we need to consider both the specific heat and the heat of transformation:

● EXAMPLE 20-5 MELTDOWN IN THE KITCHEN?

You put 650 g of water at 20°C in a pan on a 1.5-kW stove burner and forget about it. How long will it take to boil the pan dry, assuming all the burner's energy goes into the water?

Solution

First we calculate the energy needed to raise the water to its boiling point. Using the appropriate specific heat from Table 19-1 in Equation 19-4 gives

$$Q_1 = mc\,\Delta T = (0.65 \text{ kg})(4.184 \text{ kJ/kg·K})(80 \text{ K}) = 218 \text{ kJ}.$$

Equation 20-6, with data from Table 20-1, then gives the energy needed to transform all the water to the gaseous state:

$$Q_2 = mL_v = (0.65 \text{ kg})(2257 \text{ kJ/kg}) = 1467 \text{ kJ}.$$

So total energy needed is 218 kJ + 1467 kJ = 1685 kJ; at a heating rate of 1.5 kW, this takes 1685 kJ/1.5 kJ/s = 1123 s, or about 19 minutes.

> **TIP Break a Problem into Separate Parts** To work this example required solving two separate problems, one about temperature change and one about boiling. You'll often need to break a complex problem into several simpler ones, as this example suggests.

EXERCISE You put 100 g of ice at 0°C in a 625-W microwave oven. How long will it take until you have water at 90°C?

Answer: 114 s

Some problems similar to Example 20-5: 39–44 ●

● **EXAMPLE 20-6** ENOUGH ICE?

200 g of ice at $-10°C$ is added to 1.0 kg of water at $15°C$ in an insulated container. Is there enough ice to cool the water to $0°C$? If so, how much ice and water are present once equilibrium is reached?

Solution
First we determine whether there is enough ice. To bring the ice to its melting point and melt all of it requires

$$Q_1 = mc\,\Delta T + mL_f$$
$$= (0.20 \text{ kg})(2.05 \text{ kJ/kg·K})(10 \text{ K}) + (0.20 \text{ kg})(334 \text{ kJ/kg})$$
$$= 4.1 \text{ kJ} + 66.8 \text{ kJ} = 70.9 \text{ kJ},$$

where we used data from Tables 19-1 and 20-1, and where the temperature change in °C is equal to its value in kelvins. (Here we've algebraically combined the two subproblems mentioned in the tip in Example 20-5.)

Cooling the water to $0°C$ extracts an amount of heat given by

$$Q_2 = mc\,\Delta T = (1.0 \text{ kg})(4.184 \text{ kJ/kg·K})(15 \text{ K}) = 62.8 \text{ kJ}.$$

This is far more than the 4.1 kJ needed to bring the ice to $0°C$, but not quite the 70.9 kJ needed to leave it all melted. So there's enough ice to cool the water to $0°C$, with some left over. How much? Our calculation of Q_1 shows that 4.1 kJ go into raising the ice temperature. Of the 62.8 kJ extracted from the water, the remaining 58.7 kJ go to melting ice. From Equation 20-6, the amount of ice melted is then

$$m = \frac{Q}{L_f} = \frac{58.7 \text{ kJ}}{334 \text{ kJ/kg}} = 0.176 \text{ kg} = 176 \text{ g}.$$

Thus we're left with 24 g of ice in 1176 g of water, all at $0°C$.

EXERCISE Determine the final mix and temperature in Example 20-6 if the initial amount of ice is 100 g.

Answer: all water at $5.9°C$

Some problems similar to Example 20-6: 45–49, 51   ●

20-3 THERMAL EXPANSION

We have explored changes of temperature and phase that occur when matter is heated. But our *PVT* diagrams show that matter may also change in volume or pressure when heated. When an ideal gas is held at constant pressure, for example, its volume is directly proportional to its temperature.

The volume and pressure relations in the liquid and solid phases are not quite so simple. Liquids and solids are far less compressible than gases, so thermal expansion is less pronounced. This low compressibility is reflected in the steepness of the *PVT* surface in the regions of pure liquid or solid. On the microscopic level, the molecules in a liquid or solid are closely spaced; to move them requires that we do work against large (electrical) forces.

We can characterize the change in the volume of a substance with temperature in terms of its **coefficient of volume expansion,** β, defined as the fractional change in volume when the substance undergoes a small temperature change ΔT:

$$\beta = \frac{\Delta V/V}{\Delta T}. \tag{20-7}$$

This equation assumes that β is independent of temperature; if it varies significantly then we would need to define β in terms of the derivative dV/dT (see Problem 78). Our definition of β also assumes constant pressure; we could entirely inhibit thermal expansion with appropriate pressure increases.

Often we want to know how one linear dimension of a solid changes with temperature. This is especially true with long, rodlike structures where the absolute change is greatest along the long dimension (Fig. 20-14). We then speak of the **coefficient of linear expansion,** α, defined by

(a)

(b)

FIGURE 20-14 A power line sags as its length increases on a hot summer day.

FIGURE 20-15 How will the size of the hole change when the object expands?

▲ **TABLE 20-2** EXPANSION COEFFICIENTS*

SOLIDS	$\alpha(\text{K}^{-1})$	LIQUIDS AND GASES	$\beta(\text{K}^{-1})$
Aluminum	24×10^{-6}	Air	3.7×10^{-3}
Brass	19×10^{-6}	Alcohol, ethyl	75×10^{-5}
Copper	17×10^{-6}	Gasoline	95×10^{-5}
Glass (Pyrex)	3.2×10^{-6}	Mercury	18×10^{-5}
Ice	51×10^{-6}	Water, 1°C	-4.8×10^{-5}
Invar†	0.9×10^{-6}	Water, 20°C	20×10^{-5}
Steel	12×10^{-6}	Water, 50°C	50×10^{-5}

*At approximately room temperature unless noted.

† Invar, consisting of 64% iron and 36% nickel, is an alloy designed for use where thermal expansion must be minimized.

$$\alpha = \frac{\Delta L / L}{\Delta T}. \tag{20-8}$$

The volume expansion coefficient and linear expansion coefficient α are related in a simple way:

$$\beta \doteq 3\alpha, \tag{20-9}$$

as you can show in Problem 81. This relation means that either of these coefficients fully characterizes the thermal expansion of a material. However, the linear expansion coefficient α is really only meaningful with solids, because liquids and gases deform readily and therefore do not generally expand proportionately in all directions. Table 20-2 lists expansion coefficients for some common substances.

● **EXAMPLE 20-7** THERMAL EXPANSION

A steel girder is 15 m long at −20°C. It is used in the construction of a building where temperature extremes of −20°C to 40°C are expected. By how much does the girder length change between these extremes?

Solution

Table 20-2 gives the coefficient of linear expansion for steel: $\alpha = 12 \times 10^{-6}$ K^{-1}. Equation 20-8 then gives

$\Delta L = \alpha L \Delta T = (12 \times 10^{-6}$ K$^{-1})(15$ m$)(60$ K$) = 1.1$ cm.

EXERCISE A telescope assembly includes an aluminum rod 1.3 m long at 10°C. What is the maximum allowable temperature if the rod length is not to increase by more than 0.50 mm?

Answer: 26°C

Some problems similar to Example 20-7: 53–56 ●

What happens when a hollow object expands? In Fig. 20–15, you might expect that the hole would shrink as material expands into it. Actually, as Fig. 20–16 shows, *every* linear dimension expands in the same proportion, causing the interior space to expand, too.

FIGURE 20-16 A hollow solid made of rectangular slabs. As the slabs expand, so does the interior space.

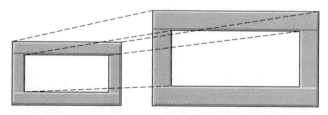

● **EXAMPLE 20-8** SPILLED GASOLINE

A car's gasoline tank is made of steel and has internal dimensions 60 cm × 60 cm × 20 cm at 10°C. It is filled with gasoline at 10°C. If the temperature now increases to 30°C, by how much does the volume of the tank increase? How much gasoline spills out of the tank?

Solution
The initial volume of the tank is

$$V = (0.60 \text{ m})(0.60 \text{ m})(0.20 \text{ m}) = 0.072 \text{ m}^3.$$

Equation 20-7 then gives the volume increases:

$$\Delta V = \beta V \Delta T,$$

so $\quad \Delta V_{\text{tank}} = (3)(12 \times 10^{-6} \text{ K}^{-1})(0.072 \text{ m}^3)(20 \text{ K})$
$$= 5.18 \times 10^{-5} \text{ m}^3,$$

and

$$\Delta V_{\text{gas}} = (95 \times 10^{-5} \text{ K}^{-1})(0.072 \text{ m}^3)(20 \text{ K}) = 1.37 \times 10^{-3} \text{ m}^3.$$

Since Table 20-2 gives the coefficient of linear expansion, α, for solids, we used $\beta = 3\alpha$ in calculating the volume change of the tank.

The expansion of the steel tank is negligible compared with that of the gasoline, so we lose 1.32×10^{-3} m³, or 1.32 L of gasoline. Don't fill your gas tank to the very top!

EXERCISE A farmer milks a cow into a 20-L steel milk pail; the milk comes from the cow at approximately 37°C. If the pail is initially full and also at 37°C, how much empty space will there be when it's been cooled to 3°C? The average thermal expansion coefficient for milk in this temperature range is 2.0×10^{-4} K^{-1}.

Answer: 112 mL

Some problems similar to Example 20-8: 60, 62, 76 ●

Thermal Expansion of Water

The entry for water at 1°C in Table 20-2 is remarkable, for the negative expansion coefficient shows that water at this temperature actually contracts on heating. Figure 20-17 shows the volume occupied by one gram of water as a function of temperature. From 0°C to 4.0°C, the volume decreases with increasing temperature, indicating a negative expansion coefficient. This curious behavior occurs only near the melting point, and is related to another anomalous property of water—the solid is less dense than the liquid. This unusual situation occurs because the structure of the ice crystal prevents H₂O molecules in the solid state from coming as close as they do in the liquid state (Fig. 20-18). Just above the melting point, the same intermolecular forces that give the ice crystal its structure still influence the molecules strongly enough to cause the liquid to contract with increasing temperature. Finally, above 4.0°C, thermal motion of the

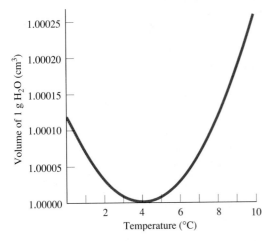

FIGURE 20-17 Volume of one gram of water near its melting point. Below 4°C water actually expands when cooled.

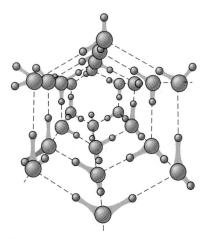

FIGURE 20-18 Water molecules in an ice crystal form an open structure, giving solid water a lower density than the liquid state.

molecules becomes more important than these intermolecular forces, and the liquid exhibits normal thermal expansion with increasing temperature.

The anomalous behavior of water has important consequences for aquatic life. Because water occupies the least volume at 4.0°C, water at this temperature sinks to the bottom of freshwater lakes. For lakes deeper than a few meters, sunlight is insufficient to change this temperature, so the deep water remains at 4.0°C year-round. In winter, the surface waters cool below 4.0°C and eventually freeze, forming an insulating layer of ice that, because of its lower density, floats on the surface. As a result, ice cover in temperate climates rarely exceeds a meter or so, and aquatic life can continue below the ice layer. If water behaved like most other liquids, the coldest water would be at the bottom and lakes would freeze from the bottom up, destroying most aquatic life.

Thermal Stresses

By enclosing a liquid or gas in a rigid container, or by tightly clamping the ends of a solid in place, we can inhibit thermal expansion. Following its PVT diagram, the material must then experience a pressure increase. If a container is unable to withstand this increase, it will burst (Fig. 20-19). If a solid is restrained along only one dimension, it will deform and perhaps crack (Fig. 20-20). Similarly, when a solid is clamped at both ends and cooled, tension forces develop that may cause it to break. Finally, if a solid is heated or cooled unevenly, part of the material expands or contracts at a different rate from the surrounding material. Again, the resulting forces may damage the material.

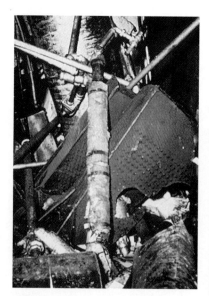

FIGURE 20-19 Industrial disaster! This boiler burst from overpressure caused by thermal expansion.

FIGURE 20-20 Thermal expansion distorted these railroad tracks, causing a derailment.

CHAPTER SYNOPSIS

Summary

1. An **ideal gas** is described by the **ideal gas law,** a relation among pressure, temperature, volume, and amount of gas.

$$PV = NkT.$$

Here N is the number of gas molecules and $k = 1.38 \times 10^{-23}$ J/K is **Boltzmann's constant.** The ideal gas law may also be written

$$PV = nRT,$$

where n is the number of moles and $R = 8.314$ J/K·mol is the **universal gas constant.**

2. **Kinetic theory** provides a description of the ideal gas in terms of the microscopic physics of large numbers of molecules. Kinetic theory relates temperature to the average translational kinetic energy of the gas molecules:

$$\tfrac{1}{2}m\overline{v^2} = \tfrac{3}{2}kT.$$

3. **Real gases** deviate from ideal behavior at high densities, where molecular size and intermolecular forces become significant. Real gases are described approximately by the **van der Waals equation:**

$$\left(P + \frac{n^2a}{V^2}\right)(V - nb) = nRT,$$

where a and b are constants describing a particular gas.

4. Most substances can exist in several **phases,** including gas, liquid, and solid. The relations among temperature, pressure, and volume in these phases are described by a **PVT surface.** Projection of the PVT surface onto the PT plane gives the **phase diagram.** Important points on this diagram are the **triple point,** where the three phases can coexist in equilibrium, and the **critical point,** where the boundary between liquid and gas phases ceases to be distinct.

5. During a phase change, energy is transferred to or from a substance without an accompanying temperature change. The energy per unit mass associated with a phase change is the **heat of transformation.** The **heat of fusion** describes the solid-liquid transition, the **heat of vaporization** describes the liquid-gas transition, and the **heat of sublimation** describes the solid-gas transition.

6. Most substances expand when heated at constant pressure. The change in volume is given by

$$\Delta V = \beta V \Delta T,$$

where β is the **coefficient of volume expansion.** The change in length of a solid is given by a similar expression:

$$\Delta L = \alpha L \Delta T,$$

where $\alpha = \beta/3$ is the **coefficient of linear expansion.** Water provides an important exception to the general rule that substances expand when heated: In the range from 0°C to 4°C, water contracts on heating.

7. If a substance is held rigidly, so that thermal expansion or contraction cannot occur, then **thermal stresses** develop that may cause damage.

Terms You Should Understand

(Pairs are closely related terms whose distinction is important; number in parentheses is chapter section where term first appears.)

ideal gas (20-1)
ideal gas law (20-1)
Boltzmann's constant (20-1)
universal gas constant (20-1)
Maxwell-Boltzmann distribution (20-1)
van der Waals equation (20-1)
phase, phase change (20-2)
isotherm (20-2)
PV diagram, PVT surface (20-2)
critical temperature, critical point (20-2)
triple point (20-2)
phase diagram (20-2)
heat of transformation; heats of fusion and vaporization (20-2)
coefficients of volume and linear expansion (20-3)

Symbols You Should Recognize

N (20-1)
k (20-1)
N_A (20-1)
R (20-1)
$\overline{v^2}$ (20-1)
L, L_f, L_v (20-2)
α, β (20-3)

Problems You Should Be Able to Solve

evaluating ideal gas properties (20-1)
relating temperature and molecular kinetic energy (20-1)
evaluating gas properties in the van der Waals approximation (20-1)
analyzing PVT diagrams (20-2)
determining energy involved in phase changes (20-2)
calculating thermal expansion (20-3)

Limitations to Keep in Mind

The ideal gas law is an approximation valid for rarefied gases where molecular size and intermolecular forces are not important. For most gases under normal conditions the law provides an excellent approximation.

Despite our everyday experience, the distinction between liquid and gas is not always clearcut. Similarly, the temperature regimes in which the different phases can exist depend on pressure and may sometimes run counter to common experience.

Equations 20-7 and 20-8 are strictly valid only for temperature changes small enough that the expansion coefficients β and α remain essentially constant.

QUESTIONS

1. If the volume of an ideal gas is increased, must the pressure drop proportionately? Explain.
2. According to the ideal gas law, what should the volume of a gas be at absolute zero? Why is this result absurd?
3. Why are you supposed to check the pressure in a tire when the tire is cold?
4. The average *speed* of the molecules in a gas increases with increasing temperature. What about the average *velocity*?
5. Suppose you start running while holding a jar of air. Do you change the average speed of the air molecules? The average velocity? The temperature?
6. Gas thermometers containing different gases at relatively high pressures do not agree exactly at all temperatures, even if they were all calibrated at the triple point of water. Why not?
7. Do all molecules in a gas have the same speed? If so, how does this come about? If not, how can the notion of temperature be meaningful?
8. The speed of sound in a gas is closely related to the thermal speed of the molecules. Why?
9. Two different gases are at the same temperature, and both are at low enough densities that they behave like ideal gases. Do their molecules have the same thermal speeds? Explain.
10. The atmosphere of a small planet such as Earth contains very little of the lighter gases like hydrogen and helium. The massive planet Jupiter has an atmosphere rich in light gases. Why might this be?
11. Is the van der Waals force attractive or repulsive? Justify your answer in terms of the sign of the term $n^2 a/V^2$ appearing in the van der Waals equation, assuming that the parameter a is positive.
12. Some people think that ice and snow must be at 0°C. Is this always true? Under what circumstances must it be true? Could you paraphrase the same arguments for another substance? Try steel, for example.
13. What is the temperature of water just under the ice layer of a frozen lake? At the bottom of the lake?
14. Deep lakes usually "turn over" twice each year. During the overturn, water from the lake bottom gets mixed with surface water, while at other times deep water and surface water do not mix. Under what conditions should such an overturn occur?
15. How is it possible to have liquid water at 0°C?

16. Some ice and water have been together in a glass for a long time. Is the water hotter than the ice?
17. Does it take more heat to melt a gram of ice at 0°C than to bring the resulting water to the boiling point? Once at the boiling point, does it take more heat to boil all the water than it did to bring it from the melting point to the boiling point?
18. Table 20-1 suggests that substances that are gases at room temperature and atmospheric pressure have relatively low heats of vaporization. Why might this be?
19. Why do we use the triple point of water for thermometer calibration? Why not just use the melting point or boiling point?
20. When the average temperature drops below freezing in winter, regions near large lakes remain warmer for some time. Why?
21. Must all substances exhibit solid, liquid, and gas phases? What might happen before a substance reaches the gas phase? Consider sugar, for example.
22. What would you do to air to liquefy it in such a way that the gas-liquid transition was obvious? In such a way that it was not obvious?
23. How is it possible to have boiling water at a temperature other than 100°C?
24. Why does the water in a car radiator boil explosively when you remove the radiator cap?
25. How does a pressure cooker work?
26. Why does a double boiler prevent food from burning?
27. Figure 20-21 shows a thin, flexible wire draped over an ice cube, with weights tied on either end of the wire. If the weights are heavy enough, the wire will gradually melt its

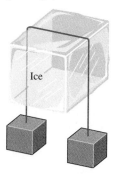

FIGURE 20-21 Question 27.

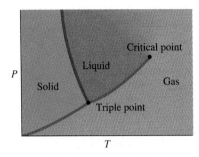

FIGURE 20-22   Question 27. Phase diagram for water. Compare with Fig. 20-23.

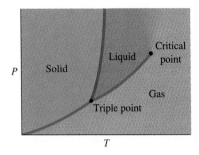

FIGURE 20-23   Question 28. Phase diagram for CO_2.

way through the ice cube, nevertheless leaving the cube in one piece. Explain this phenomenon in terms of the phase diagram for water (Fig. 20-22).

28. Figure 20-23 shows a phase diagram for carbon dioxide. Could you perform the experiment in Fig. 20-21 on a dry ice cube? Explain by comparing Figs. 20-22 and 20-23.

29. Which of Figures 20-22 and 20-23 is a more typical phase diagram? Explain.

30. Earth's inner core is solid and is surrounded by a liquid outer core made of the same material but cooler. How is this possible?

31. Suppose mercury and glass had the same coefficient of volume expansion. Could you build a mercury thermometer?

32. Why are the coefficients of volume expansion generally greater for liquids than for solids?

33. A bimetallic strip consists of thin pieces of brass and steel bonded together (Fig. 20-24). Such strips are often used in thermostats. What will happen when the strip is heated? *Hint:* Consult Table 20-2.

Brass
Steel

FIGURE 20-24   Question 33.

34. Why are power lines more likely to break in winter?

PROBLEMS

Section 20-1 Gases

1. Mars's atmosphere has a pressure only 0.0070 times that of Earth, and an average temperature of 218 K. What is the volume of 1 mole of the Martian atmosphere?

2. How many molecules are in an ideal gas sample at 350 K that occupies 8.5 L when the pressure is 180 kPa?

3. What is the pressure of an ideal gas if 3.5 moles occupy 2.0 L at a temperature of $-150°C$?

4. An ideal gas occupies a volume V at 100°C. If the gas pressure is held constant, by what factor does the volume change (a) if the Celsius temperature is doubled and (b) if the kelvin temperature is doubled?

5. If 2.0 mol of an ideal gas are at an initial temperature of 250 K and pressure of 1.5 atm, (a) what is the gas volume? (b) The pressure is now increased to 4.0 atm, and the gas volume drops to half its initial value. What is the new temperature?

6. The solar corona is an extended atmosphere of hot $(2 \times 10^6$ K) gas surrounding the cooler visible surface of the Sun. The gas pressure in the solar corona is about 0.03 Pa. What is the coronal density in particles per cubic meter? Compare with Earth's atmosphere.

7. A pressure of 1.0×10^{-10} Pa is readily achievable with laboratory vacuum apparatus. If the residual air in this "vacuum" is at 0°C, how many air molecules are in one liter?

8. A cubical metal box with thin walls measures 10 cm on each side. It is filled with 0.50 mol of an ideal gas at 180 K, and is surrounded by air at atmospheric pressure. What is the net pressure force on each side of the box?

9. A helium balloon occupies 8.0 L at 20°C and 1.0 atm pressure. The balloon rises to an altitude where air pressure is 0.65 atm and the temperature is $-10°C$. What is its volume when it reaches equilibrium at the new altitude?

10. A compressed air cylinder stands 100 cm tall and has an internal diameter of 20.0 cm. At room temperature, the pressure is 180 atm. (a) How many moles of air are in the cylinder? (b) What volume would this air occupy at 1.0 atm and room temperature?

11. An aerosol can of whipped cream is pressurized at 440 kPa when it's refrigerated at 3°C. The can warns against temperatures in excess of 50°C. What is the maximum safe pressure for the can?

12. A student's dormitory room measures 3.0 m × 3.5 × 2.6 m. (a) How many air molecules does it contain? (b) What is the total translational kinetic energy of these molecules? (c) How does this energy compare with the kinetic energy of the student's 1200-kg car going 90 km/h?

13. A 3000-ml flask is initially open while in a room containing air at 1.00 atm and 20°C. The flask is then closed, and immersed in a bath of boiling water. When the air in the flask has reached thermodynamic equilibrium, the flask is opened and air allowed to escape. The flask is then closed and cooled back to 20°C. (a) What is the maximum pressure reached in the flask? (b) How many moles escape when air is released from the flask? (c) What is the final pressure in the flask?

14. What is the thermal speed of hydrogen (H_2) molecules at 800 K?

15. At what temperature would the thermal speed of oxygen molecules in air equal the speed of sound in water at 25°C? (Consult Table 17-2.)

16. In which gas are the molecules moving faster: hydrogen (H_2) at 75 K or sulfur dioxide (SO_2) at 350 K?

17. At what temperature would the thermal speed of nitrogen molecules in air be 1% of the speed of light? Why might it not be possible to achieve this temperature in N_2 gas?

18. The principal gases in Earth's atmosphere are N_2, O_2, and Ar, with smaller amounts of CO_2 and H_2O. In air at 300 K, what are the thermal speeds of these gases?

19. The van der Waals constants for helium gas (He) are $a = 0.0341$ L^2·atm/mol^2 and $b = 0.0237$ L/mol. What is the temperature of 3.00 mol of helium at 90.0 atm pressure if the gas volume is 0.800 L? How does this result differ from the ideal gas prediction?

20. At what pressure does the van der Waals calculation of temperature for the gas of Example 20-3 differ by 10% from the ideal gas calculation?

21. Because the correction terms (n^2a/V^2 and $-nb$) in the van der Waals equation have opposite signs, there is a point at which the van der Waals and ideal gas equations predict the same temperature. For the gas of Example 20-3, at what pressure does that occur?

22. Plot the Maxwell-Boltzmann distribution of molecular speeds for samples of monatomic helium gas (He) containing 10^{18} atoms at temperatures of 50 K and 250 K.

23. In a sample of 10^{24} hydrogen (H_2) molecules, how many molecules have speeds between 900 and 901 m/s (a) at a temperature of 100 K and (b) at 450 K?

24. For the gas of Fig. 20-4, what would be the number of molecules with speeds between 1000 and 1001 m/s at temperatures of (a) 300 K; (b) 500 K?

Section 20-2 Phase Changes

25. How much energy does it take to melt a 65-g ice cube?

26. It takes 200 J to melt an 8.0-g sample of one of the substances in Table 20-1. What is the substance?

27. If it takes 840 kJ to vaporize a sample of liquid oxygen, how large is the sample?

28. Carbon dioxide sublimes (changes from solid to gas) at 195 K. The heat of sublimation is 573 kJ/kg. How much heat must be extracted from 250 g of CO_2 gas at 195 K in order to solidify it?

29. Find the energy needed to convert 28 kg of liquid oxygen at its boiling point into gas.

30. A tea kettle containing 2.1 kg of boiling water is allowed to boil dry on a stove burner. How much energy did the burner supply to the water?

31. If a 1-megaton nuclear bomb were exploded deep in the Greenland ice cap, how much ice would it melt? Assume the ice is initially at about its freezing point, and consult Appendix C for the appropriate energy conversion.

32. How much ice can a 625-W microwave oven melt in 1.0 min, if the ice is initially at 0°C?

33. What is the power of a microwave oven that takes 20 min to boil dry a 300-g cup of water initially at its boiling point?

34. At winter's end, Lake Superior's 82,000-km^2 surface is frozen to a depth of 1.3 m. The density of ice is 917 kg/m^3. (a) How much energy does it take to melt the ice? (b) If the ice disappears in 3 weeks, what is the average power supplied to melt it?

35. A refrigerator extracts energy from its contents at the rate of 95 W. How long will it take to freeze 750 g of water already at 0°C?

36. The size of industrial air conditioning and refrigeration equipment is often given in tons, meaning the number of tons of water at its melting point that the unit could freeze in one day. (This unit is a holdover from the days when ice was cut from lakes in winter.) What is the rate of heat extraction, in watts, of a 15-ton refrigeration unit?

37. At its "thaw" setting a microwave oven delivers 210 W. How long will it take to thaw a frozen 1.8-kg roast, assuming the roast is essentially water and is initially at 0°C?

38. An ice layer 50 cm thick covers a lake; soot from air pollution covers the ice and absorbs 75% of the incident sunlight, whose time-average intensity is 200 W/m^2. How long will it take to melt the ice, assuming an initial temperature of about 0°C? The density of ice is 917 kg/m^3.

39. A 100-g block of ice, initially at −20°C, is placed in a 500-W microwave oven. (a) How long must the oven be on to produce water at 50°C? (b) Make a graph showing temperature versus time during this entire interval.

40. Repeat Example 20-6 if the initial mass of ice is 50 g.

41. How much energy does it take to melt 10 kg of ice initially at −10°C? Consult Table 19-1.

42. Water is brought to its boiling point and then allowed to boil away completely. If the energy needed to raise the water to the boiling point is one-tenth of that needed to boil it away, what was the initial temperature?

43. A 250-g piece of ice at 0°C is placed in a 500-W microwave oven and the oven run for 5.0 min. What is the temperature at the end of this time?

44. During a nuclear accident 420 m^3 of emergency cooling water at 20°C are injected into a reactor vessel where the

reactor core is producing heat at the rate of 200 MW. If the water is allowed to boil at normal atmospheric pressure, how long will it take to boil the reactor dry?

45. What is the minimum amount of ice in Example 20-6 that will ensure a final temperature of 0°C?

46. What would have to be the initial ice temperature in Example 20-6 in order that the final mixture ends up with more than 200 g of ice?

47. A 500-g chunk of solid mercury at its 234 K melting point is added to 500 g of liquid mercury at room temperature (293 K). Determine the equilibrium mix and temperature.

48. Repeat the preceding problem if the initial amount of liquid is 1.0 kg.

49. A bowl contains 16 kg of punch (essentially water) at a warm 25°C. What is the minimum amount of ice at 0°C that will cool the punch to 0°C?

50. Water at 300 K is sprinkled onto 200 g of molten copper at its 1356-K melting point. The water boils away, leaving solid copper still at 1356 K. How much water does this take? Assume that the only heat loss from the copper is to the liquid water.

51. A 50-g ice cube at −10°C is placed in an equal mass of water. What must be the initial water temperature if the final mixture still contains equal amounts of ice and water?

52. A 40-kg block of aluminum is initially at 50°C. A jet of steam at 100°C hits the block and condenses, and the resulting water drops off before its temperature changes. How much steam must hit the block to raise its temperature to 100°C?

Section 20-3 Thermal Expansion

53. A Pyrex glass marble is 1.00000 cm in diameter at 20°C. What will be its diameter at 85°C?

54. At 0°C, the hole in a steel washer is 9.52 mm in diameter. To what temperature must it be heated in order to fit over a 9.55-mm-diameter bolt?

55. Suppose a single piece of welded steel railroad track stretched 5000 km across the continental United States. If the track were free to expand, by how much would its length change if the entire track went from a cold winter temperature of −25°C to a hot summer day at 40°C?

56. A glass marble 1.000 cm in diameter is to be dropped through a hole in a steel plate. At room temperature the hole diameter is 0.997 cm. By how much must the steel temperature be raised so that the marble will fit through the hole?

57. The tube in a mercury thermometer is 0.10 mm in diameter. What should be the volume of the thermometer bulb if a 1.0-mm rise is to correspond to a temperature change of 1.0°C? Neglect the expansion of the glass.

58. A 2000-mL graduated cylinder is filled with liquid at 350 K. When the liquid is cooled to 300 K, the cylinder is full only to the 1925-mL mark. Use Table 20-2 to identify the liquid.

59. A steel ball bearing is encased in a Pyrex glass cube 1.0 cm on a side. At 330 K, the ball bearing fits tightly in the cube. At what temperature will it have a clearance of 1.0 μm all around?

60. Gasoline comes from its underground tank at 10°C. On a summer day, how much gas can you put in your car's 60-L tank if the tank is not to overflow when the gas reaches the ambient temperature of 25°C?

61. Lake Erie's volume is 480 km³. If the lake temperature were uniform (it isn't!) and increased by 0.50°C, by how much would the volume change if the initial temperature were (a) 1°C and (b) 20°C?

62. A 250-ml Pyrex glass beaker is filled to the brim with ethyl alcohol; both beaker and alcohol are at 20°C. How much alcohol spills over the top when the temperature is increased to 30°C?

Paired Problems

(Both problems in a pair involve the same principles and techniques. If you can get the first problem, you should be able to solve the second one.)

63. What is the density, in moles per m³, of air in a tire whose absolute pressure is 300 kPa at 34°C?

64. Venus's atmospheric pressure is 90 times that of Earth, and its average temperature is 730 K. What is the volume of 1 mole of Venus's atmosphere?

65. What power is needed to melt 20 kg of ice in 6.0 min?

66. How long will it take a 140-W heat source to vaporize 250 g of ethyl alcohol already at its boiling point?

67. You put 300 g of water into a 500-W microwave oven and accidentally set the time for 20 min instead of 2.0 min. If the water is initially at 20°C, how much is left at the end of 20 min?

68. If 4.5×10⁵ kg of emergency cooling water at 10°C are dumped into a malfunctioning nuclear reactor whose core is producing energy at the rate of 200 MW, and if no circulation or cooling of the water is provided, how long will it be before half the water has boiled away?

69. Describe the composition and temperature of the equilibrium mixture after 1.0 kg of ice at −40°C is added to 1.0 kg of water at 5.0°C.

70. Repeat the preceding problem if the ice is initially at −80°C and if there are initially only 200 g of liquid water.

Supplementary Problems

71. How long will it take a 500-W microwave oven to vaporize completely a 500-g block of ice initially at 0°C?

72. The thermal resistance of a refrigerator's walls is 0.12 K/W. During a power failure you put a 15-kg block of ice at 0°C in the refrigerator, bringing the interior temperature to 0°C. If room temperature is 20°C, how long will the ice last?

FIGURE 20-25 A solar-heated house (Problem 73).

73. A solar-heated house (Fig. 20-25) stores energy in 5.0 tons of Glauber salt ($Na_2SO_4 \cdot 10H_2O$), a substance that melts at 90°F. The heat of fusion of Glauber salt is 104 Btu/lb, and the specific heats of the solid and liquid are, respectively, 0.46 Btu/lb·°F and 0.68 Btu/lb·°F. After a week of sunny weather, the storage medium is all liquid at 95°F. Then a cool, cloudy period sets in during which the house loses heat at an average rate of 20,000 Btu/h. (a) How long is it before the temperature of the storage medium drops below 60°F? (b) How much of this time is spent at 90°F?

74. (a) Show that Equation 20-3 implies a thermal speed given by $\sqrt{3kT/m}$. (b) By differentiating Equation 20-4, show that the most probable molecular speed is lower by a factor of $\sqrt{2/3}$.

75. Show that the coefficient of volume expansion of an ideal gas at constant pressure is just the reciprocal of its kelvin temperature.

76. A constant-volume gas thermometer is made from Pyrex glass. If the thermometer is calibrated at the triple point of water and then used to determine the boiling point of water, how much error will be introduced by ignoring the expansion of the glass?

77. Water's coefficient of volume expansion in the temperature range from 0°C to about 20°C is given approximately by $\beta = a + bT + cT^2$, where T is in Celsius and $a = -6.43 \times 10^{-5}$ °C^{-1}, $b = 1.70 \times 10^{-5}$ °C^{-2}, and $c = -2.02 \times 10^{-7}$ °C^{-3}. Show that water has its greatest density at approximately 4.0°C.

78. When the expansion coefficient varies with temperature, Equation 20-7 should be written $\beta = \dfrac{1}{V}\dfrac{dV}{dT}$. If a sample of water occupies 1.00000 L at 0°C, find its volume at 12°C. Use the information from the preceding problem, and integrate the equation above.

79. Ignoring air resistance, find the height from which you must drop an ice cube at 0°C so it melts completely on impact. Assume no heat exchange with the environment.

80. The timekeeping of an old clock is regulated by a brass pendulum 20.0 cm long. If the clock is accurate at 20°C but is in a room at 18°C, how long will it be before the clock is in error by 1 minute? Will it be too fast or too slow?

81. Prove Equation 20-9 by considering a cube of side s and therefore volume $V = s^3$ that undergoes a small temperature change dT and corresponding length and volume changes ds and dV.

CHAPTER 21

HEAT, WORK, AND THE FIRST LAW OF THERMODYNAMICS

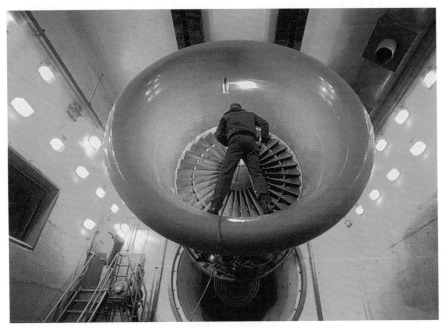

A jet aircraft engine, shown here being repaired, converts energy released in burning fuel to mechanical motion. Jet engines, gasoline and diesel engines, and electric power plants all rely on thermodynamics to effect energy conversion.

How do we change the temperature of a substance? One approach involves heat transfer between substances at different temperatures. Another approach—which you've experienced if you ever pulled a rope too quickly through your hands, or touched a drill bit just after drilling a hole, or smelled your car's brakes burning as you went down a steep hill—involves mechanical energy. Here we explore this second approach to temperature change and, in the process, come to a deeper understanding of the relation between heat and mechanical energy.

21-1 THE FIRST LAW OF THERMODYNAMICS

Heat is energy being transferred from one object to another because of a temperature difference alone. As a result of heat transfer, a substance undergoes changes in its thermodynamic state—its temperature, pressure, and volume

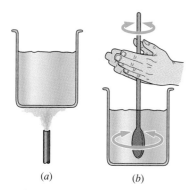

FIGURE 21-1 Either a flame or mechanical agitation will raise the water temperature.

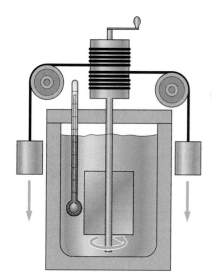

FIGURE 21-2 In Joule's apparatus, potential energy of the falling weights is converted to kinetic energy of the rotating paddle, which in turn becomes internal energy of the water.

may change, and it may undergo a change of phase. But we can accomplish these same changes with mechanical energy. As Fig. 21-1 suggests, we can raise the temperature of water by heating with a flame or by stirring violently with a spoon. For the flame to heat the water, the flame temperature must exceed the water temperature. But the spoon can have the same temperature as the water, since it is the mechanical energy of the spoon's motion that raises the water temperature. We use the term **work** to describe energy transfer that does not require a temperature difference. This sense is consistent with our earlier use of the word work to describe mechanical energy transfer. Sometimes that energy ends up as stored potential energy, but often it results in a temperature change—as when we do work against frictional forces. That both heat and mechanical work can raise temperature is what made possible Joule's quantitative identification of heat as a form of energy (Fig. 21-2).

What happens to energy transferred to a system through either heat or mechanical work? It ends up changing the **internal energy** of the system—that is, the sum of the different kinds of energy associated with all the individual particles comprising the system. For an ideal gas, internal energy is the total kinetic energy of the gas molecules. More complicated substances include potential energy as well; fuels like gasoline or uranium, for example, have considerable internal potential energy.

If we keep track of all energy entering and leaving a system—both heat and work—and monitor changes in internal energy, we find experimentally that energy is conserved. This is hardly surprising, given our emphasis on energy conservation in mechanics. But our statement of energy conservation is now broader since it includes heat as well as mechanical work. This broader statement of energy conservation is known as the **first law of thermodynamics:**

First law of thermodynamics: The change in the internal energy of a system is equal to the heat transferred to the system minus the work done by the system.

Mathematically, the first law reads

$$\Delta U = Q - W, \tag{21-1}$$

where ΔU is the change in a system's internal energy, Q the heat transferred *to* the system, and W the work done *by* the system. The minus sign arises because the first law was developed in connection with engines, which take heat from a source and give out mechanical work. Keeping track of energy transfers in an engine is easier if both the heat input and the work output are considered positive quantities; their difference is then the change in internal energy.

The first law of thermodynamics shows that the change in a system's internal energy doesn't depend on the details of the energy transfer—only on the total amount of energy transferred. We could transfer a given quantity of energy by heating, by mechanical work, or by some combination of the two, and the effect would be the same. For this reason internal energy is called a **thermodynamic state variable,** meaning a quantity whose value doesn't depend on how a system got into its particular state. Temperature and pressure are also thermodynamic state variables; heat and work are not.

We're frequently concerned with *rates* of energy flow. Differentiating the first law with respect to time gives a statement about these rates:

$$\frac{dU}{dt} = \frac{dQ}{dt} - \frac{dW}{dt},$$ (21-2)

where dU/dt is the rate of change of a system's internal energy, dQ/dt the rate of heat transfer to the system, and dW/dt the rate at which the system does work on its surroundings.

● EXAMPLE 21-1   NUCLEAR POWER, THERMAL POLLUTION

A nuclear power plant's reactor supplies energy at the rate of 3.0 GW, boiling water to produce steam that turns a turbine-generator (Fig. 21-3). The spent steam is then condensed through thermal contact with water taken from a river and sent back to the reactor. If the power plant produces electrical energy at the rate of 1.0 GW, at what rate is heat transferred to the river?

Solution
Here the system consists of the power plant and its reactor, so dU/dt in Equation 21-2 represents rate of change of internal energy in the nuclear fuel. Since that internal energy is being converted to other forms, dU/dt is negative; its value is −3.0 GW. On the other hand, dW/dt is the rate at which the plant does work—i.e., the rate at which it supplies electrical energy to the external world. Thus $dW/dt = +1.0$ GW. Solving for the heat-transfer rate then gives

$$\frac{dQ}{dt} = \frac{dU}{dt} + \frac{dW}{dt} = -3.0 \text{ GW} + 1.0 \text{ GW} = -2.0 \text{ GW}.$$

Does this make sense? Since a positive Q represents heat transferred *to* the system, the minus sign in our answer shows that heat is transferred *from* the power plant to the river at the rate of 2.0 GW. The numbers in this example—with 1 GW = 10^9 W—are typical for large nuclear and coal-burning power plants, and show that about two-thirds of the energy extracted from the fuel is wasted in heating the environment. We'll see in the next chapter just why this waste occurs.

FIGURE 21-3   A nuclear power plant, showing two reactor buildings and a large cooling tower. The plant takes in water from the river in the foreground. The water absorbs waste heat and then passes through the cooling tower before returning to the river. Example 21-1 shows that two-thirds of the energy produced in the reactor is discarded as waste heat.

> **TIP   Identify the System**   Thermodynamics problems often deal with energy flows into and out of a system. Before attacking a problem, be sure you're clear on just what constitutes the system. The first law holds no matter how you define the system, but the meanings of the various quantities change with that definition. In Example 21-1, for example, we could have excluded the nuclear fuel from the system. The net result—2 GW dumped to the river—would have been the same, but now we would have had an additional heat input of 3 GW and no change in internal energy.

EXERCISE   Gasoline burning in an automobile engine releases energy at the rate of 160 kW. Heat is exhausted through the car's radiator at the rate of 51 kW and out the exhaust at 50 kW. An additional 23 kW goes to frictional heating within the machinery of the car. What fraction of the fuel energy is available for propelling the car?

Answer: 22.5%

Some problems similar to Example 21-1:   4–6   ●

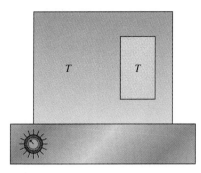

FIGURE 21-4 A quasi-static, or reversible, process. The gas sample is heated by immersion in a heat reservoir whose temperature is slowly increased. The entire gas sample is always at essentially the same temperature as the reservoir, so it is always in a well-defined thermodynamic state.

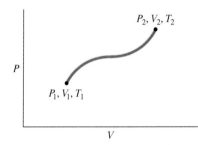

FIGURE 21-5 Because a system undergoing a quasi-static change is always in thermodynamic equilibrium, the change may be described by a succession of states—a continuous path—in its *PV* diagram.

21-2 THERMODYNAMIC PROCESSES

The first law relates two kinds of energy transfers—heat and work—that can change a system's thermodynamic state. Although the first law applies to *any* system, it is easiest to understand when applied to an ideal gas. The ideal gas law (Equation 20-2) relates the temperature, pressure, and volume of a given gas sample: $PV = nRT$. The thermodynamic state of the gas is completely determined by any two of the quantities *P*, *V*, or *T*. It's therefore convenient to represent each state by a point on the *PV* diagram for the gas.

Reversible and Irreversible Processes

Imagine heating a gas sample by immersing it in a large reservoir of water whose temperature we can control (Fig. 21-4). If the gas and water temperatures are equal, the gas will stay in thermodynamic equilibrium. If we then raise the reservoir temperature very slowly, both water and gas temperatures will rise essentially in unison, and the gas will remain in equilibrium. Such a slow change is called a **quasi-static process.** Because a system undergoing a quasi-static process is always in thermodynamic equilibrium, its overall change from one state to another is then described by a continuous sequence of points—a curve—in its *PV* diagram (Fig. 21-5).

We could reverse this heating process by very slowly lowering the reservoir temperature; the gas would then cool while remaining essentially in thermodynamic equilibrium and would go back along the same path in its *PV* diagram. For this reason, a quasi-static process is also called a **reversible process.** A process like the sudden plunging of the gas sample into boiling water is, in contrast, **irreversible** (Fig. 21-6). During an irreversible change the system is not in equilibrium, and thermodynamic variables like temperature and pressure do not have well-defined values. It therefore makes no sense to think of a path in the *PV* diagram. A process may be irreversible even though it returns a system to its original state. The distinction lies not in the end states but in the *process* that takes the system between states.

There are many ways to change the thermodynamic state of a system. Here we consider several important special cases as they apply to an ideal gas system. These special cases illustrate the physical principles behind a myriad of technological devices and natural phenomena, from the operation of a gasoline engine to the propagation of a sound wave to the oscillations of a star.

Our system consists of an ideal gas confined to a cylinder sealed with a movable piston (Fig. 21-7). The piston and cylinder walls are perfectly insulating—they block all heat transfer—while the bottom is a perfect conductor of heat. We can change the thermodynamic state of the gas mechanically, by moving the piston, or thermally, by transferring heat through the bottom. We will consider only reversible processes, which we can describe by paths in the *PV* diagram for the gas.

Work and Volume Changes

Before examining specific processes, we develop a relation between volume change and work that holds for all processes. If *A* is the cross-sectional area of our piston and cylinder and *P* the gas pressure, then $F = PA$ is the force the gas

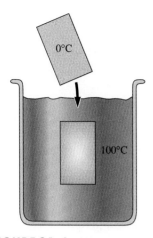

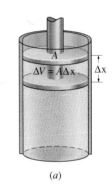

FIGURE 21-7 A gas-cylinder system. The cylinder walls are insulating, but the bottom is a good conductor of heat.

FIGURE 21-6 Plunging a gas sample at 0°C into boiling water is an irreversible process. It takes some time for the gas to reach equilibrium at the higher temperature, during which time the gas itself has no well-defined thermodynamic state.

exerts on the piston (Fig. 21-8a). If the piston moves a small distance Δx, then the work done by the gas is

$$\Delta W = F\Delta x = PA\,\Delta x = P\Delta V,$$

where $V = A\,\Delta x$ is the gas volume. The work ΔW is positive when the volume increases, indicating that the gas does work on the piston, and negative if ΔV is negative, indicating work done on the gas. Pressure may vary with volume, in which case our result is strictly valid only in the limit of very small volume changes. To find the work associated with a larger volume change, we take the limit $\Delta V \to 0$, and integrate over the volume change:

$$W = \int dW = \int_{V_1}^{V_2} PdV, \qquad (21\text{-}3)$$

where V_1 and V_2 are the initial and final volumes. Figure 21-8b shows the geometrical interpretation of this equation: the work done by the gas is simply the area under the PV curve.

Isothermal Processes

To perform a reversible **isothermal process,** we place the heat-conducting bottom of our gas cylinder in good thermal contact with a heat reservoir whose temperature is held constant (Fig. 21-9). We then move the piston to change the volume of the system, doing so slowly enough that the gas remains in equilibrium with the heat reservoir. The system then moves from its initial state to its final state along a curve of constant temperature—an **isotherm**—in the PV

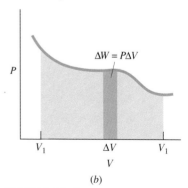

FIGURE 21-8 (a) The gas exerts a pressure force PA on the piston of area A, so the work done in moving the piston a small distance Δx is $PA\,\Delta x$, or $P\Delta V$. (b) On the PV diagram, this work corresponds to the area of the dark strip of width ΔV. The total work to take the system from volume V_1 to V_2 is the area under the PV curve between V_1 and V_2.

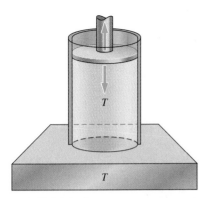

FIGURE 21-9 During an isothermal process, the piston moves slowly while the system is held in thermal contact with a heat reservoir at fixed temperature.

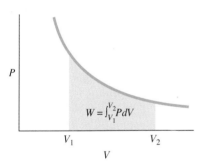

FIGURE 21-10 *PV* diagram for an isothermal process. The work done during the process is the area under the curve.

diagram (Fig. 21-10). The work done in the process is given by Equation 21-3 and is equal to the area under the isotherm.

For an ideal gas, we can relate pressure P and volume V through the ideal gas law:

$$P = \frac{nRT}{V}.$$

Then Equation 21-3 becomes

$$W = \int_{V_1}^{V_2} \frac{nRT}{V} \, dV.$$

For an isothermal process, the temperature T is constant, giving

$$W = nRT \int_{V_1}^{V_2} \frac{dV}{V} = nRT \, \ln V \Big|_{V_1}^{V_2} = nRT \, \ln\left(\frac{V_2}{V_1}\right).$$

The internal energy of an ideal gas consists only of the kinetic energy of its molecules which, in turn, depends only on the temperature. We showed this in the preceding chapter for a gas of structureless point particles and will extend it later in this chapter to gases comprised of more complicated molecules. Thus, there is no change in the internal energy of an ideal gas during an isothermal process. The first law of thermodynamics then gives

$$\Delta U = 0 = Q - W$$

so

$$Q = W = nRT \, \ln\left(\frac{V_2}{V_1}\right). \quad \text{(isothermal process)} \quad (21\text{-}4)$$

Does this result $Q = W$ make sense? Recall that Q is the heat transferred to the gas, while W is the work done by the gas. Therefore, our result says that when an ideal gas does work W on the external world with no change in temperature, it must absorb an equal amount of heat from outside. Similarly, if work is done on the gas, then the gas must reject an equal amount of heat to the outside if its temperature is not to change.

● **EXAMPLE 21-2** A DIVER EXHALES

A scuba diver (Fig. 21-11) is swimming at a depth of 25 m, where the pressure is 3.5 atm (recall Example 18-1). The air she exhales forms bubbles 8.0 mm in radius. How much work is done by each bubble as it expands on rising to the surface? Assume that the bubbles remain at the uniform 300 K temperature of the surrounding water.

Solution

Since the bubbles remain at constant temperature, the process is isothermal and Equation 21-4 applies. Just before they break the surface, the bubbles are at essentially 1 atm pressure, so their pressure has decreased by a factor of 3.5. The ideal gas law, $PV = nRT$, shows that when temperature is constant,

FIGURE 21-11 How much work do the diver's exhaled air bubbles do as they rise? (Example 21-2)

pressure and volume are inversely related. Therefore, the bubble volume has increased by a factor of 3.5. To apply Equation 21-4, we also need the quantity nRT for a bubble. We can get this from the ideal gas law using the given radius of the bubbles at 3.5 atm pressure:

$$nRT = PV = P(\tfrac{4}{3}\pi r^3),$$

so Equation 21-4 becomes

$$W = nRT \ln\left(\frac{V_2}{V_1}\right) = \tfrac{4}{3}\pi r^3 P \ln\left(\frac{V_2}{V_1}\right)$$
$$= \tfrac{4}{3}\pi (8.0 \times 10^{-3} \text{ m})^3 (3.5 \text{ atm})(1.01 \times 10^5 \text{ Pa/atm}) \ln(3.5)$$
$$= 0.95 \text{ J}.$$

Where does this energy go? As a bubble expands, it pushes water outward and, ultimately, upward. It therefore raises the gravitational potential energy of the ocean. When the bubble breaks the surface, this excess potential energy becomes kinetic energy, appearing in the form of small waves on the water surface.

EXERCISE Consider 3.2 mol of an ideal gas in contact with a heat reservoir at 320 K. Find the work done by the gas if its volume doubles.

Answer: 5.9 kJ

Some problems similar to Example 21-2: 11, 13–16

Constant-Volume Processes and Specific Heat

A **constant-volume process** (also called isometric, isochoric, or isovolumic) occurs in a rigid, closed container whose volume cannot change. In our piston-cylinder arrangement, we could tightly clamp the piston for a constant-volume process. Because the piston doesn't move, the gas does no work, and the first law becomes simply

$$Q = \Delta U. \qquad (21\text{-}5)$$

To express this result in terms of a temperature change ΔT, we introduce the **molar specific heat at constant volume,** C_V, defined by the equation

$$Q = nC_V\Delta T, \qquad (21\text{-}6)$$

where n is the number of moles. This molar specific heat is like the specific heat defined in Chapter 19, except that with a gas it's more convenient to consider the heat per mole rather than per unit mass. Introducing the definition of C_V into Equation 21-5 gives

$$nC_V\Delta T = \Delta U. \qquad (21\text{-}7)$$

Solving for C_V then gives

$$C_V = \frac{1}{n}\frac{\Delta U}{\Delta T}. \qquad (21\text{-}8)$$

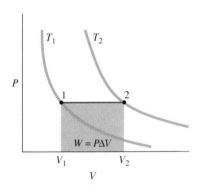

FIGURE 21-12 *PV* diagram for an isobaric process. The gas moves along the isobar from state 1 to state 2, doing work $P\Delta V$ in the process. Also shown are isotherms for the initial and final temperatures.

For an ideal gas, the internal energy is a function of temperature alone, so $\Delta U/\Delta T$ has the same value no matter what process the gas undergoes. Therefore Equation 21-7, relating the temperature change ΔT and internal energy change ΔU, applies not only to a constant-volume process, but also to *any* ideal gas process. Why, then, have we been so careful to label C_V the specific heat *at constant volume?* Although the relation $nC_V\Delta T = \Delta U$ holds for any process, it is only when no work is done that the first law allows us to write $Q = \Delta U$, and therefore only for a constant-volume process that Equation 21-6 holds.

Isobaric Processes and Specific Heat

Isobaric means constant pressure. In a reversible isobaric process, a system moves along an isobar, or curve of constant pressure, in its *PV* diagram (Fig. 21-12). The work done as the gas volume changes from V_1 to V_2 is the area under the isobar, or

$$W = P(V_2 - V_1) = P\Delta V, \qquad (21\text{-}9)$$

a result we could obtain formally by integrating Equation 21-3.

Solving the first law (Equation 21-1) for Q and using our expression for work gives

$$Q = \Delta U + W = \Delta U + P\Delta V.$$

But we've found that the change in internal energy of an ideal gas is given by $\Delta U = nC_V\Delta T$ for *any* process. Therefore

$$Q = nC_V\Delta T + P\Delta V \qquad (21\text{-}10)$$

for an isobaric process. We define the **molar specific heat at constant pressure,** C_P as the heat required to raise one mole of gas through a unit temperature change at constant pressure, or

$$Q = nC_P\Delta T.$$

Equation 21-10 can then be written

$$nC_P\Delta T = nC_V\Delta T + P\Delta V. \quad \text{(isobaric process)} \qquad (21\text{-}11)$$

This is a useful form for calculating temperature changes in an isobaric process, if we know both specific heats C_P and C_V. However, we really need only one of these specific heats, for a simple relation holds between the two. The ideal gas law, $PV = nRT$, allows us to write

$$P\Delta V = nR\,\Delta T$$

for an isobaric process. Using this expression in Equation 21-11 gives

$$nC_P\Delta T = nC_V\Delta T + nR\,\Delta T,$$

so $$C_P = C_V + R. \qquad (21\text{-}12)$$

Does this result make sense? Specific heat measures the heat needed to cause a given temperature change. In a constant-volume process, no work is done and all the heat goes into raising the internal energy and thus the temperature of an ideal gas. In a constant-pressure process, work is done and some of the added heat ends up as mechanical energy, leaving less energy available for raising the temperature. Therefore, in a constant-pressure process, *more* heat is needed to achieve a given temperature change, so the specific heat at constant pressure is larger than at constant volume, as reflected in Equation 21-12.

Why didn't we distinguish specific heats at constant volume and constant pressure much earlier? Because we were concerned mostly with solids and liquids, whose coefficients of expansion are far lower than those of gases. As a result of its relatively small expansion, the work done by a solid or liquid is much less than that done by a gas. Since work is what gives rise to the difference between C_V and C_P, the distinction is less significant for solids and liquids. As a practical matter, measured specific heats are usually at constant pressure, because enormous forces would be needed to prevent volume changes in solids or liquids.

Adiabatic Processes

In an **adiabatic process,** no heat transfer occurs between a system and its environment. The simplest way to achieve this is to surround the system with perfect thermal insulation. Even without insulation, processes that occur quickly are often approximately adiabatic because they're over before significant heat transfer has had time to occur. In a gasoline engine, for example, compression of the gasoline-air mixture and subsequent expansion of the combustion products are nearly adiabatic because they occur so rapidly that little heat flows through the cylinder walls during the process.

Since the heat Q is zero in an adiabatic process, the first law becomes simply

$$\Delta U = -W. \quad \text{(adiabatic process)} \qquad (21\text{-}13)$$

This says that if a system does work on its surroundings, then—in the absence of heat transfer—its internal energy must decrease by the same amount. Microscopically, this internal energy loss occurs in a gas-cylinder system when molecules collide with the moving piston, returning to the gas with less energy (Fig. 21-13).

As a gas expands adiabatically, its volume increases while its internal energy and temperature decrease. The ideal gas law, $PV = nRT$, then requires that the pressure decrease as well—and by more than it would in an isothermal process where T remains constant. In a PV diagram, the path of an adiabatic process—called an **adiabat**—therefore drops more steeply than the isotherms (Fig. 21-14).

Box 21-1 details the mathematics involved in finding the shape of the adiabatic path; the result is

$$PV^\gamma = \text{constant} = P_0 V_0^\gamma, \quad \text{(adiabatic process)} \qquad (21\text{-}13a)$$

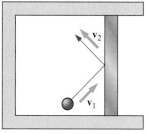

(a) Stationary piston

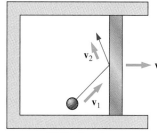

(b) Moving piston

FIGURE 21-13 (*a*) Molecules bouncing off a stationary piston rebound with no loss of speed. Here the gas is doing no work, and its internal energy remains constant. (*b*) Molecules bouncing off an outward-moving piston give some energy to the piston and rebound with lower speed. The internal energy of the gas decreases as its molecules do work on the piston. If the piston were moving inward the molecules would gain energy, and the piston would be doing work on the gas.

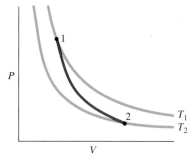

FIGURE 21-14 *PV* curve for an adiabatic process (red) shows that the pressure drops more than in an isothermal process going between the same two volumes. Two isotherms shown are for the initial and final temperatures of the adiabatic process. The adiabatic path is steeper because the gas loses internal energy as it does work.

where $\gamma = C_P/C_V$ is the ratio of the specific heats and P_0 and V_0 are the values of P and V at some point in the process.

Because $C_P = C_V + R$, the ratio $\gamma = C_P/C_V$ is always greater than 1. Therefore an adiabatic process with a given volume change results in a greater pressure change than would a comparable isothermal process, as reflected in the steeper adiabatic path in Fig. 21-14. Physically, the adiabatic path is steeper because the gas loses internal energy as it does work, so its temperature drops.

TIP   **How Can PV^γ = constant and $PV = nRT$?**   You might think these two equations are contradictory. But the ideal gas law implies PV = constant only during an isothermal process, when the temperature T doesn't change. A reversible ideal-gas process cannot be simultaneously both adiabatic (no heat exchange) and isothermal (no temperature change), so there's no contradiction. During an adiabatic process the temperature of an ideal gas *necessarily* changes.

We can find the temperature change in an adiabatic process by using the ideal gas law to eliminate pressure from Equation 21-13a. Writing that equation in the form $(PV)(V^{\gamma-1}) = (P_0V_0)(V_0^{\gamma-1})$ and substituting for PV from the ideal gas law gives $nRTV^{\gamma-1} = nRT_0V_0^{\gamma-1}$. Dividing both sides by nR gives

$$TV^{\gamma-1} = \text{constant} = T_0V_0^{\gamma-1} \qquad (21\text{-}13b)$$

for the relation between temperature and volume in an adiabatic process. We could have written the adiabatic equation in yet a third way, eliminating V in favor of T and P (see Problem 28).

BOX 21-1   DERIVING THE ADIABATIC RELATION

We derive the PV relation for an adiabatic process by integrating over a sequence of infinitesimal adiabatic changes. Equation 21-7, which holds for *any* process, shows that the internal energy change accompanying an infinitesimal temperature change is $dU = nC_V dT$. The work done in the process is $dW = P dV$ so, with $Q = 0$ for an adiabatic process, the first law becomes

$$nC_V dT = -P dV.$$

We now eliminate dT by differentiating the ideal gas law, allowing *both* P and V to change:

$$nR dT = d(PV) = P dV + V dP.$$

Solving for dT, substituting in our first-law statement, and multiplying through by R leads to

$$C_V V dP + (C_V + R)P dV = 0.$$

But $C_V + R = C_P$; substituting this and dividing through by $C_V PV$ gives

$$\frac{dP}{P} + \frac{C_P}{C_V}\frac{dV}{V} = 0.$$

Defining $\gamma \equiv C_P/C_V$ and integrating gives

$$\ln P + \gamma \ln V = \ln(\text{constant}),$$

where we've chosen to call the constant of integration $\ln$ (constant). Since $\gamma \ln V = \ln V^\gamma$, it follows by exponentiation that

$$PV^\gamma = \text{constant},$$

where the constant is the quantity $P_0 V_0^\gamma$ at a point where P and V have the known values P_0 and V_0.

● EXAMPLE 21-3 DIESEL POWER

Ignition of the fuel in a diesel engine (Fig. 21-15) occurs on contact with air heated by compression as the piston moves to the top of its stroke. (In this way a diesel differs from a gasoline engine, in which a spark ignites the fuel.) The compression occurs fast enough that very little heat flows out of the gas, so the process is essentially adiabatic. If a temperature of 500°C is required for ignition, what must be the compression ratio (ratio of maximum to minimum cylinder volume) of the diesel engine? Air has a specific heat ratio γ of 1.4, and before compression its temperature is 20°C.

Solution
Equation 21-13b gives the relation between temperature and volume in an adiabatic process. Writing T_0 and V_0 for the temperature and volume at the bottom of the piston stroke, and T_1 and V_1 for the top of the stroke, Equation 21-13b becomes

$$T_1 V_1^{\gamma-1} = T_0 V_0^{\gamma-1}.$$

Solving for the compression ratio V_0/V_1 gives

$$\frac{V_0}{V_1} = \left(\frac{T_1}{T_0}\right)^{1/(\gamma-1)} = \left(\frac{773\text{ K}}{293\text{ K}}\right)^{1/0.4} = 11.$$

Practical diesel engines have considerably higher compression ratios to ensure reliable ignition. Conversely, compressional heating places an upper limit on the compression ratios of gasoline engines, where ignition by hot air would circumvent the carefully timed spark ignition system (see Problem 26).

FIGURE 21-15 Mechanics at work on a diesel engine. ●

● EXAMPLE 21-4 ADIABATIC AND ISOTHERMAL EXPANSION

Two identical gas-cylinder systems each contain 0.060 mol of ideal gas at 300 K and 2.0 atm pressure. The specific heat ratio γ is 1.4. The gas samples are allowed to expand, one adiabatically and one isothermally, until both are at 1.0 atm pressure. What are the final temperatures and volumes of each?

Solution
The initial volume V_0 of both samples may be obtained from the ideal gas law, $PV = nRT$. Solving for V gives

$$V_0 = \frac{nRT_0}{P_0} = \frac{(0.060\text{ mol})(8.314\text{ J/K·mol})(300\text{ K})}{(2.0\text{ atm})(1.013\times10^5\text{ Pa/atm})}$$
$$= 7.39\times10^{-4}\text{ m}^3 = 0.739\text{ L}.$$

For the isothermal sample, T remains constant at 300 K. With constant temperature, $PV = \text{constant}$, so as the pressure drops in half, the volume doubles, becoming 1.48 L.

For the adiabatic expansion, solving Equation 21-13a for the volume gives

$$V = \left(\frac{P_0}{P}\right)^{1/\gamma} V_0 = \left(\frac{2.0 \text{ atm}}{1.0 \text{ atm}}\right)^{1/1.4} (0.739 \text{ L}) = 1.21 \text{ L}.$$

The temperature follows from Equation 21-13b:

$$TV^{\gamma-1} = \text{constant} = T_0 V_0^{\gamma-1},$$

or

$$T = T_0\left(\frac{V_0}{V}\right)^{\gamma-1} = (300 \text{ K})\left(\frac{0.739 \text{ L}}{1.21 \text{ L}}\right)^{0.4} = 246 \text{ K}.$$

Since we knew both P and V, we could equally well have determined the temperature from the ideal gas law.

In this example, both gas samples do work against the outside world as they expand. In the isothermal process, energy leaving the system as work is replaced by energy entering as heat, so the temperature and internal energy remain constant. But heat cannot flow during the adiabatic process, so the work done is at the expense of the internal energy of the gas. As a result, the gas temperature drops, and the volume does not increase as much. Figure 21-16 shows both processes on a PV diagram.

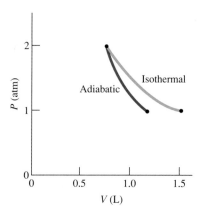

FIGURE 21-16 PV diagram showing isothermal and adiabatic processes of Example 21-4.

pressure. The process occurs rapidly enough that it is essentially adiabatic. What are the volume and temperature of the compressed air? For air, $\gamma = 1.4$.

Answers: $V = 0.11$ L, $T = 153°C$

EXERCISE A bicycle pump compresses 0.30 L of air, initially at 14°C and 1 atm pressure, to 4 times atmospheric

Some problems similar to Examples 21-3 and 21-4: 19, 22–26 ●

How much work is done during an adiabatic process? Since there's no heat transfer, $W = -\Delta U$. But we've found that $\Delta U = nC_V\Delta T$ for *any* process. Therefore in an adiabatic process

$$W = -nC_V\Delta T = nC_V(T_1 - T_2),$$

where T_1 and T_2 are the initial and final temperatures. We can express this result in terms of pressure and volume changes by solving the ideal gas law, $PV = nRT$, for temperature and using the result in the equation above:

$$W = nC_V\frac{P_1V_1 - P_2V_2}{nR}.$$

But $R = C_P - C_V$, so

$$\frac{C_V}{R} = \frac{C_V}{C_P - C_V} = \frac{1}{(C_P/C_V) - 1} = \frac{1}{\gamma - 1}.$$

Then our expression for the adiabatic work becomes

$$W = \frac{P_1V_1 - P_2V_2}{\gamma - 1}. \tag{21-14}$$

We could also have obtained this result by direct integration of Equation 21-3 for an adiabatic process (see Problem 61).

■ APPLICATION SMOG ALERT

The smog shown in Fig. 21-17a is an unfortunate manifestation of our high-energy industrial society. The presence of that smog is closely related to the adiabatic processes we've just considered.

Normally, atmospheric temperature decreases with altitude at a so-called **lapse rate** of about 6.5°C per kilometer. The main reason for the decrease is that sunlight first heats Earth's surface, which then transfers heat to the atmosphere. Now consider a parcel of air that for some reason starts moving upward. That initial rise could occur because the air is over a darker region, like pavement, that absorbs more solar energy. The heated air parcel, being less dense than the surrounding air, then rises under the influence of the buoyancy force we discussed in Chapter 18. Or the rise could occur as horizontally moving air is forced upward to clear an obstruction. Or it could be that hot air and other gases are belched out of a smokestack or automobile exhaust pipe. Whatever the reason, the air parcel starts upward. We want to know what happens next.

As the air parcel rises, the pressure of the surrounding air drops, and the parcel therefore expands to maintain pressure equilibrium. Now, air is a good thermal insulator, so a parcel of reasonable size exchanges little heat with its surroundings. Its expansion is therefore adiabatic, and thus, its temperature drops as it expands. The decrease in pressure with altitude is such that dry air cools adiabatically at the rate of about 10°C per kilometer as it rises. If the surrounding air has the typical lapse rate of 6.5°C/km, then the rising parcel cools faster. Even if it started out hotter than its surroundings, it will soon reach an altitude where it is at the same temperature (Fig. 21-18a). Then it will no longer be buoyant, and its upward motion will soon cease. Air under these conditions is therefore stable, in that a rising parcel soon reaches an equilibrium altitude. You can convince yourself that a falling parcel—falling because it's cooler than its surroundings—will also reach equilibrium.

Under stable conditions, smog-forming pollutants cannot escape to the upper atmosphere. Instead, they're trapped at an altitude where adiabatic cooling brings them to the same temperature as the surrounding air. The greater the difference between the actual lapse rate and the approximately 10°C/km adiabatic cooling rate, the stronger the "trapping" of the air and its pollutants. The most serious problems arise during a **temperature inversion,** in which the air temperature actually increases with height (Fig. 21-18b). A rising parcel still cools adiabatically, and now its temperature drops so quickly relative to its surroundings that the parcel is strongly trapped at low altitudes.

Sometimes the atmospheric lapse rate exceeds the rate of adiabatic cooling. This can occur on crisp, clear days when the ground is strongly heated, or it can occur with moist air for which the adiabatic cooling rate is only about 6°C/km. In this

(a)

(b)

FIGURE 21-17 (a) Smog over Los Angeles. The smog is trapped because adiabatic expansion cools rising smog to the same temperature as the surrounding air. (b) In the violently unstable thunderstorm, adiabatic cooling is not sufficient to bring rising or falling parcels of air to thermal equilibrium.

case a rising parcel cools more slowly than its surroundings, so its buoyancy increases and it accelerates upward (Fig. 21-18c). A falling parcel, similarly, accelerates downward. Under these conditions the air is unstable. The puffy cumulus clouds that form on an otherwise clear summer day are associated with plumes of warm air that rise through the unstable air until they're cool enough that moisture condenses to form clouds. A more dramatic example is the thunderstorm of Fig. 21-17b, in which unstable air undergoes violent up- and downdrafts driven by the heat of vaporization released as water vapor condenses.

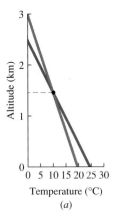

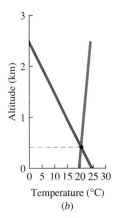

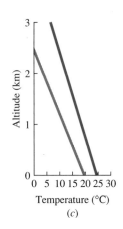

FIGURE 21-18 Conditions governing atmospheric stability. Blue lines show temperature versus altitude in the stationary atmosphere. Red lines are the rates at which rising air parcels cool adiabatically with increasing altitude. (*a*) Under normal stable conditions, adiabatic cooling occurs more rapidly and a parcel heated at ground level soon reaches equilibrium with the surrounding air, here at about 1.4 km. (*b*) A temperature inversion. Blue line shows air temperature increasing with altitude. Here equilibrium of a heated parcel occurs at low altitude, trapping pollutants near the ground. (*c*) When the lapse rate exceeds the adiabatic cooling rate, the air is unstable. A rising parcel continues to rise, without reaching equilibrium.

Cyclic Processes

Many natural and technological systems undergo **cyclic processes,** in which the system returns periodically to the same thermodynamic state. Engineering examples include engines, refrigerators, and air compressors, whose mechanical construction ensures cyclic behavior. Many natural oscillations, like those of a sound wave or a pulsating star, are essentially cyclic.

Cyclic processes often involve the four simple processes we've just outlined, which are summarized in Table 21-1. We've seen that the work done in any reversible process is just the area under the *PV* curve describing that process. A cyclic process returns to the same point in the *PV* diagram, so it generally involves both expansion and compression of the gas. During expansion the gas does work on its surroundings; during compression work gets done on the gas.

▲ **TABLE 21-1** IDEAL GAS PROCESSES

PROCESS	DEFINING CHARACTERISTIC	FIRST LAW	WORK DONE BY GAS	OTHER RELATIONS
Isothermal	T = constant	$Q = W$	$W = nRT \ln\left(\dfrac{V_2}{V_1}\right)$	PV = constant
Constant-volume	V = constant	$Q = \Delta U$	0	$Q = nC_V\Delta T$
Isobaric	P = constant	$Q = \Delta U + W$	$W = P(V_2 - V_1)$	$Q = nC_P\Delta T$ $C_P = C_V + R$
Adiabatic	$Q = 0$	$\Delta U = -W$	$W = \dfrac{P_1 V_1 - P_2 V_2}{\gamma - 1}$	PV^{γ} = constant $TV^{\gamma-1}$ = constant

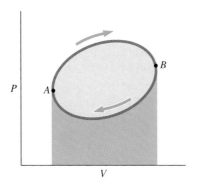

FIGURE 21-19 A cyclic path, traversed clockwise in the *PV* diagram. Area between the upper part of the curve and the *V* axis is the work done *by* the gas as it goes from state *A* to state *B*. Area between the lower part of the curve and the *V* axis is the work done *on* the gas as it returns form *B* to *A*. Net work done *by* the gas is the area enclosed by the cyclic path. If the path were traversed in the counterclockwise direction the area would instead be the work done *on* the gas.

The net work is the difference between the two. Figure 21-19 shows that this work is the area—measured as pressure times volume—enclosed by the cyclic path in the *PV* diagram.

● **EXAMPLE 21-5** A CYCLIC PROCESS

An ideal gas with $\gamma = 1.4$ occupies 4.0 L at 300 K and 100 kPa pressure. It is compressed adiabatically to one-fourth of its original volume, then cooled at constant volume to 300 K, and finally allowed to expand isothermally to its original volume. How much work is done on the gas?

Solution
Figure 21-20 shows the cyclic path *ABCA* in the *PV* diagram. To calculate the work done on the gas, we consider the work involved in moving along each of the sections *AB*, *BC*, and *CA*. The adiabatic work W_{AB} is given by Equation 21-14:

$$W_{AB} = \frac{P_A V_A - P_B V_B}{\gamma - 1}.$$

We know P_A, V_A, and V_B. To get P_B, we solve the adiabatic relation $P_B V_B^{\gamma} = P_A' V_A^{\gamma}$ to obtain

$$P_B = P_A \left(\frac{V_A}{V_B}\right)^{\gamma}.$$

Then the adiabatic work is

$$W_{AB} = \frac{P_A[V_A - (V_A/V_B)^{\gamma}V_B]}{\gamma - 1}$$

$$= \frac{(1.0 \times 10^5 \text{ Pa})[4.0 \times 10^{-3} \text{ m}^3 - (4^{1/4})(1.0 \times 10^{-3} \text{ m}^3)]}{0.4}$$

$$= -741 \text{ J}.$$

The minus sign indicates that work is done *on* the gas during this adiabatic compression.

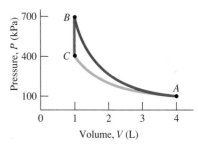

FIGURE 21-20 The cyclic process *ABCA* of Example 21-5 includes adiabatic (*AB*), constant volume (*BC*), and isothermal (*CA*) sections.

No work is done during the constant-volume process *BC*, while the isothermal work W_{CA} is given by Equation 21-4:

$$W_{CA} = nRT \ln\left(\frac{V_A}{V_C}\right).$$

The quantity nRT is given by the ideal gas law, $nRT = PV$, where we can evaluate P and V at any point along the isotherm because it's a curve of constant temperature. We know these values at point *A*, so the isothermal work becomes

$$W_{CA} = P_A V_A \ln\left(\frac{V_A}{V_C}\right)$$

$$= (1.0 \times 10^5 \text{ Pa})(4.0 \times 10^{-3} \text{ m}^3)\ln\left(\frac{4.0 \text{ L}}{1.0 \text{ L}}\right) = 555 \text{ J}.$$

The net work is then

$$W_{ABC} = W_{AB} + W_{BC} + W_{CA}$$
$$= -741 \text{ J} + 0 \text{ J} + 555 \text{ J} = -187 \text{ J}.$$

The minus sign indicates that net work is done on the gas.

We could achieve the cyclic process described here with our simple gas-cylinder system by first insulating the bottom of the cylinder and moving the piston to compress the gas to one-fourth its original volume. We would then clamp the piston in place to maintain constant volume and cool the gas to 300 K by placing the cylinder in contact with a heat reservoir whose temperature was slowly decreased. Finally, we would maintain contact with the 300 K heat reservoir while allowing the gas to expand to its original volume.

Since the system returns to its original state, its internal energy undergoes no net change. That means any work done on it must be rejected to its surroundings as heat. Since no heat flows during the adiabatic process AB, and since the gas *absorbs* heat during its isothermal expansion CA, the heat rejection takes place entirely during the constant-volume cooling period BC.

EXERCISE   The gas of this example starts at state A in Fig. 21-20 and is compressed adiabatically until its temperature is 400 K. It's then cooled while maintaining constant volume until it reaches 300 K, then allowed to expand isothermally back to state A. Find (a) the net work done on the gas and (b) the minimum volume reached.

Answers: (a) 45.7 J; (b) 1.95 L

Some problems similar to Example 21-5:   20, 21, 27, 31–33, 55, 56, 63   ●

21-3 SPECIFIC HEATS OF AN IDEAL GAS

We've found that the thermodynamic behavior of an ideal gas depends on the specific heats C_V and C_P. What are the values of those quantities?

Our ideal gas model of Chapter 20 assumed the gas molecules were structureless point particles. The only energy of such particles is their translational kinetic energy, and the internal energy U of the gas is the sum of all the molecular translational energies. But the average kinetic energy is directly proportional to the temperature:

$$\tfrac{1}{2}m\overline{v^2} = \tfrac{3}{2}kT.$$

If we have n moles of gas, the internal energy is then

$$U = nN_A\left(\tfrac{1}{2}m\overline{v^2}\right) = \tfrac{3}{2}nN_A kT,$$

where N_A is Avogadro's number. But $N_A k = R$, the gas constant, so

$$U = \tfrac{3}{2}nRT.$$

Using this result in Equation 21-8 for the molar specific heat gives

$$C_V = \frac{1}{n}\frac{\Delta U}{\Delta T} = \tfrac{3}{2}R. \tag{21-15}$$

For this simple gas of structureless particles, the adiabatic exponent γ is then

$$\gamma = \frac{C_P}{C_V} = \frac{C_V + R}{C_V} = \frac{\tfrac{5}{2}R}{\tfrac{3}{2}R} = \frac{5}{3} = 1.67. \tag{21-16}$$

Some gases, notably the inert gases helium (He), neon (Ne), argon (Ar), and others in the last column of the periodic table, have adiabatic exponents and specific heats given by these equations. But others do not. At room temperature, for example, hydrogen (H_2), oxygen (O_2), and nitrogen (N_2) obey adiabatic laws with γ very nearly $\frac{7}{5}$ ($= 1.4$). Solving Equation 21-16 for C_V shows that for these gases

$$C_V = \frac{R}{\gamma - 1} = \frac{R}{\frac{7}{5} - 1} = \tfrac{5}{2}R.$$

FIGURE 21-21 "Dumbbell" model of a diatomic molecule.

On the other hand, sulphur dioxide (SO_2) and nitrogen dioxide (NO_2) have specific heat ratios close to 1.3 and therefore volume specific heats of about $3.4R$.

What's going on here? A clue lies in the structure of individual gas molecules, reflected in their chemical formulas. The inert gas molecules are **monatomic,** consisting of single atoms. These atoms behave approximately like our model's structureless mass points. Hydrogen, oxygen, and nitrogen molecules are **diatomic,** as suggested in Fig. 21-21. Although a gas of such molecules should still obey the ideal gas law $PV = nRT$, the molecules differ in one important respect from mass points: In addition to translational kinetic energy, they can have significant rotational energy as well.

Its dumbbell shape means that a diatomic molecule has significant rotational inertia about two mutually perpendicular axes, as shown in Fig. 21-22. The molecule's rotational energy is therefore a sum of two terms, each representing rotation about one axis. In addition, of course, the molecule can also have translational energy associated with motion in any of three mutually perpendicular directions. So its total energy becomes

$$E = E_{\text{trans}} + E_{\text{rot}} = \tfrac{1}{2}mv_x^2 + \tfrac{1}{2}mv_y^2 + \tfrac{1}{2}mv_z^2 + \tfrac{1}{2}I_{x'}\omega_{x'}^2 + \tfrac{1}{2}I_{y'}\omega_{y'}^2,$$

where the I's are the rotational inertias about the axes of Fig. 21-22. The five terms in this equation represent five independent motions available to the diatomic molecule. The number of terms needed to describe the energy of a system is called the number of **degrees of freedom** of that system. A diatomic molecule has five degrees of freedom. A monatomic molecule, in contrast, has only three, corresponding to the three directions of translational motion.

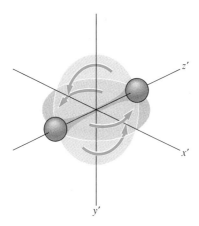

FIGURE 21-22 A diatomic molecule can rotate about two of three mutually perpendicular axes, in this case x' and y'. Rotation about the z' axis is not significant because the rotational inertia about this axis is negligible.

The Equipartition Theorem

We showed in Chapter 20 that the average translational kinetic energy associated with a gas molecule's motion in one direction is $\tfrac{1}{2}kT$. We then argued that all three directions of motion are equally probable, making the molecular kinetic energy, on average, $\tfrac{3}{2}kT$. The argument from one direction to three is statistical, based on the assumption that random collisions will "share" energy equally among the possible motions. When a molecule can rotate as well as translate, energy should be shared also among possible rotational motions.

Not only is energy shared, but it's shared equally. The nineteenth century Scottish physicist James Clerk Maxwell first proved this fact, which is now known as the **equipartition theorem:**

> **Equipartition theorem: When a system is in thermodynamic equilibrium, the average energy per molecule is $\frac{1}{2}kT$ for each degree of freedom.**

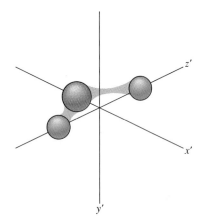

FIGURE 21-23 A triatomic molecule like NO_2 has significant rotational inertia about each of the three axes, and therefore has three rotational degrees of freedom.

We can use the equipartition theorem to calculate the specific heat of a diatomic molecule. Each molecule has five degrees of freedom, three translational and two rotational. The average energy of a molecule is then $5(\frac{1}{2}kT) = \frac{5}{2}kT$, so the total internal energy in n moles of a diatomic gas is

$$U = nN_A(\tfrac{5}{2}kT) = \tfrac{5}{2}nRT.$$

Equation 21-8 then gives the specific heat at constant volume:

$$C_V = \frac{1}{n}\frac{\Delta U}{\Delta T} = \tfrac{5}{2}R. \qquad \text{(diatomic molecule)}$$

Our result $C_P = C_V + R$ still holds, since it was derived from the first law of thermodynamics without regard to molecular structure, so

$$C_P = \tfrac{7}{2}R$$

and

$$\gamma = \frac{C_P}{C_V} = \frac{7}{5} = 1.4.$$

These results describe the observed behavior of diatomic gases like hydrogen, oxygen, and nitrogen at room temperature.

● EXAMPLE 21-6 MIXING GASES

A gas mixture consists of 2.0 mol of oxygen (O_2) and 1.0 mol of argon (Ar). What is the volume specific heat of the mixture?

Solution

Being diatomic, each oxygen molecule has 5 independent components of motion. Then the equipartition theorem gives the total internal energy of all the oxygen molecules:

$$U_{O_2} = nN_A(\tfrac{5}{2}kT) = (2.0 \text{ mol})(\tfrac{5}{2}RT) = 5.0RT,$$

where we used $N_A k = R$. Similarly, the internal energy associated with all the monatomic argon molecules is

$$U_{Ar} = nN_A(\tfrac{3}{2}kT) = (1.0 \text{ mol})(\tfrac{3}{2}RT) = 1.5RT.$$

Then the total internal energy of the gas is

$$U = U_{O_2} + U_{Ar} = 6.5RT,$$

so

$$C_V = \frac{1}{n}\frac{\Delta U}{\Delta T} = \frac{6.5R}{3.0 \text{ mol}} = 2.2R.$$

EXERCISE Find the specific heat at constant volume and the adiabatic index γ for a mixture of equal numbers of nitrogen (N_2) and helium (He) molecules.

Answers: $C_V = 2R$, $\gamma = 1.5$

Some problems similar to Example 21-6: 42–45 ●

A polyatomic molecule like NO_2 (Fig. 21-23) has significant rotational inertia about any of the three axes and therefore has three rotational degrees of freedom. With a total of six degrees of freedom, the internal energy of n moles is then

$$U = (nN_A)(6)(\tfrac{1}{2}kT) = 3nRT,$$

so

$$C_V = \frac{1}{n}\frac{dU}{dT} = 3R,$$

$$C_P = C_V + R = 4R,$$

and
$$\gamma = \frac{C_P}{C_V} = \frac{4}{3} = 1.33.$$

These values are reasonably close to the experimental values $\gamma = 1.29$ and $C_V = 3.47R$, although the agreement is noticeably poorer than for monatomic and diatomic gases.

Quantum Effects

Accounting for the effect of molecular structure on gas behavior shows how remarkably successful Newtonian mechanics can be. But at this molecular level we're pushing the limits of Newtonian mechanics, approaching the realm where a quantum-mechanical description is necessary. We can see the breakdown of Newtonian mechanics in Fig. 21-24, which shows the volume specific heat of hydrogen (H_2) as a function of temperature.

In the region from about 250 K to 750 K, the specific heat is very nearly equal to the value $\frac{5}{2}R$ that our dumbbell model predicts. But between 20 K and about 100 K, hydrogen's specific heat is close to $\frac{3}{2}R$—what we'd expect for a monatomic gas. And above about 3000 K the specific heat is $\frac{7}{2}R$, as though the molecules had acquired another two degrees of freedom.

What's going on here? The molecules seem to have different degrees of freedom, depending on temperature. And they do. In quantum mechanics, the energy associated with a periodic motion like rotation comes only in discrete multiples of some minimum amount. At low temperatures the average thermal energy $\frac{1}{2}kT$ is below this minimum, and the molecules can't rotate. So they have only three degrees of freedom, and the gas acts as if it were monatomic. As the temperature rises, molecules occasionally gain enough energy to rotate, and the specific heat increases. Eventually the temperature and thermal energy are high enough that essentially all the molecules are rotating, and the gas then acts like our diatomic model suggests it should. At still higher temperatures a new motion becomes possible, with the molecules vibrating like miniature mass-spring systems (Fig. 21-25). Energy is shared among both the kinetic and potential energies of this vibration, giving rise to two more degrees of freedom. The specific heat consequently rises to $\frac{7}{2}R$.

Are you bothered by the strange restrictions quantum mechanics imposes on the rotation and vibration of molecules? You should be! Nothing in the physics you've studied until now, and nothing in your everyday experience, suggests that a rotating object can't have any amount of energy you care to give it. But quantum mechanics deals with a realm of objects much smaller than those of our daily experience. The quantization of energy levels is only one of many unusual things that occur in the quantum realm. We will explore more quantum phenomena in Chapter 39.

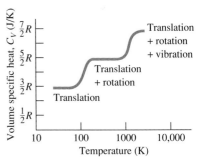

FIGURE 21-24 Volume specific heat of H_2 gas as a function of temperature. Below 20 K hydrogen is liquid and above 3200 K it dissociates into individual atoms.

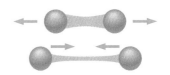

FIGURE 21-25 At high temperatures a diatomic molecule has enough energy to undergo vibrational motion.

CHAPTER SYNOPSIS

Summary

1. The **first law of thermodynamics** is a statement of energy conservation, equating the change in a system's internal energy to the difference between the heat Q added to the system and the work W done by the system:

$$\Delta U = Q - W.$$

2. A **thermodynamic process** takes a system from one thermodynamic state to another. A **quasi-static process** occurs slowly enough that the system moves through a sequence of equilibrium states, so the process can be

described by a path in the PV diagram. The work done by a system undergoing a quasistatic process that takes its volume from V_1 to V_2 is

$$W = \int_{V_1}^{V_2} P \, dV.$$

3. Four basic processes are **isothermal** (constant temperature), **constant volume, isobaric** (constant pressure), and **adiabatic** (no heat transfer). For an ideal gas, the work done and the relations among various thermodynamic variables are summarized in Table 21-1 in the text. A **cyclic process** takes a system around a closed path in its PV diagram, returning to its original state.

4. An ideal gas is characterized by its **molar specific heat at constant volume,** C_V, and its **molar specific heat at constant pressure,** C_P. The heat Q required to raise the temperature of n moles an amount ΔT is given by $Q = nC\Delta T$, where C is the specific heat appropriate to the condition— constant volume or pressure—under which the temperature change occurs. The specific heats are related by $C_P = C_V + R$, where R is the gas constant.

5. The **equipartition theorem** states that the average molecular energy per **degree of freedom** in thermodynamic equilibrium is $\frac{1}{2}kT$. Monatomic molecules have three degrees of freedom, associated with the three directions of translational motion; this leads to a volume specific heat $C_V = \frac{3}{2}R$. Diatomic molecules also undergo rotation, adding two more degrees of freedom and giving a volume specific heat of $\frac{5}{2}R$.

Terms You Should Understand

(Pairs are closely related terms whose distinction is important; number in parentheses is chapter section where term first appears.)
work (21-1)
internal energy (21-1)
first law of thermodynamics (21-1)

thermodynamic state, state variable (21-1)
quasi-static process (21-2)
reversible process, irreversible process (21-2)
isothermal process (21-2)
constant-volume process (21-2)
isobaric process (21-2)
molar specific heat at constant volume,
    . . . at constant pressure (21-2)
adiabatic process (21-2)
cyclic process (21-2)
degrees of freedom (21-3)
equipartition theorem (21-3)

Symbols You Should Recognize

ΔU (21-1)
C_V, C_P (21-2)
γ (21-2)

Problems You Should Be Able to Solve

calculating internal energy change, heat transfer, or work from the other two (21-1)
relating volume, pressure, and temperature changes for the basic ideal gas processes (21-2)
determining work done in basic ideal gas processes (21-2)
determining work involved in cyclic processes (21-2)
evaluating specific heats based on molecular structure (21-3)

Limitations to Keep in Mind

The ideal gas, with its internal energy a function of temperature alone, is an approximation. In real gases the weak forces among molecules make the internal energy depend slightly on volume as well.

Most relations developed in this chapter are valid only in the quasi-static limit; real processes may approach but not actually achieve this limit.

QUESTIONS

1. In Chapter 8 we wrote the conservation of energy principle in the form $K + U = $ constant. In what way is the first law of thermodynamics a broader statement than that?

2. The temperature of the water in a jar is raised by violently shaking the jar. Which of the terms Q and W in the first law of thermodynamics is involved in this case?

3. In Example 21-1, we considered the electric power leaving a power plant as part of the work done by the power plant. Why did we include the electric power with the work rather than with the heat?

4. What is the difference between heat and internal energy?

5. Some water is tightly sealed in a perfectly insulated container. Is it possible to change the water temperature? Explain.

6. Is the internal energy of a van der Waals gas a function of temperature alone? Why or why not?

7. Are the initial and final equilibrium states of an irreversible process describable by points in a PV diagram? Explain.

8. Why can an irreversible process not be described by a path in a PV diagram?

9. Does the first law of thermodynamics apply to an irreversible process?

10. Is it possible to have a process that is isothermal but irreversible? Explain.

11. A quasi-static process begins and ends at the same temperature. Is the process necessarily isothermal?

12. Figure 21-26 shows two processes connecting the same initial and final states. For which process is more heat added to the system?

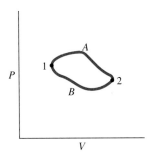

FIGURE 21-26 Question 12.

13. Two identical gas-cylinder systems are taken from the same initial state to the same final state, but by different processes. Is the work done in each case necessarily the same? The heat added? The change in internal energy?

14. When you let air out of a tire, the air seems cool. Why? What kind of process is occurring?

15. Blow on the back of your hand with your mouth wide open. You breath will feel hot. Now tighten your lips into a small opening, and blow again. Now your breath feels cool. Why?

16. How is it possible to have PV^γ = constant in an adiabatic process and yet have $PV = nRT$?

17. Water is boiled in an open pan. Of which of the four specific processes we considered is this an example?

18. Is it possible to involve an ideal gas in a process that is simultaneously isothermal and isobaric? Is there any real substance that could undergo such a process? Give an example.

19. Three identical gas-cylinder systems expand from the same initial state to final states that have the same volume. One system expands isothermally, one adiabatically, and one isobarically. Which does the most work? Which does the least work?

20. If the relation $\Delta U = nC_V\Delta T$ holds for any ideal gas process, why do we call C_V the molar specific heat at constant volume?

21. Why is the specific heat at constant pressure greater than at constant volume?

22. Imagine a gas of very complicated molecules, each with several hundred degrees of freedom. What, approximately, is the specific heat ratio γ of this gas? Is it easy or hard to change the gas temperature?

23. In what sense can a gas of diatomic molecules be considered an ideal gas, given that its molecules are not point particles?

PROBLEMS

Section 21-1 The First Law of Thermodynamics

1 In a perfectly insulated container, 1.0 kg of water is stirred vigorously until its temperature rises by 7.0°C. How much work was done on the water?

2. In a closed but uninsulated container, 500 g of water is shaken violently until its temperature rises by 3.0°C. The mechanical work required in the process is 9.0 kJ. (a) How much heat is transferred during the shaking? (b) How much mechanical energy would have been required had the container been perfectly insulated?

3. A 40-W heat source is applied to a gas sample for 25 s, during which time the gas expands and does 750 J of work on its surroundings. By how much does the internal energy of the gas change?

4. What is the rate of heat flow into a system whose internal energy is increasing at the rate of 45 W, given that the system is doing work at the rate of 165 W?

5. An engine produces useful work at the rate of 45 kW and heats its environment at the rate of 95 kW. At what rate does the engine extract energy from its fuel?

6. In a certain automobile engine, 17% of the total energy released in burning gasoline ends up as mechanical work. What is the engine's mechanical power output if its heat output is 68 kW?

7. Water flows over Niagara Falls (height 50 m) at the rate of about 10^6 kg/s. Suppose that all the water passes through a turbine connected to an electric generator producing 400 MW of electric power. If the water has negligible kinetic energy after leaving the turbine, by how much has its temperature increased between the top of the falls and the outlet of the turbine?

Section 21-2 Thermodynamic Processes

8. An ideal gas expands from the state (P_1, V_1) to the state (P_2, V_2), where $P_2 = 2P_1$ and $V_2 = 2V_1$. The expansion proceeds along the straight diagonal path AB shown in Fig. 21-27. Find an expression for the work done by the gas during this process.

9. Repeat the preceding problem for a process that follows the path ACB in Fig. 21-27.

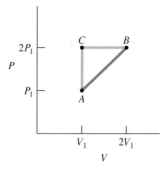

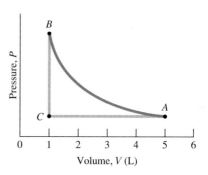

FIGURE 21-27 Problems 8, 9.

FIGURE 21-28 Problems 20, 21.

10. What are the minimum and maximum amounts of work needed to take the gas of Problem 8 between its initial and final states, if the gas pressure cannot be less than P_1 nor more than P_2? Show the appropriate paths in the PV diagram.

11. A balloon contains 0.30 mol of helium. It rises, while maintaining a constant 300 K temperature, to an altitude where its volume has expanded 5 times. How much work is done by the gas in the balloon during this isothermal expansion?

12. The balloon of the preceding problem starts at a pressure of 100 kPa and rises to an altitude where the pressure is 75 kPa, maintaining a constant 300 K temperature. (a) By what factor does its volume increase? (b) How much work does the gas in the balloon do?

13. How much work does it take to compress 2.5 mol of an ideal gas to half its original volume while maintaining a constant 300 K temperature?

14. An ideal gas expands to 10 times its original volume, maintaining a constant 440 K temperature. If the gas does 3.3 kJ of work on its surroundings (a) how much heat does it absorb and (b) how many moles of gas are there?

15. A 0.25 mol sample of an ideal gas initially occupies 3.5 L. If it takes 61 J of work to compress the gas isothermally to 3.0 L, what is the temperature?

16. A 1.5 mol sample of ideal gas occupies 30 L. If 750 J of work are done on the gas while its temperature remains constant at 273 K, what will be its new volume?

17. It takes 600 J to compress a gas isothermally to half its original volume. How much work would it take to compress it by a factor of 10 starting from its original volume?

18. A gas undergoes an adiabatic compression during which its volume drops to half its original value. If the gas pressure increases by a factor of 2.55, what is its specific heat ratio γ?

19. A gas with $\gamma = 1.4$ is at 100 kPa pressure and occupies 5.00 L. (a) How much work does it take to compress the gas adiabatically to 2.50 L? (b) What is its final pressure?

20. A gas sample undergoes the cyclic process $ABCA$ shown in Fig. 21-28, where AB lies on an isotherm. The pressure at

point A is 60 kPa. Find (a) the pressure at B and (b) the net work done on the gas.

21. Repeat the preceding problem taking AB to be on an adiabat and using a specific heat ratio of $\gamma = 1.4$.

22. A gas undergoes an adiabatic expansion in which its volume doubles. If it was originally at 160 kPa pressure and 410 K temperature, what are its new pressure and temperature? The gas has $\gamma = 1.3$.

23. A gasoline engine has a compression ratio of 8.5. If the fuel-air mixture enters the engine at 30°C, what will be its temperature at maximum compression? Assume the compression is adiabatic and that the mixture has $\gamma = 1.4$.

24. By how much must the volume of a gas with $\gamma = 1.4$ be changed in an adiabatic process if the pressure is to double?

25. By how much must the volume of a gas with $\gamma = 1.4$ be changed in an adiabatic process if the kelvin temperature is to double?

26. A B21F Volvo engine has a compression ratio of 9.3. If air at 320 K fills an engine cylinder when the piston is at the bottom of its stroke, what will be the air temperature when the piston is in its highest position? Treat the process as adiabatic, with $\gamma = 1.4$.

27. A gas expands isothermally from state A to state B, in the process absorbing 35 J of heat. It is then compressed isobarically to state C, where its volume equals that of state A. During the compression, 22 J of work are done on the gas. The gas is then heated at constant volume until it returns to state A. (a) Draw a PV diagram for this process. (b) How much work is done on or by the gas during the complete cycle? (c) How much heat is transferred to or from the gas as it goes from B to C to A?

28. Derive an equation relating pressure and temperature in an adiabatic process.

29. A 2.0 mol sample of ideal gas with molar specific heat $C_V = \frac{5}{2}R$ is initially at 300 K and 100 kPa pressure. Determine the final temperature and the work done by the gas when 1.5 kJ of heat is added to the gas (a) isothermally, (b) at constant volume, and (c) isobarically.

30. Prove that the slope of an adiabat at a given point in a PV diagram is γ times the slope of the isotherm passing through the same point.

31. An ideal gas with $\gamma = 1.67$ starts at point A in Fig. 21-29, where its pressure and volume are 1.00 m^3 and 250 kPa, respectively. It then undergoes an adiabatic expansion that triples its volume, ending at point B. It's then heated at constant volume to point C, then compressed isothermally back to A. Find (a) the pressure at B, (b) the pressure at C, and (c) the net work done on the gas.

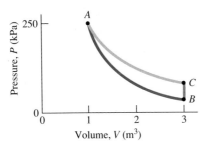

FIGURE 21-29 Problem 31.

32. The gas of Example 21-5 starts at state A in Fig. 21-20 and is compressed adiabatically until its volume is 2.0 L. It's then cooled while maintaining constant pressure until it reaches 300 K, then allowed to expand isothermally back to state A. Find (a) the net work done on the gas and (b) the minimum volume reached.

33. The gas of Example 21-5 starts at state A in Fig. 21-20 and is heated at constant volume until its pressure has doubled. It's then compressed adiabatically until its volume is one-fourth its original value, then cooled at constant volume to 300 K, and finally allowed to expand isothermally to its original state. Find the net work done on the gas.

34. A 25 L sample of an ideal gas with $\gamma = 1.67$ is at 250 K and 50 kPa. The gas is compressed isothermally to one-third of its original volume, then heated at constant volume until its state lies on the adiabatic curve that passes through its original state, and then allowed to expand adiabatically to that original state. Find the net work involved. Is work done on or by the gas?

35. A 25 L sample of an ideal gas with $\gamma = 1.67$ is at 250 K and 50 kPa. The gas is compressed adiabatically until its pressure triples, then cooled at constant volume back to 250 K, and finally allowed to expand isothermally to its original state. (a) How much work is done on the gas? (b) What is the minimum volume reached? (c) Sketch this cyclic process in a PV diagram.

36. A 25 L sample of ideal gas is at 250 K and 50 kPa. The gas is heated at constant volume until its pressure triples, then cooled at constant pressure until its temperature is back to 250 K, and finally allowed to expand isothermally to its original state. (a) How much work is done on the gas? (b) What is the minimum volume reached? (c) Sketch this cyclic process in a PV diagram.

37. A bicycle pump consists of a cylinder 30 cm long when the pump handle is all the way out. The pump contains air ($\gamma = 1.4$) at 20°C. If the pump outlet is blocked and the handle pushed until the internal length of the pump cylinder is 17 cm, by how much does the air temperature rise? Assume that no heat is lost.

38. A tightly sealed flask contains 5.0 L of air at 0°C and 100 kPa pressure. How much heat is required to raise the air temperature to 20°C? The molar specific heat of air at constant volume is 2.5R.

39. A balloon contains 5.0 L of air at 0°C and 100 kPa pressure. How much heat is required to raise the air temperature to 20°C, assuming the gas stays in pressure equilibrium with its surroundings? Neglect tension forces in the balloon. The molar specific heat of air at constant volume is 2.5R.

40. A piston-cylinder system has initial volume 1.0 L and contains ideal gas at 300 K and 1.0 atm, with $C_V = 2.5R$. The piston is held fixed and the cylinder placed in contact with a heat reservoir at 600 K. Once equilibrium is reached, the gas is allowed to expand isothermally. When the pressure reaches 1.0 atm, the gas is cooled at constant pressure to 300 K. This cyclic process is repeated once a second. (a) Draw a PV diagram for the process. (b) What is the rate at which the gas does work?

Section 21-3 Specific Heats of an Ideal Gas

41. What would be (a) the volume specific heat and (b) the specific heat ratio for a gas whose molecules had 9 degrees of freedom?

42. A gas mixture contains 2.5 mol O_2 and 3.0 mol Ar. What are the molar specific heats at constant volume and pressure for this mixture?

43. A mixture of monatomic and diatomic gases has specific heat ratio $\gamma = 1.52$. What fraction of the molecules are monatomic?

44. What should be the approximate specific heat ratio of a gas consisting of 50% NO_2 ($\gamma = 1.29$), 30% O_2 ($\gamma = 1.4$), and 20% Ar ($\gamma = 1.67$)?

45. A gas mixture contains monatomic argon and diatomic oxygen. An adiabatic expansion that doubles its volume results in the pressure dropping to one-third of its original value. What fraction of the molecules are argon?

46. You have 2.0 mol of an ideal diatomic gas whose molecules can rotate but not vibrate. Suppose you arrange to give the gas 8.0 kJ of energy in such a way that it all goes initially into translational motion of the molecules. When equilibrium is reached, what will be the gas temperature?

47. How much of a triatomic gas with $C_V = 3R$ would you have to add to 10 mol of monatomic gas to get a mixture whose thermodynamic behavior was like that of a diatomic gas?

48. By how much does the temperature of (a) an ideal monatomic gas and (b) an ideal diatomic gas (with molecular rotation but no vibration) change in an adiabatic process in which 2.5 kJ of work are done on each mole of gas?

Paired Problems

(Both problems in a pair involve the same principles and techniques. If you can get the first problem, you should be able to solve the second one.)

49. A 5.0 mol sample of ideal gas with $C_V = \frac{5}{2}R$ undergoes an expansion during which the gas does 5.1 kJ of work. If it absorbs 2.7 kJ of heat during the process, by how much does its temperature change? *Hint:* Remember that Equation 21-7 holds for *any* ideal gas process.

50. External forces compress 21 mol of ideal monatomic gas; during the process the gas transfers 15 kJ of heat to its surroundings, yet its temperature rises by 160 K. How much work was done on the gas?

51. A gas with $\gamma = \frac{5}{3}$ is at 450 K at the start of an expansion that triples its volume. The expansion is isothermal until the volume has doubled, then adiabatic the rest of the way. What is the final gas temperature?

52. A gas with $\gamma = \frac{7}{5}$ is at 273 K when it's compressed isothermally to one-third of its original volume, then further compressed adiabatically to one-fifth of its original volume. What is its final temperature?

53. An ideal gas with $\gamma = 1.4$ is initially at 273 K and 100 kPa. The gas expands adiabatically until its temperature drops to 190 K. What is its final pressure?

54. An ideal gas with $\gamma = 1.3$ is initially at 273 K and 100 kPa. The gas is compressed adiabatically to 240 kPa pressure. What is its final temperature?

55. The curved path in Fig. 21-30 lies on the 350-K isotherm for an ideal gas with $\gamma = 1.4$. (a) Calculate the net work done on the gas as it goes around the cyclic path *ABCA*. (b) How much heat flows into or out of the gas on the segment *AB*?

56. Repeat part (a) of the preceding problem for the path *ACDA* in Fig. 21-30. (b) How much heat flows into or out of the gas on the segment *CD*?

Supplementary Problems

57. An 8.5-kg rock at 0°C is dropped into a well-insulated vat containing a mixture of ice and water at 0°C. When equilibrium is reached there are 6.3 g less ice. From what height was the rock dropped?

58. A piston-cylinder arrangement containing 0.30 mol of nitrogen at high pressure is in thermal equilibrium with an ice-water bath containing 200 g of ice. The pressure of the ambient air is 1.0 atm. The gas is allowed to expand isothermally until it is in pressure balance with its surroundings. After the process is complete, the bath contains 210 g of ice. What was the original gas pressure?

59. Repeat Problem 8 for the case when the gas expands along a path given by $P = P_1 \left[1 + \left(\dfrac{V - V_1}{V_1} \right)^2 \right]$. Sketch the path in the *PV* diagram, and determine the work done.

60. A piston-cylinder arrangement contains 1.0 g of water at 50°C. The piston, which has negligible mass, is free to move and is exposed to atmospheric pressure (1.0 atm). The system is heated to 200°C. Assuming that liquid water is incompressible and that steam is an ideal gas with volume specific heat 4.3R, determine (a) the work done by the system and (b) the heat added to it.

61. Show that the application of Equation 21-3 to an adiabatic process results in Equation 21-14.

62. Find an expression for the molar specific heat at constant volume of an ideal gas in terms of its adiabatic exponent γ.

63. An ideal gas is taken clockwise around the circular path shown in Fig. 21-31. (a) How much work does the gas do? (b) If there are 1.3 moles of gas, what is the maximum temperature reached?

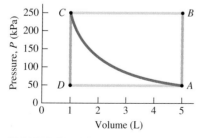

FIGURE 21-30 Problems 55, 56.

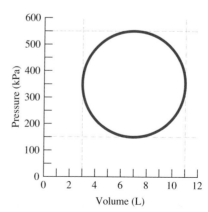

FIGURE 21-31 Problem 63.

64. A piston-cylinder system contains 0.50 mol of hydrogen (H_2) at 400 K and 300 kPa. An identical system contains 0.50 mol of helium (He) under identical conditions. Each gas undergoes an expansion that quadruples the system volume. Calculate the work done by each, if the expansion is (a) isothermal and (b) adiabatic.

65. Show that the work done by a van der Waals gas undergoing isothermal expansion from volume V_1 to V_2 is

$$W = nRT \ln\left(\frac{V_2 - nb}{V_1 - nb}\right) + an^2\left(\frac{1}{V_2} - \frac{1}{V_1}\right),$$

where a and b are the constants in Equation 20-5.

66. The compression and rarefaction in a sound wave occur fast enough that very little heat flow occurs, making the process essentially adiabatic. Show in this case that the bulk modulus (Equation 17-5) become $B = \gamma P$, and then show that Equation 17-6 leads to Equation 17-1 for the sound speed. *Hint:* Write Equation 17-5 in derivative form:

$$B = -V\frac{dP}{dV}.$$

67. A horizontal piston-cylinder system containing n mol of ideal gas is surrounded by air at temperature T_0 and pressure P_0. If the piston is displaced slightly from equilibrium, show that it executes simple harmonic motion with angular frequency $\omega = AP_0/\sqrt{MnRT_0}$, where A and M are the piston area and mass, respectively. Assume the gas temperature remains constant.

68. Repeat the preceding problem, now assuming that no heat flows into or out of the gas.

69. A cylinder of cross-sectional area A is closed by a massless piston. The cylinder contains n mol of ideal gas with specific heat ratio γ, and is initially in equilibrium with the surrounding air at temperature T_0 and pressure P_0. The piston is initially at height h_1 above the bottom of the cylinder. Sand is gradually sprinkled onto the piston until it has moved downward to a final height h_2. Find the total mass of the sand if the process is (a) isothermal and (b) adiabatic.

CHAPTER 22

THE SECOND LAW OF THERMODYNAMICS

A power plant's cooling towers transfer waste heat to the environment. In a typical power plant, two-thirds of the energy released from the fuel ends up as waste heat—a consequence of the second law of thermodynamics.

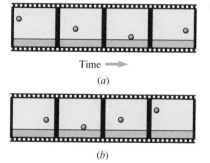

Time ⟶

(a)

(b)

FIGURE 22-1 A movie of a bouncing ball makes sense whether it's shown forward or backward.

The first law of thermodynamics relates heat and other forms of energy. Much of our world works because of this relation. Cars run on energy extracted from the heat of burning gasoline. Most of the electrical energy that powers lights and motors and computers originates in heat released in burning fuels or fissioning uranium. Our own bodies run on energy that was once released as heat deep in the Sun's core. But the first law doesn't tell the whole story. Heat and mechanical energy are not exactly the same thing, and the difference between them makes the conversion of heat to work a more subtle task than the first law alone would imply.

22-1 REVERSIBILITY AND IRREVERSIBILITY

Figure 22-1 shows a movie of a bouncing ball. We can run it forward or backward and still have a plausible sequence of physical events. Figure 22-2 shows a movie of another simple physical process. A block slides along a table, eventually coming to a stop. Friction warms the block, so we see its temperature rise. Play this one in reverse, and it makes no sense. We never see a block at rest suddenly start to move, cooling in the process. Yet energy could be conserved if it did, so the first law of thermodynamics would be satisfied.

Beat an egg, and the yolk and white quickly blend. You would be surprised if, on reversing the beater, the scrambled yolk and white separated—yet nothing in the laws of mechanics would prevent it. Or put cups of cold and hot water in thermal contact. The hot water gets colder, and the cold water gets hotter. The reverse never occurs—even though energy would be conserved if it did.

Most events are **irreversible,** in the sense of the block, the egg, and the water. What is the origin of this irreversibility? In each case we start with matter in an organized state. In the sliding block all molecules share a common motion. The egg is organized so all the yolk molecules are in one place. The hot water has a greater number of energetic molecules. Of all the possible states into which matter might arrange itself, these *organized* states are relatively rare. There are many more *chaotic* states—for instance, all the possible arrangements of molecules in a scrambled egg—that have less organization. When a system evolves, chances are it will end up in a less organized state, simply because there are far more such states available to it. It's then very unlikely to assume spontaneously a more organized state.

A key word here is *spontaneous*. We could restore a system to a more organized state—for example, by putting one cup of lukewarm water on the stove and the other in the refrigerator—but then we have to carry out a rather deliberate and energy-consuming process.

Thermodynamic processes can also be irreversible, as we discussed in the preceding chapter. Plunging a cool gas sample into boiling water, for example, creates temporarily a more organized state in which gas molecules near the container walls have higher energy (recall Fig. 21-6). Molecular collisions soon spread the energy throughout the gas, and the organization disappears. A spontaneous return to the organized state is highly improbable, and in this sense the process is irreversible.

Irreversibility is a probabilistic notion. Events that *could* occur without violating the principles of Newtonian physics nevertheless *don't* occur because they're too improbable. A practical consequence is that it's difficult to harness the considerable energy tied up as internal energy associated with random molecular motions because those motions won't spontaneously become organized. That makes much of the world's energy unavailable for doing useful work.

22-2 THE SECOND LAW OF THERMODYNAMICS

Heat Engines

It is impossible to convert *all* the internal energy of a system to useful work. But devices called **heat engines** can extract *some* of that internal energy. Examples of heat engines include gasoline and diesel engines, fossil-fueled and nuclear power plants, and jet aircraft engines.

Conceptually, we characterize a heat engine with a diagram showing heat flowing into the engine and work emerging. Figure 22-3*a* is such an energy-flow diagram for a "perfect" heat engine—one that extracts heat from a heat reservoir

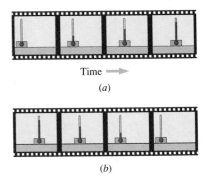

FIGURE 22-2 Movie of a block warming as friction dissipates its kinetic energy. The reverse sequence would never happen, even though it does not violate energy conservation.

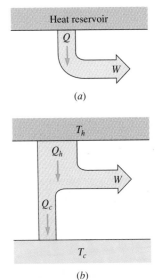

FIGURE 22-3 (*a*) Energy flow diagram for a perfect heat engine, which extracts heat Q from a reservoir and delivers an equal amount of work. (*b*) A real engine delivers as work only a fraction of the heat Q_h extracted from the high-temperature reservoir; the remainder is rejected to the low-temperature reservoir.

and converts it all to work. Such an engine would do exactly what we argued against in the preceding section—it would convert the random energy of thermal motion entirely into the ordered motion associated with mechanical work. In fact a perfect heat engine is impossible, for the same reason that we can't unscramble an egg or cause a block to accelerate spontaneously at the expense of its internal energy. This fact comprises one statement of the **second law of thermodynamics:**

> *Second law of thermodynamics (Kelvin-Planck statement): It is impossible to construct a heat engine operating in a cycle that extracts heat from a reservoir and delivers an equal amount of work.*

This particular form of the second law was formulated by Lord Kelvin and Max Planck. The phrase "in a cycle" means that a practical engine must go through a repeated sequence of steps, as in the back-and-forth motions of the pistons in a gasoline engine.

A simple heat engine consists of a gas-cylinder system and a heat reservoir, the latter kept hot, perhaps, by burning a fuel. With the gas initially at high pressure, we place the cylinder in contact with the heat reservoir. The gas expands and does work W on the piston. In this isothermal process, the gas extracts heat $Q = W$ from the reservoir. Eventually the gas reaches pressure equilibrium and stops expanding. The piston must then be returned to its original position if it's to do more work.

If we just push the piston back, we'll have to do as much work as was extracted during the expansion, and our engine will produce no net work. Instead we can cool the gas to reduce its volume, through thermal contact with a cool reservoir. But then some energy leaves the system as heat rather than work, as shown conceptually in Fig. 22-3b. Our engine extracts heat from a source and delivers mechanical work, but over a full cycle the amount of work is always less than the heat extracted. The remaining energy is rejected to the lower temperature reservoir, usually the environment. In practical terms, much of the energy released from fuels in car engines and power plants ends up as waste heat (Fig. 22-4).

The second law of thermodynamics states that we can't build a perfect heat engine. But how close can we come? Since we pay for the fuel that heats the high-temperature reservoir, it's clearly advantageous to minimize the rejected

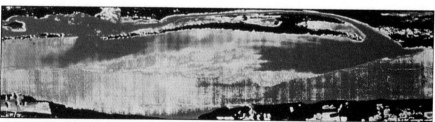

(a)   (b)

FIGURE 22-4 Real heat engines necessarily reject heat to the environment. (a) A car radiator transfers much of the waste heat to the surrounding air. (b) Power plants usually transfer waste heat to a nearby body of water. This infrared aerial photo shows hot water (red) discharged from a power plant. Such "thermal pollution" can cause serious ecological problems.

heat. We define the **efficiency** of an engine as the ratio of work W output by the engine in one cycle to the heat Q_h absorbed from the high-temperature reservoir,

$$e = \frac{W}{Q_h}.$$

Since the process is cyclic, there's no net change in internal energy over one cycle. Then the first law of thermodynamics ensures that the work done is the difference between the heat Q_h extracted from the high-temperature reservoir and the heat Q_c rejected to the cool reservoir. So the efficiency is

$$e = \frac{Q_h - Q_c}{Q_h} = 1 - \frac{Q_c}{Q_h}. \tag{22-1}$$

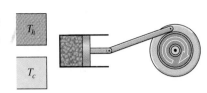

FIGURE 22-5 A simple heat engine. The gas absorbs heat from the hot reservoir, does work on the piston and wheel, and rejects heat to the cool reservoir.

Figure 22-5 shows a simple heat engine whose efficiency we can readily calculate. The engine consists of a closed cylinder containing an ideal gas. A movable piston is connected to a rod that drives a wheel. We start with the piston in its most retracted position and place the cylinder in contact with the high-temperature reservoir at T_h. The gas expands isothermally along path AB in its PV diagram (Fig. 22-6). In the process, the gas extracts heat Q_h from the reservoir and, because its internal energy does not change, delivers an equal amount of work to the wheel. At state B, we remove the cylinder from the reservoir so the expansion becomes adiabatic. We design the engine so that, when the piston reaches its maximum displacement, the gas temperature has cooled to T_c, the temperature of the cool reservoir. At this point the gas is in state C in its PV diagram. We then place the cylinder in contact with the cool reservoir. The wheel's inertia keeps it turning, and the wheel does work on the gas, compressing it isothermally from state C to state D. This work ends up as heat rejected to the cool reservoir. Finally, at state D, we separate the cylinder from the reservoir and allow the compression to continue adiabatically until the gas temperature has risen to T_h and the piston is once again fully retracted.

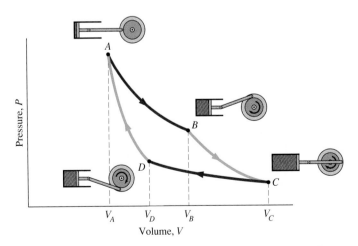

FIGURE 22-6 PV diagram for the engine of Fig. 22-5. AB and DC are isotherms at the temperatures of the hot and cool reservoirs, respectively; AD and BC are adiabatic curves.

The engine we have just described undergoes a cyclic process consisting of four reversible steps, two isothermal and two adiabatic. This cycle is called a **Carnot cycle,** and the engine a **Carnot engine,** after the French engineer Sadi Carnot (1796–1832), who explored the properties of such an engine even before the first law of thermodynamics was formulated. The particular configuration of the engine is not important, nor is the choice of an ideal gas as the engine's **working fluid.** What distinguishes the Carnot cycle from others is the sequence of thermodynamic processes and the fact that these processes are reversible. The Carnot engine is an example of a **reversible engine**—one in which thermodynamic equilibrium is maintained so that all steps could, in principle, be reversed.

What is the efficiency of a Carnot engine? To find out, we need the heats Q_h and Q_c absorbed and rejected during the isothermal parts of the cycle shown in Fig. 22-6. We developed Equation 21-4 to deal with such isothermal processes. Applying that equation gives the heat Q_h absorbed during the isothermal expansion AB:

$$Q_h = nRT_h \ln\left(\frac{V_B}{V_A}\right),$$

and the heat Q_c rejected during the isothermal compression CD:

$$Q_c = -nRT_c \ln\left(\frac{V_D}{V_C}\right) = nRT_c \ln\left(\frac{V_C}{V_D}\right).$$

We put the minus sign here because our statement of the first law describes Q as the heat *absorbed,* while Equation 22-1 for the engine efficiency requires that Q_c be the heat *rejected.* To calculate engine efficiency according to Equation 22-1, we need the ratio Q_c/Q_h:

$$\frac{Q_c}{Q_h} = \frac{T_c \ln(V_C/V_D)}{T_h \ln(V_B/V_A)}. \tag{22-2}$$

This expression can be simplified by applying Equation 21-13b to the adiabatic processes BC and DA in the Carnot cycle:

$$T_h V_B^{\gamma-1} = T_c V_C^{\gamma-1} \quad \text{and} \quad T_h V_A^{\gamma-1} = T_c V_D^{\gamma-1}.$$

Dividing these two equations gives

$$\left(\frac{V_B}{V_A}\right)^{\gamma-1} = \left(\frac{V_C}{V_D}\right)^{\gamma-1} \quad \text{or} \quad \frac{V_B}{V_A} = \frac{V_C}{V_D}$$

so Equation 22-2 becomes simply

$$\frac{Q_c}{Q_h} = \frac{T_c}{T_h}.$$

Using this result in Equation 22-1 then gives the efficiency of the Carnot engine:

$$e_{\text{Carnot}} = 1 - \frac{T_c}{T_h},\qquad(22\text{-}3)$$

where the temperatures are measured on an absolute scale (Kelvin or Rankine). Equation 22-3 tells us that the efficiency of a Carnot engine depends only on the highest and lowest temperatures of the working fluid. For a practical engine, the low temperature is usually the ambient temperature of the environment. Then to maximize the efficiency, we must make the high temperature as high as possible. Real engines represent a compromise between efficiency and the ability of materials to withstand high temperatures and pressures.

● **EXAMPLE 22-1** A CARNOT ENGINE

A Carnot engine extracts 240 J of heat from a high-temperature reservoir during each cycle. It rejects 100 J of heat to a reservoir at 15°C. How much work does the engine do in one cycle? What is its efficiency? What is the temperature of the hot reservoir?

Solution
The first law of thermodynamics requires that energy not rejected as heat be delivered as work, so the engine does

$$W = 240\text{ J} - 100\text{ J} = 140\text{ J}$$

of work. The efficiency is defined by Equation 22-1 as the ratio of work to heat extracted from the hot reservoir:

$$e = \frac{W}{Q_h} = \frac{140\text{ J}}{240\text{ J}} = 0.583 = 58.3\%.$$

Knowing the efficiency, we can solve Equation 22-3 for the high temperature to get

$$T_h = \frac{T_c}{1 - e} = \frac{288\text{ K}}{1 - 0.583} = 691\text{ K} = 418°C.$$

Note that in using Equation 22-3 we must work with absolute temperatures.

EXERCISE A Carnot engine operates between heat reservoirs at 520 K and 280 K. (a) What is its efficiency? (b) If it produces useful work at the rate of 400 W, at what rate does it reject waste heat?

Answers: (a) 46%; (b) 467 W

Some problems similar to Example 22-1: 5–8 ●

Engines, Refrigerators, and the Second Law

What's so special about the Carnot engine? Couldn't we build a different kind of engine with greater efficiency? The answer is no. The special role of the Carnot cycle is embodied in **Carnot's theorem:**

> **Carnot's theorem: All reversible Carnot engines operating between temperatures T_h and T_c have the same efficiency (given by Equation 22-3), and no other engine operating between the same two temperatures can have a greater efficiency.**

To prove Carnot's theorem, we introduce the **refrigerator.** A refrigerator is the opposite of an engine: It extracts heat from a cool reservoir and rejects it to a hotter one, taking in work in the process (Fig. 22-7). A refrigerator forces heat to flow the way it doesn't spontaneously go—from cold to hot—but to do so it requires work. A household refrigerator cools its contents and warms the house (you can feel the heat coming out the back), but in the process it uses electricity. That heat doesn't flow spontaneously from cold to hot constitutes another statement of the second law of thermodynamics, this one due to the German physicist Rudolph Clausius (1822–1888):

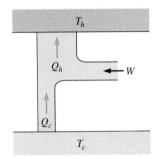

FIGURE 22-7 Energy flow diagram for a real refrigerator. The device takes in mechanical work and transfers heat from the cool to the hot reservoir.

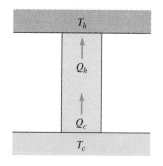

FIGURE 22-8 A perfect refrigerator would require no work to transfer heat from a cool object to a hotter one. The second law of thermodynamics rules out such a device.

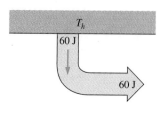

(a)

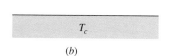

(b)

FIGURE 22-9 (a) A real heat engine combined with a perfect refrigerator. (b) The combination is equivalent to a perfect heat engine.

> **Second law of thermodynamics (Clausius statement): It is impossible to construct a refrigerator operating in a cycle whose sole effect is to transfer heat from a cooler object to a hotter one.**

The Clausius statement rules out a "perfect" refrigerator, like that shown in Fig. 22-8.

Suppose the Clausius statement were false. Then we could build the device of Fig. 22-9a, consisting of a reversible Carnot engine and a perfect refrigerator. In each cycle the engine would extract, say, 100 J from the hot reservoir, put out 60 J of useful work, and reject 40 J to the cool reservoir. The perfect refrigerator could transfer the 40 J back to the hot reservoir. The net effect would then be to extract 60 J from the hot reservoir and convert it entirely to work (Fig. 22-9b)—and we would have a perfect heat engine, in violation of the Kelvin-Planck statement of the second law. A similar argument shows that if a perfect heat engine is possible, so is a perfect refrigerator. So the Clausius and Kelvin-Planck statements of the second law of thermodynamics are equivalent, in that if one is false then so is the other.

Because the Carnot engine is reversible, we could run it backwards and reverse its path in the PV diagram of Fig. 22-6. Each process would run in reverse, and the engine would extract heat from the cool reservoir, take in work, and reject heat to the hot reservoir. It would be a refrigerator. Although real refrigerators are not designed exactly like engines, the two are, in principle, interchangeable.

We're now ready to prove Carnot's assertion that Equation 22-3 gives the maximum engine efficiency. Consider again the Carnot engine shown in Fig. 22-9a. It extracts 100 J of heat and delivers 60 J of work, so it's 60% efficient. Suppose we had another engine operating between the same two reservoirs, but with 70% efficiency. Since the Carnot engine is reversible, we can run it as a refrigerator. If we then put the two together, we get the device of Fig. 22-10a. Its net effect is to extract 10 J of heat from the cool reservoir and deliver 10 J of work—so it's a perfect heat engine, in violation of the second law (Fig. 22-10b). We could make the same argument for any combination of a reversible Carnot engine and another, more efficient engine. It's therefore impossible to make an engine that's more efficient than a reversible Carnot engine, and thus, Equation 22-3 gives the maximum possible efficiency for *any* heat engine operating between the same two fixed temperatures.

Irreversible engines, because they involve processes that dissipate organized motion, are necessarily *less* efficient. So are many reversible engines, if their heat exchange does not take place solely at the highest and lowest temperatures available (see Problem 54). The ordinary gasoline engine is a case in point; even if it could be made perfectly reversible, its efficiency would be less than that of a comparable Carnot engine (see Problem 61).

22-3 APPLICATIONS OF THE SECOND LAW OF THERMODYNAMICS

The world abounds with thermal energy, but the second law of thermodynamics imposes a fundamental limitation on our ability to turn that energy to our own uses. Any device we construct that involves the interchange of heat and work is a heat engine or refrigerator, and is therefore subject to the second law.

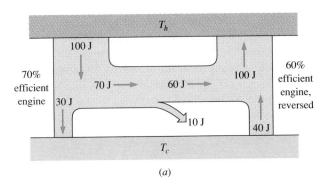

(a)

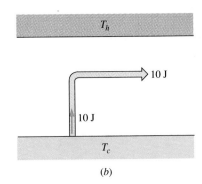

(b)

FIGURE 22-10 (a) A 60% efficient reversible engine run as a refrigerator, along with a hypothetical engine with 70% efficiency. (b) The combination is equivalent to a perfect heat engine delivering 10 J per cycle.

Limitations on Heat Engines

Most of the electricity used in the United States is produced in large power plants that are basically heat engines powered by fossil fuels like coal, oil, or natural gas, or by nuclear fission of uranium (Fig. 22-11). Figure 22-12 shows a schematic diagram of such a power plant. The working fluid is water, heated in a boiler and converted to steam at high pressure. The steam expands adiabatically against the blades of a turbine—a fanlike device that spins when struck by the steam. The turbine turns a generator that converts mechanical work into electrical energy (more on this in Chapter 31).

Steam leaving the turbine is still in the gaseous state, and is hotter than the water supplied to the boiler. Here's where the second law enters! Had the water returned to its original state, we would have extracted as work all the energy acquired in the boiler, in violation of the second law. To use the steam again, we run it through a **condenser**—a device where pipes carrying the steam are in contact with large volumes of cool water, typically from a river, lake, or ocean. The condensed steam, now cool water, is fed back into the boiler to repeat the cycle.

FIGURE 22-11 A nuclear power plant, showing reactor containment vessel at left, and large cooling towers at right.

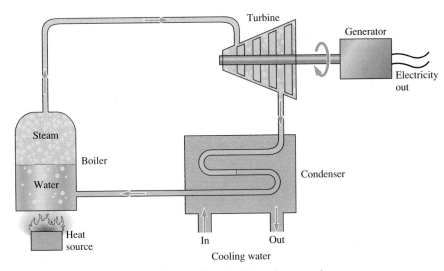

FIGURE 22-12 Schematic diagram of a typical electric power plant.

The maximum steam temperature in a power plant is limited by the materials used in its construction. For a fossil-fueled plant, current technology permits high temperatures around 650 K. Potential damage to nuclear fuel rods limits the temperature in a nuclear plant to around 570 K. Cooling-water temperature averages about 40°C (310 K), so the maximum possible efficiencies for these power plants, given by Equation 22-3, are

$$e_{\text{fossil}} = 1 - \frac{310 \text{ K}}{650 \text{ K}} = 0.52 = 52\%$$

and

$$e_{\text{nuclear}} = 1 - \frac{310 \text{ K}}{570 \text{ K}} = 0.46 = 46\%.$$

Temperature differences between the exhausted steam and the cooling water, mechanical friction, and the need to divert energy for driving pumps and pollution control devices all reduce efficiency further, to about 40% for fossil-fueled plants and 34% for nuclear plants. This means that, when we make electricity, roughly two-thirds of the fuel energy released ends up as waste heat.

A typical large power plant produces 1000 MW of electricity, so another 2000 MW of waste heat is dumped into the cooling water. This rate of energy addition can cause a large temperature rise in even a major river, and can lead to serious ecological problems. The need for power plant cooling water—imposed on us by the second law of thermodynamics—is so great that a substantial fraction of all rainwater falling on the United States eventually finds its way through the condensers of power plants (see Problem 13).

To reduce this "thermal polllution," power plants employ cooling towers—expensive devices which release waste heat to the air rather than to water. Environmental effects of the heated air are less serious than with heated water, although the increase in humidity can bring about unfortunate changes in local weather (Fig. 22-13).

FIGURE 22-13 Fog forms above the cooling towers of a large coal-fired electric power plant.

● EXAMPLE 22-2 VERMONT YANKEE

The Vermont Yankee nuclear power plant at Vernon, Vermont, produces 540 MW of electric power, while energy from nuclear fission is released as heat at the rate of 1590 MW. Steam produced in the reactor enters the turbine at a temperature of 556 K and is discharged to the condenser at 313 K. Water from the Connecticut River is pumped through the condenser at the rate of 2.27×10⁴ kg/s.

What is the maximum efficiency of the power plant, as limited by the second law of thermodynamics? What is the actual efficiency? How much does the temperature of the cooling water rise? How many houses like that of Example 19-5 could be heated with the waste heat from Vermont Yankee?

Solution

The second-law efficiency is given by Equation 22-3:

$$e = 1 - \frac{T_c}{T_h} = 1 - \frac{313 \text{ K}}{556 \text{ K}} = 0.437 = 43.7\%.$$

The actual efficiency is the ratio of electric power output to the rate of heat extraction from the nuclear fuel:

$$e = \frac{dW/dt}{dQ_h/dt} = \frac{540 \text{ MW}}{1590 \text{ MW}} = 0.340 = 34.0\%.$$

Waste heat is discharged to the river at a rate given by the difference between the thermal power output and the electric power output:

$$\frac{dQ_c}{dt} = \frac{dQ_h}{dt} - \frac{dW}{dt} = 1590 \text{ MW} - 540 \text{ MW} = 1050 \text{ MW}.$$

We can relate the waste power output to the temperature rise using Equation 19-5:

$$\Delta T = \frac{\Delta Q_c}{mc} = \frac{dQ_c/dt}{c \, dm/dt}$$

$$= \frac{1050 \times 10^6 \text{ W}}{(4184 \text{ J/kg·°C})(2.27 \times 10^4 \text{ kg/s})} = 11.1°\text{C},$$

where we used the given values of heat and mass per unit time for the ratio $\Delta Q/m$. The house of Example 19-5 uses 7926 Btu/h, or 2.3×10^{-3} MW. Then the waste heat from Vermont Yankee could heat $(1050 \text{ MW})/(2.4\times10^{-3} \text{ MW/house}) = 457{,}000$ houses. This number is unrealistic because most houses aren't as energy efficient as the one in Example 19-5; nevertheless, it shows that waste heat from power plants is a potentially valuable source for heating buildings.

EXERCISE A 40% efficient coal-fired power plant produces 1100 MW of electric power. (a) If spent steam leaves the turbine at 350 K, what is the lowest possible value for the steam temperature as it leaves the boiler? (b) If the plant is cooled with essentially the entire 4.4×10^{4}-kg/s flow of a river, by how much does the river temperature rise?

Answers: (a) 583 K; (b) 9.0°C

Some problems similar to Example 22-2: 10–12, 47, 48, 60

Gasoline and diesel engines provide another pervasive example of heat engines. A typical automobile engine has a theoretical maximum efficiency of just over 50%, but irreversible thermodynamic processes make the actual efficiency much lower. Mechanical friction dissipates additional energy, and the end result is that less than 20% of the fuel energy is available at the driving wheels. Problems 61 and 62 explore the thermodynamics of the gasoline engine.

We wouldn't be so concerned with efficiency if we didn't have to pay for fuel. Even a very small temperature difference can drive a heat engine, although with low efficiency. Surface temperatures in the tropical oceans are around 25°C (298 K), while hundreds of meters down the water temperature is only about 5°C (278 K). A heat engine operating between these temperatures would have an efficiency of only $1 - 278 \text{ K}/298 \text{ K} = 0.07$ or 7%. Nevertheless, there are large amounts of warm ocean water, and its energy—solar in origin—is free. Substantial engineering problems remain before such **ocean thermal energy conversion** (OTEC) becomes practical, but pilot plants are already being tested (Fig. 22-14).

Another solar-powered heat engine uses mirrors to concentrate sunlight, heating a fluid and driving a turbine. Several hundred megawatts of such **solar-thermal power plants** have been built in California, and are generating power at rates just a few cents above those of conventional power plants (Fig. 22-15).

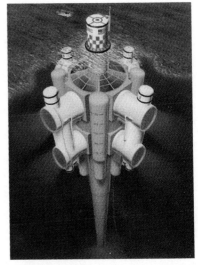

FIGURE 22-14 Artist's conception of an OTEC power plant.

Refrigerators and Heat Pumps

Reversing a heat engine gives a refrigerator, which takes in mechanical work and transfers heat from cooler to hotter (Fig. 22-16). The heat coming out the back of your home refrigerator comprises both the energy removed from the refrigerator's contents as well as the energy that was supplied as electricity to run the refrigerator. A well-designed refrigerator should minimize the amount of work (or its equivalent, electrical energy) needed to extract a given amount of heat. But again, the second law of thermodynamics imposes limitations on even the best refrigerator.

The **coefficient of performance** (COP) is the ratio of heat extracted at the lower temperature to work input by a refrigerator:

$$\text{COP} = \frac{Q_c}{W} = \frac{Q_c}{Q_h - Q_c}. \tag{22-4}$$

For a reversible Carnot engine, we found in deriving Equation 22-3 that the heat ratio Q_c/Q_h is equal to the temperature ratio T_c/T_h. The best refrigerator we can

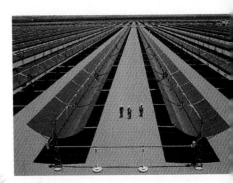

FIGURE 22-15 This solar-thermal power plant in California's Mojave Desert generates 90 MW of electricity. Sun-tracking parabolic reflectors concentrate sunlight on pipes carrying synthetic oil, which then transfers the energy to boil water that drives a steam turbine.

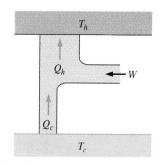

FIGURE 22-16 Energy flow diagram for a refrigerator.

build operates in a reversible Carnot cycle, so for this refrigerator the heat and temperature ratios are also equal. Then the COP becomes

$$COP = \frac{T_c}{T_h - T_c}. \qquad (22\text{-}5)$$

As the high and low temperatures become arbitrarily close, Equation 22-5 shows that the COP becomes large—meaning that the refrigerator requires relatively little work to do its job. But if we wish to effect heat transfer between two widely separated temperatures, then the COP drops and the work becomes considerable. In the limit of very large T_h, the COP approaches zero, indicating that we're simply converting mechanical work into a nearly equal amount of heat, and transferring very little additional heat from the cool reservoir. In operating a real refrigerator, we would like to minimize the work needed. But Equation 22-5—like Equation 22-3 for a heat engine—imposes a fundamental limitation on that minimum amount of work.

● EXAMPLE 22-3 A HOME FREEZER

A typical home freezer operates between a low of 0°F (-18°C) and a high of 86°F (30°C). What is the maximum possible COP of this freezer? With this COP, how much electrical energy would be required to freeze 500 g of water, initially at 0°C?

Solution
Equation 22-5 gives the COP:

$$COP = \frac{T_c}{T_h - T_c} = \frac{255 \text{ K}}{303 \text{ K} - 255 \text{ K}} = 5.3.$$

To produce 500 g of ice takes

$$Q_c = mL_f = (0.50 \text{ kg})(334 \text{ kJ/kg}) = 170 \text{ kJ},$$

where we obtained the heat of fusion from Table 20-1. This Q_c is the heat that must be removed from the low-temperature end of the freezer. The COP is the ratio of the heat removed to the work done, so

$$W = \frac{Q_c}{COP} = \frac{170 \text{ kJ}}{5.3} = 32 \text{ kJ}.$$

In a real freezer, the COP would be lower, and the work correspondingly higher, because of irreversible processes in the refrigeration cycle.

EXERCISE (a) What is the COP of a refrigerator that maintains its interior at 4°C while exhausting heat at 40°C? (b) At what average rate does the refrigerator consume electricity if it extracts heat from its contents at the rate of 300 W?

Answers: (a) 7.7; (b) 39 W

Some problems similar to Example 22-3: 16–20 ●

Refrigerators are not confined to kitchens. An air conditioner is a refrigerator designed to cool an entire room or building. A heat pump is a refrigerator that cools a building in the summer and heats it in the winter (Fig. 22-17). Heat pumps are widely used in the southern United States, where in winter they pump energy from the cool outside air to the warm house. A heat pump requires electricity, but it supplies more energy as heat than it uses in electricity. The excess energy comes from the outdoor environment. Heat pumps operating from the constant 10°C temperature found two meters below the ground have been used for years in Scandinavian countries and are becoming more popular in the northern United States.

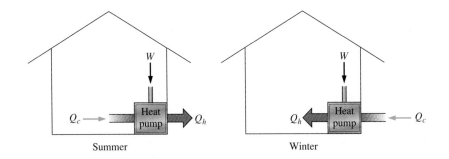

FIGURE 22-17 A heat pump. In summer the device acts as a refrigerator that cools the house interior by transferring heat to the outdoor environment. In winter its operation reverses, so it pumps heat from outdoors into the house. The work, W, is typically in the form of electricity.

● **EXAMPLE 22-4** A HEAT PUMP

A Carnot heat pump extracts energy from the ground at 10°C and transfers it to water at 70°C. The water is circulated to heat a building. (a) What is the COP of the heat pump? (b) What is its electrical power consumption if it supplies heat at the rate of 20 kW? (c) Compare the operating cost of the heat pump with that of an oil furnace if oil costs 93¢ per gallon and electricity costs 10¢ per kWh. (A gallon of oil releases about 30 kWh of heat when burned.)

Solution
Equation 22-5 gives the COP:

$$\text{COP} = \frac{T_c}{T_h - T_c} = \frac{283 \text{ K}}{343 \text{ K} - 283 \text{ K}} = 4.7.$$

Equation 22-4 shows what this means: for every kWh of electricity used, the pump transfers an additional 4.7 kWh of heat from the ground, providing a total of 5.7 kWh. The pump supplies heat at the rate of 20 kW, so the electrical power needed is

$$P_{\text{electric}} = \frac{20 \text{ kW}}{5.7} = 3.5 \text{ kW}.$$

At 10¢/kWh, it costs 35¢ per hour to run the pump. Since oil supplies 30 kWh/gallon, we would need to burn two-thirds of a gallon per hour to get 20 kW, at a cost of $(\frac{2}{3}$ gal/h$)(93$¢/gal$) = 62$¢ per hour.

Real heat pumps run at closer to half their theoretical maximum COP, and initial installation costs are much more expensive than for oil heat, so economics do not necessarily favor heat pumps.

If we're concerned about energy and not just cost, we should also consider the source of electricity. If this is a thermal power plant operating at an efficiency of, say, 33%, then the power plant uses three units of fuel energy for each unit it supplies to the heat pump. The overall efficiency of heat pump and power plant could actually be less than that of the oil furnace (see Problem 24).

EXERCISE A house requires 12 kW of heating power. If electricity costs 8.5¢/kWh, what is the minimum COP required if a heat pump is to operate for less than $5 per day?

Answer: 3.9

Some problems similar to Example 22-4: 22, 23, 63   ●

22-4 THE THERMODYNAMIC TEMPERATURE SCALE

None of the practical thermometers we introduced in Chapter 19 works at all temperatures; even helium gas thermometers become useless below about 1 K.

The operation of a reversible Carnot engine between two fixed temperatures provides another way of measuring temperature that works at any temperature. Measuring the heats Q_c and Q_h, or one of these and the work output, allows us to calculate the temperature ratio. If one temperature is known, we can then determine the other. Although this m ay seem a rather obscure way to measure temperature, it is actually used in certain experiments at very low temperature. And it provides an absolute way of defining temperature that can, in principle, be used at any temperature. The temperature scale so defined is called the **thermodynamic temperature scale.**

The zero of the thermodynamic scale—**absolute zero**—is absolute in that it represents a state of maximum order. This is reflected in the fact that a heat engine rejecting heat at absolute zero would operate at 100% efficiency, as suggested by Equation 22-3 with $T_c = 0$. Unfortunately, there are no heat reservoirs at absolute zero, and it can be shown that it is impossible to cool anything to this temperature in a finite number of steps. The statement that it is impossible to reach absolute zero in a finite number of steps is called the **third law of thermodynamics.**

22-5 ENTROPY AND THE QUALITY OF ENERGY

If offered a joule of energy, would you rather have it delivered in the form of mechanical work, heat from an object at 1000 K, or heat from an object at 300 K? Your answer might depend on what you want to do. To lift or accelerate a mass, you would be smart to take your energy as work. But if you want to keep warm, then heat from the 300 K object would be perfectly acceptable.

If, on the other hand, you're not sure what you want to do with the energy, then which should you take? The second law of thermodynamics makes the answer clear: you should take the work. Why? Because you could use it directly as mechanical energy, or you could, through friction or other irreversible processes, use it to raise the temperature of something.

If you chose 300-K heat for your joule of energy, then you could supply a full joule only to objects cooler than 300 K. You couldn't do mechanical work unless you ran a heat engine. With its maximum temperature only a little above ambient, your engine would be very inefficient, and you could only extract a small fraction of a joule of mechanical energy. Nor could you heat anything hotter than 300 K, unless you ran a refrigerator—and that would take additional work. You would be better off with 1000-K heat since you could transfer it to anything cooler than 1000 K, or could run a heat engine to produce up to 0.70 joule of mechanical energy.

Taking your energy in the form of work gives you the most options. Anything you can do with a joule of energy, you can do with the work. Heat is less versatile, with 300-K heat the least useful of the three. We're not talking here about the quantity of energy—we have exactly one joule in each case—but about **energy quality,** indicated by its ability to do a variety of useful tasks. Schematically, we can describe energy quality on a diagram like Fig. 22-18. We can readily convert an entire amount of energy from higher to lower quality, but the second law of thermodynamics prevents us from going in the opposite direction with 100% efficiency.

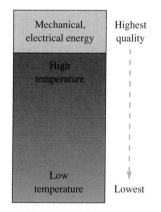

FIGURE 22-18 Energy quality measures the versatility of different energy forms. Electrical energy ranks with mechanical energy because it flows without the need for a temperature difference, thus qualifying as work.

■ APPLICATION   ENERGY QUALITY, END USE, AND COGENERATION

Energy quality has important implications for our efficient use of energy, suggesting that we try to match the quality of available energy sources to our energy needs. Figure 22-19 shows a breakdown of United States energy use by quality; it makes clear that much of our energy demand is for relatively low-quality heat. Although we can do anything we want with a high-quality energy source like flowing water or electricity, it makes sense to put our high-quality energy to high-quality uses, and use lower-quality sources to meet other needs. For example, we often heat water with electricity—a great convenience to the homeowner, but a thermodynamic folly, for in so doing we convert the highest-quality form of energy into low-grade heat. If our electricity comes from a thermal power plant, we already threw away two-thirds of the energy from fuel as waste heat. It makes little sense to run an elaborate and inefficient heat engine only to have its high-quality output of electricity used for low-temperature heating. On the other hand, if we heat our water by burning a fuel directly at the water heater, then we can, in principle, transfer all the fuel's energy to the water. Another approach to matching energy quality with energy needs is to combine the production of steam for heating with the generation of electricity. Used in Europe for years, such **cogeneration** is gaining popularity in the United States (Fig. 22-20).

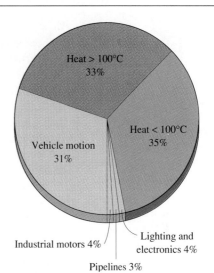

FIGURE 22-19   Energy use in the United States, organized by energy quality.

(a)

(b)

FIGURE 22-20   The Midland Cogeneration Venture in Michigan is a gas-fired facility that produces electric power for utility companies and steam—the second-law waste heat from power generation—for industrial uses at Dow Chemical Company. (a) Aerial view of MCV. The facility was originally to be a nuclear power plant, and the unfinished reactor containment buildings are the concrete structures at left. (b) Steam piping at MCV.

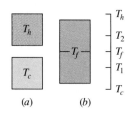

FIGURE 22-21 Two objects, initially at different temperatures, come to equilibrium at an intermediate temperature T_f. No energy is lost, but the system does lose its ability to do work.

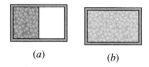

FIGURE 22-22 Adiabatic free expansion. Gas initially confined to one side of an insulated container expands to fill the entire container. The gas energy remains constant but its quality deteriorates.

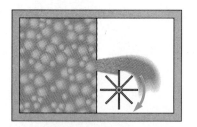

FIGURE 22-23 A means of extracting useful work from the gas in the initial state of Fig. 22-22.

Entropy

How can we quantify the notion of energy quality? Imagine bringing two identical objects, initially at different temperatures, into thermal contact (Fig. 22-21). Heat flows from the hotter to the cooler object until they reach the same temperature. If they're thermally insulated from their environment there's no energy lost in the process. But something has changed: The system has lost the ability to do useful work. In the initial state we could have run a heat engine using the two objects as its hot and cool reservoirs. In the final state there's no temperature difference and therefore no way to run a heat engine. The quantity of energy has stayed the same, but its quality has decreased. Can we find some system property that reflects this change?

Figure 22-22 shows another system whose energy quality deteriorates without any loss of total energy. We start with a gas confined to one side of an insulated container; the other side is evacuated. Remove the partition, and the gas expands freely to fill the container. Since the container is insulated, the process is adiabatic, and no heat flows out. Since the gas expands into a vacuum, it does no work either. Therefore, its energy stays constant. But again we've lost the ability to do work. Starting with the initial state, we could have made the gas run a turbine between the high pressure region and the vacuum, extracting useful work (Fig. 22-23). In the final state there's no pressure difference and therefore no way to run the turbine. Again, can we find a quantity that describes this decrease in energy quality?

To find the quantity we're after, consider an ideal gas undergoing a Carnot cycle—that is, a cycle of two isothermal and two adiabatic processes (recall Fig. 22-6). In deriving Equation 22-3 for the efficiency of this cycle we found that

$$\frac{Q_c}{Q_h} = \frac{T_c}{T_h}, \tag{22-6}$$

where Q_c was the heat *rejected* from the system to the low-temperature reservoir at T_c and Q_h the heat *added* from the reservoir at T_h.

We now change the definition of Q_c so it also means heat *added* to the system. This redefinition just changes the sign of Q_c, so now Equation 22-6 can be written

$$\frac{Q_c}{T_c} + \frac{Q_h}{T_h} = 0.$$

We can generalize this result to any reversible cycle by approximating the cycle as a sequence of adiabatic and isothermal steps (Fig. 22-24). Figure 22-24's caption shows how summing over the individual hot and cold isothermal segments then gives

$$\sum \frac{Q}{T} = 0.$$

We can approximate the closed cycle ever closer by more and more adiabatic and isothermal segments. In the limit the approximation becomes exact and the sum becomes an integral:

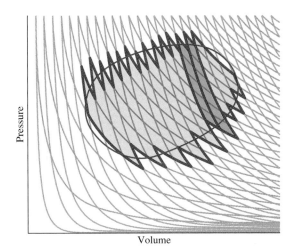

FIGURE 22-24 An arbitrary reversible cycle approximated by a sequence of adiabatic and isothermal steps. The region may be divided into elongated Carnot cycles like the two emphasized near the upper right. Heat is transferred to or from the gas only on the isothermal segments, all of which lie on the jagged approximation to the closed curve. The quantity $\dfrac{Q_c}{T_c} + \dfrac{Q_h}{T_h}$ for each individual cycle is determined from the isothermal segments only; therefore the sum of these quantities going around the entire cycle is equal to the sum of its values—namely zero—over the individual Carnot cycles.

$$\oint \frac{dQ}{T} = 0, \tag{22-7}$$

where the circle on the integral sign indicates that we integrate over all the heat transfers along a *closed* path.

Equation 22-7 tells us that there is some quantity S, a small amount of which is given by $dS = dQ/T$, that does not change when we take a system around a cyclic path. This quantity S is called **entropy.** If we take a system around a path that isn't closed, then its entropy will, in general, change. That change is given by integrating $dS = dQ/T$ over the path in question:

$$\Delta S = \int_1^2 \frac{dQ}{T}, \tag{22-8}$$

where ΔS is the change in entropy between the two thermodynamic states 1 and 2. Note that entropy has the units J/K.

What good is this concept of entropy? Equation 22-8 shows that there's no entropy change when we take a system from some initial state and eventually return it to that state. Now suppose we take a system part way around a closed path, from state 1 to state 2 shown in Fig. 22-25. The entropy change from state 2 back to state 1 must be exactly opposite the change from 1 to 2, so that the change going around the closed path is zero. But this must be true no matter how we get from state 2 to state 1, as suggested by the two possible paths in Fig. 22-25. That is, the entropy change given by Equation 22-8 is independent of path; it is

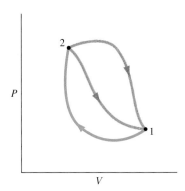

FIGURE 22-25 Entropy change is zero when the system returns to its original state. Therefore the entropy change from state 2 to state 1 is the same on both return paths shown. In fact, entropy change is always independent of path.

a property only of the initial and final states themselves. Like pressure, temperature, and volume, entropy is a thermodynamic state variable—a quantity that characterizes a given state independently of how a system got into that state.

● **EXAMPLE 22-5** ENTROPY

An object of mass m and specific heat c is heated from temperature T_1 to temperature T_2. What is the change in its entropy?

Solution

Equation 22-8 gives the entropy change. To evaluate the integral in this equation we must relate dQ and T. Here we can do so because we know the specific heat: a small temperature change dT requires a heat dQ given by

$$dQ = mc\, dT.$$

Then Equation 22-8 becomes

$$\Delta S = \int_1^2 \frac{dQ}{T} = \int_{T_1}^{T_2} mc\, \frac{dT}{T} = mc \ln\left(\frac{T_2}{T_1}\right). \tag{22-9}$$

EXERCISE What is the specific heat of a substance if 1.0 kg of it undergoes an entropy increase of 260 J/K when its temperature doubles?

Answer: 375 J/kg·K

Some problems similar to Example 22-5: 32, 33, 37, 38 ●

Equation 22-8 is meaningful only for reversible processes since an irreversible process takes a system temporarily into thermodynamic disequilibrium, in which temperature is not well defined. But because entropy doesn't depend on how a system got into its thermodynamic state, we can calculate the entropy change during an *irreversible* process by finding a *reversible* process that takes the system between the same two states, then using Equation 22-8 to calculate the entropy change for the reversible process.

At the beginning of this section we considered two irreversible processes in which energy quality deteriorated even though total energy remained constant. We can come to understand the meaning of entropy by calculating entropy changes for these two processes.

Irreversible Heat Transfer

Figure 22-21 showed two objects, initially at temperatures T_h and T_c, placed in thermal contact until they reach a common temperature T_f. The total energy of the system remains unchanged, but energy quality deteriorates. We can calculate the entropy change by considering reversible processes that take the two objects to the final temperature T_f. We could, for example, place each in contact with a heat reservoir at its initial temperature and then gradually change the temperature until it reaches T_f. The entropy change for the cool object would then be

$$\Delta S_c = \int_{T_c}^{T_f} \frac{dQ}{T}.$$

Although we might evaluate this integral (see Problem 42), note instead that there must be some intermediate temperature T_1 *that lies between T_c and T_f*, for which

$$\Delta S_c = \frac{Q_c}{T_1},$$

where Q_c is the heat absorbed by the cool object. We've marked T_1 on Fig. 22-21. Similarly, the entropy change for the hot object is

$$\Delta S_h = \int_{T_h}^{T_f} \frac{dQ}{T} = \frac{Q_h}{T_2},$$

where T_2 *lies between* T_h and T_f.

The entropy change of the system is then

$$\Delta S = \Delta S_c + \Delta S_h = \frac{Q_c}{T_1} + \frac{Q_h}{T_2}.$$

But the two objects are insulated, so any heat gained by the cool object in the actual irreversible process is equal to the heat lost by the hot object. Thus $Q_h = -Q_c$, and the entropy change becomes

$$\Delta S = \frac{Q_c}{T_1} - \frac{Q_c}{T_2} = Q_c\left(\frac{1}{T_1} - \frac{1}{T_2}\right).$$

Now T_1 lies *below* the equilibrium temperature T_f and T_2 lies *above* T_f. That makes ΔS a *positive* quantity—meaning that entropy has *increased* during the irreversible heat transfer.

Adiabatic Free Expansion

Figure 22-22 showed a gas undergoing an irreversible adiabatic free expansion. Neither the temperature nor the internal energy changes during the process, but energy quality deteriorates. We can calculate the entropy change by considering a reversible process that takes the gas between the same initial and final states. Since the final and initial temperatures are equal, one such process is an isothermal expansion. Equation 21-4 gives the heat added during such an expansion:

$$Q = nRT \ln\left(\frac{V_2}{V_1}\right).$$

Since the temperature is constant, the entropy change of Equation 22-9 becomes

$$\Delta S = \int \frac{dQ}{T} = \frac{1}{T} \int dQ = \frac{Q}{T} = nR \ln\left(\frac{V_2}{V_1}\right). \tag{22-10}$$

Since the final volume V_2 is larger than V_1, the entropy has *increased*.

In calculating this entropy change, we used the heat Q that would be required during a reversible isothermal process. Of course, no heat is transferred during the actual adiabatic free expansion. Nevertheless the entropy change is the same as for the reversible isothermal process that takes the system between the same initial and final states—and that process does involve heat transfer.

Entropy and the Availability of Work

Entropy increases during both the irreversible processes we just considered. Earlier, we argued that those processes result in a deterioration of energy quality, in that both systems lose the ability to do work. Suppose we had let the gas in Fig. 22-22 undergo a reversible isothermal expansion instead of the adiabatic free expansion. Then it would have done work given by Equation 21-4:

$$W = nRT \ln\left(\frac{V_2}{V_1}\right).$$

After the irreversible free expansion, the gas can no longer do this work, even though its energy is unchanged. Comparing with the entropy change of Equation 22-10, we see that the energy that becomes unavailable to do work is

$$E_{\text{unavailable}} = T\Delta S. \qquad (22\text{-}11)$$

Equation 22-11 is an example of a more general relation between entropy and the quality of energy:

> **Entropy and the Quality of Energy: During an irreversible process in which the entropy of a system increases by ΔS, energy E = $T_{\text{min}}\Delta S$ becomes unavailable to do work, where T_{min} is coolest temperature available to the system.**

This statement shows that entropy provides our desired measure of energy quality. Given two systems with identical energy content, the one with the lower entropy contains the higher quality energy. An entropy increase always corresponds to a degradation in energy quality, in that some energy becomes unavailable to do work.

● **EXAMPLE 22-6** ENTROPY AND ENERGY QUALITY

A cylinder contains 5.0 mol of compressed gas at 300 K, confined to 2.0 L. If the cylinder is discharged into a 150-L vacuum chamber and its temperature remains at 300 K, how much energy becomes unavailable to do work?

Solution
This is essentially an adiabatic free expansion, so Equations 22-10 and 22-11 give

$$E_{\text{unavailable}} = T\Delta S = nRT \ln\left(\frac{V_2}{V_1}\right)$$
$$= (5.0 \text{ mol})(8.314 \text{ J/K·mol})(300 \text{ K}) \ln\left(\frac{152 \text{ L}}{2.0 \text{ L}}\right)$$
$$= 54 \text{ kJ}.$$

EXERCISE Astronauts pressurize an evacuated space station by discharging 670 mol of air from a 200-L cylinder into the station's 15-m³ volume. The gas temperature remains at 290 K. How much work becomes unavailable as a result of this process?

Answer: 7.0 MJ

Some problems similar to Example 22-6: 44, 45

Entropy and the Second Law of Thermodynamics

We started this chapter by arguing that natural processes are generally irreversible, going from ordered states to disordered states. It's this loss of order—as when a hot and a cold object eventually reach the same temperature—that makes energy unavailable to do work. Entropy is a measure of this disorder. Given the tendency of systems to evolve toward disordered states, we can make a statement about entropy that is, in fact, a general statement of the second law of thermodynamics:

Second law of thermodynamics: The entropy of a closed system can never decrease.

At best, the entropy of a closed system remains constant—and this happens only in an ideal, reversible process. If anything irreversible occurs—a slight amount of friction or a deviation from exact thermodynamic equilibrium—then entropy increases. There's no going back. As entropy increases, some energy becomes unavailable to do work, and nothing within the closed system can restore that energy to its original quality. This statement of the second law in terms of entropy is equivalent to our previous statements about the impossibility of perfect heat engines and refrigerators, for the operation of either device would require a decrease in entropy.

What about a system that isn't closed? Can't we decrease its entropy? Yes—but only by supplying high-quality energy from outside. Running a refrigerator decreases the entropy of its contents (Fig. 22-26). But we must do work on the refrigerator to effect this entropy decrease—to make heat flow in the direction it doesn't normally go. If we enlarge our closed system to include the power plant or whatever else supplies the refrigerator with work, we would find that the entropy of this new closed system does not decrease. The entropy decrease at the refrigerator is offset by entropy increases elsewhere in the system (Fig. 22-27). If irreversible processes occur anywhere in the system, then there is a net entropy increase.

Any system whose entropy seems to decrease—that gets more rather than less organized—cannot be a closed system. If we enlarge the system boundaries to encompass anything that might exchange energy with the original system, then the entropy of the larger system will not decrease. Ultimately, we can enlarge the system to include the entire universe. Then we have the ultimate statement of the second law:

Second law of thermodynamics: The entropy of the universe can never decrease.

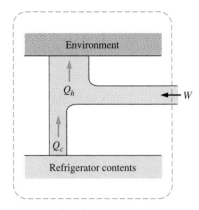

FIGURE 22-26 A refrigerator decreases the entropy of its contents, but only when work is done on the system from outside.

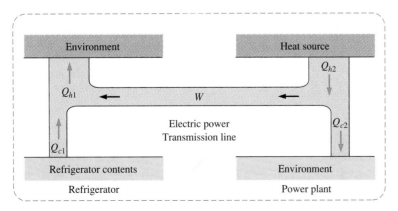

FIGURE 22-27 When the source of work for the refrigerator is included in the system, then the entropy of the entire system can at best remain constant.

As examples of this broad statement, consider the growth of a living thing from the random mix of molecules in its environment, or the construction of a skyscraper from materials that were originally dispersed about the Earth, or the appearance of ordered symbols on a printed page from a bottle of ink. All these are processes in which matter goes from near chaos to a highly organized state—akin to separating yolk and white from a scrambled egg. They are certainly processes in which entropy decreases. But Earth is not a closed system. It gets high-quality energy from the Sun, energy that is ultimately responsible for life and all its actions. If we consider the Earth-Sun system, then the entropy decrease associated with life and civilization is more than balanced by the entropy increase associated with nuclear fusion inside the Sun. We living things represent a remarkable phenomenon—the organization of matter in a universe governed by a tendency toward disorder. But we do not escape the second law of thermodynamics. Our highly organized selves and society, and the entropy decreases they represent, come into being only at the expense of greater entropy increases elsewhere in the universe.

CHAPTER SYNOPSIS

Summary

1. **Irreversible processes** decrease a system's organization or order. Because there are many more ways to arrange matter in disorganized states, the probability that a disorganized state will become spontaneously organized is impossibly small. The many equivalent statements of the **second law of thermodynamics** embody this fact.

2. A **heat engine** extracts heat from a hot reservoir and converts some of it to useful work. Applied to heat engines, the second law says that it is impossible to build a perfect heat engine that converts all the heat to work. In particular, no engine operating between temperatures T_h and T_c can be more efficient than a reversible **Carnot engine;** thus

$$e = \frac{W}{Q_h} \le e_{\text{Carnot}} = 1 - \frac{T_c}{T_h}.$$

This second-law limitation on efficiency means that real engines like automobile engines and power plants reject a great deal of waste heat to their environments.

3. A **refrigerator** extracts heat Q_c from a cool reservoir and transfers it to a hotter one. The second law of thermodynamics implies that a perfect refrigerator is impossible; real refrigerators must take in work W. The **coefficient of performance** measures the efficiency of a refrigerator:

$$\text{COP} = \frac{Q_c}{W} \le \text{COP}_{\text{Carnot}} = \frac{T_c}{T_h - T_c},$$

where the inequality shows that no refrigerator can have a higher COP than a reversible refrigerator operating on a Carnot cycle.

4. The second-law limit on the efficiency of a heat engine defines the **thermodynamic temperature scale. Absolute zero** corresponds to a state of maximum order, but the **third law of thermodynamics** states that it is impossible to reach absolute zero in a finite number of thermodynamic steps.

5. **Energy quality** describes the ability of a given quantity of energy to do work. The highest quality energy is mechanical or electrical, followed by the energy associated with large temperature differences, followed by ever smaller temperature differences. Higher quality energy can be converted to lower quality with 100% efficiency, but the second law prohibits the reverse process from occurring with 100% efficiency.

6. **Entropy** measures the relative disorder of a system, so increasing entropy corresponds to decreasing energy quality. Entropy is a thermodynamic state variable; its value depends only on the state of a system and not on how the system got into that state. The entropy difference between two states can be found by evaluating the expression

$$\Delta S = \int_1^2 \frac{dQ}{T}$$

for any reversible process connecting those states.

7. In terms of entropy, the second law of thermodynamics asserts that the entropy of the universe can never decrease. In this form, the second law is a universal statement about the tendency of systems to evolve toward states of greater disorder.

Terms You Should Understand

(Pairs are closely related terms whose distinction is important; number in parentheses is chapter section where term first appears.)

irreversible process (22-1)
second law of thermodynamics (22-2, 22-5)

heat engine, refrigerator (22-2)
efficiency (22-2), coefficient of performance (22-3)
thermodynamic temperature scale (22-4)
absolute zero (22-4)
third law of thermodynamics (22-4)
energy quality (22-5)
entropy (22-5)

Symbols You Should Recognize

T_c, T_h (22-2)   COP (22-3)
Q_c, Q_h (22-2) ΔS (22-5)
e (22-2)

Problems You Should Be Able to Solve

calculating efficiencies and waste heat of engines (22-2, 22-3)
calculating COP and related quantities for refrigerators and
   heat pumps (22-3)

calculating entropy changes for simple thermodynamic processes (22-5)
calculating loss of available work associated with entropy increase (22-5)

Limitations to Keep in Mind

The second law of thermodynamics, unlike physical laws you
   have encountered previously, is fundamentally *statistical.*
   It rules out events not because they violate basic laws of
   mechanics but because they're simply too improbable.
A completely reversible process is an idealization that cannot
   be realized in the macroscopic world where friction and
   other energy-dissipation mechanisms are present.

QUESTIONS

1. Which of the following processes is irreversible?
   (a) Stirring sugar into coffee
   (b) Building a house
   (c) Demolishing a house with a wrecking crane
   (d) Demolishing a house by taking it apart piece-by-piece
   (e) Warming a bottle of milk by transferring it directly from a refrigerator to stove top
   (f) Writing a sentence
   (g) Harnessing the energy of falling water in order to drive machinery
2. Could you cool the kitchen by leaving the refrigerator door open? Explain.
3. Could you heat the kitchen by leaving the oven open? Explain.
4. Why don't we simply refrigerate the cooling water of a power plant before it goes to the plant, thereby increasing the plant's efficiency?
5. Should a car get better mileage in the summer or the winter? Explain.
6. Is there a limit to the maximum temperature that can be achieved by focusing sunlight with a lens? If so, what is it?
7. Name some irreversible processes that occur in a real engine.
8. A power company claims that electric heat is one hundred per cent efficient. Discuss this claim.
9. Steam leaves the turbine of a power plant at 150°C and is condensed to water at the same temperature by contact with a river at 20°C. What temperature should be used as T_c in evaluating the thermodynamic efficiency?
10. A hydroelectric power plant, using the energy of falling water, can operate with an efficiency arbitrarily close to one hundred per cent. Why?

11. Viewed in isolation, a windmill is not a heat engine, for it converts mechanical energy of wind directly into electrical energy. But taking a broader view, the windmill is part of a natural heat engine. Explain.
12. To maximize the COP of a refrigerator, should you strive for a large or a small temperature difference? Explain.
13. The manufacturer of a heat pump claims that the device will heat your home using only energy already available in the ground. Is this true?
14. Why are heat pumps more widely used in warmer climates?
15. Proponents of ocean thermal energy conversion don't seem bothered by efficiencies so low as to be intolerable in fossil-fueled or nuclear power plants. Why the difference?
16. What might be done with the waste heat from power plants?
17. Does sunlight represent high- or low-quality energy? Explain; see also Question 6.
18. If new materials were developed that could withstand higher temperatures than those now encountered in power plants, how would that help us save energy?
19. The heat Q added during adiabatic free expansion is zero. Why can't we then argue from Equation 22-8 that the entropy change is zero?
20. Energy is conserved, so why can't we recycle it as we do materials?
21. A power plant provides electricity to run a heat pump. Can the heat output of the pump be greater than the heat extracted from the power plant's fuel?
22. Why does the evolution of human civilization not violate the second law of thermodynamics?

PROBLEMS

Section 22-1 Reversibility and Irreversibility

1. The egg carton shown in Fig. 22-28 has places for one dozen eggs. (a) How many distinct ways are there to arrange six eggs in the carton? (b) Of these, what fraction correspond to all six eggs being in the left half of the carton? Treat the eggs as distinguishable, so an interchange of two eggs gives rise to a new state.

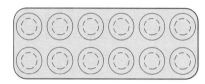

FIGURE 22-28   An egg carton (Problem 1).

2. A gas consists of four distinguishable molecules, all moving with the same speed v either to the left or to the right. A microscopic state is specified by telling which molecules are moving in which direction. (a) How many possible microscopic states are there? (b) How many of these correspond to all four molecules moving in the same direction? (c) What is the probability of finding the gas with all its molecules moving in one direction? Repeat parts (a–c) for a gas of (d) 10 molecules and (e) a more typical value, 10^{23} molecules.
3. Estimate the energy that could be extracted by cooling the world's oceans by 1°C. How does your estimate compare with humanity's yearly energy consumption of about 2.5×10^{20} J?
4. Two grains of salt and two grains of pepper are mixed. (a) If the mixture is shaken randomly and then divided into two equal parts, what is the approximate probability that all the salt will be in one part and all the pepper in another? Repeat for the case when (b) 10 grains and (c) 1000 grains of each are mixed.

Sections 22-2 and 22-3 The Second Law and Its Applications

5. What are the efficiencies of reversible heat engines operating between (a) the normal freezing and boiling points of water, (b) the 25°C temperature at the surface of a tropical ocean and deep water at 4°C, and (c) a 1000°C candle flame and room temperature?
6. A cosmic heat engine might operate between the Sun's 5600-K surface and the 2.7-K temperature of intergalactic space. What would be its efficiency?
7. A reversible Carnot engine operating between helium's melting point and its 4.25-K boiling point has an efficiency of 77.7%. What is the melting point?

8. A Carnot engine extracts 890 J from a 550-K reservoir during each cycle and rejects 470 J to a cooler reservoir. (a) How much work does it do each cycle? (b) What is its efficiency? (c) What is the temperature of the cool reservoir? (d) If the engine undergoes 22 cycles per second, what is its mechanical power output?
9. The maximum temperature in a nuclear power plant is 570 K. The plant rejects heat to a river where the temperature is 0°C in the winter and 25°C in the summer. What are the maximum possible efficiencies for the plant in these seasons? Why might the plant not achieve these efficiencies?
10. The minimum flow in the Connecticut River is 3.40×10^4 kg/s. By how much will the Vermont Yankee nuclear power plant described in Example 22-2 heat the entire river?
11. A power plant's electrical output is 750 MW. Cooling water at 15°C flows through the plant at 2.8×10^4 kg/s, and its temperature rises by 8.5°C. Assuming the plant's only energy loss is to the cooling water and that the cooling water is effectively the low-temperature reservoir, find (a) the rate of energy extraction from the fuel, (b) the plant's efficiency, and (c) its highest temperature.
12. A power plant extracts energy from steam at 250°C and delivers 800 MW of electric power. It discharges waste heat to a river at 30°C. The overall efficiency of the plant is 28%. (a) How does this efficiency compare with the maximum possible at these temperatures? (b) What is the rate of waste heat discharge to the river? (c) How many houses, each requiring 18 kW of heating power, could be heated with the waste heat from this plant?
13. The electric power output of all the thermal electric power plants in the United States is about 2×10^{11} W, and these plants operate at an average efficiency around 33%. What is the rate at which all these plants use cooling water, assuming an average 5 °C rise in cooling-water temperature? Compare with the 1.8×10^7 kg/s average flow at the mouth of the Mississippi River.
14. Consider a Carnot engine operating between temperatures T_h and T_c, where T_c is still above the ambient temperature T_0 (Fig. 22-29). It should be possible to operate a second engine between T_c and T_0. Show that the maximum overall efficiency of such a two-stage engine is the same as that of a single engine operating between T_h and T_0.
15. A Carnot engine absorbs 900 J of heat each cycle and provides 350 J of work. (a) What is its efficiency? (b) How much heat is rejected each cycle? (c) If the engine rejects heat at 10°C, what is its maximum temperature?
16. What is the COP of a reversible refrigerator operating between 0°C and 30°C?
17. How much work does a refrigerator with a COP of 4.2 require to freeze 670 g of water already at its freezing point?

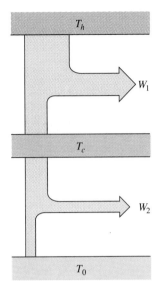

FIGURE 22-29 A two-stage engine (Problem 14).

18. An industrial freezer operates between 0°C and 32°C, consuming electrical energy at the rate of 12 kW. Assuming the freezer is perfectly reversible, (a) what is its COP? (b) How much water at 0°C can it freeze in 1 hour?

19. A 4.0 L sample of water at 9.0°C is put into a refrigerator. The refrigerator's 130-W motor then runs for 4.0 min to cool the water to the refrigerator's low temperature of 1.0°C. (a) What is the COP of the refrigerator? (b) How does this compare with the maximum possible COP if the refrigerator exhausts heat at 25°C?

20. A refrigerator maintains an interior temperature of 4°C while the temperature near its heat exhaust is 30°C. The refrigerator's insulation is imperfect, and heat leaks into the refrigerator at the rate of 340 W. Assuming the refrigerator is reversible, at what rate must it consume electrical energy to maintain a constant 4°C interior?

21. A heat pump consumes electrical energy at the rate P_e. Show that it delivers heat at the rate $(COP + 1)P_e$.

22. A store is heated by an oil furnace that supplies 30 kWh of heat from each gallon. The store's owners are considering switching to a heat pump system. Oil costs 87¢/gallon, and electricity costs 7.8¢/kWh. What is the minimum heat pump COP that will result in a savings in heating costs?

23. A heat pump transfers heat between the interior of a house and the outside air. In the summer the outside air averages 26°C, and the pump operates by chilling water to 5°C for circulation throughout the house. In winter the outside air averages 2°C, and the pump operates by heating water to 80°C for circulation throughout the house. (a) Find the coefficients of performance in summer and winter. How much work does the pump require (b) for each joule of heat

removed from the house in summer and (c) for each joule supplied to the house in winter?

24. A house is heated by a heat pump with COP = 2.3. The heat pump is run by electricity generated in a power plant whose efficiency is 28%. (a) What is the overall efficiency of this process, defined as the ratio of heat delivered to the house to heat released from fuel at the power plant? (b) Could the efficiency so defined ever exceed 100%?

25. A 0.20-mol sample of an ideal gas goes through the Carnot cycle of Fig. 22-30. Calculate (a) the heat Q_h absorbed, (b) the heat Q_c rejected, and (c) the work done. (d) Use these quantities to determine the efficiency. (e) Find the maximum and minimum temperatures, and show explicitly that the efficiency as defined in Equation 22-1 is equal to the Carnot efficiency of Equation 22-3.

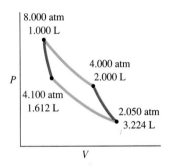

FIGURE 22-30 Problem 25 (Diagram is not to scale.)

26. Show that it is impossible for two adiabats to intersect. *Hint:* Suppose two adiabats do intersect. Connect them by a suitable isotherm to form a three-step cycle. Show that a heat engine operating in this cycle would have 100% efficiency, in violation of the second law.

27. Use appropriate energy flow diagrams to show that the existence of a perfect heat engine would permit the construction of a perfect refrigerator, thus violating the Clausius statement of the second law.

Section 22-4 The Thermodynamic Temperature Scale

28. A small sample of material is taken through a Carnot cycle between a reservoir of boiling helium at 1.76 K and an unknown lower temperature. During the process, 7.5 mJ of heat are absorbed from the helium, and 0.44 mJ rejected at the lower temperature. What is the unknown low temperature?

29. A Carnot engine operating between a vat of boiling sulfur and a bath of water at its triple point has an efficiency of 61.95%. What is the boiling point of sulfur?

30. A heat engine operating between a hotter reservoir of boiling hydrogen and a cooler one of boiling helium has 79.1%

efficiency. What is the ratio of hydrogen to helium boiling temperatures?

Section 22-5 Entropy and the Quality of Energy

31. Calculate the entropy change associated with melting 1.0 kg of ice at 0°C.
32. You heat 250 g of water from 10°C to 95°C. By how much does the entropy of the water increase?
33. A 2.0-kg sample of water is heated to 35°C. If the entropy change is 740 J/K, what was the initial temperature?
34. Melting a block of lead already at its melting point results in an entropy increase of 900 J/K. What is the mass of the lead? *Hint:* Consult Table 20-1.
35. A shallow pond contains 94,000 kg of water. In winter it's entirely frozen. By how much does the entropy of the pond increase when the ice, already at 0°C, melts and then heats to its summer temperature of 15°C?
36. Figure 22-31 shows a 500-g copper block at 80°C dropped into 1.0 kg of water at 10°C. (a) What is the final temperature? (b) What is the entropy change of the system?

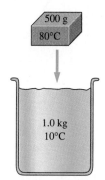

FIGURE 22-31 Problem 36.

37. The temperature of n moles of ideal gas is changed from T_1 to T_2 while the gas volume is held constant. Show that the corresponding entropy change is $\Delta S = nC_V \ln(T_2/T_1)$.
38. The temperature of n moles of ideal gas is changed from T_1 to T_2 while gas pressure is held constant. Show that the corresponding entropy change is $\Delta S = nC_P \ln(T_2/T_1)$.
39. A 5.0-mol sample of an ideal diatomic gas $(C_V = \frac{5}{2}R)$ is initially at 1.0 atm pressure and 300 K. What is the entropy change if the gas is heated to 500 K (a) at constant volume, (b) at constant pressure, and (c) adiabatically?
40. The interior of a house is maintained at 20°C while the outdoor temperature is −10°C. The house loses heat at the rate of 30 kW. At what rate does the entropy of the universe increase because of this irreversible heat flow?
41. A 250-g sample of water at 80°C is mixed with 250 g of water at 10°C. Find the entropy changes for (a) the hot water, (b) the cool water, and (c) the system.

42. Two identical objects of mass m and specific heat c at initial temperatures T_h and T_c are placed together in an insulated box and allowed to come to equilibrium. Show that the system's entropy increase is $\Delta S = mc \ln \dfrac{(T_h + T_c)^2}{4T_h T_c}$.
43. A 5.0-mol sample of ideal monatomic gas undergoes the cycle shown in Fig. 22-32, in which the process BC is isothermal. Calculate the entropy change associated with each of the three steps, and show explicitly that there is zero net entropy change over the full cycle.

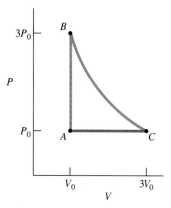

FIGURE 22-32 Problem 43.

44. In an adiabatic free expansion, 8.7 mol of ideal gas at 450 K expand 10-fold in volume. How much energy becomes unavailable to do work?
45. Ideal gas occupying 1.0 cm³ is placed in a 1.0-m³ vacuum chamber, where it expands adiabatically. If 6.5 J of energy become unavailable to do work, what was the initial gas pressure?
46. Make an argument based on Fig. 22-21 to show that a perfect refrigerator would decrease the entropy of the universe, in violation of the second law.

Paired Problems

(Both problems in a pair involve the same principles and techniques. If you can get the first problem, you should be able to solve the second one.)

47. Cooling water circulates through a reversible Carnot engine at 3.2 kg/s. The water enters at 23°C and leaves at 28°C; the average temperature is essentially that of the engine's cool reservoir. If the engine's mechanical power output is 150 kW, what are (a) its efficiency and (b) its highest temperature?
48. A reversible Carnot engine operates between 300 K and 640 K, running at 45 cycles per second. In each cycle the engine extracts 2.7 kJ from the high-temperature reservoir. Find (a) the efficiency and (b) the mechanical power output. (c) The low-temperature reservoir is provided by a flow of

300-K cooling water. If the water temperature increases 2.5°C, what is the flow rate?

49. Which would provide the greatest increase in efficiency of a Carnot engine, a 10 K increase in the maximum temperature or a 10 K decrease in the minimum temperature?

50. Which would provide the greatest increase in the COP of a reversible refrigerator, a decrease of 10 K in the maximum temperature or an increase of 10 K in the minimum temperature?

51. It costs $180 to heat a house with electricity in a typical winter month. (Electric heat simply converts all the incoming electrical energy to heat.) What would be the monthly heating bill following conversion to an electrically powered heat pump system with COP = 2.1?

52. It costs $140 each summer to operate a home air conditioner with a COP of 1.7. What will be the yearly savings in upgrading to a more efficient model with a COP of 2.4, assuming the same amount of heat is to be extracted from the house?

53. A reversible engine contains 0.20 mol of ideal monatomic gas, initially at 600 K and confined to 2.0 L. The gas undergoes the following cycle:

• Isothermal expansion to 4.0 L.
• Isovolumic cooling to 300 K.
• Isothermal compression to 2.0 L.
• Isovolumic heating to 600 K.

(a) Calculate the net heat added during the cycle and the net work done. (b) Determine the engine's efficiency, defined as the ratio of the work done to only the heat *absorbed* during the cycle.

54. (a) Determine the efficiency for the cycle shown in Fig. 22-33, using the definition given in the preceding problem. (b) Compare with the efficiency of a Carnot engine operating between the same temperature extremes. Why are the two efficiencies different?

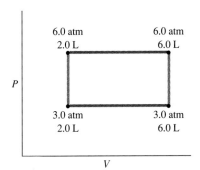

FIGURE 22-33 Problem 54.

55. You dump three 10-kg buckets of 10°C water into an empty tub, then add one 10-kg bucket of 70°C water. By how much does the entropy of the water increase?

56. Find the entropy change when a 2.4-kg aluminum pan at 155°C is plunged into 3.5 kg of water at 15°C.

Supplementary Problems

57. You're lying in a bathtub of water at 42°C. Suppose, in violation of the second law of thermodynamics, that the water spontaneously cooled to room temperature (20°C) and that the energy so released was transformed into your gravitational potential energy. Estimate the height to which you would rise above the bathtub.

58. An engine with mechanical power output 8.5 kW extracts heat from a source at 420 K and rejects it to a 1000-kg block of ice at its melting point. (a) What is its efficiency? (b) How long can it maintain this efficiency if the ice is not replenished?

59. A solar-thermal power plant is to be built in a desert location where the only source of cooling water is a small creek with average flow of 100 kg/s and an average temperature of 30°C. The plant is to cool itself by boiling away the entire creek. If the maximum temperature achieved in the plant is 500 K, what is the maximum electric power output it can sustain without running out of cooling water?

60. The McNeil generating station in Burlington, Vermont, is one of the world's largest wood-fired power plants. Here are some facts about the McNeil plant:

• The electric power output is 59 MW.
• Fuel consumption is 1.2×10^6 kg of wood per day. The wood's energy content is 1.4×10^7 J/kg, and 85% of this energy is released in burning.
• Steam enters the plant's turbine at 1000°F and leaves at 200°F.
• Cooling water circulates through the condenser at 1.8×10^5 kg/min, and its temperature increases by 18°F.

Calculate the plant's efficiency in three ways: (a) Using only the third fact above to get the maximum possible efficiency; (b) using only the first and second facts; and (c) using only the first and fourth facts. What might account for the discrepancy between the second and third parts?

61. Gasoline engines operate approximately on the **Otto cycle,** consisting of two adiabatic and two constant-volume segments. The Otto cycle for a particular engine is shown in Fig. 22-34. (a) If the gas in the engine has specific heat ratio γ, find the engine's efficiency, assuming all processes are reversible. (b) Find the maximum temperature, in terms of the minimum temperature T_{min}. (c) How does the efficiency compare with that of a Carnot engine operating between the same two temperature extremes? *Note:* Fig. 22-34 neglects the intake of fuel-air and the exhaust of combustion products, which together involve essentially no net work.

62. The compression ratio r of an engine is the ratio of maximum to minimum gas volume. For the engine of the preceding problem, Fig. 22-34 shows that the compression ratio

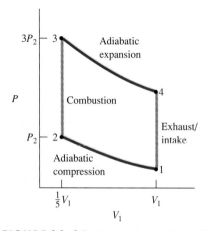

FIGURE 22-34 Otto cycle (Problems 61, 62).

is 5. (a) Find an expression for the engine efficiency as a function of compression ratio. Assume that pressure continues to triple during the combustion phase, as shown in Fig. 22-34. (b) Make a graph of efficiency versus r, and on the same plot show also the efficiency of a Carnot engine operating between the same temperature extremes.

63. A heat pump designed for southern climates extracts heat from outside air and delivers air at 40°C to the inside of a house. (a) If the average outside temperature is 5°C, what is the average COP of the heat pump? (b) Suppose the pump is used in a northern climate where the average winter temperature is −10°C. Now what is the COP? (c) Two *identical* houses, one in the north and one in the south, are heated by this pump. Both houses maintain indoor temperatures of 19°C. What is the ratio of electric power consumption in the two houses? *Hint:* Think about heat loss as well as COP!

64. The specific heat of copper in the range from a few K to about 50 K is given by $c = aT^3$, where $a = 7.68 \times 10^{-4}$ J/kg·K⁴. Find the entropy change in a 200-g piece of copper as it's heated from 10 K to 25 K.

65. A Carnot engine extracts heat from a block of mass m and specific heat c that is initially at temperature T_{h0} but which has no heat source to maintain that temperature. The engine rejects heat to a reservoir at a constant temperature T_c. The engine is operated so its mechanical power output is proportional to the temperature difference $T_h - T_c$:

$$P = P_0 \frac{T_h - T_c}{T_{h0} - T_c},$$

where T_h is the instantaneous temperature of the hot block and P_0 is the initial power output. (a) Find an expression for T_h as a function of time, and (b) determine how long it takes for the engine's power output to reach zero.

66. You have 50 kg of steam at 100°C, but no heat source to maintain it in that condition. You also have a heat reservoir at 0°C. Suppose you operate a heat engine with this system, so the steam gradually condenses and cools until it reaches 0°C. (a) Calculate the total entropy change of the steam and subsequent water. (b) Find the total amount of work the engine can do. *Hint:* This engine is reversible. But had the change in steam condition occurred irreversibly, the entropy change and thus the amount of energy that became unavailable to do work would be the same.

67. An ideal diatomic gas undergoes the cyclic process described in Fig. 22-35. Fill in the blank spaces in the table below:

	P	**V**	**T**	**U − U$_A$**	**S − S$_A$**
A	P_0	V_0	T_0	0	0
B	$3.4P_0$	V_0			
C					
D	P_0	$3.0V_0$			

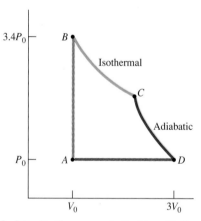

FIGURE 22-35 Cyclic process for Problem 67. Process *BC* is isothermal while *CD* is adiabatic. Diagram is not accurately scaled, and the location of point *C* is not quantitatively correct.

P A R T 3   CUMULATIVE PROBLEMS

1. Figure 1 shows the thermodynamic cycle of a diesel engine. Note that this cycle differs from that of a gasoline engine (see Fig. 22-34) in that combustion takes place isobarically. As with the gasoline engine, the compression ratio r is the ratio of maximum to minimum volume; $r = V_1/V_2$. In addition, the so-called *cutoff ratio* is defined by $r_c = V_3/V_2$. Find an expression for the engine's efficiency, in terms of the ratios r and r_c and the specific heat ratio γ. Although your expression suggests that the diesel engine might be less efficient than the gasoline engine (see Problem 61 of Chapter 22), the diesel's higher compression ratio more than compensates, giving it a higher efficiency.

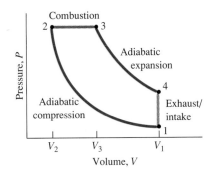

FIGURE 1   Cumulative Problem 1.

2. Deep in intergalactic space, an astronaut releases a thin-walled spherical container of radius R and negligible mass, containing a mass M of liquid water at temperature T_0 (Fig. 2). The container's surface is painted black to give it emissivity $e = 1$. The amount of radiation incident on the sphere from outside is negligible (since the temperature of intergalactic space is only 2.7 K). (a) Find an expression for the temperature of the liquid water as a function of time, assuming its heat conductivity is high enough that the water remains at essentially uniform temperature. (b) Taking $M = 100$ kg, $R = 29$ cm, and $T_0 = 300$ K, find the time from when the sphere is deployed until the water is completely frozen.

3. Equation 21-4 gives the work done by an ideal gas undergoing an isothermal expansion from volume V_1 to volume V_2. Find the analogous expression for a van der Waals gas described by Equation 20-5. Is the work done equal to the heat transferred to the gas in this case? Why or why not?

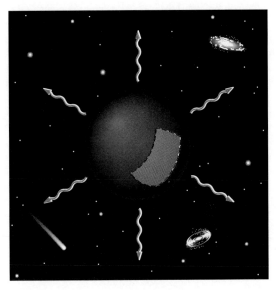

FIGURE 2   Cumulative Problem 2.

4. A solid sphere of mass m and radius R is made from a material with specific heat c, and is initially at a uniform temperature T_1. The sphere is surrounded by insulation of thermal conductivity k and thickness a (see Fig. 3). The temperature outside the insulation is fixed at T_0. (a) Assuming conduction is the only significant heat-loss mechanism, find an expression for the magnitude of the sphere's heat-loss rate as a function of its temperature T. (b) Assuming the temperature throughout the sphere remains uniform, find an expression for the temperature T as a function of time.

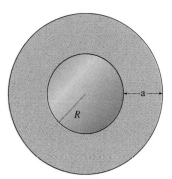

FIGURE 3   Cumulative Problem 4.

5. The ideal Carnot engine shown in Fig. 4 operates between a heat reservoir and a block of ice with mass M. An external energy source maintains the reservoir at a constant temperature T_h. At time $t = 0$ the ice is at its melting point T_0, but it is insulated from everything except the engine, so it is free to change state and temperature. The engine is operated in such a way that it extracts heat from the reservoir at a constant rate P_h. (a) Find an expression for the time t_1 at which the ice is all melted, in terms of the quantities given and any other appropriate thermodynamic parameters. (b) Find an expression for the mechanical power output of the engine as a function of time for times $t > t_1$. (c) Your expression in (b) holds only up to some maximum time t_2. Why? Find an expression for t_2.

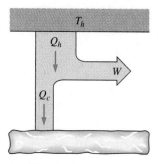

FIGURE 4   Cumulative Problem 5.

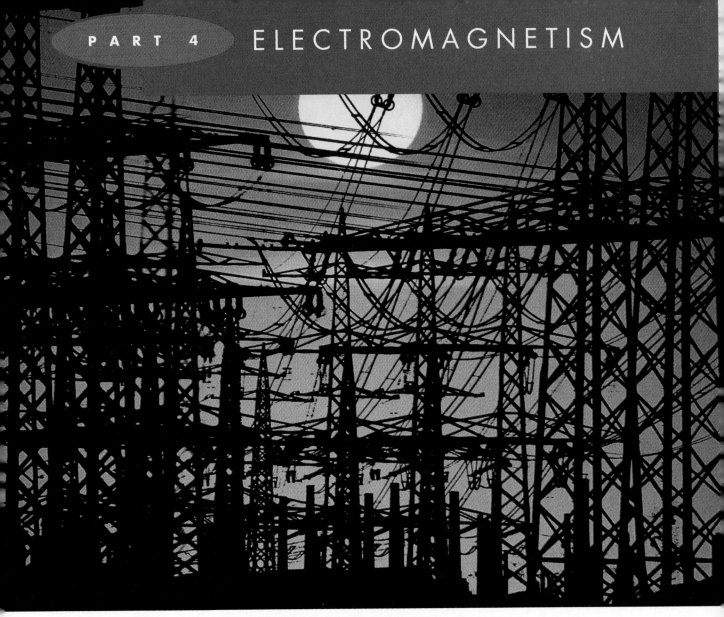

Electricity and magnetism mediate the generation, transmission, and use of energy. They also govern the structure of matter at the atomic and molecular level. Even sunlight, shown here as the Sun sets behind an electric power substation, is an electromagnetic phenomenon.

CHAPTER 23

ELECTRIC CHARGE, FORCE, AND FIELD

Lightning strikes at Kitt Peak National Observatory in Arizona relieve the buildup of electric charge in the atmosphere.

What force keeps the molecules in your body together? What force keeps a skyscraper standing or prevents a mountain from spreading into a flat blob? What force holds your car on the road as you round a turn? What force accelerates the electrons that paint the picture on your TV screen? What force underlies the awesome beauty of a thunderstorm?

Remarkably, these and all other forces except gravity that we encounter in our everyday lives—and in nearly all scientific work—are manifestations of a single force: the **electromagnetic force.** The friction, tension, and normal forces of mechanics are ultimately electromagnetic. So are interactions as diverse as the focusing of light in the lens of your eye, the extraction of information from a computer disk, or the formation of a water molecule from separate atoms. Electromagnetism is so important in science and engineering that we devote the next 12 chapters to it.

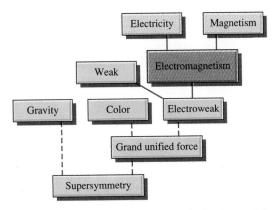

FIGURE 23-1 The place of electromagnetism among the fundamental forces. Electromagnetism comprises electricity and magnetism, once thought to be separate. Similarly, electromagnetism and the weak force are both aspects of the electroweak force. Grand unification and supersymmetry are two theories being considered for further unification of the fundamental forces.

23-1 ELECTROMAGNETISM

Electromagnetism is among the fundamental forces of nature that we introduced in Chapter 5 (see Fig. 23-1) and is the dominant force in a vast range of natural and technological phenomena. Other than gravity, electromagnetism is the only force most of us will ever deal with.

Three distinct themes motivate our study of electromagnetism:

1. The electromagnetic force is solely responsible for the structure of matter from atoms to objects of roughly human size. Much of physics, all of chemistry, and most of biology deal in this realm. Only at much smaller scales do the color and weak forces become important; only at larger scales is gravity significant.

    The wonderful diversity of chemical compounds is testimony to the rich possibilities contained in the electromagnetic interaction. Even life itself, and the DNA replicating mechanism at its heart, are manifestations of electromagnetism (Fig. 23-2). For students of physical and biological sciences, understanding electromagnetism means understanding the most fundamental basis of these disciplines.

2. We live in a technological world increasingly dominated by devices that operate on electromagnetic principles. Electric lights, motors, batteries, and generators have been essential throughout the twentieth century. More recently, electronic technology has led to the proliferation of devices for storing and processing information, for sensing and measuring, and for sophisticated control of industrial, scientific, medical, and even household systems.

3. Studying electromagnetism leads to an understanding of the nature of light and from there to the theory of relativity. Relativity profoundly alters our ideas of space and time—the very basis of physical reality.

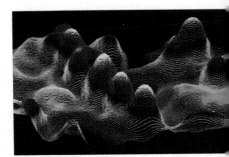

FIGURE 23-2 Electric forces govern the structure and replication of the DNA molecule, shown here in a scanning tunneling microscope image.

We've been speaking of electromagnetism, yet you're probably more familiar with electricity and magnetism separately. Although we begin with separate studies of these seemingly distinct phenomena, the relation between the two will become increasingly central. Eventually you will understand electricity and magnetism as two aspects of a single phenomenon that is basic to the workings of the universe.

23-2 ELECTRIC CHARGE

Electric charge is a fundamental property of nature. Of the three building blocks of ordinary matter—the electron, the proton, and the neutron—two carry electric charge. What is electric charge? At the most fundamental level, we don't know. We don't know what mass "really" is either, but we're familiar with it because we've spent our lives pushing objects around. By studying electrical interactions we gain a similar familiarity with charge that is as close as we can get to understanding what it "really" is.

Two Kinds of Charge

Electric charge comes in two varieties, which Benjamin Franklin designated **positive** and **negative.** These are convenient labels, but they have no physical significance. There's nothing "missing" about negative charge. Positive and negative charge are complementary properties, not the presence and absence of something. The utility in the names is mathematical since it's the algebraic sum of charges—described with positive and negative numbers—that has physical significance.

Quantities of Charge

All electrons carry the same charge, and all protons carry the same charge. The proton's charge has *exactly* the same magnitude as the electron's, but with opposite sign. Given that electrons and protons differ substantially in other properties—like mass—this electrical relation is remarkable. Problem 1 shows how dramatically different our world would be if there were even a slight difference between the magnitudes of the electron and proton charges.

The magnitude of the electron or proton charge is the **elementary charge,** e. Electric charge is **quantized**—that is, it comes only in discrete amounts. In a famous experiment in 1909, the American physicist R. A. Millikan used electric forces to suspend small oil drops. From the electric force he computed the charge on each drop and found it was always a multiple of a basic value we now know as the elementary charge (Fig. 23-3).

Modern elementary particle theories suggest that the most basic unit of charge is actually $\frac{1}{3}e$. Such "fractional charges" reside on quarks, the basic building blocks of protons, neutrons, and many other particles. Quarks always join to produce particles with integer multiples of the full elementary charge, and it seems impossible to isolate individual quarks.

The SI unit of charge is the **coulomb** (C), named for the French physicist Charles Augustin de Coulomb (1736–1806). Although the coulomb's formal definition is in terms of electric current, it's convenient to describe one coulomb as being about 6.25×10^{18} elementary charges, making the elementary charge approximately 1.60×10^{-19} C.

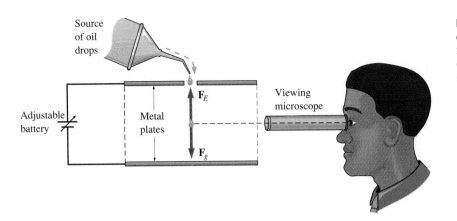

FIGURE 23-3 Millikan's oil-drop experiment. By balancing the electric force $\mathbf{F}_E$ and gravitational force $\mathbf{F}_g$ on oil drops, Millikan showed that charge is quantized.

Charge Conservation

Electric charge is a conserved quantity, meaning that the algebraic sum of the electric charges—i.e., the **net charge**—in a closed region remains constant. Charged particles may be created or annihilated, but always in pairs of equal and opposite charge (Fig. 23-4). The net charge always remains the same.

23-3 COULOMB'S LAW

You can transfer charge to a balloon by rubbing it on your clothing; you'll then find that the balloon sticks to you. Charge another balloon and the two balloons will repel each other (Fig. 23-5). You're seeing a manifestation of the fundamental fact that unlike charges attract and like charges repel. Socks clinging to other clothes when they come out of a dryer, dust attracted to the front of a TV screen or computer monitor, and the shocks you get when you cross a carpet and touch a doorknob are other common examples where you're directly aware of this electrical interaction.

But electricity would be rather unimportant if the only significant electrical interactions were these obvious ones. In fact, all interactions of everyday matter—from the motion of a car to the movement of a muscle to the growth of a tree—are dominated by the electric force. It's just that matter on the large scale is nearly perfectly neutral, so electrical effects in bulk matter are not obvious. At the molecular or even cellular level the appearance of individual charged particles makes the electrical nature of matter immediately obvious (Fig. 23-6).

The attraction and repulsion of electric charges implies that a force acts between them. The physicist Coulomb first investigated this force in the late 1700s. He found that the force between two charges acts along the line joining them, with a magnitude proportional to the product of the charges and inversely proportional to the square of the distance between them. These results are summarized in **Coulomb's law:**

$$\mathbf{F}_{12} = \frac{kq_1q_2}{r^2}\hat{\mathbf{r}} \qquad \text{Coulomb's law}. \qquad (23\text{-}1)$$

FIGURE 23-4 The creation of an electron and its antiparticle, a positron. The two particles' paths are the oppositely directed spirals that originate in a common point where the pair was created. The net charge remains zero before and after the particle creation.

FIGURE 23-5 Two balloons carrying similar electric charge repel each other.

FIGURE 23-6   (a) The salt shaker and even a single salt grain are electrically neutral, so the role of electric forces in these structures is not obvious. (b) At the atomic scale, electric forces bind individual sodium and chlorine ions together and are, in fact, responsible for the cubical shape of the salt crystal.

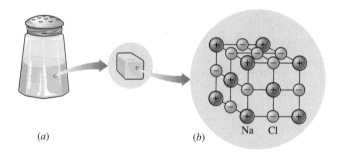

(a)   (b)   Na   Cl

$$\mathbf{F}_{12} = \frac{kq_1q_2}{r^2}\hat{\mathbf{r}}$$

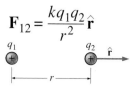

FIGURE 23-7   Quantities in Coulomb's law used in calculating the electric force of charge q_1 on q_2. The unit vector $\hat{\mathbf{r}}$ always points in the direction from q_1 toward q_2, regardless of the signs of the charges. The force $\mathbf{F}_{12}$ points in the same direction as $\hat{\mathbf{r}}$ if the two charges have the same sign, and otherwise in the opposite direction.

Here $\mathbf{F}_{12}$ is the force exerted *by* the charge q_1 *on* the charge q_2 and r is the distance between the two. The quantity k is a proportionality constant whose value in SI units is approximately 9.0×10^9 N·m²/C². The force $\mathbf{F}_{12}$ is a vector, and $\hat{\mathbf{r}}$ is a unit vector giving its direction. Figure 23-7 shows that $\hat{\mathbf{r}}$ lies on the line passing through the two charges and points in the direction *from q_1 toward q_2*.

Coulomb's law is a vector equation that covers all possible combinations of charges and the associated attractive and repulsive forces. If q_1 and q_2 have the same sign—either positive or negative—then the product q_1q_2 is positive, and the force points in the same direction as $\hat{\mathbf{r}}$. The force on q_2 is thus away from q_1, or repulsive, as it should be. But if q_1 and q_2 have opposite signs, then q_1q_2 is negative and the direction of the force is opposite that of $\hat{\mathbf{r}}$—that is, attractive.

What about the force $\mathbf{F}_{21}$ that q_2 exerts on q_1? Equation 23-1 shows it has the same magnitude as $\mathbf{F}_{12}$, but a drawing like Fig. 23-7 gives the opposite direction. Thus the electric force obeys Newton's third law.

● EXAMPLE 23-1   TWO CHARGES

A 1.0-μC charge is located at $x = 1.0$ cm, and a -1.5-μC charge at $x = 3.0$ cm. (a) What force does the positive charge exert on the negative one? (b) How would the force change if the distance between the charges were tripled?

Solution

(a) A vector from the positive charge toward the negative charge points in the $+x$ direction, so $\hat{\mathbf{r}}$ is simply the unit vector $\hat{\mathbf{i}}$. Then Equation 23-1 becomes

$$\mathbf{F} = \frac{kq_1q_2}{r^2}\hat{\mathbf{r}}$$

$$= \frac{(9.0\times10^9 \text{ N·m}^2/\text{C}^2)(1.0\times10^{-6} \text{ C})(-1.5\times10^{-6} \text{ C})}{(0.020 \text{ m})^2}\hat{\mathbf{i}}$$

$$= -34\hat{\mathbf{i}} \text{ N}.$$

The minus sign shows that the force is in the negative x direction—toward the positive charge, or attractive.

(b) If the distance were tripled the force would drop by a factor of $1/3^2$, to $-3.8\hat{\mathbf{i}}$ N.

TIP   Let the Algebra Take Care of the Signs   Equation 23-1 accounts for all aspects of the force calculation. If you correctly take $\hat{\mathbf{r}}$ to be a unit vector pointing *away* from the charge giving rise to the force—whatever its sign—and *toward* the charge being acted on—whatever its sign—then calculation of $\frac{kq_1q_2}{r^2}\hat{\mathbf{r}}$ gives correctly the magnitude and direction of the force.

EXERCISE   A 2.5-μC charge is at the origin. Find the force it exerts on a 4.1-μC charge at the point $x = 2$ m, $y = 2$ m.

Answer: $8.15\hat{\mathbf{i}} + 8.15\hat{\mathbf{j}}$ mN

Some problems similar to Example 23-1:   7, 12, 13

●

Coulomb's law for the electric force is similar to Newton's law for the gravitational force. Both show the same inverse-square decrease with distance, and both are proportional to the product of the interacting charges or masses. But there is an important difference: There's only one kind of mass (even antimatter has "positive mass") and the gravitational force it produces is always attractive. That means large concentrations of mass give rise to large gravitational forces. But charge comes in two varieties, so large concentrations of charge tend to be electrically neutral and therefore give rise to weak electric forces. The electric force between individual particles is vastly stronger than the gravitational force between the same particles (see Problem 6), and it's only because of the nearly complete cancellation of positive and negative charge in bulk matter that gravity becomes important on the macroscopic scale.

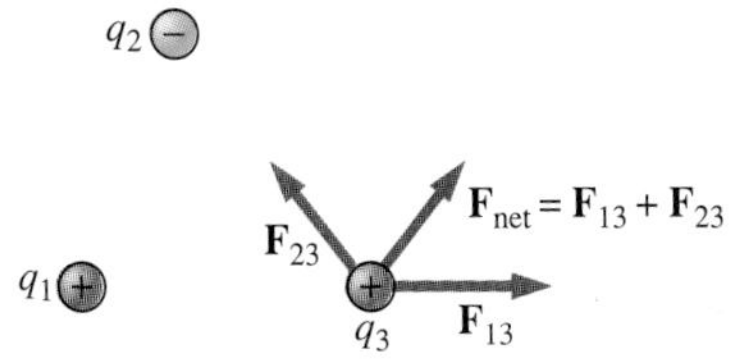

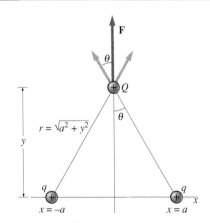

FIGURE 23-8 The superposition principle allows us to add vectorially the forces from two or more charges.

The Superposition Principle

Coulomb's law describes the force between *two* charges. But what if we have more than two charges? If we want the force on q_3 arising from two other charges q_1 and q_2, we simply calculate $\mathbf{F}_{13}$ and $\mathbf{F}_{23}$ from Equation 23-1 and add the resulting force vectors (Fig. 23-8). That is, the force that q_1 exerts on q_3 is unaffected by the presence of q_2, and the force that q_2 exerts on q_3 is unaffected by the presence of q_1. This may seem obvious, but nature need not have been that simple.

The fact that electric forces add vectorially is called the **superposition principle.** Our confidence in this principle is ultimately based on experiments that show electric and indeed electromagnetic phenomena behave according to the principle. With superposition we can solve relatively complicated problems by breaking them into simpler parts. If the superposition principle did not hold, the mathematical description of electromagnetism would be far more complicated than it is.

● **EXAMPLE 23-2** RAINDROPS

The charging of individual raindrops is ultimately responsible for the electrical activity of a thunderstorm. Suppose two drops with equal charge q are located on the x axis at $\pm a$, as shown in Fig. 23-9. Find the electric force on a third drop with charge Q located at an arbitrary point on the y axis.

Solution
Charge Q is the same distance $r = \sqrt{a^2 + y^2}$ from the two other drops, so the force from each has the same magnitude, given by Equation 23-1:

$$F = \frac{kqQ}{a^2 + y^2}.$$

But the directions of the two forces are different. It's evident from Fig. 23-9 that the x components cancel, while the y components add to give

$$F_y = 2\frac{kqQ}{a^2 + y^2}\cos\theta.$$

FIGURE 23-9 The force on Q is the vector sum of the forces from the individual charges.

But Fig. 23-9 shows that $\cos\theta = \dfrac{y}{r} = \dfrac{y}{\sqrt{a^2 + y^2}}$, so the force on Q becomes

$$\mathbf{F} = \frac{2kqQy}{(a^2 + y^2)^{3/2}}\hat{\mathbf{j}}.$$

Does this result make sense? Evaluating $\mathbf{F}$ at $y = 0$ gives zero force. Here, midway between the two charges, Q experiences equal but opposite forces from the two q's, so the net force must be zero. At very large distances such that $y \gg a$, on the other hand, we can neglect a^2 compared with y^2, and the force becomes

$$\mathbf{F} = \frac{k(2q)Q}{y^2}\hat{\mathbf{j}}. \qquad (y \gg a)$$

This is just the force we would expect from a single charge of magnitude $2q$ located a distance y from Q—showing that this system of two charges acts like a single charge $2q$ at distances large compared with the charge separation.

In drawing Fig. 23-9 we tacitly assumed that q and Q have the same signs. But our analysis holds even if they don't; in that case the product qQ is negative, and $\mathbf{F}$ therefore points in the opposite direction from $\hat{\mathbf{j}}$.

EXERCISE Find the net force on the charge located at the origin in Fig. 23-10.

Answer: $73\hat{\mathbf{i}} - 23\hat{\mathbf{j}}$ mN

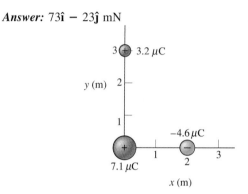

FIGURE 23-10 What is the force on the charge at the origin?

Some problems similar to Example 23-2: 19–22, 34

● EXAMPLE 23-3 BALANCING FORCES

A positive charge $+2q$ lies on the x axis at $x = -a$, and a charge $-q$ lies at $x = +a$, as shown in Fig. 23-11. Find a point where the electric force on a third charge Q would be zero.

Solution
The point must lie on the x axis since off axis the individual force vectors cannot point in opposite directions. But where on the axis? In between the two charges the repulsion of one and attraction of the other would add to give a nonzero net force. To the left of $+2q$, a third charge Q would always be closer to $+2q$ than to $-q$. The force from $+2q$ would always be greater, partly because Q would be closer to it and partly because of its greater charge. So the only place the forces from the two charges might cancel is to the right of $-q$. Even though any point in this region is closer to $-q$, the larger magnitude of $+2q$ might lead to the two forces being equal.

So suppose the third charge Q lies at some point x to the right of $-q$, so $x > a$. Then a unit vector from either $+2q$ or $-q$ toward Q points in the $+x$ direction, so is the vector $\hat{\mathbf{i}}$. Summing the two forces as given by Equation 23-1 and setting the result to zero then gives

$$\mathbf{F} = \mathbf{F}_{2q} + \mathbf{F}_{-q} = \frac{k(2q)Q}{(x + a)^2}\hat{\mathbf{i}} + \frac{k(-q)Q}{(x - a)^2}\hat{\mathbf{i}} = 0,$$

where the denominators are the distances from any point $x > a$ to the charges located at $x = -a$ and $x = +a$, respectively. Note that Q cancels from the equation, showing that the

FIGURE 23-11 Where will a third charge experience no net force (Example 23-3)?

point we're finding will be a point of zero force for *any* charge, no matter what its sign or magnitude. The quantities k and q also cancel, so the x component of our equation for the net force becomes

$$\frac{2}{(x + a)^2} = \frac{1}{(x - a)^2}.$$

Inverting and taking square roots gives

$$\frac{x + a}{\sqrt{2}} = \pm(x - a).$$

We solve separately for the two possible signs. For the $+$ sign, we have

$$x = a\frac{\sqrt{2} + 1}{\sqrt{2} - 1} = a\frac{(\sqrt{2} + 1)^2}{(\sqrt{2} - 1)(\sqrt{2} + 1)}$$

$$= (3 + 2\sqrt{2})a = 5.83a.$$

Since this value of x is greater than a, this point does lie to the right and therefore is indeed a point of zero force. Physically, we can understand the location of this point by noting that very close to $-q$ its force dominates. Far to the right of $-q$, on the other hand, we're nearly equal distances from the two charges and therefore $+2q$ dominates by virtue of its greater charge. At some intermediate point the two forces must balance; this is the point we've found.

You can verify that the solution for the minus sign lies to the left of $-q$ and is therefore inconsistent with our choice of direction for the unit vector from $-q$. This second solution is therefore not a meaningful answer.

> **TIP** **Let the Algebra Take Care of Signs** Once again, Equation 23-1 tells it all—including the direction of the force once you choose the unit vector cor-

rectly. You may be tempted in a problem like Example 23-3 simply to set the two forces equal and solve for x. That may work if you're careful, but it's safer to remember that you're really solving for a point where *the vector sum of two forces is zero*. You can't go wrong if you write those forces carefully with the correct vector directions and then sum them.

EXERCISE Repeat Example 23-3 with the charge $-q$ changed to $+q$.

Answer: $(3 - 2\sqrt{2})a = 0.17a$

Some problems similar to Example 23-3: 16–18, 32, 34 ●

Point Charges and Charge Distributions

Strictly speaking, Coulomb's law applies only to **point charges**—charged objects of arbitrarily small size. We're often interested in the forces arising from **charge distributions**—arrangements of charge spread over space. The two-charge systems of Examples 23-2 and 23-3 constitute simple charge distributions. The DNA molecule shown in Fig. 23-2 is a very complicated charge distribution. Other charge distributions include the electrodes in a TV tube, a memory cell in a computer memory chip, your heart, and a thundercloud. Ultimately these charge distributions consist of point-like electrons and protons to which Coulomb's law does apply, so we can in principle calculate the associated forces using the superposition principle.

Although the force of one point charge acting on another decreases with the inverse square of the distance, it's important to recognize that the same may or may not be true of the force arising from a charge distribution. For the distribution of two identical point charges in Example 23-2, for example, we found that the force on a third charge Q drops as the inverse square of the distance only at large distances; closer in, the dependence on distance is more complicated. This is generally true: When we're very far from a finite-sized charge distribution, it acts like a point charge with the total net charge of the distribution. But in closer the detailed arrangement of the individual charges becomes important, and the force associated with the distribution is generally no longer like that of a point charge.

23-4 THE ELECTRIC FIELD

We're often interested in the effect a charge distribution has, not on some particular charge, but on *any* charge we place in its vicinity. Accordingly, we define the **electric field:**

> *The electric field at any point is the force per unit charge experienced by a charge at that point. Mathematically,*
> $$E = \frac{F}{q}.$$ **(23-2a)**

The entire field is a set of vectors, one for each point in space, giving the magnitude and direction of the force per unit charge at that point. The units of electric field are newtons/coulomb. Fields of hundreds to thousands of N/C are commonplace, while fields of 3×10^6 N/C or more will tear electrons from air molecules. Electric fields within atoms may exceed 10^{12} N/C.

If we know the electric field **E** at a point we can compute the force on *any* charge q by rearranging Equation 23-2a:

$$\mathbf{F} = q\mathbf{E}. \tag{23-2b}$$

In Chapter 9 we gave a definition of the gravitational field as the gravitational force per unit mass. Conceptually, the gravitational field concept replaced the notion of action-at-a-distance—for example, Earth "reaching out" across empty space to pull on the moon—with the view that Earth creates a gravitational field in its vicinity and that the moon then responds to the gravitational field at its location. Defining the electric field implies the same conceptual shift; instead of thinking of one charge attracting or repelling another, we view a charge as creating an electric field throughout the space surrounding it. A second charge then responds to the field at its immediate location.

At this point you may regard the electric field as a mere mathematical construct we introduce for computational convenience or to satisfy some philosophical aversion to action-at-a-distance. But as you advance into the study of electromagnetism, you'll find that fields seem increasingly real. To a physicist, in fact, fields are every bit as real as matter itself.

There are two practical difficulties in using Equation 23-2a to measure electric fields. First, we must be sure the measured force is caused only by the electric field; if not, we must subtract the effect of other forces such as gravity. Second, the field we're trying to measure arises from one or more other charges. If the charge q is large, its own field may be strong enough to move the other charges, thereby altering the field we're trying to measure. For this reason we usually think of measuring the field with a very small "test charge."

● **EXAMPLE 23-4** A THUNDERSTORM: FORCE AND FIELD

A charged raindrop carrying 10 μC experiences an electric force of 0.30 N in the $+x$ direction. What is the electric field at its location? What would be the force on a -5.0-μC drop at the same location?

Solution

Equation 23-2a defines the electric field; here the force is $0.30\hat{\imath}$ N, so

$$\mathbf{E} = \frac{\mathbf{F}}{q} = \frac{0.30\hat{\imath}\ \text{N}}{10 \times 10^{-6}\ \text{C}} = 30\hat{\imath}\ \text{kN/C}.$$

Acting on a -5.0-μC charge, this field would give rise to a force

$$\mathbf{F} = q\mathbf{E} = (-5.0\ \mu\text{C})(30\hat{\imath}\ \text{kN/C}) = -0.15\hat{\imath}\ \text{N}.$$

TIP The Field Is Independent of the Test Charge You might wonder if the field should point in the $-x$ direction when we talk about putting a negative charge in the field. It doesn't—because the whole point of the field concept is to provide a description that's independent of the particular charge experiencing that force. The electric field in this example points in the $+x$ direction *no matter what charge* we may choose to put in the field. For a positive charge the force $q\mathbf{E}$ points in the *same* direction as the field; for a negative charge $q < 0$, and the force is *opposite* the field direction. As always, the algebra takes care of the signs.

EXERCISE An electric field $\mathbf{E} = -450\hat{\mathbf{i}}$ kN/C is used to accelerate electrons in a portion of a TV picture tube, where the x axis points from the back to the front of the tube. Find the force in this field experienced by (a) an electron and (b) an ion carrying $+2$ elementary charges. (c) Which particle will be accelerated toward the front of the tube?

Answers: (a) $7.2 \times 10^{-14}\hat{\mathbf{i}}$ N; (b) $-1.4 \times 10^{-13}\hat{\mathbf{i}}$ N; (c) the electron

Some problems similar to Example 23-4: 26–28

The Field of a Point Charge

Once we know the field of a charge distribution we can calculate its effect on other charges. The simplest charge distribution is a single point charge. Coulomb's law gives the force on a test charge q_1 located a distance r from a point charge q:

$$\mathbf{F} = \frac{kqq_1}{r^2}\hat{\mathbf{r}},$$

where $\hat{\mathbf{r}}$ is a unit vector pointing *away* from q. The electric field arising from q is the force per unit charge, or

$$\mathbf{E} = \frac{\mathbf{F}}{q_1} = \frac{kq}{r^2}\hat{\mathbf{r}}. \qquad \text{(field of a point charge)} \qquad (23\text{-}3)$$

Since it is so closely related to Coulomb's law for the electric force, we also refer to Equation 23-3 as Coulomb's law. Note that the equation contains no reference to the test charge q_1, since the field of q exists independently of any other charge. Since $\hat{\mathbf{r}}$ always points *away* from q, the direction of $\mathbf{E}$ is radially outward if q is positive and radially inward if q is negative. Figure 23-12 shows some field vectors for positive and negative point charges.

23-5 ELECTRIC FIELDS OF CHARGE DISTRIBUTIONS

The electric field is just the electric force per unit charge. Since the electric force obeys the superposition principle, so does the electric field. That means the field of a charge distribution is the vector sum of the fields of the individual point charges comprising the distribution:

$$\mathbf{E} = \mathbf{E}_1 + \mathbf{E}_2 + \mathbf{E}_3 + \cdots = \sum_i \mathbf{E}_i = \sum_i \frac{kq_i}{r_i^2}\hat{\mathbf{r}}_i, \qquad (23\text{-}4)$$

where the $\mathbf{E}_i$'s are the fields of the point charges q_i located at distances r_i from the point where we're evaluating the field, and where the $\hat{\mathbf{r}}_i$'s are unit vectors pointing *from* each point charge *toward* where we're evaluating the field. In principle, Equation 23-4 gives the electric field of *any* charge distribution. In practice, the process of summing the individual field vectors is often complicated unless the charge distribution contains relatively few charges arranged in a symmetric way.

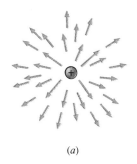

(a)

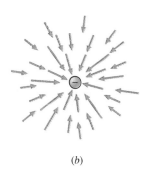

(b)

FIGURE 23-12 Field vectors for *(a)* positive and *(b)* negative point charges. Vector directions show that field vectors point radially outward or inward, and lengths show the magnitudes decreasing with the inverse square of the distance. In three dimensions the field vectors fill all space in a spherically symmetric fashion.

● EXAMPLE 23-5 TWO PROTONS

Two protons are 3.6 nm apart. (a) Find the electric field at the point P shown in Fig. 23-13. (b) Find the force on an electron at point P.

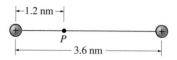

FIGURE 23-13 What is the electric field at P (Example 23-5)?

Solution

If we take the x axis along the line joining the protons, then $\hat{\mathbf{r}}_1$ is just $\hat{\mathbf{i}}$ and $\hat{\mathbf{r}}_2$ is $-\hat{\mathbf{i}}$, where the subscripts 1 and 2 refer to the left and right protons, respectively. Then Equation 23-4 gives

$$\mathbf{E} = \mathbf{E}_1 + \mathbf{E}_2 = \frac{ke}{r_1^2}\hat{\mathbf{i}} + \frac{ke}{r_2^2}(-\hat{\mathbf{i}}) = ke\left(\frac{1}{r_1^2} - \frac{1}{r_2^2}\right)\hat{\mathbf{i}}$$

$$= (9.0 \times 10^9 \text{ N·m}^2/\text{C}^2)(1.6 \times 10^{-19} \text{ C})$$

$$\times \left[\frac{1}{(1.2 \times 10^{-9} \text{ m})^2} - \frac{1}{(2.4 \times 10^{-9} \text{ m})^2}\right]\hat{\mathbf{i}}$$

$$= 750\hat{\mathbf{i}} \text{ MN/C}.$$

An electron in this field will experience a force

$$\mathbf{F} = q\mathbf{E} = (-1.6 \times 10^{-19} \text{ C})(750 \times 10^6 \hat{\mathbf{i}} \text{ N/C})$$

$$= -1.2 \times 10^{-10}\hat{\mathbf{i}} \text{ N}.$$

EXERCISE Find the magnitude of the electric field (a) 4.8 nm and (b) 200 nm to the right of the right-hand proton in Fig. 23-13. (c) Show that your answer to (b) is nearly equal to the field of a single charge $2e$ located midway between the two protons.

Answers: (a) 82.9 MN/C; (b) 70.7 kN/C

Some problems similar to Example 23-5: 31–34 ●

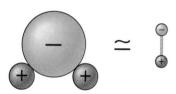

FIGURE 23-14 (left) A water molecule. Electrons spend more time in the vicinity of the single oxygen atom, giving rise to separate regions of negative and positive charge. The molecule therefore approximates the dipole shown at right, and at large distances gives rise to an electric field like that of Example 23-6.

The Electric Dipole

One of the most important charge distributions is the **electric dipole,** consisting of two point charges of equal magnitude but opposite sign held a fixed distance apart. Many molecules are essentially electric dipoles, so an understanding of the dipole helps explain molecular behavior (Fig. 23-14). During contraction the heart muscle becomes essentially a dipole, and physicians performing electrocardiography are measuring, among other things, the strength and orientation of that dipole. The dipole configuration is also used in a number of technological devices such as radio and TV antennas.

● EXAMPLE 23-6 MODELING A MOLECULE

A molecule consists of separated regions of positive and negative charge, modeled approximately as a positive charge q at $x = a$ and a negative charge $-q$ at $x = -a$, as shown in Fig. 23-15. Find a general expression for the electric field at any point on the y axis, and an approximate expression valid at large distances ($y \gg a$).

Solution

Figure 23-15 shows the individual field vectors $\mathbf{E}_-$ and $\mathbf{E}_+$, along with their sum. The y components cancel to give a net field parallel to the x axis. The x components of the two fields are clearly the same, so we have

$$E_x = E_{x-} + E_{x+} = -2\left(\frac{kq}{r^2}\sin\theta\right),$$

where the minus sign occurs because, as Fig. 23-15 shows, the net field points in the negative x direction. Fig. 23-15 also shows that the distance from both charges is $r = \sqrt{y^2 + a^2}$ and that $\sin\theta = a/r = a/\sqrt{y^2 + a^2}$. Then we have

$$\mathbf{E} = E_x\hat{\mathbf{i}} = -\frac{2kqa}{(y^2 + a^2)^{3/2}}\hat{\mathbf{i}}.$$

Does this result make sense? Midway between the charges the fields from each charge point in the same direction and have the same magnitude, so we expect a resultant field twice that of either charge alone. Setting $y = 0$ at the midpoint gives

$$\mathbf{E}(y=0) = -\frac{2kqa}{(a^2)^{3/2}}\hat{\mathbf{i}} = -\frac{k(2q)}{a^2}\hat{\mathbf{i}},$$

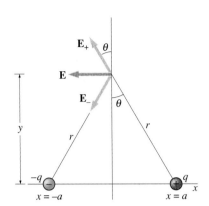

FIGURE 23-15 An electric dipole. The electric field on the dipole's perpendicular bisector (here the y axis) is parallel to the dipole axis.

which is indeed twice the field of either charge at the distance a.

We're frequently interested in the field far from a dipole, which is why this example asks for an approximate expression for $y \gg a$. (One needn't go very far from a molecule for its size to be insignificant.) Under this approximation we can neglect a^2 compared with y^2 in our expression for the field, giving

$$\mathbf{E} = -\frac{2kqa}{y^3}\hat{\mathbf{i}}, \qquad y \gg a.$$

We'll see later how forces associated with such molecular electric fields give rise to the van der Waals interaction that we considered in Chapter 20.

TIP Approximations Making approximations requires some care. Here we're basically asking for the field when y is so large that a is negligible *compared with y*. So we neglect a^2 compared with y^2 when the two are summed, but we *don't* neglect a when it appears in the numerator, where it's not summed with y.

EXERCISE Find (a) a general expression for the dipole field at points on the x axis to the right of $x = a$ in Fig. 23-15, and (b) an approximate expression valid for $x \gg a$.

Answers: (a) $\mathbf{E} = \dfrac{4kqax}{(x^2 - a^2)^2}\hat{\mathbf{i}}$; (b) $\mathbf{E} = \dfrac{4kqa}{x^3}\hat{\mathbf{i}}$

Some problems similar to Example 23-6: 31, 37, 38, 41 ●

The dipole fields in Example 23-6 and its exercise both decrease, at large distances, as the inverse *cube* of distance. Physically, this is because the dipole has zero *net* charge. Its field arises entirely from the slight separation of two opposite charges. Because of this separation the dipole field isn't exactly zero, but it is weaker and more localized than the field of a point charge. Many complicated charge distributions exhibit the essential characteristic of a dipole—namely, they're neutral but consist of separated regions of positive and negative charge—and at large distances such distributions all have essentially the same field configuration (see Problems 48 and 82).

At large distances the dipole's physical characteristics q and a enter the equations for the electric field only in the product qa. We call the product of the charge q and separation $2a$ the **dipole moment,** p. Its units are C·m. Using this definition, the fields given in Example 23-6 and its exercise for the dipole of Fig. 23-15 are

$$\mathbf{E} = -\frac{kp}{y^3}\hat{\mathbf{i}} \qquad \left(\begin{array}{c}\text{dipole field for } y \gg a, \\ \text{on perpendicular bisector}\end{array}\right) \qquad (23\text{-}5a)$$

and

$$\mathbf{E} = \frac{2kp}{x^3}\hat{\mathbf{i}} \qquad \left(\begin{array}{c}\text{dipole field} \\ \text{for } x \gg a, \text{ on axis}\end{array}\right). \qquad (23\text{-}5b)$$

Problem 41 generalizes these results to an arbitrary point not necessarily on the axis or bisector. Because the dipole isn't spherically symmetric, its field is a function of both distance and the angle between the position vector and the

FIGURE 23-16 The dipole moment vector has magnitude given by the product of the charge and separation, and it points from the negative toward the positive charge.

dipole axis. For this reason it's convenient to think of the dipole moment as a vector, **p**, whose magnitude is the product of the charge and separation and whose direction is from the negative to the positive charge (Fig. 23-16).

We emphasize that Equations 23-5 are approximations valid far from the dipole. It's useful to imagine a dipole whose separation $2a$ shrinks toward zero while the magnitude of its two charges grows to keep the product $p = 2qa$ constant. In the limit as $a \to 0$ we have a **point dipole.** Since the point dipole has zero size, Equations 23-5 become exact in this case. Although the point dipole is an idealization, it's a useful approximation when we're far from a real dipole.

Continuous Charge Distributions

Although any charge distribution ultimately consists of point-like electrons and protons, it would be impossible in practice to sum all the field vectors from the 10^{23} or so particles comprising a typical piece of matter. Instead, it's convenient to make the approximation that charge is spread continuously over the distribution. If the charge distribution extends throughout a volume, we describe it in terms of the **volume charge density,** ρ, with units of C/m^3. For charge distributions spread over surfaces or lines the corresponding quantities are the **surface charge density,** σ, and **line charge density,** λ. Their units are C/m^2 and C/m, respectively.

To calculate the field of a continuous charge distribution, we break the charged region into very many small charge elements dq, each small enough that it is essentially a point charge. Each dq then produces an electric field $d\mathbf{E}$ given by Equation 23-3:

$$d\mathbf{E} = \frac{k\, dq}{r^2}\hat{\mathbf{r}}.$$

We then form the vector sum of all the $d\mathbf{E}$'s. In the limit of infinitely many infinitesimally small dq's and their corresponding $d\mathbf{E}$'s, that sum becomes an integral and we have

$$\mathbf{E} = \int d\mathbf{E} = \int \frac{k\, dq}{r^2}\hat{\mathbf{r}}. \tag{23-6}$$

The limits of this integral are chosen to include the entire charge distribution. Figure 23-17 shows the meaning of Equation 23-6; note in particular that both the distance r and the direction specified by $\hat{\mathbf{r}}$ generally vary with position.

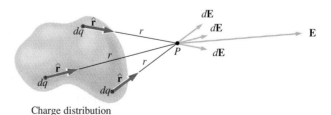

Charge distribution

FIGURE 23-17 The electric field at point P, given by Equation 23-6, is the sum of the vectors $d\mathbf{E}$ arising from the individual charge elements dq in the entire distribution, each calculated using the appropriate distance r and unit vector $\hat{\mathbf{r}}$.

● **EXAMPLE 23-7** A CHARGED ROD

Wires, antennas, and similar elongated structures can often be considered as thin rods carrying electric charge. Suppose such a rod of length ℓ carries a positive charge Q distributed uniformly over its length. Find the electric field at point P in Fig. 23-18, a distance a from the end of the rod.

Solution

Let the y axis lie along the rod, with origin at P. Consider a small length dy of the rod, containing charge dq, and located a distance y from P. A unit vector from dq toward P is $-\hat{\mathbf{j}}$, so the field at P due to dq is

$$d\mathbf{E} = -\frac{k\,dq}{y^2}\hat{\mathbf{j}}.$$

The net field at P is the sum—that is, the integral—of all the fields $d\mathbf{E}$ arising from all the dq's along the rod:

$$\mathbf{E} = \int d\mathbf{E} = -\hat{\mathbf{j}}\int_{y=a}^{y=a+\ell}\frac{k\,dq}{y^2},$$

where we've chosen the limits to cover the entire rod. To evaluate the integral we must relate dq and y. How? The rod carries total charge Q, distributed uniformly over its length ℓ. The line charge density is therefore $\lambda = Q/\ell$. This is the charge per unit length; a length dy therefore carries charge $dq = \lambda\,dy$, or $dq = Q\,dy/\ell$. Then our integral becomes

$$\mathbf{E} = -\hat{\mathbf{j}}\int_a^{a+\ell}\frac{k(Q\,dy/\ell)}{y^2} = -\frac{kQ}{\ell}\hat{\mathbf{j}}\int_a^{a+\ell}\frac{dy}{y^2}$$

$$= -\frac{kQ}{\ell}\hat{\mathbf{j}}\left[-\frac{1}{y}\right]_a^{a+\ell} = -\frac{kQ}{\ell}\hat{\mathbf{j}}\left(\frac{1}{a}-\frac{1}{a+\ell}\right) = -\frac{kQ}{a(a+\ell)}\hat{\mathbf{j}},$$

where we took k, Q, ℓ, and $\hat{\mathbf{j}}$ outside the integral because they are constants.

Does this result make sense? First consider the direction: The negative sign shows that the field is downward for positive Q and upward for negative Q, as we should expect. Now suppose P is very far from the rod, so $a \gg \ell$. Then our result becomes approximately

$$\mathbf{E} = -\frac{kQ}{a^2}\hat{\mathbf{j}}, \qquad (a \gg \ell),$$

which is just what we expect for the field of a point charge Q. In this case we're so far from the rod that its length becomes negligible, and indeed it acts like a point charge. But as we move closer, the field becomes a more complicated superposition of the fields of all the dq's at different locations along the rod, and the field no longer exhibits the inverse-square dependence of the point-charge field.

Knowing the field of the charged rod, we can use superposition to find the fields of charge distributions involving more than one rod or a combination of rods and point charges (see Problems 44–46).

FIGURE 23-18 The field at P is the sum—or integral—of the fields arising from all the infinitesimal charge elements dq along the rod.

TIP Find a Single Integration Variable

Evaluating the integral in Equation 23-6 requires that we relate the charge element dq and the position variable r so we'll have the integral expressed in terms of a single variable. The charge density provides the link needed, since it allows us to write dq as the charge density multiplied by an appropriate element of volume, area, or length:

$$dq = \rho\,dV, \quad dq = \sigma\,dA, \quad \text{or} \quad dq = \lambda\,dx.$$

Which of these we use depends on whether charge is distributed over a volume, area, or length; for the thin rod of Example 23-7, the appropriate charge density was the line charge density λ and the position variable was y, so we had $dq = \lambda\,dy$. Sometimes you'll be given the charge density explicitly, and other times you can compute it from the charge and dimensions of the charge distribution.

EXERCISE A thin rod of length ℓ lies along the y axis with its bottom end at the origin. It carries a line charge density λ that varies with position, being given by $\lambda = Q_0\dfrac{y^3}{\ell^4}$, where Q_0 is a constant and y is the distance from the origin. Find the magnitude of the electric field at $y = 0$.

Answer: $\dfrac{kQ_0}{2\ell^2}$

Some problems similar to Example 23-7: 42–45, 73, 74, 82

● **EXAMPLE 23-8** A CHARGED RING

A thin ring of radius a is centered on the origin and carries a total charge Q distributed uniformly around the ring, as shown in Fig. 23-19. Find the electric field at a point P located a distance x along the axis of the ring, and show that the result makes sense when $x \gg a$.

Solution
In Example 23-7 the magnitude but not direction of the individual field vectors $d\mathbf{E}$ from all the charge elements dq varied. Here we have the opposite situation: a point on the ring axis is equidistant from all points on the ring, so the field magnitudes dE are the same but their directions vary. Figure 23-19 shows, however, that components perpendicular to the x axis cancel for any pair of charge elements on opposite sides of the ring, leaving a net field in the x direction. Each charge element contributes an amount dE_x to the field:

$$dE_x = \frac{k\,dq}{r^2}\cos\theta = \frac{k\,dq}{x^2 + a^2}\frac{x}{\sqrt{x^2 + a^2}} = \frac{kx\,dq}{(x^2 + a^2)^{3/2}},$$

where the geometry of Fig. 23-19 gives $r = \sqrt{x^2 + a^2}$ and $\cos\theta = x/\sqrt{x^2 + a^2}$. We now need to integrate this expression over the entire ring. In this integration k, a, and x are all constants—they don't change as we move around the ring. So we have

$$E = E_x = \int_{\text{ring}} dE_x = \int_{\text{ring}} \frac{kx\,dq}{(x^2 + a^2)^{3/2}}$$

$$= \frac{kx}{(x^2 + a^2)^{3/2}} \int_{\text{ring}} dq = \frac{kx\,Q}{(x^2 + a^2)^{3/2}},$$

where the last step follows because $\int_{\text{ring}} dq$ simply means the total charge on the ring. For positive Q this field points away from the ring.

Does this result make sense? At large distances from the ring we can neglect its size a compared with the distance x, and our result reduces to $E = \dfrac{kQ}{x^2}$—just what we would expect for

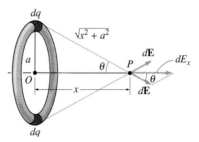

FIGURE 23-19   The electric field of a charged ring points along the ring axis since field components perpendicular to the axis cancel in pairs.

a point charge Q. As always, a finite-size charge distribution looks essentially like a point charge at distances large compared with its size.

> **TIP   How Can x Be a Constant?**   In this exercise x is constant because it's the distance from the center of the ring to the point where we're evaluating the field—that is, it's the distance from the origin to the *field point, P*. In the integration to find the field we're supposed to consider all *source points*—all points where charge is located that contributes to the field at P. Moving around the ring doesn't affect the value of x, so x is a constant for this integration. However, x is arbitrary, so our result holds for *any* value of x.

EXERCISE   Find the point on the x axis where the electric field of the ring in Example 23-8 has its greatest magnitude.

Answer: $x = a/\sqrt{2}$

Some problems similar to Example 23-8:   48–51, 83   ●

● **EXAMPLE 23-9** A POWER LINE'S FIELD

A long, straight electric power line coincides with the x axis and carries a line charge density λ C/m. What is the electric field at a point P on the y axis? Use the approximation that the line is infinitely long.

Solution
Here *both* the direction and magnitude of the field element $d\mathbf{E}$ arising from a charge element on the line vary with the position x of the charge element. But Fig. 23-20 shows that charge elements on opposite sides of the y axis give rise to electric

fields whose x components cancel. Thus the net field points in the y direction—that is, away from the line if λ is positive. So each charge element contributes an amount dE_y to the net field:

$$dE_y = \frac{k\,dq}{r^2}\cos\theta = \frac{k\lambda\,dx}{x^2 + y^2}\frac{y}{\sqrt{x^2 + y^2}} = \frac{k\lambda y\,dx}{(x^2 + y^2)^{3/2}},$$

where we've written $dq = \lambda\,dx$ and used the geometry of Fig. 23-20 to write $r = \sqrt{x^2 + y^2}$ and $\cos\theta = y/r$. To find the net field, we integrate over the entire line, from $x = -\infty$ to

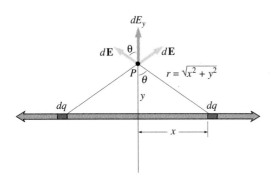

FIGURE 23-20 The field of a charged line is the vector sum of the fields $d\mathbf{E}$ from all the individual charge elements dq along the line. The x components from each pair of charge elements cancel, giving a net field that points directly away from the line.

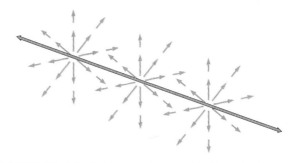

FIGURE 23-21 Field vectors for an infinite line of positive charge point radially outward from the line, with magnitude decreasing inversely with distance.

$x = +\infty$. The quantities k, λ, and y don't change as we move along the line, so we have

$$E = E_y = \int_{x=-\infty}^{x=\infty} \frac{k\lambda y \, dx}{(x^2 + y^2)^{3/2}} = k\lambda y \int_{-\infty}^{\infty} \frac{dx}{(x^2 + y^2)^{3/2}}$$

$$= k\lambda y \left[\frac{x}{y^2\sqrt{x^2 + y^2}} \right]_{-\infty}^{\infty} = k\lambda y \left[\frac{1}{y^2} - \left(-\frac{1}{y^2} \right) \right] = \frac{2k\lambda}{y}.$$

(23-7)

Here we evaluated the integral using the integral table in Appendix A and applied the limits $\pm\infty$ by noting that as $x \to \pm\infty$ the term y^2 becomes negligible compared with x^2, giving $x/\sqrt{x^2 + y^2} \to x/\sqrt{x^2} = \pm 1$. The integral could also be evaluated by rewriting it in terms of the angle θ (see Problem 87).

Since the line is infinite in both directions and has cylindrical symmetry, Equation 23-7 holds for *any* point a distance y from the line. Our result thus shows that the electric field of a long line of positive charge points radially away from the line, with magnitude that drops inversely with distance from the line (Fig. 23-21).

What about the field at large distances? Shouldn't it resemble that of a point charge, falling with the inverse square of the distance? No: The charged line is infinitely long, so no matter how far away we go it never resembles a point. Its slower dropoff reflects that fact.

Of course our infinite line is an impossibility. But many real charge distributions, including the power line of this example, have long, thin shapes that approximate an infinite line. Equation 23-7 is therefore a good approximation to the field of a *finite* line as long as we're much closer to it than its length, and not too near either end. Very far from a *finite* line, on the other hand, the field does approach that of a point charge.

EXERCISE A thin rod 2.0 m long carries 50 μC distributed uniformly over its length. Find the electric field strength (a) 1.0 cm from the rod axis, not near either end and (b) 500 m from the rod. Make suitable approximations in both cases.

Answers: (a) 45 MN/C; (b) 1.8 N/C

Some problems similar to Example 23-9: 52–55 ●

23-6 MATTER IN ELECTRIC FIELDS

We're ultimately interested in electric fields since they give rise to forces on charged particles. Because matter consists of such particles, much of the behavior of matter is fundamentally determined by electric fields.

Point Charges in Electric Fields

The motion of a single charge in an electric field is governed by the definition of the electric field (Equation 23-2):

$$\mathbf{F} = q\mathbf{E}$$

and Newton's law:
$$\mathbf{F} = m\mathbf{a}.$$

Combining these equations gives the acceleration of a particle with charge q and mass m in an electric field $\mathbf{E}$:

$$\mathbf{a} = \frac{q}{m}\mathbf{E}. \qquad (23\text{-}8)$$

This equation shows that it's the charge-to-mass ratio that determines a particle's response to an electric field. Electrons, nearly 2000 times less massive than protons but carrying the same charge, are readily accelerated by electric fields. Many practical devices, from x-ray machines to TV tubes, make use of electrons accelerated in electric fields.

When the electric field is uniform, problems involving the motion of charged particles reduce to the constant-acceleration problems we considered in Chapter 2. We'll see in the next chapter that uniform fields are easily produced by flat, uniformly charged plates.

● EXAMPLE 23-10 INSIDE A HEART MONITOR

A heart monitor used in a hospital's intensive care unit includes a cathode-ray tube that gives continuous visual display of a patient's heartbeat (Fig. 23-22). An electron beam "paints" the display on a phosphor screen at the front of the tube, being "steered" to different parts of the screen by electric forces. Suppose the beam is initially heading horizontally to the right at 4.0 Mm/s when it enters the "steering" electric field of 1.0 kN/C pointing downward. The field region extends horizontally for 2.0 cm, as shown in Fig. 23-23. In what direction is the electron moving when it leaves the field region?

Solution
As usual in two-dimensional motion, the vertical force due to the electric field does not affect the horizontal component of the electron's velocity. Therefore the electron spends a time $t = \Delta x/v_x$ in the field region. During this time it experiences a

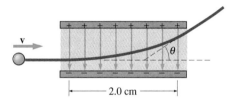

FIGURE 23-23 A uniform electric field deflects an electron from its straight-line path (Example 23-10).

vertical acceleration qE/m, and therefore gains a vertical velocity component given by

$$v_y = a_y t = \frac{qE_y}{m}\frac{\Delta x}{v_x}$$

$$= \frac{(-1.6\times10^{-19}\text{ C})(-1.0\times10^3\text{ N/C})}{9.11\times10^{-31}\text{ kg}}\left(\frac{2.0\times10^{-2}\text{ m}}{4.0\times10^6\text{ m/s}}\right)$$

$$= 8.78\times10^5\text{ m/s}.$$

Note that the field points downward, as reflected by the minus sign, and combined with the electron's negative charge thus results in an upward velocity. The electron thus leaves the field region moving at an angle of

$$\theta = \tan^{-1}\left(\frac{v_y}{v_x}\right) = \tan^{-1}\left(\frac{8.78\times10^5\text{ m/s}}{4.0\times10^6\text{ m/s}}\right) = 12.4°$$

to the horizontal.

As it traverses the field region the electron describes the parabolic trajectory of an object undergoing constant acceleration in two dimensions; once outside the field it continues in a straight line with its new velocity, as indicated in Fig. 23-23.

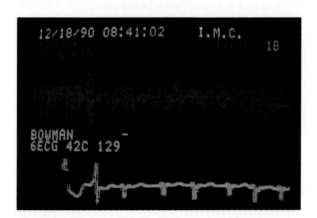

FIGURE 23-22 This heart monitor uses an electron beam deflected by electric fields to display patients' heartbeats.

Field configurations like this one are used to "steer" electron beams not only in heart monitors but also in oscilloscopes and other electronic instrumentation.

EXERCISE What electric field strength would be needed to cause the electron of Example 23-10 to leave the field region at a point 4.6 mm above where it entered? The electron's initial velocity and the length of the field region are the same as in Example 23-10.

Answer: 2.1 kN/C

Some problems similar to Example 23-10: 59, 60, 77, 78 ●

When the field is not uniform it's generally difficult to calculate particle trajectories. An important exception is the case of a particle moving at right angles to a field that points radially. In that case—the subject of the following example—the electric force changes the particle's direction but not its speed, so the motion is uniform circular.

● **EXAMPLE 23-11** AN ELECTROSTATIC ANALYZER

Two curved metal plates are used to establish an electric field given by

$$E = E_0 \frac{b}{r},$$

where $E_0 = 24$ kN/C and $b = 5.0$ cm. The field points toward the center of curvature, as shown in Fig. 23-24, and r is the distance from that center. A beam of protons with a mix of speeds is incident on the device. Find the speed v for which an incident proton will leave the analyzer moving horizontally in Fig. 23-24.

Solution
To exit with its velocity horizontal, a proton must describe a circular arc while inside the analyzer. The field provides the v^2/r acceleration required for that circular motion:

$$a = \frac{v^2}{r} = \frac{eE}{m} = \frac{e}{m} E_0 \frac{b}{r}.$$

Solving for v gives

$$v = \sqrt{\frac{eE_0 b}{m}} = \sqrt{\frac{(1.6\times10^{-19}\ \text{C})(2.4\times10^4\ \text{N/C})(0.050\ \text{m})}{1.67\times10^{-27}\ \text{kg}}}$$

$$= 3.4\times10^5\ \text{m/s}.$$

Note that it doesn't matter where the protons enter the analyzer since the $1/r$ decrease in field strength matches the $1/r$ depen-

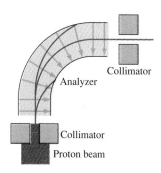

FIGURE 23-24 An electrostatic analyzer, showing the trajectories of protons in the radial electric field. Only those entering with the right speed will emerge at the top moving horizontally. Collimators block protons not moving at right angles to the field.

dence of the acceleration. Devices of this sort have been used on spacecraft to analyze charged particles in interplanetary space.

EXERCISE A proton is in circular motion centered on a long charged wire carrying uniform negative line charge density $-\lambda$. Find its speed. *Hint:* Consult Example 23-9 for the field of the charged wire.

Answer: $v = \sqrt{2k\lambda e/m}$, with m the proton mass

Some problems similar to Example 23-11: 61–65 ●

Dipoles in Electric Fields

Earlier in this chapter we calculated the field of an electric dipole, which consists of two opposite charges of equal magnitude. Here we study a dipole's response to electric fields. Since the dipole configuration provides a simple model for molecules, our results help explain molecular behavior.

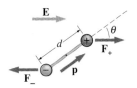

FIGURE 23-25 A dipole in a uniform electric field experiences no net force, but it does experience a torque.

Figure 23-25 shows a dipole with charges $\pm q$ separated by a distance d, located in a uniform electric field. The dipole moment vector $\mathbf{p}$ has magnitude qd and points from the negative to the positive charge (recall Fig. 23-16). Since the field is uniform it's the same at both ends of the dipole. Since the dipole charges are equal in magnitude but opposite in sign, they experience equal but opposite forces $\pm q\mathbf{E}$—and therefore there's no net force on the dipole.

However, Fig. 23-25 shows that the dipole does experience a torque that tends to align it with the field. In Chapter 13 we described torque as the cross product of the position vector with the force: $\boldsymbol{\tau} = \mathbf{r} \times \mathbf{F}$, where the magnitude of the torque vector is $rF\sin\theta$ and its direction is given by the right-hand rule. Figure 23-25 thus shows that the torque about the center of the dipole due to the force on the positive charge has magnitude

$$\tau_+ = rF\sin\theta = \frac{1}{2}d\,qE\sin\theta.$$

The torque associated with the negative charge has the same magnitude and both torques are in the same direction since both tend to rotate the dipole in Fig. 23-25 clockwise. Thus the net torque has magnitude $\tau = qdE\sin\theta$; applying the right-hand rule shows that this torque is into the page. But qd is the magnitude of the dipole moment $\mathbf{p}$, and Fig. 23-25 shows that θ is the angle between the dipole moment vector and the electric field $\mathbf{E}$; therefore we can write the torque vectorially as

$$\boldsymbol{\tau} = \mathbf{p} \times \mathbf{E}. \qquad \text{(torque on a dipole)} \qquad (23\text{-}9)$$

Because of this torque, it takes work to rotate a dipole in an electric field. If we start with the dipole oriented at right angles to the field ($\theta = \pi/2$), then Equation 12-28b gives the work required to rotate it to a new angle θ:

$$W = \int_{\pi/2}^{\theta} \tau d\theta = \int_{\pi/2}^{\theta} pE\sin\theta\,d\theta = pE\,[-\cos\theta]_{\pi/2}^{\theta} = -pE\cos\theta,$$

where the last step follows because $\cos(\pi/2) = 0$. This work ends up as stored potential energy U. Since the product of two vector magnitudes with the cosine of the angle between the vectors defines the dot product, we can write the potential energy in compact form as

$$U = -\mathbf{p} \cdot \mathbf{E}, \qquad (23\text{-}10)$$

where the zero of potential energy corresponds to the dipole at right angles to the field.

■ APPLICATION MICROWAVE COOKING AND LIQUID CRYSTALS

The torque on dipoles in electric fields forms the basis of two widespread contemporary technologies: the microwave oven and the liquid crystal display (LCD) (Fig. 23-26).

A microwave oven works by generating an electric field whose direction changes several billion times per second. Water molecules, whose dipole moment is much greater than that

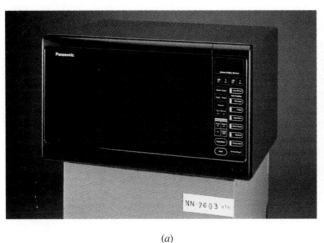

(a)   (b)

FIGURE 23-26 (a) Microwave ovens and (b) liquid crystal displays make use of the torque that dipolar molecules experience in electric fields.

of most other molecules, respond by attempting to align with the field. But the field is constantly changing, so the molecules swing rapidly back and forth. As they jostle against each other, the energy they gain from the field is dissipated as heat that cooks the food.

Calculators, laptop computers, gas pumps, and many other devices display numerical and alphabetic information with liquid crystals. This unique state of matter combines the fluidity of a liquid with the order of a solid. The liquid crystal consists of long molecules whose chemical structure gives rise to a

dipole-like charge separation. In response to each others' electric fields, the molecules tend to align with each other (Fig. 23-27). But an external electric field can rotate the liquid crystal dipoles, altering the optical properties of the material. Figure 23-28 shows how this effect is exploited to make practical displays. Liquid crystal displays have the advantage that they consume very little power. On the other hand they generate no light of their own, and therefore an external light source is needed to illuminate the display when it's dark.

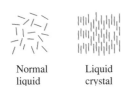

Normal	Liquid
liquid	crystal

FIGURE 23-27 Alignment of dipole-like molecules in a liquid crystal. The dipole's direction can be rotated by applying an electric field.

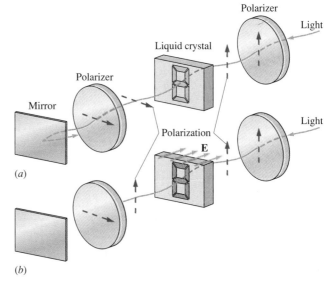

FIGURE 23-28 A typical liquid-crystal display consists of seven small bands of liquid crystal sandwiched between optical polarizers. Applying an electric field to any segment rotates its liquid crystal molecules, changing the polarization and making the segment appear dark. (We deal further with polarization in Chapter 34). (a) Electric field off. (b) Electric field on.

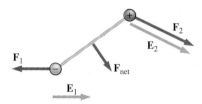

FIGURE 23-29 When the electric field differs in magnitude or direction at the two ends of the dipole, then the dipole experiences a nonzero net force as well as a torque.

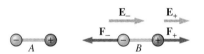

FIGURE 23-30 Dipole B is aligned with the electric field of dipole A. Since the field of A is stronger at the negative end of B, $\mathbf{F}_-$ is greater in magnitude than $\mathbf{F}_+$, so dipole B experiences a net force toward dipole A. This is the origin of the van der Waals force between gas molecules.

When the electric field is not uniform, the charges at opposite ends of the dipole experience forces that differ in magnitude and/or are not exactly opposite in direction. Then the dipole experiences a net force as well as a torque (Fig. 23-29). An important instance of this effect is the force on a dipole in the field of another dipole (Fig. 23-30). Because the dipole field falls off rapidly with distance and because the dipole responding to the field has closely spaced charges of equal magnitude but opposite sign, the dipole-dipole force is quite weak and falls extremely rapidly with distance. This weak force, which Fig. 23-30 shows to be attractive, is the basis of the van der Waals interaction between gas molecules that we considered in Chapter 20. Problems 70 and 85 deal with forces on dipoles in nonuniform fields.

Conductors, Insulators, and Dielectrics

Bulk matter consists ultimately of vast numbers of point charges, namely electrons and protons. In some matter—notably metals, ionic solutions, and ionized gases—individual charges are free to move throughout the material. In such materials—called **conductors**—the application of an electric field results in the ordered motion of electric charge that we call **electric current.** We'll consider the behavior of conductors and related materials called semiconductors in subsequent chapters.

Materials in which charge is not free to move are called **insulators,** since they do not support electric current. Insulators, however, still contain charges—it's just that their charges are bound into neutral molecules. Some molecules, like water, have intrinsic dipole moments and therefore rotate in response to an applied electric field. Even if they don't have dipole moments, molecules may respond to an electric field by stretching and acquiring **induced dipole moments** (Fig. 23-31). In either case, the application of an electric field results in the alignment of molecular dipoles with the field (Fig. 23-32). The fields of the dipoles, pointing from their positive to their negative charges, then reduce the applied electric field within the material. We'll explore the consequences of this effect further in Chapter 26. Materials in which molecules have either intrinsic dipole moments or acquire induced moments are called **dielectrics.**

If the electric field applied to a dielectric becomes too great, individual charges are ripped free, and the material then acts like a conductor. Such **dielectric breakdown** can cause severe damage in electrical equipment (Fig. 23-33). On a larger scale, lightning results from dielectric breakdown in air.

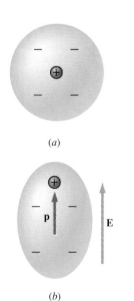

FIGURE 23-31 (*a*) A molecule with no dipole moment has negative charge concentric with positive charge. (*b*) In an applied electric field, the molecule stretches and acquires a dipole moment.

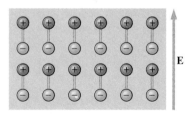

FIGURE 23-32 The alignment of molecular dipoles in a dielectric results in a reduction of the electric field within the dielectric.

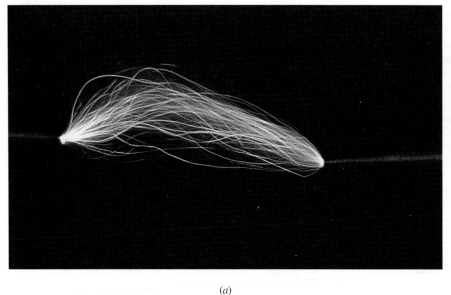

(a)

(b)

FIGURE 23-33 (a) Dielectric breakdown in air results in sparks jumping between two highly charged wires. (b) Dielectric breakdown in a solid produced this tree-like pattern.

CHAPTER SYNOPSIS

Summary

1. **Electromagnetism** is among the fundamental forces of nature. Electromagnetism comprises electricity and magnetism, seemingly distinct phenomena that are actually intimately related.

2. **Electric charge** is a fundamental property of matter. Charge comes in two types, arbitrarily called positive and negative. Charge is **quantized,** with one elementary charge—the magnitude of the electron or proton charge—equal to 1.60×10^{-19} C, where the coulomb (C) is the SI unit of charge. Charge is also **conserved,** in that the algebraic sum of the charges in a closed system never changes.

3. Charges interact via the **electric force.** This force is proportional to the product of the charges and inversely proportional to the square of the distance between them. **Coulomb's law** provides a mathematical description of the electric force between point charges:

$$\mathbf{F}_{12} = \frac{kq_1q_2}{r^2}\hat{\mathbf{r}}.$$

4. The **superposition principle** states that the electric force on a charge arising from two or more other charges is the vector sum of the force arising from each according to Coulomb's law. The superposition principle greatly simplifies the calculation of electrical effects of various **charge distributions.**

5. The electric field at a point is a vector giving the electric force per unit charge that would be experienced by a charge at that point:

$$\mathbf{E} = \frac{\mathbf{F}}{q}.$$

The electric field of a point charge q is therefore

$$\mathbf{E} = \frac{kq}{r^2}\hat{\mathbf{r}}.$$

6. The superposition principle shows that the electric field of a charge distribution is the sum of the fields of its individual point charges:

$$\mathbf{E} = \sum_i \mathbf{E}_i = \sum_i \frac{kq_i}{r_i^2}\hat{\mathbf{r}}_i.$$

A particularly important charge distribution is the **electric dipole,** consisting of two point charges with equal magnitude but opposite sign, separated by a fixed distance. At large distances from a dipole the field decreases as the inverse cube of the distance.

7. With continuous distributions of charge, the sum over all point charges becomes an integral, giving

$$\mathbf{E} = \int d\mathbf{E} = \int \frac{k\,dq}{r^2}\hat{\mathbf{r}}.$$

For any finite charge distribution with nonzero net charge, the field approaches that of a point charge at large distances.

8. A point charge in an electric field experiences a force $\mathbf{F} = q\mathbf{E}$; if this is the only force acting, the charge undergoes an acceleration $\mathbf{a} = (q/m)\mathbf{E}$ in accordance with Newton's law.

9. Because it consists of two equal but opposite point charges, an electric dipole experiences no net force in a uniform electric field. It does, however, experience a torque given by $\tau = \mathbf{p} \times \mathbf{E}$, where $\mathbf{p}$ is the dipole moment vector. A dipole in an electric field has potential energy given by $U = -\mathbf{p} \cdot \mathbf{E}$, where the zero of potential energy corresponds to the dipole oriented perpendicular to the field. In a nonuniform field, a dipole experiences both a torque and a net force.

Terms You Should Understand

(Pairs are closely related terms whose distinction is important; number in parentheses is chapter section where term first appears.)

electromagnetism (23-1)
electric charge (23-2)
coulomb (23-2)
Coulomb's law (23-3)
superposition principle (23-3)
point charge, charge distribution (23-3)
electric field (23-4)
electric dipole (23-5)
dipole moment (23-5)
volume, surface, and line charge density (23-5)
conductor, insulator, dielectric (23-6)
dielectric breakdown (23-6)

Symbols You Should Recognize

C (23-2)
e (23-2)
k (23-3)

q (23-3)
$\hat{\mathbf{r}}$ (23-3)
$\mathbf{E}$ (23-4)
$\mathbf{p}$ (23-5)
ρ, σ, λ (23-5)

Problems You Should Be Able to Solve

calculating electric forces arising from one or more charges acting on another (23-3)
calculating electric fields of distributions of discrete charges (23-5)
calculating electric fields of continuous charge distributions (23-5)
approximating electric fields at large distances from charge distributions (23-5)
evaluating forces on charges in electric fields (23-4, 23-6)
describing the motion of charged particles in uniform and radial electric fields (23-6)
evaluating torques on dipoles in electric fields (23-6)

Limitations to Keep in Mind

Coulomb's law applies strictly only to one point charge acting on another. The forces and electric fields arising from more than one point charge or from a continuous distribution must be calculated using the superposition principle.
At large distances, the field of a charge distribution approaches that of a point charge only if the distribution is both finite in size and has nonzero net charge.
The dipole field decreases with the inverse cube of the distance only for distances large compared with the dipole's charge separation.

QUESTIONS

1. How might a universe with only one kind of electric charge differ from our universe?
2. Discuss this statement: It is precisely because the electric force is so strong that the electrical nature of most everyday interactions is not obvious.
3. The gravitational force between an electron and proton is about 10^{-40} times weaker than the electrical force between the two. Since matter consists largely of electrons and protons, why is the gravitational force ever important?
4. You are given two electric charges. Could you determine whether they had the same or opposite signs? Could you determine the signs of each?
5. In Example 23-3 we found a point where the electric force on a third charge would be zero. Would a charge placed at that point be in stable equilibrium? Why or why not?
6. The gravitational force between an electron and a proton is about 10^{-40} times the electrical force between them. Does this ratio depend on how far apart they are? Explain.

7. In which of the following phenomena does electromagnetism play a dominant role?
   (a) Gasoline burns in a car engine.
   (b) The moon orbits Earth.
   (c) A nerve impulse travels from your brain to a muscle.
   (d) Protons and neutrons join to form an atomic nucleus.
   (e) A chemist synthesizes a new polymer.
   (f) You sit in a chair and the chair doesn't collapse.
8. A free neutron is unstable, and soon decays into other particles. One of the decay products is a proton. Must there be others? If so, what electrical properties must they have?
9. Where in Fig. 23-9 could you put a third charge so it would experience no net force? Would it be in stable or unstable equilibrium?
10. Why should the test charge used to measure an electric field be small?
11. Equation 23-3 gives the electric field of a point charge. Does the direction of $\hat{\mathbf{r}}$ depend on whether the charge is

positive or negative? Does the direction of **E** depend on the sign of the charge?

12. Is the electric force on a charged particle always in the direction of the field? Explain.

13. Why does a dipole produce an electric field at all? After all, the dipole has no net charge.

14. The rod in Example 23-7 carries total charge Q, and the point P is a distance a from the rod. So why isn't the electric field of the rod just kQ/a^2?

15. The field of a dipole decreases with the inverse cube of the distance. Why doesn't this violate our assertion that the field of a finite size charge distribution with nonzero net charge approaches that of a point charge at large distances?

16. A spherical balloon is initially uncharged. If you spread positive charge uniformly over the balloon's surface, would it expand or contract? What would happen if you spread negative charge instead?

17. Suppose someone argued that the force we call gravity is really an electric force, arising from a net electric charge on Earth. How could you disprove this?

18. Two cubical blocks of wood are each 10 cm on a side and carry electric charge spread over their surfaces. If they're 5 cm apart, would you be justified in writing kq_1q_2/r^2 for the force between them? How about if they're 5 m apart? Explain the difference.

19. Two charged particles are suspended in the same electric field, the electric force on each balancing the gravitational force. What quantity must they have in common?

20. A deuteron (heavy hydrogen nucleus) has twice the mass but the same charge as a normal hydrogen nucleus (a proton). Both are released from rest in the same uniform electric field. Compare the distances each goes in the same time.

21. Under what circumstances is the path of a charged particle a parabola? A circle?

22. Explain why a nonuniform field is required for a net force on a dipole.

23. Why should there be a force between two dipoles? After all, each has zero net charge.

24. Dipoles A and B are both located in the field of a point charge Q, as shown in Fig. 23-34. Does either experience a net torque? A net force? If each dipole is released from rest, describe qualitatively its subsequent motion.

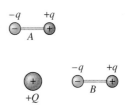

FIGURE 23-34 Question 24.

PROBLEMS

Section 23-3 Electric Charge

1. Suppose the electron and proton charges differed by one part in one billion. Estimate the net charge you would carry.

2. A typical lightning flash delivers about 25 C of negative charge from cloud to ground. How many electrons are involved?

3. Protons and neutrons are made from combinations of the two most common quarks, the u quark and the d quark. The u quark's charge is $+\frac{2}{3}e$ while the d quark carries $-\frac{1}{3}e$. How could three of these quarks combine to make (a) a proton and (b) a neutron?

4. A 2-g ping-pong ball rubbed against a wool jacket acquires a net positive charge of 1 μC. Estimate the fraction of the ball's electrons that have been removed.

Section 23-3 Coulomb's Law

5. If the charge imbalance of Problem 1 existed, what would be the approximate force between you and another person 10 m away? Treat the people as point charges, and compare the answer with your weight.

6. Compare the gravitational force between an electron and a proton with the electrical force between the two. At what distance(s) is your answer correct?

7. The electron and proton in a hydrogen atom are 52.9 pm apart. What is the magnitude of the electric force between them?

8. How far apart should an electron and proton be so the force of Earth's gravity on the electron is equal to the electric force arising from the proton? Your answer shows why gravity is unimportant on the molecular scale!

9. Two charges, one twice as large as the other, are located 15 cm apart and experience a repulsive force of 95 N. What is the magnitude of the larger charge?

10. Earth carries a net charge of -4.3×10^5 C. The force due to this charge is the same as if it were concentrated at Earth's center. How much charge would you have to place on a 1.0-g mass in order for the electrical and gravitational forces on it to balance?

11. A 6.5-μC charge is held at rest, while a small charged sphere of mass 2.3 g is released 50 cm away. Immediately after its release, the sphere accelerates toward the charge at 340 m/s^2. What is the sphere's charge?

12. A proton is at the origin and an electron is at the point $x = 0.41$ nm, $y = 0.36$ nm. Find the electric force on the proton.

13. A 9.5-μC charge is at $x = 16$ cm, $y = 5.0$ cm, and a -3.2-μC charge is at $x = 4.4$ cm, $y = 11$ cm. Find the force on the negative charge.

14. A spring of spring constant 100 N/m is stretched 10 cm beyond its 90-cm equilibrium length. If you want to keep it stretched by attaching equal electric charges to the opposite ends, what magnitude of charge should you use?

15. Two small spheres with the same mass m and charge q are suspended from massless strings of length ℓ, as shown in Fig. 23-35. Each string makes an angle θ with the vertical. Show that the charge on each sphere is $q = \pm 2\ell \sin\theta \sqrt{mg \tan\theta/k}$.

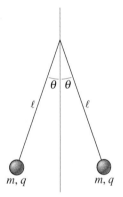

FIGURE 23-35 Problem 15.

16. A charge $3q$ is at the origin, and a charge $-2q$ is on the positive x axis at $x = a$. Where would you place a third charge so it would experience no electric force?

17. A 60-μC charge is at the origin, and a second charge is on the positive x axis at $x = 75$ cm. If a third charge placed at $x = 50$ cm experiences no net force, what is the second charge?

18. You have two charges $+4q$ and one charge $-q$. (a) How would you place them along a line so there's no net force on any of the three? (b) Is this equilibrium stable or unstable?

19. In Fig. 23-36 take $q_1 = 68$ μC, $q_2 = -34$ μC, and $q_3 = 15$ μC. Find the electric force on q_3.

20. In Fig. 23-36 take $q_1 = 25$ μC and $q_2 = 20$ μC. If the force on q_1 points in the $-x$ direction, (a) what is q_3 and (b) what is the magnitude of the force on q_1?

21. Four identical charges q form a square of side a. Find the magnitude of the electric force on any of the charges.

22. Three identical charges $+q$ and a fourth charge $-q$ form a square of side a. (a) Find the magnitude of the electric force on a charge Q placed at the center of the square. (b) Describe the direction of this force.

23. Three charges lie in the x-y plane: $q_1 = 55$ μC at $x = 0$, $y = 2.0$ m; q_2 at $x = 3.0$ m, $y = 0$; and q_3 at $x = 4.0$ m, $y = 3.0$ m. If the force on q_3 is $8.0\hat{\imath} + 15\hat{\jmath}$ N, find q_2 and q_3.

24. Two identical small metal spheres initially carry charges q_1 and q_2, respectively. When they're 1.0 m apart they experience a 2.5-N attractive force. Then they're brought together so charge moves from one to the other until they have the same net charge. They're again placed 1.0 m apart, and now they repel with a 2.5-N force. What were the original values of q_1 and q_2?

Section 23-4 The Electric Field

25. An electron placed in an electric field experiences a 6.1×10^{-10} N electric force. What is the field strength?

26. What is the magnitude of the force on a 2.0-μC charge in a 100 N/C electric field?

27. A 68-nC charge experiences a 150-mN force in a certain electric field. Find (a) the field strength and (b) the force that a 35-μC charge would experience in the same field.

28. A -1.0-μC charge experiences a $10\hat{\imath}$-N electric force in a certain electric field. What force would a proton experience in the same field?

29. The electron in a hydrogen atom is 0.0529 nm from the proton. What is the proton's electric field strength at this distance?

30. A 65-μC point charge is at the origin. Find the electric field at the points (a) $x = 50$ cm, $y = 0$; (b) $x = 50$ cm, $y = 50$ cm; (c) $x = -25$ cm, $y = 75$ cm.

Section 23-5 Electric Fields of Charge Distributions

31. In Fig. 23-37, point P is midway between the two charges. Find the electric field in the plane of the page (a) 10 cm directly above P, (b) 10 cm directly to the right of P, and (c) at P.

FIGURE 23-36 Problems 19, 20.

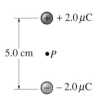

FIGURE 23-37 Problem 31.

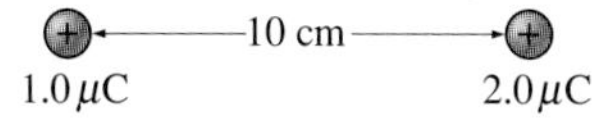

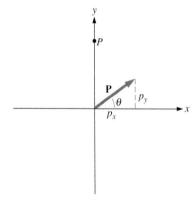

FIGURE 23-38 Problem 32.

32. A 1.0-μC charge and a 2.0-μC charge are 10 cm apart, as shown in Fig. 23-38. Find a point where the electric field is zero.

33. A proton is at the origin and an ion is at $x = 5.0$ nm. If the electric field is zero at $x = -5$ nm, what is the charge on the ion?

34. For the situation of Example 23-3, (a) write an expression for the electric field as a function of x for points to the right of the charge $-q$ shown in Fig. 23-11. (b) Taking $q = 1.0$ μC and $a = 1.0$ m, plot the field as a function of position for $x = 5$ m to $x = 25$ m.

35. (a) Find an expression for the electric field on the y axis due to the two charges q in Fig. 23-9. (b) At what point does the field on the y axis have its maximum strength?

36. Write an expression for the dipole moment vector of the dipole shown in Fig. 23-15.

37. A dipole lies on the y axis, and consists of an electron at $y = 0.60$ nm and a proton at $y = -0.60$ nm. Find the electric field (a) midway between the two charges, (b) at the point $x = 2.0$ nm, $y = 0$, and (c) at the point $x = -20$ nm, $y = 0$.

38. What is the electric field strength 10 cm from a point dipole with dipole moment 3.8 μC·m (a) on the dipole's perpendicular bisector and (b) on its axis?

39. The dipole moment of the water molecule is 6.2×10^{-30} C·m. What would be the separation distance if the molecule consisted of charges $\pm e$? (The effective charge is actually less because electrons are shared by the oxygen and hydrogen atoms.)

40. You're 1.5 m from a charge distribution whose size is much less than 1 m. You measure an electric field strength of 282 N/C. You move to a distance of 2.0 m and the field strength becomes 119 N/C. What is the net charge of the distribution? *Hint:* Don't try to calculate the charge. Determine instead how the field decreases with distance, and from that infer the charge.

41. A point dipole lies at the origin, with its dipole moment vector **p** making an angle θ with the x axis, as shown in Fig. 23-39. By resolving **p** into components and applying Equations 23-5a and 23-5b to the x and y components, respectively, show that the electric field at an arbitrary point P on the y axis is given by

$$\mathbf{E} = \frac{kp}{y^3}(-\hat{\mathbf{i}}\cos\theta + 2\hat{\mathbf{j}}\sin\theta).$$

42. Three identical charges q form an equilateral triangle of side a, with two charges on the x axis and one on the positive y axis. (a) Find an expression for the electric field at points on the y axis above the uppermost charge.

FIGURE 23-39 Problem 41. (The dipole is a true point dipole, and lies exactly at the origin.)

(b) Show that your result reduces to the field of a point charge $3q$ for $y \gg a$.

43. A 30-cm-long rod carries a charge of 80 μC spread uniformly over its length. Find the electric field strength on the rod axis, 45 cm from the end of the rod.

44. A thin rod of length ℓ carries charge Q distributed evenly over its length. A point charge with the same charge Q lies a distance b from the end of the rod, as shown in Fig. 23-40. Find a point where the electric field is zero.

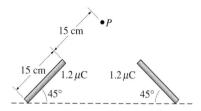

FIGURE 23-40 Problem 44.

45. The rods shown in Fig. 23-41 are both 15 cm long and both carry 1.2 μC of charge. Find the magnitude and direction of the electric field at point P.

FIGURE 23-41 Problem 45.

46. Repeat the preceding problem for the case where the right-hand rod carries -1.2 μC.

47. A thin rod of length ℓ has its left end at the origin and its right end at the $x = \ell$. It carries a line charge density given by $\lambda = \lambda_0 \frac{x^2}{\ell^2} \sin(\pi x/\ell)$, where λ_0 is a constant. Find the electric field strength at the origin.

48. Two identical rods of length ℓ lie on the x axis and carry uniform charges $\pm Q$, as shown in Fig. 23-42. (a) Find an

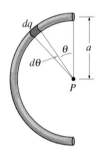

FIGURE 23-44   Problem 52.

expression for the electric field strength as a function of position x for points to the right of the right-hand rod. (b) Show that your result has the $1/x^3$ dependence of a dipole field for $x \gg \ell$. (c) What is the dipole moment of this configuration? *Hint:* See Equation 23-5b.

49. A uniformly charged ring is 1.0 cm in radius. The electric field on the axis 2.0 cm from the center of the ring has magnitude 2.2 MN/C and points toward the ring center. Find the charge on the ring.

50. Figure 23-43 shows a thin, uniformly charged disk of radius R. Imagine the disk divided into rings of varying radii r, as suggested in the figure. (a) Show that the area of such a ring is very nearly $2\pi r\, dr$. (b) If the surface charge density on the disk is σ C/m^2, use the result of (a) to write an expression for the charge dq on an infinitesimal ring. (c) Use the result of (b) along with the result of Example 23-8 to write the infinitesimal electric field dE of this ring at a point on the disk axis, taken to be the positive x axis. (d) Integrate over all such rings (that is, from $r = 0$ to $r = R$), to show that the net electric field on the disk axis is

$$E = 2\pi k\sigma\left(1 - \frac{x}{\sqrt{x^2 + R^2}}\right).$$

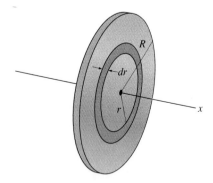

FIGURE 23-43   Problem 50.

51. Use the result of the preceding problem to show that the field of an *infinite*, uniformly charged flat sheet is $2\pi k\sigma$, where σ is the surface charge density. Note that this result is independent of distance from the sheet.

52. A semicircular loop of radius a carries positive charge Q distributed uniformly over its length. Find the electric field at the center of the loop (point P in Fig. 23-44). *Hint:* Divide the loop into charge elements dq as shown in Fig. 23-44, and write dq in terms of the angle $d\theta$. Then integrate over θ to get the net field at P.

53. The electric field 22 cm from a long wire carrying a uniform line charge density is 1.9 kN/C. What will be the field strength 38 cm from the wire?

54. What is the line charge density on a long wire if the electric field 45 cm from the wire has magnitude 260 kN/C and points toward the wire?

55. A straight wire 10 m long carries 25 μC distributed uniformly over its length. (a) What is the line charge density on the wire? Find the electric field strength (b) 15 cm from the wire axis, not near either end and (c) 350 m from the wire. Make suitable approximations in both cases.

56. Figure 23-45 shows a thin rod of length ℓ carrying charge Q distributed uniformly over its length. (a) What is the line charge density on the rod? (b) What must be the electric field direction on the rod's perpendicular bisector (taken to be the y axis)? (c) Modify the calculation of Example 23-9 to find an expression for the electric field at a point P a distance y along the perpendicular bisector. (d) Show that your result for (c) reduces to the field of a point charge Q for $y \gg \ell$.

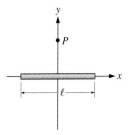

FIGURE 23-45   Problem 56.

Section 23-6 Matter in Electric Fields

57. In his famous 1909 experiment that demonstrated quantization of electric charge, R. A. Millikan suspended small oil drops in an electric field. With a field strength of 20 MN/C, what mass drop can be suspended when the drop carries a net charge of 10 elementary charges?

58. How strong an electric field is needed to accelerate electrons in a TV tube from rest to one-tenth the speed of light in a distance of 5.0 cm?

59. A proton moving to the right at 3.8×10^5 m/s enters a region where a 56 kN/C electric field points to the left. (a) How far will the proton get before its speed reaches zero? (b) Describe its subsequent motion.

60. An oscilloscope display requires that a beam of electrons moving at 8.2 Mm/s be deflected through an angle of 22° by a uniform electric field that occupies a region 5.0 cm long. What should be the field strength?

61. A uniform electric field **E** is set up between two metal plates of length ℓ and spacing d, as shown in Fig. 23-46. An electron enters the region midway between the plates moving horizontally with speed v, as shown. Find an expression for the minimum speed the electron needs to get through the region without hitting either plate. Neglect gravity.

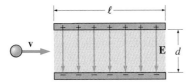

FIGURE 23-46   Problem 61.

62. An electrostatic analyzer like that of Example 23-11 has $b = 7.5$ cm. What should be the value of E_0 if the device is to select protons moving at 84 km/s?

63. An electron is moving in a circular path around a long, uniformly charged wire carrying 2.5 nC/m. What is the electron's speed?

64. Figure 23-47 shows a device its inventor claims will separate isotopes of a particular element. (Isotopes of the same element have nuclei with the same charge but different masses). Atoms of the element are first stripped completely of their electrons, then accelerated from rest through an electric field chosen to give the desired isotope exactly the right speed to pass through the electrostatic analyzer (see Example 23-11). Prove that the device won't work—that is, that it won't separate different isotopes.

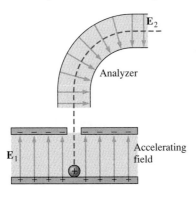

FIGURE 23-47   An isotope separator that won't work (Problem 64).

65. What is the line charge density on a long wire if a 6.8-μg particle carrying 2.1 nC describes a circular orbit about the wire with speed 280 m/s?

66. The electron in a hydrogen atom has kinetic energy 2.18×10^{-18} J. Assuming the electron is in a circular orbit around the central proton, estimate the size of the atom. (Although this problem gives a reasonable answer, the simple model of an electron orbiting a proton is not consonant with the quantum mechanical description of the atom.)

67. A dipole with dipole moment 1.5 nC·m is oriented at 30° to a 4.0-MN/C electric field. (a) What is the magnitude of the torque on the dipole? (b) How much work is required to rotate the dipole until it's antiparallel to the field?

68. A molecule has its dipole moment aligned with a 1.2-kN/C electric field. If it takes 3.1×10^{-27} J to reverse the molecule's orientation, what is its dipole moment?

69. Two identical dipoles, each of charge q and separation a, are a distance x apart as shown in Fig. 23-48. By considering forces between pairs of charges in the different dipoles, calculate the net force between the dipoles. (a) Show that, in the limit $a \ll x$, the force has magnitude $6kp^2/x^4$, where $p = qa$ is the dipole moment. (b) Is the force attractive or repulsive?

FIGURE 23-48   Problem 69.

70. A dipole with charges $\pm q$ and separation $2a$ is located a distance x from a point charge $+Q$, with its dipole moment vector perpendicular to the x axis, as shown in Fig. 23-49. Find expressions for the magnitude of (a) the net torque and (b) the net force on the dipole, both in the limit $x \gg a$. (c) What is the direction of the net force?

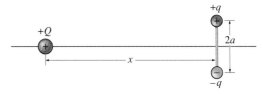

FIGURE 23-49   Problem 70.

Paired Problems

(Both problems in a pair involve the same principles and techniques. If you can get the first problem, you should be able to solve the second one.)

71. An electron is at the origin and an ion with charge $+5e$ is at $x = 10$ nm. Find a point where the electric field is zero.

72. A proton is at the origin and an ion is at $x = 5.0$ nm. If the electric field is zero at $x = -6.83$ nm, what is the charge on the ion?

73. A thin rod of length ℓ has its left end at $x = -\ell$ and its right end at the origin. It carries a line charge density given by $\lambda = \lambda_0 \dfrac{x^2}{\ell^2}$, where λ_0 is a constant. Find the electric field at the origin.

74. Repeat the preceding problem for the case when $\lambda = \lambda_0 \dfrac{x^4}{\ell^4}$.

75. A thin, flexible rod carrying charge Q spread uniformly over its length is bent into a quarter circle of radius a, as shown in Fig. 23-50a. Find the electric field strength at the point P, which is the center of the circle. *Hint:* Consult Problem 52.

76. A thin, flexible rod carrying charge Q spread uniformly over its length is bent into a circular arc of radius a, as shown in Fig. 23-50b. Find the electric field strength at the point P, which is the center of the circular arc.

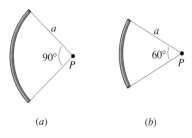

(a)   (b)

FIGURE 23-50 Problems 75, 76.

77. Ink-jet printers work by deflecting moving ink droplets with an electric field so they hit the right place on the paper. Droplets in a particular printer have mass 1.1×10^{-10} kg, charge 2.1 pC, speed 12 m/s, and pass through a uniform 97-kN/C electric field in order to be deflected through a 10° angle. What is the length of the field region?

78. If the drop speed in the printer of Problem 77 is doubled, what should be done to the electric field to have the drops hit the same point on the paper?

Supplementary Problems

79. A charge $-q$ and a charge $\frac{4}{9}q$ are located a distance a apart, as shown in Fig. 23-51. Where would you place a third charge so that all three are in static equilibrium? What should be the sign and magnitude of the third charge?

80. Two 34-μC charges are attached to the opposite ends of a spring of spring constant 150 N/m and equilibrium length 50 cm. By how much does the spring stretch?

FIGURE 23-51 Problem 79.

81. A 3.8-g particle with a 4.0-μC charge experiences a downward force of 0.24 N in a uniform electric field. Find the electric field, assuming that the gravitational force is *not* negligible.

82. A rod of length 2ℓ lies on the x axis, centered on the origin. It carries a line charge density given by $\lambda = \lambda_0 \dfrac{x}{\ell}$, where λ_0 is a constant. (a) What is the net charge on the rod? (b) Find an expression for electric field strength at all points $x > \ell$. (c) Show that your result has the $1/x^3$ dependence of a dipole field when $x \gg \ell$. *Hint:* For $\ell \ll x$, $\ln\left(\dfrac{x - \ell}{x + \ell}\right)$ becomes approximately $-\dfrac{2\ell}{x} - \dfrac{2\ell^3}{3x^3}$. (d) By comparing with Equation 23-5b, determine the dipole moment of the rod.

83. The electric field on the axis of a uniformly charged ring has magnitude 380 kN/C at a point 5.0 cm from the ring center. The magnitude 15 cm from the center is 160 kN/C; in both cases the field points away from the ring. Find the radius and charge of the ring.

84. Use the binomial theorem to show that, for $x \gg R$, the result of Problem 50 reduces to the field of a point charge whose total charge is the charge density σ times the disk area.

85. The dipole moment of a water molecule is 6.2×10^{-30} C·m. A water molecule is located 1.5 nm from a proton, with its dipole moment vector aligned as shown in Fig. 23-52. (a) Use Equation 23-5b to find the force the molecule exerts on the proton. (b) Now find the net force on the dipole in the proton's nonuniform electric field by considering that the dipole consists of two opposite charges q separated by a distance d, such that $qd = 6.2 \times 10^{-30}$ C·m. Take the limit as d becomes very small, and show that the force has the same magnitude as that of part (a), as required by Newton's third law.

FIGURE 23-52 Problem 85.

86. An *electric quadrupole* consists of two oppositely directed dipoles in close proximity. (a) Calculate the field of the quadrupole shown in Fig. 23-53 for points to the right of $x = a$, and (b) show that for $x \gg a$ the quadrupole field falls off as $1/x^4$.

87. Derive Equation 23-7 in Example 23-9 by making θ the integration variable, then evaluating the resulting integral.

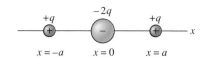

FIGURE 23-53 Problem 86.

CHAPTER 24

GAUSS'S LAW

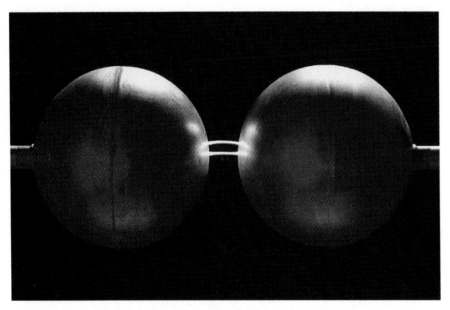

Sparks jump between two highly charged conducting spheres. Gauss's law determines the distribution of charge on these and other conductors.

W e've seen how it's possible, in principle, to calculate the electric field of any charge distribution by summing the contributions of the many individual charges comprising the distribution. But in practice that process involves a vector integration that becomes difficult for all but the simplest charge distributions. How can we hope to calculate the field of a solid ball of charge, for example, when the individual charge elements are varying distances from the field point and their field vectors point in different directions (Fig. 24-1)?

In this chapter we introduce an elegant way of describing electric fields that makes almost trivial the calculation of fields from certain charge distributions. In the process we will formulate one of the four fundamental laws of electromagnetism—a statement that is equivalent to Coulomb's law but that gives deeper insights into the behavior of the electric field.

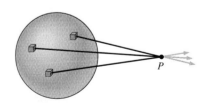

FIGURE 24-1 It would be difficult to find the field of a charged ball by summing vectorially the contributions of all the individual charge elements, three of which are shown here.

FIGURE 24-2 Electric field lines of an isolated positive point charge. The lines spread farther apart with increasing distance from the charge.

24-1 ELECTRIC FIELD LINES

The electric field is a set of vectors defined at all points in space, and we've therefore been representing fields by drawing a number of field vectors. A simpler way to visualize electric fields is with **electric field lines,** continuous lines whose direction is everywhere that of the electric field. To draw a field line, start at some point, and determine the field direction there. Move a small distance in the direction of the field, and evaluate the field direction at the new point. Extending this process in both directions from the starting point traces out an electric field line. The resulting line is a path whose direction at any point is that of the electric field at that point. Drawing many such lines gives a visualization of the overall field structure.

Tracing the field lines of a point charge is particularly simple. Starting at any point near a positive point charge, we find field vectors pointing radially outward from the charge. Move a little way outward, and the field still points in the same direction. So the field lines are straight lines, starting at the point charge extending radially outward indefinitely (Fig. 24-2).

Field lines show the direction of the field, but what about its magnitude? In Fig. 24-2 the field lines spread apart as they extend farther from the point charge. Coulomb's law tells us that the field weakens farther from the charge. So in Fig. 24-2 the field is stronger where the lines are closer and weaker where they're farther apart. This qualitative statement is always true of electric field lines, and allows us to infer relative field strength as well as field direction from field line pictures.

The relation between field strength and number of field lines is in fact quantitative. Fig. 24-3 shows a point charge field and two concentric spheres surrounding the point charge. The same number of field lines crosses the surface of each sphere, and the lines are perpendicular to the spherical surfaces. The larger sphere has twice the radius and therefore four times the surface area as the smaller one. Therefore the number of field lines *per unit area* crossing the outer sphere's surface is one-fourth that of the inner sphere. This is just the decrease in field strength given by the $1/r^2$ dependence in Coulomb's law. The number of field lines per unit area crossing a surface in the electric field is therefore proportional to the field strength. Remember in looking at two-dimensional pictures that it's the number of lines per unit *area* that counts, and that you need to consider that field lines generally spread in all three dimensions.

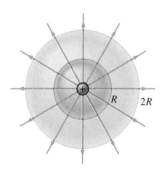

FIGURE 24-3 The outer sphere has twice the radius of the inner sphere. Since the area of a sphere is $4\pi R^2$, the outer sphere's area is four times that of the inner sphere. The same number of field lines crosses each, so the number of field lines per unit area crossing the outer sphere is one-fourth that of the inner sphere. The field strength decreases in the same way, so the number of field lines crossing a unit area is proportional to the field strength.

In tracing the field lines of charge distributions we must add vectorially the contributions from all the charges comprising the distribution. Usually the direction of the field varies as we move along a field line, so the line itself is curved (Fig. 24-4). Nevertheless, the number of field lines per unit area at right angles to the field remains proportional to the strength of the field—a fact that is ultimately grounded in Coulomb's law and the superposition principle.

You might argue that "number of field lines" is vague because we can always draw as many field lines as we want. To make the field-line picture more precise, we associate a fixed number of field lines with a charge of given magnitude. In Fig. 24-5, for example, eight field lines correspond to a charge of magnitude q. Then eight lines *begin* on the *positive* charge $+q$ (Fig. 24-5a), and 16 on $+2q$ (Fig. 24-5b). Eight lines *end* on the *negative* charge $-q$ in Fig. 24-5c. Figures 24-5d–f show the fields of some two-charge distributions, drawn with the same eight-line convention. Note that field lines always *begin* on *positive* charges and *end* on *negative* charges.

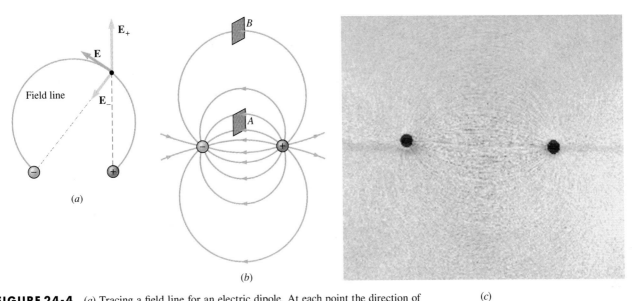

(c)

FIGURE 24-4 (a) Tracing a field line for an electric dipole. At each point the direction of the field line is that of the *net* electric field $\mathbf{E} = \mathbf{E}_+ + \mathbf{E}_-$. (b) Tracing several field lines gives an overall sense of the dipole field. Near each charge the field has the radial structure of a point-charge field, but farther away the influence of both charges becomes important and the field lines curve. Field strengh is proportional to the number of field lines per unit area crossing perpendicular to the field, so the field at A is stronger than at B. (In three dimensions the dropoff in field strength would be more dramatic.) (c) Field line pattern made visible by floating small fibers on a liquid in which two opposite charges are immersed.

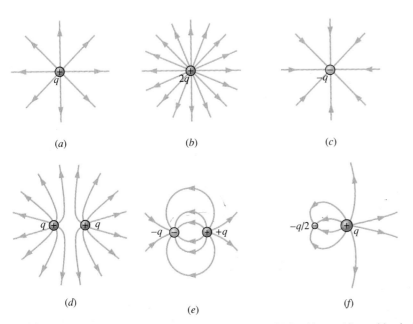

FIGURE 24-5 Field lines for (a) a positive charge q, (b) $2q$; (c) $-q$, (d) two identical charges q, (e) a dipole, consisting of two equal but opposite charges $\pm q$, and (f) opposite and unequal charges q and $-q/2$. In each drawing eight lines are used to represent a charge of magnitude q. Note in (d) and (f) that the field at large distances begins to resemble that of a single point charge.

FIGURE 24-6   In all cases, the number of field lines emerging from a closed surface is proportional to the charge enclosed.

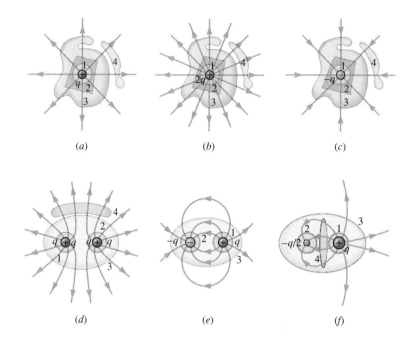

(a)   (b)   (c)

(d)   (e)   (f)

24-2 ELECTRIC FLUX

Counting Field Lines

Figure 24-6 shows the charge distributions of Fig. 24-5, each surrounded by several surfaces. (The figure shows only the two-dimensional cross section of each surface.) Each surface is closed, meaning it's impossible to get from inside to outside without crossing the surface. We now ask a simple question: How many field lines emerge from inside each surface?

In Fig. 24-6a the answer for surfaces 1 and 2 is obvious: eight. With surface 3 one field line crosses three times, twice going out and once going in. If we count a field line going inward as negative, then the algebraic sum of field lines is again eight. In fact, any *closed* surface you can draw around $+q$ will have eight field lines emerging from within. That's because eight lines begin on the charge and extend indefinitely outward, so they cross *any* closed surface surrounding the charge.

What about surface 4? Two lines cross going inward and two going outward, so the net number of field lines emerging from this surface is zero. What's different about surface 4 is that it doesn't enclose the charge. By drawing other surfaces you can convince yourself that any surface not enclosing the charge will have as many lines going in as out, and will therefore have zero net field lines emerging.

Figure 24-6b is identical except that now those surfaces enclosing the charge have 16 field lines emerging, reflecting the greater magnitude of the charge. Surfaces that don't enclose the charge still have zero net field lines emerging. Figure 24-6c is also like Fig. 24-6a, except that now the charge is negative so all field lines point inward. According to our sign convention, -8 field lines emerge from any surface enclosing the charge $-q$.

In Fig. 24-6d, surfaces 1 and 2 each enclose one of the charges q, and each has eight field lines emerging. Surface 3 encloses *both* charges, for a total enclosed charge +2q, and has 16 field lines emerging. Finally, surface 4 encloses no charge and has zero net field lines emerging.

On to Fig. 24-6e, the dipole. Surface 1 encloses charge q and has eight field lines emerging. Surface 2 encloses −q and has −8 field lines emerging. Surface 3 encloses both +q and −q, giving zero net charge enclosed. And as many field lines enter surface 3 as leave it, giving zero net field lines emerging.

Finally, in Fig. 24-6f, eight field lines emerge from surface 1—and that surface encloses +q. Surface 2 encloses −q/2, and has −4 field lines emerging. Surface 3 encloses both charges, for a net enclosed charge +q/2—and four field lines emerge from this surface. Surface 4 encloses no charge and has zero net field lines emerging.

Counting the field lines in Fig. 24-6 leads to a simple statement about how electric fields must behave:

| **The number of electric field lines emerging from any closed surface is proportional to the charge enclosed.**

This statement is very general: It doesn't matter what shape the surface is or whether the enclosed charge is a single point charge or a lot of charges adding to the same net charge. Nor does it matter how the charges are arranged, as long as they're *enclosed* by the surface in question. And the presence of charges *outside* the surface doesn't alter the conclusion about the number of field lines emerging—even though it may alter the shape of the individual lines.

We'll now rephrase our statement in a more mathematically rigorous way, obtaining one of the four fundamental laws of electromagnetism.

TIP   **Remember Fig. 24-6**   As we define new terms and write equations involving integrals, remember that the mathematics just reflects in a concise way the truth that's so obvious in Fig. 24-6—that the number of field lines emerging from a closed surface depends only on the net charge enclosed. Go back to that figure any time you begin to lose the physical significance of the mathematics.

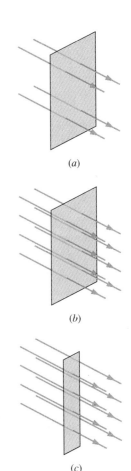

FIGURE 24-7   (a) Four field lines cross the surface shown. (b) Here the field strength has doubled, and eight lines cross the surface. (c) With the same field strength as in (b), but with half the area, only four lines cross the surface. In general, the number of field lines crossing a surface is proportional to the surface area and to the field strength.

Electric Flux

We can make the "number of field lines" more rigorous with Fig. 24-7, which shows several flat surfaces in uniform electric fields. Study the figure and its caption, and you'll see that the number of field lines crossing each surface is proportional to the surface area A and the field strength E. Figure 24-8 shows that it also depends on the orientation of the surface, specified by a vector normal to the surface. As the figure suggests, the number of field lines crossing the surface is proportional to the cosine of the angle between that normal vector **A** and the field **E**. Putting this all together, we have

$$\text{Number of field lines} \propto E\,A\cos\theta.$$

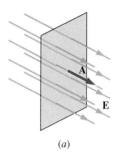

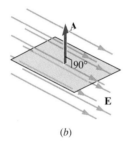

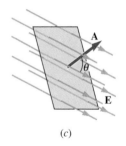

(a)   (b)   (c)

FIGURE 24-8 The number of field lines crossing a surface also depends on the surface orientation relative to the field **E**. The orientation is specified by a vector **A** normal to the surface. In (a) the surface is perpendicular to the field, so **A** and **E** are parallel; then $\theta = 0$, $\cos\theta = 1$, and the number of field lines crossing the surface is a maximum. In (b) the surface is parallel to the field, so **A** and **E** are perpendicular; here $\theta = 90°$, $\cos\theta = 0$, and no lines cross the surface. (c) In general, the number of field lines varies as $\cos\theta$, where θ is the angle between **E** and **A**.

The quantity on the right-hand side of this equation has a definite value that captures the spirit of the more vague expression "number of field lines crossing a surface." We call this quantity the **electric flux,** ϕ, through the surface. If we make the magnitude of the surface normal vector **A** equal to the surface area A, then we can define the flux compactly using the vector dot product:

$$\phi = \mathbf{E} \cdot \mathbf{A}, \qquad (24\text{-}1)$$

where the dot product, defined in Chapter 7, is the product of the two vector magnitudes with the cosine of the angle between them. Since the units of **E** are N/C, flux is measured in N·m²/C.

For the open surfaces of Fig. 24-7 and 24-8 there's an ambiguity in the sign of ϕ, since we could have taken **A** in either of the two directions along the perpendicular to the surface. But for *closed* surfaces, we unambiguously define the direction of **A** as that of the outward-pointing normal to the surface.

> **TIP**   **The Flux Is Not the Field**   The flux ϕ and field **E** are related but distinct quantities. The field is a vector defined at each point in space. Flux is a global property of the field, depending not on a single point but on how the field behaves over an extended surface. Unlike field, flux is a scalar quantity; it's simply a quantification of the "number of field lines crossing a surface."

What if a surface is curved and/or the field varies with position? Then we divide the surface into many small patches, each small enough that it's essentially flat and that the field is essentially uniform over each. If a patch has area dA, then Equation 24-1 gives the flux through it:

$$d\phi = \mathbf{E} \cdot d\mathbf{A},$$

where the vector $d\mathbf{A}$ is normal to the patch (Fig. 24-9). The total flux through the surface is then the sum over all the patches. If we make the patches arbitrarily small that sum becomes an integral, and the flux is

$$\phi = \int_{\text{surface}} \mathbf{E} \cdot d\mathbf{A}. \qquad (24\text{-}2)$$

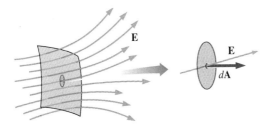

FIGURE 24-9 Even though the surface is curved and the field varies, a small enough patch of surface is essentially flat and the field is uniform over it, so the flux through the patch is $d\phi = \mathbf{E} \cdot d\mathbf{A}$.

The limits of the integral range over the entire surface, picking up contributions from all the patches $d\mathbf{A}$. Although the integral can be difficult to evaluate, we'll find it most useful in cases where its evaluation is almost trivial. Again, remember what Equation 24-2 means: The flux ϕ simply serves as a more precise measure of the "number of field lines crossing a surface."

24-3 GAUSS'S LAW

We showed in the preceding section that the number of field lines emerging from a closed surface is proportional to the charge enclosed. Now that we've developed electric flux to express more rigorously the notion "number of field lines," we can state the following:

> **The electric flux through any closed surface is proportional to the charge enclosed by that surface.**

Writing the same thing mathematically gives

$$\phi = \oint \mathbf{E} \cdot d\mathbf{A} \propto q_{\text{enclosed}},$$

where the circle on the integral sign indicates that the integral is over a *closed* surface.

To evaluate the proportionality between flux and enclosed charge, consider a positive point charge q and a spherical surface of radius r centered on the charge (Fig. 24-10*a*). The flux through this surface is given by Equation 24-2:

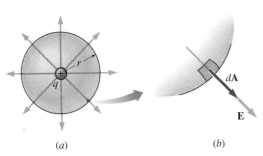

FIGURE 24-10 (*a*) The electric field of a point charge has the same magnitude everywhere on a spherical surface centered on the charge. (*b*) At each point on the surface the field and the surface normal $d\mathbf{A}$ are parallel.

(*a*)

(*b*)

$$\phi = \oint \mathbf{E} \cdot d\mathbf{A} = \oint E \, dA \cos\theta.$$

But Fig. 24-10b shows that the surface normal $d\mathbf{A}$ and the electric field $\mathbf{E}$ are parallel at any point on the sphere, so $\cos\theta = 1$. Since the electric field varies as $1/r^2$ its magnitude is the same everywhere at the fixed radius r of our sphere. Thus, we can take E outside the integral, giving

$$\phi = \oint_{\text{sphere}} E \, dA = E \oint_{\text{sphere}} dA = E(4\pi r^2),$$

where the last step follows because $\oint dA$ just means the surface area of the sphere. Now, the electric field of a point charge is given by Equation 23-3: $E = kq/r^2$. Before using this expression, we introduce the so-called permittivity constant, ε_0, defined by the relation

$$k = \frac{1}{4\pi\varepsilon_0}. \tag{24-3}$$

There's no new physics here; it's just that ε_0 will prove more convenient mathematically than k. The constant ε_0 has the value 8.85×10^{-12} C^2/N·m^2. With the definition 24-3, the electric field of the point charge q becomes $E = q/4\pi\varepsilon_0 r^2$, and our equation for flux reads

$$\phi = E(4\pi r^2) = \left(\frac{q}{4\pi\varepsilon_0 r^2}\right)(4\pi r^2) = \frac{q}{\varepsilon_0}.$$

So the proportionality constant between the flux ϕ through a closed surface and the enclosed charge is just $1/\varepsilon_0$, and our statement relating flux and charge becomes

$$\oint \mathbf{E} \cdot d\mathbf{A} = \frac{q}{\varepsilon_0}, \quad \text{(Gauss's law)} \tag{24-4}$$

where the integral is taken over *any closed surface* that *encloses* the charge q.

Equation 24-4 is **Gauss's law,** and is one of four fundamental relations that govern the behavior of electromagnetic fields throughout the universe. Whether you journey into a star in some remote galaxy, down among the strands of a DNA molecule, or into the microprocessor chip at the heart of your computer, you will find that the flux of the electric field through any closed surface depends only on the enclosed charge. In over a century of experiments, no electric field has ever been observed to violate Gauss's law.

We stress that Gauss's law, although clothed in the mathematical finery of a surface integral, is just a more rigorous way of saying what's obvious in Fig. 24-6: that the number of field lines emerging from a closed surface is proportional to the enclosed charge. This, in turn, is true because electric field lines don't begin or end in empty space, but only on point charges—a fact that reflects the inverse-square nature of the electric force.

24-4 USING GAUSS'S LAW

Gauss's law is true for *any* surface enclosing *any* charge distribution. When the charge distribution has sufficient symmetry we can choose a surface—called a **gaussian surface**—over which evaluation of the flux integral becomes simple. Then Gauss's law allows us to calculate the field far more easily than we could using Coulomb's law and superposition. We now illustrate the use of Gauss's law for three important symmetries.

Spherical Symmetry

A charge distribution is spherically symmetric if the charge density depends only on the distance from a central point. A point charge, a uniformly charged solid sphere, and a spherical surface carrying uniform surface charge density are all spherically symmetric charge distributions. Spherical symmetry implies that the magnitude of the electric field depends only on the distance r from the center of symmetry, and that the field direction is radial (Fig. 24-11).

Gauss's law applies to *any* surface we might draw around the spherically symmetric charge distribution. But the most useful surface is a sphere of arbitrary radius r centered on the center of symmetry. Then the magnitude E is the same at all points on this gaussian surface. Furthermore, **E** is everywhere in the same direction as the perpendicular to the surface, so $\cos\theta = 1$. Thus, although the direction of **E** and of the surface vary, the product $E\cos\theta$ remains constant and is simply the field magnitude E. Then the flux through our gaussian sphere becomes

$$\phi = \oint \mathbf{E} \cdot d\mathbf{A} = \oint E\cos\theta\, dA = E \oint dA = 4\pi r^2 E, \qquad (24\text{-}5)$$

where the last step follows because $\oint dA$ is just the surface area of the sphere, $4\pi r^2$. This expression for the flux does not depend on the details of the charge distribution, so long as it is spherically symmetric.

Gauss's law says that the flux through the sphere is given by q/ε_0, where q is the net charge *enclosed* by the sphere. Suppose our spherically symmetric charge distribution carries total charge Q and has radius R. That is, whatever the particular distribution of charge for $r \le R$, there is no charge at $r > R$. For any gaussian sphere with $r > R$, like the surface 1 in Fig. 24-12, the enclosed charge is the total charge Q. Equating the flux in Equation 24-5 to Q/ε_0 gives

$$4\pi r^2 E = \frac{Q}{\varepsilon_0},$$

or

$$E = \frac{1}{4\pi\varepsilon_0}\frac{Q}{r^2}. \qquad (r > R) \qquad (24\text{-}6)$$

This is just the field of a point charge! (Recall that $1/4\pi\varepsilon_0$ is the coulomb constant, k.) Equation 24-6 says that *outside* any spherically symmetric distribution of charge, the field is identical to that of a point charge located at the center of symmetry (Fig. 24-13). This is *not* an approximation—it is exactly true right up to the surface $r = R$! Imagine how hard it would have been to calculate this

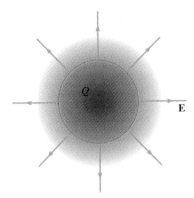

FIGURE 24-11 For a spherically symmetric charge distribution, the field vectors at a given radius all have the same magnitude and point in the radial direction—outward for a positive charge, as shown, or inward for a negative charge.

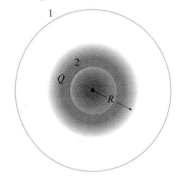

FIGURE 24-12 Two gaussian surfaces surrounding a spherical charge distribution. Surface 1 lies outside the distribution, and encloses all the charge Q. Surface 2 lies inside the distribution, and encloses only part of the charge.

FIGURE 24-13 Hairs on a highly charged person's head trace out the electric field. The essentially spherical head produces a field like that of a point charge.

field using the superposition principle! Yet somehow all the charge elements throughout the spherically symmetric distribution produce $d\mathbf{E}$'s that add vectorially to give the same field as a single point charge. Like Gauss's law itself, this result is a manifestation of the inverse-square force law.

The field *inside* the charge distribution depends on how charge is distributed. This is because a gaussian sphere with $r < R$, such as surface 2 of Fig. 24-12, does not enclose the entire charge Q. How much it encloses depends on the charge distribution. In the examples below, we consider two important special cases.

● EXAMPLE 24-1 A UNIFORMLY CHARGED SPHERE

A total charge Q is spread uniformly throughout a sphere of radius R. What is the electric field at all points in space?

Solution

This charge distribution is spherically symmetric, so the field for $r > R$ is like that of a point charge, given by Equation 24-6.

Inside the sphere, Equation 24-5 for the flux still holds, but now the charge enclosed is some fraction of Q. What fraction? The volume of the sphere is $\frac{4}{3}\pi R^3$, and it contains a total charge Q. Since charge is spread uniformly throughout the sphere, the volume charge density is given by

$$\rho = \frac{Q}{V} = \frac{Q}{\frac{4}{3}\pi R^3}.$$

The charge enclosed by a sphere of radius r is just the volume of that sphere multiplied by the volume charge density:

$$q_{\text{enclosed}} = V\rho = \frac{4}{3}\pi r^3 \frac{Q}{\frac{4}{3}\pi R^3} = Q\frac{r^3}{R^3}.$$

Equating the flux from Equation 24-5 to $q_{\text{enclosed}}/\varepsilon_0$, we have

$$4\pi r^2 E = \frac{Qr^3}{\varepsilon_0 R^3},$$

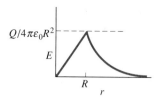

FIGURE 24-14 Field strength versus radial distance for a uniformly charged sphere of radius R. For $r > R$ the field has the inverse-square dependence of a point-charge field.

or

$$E = \frac{1}{4\pi\varepsilon_0}\frac{Qr}{R^3} = \frac{\rho r}{3\varepsilon_0}, \qquad (r < R) \qquad (24\text{-}7)$$

where we've written the field in terms of both the total charge Q and the charge density $\rho = Q/\frac{4}{3}\pi R^3$. Equation 24-7 shows that the field *inside* the charge distribution increases linearly with distance from the center. This result is entirely consistent with the inverse-square law for point charges. Although the field of each charge element decreases as $1/r^2$, in this case the amount of charge enclosed increases more rapidly—as r^3—resulting in a field that increases linearly with r. Figure 24-14 shows the combined results for the fields both inside and outside the sphere. The field direction is, of course, radial, pointing outward if Q is positive and inward if Q is negative. ●

● EXAMPLE 24-2 A THIN SPHERICAL SHELL

A thin spherical shell of radius R carries a total charge Q distributed uniformly over its surface. What is the electric field inside and outside the shell?

Solution

Since this distribution is spherically symmetric, we already know that the field outside is just the point-charge field of Equation 24-6.

For any gaussian sphere inside the shell, the enclosed charge

is zero (Fig. 24-15). Equating the flux from Equation 24-5 to this zero enclosed charge gives

$$4\pi r^2 E = 0$$

so the field is zero everywhere inside the shell! How can this be? Again, it's a manifestation of the inverse-square law. At any point inside the shell, the larger fields of nearby portions of the shell are exactly canceled by the weaker fields of more distant, but more extensive, parts of the shell (Fig. 24-16).

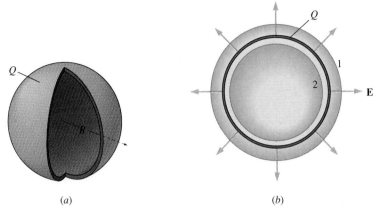

(a)

(b)

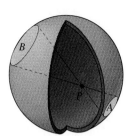

FIGURE 24-16 At any point P inside a charged shell, the field arising from the relatively few but nearby charges in region A is exactly canceled by the field arising from the more numerous but more distant charges in region B.

FIGURE 24-15 (a) A thin spherical shell carries charge Q distributed uniformly over its surface of radius R. (b) Cross-sectional view, showing two spherical gaussian surfaces. Surface 1 encloses the entire charge Q, while surface 2 encloses zero charge.

● EXAMPLE 24-3 A POINT CHARGE INSIDE A SHELL

A point charge $+q$ is at the center of a spherical shell of radius R carrying total charge $-2q$, distributed uniformly over its surface. (a) Draw the electric field lines for this configuration, using eight lines to represent a charge of magnitude q. Find expressions for the field strength for (b) $r < R$ and (c) $r > R$.

Solution
(a) The situation has spherical symmetry and so must the field. Gauss's law tells us that eight field lines must emerge from any surface surrounding $+q$ alone, and that eight field lines must go into any surface surrounding the entire distribution with its net charge $-q$. Figure 24-17 shows the only way to draw the field that's compatible with Gauss's law. Notice that a total of 16 lines end on the charged shell, consistent with its charge of $-2q$.

(b) For $r < R$ we're inside the shell, so the enclosed charge is just q. Solving Equation 24-4 for E then gives

$$E = \frac{q}{4\pi\varepsilon_0 r^2}.$$

What about the shell? Didn't we forget to take it into account? No! The shell and its charge are *irrelevant* as long as they preserve the spherical symmetry. Example 24-2 showed that the field inside a charged shell *due to the shell itself* is zero. Here we're inside the shell, so the only field we see is that of the point charge.

(c) Outside the shell, a spherical gaussian surface encloses net charge $-q$, so the field we see is that of a point charge $-q$; it has magnitude $E = q/4\pi\varepsilon_0 r^2$, and points radially inward.

What about the field at $r = R$? That's ambiguous; just inside the shell the field points outward, while just outside it points inward. In fact, the field undergoes a discontinuous jump across the infinitesimally thin surface charge layer on the shell.

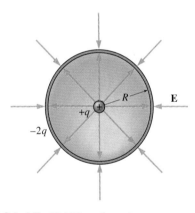

FIGURE 24-17 Field lines for a charge q surrounded by shell carrying $-2q$ (Example 24-3).

TIP Trust Gauss Gauss's law says that the flux—or number of field lines—emerging from any closed surface depends *only* on the *enclosed* charge—not on any other charge that may happen to be outside the surface. (Consult Fig. 24-6 to convince yourself of this.) *When there is enough symmetry* that often means that external charges are totally irrelevant in a field calculation, as is the shell in Example 24-3 for points with $r < R$. *But symmetry matters*: Without enough symmetry, zero net charge inside a gaussian surface is *not* sufficient to ensure that the field on the surface is zero (Fig. 24-18).

EXERCISE A solid sphere 10 cm in radius carries a uniform 40-μC charge distributed throughout its volume. It is surrounded by a concentric shell 20 cm in radius, also uniformly charged with 40 μC. Find the electric field (a) 5.0 cm, (b) 15 cm, and (c) 30 cm from the center.

Answers: (a) 18 MN/C; (b) 16 MN/C; (c) 8.0 MN/C

Some problems similar to Examples 24-1 through 24-3: 17–26, 66

FIGURE 24-18 The net charge enclosed by the gaussian sphere (gray) is zero, but the field on the sphere is not zero. Here the charge distribution—a dipole—is not spherically symmetric, so Equation 24-5 is not a valid expression for the flux.

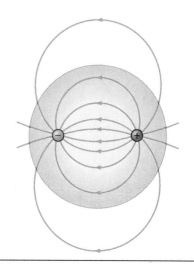

TIP **Using Gauss's Law** This section illustrated the steps needed to calculate the electric field using Gauss's law:

1. Study the symmetry to see if you can construct a gaussian surface on which the field magnitude and its direction relative to the surface are constant. (With spherical symmetry that surface was spherical.) If you can't find such a surface then Gauss's law, while still true, won't help in calculating the field.
2. Evaluate the flux. This should be easy because your choice of gaussian surface makes $E \cos \theta$ constant over the surface. This term then comes outside the integral, leaving the integral equal to the surface area.
3. Evaluate the *enclosed* charge. This is not the same as the total charge if the gaussian surface lies *within* the charge distribution.
4. Equate the flux to $q_{\text{enclosed}}/\varepsilon_0$, and solve for E. The direction of **E** should be evident from the symmetry.

Once steps 1 and 2 are done for a particular symmetry, you'll have an equation like Equation 24-5 for the flux, and you can jump right to step 3 to calculate the field in a specific case—just as we did in Examples 24-1 and 24-2.

Line Symmetry

A charge distribution has line symmetry when its charge density depends only on the perpendicular distance r from a line, called the symmetry axis (Fig. 24-19). Symmetry then requires that the field point radially and that the field magnitude depend only on distance from the axis. An appropriate gaussian surface is the cylinder of length ℓ and radius r shown in Fig. 24-19. Being radial, field lines don't cross the circular ends of the cylinder, so there's no flux through these ends. The field is everywhere perpendicular to the curved part of the surface, so

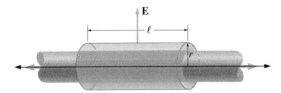

FIGURE 24-19 A positive charge distribution with line symmetry (red) extends infinitely in both directions, and its charge density depends only on the perpendicular distance from the symmetry axis. A cylindrical gaussian surface (gray) of length ℓ and radius r surrounds the charge distribution. Symmetry requires that the electric field on the surface point radially outward, and that the field magnitude E be constant over the surface. With a negative charge the field would point radially inward.

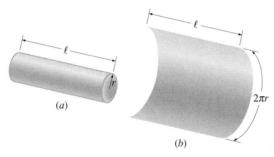

(a)

$2\pi r$

(b)

FIGURE 24-20 (a) A cylindrical gaussian surface. (b) Unrolling the cylinder gives a flat sheet of area $2\pi r \ell$.

$\cos\theta = 1$ in the expression for flux through this part. Since the field magnitude is constant at the fixed radius r of the curved surface, the flux becomes

$$\phi = \int \mathbf{E} \cdot d\mathbf{A} = \int E \, dA \cos\theta = E \int dA = 2\pi r \ell E, \qquad (24\text{-}8)$$

where the last step follows because the cylinder "unrolls" into a rectangular sheet of length ℓ and width $2\pi r$ (Fig. 24-20). Since there is no flux through the cylinder ends, Equation 24-8 gives the total flux through our entire gaussian surface. Solving for the electric field in any situation with line symmetry then amounts simply to equating the flux given in Equation 24-8 to the enclosed charge divided by ε_0, then solving for E—exactly as we did for spherical symmetry using the analogous flux equation, Equation 24-5.

● **EXAMPLE 24-4** AN INFINITE LINE OF CHARGE

Use Gauss's law to calculate the electric field of an infinite line carrying line charge density λ.

Solution
This is the same problem we solved in Example 23-9 through a tedious Coulomb's law calculation. With line charge density λ, the charge enclosed by a gaussian cylinder of length ℓ is

$q_{\text{enclosed}} = \lambda \ell$. Setting the flux given in Equation 24-8 to $q_{\text{enclosed}}/\varepsilon_0$ and solving for E then gives

$$E = \frac{\lambda \ell}{2\pi \varepsilon_0 r \ell} = \frac{\lambda}{2\pi \varepsilon_0 r}.$$

Since $1/2\pi\varepsilon_0 = 2k$, this is the same result we found in Example 23-9. The Gauss's law calculation is far simpler; symmetry and an intelligent choice of gaussian surface helped us bypass the entire integration of Example 23-9.

Although this problem dealt with an infinitesimally thin charged line, you can easily convince yourself that the same result must hold *outside* any charge distribution with line symmetry. And, as we argued in Example 23-9, the result is a good approximation for long, thin structures of finite length provided we're not too far away nor too close to the ends.

● EXAMPLE 24-5 A HOLLOW PIPE

A thin-walled pipe 3.0 m long and 2.0 cm in radius carries a net charge $q = 5.7 \ \mu C$, distributed uniformly over its surface. Find the electric field (a) 8.0 mm and (b) 8.0 cm from the pipe axis, not near either end.

Solution

Since the pipe is much longer than its diameter, we can approximate its field as that of an infinitely long charge distribution with line symmetry.

(a) A point 8.0 mm from the axis lies inside the pipe. An 8.0-mm-radius gaussian cylinder therefore encloses zero net charge; equating the flux from Equation 24-8 to zero then shows that the field at this radius—and indeed anywhere deep within the hollow pipe—is zero.

(b) For a point outside the pipe, a gaussian cylinder of length ℓ encloses charge $\lambda\ell$, where the line charge density λ is 5.7 μC/3.0 m = 1.9 μC/m. Setting the flux from Equation 24-8 to this $q_{enclosed}$ and solving for E then gives

$$E = \frac{q_{enclosed}}{2\pi\varepsilon_0 r\ell} = \frac{\lambda}{2\pi\varepsilon_0 r}$$

$$= \frac{1.9\times10^{-6} \ C/m}{(2\pi)(8.85\times10^{-12} \ C^2/N{\cdot}m^2)(8.0\times10^{-2} \ m)}$$

$$= 4.3\times10^5 \ N/C.$$

In this region the field points radially outward and falls inversely with distance from the axis.

> **TIP** **That's Distance from the Symmetry Axis** The distance r that arises in applying Gauss's law to situations with spherical or line symmetry is always the distance *from the point or line of symmetry*—as you can see from the derivations of Equations 24-4 and 24-8. Equations for the field are therefore most simply expressed in terms of that distance—*not* in terms of distance from the edge of the charge distribution.

EXERCISE Suppose the pipe in Example 24-5 were surrounded concentrically by a second pipe of the same length and 5.0 cm in diameter. What should be (a) the total charge and (b) the surface charge density (assumed uniform) on this outer pipe in order that there be no electric field outside the entire structure?

Answers: (a) $-5.7 \ \mu C$; (b) $-12 \ \mu C/m^2$

Some problems similar to Examples 24-4 and 24-5: 28, 29, 31, 32 ●

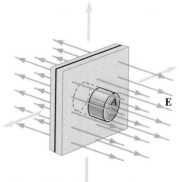

FIGURE 24-21 A charge distribution with plane symmetry. The charge density depends only on the distance from the plane of symmetry (black), and extends infinitely in both directions parallel to that plane. Also shown are the electric field and a gaussian surface.

Plane Symmetry

A charge distribution has plane symmetry when its charge density depends only on the distance from a plane. The only electric-field direction consistent with this symmetry is perpendicular to the symmetry plane (Fig. 24-21). We can evaluate the flux integral in Gauss's law using a gaussian surface whose sides are perpendicular to the symmetry plane and whose ends are parallel to it, as shown in Fig. 24-21. Our surface straddles the symmetry plane, extending equal distances on either side of it. Since no field lines cross the sides, the flux through them is zero. Symmetry of the situation implies that the field magnitude E cannot depend on position parallel to the symmetry plane. Therefore E is uniform over each end of the gaussian cylinder so, with the field perpendicular to the ends, the flux through each end is just EA, where A is the end area. Since the ends are the same distance from the symmetry plane, they must have the same field strength E. The total flux through the gaussian surface is therefore

$$\phi = 2EA. \qquad (24\text{-}9)$$

This equation holds for any charge distribution with plane symmetry; to find the field we must evaluate the charge enclosed by the gaussian surface, then apply Gauss's law.

● EXAMPLE 24-6 A SHEET OF CHARGE

An infinite sheet of charge carries a uniform surface charge density σ. What is the electric field arising from this sheet?

Solution

Since the surface charge density is uniform, this charge distribution has plane symmetry. Figure 24-22 shows the sheet and an appropriate gaussian surface. The sheet area enclosed by the gaussian surface is clearly equal to the end area A. The surface charge density—charge per unit area—is σ, so the enclosed charge is $q_{enclosed} = \sigma A$. Setting $q_{enclosed}/\varepsilon_0$ to the flux $\phi = 2EA$ given by Equation 24-9 and solving for E then gives

$$E = \frac{\sigma}{2\varepsilon_0}. \qquad (24\text{-}10)$$

This simple result says that the field strength does not depend on distance from the sheet. How can this be? By symmetry, the field must point perpendicular to the sheet. There is no charge anywhere but on the sheet, so that's the only place where field lines can begin or end. Therefore the density of field lines—the measure of field strength—is the same everywhere. Figure 24-23 shows how this result is fully consistent with Coulomb's law.

Although this example treated only an infinitesimally thin sheet of charge, we would find a uniform electric field *outside* any charge distribution with plane symmetry. The field *inside*

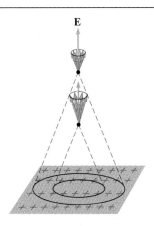

FIGURE 24-23 As we rise above an infinite sheet, the amount of charge within a given angular region increases just enough to compensate for the inverse-square decrease in field strengths of the individual charges. The result is a field that does not depend on distance from the sheet.

such a distribution would depend on how the charge density varies in the direction perpendicular to the symmetry plane (see Problem 36).

Perfect plane symmetry requires a charge distribution that is infinite in extent. But close to any large, flat, uniformly charged surface and not near an edge, the assumption of plane symmetry becomes a good approximation, and Equation 24-10 becomes reasonably accurate. We make the same approximation when we treat the acceleration of Earth's gravity as a constant; we're neglecting Earth's curvature and therefore its finite size, and approximating its gravitational field as the uniform field of an infinite sheet of mass.

EXERCISE An electron close to a large, flat sheet of charge is repelled from the sheet with a 1.8×10^{-12} N force. Find the surface charge density on the sheet.

Answer: $-200 \ \mu C/m^2$

Some problems similar to Example 24-6: 34–37 ●

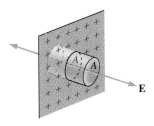

FIGURE 24-22 The area of the charged sheet enclosed by the gaussian surface is the same as the area A of its ends; the enclosed charge is therefore σA.

24-5 FIELDS OF ARBITRARY CHARGE DISTRIBUTIONS

The examples of Section 24-4 show how easy Gauss's law can sometimes make problems that would be difficult to solve using Coulomb's law. In each case the symmetry allowed us to construct a gaussian surface on which $E\cos\theta$ was constant. Only then could we take E outside the integral and solve for it. But many situations do not possess the symmetry needed to apply Gauss's law in calculating the field. Try, for example, to calculate the field of a dipole using Gauss's law. The attempt fails because it is impossible to draw an appropriate surface.

We can understand the fields associated with more complicated charge distributions by considering the fields of simpler distributions that we have already calculated using either Coulomb's law or Gauss's law. Figure 24-24 summarizes the fields of a dipole, a point charge, a uniformly charged line, and a uniformly charged plane. For the last three, note the simple relation between the number of

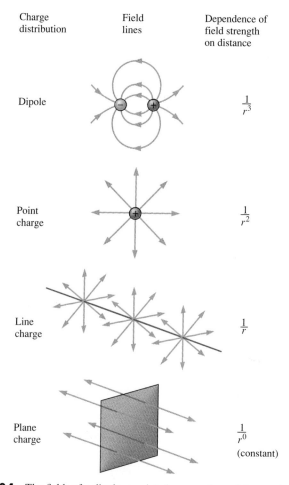

FIGURE 24-24 The fields of a dipole, a point charge, a charged line, and a charged plane.

dimensions in the charge distribution and the way the field strength depends on distance. The plane has two dimensions and its field strength is independent of distance. The line has one dimension and its field falls as $1/r$. The point has no dimensions and its field falls as $1/r^2$. In a sense, the dipole continues this progression, for it consists of two opposite point charges whose effects very nearly cancel. No wonder its field falls even faster, as $1/r^3$. In fact, one can construct a hierarchy of charge distributions whose fields fall off ever faster as dipoles nearly cancel dipoles, and so on. Such distributions are useful in the mathematical analysis of complicated charge distributions such as complex molecules or radio antennas.

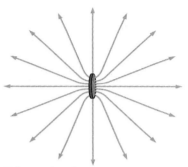

FIGURE 24-25 The field of a uniformly charged disk.

Frequently we have a charge distribution that lacks the symmetry required to make Gauss's law useful and for which a Coulomb's law calculation would be impossibly difficult. Good thinking coupled with knowledge of simpler charge distributions can go a long way toward providing a reasonable approximation to the field. Consider, for example, the uniformly charged disk shown in Fig. 24-25. For points much closer to the disk surface than to the edge, the disk looks almost like an infinite flat plane of charge. For these points the field is approximately that of an infinite plane—a field that points directly away from the plane and does not fall off with distance. Far from the disk, meanwhile, its exact size and shape are unimportant. Its field closely resembles that of a single charged point: far from the disk the field points radially outward in all directions and falls off as the inverse square of the distance from the disk. The field at intermediate distances is harder to determine. But somehow the infinite-plane field lines close to the disk must connect smoothly to the point-charge field lines far away. If we sketch these in, as in Fig. 24-25, we have a rough picture of the field everywhere. Don't underestimate the value of a simple approximation like this one! It can often tell all we need to know about a situation and may provide a much clearer understanding than would a detailed calculation.

24-6 GAUSS'S LAW AND CONDUCTORS

Electrostatic Equilibrium

In the preceding chapter we defined electrical conductors as materials containing free charges—like the free electrons in metals. Figure 24-26 shows what happens when an electric field is applied to a piece of conducting material. Free charges respond to the electric force $q\mathbf{E}$ by moving—in the direction of the field if they are positive, opposite the field if negative. The resulting charge separation gives rise to an electric field within the material that is opposite to the applied field. As more charge moves this internal field becomes stronger until its magnitude eventually equals that of the applied field. At that point free charges within the conductor experience zero net force, and the conductor is in **electrostatic equilibrium.** Although individual charges continue to move about in random thermal motion, there is no longer any net motion of charge. Once equilibrium is reached the internal and applied fields are equal but opposite, and therefore,

The electric field is zero inside a conductor in electrostatic equilibrium.

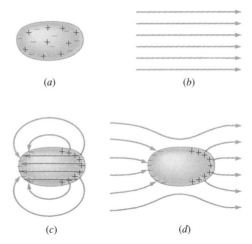

FIGURE 24-26 (*a*) A piece of conducting material contains positive and negative charge. (*b*) A uniform electric field, arising from charges outside the region shown. (*c*) Charges in the conductor separate in response to the field, resulting in an internal field that cancels the applied field. (*d*) The net field is the vector sum of the applied field and the field resulting from the redistribution of charge within the conductor. Note that field lines, as always, begin and end on charges.

It could not be otherwise: Since a conductor contains free charges, the presence of any internal electric field would result in bulk charge motion, and we would not have equilibrium. This result does not depend on the size or shape of the conductor, the magnitude or direction of the applied field, or even the nature of the material as long as it's a conductor. This ability of a conductor to cancel applied fields is the basis of shielding—the use of conductive enclosures to keep out unwanted electric fields (Fig. 24-27).

This discussion of equilibrium is a macroscopic one; it considers only overall average fields within the material. At the atomic and molecular level there are still strong electric fields near individual electrons and protons. But the *average* field,

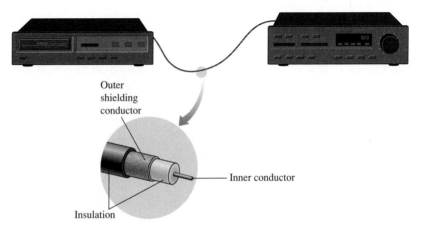

FIGURE 24-27 The shielded cable connecting a CD player to a stereo receiver keeps out stray electric fields that would otherwise introduce noise. Although the situation is not strictly static, the fields change slowly enough that charges in the shield can move to cancel them.

taken over distances many times the separation between individual charges, is zero in electrostatic equilibrium.

Charged Conductors

Although they contain free charges, conductors are normally electrically neutral since they include equal numbers of electrons and protons. But suppose we give a conductor a nonzero net charge, for example, by injecting excess electrons into its interior. There will be a mutual repulsion among the electrons and, because these are *excess* electrons, there is no compensating attraction from protons. We might expect, therefore, that the electrons will move as far apart as possible— namely to the surface of the conductor (Fig. 24-28). (Electrons might even leave the material—but that only occurs with very high charge densities.)

We now use Gauss's law to prove rigorously that excess charge *must* be at the surface of a conductor in electrostatic equilibrium. Figure 24-29 shows a piece of conducting material with a gaussian surface drawn just below the material surface. In equilibrium there is no electric field within the conductor, and thus the field is zero everywhere on the gaussian surface. The flux, $\oint \mathbf{E} \cdot d\mathbf{A}$, through the gaussian surface is therefore also zero. But Gauss's law says that the flux through a closed surface is proportional to the net charge enclosed, and therefore the net charge within our gaussian surface must be zero. This is true no matter where the gaussian surface is as long as it is *inside* the conductor. We can move the gaussian surface arbitrarily close to the conductor surface and it still encloses no net charge. If there is a net charge on the conductor it lies outside the gaussian surface, and therefore we conclude that

> If a conductor in electrostatic equilibrium carries a net charge, all excess charge resides on the conductor surface.

FIGURE 24-28 Excess charge accumulates at the surface of a charged conductor. In this elongated conductor, mutual repulsion of the excess electrons results in the greatest charge accummulation at the opposite ends.

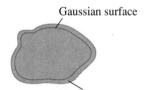

FIGURE 24-29 Since the electric field inside the conductor is zero, a gaussian surface within the conductor encloses zero net charge. Any excess charge therefore resides on the conductor surface.

● EXAMPLE 24-7 A HOLLOW CONDUCTOR

An irregularly shaped conductor has a hollow cavity, as shown in Fig. 24-30. The conductor carries a net charge of 1.0 μC. A 2.0-μC point charge is inside the cavity, not touching the conductor. Find the net charge on the cavity wall and on the outer surface of the conductor, assuming electrostatic equilibrium.

Solution
The electric field is zero everywhere within the conductor, in particular on the gaussian surface shown in Fig. 24-30. The flux through this surface is thus zero, and Gauss's law tells us that the surface therefore encloses zero net charge. But there is a 2.0-μC charge within the cavity. In order for the gaussian surface to enclose zero net charge, the cavity wall must therefore carry −2.0 μC.

Since the conductor's net charge is 1.0 μC and there is −2.0 μC on its inner wall, the outer surface of the conductor surface must carry 3.0 μC.

EXERCISE A point charge $+q$ is placed inside a hollow conducting shell carrying a net charge $-3q$. What is the total charge on the outside surface of the shell?

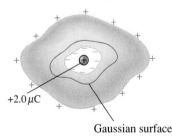

$+2.0\mu$C

Gaussian surface

FIGURE 24-30 A conductor with a hollow cavity containing a charge. The gaussian surface shown encloses no net charge, so the charge on the inner wall must be equal but opposite that of the charge within the cavity.

Answer: −2q

Some problems similar to Example 24-7: 46–49

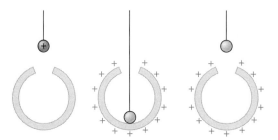

FIGURE 24-31 Experimental test of Gauss's law. When the small charged conductor contacts the interior of the hollow conductor, all its charge moves to the outside of the hollow conductor.

Experimental Tests of Gauss's Law

The fact that excess charge resides only on a conductor surface provides a very sensitive test of Gauss's law, and thus of the inverse-square law for the electric field. Figure 24-31 shows a charged conducting ball being placed inside a hollow, initially neutral conductor. When the two conductors touch, all the excess charge flows to the outer surface of the hollow conductor, leaving no net charge on the ball. In practice the experiment is often done in reverse, with an uncharged conducting ball placed within a hollow conductor. The outer conductor is then charged, and sensitive instruments used detect any charge moving to the ball. Absence of such charge motion confirms the inverse-square law. Recent experiments of this type show that the exponent 2 appearing in the inverse-square law is indeed 2 to within 3×10^{-16}. Such tests are far more sensitive than direct measurements of how the electric force varies with distance.

The Field at a Conductor Surface

There can be no electric field *within* a conductor in electrostatic equilibrium, but there *may* be a field right *at* the conductor surface. Such a field, though, cannot have a component parallel to the surface; if it did, charge would move along the surface and we would not have equilibrium. So the field at a conductor surface must be perpendicular to the surface (Fig. 24-32a).

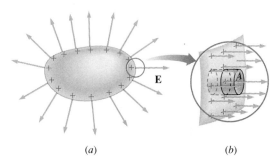

(a)   (b)

FIGURE 24-32 (a) The electric field at the surface of a charged conductor is perpendicular to the conductor surface. (b) A small gaussian surface straddles the conductor surface.

We can compute the field strength by considering a small gaussian surface that straddles the conductor surface, as shown in Fig. 24-32b. We make the gaussian surface so small that curvature of the conductor becomes negligible, and we orient the gaussian surface with its sides perpendicular and its top parallel to the conductor surface. Since the field is perpendicular to the conductor, there is then no flux through the sides of the gaussian surface. Since the field is zero inside the conductor, there is also no flux through the inner end of the gaussian surface. The only flux is through the outer end, whose area is A. Since the field is essentially uniform and perpendicular to this end, the flux is just EA. The only charge enclosed is right at the conductor surface, where it occupies the same area A. If the surface charge density is σ, then the enclosed charge is σA. Gauss's law equates the flux to $q_{enclosed}/\varepsilon_0$, so we have

$$EA = \frac{\sigma A}{\varepsilon_0}$$

or

$$E = \frac{\sigma}{\varepsilon_0}. \quad \text{(field at conductor surface)} \quad (24\text{-}11)$$

This result applies to any conductor in electrostatic equilibrium and shows that large electric fields develop where the charge density is high. Engineers designing electrical devices must avoid high charge densities whose associated fields could lead to sparks, arcing, and breakdown of electrical insulation.

● **EXAMPLE 24-8**   EARTH'S FIELD

The Earth, which is an electrical conductor, carries a net charge of -4.3×10^5 C distributed approximately uniformly over its surface. Find the surface charge density, and use Equation 24-11 to calculate the electric field at Earth's surface.

Solution

Let Q be Earth's charge and R_E its radius (which is given inside the front cover and in Appendix E). Then the surface charge density is

$$\sigma = \frac{Q}{A} = \frac{Q}{4\pi R_E^2} = \frac{-4.3 \times 10^5 \text{ C}}{(4\pi)(6.37 \times 10^6 \text{ m})^2}$$

$$= -8.43 \times 10^{-10} \text{ C/m}^2.$$

Equation 24-11 then gives the electric field at Earth's surface:

$$E = \frac{\sigma}{\varepsilon_0} = \frac{-8.43 \times 10^{-10} \text{ C/m}^2}{8.85 \times 10^{-12} \text{ C}^2/\text{N·m}^2} = -95 \text{ N/C},$$

where the minus sign indicates that the field direction is downward. This modest field is present near Earth's surface in fair weather; in thunderstorms the local field exceeds this value by many orders of magnitude.

Does our result make sense? We could also treat Earth as a spherical charge distribution, whence its surface field strength would be $E = Q/4\pi\varepsilon_0 R_E^2$. Using the symbolic form of our result for σ in Equation 24-11 gives precisely this expression, showing that these two approaches to the field are indeed consistent.

EXERCISE   Dielectric breakdown of air occurs with fields of about 3×10^6 N/C, and results in sparks jumping through the air. What is the maximum surface charge density permissible on a conductor if dielectric breakdown of the surrounding air is to be avoided?

Answer: 27 μC/m^2

Some problems similar to Example 24-8:   43, 44, 51

●

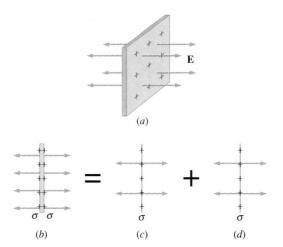

FIGURE 24-33 (*a*) An isolated, charged conducting plate. Its field points outward from both faces. (*b*) Edge-on view of the plate. If the plate is isolated, symmetry requires that there be equal charge densities on both faces. (*c*) The field anywhere is the sum of the fields of the two faces, each treated as a single charged sheet. Within the plate the fields cancel, while outside they sum to give a net field σ/ϵ_0.

Equation 24-11 gives a field that depends only on the local charge density. Does that mean the field at any point on a conductor surface arises only from the charge right at that point? No! As always, the field at any point is the vector sum of contributions from all charge elements making up the charge distribution. Remarkably, Gauss's law requires that charges on a conductor arrange themselves in such a way that the field at any point on the conductor surface depends only on the surface charge density right at that point—even though that field arises from *all* the charges on the surface (as well as from charges elsewhere if there are such)!

Consider a thin, flat, isolated, conducting sheet that has charge density σ on one of its two faces (Fig. 24-33*a*). Equation 24-11 shows immediately that the field at the surface of this plate is σ/ε_0. But if the plate is large and flat we can approximate it as an infinite sheet of charge—for which we found earlier (Equation 24-10 in Example 24-6) that the field should be $\sigma/2\varepsilon_0$. Is there a contradiction here? No! If the plate is isolated from other conductors or charges, then symmetry requires that charge spread itself evenly over *both* faces. If one face has charge density σ, so must the other—so we really have *two* charge sheets, each with density σ (Fig. 24-33*b*). Each gives rise to a field of magnitude $\sigma/2\varepsilon_0$, and *outside* the conductor those fields superpose to give the net field σ/ε_0 (Fig. 24-33*c*). *Inside* the conductor they also superpose, but here their directions are opposite and the result is that there is no field inside the conductor. Application of Equation 24-11 skips all these details. But because Equation 24-11 was derived on the assumption that the field inside the conductor is zero, it "knows" about charges everywhere on the conductor—and in this case that means on the second face of the conductor.

Equation 24-11 also applies to the pair of oppositely charged conducting plates shown in Fig. 24-34; the result, for the field between the plates, is σ/ε_0, where σ is the surface charge density on either plate. Why not $2\sigma/\varepsilon_0$? Again, Equation 24-11 always gives the field at a conductor surface—and it takes into account other charges that may be present. In this case the symmetry is broken, and charge

FIGURE 24-34 Edge view of two parallel conducting plates carrying opposite charges. Electrical attraction brings the excess charges to the inner faces. Each face constitutes a charge layer whose surface charge density has magnitude σ; between the plates their fields add to give field strength σ/ε_0.

builds up only on the inner faces of the two plates. Now each plate is a single charge layer, giving rise to a field $\sigma/2\varepsilon_0$, and between the plates the fields sum to Equation 24-11's result, σ/ε_0. Beyond the plates the fields sum to zero—a result that also follows from Equation 24-11 because now there is zero surface charge on the outer faces.

CHAPTER SYNOPSIS

Summary

1. **Electric field lines** provide a visual representation of the electric field. The direction of the field line passing through a given point is the direction of the electric field vector at that point. The number of field lines per unit area crossing an area perpendicular to the field is a measure of the field strength. Electric field lines begin and end only on charges.

2. **Electric flux** quantifies the notion "number of field lines crossing a surface." For a flat surface in a uniform field, the flux ϕ is given by $\phi = \mathbf{E} \cdot \mathbf{A}$, where $\mathbf{A}$ is a vector whose magnitude is the surface area A and whose direction is perpendicular to the surface. When the surface is curved and/or the field varies over the surface, the flux must be calculated by integration:

$$\phi = \int \mathbf{E} \cdot d\mathbf{A},$$

where $d\mathbf{A}$ is an infinitesimal vector perpendicular to the surface at each point.

3. **Gauss's law** is a fundamental relation governing the behavior of electric fields throughout the universe. Loosely, Gauss's law states that the number of field lines emerging from a closed surface depends only on the charge enclosed—itself a reflection of the inverse-square dependence of the electric force. More rigorously, Gauss's law states that the electric flux emerging from any closed surface is proportional to the charge enclosed:

$$\oint \mathbf{E} \cdot d\mathbf{A} = \frac{q_{enclosed}}{\varepsilon_0},$$

where $\varepsilon_0 = 1/4\pi k$.

4. Gauss's law is true for any surface and any distribution of charge, but it proves useful in calculating the electric field only in cases with sufficient symmetry—in particular, spherical symmetry, line symmetry, and plane symmetry. *Outside* charge distributions with these three symmetries, respectively, the field varies as $1/r^2$, as $1/r$, and exhibits no variation. The fields of more realistic charge distributions are often approximated by the fields associated with these symmetries.

5. There is no net charge motion and no electric field inside a conductor in **electrostatic equilibrium.** Gauss's law shows that any excess charge on the conductor resides on the conductor surface, and that the field at the conductor surface is perpendicular to the surface and has magnitude σ/ε_0, with σ the surface charge density.

Terms You Should Understand

(Pairs are closely related terms whose distinction is important; number in parentheses is chapter section where term first appears.)

electric field lines (24-1)

electric flux (24-2)

Gauss's law (24-3)

electrostatic equilibrium (24-6)

Symbols You Should Recognize

ϕ (24-2)

$\mathbf{A}, d\mathbf{A}$ (24-2)

$\oint$ (24-2)

$\oint \mathbf{E} \cdot d\mathbf{A}$ (24-2)

σ (24-6)

Problems You Should Be Able to Solve

drawing and interpreting field line patterns for simple charge distributions (24-1)

calculating electric flux (24-2)

using Gauss's law to calculate electric fields with spherical, line, or plane symmetry (24-4)

describing charge distributions in electrostatic equilibrium (24-6)

calculating electric fields at conductor surfaces (24-6)

Limitations to Keep in Mind

Gauss's law is universally true, but it can be used to calculate electric fields only with sufficient symmetry.

QUESTIONS

1. Can electric field lines ever cross? Why or why not?
2. If identical charged particles are placed at points A and B in Fig. 24-35, which will experience the greater force?

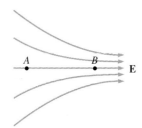

FIGURE 24-35 Question 2.

3. The electric flux through a closed surface is zero. Must the electric field be zero on that surface? If not, give an example.
4. If the flux of the gravitational field through a closed surface is zero, what can you conclude about the region interior to the surface?
5. Under what conditions can the electric flux through a surface be written as EA, where A is the surface area?
6. Eight field lines emerge from a closed surface surrounding an isolated point charge. Would this fact change if a second identical charge were brought to a point just *outside* the surface? If not, would anything change? Explain.
7. In what sense is Gauss's law equivalent to the inverse-square law?
8. If a charged particle were released from rest on a curved field line, would its subsequent motion follow the field line? Explain.
9. Gauss' law describes the flux of the electric field through a surface that may enclose charge. Must the field in Gauss's law arise only from charges within the closed surface?
10. In a certain region the electric field points to the right and its magnitude increases as you move to the right, as shown in Fig. 24-36. Does the region contain net positive charge, net negative charge, or zero net charge?

FIGURE 24-36 Question 10.

11. A spherical shell carries a nonuniform charge density. Why can't you conclude that the field inside the shell is zero?

12. A point charge is located a fixed distance from a uniformly charged sphere, outside the sphere. If the sphere shrinks in size without losing any charge, what happens to the force on the point charge?
13. In applying Equation 24-6 for the field outside a spherically symmetric charge distribution, is r the distance from the center or from the edge of the distribution?
14. The field of an infinite line of charge falls as $1/r$. How is this not a violation of the inverse-square law?
15. Why can't you use Gauss's law to determine the field of a uniformly charged cube? Why wouldn't it work to draw a cubical gaussian surface?
16. You're sitting inside an uncharged hollow spherical shell. Suddenly someone dumps a billion coulombs of charge on the shell, distributed uniformly. What happens to the electric field at your location?
17. No matter how far you get from an infinite sheet of charge, its field never changes. Why doesn't this violate our statement that far from a charge distribution its field resembles that of a point charge?
18. Why is it that the field inside a uniformly charged sphere actually increases with distance from the center? How is this consistent with Coulomb's law?
19. There is a nonzero flux through each of the surfaces in Fig. 24-7, yet there is no charge in the region shown. Why is this not a violation of Gauss's law?
20. Does Gauss's law apply to a spherical surface not centered on a point charge, as shown in Fig. 24-37? Would this be a useful surface to use in calculating the electric field?

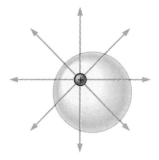

FIGURE 24-37 Question 20.

21. An insulating sphere carries charge spread uniformly throughout its volume. A conducting sphere has the same radius and net charge, but of course the charge is spread over its surface only. How do the electric fields outside these two charge distributions compare?
22. Why must the electric field be zero inside a conductor in electrostatic equilibrium?
23. Why must the electric field at the surface of a conductor in electrostatic equilibrium be perpendicular to the surface?

24. In electrostatic equilibrium, the electric field at the surface of an insulator need not be perpendicular to the insulator surface. Why not?
25. Where in Fig. 24-28 would you find the strongest electric field?
26. The electric field of a flat sheet of charge is $\sigma/2\varepsilon_0$. Yet the field of a flat conducting sheet—even a thin one, like a piece of aluminum foil—is σ/ε_0. Explain this apparent discrepancy.
27. A metal contains free electrons not bound to individual atoms. Does Gauss's law require that all these free electrons be on the metal surface?

PROBLEMS

Section 24-1 Electric Field Lines

1. What is the net charge shown in Fig. 24-38? The magnitude of the middle charge is 3 μC.

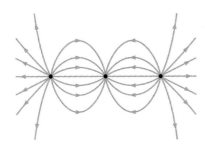

FIGURE 24-38 Problem 1.

2. A charge $+2q$ and a charge $-q$ are near each other. Sketch some field lines for this charge distribution, using the convention of eight lines for a charge of magnitude q.
3. Two charges $+q$ and a charge $-q$ are at the vertices of an equilateral triangle. Sketch some field lines for this charge distribution.
4. The net charge shown in Fig. 24-39 is $+Q$. Identify each of the charges A, B, C shown.

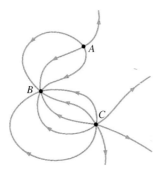

FIGURE 24-39 Problem 4.

Section 24-2 Electric Flux

5. A flat surface with area 2.0 m² is in a uniform electric field of 850 N/C. What is the electric flux through the surface

when it is (a) at right angles to the field, (b) at 45° to the field, and (c) parallel to the field?

6. What is the electric field strength in a region where the flux through 1.0 cm × 1.0 cm flat surface is 65 N·m²/C, if the field is uniform and the surface is at right angles to the field?
7. A flat surface with area 0.14 m² lies in the x-y plane, in a uniform electric field given by $\mathbf{E} = 5.1\hat{\imath} + 2.1\hat{\jmath} + 3.5\hat{k}$ kN/C. Find the flux through this surface.
8. The electric field on the surface of a 10-cm-diameter sphere is perpendicular to the sphere and has magnitude 47 kN/C. What is the electric flux through the sphere?
9. What is the flux through the hemispherical open surface of radius R shown in Fig. 24-40? The uniform field has magnitude E. *Hint:* Don't do a messy integral! Think about the flux through the open end of the hemisphere.

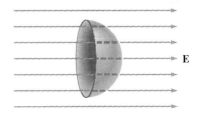

FIGURE 24-40 Problem 9.

10. The electric field shown in Fig. 24-41 is given by $\mathbf{E} = E_0\frac{y}{a}\hat{k}$, where E_0 and a are constants. Find the flux through the square of side a shown.

Section 24-3 Gauss's Law

11. What is the electric flux through each closed surface shown in Fig. 24-42?
12. A 6.8-μC charge and a -4.7 μC charge are inside an uncharged sphere. What is the electric flux through the sphere?
13. A 2.6-μC charge is at the center of a cube 7.5 cm on each side. What is the electric flux through one face of the cube? *Hint:* Think about symmetry, and don't do an integral.

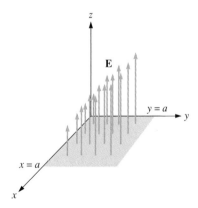

FIGURE 24-41 Problem 10.

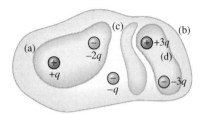

FIGURE 24-42 Problem 11.

14. If the charge in the preceding problem is still inside the cube but not at the center, (a) what is the flux through the *entire* cube? (b) Could you still calculate the flux through one face without doing an integral?

15. A dipole consists of two charges $\pm 6.1\ \mu$C located 1.2 cm apart. What is the electric flux through each surface shown in Fig. 24-43?

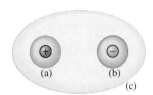

FIGURE 24-43 Problem 15.

16. The electric field in a certain region is given by $\mathbf{E} = 40x\hat{\imath}$ N/C, with x in meters. What is the volume charge density in the region? *Hint:* Apply Gauss's law to a cube 1 meter on a side.

Section 24-4 Using Gauss's Law

17. The electric field at the surface of a uniformly charged sphere of radius 5.0 cm is 90 kN/C. What would be the field strength 10 cm from the surface?

18. A solid sphere 25 cm in radius carries 14 μC, distributed uniformly throughout its volume. Find the electric field strength (a) 15 cm, (b) 25 cm, and (c) 50 cm from the sphere's center.

19. A crude model for the hydrogen atom treats it as a point charge $+e$ (the proton) surrounded by a uniform cloud of negative charge with total charge $-e$ and radius 0.0529 nm. What would be the electric field strength inside such an atom, halfway from the proton to the edge of the charge cloud?

20. Positive charge is spread uniformly over the surface of a spherical balloon 70 cm in radius, resulting in an electric field of 26 kN/C at the balloon's surface. Find the field strength (a) 50 cm from the balloon's center and (b) 190 cm from the center. (c) What is the net charge on the balloon?

21. A 10-nC point charge is located at the center of a thin spherical shell of radius 8.0 cm carrying -20 nC distributed uniformly over its surface. What are the magnitude and direction of the electric field (a) 2.0 cm, (b) 6.0 cm, and (c) 15 cm from the point charge?

22. A solid sphere 2.0 cm in radius carries a uniform volume charge density. The electric field 1.0 cm from the sphere's center has magnitude 39 kN/C. (a) At what other distance does the field have this magnitude? (b) What is the net charge on the sphere?

23. A point charge $-2Q$ is at the center of a spherical shell of radius R carrying charge Q spread uniformly over its surface. What is the electric field at (a) $r = \frac{1}{2}R$ and (b) $r = 2R$? (c) How would your answers change if the charge on the shell were doubled?

24. A spherical shell of radius 15 cm carries 4.8 μC, distributed uniformly over its surface. At the center of the shell is a point charge. (a) If the electric field at the surface of the sphere is 750 kN/C and points outward, what is the charge of the point charge? (b) What is the field just inside the shell?

25. A spherical shell 30 cm in diameter carries a total charge 85 μC distributed uniformly over its surface. A 1.0-μC point charge is located at the center of the shell. What is the electric field strength (a) 5.0 cm from the center and (b) 45 cm from the center? (c) How would your answers change if the charge on the shell were doubled?

26. The thick, spherical shell of inner radius a and outer radius b shown in Fig. 24-44 carries a uniform volume charge density ρ. Find an expression for the electric field strength

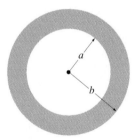

FIGURE 24-44 Problems 26, 32.

in the region $a < r < b$, and show that your result is consistent with Equation 24-7 when $a = 0$.

27. How should the charge density within a solid sphere vary with distance from the center in order that the magnitude of the electric field in the sphere be constant?

28. A long, thin wire carrying 5.6 nC/m runs down the center of a long, thin-walled, hollow pipe with radius 1.0 cm carrying -4.2 nC/m spread uniformly over its surface. Find the electric field (a) 0.50 cm from the wire and (b) 1.5 cm from the wire.

29. A long solid rod 4.5 cm in radius carries a uniform volume charge density. If the electric field strength at the surface of the rod (not near either end) is 16 kN/C, what is the volume charge density?

30. The electric field strength outside a charge distribution and 18 cm from its center has magnitude 55 kN/C. At 23 cm the field strength is 43 kN/C. Does the distribution have spherical or line symmetry?

31. An infinitely long rod of radius R carries a uniform volume charge density ρ. Show that the electric field strengths outside and inside the rod are given, respectively, by $E = \rho R^2/2\varepsilon_0 r$ and $E = \rho r/2\varepsilon_0$, where r is the distance from the rod axis.

32. Repeat Problem 26, assuming that Fig. 24-44 represents the cross section of a long, thick-walled pipe. Now the case $a = 0$ should be consistent with the result of Problem 31 for the interior of the rod.

33. A long, thin wire carries a uniform line charge density $\lambda = -6.8\ \mu\text{C/m}$. It is surrounded by a thick concentric cylindrical shell of inner radius 2.5 cm and outer radius 3.5 cm. What uniform volume charge density in the shell will result in zero electric field outside the shell?

34. A square nonconducting plate measures 4.5 m on a side and carries charge spread uniformly over its surface. The electric field 10 cm from the plate and not near an edge has magnitude 430 N/C and points toward the plate. Find (a) the surface charge density on the plate and (b) the total charge on the plate. (c) What is the electric field strength 20 cm from the plate?

35. If you "painted" positive charge on the floor, what surface charge density would be necessary in order to suspend a 15-μC, 5.0-g particle above the floor?

36. A slab of charge extends infinitely in two dimensions and has thickness d in the third dimension, as shown in Fig. 24-45. The slab carries a uniform volume charge density ρ. Find expressions for the electric field strength (a) inside and (b) outside the slab, as functions of the distance x from the center plane.

37. Figure 24-46 shows sections of three infinite flat sheets of charge, each carrying surface charge density with the same magnitude σ. Find the magnitude and direction of the electric field in each of the four regions shown.

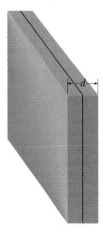

FIGURE 24-45 Section of an infinite slab of charge (Problem 36).

FIGURE 24-46 Problem 37.

Section 24-5 Fields of Arbitrary Charge Distributions

38. A rod 50 cm long and 1.0 cm in radius carries a 2.0-μC charge distributed uniformly over its length. What is the approximate magnitude of the electric field (a) 4.0 mm from the rod surface, not near either end, and (b) 23 m from the rod?

39. A nonconducting square plate 75 cm on a side carries a uniform surface charge density. The electric field strength 1 cm from the plate, not near an edge, is 45 kN/C. What is the approximate field strength 15 m from the plate?

40. Two circular plates 10 cm in diameter and 2.0 mm apart carry equal but opposite charges $\pm0.50\ \mu$C distributed uniformly over their facing surfaces. What is the electric field strength (a) between the plates but not near either edge? (b) 2.5 m from the plates on a plane passing midway between them? *Hint* for (b): See Example 23-6.

41. The electric field strength on the axis of a uniformly charged disk is given by $E = 2\pi k\sigma(1 - x/\sqrt{x^2 + a^2})$, with σ the surface charge density, a the disk radius, and x the distance from the disk center. If $a = 20$ cm, (a) for

what range of x values does treating the disk as an infinite sheet give an approximation to the field that is good to within 10%? (b) For what range of x values is the point-charge approximation good to 10%?

42. A nonconducting square 2.0 cm on a side carries a 45-nC charge spread uniformly over its surface. The x axis runs through the plate center, perpendicular to the plate, with $x = 0$ at the plate center. A -45-nC point charge is at $x = 5.0$ cm. Find approximate values for the electric field strength on the x axis at (a) $x = 1.0$ mm; (b) $x = 4.8$ cm; (c) $x = 2.5$ m. *Hint* for (c): Consult Example 23-6.

Section 24-6 Gauss's Law and Conductors

43. What is the electric field strength just outside the surface of a conducting sphere carrying surface charge density 1.4 μC/m²?

44. Calculate the acceleration of a proton at the surface of a conductor carrying surface charge density 0.60 C/m².

45. A net charge of 5.0 μC is applied on one side of a solid metal sphere 2.0 cm in diameter. After electrostatic equilibrium is reached, what are (a) the volume charge density inside the sphere and (b) the surface charge density on the sphere? Assume there are no other charges or conductors nearby. (c) Which of your answers depends on this assumption, and why?

46. A point charge $+q$ lies at the center of a spherical conducting shell carrying a net charge $\frac{3}{2}q$. Sketch the field lines both inside and outside the shell, using 8 field lines to represent a charge of magnitude q.

47. A 250-nC point charge is placed at the center of an uncharged spherical conducting shell 20 cm in radius. (a) What is the surface charge density on the outer surface of the shell? (b) What is the electric field strength at the shell's outer surface?

48. A point charge is placed at the center of an uncharged spherical conducting shell of inner radius 2.5 cm and outer radius 4.0 cm (Fig. 24-47). As a result, the outer surface of the shell acquires a surface charge density $\sigma = 71$ nC/cm². Find (a) the value of the point charge and (b) the surface charge density on the inner wall of the shell.

49. An irregular conductor containing an irregular, empty cavity carries a net charge Q. (a) Show that the electric field inside the cavity must be zero. (b) If you put a point charge inside the cavity, what value must it have in order to make the surface charge density on the outer surface of the conductor everywhere zero?

50. A neutral dime is placed in a uniform electric field of 6.2×10^5 N/C, with its faces perpendicular to the field. (a) What is the approximate charge density on the faces of the dime? (b) What is the total charge on each face? (Measure a dime!)

51. A total charge of 18 μC is applied to a thin, square metal plate 75 cm on a side. Find the electric field strength near the plate's surface.

52. Two closely spaced parallel metal plates carry surface charge densities ± 95 nC/m² on their facing surfaces, with no charge on their outer surfaces. Find the electric field strength (a) between the plates and (b) outside the plates. Treat the plates as infinite in extent.

53. A conducting sphere 2.0 cm in radius is concentric with a spherical conducting shell with inner radius 8.0 cm and outer radius 10 cm. The small sphere carries 50 nC charge and the shell has no net charge. Find the electric field strength (a) 1.0 cm, (b) 5.0 cm, (c) 9.0 cm, and (d) 15 cm from the center.

54. A coaxial cable consists of an inner wire and a concentric cylindrical outer conductor (Fig. 24-48). If the conductors carry equal but opposite charges, show that there is no surface charge density on the *outside* of the outer conductor.

FIGURE 24-48 Problem 54.

Paired Problems

(Both problems in a pair involve the same principles and techniques. If you can get the first problem, you should be able to solve the second one.)

55. A point charge $-q$ is at the center of a spherical shell carrying charge $+2q$. That shell, in turn, is concentric with a larger shell carrying charge $-\frac{3}{2}q$. Draw a cross section of this structure, and sketch the electric field lines using the convention that 8 lines correspond to a charge of magnitude q.

56. A point charge $-q$ is at the center of a spherical shell carrying charge $-\frac{3}{2}q$. That shell, in turn, is concentric with a larger shell carrying charge $+2q$. Draw a cross

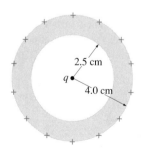

FIGURE 24-47 Problem 48.

section of this structure, and sketch the electric field lines using the convention that 8 lines correspond to a charge of magnitude q.

57. A point charge q is at the center of a spherical shell of radius R carrying charge $2q$ spread uniformly over its surface. Write expressions for the electric field strength at (a) $\frac{1}{2}R$ and (b) $2R$.

58. A point charge q is at the center of a spherical shell of radius R carrying charge $5q$. At what other distance does the electric field have the same value it does at a point halfway from the center to the shell?

59. A long, thin hollow pipe 4.0 cm in diameter carries charge at a density of $-2.6~\mu C/m$, uniformly distributed over the pipe. It is concentric with 10-cm diameter pipe carrying $+2.6~\mu C/m$, also uniformly distributed. Find the magnitude of the electric field at (a) 0.50 cm, (b) 3.5 cm, and (c) 12 cm from the axis of the pipes.

60. Two concentric hollow pipes are 5.0 cm and 12 cm in diameter, respectively. Both carry uniformly distributed electric charges. The electric field 4.0 cm from their common axis is 630 kN/C, radially outward. The field 10 cm from their common axis is 126 kN/C, radially outward. (a) Find the linear charge densities on the two pipes. (b) How would the electric field strengths at 4.0 cm and 10 cm change if the charge density on the outer pipe were doubled?

61. An early (and incorrect) model for the atom pictured its positive charge as spread uniformly throughout the spherical atomic volume. For a hydrogen atom of radius 0.0529 nm, what would be the electric field due to such a distribution of positive charge (a) 0.020 nm from the center and (b) 0.20 nm from the center?

62. A solid sphere of radius R carries a charge spread uniformly throughout its volume. At what point outside the sphere is the electric field strength equal to that at a point halfway from the center to the edge? Express your answer as a distance from the center.

63. A sphere of radius $2a$ has a hole of radius a, as shown in Fig. 24-49. The solid portion carries a uniform volume charge density ρ. Find an expression for the electric field strength within the solid portion, as a function of the distance r from the center.

64. Repeat the previous problem, now considering that the figure represents the cross section of a thick cylindrical pipe.

Supplementary Problems

65. Repeat Problem 10 for the case $\mathbf{E} = E_0\left(\dfrac{y}{a}\right)^2\hat{\mathbf{k}}$.

66. The volume charge density inside a solid sphere of radius a is given by $\rho = \rho_0 r/a$, where ρ_0 is a constant. Find (a) the total charge and (b) the electric field strength within the sphere, as a function of distance r from the center.

67. A proton is released from rest 1.0 cm from a large sheet carrying a surface charge density of -24 nC/m². How much later does it strike the sheet?

68. Fig. 24-50 shows a rectangular box with sides $2a$ and length ℓ surrounding a line of charge with uniform line charge density λ. The line passes directly through the center of the box faces. Using an expression for the field of a line charge, integrate over strips of width dx as shown to find the electric flux through one face of the box. Multiply by 4 to get the total flux through the box, and show that your result is consistent with Gauss's law.

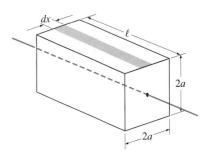

FIGURE 24-50 Problem 68.

69. Repeat Problem 36 for the case when the charge density in the slab is given by $\rho = \rho_0|x/d|$, where ρ_0 is a constant.

70. The charge density within a uniformly charged sphere of radius R is given by $\rho = \rho_0 - ar^2$, where ρ_0 and a are constants, and r is the distance from the center. Find an expression for a such that the electric field outside the sphere is zero.

71. A small object of mass m and charge q is attached by a thread of length ℓ to a large, flat, nonconducting plate carrying a uniform surface charge density σ with the same sign as q (Fig. 24-51). If the object is displaced slightly sideways from its equilibrium, show that it undergoes simple harmonic motion with period $T = 2\pi\sqrt{2\epsilon_0 m\ell/q\sigma}$. Assume the gravitational force is negligible.

72. An infinitely long nonconducting rod of radius R carries a volume charge density given by $\rho = \rho_0(r/R)$, where ρ_0 is a

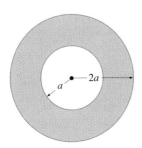

FIGURE 24-49 Problems 63, 64.

FIGURE 24-51 Problem 71.

constant. Find the electric field strength (a) inside and (b) outside the rod, as functions of the distance r from the rod axis.

73. A thick spherical shell of inner radius a and outer radius b carries a charge density given by $\rho = \dfrac{ce^{-r/a}}{r^2}$, where a and c are constants. Find expressions for the electric field strength for (a) $r < a$, (b) $a < r < b$, and (c) $r > b$.

74. A solid sphere of radius R carries a nonuniform volume charge density given by $\rho = \pi^2 \rho_0 \sin(\pi r/R)$, where r is the distance from the center and ρ_0 is a positive constant. Find the magnitude and direction of the electric field at the sphere's surface.

75. A solid sphere of radius R carries a uniform volume charge density ρ. A hole of radius $R/2$ occupies a region from the center to the edge of the sphere, as shown in Fig. 24-52. Show that the electric field everywhere in the hole points horizontally and has magnitude $\rho R/6\epsilon_0$. *Hint:* Treat the hole as a superposition of two charged spheres of opposite charge.

76. You're 5.0 m from a charge distribution and you measure an electric field strength of 850 N/C. At 2.5 m the field

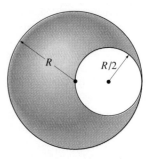

FIGURE 24-52 Problem 75.

strength has increased to about 3.4 kN/C. When you're 5.0 mm from the center of the distribution the field strength is 42.5 MN/C, and it increases to 85 MN/C at 2.5 mm. Describe the distribution as fully as you can, including its shape, any dimensions you can find, its total charge, and any appropriate charge density.

77. Two flat, parallel, closely spaced metal plates of area 0.080 m² carry total charges of $-2.1 \, \mu$C and $+3.8 \, \mu$C. Find the surface charge densities on the inner and outer faces of each plate.

78. Since the gravitational force of a point mass goes as $1/r^2$, the gravitational field **g** also obeys a form of Gauss's law. (a) Formulate this law, and (b) use it to find an expression for the gravitational field strength *within* the Earth, as a function of distance r from the center. Treat Earth as a sphere of uniform density.

ELECTRIC POTENTIAL

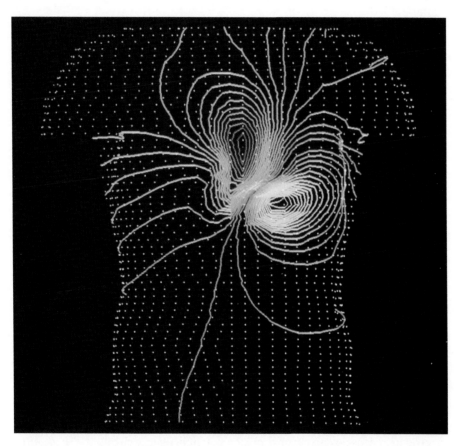

Lines of constant electric potential on the surface of the human body reflect the electric dipole structure of the heart.

You have to do work to lift your book against Earth's gravitational field. The work you do gets stored as potential energy, which is released if you lower the book. In Chapter 8 we introduced the term **conservative** to describe a force or field that, like gravity, "gives back" all the stored energy. An important property of conservative fields is path independence: The work required to move an object from one point to another does not depend on the path taken, but only on the endpoints of the path (Fig. 25-1).

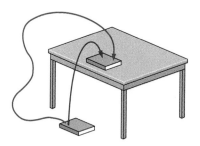

FIGURE 25-1 The gravitational field is conservative, and therefore the net work required to move between two points is independent of the path taken.

The electric field of a static charge distribution is also conservative. You can see this because Coulomb's law has essentially the same form as Newton's law of universal gravitation, and we'll soon prove path independence for the work done in moving one point charge about in the field of another. Because the electric field is conservative, it is meaningful to talk about potential energy. If you do work moving a charge against the electric force, for example, that work ends up stored as potential energy.

In this chapter we deal with electric potential, a useful and easily measured quantity that derives from the concept of potential energy. We'll see how the use of electric potential can provide a simpler approach to calculating electric fields, and how electric potential relates to the properties of everyday devices like batteries.

25-1 POTENTIAL ENERGY, WORK, AND THE ELECTRIC FIELD

In Chapter 8 we defined potential energy difference as the negative of the work $W_{A \to B}$ done by a conservative force **F** on an object moved from point A to point B:

$$\Delta U_{A \to B} = U_B - U_A = -W_{A \to B} = -\int_A^B \mathbf{F} \cdot d\boldsymbol{\ell}, \qquad (25\text{-}1)$$

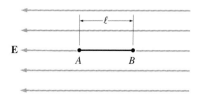

FIGURE 25-2 The work required to move a charge q from A to B in a *uniform* electric field is $qE\ell$.

where $d\boldsymbol{\ell}$ is a small element of the path from A to B.

Consider a positive charge q being moved between two points A and B a distance ℓ apart in a uniform electric field **E,** as shown in Fig. 25-2. A conservative electric force $q\mathbf{E}$ acts on the charge, and since the field is uniform that force doesn't vary as we move from A to B. Therefore, we can dispense with the integral and write

$$\Delta U_{A \to B} = -W_{A \to B} = -\mathbf{F} \cdot \boldsymbol{\ell} = -q\mathbf{E} \cdot \boldsymbol{\ell} = -qE\ell \cos(180°) = qE\ell,$$

where $\boldsymbol{\ell}$ is a vector from A to B and where the factor $\cos 180° = -1$ appears because the vectors **E** and $\boldsymbol{\ell}$ are in opposite directions. Does this result make sense? Pushing a positive charge against the electric field is like pushing a car up a hill: We do positive work, gravity does negative work, and the potential energy increases. Here we must do positive work to move the charge from A to B, the electric field does negative work, and the potential energy change—the *negative* of the work done by the conservative electric force—is positive. Let go of the charge and the field will accelerate it back toward B, changing potential to kinetic energy.

25-2 POTENTIAL DIFFERENCE

We've just found the potential energy change in moving a charge q from point A to point B in Fig. 25-2. If we had moved a charge 2q the potential energy change would have been twice as great, and it would have required twice as much work; $\frac{1}{4}q$ would have one-fourth the potential energy change and would require one-fourth the work. Since the potential energy change is directly proportional to the charge, it is convenient to consider the *potential energy change per unit charge* involved in moving between two points. We call this quantity the **electric potential difference.** Mathematically, we express the potential difference by writing $\mathbf{F} = q\mathbf{E}$ in Equation 25-1 and dividing by q:

$$\Delta V_{A \to B} = \frac{\Delta U_{A \to B}}{q} = -\frac{1}{q} \int_{A}^{B} q\mathbf{E} \cdot d\boldsymbol{\ell} = -\int_{A}^{B} \mathbf{E} \cdot d\boldsymbol{\ell}.$$

Then our definition of potential difference becomes:

> **The electric potential difference from point A to point B is the potential energy change per unit charge in moving from A to B:**
>
> $$\Delta V_{A \to B} = -\int_{A}^{B} \mathbf{E} \cdot d\boldsymbol{\ell}. \quad \text{(electric potential difference)} \quad (25\text{-}2a)$$

In other books you may see our $\Delta V_{A \to B}$ written as V_{AB} or V_{BA} or $V_B - V_A$. We use the Δ here to show explicitly that we're talking about a *change* or *difference* from one point to another, and we use the subscript $A \to B$ to make it clear that this is the potential difference going *from A to B*. In the next section we'll show how our notation is equivalent to the commonly used $V_B - V_A$, and in subsequent chapters we'll sometimes use just the symbol V for potential difference.

An external agent moving charge at constant speed between A and B would do work equal to the change in the charge's potential energy; therefore we can think of potential difference as the work per unit charge done by an external agent moving charge at constant speed between two points. We'll often say, loosely, that potential difference is the energy per unit charge involved in moving between two points; by "energy" we can mean either the work done by an external agent or the potential energy change in the electric field.

In the special case of a uniform field, Equation 25-2a reduces to

$$\Delta V_{A \to B} = -\mathbf{E} \cdot \boldsymbol{\ell}, \quad \text{(uniform field)} \quad (25\text{-}2b)$$

where $\boldsymbol{\ell}$ is a vector from A to B. Figure 25-2 shows the special case when the field $\mathbf{E}$ and path $\boldsymbol{\ell}$ are in opposite directions.

Equations 25-2 show that potential difference can be positive or negative, depending on whether the path goes against or with the field. Moving a positive charge through a positive potential difference is like going uphill: We must do work on the charge, and its potential energy increases. Moving a positive charge through a negative potential difference is like going downhill: We do negative work or, equivalently, the field does work on the charge, and its potential energy decreases. In both cases the opposite is true for a negative charge; even though the potential difference remains the same, the work and potential energy reverse because of the negative sign on the charge.

The Volt and the Electron Volt

The definition of potential difference shows that its units are joules/coulomb. Potential difference is important enough that 1 J/C has a special name—the **volt** (V). To say that a car has a 12-V battery, for example, means that the

FIGURE 25-3 Potential difference depends on *two* points. This parasailer landed on a 138,000-V power line, but he's not being electrocuted because his body is not contacting *two* points with a potential difference between them.

battery does 12 J of work on every coulomb that moves between its two terminals.

We often use the term **voltage** to speak of potential difference, especially in describing electric circuits. Strictly speaking the two terms are not synonymous, since voltage is used even in nonconservative situations that arise when fields change with time. But in common usage this subtle distinction is usually not bothersome.

> **TIP** **Potential Difference Depends on Two Points**
> Specifically, it is the energy per unit charge involved in moving *between those points*. Always think of potential difference in terms of two points. This is ultimately a very practical matter; if you forget it you won't be able to hook up a voltmeter properly, or connect jumper cables safely to your car battery! Figure 25-3 provides a dramatic illustration of this point.
>
> Sometimes we speak of "the potential (or the voltage) at point *P*." This is *always* a shorthand way of talking, and we *must* have in mind some other point. What we mean is the potential difference going from that other point to point *P*.

In molecular, atomic, and nuclear systems it's often convenient to measure energy in **electron volts** (eV), defined as follows:

> *One electron volt is the energy gained by a particle carrying one elementary charge when it moves through a potential difference of one volt.*

Since one elementary charge is 1.6×10^{-19} C, 1 eV is 1.6×10^{-19} J. Energy in eV is particularly easy to calculate when charge is given in elementary charges. However, the eV is *not* an SI unit and should be converted to joules before calculating other quantities.

● EXAMPLE 25-1 A TV PICTURE TUBE: POTENTIAL DIFFERENCE, WORK, AND ENERGY

At the back end of a TV picture tube, a uniform electric field of 600 kN/C extends over a distance of 5.0 cm and points toward the back of the tube (Fig. 25-4). (a) Find the potential difference between the back and the front end of this field region. (b) How much work would it take to move an ion with charge $+2e$ from the back to the front of the field region? (c) What would happen to an electron released at the back of the field region?

Solution
(a) With a uniform field the potential difference is given by Equation 25-2b:

$$V_{A \to B} = -\mathbf{E} \cdot \boldsymbol{\ell} = E\ell = (600 \times 10^3 \text{ N/C})(0.050 \text{ m})$$

$$= 30 \text{ kV},$$

FIGURE 25-4 A potential difference on the order of 30 kV is used to accelerate the electrons that "paint" the picture on the screen of a TV tube, shown here in a cutaway view.

where the second equality follows because a path from the back toward the front is opposite the field direction, as shown in Fig. 25-2, so $\cos\theta = -1$.

(b) Potential difference is the potential energy change per unit charge, or, equivalently, the work per unit charge an external agent must do to move charge between two points. Thus, the work needed to move the ion of charge $q = 2e$ against the 30-kV potential difference is

$$W_{ion} = q\,\Delta V = (2)(1.6\times10^{-19}\ \text{C})(30\ \text{kV}) = 9.6\times10^{-15}\ \text{J},$$

where the units work out because 1 V = 1 J/C. Since the ion carries two elementary charges, we could also express this energy as $(2e)(30\ \text{kV}) = 60\ \text{keV}$.

(c) With its negative charge, the electron would *gain* energy given by $e\,\Delta V$, or 4.8×10^{-15} J (more simply calculated as 30 keV).

EXERCISE A 1.2-μC charge is accelerated through a 3400-V potential difference. How much energy does it gain?

Answer: 4.1 mJ ●

● **EXAMPLE 25-2** POTENTIAL OF A CHARGED SHEET

An isolated, infinite charged sheet carries a uniform surface charge density σ. (a) Find an expression for the potential difference from the sheet to a point a perpendicular distance x from the sheet. (b) What is the potential difference between two points the same distance from the sheet?

Solution

Equation 24-10 gives $E = \dfrac{\sigma}{2\varepsilon_0}$ for the field of a single, isolated sheet of charge. Since the field is uniform we can apply Equation 25-2b. Moving away from the sheet in either direction is going *with* the field (assuming positive σ), so the dot product in Equation 25-2b is positive and therefore the potential difference is negative. Taking $\ell = x$ in Equation 25-2b then gives

$$\Delta V_{0\to x} = -Ex = -\frac{\sigma x}{2\varepsilon_0},$$

where the notation $0 \to x$ means we're taking the potential difference from the sheet ($x = 0$) to the point x. Thus the potential decreases linearly with distance for positive σ; for

negative σ it would increase. The important point is that the potential in a *uniform* field varies *linearly* with distance along the field direction.

If, on the other hand, we consider moving charge between two points equidistant from the sheet, then the charge moves perpendicular to the field and there is no change in its potential energy; thus, the potential difference between such points is zero.

EXERCISE Two nonconducting charged sheets carry equal but opposite surface charge densities ± 53 nC/m². The negative sheet is located at $x = 0$, the positive sheet at $x = 10$ cm. Find expressions for the potential difference from the negative sheet to the points (a) $x = 2.0$ cm, (b) $x = 5.0$ cm; (c) $x = 25$ cm, and (d) $x = -10$ cm. *Hint:* Think about the different fields *between* and *beyond* the plates.

Answers: (a) 120 V; (b) 300 V; (c) 600 V; (d) 0 V

Some problems similar to Examples 25-1 and 25-2: 5, 7–9, 18 ●

Curved Paths and Nonuniform Fields

Equations 25-2 contain a dot product that accounts for the orientation of the path relative to the field. Figure 25-5, for example, shows several straight paths of the same length ℓ in a uniform electric field. Path AB is the same path we considered in Fig. 25-2; the potential difference between its ends is just $\Delta V_{A\to B} = E\ell$ because the angle between **E** and ℓ is 180°. Path AC is at 135° to the field direction, giving a potential difference $\Delta V_{A\to C} = -E\ell\cos 135° = E\ell/\sqrt{2}$. Finally, path AD is perpendicular to the field, giving $\Delta V_{A\to D} = 0$. Quite generally, as Fig. 25-5 suggests, the potential difference depends only on the component of the path *along* the field direction. This is analogous to the situation with gravity, where the work mgh needed to lift a mass depends only on the *vertical* distance h and not on any horizontal component of the motion.

If the field is not uniform or the path is not straight, then we must use the integral form of Equation 25-2a to calculate the potential because the magnitude

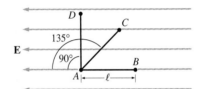

FIGURE 25-5 The potential difference depends only on the component of the path *along* the field. Mathematically, $\Delta V = -\mathbf{E}\cdot\ell = -E\ell\cos\theta$, where the quantity $-\ell\cos\theta$ can be interpreted as the path component along the field.

FIGURE 25-6 The line integral in Equation 25-2a is the sum of infinitely many infinitesimally small potential differences dV.

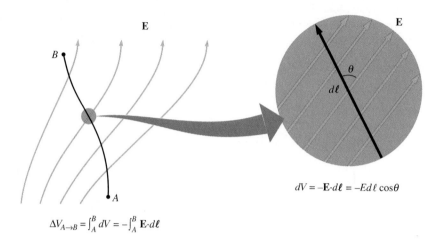

$$\Delta V_{A \to B} = \int_A^B dV = -\int_A^B \mathbf{E} \cdot d\boldsymbol{\ell}$$

$$dV = -\mathbf{E} \cdot d\boldsymbol{\ell} = -E d\ell \cos\theta$$

of $\mathbf{E}$ and/or the angle between $\mathbf{E}$ and the path is changing. Figure 25-6 shows the meaning of Equation 25-2a. If we look at a sufficiently small part of the curve in Fig. 25-6, so small that the field is essentially uniform and the path essentially straight, then Equation 25-2b for a straight path in a uniform field should apply. Describing the short path segment by a small vector $d\boldsymbol{\ell}$, we can write Equation 25-2b as

$$dV = -\mathbf{E} \cdot d\boldsymbol{\ell},$$

where dV is the potential difference from tail to head of $d\boldsymbol{\ell}$. The integral in Equation 25-2a is the sum of all the dV's over the path:

$$\Delta V_{A \to B} = \int_A^B dV = -\int_A^B \mathbf{E} \cdot d\boldsymbol{\ell}.$$

Like the work integral we introduced in Chapter 7, this **line integral** is simply a sum of scalar quantities—dot products of the vectors $\mathbf{E}$ and $d\boldsymbol{\ell}$—over some path. The limits of the integral are the endpoints of the path. Because the electrostatic field is conservative, it is not necessary to specify which of the many paths between these endpoints is taken; all such paths give the same result. Several examples in the next section illustrate the use of Equation 25-2a.

25-3 CALCULATING POTENTIAL DIFFERENCE

The Potential of a Point Charge

The electric field of a point charge q is given by Equation 23-3:

$$\mathbf{E} = \frac{kq}{r^2}\hat{\mathbf{r}},$$

where $\hat{\mathbf{r}}$ is a unit vector from the charge toward the point where the field is being evaluated. Consider two points A and B at distances r_A and r_B from a positive point charge, as shown in Fig. 25-7. What is the potential difference between these points? The distance between them is $r_B - r_A$, but we cannot just multiply this distance by the electric field because the field varies with position. Instead we integrate, following Equation 25-2a:

$$\Delta V_{A \to B} = -\int_{r_A}^{r_B} \mathbf{E} \cdot d\boldsymbol{\ell} = -\int_{r_A}^{r_B} \frac{kq}{r^2} \hat{\mathbf{r}} \cdot d\boldsymbol{\ell}.$$

As we move from r_A toward r_B, the path element vectors $d\boldsymbol{\ell}$ correspond to small increments dr in the radial direction, and therefore, we can write $d\boldsymbol{\ell} = \hat{\mathbf{r}}dr$. Then the potential becomes

$$\Delta V_{A \to B} = -\int_{r_A}^{r_B} \frac{kq}{r^2} \hat{\mathbf{r}} \cdot \hat{\mathbf{r}}dr = -kq \int_{r_A}^{r_B} r^{-2}\,dr,$$

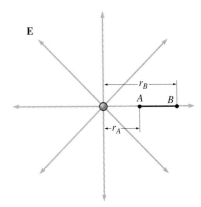

FIGURE 25-7   The potential difference between points A and B is found by integrating between r_A and r_B.

since the dot product of the unit vector $\hat{\mathbf{r}}$ with itself is simply 1. Evaluating the integral gives

$$\Delta V_{A \to B} = -kq\left[-\frac{1}{r}\right]_{r_A}^{r_B} = kq\left(\frac{1}{r_B} - \frac{1}{r_A}\right). \qquad (25\text{-}3)$$

Does this result make sense? For $r_B > r_A$ the potential difference is negative, showing that a positive test charge at r_A would "fall down" the potential "hill" toward r_B. Going the other way would require that positive work be done on a positive charge, as it's pushed "up" the potential "hill" against the repulsive force of the charge q. Although we considered q to be positive, our result holds as well for $q < 0$, in which case the sign of the potential difference changes.

Although we derived Equation 25-3 for two points on the same radial line, Fig. 25-8 shows that the result holds for *any* two points in the field of a charge q. It doesn't matter which point is at the greater distance either; if $r_B < r_A$ Equation 25-3 still gives the correct potential difference, which then becomes positive to indicate that work must be done moving a positive test charge *toward* a positive q.

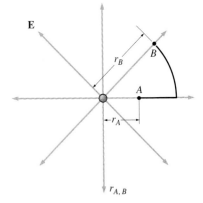

FIGURE 25-8   Here points A and B do not lie on the same radial line. But potential difference is independent of path, and one path between A and B consists of a radial segment and a circular arc. Since $\mathbf{E}$ is perpendicular to the arc, it takes no work to move a charge along the arc. The potential difference $\Delta V_{A \to B}$ therefore arises only from the radial segment, and is therefore given by Equation 25-3.

The Zero of Potential

So far we've only talked about potential differences, symbolized by the expression $\Delta V_{A \to B}$. That's because only differences in potential energy—and thus in electric potential, which is potential energy per unit charge—have physical significance. But as we did with potential energy, it's often convenient to define the **electric potential** at some point as zero and then to measure potential differences relative to that point. We then speak of "the potential at point P," designated $V(P)$ or V_P, and meaning the potential difference $V_{0 \to P}$ *from* our reference point *to* point P. Once we've defined a zero of potential, we can then write potential differences as differences in potential between two points; thus, our terminology $\Delta V_{A \to B}$ can equally well be written $V(B) - V(A)$ or $V_B - V_A$. The choice for the zero of potential is arbitrary and is usually made on the basis of mathematical or physical convenience. In electric power systems, the Earth, called "ground," is usually taken as the zero. In automobile electric systems, the car's metal structure makes a convenient zero; this is usually connected electrically to the negative battery terminal. (As we'll soon see, every point on a conductor is at the same potential, so it's appropriate to consider an entire conductor like Earth or a car's metal structure as the "point" of zero potential.)

In dealing with isolated point charges, Equation 25-3 shows that it is convenient to choose the zero of potential at infinity. If we let r_A become arbitrarily large, and drop the subscript on r_B because it can be at any radial distance r, Equation 25-3 becomes

$$V_{\infty \to r} = V(r) = \frac{kq}{r}. \qquad \text{(point-charge potential)} \qquad (25\text{-}4)$$

When we call this expression $V(r)$ "the potential of a point charge," we really mean that $V(r)$ is the potential difference going from a point very far from a charge q to a point a distance r from the charge—an interpretation that is consistent with our definition of potential difference as depending on *two* points. Because the field outside any spherically symmetric charge distribution is that of a point charge, Equation 25-4 also gives the potential outside a spherically symmetric charge distribution.

Does it bother you that potential difference can be finite over an infinite distance? The reason lies in the inverse-square dependence of the field, which drops so rapidly that the work done in moving a charge from infinity to the vicinity of a point charge remains finite. We found an analogous result in Chapter 9, where it took only a finite amount of energy—and therefore a finite "escape speed"—to escape completely from a planet's gravitational attraction. As long as a charge distribution is finite in size—so its field at large distances falls at least as fast as $1/r^2$, then it makes sense to take the zero of potential at infinity.

● **EXAMPLE 25-3** THE POTENTIAL OF A SPHERICAL CHARGE DISTRIBUTION

A charge Q is distributed in a spherically symmetric way over a sphere of radius R. (a) What is the potential at the sphere's surface, with the zero of potential taken at infinity? (b) How much work would it take to move a proton from infinity to the sphere's surface? (c) What is the potential difference from the sphere's surface to a point located $2R$ from the center?

Solution

(a) The field outside the sphere is exactly that of a point charge Q located at the sphere's center, so the potential difference between infinity and any point *outside* the sphere is given by Equation 25-4. Therefore the potential at the sphere's surface is

$$V(R) = \frac{kQ}{R}.$$

(b) This quantity is the work per unit charge needed to move a charge from infinity to the sphere's surface; since a proton carries charge e, the work involved in bringing the proton to the sphere's surface is just $W = kQe/R$.

(c) To get the potential difference from the surface ($r = R$) to $2R$, we could use Equation 25-3 with $r_B = 2R$ and $r_A = R$.

But it's perhaps easiest to keep in mind just Equation 25-4 and then find the potential difference by subtracting the potentials at the two points:

$$\Delta V_{R \to 2R} = V(2R) - V(R) = \frac{kQ}{2R} - \frac{kQ}{R} = -\frac{kQ}{2R}.$$

Does this result make sense? Yes. The negative result (for positive Q) shows that the potential energy of a positive charge would decrease as it moved away from the sphere, going in the direction of the sphere's electric field.

EXERCISE The potential at the surface of a 10-cm-radius sphere is 4.8 kV. (a) What is the charge on the sphere, assuming it's distributed in a spherically symmetric fashion? (b) What is the electric field at its surface? Assume here—and anytime it's not specified—that potential differences are taken from infinity.

Answers: (a) 53 nC; (b) 48 kN/C

Some problems similar to Example 25-3: 19–23, 25, 26, 65, 71 ●

Potentials of Arbitrary Charge Distributions

If we already know the field of a charge distribution, we can calculate potential differences by applying Equation 25-2a, as we did for the point-charge field. Example 25-4 illustrates this approach.

● **EXAMPLE 25-4**   POTENTIAL DIFFERENCE IN THE FIELD OF A LINE CHARGE

An infinite line of charge carries line charge density λ. What is the potential difference between two points at distances r_A and r_B from the line?

Solution
In the preceding chapter we used Gauss's law to obtain the result

$$\mathbf{E} = \frac{\lambda}{2\pi\varepsilon_0 r}\hat{\mathbf{r}}$$

for the field of a line charge. As we move from r_A to r_B in Fig. 25-9, we can again write $d\boldsymbol{\ell} = \hat{\mathbf{r}}\,dr$ just as we did in evaluating the point-charge potential. Then Equation 25-2a becomes

$$\Delta V_{A\to B} = -\int_{r_A}^{r_B} \mathbf{E}\cdot d\boldsymbol{\ell} = -\int_{r_A}^{r_B} \frac{\lambda}{2\pi\varepsilon_0 r}\hat{\mathbf{r}}\cdot\hat{\mathbf{r}}\,dr$$

$$= -\frac{\lambda}{2\pi\varepsilon_0}\int_{r_A}^{r_B}\frac{dr}{r} = -\frac{\lambda}{2\pi\varepsilon_0}\ln r\Big|_{r_A}^{r_B} \qquad (25\text{-}5)$$

$$= \frac{\lambda}{2\pi\varepsilon_0}\ln\left(\frac{r_A}{r_B}\right),$$

where the last step follows because $\ln x - \ln y = \ln(x/y)$. If λ is positive and $r_A < r_B$, then $\Delta V_{A\to B}$ is negative—indicating, as expected, that the electric field does work on a positive charge moving from A to B (recall that $\ln x < 0$ for $x < 1$). Conversely, moving a positive charge from B to A requires that the agent moving the charge do work since the potential difference $\Delta V_{B\to A}$ is positive.

Note that we cannot let r_A go to infinity in this case, for this would give an infinite potential difference. Physically, this reflects the fact that our charge distribution is itself of infinite

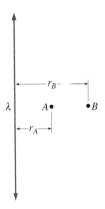

FIGURE 25-9   Two points in the field of a line charge.

extent. Mathematically, it reflects the slow $1/r$ decrease in field strength.

Although we derived Equation 25-5 for an infinitesimally thin line of charge, considerations of Section 24-4 show that this result holds outside *any* charge distribution with line symmetry.

EXERCISE   An infinitely long rod of radius R carries a uniform volume charge density ρ; Problem 24-31 shows that the electric field *inside* this rod points radially outward and has magnitude $E = \rho r/2\varepsilon_0$. Use Equation 25-2a to find the potential difference from the rod's surface to its axis.

Answer: $\rho R^2/4\varepsilon_0$

Some problems similar to Example 25-4:   28, 29, 66, 75   ●

Finding Potential Differences Using Superposition

When we don't know the field of a charge distribution, or the field is too complicated to integrate easily, we can find the potential using superposition. As we will see in the next section, this often provides an easier approach to the field as well.

Consider a charge q being brought from infinity to a point P in the vicinity of some other charges. We want to know the potential at P—by which we mean the work per unit charge required to move from infinity to P. The superposition principle states that the electric field of a charge distribution is the sum of the fields of the individual charges comprising the distribution. Therefore the work per unit charge—that is, the potential difference—between infinity and P is just the sum of the potential differences associated with the individual point charges. Mathematically, we find $V(P)$ by summing Equation 25-4 over the individual point charges q_i:

$$V(P) = \sum_i \frac{kq_i}{r_i},\qquad(25\text{-}6)$$

where the r_i's are the distances from each of the charges to the point P. Equation 25-6 has one enormous advantage over its counterpart for the electric field, Equation 23-4. Electric potential is a *scalar,* so the sum in Equation 25-6 is a scalar sum, and there's no need to consider angles or vector components.

● **EXAMPLE 25-5** THE DIPOLE POTENTIAL

The dipole of Fig. 25-10 consists of two point charges $\pm q$ separated by a distance $2a$. Find the potential at an arbitrary point P, taking the zero of potential at infinity.

Solution
We sum the potentials of the individual point charges, as Equation 25-6 suggests:

$$V(P) = \sum_i \frac{kq_i}{r_i} = \frac{kq}{r_1} + \frac{k(-q)}{r_2} = kq\left(\frac{1}{r_1} - \frac{1}{r_2}\right) = \frac{kq(r_2 - r_1)}{r_1 r_2},$$

where r_1 and r_2 are the distances from P to the positive and negative charges, respectively.

We have seen that in practical situations we're often interested in electrical effects a great distance from the dipole. If r is the distance from the dipole's center to P, then in the limit $r \gg a$ the quantity $r_1 r_2$ becomes approximately r^2, and, as Fig. 25-10 shows, $r_2 - r_1 \simeq 2a \cos\theta$. Then the dipole potential for $r \gg a$ becomes

$$V(r,\ \theta) = \frac{k(2aq)\cos\theta}{r^2} = \frac{kp\cos\theta}{r^2},\qquad(25\text{-}7)$$

with $p = 2aq$ the dipole moment.

Note that the dipole potential drops more rapidly with distance than the point-charge potential, just as the dipole field drops more rapidly with distance than the point-charge field. Note also that Equation 25-7 gives $V = 0$ along the perpendicular bisector of the dipole ($\theta = 90°$). That makes sense because a charge approaching the dipole along its bisector moves at right angles to the dipole field, so no work need be done. Since potential difference is path independent, that means it takes no

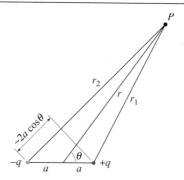

FIGURE 25-10 A dipole and a point P where its potential is to be evaluated. When P is far from the dipole, then r_2 is longer than r_1 by approximately $2a \cos\theta$. (The angle whose vertex is $-q$ is very nearly θ.)

net work to move on *any* path from infinity to a point on the dipole's bisector (see Question 20).

EXERCISE The dipole moment of a water molecule is 6.2×10^{-30} C·m. (a) Find the potential difference $V_B - V_A$ between two points on the axis of the molecular dipole, where points A and B are 8.2 nm and 5.1 nm, respectively, from the center. Both points are closer to the positive end. (b) How much work would it take to move a proton from A to B?

Answers: (a) 1.32 mV; (b) 1.32 meV $= 2.1\times10^{-22}$ J

Some problems similar to Example 25-5:
31–35, 76   ●

Continuous Charge Distributions

We can calculate the potential of a continuous charge distribution by considering it to be made up of infinitely many infinitesimal charge elements dq. Each acts like a point charge and therefore contributes to the potential at some point P an amount dV given by

$$dV = \frac{k\,dq}{r},$$

where the zero of potential is at infinity. The potential at P is the sum—in this case an integral—of the contributions dV from all the charge elements:

$$V = \int dV = \int \frac{k\,dq}{r}, \qquad (25\text{-}8)$$

where the integration is over the entire charge distribution.

● EXAMPLE 25-6 A CHARGED RING

A total charge Q is distributed uniformly around a thin ring of radius a, as shown in Fig. 25-11. What is the potential on the axis of this charged ring?

Solution

Let x be the distance from the center of the ring to some arbitrary point on the axis. The distance from each point on the ring to a point on the axis is the same, and is given by $r = \sqrt{x^2 + a^2}$. The potential on the axis is the sum of the potentials dV of all the charge elements dq around the ring, as described by Equation 25-8:

$$V = k\int_{ring} \frac{dq}{r} = \frac{k}{\sqrt{x^2 + a^2}} \int_{ring} dq,$$

where we have taken $r = \sqrt{x^2 + a^2}$ outside the integral because it's the same for all charge elements. The remaining integral is simply the total charge Q, so we have

$$V = \frac{kQ}{\sqrt{x^2 + a^2}}. \qquad (25\text{-}9)$$

Does this result make sense? At great distances from the ring ($x \gg a$), a^2 in the denominator becomes negligible, and our result becomes

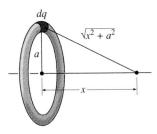

FIGURE 25-11 A charged ring (Example 25-6).

$$V = \frac{kQ}{x},$$

which is just the potential of a point charge Q—as we would expect when we're so far from the ring that its size is no longer significant. At the center of the ring, on the other hand, Equation 25-9 gives

$$V = \frac{kQ}{a}.$$

Here we're a distance a from all parts of the ring, and, since potential is a *scalar* the directions to those parts don't matter. The result is therefore the same as being a distance a from a point charge Q. ●

● EXAMPLE 25-7 A CHARGED DISK

A charged disk of radius a carries a total charge Q distributed uniformly over its surface. What is the potential at a point P on the disk axis, a distance x from the disk?

Solution

To use Equation 25-8, we must divide the disk into charge elements dq. In the preceding example we found the potential of a charged ring, so we can take the charge elements of our disk to be thin rings (see Fig. 25-12), and integrate over all the rings comprising the disk. If a ring-shaped charge element has charge dq and radius r, then Equation 25-9 gives its potential dV a distance x from the disk center:

$$dV = \frac{k\,dq}{\sqrt{x^2 + r^2}}.$$

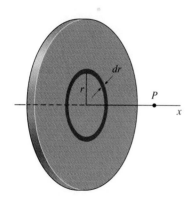

FIGURE 25-12 A charged disk, showing a ring-shaped charge element dq of radius r and width dr.

Then the potential of the entire disk is

$$V = \int_{ring} dV = \int_{r=0}^{r=a} \frac{k\, dq}{\sqrt{x^2 + r^2}}.$$

To evaluate this integral, we must relate r and dq. "Unwinding" the ring gives a strip of area $2\pi r\, dr$ (Fig. 25-13). The surface charge density σ is the total charge divided by the disk area: $\sigma = Q/\pi a^2$. Then the charge dq on our infinitesimal ring of area $2\pi r\, dr$ is

$$dq = \sigma 2\pi r\, dr = \frac{Q}{\pi a^2} 2\pi r\, dr = \frac{2Q}{a^2} r\, dr.$$

Using this result in the integral for the potential gives

$$V = \int_0^a \frac{2kQ}{a^2} \frac{r\, dr}{\sqrt{x^2 + r^2}} = \frac{kQ}{a^2} \int_0^a \frac{2r\, dr}{\sqrt{x^2 + r^2}}.$$

Note that $2r\, dr = d(r^2) = d(x^2 + r^2)$ since x is a constant with respect to the integration. The integral therefore has the form $u^{-1/2}\, du$, where $u = x^2 + r^2$, and the result is $2u^{1/2}$ or

$$V = \frac{2kQ}{a^2} \sqrt{x^2 + r^2}\, \Big|_{r=0}^{r=a} = \frac{2kQ}{a^2} \left(\sqrt{x^2 + a^2} - |x| \right).$$

Figure 25-14 shows that this complicated-looking result makes sense: Close to the sheet, the potential resembles that of an infinite sheet, while far from the disk it approaches the potential of a point charge. Example 25-9 and Problem 72 explore these limiting cases further.

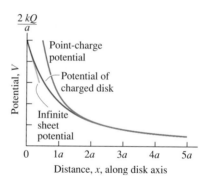

FIGURE 25-14 Charged-disk potential of Equation 25-10 approaches the potential of an infinite sheet for points close to the disk, and that of a point charge far from the disk.

EXERCISE Point P in Fig. 25-15 lies a perpendicular distance y from the end of a uniformly charged rod of length ℓ and total charge Q. Find an expression for the potential at P, taking the zero of potential at infinity.

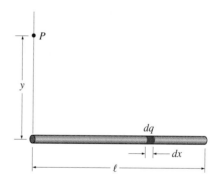

FIGURE 25-15 What is the potential at P? Figure shows a charge element dq, of length dx, to use in the integration for the potential.

Answer: $V = \dfrac{kQ}{\ell} \ln \left(\dfrac{\ell + \sqrt{\ell^2 + y^2}}{y} \right)$

Some problems similar to Examples 25-6 and 25-7: 36–39, 73, 80

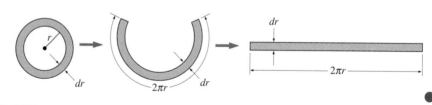

FIGURE 25-13 Unwinding the thin ring gives a strip of width dr and length $2\pi r$.

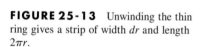

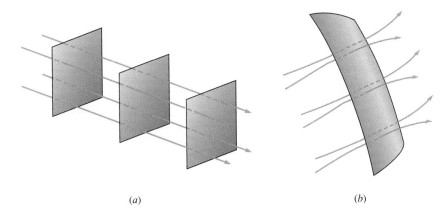

(a)   (b)

FIGURE 25-16 (a) Equipotential surfaces in a uniform electric field are planes perpendicular to the field. (b) In a nonuniform field the equipotential surfaces are curved, but are still perpendicular to the field.

25-4 POTENTIAL DIFFERENCE AND THE ELECTRIC FIELD

Equipotentials

It takes no work to move a charge at right angles to an electric field. Therefore there can be no potential difference between two points on a surface that is everywhere perpendicular to the electric field. Such surfaces are called **equipotential surfaces,** or simply **equipotentials.** Figure 25-16 shows some equipotential surfaces for both uniform and nonuniform electric fields.

Equipotentials are like contour lines used on a map to show land elevation (Fig. 25-17). A contour line is a line of constant elevation, and therefore, it takes no work to move along a contour line. Contour lines are usually spaced at even increments of elevation. Where lines are closely spaced, the elevation changes quickly. Similarly, closely spaced equipotentials indicate large potential differences between nearby points. That, in turn, means it takes a lot of work to move charge between those points—and therefore there must be a large electric field present. Figure 25-17 might just as well represent electric potential, in which case regions with closely spaced equipotentials—steep slopes on the "potential hill"—indicate large electric fields. Similarly, the equipotentials for a dipole describe the steep "hill" of the positive charge and a correspondingly deep "hole" of the negative charge (Fig. 25-18).

Calculating the Field from the Potential

Given electric field lines, we can construct equipotentials. Conversely, given equipotentials we can reconstruct the field by sketching field lines at right angles to the equipotentials. Specifying the potential at each point thus conveys all the information needed to determine the field.

We can quantify the relation between potential and field by writing the potential difference dV between two points separated by an infinitesimal displacement $d\boldsymbol{\ell}$:

$$dV = -\mathbf{E} \cdot d\boldsymbol{\ell} = -E_\ell \, d\ell,$$

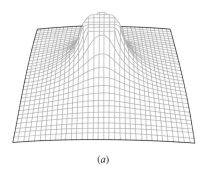

(a)

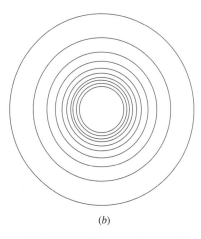

(b)

FIGURE 25-17 (a) A flat-topped hill and (b) its representation as a contour map. Closely spaced contours indicate steep slopes. Figure can also represent the potential of a uniformly charged spherical shell, whose potential is constant inside (since the electric field is zero) and falls as $1/r$ outside. Closely spaced contours then represent regions of strong electric field.

FIGURE 25-18 (*a*) Equipotentials and field lines for a dipole. The two sets of curves are everywhere perpendicular. (*b*) A plot of the potential as a function of position in the *x-y* plane shows steep "hill" for the positive charge and deep "hole" for the negative charge. The lines shown in (*b*) are *not* equipotentials.

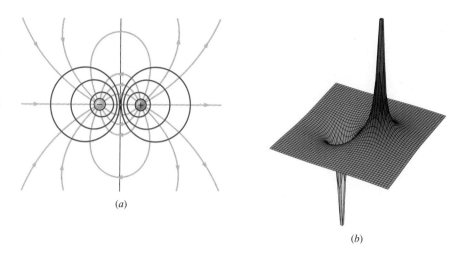

(*a*)

(*b*)

where E_ℓ designates the field component in the direction of $d\ell$. Rearranging this equation gives

$$E_\ell = -\frac{dV}{d\ell}. \tag{25-10}$$

This equation confirms our statement that the electric field is strong where potential changes rapidly with distance. The minus sign in Equation 25-10 is the same one that appears in Equation 25-2; here it tells us that if we move in the direction of *increasing* potential, then we must be moving *against* the electric field. If we want to find the field components in a chosen coordinate system, we simply choose $d\ell$ along one of the coordinate axes and apply Equation 25-10; the *x* component of the field, for example, is given by $E_x = -dV/dx$. (If *V* is a function of all three coordinates we should, strictly speaking, write E_x in terms of the *partial* derivative, $\partial V/\partial x$, that we introduced in Chapter 16.) Equation 25-11, incidentally, shows that the units of electric field can be written as V/m.

● EXAMPLE 25-8 THE FIELD OF A POINT CHARGE

Use the point-charge potential of Equation 25-4 to derive the electric field of a point charge.

Solution

The point-charge potential, $V(r) = kq/r$, depends only on *r*. Therefore the electric field points in the radial direction, and has the form $\mathbf{E} = E_r\hat{\mathbf{r}}$. Choosing $d\ell = dr$ in Equation 25-10 gives the field component:

$$E_r = -\frac{dV}{dr} = -\frac{d}{dr}\left(\frac{kq}{r}\right) = -kq\frac{d(r^{-1})}{dr} = kqr^{-2} = \frac{kq}{r^2}.$$

Thus $\mathbf{E} = \frac{kq}{r^2}\hat{\mathbf{r}}$, as expected.

●

● EXAMPLE 25-9 THE FIELD OF A CHARGED DISK

Use the result of Example 25-7 to find the electric field on the axis of a charged disk.

Solution
Symmetry shows that the field must point along the disk axis, which is the x axis in Fig. 25-12. So the field has only an x component, given by applying Equation 25-10 to the disk potential:

$$E_x = -\frac{dV}{dx} = -\frac{d}{dx}\left(\frac{2kQ}{a^2}(\sqrt{x^2 + a^2} - |x|)\right)$$

$$= \frac{2kQ}{a^2}\left(1 - \frac{|x|}{\sqrt{x^2 + a^2}}\right).$$

To see that this makes sense, consider the case $x \ll a$, for which the field becomes approximately $2kQ/a^2$. Writing Q as the surface density σ times the area πa^2 gives $E_x = 2\pi k\sigma = \sigma/2\varepsilon_0$, which is the field of an infinite charged sheet. Of course—very close to the disk it looks effectively infinite, and its field should be well approximated by that of an infinite sheet. Problem 72 shows that the field far from the disk approaches that of a point charge Q.

EXERCISE Use Equations 25-7 and 25-10 to calculate the electric field on the axis of a point dipole, and show that your result is equivalent to Equation 23-5b.

Some problems similar to Examples 25-8 and 25-9: 49–51, 80 ●

Examples 25-8 and 25-9 show that it is often much easier to calculate the electric field by first finding the potential and then differentiating, rather than doing a vector integration to get the field of a complicated charge distribution.

It's important to recognize that the *values* of the field and potential are not directly related; rather, as Equation 25-11 indicates, the field measures the *rate of change* of the potential. Field and potential are like acceleration and velocity; the *values* of the two are quite independent, with the former depending on the *rate of change* of the latter. Figure 25-19 and Example 25-10 illustrate the relation between potential and field.

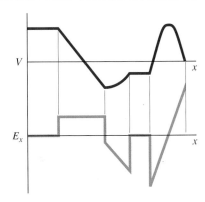

FIGURE 25-19 The x component of the electric field is the negative rate of change of the potential with respect to x.

● EXAMPLE 25-10 POTENTIAL AND FIELD

A positive charge $+2q$ lies at $x = -a$, and a charge $-q$ lies at $x = +a$. (a) Derive an expression for the potential on the x axis, and find a point on the axis in the region $x > a$ where the potential (with respect to infinity) is zero. (b) Use your expression for potential to find the electric field for $x > a$. Is the electric field zero where the potential is zero? If not, where is the field zero?

Solution
(a) The distance from any point x to the positive charge at $-a$ is $r_+ = |x - (-a)| = |x + a|$; similarly, the distance to the negative charge is $r_- = |x - a|$. Thus, the potential on the axis is

$$V(x) = \frac{k(2q)}{r_+} + \frac{k(-q)}{r_-} = kq\left(\frac{2}{|x + a|} - \frac{1}{|x - a|}\right).$$

We can find where $V = 0$ by setting the quantity in parentheses to zero. For $x > a$ both denominators are positive and we can remove the absolute value signs. For $V = 0$ we then have

$$\frac{2}{x + a} = \frac{1}{x - a}.$$

Solving for x gives $x = 3a$.

(b) The field clearly has only an x component, and Equation 25-10 then gives

$$E_x = -\frac{dV}{dx} = -kq\frac{d}{dx}\left(\frac{2}{x + a} - \frac{1}{x - a}\right)$$

$$= \frac{2kQ}{(x + a)^2} - \frac{kQ}{(x - a)^2}.$$

This result is hardly surprising: It's just the sum of two point-charge fields. We actually solved for the zero-field point in this configuration in Example 23-3; there we found that the electric force would be zero at $x = 5.83a$, which is *not* the same place where the potential is zero.

Figure 25-20 shows the potential of this charge distribution in the region $x > a$, showing clearly the point $x = 3a$ where $V = 0$. Getting a charge to this point from infinity would take

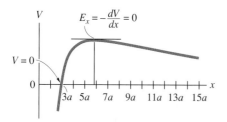

FIGURE 25-20 Potential on the x axis for a charge distribution consisting of $+2q$ at $x = -a$ and $-q$ at $x = +a$. Shown here is the region $x > a$, including the point $x = 3a$ where the potential is 0 and the point $x = 5.83a$ where the electric field—proportional to the *slope* of the potential curve—is zero. Note the deep "hole" associated with the negative charge.

no *net* work, although it would require going up and then down a potential "hill." Also clear in Fig. 25-20 is the point $x = 5.83a$ where the electric field is zero. Note that this is the point where the *slope* of the potential curve is zero, *not* where the potential itself is zero.

There are actually two points on the x axis where $V = 0$; the second lies between the charges. Both lie on an equipotential

surface of zero potential that surrounds the negative charge; Fig. 25-21 shows some equipotentials and field lines in the x-y plane.

> **TIP Field and Potential Are Not Proportional** In particular, where one is zero, the other need not be zero. You can see that in Fig. 25-20, where it clearly takes work to get from infinity to the point where $E = 0$. Just because a mountaintop is flat doesn't mean it didn't take work to climb it! Potential depends not on the field at a point but on the field over an entire path from infinity to that point. Similarly, the potential can be zero at points where the field is not, as evidenced by the steep slope of the potential curve in Fig. 25-20 at the point where it crosses zero.

EXERCISE Find the second point on the x axis in Example 25-10 where the potential is zero.

Answer: $x = a/3$

Some problems similar to Example 25-10: 48, 52, 79

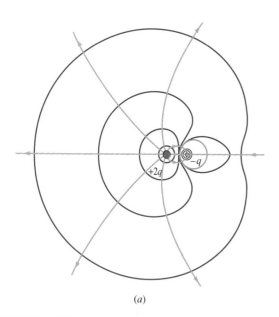

(a)

(b)

FIGURE 25-21 (a) Equipotentials and field lines for the two opposite but unequal charges of Example 25-10. Blue circle is the $V = 0$ equipotential whose x-axis intersections were calculated in Example 25-10. Values of potential inside this circle are negative; all others are positive. Note that at large distances positive equipotentials surround the entire charge distribution. (b) Three-dimensional plot of the potential over the same region shown in (a). "Hill" of the positive charge $2q$ is larger than the "hole" of the negative charge q because of the difference in charge magnitudes. Note that one would have to go "uphill" coming in from infinity before dropping into the region of negative potential near the negative charge. That means there is an equipotential with $V = 0$ surrounding the negative charge, as shown in (a).

25-5 POTENTIALS OF CHARGED CONDUCTORS

That there is no electric field inside a conductor in electrostatic equilibrium means it takes no work to move a test charge around inside the conductor; that the field at the conductor surface is perpendicular to the surface means it takes no work to move a test charge along the surface, either. Therefore the potential difference between two points in or on a conductor must be zero, and thus,

▌ *A conductor in electrostatic equilibrium is an equipotential.*

Consider an isolated, spherical conductor of radius R carrying charge Q. Since the conductor is isolated, charge is distributed uniformly over its surface, and for $r > R$ it therefore acts like a point charge. The potential at its surface is then

$$V(R) = \frac{kQ}{R},\qquad (25\text{-}11)$$

as we found in Example 25-3. The sphere itself is an equipotential, so the potential *difference* between two points *on the sphere* is zero. But that doesn't mean the potential difference between *infinity* and the sphere is zero; since the sphere is charged, it takes work to move charge to its surface from infinity, and that's what Equation 25-11 says.

Now consider two widely separated spheres of different sizes. If we connect them by a thin conducting wire, as shown in Fig. 25-22, then the system constitutes a single conductor, and charge will move through the wire until both spheres are at the same potential. But since the spheres are widely separated, each still has an essentially spherical charge distribution, so Equation 25-11 applies to each. Since the spheres have the same potential, Equation 25-11 implies that

$$\frac{kQ_1}{R_1} = \frac{kQ_2}{R_2},$$

where the subscripts label the two spheres. We can write each charge as the surface area of the sphere multiplied by the surface charge density: $Q = 4\pi R^2 \sigma$. Substituting for the Q's in the above equation and solving for the ratio of surface charge densities then gives

$$\frac{\sigma_1}{\sigma_2} = \frac{R_2}{R_1}.$$

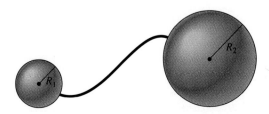

FIGURE 25-22 Two conducting spheres held at the same potential by a conducting wire. The surface charge density is greater on the smaller sphere, in inverse proportion to its radius.

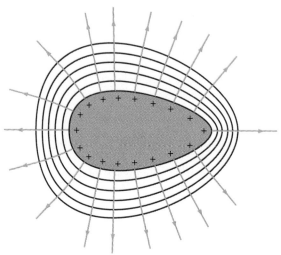

FIGURE 25-23 An irregular charged conductor. Equipotentials near the conductor have approximately its shape, and the field is strongest—and therefore the equipotentials closest—where the conductor curves most sharply.

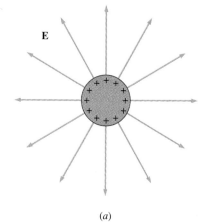

E

(a)

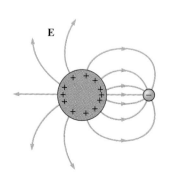

E

(b)

FIGURE 25-24 (a) An isolated conducting sphere carries a uniform surface charge density and has a spherically symmetric electric field. (b) The presence of a nearby charge distorts the surface charge distribution and therefore the electric field.

Thus the *smaller* sphere has the *larger* surface charge density. Since the electric field at a conductor surface has magnitude $E = \sigma/\varepsilon_0$, the field must be stronger at the smaller sphere.

This discussion of spherical conductors provides a qualitative description of nonspherical conductors as well. All parts of an irregularly shaped conductor must be at the same potential. Where the conductor surface curves sharply, it is like a small sphere and therefore has a higher surface charge density and a stronger electric field. In general, the field is strongest where the surface curves most sharply.*

Because a conductor surface is an equipotential and the electric field is perpendicular to the conductor surface, equipotentials just above the surface must have approximately the same shape as the surface. Because the electric field is stronger where the conductor surface curves sharply, there must be more field lines emerging from such regions. Far from a charged conductor, on the other hand, its field must resemble that of a point charge, with radial field lines and circular equipotentials. With these limiting cases in mind, we can sketch the approximate form for the field of an arbitrarily shaped conductor (Fig. 25-23).

We stress that our conclusion about surface charge density and curvature applies only to *isolated* conductors—those far from any other charges. The field of a nearby charge will modify the charge distribution of a conductor, altering the surface charge distribution (Fig. 25-24).

---

*This association of strong field and sharp curvature is only approximate and, in some unusual configurations, may not hold at all. See "The Lightning Rod Fallacy," R. H. Price and R. J. Crowley, *American Journal of Physics,* vol. 53, September 1985, p. 843.

■ **APPLICATION** CORONA DISCHARGE, POLLUTION CONTROL, AND XEROGRAPHY

The large electric fields that develop where a charged conductor is sharply curved can cause serious problems in electrical equipment; in other applications those fields are put to good use. Fields above 3 MN/C are strong enough to strip electrons from air molecules, making the air a conductor. Breakdown of air is often evidenced by a blue glow around sharply-pointed conductors (Fig. 25-25). Called **corona discharge,** this glow results from the recombination of electrons with atoms. Corona discharge causes loss of power from high-voltage transmission lines, and engineers try to avoid it by eliminating sharp edges on wires and other conducting structures.

Corona discharge is put to good use in the **electrostatic precipitator,** a pollution-control device used especially on coalburning power plants. A typical precipitator consists of parallel metal plates with thin wires running between them. Application of a high voltage between plates and wires sets up large electric fields near the wires. Exhaust gases flow between the plates and the field ionizes some gas molecules. These charged molecules, in turn, attach themselves to pollutant particles. The charged particles are driven to the collecting plates by the electric field. Every few minutes a mechanical vibrator taps the plates and the particles fall into a hopper, where they can be trucked away to use for fill or in making products like cinder blocks. A typical power plant produces some 30 pounds of particulates every second, and precipitators keep most of this out of the air. In the process, the precipitators may consume several per cent of the power plant's electrical output. Fig. 25-26 shows electrostatic precipitators at a large power plant.

Xerography (literally, "dry writing") used in copiers, laser printers, and similar devices is another widespread application of corona discharge. Copying starts as corona discharge between a charged wire and a special light-sensitive plate spreads a layer of positive charge on the plate (Fig. 25-27a). Light, imaged from the document being copied or scanned by a computer-driven laser mechanism, strikes the plate and causes charge to flow from the illuminated regions (Fig. 25-27b). Negatively charged particles called toner are then spread on the plate, where they adhere to the charged regions that correspond to dark areas in the image (Fig. 25-27c). Finally, the toner is transferred to paper and heated to fuse it into the paper, making a permanent copy.

(a)

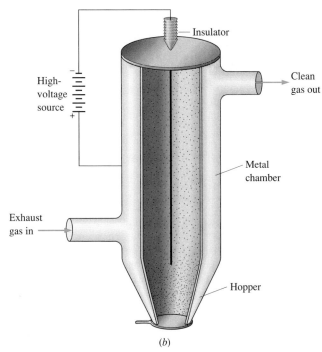

(b)

FIGURE 25-26 (a) Electrostatic precipitators at a 1300-MW coal-burning power plant. (b) In the electrostatic precipitator, high voltage applied between the metal chamber and the thin central wire results in a strong, nonuniform electric field near the wire. This field ionizes air molecules, which attach to soot particles with the result that the particles are accelerated to the chamber walls. Mechanical vibration then dislodges the particles into the hopper for eventual collection. Parallel metal plates are often used instead of the cylindrical chamber shown here.

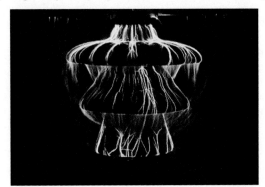

FIGURE 25-25 Corona discharge on a power-line insulator.

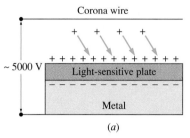
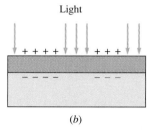
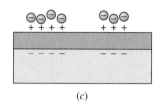

(a)   (b)   (c)

FIGURE 25-27 The xerography process, widely used in copiers and laser printers. (a) Corona discharge from a thin wire lays positive charge on a light-sensitive surface. (b) Where light hits, charge is driven into the metal substrate, leaving a charge pattern corresponding to the image being reproduced. (c) Negatively charged toner particles stick to the plate, and are later transferred and fused to paper.

CHAPTER SYNOPSIS

Summary

1. The **electric potential difference** between two points is the change in potential energy per unit charge for a charge moved between those points. Because the electric field is conservative, the potential difference between two points is independent of the path taken, and is also equal to the work per unit charge that an external agent must do in moving charge at constant speed between the two points. The potential difference between two points A and B is calculated by evaluating the **line integral** of the electric field over any path between those points:

$$\Delta V_{A \to B} = -\int_A^B \mathbf{E} \cdot d\boldsymbol{\ell}.$$

When the field is uniform, this expression reduces to

$$\Delta V_{A \to B} = -\mathbf{E} \cdot \boldsymbol{\ell}.$$

2. Defining the potential to be zero at some point allows us to speak of "the potential at a point," meaning the potential difference from the reference point to the point in question. For isolated point charges, a convenient zero is infinitely far from the charge; then the potential at an arbitrary point a distance r from the point charge q is

$$V(r) = \frac{kq}{r}. \quad \text{(point-charge potential)}$$

The potentials of charge distributions may be found by taking the line integral of the field, if the latter is known, or by summing the potentials of the point charges making up the distribution:

$$V = \sum_i \frac{kq_i}{r_i}, \begin{pmatrix} \text{discrete} \\ \text{charges} \end{pmatrix} \quad \text{or}$$

$$V = \int \frac{k\,dq}{r}. \begin{pmatrix} \text{continuous charge} \\ \text{distribution} \end{pmatrix}$$

3. **Equipotentials** are surfaces over which the potential has a constant value. Equipotentials are everywhere perpendicular to the electric field. The field is strong where equipotentials are closely spaced and vice versa. Mathematically, the field component in a given direction is related to the rate of change of potential with position in that direction:

$$E_\ell = -\frac{dV}{d\ell}.$$

4. A conductor in electrostatic equilibrium is an equipotential. The surface charge density and therefore the electric field at the conductor surface are usually greatest where the conductor curves most sharply. Very strong electric fields occur at sharp bends; if strong enough, these fields can result in **corona discharge,** in which the surrounding air becomes a conductor and charge leaks off the charged conductor.

Terms You Should Understand

(Pairs are closely related terms whose distinction is important; number in parentheses is chapter section where term first appears.)

conservative field (introduction)
potential difference (25-2)
volt, electron volt (25-2)
line integral (25-2)
equipotential (25-4)
corona discharge (25-5)

Symbols You Should Recognize

$\Delta V_{A \to B}$ (25-2)
V, eV (25-2)
$V(P)$ (25-3)
$d\ell$ (25-2)

Problems You Should Be Able to Solve

calculating the work needed to move a given charge through a given potential difference (25-2)

calculating potential differences in uniform electric fields (25-2)

calculating potential differences using the line integral of a known electric field (25-3)

evaluating potentials by summing or integrating over point charges (25-3)

finding electric field components given potential as a function of position (25-4)

sketching equipotentials of simple charge distributions (25-4)

sketching equipotentials and fields around conductors (25-5)

Limitations to Keep in Mind

Electric potential difference depends on *two points*. Phrases like "the potential at point *P*" or "the potential of a point charge" are always shorthand ways of talking about the potential difference between two points. When the second point is not specified, it is often taken to be at infinity.

QUESTIONS

1. Why can a bird perch on a high-voltage power line without getting electrocuted?

2. One proton is accelerated from rest by a uniform electric field, the other by a nonuniform electric field. If they move through the same potential difference, how do their final speeds compare?

3. Would a free electron move toward higher or lower potential?

4. The potential difference from *A* to *B* in Fig. 25-28 is zero since the two points are equidistant from the charge *Q*. How can this be, when a charge moving along the path shown clearly experiences an electric force not perpendicular to the path?

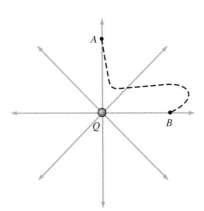

FIGURE 25-28   Question 4.

5. A proton and a positron (a particle with the electron's mass carrying charge $+e$) are accelerated through the same potential difference. How do their final energies compare? Their final speeds?

6. The electric field at the center of a uniformly charged ring is obviously zero, yet Example 25-6 shows that the potential at the center is not zero. How is this possible?

7. Must the potential be zero at any point where the electric field is zero? Explain.

8. Must the electric field be zero at any point where the potential is zero? Explain.

9. The potential is constant throughout an entire volume. What must be true of the electric field within that volume?

10. In considering the potential of an infinite flat sheet, why is it not useful to take the zero of potential at infinity?

11. The potential of a point charge is given by kq/r. Is r the distance between the two points for which this is the potential difference? Explain.

12. "Cherry picker" trucks for working in trees or power lines often carry the warning sign shown in Fig. 25-29. Explain how this hazard arises and why it might be more of a danger to someone on the ground than to a worker on the truck.

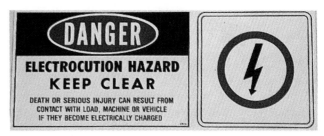

FIGURE 25-29   Question 12.

13. Two positive point charges are located a small distance apart. Are there any points, other than at infinity, where the potential is zero?

14. Is it possible for equipotential surfaces to intersect? Explain.

15. Is the potential at the center of a hollow, uniformly charged spherical shell higher, lower, or the same as at the surface?

16. A solid sphere contains charge uniformly distributed throughout its volume. Is the potential at its center higher, lower, or the same as at the surface?

17. Why do the spheres in Fig. 25-21 need to be far apart for the conclusion that surface charge density is inversely proportional to radius to hold accurately?

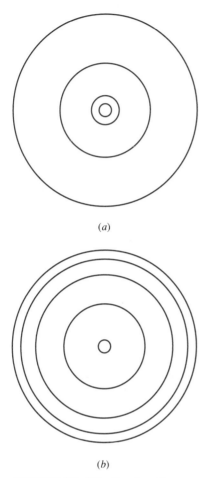

(a)

(b)

FIGURE 25-30 Question 18.

18. Figure 25-30 shows cross sections of two sets of spherical equipotentials, spaced in even increments of potential difference. Describe qualitatively how the charge distributions in the two regions differ.

19. Two equal but opposite charges form a dipole. Describe the equipotential surface on which $V = 0$.

20. Figure 25-31 shows three paths leading from infinity to a point P on the perpendicular bisector of a dipole. For each path, how much work is needed to bring a charge q from infinity to P?

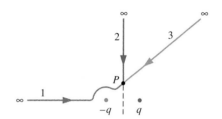

FIGURE 25-31 Question 20.

21. The electric potential in a region increases linearly with distance. What can you conclude about the electric field in this region?

22. In Fig. 25-23b the charge density on the conducting sphere is not uniform, yet the field inside must still be zero since the system is in electrostatic equilibrium. How is this possible?

23. Why is lightning likely to strike an isolated tree?

24. What is the difference between a volt and an electron volt?

PROBLEMS

Section 25-2 Potential Difference

1. How much work does it take to move a 50-μC charge against a 12-V potential difference?

2. The potential difference between the two sides of an ordinary electrical outlet is 120 V. How much energy does an electron gain when it moves from one side to the other?

3. It takes 45 J to move a 15-mC charge from point A to point B. What is the potential difference $\Delta V_{A \to B}$?

4. Show that 1 V/m is the same as 1 N/C.

5. Find the magnitude of the potential difference between two points located 1.4 m apart in a uniform 650 N/C electric field, if a line between the points is parallel to the field.

6. A charge of 3.1 C moves from the positive to the negative terminal of a 9.0-V battery. How much energy does the battery impart to the charge?

7. Two points A and B lie 15 cm apart in a uniform electric field, with the path AB parallel to the field. If the potential difference $\Delta V_{A \to B}$ is 840 V, what is the field strength?

8. Figure 25-32 shows a uniform electric field of magnitude E. Find expressions for (a) the potential difference $\Delta V_{A \to B}$ and (b) $\Delta V_{B \to C}$. (c) Use your result to determine $\Delta V_{A \to C}$.

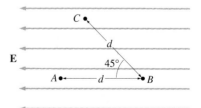

FIGURE 25-32 Problem 8.

9. A proton, an alpha particle (a bare helium nucleus), and a singly ionized helium atom are accelerated through a potential difference of 100 V. Find the energy each gains.

10. Two points A and B lie 77 cm apart in a uniform 540 V/m electric field. If $\Delta V_{A \to B} = 390$ V, what angle does a line from A to B make with the field?

11. What is the potential difference between the terminals of a battery that can impart 7.2×10^{-19} J to each electron that moves between the terminals?

12. Electrons in a TV tube are accelerated from rest through a 25-kV potential difference. With what speed do they hit the TV screen?

13. A 12-V car battery stores 2.8 MJ of energy. How much charge can move between the battery terminals before it is totally discharged? Assume the potential difference remains at 12 V, an assumption that is not realistic.

14. What is the charge on an ion that gains 1.6×10^{-15} J when it moves through a potential difference of 2500 V?

15. Two large, flat metal plates are a distance d apart, where d is small compared with the plate size. If the plates carry surface charge densities $\pm \sigma$, show that the potential difference between them is $V = \sigma d / \varepsilon_0$.

16. An electron passes point A moving at 6.5 Mm/s. At point B the electron has come to a complete stop. Find the potential difference $\Delta V_{A \to B}$.

17. A 5.0-g object carries a net charge of 3.8 μC. It acquires a speed v when accelerated from rest through a potential difference V. A 2.0-g object acquires twice the speed under the same circumstances. What is its charge?

Section 25-3 Calculating Potential Difference

Note: In these problems, the zero of potential is taken at infinity unless noted otherwise.

18. An electric field is given by $\mathbf{E} = E_0 \hat{\mathbf{j}}$, where E_0 is a constant. Find the potential as a function of position, taking $V = 0$ at $y = 0$.

19. The classical picture of the hydrogen atom has a single electron in orbit a distance 0.0529 nm from the proton. Calculate the electric potential associated with the proton's electric field at this distance.

20. Earth carries an electric charge of -4.3×10^5 C, distributed essentially uniformly over its surface. What is the potential difference between Earth's surface and the base of the ionosphere, about 80 km above the surface?

21. Points A and B lie 20 cm apart on a line extending radially from a point charge Q, and the potentials at these points are $V_A = 280$ V, $V_B = 130$ V. Find Q and the distance r between A and the charge.

22. What is the maximum potential allowable on a 5.0-cm-diameter metal sphere if the electric field at the sphere's surface is not to exceed the 3 MV/m breakdown field in air?

23. A 3.5-cm-diameter isolated metal sphere carries a net charge of 0.86 μC. (a) What is the potential at the sphere's surface? (b) If a proton were released from rest at the sphere's surface, what would be its speed far from the sphere?

24. A sphere of radius R carries a negative charge of magnitude Q, distributed in a spherically symmetric way. Find the "escape speed" for a proton at the sphere's surface—that is, the speed that would enable the proton to escape to arbitrarily large distances.

25. A thin spherical shell of charge has radius R and total charge Q distributed uniformly over its surface. What is the potential at its center?

26. A solid sphere of radius R carries a net charge Q distributed uniformly throughout its volume. Find the potential difference from the sphere's surface to its center. *Hint:* Consult Example 24-1.

27. Find the potential as a function of position in an electric field given by $\mathbf{E} = ax\hat{\mathbf{i}}$, where a is a constant and where $V = 0$ at $x = 0$.

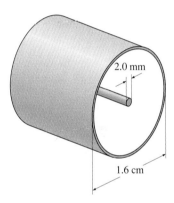

FIGURE 25-33 Problem 28.

28. A coaxial cable consists of a 2.0-mm-diameter inner conductor and an outer conductor of diameter 1.6 cm and negligible thickness (Fig. 25-33). If the conductors carry line charge densities ± 0.56 nC/m, what is the magnitude of the potential difference between them?

29. The potential difference between the surface of a 3.0-cm-diameter power line and a point 1.0 m distant is 3.9 kV. What is the line charge density on the power line?

30. Three equal charges q form an equilateral triangle of side a. Find the potential at the center of the triangle.

31. A charge $+Q$ lies at the origin, and $-3Q$ at $x = a$. Find two points on the x axis where $V = 0$.

32. Two identical charges q lie on the x axis at $\pm a$. (a) Find an expression for the potential at all points in the x-y plane. (b) Show that your result reduces to the potential of a point charge for distances large compared with a.

33. Find the potential 10 cm from a dipole of moment $p = 2.9$ nC·m (a) on the dipole axis, (b) at 45° to the axis, and (c) on the perpendicular bisector. The dipole separation is much less than 10 cm.

34. Two points A and B lie 55 cm from a dipole of moment $p = 6.4$ nC·m, whose charge separation is much less than 55 cm. A line from the dipole to A makes a 20° angle with the dipole axis, and a line to B makes a 50° angle. Find the potential difference $V_B - V_A$.

35. The potential at point P in Fig. 25-34 is 37 kV. Find the charge separation a, assuming it is much smaller than the 1.6-cm distance to P.

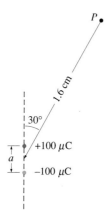

FIGURE 25-34 Problem 35.

36. A thin plastic rod 20 cm long carries 3.2 nC distributed uniformly over its length. (a) If the rod is bent into a circular ring, find the potential at its center. (b) If the rod is bent into a semicircle, find the potential at the center (i.e., at the center of the circle of which the semicircle is part).

37. A thin ring of radius R carries a charge $3Q$ distributed uniformly over three-fourths of its circumference, and $-Q$ over the rest. What is the potential at the center of the ring?

38. The potential at the center of a uniformly charged ring is 45 kV, and 15 cm along the ring axis the potential is 33 kV. Find the ring's radius and its total charge.

39. The annulus shown in Fig. 25-35 carries a uniform surface charge density σ. Find an expression for the potential at an arbitrary point P on its axis.

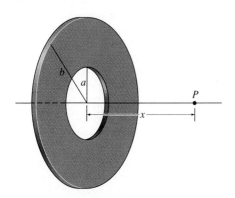

FIGURE 25-35 Problem 39.

40. A thin rod of length ℓ carries a charge Q distributed uniformly over its length. (a) Show that the potential in the plane that perpendicularly bisects the rod is given by

$$V(r) = \frac{2kQ}{\ell} \ln\left[\frac{\ell}{2r} + \sqrt{1 + \frac{\ell^2}{4r^2}}\right],$$

where r is the distance from the rod center. (b) Show that this expression reduces to an expected result when $r \gg \ell$. *Hint:* See Appendix A for a series expansion of the logarithm.

41. (a) Find the potential as a function of position in the electric field $\mathbf{E} = E_0(\hat{\mathbf{i}} + \hat{\mathbf{j}})$, where $E_0 = 150$ V/m. Take the zero of potential at the origin. (b) Find the potential difference from the point $x = 2.0$ m, $y = 1.0$ m to the point $x = 3.5$ m, $y = -1.5$ m.

Section 25-4 Potential Difference and the Electric Field

42. In a uniform electric field, equipotential planes that differ by 1.0 V are 2.5 cm apart. What is the field strength?

43. Figure 25-36 shows a plot of potential versus position along the x axis. Make a plot of the x component of the electric field for this situation.

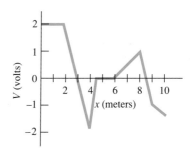

FIGURE 25-36 Problem 43.

44. Figure 25-37 shows some equipotentials in the x-y plane. (a) In what region is the electric field strongest? What are (b) the direction and (c) the magnitude of the field in this region?

45. The potential in a certain region is given by $V = axy$, where a is a constant. (a) Determine the electric field in the region. (b) Sketch some equipotentials and field lines.

46. Sketch some equipotentials and field lines for a distribution consisting of two equal point charges.

47. Figure 25-38 shows some equipotentials in the x-y plane. The equipotentials shown are 10 V apart, as indicated. Find an expression for the electric field in the region.

48. Sketch some equipotentials and field lines for a distribution consisting of two point charges $+3Q$ and $-Q$.

49. The electric potential in a region of space is given by $V = 2xy - 3zx + 5y^2$, with V in volts and the coordinates in

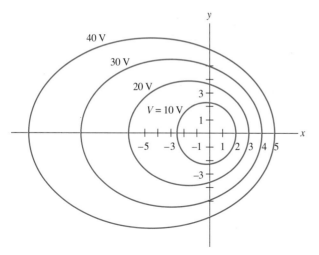

FIGURE 25-37 Problem 44.

meters. If point P is at $x = 1$ m, $y = 1$ m, $z = 1$ m, find (a) the potential at P and (b) the x, y, and z components of the electric field at P.

50. Use Equation 25-7 to calculate the electric field on the perpendicular bisector of a point dipole, and show that your result is equivalent to Equation 23-5a.

51. Use the result of Example 25-6 to determine the on-axis field of a charged ring, and verify that your answer agrees with the result of Example 23-8.

52. A charge $+4q$ is located at the origin and a charge $-q$ is on the x axis at $x = a$. (a) Write an expression for the potential on the x axis for $x > a$. (b) Find a point in this region where $V = 0$. (c) Use the result of (a) to find the electric field on the x axis for $x > a$ and (d) find a point where $\mathbf{E} = \mathbf{0}.$

53. The electric potential in a region is given by $V = -V_0(r/R)$, where V_0 and R are constants, r is the radial distance from the origin, and where the zero of potential is taken at $r = 0$. Find the magnitude and direction of the electric field in this region.

Section 25-5 Potentials of Charged Conductors

54. (a) How much charge can be placed on a metal sphere 1.0 cm in diameter before corona discharge occurs to the surrounding air? (b) What is the sphere's potential at this maximum charge?

55. The spark plug in an automobile engine has a center electrode made from wire 2.0 mm in diameter. The electrode is worn to a hemispherical shape, so it behaves approximately like a charged sphere. What is the minimum potential on this electrode that will ensure the plug sparks in air? Neglect the presence of the second electrode.

56. A large metal sphere has three times the diameter of a smaller sphere and carries three times as much charge. Both spheres are isolated, so their surface charge densities are uniform. Compare (a) the potentials and (b) the electric field strengths at their surfaces.

57. Two metal spheres each 1.0 cm in radius are far apart. One sphere carries 38 nC of charge, the other -10 nC. (a) What is the potential on each? (b) If the spheres are connected by a thin wire, what will be the potential on each once equilibrium is reached? (c) How much charge must move between the spheres in order to achieve equilibrium?

58. Sketch some equipotentials and field lines for the isolated, charged conductor shown in Fig. 25-39.

59. Two conducting spheres are each 5.0 cm in diameter and each carries 0.12 μC. They are 8.0 m apart. Determine (a) the potential on each sphere; (b) the field strength at the surface of each sphere; (c) the potential midway between the spheres; (d) the potential difference between the spheres.

60. Two small metal spheres are located 2.0 m apart. One has radius 0.50 cm and carries 0.20 μC. The other has radius 1.0 cm and carries 0.080 μC. (a) What is the potential

FIGURE 25-38 Problem 47.

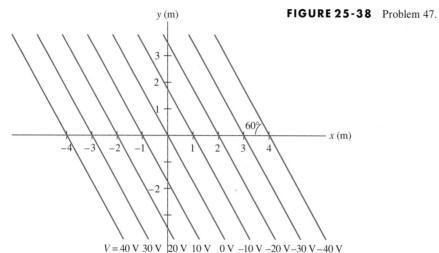

$V = 40$ V 30 V 20 V 10 V   0 V -10 V -20 V -30 V -40 V

FIGURE 25-39 Problem 58.

difference between the spheres? (b) If they were connected by a thin wire, how much charge would move along it, and in which direction?

Paired Problems

(Both problems in a pair involve the same principles and techniques. If you can get the first problem, you should be able to solve the second one.)

61. Three 50-pC charges sit at the vertices of an equilateral triangle 1.5 mm on a side. How much work would it take to bring a proton from very far away to the midpoint of one of the triangle's sides?

62. Repeat the preceding problem for the case when one of the charges is −50 pC and the proton is brought to the midpoint of the side between the two positive charges.

63. A pair of equal charges q lies on the x axis at $x = \pm a$. (a) Find expressions for the potential at points on the x axis for which $x > a$ and (b) show that your result reduces to a point-charge potential for $x \gg a$.

64. (a) For the charge distribution of the preceding problem, find an expression for the potential at *all* points on the y axis. (b) Show that your result reduces to a point-charge potential for $y \gg a$.

65. A 2.0-cm-radius metal sphere carries 75 nC and is surrounded by a concentric spherical conducting shell of radius 10 cm carrying −75 nC. (a) Find the potential difference between the shell and the sphere. (b) How would your answer change if the shell charge were changed to +150 nC?

66. A coaxial cable consists of a 2.0-mm-radius central wire carrying 75 nC/m, and a concentric outer conductor of radius 10 mm carrying −75 nC/m. (a) Find the potential difference between the outer and inner conductor. (b) How would your answer change if the outer conductor were charged to +150 nC/m?

67. On the x axis, the electric field of a certain charge distribution is given by $\mathbf{E} = a/x^4\hat{\mathbf{i}}$, where $a = 55$ V·m^3. Find the potential difference from the point $x = 1.3$ m to the point $x = 2.8$ m.

68. A sphere of radius R carries a nonuniform but spherically symmetric volume charge density that results in an electric field in the sphere given by $\mathbf{E} = E_0(r/R)^2\hat{\mathbf{r}}$, where E_0 is a constant. Find the potential difference from the sphere's surface to its center.

69. The potential as a function of position in a certain region is given by $V(x) = 3x - 2x^2 - x^3$, with x in meters and V in volts. Find (a) all points on the x axis where $V = 0$, (b) an expression for the electric field, and (c) all points on the x axis where $\mathbf{E} = \mathbf{0}$.

70. The potential in a certain region is given by $V(x) = -[2x^2 + (y - 1)^2 - 1]$. Find (a) a point where $V = 0$, (b) an expression for the electric field, and (c) a point in the x-y plane where $\mathbf{E} = \mathbf{0}$.

Supplementary Problems

71. A conducting sphere 5.0 cm in radius carries 60 nC. It is surrounded by a concentric spherical conducting shell of radius 15 cm carrying −60 nC. (a) Find the potential at the sphere's surface, taking the zero of potential at infinity. (b) Repeat for the case when the shell also carries +60 nC.

72. Show that the result of Example 25-9 approaches the field of a point charge for $x \gg a$. *Hint:* You will need to apply the binomial theorem to the quantity $1/\sqrt{x^2 + a^2}$.

73. The potential on the axis of a uniformly charged disk at 5.0 cm from the disk center is 150 V; the potential 10 cm from disk center is 110 V. Find the disk radius and its total charge.

74. A uranium nucleus (mass 238 u, charge 92e) decays, emitting an alpha particle (mass 4 u, charge 2e) and leaving a thorium nucleus (mass 234 u, charge 90e). At the instant the alpha particle leaves the nucleus, the centers of the two are 7.4 fm apart and are essentially at rest. Find their speeds when they are a great distance apart. Treat each particle as a spherical charge distribution.

75. A power line consists of two parallel wires 3.0 cm in diameter spaced 2.0 m apart. If the potential difference between the wires is 4.0 kV, what is the charge per unit length on each wire? The wires carry equal but opposite charges. *Hint:* The wires are far enough apart that they don't greatly affect each other's fields.

76. For the dipole of Example 25-5, show that the electric field at an arbitrary point far from the dipole can be written
$$\mathbf{E} = \frac{kp}{r^3}[(3\cos^2\theta - 1)\hat{\mathbf{i}} + 3\sin\theta\cos\theta\hat{\mathbf{j}}].$$

77. A thin rod of length ℓ lies on the x axis with its center at the origin. It carries a line charge density given by $\lambda = \lambda_0(x/\ell)^2$, where λ_0 is a constant. (a) Find an expression for the potential on the x axis for $x > \ell/2$. (b) Integrate the charge density to find the total charge on the rod. (c) Show that your answer for (a) reduces to the potential of a point charge whose charge is the answer to (b), for $x \gg \ell$.

78. Repeat the preceding problem for the case $\lambda = \lambda_0(x/\ell)$. Why is your answer for $x \gg \ell$ different? *Hint:* What does this charge distribution resemble at large distances?

79. For the situation of Example 25-10, find an equation for the equipotential with $V = 0$ in the *x-y* plane. Plot the equipotential, and show that it passes through the points described in Example 25-10 and its exercise.

80. A disk of radius a carries a nonuniform surface charge density given by $\sigma = \sigma_0(r/a)$, where σ_0 is a constant. (a) Find the potential at an arbitrary point on the disk axis, a distance x from the disk center. (b) Use the result of (a) to find the electric field on the disk axis, and (c) show that the field reduces to an expected form for $x \gg a$.

81. An open-ended cylinder of radius a and length $2a$ carries charge q spread uniformly over its surface. Find the potential on the cylinder axis at its center. *Hint:* Treat the cylinder as a stack of charged rings, and integrate.

CHAPTER 26

ELECTROSTATIC ENERGY AND CAPACITORS

A test firing of the Particle Beam Fusion Accelerator at Sandia National Laboratories involves the sudden release of energy stored in electric fields.

Suppose you hold two positive charges in your outstretched arms (Fig. 26-1). Bringing them closer takes work, as you move each charge against the other's electric field. That work is stored as potential energy associated with the new distribution of charge you create by moving the charges closer together. Because the static electric field is conservative, you could recover the stored energy by releasing the charges and letting them accelerate.

The example of Fig. 26-1 is trivial, but its implications are not. Energy storage in configurations of electric charge is a vital aspect of the natural and technological worlds. The energy produced in chemical reactions—including the metabolizing of food and the burning of coal, oil, and other fuels—is electrical energy released in the rearrangement of molecular charge distributions. Energy storage in systems of charged conductors is essential to the workings of electronic equipment and is important in devices that require large amounts of energy delivered in a short time. In this chapter we explore the energy of charge distributions and their electric fields, and we introduce a practical device—the capacitor—whose function is electrical energy storage.

26-1 ENERGY OF A CHARGE DISTRIBUTION

In the preceding chapter we defined the electric potential difference between two points as the change in potential energy per unit charge associated with moving charge between those points. To move charge between the two points takes work equal to the change in potential energy—equal, that is, to the electric potential difference multiplied by the charge being moved. In the simple case of two point charges, suppose charge q_1 is initially an infinite distance from a fixed charge q_2. You can assemble a new charge distribution by moving q_1 to a distance r from q_2. Equation 25-4 gives the electric potential difference from infinity to any point a distance r from q_2:

$$V_{\infty \to r} = \frac{kq_2}{r}.$$

Multiplying by q_1 gives the work you must do to bring q_1 in from infinity—that is, to assemble the new charge distribution. Since that work, W, gets stored as potential energy, U, we can write

$$W = U = \frac{kq_1q_2}{r}, \tag{26-1}$$

where we've taken the zero of potential energy when the charges are infinitely far apart. Note that W here is the work that *you* or some other agent has to do to assemble the charge distribution; it's not the work done by the electric field, which, as Chapter 8's definition of potential energy shows, is $-W$.

Equation 26-1 gives the **electrostatic potential energy** of two point charges. The equation shows that the potential energy is positive if the charges have the same sign and negative if they have opposite signs. In the latter case, it would take positive work to separate the charges. We would have obtained the same potential energy had we moved q_2 in the field of q_1, showing that the potential energy of a charge distribution is independent of how it is assembled.

These considerations hold for any charge distribution. In general, it takes work to assemble a charge distribution and that work is stored as potential energy. The potential energy can be positive, zero, or negative. Because the electric force obeys the superposition principle, the potential energy is independent of how the charge distribution is assembled: the total energy is simply the sum of the potential energies of every charge pair making up the distribution.

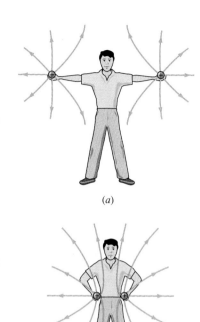

(a)

(b)

FIGURE 26-1 (a) Widely separated charges exert little force on each other. (b) Moving them together takes work, which is stored as potential energy. Note how the bending of the field lines suggests the repulsive force.

● **EXAMPLE 26-1** THE ENERGY OF A CHARGE DISTRIBUTION

Three point charges each carrying $+q$ and a fourth carrying $-q/2$ are initially infinitely far apart. They are brought together to form the square charge distribution shown in Fig. 26-2. What is the electrostatic potential energy of this charge distribution?

Solution

We can assemble the charge distribution in any order. Assume that the positive charge at the upper left is brought in first. This takes no work because no other charge is in place. Next the positive charge q at the upper right is brought to a distance a from the first positive charge. Equation 26-1 tells us that the work required is

$$W_2 = k\frac{q^2}{a}.$$

Now the third positive charge is brought to its place at the lower left. This point is a distance a from the first charge q and $\sqrt{2}a$ from the second, so the work required is

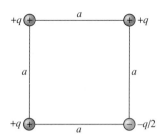

FIGURE 26-2 How much energy is stored in this square charge distribution? (Example 26-1)

$$W_3 = k\left(\frac{q^2}{a} + \frac{q^2}{\sqrt{2}a}\right).$$

Finally, the negative charge $-q/2$ is brought to a point a distance a from the second and third charges and $\sqrt{2}a$ from the first charge. The work required is again the sum of the potential differences multiplied by the charge $-q/2$:

$$W_4 = k\left(-\frac{q^2}{2a} - \frac{q^2}{2a} - \frac{q^2}{2\sqrt{2}a}\right).$$

This work is negative, indicating that it would take positive work to remove the negative charge. Adding the work required to bring in the second, third, and fourth charges gives the electrostatic potential energy of the charge distribution:

$$W = W_2 + W_3 + W_4$$

$$= k\left(\frac{q^2}{a} + \frac{q^2}{a} + \frac{q^2}{\sqrt{2}a} - \frac{q^2}{2a} - \frac{q^2}{2a} - \frac{q^2}{2\sqrt{2}a}\right)$$

$$= \frac{kq^2(2\sqrt{2} + 1)}{2\sqrt{2}a}.$$

That this is a positive quantity indicates that the work needed to assemble the three positive charges is greater than the energy gained bringing in the negative charge.

EXERCISE Repeat Example 26-1 for the case when the charge in the upper left corner is changed to $-q$.

Answer: $-3kq^2[1 - 1/(2\sqrt{2})]/a$

Some problems similar to Example 26-1:
1–4, 7, 79

●

26-2 TWO ISOLATED CONDUCTORS

An important charge distribution consists of two isolated conductors carrying equal but opposite charges. Figure 26-3 shows two such conductors, each initially uncharged. Imagine moving a small quantity of charge from one conductor to the other, giving rise to a net positive charge on one and a net negative charge on the other. This results in an electric field and therefore in a potential difference between the conductors. If we try to transfer more charge between the conductors, we must do work traversing this potential difference. The more charge we move, the harder it gets to transfer additional charge. The work it takes to transfer the charge is stored as potential energy of the charge distribution.

It is generally difficult to calculate the stored potential energy for a pair of irregularly shaped conductors like those of Fig. 26-3. An important practical case for which the potential energy may be calculated is a pair of identical, flat, parallel conducting plates whose separation is small compared with their width (Fig. 26-4a). We start with the plates uncharged, and then transfer charge Q from one plate to the other. (In practice, we would accomplish this by connecting the plates to the terminals of a battery.) Charging the plates results in an electric field between them. For closely spaced plates, this field is essentially uniform except very near the edges (Fig. 26-4b), and we may neglect this nonuniform "fringing field."

What is the electric field strength between the plates? In Chapter 24 we used Gauss's law to show that the field near the surface of a conductor carrying surface charge density σ is given by

$$E = \frac{\sigma}{\varepsilon_0}.$$

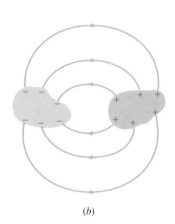

FIGURE 26-3 A pair of isolated conductors. (*a*) Initially they are uncharged, and there is no potential difference between them. (*b*) When they carry opposite charges, there is an electric field and consequently a potential difference between them.

As we discussed in Section 24.6, charge gathers entirely on the facing surfaces of the two plates, giving rise to a charge density of magnitude $\sigma = q/A$, where A is the plate area. So the electric field between the plates is

$$E = \frac{\sigma}{\varepsilon_0} = \frac{q}{\varepsilon_0 A},$$

where q is the magnitude of the charge on either plate. Shouldn't this result be doubled because there are two plates? No! Review the discussion accompanying Figs. 24-32 and 24-33 to convince yourself of this point.

The presence of the electric field means there is a potential difference between the plates. Since the field is uniform, this potential difference is a simple product of the field strength with the distance d between the plates:

$$V = Ed = \frac{qd}{\varepsilon_0 A},$$

where we're now using V rather than $\Delta V_{A \to B}$ for the potential difference.

Now imagine moving an additional very small positive charge dq from the negative to the positive plate. How much work does this take? That depends on the potential difference between the plates, which, as our expression for V shows, depends on how much charge has already been transferred. Because potential difference is work per unit charge, the work dW required to move the charge dq between the plates is

$$dW = V\,dq = \frac{qd}{\varepsilon_0 A}\,dq.$$

Suppose we start with zero net charge on either plate and gradually transfer a total charge Q from one plate to the other. Each dq that we move requires work dW as given above, so the total work is the sum of all the dW's associated with all the small quantities of charge dq that make up Q. In the limit of infinitely many infinitesimal charges dq, this sum becomes an integral, and we have

$$W = \int_0^Q dW = \int_0^Q \frac{qd}{\varepsilon_0 A}\,dq = \frac{d}{\varepsilon_0 A}\int_0^Q q\,dq.$$

That the variable q remains under the integral sign reflects the physical fact that the continually increasing charge on the plates results in an increasing potential difference and therefore makes it harder to move each additional charge dq. Continuing the integration gives

$$W = \frac{d}{\varepsilon_0 A}\frac{q^2}{2}\bigg|_0^Q = \frac{d}{2\varepsilon_0 A}Q^2.$$

Thus the work required to charge the plates increases as the square of the charge Q. The work done in charging the plates ends up as stored potential energy of the final charge distribution, so the stored energy is

$$U = W = \frac{d}{2\varepsilon_0 A}Q^2. \tag{26-2}$$

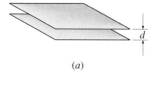

(a)

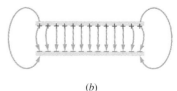

(b)

FIGURE 26-4 (a) A pair of closely spaced conducting plates. (b) Edge-on view of the plates when oppositely charged. For closely spaced plates the electric field is essentially uniform, except very near the edges, and the field outside is very small. Neglecting these small "fringing fields" is a good approximation.

The quadratic dependence of the stored energy on charge suggests that a pair of parallel plates is an excellent device for storing electrostatic energy.

26-3 ENERGY AND THE ELECTRIC FIELD

We have seen that the work required to assemble a distribution of electric charge ends up as electrostatic potential energy. Just where is the energy stored? In Example 26-1, we considered the assembly of four point charges to form a square. Surely the point charges themselves did not change; we only moved them closer together. Similarly, when we charged our pair of metal plates we did not alter the individual charges; we only moved them from one plate to another. What has changed in both these cases? The electric field has changed. In the first case we started with four isolated point charges and an electric field that looked like four isolated point-charge fields. We ended with a new charge distribution whose field did not look at all like a point-charge field.

In the case of the parallel plates, we started with uncharged plates and no electric field. As soon as we began transferring charge from one plate to the other, an electric field appeared between the plates, and this field grew in strength as more charge was transferred.

So where is the energy stored? It is stored in the electric field. As we create or alter a charge distribution, we do work and an altered electric field configuration develops. The work we do in moving the charges ultimately goes into the alteration of the field. If the work done by the applied force is positive, we have added energy to the field. If the work is negative, we have removed energy from the field.

Every electric field represents stored energy. If the field is altered, energy is either accumulated or released, depending on whether the work done is positive or negative. If the field disappears entirely, all its energy is released in some other form. Because electric forces are primarily responsible for the behavior of everyday matter, many seemingly different forms of energy storage really involve electric field energy. When you burn gasoline or metabolize food, for example, you are rearranging the charge distributions we call molecules into new configurations whose electric fields contain less energy (Fig. 26-5).

If electric fields store energy, then the amount of stored energy should depend on the field strength. Since the field strength may vary with position, we describe the stored energy in terms of **energy density,** or energy stored per unit volume. We can readily determine the energy density for our parallel plates. There we found that the field strength is given by $E = Q/\varepsilon_0 A$; solving for Q and using the result in Equation 26-2 for the stored energy gives

$$U = \frac{d}{2\varepsilon_0 A}Q^2 = \frac{d}{2\varepsilon_0 A}(\varepsilon_0 A E)^2 = \tfrac{1}{2}\varepsilon_0 E^2 A d.$$

Our assumption that the plates are very close together allowed us to conclude that the field is very nearly uniform between the plates and essentially zero outside the plates. Therefore, the energy U is stored in the region between the plates, and is distributed uniformly because the field is uniform. The volume

FIGURE 26-5 Combustion involves the rearrangement of atoms into new molecular structures. Energy released in the process comes ultimately from the electric fields associated with the charge distribution in the molecules.

between the plates is just the plate area times the separation, or Ad, and therefore the energy density u_E is given by $u_E = U/Ad$, or

$$u_E = \tfrac{1}{2}\varepsilon_0 E^2. \quad \text{(electric energy density)} \qquad (26\text{-}3)$$

Although we derived this expression for the uniform field between two parallel plates, it is in fact a universal expression that holds for *any* electric field. At any point where an electric field exists, there is stored energy whose density, in J/m^3, is given by Equation 26-3.

The deepest significance of Equation 26-3 lies in its statement that every electric field represents stored energy. As we observe a variety of physical phenomena, from everyday happenings on Earth to events in distant galaxies, we can understand that the driving energy for many of these phenomena comes from the release of energy stored in electric fields.

● EXAMPLE 26-2 ELECTRICAL ENERGY OF A THUNDERSTORM

Electric fields inside a thunderstorm have typical values of 10^5 V/m and get even higher just before electrical energy is unleashed as lightning (Fig. 26-6). The origin of these fields and hence of the energy stored in them is believed to be associated with charge transfer to rising and falling water droplets or ice crystals in the intense updrafts and downdrafts of the thunderstorm. Consider a typical thundercloud that rises to an altitude of 10 km and has a diameter of 20 km. Assuming an average field strength of 10^5 V/m, estimate the total electrostatic energy stored in the cloud. How many gallons of gasoline would you have to burn to release the same amount of energy?

Solution
The energy density is given by Equation 26-3:

$$u_E = \tfrac{1}{2}\varepsilon_0 E^2 = \tfrac{1}{2}(8.85\times10^{-12}\ \mathrm{C^2/N\cdot m^2})(10^5\ \mathrm{V/m})^2$$

$$= 4.4\times10^{-2}\ \mathrm{J/m^3}.$$

(You should verify that the units work out!) We are assuming that this energy density is the same throughout the storm, so we find the total energy by multiplying the energy density by the volume. The storm is roughly cylindrical in shape, so its volume is

$$V = \pi r^2 h = \pi(10\ \mathrm{km})^2(10\ \mathrm{km}) = 3100\ \mathrm{km^3}$$

$$= 3.1\times10^{12}\ \mathrm{m^3}.$$

Then the total stored energy is

$$U = u_E V = (4.4\times10^{-2}\ \mathrm{J/m^3})(3.1\times10^{12}\ \mathrm{m^3}) = 1.4\times10^{11}\ \mathrm{J}.$$

A gallon of gasoline contains about 10^8 J (see Appendix C), so the electrical energy stored in a thunderstorm at any given

FIGURE 26-6 Lightning is the sudden release of energy stored in atmospheric electric fields.

instant is equivalent to about 1000 gallons or 4000 L of gasoline. This comparison is not quite fair to the thunderstorm, though, because its electrical energy is continually dissipated in lightning strikes and at the same time renewed by the violent motion of the air. Problem 77 explores thunderstorm energetics in more detail.

EXERCISE In fair weather, Earth's atmospheric electric field is about 100 V/m. Find the energy stored in each km^3 of the fair-weather atmosphere.

Answer: 44 J

Some problems similar to Example 26-2: 16–18 ●

When the electric field is uniform, as in our thunderstorm example, we can find the stored energy simply by multiplying the energy density by the volume. But when the field changes with position we must resort to calculus. Consider a small volume element dV, so small that the electric field is essentially uniform over this volume. The stored energy dU in the volume element is just the energy density times the volume, or

$$dU = u_E \, dV = \tfrac{1}{2} \varepsilon_0 E^2 \, dV.$$

The total energy U is then the sum of all the dUs. In the limit of infinitesimally small volumes dV and energies dU, this sum becomes an integral:

$$U = \tfrac{1}{2} \varepsilon_0 \int E^2 \, dV, \tag{26-4}$$

where the limits on the integral are chosen to cover the entire region in which the electric field of interest exists.

We derived Equation 26-4 for the electric field energy using our previously determined expression for the work needed to assemble a simple charge distribution. We can also reverse that process, using the electric field of a charge distribution to calculate the energy density and from it the stored energy and therefore the work needed to assemble the distribution. Example 26-3 illustrates this procedure for a case when the energy density varies with position.

● **EXAMPLE 26-3**   A SHRINKING SPHERE

A sphere of radius R_1 carries a total charge Q distributed evenly over its surface (Fig. 26-7a). How much work does it take to shrink the sphere to a smaller radius R_2? Practical applications in which this question might prove important include the behavior of cell membranes, charged bubbles, and raindrops in thunderclouds.

Solution
Shrinking the sphere moves all the charge elements on its surface closer together, and therefore requires positive work. That work is equal to the change in stored electric field energy given by Equation 26-4. With all charge distributed evenly over the sphere's surface, Gauss's law tells us that there is no electric field within the sphere. Because of the spherical symmetry, Gauss's law also tells us that the field outside the sphere is identical to that of a point charge Q at the sphere's center. This means that the field at and beyond the original radius R_1 does not change as we shrink the sphere. What does change is the field between R_1 and R_2 (Fig. 26-7b). Originally this field was zero. After the sphere has shrunk, this region, too, is filled with a point-charge field. This newly created field is the site of the additional energy stored in shrinking the sphere.

Because the point-charge field between R_1 and R_2 changes with position, we must use the integral form 26-4 to calculate

the stored energy. The electric field outside the sphere is that of a point charge:

$$E = \frac{kQ}{r^2},$$

so, from Equation 26-3, the energy density as a function of r is

$$u_E(r) = \tfrac{1}{2} \varepsilon_0 E^2 = \tfrac{1}{2} \varepsilon_0 \left(\frac{kQ}{r^2} \right)^2 = \frac{kQ^2}{8 \pi r^4},$$

where we used $k = 1/4 \pi \varepsilon_0$.

To determine the total stored energy we integrate this energy density over the volume between R_2 and R_1. Because of the spherical symmetry we consider volume elements made of thin spherical shells of thickness dr. Figure 26-8 shows that the volume of the shell is very nearly $dV = 4 \pi r^2 dr$, so Equation 26-4 becomes

$$U = \int_{R_2}^{R_1} u_E \, dV = \int_{R_2}^{R_1} \frac{kQ^2}{8 \pi r^4} 4 \pi r^2 \, dr$$

$$= \frac{kQ^2}{2} \int_{R_2}^{R_1} r^{-2} \, dr = \frac{kQ^2}{2} \left(-\frac{1}{r} \right) \Big|_{R_2}^{R_1}$$

$$= \frac{kQ^2}{2} \left(\frac{1}{R_2} - \frac{1}{R_1} \right).$$

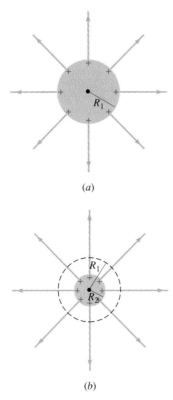

(a)

(b)

FIGURE 26-7 (a) A charged sphere and its electric field. (b) Shrinking the sphere creates new field in the region $R_2 < r < R_1$.

FIGURE 26-8 A thin spherical shell of thickness dr and radius r. Because the shell is very thin, its inner and outer surfaces have essentially the same area, namely $4\pi r^2$. Its volume is therefore $dV = 4\pi r^2\, dr$.

EXERCISE A long coaxial cable consists of an inner cylindrical conductor of radius a and an outer cylindrical conducting shell of radius b (Fig. 26-9). The conductors carry equal but opposite line charge densities $\pm\lambda$. Find the electric energy stored in a length ℓ of this cable.

This is the total energy stored in the new electric field between R_2 and R_1, and is therefore also the work done in shrinking the sphere from R_1 to R_2. If we let R_1 go to infinity, our result becomes the work required to assemble a sphere of radius R_2 carrying surface charge Q, or equivalently, the energy stored in the field of the sphere. Because the stored energy becomes infinite as R_2 approaches zero, our result suggests that the notion of a point charge is an impossible idealization. Problem 90 explores some implications of this result in the theory of elementary particles.

FIGURE 26-9 How much energy is stored in a length ℓ of the cable? (The cable is much longer than the outer conductor radius b.)

Answer: $k\lambda^2\ell\, \ln(b/a)$

Some problems similar to Example 26-3: 21–23, 26, 27

26-4 CAPACITORS

In electrical and electronic equipment, electrical energy is often stored using a pair of charged conductors separated by an insulator. Such a device is called a **capacitor.** Capacitors are typically used for short-term energy storage in situations where it is necessary to store or release electrical energy quickly. Most practical electronic devices, including radio, TV, computers, and audio equipment, would be virtually impossible to construct without capacitors. When you tune a radio, you are adjusting a capacitor. Failure of a capacitor in your car's ignition system could leave you stranded on the highway. And many high-energy experiments in physics and engineering use so much power that, were it not for capacitors, they could not be done without disrupting the supply of electric power to the rest of the world!

In an uncharged capacitor both conductors are neutral. Therefore there is no electric field in the capacitor, and no stored energy. A capacitor is charged by transferring charge (usually electrons) from one conductor to the other. The work required is generally supplied by some other source of electrical energy connected to the capacitor through wires leading to its two conductors. An electric field develops as a result of the charge separation, and the stored energy resides in this field. Although we speak of a capacitor as being charged, the capacitor remains overall electrically neutral, in that the net charge on the whole capacitor is zero. But the two individual conductors making up a charged capacitor are not neutral: one carries a charge $+Q$, the other $-Q$. When we say that the charge on a capacitor is Q, we really mean that Q is the magnitude of the charge on either conductor.

Once a capacitor is charged there is an electric field and therefore a potential difference between its two conductors. As the charge is increased the potential difference increases proportionately. Conversely, imposing a potential difference on a capacitor (by connecting it to a battery, for example) causes the capacitor to become charged in proportion to the potential difference imposed. The ratio of charge to potential difference is characteristic of a given capacitor and is called its **capacitance:**

$$C = \frac{Q}{V}.$$
(26-5)

Here Q is the magnitude of the charge on either conductor and V the potential difference (or voltage) between the conductors. Clearly the units of capacitance are coulombs/volt. One coulomb/volt is given the name **farad** (F), in honor of the nineteenth century scientist Michael Faraday. One farad is so huge a capacitance that the smaller units microfarad (10^{-6} F; abbreviated μF) and picofarad (10^{-12} F; abbreviated pF and often pronounced "puff") are widely used.

Capacitance depends on the physical construction of a capacitor—the shapes of its two conductors, their separation, and the choice of insulating material between them. Although Q and V enter the defining relation 26-5, capacitance is a constant. If V is increased, Q increases proportionately, maintaining the constant ratio C that characterizes the capacitor.

Any arrangement of two insulated conductors constitutes a capacitor. Practical capacitors are manufactured in a variety of configurations. Often they are made from two long strips of aluminum foil separated by thin layers of plastic or paper. This foil "sandwich" is then rolled into a compact cylinder, wires attached, and the whole assembly covered with a protective coating. Another common arrangement is the variable capacitor, whose configuration can be altered to change its capacitance. This change can be accomplished mechanically or, as in many modern electronic devices, electrically. Very large capacitances are achieved with so-called electrolytic capacitors, in which a thin insulating layer develops chemically under the influence of the applied voltage. Figure 26-10 shows some typical capacitors.

Calculating Capacitance

Capacitance is defined through Equation 26-5 as the ratio of charge to potential difference. To calculate the capacitance of a particular configuration of two conductors, we assume there is a charge Q on the capacitor and calculate the corresponding potential difference. Because the capacitance depends only on the physical configuration of the capacitor, it doesn't matter what value we assume for the charge—that will cancel when we take the ratio of charge to potential difference. For two irregularly shaped conductors we are back to the problem of determining the potential difference between the conductors from the distribution of charge on their surfaces. When the capacitor design includes sufficient symmetry, this calculation becomes straightforward.

By far the most important capacitor design is the parallel-plate configuration. In Section 26-2 we examined such a capacitor in some detail, although at that time we did not call the configuration a capacitor. There we found that the potential difference between plates of area A separated by a distance d is

$$V = \frac{Qd}{\varepsilon_0 A}.$$

FIGURE 26-10 Typical capacitors. The large blue unit is an 18-mF electrolytic capacitor. At top right is an air-insulated variable capacitor in which a set of metal plates rotates with respect to fixed plates in order to change the capacitance. The remaining smaller capacitors range from 43 pF to 10 μF.

Solving for the ratio $C = Q/V$ gives

$$C = \frac{Q}{V} = \frac{\varepsilon_0 A}{d}. \qquad \text{(parallel-plate capacitor)} \qquad (26\text{-}6)$$

Equation 26-6 gives the capacitance of a parallel-plate capacitor in terms of the universal constant ε_0 and factors that describe the physical configuration of the capacitor. (Strictly speaking, this expression holds only for capacitors insulated by vacuum. Later we will modify Equation 26-6 to account for other insulating materials.) Note that neither charge nor potential difference enters the final expression for capacitance, showing that the capacitance is indeed a constant. Equation 26-6 suggests that the way to make a capacitor with large capacitance is to use two plates of large area but small separation. Incidentally, Equation 26-6 shows that the units of ε_0 may be expressed as farads/meter (F/m); see Problem 32.

● **EXAMPLE 26-4** A PARALLEL-PLATE CAPACITOR

A capacitor consists of two circular metal plates of radius 10 cm separated by an air gap of 5.0 mm (Fig. 26-11). What is its capacitance? When a 12-volt battery is connected to the capacitor, how much charge appears on the plates?

Solution

Since the plate spacing is much smaller than the plate size, Equation 26-6 holds and the capacitance is

$$C = \frac{\varepsilon_0 A}{d} = \frac{\varepsilon_0 \pi r^2}{d} = \frac{(8.85 \times 10^{-12} \text{ F/m})(\pi)(0.10 \text{ m})^2}{5.0 \times 10^{-3} \text{ m}}$$

$$= 5.6 \times 10^{-11} \text{ F} = 56 \text{ pF}.$$

Equation 26-5 defines capacitance as the ratio of charge to potential difference. We can rewrite this defining relation to solve for the charge:

$$Q = CV = (56 \text{ pF})(12 \text{ V}) = 670 \text{ pC}.$$

What this really means, of course, is that the positive plate carries 670 pC and the negative plate −670 pC. Overall, the capacitor remains neutral. Note that by working with the capacitance in pF, the charge automatically comes out in pC.

EXERCISE A parallel-plate capacitor is to be made from two square pieces of aluminum foil each 8.0 cm on a side. (a) What should be the spacing between them if the capacitance is to be 47 pF? (b) What applied voltage will put ±95 nC on the plates?

Answers: (a) 1.2 mm; (b) 2.0 kV

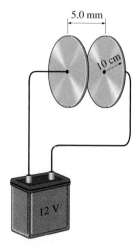

FIGURE 26-11 A parallel-plate capacitor connected to a battery. Drawing is not to scale (Example 26-4).

Some problems similar to Example 26-4: 33, 34

● **EXAMPLE 26-5** A CYLINDRICAL CAPACITOR

A capacitor consists of two long concentric metal cylinders of length L, as shown in Fig. 26-12. The inner and outer cylinders have radii a and b, respectively. What is the capacitance?

Solution
Equation 26-6 does not apply to this configuration because the field between the cylinders is not uniform. To find the capacitance, we need a relation between charge and potential difference for the cylindrical configuration. In Example 25-4, we found that the potential difference between two points outside a charge distribution with line symmetry can be written

$$V(a) - V(b) = \frac{\lambda}{2\pi\varepsilon_0} \ln\left(\frac{b}{a}\right),$$

where λ is the line charge density. Because our capacitor is long compared with its diameter, this expression is a good approximation to the potential difference due to the field of the inner conductor. What about the outer conductor? Recall (Example 24-5) that the electric field inside an empty, hollow pipe is zero; therefore, the outer conductor contributes nothing to the electric field or the potential difference between the conductors. If the magnitude of the charge on either conductor is Q, then the line charge density is $\lambda = Q/L$, and our expression for potential difference becomes

$$V = V(a) - V(b) = \frac{Q}{2\pi\varepsilon_0 L} \ln\left(\frac{b}{a}\right).$$

Capacitance is the ratio of charge to potential difference, so we have

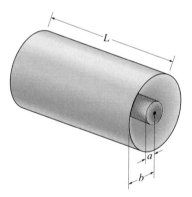

FIGURE 26-12 A cylindrical capacitor (Example 26-5).

$$C = \frac{Q}{V} = \frac{Q}{(q/2\pi\varepsilon_0 L)\ln(b/a)} = \frac{2\pi\varepsilon_0 L}{\ln(b/a)}. \quad (26\text{-}7)$$

Does this result make sense? We already found that the capacitance of a parallel-plate capacitor increases with increasing plate area or with decreasing plate separation. With the cylindrical capacitor we can increase the area of both conductors by increasing the length L of the capacitor, and indeed Equation 26-7 shows the capacitance increasing proportionately. We can decrease the spacing of the conductors by making the radii a and b more nearly equal. This makes b/a closer to one, and $\ln(b/a)$ closer to zero, again increasing the capacitance. Although the geometries of the cylindrical and parallel-plate capacitors are quite different, the same physical considerations apply to both: a large capacitance is achieved with large conductor areas and small separation. When the separation is very small, the curvature of the cylindrical capacitor cannot matter, and Equation 26-7 should reduce to Equation 26-6 for the parallel-plate capacitor (see Problem 84).

EXERCISE A conducting sphere of radius R is enclosed in a concentric spherical conducting shell of radius $\frac{3}{2}R$. What is the capacitance of this configuration?

Answer: $3R/k$

Some problems similar to Example 26-5: 35, 36, 67

26-5 ENERGY STORAGE IN CAPACITORS

In Section 26-3 we found that any electric field represents stored energy. The example that guided us to that conclusion was a parallel-plate capacitor. For that configuration, the stored energy U is given by Equation 26-2:

$$U = \frac{d}{2\varepsilon_0 A} Q^2.$$

Since $\varepsilon_0 A/d$ is the capacitance of the parallel-plate capacitor, this stored energy may be written

$$U = \frac{Q^2}{2C}. \qquad \text{(energy in a capacitor)} \qquad (26\text{-}8a)$$

It is usually easier to measure voltage than charge. To express the stored energy in terms of voltage, we can solve the equation defining capacitance, $C = Q/V$, for Q and use the result in Equation 26-8a:

$$U = \frac{Q^2}{2C} = \frac{(CV)^2}{2C} = \tfrac{1}{2}CV^2. \qquad \text{(energy in capacitor)} \qquad (26\text{-}8b)$$

Although Equations 26-8a and b were derived for a parallel-plate capacitor, they hold for any capacitor regardless of its configuration (see Problem 49).

That the stored energy depends on the *square* of the potential difference implies that more energy can be stored in a small capacitor at high voltage than in a larger one at low voltage. Practically, the difficulties of handling high voltages mitigate this conclusion somewhat, but the fact remains that the stored energy in a capacitor increases rapidly with increasing voltage.

Large capacitors can store energy for a long time. In VCRs, for example, capacitors of several farads are used to maintain the program memory in the event of power failures lasting as much as an hour or so. Energy storage in capacitors has important safety implications since large capacitors may retain dangerous voltages even after the equipment containing them is turned off (Fig. 26-13). TVs, stereos, computers, and other electronic devices use large capacitors to produce steady direct current, and before beginning work on such equipment technicians often discharge these capacitors by touching a screwdriver across their terminals. The resulting spark and loud noise are testimony to the amount of stored energy (Fig. 26-14).

FIGURE 26-13 The large black cylinders are 4700-μF electrolytic capacitors used in the power supply of a stereo amplifier.

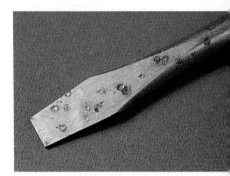

FIGURE 26-14 This screwdriver has been used repeatedly to discharge a large capacitor. Steel vaporized in each of the pitted areas.

● **E X A M P L E 2 6 - 6** WHICH CAPACITOR?

A 100-μF capacitor can tolerate a maximum potential difference of 20 V, while a 1.0-μF capacitor can tolerate 300 V. Which can store the most energy? The most charge?

Solution

Applying Equation 26-8b with V set to the maximum tolerable voltage gives

$$U_{100\,\mu F} = \tfrac{1}{2}CV^2 = \tfrac{1}{2}(100\ \mu F)(20\ V)^2 = 20\times10^3\ \mu J = 20\ mJ$$

and

$$U_{1.0\,\mu F} = \tfrac{1}{2}CV^2 = \tfrac{1}{2}(1.0\ \mu F)(300\ V)^2$$

$$= 45\times10^3\ \mu J = 45\ mJ\,.$$

Because of its higher voltage rating, the smaller capacitor can store more energy. On the other hand, the larger capacitor stores more charge, as shown by solving the defining relation $C = Q/V$:

$$Q_{100\,\mu F} = CV = (100\ \mu F)(20\ V) = 2000\ \mu C = 2.0\ mC$$

and

$$Q_{1.0\,\mu F} = CV = (1.0\ \mu F)(300\ V) = 300\ \mu C = 0.30\ mC\,.$$

Again, these numbers refer to the magnitude of the charge on each plate; overall, each capacitor remains neutral.

TIP Capacitance Isn't Capacity A capacitor is not like a bucket that can hold a fixed amount of water. Instead the charge and energy that a capacitor can hold depend on both the capacitance and on the applied voltage. In comparing the "capacities" of two capacitors one needs to know *both* the capacitance and the voltage that will be applied.

E X E R C I S E The "memory" capacitor in a VCR stores 25 J of energy with a potential difference of 3.5 V. (a) What is its capacitance? (b) What is the magnitude of the charge on each plate?

Answers: (a) 4.1 F; (b) 14 C

Some problems similar to Example 26-6: 38–41, 43, 44 ●

■ **A P P L I C A T I O N** CAMERA FLASHES, TOILET FLUSHES, AND LASER FUSION

You've probably used a camera equipped with an electronic flash. The flash unit contains a special tube filled with xenon gas. When a large potential difference is applied across the tube, dielectric breakdown occurs and the xenon suddenly ionizes. Recombination of electrons with xenon ions then results in a bright pulse of white light. After a flash picture has been taken the photographer must wait a while—typically around 10 seconds—before the flash is ready again. Why is this? Although the total energy used by the flash is small, during the short interval that the xenon is being ionized the rate of energy use far exceeds the maximum power output of the camera's battery (see Problem 45). So the battery is used to charge a capacitor at a slow rate. Once the capacitor is charged its stored energy is dumped suddenly into the flashtube, providing the short burst of high power needed to give the intense light. The flashtube cannot be used again until the capacitor is recharged.

This situation is exactly analogous to one you encounter every day in household plumbing. Flushing a toilet requires a large amount of water in a short time—far more than typical household plumbing could supply in that time. So the water is accumulated gradually in the toilet tank, then suddenly dumped when needed for flushing. In this analogy the household plumbing, with its relatively narrow pipes, corresponds to the small battery of the camera flash. The toilet tank, which gradually accumulates water, corresponds to the capacitor,

which gradually accumulates electrical energy. Of course you need to wait between flushes for the toilet tank to fill, just as you need to wait between flash pictures for the capacitor to charge.

Professional photographers needing to take flash pictures in rapid succession often carry around a large, heavy battery pack capable of supplying the flash power directly. Similarly, institutional buildings with large, high-pressure water pipes often have toilets without tanks that can be flushed in rapid succession.

The simple example of the camera flash, scaled up many times in size, shows how very large amounts of power may be obtained briefly for industrial or scientific applications. Indeed, some experiments involving high-power pulsed lasers for nuclear fusion and ballistic missile defense research require more power than all the world's electric generating stations produce (Fig. 26-15). The required energy is accumulated in huge banks of capacitors, which are suddenly discharged to provide energy to the laser. Think here about the difference between energy and power! The pulsed laser is only on for about 10^{-9} seconds, so although it consumes energy at an enormous rate while on, it does not use all that much total energy (see Problem 46). The laser is not fired very often (at least in today's test configurations), so that there is plenty of time to charge the capacitors. The *average* power consumption of the experiment is modest.

(a)

(b)

FIGURE 26-15   (a) This huge capacitor bank stores 60 MJ of energy to drive the Nova laser fusion experiment at Lawrence Livermore National Laboratory. A fraction of the stored energy can be delivered to the lasers in less than 1 ns. (b) A 2.3-F capacitor bank for a laser fusion experiment at Los Alamos National Laboratory.

26-6 CONNECTING CAPACITORS TOGETHER

The most important consideration in using a capacitor is whether its capacitance is right for the particular application. But it's also important to respect a capacitor's **working voltage,** or the maximum potential difference that should be applied across the capacitor if dielectric breakdown is to be avoided.

Large capacitances are most easily achieved using small plate separations, as Equation 26-6 suggests. But a small plate separation implies a large electric field for a given voltage. Thus in practical capacitors there is a trade-off between capacitance and working voltage. High working voltage and high capacitance together require large plate separation to keep the electric field small and avoid dielectric breakdown, while at the same time requiring large plate area to keep the capacitance up. Thus large, high-voltage capacitors are physically bulky and expensive to build. Often economics as well as physics dictates the final design of a circuit involving capacitors.

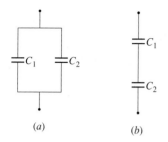

(a)   *(b)*

FIGURE 26-16 Connecting capacitors togther. (*a*) Parallel; (*b*) series. $\perp$ is the standard circuit symbol for a capacitor.

When a single capacitor with a desired combination of capacitance and working voltage is not available, we can often obtain the desired combination by connecting two or more capacitors together. There are only two ways to connect two capacitors (and indeed any electronic components with only two wires coming from them). Called **parallel** and **series,** these two possible connections are shown in Fig. 26-16. We would like to know the equivalent capacitance of each combination.

Capacitors in Parallel

Consider first the parallel combination of Fig. 26-16*a*. If we impose a potential difference *V* across the two wires coming from the combination, what will be the potential difference across each capacitor? The key to answering this question is the recognition that the wires connecting the capacitors are conductors and that in electrostatic equilibrium there can be no potential difference between any points connected to the same wire.* Thus all points connected directly together—including the top plates of each capacitor—are at the same potential. Similarly, the bottom plates and the wires connecting them are all at the same potential. Therefore, the potential differences across the two capacitors are equal. This is a very important point in the practical understanding of the electric circuits, and applies to any two circuit components that are connected in parallel:

> **The potential differences across two circuit components in parallel are equal.**

Recognizing this simple fact is essential in developing your understanding of electric circuits!

The equivalent capacitance is the ratio of the total charge on both capacitors to the voltage across the parallel combination. Solving the defining relation $C = Q/V$ for charge, we can write the charges on the two capacitors as

$$Q_1 = C_1 V \quad \text{and} \quad Q_2 = C_2 V.$$

The potential difference *V* is the *same* in both cases because the capacitors are connected in *parallel*. Thus the total charge is

$$Q = Q_1 + Q_2 = C_1 V + C_2 V = (C_1 + C_2)V.$$

Taking the ratio of total charge *Q* to the voltage *V* across the parallel combination gives the equivalent capacitance:

$$C = C_1 + C_2. \quad \text{(parallel capacitors)} \quad (26\text{-}9a)$$

Equation 26-9a is frequently stated as "capacitors in parallel add." You can understand this result physically by considering two parallel-plate capacitors with equal spacing. Connecting them in parallel amounts to adding their plate areas, giving a larger capacitance. Although we derived Equation 26-9a for two

---

* Even when we relax the equilibrium assumption, this conclusion will still hold in the approximation that the wires are perfect conductors.

parallel capacitors, the result that parallel capacitances add is easily extended to three or more capacitors (see Problem 56):

$$C = C_1 + C_2 + C_3 + \cdots \qquad \text{(parallel capacitors)} \qquad \text{(26-9b)}$$

What about the working voltage of the parallel combination? Both capacitors experience the full potential difference V, so the working voltage of the combination is that of whichever capacitor has the lower working voltage.

Capacitors in Series

Suppose we charge the series capacitor system of Fig. 26-16b, putting $+Q$ on the upper plate of C_1 and $-Q$ on the lower plate of C_2. The positive charge on the uppermost plate attracts $-Q$ to the lower plate of C_1 and the negative charge $-Q$ on the lowermost plate attracts $+Q$ to the upper plate of C_2 (Fig. 26-17). Note that this leaves the middle two plates together with zero net charge, which must be the case since these two plates are not connected to any external source of charge. With charge of magnitude Q on every plate, we can conclude that

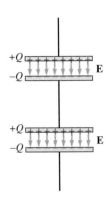

FIGURE 26-17 Capacitors in series carry the same charge.

▎**Capacitors in series carry the same charge.**

To find the equivalent capacitance, we first solve the relation $C = Q/V$ for the voltages across the two capacitors:

$$V_1 = \frac{Q}{C_1} \quad \text{and} \quad V_2 = \frac{Q}{C_2}.$$

Here there is no need to label the Q's since series capacitors carry the same charge. But now the voltages need not be the same. Since the electric fields in the two capacitors point the same way (Fig. 26-17), the voltage across the series combination is just

$$V = V_1 + V_2.$$

Inserting our expressions for the individual potential difference gives

$$V = \frac{Q}{C_1} + \frac{Q}{C_2} = Q\left(\frac{1}{C_1} + \frac{1}{C_2}\right).$$

Dividing by Q gives V/Q, which the relation $C = Q/V$ shows is the reciprocal of the equivalent capacitance:

$$\frac{1}{C} = \frac{1}{C_1} + \frac{1}{C_2}. \qquad \text{(series capacitors)} \qquad \text{(26-10a)}$$

This result is frequently described by saying that "capacitors in series add reciprocally." The result is easily extended to three or more capacitors (see Problem 56):

$$\frac{1}{C} = \frac{1}{C_1} + \frac{1}{C_2} + \frac{1}{C_3} + \cdots \qquad \text{(series capacitors)} \qquad \text{(24-10b)}$$

When there are just two capacitors, combining reciprocals over a common denominator gives

$$C = \frac{C_1 C_2}{C_1 + C_2}. \qquad \text{(2 series capacitors)} \qquad (26\text{-}10c)$$

Equations 26-10 show that the combined capacitance of two series capacitors is less than the capacitance of either. You can make physical sense of this by considering parallel-plate capacitors with equal plate areas. Putting them in series effectively adds the plate separations of the two capacitors, yielding a smaller overall capacitance.

What about the voltage rating of the series combination? The full applied voltage V is the sum of the voltages across each capacitor, so each can be rated for less than the full applied voltage. The fraction of the applied voltage that appears across each capacitor depends on the relative capacitances.

● EXAMPLE 26-7 CONNECTING CAPACITORS

(a) Find the equivalent capacitance of the combination shown in Fig. 26-18a. (b) If the maximum voltage applied between points A and B is 100 V, what should be the working voltage of C_2?

Solution

The way to handle circuit problems like this one is to reduce the circuit to a simpler one by recognizing combinations of series and parallel components, as shown in Fig. 26-18. Here C_3 and C_4 are in parallel, so they add to give an equivalent capacitance of 4.0 μF; the circuit then looks like Fig. 26-18b. This parallel combination C_{34} is in series with C_2; applying Equation 26-10c gives

$$C_{234} = \frac{C_2 C_{34}}{C_2 + C_{34}} = \frac{(12 \ \mu\text{F})(4.0 \ \mu\text{F})}{12 \ \mu\text{F} + 4.0 \ \mu\text{F}} = 3.0 \ \mu\text{F}.$$

Now the circuit looks like Fig. 26-18c, with C_1 and C_{234} in parallel. The equivalent capacitance of the entire circuit is their sum, or 7.0 μF (Fig. 26-18d).

To find the working voltage needed for C_2, we need to know the voltage across C_2 with 100 V across the entire combination. Since C_1 and the combination C_{234} are in parallel and connected to points A and B, both these capacitors experience the full 100 V. We can then use the defining relation $C = Q/V$ to find the charge on C_{234}:

$$Q_{234} = C_{234} V = (3.0 \ \mu\text{F})(100 \ \text{V}) = 300 \ \mu\text{C}.$$

But C_{234} is the series combination of C_2 and C_{34} (Fig. 26-18b,c), and the charge on series capacitors is the same. So $Q_2 = 300 \ \mu$C as well. We again use the defining relation $C = Q/V$, now to find the voltage across C_2:

$$V_2 = \frac{Q_2}{C_2} = \frac{300 \ \mu\text{C}}{12 \ \mu\text{F}} = 25 \ \text{V}.$$

This is the required working voltage for C_2.

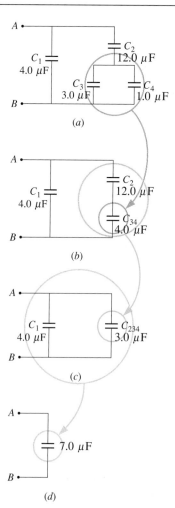

FIGURE 26-18 What is the equivalent capacitance (Example 26-7)? Analyzing the circuit involves reducing parallel and series combinations using Equations 26-9 and 26-10.

TIP Analyzing Circuits This example illustrates a very general approach to circuit analysis. You reduce the circuit to its simplest form by recognizing series and parallel combinations, then gradually build back up to find the quantity of interest. The reduction is essential, even when you're only interested in what's happening with one component. Here, for instance, we wanted to know the voltage on C_2 with 100 V across the whole combination. But to find that we had to analyze the entire circuit, and only then could we focus on C_2.

TIP Recognize Series and Parallel Parallel components have their two ends, respectively, connected *directly* together, as in Fig. 26-16a. Series components are connected in such a way that if you

imagine moving through one component, then the *only* place you can go is into the next component (Fig. 26-16b). Many connections are neither; C_2 and C_3 in Fig. 26-18a, for example, are *not* in series because after you go through C_2 the circuit splits and you could go through either C_3 or C_4. And C_1 and C_2 are *not* in parallel; even though their top plates are connected directly together, their bottom plates are not. The series and parallel formulas we've derived apply *only* to true series and parallel combinations.

EXERCISE (a) With $V_{AB} = 100$ V in Fig. 26-18, what should be C_3's working voltage? (b) What will be the charges on C_1, C_3, and C_4?

Answers: (a) 75 V; (b) $Q_1 = 400$ μC, $Q_3 = 225$ μC, $Q_4 = 75$ μC

Some problems similar to Example 26-7: 52, 54, 57, 58   ●

26-7 CAPACITORS AND DIELECTRICS

The insulating material between the plates of a capacitor serves several purposes. It maintains physical separation of the plates and minimizes charge leakage. Its molecular properties also influence the capacitance. In Chapter 23 we found that electric dipoles tend to align with an applied electric field, and we defined **dielectrics** as materials whose molecules behave as dipoles. The molecular dipole moments may be intrinsic to the molecules, or may be induced by an applied electric field, as we showed in Fig. 23-31. Essentially all insulators are dielectrics.

Suppose we have a parallel-plate capacitor charged to some voltage V_0, with air or vacuum between its plates. What happens if we insert a slab of dielectric material that fills the space between the plates? Figure 26-19 shows that molecular dipoles align with the field arising from the charge on the capacitor plates and that therefore the fields of the dipoles themselves *oppose* the capacitor field. The result is a reduction in the net electric field between the plates. How much reduction depends on details of molecular structure and molecular interactions; empirically, though, we find that a given material may be characterized by its **dielectric constant,** κ, which describes the reduction in field. If E_0 is the original field, then the field after insertion of the dielectric will be decreased by a factor $1/\kappa$: $E = E_0/\kappa$.

If the capacitor is not connected to anything, then there's no way for the charge on its plates to change. But the field between the plates has been reduced, and therefore the potential difference $V = Ed$ has decreased by the same factor $1/\kappa$; $V = V_0/\kappa$. But capacitance is the ratio of charge to voltage, so with the dielectric in place we have

$$C = \frac{Q}{V} = \frac{Q}{(V_0/\kappa)} = \frac{\kappa Q}{V_0} = \kappa C_0, \qquad (26\text{-}11)$$

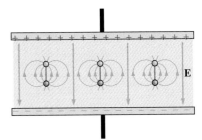

FIGURE 26-19 A capacitor with a dielectric between its plates. Molecular dipoles align with the capacitor's electric field, and their own fields point opposite to the original field. The result is a reduction in field strength.

▲ **TABLE 26-1** PROPERTIES OF SOME COMMON DIELECTRICS

DIELECTRIC MATERIAL	DIELECTRIC CONSTANT	BREAKDOWN FIELD (kV/mm)
Air	1.0006	3
Aluminum oxide	8.4	670
Glass (Pyrex)	5.6	14
Mica	5.4	100
Neoprene	6.9	12
Paper	3.5	14
Plexiglass	3.4	40
Polyethylene	2.3	50
Polystyrene	2.6	25
Quartz	3.8	8
Tantalum oxide	26	500
Teflon	2.1	60
Water	78	—

where $C_0 = Q/V_0$ is the original capacitance. Thus insertion of the dielectric increases the capacitance by a factor κ.

Capacitors are among the most difficult electronic components to miniaturize, so the ongoing revolution in microelectronics has spurred a search for suitable dielectrics with large dielectric constants, good insulating properties, and high breakdown fields. Exotic materials like tantalum oxide and strontium titanate have become widely used in recent years because of their high dielectric constants. In a few unusual applications water, with its dielectric constant of 78, is used as a dielectric. In high-energy experiments where it is necessary to store a large amount of energy for a short time, water's large dielectric constant outweighs the disadvantage of poor insulating quality. Table 26-1 lists dielectric constants and breakdown fields of selected materials.

In addition to helping build better capacitors, the relation between capacitance and dielectric constant serves as a useful probe of the structure of matter. Introducing a dielectric material between capacitor plates lowers the potential difference and therefore allows us to calculate the dielectric constant. This, in turn, gives information about the density and structure of the individual molecular dipoles. Conversely, we can use the measured dielectric constant to help identify an unknown material.

● **EXAMPLE 26-8** CAPACITORS AND DIELECTRICS

An air-insulated capacitor is charged by connecting it to a 12-V battery. The battery is then disconnected. When the space between capacitor plates is filled with an unknown plastic, the voltage between the plates drops to 4.6 V. What is the unknown material? If the plate spacing is 0.10 mm, how much voltage can the capacitor withstand with this material between its plates?

Solution

The voltage has dropped by a factor $1/\kappa = 4.6/12 = 1/2.6$. From Table 26-1, we see that a plastic with $\kappa = 2.6$ is polystyrene. With a dielectric breakdown field of 25 kV/mm,

the 0.1-mm-thick piece of polystyrene can withstand 2.5 kV. The rated working voltage would actually be lower, to allow a margin of safety.

EXERCISE An air-insulated capacitor with $C = 25 \ \mu\text{F}$ is connected to a 10-V battery and the battery is left connected as a quartz slab is inserted to fill the space between the plates. Find the charge on the capacitor (a) before and (b) after the slab is inserted. *Hint:* The battery maintains a fixed 10 V across the plates, but now charge can move from the battery to the plates.

Answers: (a) 250 μC; (b) 950 μC

Some problems similar to Example 26-8: 64–66 ●

What happens to the energy stored in a capacitor when a dielectric is inserted between its plates? The dielectric increases the capacitance by a factor κ, but it also decreases the potential difference by the same factor. If the energy is initially $U_0 = \frac{1}{2} C_0 V_0^2$, then after the dielectric is inserted it becomes

$$U = \tfrac{1}{2}(\kappa C_0)\left(\frac{V_0}{\kappa}\right)^2 = \frac{1}{2\kappa} C_0 V_0^2 = \frac{U_0}{\kappa}. \tag{26-12}$$

Since $\kappa > 1$, the energy has decreased. Where has it gone? As the dielectric moves into the capacitor, the electric field causes charge separation and rotation of the molecular dipoles. The dipoles thus gain energy from the field. In a solid the molecules interact strongly, and the energy is quickly dissipated as heat. By writing Equation 26-12 explicitly for a parallel-plate capacitor, you can show that the energy density in the presence of a dielectric is

$$u_E = \tfrac{1}{2} \kappa \varepsilon_0 E^2 \tag{26-13}$$

(see Problem 70). Here E is the field averaged over many molecules. The factor κ in Equation 26-13 reflects the presence of electric fields on the microscopic scale that are associated with the stretching and rotation of molecules.

That the energy of a capacitor is lower with a dielectric inserted suggests that a force acts to propel the dielectric slab into the capacitor. Figure 26-20 shows that this force originates in something we have heretofore intentionally ignored—the nonuniform fringing field beyond the plates of the capacitor. This nonuniform field acts on the dipoles in the dielectric to produce a net force toward the interior of the capacitor (see Problem 81).

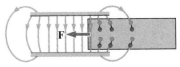

FIGURE 26-20 The nonuniform fringing field outside the capacitor is stronger nearer the plates and therefore results in a net force on the molecular dipoles in the dielectric slab. As a result, the dielectric is pulled into the capacitor.

CHAPTER SYNOPSIS

Summary

1. The work required to assemble a charge distribution is stored as the **electrostatic potential energy** of the distribution. Electrostatic potential energy resides in the electric field. Whenever an electric field is altered, energy is added to or removed from the field.
   a. The **electric field energy density** is given by

   $$u_E = \tfrac{1}{2} \varepsilon_0 E^2,$$

   where the SI units of u_E are J/m³. In a dielectric material, the energy density includes a factor of the dielectric constant κ.
   b. The electrostatic potential energy of a charge distribution may be determined either by computing the work required to assemble the individual charges of the distribution or, knowing the electric field of the distribution, by integrating the energy density over the volume containing the field:

   $$U = \int u_E \, dV = \tfrac{1}{2} \varepsilon_0 \int E^2 \, dV.$$

2. A **capacitor** is an arrangement of two conductors separated by an insulator. Transferring charge from one conductor to the other results in an electric field in the region between the conductors, and energy is stored in the field.
   a. The **capacitance** of a capacitor is the ratio of the charge to the potential difference between its conductors:

   $$C = \frac{Q}{V}.$$

   The capacitance of a parallel-plate capacitor is given by

   $$C = \frac{\varepsilon_0 A}{d},$$

   where A is the plate area and d the spacing. The capacitance of other configurations may be determined by assuming a charge, computing the associated potential difference, and taking the ratio $C = Q/V$.
   b. The energy stored in a capacitor depends on the capacitance and on the square of the potential difference:

   $$U = \tfrac{1}{2} C V^2.$$

c. Capacitors may be connected in **series** or **parallel.** The capacitances of parallel capacitors add:

$$C = C_1 + C_2 + C_3 + \cdots. \qquad \text{(parallel capacitors)}$$

The capacitances of series capacitors add reciprocally:

$$\frac{1}{C} = \frac{1}{C_1} + \frac{1}{C_2} + \frac{1}{C_3} + \cdots. \qquad \text{(series capacitors)}$$

d. The **working voltage** of a capacitor is the maximum potential difference that can be applied across the capacitor without risk of dielectric breakdown in the insulating material.

e. The **dielectric constant,** κ, of the insulating material used in a capacitor affects the capacitance, increasing it by a factor of the dielectric constant.

Terms You Should Understand

(Pairs are closely related terms whose distinction is important; number in parentheses is chapter section where term first appears.)

electrostatic potential energy (26-1)
energy density (26-3)
capacitor (26-4)
capacitance (26-4)
working voltage (26-6)
series, parallel (26-6)
dielectric constant (26-7)

Symbols You Should Recognize

U, u (26-2, 26-3)
C (26-4)
κ (26-7)

Problems You Should Be Able to Solve

calculating the work needed to assemble a distribution of discrete charges (26-1)
calculating electrostatic energy by integrating electric field energy density (26-3)
determining capacitance of simple capacitor configurations (26-4)
evaluating energy stored in capacitors (26-5)
analyzing parallel and series capacitor combinations (26-6)
determining the effect of dielectric materials in capacitors (26-7)

Limitations to Keep in Mind

Equation 26-6 for a parallel-plate capacitor is an approximation that neglects fringing fields at the plate edges; the approximation is good for capacitor plates much larger than their spacing.
Equation 26-10c applies *only* to *two* capacitors in series; with more capacitors Equation 26-10b must be used.

QUESTIONS

1. Two positive point charges are initially infinitely far apart. Is it possible, using only a finite amount of work, to move them until they are located a small distance d apart?

2. Two positive point charges are initially a small distance d apart. Is it possible, using a finite amount of work, to move them together until there is no separation between them?

3. The work required to assemble a certain charge distribution is exactly zero. Does this mean the assemblage of charges is in static equilibrium under the influence of the electric force alone? Explain.

4. How does the energy density a certain distance from a negative point charge compare with the energy density the same distance from a positive point charge of equal magnitude?

5. A dipole consists of two equal but opposite charges. Is the total energy stored in the field of the dipole zero? Why or why not?

6. Charge is spread over the surface of a balloon. The balloon is then allowed to expand. What happens to the energy of the electric field? If it is reduced, where does it go? If it is increased, where does the extra energy come from?

7. Why doesn't the superposition principle hold for electric field energy densities? That is, if you double the field strength at some point, why don't you simply double the energy density as well?

8. A student argues that the total energy associated with the electric field of a charged sphere must be infinite because its field extends throughout an infinite volume. Criticize this argument.

9. A capacitor is said to carry a charge Q. What is the net charge on the entire capacitor?

10. Does the capacitance of a capacitor describe the maximum amount of charge it can hold, in the same way that the capacity of a bucket describes the maximum amount of water it can hold? Explain and compare the meanings of capacitance and capacity.

11. A cylinder for storing compressed gas has a fixed volume, yet knowing this volume does not tell how much gas the cylinder can hold. Why not? How is the cylinder like a capacitor? Form analogies between quantities used in describing a capacitor and the amount of gas in the cylinder, the cylinder pressure, and the maximum pressure the cylinder can withstand.

12. A capacitor of capacitance C is charged to a potential difference V and carries charge Q. Why isn't the stored energy given simply by CV^2? After all, the work required to move a charge Q against a potential difference V is QV, and $C = Q/V$, so $Q = CV$.

13. Is a force needed to hold the plates of a charged capacitor in place? Explain.

14. Why do we say that capacitance depends only on the physical configuration of conductors making up a capacitor, not on the charge or potential difference, and yet we define capacitance as $C = Q/V$?

15. Why can't useful capacitors of arbitrarily large capacitance be made by simply reducing the spacing between parallel plates?

16. A solid conducting slab is inserted between the plates of a capacitor, as shown in Fig. 26-21. Does the capacitance increase, decrease, or remain the same?

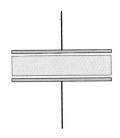

FIGURE 26-21   Question 16.

17. Why is a capacitor needed for energy storage in a camera flash? After all, the battery is the ultimate source of the

flash energy, so it should be capable of supplying the needed energy.

18. Two capacitors are connected in series. Is the equivalent capacitance more or less than that of either one?

19. Two capacitors are connected in series. Could the maximum working voltage of the combination be as great as the sum of the working voltages of both capacitors? Could it be lower than this sum? Could it be lower than the working voltage of either capacitor?

20. Explain why the potential differences across parallel capacitors must be the same.

21. Two capacitors are storing equal amounts of energy, yet one has twice the capacitance of the other. How do their voltages compare?

22. Explain why inserting a dielectric between capacitor plates increases the capacitance.

23. An air-insulated parallel-plate capacitor is connected to a battery that imposes a potential difference V across the capacitor. If a dielectric slab is inserted between the capacitor plates, what happens to (a) the potential difference; (b) the capacitor charge; and (c) the capacitance?

24. An ideal dielectric would be a material whose internal dipoles did not dissipate energy as heat as they respond to an electric field. If a slab of ideal dielectric is placed part way between the plates of a capacitor, what will be its subsequent motion? Assume that the capacitor is charged but is not connected to a battery or other external circuitry.

25. A capacitor is charged and left connected to the charging battery. If you insert a dielectric slab between the capacitor plates, do you do work on the slab, or does it do work on you? Explain.

PROBLEMS

Section 26-1 Energy of a Charge Distribution

1. Three point charges, each of $+q$, are moved from infinity to the vertices of an equilateral triangle of side ℓ. How much work is required?

2. Repeat the preceding problem for the case of two charges $+q$ and one $-q$.

3. Four 50-μC charges are brought from far apart onto a line where they are spaced at 2.0-cm intervals. How much work does it take to assemble this charge distribution?

4. Repeat Example 26-1 for the case when the negative charge is $-q$ rather than $-q/2$.

5. Suppose two of the charges in Problem 1 are held in place, while the third is allowed to move freely. If this third charge has mass m, what will be its speed when it's far from the other two charges?

6. To a very crude approximation, a water molecule consists of a negatively charged oxygen atom and two "bare" pro-

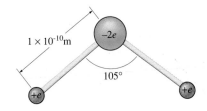

FIGURE 26-22   Problem 6.

tons, as shown in Fig. 26-22. Calculate the electrostatic energy of this configuration, which is therefore the magnitude of the energy released in forming this molecule from widely separated atoms. Your answer is an overestimate because electrons are actually "shared" among the three atoms, spending more time near the oxygen.

7. Four identical charges q, initially widely separated, are brought to the vertices of a tetrahedron of side a (Fig. 26-23). Find the electrostatic energy of this configuration.

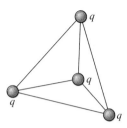

FIGURE 26-23 Problem 7 (All sides have equal length a).

8. A charge Q_0 is at the origin. A second charge, $Q_x = 2Q_0$, is brought to the point $x = a$, $y = 0$. Then a third charge Q_y is brought to the point $x = 0$, $y = a$. If it takes twice as much work to bring in Q_y as it did Q_x, what is Q_y in terms of Q_0?

Section 26-2 Two Isolated Conductors

9. Two square conducting plates 25 cm on a side and 5.0 mm apart carry charges $\pm 1.1\ \mu C$. Find (a) the electric field between the plates, (b) the potential difference between the plates, and (c) the stored energy.

10. Two square conducting plates measure 5.0 cm on a side. The plates are parallel, spaced 1.2 mm apart, and initially uncharged. (a) How much work is required to transfer 7.2 μC from one plate to the other? (b) How much work is required to transfer a second 7.2 μC?

11. (a) How much charge must be transferred between the initially uncharged plates of the preceding problem in order to store 15 mJ of energy? (b) What will be the potential difference between the plates?

12. Two parallel, circular metal plates of 15 cm radius are initially uncharged. It takes 6.3 J to transfer 45 μC from one plate to the other. How far apart are the plates?

13. A conducting sphere of radius a is surrounded by a concentric spherical shell of radius b. Both are initially uncharged. How much work does it take to transfer charge from one to the other until they carry charges $\pm Q$?

14. Show that the energy given by Equation 26-2 can be written as the product of the charge transferred with the *average* value of the potential during the transfer.

15. Two conducting spheres of radius a are separated by a distance $\ell \gg a$; since the distance is large, neither sphere affects the other's electric field significantly, and the fields remain spherically symmetric. (a) If the spheres carry equal but opposite charges $\pm q$, show that the potential difference between them is $2kq/a$. (b) Write an expression for the work dW involved in moving an infinitesimal charge dq from the negative to the positive sphere. (c) Integrate your expression to find the work involved in transferring a charge Q from one sphere to the other, assuming both are initially uncharged.

Section 26-3 Energy and the Electric Field

16. The energy density in a uniform electric field is 3.0 J/m³. What is the field strength?

17. A car battery stores about 4 MJ of energy. If all this energy were used to create a uniform electric field of 30 kV/m, what volume would it occupy?

18. Air undergoes dielectric breakdown at a field strength of 3 MV/m. Could you store energy in a uniform electric field in air with the same energy density as that of liquid gasoline? (See Appendix C.)

19. Find the electric field energy density at the surface of a proton, taken to be a uniformly charged sphere 1 fm in radius.

20. A pair of closely spaced square conducting plates measure 10 cm on a side. The electric field energy density between the plates is 4.5 kJ/m³. What is the charge on the plates?

21. The electric field strength as a function of position x in a certain region is given by $E = E_0(x/x_0)$, where $E_0 = 24$ kV/m and $x_0 = 6.0$ m. Find the total energy stored in a cube 1.0 m on a side, located between $x = 0$ and $x = 1.0$ m. (The field strength is independent of y and z.)

22. A sphere of radius R contains charge Q spread uniformly throughout its volume. Find an expression for the electrostatic energy contained within the sphere itself. *Hint:* Consult Example 24-1.

23. A sphere of radius R carries a total charge Q distributed over its surface. Show that the total energy stored in its electric field is $U = kQ^2/2R$.

24. A uranium-235 nucleus contains 92 protons and 143 neutrons, and has a diameter of 6.6 fm. Assuming that the proton charge is distributed uniformly throughout the nucleus, calculate the total electrostatic energy of this configuration. *Hint:* See the preceding two problems.

25. Two 4.0-mm-diameter water drops each carry 15 nC. They are initially separated by a great distance. Find the change in the electrostatic potential energy if they are brought together to form a single spherical drop. Assume all charge resides on the drops' surfaces.

26. A 2.1-mm-diameter wire carries a uniform line charge density $\lambda = 28\ \mu C/m$. How much energy is contained in a space 1.0 m long within one wire diameter of the wire surface?

27. A long, solid rod of radius a carries uniform volume charge density ρ. Find an expression for the electrostatic energy per unit length contained *within* the rod. *Hint:* See Problem 24-31.

Sections 26-4 Capacitors

28. A capacitor's plates hold 1.3 μC when charged to 60 V. What is its capacitance?

29. The "memory" capacitor in a VCR has a capacitance of 4.0 F and is charged to 3.5 V. What is the charge on its plates?

30. What voltage is needed to put 1.6 mC on a 100-μF capacitor?

31. Figure 26-24 shows data from an experiment in which known amounts of charge are placed on a capacitor and the resulting voltage measured. Fit a line to the data, and use it to determine the capacitance.

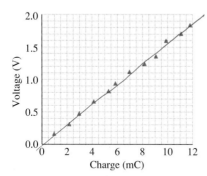

FIGURE 26-24 Problem 31 (data plot).

32. Show that the units of ε_0 may be written as F/m.

33. Find the capacitance of a parallel-plate capacitor consisting of circular plates 20 cm in radius separated by 1.5 mm.

34. A parallel-plate capacitor with 1.1-mm plate spacing has ± 2.3 μC on its plates when charged to 150 V. What is the plate area?

35. Find the capacitance of a 1.0-m-long piece of coaxial cable whose inner conductor radius is 0.80 mm and whose outer conductor radius is 2.2 mm, with air in between.

36. A capacitor consists of a conducting sphere of radius a surrounded by a concentric conducting shell of radius b. Show that its capacitance is $C = \dfrac{ab}{k(b - a)}$.

37. Figure 26-25 shows a capacitor consisting of two electrically connected plates with a third plate between them, spaced so its surfaces are a distance d from the other plates. The plates have area A. Neglecting edge effects, show that the capacitance is $2\varepsilon_0 A/d$.

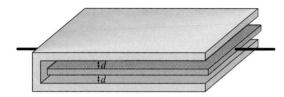

FIGURE 26-25 Problem 37.

Section 26-5 Energy Storage in Capacitors

38. The power supply of a stereo receiver contains a 2500-μF capacitor charged to 35 V. How much energy does it store?

39. Find the capacitance of a capacitor that stores 350 μJ when the potential difference across its plates is 100 V.

40. A certain capacitor stores 40 mJ of energy when charged to 100 V (a) How much would it store when charged to 25 V? (b) What is its capacitance?

41. Which can store more energy, a 1-μF capacitor rated at 250 V or a 470 pF capacitor rated at 3 kV?

42. A circuit application calls for a 10-μF capacitor that can store 12 mJ. What should be its voltage rating? The capacitors are available with voltage ratings that are multiples of 25 V.

43. A 0.01-μF, 300-V capacitor costs 25¢, a 0.1-μF, 100-V capacitor costs 35¢, and a 30-μF, 5-V capacitor costs 88¢. (a) Which can store the most charge? (b) Which can store the most energy? (c) Which is the most cost-effective energy storage device, as measured by energy stored per unit cost?

44. The charge on a capacitor is 50 mC, and it stores 2.5 J of energy. Find (a) its capacitance and (b) the voltage between its plates.

45. A camera flashtube requires 5.0 J of energy per flash. The flash duration is 1.0 ms. (a) What is the power used by the flashtube *while it is actually flashing*? (b) If the flashtube operates at 200 V, what size capacitor is needed to supply the flash energy? (c) If the flashtube is fired once every 10 s, what is its *average* power consumption?

46. The NOVA laser fusion experiment at Lawrence Livermore Laboratory in California can deliver 10^{14} W (roughly 100 times the output of all the world's power plants) of light energy when its lasers are on. But the laser pulse lasts only 10^{-9} s. (a) How much energy is delivered in one pulse? (b) The capacitor bank supplying this energy has a total capacitance of 0.26 F. Only about 0.17% (i.e., 0.0017) of the capacitor energy actually appears as light. To what voltage must the capacitor bank be charged?

47. A solid conducting slab is inserted between the plates of a charged capacitor, as shown in Fig. 26-26. The slab thickness is 60% of the plate spacing, and its area is the same as the plates'. (a) What happens to the capacitance? (b) What happens to the stored energy, assuming the capacitor is not connected to anything?

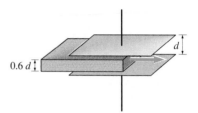

FIGURE 26-26 Problem 47.

48. Consider the two widely separated spheres of Problem 15 as a capacitor. Use energy considerations (i.e., the equation $U = \frac{1}{2}CV^2$ applies to *any* capacitor) and the answers to Problem 15 to find the capacitance.

49. The cylindrical capacitor of Example 26-5 is charged to a voltage V. Obtain an expression for the energy density as a function of radial position in the capacitor, and integrate to show explicitly that the stored energy is $\frac{1}{2}CV^2$.

Section 26-6 Connecting Capacitors Together

50. You have a 1.0-μF and a 2.0 μF capacitor. What values of capacitance could you get by connecting them in series or parallel?

51. Two capacitors are connected in series and the combination charged to 100 V. If the voltage across each capacitor is 50 V, how do their capacitances compare?

52. (a) What is the equivalent capacitance of the combination shown in Fig. 26-27? (b) If a 100-V battery is connected across the combination, what is the voltage across each capacitor? (c) What is the charge on each capacitor?

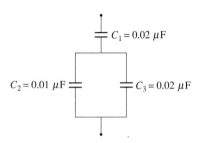

FIGURE 26-27 Problem 52.

53. You're given three capacitors: 1.0 μF, 2.0 μF, and 3.0 μF. Find (a) the maximum, (b) the minimum, and (c) two intermediate values of capacitance you could achieve with various combinations of all three capacitors.

54. What is the equivalent capacitance of the four identical capacitors in Fig. 26-28, measured between A and B?

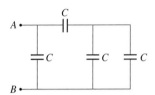

FIGURE 26-28 Problem 54.

55. You have an unlimited supply of 2.0-μF, 50-V capacitors. Describe combinations that would be equivalent to (a) a 2.0-μF, 100-V capacitor and (b) a 0.50-μF, 200-V capacitor.

56. Repeat the derivations for parallel and series capacitors, now using combinations of three capacitors.

57. What is the equivalent capacitance in Fig. 26-29?

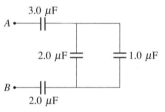

FIGURE 26-29 Problems 57, 58.

58. In Fig. 26-29, find the energy stored in the 1-μF capacitor when a 50-V battery is connected between points A and B.

59. Two capacitors C_1 and C_2 are in series, with a voltage V across the combination. Show that the voltages across the individual capacitors are

$$V_1 = \frac{C_2 V}{C_1 + C_2} \quad \text{and} \quad V_2 = \frac{C_1 V}{C_1 + C_2}.$$

60. A 0.10-μF capacitor rated at 50 V is in series with a 0.20-μF capacitor rated at 200 V. What is the maximum voltage that should be applied across the series combination? *Hint:* See the preceding problem.

61. A variable "trimmer" capacitor used to make fine adjustments has a capacitance range from 10 to 30 pF. The trimmer is in parallel with a capacitor of about 0.001 μF. Over what percentage range can the capacitance of the combination be varied?

62. Capacitors are often marked with a nominal value for the capacitance and a tolerance range within which the actual capacitance lies. For example, a 1-μF, $\pm 20\%$ capacitor has capacitance between 0.8 μF and 1.2 μF. If you connect a 0.01-μF $\pm$ 20% capacitor in series with a 0.02 μF $\pm$ 30% capacitor, in what range will the resulting capacitance lie? Express as a capacitance and its associated tolerance.

63. A. 5.0-μF capacitor is charged to 50 V, and a 2.0-μF capacitor is charged to 100 V. The two are disconnected from their charging batteries and connected in parallel, positive to positive. (a) What is the common voltage across each after they are connected? *Hint:* Charge is conserved. (b) Compare the total electrostatic energy before and after the capacitors are connected. Speculate on the discrepancy.

Section 26-7 Capacitors and Dielectrics

64. A parallel-plate capacitor has plates with 50 cm^2 area separated by a 25-μm layer of polyethylene. Find (a) its capacitance and (b) its working voltage.

65. A 470-pF capacitor consists of two circular plates 15 cm in radius, separated by a sheet of polystyrene. (a) What is the thickness of the sheet? (b) What is the working voltage?

66. An electrolytic capacitor is essentially a parallel-plate configuration in which aluminum plates are separated by a thin layer of aluminum oxide created by chemical action when a voltage is applied. If the effective plate area of a 2000-μF capacitor is 2.5 m^2, what are (a) the oxide layer thickness and (b) the working voltage?

67. Repeat Problem 35 for the more realistic case of a cable insulated with polyethylene.

68. An air-insulated parallel-plate capacitor has plate area 76 cm^2 and spacing 1.2 mm. It is charged to 900 V and then disconnected from the charging battery. A plexiglass sheet is then inserted to fill the space between the plates. What are (a) the capacitance, (b) the potential difference between the plates, and (c) the stored energy both before and after the plexiglass is inserted?

69. The capacitor of the preceding problem is connected to its 900-V charging battery and left connected as the plexiglass sheet is inserted, so the potential difference remains at 900 V. What are (a) the charge on the plates and (b) the stored energy both before and after the plexiglass is inserted?

70. Apply Equation 26-12 explicitly to a parallel-plate capacitor containing a dielectric, and show that the energy density between the plates is given by Equation 26-13.

Paired Problems

(Both problems in a pair involve the same principles and techniques. If you can get the first problem, you should be able to solve the second one.)

71. A pair of parallel conducting plates of area 0.025 m^2 carrying equal but opposite charges stores 1.6 J in its electric field. When the magnitude of the charge on both plates is increased by 5.0 μC, the stored energy increases to 2.4 J. Find the plate separation.

72. A capacitor stores 50 mJ of energy at voltage V_0. When the voltage is increased by 150 V, the stored energy increases to 75 mJ. Find the capacitance.

73. A 20-μF air-insulated parallel-plate capacitor is charged to 300 V. The capacitor is then disconnected from the charging battery, and its plate separation is doubled. Find the stored energy (a) before and (b) after the plate separation increases. Where does the extra energy come from?

74. Repeat the preceding problem, except that now the capacitor remains connected to the 300-V battery while the plates are separated.

75. In the capacitor network of Fig. 26-30, take $C = 6.0\ \mu$F. Find (a) the equivalent capacitance between A and B and (b) the charge on C when 30 V is applied between A and B.

76. Take C in Fig. 26-30 as an unknown capacitance. If 100 V is applied between A and B, the network stores 5.8 mJ of energy. Find C.

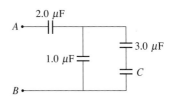

FIGURE 26-30 Problems 75, 76.

Supplementary Problems

77. A typical lightning flash transfers 30 C across a potential difference of 30 MV. Assuming such flashes occur every 5 s in the thunderstorm of Example 26-2, roughly how long could the storm continue if its electrical energy were not replenished?

78. A capacitor is constructed from a "sandwich" consisting of two long strips of aluminum foil each 2.0 cm wide and 1.6 m long, separated by two strips of 5.0-μm-thick polyethylene (Fig. 26-31). The capacitor is rolled up to make a compact cylinder. Find its capacitance. *Hint:* Because the strips are thin and closely spaced, you can treat this as a parallel-plate capacitor. But note that each foil layer in the rolled-up capacitor "sees" an oppositely charged layer on *both* sides.

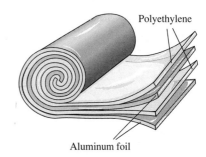

FIGURE 26-31 Problem 78.

79. Six charges $\pm q$, initially widely separated, are positioned to form a hexagon of side a, as shown in Fig. 26-32. What is the electrostatic energy of this configuration?

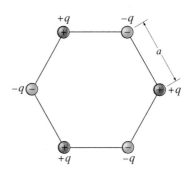

FIGURE 26-32 Problem 79.

80. Show that the result of Problem 36 reduces to that of a parallel-plate capacitor when the separation $b - a$ is much less than the radius a.

81. An air-insulated parallel-plate capacitor of capacitance C_0 is charged to voltage V_0 and then disconnected from the charging battery. A slab of material with dielectric constant κ, whose thickness is essentially equal to the capacitor spacing, is then inserted halfway into the capacitor (Fig. 26-33). Determine (a) the new capacitance, (b) the stored energy, and (c) the force on the slab in terms of C_0, V_0, κ, and the capacitor plate length L.

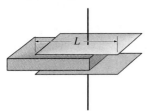

FIGURE 26-33 Problem 81.

82. Repeat parts (b) and (c) of the preceding problem, now assuming the battery remains connected while the slab is inserted.

83. We live inside a giant capacitor! Its plates are Earth's surface and the ionosphere, a conducting layer of the atmosphere beginning at about 60 km altitude. (a) What is its capacitance? *Hint:* You can treat it as either a spherical or a parallel-plate capacitor. Why? (b) The potential difference between Earth and ionosphere is about 6 MV. Find the total energy stored in this planetary capacitor.

84. Show that the result of Example 26-5 reduces to that of a parallel-plate capacitor when the separation $b - a$ is much less than the radius a. *Hint:* See Appendix A for an approximation to the logarithm.

85. Equation 26-2 gives the potential energy of a pair of oppositely charged plates. (a) Differentiate this expression with respect to the plate spacing to find the magnitude of the attractive force between the plates. (b) Compare with the answer you would get by multiplying one plate's charge by the electric field between the plates. Why do your answers differ? Which is right?

86. A solid sphere contains a uniform volume charge density. What fraction of the total electrostatic energy of this configuration is contained *within* the sphere?

87. A small dipole lies on the x axis, centered at the origin. Find an expression for the total electrostatic energy con-

tained in a thin cylindrical volume of diameter d and length ℓ, with its left end a distance ℓ from the dipole center, as shown in Fig. 26-34. Assume that ℓ is much greater than the dipole spacing. *Hint:* Since the cylinder is very thin, you can use the on-axis dipole field (Equation 23-5b) for the field throughout the cylinder.

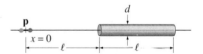

FIGURE 26-34 Problem 87.

88. A coaxial cable 15 m long consists of an inner conductor 1.0 mm in radius and an outer conductor 3.0 mm in radius, separated by polyethylene insulation. What is the electrostatic energy contained within this cable when a potential difference of 300 V is applied between its two conductors?

89. A TV antenna cable consists of two 0.50-mm-diameter wires spaced 12 mm apart. Estimate the capacitance per unit length of this cable, neglecting dielectric effects of the insulation.

90. A classical view of the electron pictures it as a purely electrical entity, whose rest mass energy mc^2 (see Section 8-7) is the energy stored in its electric field. If the electron were a sphere with charge distributed uniformly over its surface, what radius would it have to satisfy this condition? (Your answer for the electron's "size" is not consistent with modern quantum mechanics nor with experiments that suggest the electron is a true point particle.)

91. Use the fact that the static electric field is conservative to argue that there *must* be fringing field at the edges of a parallel plate capacitor. *Hint:* Remember that the plates are equipotentials, and consider the potential differences V_{AB} and V_{CD} in Fig. 26-35. What does your argument say about the strength of the fringing field relative to the field between the plates?

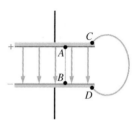

FIGURE 26-35 Problem 91.

CHAPTER 27

ELECTRIC CURRENT

Electric current in these power lines mediates the transmission of energy over long distances.

S o far our discussion of electrical phenomena has been based on the assumption of electrostatic equilibrium. We now relax that assumption, and consider situations in which charges are moving. Such motion usually occurs only in materials containing free charges, so our discussion will emphasize electrical conductors. Occasionally we will also deal with charges moving in an otherwise empty region—for example, the electron beam in a TV picture tube.

27-1 ELECTRIC CURRENT

An **electric current** is a net motion of electric charge. The measure of current is the rate at which charge crosses a given area. Accordingly, its units are coulombs per second (C/s). This unit is given the special name **ampere** (A) after the French physicist André Marie Ampère:

$$1 \text{ A} = 1 \text{ C/s}.$$

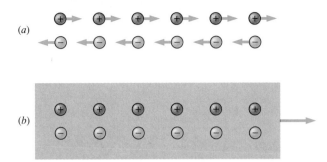

FIGURE 27-1 (*a*) Protons moving to the right constitute a current to the right. Since electrons are negative, *leftward* moving electrons also constitute a current to the *right*. The result in this case is a net current to the right. (*b*) In this bulk motion of neutral matter, the protons constitute a current to the right. But the negative electrons, also moving to the right, constitute a current to the *left*. So the net current is zero.

In electronics, biomedical applications, and many other practical situations currents are small enough that the milliampere (mA) and microampere (μA) are frequently used. When the current I is steady we write

$$I = \frac{\Delta q}{\Delta t}, \qquad \text{(steady current)} \tag{27-1a}$$

where Δq is the net charge crossing the given area in time Δt. If the current is not steady, we consider the ratio of charge to time for arbitrarily small time intervals, giving an instantaneous current that may vary with time:

$$I = \frac{dq}{dt}. \qquad \text{(instantaneous current)} \tag{27-1b}$$

The direction of current is the direction in which *positive* charge flows. If the moving charge is negative, as with electrons in a metal, then the current is opposite the charge motion. You can blame Benjamin Franklin for this confusing situation! It was Franklin who assigned the names "positive" and "negative" to the two kinds of electric charge. Had Franklin known that free charges in metals are electrons, he might well have reversed his terminology.

An electric current may consist of only one sign of charge in motion, or it may involve both positive and negative charge. It that case the current is determined by the *net* charge motion—that is, by the algebraic sum of the currents associated with both kinds of charge (Fig. 27-1).

Current—the rate at which charge crosses a given area—depends on the speed of the charge carriers, their density, and the charge carried by each. Consider a conductor containing n charges per unit volume, each carrying charge q and moving with speed v_d, as shown in Fig. 27-2. The quantity v_d is called the **drift speed.** In some cases—a beam of electrons in vacuum, for example—v_d is the actual particle speed. More commonly, v_d represents the time-average speed of the charge carriers, averaging out the effects of random thermal motion and collisions. If A is the conductor's cross-sectional area, then a length ℓ has

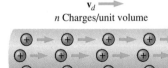

FIGURE 27-2 A conductor of cross-sectional area A containing n charge carriers per unit volume. Each moves with speed v_d and carries charge q. The total charge in a region of length ℓ is $nA\ell q$, and the current is therefore $nAqv_d$.

volume $A\ell$ and therefore contains $nA\ell$ charges. Since each carries charge q, the total charge is $\Delta Q = nA\ell q$. With drift speed v_d, the length ℓ of charge moves past a given point in time $\Delta t = \ell/v_d$, so the current is

$$I = \frac{\Delta Q}{\Delta t} = \frac{nA\ell q}{\ell/v_d} = nAqv_d. \qquad (27\text{-}2)$$

● EXAMPLE 27-1 CURRENT IN A WIRE

A copper wire with a cross-sectional area of 1.0 mm² carries a current of 5.0 A. The charge carriers in copper are electrons, and each copper atom contributes, on average, 1.3 free electrons. What is the drift speed of the electrons?

Solution
The density of copper (see inside back cover) is 8920 kg/m³, and the periodic table (also inside back cover) lists copper's atomic weight as 63.55—meaning that the mass per atom is 63.55 u. So the number density of copper atoms is

$$n = \frac{8920 \text{ kg/m}^3}{(63.55 \text{ u/atom})(1.66 \times 10^{-27} \text{ kg/u})}$$

$$= 8.46 \times 10^{28} \text{ atoms/m}^3.$$

Since each atom contributes 1.3 free electrons, the electron density is $(1.3)(8.46 \times 10^{28} \text{ m}^{-3}) = 1.10 \times 10^{29} \text{ m}^{-3}$. Solving Equation 27-2 for v_d then gives

$$v_d = \frac{I}{nAq} = \frac{5.0 \text{ A}}{(1.10 \times 10^{29} \text{ m}^{-3})(1.0 \times 10^{-6} \text{ m}^2)(1.6 \times 10^{-19}\text{C})}$$

$$= 0.284 \text{ mm/s}.$$

This remarkably small value is typical of drift speeds in metallic conductors.

EXERCISE A thin layer of gold, 0.10 mm wide and 3.2 μm thick, is used for connections to an integrated circuit. Find (a) the free electron density in gold and (b) the drift speed of electrons in the gold layer when it carries a current of 140 μA. The density of gold is 1.93×10^4 kg/m³, and each gold atom contributes, on average, 1.5 free electrons.

Answers: (a) 8.85×10^{28} m^{-3}; (b) 31 μm/s

Some problems similar to Example 27-1: 7–9, 55, 56 ●

How can the drift speed be so small? When you turn on a light switch, the light comes on immediately, not several thousand seconds later as the result of Example 27-1 might imply. Here it's important to distinguish between the speed of the electrons and that of the electrical signal in the wire. As soon as electrons at one end of the wire begin moving, their electric fields affect adjacent electrons, which also begin moving. This effect propagates down the wire at what is in fact nearly the speed of light, so the current begins everywhere almost simultaneously. The same thing happens when you turn on a garden hose full of water: Water comes out the far end even though water at the faucet has not had time to travel down the hose.

Current Density

In many cases electric currents are not so neatly confined as in a wire. Examples include currents in the oceans and solid Earth, in the atmosphere, in chemical solutions, and in the ionized gases that make up the stars and indeed much of the matter in the universe (Fig. 27-3). In these situations the current is spread over a rather ill-defined area, and its magnitude and direction may vary from point to point. It's useful to characterize such diffuse currents in terms of **current density,** defined as the current per unit area at a given point. Dividing Equation 27-2 by the area gives

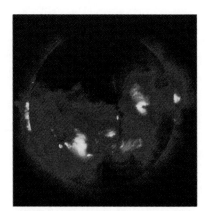

FIGURE 27-3 Strong electric currents flow in the bright yellow loops of ionized gas that arch above the Sun's surface in this image made with an orbiting x-ray telescope on the Japanese satellite Yohkoh.

$$J = \frac{I}{A} = nqv_d. \tag{27-3a}$$

The flow of charge can vary in both magnitude and direction, so current density is more generally written as a vector quantity:

$$\mathbf{J} = nq\mathbf{v}_d, \tag{27-3b}$$

where $\mathbf{v}_d$ is the drift velocity.

If current density is uniform, as in a wire, then the total current is simply the product of the current density with the wire's cross-sectional area. When the current density varies with position, then integration is necessary to calculate the total current. Problem 63 explores this case.

● EXAMPLE 27-2 AN IONIC SOLUTION

Charge carriers in salt water are positive sodium ions and negative chlorine ions, each carrying one elementary charge. A certain solution contains 6.0×10^{26} of each ion type per cubic meter. The ions drift at 2.6×10^{-5} m/s, with positive and negative ions moving in opposite directions. What is the current density in the solution? If the solution is confined to a tube of 3.0-cm^2 cross section, what is the total current?

Solution

The current density due to one type of ion—say the positive ions—is given by Equation 27-3a:

$$J = nqv_d = (6.0 \times 10^{26} \text{ m}^{-3})(1.6 \times 10^{-19} \text{ C})(2.6 \times 10^{-5} \text{ m/s})$$

$$= 2.5 \text{ kA/m}^2.$$

The negative ions drift in the opposite direction, but they also carry the opposite charge, so they provide a current density of the same magnitude and in the same direction. Therefore the net current density is 5.0 kA/m^2.

With this uniform current density through the tube's 3.0-cm^2 cross section, the total current is

$$I = JA = (5.0 \text{ kA/m}^2)(3.0 \times 10^{-4} \text{ m}^2) = 1.5 \text{ A}.$$

EXERCISE The maximum safe current density in copper wire used in household wiring is about 6 MA/m^2. (Beyond this level the wire overheats.) What is the minimum safe wire diameter in a circuit that can carry up to 15 A?

Answer: 1.8 mm

Some problems similar to Example 27-2: 2, 6, 10, 11 ●

27-2 CONDUCTION MECHANISMS

What causes electric current? Electric charges experience forces in electric fields, so applying a field to a conductor should result in a current. You might think that it would suffice to apply the field briefly to get the charges moving; Newton's law suggests they would then keep moving. In most conductors, however, charges do not move unimpeded. They bump into things—usually ions—and quickly lose any energy they've gained from the field. To sustain a current in most materials, it is therefore necessary to maintain an electric field within the material. Having such a field does not violate our conclusion that there can be no electric field in a conductor in electrostatic equilibrium since we're explicitly considering moving charges and are therefore no longer talking about equilibrium.

Although it is generally true that electric fields result in currents, the detailed relation between current and field depends on the type of conductor. In addition to metals, important conductors include ionic solutions, plasmas (ionized gases), semiconductors, and superconductors, all of which we will consider briefly in this section.

In most materials the current density and the electric field are in the same direction, and we can therefore characterize the relation between the two using the equation

$$\mathbf{J} = \sigma \mathbf{E}, \qquad (27\text{-}4a)$$

where σ is called the **conductivity** of the material. Equation 27-4a shows that conductivity is the ratio of the magnitude of the current density to the magnitude of the electric field. For many common conductors, experiment shows that this ratio is independent of field; such materials are called **ohmic,** and for them Equation 27-4 is one way of stating what is known as **Ohm's law.** (This is the *microscopic* version of Ohm's law, relating field and current density at each point. You may be familiar with the *macroscopic* version, which relates the voltage difference across a conductor to the current through it.) Ohmic materials exhibit a linear relation between current density and electric field. In **nonohmic** materials, in contrast, conductivity does depend on field, and therefore $\mathbf{J}$ and $\mathbf{E}$ are not directly proportional.

Equation 27-4a shows that the units of conductivity are A/V·m. One V/A is given the name **ohm** (symbol Ω), after the German physicist Georg Ohm (1789–1854), whose experiments clarified the relation between voltage and current. The SI unit of conductivity can therefore be written $(\Omega\cdot\text{m})^{-1}$. An equivalent way of characterizing the relation between current density and field is with the **resistivity,** ρ, defined as the inverse of the conductivity:

$$\rho = \frac{1}{\sigma},$$

so Equation 27-4a can also be written

$$\mathbf{J} = \frac{\mathbf{E}}{\rho}. \qquad (27\text{-}4b)$$

The units of resistivity are $\Omega\cdot\text{m}$.

Conductivity and resistivity vary dramatically among different materials; indeed, their measurable range is one of the broadest of any physical quantity, spanning some 24 orders of magnitude. Table 27-1 lists some typical resistivities.

● **EXAMPLE 27-3** HOUSEHOLD WIRING: THE ELECTRIC FIELD

A 1.8-mm-diameter copper wire carries 15 A to a household appliance. What is the electric field in the wire?

Solution
The current density is $J = I/A$; using this result in Equation 27-4b and solving for E gives

$$E = \frac{I\rho}{A} = \frac{(15\text{ A})(1.68\times10^{-8}\ \Omega\cdot\text{m})}{(\pi)(0.90\times10^{-3}\text{ m})^2} = 99\text{ mV/m},$$

where we found the resistivity of copper in Table 27-1. This result is much smaller than the fields we've been discussing in electrostatic situations. Because copper is such a good conductor, only a very small field is needed to drive even a substantial current. In analyzing electric circuits we will often make the approximation that the fields and therefore potential differences in copper wires are essentially zero.

EXERCISE A uniform electric field of 0.76 V/m drives a 10-A current in an iron wire. Find the wire diameter.

Answer: 1.3 mm

Some problems similar to Example 27-3: 17–19 ●

▲ **TABLE 27-1** RESISTIVITIES

MATERIAL	RESISTIVITY ($\Omega \cdot m$)
Metallic conductors (at 20°C)	
Aluminum	2.65×10^{-8}
Copper	1.68×10^{-8}
Gold	2.24×10^{-8}
Iron	9.71×10^{-8}
Mercury	9.84×10^{-7}
Silver	1.59×10^{-8}
Ionic solutions (in water at 18°C)	
1-molar copper sulfate ($CuSO_4$)	3.9×10^{-4}
1-molar hydrochloric acid (HCl)	1.7×10^{-2}
1-molar sodium chloride (NaCl)	1.4×10^{-4}
Water, pure (H_2O)	2.6×10^{5}
Sea water (typical value)	0.22
Semiconductors (pure, at 20°C)	
Germanium	0.45
Silicon	640
Insulators	
Ceramics	10^{11}–10^{14}
Glass	10^{10}–10^{14}
Polystyrene	10^{15}–10^{17}
Rubber	10^{13}–10^{16}
Wood (dry)	10^{8}–10^{14}

Conduction in Metals

Metals contain large numbers of free electrons, which respond readily to electric fields, making these materials good conductors. We can describe conduction in a semiquantitative way by considering that these electrons are in random thermal motion and undergo collisions with the ions that are fixed in the metal's crystal structure. That description is helpful in understanding why metals obey Ohm's law, but it is not strictly correct. In fact, the major cause of resistivity involves electrons scattering from imperfections in the crystal structure—an interaction whose description necessarily involves quantum mechanics.

An electric field applied in a metal accelerates the negative electrons in the direction opposite the field. In our collision model, an electron doesn't get going very fast before it collides with an ion. It then gives up whatever energy it gained from the field, and rebounds with a random velocity. But the field is still there, and the electron again accelerates (Fig. 27-4). The result is that the electrons have a very small average velocity—the drift velocity $\mathbf{v}_d$ that we introduced earlier. Their motion therefore constitutes a net current which is itself proportional to $\mathbf{v}_d$.

The drift velocity depends on two things: the acceleration of the electrons and the rate at which they undergo collisions. The electric force $-e\mathbf{E}$ gives an acceleration $\mathbf{a} = \mathbf{F}/m = -e\mathbf{E}/m$, with m the electron mass and $-e$ its charge. If we pick a random electron and ask when it last underwent a collision, the answer will be some mean time τ called the **collision time.** During that time the electron has been accelerating with acceleration $\mathbf{a} = -e\mathbf{E}/m$, and therefore its average velocity due to the presence of the electric field—i.e., the drift velocity $\mathbf{v}_d$—will be

$$\mathbf{v}_d = -\frac{e\mathbf{E}}{m}\tau.$$

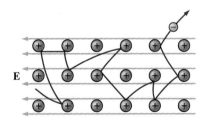

FIGURE 27-4 The path of an electron moving through a metal. The motion is almost completely random, but in the presence of an electric field there is a very slight drift antiparallel to the field.

Of course the electron also has its random thermal motion, but over many collisions this averages to zero while the velocity resulting from the electric field does not. Using our expression for the drift velocity in Equation 27-3b gives

$$\mathbf{J} = nq\mathbf{v}_d = n(-e)\mathbf{v}_d = \frac{ne^2\mathbf{E}}{m}\tau.$$

Comparing this expression with Equation 27-4a shows that the conductivity is

$$\sigma = \frac{ne^2}{m}\tau. \qquad (27\text{-}5)$$

The collision time τ depends on how fast the electrons are moving; the faster they go, the more frequent their collisions, and the lower τ should be. The electrons have two kinds of motion: their random thermal motion and the drift velocity $\mathbf{v}_d$ acquired from the electric field. In Example 27-1 we found that a typical drift speed is on the order 1 mm/s. Thermal speeds of electrons in metals, in contrast, are around 10^6 m/s. The drift speed is therefore completely negligible in determining the collision time. That means the collision time and hence the conductivity are essentially independent of the applied electric field—and that makes Equation 27-4a a linear relation for metallic conductors. In other words, metals obey Ohm's law. We stress that this conclusion is only approximate, and that Ohm's law, unlike Gauss's law, is *not* an exact, universal statement that holds everywhere and for all materials.

Although the conductivity of a metal is independent of the applied electric field, it does depend on other factors, especially temperature. From Chapter 20 we know that classical physics gives a thermal speed proportional to the square root of the temperature, so we might expect conductivity to be proportional to $1/\sqrt{T}$ and thus resistivity to $\sqrt{T}$. Experiment, however, shows that resistivity is very nearly proportional to temperature rather than to its square root (Fig. 27-5). Here we've reached the limits of classical physics, which cannot describe fully the behavior of the free electrons in a metal. Readers of the extended version of this text will explore quantum mechanics and how it deals with electrical conduction.

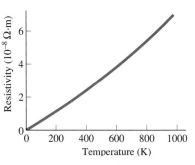

FIGURE 27-5 Resistivity of copper—the inverse of conductivity—shows a nearly linear dependence on temperature, in contrast to the classical prediction of a $\sqrt{T}$ dependence.

● **EXAMPLE 27-4** CONDUCTIVITY AND COLLISION TIME

Copper at room temperature contains 1.1×10^{29} free electrons per cubic meter. What is the collision time for these electrons?

Solution

Writing $\sigma = 1/\rho$ and solving Equation 27-5 for the collision time gives

$$\tau = \frac{m}{ne^2\rho}$$

$$= \frac{9.11\times10^{-31}\text{ kg}}{(1.1\times10^{29}\text{ m}^{-3})(1.6\times10^{-19}\text{ C})^2(1.68\times10^{-8}\ \Omega\cdot\text{m})}$$

$$= 1.9\times10^{-14}\text{ s},$$

where we found the resistivity of copper in Table 27-1. Problem 64 uses this result to estimate the mean thermal speed of the electrons.

EXERCISE Sodium contains 2.5×10^{28} free electrons per cubic meter, and the collision time is 3.4×10^{-14}s. What is the resistivity of sodium?

Answer: $4.2\times10^{-8}\ \Omega\cdot\text{m}$

Some problems similar to Example 27-4: 23, 64

●

■ **APPLICATION** NOISE IN ELECTRONIC EQUIPMENT

Although the time-average current associated with random thermal motion of charge carriers is zero, at any given instant random fluctuations may result in more charge carriers moving in one direction than in another. The result is a very small current, whose sign and magnitude fluctuate randomly. Called thermal noise, this current may disturb or even overwhelm currents of interest in sensitive electronic instruments. Thermal motion decreases at low temperatures, and sensitive circuits like the amplifiers in radio telescopes are often cooled to liquid helium temperatures (around 4 K) to reduce thermal noise (Fig. 27-6). Ultimately thermal noise limits our ability to detect and study very weak electrical signals.

FIGURE 27-6 This 12-m radio telescope at Kitt Peak National Observatory uses amplifiers cooled to 4 K with liquid helium in order to reduce electrical noise.

Ionic Solutions

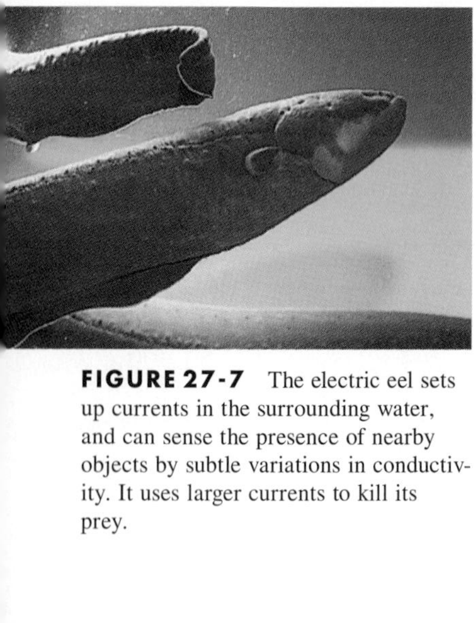

An ionic solution contains positive and negative ions that respond to an electric field by moving in opposite directions, resulting in a net current (Fig. 27-7). Conductivity of the solution is limited by collisions between the ions and neutral atoms. In addition to heating the solution, some of the energy of these collisions may go into chemical reactions that store energy. Charging a car battery, for example, involves driving a current through the battery's acid solution. Electrical energy goes into reversing the chemical reactions that normally power the battery, thus building up a supply of stored energy. Conduction in ionic solutions also plays an important role in the corrosion of metals, for example those exposed to salt solutions. And the presence of an ionic solution—sweat—increases our vulnerability to electric shock. Table 27-1 includes the resistivities of some ionic solutions. Note that these solutions are poorer conductors than metals.

FIGURE 27-7 The electric eel sets up currents in the surrounding water, and can sense the presence of nearby objects by subtle variations in conductivity. It uses larger currents to kill its prey.

Plasmas

Plasma is ionized gas that conducts because it contains free electrons and positive ions. It takes substantial energy to ionize atoms, so plasmas usually occur only in high-temperature environments. The few examples of plasmas on

Earth occur in fluorescent lamps, neon signs, devices for fusion research, the ionosphere, flames, and lightning flashes (Fig. 27-8). Yet most of the matter in the universe is probably in the plasma state; the stars, in particular, are almost entirely plasma.

The electrical properties of plasma make it so different from ordinary gas that plasma is often called "the fourth state of matter." Many plasmas are so diffuse that collisions between particles are rare. These "collisionless" plasmas are sometimes far better conductors than metals, and they can sustain large electric currents with very modest electric fields.

Semiconductors

Even in the best insulators, random thermal motions occasionally dislodge electrons, giving these materials very modest conductivity. This effect increases with temperature, but is usually insignificant at normal temperatures. But a few materials—notably the element silicon—exhibit significant conductivity at room temperature. The electrical properties of these **semiconductors** make possible the microelectronic technology that plays a pervasive role in modern civilization.

Conduction in semiconductors involves not only electrons dislodged from their places in the material structure, but also the "holes" left behind by those electrons. An adjacent electron can "fall" into the hole, with the effective result that the hole has moved in the direction of the field (Fig. 27-9). Thus holes act as positive charge carriers, and a pure semiconductor like silicon contains equal numbers of negative charge carriers (free electrons) and positive charge carriers (holes).

A pure semiconductor has rather low conductivity and is useful only in a few applications. The key to semiconductor technology lies in the control of conductivity by adding very small amounts of impurities—a process called **doping.** Adding an element with five electrons in its outermost shell—as opposed to silicon's four—results in large numbers of free electrons and a much more conductive material whose charge carriers are predominantly negative. Since its charge carriers are negative, such a material is called an **N-type** semiconductor. In contrast, doping with an element containing only three outermost electrons results in a **P-type** semiconductor, whose charge carriers are predominantly holes. The wide range of semiconductor devices in use today results from carefully engineered combinations of P- and N-type material. The application below presents one of the most important such devices.

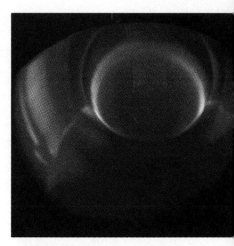

FIGURE 27-8 Plasma, or ionized gas, is an excellent electrical conductor. Photo shows glowing plasma in the Tokamak Fusion Test Reactor at the Princeton Plasma Physics Laboratory.

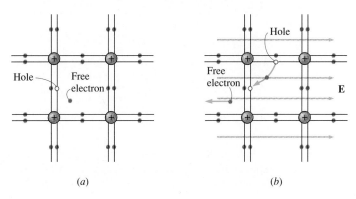

(a)   *(b)*

FIGURE 27-9 (*a*) Structure of a silicon crystal, showing each atom bound to each of its neighbors by two shared electrons. Thermal motion has dislodged one of the electrons from the crystal structure, leaving behind a hole. (*b*) In the presence of an electric field, the free electron drifts opposite the field direction. In addition, a nearby bound electron falls into the hole, effectively moving the hole in the direction of the field. Holes thus act as positive charge carriers.

■ **APPLICATION** TRANSISTORS AND INTEGRATED CIRCUITS

Few inventions have revolutionized human society as much as the transistor. Transistors are the basis of all modern electronic devices, from stereo amplifiers to automobile ignition systems to VCRs, laboratory instruments, and computers. Basically, a transistor is a semiconductor device in which one electrical signal controls another. The transistor shares that control function with its predecessor, the vacuum tube. But whereas vacuum tubes were large, fragile, expensive, unreliable, hot, slow, and consumed large amounts of power, transistors are small, rugged, cheap, reliable, cool, fast, and consume little power (Fig. 27-10).

Figure 27-11 shows a widely used type of transistor called the field effect transistor (FET). It consists of a slab of *P*-type semiconductor with two separate regions of *N*-type material on top, called the *drain* and *source.* Part of the structure is coated with silicon dioxide (SiO_2), an excellent insulator, and a metal *gate* coated on top. Now, a junction between *P*- and *N*-type semiconductors has the property that current flows readily from *P* to *N* but not from *N* to *P*. In the transistor of Fig. 27-11 current cannot flow between drain and source because either way an *N*-to-*P* junction is encountered. But suppose positive charge is placed on the gate, by connecting it to an appropriate potential. This charge will repel positive holes in the *P*-type material below the gate, and attract electrons. The result is that the "channel" between drain and source becomes temporarily *N*-type. Now there are no *N*-*P* junctions, and the transistor conducts. Thus the voltage applied to the gate can be used to control the drain–source current.

Varying the gate charge continuously varies the drain–source current in the same way; the transistor then functions as an *amplifier,* making a weak signal stronger. Audio and video systems all contain amplifying transistors; ultimately, for example, the weak signal from a cassette tape deck is amplified enough to drive a loudspeaker. In digital circuits like computers, in contrast, the transistor functions as a switch, with its drain–source channel either fully "on" (conducting) or "off" (nonconducting).

A transistor is fabricated from a single piece of silicon. By exposing various parts of the surface to dopant chemicals, oxygen, and metal atoms, the various *N* and *P* regions as well as the SiO_2 insulator and metallic gate are formed. The same process is used to produce entire circuits containing millions of transistors (Fig. 27-12). These **integrated circuits**—also called **chips**—make possible the complexity, miniaturization, and sophistication of modern electronic devices (Fig. 27-13).

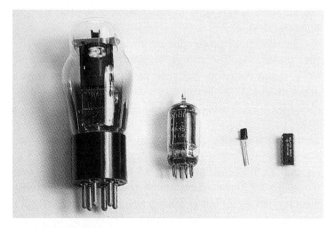

FIGURE 27-10 Vacuum tubes, a transistor, and an integrated circuit show the trend toward miniaturization in electronics. Tubes and transistor can each act as a single electronic switch, while the integrated circuit shown contains hundreds of transistors. Larger ICs now contain up to 10^8 transistors.

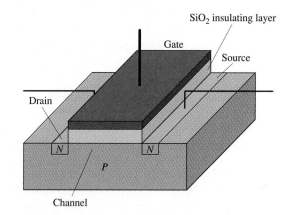

FIGURE 27-11 A field-effect transistor, or FET.

FIGURE 27-12 This PowerPC 603 microprocessor chip is at the heart of many Apple and IBM computers built in the mid-1990s. The PowerPC contains nearly 2 million transistors on a silicon chip occupying less than a square centimeter.

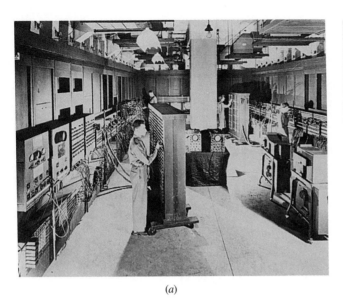

(a)

(b)

FIGURE 27-13 (a) ENIAC, one of the first computers, was built in 1946, occupied an entire room, and broke down frequently. (b) Today's laptop computers are far more powerful than ENIAC, cost orders of magnitude less, and are extremely reliable.

Superconductors

In 1911 the Dutch physicist H. Kamerlingh Onnes, studying the electrical properties of mercury at very low temperatures, found a sudden drop in resistivity at a temperature of 4.2 K. The resistivity below this temperature proved immeasurably low. Subsequent research has identified thousands of substances that become **superconductors** at sufficiently low temperatures. Currents in superconductors persist for years without any measurable decrease, suggesting that the resistivity of a superconductor is truly zero (Fig. 27-14).

For decades the known superconductors were largely metals and metal alloys that required cooling in liquid helium to achieve their superconductivity. Then, in 1986, physicists J. Georg Bednorz and K. Alex Müller of IBM's Zurich research laboratory made the stunning discovery that a class of ceramic materials becomes superconducting at temperatures around 100 K—high enough that these materials superconduct when cooled with liquid nitrogen. A flurry of research followed, soon pushing the highest superconducting temperature to well over 100 K (Fig. 27-15). Bednorz and Müller received the 1987 Nobel Prize in physics, only a year after their discovery. The search for higher temperature superconductors continues; indeed, as this book goes to press, a French team has just announced evidence for superconductivity at 250 K—about −10°F—which is warmer than a cold winter's day in the northern United States.

Superconductivity has enormous practical significance, offering loss-free transmission of electric power. Liquid-helium-cooled superconductors are used today in a number of applications, especially strong electromagnets (Fig. 27-16). Physicists and engineers are working to develop practical uses for the newer high-temperature superconductors, whose liquid nitrogen coolant is far less expensive than liquid helium. Devices based on thin films of these superconductors are already used to measure weak magnetic fields in biomedical and

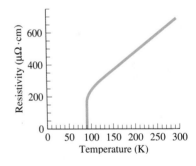

FIGURE 27-14 Resistivity versus temperature for a thin film of yttrium-barium-copper-oxide. Below the 93-K transition temperature the resistivity is truly zero.

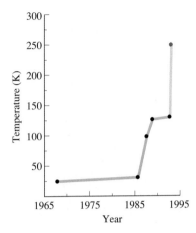

FIGURE 27-15 Since the mid-1980s, scientists have produced materials that become superconducting at increasingly high temperatures. The 250 K result is still tentative as this book goes to press.

FIGURE 27-16 Superconducting electromagnets (red structures) guide high-energy protons around the 2-km-diameter particle accelerator at Fermilab near Chicago.

geophysical applications, for measuring radiant energy, and for generating and processing microwaves. Discovery of a room-temperature superconductor would revolutionize electrical technology, and the search for such a material continues.

Although superconductivity was discovered in 1911, a satisfactory explanation of the phenomenon was not given until 1957. In that year John Bardeen, Leon Cooper, and J. Robert Schrieffer showed how superconductivity in the traditional low-temperature materials arises from a quantum-mechanical interaction among the conduction electrons. In a way that has no analog in classical physics, all the electrons move coherently through the crystal lattice, with no energy loss. Their theory earned Bardeen, Cooper, and Schrieffer the 1972 Nobel Prize in physics. A comprehensive theory of the newer, high-temperature superconducting materials has yet to be fully developed. Readers of the extended version of this text will revisit superconductivity in Chapter 41.

27-3 RESISTANCE AND OHM'S LAW

Ohm's law in the form of Equation 27-4 relates the electric field and current density within a given material. A more familiar form of Ohm's law relates voltage and current in a particular piece of material. We can relate these two—the microscopic and macroscopic forms of Ohm's law—by considering the cylindrical conductor shown in Fig. 27-17. Suppose there is a uniform electric field **E** within the conductor. Then there must be a uniform current density given by Equation 27-4b:

$$\mathbf{J} = \frac{\mathbf{E}}{\rho},$$

where ρ is the resistivity of the material. Then the total current is

$$I = JA = \frac{EA}{\rho},$$

where A is the conductor's cross-sectional area. If the conductor has length ℓ then the potential difference between its ends is

$$V = E\ell,$$

since the electric field is uniform. Note that the potential is higher (more positive) at the left, indicating that the potential drops—the energy per unit charge decreases—in the direction that charge moves through the resistor. Taking the ratio of voltage (potential difference) to current then gives

$$\frac{V}{I} = \frac{E\ell}{EA/\rho} = \frac{\rho\ell}{A}.$$

For an ohmic material, in which resistivity is independent of electric field, this equation tells us that the ratio of voltage to current depends only on the resistivity of the material and the dimensions of the particular piece. This ratio is called the electrical **resistance.** Its units are volts/ampere, or ohms (Ω). Our derivation shows that the resistance of an object with uniform cross section is just

$$R = \frac{\rho\ell}{A}. \tag{27-6}$$

Then the ratio of voltage to current becomes

$$\frac{V}{I} = R. \tag{27-7}$$

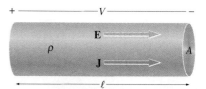

FIGURE 27-17 A cylindrical conductor made from a material with resistivity ρ. The uniform electric field **E** drives a uniform current density **J**, giving a total current $I = JA$ through the conductor's cross-sectional area A. The electric field is associated with a potential difference V across the length ℓ of the conductor.

For ohmic materials, R is constant, and Equation 27-7 is the macroscopic version of **Ohm's law.** For nonohmic materials, resistance depends on voltage, and we can consider Equation 27-7 to define the resistance at a given value of voltage and current. Again, we stress that Ohm's law is not a universal statement but an empirical law that provides a good description of ohmic materials.

Ohm's law is often written in the equivalent forms

$$V = IR \qquad \text{and} \qquad I = \frac{V}{R}.$$

This last form makes good sense, for it shows that a given voltage can push more current through a lower resistance. Two extreme cases are worth noting. An **open circuit** is a nonconducting gap with infinite resistance. No matter what the voltage across an open circuit, the current is zero. A **short circuit,** in contrast, has zero resistance. In a short circuit, currents of any magnitude can flow without requiring any potential difference or electric field. All real situations, with the exception of superconductors, lie between these two extremes.

● **EXAMPLE 27-5** RESISTANCE AND OHM'S LAW

A copper wire 0.50 cm in diameter and 70 cm long is used to connect a car battery to the starter motor. What is the wire's resistance? If the starter motor draws a current of 170 A, what is the potential difference across the wire?

Solution
Table 27-1 gives 1.68×10^{-8} $\Omega\cdot$m for the resistivity of copper. Then Equation 27-6 gives

$$R = \frac{\rho\ell}{A} = \frac{(1.68\times10^{-8}\ \Omega\cdot\text{m})(0.70\ \text{m})}{(\pi)(0.25\times10^{-2}\ \text{m})^2} = 6.0\times10^{-4}\ \Omega.$$

This low resistance is necessary because the starter motor draws such a large current. We now apply Ohm's law to find the voltage across the wire:

$$V = IR = (170\ \text{A})(6.0\times10^{-4}\ \Omega) = 0.10\ \text{V}.$$

This is small compared with the 12 volts available from the car battery, showing that the wire is well chosen for this application (Fig. 27-18). A larger voltage difference—which would occur with a thinner wire—would mean a significant reduction in power to the motor.

FIGURE 27-18 Thick jumper cables are necessary to carry the large current used by a car's starter motor.

EXERCISE What should be the diameter of an aluminum wire that carries 15 A when the voltage across 1.0 m of the wire is 0.25 V?

Answer: 1.4 mm

Some problems similar to Example 27-5: 28, 33, 34, 36

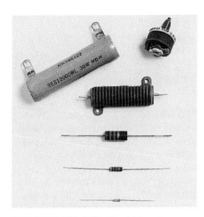

FIGURE 27-19 Typical resistors. The large unit at the upper left is a wire-wound resistor that can dissipate 30 watts. At upper right is a variable resistor, in which a rotating contact can be positioned at different points on a coil of resistance wire. Colored stripes on the smaller resistors code their resistance values. The smallest resistor shown is a carbon-film resistor that can dissipate at most $\frac{1}{4}$ watt.

Ohm's law remains valid when a conductor has nonuniform cross section, but now Equation 27-5 must be integrated to get the total resistance. Problems 69 and 70 explore this situation.

A **resistor** is a piece of conductor made to have a specific resistance. Heating elements used in electric stoves, hair dryers, irons, space heaters, and the like are all essentially resistors; so are the filaments of ordinary incandescent light bulbs. Resistors are made in a wide range of resistances for use in electronic circuits, where they help set appropriate values for voltage and current. Resistors are rated not only by their resistance but also by the maximum power they can dissipate without overheating. Figure 27-19 shows some typical resistors.

27-4 ELECTRIC POWER

We've seen that resistivity arises from collisions between electrons and ions. In these collisions, energy gained from the electric field ends up heating the conductor. Suppose a potential difference V is imposed across a conductor, driving a current I through it. The quantity V is the energy gained per unit charge as charge "falls" through the potential difference. The current I is the rate at which charge flows through the conductor. In a conductor with nonzero resistance, the energy gained from the electric field is dissipated through collisions, heating the conductor. Then the energy per unit time—that is, the power dissipated in the resistor—is the product of the energy per unit charge times the rate at which charge moves through the conductor:

$$P = IV. \qquad \text{(electric power)} \qquad (27\text{-}8)$$

Although we developed Equation 27-8 for power dissipated as heat in a resistance, it holds any time electrical energy is being converted to some other form. If, for example, we measure 5 V across an electric motor and 2 A through the motor, we can conclude that the motor is converting electrical to mechanical energy at the rate of 10 W (actually somewhat less because some of the power goes into heating).

Solving Ohm's law for V and putting the result in Equation 27-8 gives

$$P = I^2 R; \qquad (27\text{-}9a)$$

solving instead for I gives

$$P = \frac{V^2}{R}. \qquad (27\text{-}9b)$$

These are useful forms when we know the resistance and either the voltage or current. Although they may seem contradictory, both forms are equivalent and are equivalent to the more general Equation 27-8 for the case when current and voltage are related by Ohm's law.

> **TIP** **Think About What's Constant** Equation 27-9a seems to imply that power increases with increasing resistance, while Equation 27-9b seems to suggest the opposite. Both implications are correct—*if I* in Equation 27-9a and *V* in Equation 27-9b are constants. But there's no contradiction because *I* and *V* can't both be constant while the resistance *R*—the ratio of *V* to *I*—changes. In most cases we work with sources of constant voltage—for example, the power company promises to maintain 120 V between the two contacts in a household electrical outlet—and in this case the power dissipated is inversely proportional to the resistance we connect across that voltage, as shown by Equation 27-9b. See Question 22 for more on this point.

● EXAMPLE 27-6 A LIGHT BULB

The voltage in typical household wiring is 120 V. How much current does a 100-W light bulb draw? What is the bulb's resistance under these conditions?

Solution

Solving Equation 27-8 for *I* gives

$$I = \frac{P}{V} = \frac{100 \text{ W}}{120 \text{ V}} = 0.833 \text{ A}.$$

Since we know the current, we can determine the resistance directly from Ohm's law:

$$R = \frac{V}{I} = \frac{120 \text{ V}}{0.833 \text{ A}} = 144 \text{ } \Omega.$$

We could have bypassed the calculation of current and obtained *R* directly from Equation 27-9b:

$$R = \frac{V^2}{P} = \frac{(120 \text{ V})^2}{100 \text{ W}} = 144 \text{ } \Omega.$$

Finally, had we known the current but not the voltage, we could have used Equation 27-9a:

$$R = \frac{P}{I^2} = \frac{100 \text{ W}}{(0.833 \text{ A})^2} = 144 \text{ } \Omega.$$

The three approaches are equivalent. Use of Equation 27-9a or b merely amounts to solving symbolically for *V* or *I* before using Equation 27-8.

Because a light bulb filament undergoes a huge temperature change when turned on, its resistance is not independent of voltage and current. Our value 144 Ω holds when the light is on. When off, it is cool, and its resistance is much lower.

EXERCISE A power line has 0.20 Ω resistance per kilometer of length. If it carries 300 A of current, find (a) the voltage across 1.0 km of the wire and (b) the power dissipated in each km of wire.

Answers: (a) 60 V; (b) 18 kW

Some problems similar to Example 27-6: 42–44 ●

The Kilowatt-Hour

The SI unit of power is the watt (W), defined as 1 J/s and thus reflecting the definition of power as energy per time. We could equally well have defined the watt first, then defined the Joule as 1 W·s. The **kilowatt-hour** (kWh), a unit commonly used for electrical energy, is in fact defined in an analogous way. Just as 1 joule is the energy used by a device consuming 1 watt for 1 second, so a kilowatt-hour is the energy used by a device consuming 1 kW for 1 hour. Your household electric bill shows your electrical energy consumption in kWh; your cost for 1 kWh of electrical energy is typically in the range from 5¢ to 15¢. Since there are 3600 s in an hour, 1 kWh is equal to (1000 W)(3600 s) = 3.6 MJ. Burning a 100-W light bulb for 1 hour, for example, uses 100 watt-hours or

0.1 kWh. Although it is usually used only with electrical energy, the kWh is a perfectly good non-SI unit for describing any kind of energy; for example, it's useful to remember that the energy content of a gallon of oil or gasoline is about 40 kWh.

CHAPTER SYNOPSIS

Summary

1. **Electric current** is a net flow of electric charge, specified as the charge per unit time crossing a given area:

$$I = \frac{dq}{dt}.$$

If a material contains n free charges q per unit volume, moving with average speed v_d (called the **drift speed**), then the current through an area A perpendicular to the flow is

$$I = nqAv_d.$$

Current density is a vector specifying the current per unit area:

$$\mathbf{J} = nq\mathbf{v}_d.$$

2. **Conductivity** (symbol σ) is a property of a given material describing the ratio of electric field to current density in the material:

$$\mathbf{J} = \sigma\mathbf{E}.$$

For **ohmic** materials conductivity is independent of electric field and this relation constitutes the microscopic version of **Ohm's law.**

    Resistivity (symbol ρ) is the inverse of conductivity.

3. Conduction mechanisms vary with material, and include:
   a. **Metals,** in which the charge carriers are free electrons. Metals are ohmic materials in which resistivity arises from collisions of free electrons with ions.
   b. **Ionic solutions** are conductors because of the presence of negative and positive ions that can move through the solution.
   c. **Plasmas** are ionized gases, often with extremely high conductivity. Plasmas are rare on Earth but comprise much of the matter in the universe.
   d. **Semiconductors** conduct only poorly in their pure state, but their electrical properties can be radically altered by doping with impurities. Charge carriers in semiconductors can be electrons, positive "holes," or both. Semiconductors are the basis of modern electronic technology.
   e. **Superconductors** exhibit zero resistivity at sufficiently low temperatures, and consequently require no electric field or potential difference to drive a current.

4. **Resistance** is the ratio of voltage to current in a particular piece of material:

$$R = \frac{V}{I}.$$

Resistance depends on resistivity and physical dimensions. For an object of resistivity ρ, length ℓ, and uniform cross-sectional area A, the resistance is $R = \rho\ell/A$.

    For ohmic materials, the relation $R = V/I$ constitutes the macroscopic form of **Ohm's law.**

5. The rate at which electrical energy is converted to other forms is the product of the current I through a device and the potential difference V across it:

$$P = IV.$$

In a resistance the electrical energy is converted to heat, and the power can be written in the two equivalent forms

$$P = I^2R \quad \text{and} \quad P = \frac{V^2}{R}.$$

Terms You Should Understand

(Pairs are closely related terms whose distinction is important; number in parentheses is chapter section where term first appears.)

electric current, current density (27-1)
ampere (27-1)
drift speed, drift velocity (27-1)
conductivity, resistivity (27-2)
ohm (27-2)
ohmic, nonohmic materials (27-2)
Ohm's law (27-2, 27-3)
plasma (27-2)
semiconductor (27-2)
superconductor (27-2)
resistance (27-3)
open circuit, short circuit (27-3)

Symbols You Should Recognize

I (27-1)
v_d (27-1)
$\mathbf{J}$ (27-1)
σ, ρ (27-2)
Ω (27-2)
R (27-3)

Problems You Should Be Able to Solve

calculating current from drift speed and material properties (27-1)

relating current and current density (27-1)

relating electric field, current density, and conductivity or resistivity (27-2)

calculating resistance from resistivity and dimensions (27-3)

using Ohm's law to relate current, voltage, and resistance (27-3)

calculating electric power (27-4)

Limitations to Keep in Mind

Ohm's law is not a universal statement but an approximate empirical relation that holds with high accuracy for some materials, like metals.

QUESTIONS

1. If you physically move an electrical conductor, does this constitute a current?

2. In previous chapters we've stressed the absence of electric fields inside conductors in equilibrium. Why now do we allow such fields?

3. A wire carries a steady current. If the wire diameter decreases in the direction of the current, what happens to the current density?

4. When you talk on the telephone, your voice is heard almost immediately at the other end. Yet the drift speed of electrons in the telephone wire is on the order of millimeters per second. Explain the apparent discrepancy.

5. What is the difference between current and current density?

6. A constant electric field generally produces a constant drift velocity. How is this consistent with Newton's assertion that force results in acceleration, not velocity?

7. When caught in the open in a lightning storm, it is better to crouch low with the feet close together rather than lie flat on the ground. Why?

8. Why does the conductivity of a metal depend on the *square* of the electron charge?

9. Plasma physicists often use the approximation that there is no electric field in a plasma. How does this follow from the fact that plasma has a very large conductivity?

10. What are *P*- and *N*-type semiconductors? Does either carry a net electric charge?

11. Good conductors of electricity are often good conductors of heat. Why might this be?

12. Why can current persist forever in a superconductor with no applied voltage?

13. A plasma contains equal densities of free electrons and protons. Do you expect each to contribute equally to the net current? Explain.

14. Does an electric stove burner draw more current when it is first turned on or when it's fully hot?

15. A person and a cow are standing in a field when lightning strikes the ground nearby. Why is the cow more likely to be electrocuted?

16. You put a 1.5-V battery across a piece of material and a 100-mA current flows through the material. With a 9-V battery the current increases to 400 mA. Is the material ohmic or not?

17. The resistance of a metal increases with increasing temperature, while the resistance of a semiconductor decreases. Why the difference?

18. Macroscopic electric fields cannot exist inside a superconductor. Why not?

19. How does the fact that the drift speed of electrons in a metal is much less than their thermal speed imply that metals are ohmic conductors?

20. A 50-W and a 100-W light bulb are both designed to operate at 120 V. Which has the lower resistance?

21. A power line with a small but nonzero resistance is used to carry 450 MW of electric power from a nuclear power plant to a city. Is it most efficient to transmit this power at high voltage and low current or vice versa? Explain.

22. Equation 27-9a suggests that no power can be dissipated in a superconductor, since $R = 0$. But Equation 27-9b suggests the power should be infinite. Which is right, and why?

23. A motor made with superconducting wire and frictionless bearings is turning at constant speed and doing no mechanical work. Make an argument showing that the motor cannot be drawing current, even if it's connected to a battery. What would happen if the motor started to do mechanical work, like lifting a weight?

24. The resistivity of a pure semiconductor decreases with increasing temperature. Speculate on what might happen if a fixed voltage were applied across a piece of such material.

PROBLEMS

Section 27-1 Electric Current

1. A wire carries 1.5 A. How many electrons pass through the wire in each second?
2. In an ionic solution, 4.1×10^{15} ions, each carrying charge $+2e$, pass to the right each second; 3.6×10^{15} ions, each carrying $-e$, pass to the left in the same time. What is the net current?
3. A car battery is rated at 80 ampere-hours, meaning it can supply 80 A of current for 1 hour before it becomes discharged. If you accidentally leave the headlights on until the battery discharges, how much charge moves through the lights?
4. The electron beam that "paints" the image on a computer screen contains 5.0×10^6 electrons per cm of its length. If the electrons move toward the screen at 6.0×10^7 m/s, how much current does the beam carry? What is the direction of this current?
5. Electrons in the Stanford Linear Accelerator are accelerated to nearly the speed of light. These high-energy electrons are produced in pulses containing 5×10^{11} electrons each, lasting 1.6 μs. (a) Assuming an electron speed essentially that of light, what is the physical length of each pulse? (b) What is the peak current (i.e., the rate of charge flow while a pulse is going by)? (c) If the accelerator produces 180 pulses per second, what is the average current?
6. The National Electrical Code specifies a maximum current of 10 A in 16-gauge (0.129 cm diameter) copper wire. What is the corresponding current density?
7. Each atom in aluminum contributes about 3.5 conduction electrons. What is the drift speed in a 0.21-cm-diameter aluminum wire carrying 20 A?
8. What is the diameter of a copper wire carrying 15 A, if the drift speed is 0.86 mm/s?
9. What is the drift speed in a silver wire carrying a current density of 150 A/mm²? Each silver atom contributes 1.3 free electrons.
10. The filament of the light bulb in Example 27-6 has a diameter of 0.050 mm. What is the current density in the filament? Compare with the current density in a 12-gauge wire (diameter 0.21 cm) supplying current to the light bulb.
11. A gold film in an integrated circuit measures 2.5 μm thick by 0.18 mm wide. It carries a current density of 6.8×10^5 A/m². What is the total current?
12. A piece of copper wire joins a piece of aluminum wire whose diameter is twice that of the copper. The same current flows in both wires. The density of conduction electrons in copper is 1.1×10^{29} m⁻³; in aluminum it is 2.1×10^{29} m⁻³. Compare (a) the drift speeds and (b) the current densities in each.
13. A plasma used in fusion research contains 5.0×10^{18} electrons and an equal number of protons per cubic meter.

Under the influence of an electric field the electrons drift in one direction at 40 m/s, while the protons drift in the opposite direction at 6.5 m/s. (a) What is the current density? (b) What fraction of the current is carried by the electrons?

14. In Fig. 27-20, a 100-mA current flows through a copper wire 0.10 mm in diameter, a 1.0-cm-diameter glass tube containing a salt solution, and a vacuum tube where the current is carried by an electron beam 1.0 mm in diameter. The density of conduction electrons in copper is 1.1×10^{29} m⁻³. The current in the solution is carried equally by positive and negative ions with charges $\pm 2e$; the density of each ion species is 6.1×10^{23} m⁻³. The electron density in the beam is 2.2×10^{16} m⁻³. Find the drift speed in each region.

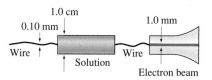

FIGURE 27-20 Problem 14.

15. The current in a wire as a function of time is given by $I(t) = 4t - 3t^2$, where t is in seconds and I in amperes. (a) Find the net charge q that has passed through the wire as a function of time, assuming $q = 0$ at $t = 0$. (b) At what time will the net charge be zero?

Section 27-2 Conduction Mechanisms

16. The electric field in an aluminum wire is 0.085 V/m. What is the current density in the wire?
17. What electric field is necessary to drive a 7.5-A current through a silver wire 0.95 mm in diameter?
18. A cylindrical tube of sea water carries a total electric current of 350 mA. If the electric field in the water is 21 V/m, what is the diameter of the tube?
19. A 1.0-cm-diameter rod carries a 50-A current when the electric field in the rod is 1.4 V/m. What is the resistivity of the rod material?
20. There is a potential difference of 2.5 V between opposite ends of a 6.0-m-long iron wire. (a) Assuming a uniform electric field in the wire, what is the current density? (b) If the wire diameter is 1.0 mm, what is the total current?
21. Use Table 27-1 to determine the conductivity of (a) copper and (b) sea water.
22. The maximum safe current in 12-gauge (0.21-cm-diameter) copper wire is 20 A. What are (a) the current density and (b) the electric field under these conditions?

23. The free-electron density in aluminum is 2.1×10^{29} m^{-3}. What is the collision time in aluminum?

24. A pure silicon crystal contains 4.9×10^{28} atoms/m^3. At room temperature, the density of electron-hole pairs is 1×10^{16} m^{-3}. In what concentration (aluminum atoms per silicon atom) must aluminum be added to give a conductivity 1000 times that of pure silicon? Assume that each aluminum atom contributes one extra hole, and that the conductivity is proportional to the density of charge carriers.

Section 27-3 Resistance and Ohm's Law

25. What is the resistance of a heating coil that draws 4.8 A when the voltage across it is 120 V?

26. What voltage does it take to drive 300 mA through a 1.2-kΩ resistance?

27. What is the current in a 47-kΩ resistor with 110 V across it?

28. The "third rail" that carries the electric power to a subway train is a rectangular iron bar whose cross section measures 10 cm × 15 cm, as shown in Fig. 27-21. What is the resistance of a 5.0-km piece of this rail?

FIGURE 27-21 A "third rail" (Problem 28).

29. What current flows when a 45-V potential difference is imposed across a 1.8-kΩ resistor?

30. A silver and an iron wire of the same length and diameter carry the same current. How do the voltages across the two compare?

31. The presence of a few ions makes air a conductor, albeit a poor one. If the total resistance between the ionosphere and Earth is 200 Ω, how much current flows as a result of a 300-kV potential difference between Earth and ionosphere?

32. A uniform wire of resistance R is stretched until its length doubles. Assuming its density and resistivity remain constant, what is its new resistance?

33. A cylindrical iron rod measures 88 cm long and 0.25 cm in diameter. (a) Find its resistance. If a 1.5-V potential difference is applied between the ends of the rod, find (b) the current, (c) the current density, and (d) the electric field in the rod.

34. You have a cylindrical piece of material 2.4 cm long and 2.0 mm in diameter. When you attach a 9-V battery to the ends of the piece, a current of 2.6 mA results. Which material from Table 27-1 do you have?

35. How must the diameters of copper and aluminum wire be related if they are to have the same resistance per unit length?

36. Extension cords are often made from 18-gauge copper wire (diameter 1.0 mm). (a) What is the resistance per unit length of this wire? (b) An electric saw that draws 7.0 A is operated at the end of an 8.0-m-long extension cord. What is the potential difference between the wall outlet and the saw?

37. Engineers call for a power line with a resistance per unit length of 50 mΩ/km. What wire diameter is required if the line is made of (a) copper or (b) aluminum? (c) If the costs of copper and aluminum wire are $1.53/kg and $1.34/kg, which material is more economical? The densities of copper and aluminum are 8.9 g/cm^3 and 2.7 g/cm^3, respectively.

38. A solid, rectangular iron bar measures 0.50 cm by 1.0 cm by 20 cm. Find the resistance between each of the three pairs of opposing faces, assuming that the faces in question are equipotentials.

39. Corrosion at battery terminals results in increased resistance, and is a frequent cause of hard starting in cars. In an effort to diagnose hard starting, a mechanic measures the voltage between the battery terminal and the wire carrying current to the starter motor. While the motor is cranking, this voltage is 4.2 V. If the motor draws 125 A, what is the resistance at the battery terminal?

40. A clear plastic trough 2.5 cm wide, 5.0 cm high, and 15 cm long has the insides of its two long sides coated with metal, as shown in Fig. 27-22. If a 60-V potential difference is applied between these sides, how much current flows when the trough contains (a) pure water and (b) sea water?

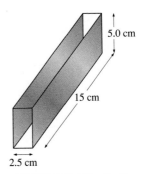

FIGURE 27-22 Problem 40.

Section 27-4 Electric Power

41. A car's starter motor draws 125 A with 11 V across its terminals. What is its power consumption?

42. A 4.5-W flashlight bulb draws 750 mA. (a) At what voltage does it operate? (b) What is its resistance?

43. A watch uses energy at the rate of 240 μW. How much current does it draw from its 1.5-V battery?

44. An electric stove burner with 35 Ω resistance consumes 1.5 kW of power. At what voltage does it operate?

45. What is the resistance of a standard 120-V, 60-W light bulb?

46. Use the numbers from Problem 31 to find the electric power dissipation in Earth's atmosphere. If we could harness this power, would it make a dent in humanity's 10^{12}-W electric power consumption?

47. If the electrons of Problem 4 are accelerated through a potential difference of 10 kV, how much power must be supplied to produce the electron beam?

48. The "instant on" feature of all the television sets in the United States requires the continuous power output of a typical large power plant—about 1000 MW. If there are 10^8 TVs in the United States, how much current does the "instant on" circuit of each draw from the 120-V power line?

49. How much total energy could the battery of Problem 3 supply?

50. During a "brown out," the power line voltage drops from 120 V to 105 V. By how much does the thermal power output of a 1500-W stove burner drop, assuming its resistance remains constant?

51. Two cylindrical resistors are made from the same material and have the same length. When connected across the same battery, one dissipates twice as much power as the other. How do their diameters compare?

52. Your author's house uses approximately 110 kWh of electrical energy each week. If that energy is supplied at 240 V, what average resistance does the house present to the power line?

53. A 2000-horsepower electric railroad locomotive gets its power from an overhead wire with 0.20 Ω/km. The potential difference between wire and track is 10 kV. Current returns through the track, whose resistance is negligible. (a) How much current does the locomotive draw? (b) How far from the power plant can the train go before 1% of the energy is lost in the wire?

54. A 100% efficient electric motor is lifting a 15-N weight at 25 cm/s. If the motor is connected to a 6.0-V battery, how much current does it draw?

Paired Problems

(Both problems in a pair involve the same principles and techniques. If you can get the first problem, you should be able to solve the second one.)

55. Electrons in a fine silver wire 20 μm in diameter drift at 0.14 mm/s. What is the current in the wire? Each silver atom contributes 1.3 free electrons.

56. A potassium slab carries a current density of 470 kA/m², and the drift speed of the electrons is 0.20 mm/s. If the density of potassium is 860 kg/m³, what is the average number of free electrons contributed by each atom?

57. What is the resistance of a column of mercury 0.75 m long and 1.0 mm in diameter?

58. An integrated circuit design calls for a 470-Ω resistor. The resistor is to be 10 μm long, 1.4 μm wide, and 0.85 μm thick, with connections made at the ends 10 μm apart. Its resistivity is to be set by the appropriate amount of doping. What should that resistivity be?

59. A power plant produces 1000 MW to supply a city 40 km away. Current flows from the power plant on a single wire of resistance 0.050 Ω/km, through the city, and returns via the ground, assumed to have negligible resistance. At the power plant the voltage between the wire and ground is 115 kV. (a) What is the current in the wire? (b) What fraction of the power is lost in transmission?

60. What should be the power line voltage in the preceding problem if the transmission loss is not to exceed 2%?

61. A 240-V electric motor is 90% efficient, meaning that 90% of the energy supplied to it ends up as mechanical work. If the motor lifts a 200-N weight at 3.1 m/s, how much current does it draw?

62. An 8.5-kN elevator is powered by a 480-V electric motor that draws 24 A. If the motor is 85% efficient, how long does it take to lift the elevator 18 m?

Supplementary Problems

63. A metal bar has a rectangular cross section 5.0 cm by 10 cm, as shown in Fig. 27-23. The bar has a nonuniform conductivity, ranging from zero at the bottom to a maximum at the top. As a result, the current density increases linearly from zero at the bottom to 0.10 A/cm² at the top. What is the total current in the bar?

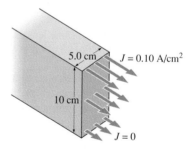

FIGURE 27-23 Problem 63.

64. Metallic copper (atomic weight 64, density 8.9 g/cm²) forms a crystal structure with copper atoms located at the corners of cubes. (a) Use the density and atomic weight to determine the distance between copper atoms. (b) Use your result, and the collision time from Example 27-4, to

estimate the mean thermal speed of the electrons in copper. Consider τ to be the mean time between collisions.

65. The electric car of Fig. 27-24 converts 70% of its electrical energy supply into mechanical energy available at the wheels. The car weighs 640 kg and has a 96-V battery. How much current does the motor draw when the car is climbing a 10° slope at 45 km/h? Neglect friction and air resistance.

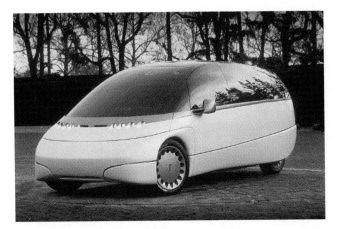

FIGURE 27-24   An electric car (Problem 65).

66. An immersion-type heating coil is connected to a 120-V outlet and immersed in a 250-ml cup of water initially at 10°C. The water comes to a boil in 85 s. What are (a) the power and (b) the resistance of the heater? Assume no heat loss, and neglect the mass of the heater.

67. A 100-Ω resistor of negligible mass is mounted inside a calorimeter. When a 12-V battery is connected for 5.0 min, the temperature inside the calorimeter rises by 26°C. What is the heat capacity of the calorimeter contents?

68. A parallel-plate capacitor has plates of 10 cm² area separated by a 1.0-mm layer of glass insulation with resistivity $\rho = 1.2 \times 10^{13}$ Ω·m and dielectric constant $\kappa = 5.6$. The capacitor is charged to 100 V and the charging battery disconnected. (a) What is the initial rate of discharge (i.e., the current through the insulation)? (b) At this rate, how long would it take the capacitor to discharge fully? (The rate does not remain constant; more on this in the next chapter.)

69. Figure 27-25 shows a resistor made from a truncated cone of material with uniform resistivity ρ. Consider the cone to be made of thin slices of thickness dx, like the one shown; Equation 27-6 shows that the resistance of each slab is

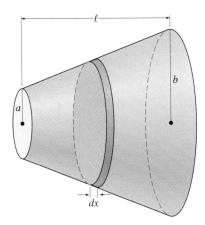

FIGURE 27-25   Problem 69.

$dR = \rho \, dx/A$. By integrating over all such slices, show that the resistance between the two flat faces is $R = \rho \ell/\pi ab$. (This method assumes the equipotentials are planes, which is only approximately true.)

70. A circular pan of radius b has a plastic bottom and metallic side wall of height h. It is filled with a solution of resistivity ρ. A metal disk of radius a and height h is placed at the center of the pan, as shown in Fig. 27-26. The side and disk are essentially perfect conductors. Show that the resistance measured from side to disk is $\rho \ln(b/a)/2\pi h$.

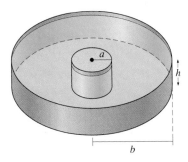

FIGURE 27-26   Problem 70.

71. At some point in a material of resistivity ρ the current density is J. Show that the power per unit volume dissipated at that point is $J^2\rho$.

72. A thermally insulated container of sea water carries a uniform current density of 75 mA/cm². How long does it take to raise its temperature from 15°C to 20°C? Use the result of the preceding problem, and assume that both the specific heat (assumed the same as pure water) and the resistivity are constant over this range.

CHAPTER 28

ELECTRIC CIRCUITS

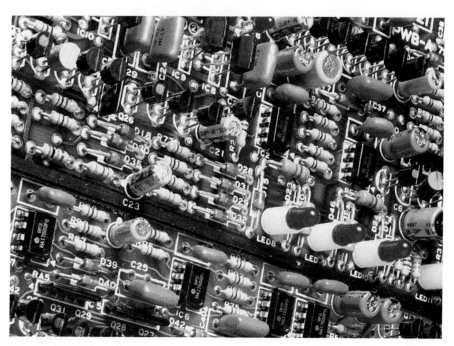

This circuit board is a complex interconnection of electronic components, including resistors, capacitors, transistors, light-emitting diodes, and integrated circuits.

28-1 CIRCUITS AND SYMBOLS

An **electric circuit** is a collection of electrical devices, called **circuit elements,** connected by conductors. A circuit usually contains a source of electrical energy, and is designed to do something useful. You are most familiar with human-made electrical circuits, which range from simple flashlights to computers, but important circuits also exist in nature. Examples include nervous systems in living organisms and Earth's global atmospheric circuit, in which thunderstorms are the batteries and the atmosphere a resistor. Your study of electric circuits should prove immensely practical, for it will help you to understand and to use effectively and safely the growing myriad of electrical and electronic devices you encounter. Basic circuit knowledge can even help you design new devices and troubleshoot old ones.

It is often helpful to represent circuits symbolically. We do so using standard symbols for circuit elements, with lines to represent the wires connecting them. We usually assume that the wires are perfect conductors, so that all points connected by wires alone are at the same potential; such points are electrically equivalent. Realizing this will greatly facilitate your interpretation of circuit diagrams! Figure 28-1 shows some common circuit symbols.

28-2 ELECTROMOTIVE FORCE

In the preceding chapter, we found that an electric field is necessary to drive a current in any conductor with nonzero resistance. But if we simply apply an electric field, say by putting excess charge on one end of the conductor, the charge will quickly redistribute itself until electrostatic equilibrium is reached and the electric field disappears. Somehow we must maintain the electric field, and with it the current, despite the tendency toward equilibrium. This requires that we compensate for the energy lost through the collisions that give the material its resistance.

What we need is a device that can maintain charge separation by converting energy from some other form into electrical energy. We call such a device a source of **electromotive force** or **emf.** (The name has historical origins; emf is not actually a force.) Most sources of emf have two electrical contacts, or **terminals,** for connection to other circuit elements. Energy conversion processes within the source move an excess of positive charge to one terminal, negative to the other, thus maintaining a potential difference between the terminals. The circuit symbol for a source of emf is shown in Fig. 28-1. The most familiar example is a battery, in which chemical energy drives electric charge to the two terminals. Other examples include electric generators, which convert mechanical energy to electrical energy; photovoltaic cells, which convert sunlight; fuel cells, which "burn" hydrogen to produce electrical energy and water; and biological cell membranes, which separate charge to control the movement of ions into the cell.

Electromotive force is quantified by the work per unit charge done by a source as it separates positive and negative charge to its two terminals. The units of emf are therefore volts, and the emf of a source is often called, loosely, its voltage. An **ideal source of emf**—one with no internal energy losses—maintains the same voltage across its terminals under all conditions. Real sources always have internal energy losses, so the terminal voltage may not equal the rated emf. We discuss this situation in the next section.

When a source of emf is not connected to any external circuit, no work is needed to maintain its terminal voltage. But current flows when the source is connected to an external circuit. This current would quickly deplete the charge at the terminals were if not for work done inside the source to separate more charge. The simple analogy of Fig. 28-2 illustrates the operation of a source of emf. The energy conversion mechanism in the emf is like the person lifting bowling balls to the table top. The balls then roll down a ramp, bumping into a series of pegs on the way down and giving up the energy they gained when lifted onto the table. Once they reach the bottom the balls roll back to where they're again lifted to repeat the cycle. An emf does the same thing: It "lifts" charge

Circuit symbols

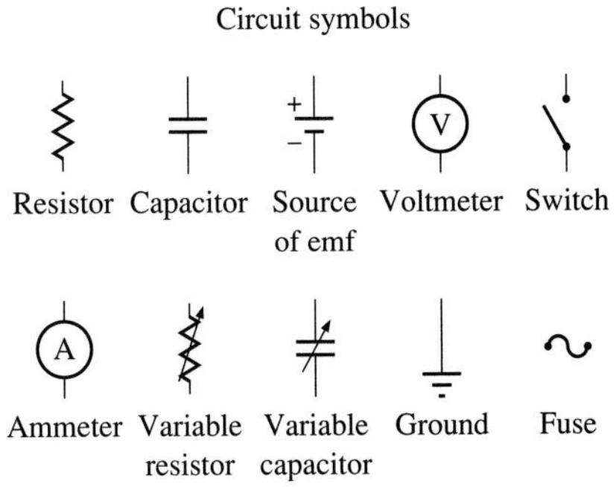

FIGURE 28-1 Common circuit symbols.

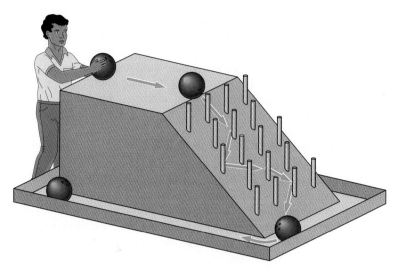

FIGURE 28-2 A gravitational analog of a source of emf. The person lifting the bowling balls represents the energy conversion mechanism in the emf, which does work on charges as it moves them against the electric field. The ramp, studded with pegs to which the balls transfer their energy, is like an external resistance connected across the emf.

against the internal electric field that points from its positive to its negative terminal. The charge then "falls" through the external circuit, dissipating its energy in the circuit resistance. When the charge returns to the source it is again "lifted," and the process continues.

28-3 SIMPLE CIRCUITS: SERIES AND PARALLEL RESISTORS

In the circuit of Fig. 28-3 a battery of emf $\mathscr{E}$ drives a current through the resistor R. How much current? The voltage across the battery is its emf $\mathscr{E}$. The battery is connected to the resistor by wires, assumed to have zero resistance. Because they have zero resistance, there is no potential difference across either wire. Therefore the voltage across the resistor is the same as the voltage across the battery, and we can immediately apply Ohm's law to find the current:

$$I = \frac{\mathscr{E}}{R}.$$

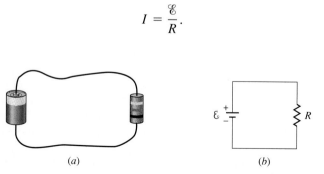

(a)   (b)

FIGURE 28-3 A circuit containing a battery and a resistor. (a) Physical circuit. (b) Schematic diagram, using the symbols of Fig. 28-1.

Energetically, this circuit is analogous to Fig. 28-2. Charge gains energy in the amount $\mathcal{E}$ joules per coulomb as it goes through the battery, then dissipates that energy in heating the resistor.

> **TIP** **Don't Get Hung Up about Wires** If the charge loses its energy flowing through the resistor, how does it then get back to the battery? If the wire is a perfect conductor, there's no problem because it takes no energy—and therefore no voltage—to drive current through it. In this case the current is determined entirely by the battery emf $\mathcal{E}$ and the resistance R. If you try to use Ohm's law to calculate the currents and voltages in ideal wires, you are needlessly complicating things! Of course real wires have some resistance, but if it's negligible compared with other resistances in the circuit, then the voltage across the wires is negligible, and we can approximate them as being ideal.

Series Resistors

Figure 28-4 shows a circuit containing two resistors in series. What is the current through these resistors? What is the voltage across each? Note that neither resistor is connected directly across the battery, so we can't argue that the voltage across either is the battery voltage. However, the full battery voltage does appear across the series combination, so if we knew the equivalent resistance we could solve for the current. What current? The current in *both* resistors. Since they are in series the only place current flowing through R_1 can go is through R_2. As long as there is no buildup of charge in the circuit, the current through both resistors—and, for that matter, through the battery as well—must be the same. This situation holds whenever circuit elements are in series:

| The current through circuit elements in series is the same.

If I is the current in the circuit of Fig. 28-4, then there must be a voltage $V_1 = IR_1$ across R_1 to drive the current through this resistor. Similarly, the voltage across R_2 is $V_2 = IR_2$. Thus, the potential difference across the two resistors together is $V_1 + V_2 = IR_1 + IR_2$. But the battery is connected directly across this series combination, so we have

$$IR_1 + IR_2 = \mathcal{E},$$

or

$$I = \frac{\mathcal{E}}{R_1 + R_2}.$$

Comparison with Ohm's law in the form $I = V/R$ shows that the two resistors in series behave like an equivalent resistance equal to the sum of their resistances. In an obvious generalization to more resistors in series, we have

$$R_{\text{series}} = R_1 + R_2 + R_3 + \cdots \qquad (28\text{-}1)$$

In other words, resistors in series add.

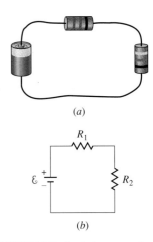

FIGURE 28-4 A battery and two resistors in series. (*a*) Physical circuit. (*b*) Schematic diagram.

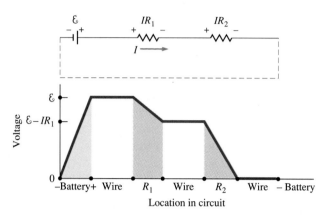

FIGURE 28-5 Voltages in the circuit of Fig. 28-4, with $V = 0$ at the negative battery terminal. Note that there is no potential difference across the wires, since they have negligible resistance, and that potential increases across the battery and decreases across the resistors. Can you tell from the graph which resistance is greater?

Given the current, we can use Ohm's law in the form $V = IR$ to solve for the voltage across each resistor:

$$V_1 = \frac{R_1}{R_1 + R_2}\mathscr{E} \tag{28-2a}$$

and

$$V_2 = \frac{R_2}{R_1 + R_2}\mathscr{E}. \tag{28-2b}$$

These expressions show that the battery voltage divides between the two resistors in proportion to their resistance. For this reason a series combination of resistors is called a **voltage divider.** Figure 28-5 depicts the voltages throughout the circuit of Fig. 28-4, and shows explicitly that the resistors divide the battery voltage.

● **EXAMPLE 28-1** DESIGNING A VOLTAGE DIVIDER

A light bulb with a resistance (when on) of 5.0 Ω is designed to operate at a current of 600 mA. To operate this lamp from a 12-V battery, what resistance should you place in series with it?

Solution

Let R_2 be the lamp and R_1 the unknown series resistor. Since resistors in series add, the current through both resistors is $I = \mathscr{E}/(R_1 + R_2)$, which is supposed to be 600 mA or 0.60 A. Solving for R_1 gives

$$R_1 = \frac{\mathscr{E} - IR_2}{I} = \frac{12\ \text{V} - (0.60\ \text{A})(5.0\ \Omega)}{0.60\ \text{A}} = 15\ \Omega.$$

You can also get this result by noting that the light bulb's proper operating voltage is $V = IR_2 = (0.60\ \text{A})(5.0\ \Omega) = 3.0$ V. This is one-fourth of the battery voltage, so the light bulb's 5-Ω resistance should be one-fourth of the total. That makes the total 20 Ω, leaving 15 Ω for R_1.

EXERCISE Suppose that in Fig. 28-4 $R_1 = 470\ \Omega$. If the voltage across R_2 is 59% of the battery voltage, find R_2.

Answer: 676 Ω

Some problems similar to Example 28-1: 23–25
●

Real Batteries

What's the difference between the two 1.5-V batteries shown in Fig. 28-6? If both were ideal there would be no difference because both would maintain 1.5 V across their terminals no matter how much current was flowing. But these are real batteries. The rate at which internal chemical reactions take place limits the amount of current each can supply. Not surprisingly, the larger battery can supply more current.

We can model a real battery by considering it to be an ideal emf in series with an **internal resistance,** as shown in Fig. 28-7. Of course this is not how batteries are made since no manufacturer can make an ideal emf! The internal resistance is intrinsic to the battery, and there is no way to circumvent it. The more powerful battery is the one with lower internal resistance; it approaches more closely the ideal of zero internal resistance and can therefore supply more current.

We can understand the effect of internal resistance by considering the circuit of Fig. 28-8. This is just the series circuit of Fig. 28-4, with R_1 the internal resistance R_{int} and R_2 the external resistance R_L. R_L is called the *load resistor* because it is the thing to which we wish to deliver electric power; it is the electrical load on the battery. From Equation 28-2b we see that if R_{int} is small compared with R_L, then the voltage across the load resistance will be very close to the battery's internal emf. In this case the battery's behavior is nearly ideal, since it has essentially $\mathcal{E}$ volts across its terminals. But if we lower R_L so it becomes comparable with R_{int}, then the voltage across R_L decreases and the battery no longer seems ideal. As we lower R_L we draw more current from the battery. It takes a higher voltage to drive this current through the fixed resistance R_{int}, so more voltage drops across R_{int}, leaving less across R_L. Even if we short-circuit the battery terminals (which is not good for the battery!) we will not get infinite current—in fact, we will simply have

$$I = \frac{\mathcal{E}}{R_{int}}. \quad \text{(battery short-circuited)}$$

We conclude that a battery or other source of emf behaves more or less ideally depending on the size of its load resistance relative to its internal resistance. A calculator, for example, has a very high resistance and draws little current. It is quite happy with a small battery whose internal resistance, while relatively high, is still small compared with the calculator's resistance. A car starter motor, on the other hand, draws a large current and thus requires a battery with very low internal resistance.

FIGURE 28-6 A 1.5-V calculator battery and a 1.5-V D-cell flashlight battery have the same voltage, but the internal resistance of the calculator battery is higher.

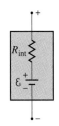

FIGURE 28-7 A real battery modeled as an ideal emf in series with an internal resistance.

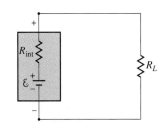

FIGURE 28-8 A real battery connected to an external load. Some voltage drops across the internal resistance, making the voltage across the battery terminals less than the battery's rated voltage.

● EXAMPLE 28-2 STARTING A CAR

Your car's starter motor draws 125 A. The car has a 12-V battery, but while the starter motor is running the voltage across the battery terminals measures only 9.5 V. What is the internal resistance of the battery?

Solution

This circuit is just that of Fig. 28-8, with the starter being the load. With 9.5 V across the starter, there must be 2.5 V left across the internal resistance to make a total of 12 V. The current is the same throughout this series circuit, so 125 A is the current through R_{int}. Knowing current and voltage, we apply Ohm's law:

$$R = \frac{V}{I} = \frac{2.5 \text{ V}}{125 \text{ A}} = 0.020 \ \Omega.$$

A battery voltage between 9 and 11 volts is typical of a car being started. A battery voltage much below 9 volts usually indicates a weak battery, a defective starter motor, or very cold weather!

EXERCISE   A 9-V battery has an internal resistance of 13 Ω. What is the maximum current that can be drawn from the battery if its terminal voltage is to remain above 8.0 V?

Answer: 77 mA

Some problems similar to Example 28-2:   12, 13, 15, 17
●

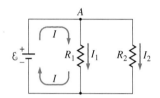

FIGURE 28-9   Parallel resistors connected across a battery. The two resistor currents sum to the battery current.

Parallel Resistors

Figure 28-9 shows two resistors in parallel, connected across an ideal battery. What is the equivalent resistance of this parallel combination? Since the two resistors are connected at top and bottom by ideal wires, the voltage across each must be the same. We made this point in Chapter 26 when we discussed parallel capacitors, and it's worth repeating here:

▌ **The voltage across circuit elements in parallel is the same.**

The parallel resistors are connected directly across the battery, so their common voltage is the battery emf $\mathcal{E}$. Applying Ohm's law then gives the current through each resistor:

$$I_1 = \frac{\mathcal{E}}{R_1}$$

and

$$I_2 = \frac{\mathcal{E}}{R_2}.$$

At the point marked A in Fig. 28-9, a current I brings in charge from the battery, while the currents I_1 and I_2 carry charge away. If charge is not to accumulate (see Problem 69), the incoming and outgoing currents must be equal; that is,

$$I = I_1 + I_2.$$

Using our expressions for the two resistor currents gives

$$I = \frac{\mathcal{E}}{R_1} + \frac{\mathcal{E}}{R_2} = \mathcal{E}\left(\frac{1}{R_1} + \frac{1}{R_2}\right).$$

Comparison with Ohm's law in the form $I = V/R$ shows that the equivalent resistance of the parallel combination is given by

$$\frac{1}{R_{\text{parallel}}} = \frac{1}{R_1} + \frac{1}{R_2}. \tag{28-3a}$$

This result is readily generalized to more than two parallel resistors:

$$\frac{1}{R_{\text{parallel}}} = \frac{1}{R_1} + \frac{1}{R_2} + \frac{1}{R_3} + \cdots. \qquad (28\text{-}3b)$$

In other words, resistors in parallel add reciprocally. Equation 28-3b shows that the resistance of a parallel combination is always lower than that of the lowest resistance in the combination. When there are only two parallel resistors, we can rewrite Equation 28-3a using a common denominator to obtain

$$R_{\text{parallel}} = \frac{R_1 R_2}{R_1 + R_2}. \qquad (28\text{-}3c)$$

Note that *parallel* resistors combine in the same way as *series* capacitors, and vice versa.

● EXAMPLE 28-3 PARALLEL AND SERIES RESISTORS

You have available three 2.0-Ω resistors. What different resistances can you make by combining all three resistors?

Solution
Figure 28-10 shows the four possible combinations. Resistors in series add, so combination (*a*) has 6.0 Ω. Resistors in parallel reciprocally add, so combination (*b*) has

$$\frac{1}{R} = \frac{1}{2.0\ \Omega} + \frac{1}{2.0\ \Omega} + \frac{1}{2.0\ \Omega} = 1.5\ \Omega^{-1},$$

for a resistance of 0.67 Ω. Combination (*c*) has two resistors in series, giving 4.0 Ω. This 4.0-Ω combination is in parallel with 2.0 Ω, so Equation 28-3c gives

$$R = \frac{(2.0\ \Omega)(4.0\ \Omega)}{2.0\ \Omega + 4.0\ \Omega} = 1.3\ \Omega.$$

Finally, combination (*d*) has two resistors in parallel, giving 1.0 Ω. This combination is in series with 2.0 Ω, for a total of 3.0 Ω. Thus you can make combinations ranging from 0.67 Ω to 6.0 Ω with these three equal resistors.

EXERCISE A 270-Ω and a 470-Ω resistor are connected in parallel, and the combination is connected in series with a 150-Ω resistor. Find the equivalent resistance of this combination.

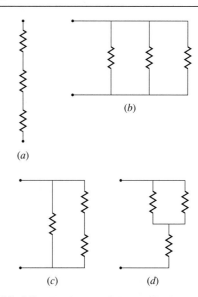

FIGURE 28-10 The four possible combinations of three equal resistors (Example 28-3).

Answer: 321 Ω

Some problems similar to Example 28-3: 16, 18, 19, 59, 60 ●

Analyzing Circuits

Many circuits contain series and parallel combinations of basic circuit elements. We can simplify these circuits by treating each series or parallel combination as a single element, often continuing the process until we can determine the voltages and currents throughout the entire circuit. Example 28-4 illustrates this procedure, which is similar to the way we dealt with capacitor combinations in Chapter 26.

● **EXAMPLE 28-4**   ANALYZING A CIRCUIT

In the circuit of Fig. 28-11, what is the current through the 2-Ω resistor?

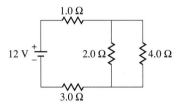

FIGURE 28-11   Circuit for Example 28-4.

Solution

We approach this problem by simplifying the circuit until we can solve for something—in this case the total current. Then we reverse the process, analyzing the circuit details until we can solve for the quantity we want. Figure 28-12 shows the steps in simplifying the circuit. We get from the original circuit, Fig. 28-12*a*, to Fig. 28-12*b* by calculating the resistance of the parallel combination of 2 Ω and 4 Ω:

$$R_\| = \frac{(2.0\ \Omega)(4.0\ \Omega)}{2.0\ \Omega + 4.0\ \Omega} = 1.33\ \Omega.$$

Figure 28-12*b* shows three resistors in series, which add to get the single-resistor circuit of Fig. 28-12*c*. From here we can calculate the total current:

$$I = \frac{\mathscr{E}}{R} = \frac{12\ \text{V}}{5.33\ \Omega} = 2.25\ \text{A}.$$

Where does this current flow? It flows from the battery through the 1-Ω resistor, then on through the parallel combination of the 2-Ω and 4-Ω resistors, then through the 3-Ω resistor and back to the battery. It does *not* all flow through the 2-Ω resistor because there are two paths the current can take when it gets to the parallel combination. However, it does all flow through the parallel combination. We already found that this combination has a resistance of 1.33 Ω, and now we know that 2.25 A flows through the combination. So the voltage across the combination is

$$V = IR = (2.25\ \text{A})(1.33\ \Omega) = 2.99\ \text{V}.$$

This same voltage appears across each of the two resistors making up the parallel combination (why?) so the current through the 2-Ω resistor is

$$I = \frac{V}{R} = \frac{2.99\ \text{V}}{2.0\ \Omega} = 1.5\ \text{A}.$$

In solving for this current we effectively reversed our original simplification of the circuit, first considering Fig. 28-12*b* to get

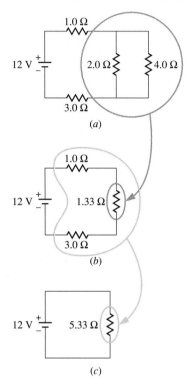

FIGURE 28-12   Simplifying the circuit by forming parallel and series combinations.

the voltage across the parallel combination, and then going to the full circuit to get the answer. At each stage we applied Ohm's law to solve for either a voltage or a current as needed.

> **TIP   Don't Abuse Ohm's Law**   Ohm's law relates the voltage *across a resistor* to the current *through that resistor*. It does *not* relate arbitrary voltages and currents anywhere in a circuit. Just because there is a 12-V battery in Fig. 28-11, for example, does *not* mean that 12 V appears across the 2-Ω resistor. Be careful, too, with series and parallel combinations. As it stands, the only individual resistors in Fig. 28-11 in either series or parallel are the parallel pair of 2 Ω and 4 Ω. The 1-Ω and 2-Ω resistors are *not* in series because current flowing through the 1-Ω resistor need not all go through the 2-Ω resistor; some can go through the 4-Ω resistor instead. Equations 28-1 and 28-3 apply *only* to combinations that are strictly series or parallel, respectively.

EXERCISE   In Fig. 28-13, find (a) the current supplied by the battery and (b) the voltage across the 180-Ω resistor.

Answers: (a) 30 mA; (b) 2.9 V

Some problems similar to Example 28-4:   26–28

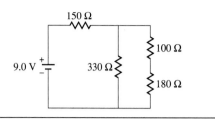

FIGURE 28-13   What is the voltage across the 180-Ω resistor?

28-4 KIRCHHOFF'S LAWS AND MULTILOOP CIRCUITS

Some circuits cannot be simplified using series and parallel combinations. This often happens when there is more than one source of emf or when circuit elements are connected in complicated ways. In Fig. 28-14, for example, are resistors R_1 and R_2 in parallel? No, because R_3 separates their lower ends. Are R_1 and R_3 in series? No, because current flowing out the bottom of R_1 can go through either R_3 or R_4. Solving circuits that aren't reducible to series and parallel combinations usually requires more general techniques than we've used so far.

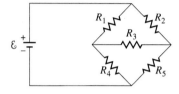

FIGURE 28-14   This circuit cannot be analyzed using series and parallel combinations.

Kirchhoff's Laws

In Fig. 28-5 we looked at changes in electric potential around the loop comprising a simple series circuit. The result was an increase in potential at the battery, followed by decreases at both resistors that summed to the gain in the battery. If we count the increase as a positive change and the decreases as negative changes, we can state the following:

| **The sum of the voltage changes across all the circuit elements around any closed loop is zero.**

This statement is known as **Kirchhoff's loop law,** and it applies not just to Fig. 28-5 but to *any* closed loop in a circuit. The loop law is ultimately about energy conservation; it says that charge moving around a loop gains as much energy from batteries or other sources as it loses in resistors or other energy-conversion devices.

In analyzing parallel resistors with Fig. 28-9 we noted that the current flowing into point A must equal the total current flowing out. This is really a statement about conservation of charge. It applies to any point in a circuit carrying steady currents since under steady-state conditions charge cannot build up or be depleted. The statement is most useful at a point where three or more wires join; such a point, like A in Fig. 28-9, is called a **node.** If we count currents flowing into a node as positive, and currents flowing out as negative, then the statement of charge conservation becomes the following:

| **The sum of the currents at any node is zero.**

This is **Kirchhoff's node law.**

Analyzing Multiloop Circuits

Even the most complex circuits can be analyzed using Kirchhoff's laws. Applying the laws amounts to writing equations expressing the loop law and node law for the distinct loops and nodes in the circuit. The number of equations needed is generally one less than the number of loops plus one less than the number of nodes; this is because the quantities in one loop and one node can be expressed entirely in terms of other loops and nodes, making one loop and one node equation each redundant. Here we give just two examples; electrical engineers take entire courses in circuit analysis using these laws.

● **EXAMPLE 28-5**   A MULTILOOP CIRCUIT

Apply Kirchhoff's laws to find the current through R_1 in Fig. 28-15a.

Solution

We don't really need Kirchhoff's laws here; since resistors R_2 and R_3 are in parallel, we could use the methods developed earlier. But this simple circuit serves to illustrate the Kirchhoff approach. In Fig. 28-15b we've identified three loops and two nodes, and we've labeled three distinct currents. In a multiloop circuit it's not always obvious which way the currents flow, and therefore we've arbitrarily assigned directions. If an answer comes out negative, that just means the current is really flowing in the opposite direction. You can probably see in this circuit that I_2 and I_3 can't really be in opposite directions, but we will assume they are for the sake of illustration.

First we apply the loop law. Going clockwise around loop 1 starting at node B, we first encounter a voltage increase $\mathcal{E}_1$, then a drop $-I_1R_1$ in resistor R_1, then a *drop* $\mathcal{E}_2$ because we traverse the second battery going from positive to negative, and finally an increase I_2R_2 in R_2. Why an increase? Because we're going *against* the indicated direction of I_2 and, therefore, from what is purportedly the lower to the higher potential end of the resistor. Never mind if the situation is really reversed; again, the algebra will take care of it. So the loop law reads

$$\mathcal{E}_1 - I_1R_1 - \mathcal{E}_2 + I_2R_2 = 0. \quad \text{(loop 1)}$$

Our choice of going clockwise is arbitrary; going counterclockwise would give the same equation since all terms would enter with the opposite sign. Loop 2 is much easier; going clockwise from B, its equation reads

$$-I_2R_2 - I_3R_3 = 0. \quad \text{(loop 2)}$$

Here both terms are negative because we go through both resistors in the indicated current direction. What about loop 3? Physically, it consists of parts of loop 1 and loop 2. Mathematically, we could get the equation for loop 3 by subtracting our previous loop equations. In other words, the equation for loop 3 contains no new information, and we need not bother with it.

Having taken care of the loops, we apply Kirchhoff's node law to the circuit nodes. The node law says that the sum of currents at a node is zero. For node A this gives

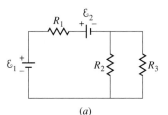

(a)

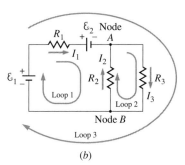

(b)

FIGURE 28-15   (a) A multiloop circuit. (b) The circuit contains 3 loops and 2 nodes. Current directions are arbitrary, and the algebraic signs of the answers will determine the actual directions (Example 28-5).

$$I_1 + I_2 - I_3 = 0, \quad \text{(node A)}$$

where I_3 enters with a minus sign since the direction we've assigned has I_3 carrying charge away from the node. Node B would give essentially the same equation. At any rate, we already have three equations in the three unknowns I_1, I_2, and I_3, and that is sufficient to solve for all three. We just need I_1, so we'll eliminate the other two. First, the node equation gives $I_3 = I_1 + I_2$; substituting in the loop 2 equation, we have $-I_2R_2 - I_1R_3 - I_2R_3 = 0$, or

$$I_2 = -\frac{I_1R_3}{R_2 + R_3}.$$

The minus sign here shows that I_1 and I_2 cannot both be in the directions indicated. Substituting this result for I_2 in the loop 1 equation and solving for I_1 then gives

$$I_1 = \frac{\mathcal{E}_1 - \mathcal{E}_2}{R_1 + \dfrac{R_2 R_3}{R_2 + R_3}}.$$

Does this result make sense? The denominator is just R_1 in series with the parallel combination of R_2 and R_3. If we inter-changed R_1 and $\mathcal{E}_2$ then we would have exactly what the equation describes: a battery of emf $\mathcal{E}_1 - \mathcal{E}_2$ connected across the resistor combination indicated by the denominator. Since the current in the series elements $\mathcal{E}_2$ and R_1 is the same, their order doesn't matter.

What about the current directions? Our answer shows that I_1 is in the direction indicated if $\mathcal{E}_1$ is the higher emf. Then I_2's direction must be opposite what we've indicated. If $\mathcal{E}_2$ is greater, these conclusions reverse. ●

● **EXAMPLE 28-6** RATE THE RESISTOR

What power dissipation must resistor R_3 of Fig. 28-16a be able to tolerate?

Solution
Again, we indicate loops, nodes, and currents (Fig. 28-16b). Here we need to know I_3 to find the power dissipation in R_3. Instead of solving algebraically as in Example 28-5, we will simply write the equations directly with their numerical values. Let's go counterclockwise around loop 1; starting at A, the loop law becomes

$$6 - 2I_1 - I_3 = 0, \qquad \text{(loop 1)}$$

where we've temporarily dropped the units. Starting from A and going clockwise around loop 2 (remember, the direction is arbitrary), we have

$$9 + 4I_2 - I_3 = 0. \qquad \text{(loop 2)}$$

Finally, the node equation at node A reads

$$-I_1 + I_2 + I_3 = 0, \qquad \text{(node } A\text{)}$$

where the first term is negative because I_1 is indicated as flowing *away* from the node. The equations for node B and loop 3 are redundant, so we're ready to solve the system. The node equation gives $I_1 = I_2 + I_3$; substituting in the loop 1 equation, we have $6 - 2I_2 - 3I_3 = 0$, or

$$I_2 = \tfrac{1}{2}(6 - 3I_3).$$

Finally, we can use this result in the loop 2 equation to get $9 + 2(6 - 3I_3) - I_3 = 0$ or

$$21 - 7I_3 = 0,$$

giving $I_3 = 3$ A. That this answer is positive indicates that I_3 is indeed upward in Fig. 28-16. The power dissipated in R_3 is therefore

$$P_3 = I_3^2 R_3 = (3 \text{ A})^2 (1 \ \Omega) = 9 \text{ W},$$

using Equation 27-9a. A 10-W resistor, the next larger size commercially available, would be just adequate.

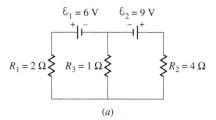

(a)

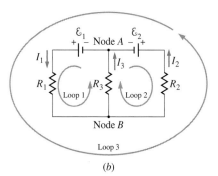

(b)

FIGURE 28-16 (a) A circuit that cannot be analyzed using series and parallel combinations. (b) There are 3 loops, 2 nodes, and 3 unknown currents whose direction assignments are, again, arbitrary (Example 28-6).

Although the circuits of Figs. 28-15 and 28-16 look very similar, that similarity is deceptive. The placement of $\mathcal{E}_2$ on the other side of node A in Fig. 28-16 means there is no way to solve this circuit using series and parallel combinations.

EXERCISE To what value should $\mathcal{E}_2$ in Fig. 28-16 be changed so that the current I_2 becomes zero?

Answer: 2.0 V

Some problems similar to Examples 28-5 and 28-6: 32–35 ●

(a)

(b)

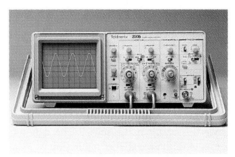
(c)

FIGURE 28-17 (a) Most voltmeters today have digital displays, although (b) analog—or moving-needle—meters are still found in older instruments. (c) An oscilloscope is essentially a pair of voltmeters in which two different voltages are indicated by the horizontal and vertical deflection of an electron beam. When the horizontal voltage varies linearly with time, the result is a plot of the vertical voltage versus time.

28-5 ELECTRICAL MEASURING INSTRUMENTS

Voltmeters

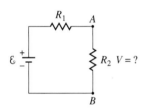

FIGURE 28-18 How to measure the voltage across R_2?

A **voltmeter** is a device that indicates the potential difference across its two terminals. The indication is usually by a digital readout, although older meters use a moving needle, and oscilloscopes use the deflection of an electron beam (Fig. 28-17). Potential difference—voltage—is a property of two points, and therefore to measure the voltage between two points we connect the two terminals of the voltmeter to those points. So to measure the voltage across resistor R_2 in Fig. 28-18, we connect the voltmeter to points A and B, as shown in Fig. 28-19a. We do *not* break the circuit and insert the meter, as in Fig. 28-19b, for then we would not be measuring the voltage *across* the resistor; in fact, we would radically alter the circuit.

How good is our voltage measurement? There are two considerations here. First, how accurately does the meter indicate the voltage across its terminals? For digital meters this is usually expressed as the number of significant figures in the digital display, while for analog (moving-needle) meters, accuracy is given as a percentage of the full-scale reading. But there is a more subtle question of accuracy. Even if the meter reads exactly the voltage across its terminals, can we be sure that voltage is the same as it was before the meter was connected? The circuit of Fig. 28-20 provides the answer. There is an open circuit between points A and B, so no current flows in the resistor R. With no current through the resistor, there is no voltage across the resistor, so the voltage between points A and B is the same as the battery voltage. We could have arrived at this result more formally using our voltage divider Equation 28-2b, with R_2 set to infinity in order to represent the open circuit.

Now connect a voltmeter between points A and B. Suppose the meter itself has a resistance R_m. Now the circuit looks like Fig. 28-20b. We have a complete circuit, with current flowing from the battery through R, through the meter, and back to the battery. A voltage across R is required to drive the current, so the voltage between points A and B is now less than it was before the meter was connected. The voltmeter reading is not the same as the voltage in the absence of the meter. This discrepancy occurs not because the meter is inherently inaccurate, but for the following reason:

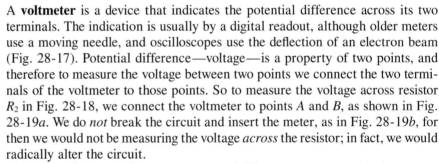
(a)

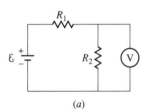

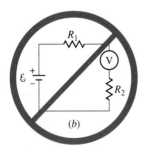
(b)

FIGURE 28-19 (a) Correct and (b) incorrect ways to measure the voltage across R_2.

The instrument affects the circuit being measured.

How far off is the meter reading? That depends on the meter resistance relative to the rest of the circuit. If the meter resistance is high, it will draw little current, and its effect on the circuit will be small. Our circuit with the meter included is identical to the voltage divider of Fig. 28-4, with R_1 replaced by R and R_2 by R_m. The meter voltage is then given by Equation 28-2b:

$$V_m = \frac{R_m}{R + R_m}\mathcal{E}.$$

As R_m becomes large compared with R, the fraction $R_m/(R + R_m)$ approaches 1 and the meter voltage becomes essentially the open-circuit voltage $\mathcal{E}$. But if R_m is not large compared with R, then the meter reading will be substantially lower. We conclude that a voltmeter should have a much higher resistance than typical resistances in the circuit being measured.

How much higher? That depends on how accurate a reading we require. For 1% accuracy, a rough rule of thumb is that the meter resistance should be 100 times the circuit resistance. If we're troubleshooting a car's electrical system, where currents are large and resistances low, we can get away with a fairly low meter resistance. But if we want to meaure the voltage developed by the electrode in a chemist's pH meter, we must use a very high resistance indeed, for the pH electrode looks like a nonideal source of emf with internal resistance as high as 10^{14} Ω.

An ideal voltmeter would draw no current and so must have infinite resistance. Older moving-needle meters have difficulty approaching this ideal because they consist fundamentally of a sensitive current meter in series with a large resistor; the current that operates the meter comes from the circuit being measured. A good meter of this type that reads 10 V full scale might have a resistance of 200 kΩ. Modern digital voltmeters, in contrast, contain amplifiers that greatly reduce the current drawn from the circuit; typical meter resistances are around 10 MΩ, and much higher values can be achieved. Today's digital meters include amplifier, digital converter, and digital display in a single integrated-circuit package, making them the most economical, reliable, and accurate variety.

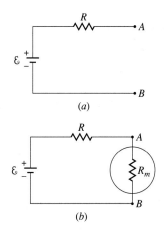

FIGURE 28-20 (*a*) Since no current flows, there is no voltage across the resistor, and the voltage between *A* and *B* is therefore the battery voltage. (*b*) A voltmeter with finite resistance R_m draws current, causing a voltage drop across R and thus lowering the voltage between *A* and *B*.

● EXAMPLE 28-7 VOLTMETERS

You wish to measure the voltage across the 40-Ω resistor of Fig. 28-21. What reading would an ideal voltmeter give? A voltmeter with a resistance of 1000 Ω?

Solution
An ideal voltmeter has infinite resistance and therefore would not alter the circuit, which is a simple voltage divider. Applying Equation 28-2b to this divider circuit gives the voltage across the 40-Ω resistor:

$$V_{40} = \frac{(40\ \Omega)(12\ \text{V})}{80\ \Omega + 40\ \Omega} = 4.0\ \text{V}.$$

Because the meter is connected in parallel with the resistor, this is also the voltage read by the meter.

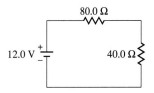

FIGURE 28-21 What is the voltage across the 40-Ω resistor (Example 28-7)?

With the nonideal voltmeter in place, the circuit becomes that of Fig. 28-22. The meter and 40-Ω resistor form a parallel combination whose resistance is given by Equation 28-3c:

$$R_{\parallel} = \frac{(40\ \Omega)(1000\ \Omega)}{40\ \Omega + 1000\ \Omega} = 38.5\ \Omega.$$

The circuit now looks like a voltage divider with $R_{\parallel}$ the lower resistor. Applying Equation 28-2b to this circuit gives

$$V_{\parallel} = \frac{(38.5\ \Omega)(12\ V)}{80\ \Omega + 38.5\ \Omega} = 3.9\ V.$$

This $V_{\parallel}$ is the voltage across the parallel combination consisting of the meter and 40-Ω resistor. Since the voltage across parallel circuit elements is the same, $V_{\parallel}$ is both the meter reading and the voltage across the 40-Ω resistor. And this value is 0.10 V—about 2.5%—lower than the value indicated by an ideal voltmeter.

EXERCISE A 100-kΩ resistor and a 150-kΩ resistor are in series, with a 250-V potential difference across the combina-

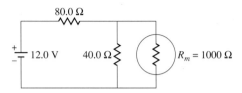

FIGURE 28-22 A nonideal voltmeter alters the circuit, lowering the overall resistance.

tion. A digital meter with a 3-significant-digit display and 1.0-MΩ resistance is used to measure the voltage across the 150-kΩ resistor. (a) What does it read? (b) By what percentage does this reading differ from that of an ideal voltmeter?

Answers: (a) 142 V; (b) 5.3%

Some problems similar to Example 28-7: 37, 38, 40, 61 ●

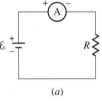

(*a*)

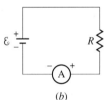

(*b*)

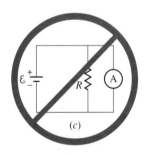

(*c*)

FIGURE 28-23 An ammeter goes in series with the circuit element whose current it is to measure. It doesn't matter whether the meter goes (*a*) before or (*b*) after the circuit element, in this case the resistor *R*. (*c*) Connecting it across the resistor is incorrect, since then the resistor current does not flow through the meter. In fact, this connection would probably destroy the meter!

Ammeters

An **ammeter** measures the current flowing *through* itself. To measure the current through a circuit element it is necessary to break the circuit and insert the ammeter in series with that element (Fig. 28-23); only then will all the current through the element also go through the meter. Connecting the meter across the resistor in Fig. 28-23 would be wrong, for then the current through the resistor would not be going through the meter.

What electrical properties should the ammeter have so it doesn't alter the circuit in which it is connected? If the meter had any resistance, then the total resistance of the circuit would increase with the meter connected in series. This in turn would decrease the current, resulting in an incorrect reading. We conclude that an ideal ammeter should have zero resistance. In practice, ammeter resistance should be much lower than typical resistances in the circuit being measured.

TIP **Watch Your Language when Connecting Meters** A voltmeter measures potential difference *between* two points; hence, we connect it *across*—i.e., in parallel with—the circuit element whose voltage we wish to measure. An ammeter measures the current *through* itself; hence, we connect it in *series* with the circuit element whose current we wish to measure. If you get used to voltages appearing *across* things and currents flowing *through* them, you'll have no trouble connecting meters. But if you insist on talking about "the voltage through" something, then you'll be unable to hook up meters accurately or safely. The ways to connect meters, and the words *across* for voltage and *through* for current, go right back to the definitions of potential difference as a property of two points and of current as a flow.

Ohmmeters and Multimeters

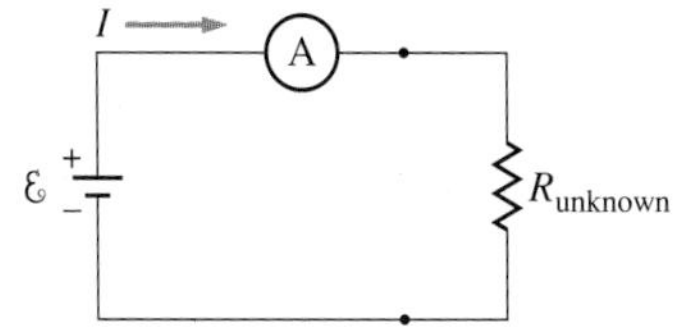

Often we would like to measure the resistance of a particular circuit element. We can do this by connecting a source of known voltage in series with an ammeter and the unknown resistance, as in Fig. 28-24. Knowing the voltage and measuring the current then allows us to calculate the unknown resistance. A meter used for this purpose can be calibrated directly in ohms even though it is really measuring current; it is then called an **ohmmeter.**

The functions of voltmeter, ammeter, and ohmmeter are often combined in a single instrument called a **multimeter.** Multimeters include switches for selecting the quantity and range to be measured, and may be either analog or digital. Figure 28-25 shows a modern digital multimeter, or DMM.

FIGURE 28-24 A simple ohmmeter consists of a known emf and an ammeter. When an unknown resistance is connected across the pair, its resistance may be determined from Ohm's law in the form $R = \mathcal{E}/I$.

Potentiometric Measurement

An elegant way to measure voltage is to compare the unknown voltage with an accurate standard, in much the same way that a pan balance weighs an unknown mass by balancing it against standard masses. Figure 28-26 shows how this scheme works. An accurately known emf $\mathcal{E}_0$ is connected across a resistor along which a sliding contact moves, forming a variable voltage divider in which the position of the sliding contact determines the voltage. The output of this voltage divider—called a **potentiometer**—is connected to the unknown voltage through a meter. The potentiometer is adjusted until the meter reads zero, at which point the potentiometer voltage—which can be read off its calibrated dial—must equal the unknown voltage. The great virtue of this method is that when the system is at null—the condition where source and unknown voltages are equal—then there is no current being drawn from the unknown regardless of the meter resistance, and therefore the method has, in principle, no effect on the circuit being measured.

In the past, potentiometric measurements were made with circuits like that of Fig. 28-26, with special batteries of precisely known emf and accurately calibrated potentiometers. Manual adjustment was used to achieve the null condition. Today, nulling is accomplished electronically or electromechanically through a process known as negative feedback. Circuits using the potentiometric technique are the basis of many powerful measurement and control devices.

FIGURE 28-25 This digital multimeter measures voltage, current, and resistance.

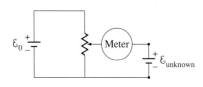

FIGURE 28-26 Potentiometric voltage measurement compares the unknown voltage with an accurately known voltage.

28-6 CIRCUITS WITH CAPACITORS

So far we have considered only circuits in which current and voltage are steady in all components. When you turn on a flashlight, for example, current starts to flow almost immediately through the bulb, batteries, and connecting metal parts. The current continues to flow steadily until you turn off the switch.

With a capacitor in a circuit, this picture changes. Circuit quantities change more gradually because of the capacitor. Why is this? Recall that a capacitor is a pair of insulated conductors that stores electrical energy when opposite charges are put on the conductors. A capacitor is characterized by its capacitance

$$C = \frac{Q}{V},$$

where Q is the magnitude of the charge on either conductor and V the voltage between the conductors. (See Section 26-4 for a review of capacitors.) Because charge and voltage are proportional in a capacitor, it is not possible to change the voltage without changing the charge. In a circuit, we change the capacitor charge by moving charge on or off the capacitor plates through wires connecting them to the rest of the circuit. This charge movement constitutes a current. The magnitude of the current is the rate at which charge is entering or leaving the capacitor. As long as the current is finite, as it is in any real circuit, the charge on the capacitor cannot change instantaneously. Because charge and voltage are proportional in a capacitor, we conclude the following:

▎**The voltage across a capacitor cannot change instantaneously.**

This simple statement is the key to understanding circuits containing capacitors.

The *RC* Circuit: Charging

Consider the circuit of Fig. 28-27. The capacitor is initially uncharged, so the voltage across it is zero. What happens when we close the switch?

The switch connects the left end of the resistor to the battery's positive terminal, so the left end of the resistor goes to $\mathcal{E}$ volts (here we take the zero of potential at the battery's negative terminal). The right end of the resistor is at the same voltage as the upper capacitor plate. But the voltage across the capacitor cannot change instantaneously, and therefore remains zero just after the switch is closed. With the capacitor plates both at zero volts, the full battery voltage $\mathcal{E}$ appears across the resistor. With $\mathcal{E}$ volts across the resistor, there must be a current $I = \mathcal{E}/R$ through the resistor. This current cannot flow "through" the capacitor but serves instead to pile positive charge on the upper plate, negative charge on the lower. The same current I flows everywhere except in the insulated gap between the capacitor plates.

Now that current is flowing, charge accumulates on the capacitor, and the capacitor voltage increases in proportion to this charge. As the capacitor voltage rises, the resistor voltage falls because the voltage across the series combination of resistor and capacitor is the battery voltage $\mathcal{E}$. But the current through the resistor is proportional to the resistor voltage, so the resistor current falls as well. This in turn decreases the *rate* at which charge accumulates on the capacitor plates, lowering the rate at which the capacitor voltage increases. The voltage across the capacitor continues to increase, and the current through the resistor to decrease, but at an ever slower rate.

What happens if we wait a long time? As the capacitor voltage approaches the battery voltage, the voltage across the resistor, hence the current through the resistor, and therefore the rate of charge buildup on the capacitor, all become very small. The whole system tends more and more slowly toward a final state in which the capacitor is charged to the full battery voltage and the current in the circuit is zero. Figure 28-28 summarizes the interplay among current, charge, and voltage, while Fig. 28-29 shows the time dependence of these quantities.

We can analyze this circuit quantitatively using the loop law. Going clockwise around the loop, we first encounter a voltage increase $\mathcal{E}$ across the battery, then a drop IR across the resistor, then a drop V_C from the upper to lower capacitor plate (Fig. 28-30). But the definition of capacitance gives $V_C = Q/C$, so the loop equation becomes

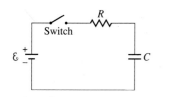

FIGURE 28-27 An *RC* circuit. The switch is closed at time $t = 0$.

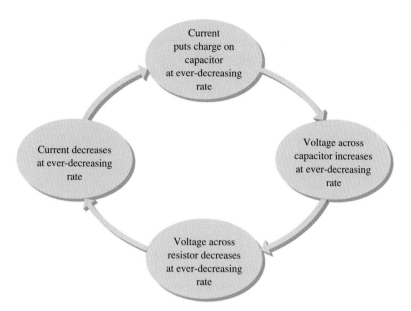

FIGURE 28-28 Interrelationships among circuit quantities in a charging *RC* circuit.

$$\mathscr{E} - IR - \frac{Q}{C} = 0.$$

This equation contains the two unknown quantities *I* and *Q*. Can we relate them? Yes—but not through a proportionality or other algebraic equation. Rather, the current is the rate at which charge is accumulating on the capacitor, or

$$I = \frac{dQ}{dt}.$$

To use this relation, we take the time derivative of the loop equation:

$$-R\frac{dI}{dt} - \frac{1}{C}\frac{dQ}{dt} = 0.$$

The battery voltage $\mathscr{E}$ does not appear in this differentiated equation because it is constant and thus its derivative is zero. Using $I = dQ/dt$ and rearranging the equation slightly gives

$$\frac{dI}{dt} = -\frac{I}{RC}. \tag{28-4}$$

This equation shows that the rate of change of current is proportional to the current itself, expressing mathematically what Fig. 28-28 shows schematically. Equations like this arise whenever the rate of change of a quantity is proportional to the quantity itself. Population growth, the increase of money in a bank account, and the decay of a radioactive element are all described by similar equations.

Like the equation for simple harmonic motion in Chapter 15, Equation 28-4 is a *differential* equation, so called because the unknown quantity *I* occurs in a

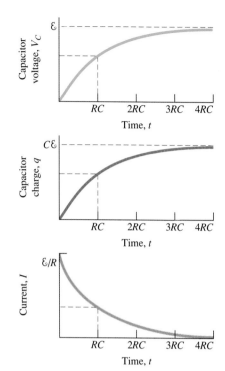

FIGURE 28-29 Time dependence of capacitor voltage, capacitor charge, and current in a charging *RC* circuit. In one time constant *RC* the capacitor voltage and charge rise to about 2/3 (actually $1 - 1/e$) of their final value, while the current drops to about 1/3 (actually $1/e$) of its initial value.

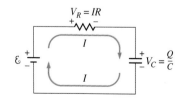

FIGURE 28-30 Voltage changes in a charging *RC* circuit. Since they sum to zero, the loop equation for this circuit is $\mathscr{E} - (IR) - (Q/C) = 0$.

derivative. The solution to a differential equation is not a single number but rather a function expressing the relation between the unknown quantity—in this case current—and the independent variable—in this case time. We can solve this particular differential equation by multiplying both sides by dt/I, in order to collect all terms involving I on one side of the equation. This gives

$$\frac{dI}{I} = -\frac{dt}{RC}.$$

We then integrate both sides, noting that RC is constant:

$$\int_{I_0}^{I} \frac{dI}{I} = -\frac{1}{RC} \int_{0}^{t} dt,$$

where $I_0 = \mathscr{E}/R$ is the initial current at the time $t = 0$ just after the switch is closed and where the integration runs from $t = 0$ to some arbitrary time t. The integral on the left is just the natural logarithm, and that on the right is just t. Then we have

$$\ln (I/I_0) = -\frac{t}{RC},$$

where we used $\ln I - \ln I_0 = \ln(I/I_0)$. To get an equation for I we exponentiate both sides, recalling that $e^{\ln x} = x$. This gives

$$\frac{I}{I_0} = e^{-t/RC},$$

or, since $I_0 = \mathscr{E}/R$,

$$I = \frac{\mathscr{E}}{R} e^{-t/RC}. \tag{28-5}$$

Thus the current in the circuit decreases exponentially with time, in agreement with our qualitative analysis.

What about the capacitor voltage? The capacitor and resistor voltages must add to the battery voltage $\mathscr{E}$, and the resistor voltage is just $V_R = IR$, or

$$V_R = \mathscr{E}e^{-t/RC}.$$

Thus the capacitor voltage is $V_C = \mathscr{E} - V_R$, or

$$V_C = \mathscr{E}(1 - e^{-t/RC}). \qquad (RC \text{ circuit, charging}) \tag{28-6}$$

Equation 28-6 shows the capacitor voltage starting at zero, and rising rapidly at first but with its rate of rise ever slowing, as it gradually approaches the battery voltage $\mathcal{E}$, again agreeing with our qualitative analysis (see Fig. 28-29, which was in fact plotted using Equations 28-5 and 28-6).

When is the capacitor fully charged? Never, according to our equations! But the rate at which it approaches full charge is determined by the quantity RC that appears in Equations 28-5 and 28-6. (Problem 45 will convince you that this quantity has the units of time.) Called the **time constant,** RC is a characteristic time for changes to occur in a circuit containing a capacitor. Equation 28-6 shows that in one time constant, the voltage rises to $\mathcal{E}(1 - 1/e)$, or to about two-thirds of the battery voltage. A practical rule of thumb says that in five time constants ($t = 5RC$) a capacitor is 99% charged (see Problem 47). The RC time constant clarifies our statement that the voltage across a capacitor cannot change instantaneously. We can now say that in times small compared with the time constant, the voltage across a capacitor cannot change appreciably. On the other hand, if we wait a long time—many time constants—we will find essentially no current flowing to the capacitor.

RC circuits with appropriate time constants are used in electronic timing applications covering microseconds to hours. In other circuits where we want voltages to change rapidly, the time constant can be annoyingly long. For example, capacitance in audio equipment can limit high-frequency response, decreasing the quality of music reproduction. You intentionally alter the response of a stereo system by adjusting bass and treble controls, which are simply variable resistors in RC circuits.

The RC Circuit: Discharging

Suppose we connect a charged capacitor across a resistor, as shown in Fig. 28-31. If the capacitor voltage is initially V_0, then when the circuit is connected this voltage will drive a current $I_0 = V_0/R$ through the resistor. This current transfers charge from the positive to the negative capacitor plate, lowering the charge on the capacitor. Since capacitor charge and voltage are proportional, the capacitor voltage drops, too. So, therefore, does the current, and therefore the rate at which the capacitor discharges. We therefore expect both the voltage and current in this circuit to decay toward zero. In terms of energy, that happens because the energy stored in the capacitor's electric field is gradually dissipated as heat in the resistor.

The loop equation for Fig. 28-31 is particularly simple; going clockwise, we have

$$\frac{Q}{C} - IR = 0,$$

where the two terms are the voltage changes across the capacitor and resistor, respectively. Since we've indicated positive current in Fig. 28-31 in the direction that would *reduce* the capacitor charge Q, the rate of change dQ/dt and the current must have opposite signs: $I = -dQ/dt$. Differentiating our loop equation and substituting this expression for I gives

$$\frac{dI}{dt} = -\frac{I}{RC}.$$

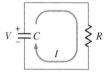

FIGURE 28-31 A discharging RC circuit.

This is Equation 28-4; the solution is therefore Equation 28-5, but with $I_0 = V_0/R$ instead of $\mathcal{E}/R$:

$$I = \frac{V_0}{R}e^{-t/RC}.\tag{28-7}$$

In this circuit the capacitor and resistor voltage are the same since the two are in parallel. Since the resistor voltage and current are proportional, the voltage across the capacitor and resistor is

$$V = V_0e^{-t/RC}.\qquad(RC\text{ circuit, discharging})\tag{28-8}$$

Equations 28-7 and 28-8 show that the capacitor discharges with the same characteristic time constant RC that governs its charging.

● EXAMPLE 28-8 A CAMERA FLASH

In Chapter 26 we considered an electronic camera flash using a capacitor to store energy. A particular camera flashtube obtains its energy from a 150-μF capacitor and requires 170 V to fire. If the capacitor is charged by a 200-V battery* through a 30-kΩ resistor, how long must the photographer wait between flashes? What is the peak power drawn from the battery? Assume the capacitor is fully discharged during a flash.

Solution
The time between flashes is the time it takes the capacitor voltage to reach 170 V. To find this time, we solve Equation 28-6 for the exponential term that contains the time:

$$e^{-t/RC} = 1 - \frac{V_C}{\mathcal{E}}.$$

We then take the natural logarithm of both sides, recalling that $\ln e^x = x$, so

$$-\frac{t}{RC} = \ln\left(1 - \frac{V_C}{\mathcal{E}}\right).$$

Solving for t and setting $V_C = 170$ V, $\mathcal{E} = 200$ V, $R = 30$ kΩ, and $C = 150$ μF gives

$$t = -RC\ln\left(1 - \frac{V_C}{\mathcal{E}}\right)$$

$$= -(30\times10^3\ \Omega)(150\times10^{-6}\ \text{F})\ln\left(1 - \frac{170\ \text{V}}{200\ \text{V}}\right) = 8.5\ \text{s}.$$

Problem 74 explores the question of power in this circuit, and shows that energy from the battery cannot all end up in the capacitor.

EXERCISE If the flash lamp in Example 28-8 has an effective resistance of 10 Ω, how long does it take the capacitor voltage to drop to 100 V as it discharges through the lamp?

Answer: 0.80 ms

Some problems similar to Example 28-8: 47–49, 51 ●

It's not always necessary to solve exponential equations in analyzing RC circuits. If we're concerned only with times short compared with the time constant, if suffices to remember that the voltage across a capacitor cannot change instantaneously. And after many time constants have passed a capacitor has essentially reached its final voltage, and there will be no current flowing to it. These two conditions are sufficient to analyze circuits on short and long time scales.

---

*Actually, much lower voltage batteries are used. But their voltage is increased using transistors and a transformer, working on principles described in Chapter 31.

EXAMPLE 28-9 LONG- AND SHORT-TIME BEHAVIOR OF AN *RC* CIRCUIT

In Fig. 28-32*a* the capacitor is initially uncharged. What is the current through R_1 the instant after the switch is closed? A long time after the switch has been closed?

Solution

The capacitor voltage cannot change instantaneously, so just after the switch is closed there can be no voltage across the capacitor and therefore none across R_2. Then the full battery voltage is across R_1, so the current in R_1 is $\mathcal{E}/R_1$. After a very long time the capacitor will be fully charged (to what voltage? —see the exercise below), and no current will flow into it. The capacitor then acts like an open circuit, and we simply have two resistors in series. The current in each is $\mathcal{E}/(R_1 + R_2)$.

How simple this example is! The uncharged capacitor has no voltage across it, so it acts instantaneously like a short circuit. The fully charged capacitor has no current into it, so it acts like an open circuit. To solve the problem we could simply redraw the circuit, once with the capacitor replaced by a wire (Fig. 28-32*b*), the second time with the capacitor simply erased from the circuit diagram (Fig. 28-32*c*). Only if we wanted to know what was happening at intermediate times would we have to resort to the solution of an equation describing the circuit.

EXERCISE When the capacitor in Example 28-9 is fully charged, what will be the voltage across it?

Answer: $\mathcal{E}R_2/(R_1 + R_2)$

Some problems similar to Example 28-9: 55–57

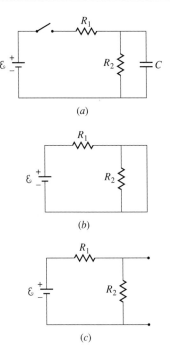

FIGURE 28-32 (*a*) A circuit with two resistors and a capacitor. (*b*) Just after the switch is closed the voltage across the capacitor is still zero, so the capacitor acts like a short circuit. (*c*) Long after the switch is closed the capacitor is fully charged. No more current flows into it, so it acts like an open circuit.

28-7 ELECTRICAL SAFETY

Whether you find yourself in a laboratory hooking up electronic equipment, or in a hospital connecting instrumentation to a patient, or on a job designing electrical devices, or simply at home plugging in appliances and tools, you should be concerned with electrical safety.

Everyone knows enough to be wary of "high voltage." People with a little more electrical sophistication are fond of saying "it isn't the voltage but the current that kills you." In fact, both points of view are partially correct. Current through the body is dangerous, but as with any resistor it takes voltage to drive that current.

Table 28-1 shows typical effects of electric currents introduced into the body through skin contact. Currents below the lethal 100 mA may result in involuntary muscle contraction that unfortunately keeps the victim in contact with the circuit. A primary danger is disturbance of the biologically generated electrical signals that pace heartbeat; this is reflected in the lethal zone of 100 to 200 mA at which currents the heart is thrown into fibrillation—uncontrolled spasms of the cardiac muscle. With electrical signals applied internally to local regions of the body, much smaller currents can be lethal. Doctors performing cardiac

▲ **TABLE 28-1** EFFECTS OF EXTERNALLY APPLIED CURRENT ON THE HUMAN ORGANISM

CURRENT RANGE	EFFECT
0.5–2 mA	Threshold of sensation
10–15 mA	Involuntary muscle contractions; can't let go
15–100 mA	Severe shock; muscle control lost; breathing difficult
100–200 mA	Fibrillation of heart; death within minutes
>200 mA	Cardiac arrest; breathing stops; severe burns

catheterization, for example, must worry about currents at the microampere level.

Above 200 mA, complete cardiac arrest may occur, breathing may stop, and there may be severe burns both internally and at the points of skin contact. Sometimes such high currents are useful: when a heart is fibrillating, doctors or emergency technicians briefly apply a high enough current to stop the heart. The heart often restarts, beating normally. The figures of Table 28-1 are rough averages, and vary widely from person to person as well as with duration of the shock and whether alternating or direct current is involved. In particular, very young children and people with heart conditions are at higher risk.

Under dry conditions, the typical human being has a resistance from one point to another on unbroken skin of about $10^5 \; \Omega$. What voltages are dangerous to such a person? At $10^5 \; \Omega$ it takes

$$V = IR = (0.1 \text{ A})(10^5 \; \Omega) = 10,000 \text{ V}$$

to drive the fatal 100 mA. But a person who is wet or sweaty has a much lower resistance and may be electrocuted by 120-V household electricity. People have been electrocuted at voltages as low as 30 V, although such cases are rare.

It takes current to harm a person, but it takes voltage to drive that current. To be dangerous, an electric circuit must have high voltage *and* be capable of driving sufficient current. For example, a car battery can deliver 300 A, but it cannot electrocute you because its 12 volts will not drive much current through you (although you could be hurt by the energy released if you accidentally short-circuit such a battery). On the other hand the 20,000 V that runs your car's spark plugs will not electrocute you either, since the internal resistance of this high-voltage circuit is so high that it cannot deliver more than a few mA.

Because potential difference is a property of two points, receiving an electric shock requires that two parts of the body be in contact with conductors at different potentials. In typical 120-V wiring used throughout North America, one of the two wires is connected physically to the ground. This ground connection is to prevent the wiring from reaching arbitrarily high potentials with respect to the ground, as might otherwise happen in a thunderstorm or if a short circuit occurred in a power line. At the same time it means that an individual contacting the "hot" side of the circuit and any grounded conductor such as the ground, a water pipe, or a bathtub will receive a shock.

A potentially dangerous situation occurs when power tools, instruments, or appliances are used by an operator who is likely to be in contact with a grounded conductor. Examples include working outdoors with an electric drill, in a kitchen with an electric mixer, or in a laboratory with an oscilloscope. Suppose you're

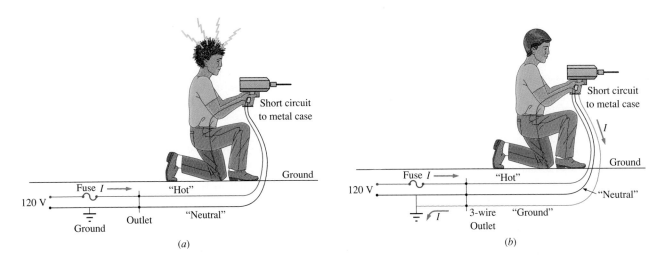

(a)

(b)

using a power tool that is plugged in through a regular two-wire cord. Normally exposed metal parts of the tool are not connected to either wire. Now suppose something goes wrong in the tool and a wire short-circuits to the metal case. If it happens to be the wire that is plugged into the grounded side of the power line there is no problem, but if it is the other wire the metal is suddenly 120 V above ground. If you are standing on the ground, or in a damp basement, or are leaning on the kitchen sink, you will receive a potentially lethal shock (Fig. 28-33a).

To avoid this danger many electrical devices are equipped with three-wire cords (Fig 28-34). The third wire runs from exposed metal parts to a grounded wire in the outlet, and normally carries no current. If a short circuit occurs the third wire provides a very low resistance path to ground (Fig. 28-33b). Large currents will flow and will blow the circuit breaker or fuse, shutting off the current. Held at ground potential by the ground wire, the operator of the device will be safe.

Because many older homes are not wired with grounded outlets, some manufacturers produce two-wire tools and other devices that are "double-insulated" to provide an extra margin of safety. Newer appliances are sometimes equipped with "polarized plugs," which can only be plugged in one way (Fig. 28-34), ensuring that exposed metal parts are most likely to end up at ground potential. Finally, electronic devices called ground-fault interrupters (Fig. 28-35) are used in kitchen, bathroom, basement, and other hazardous circuits in new homes. These devices sense a slight imbalance—5 mA or less—in current flowing in the two wires of a circuit, and shut off the circuit in less than a millisecond on the assumption that the excess current is leaking to ground—perhaps through a person. (Do ground-fault interrupters know about the node law?)

FIGURE 28-33 (a) A short circuit in an ungrounded tool could result in a lethal shock. (b) With a grounded tool, the short circuit causes a blown fuse or circuit breaker, thereby protecting the operator.

FIGURE 28-34 Plugs. (Left) grounded; (right) polarized.

FIGURE 28-35 A ground-fault interrupter protects against shock by sensing small currents leaking to ground.

CHAPTER SYNOPSIS

Summary

1. A source of **electromotive force (emf)** is a device—like a battery—that converts some form of energy into the electrical energy associated with the buildup of positive and negative charge at its two terminals. An ideal emf maintains a constant potential difference between its terminals, but energy losses in a real emf result in terminal voltage that decreases as more current is drawn from the device. When connected across a resistor, an ideal emf $\mathscr{E}$ drives a current $I = \mathscr{E}/R$ through the resistor.

2. Electric circuits may often be analyzed using series and parallel resistor combinations. Resistors in series add:

$$R_{\text{series}} = R_1 + R_2 + R_3 + \cdots,$$

while resistors in parallel add reciprocally:

$$\frac{1}{R_{\text{parallel}}} = \frac{1}{R_1} + \frac{1}{R_2} + \frac{1}{R_3} + \cdots.$$

3. More complicated circuits may be analyzed using **Kirchhoff's laws.** The **loop law** follows from conservation of energy and states that the sum of the voltage differences around any circuit loop is zero. The **node law** follows from conservation of charge and states that the sum of the currents at any circuit node is zero.

4. In using instruments to measure electrical quantities, care must be taken to ensure that the instrument does not alter the circuit being measured. An ideal **voltmeter** has infinite resistance; in practice, a voltmeter's resistance should be much higher than typical resistances in the circuit being measured. An ideal **ammeter** has zero resistance; in practice, an ammeter's resistance should be much lower than typical circuit resistances.

5. It takes time to move charge on and off the plates of a capacitor, and as a result the voltage across a capacitor cannot change instantaneously. Quantities in a resistor-capacitor (*RC*) circuit change on a characteristic time scale given by the product *RC*. When an emf $\mathscr{E}$ charges a capacitor *C* through a resistor *R*, the capacitor voltage and circuit current vary with time according to

$$V_C = \mathscr{E}(1 - e^{-t/RC})$$

$$I = \frac{\mathscr{E}}{R} e^{-t/RC}.$$

For a discharging capacitor, both voltage and current decrease exponentially with the same time constant *RC*.

6. **Electrical safety** is a matter of avoiding currents high enough to cause biological damage. The danger of electric shock depends on the current at which such damage occurs, the resistance of the organism, and the voltage available to drive the current.

Terms You Should Understand

(Pairs are closely related terms whose distinction is important; number in parentheses is chapter section where term first appears.)

circuit, circuit element (28-1)
electromotive force (emf) (28-2)
ideal, real emfs (28-2, 28-3)
voltage divider (28-3)
internal resistance (28-3)
Kirchhoff's loop and node laws (28-4)
voltmeter, ammeter, ohmmeter (28-5)
time constant (28-6)

Symbols You Should Recognize

electric circuit symbols (Fig. 28-1)
$\mathscr{E}$ (28-1)

Problems You Should Be Able to Solve

analyzing simple circuits with series and parallel resistors (28-3)
analyzing circuits with the loop and node laws (28-4)
determining the effects of nonideal measuring instruments (28-5)
analyzing charging and discharging *RC* circuits (28-6)
quickly determining short- and long-term behavior of *RC* circuits (28-6)
assessing electrical hazards (28-7)

Limitations to Keep in Mind

Real circuit elements deviate from their idealizations. Thus the voltage across a real battery may not be exactly its rated voltage, and voltage drops occur in real wires. Attention to circuit design will minimize these nonideal effects.

Electrical measuring instruments may affect the circuits they are measuring.

Currents may be far more dangerous than Table 28-1 implies if they are introduced beneath the skin.

QUESTIONS

1. In each of the circuits of Fig. 28-36, which, if any, of the resistors are in series? In parallel?

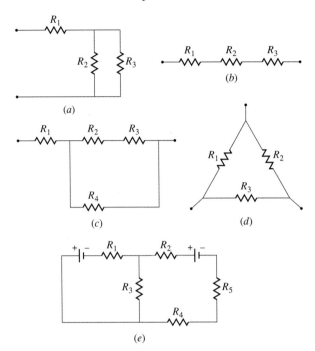

(a)

(b)

(c)

(d)

(e)

FIGURE 28-36   Question 1.

2. Are the electrical outlets in a home connected in series or parallel? How do you know?

3. In which of the circuits of Fig. 28-37 does the battery supply the same current? All the resistors have the same resistance.

4. Can the voltage across a battery's terminals differ from the rated voltage of the battery? Explain.

5. Can the voltage across a battery's terminals be higher than the rated voltage of the battery? Explain.

6. In some cities, streetlights are wired in such a way that when one light burns out, they all go out. Are the lights in series or parallel?

7. If you know the battery voltage in Fig. 28-38, can you determine the voltage between points B and C without knowing the resistance R?

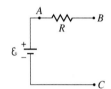

FIGURE 28-38   Questions 7, 17.

8. Must we assign zero volts to the negative terminal of a battery?

9. When the switch in Fig 28-39 is open, what is the voltage across the resistor? Across the switch?

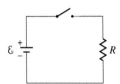

FIGURE 28-39   Question 9.

FIGURE 28-37   Question 3.

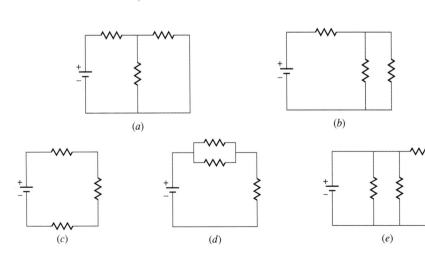

(a)

(b)

(c)

(d)

(e)

10. Two identical resistors in series dissipate equal power. How can this be, when electric charge loses energy in flowing through the first resistor?

11. What is the current through resistor R_2 in Fig. 28-40? Assume all the wires are ideal.

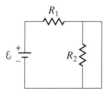

FIGURE 28-40 Question 11.

12. The resistors in Fig. 28-41 all have the same resistance. If an ideal voltmeter is connected between points A and B, what will it read?

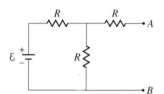

FIGURE 28-41 Question 12.

13. When a large electrical load such as a washing machine, oven, or oil burner comes on, lights throughout a house often dim. Why is this? *Hint:* Think about real wires.

14. If the node law were not obeyed at some node in an electric circuit, what would happen to the voltage at that node?

15. How would you connect a pair of equal resistors across an ideal battery in order to get the most power dissipation in the resistors?

16. You have a battery whose voltage and internal resistance are unknown. Using an ideal voltmeter and an ideal ammeter, how would you determine both these battery characteristics?

17. An ideal voltmeter is used to measure the voltages between points A and B and between C and B in the circuit of Fig. 28-38. How do the measurements compare?

18. You wish to measure the resistance R_2 in Fig. 28-42 with an ohmmeter. Can you do so while R_2 is in the circuit? Why or why not?

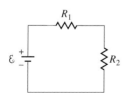

FIGURE 28-42 Question 18.

19. A student who is confused about voltage and current hooks a nearly ideal ammeter across a car battery. What happens?

20. A student who is confused about voltage and current tries to measure the voltage across a lighted light bulb by inserting a voltmeter in series with the bulb. What happens to the bulb? Explain.

21. Four identical light bulbs are connected to a battery as shown in Fig. 28-43. How does the brightness of each compare? Explain.

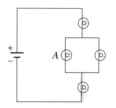

FIGURE 28-43 Questions 21, 22.

22. Suppose bulb A in Fig. 28-43 is unscrewed from its socket. How will the brightness of the three remaining bulbs change? How does the brightness of the three compare?

23. What does it mean for a capacitor to be "fully charged"?

24. Is the current into a charging capacitor in an RC circuit greatest when the capacitor voltage is greatest or when it is smallest?

25. If it takes forever to charge a capacitor fully, why is the RC time constant of any significance in describing the charging?

26. The two resistors in Fig. 28-44 have equal resistance. If the circuit has been connected for a long time, what is the voltage across the capacitor?

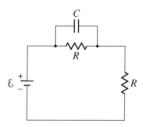

FIGURE 28-44 Question 26.

27. In one time constant, a charging capacitor reaches approximately $\frac{2}{3}$ of full charge. In one time constant, the voltage across a discharging capacitor falls to approximately $\frac{1}{3}$ of its original value. What is the origin of the approximate numerical factors $\frac{2}{3}$ and $\frac{1}{3}$ in these statements?

28. What's wrong with this news report: "A power-line worker was seriously injured when 4000 volts passed through his body"?

PROBLEMS

Section 28-1 Circuits and Symbols

1. Sketch a circuit diagram for a circuit that includes a resistor R_1 connected to the positive terminal of a battery, a pair of parallel resistors R_2 and R_3 connected to the lower-voltage end of R_1, then returned to the battery's negative terminal, and a capacitor across R_2.

2. A circuit consists of two batteries, a resistor, and a capacitor, all in series. Sketch this circuit. Does the description allow any flexibility in how you draw the circuit?

3. Resistors R_1 and R_2 are connected in series, and this series combination is in parallel with R_3. This parallel combination is connected across a battery whose internal resistance is R_{int}. Draw a diagram representing this circuit.

Section 28-2 Electromotive Force

4. What is the emf of a battery that delivers 27 J of energy as it moves 3.0 C between its terminals?

5. A 1.5-V battery stores 4.5 kJ of energy. How long can it light a flashlight bulb that draws 0.60 A?

6. If you accidentally leave your car headlights (current drain 5 A) on for an hour, how much of the 12-V battery's chemical energy is used up?

7. A battery stores 50 W·h of chemical energy. If it uses up this energy moving 3.0×10^4 C through a circuit, what is its voltage?

Section 28-3 Simple Circuits: Series and Parallel Resistors

8. A 47-kΩ resistor and a 39-kΩ resistor are in parallel, and the pair is in series with a 22-kΩ resistor. What is the resistance of the combination?

9. What resistance should be placed in parallel with a 56-kΩ resistor to make an equivalent resistance of 45 kΩ?

10. In Fig. 28-45 all resistors have the same value, R. What will be the resistance measured (a) between A and B or (b) between A and C?

11. In Fig. 28-45, take all resistors to be 1.0 Ω. If a 6.0-V battery is connected between points A and B, what will be the current in the vertical resistor?

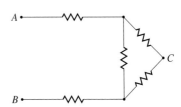

FIGURE 28-45   Problems 10, 11.

12. A defective starter motor in a car draws 300 A from the car's 12-V battery, dropping the battery terminal voltage to only 6 V. A good starter motor should draw only 100 A. What will the battery terminal voltage be with a good starter?

13. What is the internal resistance of the battery in the preceding problem?

14. Three 1.5-V batteries, with internal resistances of 0.01 Ω, 0.1 Ω, and 1 Ω, each have 1-Ω resistors connected across their terminals. To three significant figures, what is the voltage across each resistor?

15. When a 9-V battery is temporarily short-circuited, a 200-mA current flows. What is the internal resistance of the battery?

16. What possible resistance combinations can you form using three resistors whose values are 1.0 Ω, 2.0 Ω, and 3.0 Ω? (Use all three resistors.)

17. A partially discharged car battery can be modeled as a 9-V emf in series with an internal resistance of 0.08 Ω. Jumper cables are used to connect this battery to a fully charged battery, modeled as a 12-V emf in series with a 0.02-Ω internal resistance. How much current flows through the discharged battery?

18. You have a number of 50-Ω resistors, each capable of dissipating 0.50 W without overheating. How many resistors would you need, and how would you connect them, so as to make a 50-Ω combination that could be connected safely across a 12-V battery?

19. What is the equivalent resistance between A and B in each of the circuits shown in Fig. 28-46? *Hint:* In (c), think about symmetry and the current that would flow through R_2.

FIGURE 28-46   Problem 19.

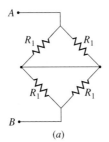

(a)

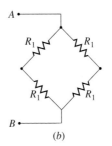

(b)

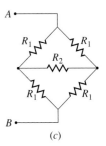

(c)

20. A 6.0-V battery has an internal resistance of 2.5 Ω. If the battery is short circuited, what is the rate of energy dissipation in its internal resistance?

21. How many 100-W, 120-V light bulbs can be connected in parallel before they blow a 20-A circuit breaker?

22. What is the current through the 3-Ω resistor in the circuit of Fig. 28-47? *Hint:* This is trivial. Can you see why?

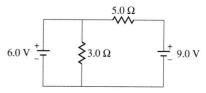

FIGURE 28-47 Problem 22.

23. Take $\mathscr{E} = 12$ V and $R_1 = 270$ Ω in the voltage divider of Fig. 28-4. (a) What should the value of R_2 in order that 4.5 V appear across R_2? (b) What will be the power dissipation in R_2?

24. A voltage divider consists of two 1.0-kΩ resistors connected in series across a 160-V emf. If a 10-kΩ resistor is connected across one of the 1.0-kΩ resistors, what will be the voltage across it?

25. In the circuit of Fig. 28-48, R_1 is a variable resistor, and the other two resistors have equal resistances R. (a) Find an expression for the voltage across R_1, and (b) sketch a graph of this quantity as a function of R_1 as R_1 varies from 0 to 10R. (c) What is the limiting value as $R_1 \rightarrow \infty$?

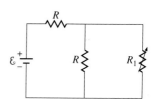

FIGURE 28-48 Problem 25.

26. In the circuit of Fig. 28-49 find (a) the current supplied by the battery and (b) the current through the 6-Ω resistor.

27. In the circuit of Fig. 28-49, how much power is being dissipated in the 4-Ω resistor?

28. A 50-Ω resistor is connected across a battery, and a 26-mA current flows. When the 50-Ω resistor is replaced with a 22-Ω resistor, a 43-mA current flows. What are the battery's voltage and internal resistance?

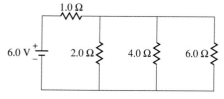

FIGURE 28-49 Problems 26, 27.

Section 28-4 Kirchhoff's Laws and Multiloop Circuits

29. In the circuit of Fig. 28-50 it makes no difference whether the switch is open or closed. What is $\mathscr{E}_3$ in terms of the other quantities shown?

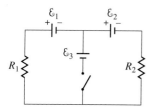

FIGURE 28-50 Problem 29.

30. What is the current through the ammeter in Fig. 28-51?

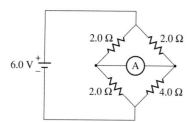

FIGURE 28-51 Problem 30.

31. In Fig. 28-52, what is the equivalent resistance measured between points A and B?

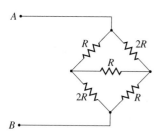

FIGURE 28-52 Problem 31.

32. Find all three currents in the circuit of Fig. 28-16, but now with $\mathscr{E}_2 = 1.0$ V.

33. Find all three currents in the circuit of Fig. 28-16 with the values given, but with battery $\mathscr{E}_2$ reversed.

34. In Fig. 28-53, take $\mathscr{E}_1 = 6.0$ V, $\mathscr{E}_2 = 1.5$ V, $\mathscr{E}_3 = 4.5$ V, $R_1 = 270$ Ω, $R_2 = 150$ Ω, $R_3 = 560$ Ω, and $R_4 = 820$ Ω. Find the current in R_3, and give its direction.

35. With all the values except $\mathscr{E}_2$ in Fig. 28-53 as given in the preceding problem, find the condition on $\mathscr{E}_2$ that will make the current in R_3 flow upward.

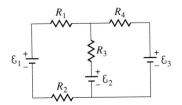

FIGURE 28-53 Problems 34, 35, 36.

36. Suppose that all resistors in Fig. 28-53 have the same value R and that $\mathcal{E}_1 = \mathcal{E}_3 = \mathcal{E}$ and $\mathcal{E}_2 = 2\mathcal{E}$. Find expressions for the currents in the four resistors, and give their directions.

Section 28-5 Electrical Measuring Instruments

37. A voltmeter with 200-kΩ resistance is used to measure the voltage across the 10-kΩ resistor in Fig. 28-54. By what percentage is the measurement in error because of the finite meter resistance?

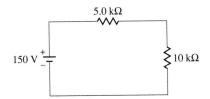

FIGURE 28-54 Problems 37, 38.

38. An ammeter with 100-Ω resistance is inserted in the circuit of Fig. 28-54. By what percentage is the measurement in error because of the nonzero meter resistance?

39. A neophyte mechanic foolishly connects an ammeter with 0.1-Ω resistance directly across a 12-V car battery whose internal resistance is 0.01 Ω. What is the power dissipation in the meter? No wonder it gets destroyed!

40. The voltage across the 30-kΩ resistor in Fig. 28-55 is measured with (a) a 50-kΩ voltmeter, (b) a 250-kΩ voltmeter, and (c) a digital meter with 10-MΩ resistance. To two significant figures, what does each read?

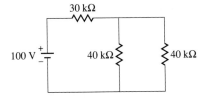

FIGURE 28-55 Problem 40.

41. You have an ammeter with 10-Ω resistance whose full-scale reading is 1.0 mA. What resistance should you put in series with it to make a voltmeter that reads 25 V full scale?

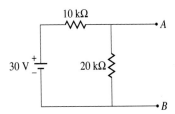

FIGURE 28-56 Problem 43.

42. Suppose you want to make a 10-A full scale ammeter out of the 1.0-mA meter of the preceding problem. What resistance should you put in parallel with it?

43. In Fig. 28-56 what are the meter readings when (a) an ideal voltmeter or (b) an ideal ammeter is connected between points A and B?

44. A resistor draws 1.00 A from an ideal 12.0-V battery. (a) If an ammeter with 0.10-Ω resistance is inserted in the circuit, what will it read? (b) If this current is used to calculate the resistance, how will the calculated value compare with the actual value?

Section 28-6 Circuits with Capacitors

45. Show that the quantity RC has the units of time (seconds).

46. If capacitance is given in μF, what will be the units of the RC time constant when resistance is given in (a) Ω, (b) kΩ, (c) MΩ? Your answers eliminate the need for tedious power-of-10 conversions.

47. Show that a capacitor is charged to approximately 99% of the applied voltage in five time constants.

48. An uncharged 10-μF capacitor and a 470-kΩ resistor are connected in series, and 250 V applied across the combination. How long does it take the capacitor voltage to reach 200 V?

49. Figure 28-57 shows the voltage across a capacitor that is charging through a 4700-Ω resistor in the circuit of Fig. 28-27. Use the graph to determine (a) the battery voltage, (b) the time constant, and (c) the capacitance.

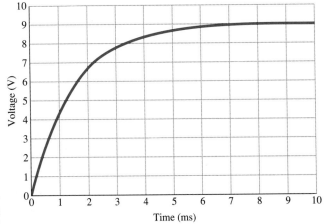

FIGURE 28-57 Problem 49.

50. The voltage across a charging capacitor in an *RC* circuit rises to $1 - 1/e$ of the battery voltage in 5.0 ms. (a) How long will it take to reach $1 - 1/e^3$ of the battery voltage? (b) If the capacitor is charging through a 22-kΩ resistor, what is its capacitance?

51. A 1.0-μF capacitor is charged to 10.0 V. It is then connected across a 500-kΩ resistor. How long does it take (a) for the capacitor voltage to reach 5.0 V and (b) for the energy stored in the capacitor to decrease to half its initial value?

52. A capacitor used to provide steady voltages in the power supply of a stereo amplifier charges rapidly to 35 V every 1/60 of a second. It must then hold that voltage to within 1.0 V for the next 1/60 s while it discharges through the amplifier circuit. If the circuit draws 1.2 A from the 35-V supply (a) what is its effective resistance and (b) what value of capacitance is needed?

53. A capacitor is charged until it holds 5.0 J of energy. It is then connected across a 10-kΩ resistor. In 8.6 ms, the resistor dissipates 2.0 J. What is the capacitance?

54. A 2.0-μF capacitor is charged to 150 V. It is then connected to an uncharged 1.0-μF capacitor through a 2.2-kΩ resistor, by closing switch *S* in Fig. 28-58. Find the total energy dissipated in the resistor as the circuit comes to equilibrium. *Hint:* Think about charge conservation.

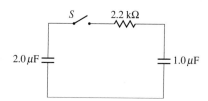

FIGURE 28-58 Problem 54.

55. For the circuit of Example 28-9, take $\mathcal{E} = 100$ V, $R_1 = 4.0$ kΩ, and $R_2 = 6.0$ kΩ, and assume the capacitor is initially uncharged. What are the currents in both resistors and the voltage across the capacitor (a) just after the switch is closed and (b) a long time after the switch is closed? Long after the switch is closed it is again opened. What are $I_1, I_2,$ and V_C (c) just after this switch opening and (d) a long time later?

56. In the circuit of Fig. 28-59 the switch is initially open and both capacitors initially uncharged. All resistors have the same value *R*. Find expressions for the current in R_2 (a) just after the switch is closed and (b) a long time after the switch is closed. (c) Describe qualitatively how you expect the current in R_3 to behave after the switch is closed.

57. In the circuit of Fig. 28-60 the switch is initially open and the capacitor is uncharged. Find expressions for the current *I* supplied by the battery (a) just after the switch is closed and (b) a long time after the switch is closed.

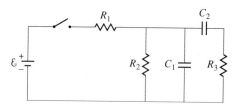

FIGURE 28-59 Problem 56.

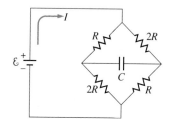

FIGURE 28-60 Problem 57.

58. Obtain an expression for the rate (dV/dt) at which the voltage across a charging capacitor increases. Evaluate your result at time $t = 0$, and show that if the capacitor continued charging steadily at this rate it would be fully charged in exactly one time constant.

Paired Problems

(Both problems in a pair involve the same principles and techniques. If you can get the first problem, you should be able to solve the second one.)

59. A 3.3-kΩ resistor and a 4.7-kΩ resistor are connected in parallel, and the pair is in series with a 1.5-kΩ resistor. What is the resistance of the combination?

60. Find the value of *R* in Fig. 28-61 that will make the resistance between points *A* and *B* equal to *R*.

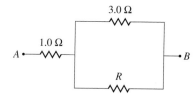

FIGURE 28-61 Problem 60.

61. A battery's voltage is measured with a voltmeter whose resistance is 1000 Ω; the result is 4.36 V. When the measurement is repeated with a 1500-Ω meter the result is 4.41 V. What are (a) the battery voltage and (b) its internal resistance?

62. An ammeter with a resistance of 1.4 Ω is connected momentarily across a battery (not the way to treat an ammeter!) and it reads 9.78 A. When the measurement is repeated with a meter whose resistance is 2.1 Ω the reading

is 7.46 A. What are (a) the battery voltage and (b) its internal resistance?

63. In Fig. 28-62, take $\mathcal{E}_1 = 12$ V, $\mathcal{E}_2 = 6.0$ V, $\mathcal{E}_3 = 3.0$ V, $R_1 = 1.0\ \Omega$, $R_2 = 2.0\ \Omega$, and $R_3 = 4.0\ \Omega$. Find the current in R_2 and give its direction.

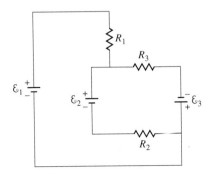

FIGURE 28-62 Problems 63, 64.

64. (a) With all values except $\mathcal{E}_2$ as given in the preceding problem, find $\mathcal{E}_2$ such that there is no current in this battery. (b) What are the currents in R_1 and R_2 under these conditions?

65. In Fig. 28-63 what are the meter readings when (a) an ideal voltmeter or (b) an ideal ammeter is connected between points A and B?

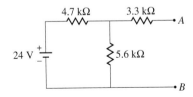

FIGURE 28-63 Problems 65, 66.

66. In Fig. 28-63 what are the meter readings when (a) a voltmeter with 50-kΩ resistance or (b) an ammeter with 150-Ω resistance is connected between points A and B?

67. An initially uncharged capacitor in an RC circuit reaches 75% of its full charge in 22.0 ms. What is the time constant?

68. Find the resistance needed in an RC circuit to bring a 20-μF capacitor from zero charge to 45% of its full charge in 140 ms.

Supplementary Problems

69. Suppose the currents into and out of a circuit node differed by 1 μA. If the node consists of a small metal sphere with diameter 1 mm, how long would it take for the electric field around the node to reach the breakdown field in air (3 MV/m)?

70. You measure the voltage across a charged 26-μF capacitor by connecting a voltmeter with 250-kΩ resistance across the capacitor. If you note the meter reading 2.0 s after connecting the meter, by what percentage does your reading differ from the initial capacitor voltage? Assume the capacitor isn't connected to any other circuitry.

71. In Fig. 28-64, what is the current in the 4-Ω resistor when each of the following circuit elements is connected between points A and B: (a) an ideal ammeter; (b) an ideal voltmeter; (c) another 4.0-Ω resistor; (d) an uncharged capacitor, right after it's connected; (e) long after the capacitor of part (d) is connected; (f) an ideal 12-V battery, with its positive terminal at A; (g) a capacitor initially charged to 12 V, right after it's connected with its positive plate at A; (h) long after the capacitor in part (g) is connected?

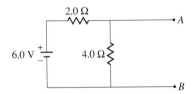

FIGURE 28-64 Problem 71.

72. A resistance R is connected across a battery with internal resistance R_{int}. Show that the maximum power dissipation in R occurs when $R = R_{int}$. *Note:* This is not the way to treat a battery! But it is the basis for matching loads in amplifiers and other devices; for example, a stereo amplifier designed to drive 8-Ω speakers has internal resistance close to 8 Ω.

73. A parallel-plate capacitor is insulated with a material of dielectric constant κ and resistivity ρ. Since the resistivity is finite, the capacitor "leaks" charge and can be modeled as an ideal capacitor in parallel with a resistor. (a) Show that the time constant of the capacitor is independent of its dimensions (provided the spacing is small enough that the usual parallel-plate approximation applies) and is given by $\varepsilon_0\kappa\rho$. (b) If the insulating material is polystyrene ($\kappa = 2.6$, $\rho = 10^{16}\ \Omega\cdot$m), how long will it take for the stored energy in the capacitor to decrease by a factor of 2?

74. Of the total energy drawn from a battery in charging an RC circuit, show that only half ends up as stored energy in the capacitor. *Hint:* What happens to the rest of it? You will need to integrate.

75. Find a general solution for the currents in Example 28-6, in terms of the symbolic quantities $\mathcal{E}_1$, $\mathcal{E}_2$, R_1, R_2, and R_3.

FIGURE 28-65 Problem 76.

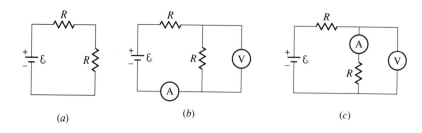

(a)   (b)   (c)

76. Two identical resistors are connected across a battery, as shown in Fig. 28-65a. Figures 28-65b and c show two ways to connect a voltmeter and ammeter in order to measure the voltage across and current through one resistor. If the voltmeter resistance is $100R$ and the ammeter resistance is $0.010R$, how would the resistances calculated from the measured voltage and current for the two connections compare with the actual resistance?

77. Write the loop and node laws for the circuit of Fig. 28-66, and show that the time constant for this circuit is $R_1 R_2 C/(R_1 + R_2)$.

78. The circuit in Fig. 28-67 extends forever to the right, and all the resistors have the same value R. Show that the equivalent resistance measured across the two terminals at left is $\frac{1}{2}R(1 + \sqrt{5})$. *Hint:* You don't need to sum an infinite series.

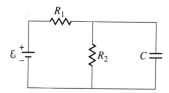

FIGURE 28-66 Problem 77.

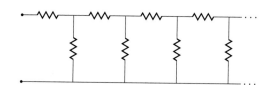

FIGURE 28-67 Problem 78.

CHAPTER 29

THE MAGNETIC FIELD

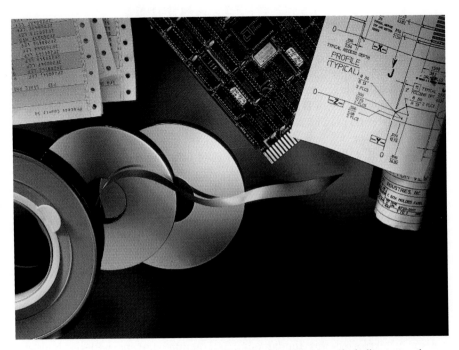

Magnetic media have replaced paper in many data storage applications, including text and graphics. An electronic circuit board controls the flow of information to and from the magnetic media.

Most people are fascinated with magnets. Magnetism—the seemingly mysterious force you feel when you try to push two magnets together in a way they don't want to go—is always intriguing. Some uses of magnets, like holding notes on refrigerators, are mundane. But others—holding gas at a temperature of one hundred million K in a nuclear fusion reactor, or converting electrical into mechanical energy in the motors of a railroad locomotive—are more impressive. And magnetism, like electricity, is at the heart of many natural phenomena and technological devices. Video and audio tape recorders, electric motors, TV picture tubes, computer disks, and electric power plants would be impossible without magnetism. Earth's magnetism helps us find our way around, provides historical evidence for the evolution of our planet, and protects us from harmful radiation. Birds, sea turtles, and some bacteria use Earth's magnetism for navigation. Without magnetism we would not even see,

FIGURE 29-1 Iron filings align with the magnetic field, tracing out the field from a pair of bar magnets. The field is strongest near the magnetic poles.

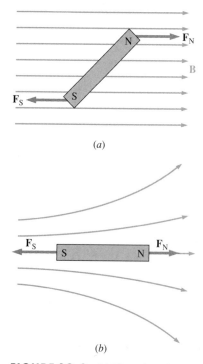

FIGURE 29-2 (*a*) The poles of a bar magnet experience opposite forces in a uniform magnetic field, giving rise to a torque that tends to align the magnet with the field. (*b*) In a nonuniform field the forces on the two poles are unequal, giving rise to a net force on the magnet.

for light itself originates in an interaction between magnetism and electricity. As with electricity, we often do not recognize the magnetic character of everyday phenomena.

In this chapter we discuss various aspects of magnetism and its relation to electricity. In subsequent chapters we will say more about the fundamental relation between magnetism and electricity, and eventually will come to understand how electricity and magnetism are manifestations of the same underlying phenomenon.

29-1 MAGNETS, POLES, AND THE MAGNETIC FIELD

An ordinary magnet has two **poles,** arbitrarily designated north and south. The interaction of magnetic poles is similar to that of electric charges: like poles repel, and unlike poles attract. This interaction may be described in terms of the **magnetic field** (symbol **B**). We say that one magnet produces a magnetic field and that a second magnet responds to the field in its immediate vicinity. As with electricity, describing magnetism in terms of fields eliminates the awkward "action at a distance" picture, in which one magnet somehow reaches across empty space to influence another.

We can trace out magnetic field lines using small bar magnets or iron filings (Fig. 29-1). The direction of the field is taken as the direction of the force on a magnetic north pole. The field is stronger—meaning it gives rise to greater forces—where field lines are closely spaced. Since magnetic poles always come in pairs, the effect of the magnetic force on a magnet is to rotate the magnet into alignment with the field (Fig. 29-2*a*). When the field is nonuniform the magnet experiences a force as well (Fig. 29-2*b*).

But what *is* a magnet? Magnets, like other matter, are composed of electrons, protons, and neutrons. The interaction of magnets and magnetic fields is fundamentally an interaction involving these particles. We begin our study of magnetism by exploring this fundamental interaction, and we will see how the phenomenon of magnetism is far more extensive than our experience with magnets would suggest. Later we'll see how magnets and magnetic materials fit into this much broader picture of magnetic phenomena.

29-2 ELECTRIC CHARGE AND THE MAGNETIC FIELD

We have mentioned several times that electricity and magnetism are closely related. One manifestation of this relation is that an *electric charge* can experience a force in a magnetic field. We defined the electric field **E** through the relation $\mathbf{F}_E = q\mathbf{E}$, where $\mathbf{F}_E$ is the electric force on a particle with charge q. Analogously, we define the magnetic field in terms of the relation between a charge q and the magnetic force $\mathbf{F}_B$ it experiences.

Consider a region in which there is no electric field, but in which a magnetic field is present. Experimentally, we find that a stationary electric charge in this region experiences no force. But if the charge is moving there may be a magnetic force on it. Experiment shows that

1. The magnetic force is always at right angles both to the velocity **v** of the charge and to the magnetic field **B** (Fig. 29-3).
2. The strength of the magnetic force is proportional to the product of the charge q, its speed v, and the magnetic field strength B.
3. The force is greatest if the charged particle is moving at right angles to the magnetic field **B** and is zero if the charge velocity **v** is parallel or antiparallel to the field. In general, the magnetic force is proportional to the sine of the angle between the vectors **v** and **B** (see Fig. 29-3).

Putting these facts together allows us to write the magnetic force compactly in terms of the vector cross product introduced in Chapter 13:

$$\mathbf{F}_B = q\mathbf{v} \times \mathbf{B}, \quad \text{(magnetic force)} \tag{29-1a}$$

where $\mathbf{F}_B$ is the magnetic force on a particle of charge q moving with velocity **v** at a point where the magnetic field is **B**. Recall from Chapter 13 that the cross product $\mathbf{A} \times \mathbf{B}$ of two vectors **A** and **B** is a vector of magnitude $AB \sin\theta$, where θ is the angle between **A** and **B**, and where the direction of $\mathbf{A} \times \mathbf{B}$ is given by the right-hand rule. The magnetic force thus has magnitude

$$F = qvB\sin\theta, \tag{29-1b}$$

while Fig. 29-4 shows how the right-hand rule applies to the vectors **v** and **B**. Because the charge q also enters Equation 29-1a, the direction of the magnetic force is that of $\mathbf{v} \times \mathbf{B}$ for a positive charge but opposite $\mathbf{v} \times \mathbf{B}$ for a negative charge.

We can regard Equation 29-1 as the definition of the magnetic field. If we put particle of charge q moving with velocity **v** in a region free of other influences (i.e., electric and gravitational fields), then the presence of a force **F** on the particle shows that there is a magnetic field **B** implied by Equation 29-1a. Equations 29-1 show that the SI units of magnetic field are N·s/C·m, a unit given the name **tesla** (T) after the Serbian-American inventor Nikola Tesla (1865–1943). One tesla is a strong magnetic field, and a smaller unit called the

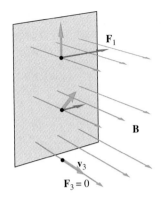

FIGURE 29-3 The magnetic force on a charged particle is perpendicular to the particle's velocity **v** and to the magnetic field **B**. The magnitude of the force depends on the angle between **v** and **B**, and is greatest when **v** and **B** are perpendicular (top). When **v** and **B** are parallel there is no magnetic force (bottom).

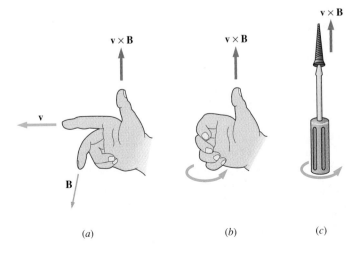

FIGURE 29-4 Three ways to remember the right-hand rule. In (b) and (c) the curved arrow represents a rotation taking the vector **v** onto the vector **B**.

FIGURE 29-5 This large superconducting magnet is 15 m long and produces a field of 6.6T. It was designed to steer charged particles in the Superconducting Super Collider, a giant particle accelerator intended to probe the fundamental structure of matter.

gauss (G), equal to 10^{-4} T, is often used. Earth's magnetic field, for example, is a little under 1 G, while the field at the poles of a toy magnet may be 100 G. Strong laboratory magnets (Fig. 29-5) produce fields ranging from several T to nearly 40 T, while neutron stars—incredibly dense, rapidly rotating objects with the mass of a star compressed into a region several km in diameter—have fields up to 10^8 T.

● **EXAMPLE 29-1** STEERING PROTONS

A magnetic field of 0.10 T is used to steer charged-particle beams in a nuclear physics experiment. The field points vertically upward. Three protons enter the field region, two moving horizontally and one vertically as shown in Fig. 29-6. All three are moving at 2.0×10^6 m/s. What is the magnetic force on each?

Solution

Proton 2 is moving parallel to the field. For it, $\mathbf{v} \times \mathbf{B} = \mathbf{0}$ and it experiences no magnetic force. Protons 1 and 3 are moving at right angles to the field, so $\sin\theta = 1$ and Equation 29-1b gives

$$F = qvB = (1.6 \times 10^{-19} \text{ C})(2.0 \times 10^6 \text{ m/s})(0.10 \text{ T})$$

$$= 3.2 \times 10^{-14} \text{ N}.$$

Since the protons carry positive charge, the direction of the force is that of the product $\mathbf{v} \times \mathbf{B}$. For proton 1, moving to the right, $\mathbf{v} \times \mathbf{B}$ is out of the page. For proton 3, moving to the left, the force is into the page. This example shows that the magnetic field alone does not determine the magnetic force. Identical particles in the same field experience different forces if their velocities are different. Had the particles been electrons the forces would have been in the opposite directions. (Why?)

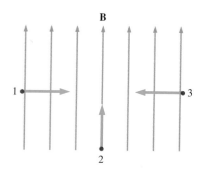

FIGURE 29-6 What is the magnetic force on each proton? (Example 29-1)

EXERCISE A 1.3-T magnetic field points in the x direction. A particle carrying 1.0 μC is moving at 20 m/s in the x-y plane and experiences a magnetic force of 1.4×10^{-5} N. What angle does the particle's velocity make with the x axis?

Answer: 33°

Some problems similar to Example 29-1: 1–11 ●

Although electricity and magnetism are related, the electric field and the magnetic field are distinct. In particular, both may be present simultaneously. In that case, a charged particle experiences both the magnetic force $\mathbf{F}_B$ of Equation 29-1 and also the electric force $q\mathbf{E}$, for a net **electromagnetic force** given by

$$\mathbf{F} = q\mathbf{E} + q\mathbf{v} \times \mathbf{B}. \qquad \text{(electromagnetic force)} \qquad (29\text{-}2)$$

● EXAMPLE 29-2 A VELOCITY SELECTOR

A region contains uniform electric and magnetic fields of magnitude E and B, respectively, and oriented at right angles as shown in Fig. 29-7. A beam of charged particles enters the region, heading straight into the page. What speed must a particle have if it is to cross the field region undeflected?

Solution
The net force on a particle must be zero if it is to be undeflected. Consider a positively charged particle heading into the page in Fig. 29-7. The electric field points to the right, so the electric force $q\mathbf{E}$ is to the right. Applying the right-hand rule with the velocity vector $\mathbf{v}$ into the page and the magnetic field $\mathbf{B}$ downward gives a magnetic force to the left; since v and B are at right angles, the magnitude of the magnetic force is qvB. So the electric and magnetic forces are in opposite directions; for them to cancel their magnitudes must be equal:

$$qE = qvB,$$

or

$$v = \frac{E}{B}.$$

Note that this condition is independent of the charge—even of its sign, since the directions of both forces would reverse for a negative charge.

The field configuration of this example is called a **velocity selector,** since it passes undeflected only those particles with speed $v = E/B$ that are moving perpendicular to both fields.

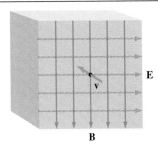

FIGURE 29-7 Crossed electric and magnetic fields give rise to oppositely-directed forces on charged particles moving perpendicular to the fields. The magnetic force depends on the particle speed, but the electric force does not; therefore, only those particles with just the right speed can traverse the field region undeflected.

Velocity selectors can be used to prepare particle beams with uniform speeds or to analyze the distribution of speeds in a charged-particle population.

EXERCISE You have a magnet that produces a 0.25-T field. What strength electric field should you apply to make a velocity selector that passes 2.8×10^5 m/s particles undeflected?

Answer: 70 kV/m

Some problems similar to Example 29-2: 12–14
●

29-3 THE MOTION OF CHARGED PARTICLES IN MAGNETIC FIELDS

Like any force, the magnetic force deflects a particle from the straight-line path Newton's first law says it would otherwise follow (Fig. 29-8). But unlike most other forces, the magnetic force is necessarily at right angles to the particle velocity. That means force and particle displacement are perpendicular. Since work is the dot product of force with displacement (Equation 7-5; $W = \mathbf{F} \cdot \Delta\mathbf{r}$), and the dot product of perpendicular vectors is zero, we can conclude that

┃ The magnetic force does no work on a charged particle.

Both the kinetic energy and thus the speed of a charged particle subject only to a magnetic force are therefore constant. The magnetic force changes only the

(a)   (b)

FIGURE 29-8 (a) In the absence of a magnetic field, an electron beam moves in a straight line. (b) A magnetic field causes the beam to deflect. If the electrons are moving from left to right, which way is the field pointing?

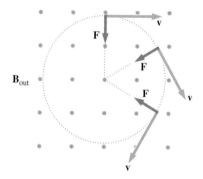

FIGURE 29-9 A charged particle moving at right angles to a uniform magnetic field describes a circular path. Note the convention that dots represent magnetic field lines emerging from the page; crosses, in contrast, would represent field lines going into the page.

direction of the particle's velocity, not its magnitude. What sort of motion results from the magnetic force? Suppose a particle with positive charge q is moving at right angles to a uniform magnetic field **B,** as shown in Fig. 29-9. At the instant the particle's velocity is to the right, the outward-pointing magnetic field gives a force $q\mathbf{v} \times \mathbf{B}$ that is downward. Since this force is at right angles to the velocity, it changes the direction but not the magnitude of the velocity. The magnetic force on the new velocity is still perpendicular to it, and since the magnitude of the velocity is unchanged, so is the magnitude of the force. What we've just described—a force of constant magnitude always at right angles to a particle's velocity—is the prescription for uniform circular motion. Thus, a charged particle moving in a plane perpendicular to a uniform magnetic field describes a circular path. The magnetic force acts exactly like the tension in a string when you tie a mass to the string and whirl it around in a circle. The tension is perpendicular to the motion and changes its direction but not its speed.

How big is the circle, and how long does it take to get around? We can answer these questions using Newton's second law. In its circular path, the charged particle undergoes an acceleration v^2/r, directed toward the center of the circle. What causes this acceleration? The magnetic force! With the field and velocity at right angles, the magnitude of the magnetic force is just

$$F = qvB,$$

and the force points toward the center of the circle. Writing Newton's law $\mathbf{F} = m\mathbf{a}$ then gives

$$qvB = m\frac{v^2}{r},$$

so

$$r = \frac{mv}{qB}. \tag{29-3}$$

This result makes sense: the larger the particle's momentum mv, the harder it is for the magnetic force to bend it out of a straight line, so the larger the radius of the orbit. On the other hand, if we make the field or charge larger, then the magnetic force increases, giving a tighter orbit.

● EXAMPLE 29-3   A TV PICTURE TUBE

In a TV picture tube, deflection of the electron beam that "paints" the TV picture is accomplished using magnetic fields. The geometry of a certain tube requires that the electron beam be bent in a circular arc with a minimum radius of 4.5 cm (Fig. 29-10). If the electrons are accelerated from rest through a 25-kV potential difference before they enter the magnetic field region, what is the magnetic field strength required? In what direction should the field point to accomplish the deflection shown in Fig. 29-10? Assume that the field is uniform over the deflecting region and zero elsewhere.

Solution

Solving Equation 29-3 for the magnetic field strength B gives

$$B = \frac{mv}{er},$$

with e the elementary charge. "Falling" through a potential difference $V = 25$ kV, the electron of charge e acquires a kinetic energy $\frac{1}{2}mv^2 = Ve$, so its speed is

$$v = \sqrt{\frac{2Ve}{m}}.$$

Our expression for the field then becomes

$$B = \frac{m}{er}\sqrt{\frac{2Ve}{m}} = \frac{1}{r}\sqrt{\frac{2mV}{e}}$$

$$= \frac{1}{0.045 \text{ m}}\left(\frac{(2)(9.1 \times 10^{-31} \text{ kg})(25 \times 10^3 \text{ V})}{1.6 \times 10^{-19} \text{ C}}\right)^{1/2}$$

$$= 0.012 \text{ T}.$$

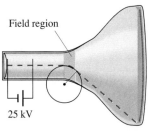

FIGURE 29-10   A TV picture tube requiring a 4.5-cm bending radius (marked with a diagonal line) for maximum deflection of the electron beam. Which way should the magnetic field point to deflect an electron in the direction indicated (Example 29-3)?

To achieve an initially downward deflection, the force $q\mathbf{v} \times \mathbf{B}$ must be initially downward. But the electron charge is negative, so $\mathbf{v} \times \mathbf{B}$ must be upward. Application of the right-hand rule to the rightward-moving electrons shows that the magnetic field $\mathbf{B}$ must be into the page in Fig. 29-10.

EXERCISE   A beam of protons moving at 2.5×10^5 m/s is deflected through a 90° turn by a 10-mT magnetic field oriented perpendicular to the beam's velocity. What is the radius of the turn?

Answer: 26 cm

Some problems similar to Example 29-3:   15, 18, 21   ●

The Cyclotron Frequency

How long does it take a charged particle to complete its circular orbit in a uniform magnetic field? The circumference of the orbit is $2\pi r$, so the period T of the circular motion is

$$T = \frac{2\pi r}{v}.$$

Using Equation 29-3 for the radius r gives

$$T = \frac{2\pi r}{v} = \frac{2\pi}{v}\frac{mv}{qB} = \frac{2\pi m}{qB}. \qquad (29\text{-}4)$$

FIGURE 29-11 The Crab Nebula, remnant of a supernova explosion observed nearly 1000 year ago. Intense radio emission occurs as electrons undergo circular motion in the Crab's magnetic field. The frequency of the emission is related to the cyclotron frequency of the electrons' motion, and its measurement therefore allows determination of the magnetic field.

This is a remarkable result, for it shows that the period of the circular motion is independent of the particle's speed and the size of the orbit. It depends only on the magnetic field and the charge-to-mass ratio of the particle. The frequency in revolutions per second is simply $1/T$, or

$$f = \frac{qB}{2\pi m}. \tag{29-5}$$

This quantity is often called the **cyclotron frequency,** since it's the frequency at which charged particles circulate in a cyclotron particle accelerator. In astrophysics, the cyclotron frequency provides a simple way to measure the magnetic fields of distant objects (Fig. 29-11), while electrons circling in a special tube called a magnetron generate the microwaves that cook your food in a microwave oven.

■ APPLICATION THE CYCLOTRON

Physicists studying the basic structure of matter need tools that can probe the atomic nucleus and its constituent particles. The only probes sufficiently small are subatomic particles themselves, accelerated to high enough energies that they can disrupt the strong nuclear force. How is this acceleration accomplished? One way is to accelerate particles of charge q through a large potential difference V, giving each particle energy qV. But there are practical problems in the generation and handling of potential differences much over a million volts (recall Fig. 25-25). To achieve higher energies, devices are used that circumvent the need for a single large potential difference. One of the earliest and most successful such devices is the **cyclotron** (Fig. 29-12). The device consists of an evacuated chamber

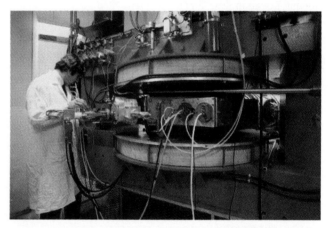

FIGURE 29-13 This cyclotron at Massachusetts General Hospital was used to produce radioisotopes for research leading to the development of the medical diagnostic procedure known as positron emission tomography (PET).

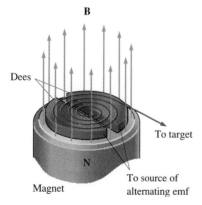

FIGURE 29-12 A cyclotron, showing one magnet pole, dees, and a typical ion trajectory. Not shown are the vacuum chamber surrounding the dees, upper pole, and frame supporting pole pieces and electromagnet windings.

between the poles of a magnet. At the center of the chamber is a source of the particles to be accelerated, usually protons or light ions. The ions undergo circular motion in the magnetic field.

Also within the evacuated chamber are two hollow conducting structures each shaped like the letter D (Fig. 29-12). A modest potential difference is applied across these two "dees," and this potential difference is made to alternate in polarity with the same frequency as the circular motion of the ions. Recall that this frequency depends only on the magnetic field strength and the charge-to-mass ratio of the ions, but not on their energy. As the ions circle around in the cyclotron, they are accelerated across the gap between the dees by the strong electric field associated with the potential difference. Because each dee is a hollow, nearly closed conducting structure, there is no electric field within the dees.

Once inside a dee, the ions simply follow a circle in the magnetic field. Halfway round they again cross the gap between dees. If the electric field were steady, the ions would be *decelerated* at this crossing. But the electric field changes direction in step with the ions' circular motion, so each time the ions cross the dee gap they are accelerated and gain more energy. They move faster and in ever larger circles, but always with the same orbital period. Eventually the ion orbits become nearly the size of the machine. At this point an electrostatic field provided by a high-voltage electrode deflects the ions out of the magnetic field and toward a target, where their interactions with target nuclei cause nuclear reactions.

In addition to providing experimental data on nuclear structure, cyclotrons are valuable in producing short-lived radioactive isotopes for a variety of purposes, particularly medical research and diagnosis. A number of large hospitals have their own cyclotrons (Fig. 29-13); in particular, the diagnostic procedure known as positron emission tomography (PET) requires cyclotron-produced radioisotopes.

At very high energies, the theory of relativity comes into play and alters our conclusion that the cyclotron frequency is independent of particle energy. As a result, the cyclotron cannot be used to achieve these relativistic energies. An alternate

FIGURE 29-14 Aerial view of Fermilab, the Fermi National Accelerator Laboratory at Batavia, Illinois. Large circle in the background is the 2-km-diameter Tevatron, a synchrotron that accelerates protons to energies of 1 TeV (10^{12} eV). Much larger synchrotrons include the Large Electron-Positron Collider at the European Center for Nuclear Research and the Superconducting Super Collider that was to have been built in Texas.

accelerator design is the **synchrotron,** in which both the magnetic field and frequency of an alternating electric field are varied to account for increasing particle energy, while the orbital radius is held constant (Fig. 29-14).

● EXAMPLE 29-4 DESIGNING A CYCLOTRON

A cyclotron is to accelerate protons to a kinetic energy of 5.0 MeV. If the magnetic field in the cyclotron is 2.0 T, what must be the radius of the cyclotron and the frequency at which the dee voltage is alternated?

Solution
The cyclotron frequency is given by Equation 29-5:

$$f = \frac{qB}{2\pi m} = \frac{(1.6\times10^{-19}\text{ C})(2.0\text{ T})}{(2\pi)(1.67\times10^{-27}\text{ kg})} = 3.0\times10^7 \text{ Hz}.$$

This is the frequency required to accelerate protons at each crossing of the dee gap; incidentally, it is about the frequency of a citizens band (CB) radio transmitter.

An energy of 5.0 MeV is equal to

$$(5.0\times10^6 \text{ eV})(1.6\times10^{-19} \text{ J/eV}) = 8.0\times10^{-13} \text{ J},$$

so the proton kinetic energy is

$$K = \tfrac{1}{2}mv^2 = 8.0\times10^{-13} \text{ J}.$$

Solving for the speed v gives

$$v = \sqrt{\frac{2K}{m}} = \sqrt{\frac{(2)(8.0\times10^{-13}\text{ J})}{1.67\times10^{-27}\text{ kg}}} = 3.1\times10^7 \text{ m/s}.$$

Equation 29-3 then gives the radius needed to accommodate 5-MeV protons:

$$r = \frac{mv}{qB} = \frac{(1.67\times10^{-27}\text{ kg})(3.1\times10^7\text{ m/s})}{(1.6\times10^{-19}\text{ C})(2.0\text{ T})} = 0.16 \text{ m}.$$

To ensure a uniform magnetic field over the particle trajectories, the radii of the magnet pole pieces would have to be somewhat larger than this value.

EXERCISE A microwave oven uses 2.4-GHz microwaves, generated by electrons circling at this frequency in the magnetic field of a magnetron tube. (a) What is the magnetic field strength? (b) If the energy associated with an electron's circular motion is 890 eV, what is the radius of its circular path?

Answers: (a) 86 mT; (b) 1.2 mm

Some problems similar to Example 29-4: 17, 25, 26   ●

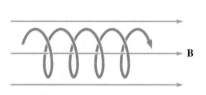

FIGURE 29-15 A particle in a uniform magnetic field describes a helical path, its motion along the field direction unaffected by the magnetic force.

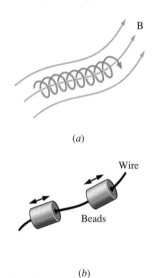

(a)

(b)

FIGURE 29-16 (a) Charged particles undergoing helical motion about the magnetic field direction are "frozen" to the field like (b) beads sliding along a wire.

FIGURE 29-17 Magnetic fields govern the behavior of matter throughout much of the universe. Here, concentrations of ionized gas trace out magnetic field loops in the Sun's atmosphere.

Particle Trajectories in Three Dimensions

What if a particle's motion is not confined to a plane perpendicular to the magnetic field? Then we can resolve its velocity into two vectors, one perpendicular and the other parallel to the magnetic field. Since the magnetic force is always perpendicular to the field, there is no component of force along the field direction—and therefore the velocity component in this direction is unaffected by the magnetic force. The force acts only on the velocity component perpendicular to the field, and here our previous analysis applies: in the plane perpendicular to the field, the particle's motion is circular with frequency given by Equation 29-5. In the absence of other forces the particle moves uniformly along the field direction while executing circular motion perpendicular to the field. The resulting trajectory is a helix, as shown in Fig. 29-15.

The absence of magnetic force along the field direction means that particles move readily along the field. But if you try to push them at right angles to the field, they simply move in larger circles about the field direction. As a result, charged particles are often described as being "frozen" to the magnetic field (Fig. 29-16). Nonuniform fields and collisions between particles make this "freezing" of particles and field less than perfect, but in many cases the particle density is low enough that the "frozen" assumption is an excellent approximation. Unlike our relatively cool planet Earth, much of the universe consists of free electrons and protons, not bound into neutral atoms. As a result, magnetic fields are a dominant influence on matter throughout much of the universe (Fig. 29-17).

An important terrestrial application of this "trapping" of particles on magnetic field lines occurs in nuclear fusion reactors. These experimental devices—whose successful development would make 1 gallon of sea water the energy equivalent of more than 300 gallons of gasoline—have to contain ionized gas (plasma) at temperatures around 100 million kelvins. They do so with "magnetic bottles," most commonly toroidal (doughnut-shaped) chambers whose circular magnetic fields never intersect the chamber walls and therefore keep the hot plasma from touching the walls (Fig. 29-18).

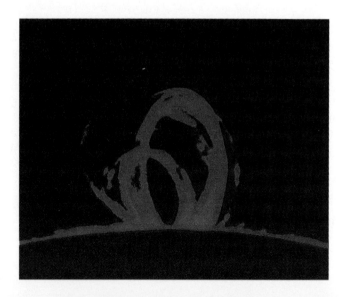

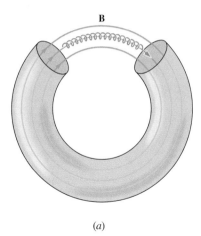

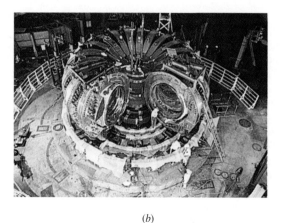

(a)

(b)

FIGURE 29-18 (*a*) Charged particles spiral about the circular field lines in a simplified fusion reactor. With its toroidal shape, the machine has no "ends" from which magnetic field lines emerge. The field therefore keeps charged particles away from the chamber walls. (*b*) The Tokamak Fusion Test Reactor at Princeton University. This view, taken during construction, shows clearly the machine's toroidal shape.

■ APPLICATION THE AURORA AND MAGNETIC MIRRORS

Earth itself possesses a magnetic field whose origin we will consider in the next chapter. Near the planet, the field resembles that of a bar magnet. A stream of electrons and protons, called the solar wind, flows outward from the Sun and is deflected by Earth's field, protecting us from potentially harmful radiation. But occasionally bursts of high-energy solar particles penetrate the magnetic field, eventually becoming trapped into helical motion along the field. Figure 29-19*a* shows that the higher field lines, where particles are most likely to be trapped, intersect Earth near the poles. High-energy particles moving along these field lines slam into the upper atmosphere near the poles, exciting oxygen and nitrogen atoms and

(b)

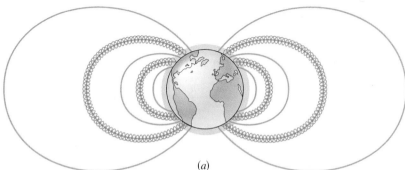

(a)

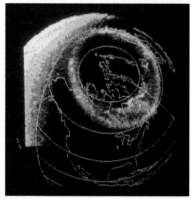

FIGURE 29-19 The aurora arises from the interaction of high-energy particles trapped in Earth's magnetic field. (*a*) Schematic diagram showing particles spiraling between the northern and southern polar regions along the magnetic field. (*b*) A colorful aurora viewed from Canada's Prince Edward Island. (*c*) The aurora appears as a circle surrounding Earth's north magnetic pole in this satellite image.

(c)

thereby producing the spectacular displays we call the aurora (Fig. 29-19b,c). These solar outbursts also buffet Earth's magnetic field, causing "magnetic storms" that can severely disrupt communications and electric power transmission.

Fig. 29-20 shows a particle spiraling into a region of stronger magnetic field, like the polar regions of Fig. 29-19a. Note that there is a component of magnetic force opposite to the general direction of motion. This force eventually reverses the particle's forward motion, with the result that the stronger field region acts like a **magnetic mirror,** reflecting particles that attempt to penetrate. This means that particles trapped in Earth's magnetic field bounce back and forth between northern and southern hemispheres, reflected by the magnetic mirrors of higher field strength near the poles. Auroral displays in opposite hemispheres are often strikingly similar since they are caused by the same particle populations mirroring back and forth.

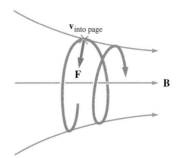

FIGURE 29-20   The magnetic mirror effect. A charged particle is spiraling into a region of stronger magnetic field. At the point shown its velocity is into the page, and the magnetic force—proportional to $\mathbf{v} \times \mathbf{B}$—has a component to the left. This force eventually reverses the rightward motion of the particle.

29-4 THE MAGNETIC FORCE ON A CURRENT

So far we have considered the magnetic force on individual charged particles. An electric current is simply a group of charged particles sharing a common motion, so we should expect a current to interact with a magnetic field.

The Force on a Straight Wire

Imagine a long straight wire carrying a current I through a uniform magnetic field **B,** as shown in Fig. 29-21. If each charge carrier has a drift velocity $\mathbf{v}_d$ along the wire, and if each carries charge q, then the magnetic force on each charge carrier is given by Equation 29-1:

$$\mathbf{F}_q = q\mathbf{v}_d \times \mathbf{B}.$$

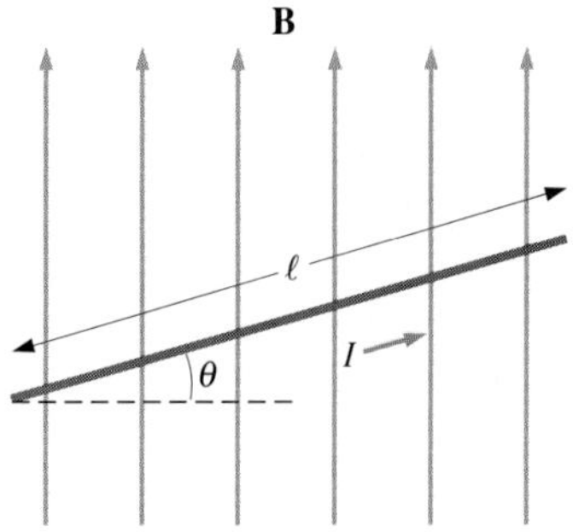

(Of course the charge carriers also have random thermal velocities. But these average to zero and therefore make no net contribution to the magnetic force.) If the wire has cross-sectional area A and contains n charge carriers per unit volume, then the net force on all the charge carriers in a length ℓ of the wire is

$$\mathbf{F} = nA\ell q\mathbf{v}_d \times \mathbf{B}.$$

The product $nAqv_d$ is just the current I, as we found in deriving Equation 27-2. If we define a vector $\boldsymbol{\ell}$ whose magnitude is the length ℓ of the wire and whose direction is along the current, then we can write

$$\mathbf{F} = I\boldsymbol{\ell} \times \mathbf{B}. \qquad \text{(magnetic force on a current)} \qquad (29\text{-}6)$$

FIGURE 29-21   A straight wire carrying current I through a uniform magnetic field. The charge carriers in the wire experience a magnetic force, in this case out of the page.

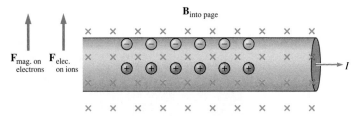

FIGURE 29-22 Electrons moving to the left in a wire are deflected upward by the magnetic force, resulting in charge separation and therefore in an upward electric force on the fixed ions. As a result the entire wire is ultimately influenced by the magnetic force on the moving electrons.

The direction of this force is at right angles to both the current and the magnetic field, or out of the page in Fig. 29-21. For a given direction of the current, the direction of the force does not depend on the sign of the charge carriers. If the current is to the right, then positive charges move to the right and the force on each is out of the page. If the charges are negative, they move to the left and the force is still out of the page because both the sign of the velocity and the sign of the charge are reversed, given the same sign for the force $q\mathbf{v} \times \mathbf{B}.$

Strictly speaking, Equation 29-6 gives the net magnetic force only on the charge carriers in the wire. But the motion of the charge carriers—typically electrons—under the influence of the magnetic force causes charge separation in the wire, and the resulting electric field exerts a force on the fixed charges—typically ions—in the wire (Fig. 29-22). Thus the entire wire experiences the force. Although its origin is not entirely magnetic, we loosely call the force given by Equation 29-6 the magnetic force on the wire.

The magnetic force on a current-carrying wire is the basis for many practical devices, including electric motors that start cars and run refrigerators, CD players, computer disk drives, subway trains, pumps, food processors, power tools, and myriad other useful instruments of modern society.

● **EXAMPLE 29-5** MAGNETIC FORCE ON A POWER LINE

A power line runs along Earth's equator, where Earth's magnetic field points horizontally from south to north and has a strength of about 0.5 G. The current in the power line is 500 A, flowing from west to east. What are the magnitude and direction of the magnetic force on 1 km of the power line?

Solution

Let eastward be the x direction, northward the y direction, and upward the z direction. After we convert gauss to tesla, Equation 29-6 gives

$$\mathbf{F} = I\boldsymbol{\ell} \times \mathbf{B} = (500 \text{ A})(1000\hat{\mathbf{i}} \text{ m}) \times (0.5 \times 10^{-4} \hat{\mathbf{j}} \text{ T})$$

$$= 25\hat{\mathbf{k}} \text{ N}.$$

This 25-N upward force is negligible compared with the weight—on the order of 2×10^4 N—of 1 km of power line.

EXERCISE A 2.0-m-long wire has a mass of 37 g. The wire extends horizontally, at 40° to a horizontal magnetic field of 0.075 T. What current in the wire will result in the magnetic force suspending it against gravity?

Answer: 3.8 A

Some problems similar to Example 29-5: 33, 35, 37 ●

■ **APPLICATION** MAGNETIC LEVITATION AND PROPULSION

Engineers throughout the world are at work on transportation systems whose operation results directly from the magnetic force on a straight current-carrying conductor. In so-called maglev vehicles, magnetic forces levitate the vehicle just a few centimeters above a conducting guideway and also provide the horizontal force that propels the vehicle (Fig. 29-23). Such vehicles should be capable of 500-km/h speeds, well above the 300 km/h maximum for high-speed rail systems. Maglev vehicles should prove especially effective for travel between cities in densely populated regions.

A related application is so-called magnetohydrodynamic (MHD) propulsion, a kind of jet propulsion in which a conducting fluid is accelerated by the magnetic force. Engineers are currently experimenting with MHD propulsion for ships. Because sea water is a reasonably good electrical conductor, passage of a current through sea water in a magnetic field results in a magnetic force on the water. Figure 29-24 shows the experimental Japanese ship Yamato-1 and its MHD propulsion system.

FIGURE 29-23 A magnetically levitated vehicle under development in Japan.

(a)

FIGURE 29-24 (a) The Japanese ship Yamato-1 uses MHD propulsion. (b) Schematic diagram of the MHD propulsion unit. Superconducting electromagnet produces a magnetic field; current flowing in the sea water between the two electrodes then experiences a magnetic force that ejects a jet of water at right. (c) One of Yamato's two MHD propulsion units.

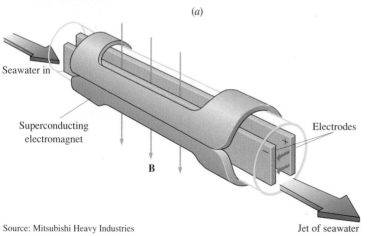

Seawater in

Superconducting electromagnet

Electrodes

B

Source: Mitsubishi Heavy Industries

Jet of seawater

(b)

(c)

■ APPLICATION THE HALL EFFECT

We have seen that the force on a current-carrying conductor is independent of the sign of the charge carriers. But there is a subtle difference. Fig. 29-25 shows two conductors, each with the same current I to the right and magnetic field B pointing into the page. In Fig. 29-25a the current is carried by electrons moving to the *left*. The product $v \times B$ is downward, but since electrons are negative the force $qv \times B$ is upward. As a result, the upper edge of the conductor is negative with respect to the lower edge. In Fig. 29-25b the current is carried by protons moving to the *right*. Again the force $qv \times B$ is upward, so now the upper edge of the conductor is positive.

This phenomenon of charge separation is the **Hall effect**. The separated charge gives rise to an electric field and therefore to a measurable potential difference—the **Hall potential**—between opposite edges of the conductor. The sign of the Hall potential depends on the sign of the charge carriers. Charge separation will stop once the electric force on the charge carriers cancels the magnetic force. The electric force has magnitude qE and, with B perpendicular to the current, the magnetic force has magnitude qv_dB, with v_d the charge-carrier drift speed. Equating these magnitudes gives $E = v_dB$. In a rectangular conductor this field will be essentially uniform, so the Hall potential becomes

$$V_H = Eh = v_dBh,$$

where h is the conductor height shown in Fig. 29-25. But Equation 27-2 gives $I = nAqv_d$, with n the number density of charge carriers, A the conductor's cross-sectional area, and q the charge on each carrier. Solving for v_d and using the result in our expression for V_H gives

$$V_H = \frac{IBh}{nAq},$$

or, since $A = ht$ with t the conductor thickness shown in Fig. 29-25,

$$V_H = \frac{IB}{nqt}. \tag{29-7}$$

The quantity $1/nq$ is the **Hall coefficient.** Measurement of V_H in a sample of known thickness carrying known current in a known magnetic field yields the value of this coefficient and, thus, gives information on the nature and density of the charge carriers. Alternatively, measurement of V_H in a material with known Hall coefficient carrying a known current provides a direct measure of the magnetic field strength (Fig. 29-26). Since V_H is inversely proportional to the charge-carrier density n, Hall-effect magnetic field measurements often use semiconductors whose carrier density n is much lower than in metals. For typical magnetic fields, Hall potentials are then in millivolts (see Problem 45).

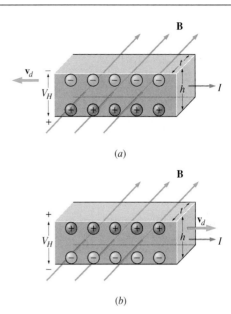

(a)

(b)

FIGURE 29-25 The Hall potential V_H arises from the deflection of charge carriers by the magnetic force when a current-carrying conductor is in a magnetic field. Although the magnetic force on the conductor is independent of the sign of the charge carriers, the sign of the Hall potential does depend on the sign of the charge carriers. In both (a) and (b) the current is to the right, carried in (a) by negative charge carriers moving to the left and in (b) by positive charge carriers moving to the right.

FIGURE 29-26 This magnetometer uses the Hall potential developed in its semiconductor probe to measure magnetic fields.

One of the most surprising discoveries of the 1980s was the so-called quantized Hall effect, where the Hall potential at low temperatures and strong magnetic fields exhibits a step-like rather than continuous increase with increasing magnetic field. This is one of the few instances where a macroscopic property manifests directly the discontinuity inherent in the quantum description of matter. Discovery of the quantized Hall effect won Klaus von Klitzing the 1985 Nobel Prize in Physics.

FIGURE 29-27 (*a*) A curved conductor in a nonuniform magnetic field.
(*b*) A small segment $d\ell$ can be treated
as a straight wire in a uniform field; the
magnetic force on the segment is
$d\mathbf{F} = I\,d\ell \times \mathbf{B}$.

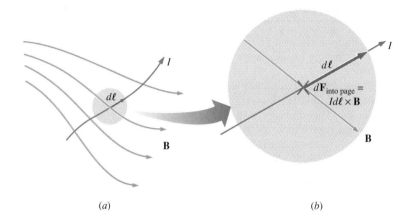

FIGURE 29-27 (*a*) A curved conductor in a nonuniform magnetic field.
(*b*) A small segment $d\ell$ can be treated
as a straight wire in a uniform field; the
magnetic force on the segment is
$d\mathbf{F} = I\,d\ell \times \mathbf{B}$.

(*a*)   (*b*)

Nonuniform Fields and Curved Conductors

Equation 29-6 applies to a straight conductor at some arbitrary angle to a
uniform magnetic field. What if the conductor bends, so its orientation relative
to the field changes? Or what if the field changes in magnitude or direction over
the length of the conductor? We can still make use of Equation 29-6 if we apply
it only to a very small segment of the conductor, so small that it is essentially
straight and the field essentially constant over its length. If the length of this
small segment is $d\ell$, then Equation 29-6 becomes

$$d\mathbf{F} = I\,d\ell \times \mathbf{B}$$

for the small force on the segment (Fig. 29-27). To find the total force on the
conductor, we sum the forces on all such segments. In the limit of very small
segments, this sum becomes an integral, and we have

$$\mathbf{F} = \int d\mathbf{F} = \int I\,d\ell \times \mathbf{B}. \qquad (29\text{-}8)$$

The integration is taken over the entire section of conductor on which we are
calculating the force.

● EXAMPLE 29-6   THE FORCE ON A CURVED CONDUCTOR

A semicircular wire connects two points C and D a horizontal
distance $2R$ apart. The wire carries a current I from C to D and
is in a uniform magnetic field **B** pointing upward, as in
Fig. 29-28a. Show that the magnetic force on this semicircular
wire is the same as the force that a straight wire from C to D
would experience if it carried the same current I.

Solution
Since the orientation of the wire relative to the field varies, we
use the integral of Equation 29-8 to calculate the force. Figure
29-28a shows an infinitesimal segment $d\ell$ of the wire, and the
angle θ that specifies its position on the semicircular arc. This

infinitesimal segment subtends an infinitesimal angle $d\theta$, as
shown. A blown-up view of the segment (Fig. 29-28b) shows
that the angle between the segment $d\ell$ and the magnetic field
B is also θ. Then the force on the segment is

$$d\mathbf{F} = I\,d\ell \times \mathbf{B}.$$

The magnitude of this force element $d\mathbf{F}$ is

$$dF = I\,d\ell B \sin\theta.$$

Application of the right-hand rule shows that the direction of
the force element is out of the page for all values of θ on the

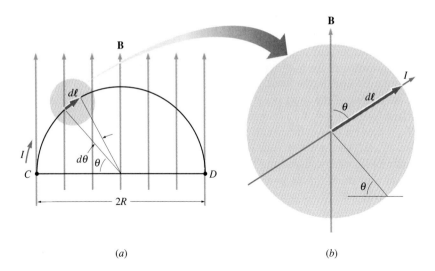

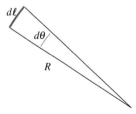

FIGURE 29-28 (*a*) A semicircular wire carries current *I* between points *C* and *D*. (*b*) Blow-up of an infinitesimal segment *d*ℓ of the wire, showing that the segment makes an angle θ with the magnetic field.

semicircle—that is, for $0 \leq \theta \leq \pi$. Therefore, the integral in Equation 29-8 describes a sum of vectors all of which point in the same direction, so the magnitude of the net force is

$$F = \int dF = \int_0^\pi IB \sin \theta \, d\ell.$$

But $d\ell$ subtends the angle $d\theta$, so $d\ell = R\,d\theta$ (Fig. 29-29). Then, taking the constants *I*, *B*, and *R* outside the integral, we have

$$F = IBR \int_0^\pi \sin \theta \, d\theta = IBR \left.(-\cos \theta)\right|_0^\pi$$

$$= IBR[-(-1 - 1)] = IB(2R).$$

This is just the force we would get from Equation 29-6 for a straight wire of length $2R$ carrying current *I* perpendicular to a magnetic field **B**.

EXERCISE A 15-cm-long wire carrying 4.8 A is bent into a quarter circle and oriented like the left half of the semicircular

wire in Fig. 29-28a (i.e., from point *C* to the top of the semicircle). If the magnetic field shown in Fig. 29-28a has magnitude 0.65 T, what is the magnetic force on the wire?

Answer: 0.30 N

Some problems similar to Example 29-6: 40–43

FIGURE 29-29 An angle in radians is defined as the ratio of subtended arc to radius. Here the angle $d\theta$ is so small that the difference between a circular arc and the straight segment $d\ell$ is insignificant. Therefore $d\ell = R\,d\theta$.

If we closed the semicircular loop of Example 29-6, as in Fig. 29-30, the force on the straight section would have the same magnitude $IB(2R)$ but the opposite direction from the force on the semicircle; thus, the net force on the closed loop would be zero. Problem 43 shows that this result is in fact true for *any* closed current loop in a *uniform* magnetic field.

29-5 A CURRENT LOOP IN A MAGNETIC FIELD

We've just seen that a closed, current-carrying loop experiences no net force in a uniform magnetic field. But such a loop generally does experience a torque, as we now show. Torques on current loops play important roles in a wide range of natural and technological systems.

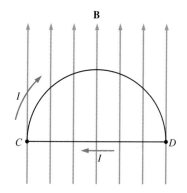

FIGURE 29-30 A closed current loop consisting of a semicircle and a straight wire. Since both carry the same current, but in opposite directions, the result of Example 29-6 shows that they experience forces of equal magnitude but opposite directions. The net force on the loop is therefore zero.

Consider a rectangular current loop in a uniform magnetic field, with the normal to the loop making an angle θ with the magnetic field, as shown in Fig. 29-31a. Since current flows in opposite directions on opposite sides and since the field is uniform, the magnetic forces on opposite sides cancel and there is no net magnetic force on the loop. The right-hand rule shows that the forces on the top and bottom of the loop point directly upward and downward, respectively, and therefore result in no net torque.

The forces on the vertical sections of the loop, however, do produce a torque. Figure 29-31b shows a top view of the loop, with the forces resulting from the upward- and downward-flowing currents in the vertical sides. From the figure, it is clear that these forces give rise to a net torque twisting the loop clockwise. Each vertical side is a straight wire of length a carrying current I at right angles to the horizontally directed magnetic field **B,** so the magnitude of the force on each side is simply

$$F_{side} = IaB.$$

The distance from the loop's central axis to the sides where this force acts is half the loop width, or $b/2$. The geometry of Fig. 29-31b shows that the angle used in calculating the torque is the same as the angle θ between the loop normal and the magnetic field. Therefore, the torque due to the force on each side is

$$\tau_{side} = F\frac{b}{2}\sin\theta = Ia\frac{b}{2}B\sin\theta.$$

Accounting for the contributions from both sides gives the magnitude of the net torque on the loop:

$$\tau = IabB\sin\theta.$$

We can express the torque in vector notation if we define a vector **A** whose magnitude is the area ab of the loop and which is perpendicular to the loop. We

FIGURE 29-31 (a) A rectangular current loop in a uniform magnetic field. The angle between the loop normal and the magnetic field is θ. (b) Top view of the current loop, showing that magnetic forces on the vertical sides result in a net torque. The angles used in the torque calculation are the same as the angle θ between the loop normal and the magnetic field.

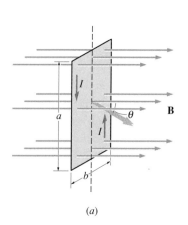

(a)

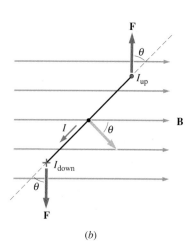

(b)

choose the direction of **A** by the right-hand rule: wrap your fingers around the loop in the direction of the current and your thumb points in the direction of **A.** Then we can write

$$\boldsymbol{\tau} = I\mathbf{A} \times \mathbf{B}. \tag{29-9}$$

Although we derived this equation for a rectangular loop, it holds in fact for any current loop. The torque on a current loop depends on the current, the loop area, the magnetic field, and the orientation between loop and field.

Equation 29-9 should remind you of Equation 23-9 for the torque on an electric dipole in an electric field. There we had

$$\boldsymbol{\tau} = \mathbf{p} \times \mathbf{E},$$

where **p** is the electric dipole moment and **E** the electric field. Comparison with Equation 29-9 suggests that a current loop in a magnetic field behaves analogously to an electric dipole in an electric field. As far as its response to magnetic fields is concerned, therefore,

▌ A current loop constitutes a magnetic dipole.

The quantity $I\mathbf{A}$ in Equation 29-9 plays the same role as the electric dipole moment in the equation $\boldsymbol{\tau} = \mathbf{p} \times \mathbf{E}.$ We therefore call this quantity the **magnetic dipole moment,** $\boldsymbol{\mu}$:

$$\boldsymbol{\mu} = I\mathbf{A}. \quad \text{(single-turn loop)}$$

The direction of the vector $\boldsymbol{\mu}$ is the same as the direction we defined for **A:** curl your right fingers in the direction of the loop current, and your thumb points in the direction of $\boldsymbol{\mu}$ (Fig. 29-32). More generally, a loop may consist of N turns of conducting wire; then each contributes $I\mathbf{A}$ to give the total magnetic moment:

$$\boldsymbol{\mu} = NI\mathbf{A}. \quad \text{(magnetic dipole moment)} \tag{29-10}$$

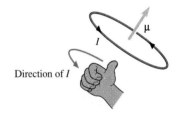

FIGURE 29-32 Using the right-hand rule to find the direction of a current loop's magnetic moment.

Clearly the units of magnetic moment are A·m². Using the magnetic moment vector, Equation 29-9 then becomes

$$\boldsymbol{\tau} = \boldsymbol{\mu} \times \mathbf{B}, \quad \text{(torque on a current loop)} \tag{29-11}$$

in analogy with the electric case. The magnetic moment vector of a current loop is perpendicular to the plane of the loop, and the torque given in Equation 29-11 tends to align the magnetic moment with the field (Fig. 29-33). It takes work to twist the loop's magnetic moment vector out of alignment with the field. In exact analogy with the electric-dipole case summarized in Equation 23-10, we express the associated potential energy as

$$U_{\text{magnetic}} = -\boldsymbol{\mu} \cdot \mathbf{B}. \tag{29-12}$$

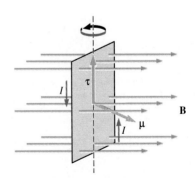

FIGURE 29-33 The torque on a current loop tends to align the loop's magnetic moment vector with the magnetic field.

● **EXAMPLE 29-7**   POINTING A SATELLITE

Some small satellites use the magnetic torque on current loops to point themselves in space. Earth itself provides the magnetic field, while electricity from solar panels powers coils that constitute the current loops; unlike rocket-based pointing systems, there's no fuel to run out. Three mutually perpendicular coils allow the satellite to point itself in any direction. One satellite uses 30-cm-diameter circular coils, each with 1000 turns. (a) If the satellite orbits at an altitude where Earth's magnetic field is 0.24 G, and if specifications call for a maximum torque of 1.3 mN·m from each coil, what should be the coil current? (b) How much work is done on the coil to rotate it from its maximum-torque orientation to an angle of 120° to the field?

Solution

(a) The maximum torque occurs when the coil's magnetic moment is perpendicular to the field; Equation 29-11 shows that this maximum torque has magnitude $\tau_{max} = \mu B$. The magnetic moment of the coil is the product of the number of turns, the current, and the coil area πR^2, as given by Equation 29-10. Thus the maximum torque is

$$\tau_{max} = NI\pi R^2 B.$$

Solving for the current gives

$$I = \frac{\tau_{max}}{N\pi R^2 B} = \frac{1.3\times10^{-3}\ \text{N·m}}{(1000)(\pi)(0.15\ \text{m})^2(0.24\times10^{-4}\ \text{T})}$$

$$= 0.766\ \text{A}.$$

(b) The loop's potential energy is given by Equation 29-12: $U = -\boldsymbol{\mu} \cdot \mathbf{B} = -\mu B \cos\theta$. Maximum torque occurs when $\theta = 90°$, so the work in rotating the coil—equal to the difference in its potential energies in the two orientations—is

$$W = U_{120°} - U_{90°} = -\mu B \cos 120° - (-\mu B \cos 90°)$$

$$= -\tau_{max} \cos 120° - 0 = (-1.3\times10^{-3}\ \text{N·m})(\cos 120°)$$

$$= 0.65\ \text{mJ},$$

where we recognized the quantity μB as the maximum torque defined earlier.

EXERCISE   An electric motor consists of a 550-turn loop with area 100 cm² in a 0.23-T magnetic field. What loop current is necessary if the motor is to develop a maximum torque of 3.7 N·m?

Answer: 2.9 A

Some problems similar to Example 29-7:   47, 48, 50–52   ●

■ **APPLICATION**   ELECTRIC MOTORS

Electric motors are so much a part of our lives that we hardly think of them. But CD players, car starters, refrigerators, vacuum cleaners, power saws, subway trains, computer disk drives, food processors, washing machines, fans, hair dryers, water pumps, oil burners, and most industrial machinery would be difficult or impossible to build without electric motors (Fig. 29-34).

At the heart of every electric motor is a current loop in a

(a)

(b)

FIGURE 29-34   Electric motors are made in a wide range of sizes and power outputs. (a) A small motor, disassembled to show the armature (rotating coil assembly). (b) This large motor drives the world's largest fully steerable radio telescope.

magnetic field. But instead of a steady current, the loop carries a current that reverses periodically. In direct-current (DC) motors, this reversal is achieved through the electrical contacts that provide current to the loop. Figure 29-35 shows a simple motor consisting of a loop in the field of a permanent magnet. Current from an external battery reaches the loop through a set of stationary "brushes," which make contact with a pair of semicircular conductors called the "commutator." The commutator is attached rigidly to the loop, which rotates to align its dipole moment with the field. Just as it reaches alignment, however, the brushes cross the gaps in the commutator. This crossing reverses the connections of loop to battery, resulting in a reversal of the loop current. This in turn reverses the magnetic moment of the loop, so it is no longer aligned with the field. The loop then rotates another 180° to align its new magnetic moment vector with the field. Just as it reaches alignment, the current again reverses. This process repeats, resulting in continuous rotation of the loop. A rigid shaft through loop and commutator delivers mechanical work to the device powered by the motor. The source of this work is the battery or whatever supplies electrical energy to the motor. The motor itself is a device that converts electrical to mechanical energy; the magnetic field is an intermediary in this conversion.

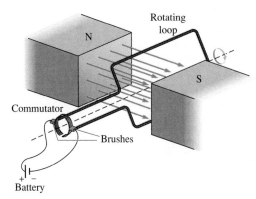

FIGURE 29-35 A simple electric motor. Current flows from the battery through the stationary brushes to the rotating commutator and loop. The direction of the loop current reverses as the two halves of the commutator rotate to contact different brushes.

What determines the rotational speed of the motor? Why doesn't it undergo a constant rotational acceleration, reaching ever greater speeds? The answer to these questions lies in another deep interaction between electricity and magnetism, an interaction that we will explore in Chapter 31.

CHAPTER SYNOPSIS

Summary

1. **Magnetism** is a fundamental interaction described in terms of **magnetic fields.** Magnetic fields interact with **moving electric charges.** A moving charged particle experiences a force that depends on its charge, its velocity, and the magnetic field. The force is at right angles to both velocity and field, and is given by

$$\mathbf{F} = q\mathbf{v} \times \mathbf{B}.$$

2. An electric current is made up of moving electric charges, so there is a net magnetic force on an electric current. The force on small current elements of length $d\ell$ is

$$d\mathbf{F} = I\,d\ell \times \mathbf{B},$$

where I is the current, $\mathbf{B}$ the magnetic field, and $d\ell$ an infinitesimal vector pointing in the local direction of the current. For a straight wire of length ℓ in a magnetic field, this equation becomes

$$\mathbf{F} = I\ell \times \mathbf{B};$$

for other cases the infinitesimal force $d\mathbf{F}$ can be integrated over a current to obtain the force on the entire current.

3. A particularly important case of a current in a magnetic field is a closed current loop. Such a loop behaves like a **magnetic dipole** with magnetic dipole moment

$$\mu = NI\mathbf{A},$$

where N is the number of turns in the loop, I the loop current, and $\mathbf{A}$ a vector perpendicular to the plane of the loop and whose magnitude is the loop area.

a. A current loop in a magnetic field experiences a torque given by

$$\tau = \mu \times \mathbf{B}.$$

If the field is nonuniform, the loop experiences a net force as well.

b. The potential energy associated with a current loop in a magnetic field is

$$U_B = -\mu \cdot \mathbf{B},$$

where the zero of potential energy is taken when the loop's magnetic moment vector is perpendicular to the field.

Terms You Should Understand

(Pairs are closely related terms whose distinction is important; number in parentheses is chapter section where term first appears.)

magnetic field (29-1, 29-2)

magnetic force, electromagnetic force (29-2)

tesla, gauss (29-2)
cyclotron frequency (29-3)
cyclotron, synchrotron (29-3)
magnetic mirror (29-3)
Hall effect, Hall potential (29-4)
current loop (29-5)
magnetic dipole moment (29-5)
electric motor (29-5)

Symbols You Should Recognize

$\mathbf{B}$ (29-1, 29-2)
T (29-2)
μ (29-5)

Problems You Should Be Able to Solve

calculating magnetic force vectors given charge, velocity, and magnetic field (29-2)

analyzing charged-particle trajectories in uniform magnetic fields (29-3)
finding magnetic forces on straight conductors in uniform fields (29-4)
using integration to determine magnetic forces with curved conductors and/or nonuniform fields (29-4)
determining magnetic dipole moments of current loops (29-5)
evaluating torque and potential energy for current loops in magnetic fields (29-5)

Limitations to Keep in Mind

Magnetic forces arise only when charges are *moving*.
The cyclotron frequency is independent of particle energy only for speeds much less than the speed of light.

QUESTIONS

1. A stationary charged particle experiences no force in a certain region. Can you conclude from this observation that there is no magnetic field in this region?
2. A charged particle moves through a region containing only a magnetic field. Under what condition will the particle experience no force?
3. An electron moving with velocity $\mathbf{v}$ through a magnetic field $\mathbf{B}$ experiences a magnetic force $\mathbf{F}$. Which of the vectors $\mathbf{F}$, $\mathbf{v}$, and $\mathbf{B}$ must be at right angles?
4. In Fig. 29-36 a high-energy gamma ray has decayed into an electron and its positively charged antiparticle, a positron. A magnetic field points out of the plane of the photograph, and the electron and positron spiral in this field. Which path belongs to which particle? Why might the paths be spirals rather than circles of constant radius?

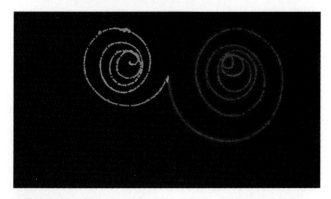

FIGURE 29-36 Creation of an electron-positron pair in the decay of a gamma ray incident from the top of the figure. Trajectories were imaged in the detector of a high-energy particle accelerator (Question 4).

5. A magnetic field points out of this page. Will a positively charged particle moving in the plane of the page circle clockwise or counterclockwise as viewed from above?
6. An electron moves through a region in a straight line at constant speed. Can you conclude that there is no magnetic field in the region? Could you so conclude if you knew that there were no electric or gravitational forces on the electron? Explain.
7. High-resolution color TV monitors sometimes have a built-in circuit that compensates for changes in the orientation of Earth's magnetic field relative to the TV picture tube as the monitor is moved from one place to another. Why is this necessary?
8. What is meant by the statement that charged particles can be "trapped" by magnetic fields?
9. An electron beam comes straight to the center of a TV screen, where it makes a spot of light. If you hold the north pole of a bar magnet on the left side of the picture tube, which way will the spot move?
10. Do particles in a cyclotron gain energy from the electric field, the magnetic field, or both? Explain.
11. A cyclotron is designed to accelerate either hydrogen or deuterium nuclei. If the magnetic field is unchanged, how must the frequency of the alternating dee voltage be changed in order to switch from hydrogen to deuterium?
12. An electron and a proton moving at the same speed enter a region containing a uniform magnetic field. Which is deflected more from its original path?
13. An electron and a proton with the same kinetic energy enter a region containing a uniform magnetic field. Which is deflected more from its original path?

14. For what orientation of electric and magnetic fields could the net force on a particle be zero?
15. Will the velocity selector of Example 29-2 work for particles coming out of the page in Fig. 29-7? Why or why not?
16. What does magnetism have to do with the fact that auroras are seen near Earth's poles?
17. In a certain region uniform electric and magnetic fields are at right angles to one another. A positively charged particle is released from rest in this region. Describe qualitatively its subsequent motion.
18. How do the period and radius of an electron's orbit in a magnetic field depend on its velocity? Assume that the electron is moving at right angles to the field.

19. Current in a certain ionic solution is carried equally by positive and negative ions. Would you expect the Hall effect to occur in this solution?
20. Two identical particles carrying equal charge are moving in opposite directions along a magnetic field, when they collide elastically head-on. Describe their subsequent motion.
21. Repeat the above question for the case when the two particles are moving instantaneously perpendicular to the field when they collide.
22. Under what conditions will a current loop in a magnetic field experience zero force? Zero torque?
23. What would happen to a motor with no commutator?

PROBLEMS

Section 29-2 Electric Charge and the Magnetic Field

1. (a) What is the minimum magnetic field needed to exert a 5.4×10^{-15}-N force on an electron moving at 2.1×10^7 m/s? (b) What magnetic field strength would be required if the field were at $45°$ to the electron's velocity?
2. An electron moving at right angles to a 0.10-T magnetic field experiences an acceleration of 6.0×10^{15} m/s². (a) What is the electron's speed? (b) By how much does its *speed* change in 1 ns ($= 10^{-9}$ s)?
3. What is the magnitude of the magnetic force on a proton moving at 2.5×10^5 m/s (a) at right angles; (b) at $30°$; (c) parallel to a magnetic field of 0.50 T?
4. A magnetic field of 0.10 T points in the x direction. A charged particle carrying 1.0 μC enters the field region moving at 20 m/s. What are the magnitude and direction of the force on the particle when it first enters the field region if it does so moving (a) along the x axis; (b) along the y axis; (c) along the z axis; (d) at $45°$ to both x and y axes?
5. A particle carrying a 50-μC charge moves with velocity $\mathbf{v} = 5.0\hat{\imath} + 3.2\hat{\mathbf{k}}$ m/s through a uniform magnetic field $\mathbf{B} = 9.4\hat{\imath} + 6.7\hat{\jmath}$ T. (a) What is the force on the particle? (b) Form the dot products $\mathbf{F} \cdot \mathbf{v}$ and $\mathbf{F} \cdot \mathbf{B}$ to show explicitly that the force is perpendicular to both $\mathbf{v}$ and $\mathbf{B}$.
6. Moving in the x direction, a particle carrying 1.0 μC experiences no force. Moving with speed v at $30°$ to the x axis, the particle experiences a force of 2.0 N. What is the magnitude of the force it would experience if it moved along the y axis with speed v?
7. A proton moving with velocity $\mathbf{v}_1 = 3.6 \times 10^4 \hat{\jmath}$ m/s experiences a force of $7.4 \times 10^{-16} \hat{\imath}$ N. A second proton moving on the x axis experiences a magnetic force of $2.8 \times 10^{-16} \hat{\jmath}$ N. Find the magnitude and direction of the magnetic field, and the velocity of the second proton.
8. The magnitude of Earth's magnetic field is a little less than 1 G near Earth's surface. What is the maximum possible magnetic force on an electron with kinetic energy of 1 keV? Compare with the gravitational force on the same electron.

9. An alpha particle (2 protons, 2 neutrons) is moving with velocity $\mathbf{v} = 150\hat{\imath} + 320\hat{\jmath} - 190\hat{\mathbf{k}}$ km/s at a point where the magnetic field is $\mathbf{B} = 0.66\hat{\imath}$ T. Find the magnitude of the force on the particle.
10. An electron is moving with velocity $\mathbf{v} = 5.1\hat{\imath} + 2.3\hat{\jmath}$ Mm/s at a point where the magnetic field is $\mathbf{B} = 0.29\hat{\imath} - 0.62\hat{\mathbf{k}}$ T. Find the magnetic force on the electron.
11. A 1.4-μC charge moving at 185 m/s experiences a magnetic force $\mathbf{F}_B = 2.5\hat{\imath} + 7.0\hat{\jmath}$ μN in a magnetic field $\mathbf{B} = 42\hat{\imath} - 15\hat{\jmath}$ mT. What is the angle between the particle's velocity and the magnetic field?
12. A velocity selector uses a 60-mT magnetic field and a 24 kN/C electric field. At what speed will charged particles pass through the selector undeflected?
13. A region contains an electric field $\mathbf{E} = 7.4\hat{\imath} + 2.8\hat{\jmath}$ kN/C and a magnetic field $\mathbf{B} = 15\hat{\jmath} + 36\hat{\mathbf{k}}$ mT. Find the electromagnetic force on (a) a stationary proton, (b) an electron moving with velocity $\mathbf{v} = 6.1\hat{\imath}$ Mm/s.
14. A charged particle is moving at right angles to both a 1.1 kN/C electric field and a 0.75-T magnetic field. If the magnitude of the electric force on the particle is twice that of the magnetic force, what is the particle's speed?

Section 29-3 The Motion of Charged Particles in Magnetic Fields

15. What is the radius of the circular path described by a proton moving at 15 km/s in a plane perpendicular to a 400-G magnetic field?
16. How long does it take an electron to complete a circular orbit at right angles to a 1.0-G magnetic field?
17. Radio astronomers detect electromagnetic radiation at a frequency of 42 MHz from an interstellar gas cloud. If this radiation is caused by electrons spiraling in a magnetic field, what is the field strength in the gas cloud?
18. A beam of electrons moving in the x direction at 8.7×10^6 m/s enters a region where a uniform magnetic

field of 180 G points in the y direction. How far into the field region does the beam penetrate?

19. Electrons and protons with the same kinetic energy are moving at right angles to a uniform magnetic field. How do their orbital radii compare?

20. The Van Allen belts are regions in space where high-energy charged particles are trapped in Earth's magnetic field. If the field strength at the Van Allen belts is 0.10 G, what are the period and radius of the helical path described (a) by a proton with a 1.0-MeV kinetic energy? (b) by a 10-MeV proton?

21. Typical particle energies in a nuclear fusion reactor are on the order of 10 keV. If the smallest dimension of the reactor is on the order of 1 m, estimate the minimum magnetic field strength needed to ensure that protons have orbits smaller than the size of the reactor. Will this field be sufficient for electrons of the same energy?

22. Show that the orbital radius of a charged particle moving at right angles to a magnetic field B can be written

$$r = \frac{\sqrt{2Km}}{qB},$$

where K is the kinetic energy in joules, m the particle mass, and q its charge.

23. Two protons, moving in a plane perpendicular to a uniform magnetic field of 500 G, undergo an elastic head-on collision. How much time elapses before they collide again? *Hint:* Draw a picture.

24. Repeat the preceding problem for the case of a proton and an antiproton colliding head-on (a) if they have the same speed and (b) if they have different speeds. (An antiproton has the same mass as a proton, but carries the opposite charge.)

25. A cyclotron is designed to accelerate deuterium nuclei. (Deuterium has one proton and one neutron in its nucleus.) (a) If the cyclotron uses a 2.0-T magnetic field, at what frequency should the dee voltage be alternated? (b) If the vacuum chamber has a diameter of 0.90 m, what is the maximum kinetic energy of the deuterons? (c) If the magnitude of the potential difference between the dees is 1500 V, how many orbits do the deuterons complete before achieving the energy of part (b)?

26. Without changing the magnetic field, how could the cyclotron of the preceding problem be modified to accelerate (a) protons and (b) alpha particles (two protons and two neutrons)? What would be the maximum energy achievable with (c) protons and (d) alpha particles?

27. Figure 29-37 shows a simple mass spectrometer, designed to analyze and separate atomic and molecular ions with different charge-to-mass ratios. In the design shown, ions are accelerated through a potential difference V, after which they enter a region containing a uniform magnetic field. They describe semicircular paths in the magnetic field, and land on a detector a lateral distance x from where

they entered the field region, as shown. Show that x is given by

$$x = \frac{2}{B}\sqrt{\frac{2V}{(q/m)}},$$

where B is the magnetic field strength, V the accelerating potential, and q/m the charge-to-mass ratio of the ion. By counting the number of ions accumulated at different positions x, one can determine the relative abundances of different atomic or molecular species in a sample.

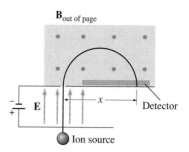

FIGURE 29-37 A mass spectrometer (Problem 27).

28. A mass spectrometer like that of the preceding problem has $V = 2000$ V and $B = 1000$ G. It is used to analyze a gas sample suspected of containing Ne, O_2, CO, SO_2, and NO_2. Ions are detected at distances of 58 cm, 68 cm, and 87 cm from the entrance to the field region. Which gases are actually present? Assume that all molecules are singly ionized.

29. A mass spectrometer is used to separate the fissionable uranium isotope U-235 from the much more abundant isotope U-238. To within what percentage must the magnetic field be held constant if there is to be no overlap of these two isotopes? Both isotopes appear as constituents of uranium hexafluoride gas (UF_6), and the gas molecules are all singly ionized.

30. An electron is moving in a uniform magnetic field of 0.25 T; its velocity components parallel and perpendicular to the field are both equal to 3.1×10^6 m/s. (a) What is the radius of the electron's helical path? (b) How far does it move along the field direction in the time it takes to complete a full orbit about the field direction?

31. An electron moving at 3.8×10^6 m/s enters a region containing a uniform magnetic field $\mathbf{B} = 18\hat{\mathbf{k}}$ mT. The electron is moving at 70° to the field direction, as shown in Fig. 29-38. Find the radius r and pitch p of its helical path, as indicated in the figure.

32. A proton in interstellar space describes a helical path about a 15-mG magnetic field, with velocity component 40 km/s perpendicular to the field. If the pitch of the helix (see Fig. 29-38) is 8.7 km, what is the proton's velocity component parallel to the field?

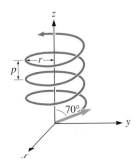

FIGURE 29-38   Problems 31, 32.

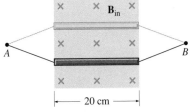

FIGURE 29-40   Problem 38.

Section 29-4 The Magnetic Force on a Current

33. What is the magnitude of the force on a 50-cm-long wire carrying 15 A at right angles to a 500-G magnetic field?

34. A wire coincides with the x axis, carrying 2.4 A in the $+x$ direction. The wire passes through a region containing a uniform magnetic field $\mathbf{B} = 0.17\hat{\mathbf{i}} + 0.32\hat{\mathbf{j}} - 0.21\hat{\mathbf{k}}$ T. Find a vector expression for the force per unit length on the wire in the magnetic field region.

35. A wire carrying 15 A makes a 25° angle with a uniform magnetic field. The magnetic force per unit length of wire is 0.31 N/m. (a) What is the magnetic field strength? (b) What is the maximum force per unit length that could be achieved by reorienting the wire in this field?

36. A wire of negligible resistance is bent into a rectangle as shown in Fig. 29-39, and a battery and resistor are connected as shown. The right-hand side of the circuit extends into a region containing a uniform magnetic field of 38 mT pointing into the page. Find the magnitude and direction of the net force on the circuit.

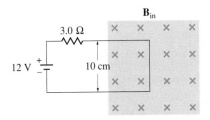

FIGURE 29-39   Problem 36.

37. In a high-magnetic-field experiment, a conducting bar carrying 7.5 kA passes through a 30-cm-long region containing a 22-T magnetic field. If the bar makes a 60° angle with the field direction, what force is necessary to hold it in place?

38. A 20-cm-long conducting rod with mass 18 g is suspended by wires of negligible mass, as shown in Fig. 29-40. The rod is in a region containing a uniform magnetic field of 0.15 T pointing horizontally into the page, as shown. An external circuit supplies current between the support points A and B. (a) What is the minimum current necessary to move the bar to the upper position shown? (b) Which direction should the current flow?

39. A piece of wire with mass per unit length 75 g/m runs horizontally at right angles to a horizontal magnetic field. A 6.2-A current in the wire results in its being suspended against gravity. What is the magnetic field strength?

40. A nonuniform magnetic field points out of the page, as shown in Fig. 29-41. The field strength increases at the rate of 2.0 mT/cm as you move to the right. A square wire loop 15 cm on a side lies in a plane perpendicular to the field, and a 2.5-A current circles the loop in the counterclockwise direction. What are the magnitude and direction of the net magnetic force on the loop?

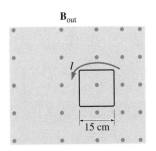

FIGURE 29-41   Problem 40.

41. A wire carrying 1.5 A passes through a region containing a 48-mT magnetic field. The wire is perpendicular to the field and makes a quarter-circle turn of radius 21 cm as it passes through the field region, as shown in Fig. 29-42. Find the magnitude and direction of the magnetic force on this section of wire.

42. A wire coincides with the x axis, and carries a current $I = 2.0$ A in the $+x$ direction. A nonuniform magnetic field points in the y direction, given by $\mathbf{B} = B_0(x/x_0)^2\hat{\mathbf{j}}$, where $B_0 = 0.22$ T, $x_0 = 1.0$ m, and x is the x coordinate. Find the force on the section of wire between $x = 1.0$ m and $x = 3.5$ m.

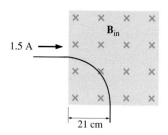

FIGURE 29-42 Problem 41.

43. Apply Equation 29-8 to a closed current loop of arbitrary shape in a *uniform* magnetic field, and show that the net force on the loop is zero. *Hint:* Both I and B are constant as you go around the loop, so you can take them out of the integral. What is the remaining vector integral?

44. A rectangular copper strip measures 1.0 mm in the direction of a uniform 2.4-T magnetic field. When the strip carries a 6.8-A current at right angles to the field, the Hall voltage across the strip is 1.2 μV. Find the number density of free electrons in the copper.

45. The probe in a Hall-effect magnetometer uses a semiconductor doped to a charge-carrier density of 7.5×10^{20} m^{-3}. The probe measures 0.35 mm thick in the direction of the magnetic field being measured, and carries a 2.5-mA current perpendicular to the field. If its Hall potential is 4.5 mV, what is the magnetic field strength?

Section 29-5 A Current Loop in a Magnetic Field

46. Show that the units of magnetic moment (A·m^2) can also be written as J/T.

47. A single-turn square wire loop 5.0 cm on a side carries a 450-mA current. (a) What is the magnetic moment of the loop? (b) If the loop is in a uniform 1.4-T magnetic field with its dipole moment vector at 40° to the field direction, what is the magnitude of the torque it experiences?

48. An electric motor contains a 250-turn circular coil 6.2 cm in diameter. If it is to develop a maximum torque of 1.2 N·m at a current of 3.3 A, what should be the magnetic field strength?

49. A bar magnet experiences a 12-mN·m torque when it is oriented at 55° to a 100-mT magnetic field. What is the magnitude of its magnetic dipole moment?

50. A single-turn wire loop 10 cm in diameter carries a 12-A current. It experiences a torque of 0.015 N·m when the normal to the loop plane makes a 25° angle with a uniform magnetic field. What is the magnetic field strength?

51. A simple electric motor like that of Fig. 29-35 consists of a 100-turn coil 3.0 cm in diameter, mounted between the poles of a magnet that produces a 0.12-T field. When a 5.0-A current flows in the coil, what are (a) its magnetic dipole moment and (b) the maximum torque developed by the motor?

52. A satellite with rotational inertia 20 kg·m^2 is in orbit at a height where Earth's magnetic field strength is 0.18 G. It

has a magnetic torquing system, as described in Example 29-7, that uses a 1000-turn coil 30 cm in diameter. What should be the current in the coil if the magnetic torque is to give the satellite a maximum angular acceleration of 0.0015 s^{-2}?

53. Nuclear magnetic resonance (NMR) is a technique for analyzing chemical structures and is also the basis of magnetic resonance imaging used for medical diagnosis. The NMR technique relies on sensitive measurements of the energy needed to flip atomic nuclei upside-down in a given magnetic field. In an NMR apparatus with a 7.0-T magnetic field, how much energy is needed to flip a proton ($\mu = 1.41 \times 10^{-26}$ A·m^2) from parallel to antiparallel to the field?

54. A wire of length ℓ carries a current I. (a) Find an expression for the magnetic dipole moment that results when the wire is wound into an N-turn circular coil. (b) For what integer value of N is this moment a maximum?

Paired Problems

(Both problems in a pair involve the same principles and techniques. If you can get the first problem, you should be able to solve the second one.)

55. Find the magnetic force on an electron moving with velocity $\mathbf{v} = 8.6 \times 10^5 \hat{\mathbf{i}} - 4.1 \times 10^5 \hat{\mathbf{j}}$ m/s in a magnetic field $\mathbf{B} = 0.18 \hat{\mathbf{j}} + 0.64 \hat{\mathbf{k}}$ T.

56. A proton moving with velocity $\mathbf{v} = 2.0 \times 10^5 \hat{\mathbf{i}} + 4.0 \times 10^5 \hat{\mathbf{j}}$ m/s experiences a magnetic force $\mathbf{F} = 10 \hat{\mathbf{i}} - 5.0 \hat{\mathbf{j}} + 21 \hat{\mathbf{k}}$ fN. What is the z component of the magnetic field?

57. Proponents of space-based particle-beam weapons have to confront the effect of Earth's magnetic field on their beams. If a beam of protons with kinetic energy 100 MeV is aimed in a straight line perpendicular to Earth's magnetic field in a region where the field strength is 48 μT, what will be the radius of the protons' circular path?

58. Electrons are accelerated through a 30-kV potential difference at the rear of a TV tube. The electron beam is initially headed straight toward the center of the tube. The TV is oriented so the beam is perpendicular to Earth's magnetic field, in a location where the field strength is 62 μT. What will be the radius of the electron beam's curved path?

59. A 170-mT magnetic field points into the page, confined to a square region as shown in Fig. 29-43. A square conduct-

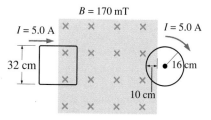

FIGURE 29-43 Problems 59, 60.

ing loop 32 cm on a side carrying a 5.0-A current in the clockwise sense extends partly into the field region, as shown. Find the magnetic force on the loop.

60. Find the force on the circular current loop shown at the right of Fig. 29-43. The loop carries 5.0 A clockwise, has radius 16 cm, and extends 10 cm into the field region.

61. An old-fashioned analog meter uses a wire coil in a magnetic field to deflect the meter needle. If the coil is 2.0 cm in diameter and consists of 500 turns of wire, what should be the magnetic field strength if the maximum torque is to be 1.6 μN·m when the current in the coil is 1.0 mA?

62. A circular wire coil 15 cm in diameter carries a 460-mA current and experiences a 0.020-N·m torque when the normal to the coil makes a 27° angle with a 42-mT magnetic field. How many turns are in the coil?

Supplementary Problems

63. Electrons in a TV picture tube are accelerated through a 30-kV potential difference and head straight for the center of the tube, 40 cm away. If the electrons are moving at right angles to Earth's 0.50-G magnetic field, by how much do they miss the screen's exact center?

64. A certain region contains both an electric field and a magnetic field. An electron moving with velocity $2.0 \times 10^4 \hat{\jmath}$ m/s experiences a force of $2.2\hat{\imath} + 4.8\hat{\jmath}$ fN. Another electron, moving with velocity $2.0 \times 10^4 \hat{\imath}$ m/s, experiences a force of $2.6\hat{\jmath} + 2.2\hat{k}$ fN. A third electron, moving with velocity $1.0 \times 10^4 \hat{k}$ m/s, experiences a force $-1.1\hat{\imath} + 4.8\hat{\jmath}$ fN. Find a combination of uniform electric and magnetic fields that could be responsible for these forces.

65. A conducting bar with mass 15.0 g and length 22.0 cm is suspended from a spring in a region where a 0.350-T magnetic field points into the page, as shown in Fig. 29-44. With no current in the bar, the spring length is 26.0 cm. The bar is supplied with current from outside the field region, using wires of negligible mass. When a 2.00-A current flows from left to right in the bar, it rises 1.2 cm from its equilibrium position. Find (a) the spring constant and (b) the unstretched length of the spring.

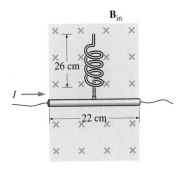

FIGURE 29-44   Problem 65.

66. In 2.0 μs, an electron moves 15 cm in the direction of a 0.10-T magnetic field. If the electron's velocity components perpendicular and parallel to the field are equal, (a) what is the length of its actual helical trajectory and (b) how many orbits about the field direction does it complete?

67. A square loop of side a is free to pivot about a horizontal rod passing through the centers of two sides. The loop carries a current I and is in a uniform vertical magnetic field **B**, as shown in Fig. 29-45. A mass m hangs from one side of the loop, as shown. Find an expression for the angle θ between the loop plane and the horizontal for which this system will be in equilibrium.

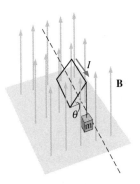

FIGURE 29-45   Problem 67.

68. A disk of radius R carries a uniform surface charge density σ and is rotating with angular frequency ω. Show that its magnetic dipole moment is $\frac{1}{4}\pi\sigma\omega R^4$. *Hint:* Divide the disk into concentric rings. Treat each as a loop carrying an infinitesimal current, and integrate over all the loops.

69. A 10-turn wire loop measuring 8.0 cm by 16 cm carrying 2.0 A lies in a horizontal plane but is free to rotate about the axis shown in Fig. 29-46. A 50-g mass hangs from one side of the loop, and a uniform magnetic field points horizontally, as shown. What magnetic field strength is required to hold the loop in its horizontal position?

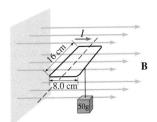

FIGURE 29-46   Problem 69.

70. A closed current loop is made from two semicircular wire arcs of radius R, joined at right angles as shown in Fig. 29-47. The loop carries a current I and is oriented with the

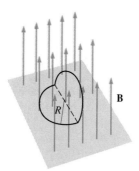

FIGURE 29-47 Problem 70.

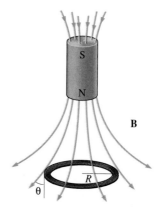

FIGURE 29-48 Problem 71.

plane of one semicircle perpendicular to a uniform magnetic field **B,** as shown. Find (a) the magnetic moment of this nonplanar loop and (b) the torque on the loop. *Hint:* You can think of the loop as a superposition of two semicircular loops, each closed along the dashed line shown (why?).

71. A circular wire loop of mass m and radius R carries a current I. The loop is hanging horizontally below a cylindrical bar magnet, suspended by the magnetic force, as shown in Fig. 29-48. If the field lines crossing the loop make an angle θ with the vertical, show that the strength of the magnet's field at the loop's position is $B = mg/2\pi RI \sin\theta$.

72. A square wire loop of mass m carries a current I. It is initially in equilibrium, with its magnetic moment vector aligned with a uniform magnetic field **B.** The loop is ro-

tated slightly out of equilibrium about an axis through the centers of two opposite sides and then released. Show that it executes simple harmonic motion with period given by $T = 2\pi\sqrt{m/6IB}$.

73. Early models pictured the electron in a hydrogen atom as being in a circular orbit of radius 5.29×10^{-11} m about the stationary proton, held in orbit by the electric force. Find the magnetic dipole moment of such an atom. This quantity is called the *Bohr magnetron* and is typical of atomic-sized magnetic moments. *Hint:* The full electron charge passes any given point in the orbit once per orbital period. Use this fact to calculate the average current.

CHAPTER 30

SOURCES OF THE MAGNETIC FIELD

Moving electric charge produces magnetic fields. Here a large electromagnet lifts scrap metal in a recycling operation.

The preceding chapter introduced the magnetic field and its effect on matter. Here we consider the opposite question: How does matter produce magnetic fields? We've seen that magnetic fields exert forces on moving electric charges—a fact that implies a deep relation between electricity and magnetism. That relation in fact goes two ways: Not only do magnetic fields affect moving electric charges, but moving charges themselves produce magnetic fields.

30-1 THE BIOT-SAVART LAW

The first inkling of a relation between electricity and magnetism came in 1820 when the Danish scientist Hans Christian Oersted discovered that a compass needle is deflected by an electric current (Fig. 30-1). A mere month after Oersted's discovery became known in Paris, the French scientists Jean Baptiste Biot and Félix Savart had experimentally determined the form of the magnetic

FIGURE 30-1 Oersted's experiment linking electricity and magnetism is commemorated on the Oersted medal, awarded by the American Association of Physics Teachers to honor "notable contributions to the teaching of physics."

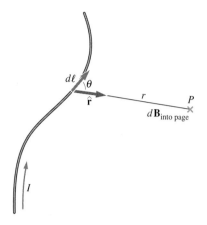

FIGURE 30-2 The Biot-Savart law gives the magnetic field $d\mathbf{B}$ at point P arising from the current I flowing along the infinitesimal vector $d\ell$. The unit vector $\hat{\mathbf{r}}$ points from the current element $I\,d\ell$ toward the field point P.

field arising from a steady current. Biot and Savart considered the contribution $d\mathbf{B}$ to the magnetic field at some point P due to a small segment $d\ell$ in the path of a steady current I. They found, in the geometry of Fig. 30-2, that

1. The field contribution $d\mathbf{B}$ is perpendicular to both the current element $I\,d\ell$ and the vector $\mathbf{r}$ from the current element to the point P.
2. The field strength dB is proportional to the current I and to the length $d\ell$ of the current element, and inversely proportional to the square of the distance r from the current element to the field point P.
3. The field strength depends on the orientation between the vectors $d\ell$ and $\mathbf{r}$, being proportional to the sine of the angle θ between these two vectors.

Mathematically, these results can be summarized in a compact vector equation called the **Biot-Savart law:**

$$d\mathbf{B} = \frac{\mu_0}{4\pi}\frac{I\,d\ell \times \hat{\mathbf{r}}}{r^2}, \quad \text{(Biot-Savart law)} \tag{30-1}$$

where $\hat{\mathbf{r}}$ is a unit vector pointing from the current element toward the field point P and where μ_0 is a constant called the permeability constant, with the exact value $4\pi \times 10^{-7}\ \text{N/A}^2$.*

Compare the Biot-Savart law with the Coulomb expression for the electric field of an infinitesimal point charge dq:

$$d\mathbf{E} = \frac{dq}{4\pi \varepsilon_0 r^2}\,\hat{\mathbf{r}}.$$

Both give fields whose strengths depend on their sources—current and charge for the magnetic and electric fields, respectively. Both fields decrease as the inverse square of the distance from those sources. But here the similarity ends. Unlike electric charge, the current element $I\,d\ell$ has direction as well as magnitude. As a result the magnetic field of a current element is not symmetric about the element but depends on the direction of the field point relative to the direction of the current element. This directional character is expressed by the cross product in Equation 30-1, which reflects Biot and Savart's discovery that the field contribution $d\mathbf{B}$ is perpendicular both to the current element $I\,d\ell$ and the vector $\hat{\mathbf{r}}$ (see Fig. 30-2).

There is another distinction between the Biot-Savart law and Coulomb's law. Both describe the fields of localized structures—current elements and point charges—that are sources of the fields. It makes sense to talk about the electric field of an isolated point charge. But can we have an isolated current element? Not in a steady-state situation, where the current flowing into a current element must be the same as the current flowing out. Thus any Biot-Savart calculation necessarily involves the fields produced by many small current elements from an entire circuit. Experimentally, we find that the magnetic field obeys the superpo-

---

* That this constant has an exact value is a consequence of the definition of the ampere in terms of magnetic forces, as discussed later in this chapter.

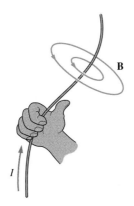

FIGURE 30-3 Magnetic field lines generally encircle a current. The field direction is given by the right-hand rule: Point your right thumb in the direction of the current, and your fingers curl in the direction of the field.

sition principle, so the net field at any point is the vector sum—or integral—of the fields of individual current elements:

$$\mathbf{B} = \int d\mathbf{B} = \int \frac{\mu_0}{4\pi} \frac{I\, d\boldsymbol{\ell} \times \hat{\mathbf{r}}}{r^2}, \qquad (30\text{-}2)$$

where the integration is taken over the entire circuit in which the current I flows. The field given by Equation 30-2 depends on the details of the current configuration, but the directionality associated with the cross product means that, quite generally, magnetic field lines encircle the current that is their source (Fig. 30-3).

● **EXAMPLE 30-1** THE FIELD OF A CURRENT LOOP

Find the magnetic field at an arbitrary point P on the axis of a circular loop of radius a carrying current I, as shown in Fig. 30-4a.

Solution

Let the loop axis be the x axis, with the origin at the loop center. Figure 30-4b shows the field contribution $d\mathbf{B}$ at P arising from a current element $I\, d\boldsymbol{\ell}$ at the top of the loop, with its direction given by the cross product $d\boldsymbol{\ell} \times \hat{\mathbf{r}}$. Figure 30-4c adds the contribution from a current element at the bottom of the loop, and shows that the net field points along the axis. To find that net field we therefore need sum only the x components of the field contributions from around the loop.

Figures 30-4b and c also show that the vectors $d\boldsymbol{\ell}$ and $\hat{\mathbf{r}}$ are perpendicular; therefore, the product $d\boldsymbol{\ell} \times \hat{\mathbf{r}}$ has magnitude $|d\boldsymbol{\ell}||\hat{\mathbf{r}}|\sin 90° = d\ell$ since $\hat{\mathbf{r}}$ is a unit vector. Then the magnitude of the field contribution $d\mathbf{B}$, as given by the Biot-Savart law, is

$$dB = \frac{\mu_0 I}{4\pi} \frac{|d\boldsymbol{\ell} \times \hat{\mathbf{r}}|}{r^2} = \frac{\mu_0 I}{4\pi} \frac{d\ell}{x^2 + a^2}.$$

To find the net field, we need to sum the x components $dB_x = dB\cos\theta$. Figure 30-4b shows that $\cos\theta = a/r = a/\sqrt{x^2 + a^2}$, so the net field becomes

$$B = \int_{\text{loop}} dB\cos\theta$$

$$= \int_{\text{loop}} \left(\frac{\mu_0 I}{4\pi}\frac{d\ell}{x^2 + a^2}\right)\left(\frac{a}{\sqrt{x^2 + a^2}}\right)$$

$$= \frac{\mu_0 I a}{4\pi(x^2 + a^2)^{3/2}} \int_{\text{loop}} d\ell,$$

where the integral reduces to such a simple form because the distance x is the same for all points on the loop. The remaining integral just means the sum of all the infinitesimal segments $d\ell$ around the loop—or the loop circumference $2\pi a$. Thus, the magnitude of the magnetic field on the loop axis becomes

$$B = \frac{\mu_0 I a^2}{2(x^2 + a^2)^{3/2}}. \qquad (30\text{-}3)$$

EXERCISE Earth's magnetic field at temperate latitudes is about 50 μT. You wish to produce a point where the net field is zero by orienting a 12-cm-diameter wire loop so the field at its center just cancels Earth's field. What loop current is required?

Answer: 4.8 A

Some problems similar to Example 30-1: 1–3, 5, 15, 17

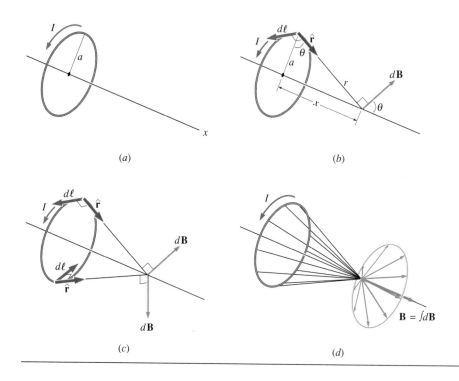

FIGURE 30-4 (*a*) A current loop whose axis is the *x* axis. (*b*) The field element *d***B** arising from the current element *I dℓ* at the top of the loop. (*c*) Field components perpendicular to the loop axis cancel in pairs. (*d*) The result is a net field along the axis.

FIGURE 30-5 The field of a current loop, as traced by iron filings. Note that field lines encircle the current-carrying wire.

We found in the preceding chapter that a current loop behaves like a magnetic dipole when it's in an external magnetic field. Does the loop also produce a dipole-like field? At points far from the loop ($x \gg a$), we can neglect a^2 compared with x^2 in the denominator of Equation 30-3, which then becomes

$$B = \frac{\mu_0 I a^2}{2x^3} = \frac{\mu_0}{2\pi}\frac{\mu}{x^3}, \qquad (x \gg a) \qquad (30\text{-}4)$$

where the second equality follows by introducing the loop's magnetic dipole moment $\mu = I\pi a^2$ as defined in Equation 29-10. Equation 30-4 shows the inverse-cube behavior we found earlier for the field of an *electric* dipole. Although we derived it for a circular loop, Equation 30-4 in fact holds at large distances from a current loop of *any* shape, suggesting that the field is essentially that of a magnetic dipole. A much more difficult calculation would confirm that the distant field off the axis also shows an inverse-cube dependence and exhibits the angular dependence typical of a dipole. We conclude that, in both its response to magnetic fields and its production of magnetic fields:

❙ A current loop constitutes a magnetic dipole.

Figure 30-5 shows the field of a current loop as traced by iron filings.

■ **APPLICATION** MAGNETIC FIELDS OF EARTH AND SUN

Earth, Sun, and many other astronomical objects possess magnetic fields. Reasonably close to Earth, the field approximates that of a magnetic dipole of dipole moment $\mu = 8.0 \times 10^{22}$ A·m² (Fig. 30-6). The direction of the dipole moment vector differs by about 11° from that of Earth's rotation axis, and this accounts for the difference between magnetic and true north. Locally, the field often deviates significantly from a pure dipole form, and these deviations provide geologists with clues to the detailed structure of the planet. The field is not constant; locally, its direction varies significantly over times as short as a few years, and the overall field reverses about every half million years or so (Fig. 30-7). At substantial distances from Earth the field is distorted from its dipole form by the solar wind, a flow of high-speed particles from the Sun.

What causes Earth's magnetic field? We know that electric currents produce magnetic fields. Deep inside Earth are a solid inner core and a liquid outer core, both rich in iron. Through an interaction not yet fully understood, the planet's rotation combined with convective motions due to internal heating produces electric currents in the liquid core. For reasons even less well understood, those currents and the resulting magnetic field undergo reversals on a time scale of approximately a million years. Problems 5 and 6 deal with Earth's magnetic field and its origin.

Earth's magnetic field is crucial to our well-being. As we saw in the preceding chapter, high-energy particles from the Sun and elsewhere are trapped in the magnetic field and have difficulty reaching Earth's surface. Thus the field protects us from harmful particulate radiation. During the field reversals, which last about 10,000 years, the magnetic field is significantly reduced and surface exposure to high-energy particulate radiation is accordingly increased. Some scientists speculate that evolutionary changes due to radiation-induced mutations may accelerate at these times.

The Sun's magnetic field probably arises in the same way as does Earth's, although the gaseous nature of the star and the intense energy flow resulting from nuclear fusion in the Sun's core make its magnetic field much more dynamic. The Sun's field reverses every 11 years, giving rise to the solar activity cycle whose best-known indicator is the count of sunspots—regions of intense magnetic field at the solar surface (Fig. 30-8).

The complex behavior of astrophysical magnetic fields is governed entirely by Newton's laws and the laws of electromagnetism. That we do not yet fully understand these magnetic fields is testimony to the rich variety of phenomena subsumed under those laws.

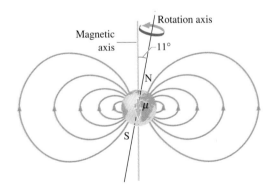

FIGURE 30-6 Earth's magnetic field approximates that of a magnetic dipole located near the center of the Earth and inclined at 11° to the rotation axis. Note from the field direction that Earth's "north" pole is really a magnetic south pole.

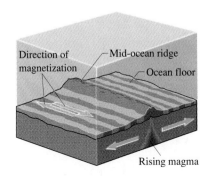

FIGURE 30-7 Magma emerges at the mid-ocean ridges, pushing older portions of the ocean floor farther apart. As the magma solidifies, it becomes magnetized in the direction of Earth's magnetic field. Analysis of the ocean floor reveals bands of alternating magnetic polarity, showing that Earth's magnetic field reverses on a time scale of 10^5 to 10^6 years. Dark bands represent regions with Earth's present magnetic polarity.

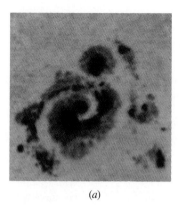

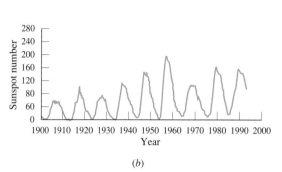

(a)

(b)

(c)

FIGURE 30-8 (a) A group of sunspots, each a region of intense magnetic field at the Sun's surface. (b) The solar magnetic activity cycle is clearly evident in this plot of the number of sunspots. (c) The solar corona traces the overall solar magnetic field in this photo taken during the 1991 eclipse in Hawaii.

Any steady current ultimately flows in a complete circuit, and therefore at great distances its magnetic field is dipolar. But closer in the field configuration depends on the details of the current. An important case is the field near a straight stretch of current. We can approximate this field as that of an infinite line of current, which we calculate in Example 30-2.

● **EXAMPLE 30-2** THE FIELD OF A STRAIGHT WIRE

What is the magnetic field produced by an infinitely long straight wire carrying a steady current I?

Solution
Let the wire coincide with the x axis, and consider an arbitrary point P on the y axis. Figure 30-9 shows a current element $I\,d\ell$ on the wire. At point P the field $d\mathbf{B}$ of this element is perpendicular to both the wire and the vector $\hat{\mathbf{r}}$—that is, $d\mathbf{B}$ points out of the page. Clearly this is true for all current elements along the wire, so we can find the net field by integrating the field contributions dB without needing a vector integration. Using the Biot-Savart law, we have

$$dB = \frac{\mu_0 I}{4\pi} \frac{|d\boldsymbol{\ell} \times \hat{\mathbf{r}}|}{r^2} = \frac{\mu_0 I}{4\pi} \frac{d\ell \sin\theta}{r^2},$$

where we used the fact that $\hat{\mathbf{r}}$ is a unit vector making an angle θ with $d\boldsymbol{\ell}$. Figure 30-9 shows that $r^2 = x^2 + y^2$ and $\sin\theta = y/r = y/\sqrt{x^2 + y^2}$. And, since the segment $d\ell$ lies along the x axis, $d\ell = dx$. Making these substitutions and integrating over the entire wire (from $x = -\infty$ to $x = +\infty$) gives the net field at point P:

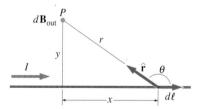

FIGURE 30-9 Geometry for calculating the field at P due to an infinite straight wire carrying a steady current I along the x axis. All current elements along the wire contribute fields at P that point out of the page.

$$B = \int dB = \int_{-\infty}^{\infty} \frac{\mu_0 I}{4\pi} \frac{y\,dx}{(x^2 + y^2)^{3/2}}$$

$$= \frac{\mu_0 Iy}{4\pi} \frac{x}{y^2 \sqrt{x^2 + y^2}}\bigg|_{-\infty}^{\infty},$$

where we found the integral in the table of integrals in Appendix A. (See Problem 73 for another approach.) At the limits

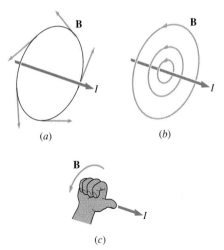

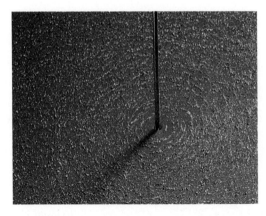

FIGURE 30-11   Iron filings trace out the circular magnetic field surrounding a current-carrying wire.

FIGURE 30-10   (*a*) Some magnetic field vectors associated with a straight wire carrying a steady current. (*b*) The corresponding magnetic field lines are circles, concentric with the wire. (*c*) The right-hand rule gives their direction.

$x = \pm\infty$ the expression $x/\sqrt{x^2 + y^2}$ takes on the values ± 1, so we have

$$B = \frac{\mu_0 I}{4\pi y}[1 - (-1)] = \frac{\mu_0 I}{2\pi y}. \qquad (30\text{-}5)$$

Since the wire has cylindrical symmetry this result must hold anywhere, and the result is circular field lines encircling the wire, as shown in Fig. 30-10.

The field given by Equation 30-5 falls as the inverse of the distance from the wire. This should not be surprising: we found

the same dependence on distance for the *electric* field of an infinitely long charged line. The field patterns differ, though, in that the electric field points radially from the charged line while the magnetic field encircles the current-carrying wire. Although Equation 30-5 applies strictly only to an infinitely long wire, it is a good approximation for a finite wire at distances small compared with the wire length (Fig. 30-11).

EXERCISE   A power line carries 450 A. How close would you have to be to the line for its magnetic field to equal Earth's 50-μT field?

Answer: 1.8 m

Some problems similar to Example 30-2:   4, 11, 59, 60 ●

30-2 THE MAGNETIC FORCE BETWEEN TWO CONDUCTORS

In the preceding chapter we found the force on a straight wire of length ℓ carrying a current I through a magnetic field **B**:

$$\mathbf{F} = I\boldsymbol{\ell} \times \mathbf{B},$$

where the vector $\boldsymbol{\ell}$ is in the direction of the current flow. Through Equation 30-5 we now know the magnetic field produced by a long wire. If two long, parallel wires carry current in the same direction, as shown in Fig. 30-12, then each will experience a force arising from the other's field. We now determine that force.

If d is the distance between the wires, then at wire 2 the field magnitude B_1 due to the current I_1 is, from Equation 30-5,

$$B_1 = \frac{\mu_0 I_1}{2\pi d}.$$

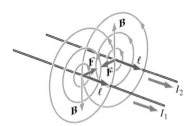

FIGURE 30-12   The magnetic force between two parallel wires carrying current in the same direction is attractive.

FIGURE 30-13 This superconducting electromagnet was torn apart by the magnetic forces from a current of 50 kA flowing in its coils. The magnetic field reached 55 T—over one million times Earth's field. Coils were made from copper/niobium composite with the strength of steel and were reinforced with Kevlar and a steel casing. The coils originally encircled the channel where a pencil has been placed to show the scale.

The field is perpendicular to wire 2, so the magnitude of the force on a length ℓ of wire 2 is

$$F_2 = I_2 \ell B_1 = \frac{\mu_0 I_1 I_2 \ell}{2\pi d}. \qquad (30\text{-}6)$$

Calculating the force on a length ℓ of wire 1 would amount to interchanging the subscripts 1 and 2, giving a force of the same magnitude.

What is the direction of the force? Evaluating the cross product of ℓ and **B** in Fig. 30-12 shows that the force on wire 2 is toward wire 1, and vice versa. By using the right-hand rule, you can convince yourself that reversing one of the currents would reverse the directions of both forces. We therefore conclude the following:

> **Conductors carrying currents in the same direction experience an attractive magnetic force, while conductors carrying currents in opposite directions repel each other.**

The force between nearby conductors must be considered in the construction of electrical devices carrying large currents. In electromagnets, particularly, conductors must have enough support that magnetic forces do not destroy the device (Fig. 30-13). The hum you often hear around electrical equipment comes from the mechanical vibration of tightly wound conductors in transformers and other electrical components. This vibration results from the changing magnetic force associated with the 60-Hz alternating current.

The magnetic force between conductors is the basis for the definition of the ampere and, consequently, the coulomb. One ampere is defined as follows:

> **If two long, parallel conductors 1 meter apart carry equal currents and experience a force of 2×10^{-7} N, then the current in each is, by definition, 1 A.**

It then follows that

> **1 C is, by definition, the amount of charge passing in 1 s through a wire carrying 1 A.**

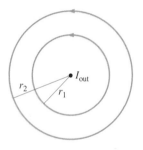

FIGURE 30-14 Two magnetic field lines surrounding a wire carrying current out of the page.

30-3 AMPÈRE'S LAW

In Chapter 23 we did several relatively cumbersome electric field calculations using Coulomb's law. For example, we found the field of an infinite line charge by integrating over all the charge elements along the line. Then in Chapter 24 we introduced Gauss's law and showed how it made electric field calculations much easier in situations with sufficient symmetry.

Can we make a statement that would enable us to calculate magnetic fields with comparable ease? Figure 30-14 shows two of the circular magnetic field lines surrounding a long wire carrying current I out of the page. Imagine moving around the inner circle. As you move a short way, form the product of the distance $d\ell$ that you travel with the component of the magnetic field in the direction you're going. Here the field is entirely in the direction of your path, so that product is simply $B\,d\ell$, where B is the field magnitude; more generally, it would be the dot product $\mathbf{B} \cdot d\ell$ to account for an arbitrary angle between the field and your path. Now consider adding up all these products as you go around a complete circle. Formally, the result is the line integral $\oint \mathbf{B} \cdot d\ell$, or, in this case when the path and field coincide, just $\oint B\,d\ell$. Here the circle indicates that the integration is done around a *closed* path. The magnitude of $\mathbf{B}$ is given by Equation 30-5: $B = \dfrac{\mu_0 I}{2\pi r_1}$, where we've replaced y by the radius r_1 since Equation 30-5 applies for any orientation of the y axis perpendicular to the wire. Evaluating $\oint \mathbf{B} \cdot d\ell$ then gives

$$\oint \mathbf{B} \cdot d\ell = \frac{\mu_0 I}{2\pi r_1} \oint d\ell = \frac{\mu_0 I}{2\pi r_1}(2\pi r_1) = \mu_0 I, \qquad (30\text{-}7)$$

where we took $B = \mu_0 I/2\pi r_1$ outside the integral because r_1 is the constant radius of the circular field line. Our answer, $\mu_0 I$, is independent of the radius r_1. We would get the same answer going around the outer circle, or indeed any circular path. On a larger path the distance is greater, but the field—dropping as $1/r$—is correspondingly weaker, making the value of $\oint \mathbf{B} \cdot d\ell$ the same. Thus the line integral $\oint \mathbf{B} \cdot d\ell$ does not depend on the radius of the circular path, but only on the current I encircled by that path.

What if our path does not coincide with a field line? Figure 30-15 shows a closed loop that encircles the wire but does not coincide with a single field line. We can evaluate $\oint \mathbf{B} \cdot d\ell$ by considering each of the four segments AB, BC, CD, and DA. The radial segments BC and DA are perpendicular to the field, so here $\mathbf{B} \cdot d\ell = 0$ and these segments make no contribution to $\oint \mathbf{B} \cdot d\ell$. Now consider the arc AB that lies on our path and the arc DC that does not. We know that $\oint \mathbf{B} \cdot d\ell$ has the same value, namely $\mu_0 I$, if we go around a path coinciding with either the inner or the outer field line. Arcs AB and DC occupy the same fractions of their respective circles, and, therefore, the contribution from AB to the line integral $\oint \mathbf{B} \cdot d\ell$ is the same as we would get going instead along DC. But then since the radial segments contribute nothing, the value of $\oint \mathbf{B} \cdot d\ell$ around the irregular path $ABCDA$ is the same as its value around the outer circle—which Equation 30-7 shows to be $\mu_0 I$.

We can approximate an arbitrary loop as a sequence of concentric arcs joined by radial segments, as shown in Fig. 30-16. Applying the arguments we used with Fig. 30-15, we conclude that the line integral $\oint \mathbf{B} \cdot d\ell$ around *any* loop has the value $\mu_0 I$. Although we have considered only the field of a single straight current, this result is in fact a universal statement about the relation between current and magnetic field:

The value of the line integral $\oint \mathbf{B} \cdot d\ell$ around any closed loop is proportional to the current encircled by that loop. Mathematically,

$$\oint \mathbf{B} \cdot d\ell = \mu_0 I_{\text{encircled}}. \qquad (30\text{-}8)$$

This statement is a form of **Ampère's law,** one of the four fundamental laws of electromagnetism. Ampère's law says that whatever arrangement of currents we

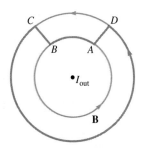

FIGURE 30-15 A closed loop that does not coincide with a single field line. The line integral of the magnetic field around this loop has the same value—$\mu_0 I$—that it has around a circular loop coinciding with a field line.

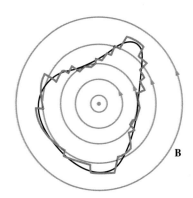

FIGURE 30-16 Approximating an irregular loop as a series of arcs and radial segments. The line integral of the magnetic field is still $\mu_0 I$, as it is for *any* closed loop.

(a)

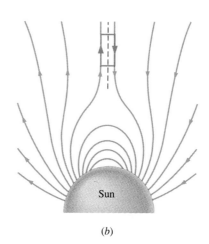

(b)

FIGURE 30-17 (a) A coronal streamer in the Sun's outer atmosphere contains oppositely directed magnetic fields in close proximity. (b) A model calculation of the coronal magnetic field. Since $\oint \mathbf{B} \cdot d\boldsymbol{\ell}$ is clearly nonzero around the loop shown, there must be current in the region encircled by the loop.

might have, and however complicated the resulting magnetic field, that field is such that the line integral $\oint \mathbf{B} \cdot d\boldsymbol{\ell}$ around any closed loop will have the value $\mu_0 I$, where I is the net current *encircled* by that loop. Compare this with Gauss's law, which says that whatever arrangement of electric charges we might have, the resulting electric field is such that the surface integral $\oint \mathbf{E} \cdot d\mathbf{A}$ has the value q/ε_0, where q is the *enclosed* charge.

Our statement of Ampère's law is true for any closed loop whatsoever, as long as the encircled current is steady—that is, never changing with time. It does not matter whether the current is in a single wire or in a number of wires or distributed throughout space; in any case, we simply add the currents to obtain the net current encircled by the loop. If there are currents flowing in opposite directions we add them with appropriate algebraic signs. A current counts as positive if, when you curl the fingers of your right hand in the direction of the loop, your right thumb points in the general direction of the current.

Although Ampère's law describes the relation between the magnetic field on a loop and the current *encircled* by that loop, we emphasize that the field **B** in Ampère's law is the *net field* from all currents, whether inside the loop or not. We found the same thing with Gauss's law, which relates the electric field **E** on a closed surface to the enclosed charge; **E** itself, however, is the *net field* arising from all sources, whether enclosed or not.

Ampère's law, like Gauss's, is a truly universal statement describing the relation between magnetic field and electric current. It holds in the electromagnetic devices we build, in atomic and molecular systems, in the interaction of fluid motion and electric charge that gives rise to Earth's magnetic field, and in distant astrophysical objects (Fig. 30-17). Although it is difficult to show mathematically, the Biot-Savart law follows logically from Ampère's law in the same sense that Coulomb's law follows from Gauss's.

● EXAMPLE 30-3   SOLAR CURRENTS

The long dimension of the rectangular loop in Fig. 30-17b is 4×10^8 m, and the magnetic field strength in the vicinity of the loop is essentially constant at 2 mT. Find the total current encircled by the rectangle. In what direction is this current flowing?

Solution

We can't insert an ammeter into the Sun's atmosphere, but we can determine the current using Ampère's law. The short ends of the rectangle in Fig. 30-17b are perpendicular to the field, so here $\mathbf{B} \cdot d\boldsymbol{\ell} = 0$ and there is no contribution to the line integral around the loop. Moving around the loop in the direction shown means going *with* the field along both long dimensions (note that this is because the field direction reverses across the dashed line). Since the field is uniform, the contribution to the line integral from each side is just $B\ell$. Thus $\oint \mathbf{B} \cdot d\boldsymbol{\ell} = 2B\ell$. Ampère's law tells us that the line integral of the magnetic field around a closed loop has the value $\mu_0 I$, so

$$2B\ell = \mu_0 I,$$

or

$$I = \frac{2B\ell}{\mu_0} = \frac{(2)(2 \times 10^{-3} \text{ T})(4 \times 10^8 \text{ m})}{4\pi \times 10^{-7} \text{ N/A}^2} = 1 \times 10^{12} \text{ A}.$$

This is a colossal current by terrestrial standards, but is typical of large-scale currents on the Sun.

Since the line integral is positive, we curl our right fingers around the loop in the direction shown by the arrows and our thumb must point in the direction of the current—that is, into the page in Fig. 30-17. In three dimensions this current flows around the Sun in approximately the equatorial plane.

> **TIP   Amperian Loops**   The loop used with Ampère's law is truly arbitrary. It need not coincide with a field line, as Fig. 30-16 showed. Example 30-3 showed this, too; here, the rectangular loop coincided with the straight field lines over its long sides but not along its shorter ends. The loop used with Ampère's law is called an **amperian loop.** Don't confuse amperian loops with field lines!

EXERCISE   Fig. 30-18 shows a cross-sectional view of three wires carrying current perpendicular to the plane of the page. Wire A carries 20 A out of the page. $\oint \mathbf{B} \cdot d\boldsymbol{\ell}$ around loop 1 in the direction shown has the value 1.26×10^{-5} T·m, and the line integral around loop 2 has the value -6.28×10^{-6} T·m. Find the currents in wires B and C, and the directions of those currents.

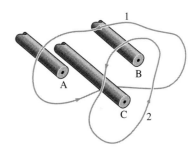

FIGURE 30-18   What are the currents in wires B and C?

Answers: B: 10 A into page; *C*: 5 A out of page
Some problems similar to Example 30-3:   28–30   ●

30-4 USING AMPÈRE'S LAW

For charge distributions with sufficient symmetry we used Gauss's law to solve for the electric field in a simple and elegant way. Similarly, for current distributions with sufficient symmetry we can use Ampère's law to solve for the magnetic field. Doing so requires finding suitable amperian loops over which we can evaluate the line integral of the magnetic field.

The Field of a Straight Wire

Here we use Ampère's law to determine the field of an infinite straight wire. This is the same calculation we did in Example 30-2 using the Biot-Savart law.

Figure 30-19 shows an end view of the wire, which carries current I out of the page. We know that magnetic field lines generally encircle the currents that are their sources. Here cylindrical symmetry requires that those field lines be circles

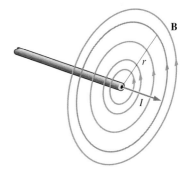

FIGURE 30-19   Circular field lines surrounding an infinitely long wire. Gray circle is an amperian loop for evaluating the line integral of the magnetic field. Although we've drawn only four field lines, the field exists everywhere, including on the amperian loop.

centered on the wire and that the magnetic field strength cannot depend on angular position around the wire. Thus B must be constant on a field line. Applying the right-hand rule—thumb along the current, curling the fingers of the right hand—shows that the magnetic field circles counterclockwise around the wire. And since the wire is infinitely long, Fig. 30-19 represents the field anywhere along the wire.

Given the symmetry, an appropriate amperian loop for evaluating the line integral is itself a circle, also shown in Fig. 30-19. The magnetic field everywhere on this amperian loop is in the same direction as the loop, so the dot product $\mathbf{B} \cdot d\boldsymbol{\ell}$ becomes simply $B \, d\ell$. Then the line integral around the loop is

$$\oint_{\text{loop}} \mathbf{B} \cdot d\boldsymbol{\ell} = \oint_{\text{loop}} B \, d\ell = B \oint_{\text{loop}} d\ell = 2\pi r B.$$

Here we can take the field magnitude B outside the integral because it's constant on the circular amperian loop; the remaining integral is then the loop circumference $2\pi r$. Ampère's law equates this line integral to μ_0 times the encircled current, so

$$2\pi r B = \mu_0 I,$$

or
$$B = \frac{\mu_0 I}{2\pi r}. \tag{30-9}$$

This result is the same as Equation 30-5, but we derived it with much less difficulty.

> **TIP** **Symmetry Is Crucial** Our use of Ampère's law to derive the field of a long wire depends crucially on symmetry. We cannot arbitrarily pull B outside the integral unless we know—as we do here from symmetry—that it is constant in magnitude and in direction relative to our amperian loop.

Our calculation made no assumptions about the diameter of the wire. Therefore Equation 30-9 holds for any long cylindrical wire, thick or thin, as long as we restrict ourselves to points *outside* the wire so that our amperian loop encircles the *entire* current. In fact, you can easily convince yourself that Equation 30-9 must hold *outside* any current distribution with cylindrical symmetry. To calculate the field *inside* a current distribution, however, we must be careful to use only the actual current encircled by our amperian loop. Example 30-4 illustrates this point.

● EXAMPLE 30-4   INSIDE A WIRE

A long, straight wire of radius R carries a current I uniformly distributed over its cross-sectional area. What is the magnetic field as a function of position within the wire?

Solution

All the symmetry arguments we used to find the field outside a wire still apply here, so the line integral around a circular

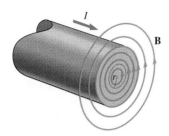

FIGURE 30-20 Cross section of a long cylindrical wire carrying current out of the page. Symmetry requires that the magnetic field be circular inside as well as outside the wire. Current is distributed uniformly over the wire; in applying Ampère's law to the amperian loop shown in gray we need the fraction of the total current encircled by the loop.

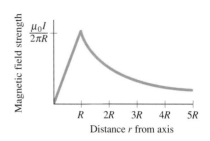

FIGURE 30-21 The magnetic field inside a wire increases linearly with radial distance, while beyond the wire radius R it drops as $1/r$.

amperian loop of radius r is still $2\pi rB$. What's different is that an amperian loop within the wire no longer encircles the entire current (Fig. 30-20). How much current is encircled? The current is distributed uniformly, giving a current density (current per unit area) $J = I/A = I/\pi R^2$. The encircled current is then the current density times the loop area, or

$$I_{\text{encircled}} = \left(\frac{I}{\pi R^2}\right)(\pi r^2) = I\frac{r^2}{R^2}.$$

Equating the line integral $2\pi rB$ to this encircled current gives

$$2\pi rB = I\frac{r^2}{R^2},$$

or

$$B = \frac{\mu_0 Ir}{2\pi R^2}. \tag{30-10}$$

Does this result make sense? The field increases linearly with distance from the axis—just as we found for the *electric* field *inside* a uniformly charged cylinder. Here the increase occurs

because we encircle more and more current—with $I_{\text{encircled}}$ growing as r^2—as long as we're inside the wire. Once we reach the surface, of course, the encircled current remains constant and the field begins to decrease inversely with distance, as described in Equation 30-9. Figure 30-21 plots the field strengths both inside and outside the wire.

This example is not merely academic; in superconducting wires, for example, strong magnetic fields can destroy superconductivity, so it's important to know the field throughout the wire.

EXERCISE A power line 4.0 cm in diameter carries 1.5 kA. Find the magnetic field (a) 1.0 cm from the axis of the wire and (b) 10 cm from the axis.

Answers: (a) 7.5 mT; (b) 3.0 mT

Some problems similar to Example 30-4: 32, 33, 37, 38 ●

● **EXAMPLE 30-5** A CURRENT SHEET

An infinite flat sheet carries a current out of the page, as shown in Fig. 30-22. The current is distributed uniformly along the sheet, with the current per unit width along the sheet given by J_s. Find the magnetic field of this sheet.

Solution

What might the field of this sheet look like? Figure 30-23 suggests that we can consider the sheet to be made of many parallel wires. The vector sum of the fields of these wires gives a net field to the left above and to the right below the sheet. We can also argue from symmetry that the fields must point horizontally and with magnitude that is independent of position parallel to the sheet; since the sheet extends infinitely to the right and left, there's nothing to favor an upward or downward deflection of the field lines or any variation in field magnitude parallel to the sheet. We've drawn these horizontal field lines, and an appropriate amperian loop, in Fig. 30-22.

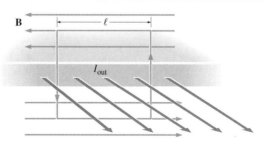

FIGURE 30-22 A current sheet (gray) extends infinitely to the left and right, as well as in and out of the page. Field lines and a rectangular amperian loop are also shown.

We already evaluated $\oint \mathbf{B} \cdot d\boldsymbol{\ell}$ for a similar geometry in Example 30-3; the result is simply $2B\ell$, with ℓ the width of the amperian rectangle. The current per unit width of the sheet is J_s, so our rectangular loop of width ℓ encircles a current

$I = J_s\ell$. Equating μ_0 times this encircled current to the line integral $\oint \mathbf{B} \cdot d\ell = 2B\ell$ gives

$$2B\ell = \mu_0 J_s \ell,$$

or
$$B = \tfrac{1}{2}\mu_0 J_s. \qquad (30\text{-}11)$$

Thus the magnetic field of an infinite current sheet is independent of distance from the sheet. We found the same result for the electric field of an infinite charged plane. Although the infinite sheet is an idealization, Equation 30-11 is a good approximation to the field near long, wide, flat conductors. Thin current sheets also form in conducting plasmas like those in fusion reactors and in the Sun's atmosphere, where they can lead to sudden dissipation of energy.

EXERCISE Current in a printed circuit board is carried in a long copper strip 2.1 mm wide and much thinner than its width. If the current in the strip is 35 mA, spread uniformly over its cross section, find the magnetic field strength near the strip's surface.

Answer: 10 μT

Some problems similar to Example 30-5: 39–41, 44

FIGURE 30-23 A current sheet approximated by closely spaced parallel wires. The magnetic field of the sheet is the vector superposition of the fields of the individual wires. In the limit of infinitesimally spaced wires, the field lines become straight and strictly horizontal.

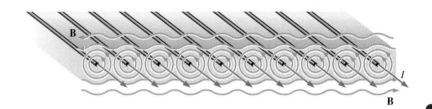

Fields of Simple Current Distributions

We have used Ampère's law to calculate the magnetic fields of two simple, symmetric current distributions. Although those magnetic fields may look quite different from the electric fields of correspondingly symmetric charge distributions, they exhibit the same dependences on distance. Table 30-1 summarizes the electric and magnetic fields of several simple charge and current distributions. Real distributions are usually more complicated, but often they can be approximated by these simple cases. Far from any current loop, for example, the magnetic field is that of a dipole. Very near *any* wire, its magnetic field is essentially that of a long straight wire. For situations that lack symmetry and in which approximations are not adequately accurate, we can always calculate the magnetic field of a steady current distribution from the Biot-Savart law, just as we can always calculate the electric field of an arbitrary charge distribution from Coulomb's law.

30-5 SOLENOIDS AND TOROIDS

We found in Chapter 26 that we can produce a uniform electric field between the two closely spaced, charged conducting plates of a capacitor. Is there an analogous device that will produce a uniform magnetic field?

Fig. 30-24 shows a coil of wire carrying current I. Close to the wire are magnetic field lines encircling the wire. We show these field lines in Fig. 30-24 at the top and bottom of the coil, where they cross the plane of the page. The net field anywhere is, of course, just the vector sum of the field contributions from all parts of the coil. You can see that inside the coil, the fields from current elements at the top and bottom all have a component to the right and so tend to

▲ **TABLE 30-1** FIELDS OF SOME SIMPLE CHARGE AND CURRENT DISTRIBUTIONS

DISTRIBUTION	FIELD DEPENDENCE ON DISTANCE*	ELECTRIC **OR** MAGNETIC **FIELD**
Electric dipole	$\dfrac{1}{r^3}$	
Magnetic dipole	$\dfrac{1}{r^3}$	
Spherically symmetric charge distribution	$\dfrac{1}{r^2}$	
Spherically symmetric current distribution	Impossible for steady current	
Charge distribution with line symmetry	$\dfrac{1}{r}$	
Current distribution with line symmetry	$\dfrac{1}{r}$	
Infinite flat sheet of change	Uniform field; no variation	
Current sheet	Uniform field; no variation	

*For field *outside* distribution.

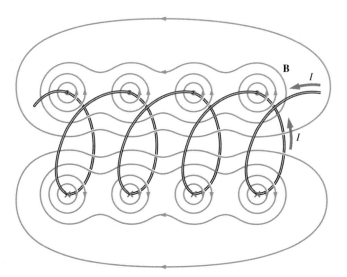

FIGURE 30-24 The magnetic field arising from the current in a loosely wound coil of wire is strongest within the coil and weaker outside. Field is shown only in the plane of the page; dots show current emerging from the page and crosses mark current going into the page.

reinforce. Outside the coil, though, field components from top and bottom are in opposite directions and thus reduce the net field, which points to the left as shown in the figure.*

Imagine winding to coil more tightly and making it longer. Such a long, tightly wound coil is a **solenoid.** Figure 30-25 shows what happens as the solenoid gets longer: the interior field stays essentially the same, but exterior field lines spread into more distant regions before closing back on themselves. Therefore, the field strength outside the solenoid—as evidenced by the density of field lines—decreases. In the limit of an infinitely long solenoid, the interior field becomes perfectly straight and the exterior field in the plane of Figs. 30-24 and 30-25 goes to zero. A real solenoid approaches this ideal limit when its length is much greater than its diameter. Figure 30-26 shows the field of a solenoid as traced by iron filings.

In the long-solenoid limit, we can use Ampère's law to find the field in the solenoid. Figure 30-27 shows a cross section through a long solenoid, with a rectangular amperian loop. Since the exterior field is zero, there is no contribution to the line integral $\oint \mathbf{B} \cdot d\boldsymbol{\ell}$ from the top of the amperian loop. The loop's vertical sides are at right angles to the field, so they, too, contribute nothing to the line integral. The only contribution is from the bottom of the loop. Going counterclockwise around the loop, we move with the field on this segment. Since the field cannot vary with position along the infinite solenoid, the contribution to the line integral is just $B\ell$; since the other three sides contribute nothing, this is the value of $\oint \mathbf{B} \cdot d\boldsymbol{\ell}$.

To apply Ampère's law we need the encircled current. If there are n turns per unit length of solenoid, the amperian loop encircles $n\ell$ turns. The same current

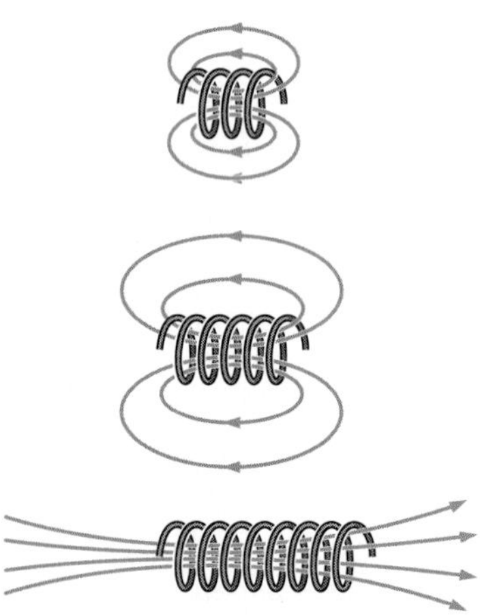

FIGURE 30-25 As a solenoid gets longer, the interior field stays nearly constant but the exterior field weakens as the field lines spread ever farther apart.

---

* Because the coil carries a net current from right to left, there is also a weak field component outside that encircles the coil. This component plays no role in the discussion that follows, and we ignore it.

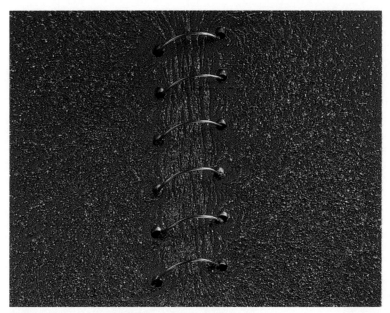

FIGURE 30-26 Iron filings trace the magnetic field of a solenoid. Note that the field is strong inside the solenoid and relatively weak outside, except near the ends.

I flows in each turn (why?), so the encircled current is $n\ell I$. Applying Ampère's law then gives

$$B\ell = \mu_0 n\ell I,$$

or $$B = \mu_0 nI. \quad \text{(solenoid)} \tag{30-12}$$

Since the vertical dimension of the amperian rectangle never entered our calculation, this field magnitude is the same anywhere inside the solenoid. The magnetic field in the solenoid is therefore uniform. Although it looks very different, the long solenoid with its uniform magnetic field is the magnetic analog of a closely spaced parallel-plate capacitor, which produces a uniform electric field.

Although Figs. 30-24 and 30-25 depict circular coils, the derivation of Equation 30-12 is based only on Fig. 30-27, which could represent a solenoid whose coils are circular, square, or any other shape. Equation 30-12 therefore holds for coils of any shape, as long as that shape is the same over the length of the solenoid.

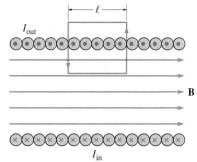

FIGURE 30-27 Cross section of a long solenoid, showing a rectangular amperian loop straddling the region where the solenoid coils emerge from the plane of the page.

■ **APPLICATION** EVERYDAY SOLENOIDS, MEDICAL SOLENOIDS

Solenoids of all sizes are used in a wide variety of experimental and practical devices. Because a solenoid is hollow, magnetic materials like iron will be pulled into the solenoid by the nonuniform magnetic field near its ends, and solenoids are therefore used to produce straight-line motion in mechanical devices. For example, turning the key to start your car sends current through a solenoid in the car's starter motor. A magnetic field develops in the solenoid, pulling in a steel plunger. Movement of the plunger joins a set of electrical contacts allowing current to flow to the starter motor (it would be difficult to switch the roughly 100-A starter current directly from inside the car). At the same time, the rod moves a small gear to the end of the motor shaft, engaging a gear on the engine's flywheel so that the starter motor can turn the engine (Fig. 30-28).

Running a dishwasher or washing machine also involves solenoids. The valves that control the flow of water in these machines are solenoid valves, opened and closed as a steel rod moves in or out of a solenoid (Fig. 30-29).

Finally, huge solenoids with human-size interiors are used to produce the strong, uniform magnetic fields needed for magnetic resonance imaging (MRI). This high-tech medical technique images the body's interior without the need for x rays or other radiation (Fig. 30-30). The MRI solenoids use superconducting wire cooled with liquid helium to achieve high currents and correspondingly high magnetic fields.

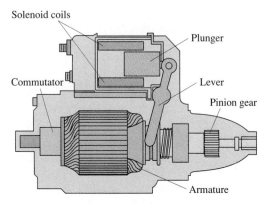

FIGURE 30-28 Cross section of a car starter motor, showing solenoid coil and mechanical linkage that engages the starter's pinion gear to turn the engine.

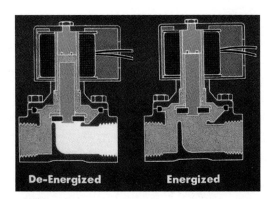

FIGURE 30-29 Cutaway diagram of a solenoid valve, widely used in washing machines, dishwashers, and many industrial applications. Solenoid coils are the red structures at top. A steel plunger extending part of the way into the coils connects to a diaphragm. At left, there is no current in the coils, and the diaphragm blocks the flow of liquid. At right, the coils are energized, producing a magnetic field. As a result, the plunger and diaphragm are pulled upward, allowing liquid to flow through the valve.

FIGURE 30-30 Patient being inserted into a superconducting solenoid used for magnetic resonance imaging (MRI), a powerful and relatively new medical diagnostic technique.

● EXAMPLE 30-6 AN MRI SOLENOID

A solenoid used in magnetic resonance imaging is 2.4 m long and 95 cm in diameter. It is wound from a niobium-titanium superconducting wire 2.0 mm in diameter, with adjacent turns separated by an insulating layer of negligible thickness. What current is necessary to produce a 1.5-T magnetic field inside the solenoid?

Solution

The magnetic field is given by Equation 30-12, $B = \mu_0 nI$, so if we know n—the number of turns per unit length—we can solve for I. Since the wire has diameter $d = 2.0$ mm, 5 turns will occupy 1 cm and therefore $n = 500$ turns/meter. Solving Equation 30-12 for I then gives

$$I = \frac{B}{\mu_0 n} = \frac{(1.5 \text{ T})}{(4\pi \times 10^{-7})(500 \text{ m}^{-1})} = 2.4 \text{ kA}.$$

This is a large current, but readily handled by the niobium-titanium superconductor.

EXERCISE Copper wire 0.40 mm in diameter is tightly wound into a long solenoid, and a 3.6-A current passed through it. What is the magnetic field in the solenoid?

Answer: 11 mT

Some problems similar to Example 30-6: 45–47, 50

Toroids

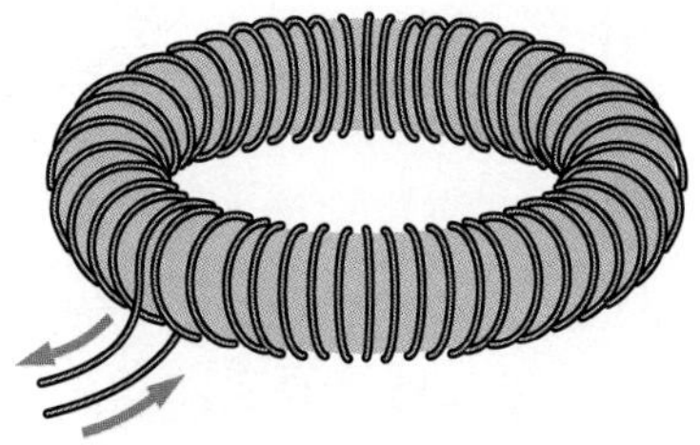

A **toroid** is a solenoid bent into a doughnut shape (Fig. 30-31). Toroidal geometry is widely used in fusion reactors, as discussed in the preceding chapter, and toroidal coils help produce the magnetic field that confines the fusion plasma.

Symmetry requires that the toroid's field lines be circular, with constant magnitude on any line. We can readily calculate the line integral of this magnetic field around a circular amperian loop, like that shown in Fig. 30-32. This loop coincides with a field line, and symmetry ensures that the magnetic field has the same magnitude over the loop. Therefore the line integral becomes

$$\oint_{\text{loop}} \mathbf{B} \cdot d\ell = \oint_{\text{loop}} B\, d\ell = B \oint_{\text{loop}} d\ell = 2\pi r B.$$

To apply Ampère's law we equate this quantity to μ_0 times the encircled current. If the toroid consists of N turns and carries a current I, then an amperian loop inside the toroid encircles a total current NI. Then Ampère's law becomes

$$2\pi r B = \mu_0 N I,$$

or

$$B = \frac{\mu_0 N I}{2\pi r}. \qquad \text{(toroid)} \qquad (30\text{-}13)$$

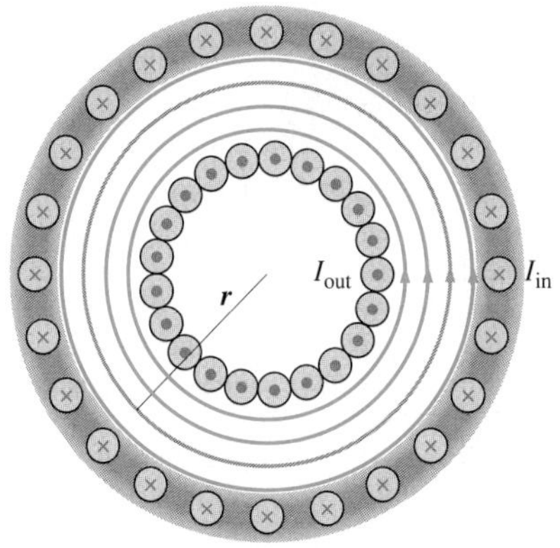

This result holds when the amperian loop is within the toroid itself. But if it's inside the inner edge there's no encircled current and, therefore, no magnetic field. And beyond the outer edge, the amperian loop encircles equal but opposite currents, giving zero net current and again no magnetic field.* As with a solenoid, the individual turns of a toroid need not be circular.

The toroidal field of Equation 30-13 is not uniform but exhibits a $1/r$ decrease. This nonuniformity causes problems with plasma confinement in fusion reactors, where it causes a drift of particles perpendicular to the field and therefore toward the walls of the machine.

30-6 MAGNETIC MATTER

We began our study of magnetism in Chapter 29 with a discussion of magnets. But we quickly moved on to concentrate on the behavior of electric charges in magnetic fields. In this chapter we've further developed the relation between magnetism and electric charge, as we've seen that moving electric charge is the source of magnetic fields. What does all this business with electric charge have to do with the familiar magnets we used to introduce magnetism?

In fact the two are one and the same phenomenon. When an electron orbits an atomic nucleus, its circular motion and charge make a miniature current loop, and it therefore constitutes a magnetic dipole (Fig. 30-33). In addition the electron possesses an intrinsic magnetic dipole moment associated with an intrinsic angular momentum called "spin." Interactions among these magnetic dipole moments determine the magnetic properties of individual atoms and of

FIGURE 30-31 A toroidal coil.

FIGURE 30-32 Cross section of a toroid showing current emerging at inner edge of coils and descending into plane of page at outer edge. Also shown are circular field lines and an amperian loop of radius r (gray) for use in calculating the field.

FIGURE 30-33 In the classical model of the atom, the circling electron constitutes a miniature current loop. The atom is therefore a magnetic dipole. Here the current—carried by the *negative* electron and therefore opposite the electron's motion—is counterclockwise when viewed from above, so the magnetic moment points upward.

---

*This neglects a weak field associated with the fact that the current has a component *around* the toroid in the plane of Fig. 30-32.

FIGURE 30-34 Magnetic domains in a thin film of ferromagnetic material. The dark and colored areas represent regions where the magnetic moments point either into or out of the plane of the figure; the different colors emphasize that regions of oppositely directed magnetic moments tend to organize into parallel structures.

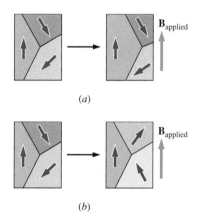

FIGURE 30-35 Domain changes in a ferromagnetic material occur through (a) domain growth and (b) domain realignment. These changes occur when a magnetic field is applied, and result in the material acquiring a net magnetic moment.

bulk matter. Although an accurate description of magnetism in matter necessarily involves quantum mechanics, we can nevertheless use our knowledge of magnetic dipoles to gain a qualitative understanding of magnetic matter.

Ferromagnetism

Experiment reveals three types of magnetic behavior in bulk matter. **Ferromagnetism,** the most familiar, is actually limited to a few substances, notably the elements iron, nickel, and cobalt, and some compounds. In a ferromagnetic material, a quantum-mechanical interaction among nearby atomic magnetic moments results in regions—called **magnetic domains**—in which all the atomic magnetic moments point in the same direction. A typical domain contains 10^{17} to 10^{21} atoms and occupies a volume on the order of 10^{-12} to 10^{-8} m^3 (Fig. 30-34). The magnetic moment of a single domain can be large since it is the sum of many atomic magnetic moment vectors all pointing in the same direction. A typical piece of ferromagnetic material, however, contains many domains with their moments in random directions and, therefore, exhibits no net magnetic moment. But when an external magnetic field is applied, a net magnetic moment develops. This occurs because domains that already happen to be aligned with the field can grow by realignment of individual atomic moments in adjacent domains (Fig. 30-35a). In addition, the magnetic moments of entire domains can rotate (Fig. 30-35b).

Since any dipole experiences a force in a nonuniform field, a piece of ferromagnetic material in such a field experiences a net force; this is why iron and other ferromagnetic materials are attracted to magnets even though they themselves are not magnets.

Removal of the applied magnetic field does not entirely destroy the overall alignment that gives the net magnetic moment in a ferromagnetic substance. The remanent magnetization is what makes permanent magnets. In so-called hard ferromagnetic materials, the remanent magnetism is strong; these materials are used specifically for making permanent magnets. Remanent magnetism is relatively weak in soft ferromagnetic materials; these are used in applications like heads for VCRs and computer disk drives, where permanent magnetization is undesirable (tapes and disks themselves use harder materials for long-term information storage).

Random thermal motions tend to disrupt the alignment of individual magnetic moments. Thus ferromagnetic effects weaken with increasing temperature. Above the so-called **Curie temperature,** ferromagnetism ceases altogether. Curie temperatures for the common ferromagnetic elements nickel, iron, and cobalt are 631, 1043, and 1395 K, respectively. The rarer ferromagnetic elements dysprosium and gadolinium have much lower Curie temperatures of 85 and 289 K, respectively. The disappearance of ferromagnetism at the Curie temperature is an example of a phase transition, analogous to the solid/liquid/gas transitions we studied in Chapter 20.

Paramagnetism

Many substances that are not ferromagnetic nevertheless consist of atoms or molecules with permanent magnetic dipole moments. What distinguishes these **paramagnetic** materials from ferromagnetic substances is the absence of a strong interaction that tends to align nearby moments. As a result, individual atomic moments are not organized into domains, but point in random directions. An applied magnetic field still brings the atomic magnetic moments into some degree of alignment, but at all but the coldest temperatures this alignment is far less complete than in a ferromagnetic material. Therefore paramagnetic materials are attracted only weakly to magnets (Fig. 30-36).

Diamagnetism

Even materials with no intrinsic magnetic moments can have magnetic moments induced when a magnet approaches or when an applied magnetic field otherwise changes. Such materials are termed **diamagnetic.** In contrast to paramagnetic and ferromagnetic materials, diamagnetic materials are repelled by magnets. We will explore the origins of diamagnetism in the next chapter.

Magnetic Susceptibility

We can characterize the effect of atomic magnetic dipoles on bulk matter just as we did in Chapter 23 for electric dipoles in dielectrics. There we found that atomic electric dipoles align with an electric field to reduce the field within the material, and we introduced the dielectric constant κ as the factor quantifying that reduction.

In a magnetic material, alignment of atomic magnetic dipoles has the opposite effect: the field in the material is increased. Figure 30-37 shows why. An electric dipole consists of two separated point charges. The strongest field associated with this dipole is the internal field pointing straight from the positive to the negative charge. When the dipole aligns with an applied electric field, this internal field is *opposite* the applied field and, therefore, *reduces* the net electric field (Fig. 30-37a). But, as we've seen, magnetic field lines always *encircle* a current. As Fig. 30-37b shows, the field inside the current loop therefore points in the *same* direction as the applied field and, therefore, *increases* the net magnetic field.

FIGURE 30-36 Liquid oxygen (O_2) is one of the more strongly paramagnetic materials. Here liquid oxygen is suspended between the poles of a magnet.

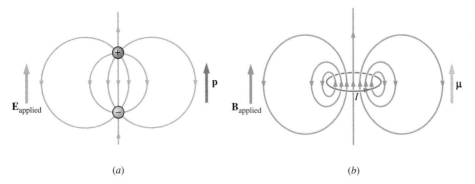

$E_{applied}$ p $B_{applied}$ I μ

(a)   (b)

FIGURE 30-37 Although the electric fields far from electric and magnetic dipoles have the same form, the fields *within* the atomic structure giving rise to the different dipole moments have opposite directions. This is because electric dipoles consist of separated point charges, while magnetic dipoles are current loops. When external electric and magnetic fields are applied, respectively, to these two structures, the result is a reduction in the net electric field but an increase in the net magnetic field.

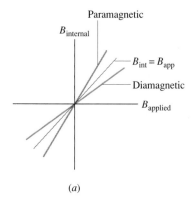

(a)

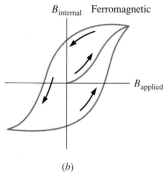

(b)

FIGURE 30-38 Internal versus applied magnetic field for different types of magnetic materials. (a) In diamagnetic and paramagnetic materials the relationship is linear; paramagnetic materials strengthen the applied field slightly, while diamagnetic materials weaken it slightly. (b) In ferromagnetic materials, the relationship depends on the strength of the applied field and on the past history of the material. In (b) field strengths along the vertical axis are much greater than along the horizontal axis, reflecting the large susceptibilities of ferromagnetic materials. Arrows indicate the direction in which fields are changed.

In analogy with the dielectric constant, we introduce the quantity κ_M, called the **relative permeability,** as the factor by which the magnetic field within a material increases as a result of the alignment of atomic magnetic dipoles. For paramagnetic materials, κ_M is slightly greater than 1; for ferromagnetic materials it is much greater than 1. In diamagnetic materials the dipoles align antiparallel to the field; this makes $\kappa_M < 1$ for these materials. Because the relative permeabilities of paramagnetic and diamagnetic materials are very close to 1, it's more convenient to work with the **magnetic susceptibility,** defined by

$$\chi_M = \kappa_M - 1.$$

The internal field in the material is then given by

$$B_{\text{int}} = \kappa_M B_{\text{applied}} = (\chi_M + 1)B_{\text{applied}}. \tag{30-14}$$

Equation 30-14 is most useful for paramagnetic and diamagnetic materials. In a ferromagnetic material, the relative permeability and susceptibility themselves depend on the applied field, and in fact on the past history of the material's exposure to magnetic fields (Fig. 30-38). This phenomenon of **hysteresis,** in which a ferromagnetic material "remembers" past fields, is what makes permanent magnets possible. Table 30-2 lists some magnetic susceptibilities for all three types of magnetic materials. Note the entry for superconductors, which are perfectly diamagnetic. Their susceptibility of -1 implies that $\kappa_M = 0$, showing that they completely exclude magnetic fields. We will explore the reasons for this in the next chapter.

▲ **TABLE 30-2** MAGNETIC SUSCEPTIBILITIES*

MATERIAL	MAGNETIC SUSCEPTIBILITY, χ_M
Diamagnetic materials:	
Copper	-9.6×10^{-6}
Lead	-1.6×10^{-5}
Mercury	-2.8×10^{-5}
Nitrogen (gas, 293 K)	-6.7×10^{-9}
Sodium chloride	-1.4×10^{-5}
Any superconductor	-1
Water	-9.1×10^{-6}
Paramagnetic materials	
Aluminum	2.1×10^{-5}
Chromium	3.1×10^{-4}
Oxygen (gas, 293 K)	1.9×10^{-6}
Oxygen (liquid, 90 K)	3.5×10^{-3}
Sodium	8.5×10^{-6}
Ferromagnetic materials (field and history dependent; maximum value listed)	
Iron (annealed)	5.5×10^{3}
Permalloy (55% Fe, 45% Ni)	2.5×10^{4}
Supermalloy (15.7% Fe, 79% Ni, 5.0% Mo; 0.30% Mn)	8.0×10^{5}
μ-metal (77% Ni, 16% Fe, 5% Cu, 2% Cr)	1.0×10^{5}

*At 300 K unless noted.

30-7 MAGNETIC MONOPOLES AND GAUSS'S LAW

Electric and magnetic fields, at least as we've studied them so far, have very different configurations. Electric fields begin and end on their sources—namely electric charges. Magnetic fields, in contrast, encircle their sources—namely *moving* electric charges. Magnetic fields generally form closed loops, while the electric fields we've encountered so far do not (Fig. 30-39).

Are there particles analogous to electric charges, from which magnetic field lines might originate? There might be. Symmetry arguments based on the existence of electric charge and the many similarities between electricity and magnetism have long suggested to physicists that such **magnetic monopoles**—isolated magnetic north and south poles—might exist. Furthermore, theories of elementary particles suggest that magnetic monopoles should have been created in the Big Bang event with which the universe began. However, the most recent version of the Big Bang theory suggests that monopoles should be spread so thinly as to make the chance of detecting one completely negligible.

If they existed, magnetic monopoles would be to the magnetic field what electric charges are to the electric field. Isolated monopoles would give rise to radial magnetic fields like the electric fields of point charges. Monopoles would experience forces in magnetic fields, forces given simply by the product of the monopole's "magnetic charge" and the magnetic field. And *moving* magnetic monopoles would produce electric fields—just as moving electric charges produce magnetic fields.

No one has yet found a magnetic monopole, however, although serious experimental searches continue. Every magnetic field we've ever seen has its origin in moving electric charge. The observed absence of magnetic monopoles means that the most fundamental magnetic field configuration is that of the dipole—in contrast to the simpler electric case, where the spherically symmetric point-charge field is possible. That's why Table 30-1 contains no magnetic entry under spherical geometry.

In the absence of magnetic monopoles, there is no place where magnetic field lines begin or end. That means there is no closed surface through which a nonzero net number of magnetic field lines emerges. In complete analogy with our discussion of Gauss's law in Chapter 24, we can say that the **magnetic flux**—mathematically, the surface integral $\oint \mathbf{B} \cdot d\mathbf{A}$—is zero for every closed surface. This statement is **Gauss's Law for magnetism:**

$$\oint \mathbf{B} \cdot d\mathbf{A} = 0. \qquad (30\text{-}15)$$

Equation 30-15 is another of the four fundamental laws of electromagnetism. We've now seen three of them: Gauss's law for electricity, Ampère's law (still incomplete because of its restriction to *steady* currents), and Gauss's law for magnetism. In the next chapter we'll encounter the fourth of these fundamental laws.

Although Gauss's law for magnetism has zero on its right-hand side, the law is far from being devoid of content. It says what we've already found to be the case: that magnetic fields must configure themselves so they have no beginnings or endings.

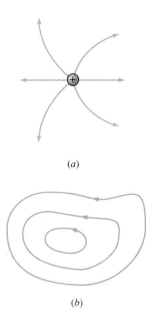

(a)

(b)

FIGURE 30-39 At this point in our study of electromagnetism, electric and magnetic field configurations are clearly distinguished. Electric fields (a) begin and end on electric charges, while magnetic fields (b) generally form closed loops.

CHAPTER SYNOPSIS

Summary

1. Magnetic fields arise from **moving electric charges.** The field at some point P arising from a steady electric current is described by the **Biot-Savart law:**

$$d\mathbf{B} = \frac{\mu_0}{4\pi} \frac{Id\boldsymbol{\ell} \times \hat{\mathbf{r}}}{r^2},$$

where $d\mathbf{B}$ is the contribution to the field from a current element flowing along the infinitesimal vector $d\boldsymbol{\ell}$ a distance r from the point P, and $\hat{\mathbf{r}}$ is a unit vector from the current element toward P. The field at P of an entire current distribution is found by integrating all its contributions $d\mathbf{B}$.

   Important special cases include a current loop, which at large distances produces a dipole field, and an infinite straight current, whose field encircles the current and decreases as $1/r$.

2. The field arising from a current exerts a magnetic force on nearby currents; as a result, parallel currents in the same direction attract and those in opposite directions repel.

3. **Ampère's law** relates the line integral of the magnetic field around an arbitrary closed loop to the current encircled by that loop:

$$\oint_{\text{loop}} \mathbf{B} \cdot d\boldsymbol{\ell} = \mu_0 I_{\text{encircled}}.$$

This form holds for all steady currents. It may be used to calculate the magnetic field in cases with sufficient symmetry, including line symmetry (straight wires), plane symmetry (current sheets), solenoids, and toroids. The Biot-Savart law follows logically from Ampère's law.

4. Individual elementary particles and orbiting atomic electrons constitute miniature current loops, which are responsible for magnetic effects in bulk matter. The **relative permeability,** κ_M, gives the ratio of the internal magnetic field in a material to the applied field; this relation is also described by the **magnetic susceptibility,** $\chi_M = \kappa_M - 1$.

   a. In **ferromagnetic** materials, atomic magnetic moments group into domains with net magnetic moments. These domains align with an applied field, greatly increasing the field within the material. Ferromagnetic materials retain a remanent magnetization even after the applied field is removed; this phenomenon accounts for permanent magnets. $\kappa_M \gg 1$ for ferromagnetic materials, although the internal field depends on both the applied field and the material's history.

   b. In **paramagnetic** materials, individual atomic magnetic moments become partially aligned with an applied field, but there is no cooperative interaction among the individual moments. $\kappa_M > 1$ for paramagnetic materials.

   c. Diamagnetic materials have no intrinsic magnetic moments. They have $\kappa_M < 1$ and are weakly repelled from magnets.

5. No **magnetic monopoles,** or magnetic analogs of electric charge, have ever been found. Magnetic fields—originating from moving electric charge—therefore form closed loops without beginnings or ends. In the absence of magnetic monopoles, **Gauss's law for magnetism** says that the flux of the magnetic field through any closed surface is zero:

$$\oint \mathbf{B} \cdot d\mathbf{A} = 0.$$

Terms You Should Understand

(Pairs are closely related terms whose distinction is important; number in parentheses is chapter section where term first appears.)

Biot-Savart law (30-1)
Ampère's law (30-3)
solenoid, toroid (30-5)
ferromagnetism, paramagnetism, diamagnetism (30-6)
permeability, susceptibility (30-6)
magnetic monopole (30-7)
Gauss's law for magnetism (30-7)

Symbols You Should Recognize

μ_0 (30-1)
$\oint \mathbf{B} \cdot d\boldsymbol{\ell}$ (30-3)
κ_M, χ_M (30-6)

Problems You Should Be Able to Solve

calculating magnetic fields of simple current distributions by integration using the Biot-Savart law (30-1)
evaluating forces between adjacent conductors (30-2)
evaluating magnetic fields using Ampère's law in situations with sufficient symmetry (30-4)
applying results derived in the text for the fields of straight wires, current loops, solenoids, and toroids (30-1–30-5)
relating internal and applied magnetic fields given magnetic susceptibility (30-6)

Limitations to Keep in Mind

The Biot-Savart law, and Ampère's law as expressed in this chapter, are exactly valid only for *steady* currents—those that never change with time.

QUESTIONS

1. In what two senses does a current loop behave like a magnetic dipole?

2. The electric field far from a pair of equal but opposite charges has the same configuration as the magnetic field far from a current loop. Yet inside the structure giving rise to the fields, the two point in opposite directions. Why?

3. The Biot-Savart law shows that the magnetic field of a current element decreases as $1/r^2$. Could you put together a complete circuit whose field exhibits this $1/r^2$ decrease? Why or why not?

4. Do currents in the same direction attract or repel? Explain.

5. If a current is passed through an unstretched spring, will the spring contract or expand? Explain.

6. Why is it advantageous to define the ampere in terms of magnetic force rather than a standard ammeter?

7. One way to heat a plasma is to initiate a large current through it. In response the plasma compresses or "pinches" in the direction perpendicular to the current, and this adiabatic compression heats the plasma. Explain the origin of this "pinch effect."

8. The field of a long, straight wire consists of circular field lines. Does Ampère's law hold for a square loop surrounding the wire? For a circular loop not concentric with the wire? See Fig. 30-40.

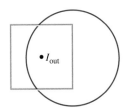

FIGURE 30-40   Question 8.

9. If the line integral around a closed loop is zero, does that mean the magnetic field on the loop is zero?

10. Must the integration path in Ampère's law coincide with a field line?

11. A short, straight length of wire has cylindrical symmetry. Why can't Ampère's law be used to find the magnetic field of a short wire? *Hint:* Think about steady currents and a short wire.

12. What must be going on inside Earth in order that it have a magnetic field?

13. Figure 30-41 shows some magnetic field lines associated with two parallel wires carrying equal currents perpendicular to the page. Are the currents in the same or opposite directions? How can you tell?

14. A solid cylinder and a hollow pipe of the same outer diameter carry the same current along their long dimensions, with the current distributed symmetrically about their

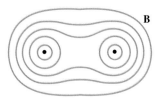

FIGURE 30-41   Question 13.

axes. How do the magnetic fields at their surfaces compare?

15. Why is a piece of iron attracted into a solenoid?

16. Would there be a net magnetic force on a piece of iron deep inside a long solenoid? Explain.

17. In what sense is a long solenoid the magnetic analog of a parallel-plate capacitor?

18. What would happen to the magnetic field inside a solenoid if you (a) doubled the solenoid length without changing the number of turns per unit length or (b) doubled the length without changing the total number of turns?

19. Identify *three* regions in the magnetic field of Fig. 30-42 where there *must* be a current. Which way is the current flowing in each region?

FIGURE 30-42   Question 19.

20. An unmagnetized piece of iron has no net magnetic dipole moment. Yet it is attracted to either pole of a bar magnet. Why?

21. How would you determine experimentally whether a substance was paramagnetic or diamagnetic?

22. Would permanent magnets be possible if the relative permeability of ferromagnetic materials were strictly constant? Explain.

23. Why do paramagnetic and ferromagnetic effects weaken with increasing temperature?

24. If magnetic monopoles existed, then a net motion of monopoles would constitute a "magnetic current." What sort of field should arise from such a "current?"

25. Figure 30-43 shows four different time-independent fields. Which are electric and which magnetic?

26. In the absence of magnetic monopoles, why do we even bother to write Gauss's law for magnetism? Does the law have any physical significance?

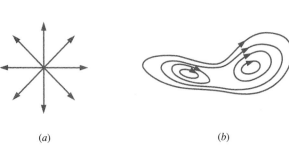

(a)   (b)

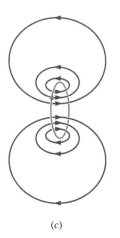

(c)

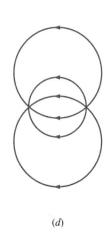

(d)

FIGURE 30-43 Question 25.

PROBLEMS

Section 30-1 The Biot-Savart Law

1. A wire carries 15 A. You form the wire into a single-turn circular loop with magnetic field 80 μT at the loop center. What is the loop radius?

2. A single-turn wire loop is 2.0 cm in diameter and carries a 650-mA current. Find the magnetic field strength (a) at the loop center and (b) on the loop axis, 20 cm from the center.

3. A 2.2-m-long wire carrying 3.5 A is wound into a tight, loop-shaped coil 5.0 cm in diameter. What is the magnetic field at its center?

4. What is the current in a long wire if the magnetic field strength 1.2 cm from the wire's axis is 67 μT?

5. Suppose Earth's magnetic field arose from a single loop of current at the outer edge of the planet's liquid core (core radius 3000 km). How large must the current be to give the observed magnetic dipole moment of 8.0×10^{22} A·m²?

6. Earth's magnetic dipole moment is 8.0×10^{22} A·m². What is the magnetic field strength on Earth's surface at either pole?

7. A single-turn current loop carrying 25 A produces a magnetic field of 3.5 nT at a point on its axis 50 cm from the loop center. What is the loop area, assuming the loop diameter is much less than 50 cm?

8. Two identical current loops are 10 cm in diameter and carry 20-A currents. They are placed 1.0 cm apart, as shown in Fig. 30-44. Find the magnetic field strength at the center of either loop when their currents are in (a) the same and (b) opposite directions.

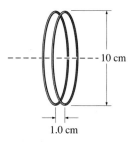

FIGURE 30-44 Problem 8.

9. You have a spool of thin wire that can handle a maximum current of 0.50 A. If you wind the wire into a loop-like coil 20 cm in diameter, how many turns should the coil have if the magnetic field at its center is to be 2.3 mT at this maximum current?

10. A single piece of wire is bent so that it includes a circular loop of radius a, as shown in Fig. 30-45. A current I flows in the direction shown. Find an expression for the magnetic field at the center of the loop.

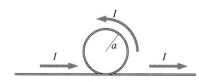

FIGURE 30-45 Problem 10.

11. Two long, parallel wires are 6.0 cm apart. One carries 5.0 A and the other 10 A, with both currents in the same direction. Where on a line perpendicular to both wires is the magnetic field zero?

12. Four long, parallel wires are located at the corners of a square 15 cm on a side. Each carries a current of 2.5 A, with the top two currents into the page in Fig. 30-46 and the bottom two out of the page. Find the magnetic field at the center of the square.

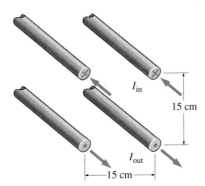

FIGURE 30-46 Problems 12, 23.

13. A power line carries a 500-A current toward magnetic north and is suspended 10 m above the ground. The horizontal component of Earth's magnetic field at the power line's latitude is 0.24 G. If a magnetic compass is placed on the ground directly below the power line, in what direction will it point?

14. An electron is moving at 3.1×10^6 m/s parallel to a 1.0-mm-diameter wire carrying 20 A. If the electron is 2.0 mm from the center of the wire, with its velocity in the same direction as the current, what are the magnitude and direction of the force it experiences?

15. Part of a long wire is bent into a semicircle of radius a, as shown in Fig. 30-47. A current I flows in the direction shown. Use the Biot-Savart law to find the magnetic field at the center of the semicircle (point P).

FIGURE 30-47 Problem 15.

16. Use the result given in Problem 59 to find the magnetic field strength at the center of a square loop of side a carrying current I.

17. Figure 30-48 shows a conducting loop formed from concentric semicircles of radii a and b. If the loop carries a current I as shown, find the magnetic field at point P, the common center.

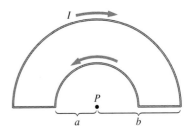

FIGURE 30-48 Problem 17.

Section 30-2 The Magnetic Force Between Two Conductors

18. In standard household wiring, parallel wires about 1 cm apart carry currents around 15 A. What is the magnitude of the force per unit length between such wires?

19. It would take a rather large apparatus to implement the definition of the ampere given in Section 30-2. Suppose you wanted to use a smaller apparatus, with wires 50 cm long separated by 2.0 cm. What force would correspond to a current of 1 A?

20. Two parallel copper rods supply power to a high-energy experiment, carrying the same current in opposite directions. The rods are held 8.0 cm apart by insulating blocks mounted every 1.5 m. If each block can tolerate a maximum tension force of 200 N, what is the maximum allowable current?

21. The structure shown in Fig. 30-49 is made from conducting rods. The upper horizontal rod is free to slide vertically on the uprights, while maintaining electrical contact with them. The upper rod has mass 22 g and length 95 cm. A battery connected across the insulating gap at the bottom of the left-hand upright drives a 66-A current through the structure. At what height h will the upper wire be in equilibrium?

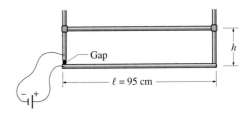

FIGURE 30-49 Problem 21.

22. Three parallel wires 4.6 m long each carry 20 A in the same direction. They are spaced at the vertices of an equilateral triangle 3.5 cm on a side. Find the magnitude of the force on each wire.

23. The wires in Fig. 30-46 carry 2.5-A currents in the directions indicated. Find the net force per unit length on the wire at lower left.

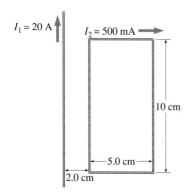

FIGURE 30-50 Problem 24.

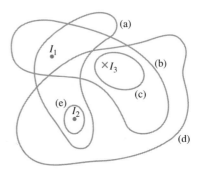

FIGURE 30-52 Problem 28.

24. A long, straight wire carries 20 A. A 5.0-cm by 10-cm rectangular wire loop carrying 500 mA is located 2.0 cm from the wire, as shown in Fig. 30-50. Find the net magnetic force on the loop.

25. A solenoid 10 cm in diameter is made with 2.1-mm-diameter copper wire wound so tightly that adjacent turns touch, separated only by enamel insulation of negligible thickness. The solenoid carries a 28-A current. In the long, straight wire approximation, what is the net force between two adjacent turns of the solenoid?

26. A long, flat conducting ribbon of width w is parallel to a long straight wire, with its near edge a distance a from the wire (Fig. 30-51). Wire and ribbon carry the same current I; this current is distributed uniformly over the ribbon. Use integration to show that the force per unit length between the two has magnitude $\dfrac{\mu_0 I^2}{2\pi w}\ln\left(\dfrac{a+w}{a}\right)$.

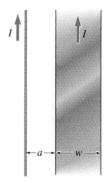

FIGURE 30-51 Problem 26.

Section 30-3 Ampère's Law

27. The line integral of the magnetic field on a closed path surrounding a wire has the value 8.8 μT·m. What is the current in the wire?

28. In Fig. 30-52, $I_1 = 2$ A flowing out of the page; $I_2 = 1$ A, also out of the page, and $I_3 = 2$ A, into the page. What is the line integral of the magnetic field taken counterclockwise around each loop shown?

29. The magnetic field shown in Fig. 30-53 has uniform magnitude 75 μT, but its direction reverses abruptly. How much current is encircled by the rectangular loop shown?

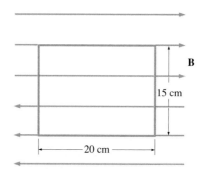

FIGURE 30-53 Problems 29, 30.

30. On the nighttime side of Earth, the solar wind draws some of the planet's magnetic field into a long "magnetotail," where field lines on opposite sides of the equatorial plane are straight but antiparallel (as in the configuration of Fig. 30-53). If the field strength is 2×10^{-4} G, what is the current per unit length flowing in the magnetotail?

31. Figure 30-54 shows a magnetic field pointing in the x direction. Its strength, however, varies with position in the

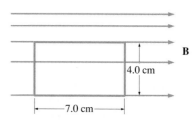

FIGURE 30-54 Problems 31, 68.

y direction. At the top and bottom of the rectangular loop shown the field strengths are 3.4 μT and 1.2 μT, respectively. How much current flows through the area encircled by the loop?

Section 30-4 Using Ampère's Law

32. (a) What is the magnetic field strength 0.10 mm from the axis of a 1.0-mm-diameter wire carrying 5.0 A distributed uniformly over its cross section? (b) What is the field strength at the surface of the wire?

33. A solid wire 2.1 mm in diameter carries a 10-A current with uniform current density. What is the magnetic field strength (a) at the axis of the wire, (b) 0.20 mm from the axis, (c) at the surface of the wire, and (d) 4.0 mm from the wire axis?

34. Show that Equations 30-9 and 30-10 give the same results when evaluated at the surface of the wire.

35. A long conducting rod of radius R carries a nonuniform current density given by $J = J_0 r/R$, where J_0 is a constant and r is the radial distance from the rod's axis. Find expressions for the magnetic field strength (a) inside and (b) outside the rod.

36. A long, hollow conducting pipe of radius R carries a uniform current I along the pipe, as shown in Fig. 30-55. Use Ampère's law to find the magnetic field strength (a) inside and (b) outside the pipe.

FIGURE 30-55 Problem 36.

37. Typically, cylindrical wires made from yttrium-barium-copper-oxide superconductor can carry a maximum current density of 6.0 MA/m² at a temperature of 77 K, as long as magnetic field at the conductor surface does not exceed 10 mT. Suppose such a wire is to carry the maximum current density. (a) At what wire diameter would the surface magnetic field equal the 10-mT limit? (b) Is this a maximum or minimum value for the diameter if the field is not to exceed the limit? (c) What current would a wire with this diameter carry?

38. A long, hollow conducting pipe of radius R and length ℓ carries a uniform current I flowing around the pipe, as shown in Fig. 30-56. Find expressions for the magnetic field (a) inside and (b) outside the pipe. *Hint:* What configuration does this pipe resemble?

39. A copper ribbon 1.0 cm wide and 0.15 mm thick is rated for a maximum safe current density of 8.8×10^6 A/m². What is the maximum magnetic field strength achievable at the surface of this ribbon?

FIGURE 30-56 Problem 38.

40. Two large, flat conducting plates lie parallel to the x-y plane. They carry equal currents, one in the $+x$ and the other in the $-x$ direction. In each plate the current per meter of width in the y direction is J_s. Find the magnetic field strength (a) between and (b) outside the plates.

41. Repeat the preceding problem for the case when one current flows in the $+x$ direction and the other in the $+y$ direction.

42. The coaxial cable shown in Fig. 30-57 consists of a solid inner conductor of radius a and a hollow outer conductor of inner radius b and thickness c. The two carry equal but opposite currents I, uniformly distributed. Find expressions for the magnetic field strength as a function of radial position r (a) within the inner conductor, (b) between the inner and outer conductors, and (c) beyond the outer conductor.

FIGURE 30-57 Problems 42, 63.

43. A hollow conducting pipe of inner radius a and outer radius b carries a current I parallel to its axis and distributed uniformly through the pipe material (Fig. 30-58). Find expressions for the magnetic field for (a) $r < a$, (b) $a < r < b$, and (c) $r > b$, where r is the radial distance from the pipe axis.

FIGURE 30-58 Problem 43.

44. A conducting slab extends infinitely in the x and y directions and has thickness h in the z direction. It carries a uniform current density $\mathbf{J} = J\hat{\imath}$. Find the magnetic field strength (a) inside and (b) outside the slab, as functions of the distance z from the center plane of the slab.

Section 30-5 Solenoids and Toroids

45. A superconducting solenoid has 3300 turns per meter and can carry a maximum current of 4.1 kA. What is the magnetic field strength in the solenoid?

46. A solenoid used in a plasma physics experiment is 10 cm in diameter, 1.0 m long, and carries a 35-A current to produce a 100-mT magnetic field. (a) How many turns are in the solenoid? (b) If the solenoid resistance is 2.7 Ω, how much power does it dissipate?

47. You have 10 m of 0.50-mm-diameter copper wire and a battery capable of passing 15 A through the wire. What magnetic field strengths could you obtain (a) inside a 2.0-cm-diameter solenoid wound with the wire as closely spaced as possible and (b) at the center of a single circular loop made from the wire?

48. A toroidal coil of inner radius 15 cm and outer radius 17 cm is wound from 1200 turns of wire. What are (a) the minimum and (b) the maximum magnetic field strengths within the toroid when it carries a 10-A current?

49. A toroidal fusion reactor requires a magnetic field that varies by no more than 10% from its central value of 1.5 T. If the minor radius of the toroidal coil producing this field is 30 cm, what is the minimum value for the major radius of the device?

50. A long solenoid with n turns per unit length carries a current I. The current returns to its driving battery along a wire of radius R that passes through the solenoid, along its axis. Find expressions for (a) the magnetic field strength at the surface of the wire and (b) the angle the field at the wire surface makes with the solenoid axis.

51. We noted that there is a nonzero magnetic field component outside a solenoid, encircling the device, associated with the component of current flow parallel to the solenoid axis. For a long solenoid of radius R, find an expression for the ratio of this external encircling field just outside the solenoid to the field inside, and show explicitly that this ratio tends to zero as the number of turns per unit length becomes large.

52. Derive Equation 30-12 for the field of a solenoid by considering the solenoid to be made of a large number of adjacent current loops. Use Equation 30-3 for the field of a current loop, and integrate over all loops.

Section 30-6 Magnetic Matter

53. When a sample of a certain substance is placed in a 250.0-mT magnetic field, the field inside the sample is 249.6 mT. Find the magnetic susceptibility of the substance. Is it ferromagnetic, paramagnetic, or diamagnetic?

54. A container of liquid oxygen at 90 K is placed in a 500.0-G magnetic field. What is the field strength within the liquid oxygen?

55. A ferromagnetic material is placed in a 2.5-G magnetic field and the field within the material is determined to be 1.8 T. What is the magnetic susceptibility of this material?

56. Figure 30-59 shows the hysteresis curve for the ferromagnetic alloy Alnico V, commonly used in permanent magnets. Use the graph to find approximate values for the maximum field strength obtainable inside this material, (a) in the absence and (b) in the presence of an externally applied field.

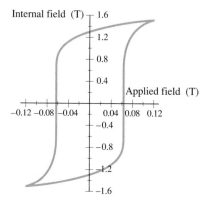

FIGURE 30-59 Problem 56.

Paired Problems

(Both problems in a pair involve the same principles and techniques. If you can get the first problem, you should be able to solve the second one.)

57. Two concentric, coplanar circular current loops have radii a and $2a$. If the magnetic field is zero at their common center, how does the current in the outer loop compare with that in the inner loop?

58. A thin conducting washer of inner radius a and outer radius $2a$ carries a current I distributed uniformly with radial position, as suggested in Fig. 30-60. Find an expression for the magnetic field strength at its center.

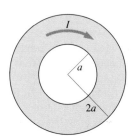

FIGURE 30-60 Problem 58.

59. Figure 30-61 shows a wire of length ℓ carrying current fed by other wires that are not shown. Point A lies on the perpendicular bisector, a distance y from the wire. Adapt the calculation of Example 30-2 to show that the magnetic field at A due to the straight wire alone has magnitude $\dfrac{\mu_0 I \ell}{2\pi y \sqrt{\ell^2 + 4y^2}}$. What is the field direction?

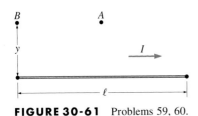

FIGURE 30-61   Problems 59, 60.

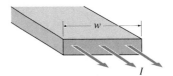

FIGURE 30-62   Problem 66.

60. Point B in Fig. 30-62 lies a distance y perpendicular to the end of the wire. Show that the magnetic field at B has magnitude $\dfrac{\mu_0 I \ell}{4\pi y \sqrt{\ell^2 + y^2}}$. What is the field direction?

61. The largest lightning strikes have peak currents around 250 kA, flowing in essentially cylindrical channels of ionized air. How far from such a flash would the resulting magnetic field be equal to Earth's magnetic field strength, about 50 μT?

62. A particle-beam weapon being tested for ballistic missile defense delivers a 2.5-cm-diameter electron beam carrying a current of 10 kA at an altitude where Earth's magnetic field is 30 μT. How close to the beam would an adversary's surveillance apparatus need to be in order to detect the beam by observing a 1% change in the local magnetic field strength? Assume the beam's magnetic field is parallel to Earth's at the point of detection.

63. A coaxial cable like that shown in Fig. 30-57 consists of a 1.0-mm-diameter inner conductor and an outer conductor of inner diameter 1.0 cm and 0.20 mm thickness. A 100-mA current flows down the center conductor and back along the outer conductor. Find the magnetic field strength (a) 0.10 mm, (b) 5.0 mm, and (c) 2.0 cm from the cable axis.

64. Repeat the preceding problem if the current in the outer conductor only is increased to 200 mA.

Supplementary Problems

65. A circular wire loop of radius 15 cm and negligible thickness carries a 2.0-A current. Use suitable approximations to find the magnetic field of this loop (a) in the loop plane, 1.0 mm outside the loop, and (b) on the loop axis, 3.0 m from the loop center.

66. A long, flat conducting bar of width w carries a total current I distributed uniformly, as shown in Fig. 30-62. Use suitable approximations to write expressions for the magnetic field strength (a) near the conductor surface ($r \ll w$) and (b) very far from the conductor ($r \gg w$).

67. (a) Use the result of Problem 59 to find an expression for the magnetic field strength on the axis of a square loop of

side a carrying current I. (b) Show that your result reduces the field of a dipole—Equation 30-4—in the limit $x \gg a$.

68. The magnetic field in Fig. 30-54 is given by $\mathbf{B} = by\hat{\imath}$, where $b = 55$ μT/m and y is the vertical coordinate in Fig. 30-54, measured in meters. Find the current density in the region.

69. A wide, flat conducting spring of spring constant $k = 20$ N/m and negligible mass consists of two 6.0-cm-diameter turns, as shown in Fig. 30-63. In its unstretched configuration the coils are nearly touching. A 10-g mass is hung from the spring, and at the same time a current I is passed through it. The spring stretches 2.0 mm. Find I, assuming the coils remain close enough to be treated as parallel wires.

FIGURE 30-63   Problem 69.

70. A solid conducting wire of radius R runs parallel to the z axis and carries a current density given by $\mathbf{J} = J_0(1 - r/R)\hat{\mathbf{k}}$, where J_0 is a constant and r the radial distance from the wire axis. Find expressions for (a) the total current in the wire, (b) the magnetic field strength for $r > R$, and (c) the magnetic field strength for $r < R$.

71. A disk of radius a carries a uniform surface charge density σ, and is rotating with angular speed ω about the central axis perpendicular to the disk. Show that the magnetic field at the disk's center is $\frac{1}{2}\mu_0\sigma\omega a$.

72. Calculate the magnetic field of an infinite current sheet by considering the sheet to be made up of infinitesimal line currents, as suggested in Fig. 30-23, and integrating the fields of these line currents.

73. Work Example 30-2 by expressing all variables in terms of the angle θ and integrating over the appropriate range in θ.

ELECTROMAGNETIC INDUCTION

Generators use electromagnetic induction to convert mechanical energy into electral energy. These generators at Arizona's Glen Canyon Dam produce 1300 MW of electric power from the energy of falling water.

A ll the electric and magnetic fields we encountered in the previous chapters had their ultimate origins in electric charge, either stationary or moving. We stressed a relation between electricity and magnetism, whereby electric charge gives rise to and interacts with both the electric field and the magnetic field. The remainder of our study of electromagnetism is devoted to a much more intimate relation between electricity and magnetism, a relation in which the fields themselves interact directly. This interaction forms the basis of new electromagnetic technologies, leads toward an understanding of the nature of light, and points the way to the theory of relativity.

31-1 INDUCED CURRENTS

In 1831, the English scientist Michael Faraday and the American Joseph Henry independently carried out experiments in which electric currents arose in circuits subjected to changing magnetic fields. Figure 31-1 shows one such exper-

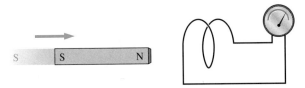

FIGURE 31-1 When a magnet is moved near a closed circuit, current flows in the circuit.

iment, in which a magnet is moved in the presence of a circuit consisting of a loop of wire and an ammeter. There is no battery or other obvious source of emf in the circuit. As long as the magnet is held still, there is no current. But while the magnet is moving, the ammeter registers a current—an **induced current.** If we modify the experiment, holding the magnet still but moving the coil, we again observe an induced current. Apparently only the relative motion matters. If we replace the magnet with another circuit in which a battery drives a current, and move the two circuits relative to each other, we get an induced current in the circuit without the battery (Fig. 31-2). If we hold the two circuits still we get no induced current. But now if we close—or open—a switch in the circuit with the battery, we find that the ammeter indicates a momentary induced current (Fig. 31-3).

The one common feature in all these experiments is a *changing magnetic field*. It does not matter whether the field changes because a magnet is moved, or because a circuit is moved near a magnet, or because the current giving rise to the field changes. In each case, an induced current appears in a circuit subjected to a changing magnetic field. We are observing here a new phenomenon—**electromagnetic induction**—whereby electrical effects arise from *changing* magnetic fields.

31-2 FARADAY'S LAW

We know from Chapter 27 that a source of electromotive force—something like a battery that supplies energy to electric charges—is needed to establish a current in a circuit. When an induced current flows in a circuit, an emf is similarly present. This **induced emf** is usually not localized at one point in the circuit, as in a battery, but may be spread throughout the conductors making up the circuit.

Experimentally, we find that the induced emf in a circuit depends on the rate of change of magnetic flux through the circuit. Before quantifying this relationship, we show how to calculate magnetic flux.

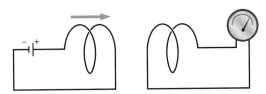

FIGURE 31-2 An induced current also arises when the magnet is replaced by a current-carrying circuit.

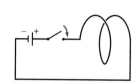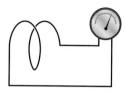

FIGURE 31-3 A current is also induced—even in the absence of any motion—when the current in an adjacent circuit changes.

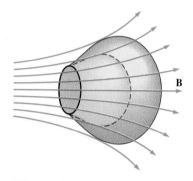

FIGURE 31-4 The flux through a circuit is the same through *any* open surface bounded by the circuit. Three such surfaces are shown here; they include the flat circular surface in the plane of the circular loop circuit, and the two bubble-like surfaces. If the fluxes through any two of these surfaces were not the same, then the flux through a *closed* surface bounded by the two would not be zero, in violation of Gauss's law for magnetism.

Magnetic Flux

We define magnetic flux in analogy with Chapter 24's definition of electric flux as the integral of the field over a surface:

$$\phi_B = \int \mathbf{B} \cdot d\mathbf{A}, \tag{31-1}$$

where $d\mathbf{A}$ is an infinitesimal vector normal to the surface. As with electric flux, magnetic flux is proportional to the number of field lines passing through a surface. With electromagnetic induction, we're interested in the flux through an *open* surface bounded by the circuit in question. Because the magnetic flux through any *closed* surface is zero, we can in fact use *any* open surface bounded by our circuit to calculate the flux through the circuit (see Fig. 31-4).

For a flat surface in a uniform magnetic field, Equation 31-1 reduces to the simple expression $\phi_B = BA\cos\theta$, with θ the angle between the field and the normal to the surface. For curved surfaces and nonuniform fields, we add—that is, integrate—all the infinitesimal flux elements $d\phi_B = \mathbf{B} \cdot d\mathbf{A}$ over the surface to get the total flux. Note that the units of magnetic flux are T·m².

● EXAMPLE 31-1 THE MAGNETIC FLUX THROUGH A SOLENOID

A solenoid of circular cross section has radius R, consists of n turns per unit length, and carries a current I. What is the magnetic flux through each turn of the solenoid?

Solution
Away from the solenoid ends, the field is uniform and parallel to the solenoid axis, with magnitude given by Equation 30-12:

$$B = \mu_0 nI.$$

A flat surface bounded by one turn of the solenoid lies at right angles to this uniform field, so the flux is simply the product of the magnetic field and the area:

$$\phi_B = \int \mathbf{B} \cdot d\mathbf{A} = BA = \mu_0 nI\pi R^2.$$

We are being a little loose here in thinking of a single turn of the solenoid as a closed loop, but if the solenoid is tightly wound this is an excellent approximation. We will be concerned frequently with the flux through a multiturn coil, and in calculating this flux it is convenient to view each turn as an individual loop. The flux through an N-turn coil in a uniform magnetic field is just N times the flux through each turn.

EXERCISE A rectangular wire loop measuring 10 cm by 15 cm is oriented so the normal to the loop makes a 30° angle with a uniform 50-mT magnetic field. Find the magnetic flux through the loop.

Answer: 6.5×10^{-4} T·m²

Some problems similar to Example 31-1: 3, 4 ●

● EXAMPLE 31-2 MAGNETIC FLUX IN A NONUNIFORM FIELD

A long, straight wire carries a current I. A rectangular wire loop of dimensions ℓ by w lies with its closest edge a distance a from the wire, as shown in Fig. 31-5. What is the magnetic flux through the loop?

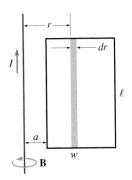

FIGURE 31-5 A rectangular loop in the magnetic field of a long wire. The field to the right of the wire points into the plane of the page, and its magnitude drops inversely with distance from the wire. Area elements for the flux calculation are strips of length ℓ and infinitesimal width dr.

Solution

The magnetic field of the wire is given by Equation 30-9:

$$B = \frac{\mu_0 I}{2\pi r},$$

where r is the distance from the wire. At the site of the loop, this field points straight into the page, perpendicular to the plane of the loop. However, the field varies with distance from the straight wire, so we cannot simply multiply the field by the loop area to get the flux. Instead, we divide the loop into thin strips of width dr and area $dA = \ell\, dr$, as shown in Fig. 31-5. With the field at right angles to each strip, $\mathbf{B} \cdot d\mathbf{A} = B\, dA$, and the flux through any strip is

$$d\phi = B\, dA = B\ell\, dr = \frac{\mu_0 I}{2\pi r} \ell\, dr.$$

Then the total flux through the loop is the integral over all such strips contained within the loop, that is, over all strips between $r = a$ and $r = a + w$:

$$\phi = \int_a^{a+w} \frac{\mu_0 I}{2\pi r} \ell\, dr = \frac{\mu_0 I \ell}{2\pi} \int_a^{a+w} \frac{dr}{r} = \frac{\mu_0 I \ell}{2\pi} \ln r \Big|_a^{a+w}$$

$$= \frac{\mu_0 I \ell}{2\pi} \ln\left(\frac{a + w}{a}\right).$$

EXERCISE A nonuniform magnetic field points in the z direction. The field strength is independent of y but varies linearly from $B = 0$ at $x = 0$ to $B = 2.0$ T at $x = 1.0$ m. A square wire loop 45 cm on a side lies in the x-y plane with one corner at the origin and two sides extending along the positive x and y axes. Find the magnetic flux through this loop.

Answer: 91 mT·m²

Some problems similar to Example 31-2: 45, 46 ●

Flux and Induced EMF

Having quantified the notion of magnetic flux, we are now ready to state rigorously the experimental fact that changing magnetic flux induces an emf in a circuit. Our statement is a special case of **Faraday's law of induction,** which constitutes another of the four basic laws of electromagnetism:

> **The induced emf in a circuit is proportional to the rate of change of magnetic flux through any surface bounded by that circuit.**

This statement is a special case; we will later broaden its scope to include situations where no circuit is present. In SI units the proportionality constant between emf and rate of change of flux is just -1, so mathematically Faraday's law is

$$\mathcal{E} = -\frac{d\phi_B}{dt}, \tag{31-2}$$

where $\mathcal{E}$ is the emf induced in a circuit and ϕ_B the magnetic flux through any surface bounded by that circuit. Problem 1 will help convince you that the units of the rate of change of magnetic flux are indeed volts, the same as the units of emf. The minus sign in Equation 31-2 is essential, and we will soon have a great deal more to say about it. We stress that Faraday's law, Equation 31-2, applies whenever the magnetic flux through a circuit changes. We could change that flux by moving a magnet near the circuit or the circuit near a magnet. Since the flux $\mathbf{B} \cdot \mathbf{A} = BA\cos\theta$ depends on the field strength B, the area A, and the angle θ between the field and the normal to the area, we could also change the flux by changing any of these three quantities. The following examples explore these possibilities.

● EXAMPLE 31-3 A CHANGING MAGNETIC FIELD

A wire loop of radius 10 cm has a resistance of 2.0 Ω. The loop is at right angles to a uniform magnetic field $\mathbf{B}$, as shown in Fig. 31-6. The field strength is increasing at 0.10 tesla/second. What is the magnitude of the induced current in the loop?

Solution

To find the induced current, we need to know the induced emf. Faraday's law tells us that the induced emf is related to the rate of change of magnetic flux through the circuit. With a uniform field at right angles to the loop, the flux is just the field strength times the loop area, or

$$\phi_B = \int \mathbf{B} \cdot d\mathbf{A} = \pi r^2 B.$$

Even though the magnetic field is changing with *time,* at any given instant it is uniform in *space,* which is why the integration was trivial.

We don't know B, but this doesn't matter because we are really interested in the *rate of change* of the flux, not in the flux itself. With the loop area constant, the rate of change of flux is

$$\frac{d\phi_B}{dt} = \pi r^2 \frac{dB}{dt} = (\pi)(0.10 \text{ m})^2(0.10 \text{ T/s}) = 3.14\times10^{-3} \text{ V}.$$

By Faraday's law, this is the magnitude of the induced emf. We then calculate the current using Ohm's law, as we would for any emf:

$$I = \frac{\mathscr{E}}{R} = \frac{3.14\times10^{-3} \text{ V}}{2.0 \text{ }\Omega} = 1.6\times10^{-3} \text{ A} = 1.6 \text{ mA}.$$

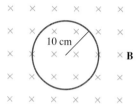

FIGURE 31-6 A circular conducting loop at right angles to a uniform magnetic field (Example 31-3).

TIP It's the Change That Counts Faraday's law relates induced emf to the *rate of change* of magnetic flux. The actual value of the flux—or of the magnetic field if that is what's changing—doesn't matter. You may be troubled by induction problems where the value of the field is not given. You don't need it; what you do need or may be asked to find is the *rate of change* of the field.

EXERCISE A square conducting loop 25 cm on a side lies at right angles to a uniform magnetic field. The loop's resistance is 8.0 Ω, and it carries a current of 14 mA. At what rate is the magnetic field changing?

Answer: 1.8 T/s

Some problems similar to Example 31-3: 4–8 ●

● EXAMPLE 31-4 A CHANGING AREA

A circuit consists of two parallel conducting rails a distance ℓ apart connected at one end by a resistance R. A conducting bar slides along the rails. The whole circuit is in a constant, uniform magnetic field $\mathbf{B}$ at right angles to the plane of the circuit, as shown in Fig. 31-7. The bar is pulled to the right with constant speed v. What is the current in the circuit? Assume the bar and rails are ideal conductors, so the total circuit resistance is R.

Solution

Here the current is driven by an induced emf arising from the change in magnetic flux that occurs as the circuit area increases. We determine the emf using Faraday's law,

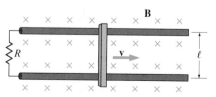

FIGURE 31-7 Pulling the bar to the right increases the circuit area, thereby increasing the magnetic flux and inducing an emf that drives a current (Example 31-4).

$\mathscr{E} = -d\phi_B/dt$. The circuit area is the rail spacing ℓ times the distance x from resistor to bar. With a uniform field perpendicular to the circuit, the flux integral $\int \mathbf{B} \cdot d\mathbf{A}$ reduces to the product of field strength with the area:

$$\phi_B = \int \mathbf{B} \cdot d\mathbf{A} = BA = B\ell x.$$

Never mind that we don't know x—we do know its rate of change and that is all we need. With the field and the rail spacing constant, the rate of change of flux is

$$\frac{d\phi_B}{dt} = B\ell \frac{dx}{dt} = B\ell v,$$

since dx/dt is just the bar velocity v. Faraday's law tells us that the magnitude of the induced emf is equal to the rate of change of flux, so

$$|\mathscr{E}| = B\ell v.$$

This emf drives a current I around the circuit:

$$|I| = \frac{|\mathscr{E}|}{R} = \frac{B\ell v}{R}.$$

EXERCISE A square wire loop 25 cm on a side moves into a region containing a uniform 1.2-T magnetic field oriented at right angles to the loop, as shown in Fig. 31-8. The loop wire is essentially resistanceless, but inserted in the loop is a 3-V flashlight bulb. How fast must the loop move to keep the bulb lit at its normal brightness?

Answer: 10 m/s

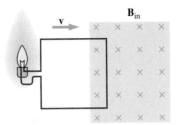

FIGURE 31-8 How fast must the loop move to light the 3-V bulb?

Some problems similar to Example 31-4: 9, 10, 16, 27, 35, 36   ●

● **EXAMPLE 31-5** A CHANGING ORIENTATION

A circular wire loop of radius a and resistance R is initially perpendicular to a constant, uniform magnetic field B. The loop rotates with angular velocity ω about an axis through a diameter, as shown in Fig. 31-9. What is the current in the loop?

Solution
Again, we must find the rate of change of magnetic flux, from which we can get the induced emf and then the current. Here the field is uniform over the area, but the orientation of the field relative to the area is changing. The definition of magnetic flux contains a dot product to account for this orientation. Evaluating the flux, we have

$$\phi_B = \int \mathbf{B} \cdot d\mathbf{A} = \int B \, dA \cos\theta = B\cos\theta \int dA = \pi a^2 B \cos\theta,$$

where θ is the angle between the field and a perpendicular to the loop area. Here we could take the field magnitude B outside the integral because the field is uniform over the loop. And even though the orientation between loop and field changes with *time*, it does not change with *position* at a given instant, so we could also take $\cos\theta$ outside the integral.

The changing orientation is described by giving θ as a function of time. Since the loop rotates with constant angular velocity ω, we can write simply $\theta = \omega t$, where we take the zero of time when $\theta = 0$. Then the flux is

$$\phi_B = \pi a^2 B \cos\omega t,$$

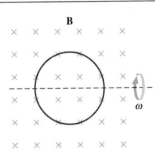

FIGURE 31-9 A wire loop rotating in a uniform magnetic field. The flux changes because of the changing orientation of the loop relative to the field, thereby inducing an emf that drives a current around the loop.

so

$$\frac{d\phi_B}{dt} = \pi a^2 B \frac{d}{dt}(\cos\omega t) = -\pi a^2 B\omega \sin\omega t.$$

By Faraday's law, the emf is then

$$\mathscr{E} = -\frac{d\phi_B}{dt} = \pi a^2 B\omega \sin\omega t,$$

giving a current

$$I = \frac{\mathscr{E}}{R} = \frac{\pi a^2 B\omega}{R} \sin\omega t.$$

(Check the units!)

Unlike the current in the previous two examples, this one changes with time. Its sinusoidal time dependence is in fact just like that of standard alternating current used for electric power—and with good reason: our rotating loop constitutes a simple alternating-current generator. Sinusoidally varying emf's and currents occur whenever conducting loops are rotated in uniform magnetic fields.

TIP   Peak Values   You'll often be asked for the peak value of voltage or current when either quantity is changing with time (more on this in Chapter 33). You've derived a formula like $\mathcal{E} = \pi a^2 B\omega \sin \omega t$ in Example 31-5, but you don't have a value for the time t.

It doesn't matter! The peak value occurs when $\sin \omega t = 1$, so just replace $\sin \omega t$ by 1. In Example 31-5, for example, the peak emf is just $\mathcal{E}_p = \pi a^2 B\omega$.

E X E R C I S E   Take the radius $a = 11$ cm in Example 31-5 and the magnetic field strength $B = 0.63$ T. With what angular speed ω must the loop rotate in order to produce a peak emf of 6.0 V?

Answer: 250 s^{-1}

Some problems similar to Example: 31-5:   12, 60, 66

You might wonder about the direction of the induced emf and current in these examples since we were concerned only with magnitudes. Although that direction follows mathematically from the minus sign in Faraday's law, we will find in the next section a physical justification for the minus sign that makes it clear how to determine the direction of induced emf's and currents.

31-3 INDUCTION AND THE CONSERVATION OF ENERGY

Move a bar magnet toward a wire loop, as shown in Fig. 31-10. Without the loop present, it would take no work to move the magnet horizontally at constant velocity. But with the loop present, moving the magnet induces a current in the loop. Electrical energy is dissipated in the loop's resistance, heating the loop. Where does the energy come from?

There's only one source for the energy—the agent moving the magnet. As you move the magnet, you must be doing positive work. Otherwise you would get energy—ultimately heating the loop—for nothing. So you must have to push the magnet against some force. What force? A magnetic force, caused by the interaction of the magnet with the magnetic field produced by the induced current. Think about this! The induced current, like any current, produces a magnetic field. The magnet experiences a force in this field, and you must do work to move the magnet through that field. With the magnetic field as intermediary, the work you do ultimately ends up heating the loop.

FIGURE 31-10   As the bar magnet moves toward the conducting loop, the changing magnetic flux through the loop induces an emf that drives a current around the loop. Conservation of energy requires that the magnetic field produced by this current *oppose* the motion of the magnet, so the agent moving the magnet does work that ends up heating the loop.

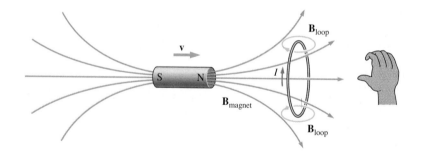

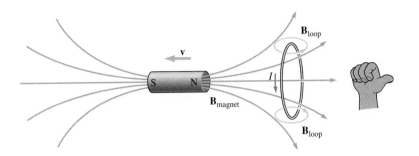

FIGURE 31-11 When the bar magnet is pulled away from the loop, the direction of the induced current is such that the magnetic force opposes the magnet's withdrawal.

As you push the magnet toward the loop, the force the loop's field exerts on the magnet must be repulsive in order that you do positive work on the magnet. In Fig. 31-10 the bar magnet's north pole is toward the loop. To provide a repulsive force the loop must therefore present a north pole to the approaching magnet; that is, field lines must emerge from the loop's interior pointing to the left. What current direction will provide such a field? Application of the right-hand rule gives the answer: Wrap the fingers of your right hand around the loop in the direction of the current, and your right thumb points in the direction of the field in the loop's interior. Thus the current direction is as shown in Fig. 31-10.

This analysis leading to the direction of the induced current was based on one simple principle: conservation of energy. For electromagnetic induction, this universal principle requires the following:

> **The direction of the induced emf and current is such that the magnetic field created by the induced current opposes the change in magnetic flux.**

This statement, called **Lenz's law,** is represented mathematically in the minus sign on the right-hand side of Faraday's law (Equation 31-2).

What happens, for example, if you pull the bar magnet away from the loop in Fig. 31-10? Now the loop must present a south pole to the receding magnet, creating an attractive force opposing the magnet's withdrawal (Fig. 31-11). The loop current must be opposite its direction in Fig. 31-10, as again you do work to overcome the magnetic force.

Motional EMF and Lenz's Law

When a conductor moves through a magnetic field, we can understand the origin of the induced emf in terms of the magnetic force on charge carriers in the wire; this emf is called **motional emf.** In the case of motional emf we can show explicitly that Lenz's law requires energy conservation.

Consider a square conducting loop of side ℓ and resistance R being pulled with constant speed v out of a uniform magnetic field **B,** as shown in Fig. 31-12. We will show that the rate of joule heating in the loop is equal to the rate at which the agent pulling the loop does work.

In Chapter 29 we found that the magnetic force on a charge is given by $\mathbf{F} = q\mathbf{v} \times \mathbf{B}.$ Pulling the loop to the right moves its free electrons through the magnetic field; the magnetic force $q\mathbf{v} \times \mathbf{B}$ on these electrons is downward in Fig. 31-12 (opposite $\mathbf{v} \times \mathbf{B}$ since electrons are negative). The resulting downward motion of the negative electrons in the left-hand side of the loop constitutes

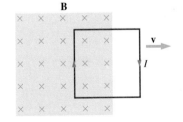

FIGURE 31-12 A conducting loop being withdrawn from a magnetic field. The magnetic force $q\mathbf{v} \times \mathbf{B}$ on charge carriers in the left side of the loop drives a current clockwise around the loop. The current direction also follows from Lenz's law: The magnetic field produced by the loop current acts to oppose the decrease in flux through the loop, and therefore points in the same direction as the original field. This requires a clockwise loop current.

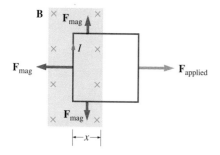

FIGURE 31-13 Forces on the loop. The total magnetic force is that on the left side alone, and the agent pulling the loop must exert an equal but opposite force to maintain constant velocity.

an upward current. This current continues clockwise around the loop, driven by an electric field associated with the separation of charge in the loop's left side.

Now we have a current I in the loop. We found in Chapter 29 that the magnetic force on a current-carrying conductor of length ℓ is $\mathbf{F} = I\boldsymbol{\ell} \times \mathbf{B}$. Applying this expression to the conducting loop in Fig. 31-12 shows that there is no magnetic force on the right-hand side (since $\mathbf{B} = \mathbf{0}$ there) and that the forces on top and bottom cancel (Fig. 31-13). So the total magnetic force on the loop is that on the left side alone; since the current is upward while the magnetic field is into the page, the magnitude of this force is $I\ell B$, and the right-hand rule shows that it points to the left. Since the loop is being pulled with constant velocity, the net force on it—the sum of the magnetic force and the force applied by the agent pulling the loop—must be zero. The direction of the magnetic force, and therefore of the loop current, is consistent with this requirement.

We could equally well determine the current direction from magnetic flux considerations. As the loop leaves the field region, the flux through it decreases. The direction of the induced current is such as to oppose this decrease in flux. Therefore the magnetic field of the induced current points into the page, as the induced current tries to maintain the flux. By the right-hand rule, a field within the loop and into the page requires that the induced current flow in the clockwise direction.

To calculate the current, we must find the induced emf, which in turn is related to the rate of change of magnetic flux through the loop. With the field and loop perpendicular, and with the field uniform in the region where it is nonzero, the magnetic flux is the product of the magnetic field strength and the loop area that lies within the field:

$$\phi_B = B\ell x,$$

where x is the distance between the left edge of the loop and the right edge of the magnetic field region. The magnetic field remains constant, but as the loop moves the distance x decreases at the rate $dx/dt = -v$ (the minus sign indicates a decrease). Then the rate of change of flux is

$$\frac{d\phi_B}{dt} = \frac{d(B\ell x)}{dt} = B\ell\frac{dx}{dt} = -B\ell v,$$

so Faraday's law gives $\qquad \mathscr{E} = -\frac{d\phi_B}{dt} = B\ell v.$

This induced emf drives a current I around the loop, where

$$I = \frac{\mathscr{E}}{R} = \frac{B\ell v}{R}.$$

The rate of energy dissipation in the loop is the product of the emf and the current (Equation 27-8):

$$P = I\mathscr{E} = \frac{B\ell v}{R}B\ell v = \frac{B^2\ell^2 v^2}{R}. \qquad \left(\begin{array}{c}\text{power dissipated}\\\text{in loop}\end{array}\right)$$

We've found that the magnetic force on the loop has magnitude $I\ell B$; since the loop is moving with constant velocity, this is also the magnitude of the applied force. Equation 7-21 shows that the rate at which work is done by a force **F** acting on an object moving with velocity **v** is $P = \mathbf{F} \cdot \mathbf{v}$; here, with the applied force and velocity in the same direction, we have

$$P = Fv = I\ell Bv = \frac{B\ell v}{R}\,\ell Bv = \frac{B^2\ell^2 v^2}{R}, \qquad \begin{pmatrix} \text{power supplied} \\ \text{to pull loop} \end{pmatrix}$$

in agreement with our expression for the electric power dissipated in the loop. Thus, the work done by the agent pulling the loop ends up as electrical energy, which is dissipated in the resistor.

Electromagnetic induction is the principle behind many important technological devices. Induction permits us to transform mechanical into electrical energy, and provides great flexibility in the handling of electric power. The applications that follow explore some uses of induction.

■ APPLICATION ELECTRIC GENERATORS

Probably the most important technological application of induction occurs in the electric generator. The world currently uses electrical energy at the phenomenal rate of about 10^{13} watts—roughly equal to the power output of 100 billion human bodies—and virtually all of this power comes from generators. A generator is just a system of conductors in a magnetic field (Fig. 31-14). Mechanical energy is supplied to rotate the conductors, resulting in a changing magnetic flux. An emf is induced and current flows through the generator and on to whatever electrical loads are connected to it. Any source of mechanical energy can power the generator, but the most common is steam from burning fossil fuels or from nuclear fission (Fig. 31-15a). Electrical energy may be generated from the kinetic energy of water or wind. (Fig. 31-15b). A small electric generator, often called an alternator, is used to recharge the battery in an automobile while the engine is running.

Lenz's law, the conservation of energy in electromagnetic induction, is very much applicable to electric generators. Were

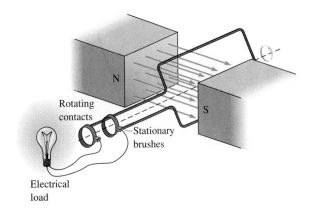

FIGURE 31-14 Simplified diagram of an electric generator. As the loop rotates in the magnetic field, the changing flux induces an emf that drives a current through the rotating contacts and stationary brushes and on through the electrical load.

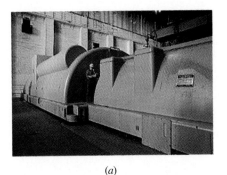

(a)

(b)

FIGURE 31-15 (a) This large power-plant generator produces over 1000 MW of electric power. (b) Wind-driven electric generators at a "wind farm" in California.

it not for Lenz's law, which requires that induced currents *oppose* the changes giving rise to them, generators would turn on their own and happily supply electricity without the need for coal, oil, or uranium. The voluminous quantities of fuel (Fig. 31-16) consumed by power plants are dramatic testimony to the minus sign appearing on the right-hand side of Equation 31-2!

An instructive introduction to Lenz's law comes about if you have access to a hand-cranked electric generator. Without any electrical load across the generator, it is easy to turn. But as you switch on increasingly heavy loads—by *lowering* the electrical load resistance—the generator gets harder to turn (Fig. 31-17). Most people find they can just sustain a 100-W light bulb with a hand generator. Think about this next time you leave a light on! You also experience Lenz's law when you turn on the headlights of a car that is idling slowly. You can hear the engine speed drop, and the car may even stall, as the car's generator gets harder for the engine to turn.

FIGURE 31-16 A 110-car trainload of coal arriving at a power plant in Texas. Some fourteen such trains arrive at the plant each week—a testimony to the minus sign in Equation 31-2!

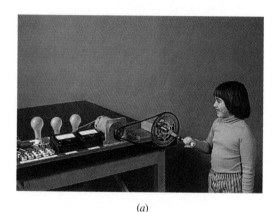

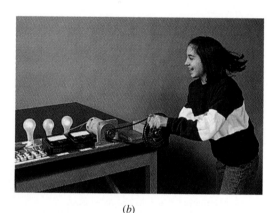

(a)   (b)

FIGURE 31-17 (a) With no electrical load, a hand-cranked generator is easy to turn. (b) With 200 watts of light bulbs connected, turning the generator becomes much more difficult. The generator emf is the same in both cases, because the generator is being turned at the same rate. But only in (b) is current flowing, giving nonzero power $P = IV$ that must be supplied by turning the generator.

● **EXAMPLE 31-6** DESIGNING A GENERATOR

An electric generator consists of a 10-turn square wire loop 50 cm on a side. The loop is turned at 60 revolutions per second, to produce standard 60-Hz alternating current like that used throughout the United States and Canada. How strong must the magnetic field be for the peak output voltage of the generator to be 170 V? (This is actually the peak voltage of standard 120-V household wiring; 120 V is an appropriate average value.)

Solution
We need to evaluate the induced emf as a function of magnetic field strength. With a uniform magnetic field, the flux through

one turn of the loop is $\int \mathbf{B} \cdot d\mathbf{A} = BA \cos\theta$, where θ is the angle between the field and the normal to the loop. But the loop rotates with angular frequency $\omega = 2\pi f$, so $\theta = 2\pi ft$. The loop area A is s^2, with s the length of the loop side, so the flux through N turns of the loop is

$$\phi_B = NBs^2 \cos(2\pi ft).$$

To find the induced emf, we take the time rate of change of this flux:

$$\mathcal{E} = -\frac{d\phi_B}{dt} = -NBs^2 \left[-2\pi f \sin(2\pi ft)\right]$$

$$= 2\pi NfBs^2 \sin(2\pi ft).$$

The peak emf is the quantity multiplying the sine; we want this to be 170 V. Solving for the unknown magnetic field B then gives

$$B = \frac{\mathscr{E}_{peak}}{2\pi Nfs^2} = \frac{170 \text{ V}}{(2\pi)(10)(60 \text{ Hz})(0.50 \text{ m})^2} = 0.18 \text{ T}.$$

This is a typical field strength near the poles of a strong permanent magnet.

EXERCISE A generator includes a circular coil 30 cm in diameter, spinning at 3600 rpm in a uniform 0.50-T magnetic field. How many turns should it have if the peak output voltage is to be 2400 V?

Answer: 180

Some problems similar to Example 31-6: 24, 26, 51, 52, 61 ●

■ APPLICATION MAGNETIC RECORDING

Magnetic materials are widely used as information storage media. Examples include audio and video cassette tapes and computer disks of both the hard and floppy variety. Retrieving the stored information involves electromagnetic induction.

In a typical magnetic recording system the magnetic medium—usually a long plastic tape or a circular disk coated with ferromagnetic oxides—is made to move past a small coil called the **head.** To record information, current is passed through the head coil to impress a magnetization pattern on the tape or disk (Fig. 31-18). This pattern reflects the information contained in the time variation of the current supplied to the head. With analog systems, like most audio and videotapes, the current varies continuously and produces a smoothly varying magnetization pattern. (The current in audio systems may itself derive from electromagnetic induction, using a microphone in which a ferromagnetic diaphragm vibrates in response to sound waves, altering the magnetic field in a nearby coil and thus inducing a current.) With digital systems, including computer data storage and the newer digital audio tape (DAT) and video formats, coil currents of the same magnitude but opposite polarity produce regions of oppositely directed magnetization; these represent the 1's and 0's in which digital information is coded.

FIGURE 31-19 Mechanism of a computer disk drive includes a head (at bottom) that literally flies as close as 0.25 μm above a spinning aluminum disk coated with ferromagnetic material (red).

To retrieve the stored information, the magnetic medium is again moved past a coil, which may be the same one used for recording. The changing magnetization on the medium results in a varying magnetic flux through the coil, which induces an emf that is amplified and processed to produce images, sound, or digital data. (In audio systems, the final conversion to sound is again usually magnetic, with the amplified signal fed to the coil of a loudspeaker, where magnetic forces result in movement of the speaker cone and the generation of sound waves.)

The rate at which information can be stored and extracted depends on how densely the regions of varying magnetization can be packed without interfering and on how fast the medium moves in relation to the heads. Audio cassette tapes move past the heads at a mere 4.8 cm/s, a speed that limits these tapes' ability to record high-frequency sound faithfully. The much higher information content of video images requires higher tape-to-head speeds; in standard videocassette recorders a speed of 39.52 m/s is achieved by moving the tape past rapidly spinning heads. High-speed computer disks boast even greater speeds, allowing rapid data storage and retrieval (Fig. 31-19).

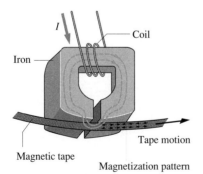

FIGURE 31-18 Recording on magnetic tape. The iron body of the head confines the magnetic field of the coil except at the gap, where a fringing field emerges to impose a magnetization pattern on the moving tape. When the tape is played back, its changing magnetization pattern imposes a changing magnetic flux at the head, inducing an emf in the head coil.

■ **APPLICATION** EDDY CURRENTS

Our discussion of induced currents has centered on conducting loops. But induced currents also appear in solid conductors through which magnetic flux is changing. The resistance of a solid piece of conductor is low, so the induced currents are large, resulting in substantial energy dissipation. The presence of these **eddy currents** can make it difficult to move a conductor rapidly through a magnetic field. For example, if you try to push a piece of metal—it need not be a ferrous metal like iron—between the poles of a magnet, you will find yourself working against a magnetic force.

A common demonstration of eddy currents consists of a pendulum with a metal bob that swings between the poles of a magnet (Fig. 31-20). As it swings toward the magnet, the pendulum experiences an increasing magnetic flux that induces eddy currents. The energy dissipated by these currents comes ultimately from the kinetic energy of the pendulum, which therefore stops abruptly between the magnet poles.

Eddy currents provide an alternative to friction brakes for stopping moving machinery. Rapidly rotating saw blades, for example, can be stopped abruptly by an electromagnet activated next to the blade. Similarly, eddy-current brakes are sometimes used on trains and in other applications involving rotating conductors.

In some instances eddy currents are a nuisance, acting just like friction in reducing the efficiency of machinery. To solve this problem, slots are often cut into moving conductors to make the current paths longer, thus increasing the electrical resistance and reducing the eddy currents. For example, if the solid pendulum bob in Fig. 31-20 is replaced by a slotted piece, it then swings more freely through the magnet.

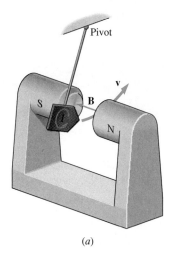

(a)

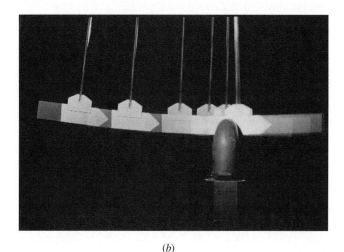

(b)

FIGURE 31-20 (a) An eddy-current pendulum consists of a conductor swinging between magnet poles. As it enters the field region, the conductor experiences a changing magnetic flux that induces a current. You should convince yourself that the current is in the direction shown as the bob begins to enter the field region. (b) The current dissipates energy in the resistance of the conductor, at the expense of its kinetic energy. Strobe photo shows rapid deceleration as the bob passes between the magnet poles.

Lenz's Law and Changing Magnetic Fields

Lenz's law—conservation of energy applied to electromagnetic induction—determines the direction of the induced current even when no motion is involved. Figure 31-21 shows a conducting loop in a magnetic field that points into the page. Suppose the field strength—and, therefore, the magnetic flux through the loop—is *decreasing*. Then the direction of the induced current must be such that the magnetic field it creates *opposes* this decrease. Therefore, the loop's field must reinforce the existing field, which means that the loop's field points into the

FIGURE 31-21 A uniform magnetic field is decreasing in strength, causing a decrease in magnetic flux through the conducting loop. The induced current in the loop must *oppose* this *decrease,* and therefore loop current goes clockwise in order to produce a magnetic field within the loop that reinforces the original field.

page. Applying the right-hand rule then shows that the loop current is clockwise in Fig. 31-21.

What if the magnetic field in Fig. 31-21 had been increasing in strength? Then the loop current would have been in the opposite direction, to create a magnetic field opposite the existing field in order to oppose its increase.

TIP **Induction Opposes Change** Faraday's law relates the induced emf to the *rate of change* of magnetic flux. Lenz's law says that the induced emf and current are in such a direction as to oppose that *change* in flux—not the flux or the field itself. Figure 31-21 is a case in point; here the induced current is in a direction that actually reinforces the original field and flux—precisely because they are both *decreasing.* The same thing happened in Fig. 31-12, where movement of the loop decreased the flux, giving an induced current that reinforced the existing field.

● **EXAMPLE 31-7** LENZ'S LAW

Two coils are arranged as shown in Fig. 31-22. If the resistance of the variable resistor is being increased, what is the direction of the induced current in the fixed resistor R?

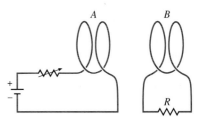

FIGURE 31-22 As the variable resistance is increased, the current in coil A decreases. This decreases the magnetic flux through coil B, resulting in an induced current whose direction is such that the magnetic field it creates in the interior of coil B reinforces the field from coil A, thereby opposing the decrease in flux through B.

Solution
Applying the right-hand rule, we find that the magnetic field of coil A emerges from the right side of the coil, pointing toward coil B. As the resistance increases, the current in coil A decreases, and with it the strength of coil A's magnetic field. This results in a decrease in the magnetic flux through coil B. The induced current in coil B acts to oppose this decrease in flux, so the magnetic field resulting from the induced current reinforces the field from coil A. Thus the field of coil B emerges from the

right end of the coil and enters on the left end. By the right-hand rule, this requires a current from right to left in the fixed resistor.

EXERCISE A circular conducting loop lies next to a long, straight wire carrying current in the direction indicated in Fig. 31-23. In which direction is the induced current in the loop when the current in the wire is (a) increasing and (b) decreasing?

Answers: (a) counterclockwise; (b) clockwise

Some problems similar to Example 31-7: 4, 10, 19, 27, 32

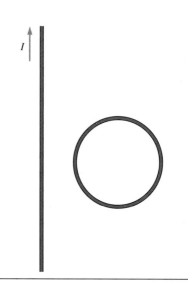

FIGURE 31-23 Which way is the current in the circular loop?

Induction in Open Circuits

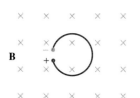

FIGURE 31-24 An open loop. In a changing magnetic field, there is an induced emf that results in charge buildup at the gap. The polarity shown results when the magnetic field strength decreases.

An induced emf also arises in an open circuit, but there it cannot drive a steady current. Figure 31-24 shows the loop and magnetic field of Fig. 31-21, but now with a small gap in the loop. Although the loop is not quite closed, there is still an induced emf. In response to this emf, charge piles up on either side of the gap, creating a voltage across it. Once the gap voltage equals the induced emf, the electric field associated with the separated charge opposes the emf's tendency to move charge, and a steady state is reached. In the open-circuit case, the entire emf implied by Equation 31-2 is available at the gap. The polarity can be determined by considering what would happen if current did flow. Since current flowed clockwise in the closed loop of Fig. 31-21, positive charge will accumulate at the bottom of the gap, as shown in Fig. 31-24.

Another example of an induced emf in an open circuit is the motional emf arising as a single conductor moves through a magnetic field. Example 31-8 treats this case quantitatively.

● EXAMPLE 31-8 THE TETHERED SATELLITE

A 1992 flight of the space shuttle Atlantis attempted to demonstrate a new method for supplying electric power to orbiting spacecraft. Once in orbit, Atlantis deployed a 520-kg satellite attached to the shuttle through a conducting tether (Fig. 31-25). With the tethered satellite flying vertically above Atlantis the two moved at approximately right angles to Earth's magnetic field, whose strength was about 30 μT at the orbital position. Find the motional emf that would have developed between the tethered satellite and the shuttle had the tether been extended its full 20-km length. (Difficulties with the tether's reel mechanism prevented full deployment from the shuttle, but a subsequent rocket-borne tether was successful.)

Solution

Figure 31-26 shows the pair flying at right angles to the magnetic field. We can imagine forming a closed circuit by letting the satellite and shuttle slide along a system of conducting rails, just like that of Example 31-4. The system is then identical to that example, and we have

$$\mathcal{E} = B\ell v = (30 \times 10^{-6} \text{ T})(20 \times 10^3 \text{ m})(7.8 \times 10^3 \text{ m/s})$$

$$= 4.7 \text{ kV},$$

where we calculated the speed in low-Earth orbit in Example 9-3.

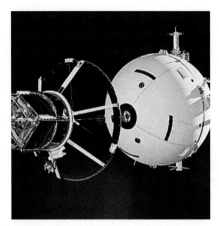

FIGURE 31-25 Tethered satellite being deployed from the space shuttle Atlantis.

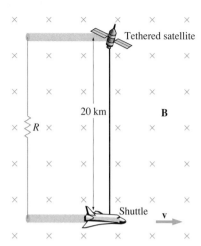

FIGURE 31-26 The tethered satellite generates electric power from the motional emf developed between the spacecraft and a satellite connected by a conducting tether (Example 31-8).

Since the magnetic flux through the rail arrangement would be increasing, the current would flow counterclockwise so that the magnetic field of the induced current would oppose this increase. That means positive charge will accumulate on the satellite and negative charge on the shuttle—a result that can also be obtained by considering the magnetic force on free charges in the tether (see Problem 57).

At the shuttle's orbital altitude there are enough free electrons from ionization of the upper atmosphere for the tenuous gas surrounding the shuttle to carry a current, completing a circuit and allowing exploitation of the induced emf for electric power generation.

EXERCISE An airplane with a wingspan of 44 m is in level flight at 300 m/s over the north pole, where Earth's magnetic field has magnitude 0.62 G and points vertically downward. Find the induced emf between the wing tips.

Answer: 0.82 V

Some problems similar to Example 31-8: 10, 13, 57, 58 ●

31-4 INDUCED ELECTRIC FIELDS

What do we really mean by an induced emf? In a circuit containing a battery, the notion of emf is clear—the emf arises in a specific device where chemical energy is converted to electrical energy associated with charge separation. This charge separation sets up an electric field that drives current in an external circuit. In the case of motional induced emf we also have a clear picture: the emf arises from the separation of charge associated with magnetic forces on the free charges in a moving conductor; again, the electric field associated with this charge separation may drive a current.

Now consider the current induced in a conducting loop by a magnetic field that changes with time. No motion is involved, yet there must be a force on the free charges in the conductor. The one force we know that acts on stationary charges is the electric force. Electric forces arise from electric fields, so there must be an electric field in a conducting loop in a changing magnetic field. This field is called an **induced electric field.** It has the same effect on charges—exerting a force $q\mathbf{E}$—as do the electric fields we considered earlier. The field itself, however, originates not in electric charges but in changing magnetic fields.

An induced electric field arises whenever a magnetic field changes with time—whether or not an electric circuit is present. When a circuit is present, then the induced field may drive a current. But the induced field, not the current, is fundamental. A single, stationary electron placed in a changing magnetic field will experience an *electric* force—clear evidence for the existence of an electric field.

We have been thinking of Faraday's law as a relation between the emf induced in a circuit and the rate of change of magnetic flux through that circuit. We now know that induced electric fields are the fundamental manifestation of changing magnetic flux, and that these fields arise whether or not circuits are present. We need to reformulate Faraday's law to describe induced electric fields without reference to circuits. The induced emf $\mathscr{E}$ that we have been writing on the left-hand side of Faraday's law means simply the work per unit charge gained by a test charge moved around a circuit. Since work is the line integral of force over distance, and electric field is the force per unit charge, we can write the emf as the line integral of the electric field. Then Faraday's law becomes

$$\oint \mathbf{E} \cdot d\boldsymbol{\ell} = -\frac{d\phi}{dt} = -\frac{d}{dt} \int \mathbf{B} \cdot d\mathbf{A}. \quad \text{(Faraday's law)} \qquad (31\text{-}3)$$

Here the line integral on the left-hand side is taken over *any* closed loop, which need not coincide with a circuit. The flux on the right-hand side is the surface integral of the magnetic field over *any* open surface bounded by the loop on the left-hand side.

Faraday's law in the form 31-3 makes no reference to wires or other circuits. It simply describes induced electric fields, which occur whenever there are changing magnetic fields. If electric circuits are present, then induced currents occur as well—but it is the induced electric fields that are fundamental. We can state Faraday's law loosely but powerfully by saying the following:

▌ A changing magnetic field creates an electric field.

This direct interaction between the fields is the basis for numerous practical devices and, as we shall see in Chapter 34, is essential to the existence of light.

Note the similarity between Faraday's law and Ampère's law (Equation 30-8). Faraday's law gives the line integral of the electric field around a closed loop in terms of the rate of change of magnetic flux through the loop. Ampère's law gives the line integral of the magnetic field around a closed loop in terms of the current through the loop. Both give fields that *encircle* their sources—current for the source of magnetic field and changing magnetic field for the induced electric field. That means the configuration of an induced electric field is very different from that of an electric field originating in electric charge. Induced fields have no beginning or end; their field lines form closed loops encircling regions of changing magnetic field.

When a changing magnetic field has sufficient symmetry, we can evaluate the induced electric field in the same way we did the magnetic field of a symmetric current distribution. Example 31-9 illustrates this procedure.

● EXAMPLE 31-9 AN INDUCED ELECTRIC FIELD

A long solenoid has circular cross section of radius R. The current in the solenoid is increasing and, as a result, the uniform magnetic field within the solenoid increases with time; the field magnitude is given by $B = bt$, with b a constant and t the time. Find the induced electric field outside the solenoid, a distance r from the solenoid axis.

Solution
The induced electric field has no beginning or end, so the field lines must encircle the solenoid. The only field consistent with the symmetry consists of circular field lines centered on the solenoid axis, as suggested in Fig. 31-27. Since the solenoid field points into the page and is increasing in strength, the direction of the induced electric field must be such that any current it might drive would produce a magnetic field *opposing* the increase in the solenoid field. Applying the right-hand rule then shows that the induced electric field runs counterclockwise.

Faraday's law relates the line integral of the electric field to the rate of change of the encircled magnetic flux. Here a suitable integration loop is itself a circle centered on the solenoid axis. Symmetry shows that the field magnitude is constant over this loop, and if we circle the loop in the direction of the field then $\cos\theta$ in the dot product $\mathbf{E} \cdot d\boldsymbol{\ell}$ is 1. Thus

$$\oint \mathbf{E} \cdot d\boldsymbol{\ell} = \oint E\, d\ell = E \oint d\ell = 2\pi r E,$$

since $\oint d\ell$ is just the loop circumference. Faraday's law relates this quantity to the rate of change of the encircled magnetic flux. Here the flux is due to the uniform magnetic field inside the solenoid, perpendicular to our loop and confined to a circular area of radius R. (We use the long-solenoid approximation, neglecting the small magnetic field that must exist outside a solenoid of finite length.) So the encircled flux is just $\phi_B = \pi R^2 B = \pi R^2 bt$, and its rate of change is

$$\frac{d\phi_B}{dt} = \frac{d}{dt}(\pi R^2 bt) = \pi R^2 b.$$

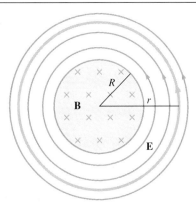

FIGURE 31-27 Cross section of a solenoid of radius R whose magnetic field is increasing with time. Field lines of the induced electric field are circles concentric with the solenoid axis. Also shown is a circular loop (gray) of radius r for evaluating the line integral in Faraday's law.

Then Faraday's law gives

$$2\pi r E = -\pi R^2 b,$$

or

$$E = -\frac{R^2 b}{2r}.$$

We've already accounted for the minus sign in arguing that the field circles counterclockwise. The $1/r$ dependence of the field strength on distance should come as no surprise; points with $r > R$ are outside a cylindrically symmetric distribution, in this case a distribution of changing magnetic flux. We found the same $1/r$ dependence for the electric and magnetic fields outside, respectively, a cylindrically symmetric charge distribution and a cylindrically symmetric current distribution.

EXERCISE Show that the electric field at points *inside* the solenoid has magnitude $br/2$.

Some problems similar to Example 31-9: 37–39, 41, 43, 44 ●

In Chapter 30 we derived the magnetic field of a solenoid using Ampère's law—which, as we've formulated it so far, applies only to steady currents. But a solenoid whose field is changing—as in Example 31-9—must have a changing current, so how can we talk about it? In fact, Ampère's law gives a good approximation to the field produced by changing current, provided that change is sufficiently slow. In the examples and problems of this chapter we assume that to be the case. In Chapter 34 we will explore what happens if we relax the assumption of slowly changing currents.

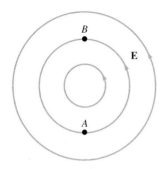

FIGURE 31-28 Two points in an induced electric field. If a positively charged particle moves counterclockwise along the field line from A to B the field does work on it. But if it moves from A to B in the clockwise direction the field does negative work; that is, an external force must do work on the charge. The work done is not path independent, and the induced electric field is not conservative.

FIGURE 31-29 This induction cooktop has no burners. Instead, coils beneath the cooktop produce time-varying magnetic fields that induce currents in the iron cookware. Resistance in the cookware then results in heating.

Conservative and Nonconservative Electric Fields

We have seen that static electric fields—those beginning and ending on stationary charge distributions—are conservative, meaning that the work required to move a charge between two points is independent of the path taken. A consequence is that it takes no work to move around a closed path in an electrostatic field; mathematically, we express this by writing

$$\oint \mathbf{E} \cdot d\boldsymbol{\ell} = 0. \quad \text{(electrostatic field)}$$

In contrast, induced electric fields generally form closed loops, and Faraday's law shows that the line integral of the electric field around a closed path in such a field is decidedly not zero. That means the induced electric field does work on a charge moved around a *closed* path and that the work done in moving between two points cannot be independent of the path taken (Fig. 31-28). The induced electric field is therefore not conservative.

The work done on charges moving around closed paths in nonconservative electric fields can be useful. A simple example is the induction cooktop (Fig. 31-29), where food cooks in a special conducting pan heated by currents in the pan itself. Those currents are driven by an induced electric field originating from a changing magnetic field in coils just below the cooktop's surface. Much more generally, any electrical device that derives its power from a generator—and that includes almost every electrical thing you use—is part of a circuit in which a nonconservative electric field drives the current.

■ APPLICATION THE TOKAMAK

In Chapter 29 we described how magnetic fields are used to confine very hot plasma in fusion reactors and showed why most reactors have a toroidal shape. The most promising of the toroidal reactor designs is the **tokamak,** in which the confining magnetic field arises in part from current within the plasma itself. Discharging a large capacitor bank through a coil produces a rapidly changing magnetic field in the "hole" of the toroidal "doughnut," resulting in an induced electric field within the torus. This field drives the plasma current that produces a magnetic field component around the minor radius of the device. Other coils around the torus provide a magnetic field component around the long dimension of the torus, giving a net magnetic field that spirals around the plasma (Fig. 31-30).

As in any conductor, particle collisions dissipate energy gained from the electric field, in this case heating the plasma. Unfortunately the resistance of a plasma drops at high temperature, making it difficult to achieve the temperature needed for fusion with resistive heating alone. Nevertheless, the induced electric field helps bring the plasma close to the fusion temperature, as well as driving currents that produce the confining magnetic field.

Since the tokamak uses an induced electric field, its operation requires a changing magnetic field. Therefore, the tokamak cannot be run continuously but operates instead in a

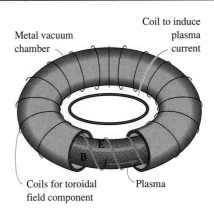

FIGURE 31-30 A tokamak fusion reactor. Current in the plasma is driven by an electric field induced by the magnetic field change associated with a rapid change in current through the central coil. Additional coils wound around the toroid give a steady magnetic field in the same direction as the plasma current. Combining with the field from the plasma current gives a net field that spirals around the plasma.

pulsed mode as current builds up rapidly and then decays in preparation for another pulse. Today's experimental tokamaks can achieve pulse durations of 10 s or more, during which they produce significant fusion energy.

■ FOR FURTHER THOUGHT NONCONSERVATIVE ELECTRIC FIELDS

The nonconservative nature of the induced electric field is strikingly demonstrated if you attempt to measure potential differences in nonconservative fields. In Chapter 25, we defined potential difference as the work required to move a unit charge between two points, and stressed that this work is independent of path for a conservative field. But when the field is nonconservative, the work is not independent of path, and the concept of potential becomes ambiguous.

Figure 31-31 shows an end view of a long solenoid surrounded by three identical resistors bent into circular arcs. If the solenoid current is increasing, an induced electric field appears in the resistors, and drives a current I in the counterclockwise direction.

Because they have the same resistance and carry the same current, the potential difference across each resistor should be the same. We could try to measure the potential difference across one resistor, for example, by connecting a voltmeter as shown in Fig. 31-31b. With current I flowing through the resis-

tance R, this meter reads IR. Since the current flows counterclockwise, we must connect the positive voltmeter terminal to point B.

But now try to measure the potential difference across the other two resistors together, as in Fig. 31-31c. With the current I flowing through the total resistance $2R$, the meter now reads $2IR$. We have two voltmeters with their terminals connected to the same points, and yet they indicate different voltages. Not only are the magnitudes of the voltages different, but even their polarities differ. How can this be? We are experiencing the nonconservative nature of the induced electric field. The two voltmeters are positioned differently with respect to the changing magnetic flux, so they sample different regions of the induced electric field. Even though the meters are connected to the same points, they measure the line integral of the induced electric field over *different* paths, and so they do not read the same voltage.*

---

* For a fascinating discussion of voltage measurement in induced fields, see R. Romer, "What do 'Voltmeters' Measure?: Faraday's Law in a Multiply Connected Region," *American Journal of Physics,* vol. 50, no. 12, pp. 1089–1091 (December 1982).

FIGURE 31-31 (*a*) End view of a long solenoid surrounded by three resistors in series. A changing magnetic field in the solenoid induces an emf in the resistors, and the same current *I* flows through each. (*b*) A voltmeter connected between points *A* and *B* indicates the voltage *IR* across one resistor. (*c*) A second voltmeter connected to the *same points* indicates voltage 2*IR* with the opposite sign.

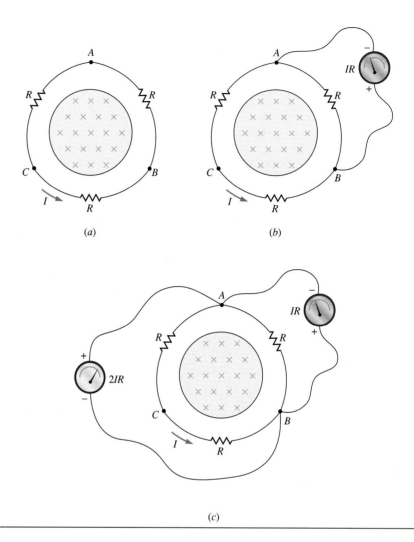

(*a*)

(*b*)

(*c*)

31-5 DIAMAGNETISM

In Section 30-6 we discussed paramagnetic and ferromagnetic materials, in which atomic magnetic dipoles align with an applied magnetic field, causing an attractive interaction between the material and a magnet. We also mentioned diamagnetic materials, in which induced magnetic dipoles align antiparallel to the applied field, causing a repulsive force. We are now ready to understand diamagnetism as a manifestation of Faraday's law at the microscopic level.

In a purely diamagnetic material, current loops associated with pairs of atomic electrons exactly cancel, leaving atoms with no intrinsic magnetic moments. Fig. 31-32*a* shows a simplified model to describe such an atom. The picture should not be taken too literally, for it uses classical physics to describe a phenomenon properly within the domain of quantum mechanics. Nevertheless, it shows qualitatively how diamagnetism is an electromagnetic induction effect.

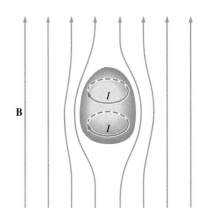

FIGURE 31-32 Simplified, classical model for a diamagnetic atom. (*a*) In the absence of an applied field, magnetic moments (cross and dot) associated with a pair of electrons exactly cancel. (*b*) The changing flux associated with an increasing magnetic field induces an electric field that speeds up one electron and slows down the other, giving rise to a net magnetic moment. (In applying the right-hand rule, remember that the electrons are negative.)

FIGURE 31-33 Induced currents in a superconductor completely cancel an applied magnetic field. Net result is that a magnetic field cannot penetrate the superconductor.

The electron orbiting clockwise in Fig. 31-32*a* has a magnetic moment pointing out of the page (apply the right-hand rule and remember that the electron carries a negative charge). In the absence of an applied magnetic field, this moment is exactly cancelled by the equal but oppositely directed moment of the other electron. But now suppose a magnetic field is applied, for example by moving the north pole of a magnet down toward the plane of Fig. 31-32. This produces a changing magnetic field in the plane of the figure, resulting in an induced electric field that alters the speeds of the orbiting electrons. The direction of this effect must be such as to oppose the increase in field. Thus the electron whose magnetic moment points out of the page has its moment increase, while the moment pointing into the page decreases. Figure 31-32*b* shows how these changes result from the alteration of electron speeds by the induced electric field.

Once the field is applied and the orbital speeds of the two electrons are no longer identical, the atom now has a net magnetic moment that points out of the page in Fig. 31-32, opposing the incoming magnet and resulting in a repulsive force. This repulsion is the distinguishing characteristic of diamagnetism. We listed a number of diamagnetic materials in Table 30-2.

A material that is entirely superconducting is perfectly diamagnetic, with magnetic susceptibility -1. This means that the magnetic field resulting from induced currents within the material completely cancels any applied field. Since these induced currents persist in the zero-resistance superconductor, the material completely excludes magnetic fields from its interior, a phenomenon known as the Meissner effect (Fig. 31-33). The repulsive force associated with the magnetic moments of a permanent magnet and a nearby superconductor results in the widely publicized phenomenon of magnetic levitation (Fig. 31-34). Readers of the extended version of this text will explore superconductivity further in Chapter 42.

FIGURE 31-34 A small magnet is levitated above a wafer of high-temperature superconductor in a bath of liquid nitrogen at 77 K.

CHAPTER SYNOPSIS

Summary

1. **Electromagnetic induction** is a fundamental phenomenon linking magnetism and electricity. Induction is described by **Faraday's law,** which states that a **changing magnetic field** produces an **induced electric field.** Unlike the conservative electrostatic field of an electric charge, this induced field is **nonconservative,** meaning it can do work on charges as they move around a closed loop. Faraday's law relates the line integral of this nonconservative electric field around an arbitrary loop to the rate of change of magnetic flux through a surface bounded by that same loop:

$$\oint \mathbf{E} \cdot d\boldsymbol{\ell} = -\frac{d\phi_B}{dt} = -\frac{d}{dt} \int \mathbf{B} \cdot d\mathbf{A}.$$

2. In order for energy to be conserved, the induced electric field is in such a direction as to oppose the change in flux that gives rise to it. This energy-conserving aspect of Faraday's law is called **Lenz's law** and is reflected mathematically in the minus sign on the right-hand side of Faraday's law.

3. When a conductor is present, the nonconservative electric field manifests itself as an **induced emf:**

$$\mathcal{E} = -\frac{d\phi_B}{dt}.$$

This emf drives an **induced current** in any circuit with finite resistance. It does not matter whether the magnetic flux is changed by moving a conductor in a magnetic field, or by moving a magnetic field near a conductor, or by altering the shape or orientation of the conductor. The generation of electric power by moving conducting loops in magnetic fields is an important technological example of induced currents.

4. **Diamagnetism** is a manifestation of electromagnetic induction on the atomic scale. Application of a magnetic field to a diamagnetic material results in induced atomic magnetic moments that cause the material to be repelled from a magnet.

Terms You Should Understand

(Pairs are closely related terms whose distinction is important; number in parentheses is chapter section where term first appears.)

electromagnetic induction (31-1)
induced current, induced emf (31-1, 31-2)
magnetic flux (31-2)
Faraday's law, Lenz's law (31-2, 31-3)
motional emf (31-3)
generator (31-3)
eddy currents (31-3)
induced electric field (31-4)

Symbols You Should Recognize

ϕ_B (31-2)
$\oint \mathbf{E} \cdot d\boldsymbol{\ell}$ (31-4)

Problems You Should Be Able to Solve

calculating magnetic flux in uniform and nonuniform fields (31-2)
calculating induced emf's from changing magnetic flux (31-2)
determining the direction of induced emf's and currents (31-3)
calculating induced electric fields in symmetric situations (31-4)

Limitations to Keep in Mind

Electromagnetic induction requires a *change* in magnetic flux. Values of magnetic field or flux alone do not matter; only the rate of change of those quantities is important.

QUESTIONS

1. A copper penny falls on a vertical path that takes it between the poles of a magnet. Does it hit the ground faster or slower than if no magnet were present?

2. A bar magnet is moved toward a conducting ring, as shown in Fig. 31-35. What is the direction of the induced current in the ring?

3. Figure 31-36 shows two concentric conducting loops, the outer connected to a battery and a switch. The switch is initially open. It is then closed, left closed for a while, then opened again. Describe the currents in the inner loop during the entire procedure.

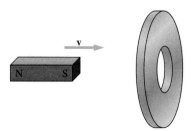

FIGURE 31-35 Question 2.

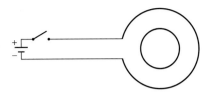

FIGURE 31-36 Question 3.

4. An electric generator is being turned at constant speed. A load resistor R is connected across the generator terminals. If the electrical resistance of the load is lowered, does the generator get easier or harder to turn?
5. Service manuals for cars often tell you to set the idle speed of the engine with the headlights on. Why? What does this have to do with electromagnetic induction?
6. Figure 31-37 shows two square wire loops, the first containing a battery and variable resistor. The resistor is initially at the midpoint of its resistance range. Should its resistance be lowered or raised in order to induce a clockwise current in the right-hand loop?

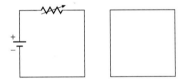

FIGURE 31-37 Question 6.

7. Consider the simple motor shown in Fig. 29-35. What happens if you connect a resistor across the motor terminals, and turn the motor by hand? What is the difference between a motor and a generator?
8. Figure 31-38 shows an open wire loop in a magnetic field. The field is changing, and charge has piled up at the loop gap with the polarity indicated. Is the magnetic field strength increasing or decreasing?

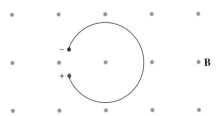

FIGURE 31-38 Question 8.

9. A constant, uniform magnetic field points into the page. A flexible, circular conducting ring lies in the plane of the page. If the ring is stretched, maintaining its circular shape, what is the direction of the induced current?

10. A student argues that it takes work to stretch the ring in the preceding question, and that therefore the ring should release all the work as energy if allowed to shrink. Is this right? Why or why not?
11. When a magnet is moved near a superconductor, the magnetic field lines never enter the superconductor. Why not?
12. Is it possible to produce an induced current that never changes? How or why not? Could you produce an induced current that was steady for some finite time? How or why not?
13. When you push a bar magnet into a conducting loop, you do work. What happens when you pull it out the other side?
14. You are turning a generator in such a way that the current it delivers remains constant. As you lower the load resistance across the generator, does the generator get easier or harder to turn?
15. Devise a way of measuring a magnetic field using Faraday's law.

FIGURE 31-39 Question 16.

16. In Fig. 31-39, a copper ring was originally resting on the wooden structure, surrounding the coil. When a rapidly changing current was applied to the coil, the ring was ejected into the air. Explain this phenomenon.
17. Fluctuations in Earth's magnetic field due to changing solar activity can wreak havoc with communications, even those using underground cables. How is this possible?
18. Why is it not possible to run a tokamak on a continuous basis?
19. Conventional brakes on a car need large surface areas to dissipate the heat of friction when the brakes are applied. Would eddy-current brakes have the same problem?
20. Which way would the eddy currents flow in Fig. 31-20a as the bob continues its swing and begins to emerge from the field?
21. In Chapter 29, we pointed out that a static magnetic field cannot change the energy of a charged particle. Is this true of a changing magnetic field? Discuss.

22. A long solenoid of circular cross section is oriented so that its magnetic field points out of the page, as shown in Fig. 31-40. The solenoid current is increasing. (a) What is the direction of the induced electric field at points A and B in the figure? (b) What is the magnitude of the induced electric field in the center of the solenoid? (Don't calculate! Argue from symmetry.)

23. Is the concept of electric potential (Chapter 25) useful in a nonconservative electric field? Give an example to substantiate your answer.

24. Can an induced electric field exist in the absence of a conductor?

25. Could you tell whether a given electric field arises from electric charge or from a changing magnetic field? How or why not?

26. Does a diamagnetic material experience a force in a uniform magnetic field?

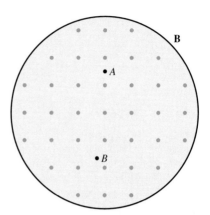

FIGURE 31-40   Question 22.

PROBLEMS

Sections 31-2 and 31-3 Faraday's Law and Induction and the Conservation of Energy

1. Show that the volt is the correct SI unit for the rate of change of magnetic flux, making Faraday's law dimensionally correct.

2. A bar magnet is moved steadily through a wire loop, as shown in Fig. 31-41. Sketch qualitatively the current and power dissipation in the loop as functions of time. Take as positive a current flowing out of the plane of the page at the top of the loop, and indicate the position of the magnet on your time axis.

3. Find the magnetic flux through a circular loop 5.0 cm in diameter oriented with the loop normal at 30° to a uniform 80-mT magnetic field.

4. A circular wire loop 40 cm in diameter has 100-Ω resistance and lies in a horizontal plane. A uniform magnetic field points vertically downward, and in 25 ms it increases linearly from 5.0 mT to 55 mT. Find the magnetic flux through the loop at (a) the beginning and (b) the end of the 25 ms period. (c) What is the loop current during this time? (d) Which way does this current flow?

5. A conducting loop of area 240 cm² and resistance 12 Ω lies at right angles to a spatially uniform magnetic field. The loop carries an induced current of 320 mA. At what rate is the magnetic field changing?

6. A conducting loop of area A and resistance R lies at right angles to a spatially uniform magnetic field. At time $t = 0$ the magnetic field and loop current are both zero. Subsequently, the current increases according to $I = bt^2$, where b is a constant with the units A/s². Find an expression for the magnetic field strength as a function of time.

7. A conducting loop with area 0.15 m² and resistance 6.0 Ω lies in the x-y plane. A spatially uniform magnetic field points in the z direction. The field varies with time according to $B_z = at^2 - b$, where $a = 2.0$ T/s² and $b = 8.0$ T. Find the loop current (a) when $t = 3.0$ s and (b) when $B_z = 0$.

8. The magnetic field inside a 20-cm-diameter solenoid is increasing at the rate of 2.4 T/s. How many turns should a coil wrapped around the outside of the solenoid have in order that the emf induced in the coil be 15 V?

9. A square wire loop of side ℓ and resistance R is pulled with constant speed v from a region of no magnetic field until

FIGURE 31-41   Problem 2.

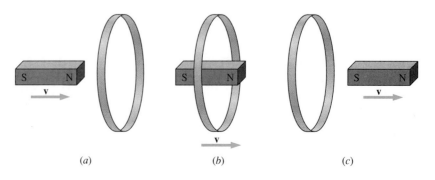

(a)   (b)   (c)

it is fully inside a region of constant, uniform magnetic field **B** perpendicular to the loop plane. The boundary of the field region is parallel to one side of the loop. Find an expression for the total work done by the agent pulling the loop.

10. A 1.8-m high runner sprints eastward at 9.5 m/s along the equator, where Earth's magnetic field points horizontally with a strength of 31 μT. (a) What is the magnitude of the emf induced between the runner's head and feet? (b) Which end is positive? *Hint:* See Fig. 30-6.

11. In Fig. 31-24 the loop radius is 15 cm, and the magnetic field is decreasing at the rate of 550 T/s. If the gap width is small compared with the loop circumference, what is the voltage across the gap?

12. A 5-turn coil 1.0 cm in diameter is rotated at 10 rev/s about an axis perpendicular to a uniform magnetic field. A voltmeter connected to the coil through rotating contacts reads a peak value of 360 μV. What is the magnetic field strength?

13. The wingspan of a 747 jetliner is 60 m. If the plane is flying at 960 km/h in a region where the vertical component of Earth's magnetic field is 0.20 G, what emf develops between the plane's wingtips?

14. A square wire loop 3.0 m on a side is perpendicular to a uniform magnetic field of 2.0 T. A 6-V light bulb is in series with the loop, as shown in Fig. 31-42. The magnetic field is reduced steadily to zero over a time Δt. (a) Find Δt such that the light will shine at full brightness during this time. (b) Which way will the loop current flow?

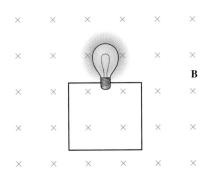

FIGURE 31-42 Problem 14.

15. In Example 31-2 take $a = 1.0$ cm, $w = 3.5$ cm, and $\ell = 6.0$ cm. Suppose the rectangular loop is a conductor with resistance 50 mΩ and that the current I in the long wire is increasing at the rate of 25 A/s. Find the induced current in the loop. In what direction does it flow?

16. A windmill with conducting blades of length ℓ is rotating with angular speed ω about a horizontal axis; the horizontal component of Earth's magnetic field at its location is B_x. Find an expression for the emf developed between the blade tips and the central axis.

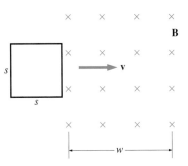

FIGURE 31-43 Problem 17.

17. A square conducting loop of side $s = 0.50$ m and resistance $R = 5.0$ Ω moves to the right with speed $v = 0.25$ m/s. At time $t = 0$ its rightmost edge enters a uniform magnetic field $B = 1.0$ T pointing into the page, as shown in Fig. 31-43. The magnetic field covers a region of width $w = 0.75$ m. Plot (a) the current and (b) the power dissipation in the loop as functions of time, taking a clockwise current as positive and covering the time until the entire loop has exited the field region.

18. A 2.0-m-long solenoid is 15 cm in diameter and consists of 2000 turns of wire. The current in the solenoid is increasing at the rate of 1.0 kA/s. (a) Find the current in a wire loop with diameter 10 cm and resistance 5.0 Ω, lying inside the solenoid in a plane perpendicular to the loop axis. (b) Repeat for a similarly oriented loop with diameter 25 cm, lying entirely outside the solenoid.

19. A solenoid 2.0 m long and 30 cm in diameter consists of 5000 turns of wire. A 5-turn coil with negligible resistance is wrapped around the solenoid and connected to a 180-Ω resistor, as shown in Fig. 31-44. The direction of the current in the solenoid is such that the solenoid's magnetic field points to the right. At time $t = 0$ the solenoid current begins to decay exponentially, being given by $I = I_0 e^{-t/\tau}$, where $I_0 = 85$ A, $\tau = 2.5$ s, and t is the time in seconds. (a) What is the direction of the current in the resistor as the solenoid current decays? What is the value of the resistor current at (b) $t = 1.0$ s and (c) $t = 5.0$ s?

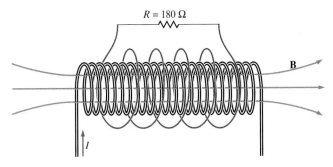

FIGURE 31-44 Problems 19–21.

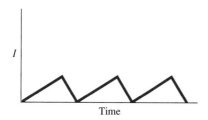

FIGURE 31-45 Problem 20.

20. Make a qualitative plot of the resistor current in the preceding problem as a function of time if the solenoid current has the form shown in Fig. 31-45. Take a left-to-right resistor current as positive.

21. (a) Find an expression for the resistor current in Problem 19 if the solenoid current is given by $I = I_0 \sin \omega t$, where $I_0 = 85$ A and $\omega = 210$ s^{-1}. (b) What is the peak current in the resistor? (c) What is the resistor current when the solenoid current is a maximum?

22. A magnetic field is described by $\mathbf{B} = B_0 \sin \omega t\, \hat{\mathbf{k}}$, where $B_0 = 2.0$ T and $\omega = 10$ s^{-1}. A conducting loop with area 150 cm^2 and resistance 5.0 Ω lies in the x-y plane. Find the induced current in the loop (a) at $t = 0$ and (b) at $t = 0.10$ s.

23. In the preceding problem, what is the first time after $t = 0$ when the loop current will be zero?

24. A car alternator consists of a 250-turn coil 10 cm in diameter in a magnetic field of 0.10 T. If the alternator is turning at 1000 revolutions per minute, what is its peak output voltage?

25. A credit-card reader extracts information from the card's magnetic stripe as it is pulled past the reader's head. At some instant the card motion results in a magnetic field at the head that is changing at the rate of 450 μT/ms. If this field passes perpendicularly through a 5000-turn head coil 2.0 mm in diameter, what will be the induced emf?

26. A generator consists of a rectangular coil 75 cm by 1.3 m, spinning in a 0.14-T magnetic field. If it is to produce a 60-Hz alternating emf (i.e., $\mathscr{E} = \mathscr{E}_0 \sin 2\pi ft$, where $f = 60$ Hz) with peak value 6.7 kV, how many turns must it have?

27. Figure 31-46 shows a pair of parallel conducting rails a distance ℓ apart in a uniform magnetic field **B**. A resistance R is connected across the rails, and a conducting bar of negligible resistance is being pulled along the rails with

velocity **v** to the right. (a) What is the direction of the current in the resistor? (b) At what rate must work be done by the agent pulling the bar?

28. The resistor in the preceding problem is replaced by an ideal voltmeter. (a) To which rail should the positive meter terminal be connected? (b) At what rate must work be done by the agent pulling the bar?

29. A battery of emf $\mathscr{E}$ is inserted in series with the resistor in Fig. 31-46, with its positive terminal toward the top rail. The bar is initially at rest, and now no agent pulls it. (a) Describe the bar's subsequent motion. (b) The bar eventually reaches a constant speed. Why? (c) What is that constant speed? Express in terms of the magnetic field, the battery emf, and the rail spacing ℓ. Does the resistance R affect the final speed? If not, what role does it play?

30. A toroidal coil of square cross section has inner radius a and outer radius b. It consists of N turns of wire and carries a time-varying current $I = I_0 \sin \omega t$. A single-turn wire loop encircles the toroid, passing through its center hole as shown in Fig. 31-47. Find an expression for the peak emf induced in the loop.

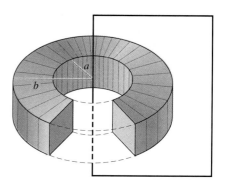

FIGURE 31-47 Problem 30.

31. A pair of parallel conducting rails 10 cm apart lie at right angles to a uniform magnetic field **B** of magnitude 2.0 T, as shown in Fig. 31-48. A 5.0-Ω and a 10-Ω resistor lie across the rails and are free to slide along them. (a) The 5-Ω resistor is held fixed, and the 10-Ω resistor is pulled to the right at 50 cm/s. What are the direction and magnitude of the induced current? (b) Now the 10-Ω resistor is held

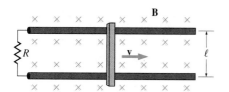

FIGURE 31-46 Problems 27–29, 33, 63.

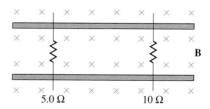

FIGURE 31-48 Problems 31, 32, and 64.

fixed, and the 5-Ω resistor is pulled to the left at 50 cm/s. What are the direction and magnitude of the induced current? (c) What is the power dissipation in the 10-Ω resistor in both cases?

32. In Fig. 31-48 the 10-Ω resistor is being moved to the right at a constant 50 cm/s. The 5-Ω resistor, initially at rest, is placed across the conducting rails. Describe qualitatively its subsequent motion, and determine its final speed.

33. In Fig. 31-46, take ℓ = 10 cm, B = 0.50 T, R = 4.0 Ω, and v = 2.0 m/s. Find (a) the current in the resistor, (b) the magnetic force on the bar, (c) the power dissipation in the resistor, and (d) the mechanical work done by the agent pulling the bar. Compare your answers to (c) and (d).

34. A square conducting loop of side ℓ and resistance R lies in the x-y plane and is being moved with constant velocity $\mathbf{v} = v\hat{\mathbf{i}}$, its sides parallel to the x and y axes. For $x < 0$ there is a uniform magnetic field $\mathbf{B} = B_0\hat{\mathbf{k}}$; for $x > 0$ the field is nonuniform and is given by $\mathbf{B} = (B_0 + bx)\hat{\mathbf{k}}$, where B_0 and b are positive constants. At time $t = 0$ the trailing side of the loop crosses the y axis, so the loop is entirely in the nonuniform field region. (a) Find an expression for the loop current for times $t \geq 0$. (b) Which way does the current flow, as viewed from the positive z axis?

35. A circular loop 40 cm in diameter is made from a flexible conductor and lies at right angles to a uniform 12-T magnetic field. At time $t = 0$ the loop starts to expand, its radius increasing at the rate of 5.0 mm/s. Find the induced emf in the loop (a) at $t = 1.0$ s and (b) at $t = 10$ s.

36. A spherical balloon initially has radius 20 cm. A conducting stripe with resistance 2.4 Ω is painted around the balloon's "equator," and a uniform 1.5-T magnetic field points perpendicular to the "equatorial plane." The balloon is deflated so its radius decreases at the constant rate of 1.0 cm/s. Find the current in the conducting stripe when the radius has been reduced to 10 cm, assuming the stripe's resistance stays constant.

Section 31-4 Induced Electric Fields

37. The induced electric field 12 cm from the axis of a solenoid with 10 cm radius is 45 V/m. Find the rate of change of the solenoid's magnetic field.

38. Find the electric force on a 50-μC charge inside the solenoid of Problem 18, if the charge is 5.0 cm from the solenoid axis.

39. Figure 31-49 shows a top view of a tokamak. The magnetic field in the center is confined to a circular area of radius 50 cm, and during a pulse it increases at the rate of 5.1 T/ms. (a) What is the magnitude of the induced electric field in the tokamak, 1.2 m from the center of the field region in Fig. 31-49? (b) What is the field direction? (c) If a proton circles the tokamak once at this radius, going with the electric field, how much energy does it gain?

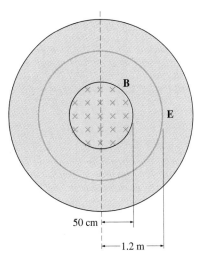

FIGURE 31-49 Top view of a tokamak (Problem 39).

40. A uniform magnetic field points into the page in Fig. 31-50. In the same region an electric field points straight up, but increases with position at the rate of 10 V/m² as you move to the right. Apply Faraday's law to a rectangular loop to show that the magnetic field must be changing with time, and calculate the rate of change.

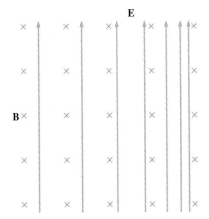

FIGURE 31-50 The magnetic field must be changing. Why? (Problem 40)

41. In Example 31-9, take the solenoid radius R = 10 cm and suppose the magnetic field inside the solenoid is given by $B = 0.10t^3 - 1.1t^2 + 2.8t$, with B in T and t in ms. (a) Find the electric field strength 14 cm from the solenoid axis at t = 1.0 ms. (b) Find a time when the induced electric field in and around the solenoid is zero.

42. Use Faraday's law to show that the electric field produced by charges on the plates of a parallel-plate capacitor cannot

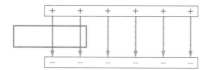

FIGURE 31-51 Problem 42.

end abruptly at the edges of the plates. *Hint:* Consider the loop shown in Fig. 31-51.

43. Figure 31-52 shows a magnetic field pointing into the page; the field is confined to a layer of thickness h in the vertical direction but extends infinitely to the left and right. The field strength is increasing with time: $B = bt$, where b is a constant. Find an expression for the electric field at all points outside the field region. *Hint:* Consult Example 30-5.

FIGURE 31-52 Problem 43.

44. The magnetic field inside a solenoid of circular cross section is given by $\mathbf{B} = bt\,\hat{\mathbf{k}}$, where $b = 2.1$ T/ms. At time $t = 0.40\,\mu$s a proton is inside the solenoid at the point $x = 5.0$ cm, $y = 0$, $z = 0$ and is moving with velocity $\mathbf{v} = 4.8\times10^6\,\hat{\mathbf{j}}$ m/s. Find the net electromagnetic force on the proton.

Paired Problems

(Both problems in a pair involve the same principles and techniques. If you can get the first problem, you should be able to solve the second one.)

45. A magnetic field is given by $\mathbf{B} = B_0(x/x_0)^2\,\hat{\mathbf{k}}$, where B_0 and x_0 are constants. Find an expression for the magnetic flux through a square of side $2x_0$ that lies in the x-y plane with one corner at the origin and two sides coinciding with the positive x and y axes.

46. A circular region of radius R lies in the x-y plane and contains a magnetic field given by $\mathbf{B} = B_0\dfrac{r}{R}\hat{\mathbf{k}}$, where r is the radial distance from the central axis of the field region. Find an expression for the magnetic flux through this region.

47. A uniform magnetic field is given by $\mathbf{B} = bt\,\hat{\mathbf{k}}$, where $b = 0.35$ T/s. Find the current in a conducting loop with area 240 cm² and resistance 0.20 Ω that lies in the x-y plane. In what direction is the current, as viewed from the positive z axis?

48. A uniform magnetic field is given by $\mathbf{B} = bt^3\,\hat{\mathbf{k}}$. A square conducting loop 15 cm on a side has 0.32-Ω resistance and lies in the x-y plane. At time $t = 2.5$ s, the current in the loop is 4.1 A. Find b.

49. A pair of vertical conducting rods are a distance ℓ apart and are connected at the bottom by a resistance R. A conducting bar of mass m runs horizontally between the rods and can slide freely down them while maintaining electrical contact. The whole apparatus is in a uniform magnetic field **B** pointing horizontally and perpendicular to the bar. When the bar is released from rest it soon reaches a constant speed. Find this speed.

50. A conducting bar of mass m slides down the conducting wedges shown in Fig. 31-53. The wedges are separated by a distance ℓ, connected at the top by a resistance R, and make an angle θ with the vertical. A uniform magnetic field **B** points horizontally, as shown. When released from rest the bar soon reaches a constant speed. Find an expression for this speed.

FIGURE 31-53 Problem 50.

51. Figure 31-54 shows an unusual design for a generator, consisting of a conducting bar that rotates about a central axis while making contact with a conducting ring of radius R. A uniform magnetic field is perpendicular to the ring. Wires from the axis and ring carry power to a load. Find an expression for the emf induced in this generator when the bar rotates with angular speed ω.

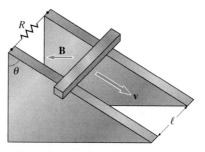

FIGURE 31-54 Problem 51.

52. A copper disk 90 cm in diameter is spinning at 3600 rpm about a conducting axle through its center, as shown in Fig. 31-55. A uniform 1.5-T magnetic field is perpendicular to the disk, as shown. A stationary conducting brush maintains contact with the disk's rim, and a voltmeter is

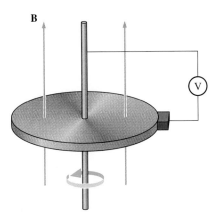

FIGURE 31-55   Problem 52.

connected between the brush and the axle. (a) What does the voltmeter read? (b) Which voltmeter lead is positive?

53. An electron is inside a solenoid, 28 cm from the solenoid axis. It experiences an electric force of magnitude 1.3 fN. At what rate is the solenoid's magnetic field changing?

54. It takes 0.43 J to push a 84-μC charge around a closed path surrounding a 1.5-m-diameter solenoid. At what rate is the solenoid's magnetic field changing?

Supplementary Problems

55. At times prior to $t = 0$, there is no current in either the solenoid or the small coil of Problem 19. Subsequently, the current in the small coil is observed to increase at 10 μA/s. What is the solenoid current as a function of time?

56. A conducting loop of area A and resistance R lies perpendicular to a uniform magnetic field B. The loop is then rotated at a uniform rate until it is upside down; this takes time Δt. Find an expression for the work done in flipping the loop.

57. A conducting rod of length ℓ moves at speed v in a plane perpendicular to a uniform magnetic field **B,** as shown in Fig. 31-56. The magnetic force on charge carriers in the rod causes charge separation, which creates an electric

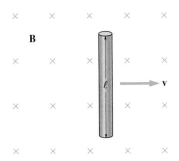

FIGURE 31-56   Problem 57.

field. Charge motion stops when the electric and magnetic forces on the charge carriers are equal. Show that this condition results in an electric field of magnitude vB and, therefore, in a potential difference $B\ell v$ between the rod ends, as we found from flux considerations in Example 31-8.

58. The tethered satellite system of Example 31-8 is flown in a low circular orbit about a planet of mass M and radius R. If the tether has length ℓ and generates an emf $\mathcal{E}$, find an expression for the planet's magnetic field strength.

59. Clever farmers whose lands are crossed by large power lines have been known to steal power by stringing wire near the power line and making use of the induced current—a practice that has been ruled legally to be theft. The scene of a particular crime is shown in Fig. 31-57. The power line carries 60-Hz alternating current with a peak current of 10 kA (that is, the current is given by $I = I_0 \sin \omega t$, where $I_0 = 10$ kA and $\omega = 2\pi f$, with $f = 60$ Hz). (a) If the farmer wants a peak voltage of 170 V, what should be the length ℓ of the loop shown in Fig. 31-57? (170 V is the peak of standard 120-V AC power.) (b) If all the equipment the farmer connects to the loop has an equivalent resistance of 5.0 Ω, what is the farmer's average power consumption? *Note:* The *average* power consumption is half the product of the peak voltage and peak current. (c) If the power company charges 10¢ per kWh, what is the monetary value of the energy stolen each day? (d) Without examining the farmer's lands, how, in principle, could the power company know that a crime is being committed?

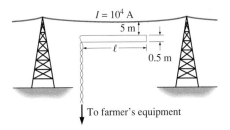

FIGURE 31-57   Problem 59.

60. A circular wire loop of resistance R and radius a lies with its plane perpendicular to a uniform magnetic field. The field strength changes from an initial value B_1 to a final value B_2. Show, by integrating the loop current over time, that the total charge that moves around the ring is

$$q = \frac{\pi a^2}{R}(B_2 - B_1).$$

Note that this result is independent of how the field changes with time.

61. A generator like that shown in Fig. 31-14 has an N-turn coil of area A spinning with angular speed ω in a uniform magnetic field B. A resistor R is connected across the generator. (a) Find an expression for the power dissipated in the resistor as a function of time. (b) Find an expression for the magnetic torque on the generator coil as a function of time. (c) Study the discussion associated with Equations 12-28a and b, and use it to show that the rate at which the agent turning the generator at constant angular speed does work is equal to the power dissipation in the resistor. Assume the coil's magnetic moment is aligned with the field a time $t = 0$.

62. A conducting disk with radius a, thickness h, and resistivity ρ is inside a solenoid of circular cross section. The disk axis coincides with the solenoid axis. The magnetic field in the solenoid is given by $B = bt$, with b a constant. Find expressions for (a) the current density in the disk as a function of the distance r from the disk center and (b) the rate of power dissipation in the entire disk. *Hint:* Consider the disk to be made up of infinitesimal conducting loops.

63. The bar in Problem 27 has mass m and is initially at rest. A constant force F is applied to the bar, pulling it to the right. (a) Formulate Newton's second law for the bar as an equation involving both v and $a = dv/dt$. (b) Use your equation to show that the bar's acceleration becomes zero when its speed reaches the value $FR/B^2\ell^2$. (c) Show by direct substitution that your equation is satisfied if v as a function of time is given by $v(t) = \dfrac{FR}{B^2\ell^2}(1 - e^{-B^2\ell^2 t/mR})$.

64. Find an expression for the speed of the left-hand resistor in Problem 32 as a function of time, in terms of its mass m, the field strength B, the speed v of the right-hand bar, the time t, and the resistance R_{left} and R_{right}.

65. A pendulum consists of a mass m suspended from two identical copper wires of negligible mass. At equilibrium the mass is a vertical distance ℓ below its supports, and the wires make 45° angles with the vertical, as shown in Fig. 31-58. A uniform magnetic field **B** points into the page. The pendulum is displaced from the plane of the page by a small angle θ_0, and at time $t = 0$ it is released. Find an expression for the voltmeter reading as a function of time.

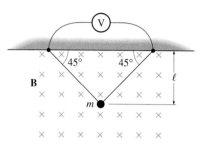

FIGURE 31-58 Problem 65.

66. A *flip coil* consists of a small coil used to measure magnetic fields. The flip coil is placed in a magnetic field with its plane perpendicular to the field, and then rotated abruptly through 180° about an axis in the plane of the coil. The coil is connected to instrumentation to measure the total charge Q that flows during this process. If the coil has N turns of area A and if its rotation axis is perpendicular to the magnetic field, show that the field strength is given by $B = QR/2NA$, where R is the coil resistance.

CHAPTER 32

INDUCTANCE AND MAGNETIC ENERGY

This transformer at an electrical substation uses mutual inductance to change the voltage at which power is transmitted.

F araday's law implies that a changing magnetic flux through a circuit produces an induced emf in the circuit. In this chapter we consider the special
case when that changing flux is itself caused by a changing current in an
electric circuit. We then speak of the **inductance** of the circuit.

32-1 MUTUAL INDUCTANCE

Consider two coils arranged so that some of the magnetic flux associated with
current in one coil also passes through the second coil, as in Fig. 32-1. If we
change the current I_1 in the first coil, an induced emf $\mathcal{E}_2$ appears in the second.
As we discussed in the preceding chapter, $\mathcal{E}_2$ depends on the rate of change of
magnetic flux through the second coil. The magnetic flux depends, in turn, on
the current in the first coil and on the geometrical arrangement of the two coils
that determines how much flux from the first coil actually links the second. We

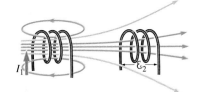

FIGURE 32-1 The mutual inductance between two coils is the ratio of
magnetic flux through one coil to the
current in the other coil A changing current in one coil induces an emf in the
other.

characterize this geometrical arrangement by the ratio of the total magnetic flux through the second coil to the current in the first coil. This ratio defines the **mutual inductance,** M, of the two coils:

$$M = \frac{\phi_2}{I_1}. \tag{32-1}$$

Solving Equation 32-1 for ϕ_2 and differentiating, we obtain

$$\frac{d\phi_2}{dt} = M\frac{dI_1}{dt}.$$

Faraday's law says that $d\phi_2/dt$ is $-\mathscr{E}_2$, the induced emf in coil 2. So we have

$$\mathscr{E}_2 = -M\frac{dI_1}{dt}, \tag{32-2}$$

where the minus sign describes the polarity of the induced emf.

We could equally well have considered the case where a changing current in the second coil induces an emf in the first. Although it is not at all obvious, the same value of mutual inductance applies in this case even when the arrangement of the two coils is far from symmetric.

From Equation 32-2, we see that the unit of mutual inductance is the volt-second/ampere. This unit is given the name henry (H) in honor of the American scientist Joseph Henry (1797–1878), who was also the first secretary of the Smithsonian Institution. Mutual inductances found in common electronic circuits usually range from microhenrys (μH) on up to several henrys.

● **EXAMPLE 32-1** MUTUAL INDUCTANCE

A 2-turn coil is wrapped around a long solenoid with cross-sectional area $A = 26$ cm², wound with $n = 3500$ turns per meter of length (Fig. 32-2). Find the mutual inductance of this arrangement.

Solution
Since the solenoid field is uniform and confined to the solenoid interior, the flux through each turn of the small coil is just BA. Accounting for the two turns and using Equation 30-12 for the solenoid's magnetic field, the total magnetic flux through the small coil becomes

$$\phi_2 = 2BA = 2\mu_0 nIA,$$

where I is the solenoid current. Equation 32-1 shows that mutual inductance is the ratio of the flux in one coil to the current in the other, or

$$M = \frac{\phi_2}{I} = \frac{2\mu_0 nIA}{I} = 2\mu_0 nA$$

$$= (2)(4\pi\times10^{-7} \text{ N/A}^2)(3500)(26\times10^{-4} \text{ m}^2)$$

$$= 23 \ \mu\text{H}.$$

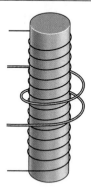

FIGURE 32-2 What is the mutual inductance of the two coils? (Example 32-1)

EXERCISE Suppose the solenoid of Example 32-1 has circular cross section, and that the smaller coil is *inside* the solenoid, with its diameter half that of the solenoid. Find the mutual inductance.

Answer: 5.7 μH

Some problems similar to Example 32-1: 6, 7, 10

●

● EXAMPLE 32-2 AN IGNITION COIL

Your car's spark plugs ignite the gasoline-air mixture that burns to power the engine. The high voltage that causes sparks to jump across the plug gaps is provided by the ignition coil—actually an arrangement of two tightly wound coils (Fig. 32-3). Current from the car's 12-V battery flows through the coil with fewer turns and is interrupted periodically by a switch in the distributor. The sudden change in current induces a large emf in the coil with more turns, and this emf drives the spark. Interruption of the current is carefully timed so that the spark occurs at exactly the right point in the engine cycle. An important part of a "tune-up" is the precise adjustment of this ignition timing.

A typical ignition coil draws 3.0 A and supplies 20 kV to the spark plugs. If the current decays in 0.10 ms when the switch opens, what is the mutual inductance of the ignition coil?

Solution

The rate of change of current is

$$\frac{dI}{dt} = \frac{3.0 \text{ A}}{0.10 \times 10^{-3} \text{ s}} = 3.0 \times 10^4 \text{ A/s}.$$

Solving Equation 32-2 for M then gives

$$M = \frac{|\mathscr{E}|}{|dI/dt|} = \frac{20 \times 10^3 \text{ V}}{3.0 \times 10^4 \text{ A}} = 0.67 \text{ H}.$$

EXERCISE An electric toothbrush has no electrical connection to the power line (Fig. 32-4). But when the toothbrush is in its stand, a coil inside the toothbrush itself rests inside another coil in the stand, and alternating current from the power line flows in the stand coil. The current thus induced in the toothbrush coil is used to charge the batteries that power the toothbrush. Suppose the mutual inductance of this arrangement is 100 mH. At an instant when the current in the stand coil is changing at the rate of 40 A/s, what is the emf in the toothbrush coil?

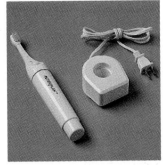

FIGURE 32-4 The batteries in this electric toothbrush are charged with energy transferred via mutual induction of coils located in the base unit, which is connected to the AC power line, and in the bottom of the brush unit. There is no direct electrical connection to the brush unit.

Answer: 4.0 V

Some problems similar to Example 32-2: 1–4 ●

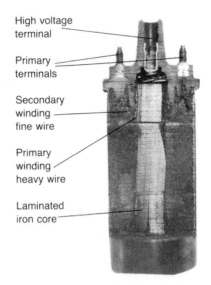

High voltage terminal

Primary terminals

Secondary winding fine wire

Primary winding heavy wire

Laminated iron core

FIGURE 32-3 An automobile ignition coil uses electromagnetic induction to produce high voltage that drives sparks to ignite the gasoline.

32-2 SELF-INDUCTANCE

So far we've considered emf and current induced in a circuit by changes—like moving a magnet or varying the current in another circuit—that were external to the circuit in question. But the changing magnetic field associated with the changing current in a circuit also affects that same circuit.

Consider a circular loop carrying current I, as shown in Fig. 32-5. Magnetic field lines arising from this current loop pass through the loop, making a magnetic flux through the loop. If the current is steady, this flux is constant, and

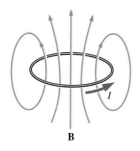

FIGURE 32-5 Magnetic flux from a circular current loop passes through the loop itself.

FIGURE 32-6 Typical inductors. At left is a 20-H unit, wound on an iron core for increased inductance. The others have inductances in the mH range.

there is no induced electric field. But if we change the loop current, then the flux changes, and an induced electric field arises. In order to conserve energy this field opposes the change that causes it—in this case the change in loop current. If the current is counterclockwise and we increase its strength, an induced electric field will appear in the clockwise direction to oppose the current increase. If we decrease the current, the induced electric field will have the opposite sense, now trying to maintain the current. The induced electric field therefore makes it difficult to change the current in the circuit.

This property of a circuit whereby its own magnetic field opposes changes in current is termed **self-inductance.** All circuits possess self-inductance, but this inductance is important only when the circuit encircles a great many of its own magnetic field lines or when current changes very rapidly. A simple piece of wire exhibits very little opposition to current changes in the 60-Hz alternating current used in electric power systems. But in a TV set or computer, where currents change on time scales of billionths of a second, self-inductance of the wires themselves must be taken into account.

An **inductor** is a device designed specifically to exhibit self-inductance. A typical inductor consists of a coil of wire, constructed so that a great deal of its own magnetic flux is encircled. Some inductors are wound on iron cores to promote flux concentration (Fig. 32-6). Ideally, the only electrical property of an inductor is its inductance. But real inductors are made from wire, so they have resistance as well.

As long as the current in an inductor is steady, the inductor acts just like a piece of wire. But when the current changes, the changing magnetic flux induces an emf that opposes the change in current. The more rapidly the current changes, the greater the rate of change of flux and so the greater the emf. We characterize the inductor by its self-inductance, L, defined as the ratio of magnetic flux through the inductor to current in the inductor:

$$L = \frac{\phi_B}{I}. \tag{32-3}$$

The unit of self-inductance, like that of mutual inductance, is the henry. Inductance is a constant determined by the physical design of an inductor. In principle, we can calculate the inductance of any inductor, but in practice this is difficult unless the geometry is particularly simple. Inductors for use in electronic circuits are available commercially in a wide range of inductance values.

● **EXAMPLE 32-3** INDUCTANCE OF A SOLENOID

A long solenoid of cross-sectional area A and length ℓ has n turns per unit length. What is its self-inductance?

Solution

Equation 32-3 shows that self-inductance is the ratio of magnetic flux to current. In Chapter 30 we used Ampère's law to find the magnetic field of a long solenoid: $B = \mu_0 nI$. Since the field is uniform and perpendicular to the solenoid coils, the flux

through each turn is just BA. With n turns per unit length, our solenoid has $N = n\ell$ turns. The total flux through the solenoid is then

$$\phi_B = NBA = (n\ell)(\mu_0 nI)(A) = \mu_0 n^2 A\ell I.$$

Then Equation 32-3 gives

$$L = \frac{\phi_B}{I} = \mu_0 n^2 A\ell. \tag{32-4}$$

Does this result make sense? As the area increases, so does the flux and therefore the inductance. As the length increases so does the number of turns, and therefore the flux and the inductance increase. Equation 32-4 reflects these trends. Can you see why the inductance should be proportional to the *square* of the number of turns per unit length?

EXERCISE What is the self-inductance of the MRI solenoid in Example 30-6 (page 768)?

Answer: 0.53 H

Some problems similar to Example 32-3: 11, 15, 20 ●

The induced emf in an inductor is determined by Faraday's law, which relates the emf to the rate of change of magnetic flux:

$$\mathcal{E} = -\frac{d\phi_B}{dt}.$$

Differentiating Equation 32-3, the definition of inductance, gives

$$\frac{d\phi_B}{dt} = L\frac{dI}{dt}.$$

Then Faraday's law becomes

$$\mathcal{E} = -L\frac{dI}{dt}. \tag{32-5}$$

This equation gives the emf $\mathcal{E}$ induced in an inductor L when the current in the inductor is changing at the rate dI/dt. The minus sign again tells us that the emf *opposes* the change in current. For this reason the inductor emf is often called a **back emf**; it works *against* changes brought about by an externally applied emf.

When the current in an inductor is steady, then $dI/dt = 0$ and there is no emf in the inductor. In this case, the inductor acts just like a piece of wire. But when the current changes the inductor responds by producing a back emf that opposes the change in current. Now the inductor acts very much like a battery, with the magnitude of its emf dependent on how fast the current changes. If we try to start or stop current suddenly, dI/dt is very large and a very large back emf appears. This is not merely mathematics! Rapid switching of inductive devices such as solenoids, solenoid valves, or motors can result in destruction of delicate electronic devices by induced currents. And people have been killed opening switches in circuits containing large inductors.

● **EXAMPLE 32-4** A DANGEROUS INDUCTOR

A current of 5.0 A is flowing in a 2.0-H inductor. The current is reduced steadily to zero in 1.0 ms. What is the magnitude of the emf in the inductor while the current is being turned off?

Solution
Because the current changes steadily, its time rate of change has magnitude

$$\frac{dI}{dt} = \frac{5.0 \text{ A}}{1.0 \text{ ms}} = 5000 \text{ A/s},$$

so

$$|\mathcal{E}| = L\frac{dI}{dt} = (2.0 \text{ H})(5000 \text{ A/s}) = 10,000 \text{ V},$$

enough to produce a lethal shock. Note that this voltage is quite unrelated to the voltage of the battery or whatever else was supplying the inductor current. We could have a 6-volt battery and still be electrocuted trying to open the circuit rapidly when a large inductance is present.

EXERCISE A neon lamp that glows only when the voltage across it exceeds 90 V is connected across a 1.2-H inductor

carrying 750 mA. When the current is interrupted the lamp flashes. Find the maximum time over which the current could have dropped to zero, assuming a steady decrease.

Answer: 10 ms

Some problems similar to Example 32-4: 12–14 ●

32-3 INDUCTORS IN CIRCUITS

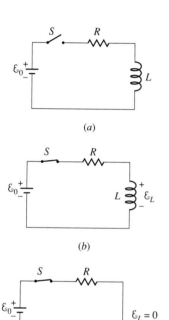

(a)

(b)

(c)

FIGURE 32-7 (a) An *RL* circuit. With the switch open no current flows. (b) Immediately after the switch is closed there is still no current. The inductor produces a back emf equal but opposite the battery emf, and therefore the rate of change of current is not zero. (c) After a long time, the rate of change of current and therefore the inductor emf approach zero. The inductor then acts like a piece of wire.

Here we examine circuits containing batteries, resistors, and inductors, analogous to the *RC* circuits of Chapter 28. In the qualitative analysis of *RC* circuits we found a useful guiding principle: The voltage across a capacitor cannot change instantaneously. We can make an analogous statement for inductors. Because the inductor emf is proportional to the rate of change of current in the inductor and because an infinite emf is physically impossible, we conclude the following:

▌ **The current through an inductor cannot change instantaneously.**

Thus, the effect an inductor has on current is analogous to the effect a capacitor has on voltage. Much of our understanding of capacitors can be applied to inductors if we interchange the words *voltage* and *current*.

Building Up the Current

Figure 32-7a shows a circuit containing a battery, switch, resistor, and inductor (symbol ⌇⌇⌇). What happens when we close the switch? Initially the inductor current is zero; since it can't change instantaneously, it must remain zero immediately after the switch is closed. But this is a series circuit, so the inductor and resistor currents are equal. With zero current immediately after the switch is closed, there must be no voltage across the resistor. Therefore the inductor must produce a back emf equal to that of the battery, with the polarity shown in Fig. 32-7b. Even though there is, at this instant, no current in the inductor, the presence of an emf indicates that the *rate of change* of the current is not zero. Going around the circuit in the direction that the battery would drive a current, the inductor polarity indicated in Fig. 32-7b shows that the inductor emf is *negative*—and Equation 32-5 therefore shows that dI/dt is *positive*—i.e., the current is *increasing*.

As the current rises, so does the voltage across the resistor (since $V_R = IR$). Since the battery emf is constant, that means the inductor emf goes *down*—and that means the rate of change of current goes down. Thus, the current in the circuit builds up, but at an ever-decreasing rate. Concurrently, the inductor emf goes down. Eventually the current reaches a steady value, at which point dI/dt and, therefore, the inductor emf are zero. In this ultimate steady state the inductor acts like a piece of wire, and the circuit looks like Fig. 32-7c. The steady-state current is just $\mathcal{E}_0/R$, where $\mathcal{E}_0$ is the battery emf. Figure 32-8 summarizes graphically this qualitative analysis of the *RL* circuit.

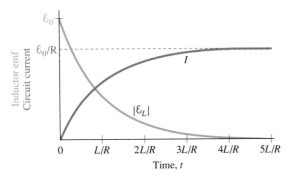

FIGURE 32-8 Inductor current and magnitude of the inductor emf as functions of time. Vertical axis represents either current or voltage.

To analyze the *RL* circuit quantitatively, we apply the loop law. Going clockwise from the negative terminal of the battery, we first encounter a voltage increase $\mathcal{E}_0$ due to the battery. Then the voltage decreases by *IR* in the resistor. Finally, there is a voltage change $\mathcal{E}_L$ in the inductor. This change is actually a decrease, because the inductor emf opposes the increase in current. However, we will simply call the inductor emf $\mathcal{E}_L$ and let Equation 32-5 take care of the signs. Then the loop law becomes

$$\mathcal{E}_0 - IR + \mathcal{E}_L = 0. \qquad (32\text{-}6a)$$

If we differentiate this equation with respect to time, the battery emf $\mathcal{E}_0$ drops out because it is constant, giving

$$\frac{d\mathcal{E}_L}{dt} = R\frac{dI}{dt}.$$

Equation 32-5 shows that $dI/dt = -\mathcal{E}_L/L$, so our differentiated loop equation becomes

$$\frac{d\mathcal{E}_L}{dt} = -R\frac{\mathcal{E}_L}{L}. \qquad (32\text{-}6b)$$

This differential equation describes a quantity—$\mathcal{E}_L$—whose rate of change is proportional to itself. We discussed such equations in Chapter 28 when we considered the *RC* circuit. Equation 32-6b is similar to Equation 28-4, but with current *I* replaced by the inductor emf $\mathcal{E}_L$, capacitance *C* by *L*, and *R* by $1/R$. The solution to Equation 32-6b is that of Equation 28-4, provided we make the appropriate substitutions for *I*, *C*, and *R*:

$$\mathcal{E}_L = -\mathcal{E}_0 e^{-Rt/L}. \qquad (32\text{-}7)$$

This equation shows that the inductor emf decays exponentially to zero, starting from an initial value of $-\mathcal{E}_0$ (negative because the inductor emf *opposes* the battery emf). We can now solve for the current using Equation 32-6a:

$$I = \frac{\mathcal{E}_0 + \mathcal{E}_L}{R} = \frac{\mathcal{E}_0 + (-\mathcal{E}_0 e^{-Rt/L})}{R} = \frac{\mathcal{E}_0}{R}(1 - e^{-Rt/L}). \qquad (32\text{-}8)$$

With a capacitor, we characterized the exponentially changing quantities in terms of the capacitive time constant RC. With an inductor, we have an **inductive time constant** L/R. Significant changes in current cannot occur on time scales much shorter than L/R. On the other hand, an RL circuit will approach a steady state, with zero $\mathcal{E}_L$, only after many inductive time constants.

Why is the inductive time constant a quotient of L and R rather than a product, as in the capacitor case? In Problem 23 you will convince yourself mathematically that L/R does indeed have the units of seconds. But you can also understand this physically. The larger L, the larger the back emf and the longer it takes the current to build up. The larger R, the smaller the final current and so, all else being equal, the smaller the rate of change of current, and, therefore, the smaller the inductive effects.

● EXAMPLE 32-5 FIRING UP AN ELECTROMAGNET

A large electromagnet used for lifting scrap metal has a self-inductance of 56 H. It is connected through a switch to a constant 440-V power source; the total resistance of the circuit is 2.8 Ω. When the switch is closed, how long does it take to bring the magnet current to 75% of its final value?

Solution

Letting $t \to \infty$ in Equation 32-8 shows that the final current is $I_{\text{final}} = \mathcal{E}_0/R$, as we argued in our qualitative analysis. Setting the current I in Equation 32-8 to $0.75\mathcal{E}_0/R$ gives

$$0.75 = 1 - e^{-Rt/L},$$

or

$$e^{-Rt/L} = 0.25.$$

Taking the natural logarithm of both sides (recall that $\ln e^x = x$) gives

$$-Rt/L = \ln(0.25),$$

or

$$t = -\frac{L}{R}\ln(0.25) = -\frac{56\text{ H}}{2.8\text{ Ω}}\ln(0.25) = 28\text{ s}.$$

(The minus sign canceled since the logarithm of a number less than 1 is negative.) Our answer is approximately one time constant ($L/R = 20$ s). This should not be surprising since we found with RC circuits that quantities following equations like Equation 28-6 (and, therefore, its analog, Equation 32-8) reach $1 - 1/e$, or about two-thirds, of their final value in one time constant.

EXERCISE A 1.0-kΩ resistor is in series with an inductor, and a 12-V battery is connected across the pair. The current rises to 8.5 mA in 21 μs. Find the inductance.

Answer: 61 mH

Some problems similar to Example 32-5: 27–29, 32 ●

The Current Decays

Figure 32-9a shows an RL circuit with a two-way switch. Throwing the switch to position A allows current to build up in the inductor as we've already described. Then, at time $t = 0$, we throw it to position B. This disconnects the battery, leaving a circuit electrically equivalent to Fig. 32-9b. Just prior to $t = 0$ there was some current I_0 flowing downward in the inductor. Since the inductor current cannot change instantaneously, that current must continue just after the switch is closed, as shown in Fig. 32-9b. To drive this current, the inductor must develop an emf in the direction shown. Now the inductor emf is positive, so Equation 32-5 shows that the current is *decreasing,* as we might well expect since the battery has been disconnected. As the current decreases, so does the voltage across the resistor. So, therefore, does the inductor emf and, therefore, the rate of change of current. We thus expect both the current and the inductor emf to decrease, but at an ever-decreasing rate.

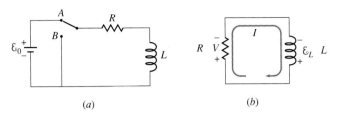

(a)   (b)

FIGURE 32-9 (a) Throwing the switch to position A allows current to build up in the inductor. (b) Throwing it to position B gives a circuit containing only the inductor and resistor. The current then decays exponentially.

Note that the inductor emf here is like any other induced emf: it *opposes* the change giving rise to it. In this case that change is the decrease in current caused by disconnecting the battery. The inductor responds with an emf in such a direction as to keep that current flowing.

Applying the loop law to Fig. 32-9b gives

$$\mathcal{E}_L - IR = 0.$$

Using $\mathcal{E}_L = -L \, dI/dt$ from Equation 32-5, the loop equation becomes

$$\frac{dI}{dt} = -\frac{R}{L} I.$$

This is just like Equation 32-6b, but with I replacing $\mathcal{E}_L$. The solution follows by analogy with Equation 32-7:

$$I = I_0 e^{-Rt/L}, \tag{32-9}$$

where I_0 is the inductor current when the switch is thrown from A to B. Equation 32-9 shows that the current decays with the same exponential time constant L/R that described its buildup (Fig. 32-10). The resistor voltage IR and therefore the inductor emf also decay in the same way.

It is not always necessary to use Equations 32-8 and 32-9 in describing RL circuits. For times very short compared with the time constant L/R, it suffices to use the fact that inductor currents cannot change instantaneously. And after many time constants, inductors in a circuit containing only steady sources will act like wires. Example 32-6 explores this situation.

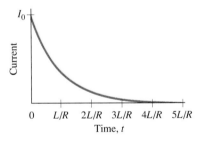

FIGURE 32-10 Exponential decay of the current in the circuit of Fig. 32-9b.

● **EXAMPLE 32-6** SHORT TIMES, LONG TIMES

In the circuit of Fig. 32-11a, the switch is initially open. What is the current in resistor R_2 immediately after the switch is closed? A long time after the switch is closed? Long after the switch is closed, it is again opened. What is the current in R_2 just after it is opened? A long time after?

Solution

Just before we close the switch, the current in the inductor is zero. The current cannot change instantaneously, so it remains zero just after the switch is closed. At this instant the inductor might as well be an open circuit, giving the circuit shown in Fig. 32-11b. Then all the current from R_1 flows through R_2, so

$$I = \frac{\mathcal{E}_0}{R_1 + R_2}.$$

If we wait long enough, the circuit will reach a steady state in which $dI/dt = 0$. So then there is no inductor emf, and the

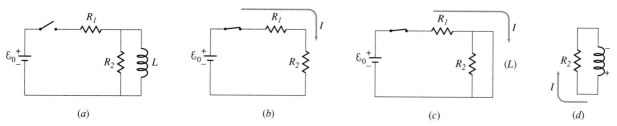

FIGURE 32-11 (a) Circuit for Example 32-6. (b) Just after the switch is closed the inductor acts like an open circuit. (c) A long time later the inductor emf is zero, so it acts like a wire. (d) When the switch opens again, current continues to flow in the inductor and on through R_2.

inductor acts like a wire. We can redraw the circuit as Fig. 32-11c, in which all the current from R_1 goes through L, and none through R_2. The resulting current in R_1 and L is just $\mathcal{E}_0/R_1$.

Now the switch is opened again. Current in R_1 stops abruptly since there is no way charge can get through the open switch. But the current through the inductor, which was $\mathcal{E}_0/R_1$ just before the switch was opened, remains $\mathcal{E}_0/R_1$ the instant after the switch is opened. There is only one place this current can go—through R_2, from bottom to top (Fig. 32-11d). So just after the switch is opened, the current in R_2 is $\mathcal{E}_0/R_1$. Notice that the value of R_2 has no effect on this current, which is determined entirely by the battery emf and the resistance R_1.

What about the voltage across R_2? This is given by Ohm's law:

$$V_2 = I_2 R_2 = \frac{\mathcal{E}_0 R_2}{R_1}.$$

The larger R_2, the larger the voltage that appears when the switch is opened. If R_2 has infinite resistance, or is not in the circuit, the voltage will be arbitrarily large as the inductor seeks at all cost to keep the current flowing. This dangerous situation can result in arcing and vaporization of circuit conductors, and even in electric shock. In circuits with large inductance, resistors are often placed in parallel with inductors to alleviate these dangers.

Finally, the current in Fig. 32-11d decays exponentially to zero. Plots of the currents in R_2 and L as functions of time are shown in Fig. 32-12.

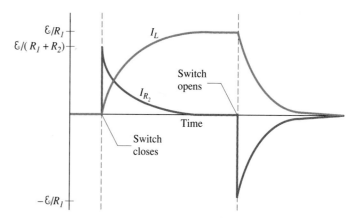

FIGURE 32-12 Currents in R_2 and L for Example 32-6. Note that the time constants before and after the switch opens are different; this reflects the fact that R_1 is out of the circuit after the switch opens.

EXERCISE In Fig. 32-11 take $\mathcal{E}_0 = 12$ V, $R_1 = 56\ \Omega$, $L = 48$ mH, and suppose the switch has been closed for a long time. What is the maximum value of R_2 for which the inductor emf will not exceed 100 V when the switch is opened?

Answer: 467 Ω

Some problems similar to Example 32-6: 36, 37, 61, 62 ●

32-4 MAGNETIC ENERGY

In the situations we considered in Figs. 32-9b and 32-11d, current flows in circuits containing only a resistor and an inductor. Energy is dissipated, heating the resistor. Where does this energy come from?

Because there is a current in the inductor, there is also a magnetic field. The change in that magnetic field is what produces the emf that drives current around the circuit. As the current decreases, so does the inductor's magnetic field. Eventually the circuit reaches a state where there is no current, no magnetic field—and a hot resistor. So where did the resistor's thermal energy come from? It came from the magnetic field.

Like the electric field, the magnetic field contains stored energy. Our decaying *RL* circuit is analogous to a discharging *RC* circuit, in which the electric field between the capacitor plates disappears as thermal energy appears in the resistor. As in the electric case, magnetic energy is not limited to circuits. *Any* magnetic field contains energy. Release of magnetic energy drives a number of practical devices and also powers violent events throughout the universe (Fig. 32-13).

We can reinterpret the *RL* circuit of Fig. 32-9 in terms of energy. With the switch in position *A*, the battery supplies energy to the resistor and inductor. In the resistor the energy is dissipated as heat, but in the inductor it goes into the magnetic field. When the switch is thrown to position *B* the battery—the ultimate energy source for the circuit—is disconnected. Energy dissipated in the resistor now comes from the decaying magnetic field of the inductor. That energy came originally from the battery but was stored temporarily in the magnetic field. Figure 32-14 outlines these energy transfers.

Magnetic Energy in an Inductor

How much energy is stored in an inductor's magnetic field? We can answer this by considering first the *rate* of energy storage. If we multiply Equation 32-6a—the loop law for an *RL* circuit—by the current *I*, the result is

$$I\mathscr{E}_0 - I^2R + I\mathscr{E}_L = 0,$$

or, using Equation 32-5 for $\mathscr{E}_L$,

$$I\mathscr{E}_0 - I^2R - LI\frac{dI}{dt} = 0.$$

What do the terms in this equation mean? The first is the product of the battery's current and emf—a product we know gives electrical power. That this term is positive means the battery supplies energy *to* the circuit at the rate $I\mathscr{E}_0$. The second term, I^2R, is the rate of energy dissipation in the resistor (recall Equation 27-9). The negative sign means that this energy is taken *from* the circuit. The third term is also negative (since the current is increasing, *dI/dt* is positive) and represents the rate at which the inductor takes energy from the circuit. But the

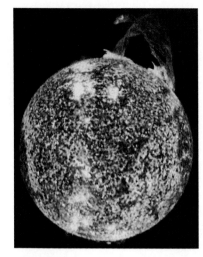

FIGURE 32-13 This eruption of a huge prominence from the Sun's surface involves the release of energy stored in magnetic fields.

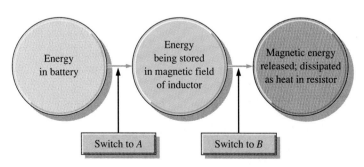

FIGURE 32-14 Energy transfers in the circuit of Fig. 32-9.

inductor does not dissipate this energy; instead, it stores the energy in its growing magnetic field. The rate at which the inductor stores energy is thus

$$P = LI\frac{dI}{dt}.$$

Suppose we increase the current in an inductor by some small amount dI over a small time interval dt. Since the power is the rate of energy storage, the energy dU stored during this time is thus

$$dU = P \, dt = LI \, dI$$

We find the total energy stored in bringing the inductor current from zero to some final value I by summing—that is, integrating—all the dU's:

$$U = \int dU = \int P \, dt = \int_0^I LI \, dI = \tfrac{1}{2}LI^2\big|_0^I$$

Evaluating at the two limits then gives the stored energy:

$$U = \tfrac{1}{2}LI^2. \tag{32-10}$$

This much energy is therefore released when the magnetic field decays.

● EXAMPLE 32-7   QUENCHING A SUPERCONDUCTING MAGNET

Loss of coolant is a danger in superconducting electromagnets. The current is suddenly left without its zero-resistance path, and energy stored in the magnetic field is rapidly released. To prevent explosive energy release, copper wire is incorporated into the conducting system to lengthen the time constant L/R in the event of such a "quench." The superconducting MRI solenoid of Example 30-6 carries a 2.4-kA current and has 0.53-H inductance. In its nonsuperconducting state, the total resistance is 31 mΩ. (a) How much energy is stored in the solenoid's magnetic field? (b) If the coils suddenly lose their superconductivity, what is the initial rate of energy release?

Solution
(a) Equation 32-10 gives the magnetic energy:

$$U = \tfrac{1}{2}LI^2 = (\tfrac{1}{2})(0.53 \text{ H})(2.4 \times 10^3 \text{ A})^2 = 1.53 \text{ MJ}.$$

(b) When superconducting ceases, the current decays according to Equation 32-9. At the instant the decay starts the current still has its original value. Since the magnetic energy released is dissipated in the resistor, the rate of energy release is just I^2R, or

$$P = I^2R = (2.4 \times 10^3 \text{ A})^2(0.031 \ \Omega) = 180 \text{ kW}.$$

This is a substantial power; equivalent to 1,800 100-W light bulbs burning in the space of this roughly human-size device. The following exercise explores the duration of this power surge.

EXERCISE   How long is it before 90% of the magnetic energy in Example 32-7 has been dissipated?

Answer: 20 s

Some problems similar to Example 32-7:   42, 44, 47   ●

Magnetic Energy Density

A long solenoid is a particularly simple inductor in which the magnetic field is essentially uniform. We can readily evaluate the energy density using this magnetic field, just as we found the electric field energy density using a parallel-plate capacitor.

In Example 32-3 we found that the inductance of a long solenoid of length ℓ and cross-sectional area A is $L = \mu_0 n^2 A \ell$, with n the number of turns per unit length. Equation 32-10 then gives the magnetic energy stored in the solenoid when it carries current I:

$$U = \tfrac{1}{2}LI^2 = \tfrac{1}{2}\mu_0 n^2 A \ell I^2 = \frac{1}{2\mu_0}(\mu_0 n I)^2 A \ell = \frac{B^2}{2\mu_0}A\ell,$$

where we recognized the quantity $\mu_0 n I$ as B, the magnetic field in the solenoid (Equation 30-12). The quantity $A\ell$ is the volume containing this field, so the energy per unit volume—the **magnetic energy density**—is

$$u_B = \frac{B^2}{2\mu_0}. \qquad \text{magnetic energy density} \qquad (32\text{-}11)$$

Although we derived this expression for the field of a solenoid, it is, in fact, a universal expression for the local magnetic energy density. Wherever there is a magnetic field, there is stored energy.

Compare Equation 32-11 with Equation 26-3 for the energy density in an electric field:

$$u_E = \tfrac{1}{2}\varepsilon_0 E^2.$$

The expressions for electric and magnetic energy densities are similar. Each is proportional to the *square* of the field strength, and each contains the appropriate constant. That the constant appears in the numerator in one case and the denominator in the other has no deep significance; it is merely a consequence of the way SI units are defined.

● **EXAMPLE 32-8** ENERGY IN EARTH'S MAGNETIC FIELD

The magnetic field strength near Earth's surface is about 50 μT. (a) How much energy is contained in 1 cubic kilometer of this field? (b) How does this compare with the electrical energy in the same volume, given a fair-weather electric field of 100 V/m?

Solution

(a) The magnetic field is essentially constant over this volume, so the total energy is the product of the energy density with the volume V:

$$U = u_B V = \frac{B^2}{2\mu_0}V = \frac{(50 \times 10^{-6}\ \text{T})^2}{(2)(4\pi \times 10^{-7}\ \text{N/A}^2)}(1 \times 10^3\ \text{m})^3$$

$$= 1\ \text{MJ}.$$

This is not a particularly large energy; a mere gallon of gasoline, for example, stores about 100 times as much.

(b) Since we're considering equal volumes, it suffices to compare energy densities. Then

$$\frac{u_B}{u_E} = \frac{B^2/2\mu_0}{\tfrac{1}{2}\varepsilon_0 E^2} = \frac{B^2}{\mu_0 \varepsilon_0 E^2}$$

$$= \frac{(50 \times 10^{-6}\ \text{T})^2}{(4\pi \times 10^{-7}\ \text{N/A}^2)(8.85 \times 10^{-12}\ \text{F/m})(100\ \text{V/m})^2}$$

$$= 2 \times 10^4$$

Thus the electrical energy is even smaller, by a factor of 20,000. The energy stored in both of Earth's electromagnetic fields is very small compared, for example, with the gravitational energy of an equivalent volume of air or the amount of solar energy incident each day on a square kilometer. In other systems, though, magnetic energy may dominate.

EXERCISE A typical sunspot is about 50,000 km in diameter, and extends into the Sun as a cylinder about 30,000 km long. Its magnetic field is about 0.2 T. What is the magnetic energy in such a spot?

Answer: 9×10^{26} J

Some problems similar to Example 32-8: 51, 52 ●

When the magnetic field varies with position it is necessary to divide the volume into infinitesimal elements and integrate to find the total magnetic energy. Example 32-9 illustrates this process:

● **EXAMPLE 32-9** MAGNETIC ENERGY IN A TOROID

A toroidal coil of square cross section has inner radius R and side ℓ as shown in Fig. 32-15. The coil consists of N turns, and carries a current I. What is the total magnetic energy stored in the toroid?

Solution
In Chapter 30, we found that the magnetic field in a toroid is given by

$$B = \frac{\mu_0 NI}{2\pi r},$$

with r the distance from the central axis of the toroid. The magnetic energy density in the toroid is then

$$u_B = \frac{B^2}{2\mu_0} = \frac{\mu_0 N^2 I^2}{8\pi^2 r^2}.$$

To calculate the total energy, consider a thin ring of thickness dr and height ℓ located at a distance r from the axis of the toroid, as shown in Fig. 32-15. The volume of this ring is

$$dV = 2\pi r \ell\, dr,$$

so the magnetic energy in the ring is

$$dU = u_B\, dV = \frac{\mu_0 N^2 I^2}{8\pi^2 r^2} 2\pi r \ell\, dr = \frac{\mu_0 N^2 I^2 \ell}{4\pi r} dr.$$

To find the total energy, we integrate over all such rings within the toroid; that is, from $r = R$ to $r = R + \ell$:

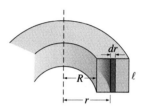

FIGURE 32-15 Cross section of a toroidal coil (Example 32-9). Also shown is a part of a volume element, in the shape of a ring of thickness dr.

$$U = \int_R^{R+\ell} dU = \frac{\mu_0 N^2 I^2 \ell}{4\pi} \int_R^{R+\ell} \frac{dr}{r} = \frac{\mu_0 N^2 I^2 \ell}{4\pi} \ln\left(\frac{R + \ell}{R}\right).$$

Comparing this result with the expression $U = \frac{1}{2}LI^2$ would allow us to calculate the inductance of the toroid (see Problem 65).

EXERCISE A long straight wire of radius R carries current I. Find an expression for the magnetic energy stored in a region of length ℓ, between the wire surface and a distance $2R$ from the wire's axis.

Answer: $(\mu_0 I^2 \ell \ln 2)/4\pi$

Some problems similar to Example 32-9: 54–56, 63, 64 ●

CHAPTER SYNOPSIS

Summary

1. The **mutual inductance** of a pair of coils is defined as the ratio of the total flux in the second coil to the current in the first:

$$M = \frac{\phi_2}{I_1}.$$

By Faraday's law, the emf in the second coil is proportional to the rate of change of current in the first:

$$\mathcal{E}_2 = -M\frac{dI_1}{dt}.$$

The same mutual inductance M describes the emf developed in the first coil as a result of changing current in the second coil.

2. A changing current in a coil or circuit gives rise to a changing magnetic flux through that same circuit. This changing flux, in turn, gives rise to an induced electric field that opposes the original change in current. A device constructed to exploit this property of **self-inductance** is called an **inductor.** The self-inductance L of an inductor is the ratio of magnetic flux to current:

$$L = \frac{\phi}{I}.$$

Faraday's law relates the emf in an inductor to the rate of change of current:

$$\mathscr{E} = -L\frac{dI}{dt}.$$

The direction of the emf is such as to oppose changes in the inductor current. Self-inductance prevents the inductor current from changing instantaneously.

3. In a circuit containing a resistor R and inductor L, changes occur exponentially with an **inductive time constant** L/R. The rising current in a series RL circuit is given by

$$I = \frac{\mathscr{E}_0}{R}(1 - e^{-Rt/L}),$$

where $\mathscr{E}_0$ is the battery emf and t the time. If the current is subsequently allowed to decay, it goes exponentially to zero:

$$I = I_0 e^{-Rt/L},$$

where t is measured from the start of the decay.

4. Electrical energy supplied to an inductor ends up stored in the inductor's magnetic field. When an inductance L carries current I, the stored magnetic energy is

$$U = \tfrac{1}{2}LI^2.$$

5. All magnetic fields, not only those of inductors, contain stored energy, with the local **magnetic energy density** given by

$$u_B = \frac{B^2}{2\mu_0}.$$

Terms You Should Understand

(Pairs are closely related terms whose distinction is important; number in parentheses is chapter section where term first appears.)

mutual inductance, self-inductance (32-1, 32-2)
inductor (32-2)
back emf (32-2)
inductive time constant (32-3)
magnetic energy, magnetic energy density (32-4)

Symbols You Should Recognize

M, L (32-1, 32-2)
u_B (32-4)

Problems You Should Be Able to Solve

calculating mutual inductance for simple coil configurations (32-1)
finding induced emf in a second coil given mutual inductance and rate of change of current in the first coil (32-1)
calculating self-inductance for simple configurations (32-2)
calculating back emf given self-inductance and rate of change of current (32-2)
analyzing time-dependent behavior of RL circuits (32-3)
calculating magnetic energy (32-4)

Limitations to Keep in Mind

Real inductors, other than those made from superconductors, have resistance as well as inductance.

QUESTIONS

1. Figure 32-16 shows two pairs of identical coils in different geometrical arrangements. For which arrangement is the mutual inductance greatest? Why?

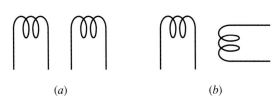

(a)   *(b)*

FIGURE 32-16 Question 1.

2. A car battery has an emf of 12 V, yet energy from the battery provides the 20,000-V spark that ignites the gasoline. How is this possible?

3. When two coils are connected in series but are physically far apart, they behave as a single inductor whose inductance is the sum of the individual inductances. Why might this not be true if they are close together?

4. You have a fixed length of wire to wind into an inductor. Will you get more inductance if you wind a short coil with large diameter, or a long coil with small diameter?

5. You have a fixed length of wire of resistance R. You want to wind the wire into a small space and use it as a resistor. How would you wind it so as to minimize its self-inductance?

6. In wiring circuits that operate at high frequencies, like TV sets or computers, it is important to avoid extraneous loops in wires. Why?

7. In a popular demonstration of induced emf, a light bulb is connected across a large inductor in an *LR* circuit, as shown in Fig. 32-17. When the switch is opened, the bulb flashes brightly and burns out. Why?

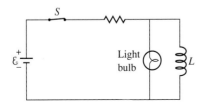

FIGURE 32-17 Question 7.

8. In the *RL* circuit of Fig. 32-7a, can the inductor emf exceed the battery emf (a) when the switch is first closed? (b) When the switch is opened after being closed for a long time?

9. What is the time constant for an inductive circuit made entirely from a superconductor?

10. Does it take more or less than one time constant for current in an *RL* circuit to build up to half its steady-state value?

11. If you increase the resistance in an *RL* circuit, what effect does this have on the inductive time constant?

12. Speculate on what would happen if you connected an ideal battery directly across an ideal inductor, with no resistance anywhere in the circuit.

13. How could you modify the simple *RL* circuit of Fig. 32-7a to prevent dangerous voltages from developing when the switch is opened?

14. List some similarities and differences between inductors and capacitors.

15. A 1-H inductor carries 10 A, and a 10-H inductor carries 1 A. Which inductor contains more stored energy?

16. Does the energy density in a magnetic field depend on the direction of the field?

17. The field of a magnetic dipole extends to infinity. Is there an infinite amount of energy stored in the dipole field? Why or why not?

18. It takes work to push two bar magnets together with like poles facing each other. Where does this energy go?

PROBLEMS

Section 32-1 Mutual Inductance

1. Two coils have a mutual inductance of 2.0 H. If current in the first coil is changing at the rate of 60 A/s, what is the emf in the second coil?

2. A 500-V emf appears in a coil when the current in an adjacent coil changes at the rate of 3.5 A/ms. What is the mutual inductance of the coils?

3. The current in one coil is given by $I = I_p \sin 2\pi f t$, where $I_p = 75$ mA, $f = 60$ Hz, and $t =$ time. Find the peak emf in a second coil if the mutual inductance between the coils is 440 mH.

4. Two coils have a mutual inductance of 580 mH. One coil is supplied with a current given by $I = 3t^2 - 2t + 4$, where I is in amperes and t in seconds. What is the induced emf in the other coil at time $t = 2.5$ s?

5. An alternating current given by $I_p \sin 2\pi f t$ is supplied to one of two coils whose mutual inductance is M. (a) Find an expression for the emf in the second coil. (b) When $I_p = 1.0$ A and $f = 60$ Hz, the peak emf in the second coil is measured at 50 V. What is the mutual inductance?

6. Find the mutual inductance of the two-coil system described in Problem 19 of Chapter 31.

7. Two long solenoids of length ℓ both have n turns per unit length. They have circular cross sections with radii R and $2R$, respectively. The smaller solenoid is mounted inside the larger one, with their axes coinciding. Find the mutual inductance of this arrangement, neglecting any nonuniformity in the magnetic field near the ends.

8. Coils A and B have mutual inductance 25 mH. At time $t = 0$ the current in coil A is zero. Subsequently a time-varying current is supplied to A, and the induced emf in coil B is given by $\mathscr{E} = 50 + 0.2t$, with $\mathscr{E}$ in V and t in ms. Find an expression for the time-varying current in coil A.

9. A rectangular loop of length ℓ and width w is located a distance a from a long, straight wire, as shown in Fig. 32-18. What is the mutual inductance of this arrangement?

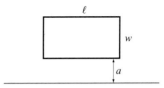

FIGURE 32-18 Problem 9.

10. Two wire loops of radii a and b lie in the same plane and have a common center. Find the mutual inductance of this arrangement, assuming $b \gg a$. *Hint:* With $b \gg a$ the magnetic field will be essentially uniform over the smaller loop. See Example 30-1.

Section 32-2 Self-Inductance

11. What is the self-inductance of a solenoid 50 cm long and 4.0 cm in diameter that contains 1,000 turns of wire?

12. The current in an inductor is changing at the rate of 100 A/s, and the inductor emf is 40 V. What is its self-inductance?

13. A 2.0-A current is flowing in a 20-H inductor. A switch is opened, interrupting the current in 1.0 ms. What emf is induced in the inductor?

14. A 60-mA current is flowing in a 100-mH inductor. Over a period of 1.0 ms the current is reversed, going steadily to 60 mA in the opposite direction. What is the inductor emf during this time?

15. A cardboard tube measures 15 cm long by 2.2 cm in diameter. How many turns of wire must be wound on the full length of the tube to make a 5.8-mH inductor?

16. The current in a 2.0-H inductor is given by $I = 3t^2 + 15t + 8$, where t is in seconds and I in amperes. Find an expression for the magnitude of the inductor emf.

17. The emf in a 50-mH inductor has magnitude $|\mathscr{E}| = 0.020t$, with t in seconds and $\mathscr{E}$ in volts. At $t = 0$ the inductor current is 300 mA. (a) If the current is increasing, what will be its value at $t = 3.0$ s? (b) Repeat for the case when the current is decreasing.

18. The current in a 40-mH inductor is given by $I = I_0 e^{-bt}$, where $I_0 = 10$ A and $b = 20$ s^{-1}. What is the magnitude of the inductor emf at (a) $t = 0$ (b) $t = 25$ ms, and (c) $t = 50$ ms?

19. A 2,000-turn solenoid is 65 cm long and has cross-sectional area 30 cm². What rate of change of current will produce a 600-V emf in this solenoid?

20. You have a plastic rod 20 cm long and 1.5 cm in diameter. What inductance will you get if you wind the entire rod with a single layer of (a) 22-gauge (0.64-mm-diameter) and (b) 34-gauge (0.16-mm-diameter) wire? Assume adjacent turns are touching, separated only by a negligible thickness of enamel insulation.

21. The emf in a 50-mH inductor is given by $\mathscr{E} = \mathscr{E}_p \sin \omega t$, where $\mathscr{E}_p = 75$ V and $\omega = 140$ s^{-1}. What is the peak current in the inductor? (Assume the current swings symmetrically about zero.)

22. A coaxial cable consists of an inner conductor of radius a and outer conductor of radius b, as shown in Fig. 32-19. Current flows along one conductor and back along the other. Show that the inductance per unit length of the cable is $\frac{\mu_0}{2\pi} \ln(b/a)$.

FIGURE 32-19 Problems 22, 68.

Section 32-3 Inductors in Circuits

23. Show that the inductive time constant has the units of seconds.

24. What inductance should you put in series with a 100-Ω resistor to give a time constant of 2.2 ms?

25. The current in a series RL circuit rises to 20% of its final value in 3.1 µs. If $L = 1.8$ mH, what is the resistance R?

26. The current in a series RL circuit rises to half its final value in 7.6 s. What is the time constant?

27. A 10-H inductor is wound of wire with resistance 2.0 Ω. If the inductor is connected across an ideal 12-V battery, how long will it take the current to reach 95% of its final value?

28. In a series RL circuit like Fig. 32-7a, $\mathscr{E}_0 = 45$ V, $R = 3.3$ Ω, and $L = 2.1$ H. If the current is 9.5 A, how long has the switch been closed?

29. In Fig. 32-7a, take $R = 2.5$ kΩ and $\mathscr{E}_0 = 50$ V. When the switch is closed, the current through the inductor rises to 10 mA in 30 µs. (a) What is the inductance? (b) What will be the current in the circuit after many time constants?

30. A series RL circuit like Fig. 32-7a has $\mathscr{E}_0 = 60$ V, $R = 22$ Ω, and $L = 1.5$ H. Find the rate of change of the current (a) immediately after the switch is closed and (b) 0.10 s later.

31. In Fig. 32-7a, take $R = 100$ Ω, $L = 2.0$ H, and $\mathscr{E}_0 = 12$ V. At 20 ms after the switch is closed, what are (a) the circuit current, (b) the inductor emf, (c) the resistor voltage, (d) the rate of change of the circuit current, and (e) the power dissipation in the resistor?

32. Show that a series RL circuit reaches 99% of its final current in approximately 5 time constants.

33. Resistor R_2 in Fig. 32-20 is to limit the emf that develops when the switch is opened. What should be its value in order that the inductor emf not exceed 100 V?

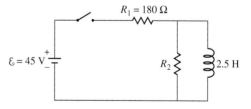

FIGURE 32-20 Problem 33.

34. In Fig. 32-9a take $\mathscr{E}_0 = 12$ V, $R = 2.7$ Ω, and $L = 20$ H. Initially the switch is in position B and there is no current anywhere. At $t = 0$ the switch is thrown to position A, and at $t = 10$ s it is thrown back to position B. Find the inductor current at (a) $t = 5.0$ s and (b) $t = 15$ s.

35. A 5.0-A current is flowing through a nonideal inductor with $L = 500$ mH. If the inductor is suddenly short circuited, the inductor current drops to 2.5 A in 6.9 ms. What is the resistance of the inductor?

36. In Fig. 32-21, take $\mathscr{E}_0 = 12$ V, $R_1 = 4.0$ Ω, $R_2 = 8.0$ Ω, $R_3 = 2.0$ Ω, and $L = 2.0$ H. What is the current I_2

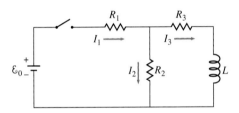

FIGURE 32-21   Problem 36.

(a) immediately after the switch is first closed and (b) a long time after the switch is closed? (c) After a long time the switch is again opened. Now what is I_2?

37. In Fig. 32-22, take $\mathcal{E}_0 = 20$ V, $R_1 = 10 \ \Omega$, $R_2 = 5.0 \ \Omega$, and assume the switch has been open for a long time. (a) What is the inductor current immediately after the switch is closed? (b) What is the inductor current a long time after the switch is closed? (c) If after a long time the switch is again opened, what will be the voltage across R_1 immediately afterwards?

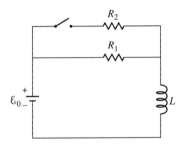

FIGURE 32-22   Problem 37.

Section 32-4 Magnetic Energy

38. How much energy is stored in a 5.0-H inductor carrying 35 A?

39. What is the current in a 10-mH inductor when the stored energy is 50 μJ?

40. A 220-mH inductor carries 350 mA. How much energy must be supplied to the inductor in raising the current to 800 mA?

41. A 12-V battery, 5.0-Ω resistor, and 18-H inductor are connected in series and allowed to reach a steady state. (a) What is the energy stored in the inductor? (b) Once in the steady state, over what time interval is the energy dissipated in the resistor equal to that stored in the inductor?

42. A battery, switch, resistor, and inductor are connected in series. When the switch is closed the current rises to half its steady-state value in 1.0 ms. How long does it take for the magnetic energy in the inductor to rise to half its steady-state value?

43. The current in a 2.0-H inductor is decreased linearly from 5.0 A to zero over 10 ms. (a) What is the average rate at

which energy is being extracted from the inductor during this time? (b) Is the instantaneous rate constant?

44. When a nonideal 1.0-H inductor is short-circuited, its magnetic energy drops to one-fourth of its original value in 3.6 s. What is its resistance?

45. The current in a 2.0-H inductor is increasing. At some instant, the current is 3.0 A and the inductor emf is 5.0 V. At what rate is the inductor's magnetic energy increasing at this instant?

46. A 500-turn solenoid is 23 cm long, 1.5 cm in diameter, and carries 65 mA. How much magnetic energy does it contain?

47. A superconducting solenoid with inductance $L = 3.5$ H carries 1.8 kA. Copper is embedded in the coils to carry the current in the event of a quench (see Example 32-7). (a) What is the magnetic energy in the solenoid? (b) What is the maximum resistance of the copper that will limit the power dissipation to 100 kW immediately after a loss of superconductivity? (c) With this resistance, how long will it take the power to drop to 50 kW?

48. Show that the quantity $B^2/2\mu_0$ has the units of energy density (J/m^3).

49. The Alcator fusion experiment at MIT has a 50-T magnetic field. What is the magnetic energy density in Alcator?

50. What is the magnetic field strength in a region where the magnetic energy density is 7.8 J/cm^3?

51. The magnetic field of a neutron star is about 10^8 T. How does the energy density in this field compare with the energy density stored in (a) gasoline and (b) pure uranium-235 (mass density 19×10^3 kg/m^3)? Consult Appendix C.

52. A loop of magnetic field arches above the Sun's surface, forming a tube approximately 10^5 km long and 10^4 km in diameter. If the magnetic field strength in the tube is 50 G, what is its magnetic energy content?

53. A single-turn loop of radius R carries current I. How does the magnetic energy density at the loop center compare with that of a long solenoid of the same radius, carrying the same current, and consisting of n turns per unit length?

54. A magnetic field is given by $\mathbf{B} = B_0(x/a)^2\hat{\mathbf{j}}$, where B_0 and a are constants. Find an expression for the magnetic energy in a cube of side a with one corner at the origin and sides extending along the coordinate axes.

55. Estimate the total energy in Earth's magnetic field by integrating the magnetic energy density over the entire volume outside the planet, assuming that the field strength at Earth's surface is a constant 0.5 G and drops as $1/r^3$, with r the distance from Earth's center. (This estimate neglects the factor-of-two variation from pole to equator, any field inside the planet, and distortions of Earth's magnetic field by the solar wind.)

56. A toroid of inner radius 1.5 m and square cross section is wound with 2,500 turns. What must be the length ℓ of its cross-sectional square if the toroid contains 80 J of magnetic energy at a current of 63 A?

Paired Problems

(Both problems in a pair involve the same principles and techniques. If you can get the first problem, you should be able to solve the second one.)

57. Two coils have mutual inductance M. The current supplied to coil A is given by $I = bt^2$. Find an expression for the magnitude of the induced emf in coil B.

58. Two coils have mutual inductance M, and a time-varying current is supplied to coil A; at time $t = 0$ that current is zero. The magnitude of the induced emf in coil B is given by $\mathcal{E} = b\sqrt{t}$. Find an expression for the current in coil A.

59. In the circuit of Fig. 32-7a, take $\mathcal{E}_0 = 5.0$ V and $R = 1.8\ \Omega$. At 2.5 s after the switch is closed, the circuit current is 250 mA. Find the inductance.

60. In the circuit of Fig. 32-7a, take $\mathcal{E}_0 = 5.0$ V and $R = 1.8\ \Omega$. At 2.5 s after the switch is closed, the inductor emf is 2.1 V. Find the inductance.

61. In Fig. 32-11a, take $\mathcal{E}_0 = 25$ V, $R_1 = 1.5\ \Omega$, and $R_2 = 4.2\ \Omega$. What is the voltage across R_2 (a) immediately after the switch is first closed and (b) a long time after the switch is closed? (c) Long after the switch is closed it is again opened. Now what is the voltage across R_2?

62. In Fig. 32-11, take $R_2 = 5R_1$. If the maximum possible value for the inductor emf in this circuit is 300 V, what is the battery emf $\mathcal{E}_0$?

63. A wire of radius R carries a current I distributed uniformly over its cross section. Find an expression for the magnetic energy per unit length in the region from R to $100R$.

64. A wire of radius R carries a current I distributed uniformly over its cross section. Find an expression for the total magnetic energy per unit length *within* the wire.

Supplementary Problems

65. (a) Use the result of Example 32-9 to determine the inductance of a toroid. (b) Show that your result reduces to the inductance of a long solenoid when $R \gg \ell$.

66. Two long, flat parallel bars of width w and spacing d carry equal but opposite currents I, as shown in Fig. 32-23. (a) Use Ampère's law to find the magnetic field between the bars. Take $d \ll w$ so you can neglect fringing fields. (b) Use your result to find the magnetic energy per unit length stored between the bars. (c) Compare your result in (b) with the expression $U = \frac{1}{2}LI^2$ to find an expression for the inductance per unit length of the bars.

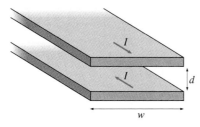

FIGURE 32-23 Problem 66.

67. (a) Use Equation 32-9 to write an expression for the power dissipation in the resistor as a function of time, and (b) integrate from $t = 0$ to $t = \infty$ to show that the total energy dissipated is equal to the energy initially stored in the inductor, namely, $\frac{1}{2}LI_0^2$.

68. (a) Find the magnetic energy density as a function of radial distance for the coaxial cable of Problem 22. (b) Integrate over the volume between the cable conductors to show that the total energy per unit length of the cable is

$$U = \frac{\mu_0}{4\pi} I^2 \ln(b/a).$$

Hint: Your volume element should be a cylindrical shell of radius r, thickness dr, and length ℓ. What is its volume dV? (c) Use the expression $U = \frac{1}{2}LI^2$ to find the inductance per unit length, and show that your result agrees with that of Problem 22.

69. An electric field and a magnetic field have the same energy density. Obtain an expression for the ratio E/B, and evaluate this ratio numerically. What are its units? Is your answer close to any of the fundamental constants listed inside the front cover?

70. Two long, straight, parallel wires are a distance d apart. The wires have radius a, where $a \ll d$. Current flows down one wire and back along the other. Find the inductance per unit length of the parallel wires. Assume the wire radius is so small that you can neglect the magnetic flux within the wires themselves.

71. The switch in the circuit of Fig. 32-24 is closed at time $t = 0$, at which instant the inductor current is zero. Write the loop and node laws for this circuit, and show that they are satisfied if the inductor current is given by $I = (\mathcal{E}_0/R_1)(1 - e^{-R_\parallel t/L})$, where $R_\parallel$ is the resistance of R_1 and R_2 were they connected in parallel.

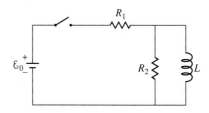

FIGURE 32-24 Problem 71.

72. Earth's magnetic field ends abruptly on the sunward side at approximately the point where the magnetic energy density has dropped to the same value as the kinetic energy density in the solar wind. Near Earth, the solar wind contains about 5 protons and 5 electrons per cubic centimeter, and flows at 400 km/s. Treating Earth's field as that of a dipole with dipole moment 8×10^{22} J/T, estimate the distance to the point above the equator where the field ends.

CHAPTER 33

ALTERNATING-CURRENT CIRCUITS

Time-varying voltages and currents are essential in electronic circuits and in electric power transmission. This oscilloscope displays time-varying voltages measured in a computer circuit.

hapter 28 considered direct-current (DC) circuits, in which the source of electrical energy is a battery or other device whose emf does not change with time. When we turn on a circuit containing only resistors and a DC emf, currrent starts to flow immediately and remains steady until the circuit is turned off. Even when we add capacitance, as in Section 28-6, or inductance, as in Section 32-3, all currents and voltages eventually reach steady values.

We now turn our attention to alternating-current (AC) circuits, in which sources of electrical energy vary with time. A familiar AC circuit is standard household wiring. Alternating current with a frequency of 60 Hz is used almost universally for electric power generation and transmission, for reasons we will discuss in Section 33-6. Devices such as stereos, TVs, radios, and microwave ovens involve more rapidly varying alternating currents.

33-1 ALTERNATING CURRENT

In describing time-varying electrical quantities, we will consider only sinusoidal variations. More complicated variations can be analyzed as superpositions of sinusoidal functions, as we described in Section 16-5. A sinusoidal AC voltage or current is characterized by its amplitude, frequency, and phase constant—the same quantities we developed in Chapter 15 to describe simple harmonic motion. Amplitude is specified by giving the peak value (V_p, I_p) or the **root-mean-square** value (V_{rms}, I_{rms}). The rms value is an average obtained by squaring the signal, taking the time average, and then taking the square root. This procedure is used because the direct average of an AC signal is zero since it spends as much time below zero as above. Use of rms values also facilitates the calculation of power in AC circuits. For a sine wave, rms and peak values are related by

$$V_{rms} = \frac{V_p}{\sqrt{2}} \quad \text{and} \quad I_{rms} = \frac{I_p}{\sqrt{2}}, \tag{33-1}$$

as you can show in Problem 7. When we speak of 120-V household wiring, for example, we are giving the rms voltage.

In practical and engineering situations we usually describe frequency f in cycles per second, or hertz (Hz). In mathematical analysis of alternating current, it is usually more convenient to use the angular frequency ω, measured in radians per second or, equivalently, inverse seconds (s^{-1}). The relation between the two,

$$\omega = 2\pi f, \tag{33-2}$$

is the same as for rotational and simple harmonic motion, and for the same reason: a full cycle contains 2π radians.

Sometimes we are interested in the phase constant ϕ of an AC signal, which describes when the sine curve crosses zero. A full mathematical description of an AC voltage or current then includes its amplitude (V_p, I_p), frequency (ω), and phase constant (ϕ):

$$V = V_p \sin(\omega t + \phi_V) \quad \text{and} \quad I = I_p \sin(\omega t + \phi_I). \tag{33-3}$$

Here we've labeled the phase constants with subscripts V and I to indicate that voltage and current—even in the same circuit element—need not have the same phase constant. We will often take one phase constant to be zero; then the other describes the *phase difference* between the voltage and current. Phase difference is an important quantity in AC circuits.

Figure 33-1 plots a typical AC voltage, showing the relation between peak and rms amplitudes.

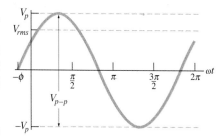

FIGURE 33-1   A sinusoidally varying AC voltage, showing peak, rms, and peak-to-peak amplitudes; the latter is just the difference between the extreme values, here equal to $2V_p$. The waveform completes a full cycle as ωt increases by 2π. The phase constant ϕ is $\pi/6$ or 30°; note that a positive phase constant shifts the curve to the left.

● **EXAMPLE 33-1** HOUSEHOLD VOLTAGE

Standard household wiring in the United States supplies 120 V rms at 60 Hz. Express this voltage mathematically in the form of Equation 33-3, assuming that the voltage is rising through zero at time $t = 0$.

Solution

The rms and peak voltages are related by Equation 33-1, so

$$V_p = \sqrt{2}V_{rms} = (\sqrt{2})(120 \text{ V}) = 170 \text{ V}.$$

The angular frequency is 2π times the frequency in Hz, so

$$\omega = 2\pi f = (2\pi)(60 \text{ Hz}) = 377 \text{ s}^{-1}.$$

When $t = 0$ we want Equation 33-3 to give $V = 0$; that requires $\phi = 0$ or $\phi = \pi$. But only with $\phi = 0$ will the curve be *rising* at $t = 0$, so Equation 33-3 becomes

$$V = 170\sin(377t),$$

with V in volts and t in seconds. We will frequently take the phase constant to be zero as in this example; only when we're comparing signals with different phase does the value of ϕ become significant.

EXERCISE A 1.0-kHz sinusoidal current with rms amplitude 1.5 A drives a loudspeaker in a test of a stereo system. Express this current in the form of Equation 33-3, assuming zero phase constant.

Answer: $I = 2.12\sin(6.28 \times 10^3 t)$, with I in A and t in s.

Some problems similar to Example 33-1: 1, 4, 5 ●

33-2 CIRCUIT ELEMENTS IN AC CIRCUITS

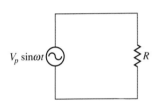

FIGURE 33-2 A resistor connected across an AC generator (symbol ⊝).

Here we examine separately the AC behavior of resistors, capacitors, and inductors so we can subsequently understand what happens when we combine these elements in AC circuits.

Resistors

An ideal resistor is a device whose current and voltage are always proportional:

$$I = \frac{V}{R}.$$

Figure 33-2 shows a resistor R connected across an AC generator, making the voltage across the resistor equal to the generator voltage. The generator voltage is described by Equation 33-3, where we take $\phi_V = 0$. Then the current is

$$I = \frac{V}{R} = \frac{V_p \sin \omega t}{R} = \frac{V_p}{R}\sin \omega t.$$

The current has the same frequency as the voltage, and, since its phase constant is also zero, the voltage and current are *in phase*—that is, they peak at the same time. The peak current is simply the peak voltage divided by the resistance: $I_p = V_p/R$. Because voltage and current are both sinusoidal, their rms values are in the same ratio as their peak values; thus $I_{rms} = V_{rms}/R$.

Capacitors

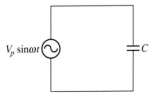

FIGURE 33-3 A capacitor across an AC generator.

Figure 33-3 shows a capacitor connected across an AC generator. In Chapter 26, we defined a capacitor as a device in which voltage and charge are directly proportional:

$$q = CV.$$

Differentiating this relation gives

$$\frac{dq}{dt} = C\frac{dV}{dt}.$$

But dq/dt is the capacitor current I, so

$$I = C\frac{dV}{dt}.$$

The generator voltage $V_p \sin \omega t$ appears directly across the capacitor, so the current is

$$I = C\frac{d}{dt}(V_p \sin \omega t)$$

$$= \omega C V_p \cos \omega t = \omega C V_p \sin(\omega t + \pi/2). \qquad (33\text{-}4)$$

This equation shows clearly the phase and amplitude relations between current and voltage in a capacitor. Because the cosine curve is just a sine curve shifted left by $\pi/2$ or $90°$, Equation 33-4 tells us that:

❙ The current in a capacitor leads the voltage by 90°.

Figure 33-4 shows graphically this relation between current and voltage in a capacitor.

The term $\omega C V_p$ multiplying the cosine in Equation 33-4 is the amplitude of the current, so we can write

$$I_p = \omega C V_p,$$

or, in a form resembling Ohm's law,

$$I_p = \frac{V_p}{1/(\omega C)} = \frac{V_p}{X_C}, \qquad (33\text{-}5)$$

where we have defined $X_C = 1/\omega C$.

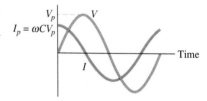

FIGURE 33-4   Current in a capacitor leads the voltage by $\pi/2$ or $90°$.

Equation 33-5 shows that the capacitor acts somewhat like a resistor of resistance $X_C = 1/\omega C$. But not quite! This "resistance" does give the relation between the peak voltage and peak current, but it doesn't tell the whole story. The capacitor also introduces a phase difference between voltage and current. This phase difference reflects a fundamental physical difference between resistors and capacitors. A resistor dissipates electrical energy as heat. A capacitor stores and releases electrical energy. Over a complete cycle, the agent turning the generator in Fig. 33-3 does no net work, while the agent turning the generator with the resistive load of Fig. 33-2 continually does work that gets dissipated as heat in the resistor. Because the quantity X_C in Equation 33-5 does not act quite like a resistance, we give it the special name **capacitive reactance**. Like resistance, reactance is measured in ohms (Ω).

Does it make sense that X_C depends on frequency? Yes—as frequency goes to zero, X_C goes to infinity. At zero frequency nothing is changing, there is no need to move charge on or off the plates, so no current flows, and the capacitor might as well be an open circuit. As frequency increases, larger currents flow to move charge on and off the capacitor in ever shorter times, so the capacitor looks increasingly like a short circuit. We often summarize this behavior qualitatively by saying that a capacitor at low frequencies acts like an open circuit, while at high frequencies it acts like a short circuit.

Why does the capacitor current *lead* the voltage? Because the capacitor voltage is proportional to its charge, and it takes current to move charge onto the capacitor plates. Therefore current must flow *before* the voltage can change significantly. We found the same thing with the *RC* circuit of Chapter 28; there current flowed as soon as the switch was closed, but it took time to build up the capacitor voltage.

Inductors

Figure 33-5 shows an inductor connected across an AC generator. The loop law for this circuit is

$$V_p \sin \omega t + \mathcal{E}_L = 0.$$

From the preceding chapter, we know that the inductor emf is given by

$$\mathcal{E}_L = -L\frac{dI}{dt},$$

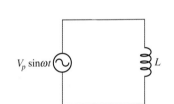

FIGURE 33-5 An inductor across an AC generator.

so the loop law becomes

$$V_p \sin \omega t = L\frac{dI}{dt}.$$

To obtain a relation involving the current I rather than its derivative, we integrate this equation:

$$\int V_p \sin \omega t\, dt = \int L\frac{dI}{dt}dt = \int L\, dI,$$

or

$$-\frac{V_p}{\omega}\cos \omega t = LI.$$

Here we have set the integration constants to zero because nonzero values would represent a DC emf and current that are absent in this circuit. Solving for I then gives

$$I = -\frac{V_p}{\omega L}\cos \omega t = \frac{V_p}{\omega L}\sin(\omega t - \pi/2), \quad (33\text{-}6)$$

where the last step follows because $\sin(\alpha - \pi/2) = -\cos\alpha$ for any α.

Equation 33-6 shows that the current in the inductor lags the applied voltage by $\pi/2$ or 90° (i.e., Equation 33-6 is Equation 33-3 with $\phi = -\pi/2$). Equivalently:

The voltage across an inductor leads the inductor current by 90°.

Figure 33-6 plots this phase relation.

Equation 33-6 also shows that the peak current is

$$I_p = \frac{V_p}{\omega L} = \frac{V_p}{X_L}.$$ (33-7)

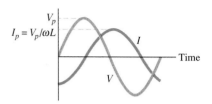

FIGURE 33-6 Voltage across an inductor leads the current by $\pi/2$ or 90°.

Again, this equation resembles Ohm's law, with a "resistance" $X_L = \omega L$. But as with the capacitor, no power is dissipated in the inductor. Instead, energy is alternately stored and released as the inductor's magnetic field builds up, then decays. To distinguish it from dissipative resistance, we call X_L the **inductive reactance.** Inductive reactance, too, is measured in ohms.

Does it make sense that the inductive reactance increases with increasing ω and increasing L? An inductor is a device that, through its induced back emf, opposes changes in current. The greater the inductance, the greater the opposition to changing current. And the more rapidly the current is changing, the more vigorously the inductor opposes the change, so the inductive reactance increases at high frequencies. In the extreme case of very high frequencies, an inductor looks like an open circuit. But at very low frequencies it looks more and more like a short circuit, until with direct current (zero frequency), an inductor exhibits zero reactance because there is no change in current.

Why does the inductor voltage *lead* the current? Because a changing current in an inductor induces an emf. *Before* the current can build up significantly there must first, therefore, be voltage across the inductor.

Table 33-1 summarizes the phase and amplitude relations in resistors, capacitors, and inductors.

● **EXAMPLE 33-2** INDUCTORS AND CAPACITORS

A capacitor is connected across the 60-Hz, 120-V rms power line, and an rms current of 200 mA flows. What is the capacitance? What inductance would have to be connected across the power line for the same current to flow? Would there be anything different about the circuit containing the inductor?

Solution

The peak current and voltage are related through Equation 33-5:

so

$$I_p = \frac{V_p}{1/\omega C},$$

$$C = \frac{I_p}{\omega V_p}.$$

We are given the rms voltage and current, but since only the ratio of these quantities appears in our equation, it doesn't

▲ **TABLE 33-1** PHASE AND AMPLITUDE RELATIONS
IN CIRCUIT ELEMENTS

CIRCUIT ELEMENT	PEAK CURRENT/VOLTAGE	PHASE RELATION
Resistor	$I_p = \dfrac{V_p}{R}$	V, I in phase
Capacitor	$I_p = \dfrac{V_p}{X_C} = \dfrac{V_p}{1/\omega C}$	I leads V by 90°
Inductor	$I_p = \dfrac{V_p}{X_L} = \dfrac{V_p}{\omega L}$	V leads I by 90°

matter whether we use rms or peak values. With $f = 60$ Hz, $\omega = 2\pi f$ or 377 s^{-1}, so

$$C = \frac{I}{\omega V} = \frac{0.20 \text{ A}}{(377 \text{ s}^{-1})(120 \text{ V})} = 4.4 \ \mu\text{F}.$$

An inductor that passes the same current must have the same reactance, so

$$\omega L = \frac{1}{\omega C},$$

or

$$L = \frac{1}{\omega^2 C} = \frac{1}{(377 \text{ s}^{-1})^2(4.4 \times 10^{-6} \text{ F})} = 1.6 \text{ H}.$$

Although the currents are the same, the two situations differ in that current leads voltage by 90° in the capacitor and lags by 90° in the inductor.

EXERCISE Inductors (called "ballast") are often used to limit the current in fluorescent lamps. In a particular lamp operating at 60 Hz, 80 V (rms) appears across a 0.53-H ballast inductor. (a) What is the rms inductor current? (b) What capacitance could be used in place of the inductor to provide the same current?

Answers: (a) 400 mA; (b) 13 μF

Some problems similar to Example 33-2: 13, 14, 17, 20

Phasor Diagrams

Phase and amplitude relations in AC circuits may be summarized graphically in **phasor diagrams.** A phasor is an arrow whose fixed length represents the amplitude of an AC voltage or current. The phasor rotates counterclockwise about the origin with the angular frequency ω of the AC quantity. The component of the phasor on the vertical axis then represents the sinusoidally varying AC signal. Figure 33-7a shows phasors for the current and voltage in a resistor. The lengths of the phasors are related by Ohm's law, $V_p = I_p R$. The current and voltage phasors always point in the same direction, showing that current and voltage in the resistor are in phase. Figures 33-7b and 33-7c show phasor diagrams for a capacitor and an inductor. In each, the lengths of the phasors are related by the appropriate reactance, so $V_p = I_p X$. As the phasors rotate, they remain at right angles, indicating the phase relation between current and voltage in these reactive circuit elements. You should convince yourself that all the relationships of Table 33-1 are correctly described by the phasor diagrams of Fig. 33-7. Although phasor diagrams do not add much to our understanding of AC circuits containing only one circuit element, they will greatly simplify the analysis of more complicated circuits.

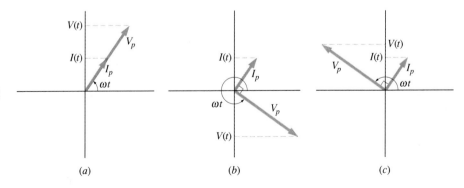

FIGURE 33-7 Phasor diagrams showing voltage and current in (a) a resistor, (b) a capacitor, and (c) an inductor. Lengths of phasors correspond to amplitudes V_p and I_p, while projection on the vertical axis gives the instantaneous value. Angle between voltage and current phasors gives their relative phase.

(a)   (b)   (c)

▲ **TABLE 33-2** CAPACITORS AND INDUCTORS

	CAPACITOR	**INDUCTOR**
Defining relation	$C = \dfrac{q}{V}$	$L = \dfrac{\phi_B}{I}$
Defining relation, differential form	$I = C\dfrac{dV}{dt}$	$\mathcal{E} = -L\dfrac{dI}{dt}$
Opposes changes in	Voltage	Current
Energy storage	In electric field $U = \frac{1}{2}CV^2$	In magnetic field $U = \frac{1}{2}LI^2$
Behavior in low-frequency limit	Open circuit	Short circuit
Behavior in high-frequency limit	Short circuit	Open circuit
Reactance	$X_C = 1/\omega C$	$X_L = \omega L$
Phase	Current leads by 90°	Voltage leads by 90°

Capacitors and Inductors: A Comparison

Here and in previous chapters, we have considered separately the behavior of capacitors and inductors. Many of the properties of these devices are analogous. A capacitor opposes instantaneous changes in voltage, while an inductor opposes instantaneous changes in current. In an *RC* circuit with a DC emf, voltage builds up exponentially across the capacitor, with time constant *RC*. In the analogous *RL* circuit, current builds up exponentially in the inductor, with time constant *L/R*. A capacitor stores electrical energy given by $\frac{1}{2}CV^2$. An inductor stores magnetic energy given by $\frac{1}{2}LI^2$. A capacitor acts like an open circuit at low frequencies, an inductor like a short circuit at low frequencies. Each exhibits the opposite behavior at high frequencies.

Capacitors and inductors are complementary devices, reflecting a deeper complementarity between electric and magnetic fields. Any verbal description of a capacitor applies to an inductor if we replace the words "capacitor" with "inductor," "electric" with "magnetic," and "voltage" with "current." Table 33-2 summarizes the complementary aspects of capacitors and inductors.

■ **APPLICATION** LOUDSPEAKER SYSTEMS

Loudspeakers in high-quality audio systems invariably contain two or more individual *drivers*—devices for converting electrical energy to sound—within the same enclosure (Fig. 33-8a). The most common form of driver includes a wire coil attached to a diaphragm and suspended in the field of a permanent magnet (Fig. 33-8b). Varying current in the coil then leads to a varying force that drives the diaphragm back and forth to produce sound.

Faithful reproduction of low-frequency sound requires a large driver, called a *woofer*. The woofer is large both because much of the sound power in music lies in the low-frequency range (think of a drum versus a flute!) and because the human

ear is much less sensitive to low frequencies (recall Fig. 17-4). But the woofer's large size gives it a large mechanical inertia, which means it cannot respond effectively with the rapid movements needed to reproduce high-frequency sound. Consequently, a much smaller driver, the *tweeter,* is used for high frequencies.

Power from an amplifier does not "know" about the mechanical properties of the drivers, so connecting both drivers directly to the same amplifier would result in low-frequency power being dissipated ineffectually in the tweeter and high-frequency power in the woofer. To prevent this inefficiency, a speaker system contains a *crossover network* to "steer" power to the appropriate drivers. Figure 33-8*c* shows a simple crossover network that exploits the frequency-dependent behavior of capacitors and inductors. At low frequencies the inductor reactance ωL is low, and current flows readily through the inductor to the woofer coil. But at high frequencies the inductor reactance is high, and little high-frequency power reaches the woofer. The capacitor's behavior is the opposite: It blocks low-frequency power from reaching the tweeter, while passing high-frequency power. Many speaker systems also employ a *midrange* driver, with a capacitor and inductor in series to block power at both high- and low-frequency extremes. Example 33-4, later in this chapter, explores quantitatively the behavior of the midrange circuitry.

The circuits of Fig. 33-8*c* are examples of *filters,* widely used in electronic systems to pass preferentially a range of frequencies. Problem 74 explores a simple filter.

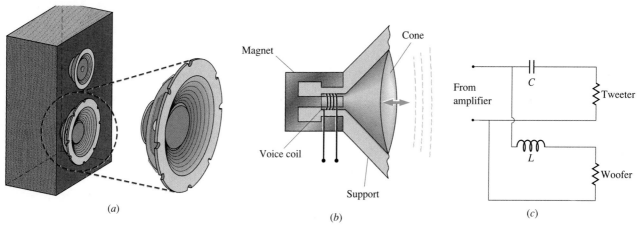

FIGURE 33-8 (*a*) A loudspeaker system, showing the woofer and tweeter drivers for low- and high-frequency sound production. (*b*) Cutaway view of a moving-coil driver. (*c*) The crossover network "steers" power at the appropriate frequencies to the woofer and tweeter. Resistances represent the two drivers.

33-3 *LC* CIRCUITS

In this section we consider circuits containing both inductors and capacitors. The properties of these circuits reflect directly the complementary nature of the two devices.

LC Oscillations

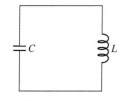

FIGURE 33-9 An *LC* circuit.

Figure 33-9 shows a circuit with a capacitor C and inductor L. Suppose the capacitor is initially charged to some voltage V_p and corresponding charge q_p, then connected to the inductor. What happens?

Initially, the capacitor is fully charged, while the inductor current is zero. There is electrical energy stored in the capacitor, but no energy in the inductor. This initial state is shown in Fig. 33-10*a*. Then the capacitor begins to discharge through the inductor. It cannot do so all at once, because the inductor opposes changes in current. So current in the inductor rises gradually, and with it the magnetic energy stored in the inductor. At the same time the capacitor voltage, charge, and stored electrical energy decrease. Some time later, the initial energy is divided equally between the capacitor and inductor, as in Fig. 33-10*b*. But the capacitor keeps discharging, eventually reaching zero charge, as in Fig. 33-10*c*. Now there is no voltage across the capacitor and no stored electrical energy. All the energy that was initially in the electric field of the capacitor is in the magnetic field of the inductor.

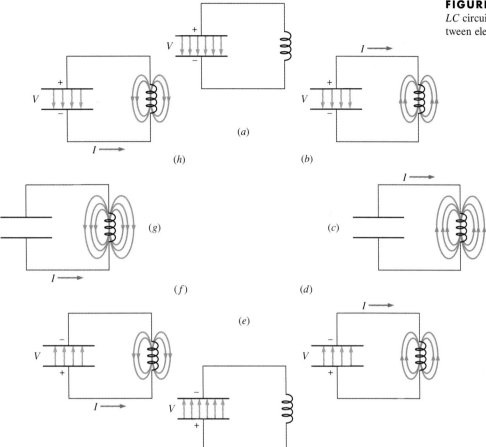

FIGURE 33-10 Oscillation in an *LC* circuit, showing energy transfer between electric and magnetic fields.

Does everything stop at this point? No, because current is flowing in the inductor. Current in an inductor cannot change abruptly, so the current keeps flowing and piles positive charge on the bottom plate of the capacitor. Stored electrical energy increases as the capacitor charges, and the inductor current and stored magnetic energy decrease. Eventually (Fig. 33-10*e*), the capacitor is fully charged in the opposite direction from its initial state. Again all the energy is in the capacitor. Again the capacitor begins to discharge, and the process repeats, now with a counterclockwise current. All the energy is transferred to the inductor (Fig. 33-10*g*), and then back to the capacitor, which again attains its initial state (Fig. 33-10*a*). Provided there is no energy loss, the oscillation repeats indefinitely.

This *LC* oscillation should remind you of the mass-spring system we studied in Chapter 15. There, energy was transferred back and forth between kinetic energy of the mass and potential energy of the spring. Here, energy is transferred back and forth between magnetic energy of the inductor and electrical energy of the capacitor. The mass-spring system oscillates with frequency determined by the mass m and spring constant k. Similarly, the *LC* circuit oscillates with frequency determined by the inductance L and capacitance C. Table 33-3 shows some analogies between mass-spring systems and *LC* circuits. We will develop these analogies more rigorously in the next section.

Analogies with *LC* circuits are so useful that engineers sometimes simulate complicated systems, such as bridges, automobile suspensions, or world energy

▲ **T A B L E 3 3 - 3** *LC* CIRCUITS AND MASS-SPRING SYSTEMS

LC CIRCUIT	MASS-SPRING
Charge q	Displacement x
Current $I = dq/dt$	Velocity $v = dx/dt$
Inductance L	Mass m
Capacitance C	$1/k$ (k = spring constant)
Magnetic energy $U_B = \frac{1}{2}LI^2$	Kinetic energy $U_K = \frac{1}{2}mv^2$
Electric energy $U_E = \frac{1}{2}(1/C)q^2$	Potential energy $U = \frac{1}{2}kx^2$
Frequency $\omega = 1/\sqrt{LC}$	Frequency $\omega = \sqrt{k/m}$
Resistance	Friction

usage, with networks of *LC* circuits. Such a network is called an analog computer because its behavior is analogous to that of the system under study.

Analyzing the *LC* Circuit

We described the *LC* circuit qualitatively in terms of transfer between electric and magnetic energy. This description suggests a way to analyze the circuit quantitatively. The total energy in the circuit is the sum of the magnetic and electric energy:

$$U = U_B + U_E = \frac{1}{2}LI^2 + \frac{1}{2}\frac{q^2}{C}.$$

The time derivative of this equation is

$$\frac{dU}{dt} = \frac{d}{dt}\left(\frac{1}{2}LI^2 + \frac{1}{2}\frac{q^2}{C}\right).$$

But since the total energy does not change, $dU/dt = 0$. Carrying out the differentiations in our expression for dU/dt, we then have

$$LI\frac{dI}{dt} + \frac{q}{C}\frac{dq}{dt} = 0. \tag{33-8}$$

Substituting $I = dq/dt$ and $dI/dt = d^2q/dt^2$ gives

$$L\frac{d^2q}{dt^2} + \frac{1}{C}q = 0. \tag{33-9}$$

Equation 33-9 is a differential equation describing the capacitor charge as a function of time. We encountered a similar equation in Chapter 15 when we studied the mass-spring system:

$$m\frac{d^2x}{dt^2} + kx = 0. \tag{15-4}$$

We found that Equation 15-4 could be satisfied by a sinusoidal function of time, with frequency given by

$$\omega = \sqrt{k/m}. \tag{15-11}$$

Equation 33-9 is identical to Equation 15-4 except that q replaces x, L replaces m, and $1/C$ replaces k. Therefore the solution of Equation 33-9 is a sinusoidal oscillation whose frequency is given by Equation 15-11 with L replacing m and $1/C$ replacing k:

$$q = q_p \cos \omega t, \tag{33-10}$$

where

$$\omega = \frac{1}{\sqrt{LC}}. \tag{33-11}$$

Here we chose cosine rather than sine since our qualitative description (Fig. 33-10) started with the capacitor charge at its peak value. We could equally well have used a sine function with phase constant $\pi/2$. Differentiating the charge (Equation 33-10) gives the current:

$$I = \frac{dq}{dt} = \frac{d}{dt}(q_p \cos \omega t) = -\omega q_p \sin \omega t. \tag{33-12}$$

All other circuit quantities follow from Equations 33-10 and 33-12. The capacitor voltage, obtained from the relation $q = CV$, is

$$V_C = \frac{q}{C} = \frac{q_p}{C} \cos \omega t.$$

The electrical energy in the capacitor is, therefore,

$$U_E = \tfrac{1}{2}CV^2 + (\tfrac{1}{2}C)\left(\frac{q_p}{C}\cos \omega t\right)^2 = \frac{q_p^2}{2C}\cos^2 \omega t,$$

while the magnetic energy in the inductor is

$$U_B = \tfrac{1}{2}LI^2 = \tfrac{1}{2}L(-\omega q_p \sin \omega t)^2 = \tfrac{1}{2}L\omega^2 q_p^2 \sin^2 \omega t.$$

We can verify that our solution conserves energy by adding the electric and magnetic energies:

$$U_{\text{total}} = U_E + U_B = \frac{q_p^2}{2C}\cos^2 \omega t + \tfrac{1}{2}L\omega^2 q_p^2 \sin^2 \omega t.$$

But Equation 33-11 shows that $\omega^2 = 1/LC$, so we have

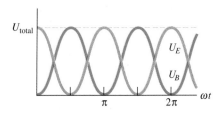

FIGURE 33-11 Electric and magnetic energies in an *LC* circuit. Their sum is constant.

$$U_{total} = \frac{q_p^2}{2C}\cos^2 \omega t + \frac{1}{2}\frac{L}{LC}q_p^2 \sin^2 \omega t$$

$$= \frac{q_p^2}{2C}(\cos^2 \omega t + \sin^2 \omega t) = \frac{q_p^2}{2C},$$

since $\cos^2 \omega t + \sin^2 \omega t = 1$. Thus the total energy is independent of time and is equal to the initial energy stored in the capacitor. Figure 33-11 is a plot of the electric and magnetic energies as functions of time, showing that the two always sum to a constant.

● **EXAMPLE 33-3** A PIANO TUNER

You wish to make an *LC* circuit oscillate at 440 Hz (A above middle C) to assist in tuning pianos. You have available a 2.0-H inductor. What value of capacitance should you use? If you initially charge the capacitor to 5.0 V, what will be the peak charge on the capacitor and the peak current in the circuit?

Solution
The oscillation frequency is given by Equation 33-11. Solving for *C* gives

$$C = \frac{1}{\omega^2 L} = \frac{1}{(2\pi f)^2 L} = \frac{1}{[(2\pi)(440\ \text{Hz})]^2(2.0\ \text{H})} = 65.4\ \text{nF}.$$

The capacitor charge and voltage are related through $C = q/V$, so

$$q_p = CV_p = (65.4\ \text{nF})(5.0\ \text{V}) = 327\ \text{nC}.$$

Equation 33-12 then shows that the peak current is

$$I_p = \omega q_p = 2\pi f q_p = (2\pi)(440\ \text{Hz})(327\ \text{nC}) = 0.90\ \text{mA}.$$

Problem 38 shows how the current also follows from energy considerations.

EXERCISE An FM radio transmitter requires an *LC* circuit that oscillates at the 89.5-MHz transmitter frequency. What should be the inductance if the circuit capacitance is 47 pF?

Answer: 67 nH

Some problems similar to Example 33-3: 27, 28, 31, 32
●

Resistance in *LC* Circuits—Damping

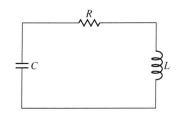

FIGURE 33-12 An *RLC* circuit.

Real inductors, capacitors, and wires have resistance. Both this intrinsic resistance and any external resistance are represented by the resistor *R* in Fig. 33-12. What happens in such a resistive *LC* circuit?

Provided the resistance is small—small enough that only a small fraction of the energy is lost in one cycle—then our analysis of the preceding section applies. The circuit oscillates at a frequency given very nearly by Equation 33-11. But as current flows back and forth through the resistor, energy is dissipated as heat. On each successive cycle, the total energy decreases. Consequently, the amplitude of the oscillations decreases with time.

We can analyze this *RLC* circuit by starting with Equation 33-8, but now setting dU/dt equal to the rate of energy dissipation in the resistor:

$$LI\frac{dI}{dt} + \frac{q}{C}\frac{dq}{dt} = -I^2R,$$

where the minus sign indicates that energy is *lost* in the resistor. Writing $I = dq/dt$ as we did in the preceding section leads to

$$L\frac{d^2q}{dt^2} + R\frac{dq}{dt} + \frac{q}{C} = 0.$$

This equation is mathematically identical to Equation 15-22 for damped simple harmonic motion, showing that our analogies of Table 33-3 continue to hold when resistance is present. Using Equation 15-23, which is the solution to Equation 15-22, and the appropriate analogies from Table 33-3, we can construct the solution for our decaying *RLC* circuit:

$$q = q_p e^{-Rt/2L} \cos \omega t. \qquad (33\text{-}13)$$

Other quantities show similar behavior, with oscillation amplitude decaying exponentially with time constant $2L/R$. Figure 33-13 shows an oscilloscope trace of the capacitor voltage in a circuit undergoing damped oscillations.

Equations 15-23 and 33-13 are correct only when the energy dissipation is small. As the electrical resistance increases, the oscillations decay more rapidly and the frequency of oscillation decreases. Finally, when the exponential time constant $2L/R$ equals the inverse of the natural frequency given by Equation 33-11, much of the energy is lost in the time of one undamped oscillation period. This situation is termed **critical damping**, and at this value of R circuit quantities decay exponentially to zero, in analogy with a critically damped mechanical system (Section 15-6). For greater values of R, the circuit is **overdamped** and also exhibits no oscillation.

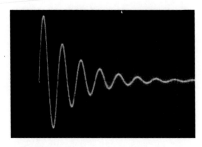

FIGURE 33-13 An oscilloscope displays the capacitor voltage in a damped *RLC* circuit. Note the exponential decline in amplitude of the oscillations.

33-4 DRIVEN *RLC* CIRCUITS AND RESONANCE

What happens if we connect an *RLC* circuit to an AC generator, as shown in Fig. 33-14? Because the circuit is analogous to a mass-spring system, we might expect it to exhibit resonant behavior analogous to the mechanical resonance discussed in Section 15-7.

Resonance in the *RLC* Circuit

Suppose we vary the generator frequency in Fig. 33-14 while keeping the peak voltage constant. How much current flows in the *RLC* circuit? At low frequencies the capacitor acts almost like an open circuit (its reactance $X_C = 1/\omega C$ is large), so little current flows. At high frequencies the inductor acts almost like an open circuit (its reactance $X_L = \omega L$ is large), so little current flows. At some intermediate frequency the current must be a maximum. We now show that this **resonant frequency,** ω_r, is in fact the undamped natural frequency $1/\sqrt{LC}$.

Figure 33-14 is a series circuit, so the *same* current flows through all the components. We know that the voltage in a capacitor lags the current by 90°,

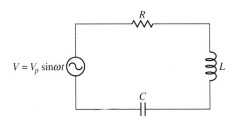

FIGURE 33-14 A series *RLC* circuit driven by an AC generator.

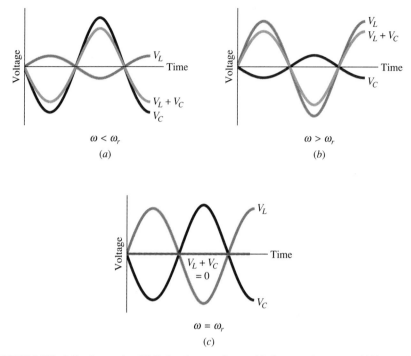

FIGURE 33-15 In a series *RLC* circuit, capacitor and inductor voltages are 180° out of phase at all frequencies. (*a*) At low frequencies, the voltage is greatest across the capacitor. (*b*) At high frequencies, the voltage is greatest across the inductor. (*c*) At resonance, capacitor and inductor voltages have equal amplitude, and therefore they cancel completely.

while the voltage in an inductor leads by 90°. Since the same current flows through the inductor and capacitor, the inductor and capacitor voltages are therefore 180° out of phase. This phase relation holds at *any* frequency. At low frequencies, however, the capacitor's reactance is greatest and the capacitor voltage is therefore greater than the inductor voltage (Fig. 33-15*a*). At high frequencies the opposite is true, with the inductor voltage being greatest (Fig. 33-15*b*). Since the two voltages are 180° out of phase, they tend to cancel, but at high or low frequencies Figs. 33-15*a* and *b* show that their different amplitudes mean this cancellation is not complete.

Is there a frequency at which the capacitor and inductor voltages exactly cancel? The peak current and voltage in these two components are related by Equations 33-5 and 33-7, respectively. Since the current is the same, comparison of these equations shows that the voltages will be the same when the capacitive reactance $1/\omega C$ is equal to the inductive reactance ωL, which gives

$$\omega_r^2 = \frac{1}{LC}.$$

This is precisely Equation 33-11's condition for the undamped natural frequency.

At resonance, then, the capacitor and inductor voltages completely cancel. The voltage across the pair together is zero, and—at the resonant frequency only—they might as well be replaced by a wire. The circuit current at resonance is determined entirely by the resistance. At any other frequency the effects of capacitance and inductance do not cancel completely, and the current is lower.

Frequency Response of the *RLC* Circuit

Here we derive a general expression for the current as a function of frequency in the series *RLC* circuit, using the phasor diagrams introduced in Section 33-2. Since the same current flows through all components of the series circuit, we represent this current by a single phasor of length I_p in Fig. 33-16. Also shown are phasors for the resistor, capacitor, and inductor voltages. The resistor voltage phasor is in the same direction as the current because these two are in phase, but the capacitor and inductor voltages are at 90° relative to the current, representing the phase relations in these components.

Applying the loop law to Fig. 33-14 gives $V - V_R - V_L - V_C = 0$; thus, the applied voltage V is the sum of the voltages across the other three circuit elements. This statement is true for the instantaneous values of the voltages at any time. But because they have different phase constants, the voltages V_R, V_L, and V_C peak at different times—and, thus, the peak applied voltage is *not* the sum of the peak voltages across the resistor, inductor, and capacitor. We can find the relation among the peak voltages by adding their phasors *vectorially*—a process that accounts for both magnitude and phase. Figure 33-16*b* shows the result of this phasor addition. Applying the Pythagorean theorem, we see that the magnitude of the applied voltage phasor—that is, the peak applied voltage—is given by

$$V_p^2 = V_{Rp}^2 + (V_{Cp} - V_{Lp})^2.$$

Expressing this in terms of the current and the resistances and reactances gives

$$V_p^2 = I_p^2 R^2 + (I_p X_c - I_p X_L)^2,$$

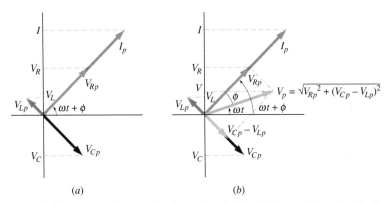

(a) $\qquad\qquad\qquad\qquad$ (b)

FIGURE 33-16 Phasor diagrams for the series *RLC* circuit. As usual, the length of each phasor indicates the peak value (subscript *p*) of the associated quantity. (*a*) The current in the three circuit elements is the same, but the voltages are out of phase. The whole system rotates with angular speed ω, and instantaneous values of voltages and current are given by the projections on the vertical axis. (*b*) The three voltage phasors sum vectorially to the applied voltage. Note that for this case the current leads the applied voltage; the phase difference ϕ is given by Equation 33-16.

or
$$I_p = \frac{V_p}{\sqrt{R^2 + (X_C - X_L)^2}} = \frac{V_p}{Z}, \qquad (33\text{-}14)$$

where we have defined

$$Z = \sqrt{R^2 + (X_C - X_L)^2}.$$

Equation 33-14 has the form of Ohm's law, with Z playing the role of resistance. We call Z the **impedance** of the circuit. Impedance is a generalization of resistance to include the frequency-dependent effects of capacitance and inductance. Putting in our expressions for the reactances gives

$$Z = \sqrt{R^2 + \left(\frac{1}{\omega C} - \omega L\right)^2}. \qquad (33\text{-}15)$$

In agreement with our earlier analysis, this equation shows that the circuit impedance becomes very large at high and low frequencies, and has its lowest value, R, at resonance.

Figure 33-17 is a plot of Equation 33-14, showing peak current versus frequency for several values of resistance. As we lower the resistance, the peak current at resonance rises. Although the current at other frequencies rises, too, it does so to a much lower extent than at resonance. This is because the impedance at resonance depends only on the resistance, but includes reactive effects at other frequencies. As a result, the resonance curve becomes more sharply peaked as the resistance drops. For a circuit with very low resistance, the current at resonance is dramatically different from that at even a slightly different frequency. Such a circuit, called a **high-Q** (for high-quality) circuit, does a good job of distinguishing its resonant frequency from nearby frequencies. A rigorous definition of Q can be given in terms of the width of the resonance curve (see Problem 81). When you tune an older radio, you are adjusting the variable capacitor in a high-Q LC circuit that selects the desired station from among the hodgepodge of radio signals reaching the antenna (Fig. 33-18).

Equation 33-14 relates the peak current and the applied voltage in the RLC circuit, but it does not tell the whole story. In general, current and voltage are

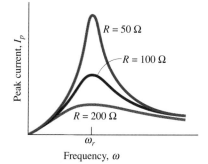

FIGURE 33-17 Resonance curves for an RLC circuit with $L = 5.0$ mH, $C = 0.22$ μF, and several resistance values.

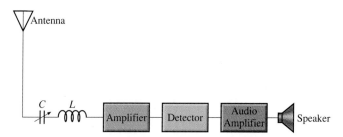

FIGURE 33-18 Simplified diagram of an older radio, using an LC circuit with a variable capacitor to select the desired station frequency. The signal is subsequently amplified, the audio information is extracted in the detector, and the amplified audio signal drives the loudspeaker. Modern radios often use sophisticated digital circuitry in place of the LC frequency selector.

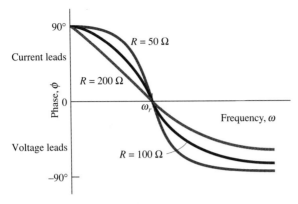

FIGURE 33-19 Phase relations for the *RLC* circuits whose resonance curves are shown in Fig. 33-17.

not in phase. Figure 33-16*b* shows that the phase difference ϕ between the two is

$$\tan \phi = \frac{V_{Cp} - V_{Lp}}{V_{Rp}}.$$

Because the voltages are proportional to the reactances and resistance, this expression may be written

$$\tan \phi = \frac{X_C - X_L}{R} = \frac{1/\omega C - \omega L}{R}. \qquad (33\text{-}16)$$

Does this equation make sense? At resonance, $X_L = X_C$ and the phase difference is zero. Here the effects of capacitance and inductance cancel, and the circuit behaves like a pure resistance. At frequencies below resonance, capacitive reactance is greatest and the phase difference is positive, indicating that current leads voltage—as we expect in a circuit dominated by capacitance. Above resonance $X_L > X_C$ and the phase difference is negative, indicating that voltage leads current—as we expect in a circuit dominated by inductance. At high- and low-frequency extremes, $\tan \phi$ becomes arbitrarily large and the phase differences approach 90°, as shown in Fig. 33-19.

TIP **Phase Matters** You can't analyze an AC circuit by treating resistors, capacitors, and inductors all as "resistors" with resistances R, X_C, and X_L. Why not? Because associated with each component is a different phase relation between current and voltage. Only a phasor diagram—or mathematical analysis that includes phase relations—correctly characterizes the circuit. The phase relations in a series *RLC* circuit show up in the minus sign joining the capacitive and inductive reactance, and in the Pythagorean-like addition of the resistance and reactances in Equations 33-14 and 33-15.

● **EXAMPLE 33-4** DESIGNING A LOUDSPEAKER SYSTEM

Current flows to the midrange speaker in a loudspeaker system through a 2.2-mH inductor and a capacitor, in series. (See *Application: Loudspeaker Systems*, earlier in this chapter.) (a) What should be the capacitance in order that a given applied voltage produces the maximum current at a frequency of 1.0 kHz? (b) What should be the speaker's resistance in order that the same voltage produces a current with half the maximum value at 600 Hz? (c) If the peak output voltage of the amplifier is 20 V, what should be the capacitor's peak voltage rating?

Solution

(a) Inductor, capacitor, and resistor comprise a series *RLC* circuit, with resonant frequency $f_r = \omega_r/2\pi = 1/2\pi\sqrt{LC}$. Solving for *C* gives

$$C = \frac{1}{4\pi^2 f_r^2 L} = \frac{1}{(4\pi^2)(1.0\times10^3 \text{ Hz})^2(2.2\times10^{-3} \text{ H})} = 11.5 \text{ }\mu\text{F}.$$

(b) Equation 33-14 gives the current in an *RLC* circuit. The denominator in this equation—the impedance $Z = \sqrt{R^2 + (X_C - X_L)^2}$—has the value *R* at resonance, when $X_L = X_C$. Thus the current will have half its maximum value where the $Z = 2R$. Using $X_C = 1/\omega C$ and $X_L = \omega L$, we want *R* such that

$$Z = \sqrt{R^2 + \left(\frac{1}{\omega C} - \omega L\right)^2} = 2R$$

when $f = \omega/2\pi = 600$ Hz, or $\omega = 3.77\times10^3$ s⁻¹. Squaring and solving for *R* gives

$$R = \frac{1}{\sqrt{3}}\left(\frac{1}{\omega C} - \omega L\right) = 8.53 \text{ }\Omega,$$

where we used the 11.5-μF answer to part (a) for the capacitance *C*. This resistance value is typical for a loudspeaker. The exercise below shows that the current also has half its maximum value at 1.7 kHz, so the midrange speaker gets at least half its maximum current in the range from about 600 to 1700 Hz.

(c) The maximum current flows at the resonant frequency, so the Ohm's-law-like Equation 33-5 shows that the maximum capacitor voltage also occurs at this frequency and is given by

$$V_{Cp} = I_p X_C = \left(\frac{V_p}{R}\right)\left(\frac{1}{\omega C}\right)$$

$$= \left(\frac{20 \text{ V}}{8.53 \text{ }\Omega}\right)\left(\frac{1}{(2\pi)(1000 \text{ Hz})(11.5\times10^{-6} \text{ F})}\right)$$

$$= 32 \text{ V}.$$

How can this value be *greater* than the peak applied voltage? Remember that there is another source of emf in the circuit—the inductor, whose emf depends on the rate of change of current and may therefore exceed the applied voltage. In this relatively low-*Q* circuit the peak capacitor voltage is not too much greater than the peak applied voltage, but in high-*Q* circuits like those used in radio transmitters capacitors may have to withstand voltages hundreds of times those supplied to the circuit.

EXERCISE Find a second frequency where the speaker current in Example 33-4 has half its maximum value. *Hint:* Square Equation 33-15, isolate the frequency-dependent terms, take a square root, and solve a quadratic equation. Or use graphical techniques.

Answer: 1.7 kHz

Some problems similar to Example 33-4: 43, 44, 46, 69, 70, 76 ●

33-5 POWER IN AC CIRCUITS

In Section 33-2, we noted that average power dissipation is zero in a circuit containing only a capacitor or an inductor. We can understand this physically because the reactive element alternately stores and releases energy rather than dissipating it as heat. Mathematically, we can see this from Fig. 33-20*a*, which shows the current, voltage, and instantaneous power in a capacitor. The power is the product of the current and voltage. Because these two are out of phase, the power is positive half the time, and negative half the time. When the power is positive, the capacitor is absorbing energy from the source of emf that drives the current. When the power is negative, the capacitor is returning energy to the driving source. The net energy transferred to the capacitor over one cycle is $\int P \, dt$, or the area under the power versus time curve, and is zero in this case.

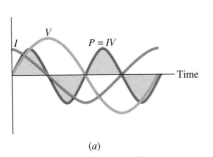

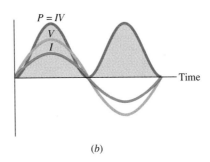

FIGURE 33-20   Current I, voltage V, and power consumption IV. (*a*) Current and voltage in a capacitor are out of phase by 90°. The power is alternately positive (absorbing energy from the source of emf) and negative (returning energy to the source), and the net energy consumption (shaded area) over one cycle is zero. (*b*) Current and voltage are in phase in a resistor and therefore the power is always positive, meaning that the resistor is a net energy consumer.

Figure 33-20*b*, in contrast, shows current, voltage, and instantaneous power in a resistor. Since current and voltage are always in phase, the power is always positive, and the resistor always takes energy from the source. Comparison of Figs. 33-20*a, b* suggests that the phase difference between current and voltage is important in determining the average power consumption of an AC circuit. We can see this more clearly if we imagine slipping the current and voltage just slightly out of phase, as in Fig. 33-21. Then there are narrow regions where the power is negative, so the average power over one cycle is slightly less than in the resistor case. As the phase difference increases, so does the time that the power is negative, until at 90° phase difference, the time-average power is zero.

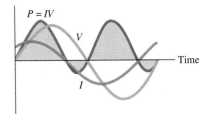

FIGURE 33-21   Average power consumption decreases as current and voltage go out of phase.

We can develop a general expression for power in AC circuits by considering the time-average product of voltage and current with arbitrary phase difference ϕ:

$$\langle P \rangle = \langle [I_p \sin(\omega t + \phi)][V_p \sin \omega t] \rangle,$$

where $\langle \, \rangle$ indicates a time average over one cycle. Expanding the current term using a trig identity (see Appendix A) gives

$$\langle P \rangle = I_p V_p \langle (\sin^2 \omega t)(\cos \phi) + (\sin \omega t)(\cos \omega t)(\sin \phi) \rangle.$$

The average of $(\sin \omega t)(\cos \omega t)$ is zero, as we've just shown for two signals 90° out of phase. The quantity $\sin^2 \omega t$ swings from 0 to 1, and is symmetric about $\frac{1}{2}$, so its average value is $\frac{1}{2}$ (Fig. 33-22). Then we have

$$\langle P \rangle = \tfrac{1}{2} I_p V_p \cos \phi.$$

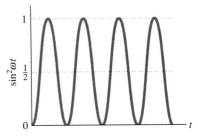

FIGURE 33-22   The average of $\sin^2 \omega t$ is $\frac{1}{2}$.

Writing the peak values as $\sqrt{2}$ times the rms values gives

$$\langle P \rangle = \tfrac{1}{2}\sqrt{2}I_{rms} \sqrt{2}V_{rms} \cos \phi = I_{rms} V_{rms} \cos \phi. \qquad (33\text{-}17)$$

This equation confirms our earlier graphical arguments. When the voltage and current are in phase, the average power is just the product $I_{rms} V_{rms}$. (This, in fact, is a principal reason for using rms values: with them, the expression for average power is the same as in the DC case.) But with current and voltage out of phase, the average power is smaller; at 90° phase difference it is zero.

The factor cos ϕ is called the **power factor.** A purely resistive circuit has a power factor of 1, while a circuit with only inductance and capacitance has a power factor of zero. In circuits containing resistance along with capacitance and/or inductance, the power factor generally depends on frequency; in the series *RLC* circuit, for example, it is 1 at resonance but lower at other frequencies.

33-6 TRANSFORMERS AND POWER SUPPLIES

A **transformer** is a pair of wire coils in close proximity. Supplying an alternating current to one coil—called the **primary**—results in a changing magnetic flux through the other coil, or **secondary.** According to Faraday's law, an induced emf then appears in the secondary coil. This emf can drive a current in circuitry connected to the secondary coil. Thus, the transformer transfers electrical energy from the primary circuit to the secondary circuit, even though there is no direct electrical connection between the two.

Practical transformers are often wound on iron cores that concentrate flux, ensuring that essentially all the magnetic flux produced by the primary coil goes through the secondary. Figure 33-23 shows a simplified diagram of a transformer, along with its circuit symbol.

The transformer in Fig. 33-23a has two turns in its primary and four in its secondary. Since the same changing flux passes through each turn of each coil, the total emf induced the secondary must be twice that of the primary. The transformer is therefore a **step-up-transformer;** it steps up the voltage by a factor of 2. Interchanging primary and secondary would give a **step-down** transformer, with the secondary voltage half that of the primary. In general, the ratio of the peak (or rms) primary voltage V_1 to the peak (or rms) secondary voltage V_2 is the same as the ratio of the numbers of turns in the two coils:

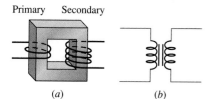

Primary   Secondary

$$V_2 = \frac{N_2}{N_1} V_1. \tag{33-18}$$

FIGURE 33-23 (*a*) A transformer, consisting of two coils wound on an iron core. (*b*) Transformer circuit symbol.
58,7.5
71,5.02
0

Aren't we getting something for nothing with a step-up transformer? No—a step-up transformer increases voltage but not power. In an ideal transformer, all the power supplied to the primary is transferred to the secondary; therefore,

$$I_1 V_1 = I_2 V_2. \tag{33-19}$$

If the voltage goes up, the current goes down, and vice versa. Real transformers have losses associated with resistance in their windings and heating of their iron cores, but good engineering holds these losses to a few per cent of the total power.

■ APPLICATION ELECTRIC POWER DISTRIBUTION

The power loss in a resistance R carrying current I is given by $I^2 R$. That means it's most efficient to transmit electric power at high voltage and low current. Lower voltages, on the other hand, are safer and easier to handle. How can we satisfy both these considerations in practical power systems? With transformers!

Power-plant generators operate at about 20 kV. Transformers at the power plant then step this up to several hundred kilovolts for long-distance transmission. At a city or town the voltage is dropped to several kilovolts for distribution within the municipality. Transformers near each building reduce the voltage further, for example, to 120 V and 240 V for household use. Individual electrical devices within a building use transformers to meet their particular voltage requirements. Figure 33-24 outlines the voltage transformations in power transmission.

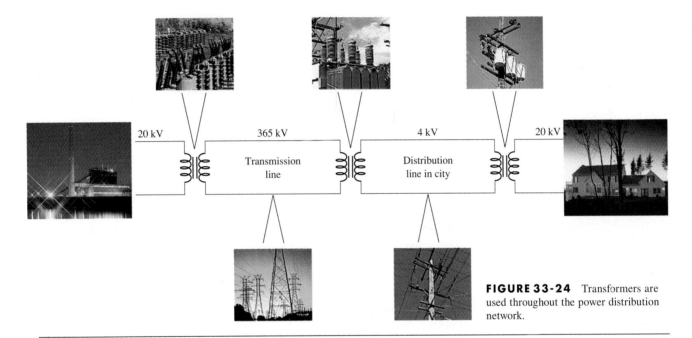

FIGURE 33-24 Transformers are used throughout the power distribution network.

Transformers are widely used to provide voltages different from those of available sources. Most transistor circuits, for example, require a few tens of volts or less; step-down transformers bring the 120-V AC power line down to this level. Television tubes and computer monitors require tens of kV, calling for step-up transformers.

Transformers exploit electromagnetic induction, and therefore, they work only with time-varying current—that is, AC. A transformer cannot be used to step up the DC voltage from a battery, unless current from the battery is somehow interrupted periodically—as it is, for example, by the distributor in an automobile ignition system, whose ignition coil is really a transformer (see Example 32-2). In fact, one of the main reasons for using AC power is that it is readily transformed from one voltage level to another.

FIGURE 33-25 Circuit symbol for a diode, with preferred direction of current indicated.

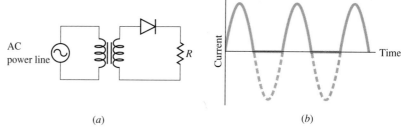

(a)

(b)

FIGURE 33-26 (a) A simple DC power supply consists of a transformer and diode, which supply current to the load resistance. The diode passes current in only one direction, cutting off the negative half of each cycle and giving the load current shown in (b).

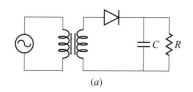

(a)

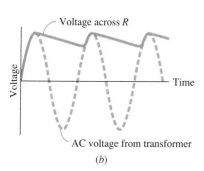

(b)

FIGURE 33-27 (a) Adding a capacitor reduces variations in the load voltage and therefore in the load current. This circuit is a complete DC power supply. (b) Load voltage as a function of time.

DC Power Supplies

Devices like light bulbs and electric heaters work equally well on AC or DC. But others, especially electronic circuits, require DC power. How do we get DC from the AC power line?

In Chapter 27's discussion of semiconductors, we mentioned that a junction between P-type and N-type semiconductors passes current in one direction but not the other. A **diode** is a PN junction designed specifically to be such a "one-way valve" for electric current. An ideal diode acts like a short circuit to current flowing in the preferred direction and like an open circuit in the opposite direction (Fig. 33-25).

Figure 33-26a shows a diode and resistor connected to the secondary of a transformer. The resistor represents the load—that is, whatever circuit we wish to supply with DC power. The transformer brings the 120-V AC power line to a level appropriate to the load. The diode passes current in only one direction, resulting in the load current shown in Fig. 33-26b.

Although the load current in Fig. 33-26 flows in one direction, it still varies drastically with time. If the load were a stereo amplifier, for example, its speakers would emit a loud hum. So we smooth, or **filter,** the output of the diode. The simplest filter is a capacitor, as shown in Fig. 33-27a. As the voltage on the left side of the diode rises, the capacitor voltage rises rapidly because of the short time constant associated with the low resistance of the diode in its preferred direction. But then the AC voltage begins to fall, and the diode "turns off." The capacitor cannot discharge from right to left through the diode but only through the load resistance. If the time constant RC is long enough, the capacitor voltage will drop only slightly before the AC voltage again rises and sends a new surge of current through the diode to bring the capacitor to its maximum charge (Fig. 33-27b). By making the capacitance large enough—so large that the time constant is much longer than the period of the AC power—we can make load current and voltage arbitrarily smooth. Large capacitors are expensive, so in practice additional filters—often using transistors or integrated circuits—are added to the simple capacitive filter in critical applications. High-quality power supplies for audio equipment or electronic instrumentation achieve **ripple factors**—the ratio of the fluctuation amplitude to the DC level—of 10^{-5} or better. More complicated diode circuits can also be used to produce signals requiring less filtering (see Problem 80).

CHAPTER SYNOPSIS

Summary

1. **Alternating current** varies with time. A sinusoidal AC signal is characterized by its peak amplitude, frequency, and phase constant; for example:

$$I = I_p \sin(\omega t + \phi).$$

2. In a resistor, the ratio of voltage to current is always constant, so

$$I_p = \frac{V_p}{R},$$

and the current and voltage are in phase.

3. In a capacitor the ratio of peak voltage and current is determined by the **capacitive reactance:**

$$I_p = \frac{V_p}{X_C}, \quad \text{where} \quad X_C = \frac{1}{\omega C}.$$

The current in the capacitor **leads** the voltage by 90°.

4. In an inductor the ratio of peak voltage and current is determined by the **inductive reactance:**

$$I_p = \frac{V_p}{X_L}, \quad \text{where} \quad X_L = \omega L.$$

The current in the inductor **lags** the voltage by 90°.

5. **Phasors** are vector-like arrows showing the amplitude and phase of AC signals, and are useful in analyzing AC circuits.

6. In an undriven LC circuit, energy oscillates between electric and magnetic forms at the **resonant frequency** $\omega_r = 1/\sqrt{LC}$. The amplitude of the oscillation decays exponentially as energy is dissipated in the circuit resistance.

7. In a series RLC circuit, the effects of inductance and capacitance exactly cancel at the resonant frequency. At this frequency the circuit exhibits the minimum **impedance,** and therefore passes the maximum current. At resonance the current and voltage are in phase. At lower frequencies the capacitor dominates and current leads voltage, while at higher frequencies the inductor dominates and voltage leads current. The impedance of a series RLC circuit is

$$Z = \sqrt{R^2 + (X_C - X_L)^2},$$

while the phase difference between current and voltage is given by

$$\cos \phi = \frac{R}{Z}.$$

8. The power dissipated in an AC circuit depends on the relative effects of resistance and reactance. In a purely reactive circuit, current and voltage are 90° out of phase, and no power is dissipated. In a purely resistive circuit, the average power dissipation is $I_{rms} V_{rms}$. When both resistance and reactance are present, the power dissipation depends on the **power factor,** $\cos \phi$, where ϕ is the phase difference between current and voltage. In general, the time-average power consumed in an AC circuit is

$$P = I_{rms} V_{rms} \cos \phi.$$

9. A **transformer** uses electromagnetic induction to transfer electric power between two circuits. The ratio of the peak or rms voltages in the transformer's two windings is the same as the ratio of the numbers of turns in the windings:

$$V_2 = \frac{N_2}{N_1} V_1,$$

while the power VI is the same in each winding of an ideal transformer.

10. DC **power supplies** use **diodes** to change alternating to direct power. Capacitors then smooth out the remaining time variation to produce the steady voltages and currents needed to power electronic equipment.

Terms You Should Understand

(Pairs are closely related terms whose distinction is important; number in parentheses is chapter section where term first appears.)

AC, DC (introduction)
peak, rms amplitudes (33-1)
frequency, angular frequency (33-1)
capacitive reactance, inductive reactance (33-2)
phasor (33-2)
resonant frequency (33-4)
power factor (33-5)
transformer (33-6)
diode (33-6)

Symbols You Should Recognize

$I_p, V_p; I_{rms}, V_{rms}$ (33-1)
f, ω (33-1)
ϕ (33-1)
X_C, X_L (33-2)
Z (33-4)

Problems You Should Be Able to Solve

calculating current and voltage in AC circuits involving individual resistors, capacitors, and inductors (33-2)
analyzing oscillating LC circuits (33-3)
analyzing driven RLC circuits (33-4)

calculating power in AC circuits (33-5)
designing simple power supplies using transformers, diodes, and capacitive filters (33-6)

Limitations to Keep in Mind

This chapter considers only sinusoidally varying AC signals. More complex time variations must first be analyzed into sums of sinusoidal terms, using techniques mentioned in Chapter 16.

Care must be taken in adding AC signals with different phases. Phasor diagrams provide a convenient way of doing this.

QUESTIONS

1. Two AC signals have the same amplitude but different frequencies. Are their rms amplitudes the same?
2. Does it make sense to talk about the phase difference between two AC signals of different frequencies? Sketch a diagram to confirm your answer.
3. What is meant by the statement "a capacitor is a DC open circuit"?
4. How can current keep flowing in an AC circuit containing a capacitor? After all, a capacitor contains a gap between two conductors, and no charge can cross this gap.
5. Why does it make sense that inductive reactance increases with frequency?
6. The same AC voltage appears across a capacitor and a resistor, and the same rms current flows in each. Is the power dissipation the same in each?
7. When a particular inductor and capacitor are connected across the same AC voltage, the current in the inductor is larger than in the capacitor. Will this be true at all frequencies?
8. An inductor and capacitor are connected in series across an AC generator, and the rms voltage across the inductor is found to be larger than that across the capacitor. Is the generator frequency above or below resonance?
9. When the capacitor voltage in an undriven LC circuit reaches zero, why don't the oscillations stop?
10. Why is the quantity ωL not called the resistance of an inductor?
11. Why is Equation 33-5 not a full description of the relation between voltage and current in a capacitor? What equation does give the full relation?
12. If you double both the capacitance and inductance in an LC circuit, what effect does this have on the resonant frequency?
13. In a series RLC circuit, the applied voltage lags the current. Is the frequency above or below resonance?
14. In a series RLC circuit, the applied voltage leads the current. Is the peak voltage greater across the capacitor or the inductor?

15. At a certain frequency, the impedance of a series RLC circuit is twice the resistance of the circuit. Can you tell whether the frequency is above or below resonance? Which is it, or why can't you tell?
16. What does it mean to say that the capacitor in an RLC circuit dominates at low frequencies?
17. The voltage across two circuit elements in series is zero. Is it possible that the voltages across the individual elements are nevertheless not zero? Give an example.
18. If you measure the rms voltages across the resistor, capacitor, and inductor in a series RLC circuit, will they add to the rms value of the generator voltage? Reconcile your answer with the loop law. (See also Problem 47.)
19. In a fluorescent light fixture an inductor, called the ballast, is used to limit current to the lamp. Why is an inductor preferable to a resistor?
20. What is the power factor in a circuit containing only a resistor? Does this power factor change with frequency?
21. To save electrical energy, should you strive for a large or small power factor?
22. When an AC motor runs with no significant mechanical load, its power factor is very nearly zero. What must happen when the motor begins doing mechanical work?
23. A step-up transformer increases voltage, or energy per unit charge. How is this possible without violating conservation of energy?
24. The iron cores of transformers are often made by laminating together thin sheets of iron separated by an insulating substance. Why is this preferable to making the cores out of solid pieces of iron? *Hint:* Think about eddy currents.
25. A battery charger runs off the 120-V AC power line. It supplies up to 30 A to recharge a 12-V car battery, yet it can be plugged into a 15-A circuit without blowing the circuit breaker. How is this possible?
26. Manuals for electronic instruments that run off the AC power line, including stereo amplifiers, often caution against connecting the instrument to DC power sources. Why? *Hint:* What would happen if a DC voltage were imposed across the transformer in such an instrument?

PROBLEMS

Section 33-1 Alternating Current

1. Much of Europe uses AC power at 230 V rms and 50 Hz. Express this AC voltage in the form of Equation 33-3, taking $\phi = 0$.
2. An rms voltmeter connected across the filament of a TV picture tube reads 6.3 V. What is the peak voltage across the filament?
3. An oscilloscope displays a sinusoidal signal whose peak-to-peak voltage (see Fig. 33-1) is 28 V. What is the rms voltage?
4. An industrial electric motor runs at 208 V rms and 400 Hz. What are (a) the peak voltage and (b) the angular frequency?
5. An AC current is given by $I = 495 \sin(9.43t)$, with I in milliamperes and t in milliseconds. Find (a) the rms current and (b) the frequency in Hz.
6. What are the phase constants for each of the signals shown in Fig. 33-28?

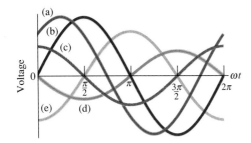

FIGURE 33-28   Problem 6.

7. The rms amplitude is defined as the square root of the average of the square of the signal. For a periodic function the time average is the integral over one period, divided by the period. For a sinusoidal voltage given by $V = V_p \sin \omega t$, show explicitly that $V_{\text{rms}} = V_p/\sqrt{2}$.
8. How are the rms and peak voltages related for the square wave shown in Fig. 33-29? See Problem 7.

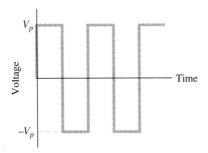

FIGURE 33-29   A square wave (Problem 8).

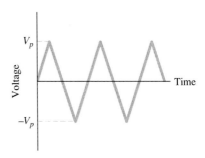

FIGURE 33-30   A triangle wave (Problem 9).

9. How are the rms and peak voltages related for the triangle wave in Fig. 33-30? See Problem 7.
10. A 10-V rms AC signal with frequency 1.0 kHz is 90° ahead of another signal of the same frequency but with peak voltage 7.1 V. On the same graph, plot voltage versus time for both these signals. Choose scales appropriate to display one full cycle.
11. The most general expression for a sinusoidal AC current may be written either $I = I_1 \sin \omega t + I_2 \cos \omega t$ or $I = I_p \sin(\omega t + \phi)$. Find relations between $I_1, I_2, I_p,$ and ϕ that make these expressions equivalent. (See Appendix A for trig identities.)

Section 33-2 Circuit Elements in AC Circuits

12. Show that the unit of both capacitive and inductive reactance is the ohm.
13. What is the rms current in a 1.0-μF capacitor connected across the 120-V rms, 60-Hz AC line?
14. A 470-Ω resistor, 10-μF capacitor, and 750-mH inductor are each connected across 6.3-V rms, 60-Hz AC power sources. Find the rms current in each.
15. Find the reactance of a 3.3-μF capacitor at (a) 60 Hz, (b) 1.0 kHz, and (c) 20 kHz.
16. A 15-μF capacitor carries an rms current of 1.4 A. What is the minimum safe voltage rating for the capacitor if the frequency is (a) 60 Hz or (b) 1.0 kHz?
17. A capacitor and a 1.8-kΩ resistor pass the same current when each is separately connected across a 60-Hz power line. What is the capacitance?
18. (a) A 2.2-H inductor is connected across the 120-V rms, 60-Hz power line. What is the rms inductor current? (b) Repeat if the same inductor is connected across the 230-V rms, 50-Hz power line commonly used in Europe.
19. A 50-mH inductor is connected across a 10-V rms AC generator, and an rms current of 2.0 mA flows. What is the generator frequency?
20. A 2.0-μF capacitor has a capacitive reactance of 1.0 kΩ. (a) What is the frequency of the applied voltage? (b) What

inductance would give the same value for inductive reactance at this frequency? (c) How would the two reactances compare if the frequency were doubled?

21. A 1.2-μF capacitor is connected across a generator whose output is given by $V = V_p \sin 2\pi ft$, where $V_p = 22$ V, $f = 60$ Hz, and t is in seconds. (a) What is the peak current? (b) What are the magnitudes of the voltage and (c) the current at $t = 6.5$ ms?

22. A voltage $V = V_p \sin \omega t$ is applied across a capacitor. What is the minimum frequency for which the current will be zero at time $t = 20 \ \mu s$?

23. What is the maximum charge on the plates of a 16-μF capacitor connected across the 120-V rms, 60-Hz AC power line?

24. At 10 kHz an inductor has 10 times the reactance of a capacitor. At what frequency will their reactances be equal?

25. A 0.75-H inductor is in series with a fluorescent lamp, and the series combination is across the 120-V rms, 60-Hz power line. If the rms inductor voltage is 90 V, what is the rms lamp current?

26. A 2.2-nF capacitor and a capacitor of unknown capacitance are connected in parallel across a 10-V rms sine-wave generator. At 1.0 kHz, the generator supplies a total current of 3.4 mA rms. The generator frequency is then decreased until the rms current has dropped to 1.2 mA. Find (a) the unknown capacitance and (b) the lower frequency.

Section 33-3 LC Circuits

27. Find the resonant frequency of an LC circuit consisting of a 0.22-μF capacitor and a 1.7-mH inductor.

28. An LC circuit with $C = 18$ mF undergoes LC oscillations with period 2.4 s. What is the inductance?

29. You have a 2.0-mH inductor and wish to make an LC circuit whose resonant frequency spans the AM radio band (550 kHz to 1600 kHz). What range of capacitance should your variable capacitor cover?

30. The FM radio band covers the frequency range from 88 MHz to 108 MHz. If the variable capacitor in an FM receiver ranges from 10.9 pF to 16.4 pF, what inductor should be used to make an LC circuit whose resonant frequency spans the FM band?

31. You want to use an LC circuit in a timing application. The circuit is to start with the capacitor fully charged, and the voltage should drop to zero in 15 s. You have available a 25-H inductor. What capacitance should you use?

32. An LC circuit includes a 20-μF capacitor and has a period of 5.0 ms. The peak current is 25 mA. Find (a) the inductance and (b) the peak voltage.

33. An LC circuit includes a 0.025-μF capacitor and a 340-μH inductor. (a) If the peak voltage on the capacitor is 190 V, what is the peak current in the inductor? (b) How long after the voltage peak does the current peak occur?

34. If the capacitor in the preceding problem is rated at 600 V, what is the maximum inductor current that will keep the capacitor within its rating?

35. At the instant when the electric and magnetic energies are equal in the LC circuit of Problem 33, the current is 540 mA. (a) What is the instantaneous voltage? Find (b) the peak voltage, (c) the peak current, and (d) the total energy.

36. In an LC circuit, what fraction of a cycle passes before the energy in the capacitor falls to one-fourth of its peak value?

37. One-eighth of a cycle after the capacitor in an LC circuit is fully charged, what are each of the following as fractions of their peak values: (a) capacitor charge, (b) energy in the capacitor, (c) inductor current, (d) energy in the inductor?

38. Show from conservation of energy that the peak voltage and current in an LC circuit are related by $I_p = V_p \sqrt{\dfrac{C}{L}}$.

39. The 2000-μF capacitor in Fig. 33-31 is initially charged to 200 V. Describe how you would manipulate switches A and B to charge the 500-μF capacitor to 400 V. Include the times you would throw the switches.

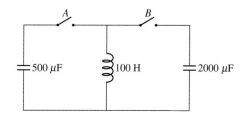

FIGURE 33-31 Problem 39.

40. A damped LC circuit consists of a 0.15-μF capacitor and a 20-mH inductor with resistance 1.6 Ω. How many cycles will the circuit oscillate before the peak voltage on the capacitor drops to half its initial value?

41. A damped RLC circuit includes a 5.0-Ω resistor and a 100-mH inductor. If half the initial energy is lost after 15 cycles, what is the capacitance?

Section 33-4 Driven RLC Circuits and Resonance

42. A series RLC circuit has $R = 75$ Ω, $L = 20$ mH, and a resonant frequency of 4.0 kHz. (a) What is the capacitance? (b) What is the impedance of the circuit at resonance? (c) What is the impedance at 3.0 kHz?

43. If the speaker system of Example 33-4 is driven by a 10-V peak, 1.0-kHz sine wave, what will be the peak voltage across the capacitor?

44. An RLC circuit includes a 1.5-H inductor and a 250-μF capacitor rated at 400 V. The circuit is connected across a sine-wave generator whose peak voltage is 32 V. What minimum resistance must the circuit have to ensure that the

capacitor voltage does not exceed its rated value when the generator is at the resonant frequency?

45. TV channel 2 occupies the frequency range from 54 MHz to 60 MHz. A series *RLC* tuning circuit in a TV receiver includes an 18-pF capacitor and resonates in the middle of the channel 2 band. (a) What is the inductance? (b) To let the whole signal in, the resonance curve must be broad enough that the current throughout the band be no less than 70% of the current at the resonant frequency. What constraint does this place on the circuit resistance?

46. An *RLC* circuit includes a 10-Ω resistor, 1.5-μF capacitor, and 50-mH inductor. The capacitor is rated at 1200 V. The circuit is driven by an AC source whose peak voltage is 100 V. (a) What would be the peak capacitor voltage at resonance? (b) Make a graph of the peak capacitor voltage as a function of frequency, and from it determine the frequency range that should be avoided for the capacitor to stay within its voltage rating.

47. A 2.0-H inductor and a 3.52-μF capacitor are connected in series with a 50-Ω resistor, and the combination is connected to an AC generator supplying 24 V peak at 60 Hz. (a) At the instant the generator voltage is at its peak, what is the instantaneous voltage across each circuit element? Show explicitly that these sum to the generator voltage. (b) If rms voltmeters are connected across each of the three components, what will they read? Do their readings sum to the rms generator voltage? Does this contradict the loop law?

48. Show that the impedance of an *RLC* circuit driven at frequency ω can be written

$$Z = \sqrt{R^2 + \omega^2 L^2 (1 - \omega_r^2/\omega^2)^2},$$

where ω_r is the resonant frequency.

49. Figure 33-32 shows the phasor diagram for an *RLC* circuit. (a) Is the driving frequency above or below resonance? (b) Complete the diagram by adding the applied voltage phasor, and from your diagram determine the phase difference between applied voltage and current.

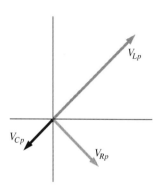

FIGURE 33-32 Problem 49.

50. An AC voltage of fixed amplitude is applied across a series *RLC* circuit. The component values are such that the cur-

rent at half the resonant frequency is half the current at resonance. (a) Show that the current at twice the resonant frequency is also half that at resonance. (b) Sketch phasor diagrams for both off-resonance cases.

51. For the circuit of Problem 46, find the phase relation between applied voltage and current at frequencies of (a) 550 Hz and (b) 700 kHz.

52. For the circuit of Problem 46, find two frequencies at which the voltage and current will be 60° out of phase. Which leads in each case?

Section 33-5 Power in AC Circuits

53. An electric drill draws 4.6 A rms at 120 V rms. If the current lags the voltage by 25°, what is the drill's power consumption?

54. A series *RLC* circuit has resistance 100 Ω and impedance 300 Ω. (a) What is the power factor? (b) If the rms current is 200 mA, what is the power dissipation?

55. A series *RLC* circuit has power factor 0.80 and impedance 100 Ω at 60 Hz. (a) What is the circuit resistance? (b) If the inductance is 0.10 H, what is the resonant frequency?

56. An *RLC* circuit with $R = 10$ Ω, $C = 2.0$ μF, and $L = 500$ mH is connected to an AC generator supplying 80 V rms. (a) Find the power factor when the generator frequency is half the resonant frequency. Find the power dissipation (b) at half the resonant frequency and (c) at the resonant frequency.

57. A power plant produces 60-Hz power at 365 kV rms and 200 A rms. The plant is connected to a small city by a transmission line with total resistance 100 Ω. What fraction of the power is lost in transmission if the city's power factor is (a) 1.0 or (b) 0.60? (c) Is it more economical for the power company if the load has a large power factor or a small one? Explain.

Section 33-6 Transformers and Power Supplies

58. A rural power line carries 2.3 A rms at 4000 V. A step-down transformer reduces this to 235 V rms to supply a house. Find (a) the turns ratio of the transformer and (b) the current in the 235-V line to the house.

59. A transformer steps up the 120-V rms AC power line voltage to 23 kV rms for a TV picture tube. If the rms current in the primary is 1.0 A, and the transformer is 95% efficient, what is the secondary current?

60. A car battery charger runs off the 120-V rms AC power line, and supplies 10 A DC at 14 V. (a) If the charger is 80% efficient in converting the line power to the DC power it supplies to the battery, how much current does it draw from the AC line? (b) If electricity costs 9.5¢/kWh, how much does it cost to run the charger for 10 hours? Assume the power factor is 1.

61. The transformer in the power supply of Fig. 33-27*a* has an output voltage of 6.3 V rms at 60 Hz, and the capacitance

is 1200 μF. (a) With an infinite load resistance, what would be the output voltage of the power supply? (b) What is the minimum load resistance for which the output would not drop more than 1% from this value? Assume that the discharge time in Fig. 33-27b is essentially a full cycle.

62. A power supply like that in Fig. 33-27a is supposed to deliver 22 V DC at a maximum current of 150 mA. The transformer's peak output voltage is appropriate to charge the capacitor to a full 22 V, and the primary is supplied with 60 Hz AC. What value of capacitance will ensure that the output voltage stays within 3% of the rated 22 V?

Paired Problems

(Both problems in a pair involve the same principles and techniques. If you can get the first problem, you should be able to solve the second one.)

63. A sine-wave generator delivers a signal whose peak voltage is independent of frequency. Two identical capacitors are connected in parallel across the generator, and the generator supplies a peak current I_p at frequency f_1. The capacitors are then connected in series across the generator. To what frequency should the generator be tuned to bring the current back to I_p?

64. A 1.0-μF capacitor and a 2.0-μF capacitor are connected in parallel across a sine-wave generator, and the generator supplies a total rms current of 25 mA at 1.0 kHz. Assuming the generator voltage is independent of frequency, at what frequency will it supply the same current when the capacitors are in series?

65. The peak current in an oscillating LC circuit is 850 mA. If $L = 1.2$ mH and $C = 5.0$ μF, what is the peak voltage?

66. An LC circuit with $C = 470$ pF is oscillating at 7.3 MHz. If the peak voltage is 95 V, what is the peak current?

67. An RLC circuit includes a 3.3-μF capacitor and a 27-mH inductor. The capacitor is charged to 35 V, and the circuit begins oscillating. Ten full cycles later the capacitor voltage peaks at 28 V. What is the resistance?

68. An RLC circuit with $R = 1.2$ Ω and $C = 10$ μF loses 2% of its initial energy in one oscillation cycle. What is its inductance?

69. A series RLC circuit with $R = 5.5$ Ω, $L = 180$ mH, and $C = 0.12$ μF is connected across a sine-wave generator. If the inductor can handle a maximum current of 1.5 A, what is the maximum safe value for the generator's peak output voltage when it is tuned to resonance?

70. A series RLC circuit with $R = 1.3$ Ω, $L = 27$ mH, and $C = 0.33$ μF is connected across a sine-wave generator. If the capacitor's peak voltage rating is 600 V, what is the maximum safe value for the generator's peak output voltage when it is tuned to resonance?

Supplementary Problems

71. Two capacitors are connected in parallel across a 10-V rms, 10-kHz sine-wave generator, and the generator sup-

plies a total rms current of 30 mA. When the capacitors are rewired in series, the rms generator current drops to 5.5 mA. Find the values of the two capacitances.

72. An LC circuit starts at $t = 0$ with its 2000-μF capacitor at its peak voltage of 14 V. At $t = 35$ ms the voltage has dropped to 8.5 V. (a) What will be the peak current? (b) When will the peak current occur?

73. An undriven RLC circuit with inductance L and resistance R starts oscillating with total energy U_0. After N cycles the energy is U_1. Find an expression for the capacitance, assuming the circuit is not heavily damped.

74. Figure 33-33 shows a **low-pass filter.** When an alternating voltage is applied at the V_{in} terminals, the output voltage V_{out} depends on frequency. (a) Show that $V_{out} = V_{in}/\sqrt{1 + (RC\omega)^2}$, where the voltages are either peak or rms values (there is also a phase difference). (b) At what frequency is the output voltage down from the input voltage by a factor of $1/\sqrt{2}$? (This is called the *half-power point* since the power—proportional to V^2—is down by a factor of $\frac{1}{2}$.) *Hint:* You can repeat the phasor analysis of Section 33-4, but without the inductor. Or you can start from Equation 33-14, with X_L set to zero.

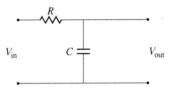

FIGURE 33-33 Problem 74.

75. You wish to make a "black box" with two input connections and two output connections, as shown in Fig. 33-34. When you put a 12-V rms, 60-Hz sine wave across the input, a 6.0-V, 60-Hz signal should appear at the output, with the output voltage leading the input voltage by 45°. Design a circuit that could be used in the "black box."

FIGURE 33-34 Problem 75.

76. A series RLC circuit with $R = 47$ Ω, $L = 250$ mH, and $C = 4.0$ μF is connected across a sine-wave generator whose peak output voltage is independent of frequency. Find the frequency range over which the peak current will exceed half its value at resonance. *Hint:* You can solve this problem graphically or, with appropriate algebraic manipulations, using quadratic equations.

77. A sine-wave generator with peak output voltage of 20 V is applied across a series RLC circuit. At the resonant fre-

quency of 2.0 kHz the peak current is 50 mA, while at 1.0 kHz it is 15 mA. Find R, L, and C.

78. Use phasor analysis to show that the parallel RLC circuit of Fig. 33-35 has impedance $Z = \left[\dfrac{1}{R^2} + \left(\dfrac{1}{X_C} - \dfrac{1}{X_L} \right)^2 \right]^{-1/2}$

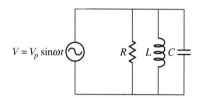

FIGURE 33-35 Problem 78.

79. A 2.5-H inductor is connected across a 1500-μF capacitor. A 5.0-kg mass is connected to a spring. What should be the spring constant if the mechanical and electrical systems have the same resonant frequency?

80. Figure 33-36 shows a diode circuit called a *full-wave bridge,* whose output requires less filtering than the single-diode circuit of Fig. 33-26. (a) Sketch a graph of the resistor current as a function of time, covering two full cycles of the AC generator, and (b) explain why less filtering is needed.

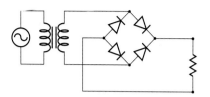

FIGURE 33-36 A full-wave diode bridge (Problem 80).

81. For RLC circuits in which the resistance is not too large, the Q factor may be defined as the ratio of the resonant frequency to the difference between the two frequencies where the power dissipated in the circuit is half that dissipated at resonance. Show, using suitable approximations, that this definition leads to the expression $Q = \omega_r L/R$, with ω_r the resonant frequency.

82. Consider a series circuit containing an AC generator, a resistor, and a capacitor. Construct a phasor diagram, and derive expressions for the circuit impedance and the phase angle between the applied voltage and the current. Show that the current always leads the voltage.

83. Consider a series circuit containing an AC generator, a resistor, and an inductor. Construct a phasor diagram, and derive expressions for the circuit impedance and the phase angle between the applied voltage and the current. Show that the voltage always leads the current.

CHAPTER 34

MAXWELL'S EQUATIONS AND ELECTROMAGNETIC WAVES

Electromagnetic waves include light and the radio waves essential to modern communications. This antenna communicates with orbiting satellites.

A t this point we have introduced the four fundamental laws of electromagnetism—Gauss's law for electricity, Gauss's law for magnetism, Ampère's law, and Faraday's law—that govern the behavior of electric and magnetic fields throughout the universe. We have seen how these laws describe the electric and magnetic interactions that make matter act as it does and have explored many practical devices that exploit the laws of electromagnetism. Here we extend the fundamental laws to their most general form and show how they predict the existence of electromagnetic waves.

34-1 THE FOUR LAWS OF ELECTROMAGNETISM

Table 34-1 summarizes the four laws as we introduced them in earlier chapters. As you look at these four laws together, you can't help noticing some strong similarities. On the left-hand sides of the equations, the two laws of Gauss are identical but for the interchanging of **E** and **B**. Similarly, the laws of Ampère and Faraday have left-hand sides that differ only in the interchange of **E** and **B**.

On the right-hand sides, things are more different. Gauss's law for electricity involves the charge enclosed by the surface of integration, while Gauss's law for magnetism has zero on the right-hand side. Actually, though, these laws are similar. Since we have no experimental evidence for the existence of isolated magnetic charge, the enclosed magnetic charge on the right-hand side of Gauss's law for magnetism is zero. If and when magnetic monopoles are discovered, then the right-hand side of Gauss's law for magnetism would be nonzero for any surface enclosing net magnetic charge.

The right-hand sides of Ampère's and Faraday's laws are distinctly different. In Ampère's law we find the current—the flow of electric charge—as a source of magnetic field. We can understand the absence of a similar term in Faraday's law because we have never observed a flow of magnetic monopoles. If we had such a flow, then we would expect this magnetic current to produce an electric field encircling the magnetic current.

Two of the differences among the four laws of electromagnetism would be resolved if we knew for sure that magnetic monopoles exist. That current theories of elementary particles suggest the existence of monopoles is a tantalizing hint that there may be a fuller symmetry between electric and magnetic phenomena. The search for symmetry, based not on logic or experimental evidence but on an intuitive sense that nature should be simple, has motivated some of the most important discoveries in physics.

34-2 AMBIGUITY IN AMPÈRE'S LAW

There remains one difference between the equations of electricity and magnetism that would not be resolved by the discovery of magnetic monopoles. On the right-hand side of Faraday's law we find the term $d\phi_B/dt$ that describes

▲ TABLE 34-1 FOUR LAWS OF ELECTROMAGNETISM (STILL INCOMPLETE)

LAW	MATHEMATICAL STATEMENT	WHAT IT SAYS
Gauss for **E**	$\oint \mathbf{E} \cdot d\mathbf{A} = \dfrac{q}{\varepsilon_0}$	How charges produce electric field; field lines begin and end on charges
Gauss for **B**	$\oint \mathbf{B} \cdot d\mathbf{A} = 0$	No magnetic charge; magnetic field lines do not begin or end
Faraday	$\oint \mathbf{E} \cdot d\boldsymbol{\ell} = -\dfrac{d\phi_B}{dt}$	Changing magnetic flux produces electric field
Ampère (steady currents only)	$\oint \mathbf{B} \cdot d\boldsymbol{\ell} = \mu_0 I$	Electric current produces magnetic field

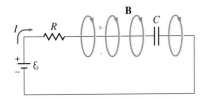

FIGURE 34-1 A charging *RC* circuit, showing some magnetic field lines surrounding the current-carrying wire.

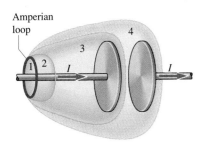

FIGURE 34-2 Ampère's law relates the line integral around the loop to the current through *any* surface bounded by the loop. There is no current through surface 3, so Ampère's law is ambiguous.

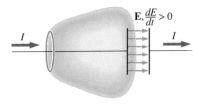

FIGURE 34-3 There is a changing electric field in the charging capacitor and, therefore, a changing electric flux through surface 3 of Fig. 34-2.

changing magnetic flux as a source of electric field. We find no comparable term in Ampère's law. Are we missing something? Is it possible that a changing electric flux produces a magnetic field? So far, we have described no experimental evidence for such a conjecture. It is suggested only by our sense that the near symmetry between electricity and magnetism is not a coincidence. If a changing electric flux did produce a magnetic field, just as a changing magnetic flux produces an electric field, then we would expect a term $d\phi_E/dt$ on the right-hand side of Ampère's law.

In our statement of Ampère's law in Chapter 30, we emphasized that the law applied only to steady currents—those that never change. The reason for this restriction is suggested in Fig. 34-1, which shows a simple *RC* circuit. While the capacitor charges, there is a current in the circuit that decreases with time. This current produces a magnetic field, as suggested in the figure. Let us apply Ampère's law to calculate this field.

Ampère's law tells us that the line integral around any closed loop is proportional to the encircled current:

$$\oint \mathbf{B} \cdot d\boldsymbol{\ell} = \mu_0 I.$$

By the encircled current, we mean the current through *any* surface bounded by the loop. Figure 34-2 shows four such surfaces. The same current flows through surfaces 1, 2, and 4. But there is no current through surface 3, which passes between the capacitor plates. So Ampère's law is ambiguous in that surface 3 would not give the same answer for the magnetic field.

This ambiguity does not arise with steady currents. In an *RC* circuit the steady-state current is everywhere zero, and thus the right-hand side of Ampère's law is zero for *any* surface. It's only when currents are changing with time that there may be situations like that of Fig. 34-2 where Ampère's law becomes ambiguous.* That is why the form of Ampère's law we have used until now is strictly valid only for steady currents.

Can we salvage Ampère's law, extending it to cover unsteady currents without affecting its validity in the steady case? Symmetry between Ampère's and Faraday's laws has already suggested that a changing electric flux might produce a magnetic field. Between the plates of a charging capacitor there is an electric field whose magnitude is increasing (Fig. 34-3). That means there is a changing electric flux through surface 3 of Figure 34-2.

It was the Scottish physicist James Clerk Maxwell who, about 1860, suggested that a changing electric flux should give rise to a magnetic field. Since that time many experiments, including direct measurement of the magnetic field inside a charging capacitor, have confirmed Maxwell's remarkable insight. Maxwell quantified his idea by introducing a new term into Ampère's law:

$$\oint \mathbf{B} \cdot d\boldsymbol{\ell} = \mu_0 I + \mu_0 \varepsilon_0 \frac{d\phi_E}{dt}. \tag{34-1}$$

* You might argue that we could produce a nonzero steady current in the *RC* circuit by steadily increasing the applied emf, again making Ampère's law ambiguous. But we could not do so forever since that would require infinite energy. Steady current means current that *never* changes. A current that is steady over a finite time interval must be started and stopped, so it isn't really steady.

Now there is no ambiguity. The integral is taken around any loop, I is the current through *any* surface bounded by the loop, and ϕ_E is the electric flux through that surface. With our charging capacitor, Equation 34-1 gives the same magnetic field no matter which surface we choose. For surfaces 1, 2, and 4 of Fig. 34-3, the current I makes all the contribution to the right-hand side of the equation (here we assume that the electric field outside the capacitor is zero). For surface 3, through which no current flows, the right-hand side of Equation 34-1 comes entirely from the changing electric flux. You can readily verify that the term $\varepsilon_0 d\phi_E/dt$ has the units of current, and that, for the charging capacitor, this term is numerically equal to the current I (see Problem 3). Although the changing electric flux is not an electric current, it has the same effect as a current in producing a magnetic field. For this reason Maxwell called the term $\varepsilon_0 d\phi_E/dt$ the **displacement current.** The word "displacement" has historical roots that do not provide much physical insight. But the word "current" is meaningful in that the effect of displacement current is indistinguishable from that of real current in producing magnetic fields.

● **EXAMPLE 34-1** DISPLACEMENT CURRENT PRODUCES MAGNETIC FIELD

A parallel-plate capacitor with circular plates a distance d apart is charged through long, straight wires as shown in Fig. 34-4. The potential difference between the plates is increasing at the rate dV/dt. Find an expression for the magnetic field as a function of position between the plates.

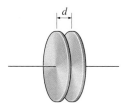

FIGURE 34-4 A circular capacitor (Example 34-1).

Solution

With long, straight feed wires, the situation has cylindrical symmetry. The only magnetic field with this symmetry has circular field lines and a magnitude that depends only on the radial distance r from the symmetry axis, as shown in Fig. 34-5. A magnetic field line within the capacitor encircles no conduction current—no flow of charge—but it does encircle a changing electric field and therefore a displacement current. If the field line has radius r, the encircled electric flux is

$$\phi_E = \int \mathbf{E} \cdot d\mathbf{A} = \pi r^2 E = \pi r^2 \frac{V}{d},$$

where the uniformity of the field allows us to calculate the field as the ratio of potential difference to plate spacing and the flux as a simple product of field and area. Then the displacement current is

$$I_D = \varepsilon_0 \frac{d\phi_E}{dt} = \frac{\varepsilon_0 \pi r^2}{d} \frac{dV}{dt}.$$

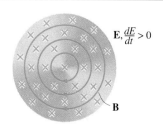

$$\mathbf{E}, \frac{dE}{dt} > 0$$

$\mathbf{B}$

FIGURE 34-5 Electric and magnetic fields between the circular capacitor plates. The electric field strength is increasing, so the displacement current is in the same direction as the electric field. Pointing the right thumb in this direction then shows that the magnetic field circles clockwise.

With cylindrical symmetry, the line integral on the left-hand side of Ampère's law becomes

$$\oint \mathbf{B} \cdot d\boldsymbol{\ell} = 2\pi r B.$$

Equating this quantity to μ_0 times the encircled displacement current gives

$$2\pi r B = \frac{\mu_0 \varepsilon_0 \pi r^2}{d} \frac{dV}{dt},$$

so

$$B = \frac{\mu_0 \varepsilon_0 r}{2d} \frac{dV}{dt}.$$

This field, with its magnitude increasing linearly with r, should remind you of the magnetic field inside a cylindrical wire (see Example 30-4). Problem 4 extends this calculation to the field outside the capacitor.

We can find the direction of the induced magnetic field just as we did for the fields of ordinary conduction currents: Point your right thumb in the direction of the current and your right fingers curl in the direction of the magnetic field. But which way does the displacement current go? In this example the electric field strength is increasing, so $d\phi_E/dt$ is positive, and the displacement current is in the direction of the electric field (see Fig. 34-5). If the electric field strength were decreasing, the displacement current would be opposite the field.

The induced magnetic field in a practical capacitor is minuscule, as the following exercise illustrates. We'll soon see, however, that the significance of displacement-current-induced magnetic fields is vastly greater than in this simple example.

EXERCISE In 1984 D. F. Bartlett and T. R. Corle of the University of Colorado first measured the magnetic field inside a charging capacitor using a sensitive magnetometer called a superconducting quantum interference detector (SQUID). They used a capacitor with circular plates spaced 1.22 cm, connected across a 340-V peak sine-wave generator operating at 1.25 kHz. What was the peak magnetic field strength 3.0 cm from the capacitor axis?

Answer: 3.65×10^{-11} T, less than one millionth of Earth's magnetic field

Some problems similar to Example 34-1: 4–6 ●

34-3 MAXWELL'S EQUATIONS

It was Maxwell's genius to recognize that Ampère's law should be modified to reflect the symmetry suggested by Faraday's law. The consequences of Maxwell's discovery go far beyond anything he could have imagined. To honor Maxwell, the four complete laws of electromagnetism are given the collective name **Maxwell's equations.** This full and complete set of equations, first published in 1864, governs the behavior of electric and magnetic fields everywhere. Table 34-2 summarizes Maxwell's equations.

These four simple, compact statements are all it takes to describe classical electromagnetic phenomena. Everything electric or magnetic that we have considered and will consider—from polar molecules to electric current; resistors, capacitors, inductors, and transistors; solar flares and cell membranes; electric generators and thunderstorms; computers and TV sets; the northern lights and fusion reactors—all these can be described using Maxwell's equations. And despite this wealth of phenomena, we have yet to discuss a most important manifestation of electromagnetic fields.

Maxwell's Equations in Vacuum

Consider Maxwell's equations in a region free of any matter—in vacuum. We have learned enough about electromagnetism to anticipate that the fields them-

▲ **TABLE 34-2** MAXWELL'S EQUATIONS

LAW	MATHEMATICAL STATEMENT	WHAT IT SAYS	EQUATION NUMBER
Gauss for **E**	$\oint \mathbf{E} \cdot d\mathbf{A} = \dfrac{q}{\varepsilon_0}$	How charges produce electric field; field lines begin and end on charges	(34-2)
Gauss for **B**	$\oint \mathbf{B} \cdot d\mathbf{A} = 0$	No magnetic charge; magnetic field lines do not begin or end	(34-3)
Faraday	$\oint \mathbf{E} \cdot d\boldsymbol{\ell} = -\dfrac{d\phi_B}{dt}$	Changing magnetic flux produces electric field	(34-4)
Ampère	$\oint \mathbf{E} \cdot d\boldsymbol{\ell} = \mu_0 I + \varepsilon_0 \mu_0 \dfrac{d\phi_E}{dt}$	Electric current and changing electric flux produce magnetic field	(34-5)

selves will still be able to interact, change, and carry energy even in the absence of matter. To express Maxwell's equations in vacuum, we simply remove all reference to matter—that is, to electric charge:

$$\oint \mathbf{E} \cdot d\mathbf{A} = 0 \tag{34-6}$$

$$\oint \mathbf{B} \cdot d\mathbf{A} = 0 \tag{34-7}$$

$$\oint \mathbf{E} \cdot d\boldsymbol{\ell} = -\frac{d\phi_B}{dt} \tag{34-8}$$

$$\oint \mathbf{B} \cdot d\boldsymbol{\ell} = \mu_0 \varepsilon_0 \frac{d\phi_E}{dt}. \tag{34-9}$$

In vacuum the symmetry is complete, in that electric and magnetic fields enter on an equal footing.* The only source of each field is a change in the other field.

34-4 ELECTROMAGNETIC WAVES

Faraday's law—Equation 34-8—shows that a changing magnetic field induces an electric field. In general, this induced electric field is itself changing. Ampère's law—Equation 34-9—shows that a changing electric field induces a magnetic field, which itself may be changing. The two laws together suggest the possibility of **electromagnetic waves,** in which each type of field continually produces the other in an electromagnetic structure that propagates through space. We will now show, directly from Maxwell's equations, that such waves are indeed possible. In the process we will discover the properties of electromagnetic waves and will come to a deep understanding of the nature of light.

In Chapter 16 we found that a sinusoidal wave propagating in the x direction can be represented by a function of the form $A \sin(kx - \omega t)$, where A is the wave amplitude, k the wave number, and ω the angular frequency. We now demonstrate that electric and magnetic fields of this form satisfy Maxwell's equations. Our demonstration is highly mathematical, but its physical significance is profound, for it reveals a rich new realm of electromagnetic phenomena that play a vital role in the workings of our universe.

We propose an electromagnetic wave consisting of electric and magnetic fields oriented at right angles to each other and to the direction of wave motion, as suggested in Fig. 34-6. The electric field points in the y direction, the magnetic field points in the z direction, and the wave travels in the x direction. In Fig. 34-6b the field lines are pictured only on the surface of a rectangular box. We must imagine these lines continuing straight forever in the y and z directions, giving a wave whose properties do not vary with y or z. Such a wave, whose

---

* The appearance of the constants ε_0 and μ_0 in Ampère's law but not in Faraday's law is an accident of our choice of units. That Faraday's law contains a minus sign, while Ampère's does not, is actually a symmetry, reflecting the complementary way in which electric and magnetic fields produce each other.

FIGURE 34-6 Fields of a plane electromagnetic wave. (*a*) Field vectors (*not* field lines), showing sinusoidal variation that indicates that this is a simple harmonic wave. The electric and magnetic fields are perpendicular and in phase. (*b*) Fields in the plane wave do not depend on *y* or *z*, and thus the fields actually extend forever in the *y* and *z* directions. This figure shows some field lines in a rectangular slab of the wave. Note that their spacing reflects the orientation and the sinusoidal variation shown in (*a*).

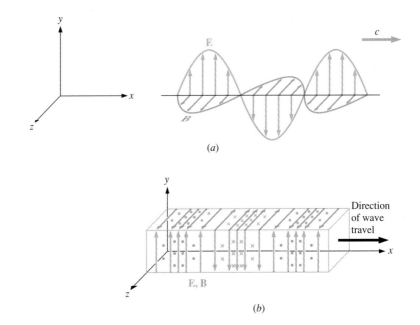

properties are independent of position in planes perpendicular to the propagation direction, is called a **plane wave.** Mathematically, our plane wave fields are described by

$$\mathbf{E}(x,\, t) = E_p \sin(kx - \omega t)\hat{\mathbf{j}} \qquad (34\text{-}10)$$

and

$$\mathbf{B}(x,\, t) = B_p \sin(kx - \omega t)\hat{\mathbf{k}}, \qquad (34\text{-}11)$$

where the peak amplitudes E_p and B_p are constants and where $\hat{\mathbf{j}}$ and $\hat{\mathbf{k}}$ are unit vectors in the *y* and *z* directions. In Fig. 34-6*b* the variation in spacing of the field lines, and their reversal from one region of densely spaced lines to another, reflect the sinusoidal dependence of the wave fields on position in the *x* direction. We chose a sinusoidal wave shape because we are familiar with such waves from Chapter 16 and because a specific mathematical form will make our derivation more concrete. But we emphasize here, as in Chapter 16, that *any* functions of $x \pm vt$ or, equivalently, $kx \pm \omega t$, are admissable waveforms (see Problem 74 here and Problem 54 of Chapter 16).

We now show that the electric and magnetic fields pictured in Fig. 34-6 and described by Equations 34-10 and 34-11 do indeed satisfy Maxwell's equations. Note first that our field lines continue forever, with no beginnings or ends; therefore Gauss's laws for electricity and magnetism in vacuum (Equations 34-6 and 34-7) are satisfied.

Faraday's Law

To see that Faraday's law is satisfied, consider an observer looking directly toward the *x-y* plane in Fig. 34-6. Such an observer would see electric field lines going up and down and magnetic field lines coming straight in and out, as shown in Fig. 34-7. Consider the small rectangular loop of height h and infinitesimal width dx shown in the figure. Evaluating the line integral of the electric field $\mathbf{E}$ around this loop, we get no contribution from the short ends because they are at right angles to the field. Going around counterclockwise, we get a contribution $-Eh$ as we go down the left side against the field direction. Then we get a positive contribution going up the right side. Because of the variation in field strength with position, the field strength on the right side of the loop is different from that on the left. Let the change in field be dE, so the field on the right side of the loop is $E + dE$, giving a contribution of $(E + dE)h$ to the line integral. Then the line integral of $\mathbf{E}$ around the loop is

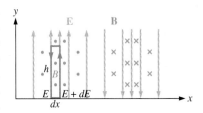

FIGURE 34-7   Cross section of Fig. 34-6 in the x-y plane. Also shown is a rectangular loop for evaluating the line integral in Faraday's law.

$$\oint \mathbf{E} \cdot d\boldsymbol{\ell} = -Eh + (E + dE)h = h\,dE.$$

Physically, this nonzero line integral means that we are dealing with an induced electric field. Induced by what? By a changing magnetic flux through the loop. The electric field of the wave arises because of the changing magnetic field of the wave. The area of the loop is $h\,dx$, and the magnetic field $\mathbf{B}$ is at right angles to this area, so the magnetic flux through the loop is just

$$\phi_B = Bh\,dx.$$

The rate of change of flux through the loop is then

$$\frac{d\phi_B}{dt} = h\,dx\,\frac{dB}{dt}.$$

Faraday's law relates the line integral of the electric field to the rate of change of flux:

$$\oint \mathbf{E} \cdot d\boldsymbol{\ell} = -\frac{d\phi_B}{dt},$$

or, using our expressions for the line integral and the rate of change of flux,

$$h\,dE = -h\,dx\,\frac{dB}{dt}.$$

Dividing through by $h\,dx$, we have

$$\frac{dE}{dx} = -\frac{dB}{dt}. \tag{34-12a}$$

In deriving this equation, we considered changes in E with position at a fixed instant of time, as pictured in Fig. 34-7, so our derivative dE/dx means the rate of change of E with position while time is held fixed. Similarly, in evaluating the derivative of magnetic flux, we were concerned only with the time rate of change at the fixed position of our loop. Both our derivatives represent rates of change with respect to one variable while the other variable is held fixed, and are therefore partial derivatives (see Tip box on p. 383 if you're not familiar with partial derivatives). Equation 34-12a should then be written more properly:

$$\frac{\partial E}{\partial x} = -\frac{\partial B}{\partial t}. \qquad (34\text{-}12b)$$

Equation 34-12b—which is just Faraday's law applied to our electromagnetic wave—tells us that the rate at which the electric field changes with *position* is related to the rate at which the magnetic field changes with *time*.

Ampère's Law

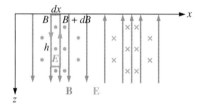

FIGURE 34-8 Cross section of Fig. 34-6 in the *x-z* plane, showing a rectangular loop for evaluating the line integral in Ampère's law.

Now imagine an observer looking down on Fig. 34-6 from above. This observer sees the magnetic field lines lying in the *x-z* plane, and electric field lines emerging perpendicular to the *x-z* plane as shown in Fig. 34-8. We can apply Ampère's law (Equation 34-9) to the infinitesimal rectangle shown, just as we applied Faraday's law to a similar rectangle in the *x-y* plane. Going counterclockwise around the rectangle, we get no contribution to the line integral of the magnetic field on the short sides, since they lie perpendicular to the field. Going down the left side, we get a contribution Bh to the line integral. Going up the right side, against the field, we get a negative contribution $-(B + dB)h$, where dB is the change in B from one side of the rectangle to the other. So the line integral in Ampère's law is

$$\oint \mathbf{B} \cdot d\boldsymbol{\ell} = Bh - (B + dB)h = -h\,dB.$$

The electric flux through the rectangle is simply $Eh\,dx$, so the rate of change of electric flux is

$$\frac{d\phi_E}{dt} = h\,dx\,\frac{dE}{dt}.$$

Ampère's law relates the line integral of the magnetic field to this time derivative of the electric flux, giving

$$-h\,dB = \varepsilon_0\mu_0 h\,dx\,\frac{dE}{dt}.$$

Dividing through by $h\,dx$ and noting again that we are really dealing with partial derivatives, we have

$$\frac{\partial B}{\partial x} = -\varepsilon_0\mu_0\frac{\partial E}{\partial t}. \qquad (34\text{-}13)$$

Equations 34-12 and 34-13—derived from Faraday's and Ampère's laws—express fully the requirements that Maxwell's universal laws of electromagnetism pose on the field structure postulated in Fig. 34-6. The two equations are remarkable in that each describes an induced field that arises from the changing of the other field. That other field, in turn, arises from the changing of the first field. Thus we have a self-perpetuating electromagnetic structure, whose fields exist and change without the need for charged matter. If Equations 34-10 and 34-11, which describe the fields in Fig. 34-6, can be made consistent with Equations 34-12 and 34-13, then we will have demonstrated that our electromagnetic wave does indeed satisfy Maxwell's equations and is thus a possible configuration of electric and magnetic fields.

Faraday, Ampère, and the Wave Fields

To see that Equation 34-12 is satisfied, we differentiate the electric field of Equation 34-10 with respect to x, and the magnetic field of Equation 34-11 with respect to t:

$$\frac{\partial E}{\partial x} = \frac{\partial}{\partial x}[E_p \sin(kx - \omega t)] = kE_p \cos(kx - \omega t)$$

and
$$\frac{\partial B}{\partial t} = \frac{\partial}{\partial t}[B_p \sin(kx - \omega t)] = -\omega B_p \cos(kx - \omega t).$$

Putting these expressions in for the derivatives in Equation 34-12 gives

$$kE_p \cos(kx - \omega t) = -[-\omega B_p \cos(kx - \omega t)].$$

The cosine term cancels from this equation, showing that the equation holds if

$$kE_p = \omega B_p. \tag{34-14}$$

To see that Equation 34-13 is also satisfied, we differentiate the magnetic field of Equation 34-11 with respect to x and the electric field of Equation 34-10 with respect to t:

$$\frac{\partial B}{\partial x} = kB_p \cos(kx - \omega t)$$

and
$$\frac{\partial E}{\partial t} = -\omega E_p \cos(kx - \omega t).$$

Using these expressions in Equation 34-13 then gives

$$kB_p \cos(kx - \omega t) = -\varepsilon_0 \mu_0[-\omega E_p \cos(kx - \omega t)].$$

Again, the cosine term cancels, showing that this equation is satisfied if

$$kB_p = \varepsilon_0 \mu_0 \omega E_p. \tag{34-15}$$

Our analysis has shown that electromagnetic waves whose form is given by Fig. 34-6 and Equations 34-10 and 34-11 can exist, provided that the amplitudes E_p and B_p, and the frequency ω and wave number k, are related by Equations 34-14 and 34-15. Physically, the existence of these waves is possible because a change in either kind of field—electric or magnetic—induces the other kind of field, giving rise to a self-perpetuating electromagnetic field structure. Maxwell's theory thus leads to the prediction of an entirely new phenomenon—the electromagnetic wave. We will now explore some properties of these waves.

34-5 THE SPEED OF ELECTROMAGNETIC WAVES

In Chapter 16 we found that the speed of a sinusoidal wave is given by the ratio of the angular frequency and wave number:

$$\text{wave speed} = \frac{\omega}{k}.$$

To determine the speed of our electromagnetic wave, we solve Equation 34-14 for E_p:

$$E_p = \frac{\omega B_p}{k},$$

and use this expression in Equation 34-15:

$$kB_p = \varepsilon_0\mu_0\omega E_p = \frac{\varepsilon_0\mu_0\omega^2 B_p}{k}.$$

Solving for the wave speed ω/k then gives

$$\text{wave speed} = \frac{\omega}{k} = \frac{1}{\sqrt{\varepsilon_0\mu_0}}. \qquad (34\text{-}16)$$

This remarkably simple result shows that the speed of an electromagnetic wave in vacuum depends only on the electric and magnetic constants ε_0 and μ_0. All such electromagnetic waves, regardless of frequency or amplitude, share this speed. Although we derived this result for sinusoidal waves, superposition considerations of Section 16-5 show that it holds for any wave shape.

We can easily calculate the speed given in Equation 34-16:

$$\frac{1}{\sqrt{\varepsilon_0\mu_0}} = \frac{1}{[(8.85\times10^{-12}\text{ F/m})(4\pi\times10^{-7}\text{ H/m})]^{1/2}} = 3.00\times10^8\text{ m/s}.$$

But this is precisely the speed of light! As early as 1600, Galileo had tried to measure the speed of light by uncovering lanterns on different mountain tops. He was able to conclude only that "If not instantaneous, it is extraordinarily rapid." In 1728, James Bradley, in England, used changes in the apparent positions of stars resulting from Earth's orbital motion to calculate a value of 2.95×10^8 m/s for the speed of light. Bradley's result differs by less than 2% from the value 2.99792458×10^8 m/s used in the 1983 definition of the meter in terms of the speed of light. Furthermore, the Dutch physicist Christian Huygens had suggested in 1678—again, about 200 years before Maxwell—that light is a wave. A substantial body of optical experiments had confirmed Huygens' theory, although neither theory nor experiment could say what sort of wave light might be. Now, in the 1860s, came Maxwell. Using a theory developed from laboratory experiments involving electricity and magnetism, with no reference whatever to optics or light, Maxwell showed how the interplay of electric and magnetic fields could result in electromagnetic waves. The speed of those waves—calculated from the quantities ε_0 and μ_0 that were determined in laboratory experiments having nothing to do with light—was precisely the known speed of light. Maxwell was led inescapably to the conclusion that light is an electromagnetic wave.

Maxwell's identification of light as an electromagnetic phenomenon is a classic example of the unification of knowledge toward which science is ever striving. With one simple calculation, Maxwell brought the entire science of optics under the umbrella of electromagnetism. Maxwell's work stands as a crowning intellectual triumph, an achievement whose implications are still expanding our view of the universe.

34-6 PROPERTIES OF ELECTROMAGNETIC WAVES

Our demonstration that electromagnetic waves satisfy Maxwell's equations places definite constraints on the properties of those waves. The wave frequency ω and wave number k are not both arbitrary, but must be related through

$$\frac{\omega}{k} = c,\tag{34-17a}$$

where $c = 1/\sqrt{\varepsilon_0\mu_0}$ is the speed of light. In Chapter 16 we related the angular frequency ω and wave number k to the more familiar frequency f and wavelength λ through the equations $\omega = 2\pi f$ and $k = 2\pi/\lambda$. Therefore we can also write Equation 34-17a in the form

$$f\lambda = c.\tag{34-17b}$$

Furthermore, Equation 34-14 shows that

$$E = \frac{\omega}{k}B = cB.\tag{34-18}$$

Thus, the field magnitudes in the wave are not independent but are in the ratio of the speed of light. Also, equations 34-10 and 34-11 require that the electric and magnetic fields be in phase. (This is why we wrote E and B in Equation 34-18 even though Equation 34-14 relates only the peak values E_p and B_p.) Finally, Fig. 34-6 has the electric and magnetic fields perpendicular to each other and to the direction of wave propagation. Although it is not clear that we had to start with a wave of this form, it is in fact the case that only waves with **E** and **B** in phase and with **E, B,** and the propagation direction all perpendicular to each other can satisfy Maxwell's equations in vacuum (see Problem 11).

● EXAMPLE 34-2 LASER LIGHT

A laser beam with wavelength 633 nm is propagating in the $+z$ direction. Its electric field is parallel to the x axis and has amplitude 6.0 kV/m. Find the wave frequency, and the direction and amplitude of the magnetic field.

Solution

Equation 34-17b relates the wavelength and frequency to the speed of light. Solving for f gives

$$f = \frac{c}{\lambda} = \frac{3.00 \times 10^8 \text{ m/s}}{633 \times 10^{-9} \text{ m}} = 4.74 \times 10^{14} \text{ Hz}.$$

If we imagine reorienting the wave of Fig. 34-6 so it propagates along the z direction, then rotate it about the z direction so the electric field is parallel to the x axis, we find that the magnetic field is parallel to the y axis. The magnetic field amplitude follows from Equation 34-18:

$$B_p = \frac{E_p}{c} = \frac{6.0 \times 10^3 \text{ V/m}}{3.00 \times 10^8 \text{ m/s}} = 2.0 \times 10^{-5} \text{ T}.$$

EXERCISE An electromagnetic wave is propagating in the $-y$ direction, with its magnetic field parallel to the x axis. The magnetic field amplitude is 8.0 μT. Write an expression for the wave's electric field vector at the point where the magnetic field points in the $+x$ direction and is at its peak value.

Answer: $\mathbf{E} = -2.4\hat{\mathbf{k}}$ kV/m

Some problems similar to Example 34-2: 24–27

●

34-7 THE ELECTROMAGNETIC SPECTRUM

Although an electromagnetic wave's frequency and wavelength must be related by Equation 34-17b, one or the other of these quantities is completely arbitrary. That means we can have electromagnetic waves of any frequency, or, equivalently, any wavelength. Direct measurement shows that visible light occupies a wavelength range from about 400 nm to 700 nm, corresponding to frequencies from 7.5×10^{14} Hz to 4.3×10^{14} Hz. The different wavelengths or frequencies correspond to different colors, with red at the long-wavelength, low-frequency end of the visible region and the blue at the short-wavelength, high-frequency end (Fig. 34-9).

The range of frequencies occupied by visible light is rather limited. What about electromagnetic waves whose frequencies lie above and below the visible range? Such invisible electromagnetic waves were unknown in Maxwell's time. A brilliant confirmation of Maxwell's theory came in 1888, when the German physicist Heinrich Hertz succeeded in generating and detecting electromagnetic

FIGURE 34-9 Visible light occupies a wavelength range from about 400 nm to 700 nm. But electromagnetic waves come in a much broader range of wavelengths.

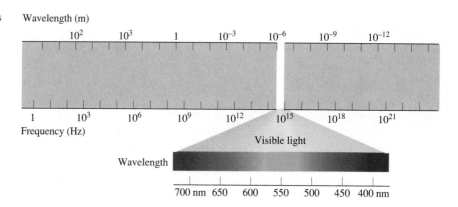

waves of much lower frequency than visible light. Hertz intended his work only to verify Maxwell's modification of Ampère's law, but the practical consequences have proven enormous. In 1896, the Italian scientist Guglielmo Marconi demonstrated that he could generate and detect the so-called "Hertzian waves." In 1901, he transmitted electromagnetic waves across the Atlantic Ocean, creating a public sensation. From the pioneering work of Hertz and Marconi, spurred by the theoretical efforts of Maxwell, came the entire technology of radio, television, and microwaves that so dominates modern society. We now consider all electromagnetic waves in the frequency range from a few Hz to about 3×10^{11} Hz as radio waves, with ordinary AM radio at about 10^6 Hz, FM radio at 10^8 Hz, and microwaves used for radar, cooking, and satellite communications at 10^9 Hz and above.

Between radio waves and visible light lies the infrared frequency range. Electromagnetic waves in this region are emitted by warm objects, even when they are not hot enough to glow visibly. For this reason, infrared detectors are used to determine subtle body temperature differences in medical diagnosis, to examine buildings for heat loss, and to study the birth of stars in clouds of interstellar gas and dust (Fig. 34-10).

Beyond the visible region are the ultraviolet rays responsible for sunburn, then the highly penetrating x rays, and finally the gamma rays whose primary terrestrial source is radioactive decay. All these phenomena, from radio to gamma rays, are fundamentally the same: They are all electromagnetic waves, differing only in frequency and wavelength. All travel with speed c, and all consist of electric and magnetic fields produced from each other through the induction processes described by Faraday's and Ampère's laws. Naming the different types of electromagnetic waves is just a matter of convenience; there are no gaps in the continuous range of allowed frequencies and wavelengths. Practical differences arise because waves of different wavelengths interact differently with matter; in particular, shorter wavelengths tend to be generated and absorbed most efficiently by smaller systems. Figure 34-11 shows the range of electromagnetic waves—the **electromagnetic spectrum**—displayed in a single diagram.

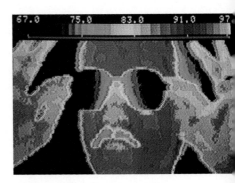

FIGURE 34-10 Subtle variations in body temperature are color coded in this infrared image of a human face.

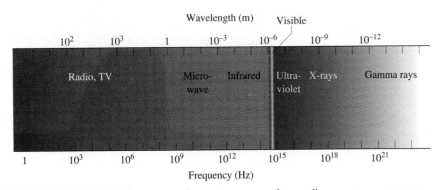

FIGURE 34-11 The electromagnetic spectrum ranges from radio waves to gamma rays. Note the logarithmic scale, in which equal intervals on the diagram correspond to factors-of-10 changes in frequency and wavelength.

■ **APPLICATION** THE NEW ASTRONOMY

FIGURE 34-12 Earth's atmosphere is opaque to most electromagnetic waves, although "windows" of transparency exist in several wavelength ranges—especially the visible. Bottom of diagram represents Earth's surface, with top of the gray curve the height to which waves of a given wavelength penetrate from outer space.

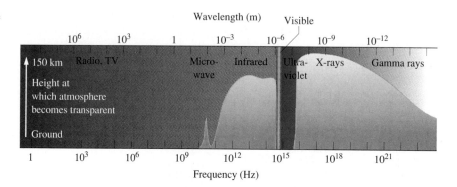

A glance at Fig. 34-11 shows that visible light occupies a small fraction of the electromagnetic spectrum. For centuries our only information about the universe beyond Earth—except for an occasional meteorite—came from visible light. Processes like those occurring on the visible surface of our Sun and many other stars produce predominantly visible light. Optical astronomy, utilizing visible light, gave a good picture of the universe to the extent that it consists of objects not too different from the visible part of the Sun. The restriction to optical astronomy was in part imposed by Earth's atmosphere. Transparent to visible light, the atmosphere is largely opaque to other forms of electromagnetic radiation, although "windows" of relative transparency exist in parts of the radio and infrared bands (Fig. 34-12). The discovery by Bell Telephone Laboratories electrical engineer Karl Jansky in 1931 that radio waves from outer space can be detected on Earth led to the development of radio astronomy. For decades, radio astronomy has given a picture of the universe that complements the optical view, showing phenomena that are simply not detectable by optical means (Fig. 34-13).

The onset of the space age in the late 1950s finally opened the entire electromagnetic spectrum to astronomers. Before this time there were surprisingly few suggestions that anything interesting might be found beyond the visible range. But satellites carrying infrared, ultraviolet, x-ray, and gamma-ray detectors have literally revolutionized our view of the universe (Fig. 34-14). Exotic objects like neutron stars—with the mass of the Sun crushed to a diameter of a few kilometers—and black holes, whose gravity is so strong that not even light can escape—are now objects of astronomical study. The opening of the entire electromagnetic spectrum has brought a new richness to astronomy, showing that our universe contains some of the most unusual objects that the laws of physics permit. Phenomena that were once bizarre conjectures of theoreticians are now observed regularly. Closer to home, observations of the Sun with ultraviolet and x-ray instruments have brought new understandings of the star that sustains us. And by turning space-borne infrared detectors toward Earth, we have learned much about the structure and resources of our own planet (Fig. 34-15).

(a)

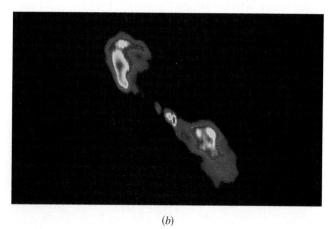

(b)

FIGURE 34-13 (a) Some of the 27 dish antennas that comprise the Very Large Array (VLA) radio telescope in New Mexico. (b) The galaxy Centaurus A, a powerful radio emitter, imaged with the VLA. The two lobes are jets of material ejected from the galaxy's central core. The VLA was tuned to a wavelength of 20 cm for this observation.

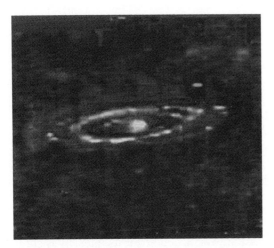

FIGURE 34-14 This false-color infrared image of the Andromeda galaxy was taken with the Infrared Astronomical Satellite. Yellow indicates regions of brightest infrared emission, corresponding to places where new stars are probably forming.

FIGURE 34-15 Infrared image of the New York City area, taken from a Landsat satellite. Areas of vegetation show reddish in this false-color image; note Central Park on Manhattan Island. Landsat images resolve objects as small as 30 m; here you can see individual ships near the Verrazano Bridge at lower center.

34-8 POLARIZATION

The fields of an electromagnetic wave in vacuum (and in most materials) are perpendicular to the propagation direction, but within the plane perpendicular to the wave propagation the orientation of one field is still arbitrary. **Polarization** is a wave property that specifies the electric field direction; since the two fields are perpendicular, polarization also determines the magnetic field direction (Fig. 34-16).

Electromagnetic waves used in radio, TV, and radar are generated in such a way that they have a definite polarization. So are the light waves produced by most lasers. In contrast, visible light from hot sources like the Sun or a light bulb is **unpolarized,** consisting of a mixture of electromagnetic waves with random field orientations.

Unpolarized light may be polarized either by reflection off surfaces or when it passes through substances whose molecular or crystal structure has a preferred direction called the **transmission axis.** Many crystals and synthetic materials like the plastic Polaroid have this property. For example, sunlight reflecting off the hood of a car becomes partially polarized in the horizontal direction. Polaroid sunglasses, with their transmission axis vertical, block this reflected glare without significantly reducing overall light intensity.

FIGURE 34-16 Some field vectors for an electromagnetic wave polarized in the y direction.

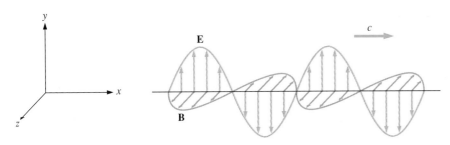

FIGURE 34-17 Two pairs of Polaroid sunglasses with their transmission axes at right angles. Where they overlap, no light can get through.

A polarizing material passes unattenuated only that component of the wave's electric field that lies along its preferred direction. If θ is the angle between the field and the polarizer's preferred direction, then the field component in the preferred direction is $E\cos\theta$. As we will show shortly, the intensity S of an electromagnetic wave is proportional to the square of the field strength; as a result a wave of intensity S_0 incident on a polarizer emerges with intensity given by the so-called **Law of Malus:**

$$S = S_0 \cos^2\theta. \tag{34-19}$$

This equation shows that electromagnetic waves will be blocked completely if $\theta = 90°$, a situation that occurs when unpolarized light passes through one polarizer to give it a definite polarization, then through another oriented at $90°$ to the first (Fig. 34-17).

When unpolarized light passes through a polarizer its intensity is cut in half. You can see this from Equation 34-19 because the unpolarized light includes a mix of waves with random polarization angles θ. Averaging over all possible angles in Equation 34-19 amounts to taking the average of $\cos^2\theta$ over a full cycle. We've seen on several occasions that the average of the square of a sinusoidal function is $\frac{1}{2}$, so a polarizer does indeed cut the intensity of unpolarized light in half.

● EXAMPLE 34-3 MULTIPLE POLARIZERS

Unpolarized light with intensity S_0 is incident on a "stack" of three polarizers. The first has its polarization axis vertical, the second is at $25°$ to the vertical, and the third is at $70°$ to the vertical (Fig. 34-18). What is the intensity of light emerging from this stack?

Solution
We've just seen that the first polarizer cuts the unpolarized intensity in half, giving $\frac{1}{2}S_0$ for the intensity incident on the second polarizer. Equation 34-19 shows that the second and third polarizers each reduce the intensity by a factor $\cos^2\theta$, where θ is the angle between the incident polarization direction—established by one polarizer—and the next polarizer's axis. For the second polarizer this angle is $25°$; for the third it is $70° - 25° = 45°$. Thus light emerges from the stack with intensity

$$S = (\tfrac{1}{2})(\cos^2 25°)(\cos^2 45°)S_0 = 0.205 S_0.$$

Interestingly, this is greater than the intensity we would get passing light through a vertical polarizer followed by a single polarizer at $70°$. And, as the exercise below shows, it is much

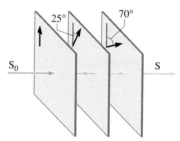

FIGURE 34-18 A stack of polarizers. Arrows on the sheets indicate directions of the polarization axes.

greater than what we would get by interchanging the second and third polarizers. Can you see why?

EXERCISE Rework Example 34-3 with the second and third polarizers interchanged.

Answer: $S = 0.029 S_0$

Some problems similar to Example 35-3: 35, 36, 72 ●

Polarization can tell us much about sources of electromagnetic waves or about materials through which the waves travel. Many astrophysical processes produce polarized waves; measuring the polarization then gives clues to the mechanisms operating in distant objects. Polarization of light as it passes

FIGURE 34-19 Photomicrograph of a thin section of rock placed between crossed polarizers. Individual mineral crystals within the rock rotate the light's electric field, altering the transmitted light intensity.

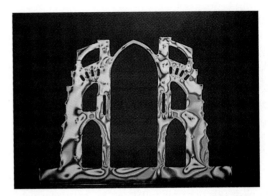

FIGURE 34-20 Plastic model of a Gothic cathedral, photographed between polarizing sheets. The resulting patterns reveal stresses, helping architects and engineers understand the response of the building to wind and weight loading.

through materials helps geologists to understand the composition and formation of rocks (Fig. 34-19) and helps engineers to locate stresses in structures (Fig. 34-20).

■ **APPLICATION** ELECTRO-OPTIC MODULATION

The atomic structure of some materials causes them to rotate the polarization direction of incident light, an effect demonstrated in Figs. 34-19 and 34-20. Applying an electric field may alter the structure and, therefore, the material's effect on light passing through it. This is called the **electro-optic** effect.

An **electro-optic modulator** (EOM) is a device that uses the electro-optic effect to control the intensity of light. Figure 34-21 shows an EOM consisting of an electro-optic crystal between crossed polarizers. Electrodes coated on the ends of the crystal allow application of a voltage that produces an

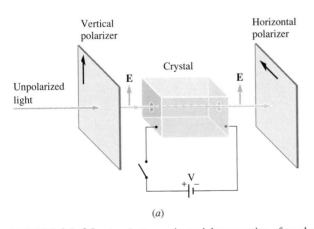

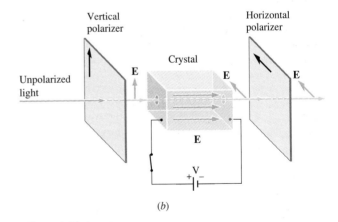

FIGURE 34-21 An electro-optic modulator consists of an electro-optic crystal between crossed polarizers. Electrodes permit application of a voltage across the crystal, and "windows" in the electrodes let light through. (*a*) With no applied voltage there is no rotation of the polarization, and the crossed polarizers therefore prevent light transmission through the device. (*b*) An applied voltage produces an electric field in the crystal. This alters the crystal structure and causes it to rotate the polarization of the incident light. A 90° rotation, shown here, gives maximum light transmission.

electric field in the crystal. With the proper design and orientation, the crystal in Fig. 34-21 has no effect on the polarization direction of light when there is no applied voltage. Since the polarizers are at right angles, light polarized by the first is then blocked by the second and so no light gets through the device (Fig. 34-21a).

When a voltage is applied across the crystal the polarization direction of the incident light is now rotated. At some value of the voltage—typically a kilovolt or more for most crystals—the polarization rotates through 90°. Then, light emerging from the crystal is polarized in the direction of the second polarizer and passes unattenuated through that polarizer. Thus the entire EOM passes the maximum light at this particular voltage.

Applying a time-varying voltage causes the EOM to vary, or *modulate,* the light intensity. Voice or music, for example, can

be transmitted on a laser beam by applying the amplified signal from a microphone to an EOM and using a light-sensitive detector to extract the audio information at the other end.

Turning on and off the voltage that gives maximum light transmission makes the EOM into an electronic shutter. Typical EOMs achieve on/off times on the order of a nanosecond, making them much faster than mechanical or even conventional electronic shutters and flash lamps. Prior to the 1983 redefinition of the meter in terms of the speed of light, EOM-based systems provided some of the most accurate measurements of c. Today EOM's enjoy a wide variety of scientific and commercial applications, especially those requiring rapid control of laser beams. An increasingly important use is the conversion of electrical signals to light pulses on the highest speed fiber optic communications systems.

34-9 PRODUCING ELECTROMAGNETIC WAVES

We have shown that electromagnetic waves can exist, and have explored some of their properties. But how do these waves originate?

All that's necessary is to produce a changing electric or magnetic field. Once a changing field of either type exists, Faraday's and Ampère's laws ensure the production of the other type—and so on, to give a propagating electromagnetic wave. Ultimately, changing fields of either type occur when we alter the motion of electric charge. Therefore:

| *Accelerated* electric charge is the source of electromagnetic waves.

In a radio transmitter, the accelerated charges are electrons moving back and forth in an antenna, driven by an alternating voltage from an *LC* circuit (Fig. 34-22). In an x-ray tube, high-energy electrons decelerate rapidly as they slam into a target; their deceleration is the source of the electromagnetic waves, now in the x-ray region of the spectrum. In the hot plasma of a fusion reactor, high-speed electrons veer around ions under the influence of the attractive electric force; their change in direction represents an acceleration that is responsible for the production of electromagnetic waves—a detrimental effect because the waves sap the plasma's energy, thereby cooling it. In the magnetron tube of a microwave oven, electrons circle in a magnetic field; their centripetal acceleration is the source of the microwaves that cook your food. And the altered movement of electrons in atoms—although described accurately only by quantum mechanics—is the source of most visible light. If the motion of the accelerated charges is periodic, then the wave frequency is that of the motion; more generally, systems are most efficient at producing (and receiving) electromagnetic waves whose wavelength is comparable to the size of the system. That's why TV antennas are on the order of 1 m in size, while nuclei—some 10^{-15} m in diameter—produce gamma rays.

Calculation of electromagnetic waves emitted by accelerated charges presents challenging but important problems for physicists and communications engineers. Figure 34-23a shows a "snapshot" of the electric field produced by a single point charge undergoing simple harmonic motion, while Fig. 34-23b

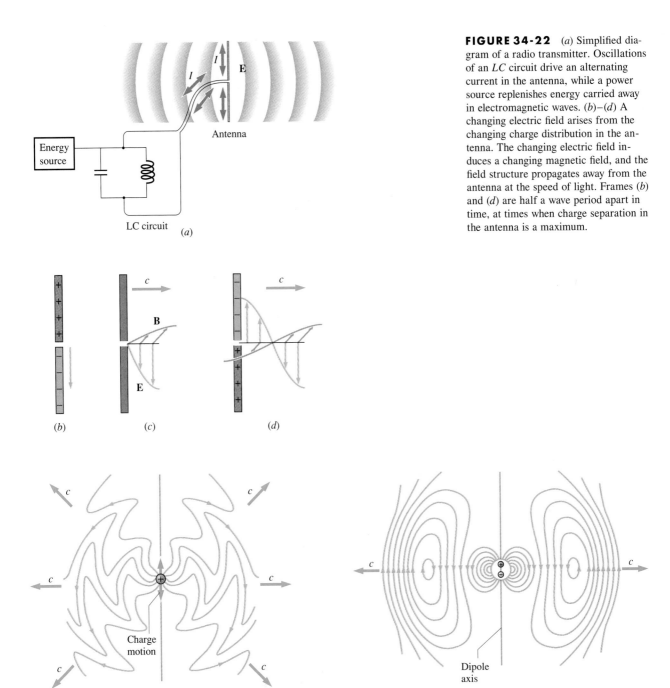

FIGURE 34-22 (*a*) Simplified diagram of a radio transmitter. Oscillations of an *LC* circuit drive an alternating current in the antenna, while a power source replenishes energy carried away in electromagnetic waves. (*b*)–(*d*) A changing electric field arises from the changing charge distribution in the antenna. The changing electric field induces a changing magnetic field, and the field structure propagates away from the antenna at the speed of light. Frames (*b*) and (*d*) are half a wave period apart in time, at times when charge separation in the antenna is a maximum.

FIGURE 34-23 "Snapshots" showing the electric fields of oscillating charge distributions. (*a*) A single point charge executing simple harmonic motion in the horizontal plane. Note that the field close to the charge approximates the radial field of a stationary point charge, but that farther out the "kinks" in the field—arising from the accelerated motion—are more prominent and become essentially perpendicular to the radial direction. These transverse kinks move outward at the speed of light, and constitute the wave fields. (*b*) The electric field of an oscillating dipole. Note that the field forms closed loops, detached from the dipole. These are the outward propagating wave fields. The larger field loops shown were formed when the oscillating dipole had the opposite orientation. Not shown in either figure are the equally important magnetic fields.

shows the field of an oscillating dipole—a configuration approximated by many systems from antennas to atoms and molecules. Both figures show that the waves are strongest in the direction at right angles to the acceleration of the charge distribution and that there is no radiation in the direction of the acceleration. This accounts for, among other phenomena, the directionality of radio and TV antennas, which transmit and receive most effectively perpendicular to the long direction of the antenna.

The fields shown in Fig. 34-23 seem to bear little resemblance to the plane-wave fields of Fig. 34-7 that we used to demonstrate the possibility of electromagnetic waves. We could produce true plane waves only with an infinite sheet of accelerated charge—an obvious impossibility. But far from the source, the curved field lines evident in Fig. 34-23b, for example, would appear straight, and the wave would begin to approximate a plane wave. So our plane-wave analysis is a valid approximation at great distances—typically many wavelengths—from a localized wave source. Closer to the source more complicated expressions for the wave fields apply, but these, too, satisfy Maxwell's equations.

34-10 ENERGY IN ELECTROMAGNETIC WAVES

We showed in previous chapters that electric and magnetic fields contain energy. Here we have considered electromagnetic waves, in which a combination of electric and magnetic fields propagates through space. As the wave moves, it must transport the energy contained in those fields.

We define the **intensity,** S, as the rate at which an electromagnetic wave transports energy across a unit area. This is the same definition we used for wave intensity in Chapter 16, and the units are also the same: power per unit area, or W/m². We can calculate the intensity of a plane electromagnetic wave by considering a rectangular box of thickness dx and cross-sectional area A with its face perpendicular to the wave propagation (Fig. 34-24). Within this box are wave fields **E** and **B** whose energy densities are given by Equations 26-3 and 32-11:

$$u_E = \tfrac{1}{2} \varepsilon_0 E^2$$

$$u_B = \frac{B^2}{2\mu_0}.$$

If dx is sufficiently small, the fields don't change much over the box, so the total energy in the box is just the sum of the electric and magnetic energy densities multiplied by the box volume $A\,dx$:

$$dU = (u_E + u_B)\,A\,dx = \frac{1}{2}\left(\varepsilon_0 E^2 + \frac{B^2}{\mu_0}\right) A\,dx.$$

This energy moves with speed c, so all the energy contained in the box length dx moves out of the box in a time $dt = dx/c$. The rate at which energy moves through the cross-sectional area A is then

$$\frac{dU}{dt} = \frac{1}{2}\left(\varepsilon_0 E^2 + \frac{B^2}{\mu_0}\right)\frac{A\,dx}{dx/c} = \frac{c}{2}\left(\varepsilon_0 E^2 + \frac{B^2}{\mu_0}\right) A.$$

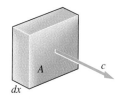

FIGURE 34-24 A box of length dx and cross-sectional area A at right angles to the propagation of an electromagnetic wave.

So the intensity S, or rate of energy flow per unit area, is

$$S = \frac{c}{2}\left(\varepsilon_0 E^2 + \frac{B^2}{\mu_0}\right).$$

We can recast this equation in simpler form by noting that, for an electromagnetic wave, $E = cB$ and $B = E/c$. Using these expressions to replace one of the E's in the term E^2 with B and similarly one of the B's in the term B^2 with E, we have

$$S = \frac{c}{2}\left(\varepsilon_0 cEB + \frac{EB}{\mu_0 c}\right) = \frac{1}{2\mu_0}(\varepsilon_0 \mu_0 c^2 + 1)\, EB.$$

But $c = 1/\sqrt{\varepsilon_0 \mu_0}$, so $\varepsilon_0 \mu_0 c^2 = 1$, giving

$$S = \frac{EB}{\mu_0}. \qquad\qquad (34\text{-}19a)$$

Although we derived Equation 34-19a for an electromagnetic wave, it is in fact a special case of the more general result that nonparallel electric and magnetic fields are accompanied by a flow of electromagnetic energy. In general, the rate of energy flow per unit area is given by

$$\mathbf{S} = \frac{\mathbf{E} \times \mathbf{B}}{\mu_0}. \qquad\qquad (34\text{-}19b)$$

Here a vector $\mathbf{S}$ is used to signify not only the magnitude of the energy flow, but also its direction. For an electromagnetic wave in vacuum, in which $\mathbf{E}$ and $\mathbf{B}$ must be at right angles, Equation 34-19b reduces to Equation 34-19a, with the direction of energy flow the same as the direction of wave travel. The vector intensity $\mathbf{S}$ is called the **Poynting vector** after the English physicist J. H. Poynting, who suggested it in 1884. Poynting's name is especially fortuitous, for the Poynting vector points in the direction of energy flow. Problem 76 explores an important application of the Poynting vector to fields that do not constitute an electromagnetic wave.

Equations 34-19 give the intensity at the instant when the fields have magnitudes E and B. In an electromagnetic wave the fields oscillate, and so does the intensity. We're usually not interested in this rapid oscillation. For example, an engineer designing a solar collector doesn't care that sunlight intensity oscillates at about 10^{14} Hz. What she really wants is the *average* intensity, $\bar{S}$. Because the instantaneous intensity of Equation 34-19a contains a product of sinusoidally varying terms, which are in phase, the average intensity is just half the peak intensity:

$$\bar{S} = \frac{\overline{EB}}{\mu_0} = \frac{E_p B_p}{2\mu_0} \quad \text{(average intensity)} \qquad (34\text{-}20a)$$

(This follows because, as we've seen on several occasions, the average of $\sin^2 \omega t$ over one cycle is $\frac{1}{2}$.) We wrote Equation 34-20a in terms of both the electric and magnetic fields, but we can use the wave condition $E = cB$ to eliminate either field in terms of the other:

$$\bar{S} = \frac{E_p{}^2}{2\mu_0 c} \tag{34-20b}$$

and

$$\bar{S} = \frac{cB_p{}^2}{2\mu_0}. \tag{34-20c}$$

● EXAMPLE 34-4 SOLAR ENERGY

The average intensity of sunlight on a clear day at noon is about 1 kW/m². What are the electric and magnetic fields in sunlight? How many solar collectors would you need to replace a 4.8-kW electric water heater in noonday sun, if each collector has an area of 2.0 m² and converts 40% of the incident sunlight to heat?

Solution

Solving Equation 34-20b for the electric field gives

$$E_p = \sqrt{2\mu_0 c \bar{S}}$$

$$= [(2)(4\pi \times 10^{-7}\ \text{H/m})(3.0 \times 10^8\ \text{m/s})(1 \times 10^3\ \text{W/m}^2)]^{1/2}$$

$$= 0.87\ \text{kV/m}.$$

The peak magnetic field is then given by $B_p = E_p/c$, so

$$B_p = \frac{E_p}{c} = \frac{870\ \text{V/m}}{3.0 \times 10^8\ \text{m/s}} = 3 \times 10^{-6}\ \text{T}.$$

At 1 kW/m², we would then need 4.8 m² of collector area if the collectors were 100% efficient. At 40% efficiency, we therefore need 4.8 m²/0.40 = 12 m², for a total of 6 collectors.

EXERCISE A laser produces an average power of 7.0 W in a light beam 1.0 mm in diameter. Find (a) the average intensity and (b) the peak electric field of the laser light.

Answers: (a) 8.9 MW/m²; (b) 82 kV/m

Some problems similar to Example 34-4: 38–41, 43

Waves from Localized Sources

As an electromagnetic wave propagates through empty space, its total energy does not change. With plane waves the intensity—power per unit area—does not change either. But when a wave originates in a localized source like an atom, a radio transmitting antenna, a light bulb, or a star, its wavefronts are not planes but expanding spheres (recall Fig. 16-18). The wave's total energy remains the same, but as it expands that energy is spread over the area of an ever larger sphere—whose area increases as the square of the distance from the source. Therefore, as we found in Chapter 16, the power per unit area—the intensity—decreases as the inverse square of the distance:

$$S = \frac{P}{4\pi r^2}. \tag{34-21}$$

Here S and P can be either the peak or average intensity and power, respectively, and r is the distance from a localized source. This intensity decrease occurs not because electromagnetic waves "weaken" and lose energy but because that energy gets spread ever more thinly.

Because the intensity of an electromagnetic wave is proportional to the *square* of the field strengths (Equations 34-20), Equation 34-21 shows that the *fields* of a spherical wave decrease as $1/r$. Contrast that with the $1/r^2$ decrease in the electric field of a stationary point charge, and you can see why the electromagnetic wave fields associated with an accelerated charge dominate in all but the immediate vicinity of the charge (see Fig. 34-23a).

● **EXAMPLE 34-5**   A GARAGE-DOOR OPENER

A radio-activated garage-door opener responds to signals with average intensity as weak as 20 μW/m^2. If the transmitter unit produces a 240-mW signal, broadcast in all directions, what is the maximum distance at which the transmitter will activate the door opener? What is the minimum value for the peak electric field to which the unit responds?

Solution

Since the waves spread out in all the directions, Equation 34-21 applies. Solving for r gives

$$r = \sqrt{\frac{P}{4\pi S}} = \sqrt{\frac{240\times10^{-3}\text{ W}}{(4\pi)(20\times10^{-6}\text{ W/m}^2)}} = 31\text{ m}.$$

Solving Equation 34-20b gives the electric field corresponding to the unit's 20-μW/m^2 sensitivity:

$E_p = \sqrt{2\mu_0 c\bar{S}}$

$= \sqrt{(2)(4\pi\times10^{-7}\text{ H/m})(3.00\times10^8\text{ m/s})(20\times10^{-6}\text{ W/m}^2)}$

$= 0.12$ V/m.

The sensitivity of radio receiving equipment is often expressed in terms of the minimum electric field strength.

EXERCISE   A stereo receiver's AM tuner section has a rated sensitivity of 2.1 mV/m. What is the maximum distance at which this unit can receive broadcasts from a radio station's 5.0-kW transmitter, assuming the signal is broadcast in all directions?

Answer: 261 km

Some problems similar to Example 34-5:   49, 51, 52, 61, 62   ●

34-11   WAVE MOMENTUM AND RADIATION PRESSURE

We know from mechanics that moving objects carry both energy and momentum. The same is true for electromagnetic waves. Maxwell showed that the wave energy U and momentum p are related by

$$p = \frac{U}{c}. \tag{34-22}$$

If an electromagnetic wave is incident on an object and the object absorbs the wave energy (as, for example, a black object exposed to sunlight), then the object also absorbs the momentum given by Equation 34-22. If the wave's average intensity is $\bar{S}$, then it carries energy per unit area at the average rate $\bar{S}$ J/s/m^2. According to Equation 34-22 it therefore carries momentum per unit area at the rate $\bar{S}/c$. Newton's law in its general form $\mathbf{F} = d\mathbf{p}/dt$ tells us that the rate of change of an object's momentum is equal to the net force on the object. Therefore, if an object absorbs electromagnetic wave momentum $\bar{S}/c$ per unit area per unit time, it experiences a force per unit area of this magnitude. Since force per unit area is pressure, we call this quantity the **radiation pressure:**

$$P_{\text{rad}} = \frac{\bar{S}}{c}. \tag{34-23}$$

The radiation pressure is doubled if an object reflects electromagnetic waves, in the same way that bouncing a basketball off a backboard changes the ball's momentum by $2mv$ and, therefore, delivers momentum $2mv$ to the backboard.

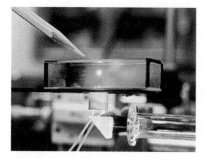

FIGURE 34-25 The star-like image is a 20-micron particle levitated by a laser beam reflected upward by the prism shown at the bottom. The star-like rays are due to diffraction (see Chapter 37) inside the camera.

FIGURE 34-26 Two-stage sailing spacecraft proposed for interstellar travel. At the target star, laser light from the solar system would reflect from the large sail to the smaller one, bringing the latter to a stop. The diameter of the large sail is 1000 km.

The pressure exerted by ordinary light is very small, but Dartmouth College physicists E. F. Nichols and G. F. Hull demonstrated its existence in a sensitive experiment performed in 1903. With high-energy laser light or with objects of low mass and large area, radiation pressure can be appreciable. Lasers exert enough light pressure to levitate small particles (Fig. 34-25), and the pressure of sunlight has been suggested as a means of driving interplanetary "sailing ships" (Fig. 34-26). Finally, the idea that electromagnetic waves carry momentum played a crucial role in Einstein's development of his famous equation $E = mc^2$.

● EXAMPLE 34-6 STAR WARS

A proposed ballistic missile defense system calls for a laser that can focus 25 MW of light on an attacking warhead. The weapon works by heating the warhead to the point of destruction, but it also delivers momentum that alters the warhead's trajectory. If the beam dwells on a 200-kg warhead for 15 s, what velocity change does it impart to the warhead? Estimate the distance by which the warhead will be knocked off course over its remaining 30 minutes of flight.

Solution

The energy delivered in a 25-MW beam acting for 15 s is (25 MW)(15 s) = 375 MJ. According to Equation 34-22, the associated momentum is

$$p = \frac{U}{c} = \frac{375 \times 10^6 \text{ J}}{3.00 \times 10^8 \text{ m/s}} = 1.25 \text{ kg} \cdot \text{m/s}.$$

The change in the warhead's velocity is given by $m\Delta v = \Delta p$, where Δp is the change in its momentum. Assuming the warhead absorbs all the beam's momentum, we then have

$$\Delta v = \frac{\Delta p}{m} = \frac{1.25 \text{ kg} \cdot \text{m/s}}{200 \text{ kg}} = 0.00625 \text{ m/s}.$$

This is insignificant compared with a typical warhead speed of 7 km/s. Even though we don't know the direction of the velocity change, we can estimate crudely the error Δx in the impact point by multiplying this change by the flight time:

$$\Delta x = \Delta v \, t = (0.00625 \text{ m/s})(30 \text{ min})(60 \text{ s/min}) = 11 \text{ m}.$$

This is totally negligible, especially for a nuclear warhead. Even with this enormously powerful laser, radiation pressure has an insignificant effect.

E X E R C I S E   A laser delivers 5.0 MW/m². If the beam is directed upward, what is the maximum mass for a 100-μm-diameter particle to be suspended in the beam?

Answer: 1.3×10^{-11} kg

Some problems similar to Example 34-6:   56–59 ●

CHAPTER SYNOPSIS

Summary

1. Maxwell's modification of Ampère's law adds a **displacement current** term $\varepsilon_0 d\phi_E/dt$, showing that changing electric flux is a source of magnetic field. This modified law completes the set of **Maxwell's equations**—the four equations that govern the behavior of electromagnetic fields.
2. The interplay of electric and magnetic fields described by Faraday's and Ampère's laws gives rise to **electromagnetic waves.** When they propagate through vacuum, these waves
   (a) Travel at the speed of light, $c = 1/\sqrt{\varepsilon_0 \mu_0} = 3.00 \times 10^8$ m/s.
   (b) Have their electric and magnetic fields at right angles to each other and to the direction of wave propagation.
   (c) Have their field magnitudes related by $E = cB$.
   (d) Can have any frequency or wavelength, provided the two are related by the equivalent expressions $f\lambda = c$ or $\omega/k = c$.
3. Radio waves, television, microwaves, infrared, visible light, ultraviolet, x rays, and gamma rays are all forms of electromagnetic radiation. They differ only in frequency and wavelength, and together comprise the **electromagnetic spectrum.**
4. **Polarization** describes the orientation of a wave's electric field in the plane perpendicular to the propagation direction. When a polarized wave passes through a polarizing material, its intensity is reduced by a factor $\cos^2 \theta$, where θ is the angle between the wave polarization and the preferred axis of the material.
5. Electromagnetic waves are produced by accelerated electric charges, as in the alternating current of a radio antenna.
6. Electromagnetic waves carry energy. The rate at which energy is transported per unit area is the wave **intensity.** The **Poynting vector,**

$$\mathbf{S} = \frac{\mathbf{E} \times \mathbf{B}}{\mu_0},$$

describes this energy transport for any configuration of electromagnetic fields. The average intensity has half the peak value, or

$$\bar{S} = \frac{E_p B_p}{2\mu_0}.$$

The intensity of a plane wave remains constant, while the intensity from a localized source decreases as the inverse square of the distance from the source.

7. An electromagnetic wave with energy U also carries momentum $p = U/c$. As a result, it exerts a **radiation pressure** $P_{rad} = \bar{S}/c$ on an object that absorbs the wave and twice this pressure on a reflecting object.

Terms You Should Understand

(Pairs are closely related terms whose distinction is important; number in parentheses is chapter section where term first appears.)

displacement current (34-2)
Maxwell's equations (34-3)
electromagnetic wave (34-4)
electromagnetic spectrum (34-7)
polarization (34-8)
intensity, Poynting vector (34-10)
radiation pressure (34-11)

Symbols You Should Recognize

$\varepsilon_0 d\phi_E/dt$ (34-2)	$S, \mathbf{S}, \bar{S}$ (34-10)
c (34-5)	P_{rad} (34-11)

Problems You Should Be Able to Solve

evaluating induced magnetic fields in symmetric situations (34-2)
relating frequency and wavelength of electromagnetic waves (34-6)
relating electric and magnetic field strengths in electromagnetic waves (34-6)
calculating light intensity emerging from one or more polarizers (34-8)
relating wave intensity and fields (34-10)
evaluating wave intensity and fields as a function of distance from localized sources (34-10)
calculating radiation pressure and its effects (34-11)

Limitations to Keep in Mind

The description of electromagnetic waves developed in this chapter applies strictly only in vacuum.

QUESTIONS

1. Why is Maxwell's modification of Ampère's law essential to the existence of electromagnetic waves?
2. The presence of magnetic monopoles would require modification of Gauss's law for magnetism. Which of the other Maxwell equations would also need modification?
3. There is displacement current between the plates of a charging capacitor, yet no charge is moving between the plates. In what sense is the word "current" appropriate here?
4. Is there displacement current in an electromagnetic wave? Is there ordinary conduction current?
5. List some similarities and differences between electromagnetic waves and sound waves.
6. What aspect of the electromagnetic wave considered in Section 34-4 ensures that Gauss's laws for electricity and magnetism are satisfied?
7. Explain why parallel electric and magnetic fields in vacuum could not constitute an electromagnetic wave.
8. The speed of an electromagnetic wave is given by $c = \lambda f$. How does the speed depend on frequency? On wavelength?
9. When astronomers observe a supernova explosion in a distant galaxy, they see a sudden, simultaneous rise in visible light and other forms of electromagnetic radiation. How is this evidence that the speed of light is independent of frequency?
10. Turning a TV antenna so its rods point vertically may change the quality of your TV reception. Why? Think about polarization.
11. Unpolarized light is incident on two sheets of Polaroid with their polarization directions at right angles, and no light gets through. A third sheet is inserted between the other two, and now some light gets through. How can this be?
12. Why is it not possible to define exactly where the visible region of the spectrum ends?
13. Why did the field of x-ray astronomy flourish only after the advent of space flight?
14. The Sun emits most of its electromagnetic wave energy in the visible region of the spectrum, with the peak in the yellow-green. Our eyes are sensitive to the same range, with peak sensitivity in the yellow-green. Is this a coincidence?
15. Suppose your eyes were sensitive to radio waves rather than light. What things would look bright?
16. An LC circuit is made entirely from superconducting materials, yet its oscillations eventually damp out. Why?
17. If you double the field strength in an electromagnetic wave, what happens to the intensity?
18. The intensity of light falls off as the inverse square of the distance from the source. Does this mean that electromagnetic wave energy is lost? Explain.
19. When your picture is taken with a flash camera, why doesn't the momentum of the light flash knock you over?
20. Some long-distance power transmission lines use DC rather than AC, despite the need to convert between DC amd AC at either end. Why might this be? What energy loss mechanism occurs with AC but not DC?
21. Electromagnetic waves do not readily penetrate metals. Why might this be?

PROBLEMS

Section 34-2 Ambiguity in Ampère's Law

1. A uniform electric field is increasing at the rate of 1.5 V/m·µs. What is the displacement current through an area of 1.0 cm² at right angles to the field?
2. A parallel-plate capacitor has square plates 10 cm on a side and 0.50 cm apart. If the voltage across the plates is increasing at the rate of 220 V/ms, what is the displacement current in the capacitor?
3. A parallel-plate capacitor of plate area A and spacing d is charging at the rate dV/dt. Show that the displacement current in the capacitor is equal to the conduction current flowing in the wires feeding the capacitor.
4. A capacitor with circular plates is fed with long, straight wires along the axis of the plates. Show that the magnetic field *outside* the capacitor, in a plane that passes through the interior of the capacitor and is perpendicular to the

axis, is given by $B = \dfrac{\mu_0 \varepsilon_0 R^2}{2rd} \dfrac{dV}{dt}$. Here R is the plate radius, d the spacing, dV/dt the rate of change of the capacitor voltage, and r the distance from the axis.
5. A parallel-plate capacitor has circular plates with radius 50 cm and spacing 1.0 mm. A uniform electric field between the plates is changing at the rate 1.0 MV/m·s. What is the magnetic field between the plates (a) on the symmetry axis, (b) 15 cm from the axis, and (c) 150 cm from the axis?
6. An electric field points into the page and occupies a circular region of radius 1.0 m, as shown in Fig.34-27. There are no electric charges in the region, but there is a magnetic field forming closed loops pointing clockwise, as shown. The magnetic field strength 50 cm from the center of the region is 2.0 µT. (a) What is the rate of change of the electric field? (b) Is the electric field increasing or decreasing?

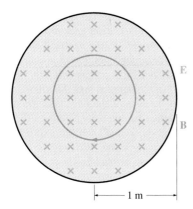

FIGURE 34-27 Problem 6.

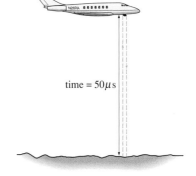

$time = 50 \mu s$

FIGURE 34-28 Problem 14.

Section 34-4 Electromagnetic Waves

7. At a particular point the instantaneous electric field of an electromagnetic wave points in the $+y$ direction, while the magnetic field points in the $-z$ direction. In what direction is the wave propagating?

8. The fields of an electromagnetic wave are $\mathbf{E} = E_p \sin(kz + \omega t)\hat{\mathbf{j}}$ and $\mathbf{B} = B_p \sin(kz + \omega t)\hat{\mathbf{i}}$. Give a unit vector in the direction of propagation.

9. The electric field of a radio wave is given by $\mathbf{E} = E \sin(kz - \omega t)(\hat{\mathbf{i}} + \hat{\mathbf{j}})$. (a) What is the peak amplitude of the electric field? (b) Give a unit vector in the direction of the magnetic field at a place and time where $\sin(kz - \omega t)$ is positive.

10. Show by differentiation and substitution that Equations 34-12b and 34-13 can be satisfied by fields of the form $E(x, t) = E_p f(kx \pm \omega t)$ and $B(x, t) = B_p f(kx \pm \omega t)$, where f is any function of the argument $kx \pm \omega t$.

11. Show that it is impossible for an electromagnetic wave in vacuum to have a time-varying component of its electric field in the direction of its magnetic field. *Hint:* Assume $\mathbf{E}$ does have such a component, and show that you cannot satisfy both Gauss and Faraday.

Section 34-5 The Speed of Electromagnetic Waves

12. A light-minute is the distance light travels in one minute. Show that the Sun is about 8 light-minutes from Earth.

13. Your intercontinental telephone call is carried by electromagnetic waves routed via a satellite in geosynchronous orbit at an altitude of 36,000 km. Approximately how long does it take before your voice is heard at the other end?

14. An airplane's radar altimeter works by bouncing radio waves off the ground and measuring the round-trip travel time (Fig. 34-28). If that time is 50 μs, what is the altitude?

15. Roughly how long does it take light to go 1 foot?

16. If you speak via radio from Earth to an astronaut on the moon, how long is it before you can get a reply?

17. "Ghosts" on a TV screen occur when part of the signal goes directly from transmitter to receiver, while part takes a longer route, reflecting off mountains or buildings (Fig. 34-29). The electron beam in a 50-cm-wide TV tube "paints" the picture by scanning the beam from left to right across the screen in about 10^{-4} s. If a "ghost" image appears displaced about 1 cm from the main image, what is the difference in path lengths of the direct and indirect signals?

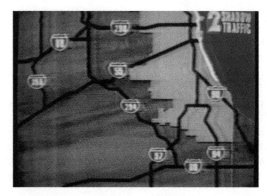

FIGURE 34-29 Ghost images of highways appear on this TV traffic report (Problem 17).

18. A computer can fetch information from its memory in 3.0 ns, a process that involves sending a signal from the central processing unit (CPU) to memory and awaiting the return of the information. If signals in the computer's wiring travel at $0.60c$, what is the maximum distance between the CPU and the memory? Your answer shows why high-speed computers are necessarily compact.

19. The speed of electromagnetic waves in a transparent dielectric is given by $1/\sqrt{\kappa \varepsilon_0 \mu_0}$, where κ is the dielectric

constant described in Chapter 26 (see Problem 69). An experimental measurement gives 1.97×10^8 m/s for the speed of light in a piece of glass. What is the dielectric constant of this glass at optical frequencies?

Section 34-6 Properties of Electromagnetic Waves

20. What are the wavelengths of (a) a 100-MHz FM radio wave, (b) a 3.0-GHz radar wave, (c) a 6.0×10^{14}-Hz light wave, and (d) a 1.0×10^{18}-Hz x ray?

21. A 60-Hz power line emits electromagnetic radiation. What is the wavelength?

22. Antennas for transmitting and receiving electromagnetic radiation usually have typical dimensions on the order of half a wavelength. Look at a TV antenna, and estimate the wavelength and frequency of a TV signal.

23. A CB radio antenna is a vertical rod 2.75 m high. If this length is one-fourth of the CB wavelength, what is the CB frequency?

24. A microwave oven operates at 2.4 GHz. What is the distance between wave crests in the oven?

25. What would be the electric field strength in an electromagnetic wave whose magnetic field equalled that of Earth, about 50 μT?

26. Dielectric breakdown in air occurs at an electric field strength of about 3×10^6 V/m. What would be the peak magnetic field in an electromagnetic wave with this value for its peak electric field?

27. A radio receiver can detect signals with electric fields as low as 320 μV/m. What is the corresponding magnetic field?

Section 34-8 Polarization

28. An electromagnetic wave is propagating in the z direction. What is its polarization direction, if its magnetic field is in the y direction?

29. Polarized light is incident on a sheet of polarizing material, and only 20% of the light gets through. What is the angle between the electric field and the polarization axis of the material?

30. Vertically polarized light passes through a polarizer whose polarization axis is oriented at 70° to the vertical. What fraction of the incident intensity emerges from the polarizer?

31. A polarizer blocks 75% of a polarized light beam. What is the angle between the beam's polarization and the polarizer's axis?

32. An electro-optic modulator (see Fig. 34-21) is supposed to switch a laser beam between fully off and fully on, as its crystal rotates the beam polarization by 90° when a voltage is applied. But a power-supply failure results in only enough voltage for a 72° beam rotation. What fraction of the laser light is transmitted when it is supposed to be fully on?

33. Unpolarized light of intensity S_0 passes first through a polarizer with its polarization axis vertical, then through one with its axis at 35° to the vertical. What is the light intensity after the second polarizer?

34. Vertically polarized light passes through two polarizers, the first at 60° to the vertical and the second at 90° to the vertical. What fraction of the light gets through?

35. Unpolarized light with intensity S_0 passes through a stack of five polarizing sheets, each with its axis rotated 20° with respect to the previous one. What is the intensity of the light emerging from the stack?

36. Unpolarized light of intensity S_0 is incident on a "sandwich" of three polarizers. The outer two have their transmission axes perpendicular, while the middle one has its axis at 45° to the others. What is the light intensity emerging from this "sandwich?"

37. Polarized light with average intensity S_0 passes through a sheet of polarizing material which is rotating at 10 rev/s. At time $t = 0$ the polarization axis is aligned with the incident polarization. Write an expression for the transmitted intensity as a function of time.

Section 34-10 Energy in Electromagnetic Waves

38. A typical laboratory electric field is 1000 V/m. What is the average intensity of an electromagnetic wave with this value for its peak field?

39. What would be the average intensity of a laser beam so strong that its electric field produced dielectric breakdown of air (which requires $E_p = 3 \times 10^6$ V/m)?

40. Estimate the peak electric field inside a 625-W microwave oven under the simplifying approximation that the microwaves propagate as a plane wave through the oven's 750-cm² cross-sectional area.

41. A radio receiver can pick up signals with peak electric fields as low as 450 μV/m. What is the average intensity of such a signal?

42. Show that the electric and magnetic energy densities in an electromagnetic wave are equal.

43. A laser blackboard pointer delivers 0.10 mW average power in a beam 0.90 mm in diameter. Find (a) the average intensity, (b) the peak electric field, and (c) the peak magnetic field.

44. The laser of Example 34-6 produces a spot 80 cm in diameter at its target. What are the rms electric and magnetic fields at the target?

45. The United States' safety standard for continuous exposure to microwave radiation is 10 mW/cm². The glass door of a microwave oven measures 40 cm by 17 cm and is covered with a metal screen that blocks microwaves. What fraction of the oven's 625-W microwave power can leak through the door window without exceeding the safe exposure to someone right outside the door? Assume the power leaks uniformly through the window area.

46. A 1.0-kW radio transmitter broadcasts uniformly in all directions. What is the intensity of its signal at a distance of 5.0 km from the transmitter?

47. Use the fact that sunlight intensity at Earth's orbit is 1368 W/m^2 to calculate the Sun's total power output.

48. About two-thirds of the solar energy at Earth's orbit reaches the planet's surface. At what rate is solar energy incident on the entire Earth? See the previous problem, and compare your result with the roughly 10^{13} W rate at which humanity consumes energy.

49. During its 1989 encounter with Neptune, the Voyager 2 spacecraft was 4.5×10^9 km from Earth (Fig. 34-30). Its images of Neptune were broadcast by a radio transmitter with a mere 21-W average power output. What would be (a) the average intensity and (b) the peak electric field received at Earth if the transmitter broadcast equally in all directions? (The received signal was actually somewhat stronger because Voyager used a directional antenna.)

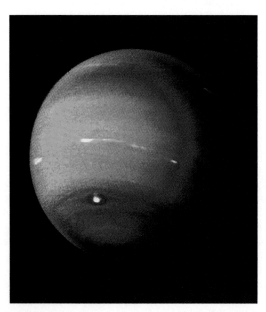

FIGURE 34-30 Neptune, photographed by the Voyager II spacecraft when it was 4.5×10^9 km from Earth. How long did it take the radio signal carrying this image to reach Earth? (Problem 49)

50. A quasar 10 billion light-years from Earth appears the same brightness as a star 50,000 light-years away. How does the power output of the quasar compare with that of the star?

51. At 1.5 km from the transmitter, the peak electric field of a radio wave is 350 mV/m. (a) What is the transmitter's power output, assuming it broadcasts uniformly in all directions? (b) What is the peak electric field 10 km from the transmitter?

52. The peak electric field at a point 25 m from a point source of electromagnetic waves is 4.2 kV/m. What is the peak magnetic field 1.0 m from the source?

53. A typical fluorescent lamp is a little over 1 m long and a few cm in diameter. How do you expect the light intensity to vary with distance (a) near the lamp but not near either end and (b) far from the lamp?

Section 34-11 Wave Momentum and Radiation Pressure

54. A camera flash delivers 2.5 kW of light power for 1.0 ms (Fig. 34-31). Find (a) the total energy and (b) the total momentum carried by the flash.

FIGURE 34-31 How much energy and momentum are in light from the camera flash? (Problem 54)

55. What is the radiation pressure exerted on a light-absorbing surface by a laser beam whose intensity is 180 W/cm^2?

56. A laser beam shines vertically upward. What laser power is necessary for this beam to support a flat piece of aluminum foil with mass 30 μg and diameter equal to that of the beam? Assume the foil reflects all the light.

57. The average intensity of noonday sunlight is about 1 kW/m^2. What is the radiation force on a solar collector measuring 60 cm by 2.5 m if it is oriented at right angles to the incident light and absorbs all the light?

58. Serious proposals have been made to "sail" spacecraft to the outer solar system using the pressure of sunlight. How much sail area must a 1000-kg spacecraft have if its accel-

eration at Earth's orbit is to be 1 m/s²? Assume the sails are made from reflecting material. *Hint:* Can you neglect the Sun's gravity?

59. A 65-kg astronaut is floating in empty space. If the astronaut shines a 1.0-W flashlight in a fixed direction, how long will it take the astronaut to accelerate to a speed of 10 m/s?

60. A "photon rocket" emits a beam of light instead of the hot gas of an ordinary rocket. How powerful a light source would be needed for a photon rocket with thrust equal to that of a space shuttle (35 MN)? Compare your answer with humanity's total electric power generating capability, about 10^{12} W.

Paired Problems

(Both problems in a pair involve the same principles and techniques. If you can get the first problem, you should be able to solve the second one.)

61. Find the peak electric and magnetic fields 1.5 m from a 60-W light bulb that radiates equally in all directions.

62. At 4.6 km from a radio transmitter, the peak electric field in the radio wave measures 380 mV/m. What is the transmitter's power, assuming it broadcasts equally in all directions?

63. Unpolarized light is incident on two polarizers with their axes at 45°. What fraction of the incident light gets through?

64. Find the angle between two polarizers if unpolarized light incident on the pair emerges with 10% of its incident intensity.

65. What is the radiation force on the door of a microwave oven if 625 W of microwave power hits the door at right angles and is reflected?

66. What is the power output of a laser whose beam exerts a 55-mN force on an absorbing object oriented at right angles to the beam? The object is larger than the beam's cross section.

67. A 60-W light bulb is 6.0 cm in diameter. What is the radiation pressure on an opaque object at the bulb's surface?

68. A white dwarf star is approximately the size of Earth but radiates about as much energy as the Sun. Estimate the radiation pressure on an absorbing object at the white dwarf's surface.

Supplementary Problems

69. Maxwell's equations in a dielectric resemble those in vacuum (Equations 34-6 through 34-9), but with ϕ_E in Ampère's law replaced by $\kappa \phi_E$, where κ is the dielectric con-

stant introduced in Chapter 26. Show that the speed of electromagnetic waves in such a dielectric is $c/\sqrt{\kappa}$.

70. Use appropriate data from Appendix E to calculate the radiation pressure on a light-absorbing object at the Sun's surface.

71. A radar system produces pulses consisting of 100 full cycles of a sinusoidal 70-GHz electromagnetic wave. The average power while the transmitter is on is 45 MW, and the waves are confined to a beam 20 cm in diameter. Find (a) the peak electric field, (b) the wavelength, (c) the total energy in a pulse, and (d) the total momentum in a pulse. (e) If the transmitter produces 1000 pulses per second, what is its average power output?

72. In a stack of polarizing sheets, each sheet has its polarization axis rotated 14° with respect to the preceding sheet. If the stack passes 37% of the incident, unpolarized light, how many sheets does it contain?

73. The peak electric field measured at 8.0 cm from a light source is 150 W/m², while at 12 cm it measures 122 W/m². Describe the shape of the source.

74. Show that Equations 34-12b and 34-13 may be combined to yield a wave equation like Equation 16-13. *Hint:* Take the partial derivative of one equation with respect to x and of the other with respect to t, and use the fact that

$$\frac{\partial}{\partial x}\left(\frac{\partial f}{\partial t}\right) = \frac{\partial}{\partial t}\left(\frac{\partial f}{\partial x}\right)$$

for any well-behaved function $f(x, t)$.

75. Studies of the origin of the solar system suggest that sufficiently small particles might be blown out of the solar system by the force of sunlight. To see how small such particles must be, compare the force of sunlight with the force of gravity, and solve for the particle radius at which the two are equal. Assume the particles are spherical and have density 2 g/cm³. Why do you not need to worry about the distance from the Sun?

76. A cylindrical resistor of length ℓ, radius a, and resistance R carries a current I. Calculate the electric and magnetic fields at the surface of the resistor, assuming the electric field is uniform throughout, including at the surface. Calculate the Poynting vector, and show that it points into the resistor. Calculate the flux of the Poynting vector (that is, $\int \mathbf{S} \cdot d\mathbf{A}$) over the surface of the resistor to get the rate of electromagnetic energy flow into the resistor, and show that the result is just I^2R. Your result shows that the energy heating the resistor comes from the fields surrounding it. These fields are sustained by the source of electrical energy that drives the current.

P A R T 4 CUMULATIVE PROBLEMS

These problems combine material from chapters throughout the entire part or, in addition, from chapters in earlier parts, or they present special challenges.

1. An air-insulated parallel-plate capacitor has plate area 100 cm² and spacing 0.50 cm. The capacitor is charged to a certain voltage and then disconnected from the charging battery. A thin-walled, nonconducting box of the same dimensions as the capacitor is filled with water at 20.00°C. The box is released at the edge of the capacitor and moves without friction into the capacitor (Fig. 1). When it reaches equilibrium the water temperature is 21.50°. What was the original voltage on the capacitor?

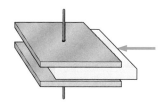

FIGURE 1 Cumulative Problem 1.

2. A wire of length ℓ and resistance R is formed into a closed rectangular loop twice as long as it is wide. It is mounted on a nonconducting horizontal axle parallel to its longer dimension, as shown in Fig. 2. A uniform magnetic field **B** points into the page, as shown. A string of negligible mass is wrapped around a drum of radius a attached to the axle, and a mass m is attached to the string. The string is many times longer than the drum circumference; thus, many turns are wrapped around the drum. When the mass is released it falls and eventually reaches a speed that, averaged over one cycle of the loop's rotation, is constant from one rotation to the next. Find an expression for that average terminal speed.

3. Five wires of equal length 25 cm and resistance 10 Ω are connected to form two equilateral triangles that share a common side, as shown in Fig. 3. Two solenoids are perpendicular to the plane of the figure, as shown. Both solenoids have diameter 10 cm, and both extend a long way in and out of the page. The magnetic fields of both solenoids point out of the page; the field strength in the left-hand solenoid is increasing at 50 T/s while that in the right-hand solenoid is decreasing at 30 T/s. Find the current in the resistance wire shared by both triangles. Which way does the current flow?

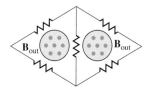

FIGURE 3 Cumulative Problem 3.

4. A long solenoid of length ℓ and radius R has a total of N turns. The solenoid current is increasing linearly with time: $I(t) = bt$, where b is a constant. (a) Find an expression for the rate at which the magnetic energy in the solenoid is increasing. (b) Find an expression for the induced electric field at the inner edge of the solenoid coils. (c) Evaluate the Poynting vector at the inner edge of the coils, and show by integration that electromagnetic energy is flowing into the solenoid at a rate equal to the buildup of magnetic energy. Use the long-solenoid approximation throughout, neglecting any variations along the direction of the solenoid axis.

5. A coaxial cable consists of an inner conductor of radius a and an outer conductor of radius b; the space between the conductors is filled with insulation of dielectric constant κ (Fig. 4). The cable's axis is the z axis. The cable is used to carry electromagnetic energy from a radio transmitter to a broadcasting antenna. The electric field between the conductors points radially from the axis,

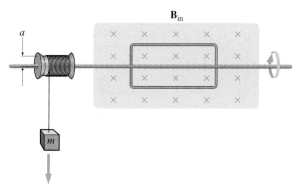

FIGURE 2 Cumulative Problem 2.

and is given by $E = E_0 \dfrac{a}{r} \cos(kz - \omega t)$. The magnetic field encircles the axis, and is given by $B = B_0 \dfrac{a}{r} \cos(kz - \omega t)$. Here E_0, B_0, k, and ω are constants. (a) Show, using appropriate closed surfaces and loops, that these fields satisfy Maxwell's equations. Your result shows that the cable acts as a "waveguide," confining an electromagnetic wave to the space between the conductors. (b) Find an expression for the speed at which the wave propagates along the cable.

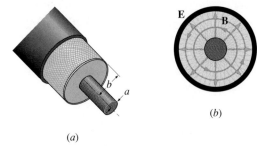

(a)

(b)

FIGURE 4 Cumulative Problem 5. (a) A coaxial cable. (b) Cross section, showing the electric and magnetic fields. The fields also vary with position z along the cable axis, according to the equations given.

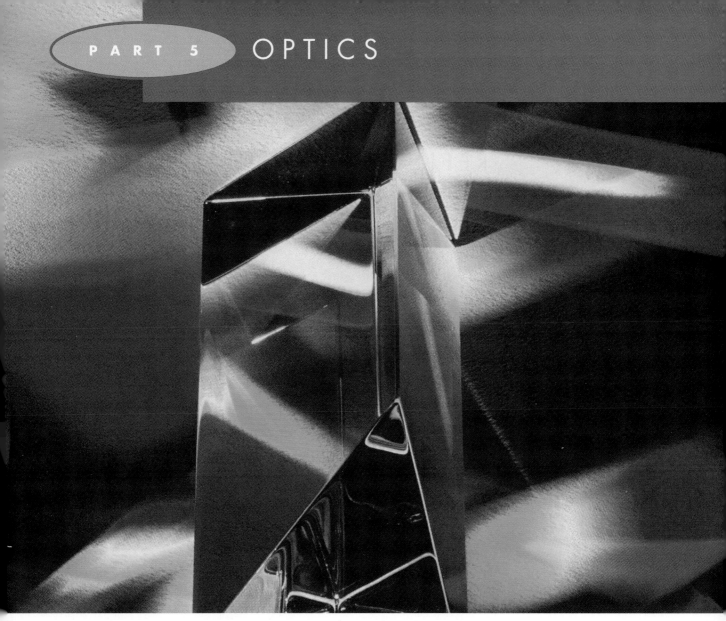

In "Prismatic Abstract," multiple light beams converge on a prism to create a dazzlingly color-ful effect. Photographic artist Pete Saloutous exploited the principles of optics in creating this work.

CHAPTER 35

REFLECTION AND REFRACTION

Reflection and refraction guide light in a wide range of natural and technological applications. Here the image of a bee's head appears in a large-scale version of the optical fibers that are essential to modern communications.

Maxwell's brilliant work shows that the phenomena of **optics**—that is, the behavior of light—are manifestations of the laws of electromagnetism. Except in the atomic realm, where it is necessary to use quantum mechanics, all optical phenomena can be understood in terms of electromagnetic wave fields as described by Maxwell's equations. But in many cases we need not resort to the full electromagnetic or even wave description of light to understand optics. When the objects with which light—or, for that manner, any other wave—interacts are much larger than the wavelength, then to a good approximation the light travels in straight lines called **rays** (Fig. 35-1), except when it actually hits something. **Geometrical optics** is the study of light under conditions when the ray approximation is valid. In this chapter we show how geometrical optics describes the behavior of light at interfaces between two different materials. In the next chapter we will show how geometrical optics explains optical systems including mirrors, lenses, the human eye, and many optical instruments.

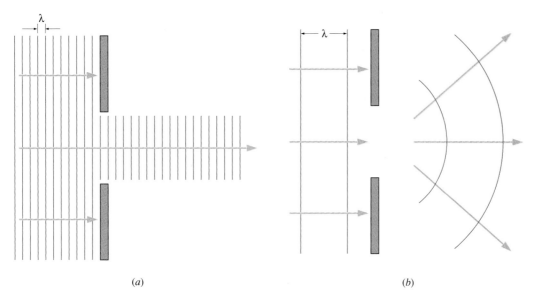

FIGURE 35-1 Light waves incident on an opaque barrier with a hole in it. (*a*) When the hole diameter is much greater than the wavelength λ, light emerges from the hole in essentially a straight line. In this case it is appropriate to treat the light as traveling in straight rays except where it actually hits something. (*b*) When the hole diameter is comparable to or smaller than the wavelength, then interference causes the wavefronts to bend, and the ray approximation is no longer adequate.

35-1 REFLECTION AND TRANSMISSION

When a light ray propagating in one medium strikes the interface with a second medium, some light may be reflected back into the original medium and some transmitted into the new medium. Properties of the two materials, and the propagation direction of the incident light, determine the details of the reflection and transmission. It is also possible for light energy to be absorbed as it propagates through a medium, but here we will neglect this absorption.

Some materials—notably metals—reflect nearly all the light incident on them. It's no coincidence that these materials are also good electrical conductors. The oscillating electric field of a light wave drives the metal's free electrons into oscillatory motion, and this accelerated motion, in turn, produces electromagnetic waves. The effect of all the oscillating electrons together is simply to reradiate the wave back into the original medium. Ultimately, that's why metals appear shiny.

In other materials, the atomic and molecular configurations are such that the material absorbs most of the incident light energy. Such materials therefore appear dark and opaque. Since opaque materials basically destroy light by converting it to other forms of energy, they will play little role in our exploration of optics.

Still other materials are essentially transparent, allowing light to propagate with little energy loss. These materials are generally insulators whose electrons, bound to individual atoms, cannot respond freely to the fields of an incident wave. The electrons do respond in a more limited way, however, effectively producing oscillating molecular dipoles. Although a microscopic description of

FIGURE 35-2 A laser beam incident on a glass prism shows partial reflection and partial transmission at the interfaces between air and glass.

this process is complicated, the net effect in most simple dielectric materials is a reduction in the propagation speed of electromagnetic waves, as wave energy is absorbed and reradiated by the molecular dipoles.

In this chapter we explore the processes that occur at the interfaces between transparent and reflective materials and between different transparent materials. Reflection is the significant process in the former case, while in the latter both reflection and transmission generally occur (Fig. 35-2).

35-2 REFLECTION

Reflection returns some or all of the light incident on an interface to its original medium. Whether the reflection is essentially complete, as from a metal, or partial, as from a transparent material, it satisfies the same geometrical condition. Experiment—as well as analysis based on Maxwell's equations—shows that the incident ray, the reflected ray, and the normal to the interface between the two materials all lie in the same plane, and that the angle θ_1' that the reflected ray makes with the normal is equal to the angle θ_1 made by the incident ray (Fig. 35-3a). That is,

$$\theta_1' = \theta_1, \tag{35-1}$$

where the subscript 1 designates angles in the first medium. These angles are designated the **angle of reflection** and **angle of incidence,** respectively.

When a beam of parallel light rays reflects off a smooth surface, each ray in the beam reflects at the same angle, and the entire beam is thus reflected without distortion (Fig 35-3b). This process is called **specular reflection.** But if the surface is rough, then individual rays, while still obeying Equation 35-1, reflect from differently oriented pieces of the surface. As a result the reflection is **diffuse,** with the original beam spreading in all directions (Fig. 35-3c). White wall paint is a good example of a diffuse reflector, while the aluminum or silver coating of a mirror is an excellent specular reflector. When we speak of reflection we generally mean specular reflection.

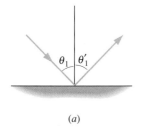

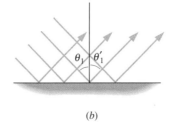

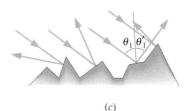

(a)   *(b)*   *(c)*

FIGURE 35-3 (*a*) The angles of reflection and incidence are equal. (*b*) In specular reflection, a smooth surface reflects a light beam undistorted. (*c*) A rough surface results in diffuse reflection. The law of reflection still holds for each individual ray.

● **EXAMPLE 35-1** THE CORNER REFLECTOR

Two mirrors stand vertically, at right angles to each other as shown in Fig. 35-4a. Show that any light ray incident in the horizontal plane will return antiparallel to its incident direction.

Solution

Figure 35-4b shows a top view of the two mirrors. The law of reflection ensures that the two angles marked θ are equal, as are the two marked ϕ. But the mirrors make a 90° angle, and therefore $\phi = 90° - \theta$. The incident and first reflected ray make an angle 2θ, while the first reflected ray and the outgoing ray make an angle $2\phi = 180° - 2\theta$. Thus these two angles sum to 180°, which shows that the incident and outgoing rays are antiparallel. This result holds regardless of the actual value of the incidence angle θ.

Adding a third mirror at right angles to the other two allows the corner reflector to return any beam in the direction from which it came—regardless of the orientation of the beam or the reflector. Corner reflectors, often made with prisms rather than mirrors, are widely used in optics. A corner reflector left on the moon allows laser-based measurements of the moon's distance to within about 15 cm (Fig. 35-5).

EXERCISE Two mirrors make an angle of 135°. A light ray is incident on the first mirror at an angle of 60° to the mirror's normal. Find (a) the angle of incidence on the second mirror and (b) the angle between the original incident beam and the final outgoing beam.

Answers: (a) 75°; (b) 90°

Some problems similar to Example 35-1: 2, 4, 5

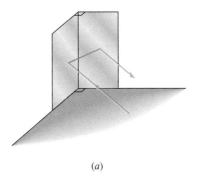

(a)

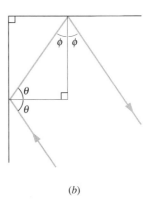

(b)

FIGURE 35-4 (a) Two-dimensional corner reflector, made from two mirrors at right angles. (b) Any ray incident in the horizontal plane returns parallel to its original direction.

FIGURE 35-5 Astronauts left this array of corner reflectors on the moon. Reflecting laser beams off the array allows determination of the moon's position to within 15 cm.

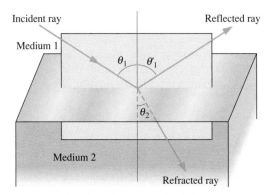

FIGURE 35-6 Reflection and refraction at an interface. This case, where the refracted ray is bent *toward* the normal, occurs when medium 2 has a higher refractive index. Angles θ_1, θ_1', and θ_2 are, respectively, the angles of incidence, reflection, and refraction. Light-colored surface is a plane perpendicular to the interface.

35-3 REFRACTION

Figures 35-2 and 35-6 show that light passing between two transparent media is partially reflected and partially transmitted at their interface. In addition, the transmitted beam changes direction at the interface—a phenomenon known as **refraction.**

Physically, refraction occurs because the wave speeds in the two media are different. That difference, in turn, results from delays introduced by the interaction of the electromagnetic wave fields with atomic electrons. But the wave frequency f—and thus the wave period $T = 1/f$—must be the same on both sides of the interface; otherwise, on one side of the interface more wave crests would pass in a given time than on the other side, implying the creation or destruction of waves at the interface. In Fig. 35-7a waves in medium 1 travel at speed v_1, and the wavelength—the distance between wave crests—is therefore $v_1 T$. In medium 2 the speed has some lower value v_2, and the wavelength $v_2 T$ is correspondingly shorter. The shaded triangles in Fig. 35-7a are right triangles with a common hypotenuse, and the ratio of the opposite side to the length of this hypotenuse defines the sines of the angles θ_1 and θ_2. Equating expressions for the hypotenuse length (opposite side divided by sine) in terms of these two angles gives

$$\frac{v_1 T}{\sin \theta_1} = \frac{v_2 T}{\sin \theta_2},$$

or

$$\frac{\sin \theta_2}{\sin \theta_1} = \frac{v_2}{v_1}.$$

We characterize the effect of a transparent medium on light through its **index of refraction**, n, defined as the ratio of the speed c of light in vacuum to the speed v of light in the medium:

$$n = \frac{c}{v}. \qquad \text{(index of refraction)} \qquad (35\text{-}2)$$

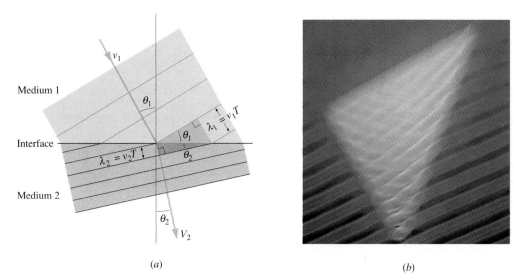

(a)   (b)

FIGURE 35-7 (a) Waves refract at an interface because wave speeds—and therefore the wavelengths—are different in the two media. Two shaded triangles have a common hypotenuse, whose length is $vT/\sin\theta$, where v and θ are taken on either side of the interface. (b) Water waves refract as they pass over a triangular patch of shallower water. Note the decreased wavelength and different orientations of the waves within the shallow region. Reflection at the interfaces has resulted in waves propagating in several directions.

(Again there's an electromagnetic connection: the index of refraction is just the square root of the dielectric constant that we introduced in Chapter 26. But don't try to calculate refractive indices from the dielectric constants of Table 26-1; the dielectric constant is frequency dependent, and its value at optical frequencies is significantly different from the DC values of Table 26-1.) Table 35-1 lists some indices of refraction. Note that for gases the index of refraction is essentially 1, the same as that of vacuum.

▲ **TABLE 35-1** INDICES OF REFRACTION*

SUBSTANCE	INDEX OF REFRACTION, n
Gases	
Air	1.000293
Carbon dioxide	1.00045
Liquids	
Water	1.333
Ethyl alcohol	1.361
Glycerine	1.473
Benzene	1.501
Diiodomethane	1.738
Solids	
Ice (H_2O)	1.309
Fused quartz (SiO_2)	1.458
Polystyrene	1.49
Glass (crown)	1.52
Sodium chloride (NaCl)	1.544
Glass (flint)	1.6–1.9
Diamond (C)	2.419
Rutile (TiO_2)	2.62

*At 1 atm pressure and temperatures ranging from 0°C to 20°C, measured at a wavelength of 589 nm (the yellow line of sodium).

Using the definition 35-2 in our ratio of the sines of the angles of refraction and incidence, and then cross multiplying, gives **Snell's law:**

$$n_1 \sin \theta_1 = n_2 \sin \theta_2. \quad \text{(Snell's law)} \quad (35\text{-}3)$$

This law was developed geometrically in 1621 by Willebrord van Roijen Snell of the Netherlands and described analytically in the 1630s by René Descartes of France, where it is known to this day as Descartes' law. It allows us to predict what will happen to light at an interface, provided we know the refractive indices of the two media.

● EXAMPLE 35-2   CD MUSIC

The laser beam that "reads" information from a compact disc is 0.737 mm wide at the point where it strikes the underside of the disc and forms a converging cone with half-angle 27°, as shown in Fig. 35-8. It then travels through a 1.2-mm-thick layer of transparent plastic with refractive index 1.55 before reaching the very thin, reflective information layer near the disc's top surface. What is the beam diameter (d in Fig. 35-8) at the information layer?

Solution
Figure 35-8 shows that $d = D - 2x$, where $D = 0.737$ mm is the beam diameter as it hits the disc. From the figure we also see that $x = t \tan \theta_2$, where $t = 1.2$ mm is the thickness of the plastic. Finally, Snell's law gives $\theta_2 = \sin^{-1}\left(\dfrac{\sin \theta_1}{n}\right)$, where we set the refractive index of air to 1. Putting these relations together gives

$$d = D - 2t \tan\left[\sin^{-1}\left(\frac{\sin \theta_1}{n}\right)\right]$$

$$= 737 \ \mu\text{m} - (2)(1200 \ \mu\text{m}) \tan\left[\sin^{-1}\left(\frac{\sin 27°}{1.55}\right)\right]$$

$$= 1.8 \ \mu\text{m},$$

which is a little larger than the "pits" cut into the CD to store its information. This narrowing of the beam plays a crucial role in keeping CDs noise free. The tiniest dust speck would blot out information at the μm-scale information layer, but at the point where the laser beam actually enters the disc—the closest dust can get to the information layer—it would take mm-size dust to cause problems.

EXERCISE Figure 35-9 shows a polystyrene cylinder whose height is equal to its diameter. What is the maximum incidence angle θ_1 at which a light ray striking the top center of the cylinder will emerge through the bottom without first striking the side? The medium surrounding the cylinder is air.

Answer: 42°

Some problems similar to Example 35-2: 14, 15, 17, 18

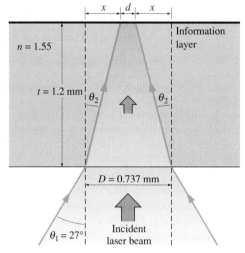

FIGURE 35-8 Section through a compact disc, showing convergence of the laser beam to a narrow spot at the information layer. All CDs share a common refractive index of 1.55 (Example 35-2).

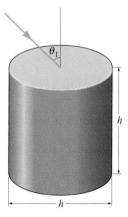

FIGURE 35-9 What is the maximum θ_1 for which the beam will emerge through the bottom of the polystyrene cylinder?   ●

Snell's law applies whether light goes from a medium of lower to higher refractive index or the reverse, as you can see by reversing the path of the light in Fig. 35-7. In the former case the refracted ray bends *toward* the normal, while in the latter case it bends *away* from the normal.

● **EXAMPLE 35-3** IN AND OUT: PARALLEL RAYS

In Fig. 35-10, a light ray propagating in air strikes a transparent slab of thickness d and refractive index n with incidence angle θ_1. Show that the ray emerges from the slab propagating parallel to its original direction.

Solution
Applying Snell's law to the upper interface gives

$$\sin\theta_2 = \frac{\sin\theta_1}{n},$$

where we've taken $n_1 = 1$ for air and $n_2 = n$ for the slab. At the lower interface, with the light going from the slab to air, we have $n_1 = n$ and $n_2 = 1$, so Snell's law gives

$$\sin\theta_4 = n\sin\theta_3.$$

But the slab faces are parallel so, as the figure suggests, $\theta_3 = \theta_2$. Combining our two versions of Snell's law then gives

$$\sin\theta_4 = n\left(\frac{\sin\theta_1}{n}\right) = \sin\theta_1,$$

showing that the incident and outgoing rays are indeed parallel. They are, however, displaced by the distance x shown in the figure. You can find that displacement in Problem 60.

EXERCISE A piece of crown glass is immersed in diiodomethane. A light ray strikes the glass with incidence angle

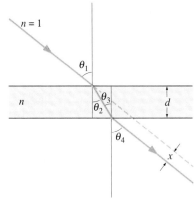

FIGURE 35-10 A light beam incident on a transparent slab emerges with its original direction, but is displaced from its original path.

40°. Find the angle the refracted beam in the glass makes with the normal to the interface.

Answer: 47.3°

Some problems similar to Example 35-3: 11, 13, 60

●

We derived Snell's law using Fig. 35-7a, which shows the wavelength change that occurs as the wave speed changes between two media. We argued that the frequencies in the two media must be the same, and since wavelength and frequency are related by $\lambda f = v$, Equation 35-2 then shows that the wavelengths in the two media are related by

$$\frac{\lambda_2}{\lambda_1} = \frac{n_1}{n_2}. \tag{35-4}$$

● **EXAMPLE 35-4** LIGHT AND DIAMOND

Light with wavelength 589 nm in vacuum enters a diamond. Find the light speed and wavelength in the diamond.

Solution
Equation 35-2 gives

$$v = \frac{c}{n} = \frac{3.00\times10^8 \text{ m/s}}{2.419} = 1.24\times10^8 \text{ m/s}.$$

The refractive index of vacuum is 1, so Equation 35-4 gives

$$\lambda_{\text{diamond}} = \frac{\lambda_{\text{vacuum}}}{n} = \frac{589 \text{ nm}}{2.419} = 243 \text{ nm}.$$

The wavelength and speed are reduced by the same factor, as they must be to keep the wave frequency unchanged.

EXERCISE Microwaves propagate through glass at about 1.4×10^8 m/s. What is the refractive index of glass at microwave frequencies?

Answer: 2.14

Some problems similar to Example 35-4: 8, 9, 19

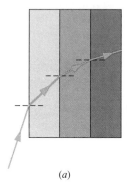

(a)

(b)

FIGURE 35-11 (a) Light propagating through a series of slabs with increasing refractive indices. At each interface the light bends more toward the normal. (b) Light follows a curved path in a medium with a continuously increasing refractive index.

Multiple Refractions

Engineered optical systems often use several layers of refractive material to minimize reflective losses and certain types of distortion. To describe the path of a light ray in such a system, we need only apply Snell's law at each of the interfaces using the appropriate pair of refractive indices (Fig. 35-11a). When the layers are parallel the angular deflection of a light ray is the same as if it had gone through a single interface from the first medium to the last (see Problem 54). In many natural situations, including the human eye and Earth's atmosphere, the index of refraction is a continuously varying function of position (Fig. 35-11b). We can approximate such a case as a sequence of thin layers, each with a different refractive index. Going toward the limit of infinitely many infinitesimal layers gets us arbitrarily close to the exact solution, in which the light follows a curved path. The mirage shown in Fig. 35-12 results from the fact that the refractive index of air is temperature dependent, causing the path of light rays to bend continuously.

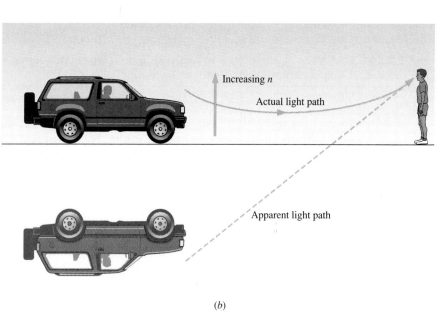

Increasing n

Actual light path

Apparent light path

(b)

(a)

FIGURE 35-12 (a) Mirages on a hot road. (b) The mirage occurs because hot air near the highway surface has a lower refractive index, resulting in a curved path for the light. The apparent position of the vehicle is not its actual position. Can you see why it appears upside down?

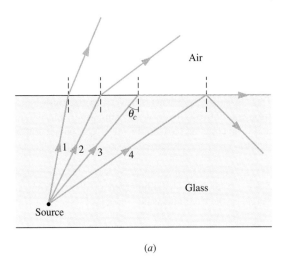

(a)

(b)

FIGURE 35-13   (a) Light propagating inside a glass block is refracted away from the normal at the glass-air interface. Ray 3, incident at the critical angle, just skims along the interface. At higher incidence angles, as with ray 4, the light undergoes total internal reflection at the interface. Not shown are weak reflected rays at angles less than the critical angle. (b) Three light beams strike a lens and are refracted at the air-glass interface. The blue beam strikes the subsequent glass-air interface nearly perpendicularly, and essentially all of it is transmitted. The yellow beam strikes more obliquely but is still largely transmitted. The red beam strikes at an incidence angle greater than the critical angle and undergoes total internal reflection.

35-4 TOTAL INTERNAL REFLECTION

Figure 35-13 shows light propagating inside a glass block and striking the interface with the surrounding air. Since air's refractive index is lower than that of glass, rays are bent *away* from the normal as they leave the glass. As the incidence angle increases, so does the angle of refraction—and Snell's law shows that the latter is always greater than the incidence angle. So at some incidence angle (see ray 3 in Fig. 35-13) the angle of refraction reaches 90°. What then?

If the incidence angle is increased further, we find that the light is *totally* reflected at the interface. (There is always *some* reflection back into the glass, but now it's *all* reflected.) This phenomenon is called **total internal reflection,** and the incidence angle at which it first occurs is the **critical angle,** θ_c. We can find θ_c by setting $\theta_2 = 90°$ (i.e., $\sin\theta_2 = 1$) in Snell's law (Equation 35-3). The critical angle is then θ_1, and we have

$$\sin\theta_c = \frac{n_2}{n_1}. \qquad (35\text{-}5)$$

Since the sine of an angle cannot exceed 1, we must have $n_2 \leq n_1$ in order for this equation to have a solution. Thus, total internal reflection occurs only when light propagating in one medium strikes an interface with a medium of lower refractive index, and it occurs whenever the incidence angle exceeds the critical angle.

Total internal reflection makes uncoated glass an excellent reflector when it's oriented appropriately (Fig. 35-14). Corner reflectors (Example 35-1) actually use total internal reflection in solid cubes of glass, rather than individual mirrors at right angles. For an observer inside a medium of higher refractive index, the existence of the critical angle affects the view of the outside world, as the example below shows. Finally, total internal reflection is the basis of the optical fibers now widely used in communications.

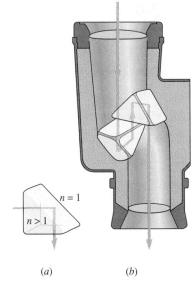

(a)   (b)

FIGURE 35-14   (a) A glass prism redirects light through total internal reflection. (b) Binoculars use a pair of prisms to "fold" the light path, allowing a more compact design.

● EXAMPLE 35-5 WHALE WATCH

Planeloads of whale watchers fly above the ocean, as shown in Fig. 35-15. A whale looks upward, watching the planes. Within what range of viewing angles can the whale see the planes?

Solution

If the whale emitted light, that light would emerge from the ocean surface only for incidence angles less than the critical angle—that is, those angles within the cone shown in Fig. 35-15. Since the path of light is reversible, the whale can only see objects above the surface when it looks within this cone. For the water-air interface Equation 35-5 gives

$$\theta_c = \sin^{-1}\left(\frac{1}{1.333}\right) = 48.6°,$$

where we found water's refractive index in Table 35-1. The geometry of Fig. 35-15 then shows that this is also the half-angle of the cone, so the whale must look within 48.6° of the vertical to see out.

Can the whale see a plane only if it's *actually* within this angle? No. As the strongly refracted ray from the right-hand plane suggests, the entire outside world appears to the whale compressed into a cone of half-angle θ_c. If the whale looks beyond this cone, it will see instead reflections of objects below the surface.

EXERCISE A diamond is submerged in water. What is the critical angle at the diamond-water interface?

Answer: 33.4°

Some problems similar to Example 35-5: 22, 23, 26, 28, 33

FIGURE 35-15 The whale sees the entire world above the surface in a cone of half-angle θ_c. Looking beyond this cone, it sees reflections of objects below the surface.

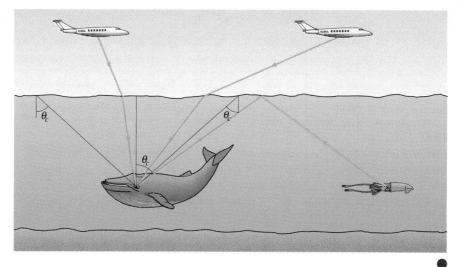

■ APPLICATION OPTICAL FIBERS AND LIGHTWAVE COMMUNICATION

Refraction and total internal reflection are the basis for **optical fibers,** which revolutionized communications in the 1980s. Some three *billion* km of optical fiber were installed in the United States alone during that decade, and today they carry telephone conversations, television signals, computer data, and other information. Undersea fibers link the entire planet in a fiber-optic network.

A typical fiber consists of a glass core only 8 μm in diameter, surrounded by a so-called cladding consisting of glass with a lower refractive index than the core. The propagation of light in the fiber is a process whereby the core-cladding interface guides the light along the fiber. In some fibers this takes place

by abrupt internal reflection at the interface (Fig. 35-16), while in others gradual refraction in the cladding guides the light. The glass used in optical fibers is so pure that a 1-km thick slab would appear as transparent as an ordinary window pane. Today's fibers use infrared light at 1.31 μm wavelength, although the anticipated development of 1.55-μm systems should result in lower losses. And fibers are under study that actually regenerate signals as they travel, extracting energy from an additional light beam.

Optical fibers offer several advantages over copper wire or open-air transmission of electromagnetic waves. The main advantage is their very high **bandwidth**—the rate at which they

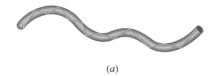

(a)

(b)

FIGURE 35-16 (a) A beam of light undergoes a series of total internal reflections that guide it along an optical fiber. (b) A bundle of actual fibers.

FIGURE 35-17 The thin fiber-optic cable and the much bulkier cable of copper wires carry information at the same rate.

can carry information. This bandwidth arises because the information carried on fiber is encoded on infrared radiation with a frequency on the order of 10^{14} Hz—much higher than the microwave frequencies used in conventional communication systems. A single fiber, for example, can carry tens of thousands of telephone conversations (Fig. 35-17).

Fibers are lighter and more rugged than copper cables and are less easy to "tap" illicitly. Because they are insulators, optical fibers are also less susceptible to electrical "noise" and are therefore used to carry information in electrically noisy environments like power plants or high-energy physics laboratories.

Optical fibers play a key role in the network that today links nearly all the world's computers. Computers on one floor or within a building may be connected in a small, relatively low-bandwidth network with copper wire. But connections between buildings are nearly always made with optical fibers, linking individual local networks into larger institution-wide networks whose optical links can transfer large amounts of information with reasonable speed. Entire institutions—corporations, universities, government laboratories, and the like—are then linked by very high speed fiber-optic "superhighways" that carry enormous loads of information "traffic."

Optical fiber technology requires more than just fiber: appropriate opto-electronic devices are needed at each end to convert information from light to electrical forms and vice versa. Such devices include miniature lasers made from semiconductors, and high-speed light-sensitive diodes. As the cost of these devices drops, high-speed optical fiber will soon connect to individual homes and offices, allowing nearly instant access to telecommunications, video resources, libraries, and databases.

35-5 DISPERSION

Refraction ultimately occurs because of the interaction of electromagnetic wave fields with atomic electrons. Although the details of that interaction require a quantum description, it should not be surprising that the index of refraction depends on frequency. After all, as we found in Chapter 15, the response of a system (here an atom) generally depends on the frequency of the driving force (here the oscillating fields of a wave).

Because the index of refraction depends on frequency, different frequencies—different colors for visible light—will be refracted through different angles. A light beam containing many colors will therefore be spread into its constituent colors, a process called **dispersion** (Fig. 35-18). The classic example of dispersion is Newton's demonstration that white light is really a mixture of all the colors in the visible spectrum. Newton not only broke white light into its

FIGURE 35-18 Dispersion of white light by a glass prism. The refractive index of the glass is greater at higher frequencies—shorter wavelengths—and therefore results in greater refraction at the blue end of the spectrum. The incident beam is at bottom left, and a white reflected beam leaves the prism going straight downward. This beam has undergone an additional refraction that has again combined its colors.

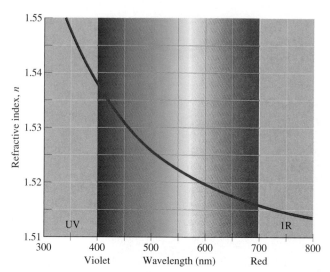

FIGURE 35-19 Index of refraction as a function of wavelength for high-dispersion crown glass.

constituent colors, as in Fig. 35-18, but he also recombined the colors to produce the original white light.

In most materials in most frequency ranges, the index of refraction increases with increasing frequency and therefore decreases with increasing wavelength (Fig. 35-19). That means colors toward the violet end of the spectrum are refracted through the greatest angles, as is evident in Fig. 35-18.

Dispersion is the basis for the widely used technique of **spectral analysis** or **spectroscopy,** in which substances or processes are characterized by analyzing the frequencies of electromagnetic radiation they emit, transmit, or absorb. Hot, dense objects, for example, emit a continuous band of radiation, while diffuse gases radiate at only a few specific wavelengths (Fig. 35-20). The existence of those discrete wavelengths provided some of the strongest evidence for the nature of the atom, and today spectral analysis allows astronomers to identify

FIGURE 35-20 (Top) The solar spectrum is an essentially continuous band of wavelengths, produced by the hot, dense gases of the Sun's visible surface. Dark lines are discrete wavelengths absorbed by overlying gases. (Bottom) Spectrum of a diffuse gas—in this case hydrogen—consists of light at discrete wavelengths. The pattern of lines allows identification of the emitting material.

and measure the abundances of elements in distant astrophysical objects. Geologists use spectral analysis to identify minerals, and chemists use infrared spectra to study molecules. Spectral analysis is a powerful tool in nearly every branch of science. Although early spectroscopes used prisms, most modern instruments use instead a device called a diffraction grating, whose operation we describe in Chapter 37.

Dispersion can be a nuisance in optical systems. Glass lenses, for example, focus different colors at different points, resulting in a distortion known as chromatic aberration. This effect can be minimized by making composite lenses of materials with different refractive indices.

● EXAMPLE 35-6   HOW MUCH DISPERSION?

White light strikes the prism in Fig. 35-21 normal to one surface. The prism is made of the glass whose refractive index is plotted in Fig. 35-19. Find the angle between outgoing red (700 nm) and violet (400 nm) light.

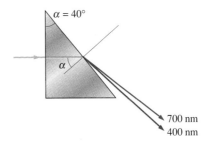

FIGURE 35-21   Example 35-6.

Solution
There is no refraction at the air-glass interface since the incident ray is normal to the surface. At the second interface medium 2 is air with $n_2 = 1$, and medium 1 is the glass whose indices at the two wavelengths we can read from Fig. 35-19: $n_{400} = 1.538, n_{700} = 1.516$. The geometry of Fig. 35-21 shows that the incidence angle at the second interface is the angle $\alpha = 40°$ at the top of the prism. The angles of refraction are then given by solving Snell's law for the two values of θ_2:

$$\theta_{400} = \sin^{-1}(n_{400} \sin \alpha) = \sin^{-1}[(1.538)(\sin 40°)] = 81.34°,$$

and

$$\theta_{700} = \sin^{-1}(n_{700} \sin \alpha) = \sin^{-1}[(1.516)(\sin 40°)] = 77.02°.$$

The angle between the two outgoing beams is therefore $\theta_{400} - \theta_{700} = 4.32°$, with the violet beam (400 nm) experiencing the greatest deflection. This 4.32° spread is called the *angular dispersion* of the beam.

EXERCISE   If the angle α in Fig. 35-21 is increased, what is the maximum angular dispersion that can be achieved by refracting both the 400 nm and 700 nm beams? At what α does this occur?

Answer: $\Delta\theta = 9.7°$ at $\alpha = \sin^{-1}(1/n_{400}) = 40.556°$

Some problems similar to Example 35-6:   34–38 ●

■ APPLICATION   THE RAINBOW

Nature provides a beautiful application of dispersion and internal reflection in the rainbow (Fig. 35-22), which occurs when sunlight strikes rain or other water droplets in the air. An observer standing between the Sun and the rain then sees the circular arc of colored bands. Figure 35-23 shows that the center of that arc lies on the line joining the Sun to the observer's eye. That means each observer sees a different rainbow! Furthermore, the rainbow's arc always subtends an angle of approximately 42°. How does the rainbow form, and why does it have this geometry?

Theories of the rainbow date back many centuries. By 1635 Descartes had produced a nearly complete explanation of the

FIGURE 35-22   Reflection, refraction, and dispersion all act to produce a rainbow.

rainbow's shape and apparent location, but because he did not know about dispersion he could not account for the colors. Some years later Newton, in his *Optics,* produced a full explanation.

Figure 35-24 shows a light ray passing through a spherical raindrop. The incidence angle θ in Fig. 35-24 is arbitrary, and parallel light rays striking the curved surface of the drop will experience a range of values for θ. There will, therefore, be a range of angles ϕ between the incident and outgoing rays. As Fig. 35-25 shows, however, there is a maximum angle* ϕ_{max} of about 42° and more light is returned at angles close to ϕ_{max} than at other angles. That is why a bright band—the rainbow—appears in the sky at an angle of about 42° to the direction of the Sun's rays. Problems 64 and 65 show how to find ϕ_{max}.

The "bunching" of light rays near ϕ_{max} explains why a bright band should appear, but why should it be colored? Because the refractive index varies with wavelength, so does ϕ_{max}. Each color will therefore appear brightest at a slightly different angle. For water, the refractive index in the visible region ranges from $n_{red} = 1.330$ to $n_{violet} = 1.342$. Using these values with the results of Problems 64 and 65 yields $\phi_{red} = 42.53°$ and $\phi_{violet} = 40.78°$. Thus the rainbow is seen as a band, itself subtending an angle of about 1.75°, with red at the top.

One occasionally sees a fainter and larger arc above the primary rainbow. This secondary rainbow results from two internal reflections, and as a result the order of its colors is reversed. Problem 66 explores the secondary rainbow.

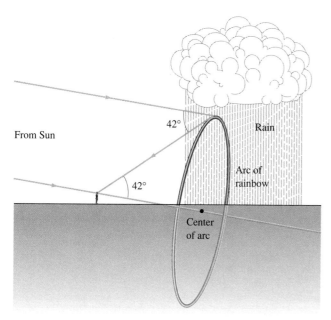

FIGURE 35-23 The rainbow is a circular arc located at 42° from the line that includes the Sun, the observer, and the center of the arc. The Sun is so far away that its rays are essentially parallel. Here part of the rainbow is blocked by the ground; however, observers in aircraft can sometimes see the entire circle.

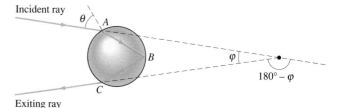

FIGURE 35-24 A light ray passing through a spherical raindrop undergoes refraction at each interface and internal reflection at the back of the drop. (Not shown are the reflected rays at interfaces A and C and the transmitted ray at interface B.) The ray undergoes an overall deflection of $180° - \phi$.

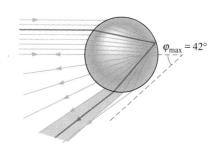

FIGURE 35-25 Parallel rays striking a water drop. The minimum angle through which a ray is deflected is about 138°, corresponding to a maximum of about 42° in the angle ϕ of Fig. 35-24. Furthermore, rays tend to "bunch" at this angle, as suggested by the broad outgoing beam defined by three nearly parallel rays. This "bunching" causes an observer to see a bright arc at 42° from the antisolar direction in Fig. 35-23. The thicker ray undergoes the minimum deviation and has, correspondingly, the maximum ϕ.

---

* This angle actually corresponds to a *minimum* deflection in which the ray is turned through an angle of $180° - 42° = 138°$. Other rays suffer more deflection; one hitting normal to the drop surface, for example, is turned through 180°, corresponding to $\phi = 0$.

35-6 REFLECTION COEFFICIENTS AND THE POLARIZING ANGLE

The laws of geometrical optics describe the paths of light rays at interfaces between different media, but they cannot tell how much of the light is reflected and how much transmitted at an interface. But application of Ampère's and Faraday's laws to electromagnetic wave fields at an interface does give that information. Such a calculation is beyond the scope of this book; here we look briefly at the results.

Normal Incidence

When light is incident normally on an interface, application of Ampère's and Faraday's laws yields a fairly simple result:

$$R = \left(\frac{n_2 - n_1}{n_2 + n_1}\right)^2, \tag{35-6}$$

where n_1 and n_2 are the refractive indices of the two media, and the **reflection coefficient** R is the fraction of the incident intensity that is reflected back into the medium of incidence. Since the difference $n_2 - n_1$ appears squared in Equation 35-6, the fraction of light reflected is the same for light crossing the interface in either direction. Energy conservation requires that light not reflected must be transmitted into the second medium, so the **transmission coefficient** is therefore

$$T = 1 - R = \frac{4n_1 n_2}{(n_1 + n_2)^2}. \tag{35-7}$$

Note that these results give $R = 0$ and $T = 1$ for $n_1 = n_2$, as we would expect for this case when there is really no interface. Equations 35-6 and 35-7 are closely related to Equations 17-7 and 17-8 for the reflection of waves on strings, although there we deal with amplitudes and here with intensities.

● **EXAMPLE 35-7** A TRIPLE-GLAZED WINDOW

In an effort to cut heat loss, a homeowner replaces single-glazed windows with triple-glazed windows. Both use the same type of glass, whose refractive index is 1.5. Unfortunately the new windows not only conserve heat, but they also let in less light. Find the reduction in intensity for light incident normally on the window as a result of the change from single to triple glazing.

Solution

Equation 35-7 shows that the fraction of light transmitted at a single air-glass or glass-air interface is

$$T = \frac{4n_1 n_2}{(n_1 + n_2)^2} = \frac{(4)(1)(1.5)}{(1 + 1.5)^2} = 0.96.$$

A single sheet of glass has two such interfaces, each cutting the intensity by this factor 0.96. So the transmitted intensity is down by a factor $0.96^2 = 0.922$ from its incident value. A triple-glazed window, on the other hand, has a total of six interfaces, cutting the intensity to $0.96^6 = 0.783$ of its incident value. Comparing the two results gives an intensity ratio of $0.783/0.992 = 0.79$ for the triple as compared with the single glazing.

Our calculation here is approximate for a subtle reason. Both reflection and transmission occur at each interface. Of the light entering a slab of glass, some is reflected back into the glass at the exit side, and of that a small fraction is reflected again at the first side, and most of that eventually joins the first exiting beam. "Higher order" effects—associated with more reflections—may also affect the outgoing beam. In our window example these effects are small. But by carefully matching the thicknesses of transparent slabs to appropriate multiples of the light wavelength, it is possible to use wave interference to enhance or diminish the transmitted or reflected intensity. Anti-reflection coatings on lenses use this effect, which we will study in Chapter 37.

EXERCISE A light beam propagating through air is incident normally on a transparent material, and 17.2% of the incident intensity is reflected. Find the refractive index and use Table 35-1 to identify the material.

Answer: 2.42, diamond

Some problems similar to Example 35-7: 40–42

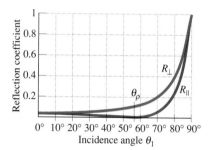

FIGURE 35-26 Reflection coefficients as a function of incidence angle for the two cases of light polarized with its electric field parallel and perpendicular to the plane defined by the incident and reflected rays, for an interface between air and glass with $n = 1.5$. At 56° incidence the parallel polarization is entirely transmitted ($R_\parallel = 0$), so an unpolarized incident beam will become polarized on reflection.

Oblique Incidence

It should be no surprise that the reflection and transmission coefficients for light incident on an interface at an oblique angle depend on that angle. Less obvious is the fact that they also depend on the polarization of the light. The equations for these coefficients are quite complicated, and we give only their graphical representation (Fig. 35-26). The graph shows two curves for the reflection coefficients $R_\perp$ and $R_\parallel$ at an air-glass interface. These correspond to polarizations with the wave electric field perpendicular and parallel, respectively, to the plane of Fig. 35-6. Light of arbitrary polarization can be expressed as a combination of these two.

At normal incidence ($\theta_1 = 0$), both curves start from the value $R = 0.04$ implied in our calculation giving $T = 0.96$ in Example 35-7. At large incidence angles both coefficients approach 1. You've undoubtedly experienced this effect when standing in front of a large plate-glass window; as you look along the window at an oblique angle, the reflection of objects on your side of the window becomes more prominent (Fig. 35-27).

The Polarizing Angle

Figure 35-26 shows that light with the perpendicular polarization reflects more strongly than does light with the parallel polarization, except at normal and grazing ($\theta_1 = 90°$) incidence. That means unpolarized light reflecting from the surface will come off partially polarized. At one particular angle, marked θ_p in Fig. 35-26, the parallel polarization does not reflect at all. Unpolarized light incident at this special angle—called the **polarizing angle** or the **Brewster angle**—comes off entirely polarized. Orienting a piece of glass or other transparent material at its polarizing angle therefore offers one way of producing light polarized perpendicular to the plane of the incident and reflected rays. Alternatively, successive transmissions at the polarizing angle can enhance the parallel component of polarization.

The reason there is no reflection for the parallel polarization at a particular angle follows from the fact that electromagnetic radiation ultimately arises from accelerated charges and that, as we saw in Fig. 34-23, there is no radiation in the direction of the acceleration. Figure 35-28 shows what happens when light with parallel polarization is incident at the polarizing angle, 56.3° in glass with $n = 1.5$. Snell's law gives an angle of refraction $\theta_2 = \sin^{-1}(\sin\theta_p/n) = 33.7°$. Now $56.3° + 33.7° = 90°$ and that means, as Fig. 35-28 shows, that any reflected ray would be perpendicular to the refracted ray. But the reflected ray really arises from the acceleration of the electrons in the glass—acceleration

FIGURE 35-27 Reflection in a plate-glass window is especially prominent at oblique incidence angles.

which, according to Newton's law of motion, is in the direction of the driving force. But here the driving force comes from the wave's electric field, and with polarization in the plane of Fig. 35-28, that field is itself parallel to the direction the reflected ray would take. But we've seen that there is no radiation in the direction of the acceleration, and hence for this special geometrical condition there is no reflected ray.

Figure 35-28 shows that, quite generally, this condition of no reflection of the parallel polarization will be met when the angles of incidence and refraction sum to 90°: $\theta_p + \theta_2 = 90°$, where we've set the incidence angle equal to the polarizing angle. This, in turn, implies that $\sin \theta_2 = \cos \theta_p$. But with the incidence angle $\theta_1 = \theta_p$, Snell's law gives $\sin \theta_2 = (n_1/n_2)\sin\theta_p$. Applying our condition $\sin \theta_2 = \cos \theta_p$, we then have $\cos \theta_p = (n_1/n_2)\sin \theta_p$. Multiplying both sides by $n_2/n_1 \cos \theta_p$ then gives

$$\tan \theta_p = \frac{n_2}{n_1}. \tag{35-8}$$

For the air-glass interface whose reflection coefficients are plotted in Fig. 35-26, with $n_1 = 1$ and $n_2 = 1.5$, this equation indeed gives $\theta_p = 56.3°$.

A phenomenon similar to the polarizing angle holds for reflection from metals and other surfaces, although here the parallel polarization is minimized but not completely eliminated. Reduction of glare using Polaroid sunglasses works by eliminating the polarized light reflected off surfaces (Fig. 35-29).

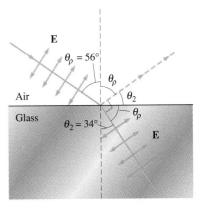

FIGURE 35-28 An electromagnetic wave polarized in the plane of the figure is incident on an interface. When the angles of incidence and refraction sum to 90°, the reflected ray would be in the direction of the electric field in the lower medium. Since there is no radiation in the direction of charged-particle acceleration, there is no reflected ray in this special case. Numbers shown are for an air-glass interface.

(a)

(b)

FIGURE 35-29 Identical views of a storefront, photographed (a) without and (b) with a polarizing filter. Light reflecting obliquely from the glass is partially polarized. The polarizing filter eliminates this component, allowing the inside of the store to show much more clearly.

CHAPTER SYNOPSIS

Summary

1. **Geometrical optics** treats the behavior of light under the approximation that it travels in straight lines—**rays**—except at interfaces between different materials. This approximation is valid whenever waves interact with systems that are much larger than their wavelength.

2. **Reflection** occurs to some extent at nearly all interfaces. When a light ray reflects, the **angle of incidence** and **angle of reflection**—both measured from the normal to the interface—are equal. **Specular reflection** occurs from smooth surfaces, while rough surfaces produce **diffuse reflection.**

3. **Refraction** is the bending of light at an interface between transparent materials. The directions of the incident and refracted rays are related by **Snell's law:**

$$n_1 \sin\theta_1 = n_2 \sin\theta_2,$$

where θ_1 and θ_2 are the angles of incidence and refraction, respectively. The **indices of refraction** are n_1 and n_2, defined as the ratios of the speed of light in vacuum to the speed in a given medium.

4. **Total internal reflection** occurs when light propagating in a medium with refractive index n_1 is incident on an interface at an angle greater than the **critical angle** θ_c given by

$$\sin\theta_c = \frac{n_2}{n_1}.$$

Total internal reflection is possible only when $n_2 < n_1$.

5. The index of refraction generally depends on frequency. This results in **dispersion** of the different colors as they refract through different angles. Dispersion is the basis of **spectroscopy.**

6. For normal incidence, the fraction of light intensity reflected at an interface is given by the **reflection coefficient:**

$$R = \left(\frac{n_2 - n_1}{n_2 + n_1}\right)^2.$$

The **transmission coefficient** is $T = 1 - R$. For oblique incidence the reflection and transmission coefficients depend on both the incidence angle and the polarization. When the angle of incidence is equal to the **polarizing angle** (also called the **Brewster angle**) there is no reflection of light polarized in the plane of the incident and refracted rays.

Terms You Should Understand

(Pairs are closely related terms whose distinction is important; number in parentheses is chapter section where term first appears.)

geometrical optics (introduction)
ray (introduction)
angles of incidence, reflection, refraction (35-2, 35-3)
specular reflection, diffuse reflection (35-2)
refraction (35-3)
index of refraction (35-3)
Snell's law (35-3)
total internal reflection (35-4)
critical angle (35-4)
optical fiber (35-4)
dispersion (35-5)
spectroscopy (35-5)
reflection coefficient, transmission coefficient (35-6)
polarizing angle (35-6)

Symbols You Should Recognize

θ_1, θ_1', θ_2 (35-2, 35-3)
n (35-3)
θ_c (35-4)
θ_p (35-6)

Problems You Should Be Able to Solve

determining the directions of reflected and refracted rays at interfaces (35-2 and 35-3)
evaluating conditions for total internal reflection (35-4)
analyzing dispersion (35-5)
determining reflected and transmitted intensities at normal incidence (35-6)
evaluating the polarizing angle (35-6)

Limitations to Keep in Mind

Geometrical optics is an approximation valid only when light interacts with systems much larger than its wavelength.
Characterizing the optical properties of a material by a single number, the index of refraction, is a simplification. In many materials optical properties depend on direction, and include absorption as well as transmission.

QUESTIONS

1. Are light rays real? Discuss.
2. It's usually inappropriate to consider low-frequency sound waves as traveling in rays. Why? Why is it more appropriate for high-frequency sound and for light?
3. Describe why a spoon appears bent when it's in a glass of water.
4. Why do a diamond and an identically shaped piece of glass sparkle differently?
5. Specular reflection occurs with "smooth" surfaces. But on the microscopic scale all surfaces are rough. What do you suppose should be the criterion for "smoothness" in dealing with reflection?
6. White light goes from air through a glass slab with parallel surfaces. Will its colors be dispersed when it emerges from the glass?
7. Would light behave as in Fig. 35-14a if the refractive index of the prism were 1.25? Explain.
8. You send white light through two identical glass prisms, oriented as shown in Fig. 35-30. Describe the beam that emerges from the right-hand prism.

White light

FIGURE 35-30   Question 8.

9. Why can optical fibers carry much more information than copper wires?
10. What would happen if you scratched the outside of the reflecting surface of a prism used for total internal reflection?
11. Lightning produces a sudden burst of static in a nearby radio receiver. The static comprises a broad band of radio frequencies. But in a very distant receiver the "noise" from the flash arrives over an extended time. What does this say about the refractive index of Earth's atmosphere for radio waves?
12. What does a fish see as it looks around in directions above the horizontal? Explain.
13. In glass, which end of the visible spectrum has the lowest critical angle for total internal reflection?
14. Looking out the window of a lighted room at night, you see clear reflections of the room's interior. In the daytime those reflections would be much less obvious, yet the reflection coefficient has not changed. Explain.
15. Why can't you walk to the end of the rainbow?
16. What is wrong with the painting in Fig. 35-31? *Hint:* The rainbow subtends a half-angle of 42°.

FIGURE 35-31   (What's wrong with this painting (*Niapra*, by Harry Fenn)? (Question 16.)

17. Suppose the refractive index of water were not frequency dependent. Would anything like the rainbow occur?
18. Figure 35-32 shows a ball inside a transparent sphere inside an aquarium tank. The transparent sphere and the tank can each be filled with water. Explain which combination is occurring in each frame.

FIGURE 35-32   Question 18.

19. Why are polarizing sunglasses better than glasses that simply cut down on the total amount of light?
20. Does the transmitted intensity always exceed the reflected intensity at an air-glass interface?
21. Under what conditions will the polarizing angle be less than 45°?

PROBLEMS

Section 35-2 Reflection

1. Through what angle should you rotate a mirror in order that a reflected ray rotate through 30°?
2. The mirrors in Fig. 35-33 make a 60° angle. A light ray enters parallel to the symmetry axis, as shown. (a) How many reflections does it make? (b) Where and in what direction does it exit the mirror system?

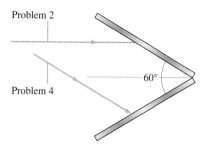

FIGURE 35-33   Problems 2, 4, 5.

3. To what angular accuracy must two ostensibly perpendicular mirrors be aligned in order that an incident ray returns on a path within 1° of its incident direction?
4. If a light ray enters the mirror system of Fig. 35-33 propagating in the plane of the page and parallel to one mirror, through what angle will it be turned?
5. Suppose the angle in Fig. 35-33 is changed to 75°. A ray enters the mirror system parallel to the axis. (a) How many reflections does it make? (b) Through what angle is it turned when it exits the system?
6. Two plane mirrors make an angle ϕ. For what value of ϕ will a light ray reflecting once off each mirror be turned through the same angle ϕ?
7. Two plane mirrors make an angle ϕ. A light ray enters the system and is reflected once off each mirror. Show that the ray is turned through an angle $360° - 2\phi$.

Section 35-3 Refraction

8. In which substance in Table 35-1 does the speed of light have the value 2.292×10^8 m/s?
9. Information in a compact disc is stored in "pits" whose depth is essentially one-fourth of the wavelength of the laser light used to "read" the information. That wavelength is 780 nm in air, but the wavelength on which the pit depth is based is measured in the $n = 1.55$ plastic that makes up most of the disc. Find the pit depth.
10. Light is incident on an air-glass interface, and the refracted light in the glass makes a 40° angle with the normal to the interface. The glass has refractive index 1.52. Find the incidence angle.
11. A light ray propagates in a transparent material at 15° to the normal to the surface. When it emerges into the sur-

rounding air, it makes a 24° angle with the normal. What is the refractive index of the material?
12. Light propagating in the glass ($n = 1.52$) wall of an aquarium tank strikes the interior edge of the wall with incidence angle 12.4°. What is the angle of refraction in the water?
13. A block of glass with $n = 1.52$ is submerged in one of the liquids listed in Table 35-1. For a ray striking the glass with incidence angle 31.5°, the angle of refraction is 27.9°. What is the liquid?
14. A meter stick lies on the bottom of the rectangular trough in Fig. 35-34, with its zero mark at the left edge of the trough. You look into the long dimension of the trough at a 45° angle, with your line of sight just grazing the top edge of the tank, as shown. What mark on the meter stick do you see if the trough is (a) empty, (b) half full of water, and (c) full of water?

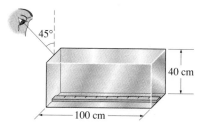

FIGURE 35-34   Problem 14.

15. You look at the center of one face of a solid cube of glass, on a line of sight making a 55° angle with the normal to the cube face. What is the minimum refractive index of the glass for which you will see through the opposite face of the cube?
16. The cylindrical tank in a public aquarium is 10 m deep, 11 m in diameter, and is full to the brim with water. If a flashlight shines on the tank from above, what is the minimum angle its beam can make with the horizontal if it is to illuminate part of the tank bottom?
17. You're standing 2.3 m horizontally from the edge of a 4.5-m-deep lake, with your eyes 1.7 m above the water surface. A diver holding a flashlight at the lake bottom shines the light so you can see it. If the light in the water makes a 42° angle with the vertical, at what horizontal distance is the diver from the edge of the lake?
18. You've dropped your car keys at night off the end of a dock into water 1.6 m deep. A flashlight held directly above the dock edge and 0.50 m above the water illuminates the keys when it's pointed at 40° to the vertical, as shown in Fig. 35-35. What is the horizontal distance x from the edge of the dock to the keys?
19. A light ray is propagating in a crystal where its wavelength is 540 nm. It strikes the interior surface of the crystal with an incidence angle of 34° and emerges into the surrounding

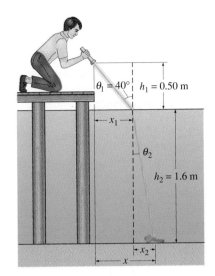

FIGURE 35-35 What is the horizontal distance x from the edge of the dock to the lost keys? (Problem 18)

air at 76° to the surface normal. Find (a) the light's frequency and (b) its wavelength in air.

20. The prism in Fig. 35-36 has $n = 1.52$, $\alpha = 60°$, and is surrounded by air. A light beam is incident at $\theta_1 = 37°$. Find the angle δ through which the beam is deflected.

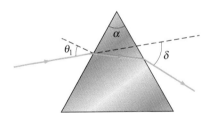

FIGURE 35-36 Problems 20, 57, 61.

Section 35-4 Total Internal Reflection

21. Find the critical angle for total internal refraction in (a) ice, (b) polystyrene, and (c) rutile. Assume the surrounding medium is air.

22. A drop of water is trapped in a block of ice. What is the critical angle for total internal reflection at the water-ice interface?

23. What is the critical angle for light propagating in crown glass when the glass is immersed in (a) water, (b) benzene, and (c) diiodomethane?

24. Total internal reflection occurs at an interface between a plastic and air at incidence angles greater than 37°. What is the refractive index of the plastic?

25. Light propagating in a medium with refractive index n_1 encounters a parallel-sided slab with index n_2. On the other side is a third medium with index $n_3 < n_1$. Show that the condition for avoiding internal reflection at *both* interfaces

is that the incidence angle at the n_1-n_2 interface be less than the critical angle for an n_1-n_3 interface. In other words, the index of the intermediate material doesn't matter.

26. An aquarium measures 30 cm front to back, as shown in Fig. 35-37. It is made of glass with thickness much less than the size of the aquarium and is full of water with $n = 1.333$. You put your eye right up to the center of the aquarium's front wall and can still see the entire back wall. What is the maximum value of the aquarium's width w? *Hint:* You can ignore the glass; see the preceding problem.

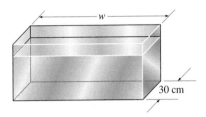

FIGURE 35-37 Problem 26.

27. What is the minimum refractive index for which total internal reflection will occur as shown in Fig. 35-14a? Assume the surrounding medium is air and that the prism is an isosceles right triangle.

28. Where and in what direction would the main beam emerge if the prism in Fig. 35-14a were made of ice?

29. What is the speed of light in a material for which the critical angle at an interface with air is 61°?

30. The prism of Fig. 35-14a has $n = 1.52$. When it is immersed in a liquid, a beam incident as shown in the figure ceases to undergo total reflection. What is the minimum value for the liquid's refractive index?

31. A compound lens is made from crown glass ($n = 1.52$) bonded to flint glass ($n = 1.89$). What is the critical angle for light incident on the flint-crown interface?

32. Find a simple expression for the speed of light in a material in terms of the critical angle at an interface between the material and vacuum.

33. A scuba diver sets off a camera flash a distance h below the surface of water with refractive index n. Show that light emerges from the water surface through a circle of diameter $2h/\sqrt{n^2 - 1}$.

Section 35-5 Dispersion

34. Laser beams with wavelengths 650 nm (red) and 410 nm (blue) strike an air-glass interface with incidence angle 50°. If the glass has refractive indices of 1.680 and 1.621 for the blue and red light, respectively, what will be the angle between the two beams in the glass?

35. Suppose the red and blue beams of the preceding problem are now propagating in the same direction *inside* the glass. For what range of incidence angles on the glass-air interface will one beam be totally reflected and the other not?

36. White light propagating in air is incident at 45° on the equilateral prism of Fig. 35-38. Find the angular dispersion γ of the outgoing beam, if the prism has refractive indices $n_{red} = 1.582$, $n_{violet} = 1.633$.

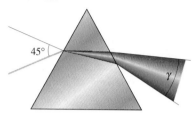

FIGURE 35-38 Problem 36 (angles of dispersed rays are not accurate.)

37. Two of the prominent spectral lines—discrete wavelengths of light—emitted by glowing hydrogen are hydrogen-α at 656.3 nm and hydrogen-β at 486.1 nm. Light from glowing hydrogen passes through a prism like that of Fig. 35-21, then falls on a screen 1.0 m from the prism. How far apart will these two spectral lines be? Use Fig. 35-19 for the refractive index.

38. Light from glowing sodium contains the two discrete wavelengths 589.0 nm and 589.6 nm. This light is passed through a prism like that of Fig. 35-21 and then allowed to fall on a screen 2.0 m distant. For wavelengths near 600 nm, the refractive index of the prism is $n = 1.546 - 4.47 \times 10^{-5}\lambda$, with λ the wavelength in nm. What must be the prism's apex angle α in order that the two sodium wavelengths be separated on the screen by 1.5 mm?

Section 35-6 Reflection Coefficients and the Polarizing Angle

39. Light is normally incident on ice. What fraction of the intensity is transmitted?

40. An aquarium's walls are made from crown glass with $n = 1.52$. What fraction of the intensity is reflected for light incident normally (a) from inside the water-filled aquarium and (b) from outside?

41. What is the refractive index of a material that transmits 92.4% of the light normally incident on it from air?

42. Light is incident normally on the outside of a glass-walled aquarium. The refractive indices of the glass and water are 1.52 and 1.333, respectively. What fraction of the incident intensity is transmitted into the water?

43. When a crystal is submerged in water ($n = 1.333$) the transmission coefficient for light incident normally on the crystal increases by 10% over its value in air. What is the refractive index of the crystal?

44. What would be the refractive index of a material for which normally incident light was half reflected, half transmitted? Assume the light is incident from air.

45. The reflection coefficient for normally incident light is the same when a block of plastic is submerged in water and in diiodomethane. What is the refractive index of the plastic?

46. Find the polarizing angle for diamond when light is incident from air.

47. What is the refractive index of a material for which the polarizing angle in air is 62°?

48. What is the polarizing angle for light incident from below on the surface of a pond?

Paired Problems

(Both problems in a pair involve the same principles and techniques. If you can get the first problem, you should be able to solve the second one.)

49. Light propagating in air strikes a transparent crystal at incidence angle 35°. If the angle of refraction is 22°, what is the speed of light in the crystal?

50. A laser beam with wavelength 633 nm is propagating in air when it strikes a transparent material at incidence angle 50°. If the angle of refraction is 27°, what is the wavelength in the material?

51. A cylindrical tank 2.4 m deep is full to the brim with water. Sunlight first hits part of the tank bottom when the rising Sun makes a 22° angle with the horizon. Find the tank's diameter.

52. For what diameter tank in the preceding problem will sunlight strike some part of the tank bottom whenever the Sun is above the horizon?

53. Light is incident from air on the flat wall of a polystyrene water tank. If the incidence angle is 40°, what angle does the light make with the tank normal in the water?

54. A parallel-sided slab with refractive index n_2 separates two media with indices n_1 and n_3. Show that a light ray incident on the n_1-n_2 interface then enters the third medium with the same angle of refraction it would have had if the slab had not been present. (Assume total internal reflection does not occur at either interface.)

55. Light strikes a right-angled glass prism ($n = 1.52$) in a direction parallel to the prism's base, as shown in Fig. 35-39. The point of incidence is high enough that the refracted ray hits the opposite sloping side. (a) Through which side of the prism does the beam emerge? (b) Through what angle has it been deflected?

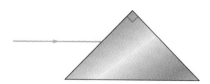

FIGURE 35-39 Problems 55, 56.

56. Repeat the preceding problem if the prism has refractive index 1.15.

57. Repeat Problem 20 for the case $n = 1.75$, $\alpha = 40°$, and $\theta_1 = 25°$.

58. The surfaces of a glass sheet are not quite parallel, but rather make an angle of 10°. The glass has refractive index $n = 1.52$. A light beam strikes one side of the sheet at incidence angle 35°, coming in on the thicker side of the normal. Through what angle is the beam direction changed when it emerges on the opposite side of the glass sheet?

Supplementary Problems

59. A cubical block is made from two equal-size slabs of materials with different refractive indices, as shown in Fig. 35-40. Find the index of the right-hand slab if a light ray is incident on the center of the left-hand slab and then describes the path shown.

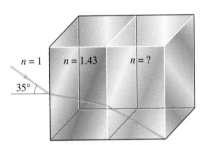

FIGURE 35-40 Problem 59.

60. Find an expression for the displacement x in Fig. 35-10, in terms of θ_1, d, and n.

61. Light is incident with incidence angle θ_1 on a prism with apex angle α and refractive index n, as shown in Fig. 35-36. Show that the angle δ through which the outgoing beam deviates from the incident beam is given by

$$\delta = \theta_1 - \alpha + \sin^{-1}\left\{n \sin\left[\alpha - \sin^{-1}\left(\frac{\sin\theta_1}{n}\right)\right]\right\}.$$

Assume the surrounding medium has $n = 1$.

62. Taking $n = 1.5$ and $\alpha = 60°$, plot the deviation δ of the preceding problem over the range $45° < \theta_1 < 50°$, and use your plot to find the incidence angle for minimum deviation. Trace the incident beam for this value of θ_1. Your result should be the symmetric path shown in Fig. 35-41; in fact, the minimum deviation always occurs with the incidence angle that gives this path, for any n and α.

63. Show that a three-dimensional corner reflector (three mirrors in three mutually perpendicular planes, or a solid cube in which total internal reflection occurs) turns an incident light ray through 180°, so it returns in the direction from which it came. *Hint:* Let $\mathbf{q} = q_x\hat{\mathbf{i}} + q_y\hat{\mathbf{j}} + q_z\hat{\mathbf{k}}$ be a vector in the direction of propagation. How does this vector get

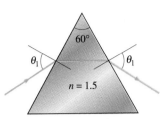

FIGURE 35-41 Minimum deviation through an equilateral prism occurs when the path is symmetric (Problem 62).

changed on reflection by a mirror in a plane defined by two of the coordinate axes?

64. Show that the angle ϕ that appears in Fig. 35-24 is given by $\phi = 4 \sin^{-1}\left(\frac{\sin\theta}{n}\right) - 2\theta$, where θ is the angle of incidence.

65. (a) Differentiate the result of the preceding problem to show that the maximum value of ϕ occurs when the incidence angle θ is given by $\cos^2\theta = \frac{1}{3}(n^2 - 1)$. (b) Use this result and that of the preceding problem to find ϕ_{max} in water with $n = 1.333$.

66. Figure 35-42 shows the approximate path of a light ray that undergoes internal reflection twice in a spherical water drop. Find the maximum angle ϕ for this case, taking $n = 1.333$. This is the angle at which the secondary rainbow appears.

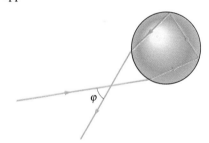

FIGURE 35-42 Problem 66.

67. *Fermat's principle* states that the path of a light ray between two points is such that the time to traverse that path is an extremum (either a minimum or a maximum) when compared with the times for nearby paths. Consider two points A and B on the same side of a reflecting surface, and show that a light ray traveling from A to B via a point on the reflecting surface will take the least time if its path obeys the law of reflection. Thus, the law of reflection (Equation 35-1) follows from Fermat's principle.

68. Use Fermat's principle (see preceding problem) to show that a light ray going from point A in one medium to point B in a second medium will take the least time if its path obeys Snell's law. Thus, Snell's law follows from Fermat's principle.

CHAPTER 36

IMAGE FORMATION AND OPTICAL INSTRUMENTS

This mirror directs sunlight into the McMath–Pierce Solar Telescope in Arizona, which uses optical technology to form a detailed image of the Sun's surface.

Reflection and refraction alter the direction of light propagation, following laws we developed in the preceding chapter. A wide variety of natural and technological systems use this effect to form images. In this chapter we study image formation using the approximation of geometrical optics—an approximation that remains valid as long as we consider only length scales much greater than the wavelength of light.

When we view an object directly, light comes to our eyes straight from the object. When we view an object with an optical system, our eyes perceive light that seems to come straight from the object but whose path has actually been altered. As a result we see an **image** that may be different in size, orientation, or apparent position from the actual object. In some cases light actually comes from the image to our eyes; the image is then called a **real image.** In other cases light only apparently comes from the image location; the image is then called a **virtual image.**

3 6 - 1   P L A N E   M I R R O R S

When you look at yourself in a flat mirror, you see an image that appears to be behind the mirror by the same distance that you are in front of it. The image is upright and the same size as you are, but appears reversed. Why?

Figure 36-1 shows how the image in a plane mirror comes about. In Fig. 36-1a we concentrate on a small object, in this case an arrowhead. (We'll frequently use arrows to represent objects in image-forming situations because they're both simple and sufficiently asymmetric that we can see whether images are inverted or upright.) We have drawn three light rays that leave the object, reflect off the mirror, and enter the observer's eye. The rays reflect at the mirror with equal angles of incidence and reflection. As Fig. 36-1a shows, light looks to the observer like it's coming from a point behind the mirror. That point is the location of the arrowhead's image. In this case the image is virtual because no light actually comes from behind the mirror.

Since two nonparallel lines define a point, we need only two rays to locate the arrowhead in Fig. 36-1a. We've repeated this image-location process in Fig. 36-1b, using as one of the rays the ray that reflects normally. The same procedure also locates the bottom of the arrow, and we could obviously fill in additional points to locate the entire arrow; the resulting image is shown in Fig. 36-1b.

Note that the triangles OPQ and $O'PQ$ in Fig. 36-1b share a common side and that the angles OPQ and $O'PQ$ are both right angles. And because the angles of incidence and reflection are equal, so are the angles OQP and $O'QP$. Therefore triangles OPQ and $O'PQ$ are congruent, showing that the distance PQ' that the arrowhead's image lies behind the mirror is equal to the distance OP from the actual arrowhead to the mirror. A similar analysis applies to the rays from the bottom of the arrow, so we can conclude that the image appears as far behind the mirror as the object is in front. Furthermore, since extensions of the rays from the top and bottom of the arrow normal to the mirror pass through the top and bottom of the image, respectively, the image must be the same height as the object itself.

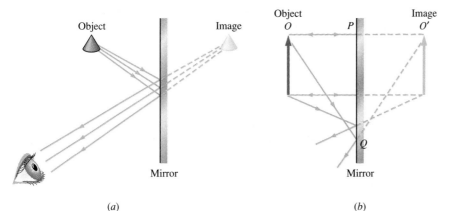

(a)

(b)

FIGURE 36-1 Image formation in a plane mirror. (a) Light reflected off the mirror seems to come from an image behind the mirror. Since there is really no light coming from behind the mirror, this is a virtual image. (b) Two rays from each point on an object serve to locate that point's image. Triangles OPQ and $O'PQ$ are congruent, showing that object and image are equidistant from the mirror.

● **EXAMPLE 36-1** WHAT SIZE MIRROR TO BUY?

You want the smallest mirror that will show your full image. How tall must it be?

Solution

Figure 36-2 shows the situation. Because the angles of incidence and reflection are equal, light from your foot reflects from the mirror at a point that's vertically halfway between your eye and the floor. Similarly, light from the top of your head reflects midway between that point and your eye. The total distance from top to bottom of the mirror then needs to be half your eye-foot distance plus half the distance from your eye to the top of your head—for a total of half your height. Note that this result does not depend on how far from the mirror you stand. It does, however, require that you fix the mirror to the wall at just the right height.

EXERCISE You buy a mirror that's half your height, but you affix it to the wall 2 cm below its optimum location. How much of your image will be cut off?

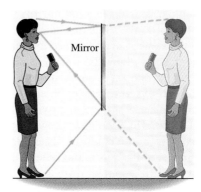

FIGURE 36-2 A mirror half your height shows your entire image.

Answer: Top 4 cm

Some problems similar to Example 36-1: 1–3 ●

Image Reversal

FIGURE 36-3 The image of a right hand is the image's left hand—but it's still the image of the *right* hand.

Figure 36-3 shows that the image in a plane mirror appears reversed left to right. So why isn't it inverted top to bottom as well?

The image does indeed appear reversed, but in the sense that its left hand is the image of the object's *right* hand. It's not that the image of the object's left hand appears opposite the right hand, any more than the image of the head appears opposite the feet. A more accurate description is that the mirror reverses front to back. Objects lying parallel to the mirror are not altered at all, but an object pointing perpendicular to the mirror is reversed. In Fig. 36-4 the effect is to alter only one of the three coordinate axes, and that alters handedness, rotation, and all other phenomena connected with the right-hand rules we've been using.

Although image formation in a single plane mirror is straightforward, multiple reflections with more than one mirror can produce many—in some cases infinitely many—images; see Fig. 36-5 and Problem 2.

36-2 CURVED MIRRORS

In contrast to plane mirrors, curved mirrors form images that may be upright or inverted, virtual or real, and whose sizes need not be those of the original objects.

Parabolic Mirrors

A parabola—the curve generated by a quadratic function like $y = x^2$—has an interesting geometrical property. For any point on the parabola, a line drawn parallel to the parabola's axis makes the same angle to the normal as does a

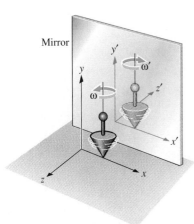

FIGURE 36-4 Actually, the mirror reverses front to back. Here the *x* and *y* axes of a normal right-handed coordinate system are unchanged in the mirror, but the image of the *z*-axis—which runs perpendicular to the mirror—is reversed. As a result the image is a left-handed coordinate system. The rotation of the spinning top, for example, is reversed in the mirror. But its angular velocity vector still points upward because a left-hand rule applies in the mirror world.

FIGURE 36-5 Multiple images formed by several plane mirrors.

second line drawn to a special point called the **focus** or **focal point** (Fig. 36-6*a*). That means a concave mirror of parabolic shape will reflect rays parallel to the parabola's axis so they converge at the focus (Fig. 36-6*b*). This effect can be used to concentrate light to very high intensities (Fig. 36-7*a*). Conversely, a point source of light at the focus will emerge from the mirror in a beam of parallel rays (Fig. 36-7*b*).

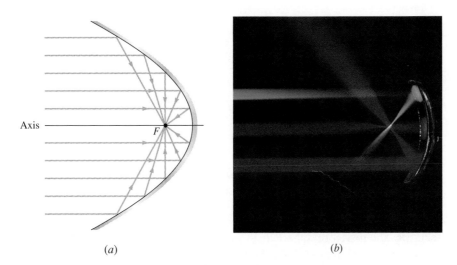

(*a*)   (*b*)

FIGURE 36-6 Drawing and photo show that a parabolic mirror reflects rays parallel to its axis to a common focus. (*a*) Ray diagram. (*b*) Photo shows rays reflecting from an actual mirror.

(a)

(b)

FIGURE 36-7 (a) This solar furnace uses a huge parabolic reflector to concentrate sunlight. (b) A flashlight has a light bulb near the focus of a parabolic mirror and thus produces a beam of nearly parallel light rays.

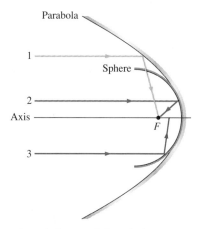

FIGURE 36-8 A sphere and parabola essentially coincide over a limited area. Rays 1 and 2 strike the parabolic mirror and reflect to the focus F, and 2 essentially reaches F from the spherical mirror as well. But ray 3, striking the spherical mirror farther from the axis, does not.

Spherical Concave Mirrors

It's much easier to make a spherical mirror than a parabolic one. Figure 36-8 shows that a parabola and a sphere essentially coincide over a limited range of the sphere's surface, and therefore over this range we can expect a sphere, too, to focus parallel light with a high degree of accuracy. Most mirrors used for focusing are therefore spherical in shape. Because a spherical mirror is not exactly parabolic, parallel rays do not converge at exactly the same point. The resulting distortion, called **spherical aberration,** is minimized by making the actual mirror only a tiny fraction of the entire sphere.

■ **APPLICATION** HUBBLE TROUBLE

The Hubble Space Telescope was launched in 1990 as the flagship of a new generation of space-based astronomical observatories. Although Hubble is smaller than many ground-based telescopes, its vantage point above the atmosphere was to make it optically superior in resolving astronomical objects. Furthermore, Hubble can observe in infrared, visible, and ultraviolet wavelengths.

In the months following Hubble's launch, engineers and scientists checking the telescope were frustrated by their inability to achieve a clear focus. They came reluctantly to the conclusion that Hubble's 2.4-m-diameter primary mirror was flawed. Subsequent investigation showed that an error of 1.3 mm in the placement of instruments used during manufacture of the mirror had resulted in its being ground to the wrong curvature. The mirror itself is off by only 2.3 μm at its edge, but this is an enormous flaw in an optical system designed to be accurate to a small fraction of the wavelength of light. Hubble specifications called for 70% of the light energy from a point source to fall within an angular diameter of 0.1 second of arc; because of its spherical aberration, at best 16% of the light fell in the prescribed zone.

Although the Hubble mirror could not be replaced or repaired, in 1993 astronauts managed to install corrective lenses that restored Hubble's optical system to better than its original design specifications. The result is a superb astronomical instrument with image quality greatly superior to that of ground-based telescopes (Fig. 36-9).

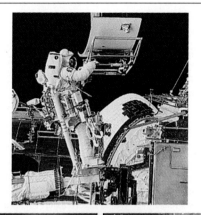

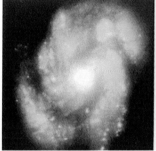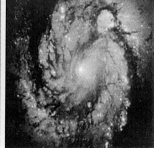

FIGURE 36-9 (Top) An astronaut installing new optical equipment during the 1993 repair of the Hubble Space Telescope. (Bottom) Hubble images of M-100, a galaxy 50 million light-years from Earth, taken before and after the repair.

We can see how spherical mirrors form images by tracing two rays from each of several points on the object, just as we did for plane mirrors. There are some special rays that make this process much simpler. First is any ray parallel to the axis; under the assumption that the spherical mirror approximates a parabola, we know that this ray reflects through the focal point *F*. A second useful ray lies along the line from the focus to the point on the object; under the parabolic approximation we know that this ray will emerge parallel to the axis. Two other useful rays are the ray that strikes the very center of the mirror, which reflects with equal angles on either side of the mirror axis, and the ray through the center of curvature, which, because it strikes normal to the surface of the spherical mirror, returns on itself. Any two of these rays suffice to locate the image.

Figure 36-10 shows the results of these ray tracings, using the two rays involving the focus, to find the image location in three cases. In all three cases symmetry ensures that the bottom of the image arrow will be on the axis, so we haven't bothered to trace it. In the more general case where the object isn't sitting on the axis we would also trace to find the bottom of the image arrow. In Fig. 36-10*a* we see that a distant object—beyond the mirror's center of curvature (*C*)—forms a smaller, inverted image. Light actually emerges from this image, so it's a *real* image. If you looked from the left in Fig. 36-10*a* you would actually see the image in space in front of the mirror (Fig. 36-11).

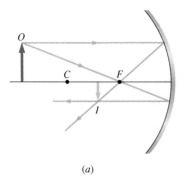

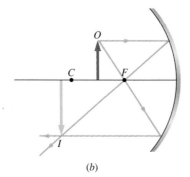

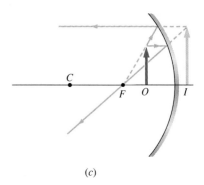

| (a) | (b) | (c) |

FIGURE 36-10 Image formation in a concave spherical mirror, found by exploiting the fact that light from the focus reflects paraxially from the mirror, and vice versa. *O* denotes the object and *I* its image. (*a*) With the object beyond the mirror's center of curvature *C*, the mirror forms a real image that is inverted and reduced in size. (*b*) With the object between the center of curvature and focus, the image is real, inverted, and magnified. (*c*) When the object is closer than the focus, the image is enlarged, upright, and virtual. Note, in this case, to get the ray including the object and the focus we had to extend the actual light path back toward the focus.

FIGURE 36-11 A bear meets its real image, formed by the concave mirror at rear. Note that the bear and its image are both in *front* of the mirror. From the fact that they're about the same size, what can you conclude about their position relative to the focal point?

As the object moves closer to the mirror the real image grows; with the object between the center of curvature and the focus, the image is larger than the object, and farther from the mirror (Fig. 36-10*b*). As the object moves toward the focus the image grows larger and moves rapidly away from the mirror. With the object right at the focus the rays emerge in a parallel beam and there is no image. Finally, rays from an object closer to the mirror than the focus diverge after reflection. To an observer they appear to come from a point behind the mirror. Thus there is a virtual image, in this case upright and enlarged (Fig. 36-10*c*).

Convex Mirrors

A convex mirror reflects on the outside of its spherical curvature, causing light to diverge rather than focus. Therefore, there is no possibility of forming a real image with a convex mirror. But Fig. 36-12 shows that the mirror can still form a virtual image. Although the focus has less obvious physical significance in this case, its location still controls the geometry of reflected rays. As Fig. 36-12 shows, we can still draw a ray parallel to the axis and another ray that would go through the focus if the mirror weren't in its way. By tracing these rays through the mirror we can see the directions in which they reflect. The reflected rays appear to diverge from a common point to the right, showing that there is a virtual image, upright and reduced in size, to the right of the mirror. You can convince yourself that for a convex mirror the image always has these characteristics. Convex mirrors are widely used where an image of a broad region needs to be captured in a small space (Fig. 36-13).

Curved and Plane Mirrors

You can understand curved mirrors qualitatively by thinking about bending a plane mirror. A plane mirror produces a virtual image of the same size as the object. If you bend it so it becomes slightly convex, the rays diverge more and the virtual image shrinks a little. If you bend it so it becomes slightly concave, the rays converge more and the virtual image grows. Figure 36-14 shows the effect of both these changes. Making the mirror concave also moves the focus inward from infinitely far away. If you bend it enough that the focus moves closer to the mirror than the object, then you're in a whole new realm where the mirror produces a real image, as in Figs. 36-10a and b.

The Mirror Equation

Drawing ray diagrams is a useful way to get an intuitive feel for image formation with curved mirrors. However, more precise image locations are obtained from the **mirror equation,** which we now derive.

Figure 36-15 shows a portion of a concave spherical mirror whose **focal length**—the distance from the mirror to the focal point—is f. As usual, we assume that the mirror is such a small portion of a sphere that it's essentially parabolic. Mathematically, that means the mirror's extent in the horizontal direction of Fig. 36-15 is much smaller than its vertical extent. As a result, we can approximate horizontal distances to points on the mirror as being measured to a vertical line through the mirror's apex, point A in Fig. 36-15. Under this approximation the base of the green triangle is *approximately* the focal length f. Using this triangle we can then write

$$\tan \alpha = \frac{h}{f},$$

with h the object height. Using the yellow triangle, we can equally well write

$$\tan \alpha = \frac{-h'}{\ell' - f},$$

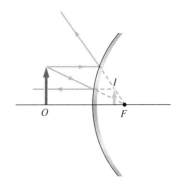

FIGURE 36-12 Image formation using a convex mirror. The image is always virtual, upright, and reduced in size.

FIGURE 36-13 Reflecting spheres produce reduced, upright, virtual images.

FIGURE 36-14 Funhouse mirrors are alternately convex and concave, producing stretched and shrunken images.

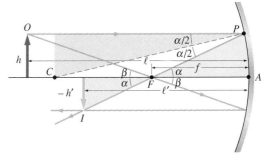

FIGURE 36-15 Ray diagram for derivation of the mirror equation. Horizontal distances are approximated as being measured to the horizontal position of the mirror apex A.

where ℓ' is the distance from the image to the mirror. We write $-h'$ so h' can be a *negative* number to indicate an *inverted* image. Equating our two expressions for $\tan\alpha$ and rearranging slightly gives

$$\frac{h'}{h} = -\frac{\ell'-f}{f} \qquad (36\text{-}1)$$

We can do a similar analysis for the two triangles with vertex angles β, giving $\tan\beta = -h'/f$ and $\tan\beta = h/(\ell-f)$, where ℓ is the distance from mirror to object. Equating these two expressions gives

$$\frac{h'}{h} = -\frac{f}{\ell-f}. \qquad (36\text{-}2)$$

Equating the right-hand sides of Equations 36-1 and 36-2 and carrying out some algebra then gives the mirror equation:

$$\frac{1}{\ell} + \frac{1}{\ell'} = \frac{1}{f}. \qquad \text{mirror equation} \qquad (36\text{-}3)$$

We define the ratio h'/h as the mirror's **magnification,** M. Given the focal length and the image or object distance, M can be found from Equation 36-1 or 36-2, respectively. Or, by solving Equation 36-3 for f, the magnification is minus the ratio of image to object distance:

$$M = \frac{h'}{h} = -\frac{\ell'}{\ell}. \qquad (36\text{-}4)$$

We can extract one more useful fact from Fig. 36-15 by considering a line from the point P where the parallel ray hits the mirror to point C, the mirror's center of curvature. Since it's a radius, this line is normal to the mirror, so the law of reflection therefore ensures that it bisects the angle between two rays at P, which is equal to the angle α. By considering the light gray triangle we can then write $\tan(\alpha/2) = h/R$, where R is the mirror's curvature radius. Now the angle α in Fig. 36-15 must be very small under our assumption that the mirror curves only slightly. We can therefore use the approximations $\tan\alpha \simeq \alpha$ and $\tan(\alpha/2) \simeq \alpha/2$. Rewriting the expressions $\tan\alpha = h/f$ and $\tan(\alpha/2) = h/R$ using this small-angle approximation then gives $h/R = h/2f$, or

$$f = \frac{R}{2}. \qquad (36\text{-}5)$$

Thus, the focal length of a spherical mirror is half its curvature radius.

We emphasize that the formulas derived in this section are approximations based on the assumption that the spherical mirror is a good approximation to a parabola. Equivalent ways of stating this same approximation are that the curvature radius and focal length of the mirror are large compared with the mirror's actual diameter, and that all rays reflecting from the mirror make only small angles with the mirror axis.

● EXAMPLE 36-2 OPTICAL ILLUSION

A popular "optical illusion" gadget consists of a tiny model car that seems to "float" in the top of a bowl-shaped depression (Fig. 36-16). The "floating" car is actually the real image of a solid model mounted just above a concave mirror at the bottom of the device. The mirror's focal length is 2.0 cm. If the image is 11 cm above the mirror and 3.2 cm long, where and what size is the actual model car?

Solution
We solve Equation 36-3 for the object distance:

$$\ell = \left(\frac{1}{f} - \frac{1}{\ell'}\right)^{-1} = \left(\frac{1}{2.0 \text{ cm}} - \frac{1}{11 \text{ cm}}\right)^{-1} = 2.44 \text{ cm}.$$

Then the object size follows from Equation 36-4:

$$h = -\frac{\ell h'}{\ell'} = -\frac{(2.44 \text{ cm})(-3.2 \text{ cm})}{11 \text{ cm}} = 0.71 \text{ cm},$$

where we use a negative value for h' because the real image is inverted. We could also have obtained this result from Equation 36-1.

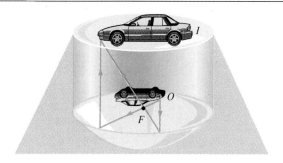

FIGURE 36-16 Optical illusion: A real image of the model car floats above the bottom of the bowl.

EXERCISE You stand 7.8 m from a concave spherical mirror, and, looking toward the mirror, you see an upside-down image of yourself with half your height. (a) How far apart are you and your image? (b) What is the mirror's curvature radius?

Answers: (a) 3.9 m; (b) 5.2 m

Some problems similar to Example 36-2: 4–8 ●

Although we derived the mirror equation in a situation involving a real image, the equation applies as well to virtual-image formation if we adopt the convention that a *negative* image distance puts the image *behind* the mirror—in the realm where we find virtual images. And we can handle *convex* mirrors as well if we take the focal length and therefore the curvature radius as *negative* quantities.

> **TIP** **Sign Conventions—Positive and Real, Negative and Virtual** The sign conventions for mirrors make sense: Objects and real images have positive distances. Images that aren't real (i.e., virtual images) are more absurd: they have negative distances. Mirrors that can make real images have positive focal lengths; mirrors that cannot, have negative focal lengths. On the other hand, real images may have negative height—but that's understandable because it simply means they're inverted.

● EXAMPLE 36-3 SIZING UP HUBBLE

Figure 36-17 shows a technician standing in front of the Hubble Space Telescope's primary mirror, whose focal length is 5.52 m. Use the photo to estimate the locations of the technician and his image relative to the mirror.

Solution
Direct measurement of the photo shows that the image of the technician's head appears about 3.3 times the size of his actual head. (This is an underestimate because the image is farther from the camera and thus appears smaller in relation to the technician than it really is.) The image is upright, so its height h' is positive. With $h'/h = 3.3$, Equation 36-4 then gives $\ell' = -3.3\ell$. The negative sign here indicates that the image is *behind* the mirror and is, therefore, virtual. Using this result in the mirror equation 36-3 gives

FIGURE 36-17 A technician standing in front of the Hubble Space Telescope mirror. How far are the technician and his image from the mirror (Example 36-3)?

$$\frac{1}{\ell} + \frac{1}{-3.3\ell} = \frac{1}{f}.$$

Solving for ℓ we have

$$\ell = f\left(1 - \frac{1}{3.3}\right) = 0.697f = (0.697)(5.52 \text{ m}) = 3.85 \text{ m}.$$

Thus, the technician is 3.85 m in front of the mirror, and his image is 3.3 times this, or about 13 m, behind the mirror.

Does this result make sense? The 3.8-m object distance is *less* than the focal length, as Fig. 36-10 showed it must be for the concave mirror to make a virtual image.

EXERCISE You're scrutinizing your nose using a hand-held concave vanity mirror with curvature radius 2.2 m. How far from your face should you hold the mirror to see your nose doubled in size?

Answer: 55 cm

Some problems similar to Example 36-3: 9–11 ●

■ APPLICATION NONIMAGING OPTICS AND THE QUEST FOR SOLAR ENERGY

Sometimes we want to concentrate light energy without necessarily forming an image. In solar energy systems producing heat, for example, concentrated sunlight leads to higher temperature and thus to higher efficiency; with photovoltaic systems inexpensive reflectors can be used to concentrate solar energy onto more expensive photovoltaic cells, reducing the cell area needed.

Nonimaging optics is a new field, and the design of concentrators is an ongoing challenge for optical engineers. Figure 36-18 shows a popular concentrator design, called a compound parabolic concentrator. Unlike a parabolic mirror designed for image formation, this concentrator is made from parabolic segments that do not form a single complete parabola. As Fig. 36-18a shows, all light entering the concentrator from a broad range of angles ends up at the narrow exit end. The light doesn't converge to form an image, but its intensity has nevertheless been increased because its energy is spread over a smaller area. Thus, the concentrator acts as a "funnel" for light.

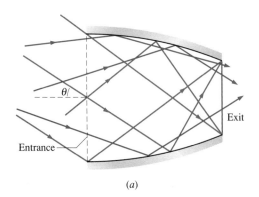

(a)

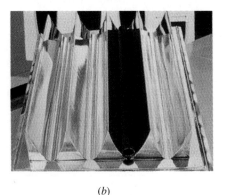

(b)

FIGURE 36-18 (a) A compound parabolic concentrator. Rays (blue) entering at the angle θ to the concentrator axis are reflected to the edge of the exit aperture. Rays incident at smaller angles (red) pass through the exit aperture after at most one reflection; thus all light entering over the angular range 2θ is concentrated at the exit. (b) A section of parabolic concentrators used to concentrate solar energy on a black tube carrying a heat-transfer fluid. A piece of pipe lies at the exit of one concentrator; since light paths are reversible, the fact that the entire concentrator appears black confirms that the path of an incident light ray ends up on the black tube.

Figure 36-18b shows a section of a compound parabolic reflector used to concentrate light onto a tube carrying heat-transfer fluid in a solar thermal power plant. Because the concentrator accepts light from a broad angular range, it works without the need for expensive tracking equipment that would keep it facing the Sun.

Nonimaging concentrators have other applications as well. The first compound parabolic concentrator, built in 1965, was designed to collect light from a high-energy physics experiment. Nonimaging optics in the infrared were used on the Cosmic Background Explorer satellite that, in 1992, provided crucial evidence in support of the Big Bang theory of the origin of the universe. And under development are solar-powered lasers that use nonimaging concentrators as the "pump" that supplies energy to run the laser.

36-3 LENSES

A **lens** is a piece of transparent material shaped in such a way that parallel light rays are refracted toward a single point, again called the **focus** or **focal point.** The focus may be real or only apparent—i.e., virtual— depending on whether the lens is convex (**converging lens;** see Fig. 36-19) or concave (**diverging lens;** see Fig. 36-20). We will see in the next section that a lens made with spherical surfaces approximates this focusing behavior and will show how the focal length is related to the curvatures of the lens surfaces and to its index of refraction. (An alternative lens design—the **graded-index lens**—is flat but has a refractive index that varies with position.) For now we assume we have working lenses and show how they form images.

We will use the **thin-lens** approximation, in which a lens's thickness is much less than the curvature radii of its surfaces, its focal length, and the distances to any objects and their images. Unlike a mirror, a lens works in both directions and, therefore, has two focal points. We will show, again in the next section, that the two focal lengths of a thin lens are the same.

A lens works by refracting light as it enters the first lens surface, and again as it exits the second. Because the two surfaces are not parallel, there is a net change in the direction of a light ray. We will consider the details of those refractions in the next section. In the thin-lens approximation, however, it suffices to consider that light simply bends when it crosses the center plane of the lens. Although that isn't really what happens, the lens is so thin compared with other lengths of interest that the distance between its two surfaces is negligible. Under this approximation a light ray crossing the lens at its axis suffers no effect whatsoever since for it the refractive effects of the opposite surfaces cancel (recall Example 35-3).

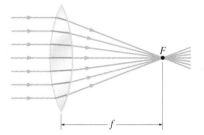

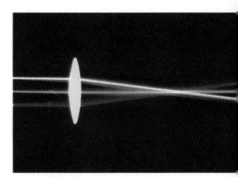

FIGURE 36-19 A convex lens brings parallel light rays to a real focus *F*. The focal length *f* is the distance from the lens to the focal point.

Lens Images by Ray Tracing

We can use ray tracing to find lens images, just as we did with mirrors. Again, two rays from any point serve to fix its image. It's convenient with lenses to use the ray parallel to the lens axis—which gets refracted through the focus—and the ray that passes undeviated through the lens center. Figure 36-21 shows the results for different object placements in relation to a converging lens. In Fig. 36-21a we see that an object farther out than two focal lengths produces a smaller, inverted real image on the other side of the lens. Since light really emanates from this image, you could see it without actually looking through the lens. As the object moves in, the image grows until, when the object lies between one and two focal lengths, the image is farther than 2 *f* from the lens and is larger

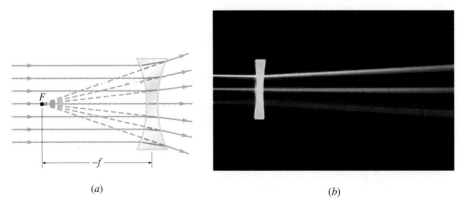

(a) (b)

FIGURE 36-20 A concave lens has a virtual focus, since light rays only appear to meet at the focus. The focal length is negative, indicating that the lens causes light to diverge.

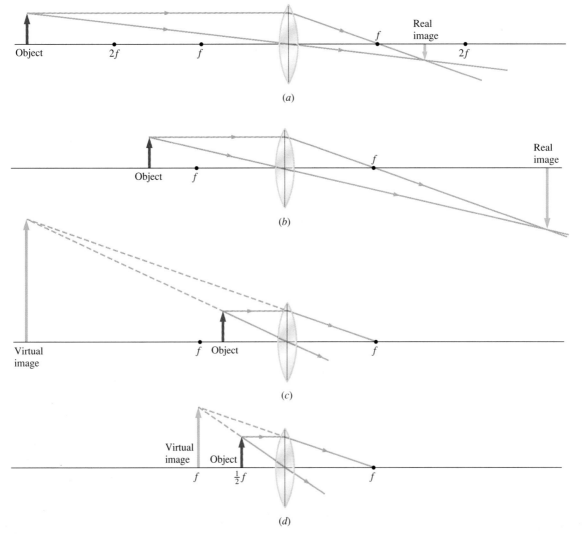

FIGURE 36-21 Image formation with a converging lens. Marks on axis are distances from the lens center, in units of the focal length *f*. In (*a*) and (*b*), an object distance greater than the focal length results in an inverted real image. In (*b*) the image is enlarged if the object distance is between one and two focal lengths. In (*c*) and (*d*), a virtual image forms when the object is within the focal length. When the object moves within half the focal length the image moves within the focal length, as shown in (*d*).

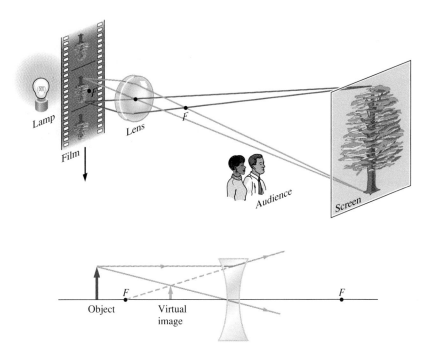

FIGURE 36-22 Simplified diagram of a movie projector. The film lies just beyond the focal length, and thus the lens produces an enlarged real image on the distant screen. The audience normally views the image from the projector side of the screen. Note that the film is upside down.

FIGURE 36-23 A diverging lens always forms a reduced, upright, virtual image, visible only through the lens.

than the object. The image on a movie screen is formed in this way (Fig. 36-22). Moving the object closer than the focal point produces an enlarged virtual image that can be seen only by an observer looking *through* the lens (Fig. 36-21*c*). Moving the object still closer causes the virtual image to move inward and shrink; when the object is within half the focal length of the lens, then the image is within one focal length (Fig. 36-21*d*). The virtual image remains always larger than the object.

Figure 36-23 shows ray tracings for a diverging lens. Like a convex mirror, this lens produces only virtual images that are upright and reduced in size. Like the virtual images of Fig. 36-21*c* and *d*, these virtual images are visible only through the lens. You should convince yourself that the basic geometry of Fig. 36-23 does not change even if the object moves within the focal length.

> **TIP** **Understanding Lenses** Before going on to the mathematics of lenses, try answering the following simple questions about the image shown in Fig. 36-24. Then check your answers against those in the footnote below.*
> (a) What would happen if the bottom half of the lens were covered?
> (b) What would happen if the lens were removed?
> (c) What would happen if the screen's distance from the lens were doubled?
> (d) Would there be an image in the absence of the screen?

*Answers: (a) You might think that half the image would disappear. But Fig. 36-21 shows complete images forming from rays that travel only through the top half of the lens. We could equally well have formed the images with rays from the bottom half, or any other portion of the lens. Covering part of the lens—any part—diminishes the intensity of the image but does not block *any* of it. (b) Did you think the image would now appear upright? No! The lens is *essential* to the image formation; it's what brings rays leaving each point of the candle flame together again at a single point. Absent the lens, the screen would appear diffusely lit by the candle's light, but there would
(*continued*)

FIGURE 36-24 A simple optical system, consisting of an object (the candle), a converging lens, and a white screen on which the image appears. This system is a realization of Fig. 36-21*b*.

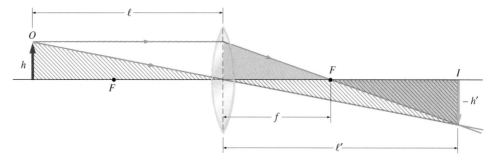

FIGURE 36-25   Ray diagram for deriving the lens equation. The two colored triangles are similar, as are the hatched triangles.

A Lens Equation

We can locate lens images quantitatively by deriving an equation like we did for mirrors, now using Fig. 36-25. Note that the two hatched (////) triangles in Fig. 36-25 are similar, and therefore the magnification is

$$M = \frac{h'}{h} = -\frac{\ell'}{\ell}, \tag{36-6}$$

where ℓ and ℓ' are the object and image distances, respectively, and where we again take a negative height to signify an inverted image. Thus, the magnitude of the magnification is the ratio of object to image distance, as it was with mirrors. The two colored triangles are also similar, and therefore

$$\frac{-h'}{\ell' - f} = \frac{h}{f}.$$

Combining this result with Equation 36-6 and doing some algebra then gives

$$\frac{1}{\ell} + \frac{1}{\ell'} = \frac{1}{f}, \quad \text{lens equation} \tag{36-7}$$

which is identical to the mirror equation 36-3. Note that putting the object at infinity ($\ell = \infty$) gives $\ell' = f$ and that putting the object at the focus ($\ell = f$)

be no image. (c) Did you think the image size would double? No! The image exists only at the one place where rays from each point on the candle converge. Again, the screen would be diffusely lit. Moving it very slightly from its position, in contrast, would result in an blurred rendition of the candle. (Moving the lens also would allow refocusing the image at the new screen location.) (d) The screen helps make the image visible to observers looking from various directions. But it plays no role in the actual image formation; absent the screen, there would still be an image at the point where the screen is now located. An observer looking toward the lens from the right, into the cone of light diverging from the image, would see an upside-down candle flame apparently in "thin air."

For more intuition on image formation, see Fred M. Goldberg and Lillian C. McDermott, "An Investigation of Student Understanding of the Real Image Formed by a Converging Lens or Concave Mirror," *American Journal of Physics,* vol. 55, no. 2, pp. 108–119 (February 1987).

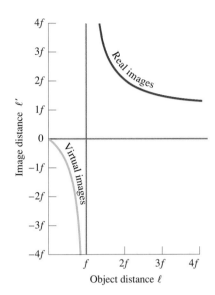

FIGURE 36-26 Image versus object distance for converging lenses. Object distances less than the focal length f correspond to virtual images (negative ℓ'). The image moves toward infinity as the object distance approaches the focal length from either direction, and concurrently the image grows arbitrarily large.

gives $\ell' = \infty$. Thus, Equation 36-7 reflects our ray-tracing diagrams that show parallel rays (such as those from an object at a great distance) converging at the focal point, and vice versa.

Although we derived Equation 36-7 for the case of a real image, the equation holds for virtual images if we consider the image distance negative; in that case the image is on the same side of the lens as is the object. And it holds for diverging lenses if we consider the focal length negative. Again these conventions make sense, as suggested in the tip before Example 36-4. Figure 36-26 is a graphical summary of the lens equation for a converging lens.

TIP   **Sign Conventions**   As with mirrors, the sign conventions for lenses make sense. Objects and real images have positive distances; virtual images "aren't really there" and have negative distances. Converging lenses—those that can form real images—have positive focal lengths; diverging lenses can form only virtual images, and they have negative focal lengths.

● **EXAMPLE 36-4** SLIDE SHOW

You're projecting slides onto a wall 2.6 m from a slide projector whose single lens has focal length 12.0 cm. (a) How far should the slides be from the lens? (b) How big will be the image of a 35-mm slide?

Solution
We want the image focused on the screen, so the image distance ℓ' is 2.6 m. Solving Equation 36-7 for the object distance ℓ then gives

$$\ell = \left(\frac{1}{f} - \frac{1}{\ell'} \right)^{-1} = \left(\frac{1}{12 \text{ cm}} - \frac{1}{260 \text{ cm}} \right)^{-1} = 12.58 \text{ cm}.$$

This is just beyond the focal length, as Figs. 36-21 and 36-26 suggest it should be to get an enlarged, distant, real image.

Equation 36-6 then gives the height h' of the image formed of a 35-mm slide:

$$h' = -h\frac{\ell'}{\ell} = -(3.5 \text{ cm})\left(\frac{260 \text{ cm}}{12.58 \text{ cm}} \right) = -72 \text{ cm}.$$

The minus sign indicates inversion, showing that the slide must go in the projector upside down to get an upright image.

EXERCISE   If the slide in Example 36-4 moves 1 mm farther from the lens, what will happen to the position of the image?

Answer: moves 36 cm toward projector

Some problems similar to Example 36-4: 16–18 ●

● **E X A M P L E 3 6 - 5** FINE PRINT

You're using a magnifying glass (a converging lens) with 30 cm focal length to read a telephone book (Fig. 36-27). How far from the page should you hold the lens in order to see the print enlarged 3 times?

Solution

Here the image is virtual, so ℓ' is negative and $3\times$ magnification then corresponds to $\ell' = -3\ell$. So Equation 36-7 becomes

$$\frac{1}{\ell} - \frac{1}{3\ell} = \frac{2}{3\ell} = \frac{1}{f} = \frac{1}{30 \text{ cm}},$$

whence $\ell = (2)(30 \text{ cm})/3 = 20$ cm. Figure 36-27 confirms that the image appears further from the lens than the actual page.

E X E R C I S E A magnifying glass enlarges print by 50% when it's held 9.0 cm from a page. What is its focal length?

FIGURE 36-27 A magnifying glass is a converging lens used in its virtual-image mode.

Answer: 27 cm

Some problems similar to Example 36-5: 19, 25 ●

36-4 THE LENSMAKER'S FORMULA

We now examine in detail the refraction effects that cause lenses to form images. Every lens has two refracting surfaces, at least one, and usually both, of which are curved. We therefore look first at image formation due to refraction at a *single* curved interface. To understand the behavior of a lens we will then consider two such interfaces, using the image formed by the first as the object for the second.

Refraction at a Curved Surface

Figure 36-28 shows part of a spherically curved piece of material (typically glass or plastic) with curvature radius R and refractive index n_2, surrounded by a medium with index n_1 (typically air). A point-like object O lies on an axis through the center of curvature of the refracting surface. We consider only light rays from O that are nearly parallel to the axis. Under this restriction the labeled angles in Fig. 36-28 are all small, and we can use the approximation $\sin x \simeq \tan x \simeq x$, where $x \ll 1$ with x an angle measured in radians.

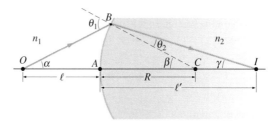

FIGURE 36-28 Refraction at a curved surface produces an image at I of the object at O.

In Fig. 36-28 a light ray from O propagating along the axis suffers no refraction because it is normal to the interface. An image of O should therefore form at a point where other rays from O intersect the axis. The figure shows one such ray, which hits the interface at B with some small incidence angle θ_1, and propagates in the second medium at angle θ_2 to the normal. Under the small-angle approximation, Snell's law—$n_1 \sin \theta_1 = n_2 \sin \theta_2$—becomes simply

$$n_1 \theta_1 = n_2 \theta_2.$$

From triangles BCI and OBC, we see that $\theta_2 = \beta - \gamma$ and $\theta_1 = \alpha + \beta$. Using these results in our small-angle version of Snell's law gives

$$n_1(\alpha + \beta) = n_2(\beta - \gamma).$$

Furthermore, in the small-angle approximation the arc BA is so close to a straight line that we can write $\alpha \simeq \tan \alpha \simeq BA/\ell$, with $\ell = OA$ the object's distance from the refracting surface. Similarly, $\beta \simeq BA/R$, and $\gamma \simeq BA/\ell'$. Thus, our expression of Snell's law becomes

$$n_1 \left(\frac{BA}{\ell} + \frac{BA}{R} \right) = n_2 \left(\frac{BA}{R} - \frac{BA}{\ell'} \right)$$

or, on canceling the arc length BA and rearranging,

$$\frac{n_1}{\ell} + \frac{n_2}{\ell'} = \frac{n_2 - n_1}{R}. \tag{36-8}$$

Notice that the angle α does not appear in this equation. Therefore, this relation between the object and image distances holds for *all* rays as long as they satisfy the small-angle approximation. That means all such rays come to a common focus at I in Fig. 36-28, so an image must appear at this point.

Although we derived Equation 36-8 for a case of real image formation, as usual it holds for virtual images as well if we consider the image distance negative. It also works for surfaces that are concave toward the object—the opposite of Fig. 36-28—if we take R as a negative number.

● **EXAMPLE 36-6** A CYLINDRICAL AQUARIUM

An aquarium is made from a thin-walled tube of transparent plastic 70 cm in diameter (Fig. 36-29a). For a cat looking directly into the aquarium, what is the apparent distance to a fish 15 cm from the aquarium wall?

Solution

We can neglect the aquarium wall; since it's thin and has essentially parallel faces, light suffers no net deflection in passing through the wall. The surface is concave toward the object, so $R = -35$ cm. With $\ell = 15$ cm, $n_1 = 1.333$ for water, and $n_2 = 1$ for air, we can solve Equation 36-8 for the image distance ℓ':

$$\ell' = \frac{n_2}{\left(\dfrac{n_2 - n_1}{R} - \dfrac{n_1}{\ell} \right)} = \frac{1}{\left(\dfrac{1 - 1.333}{-35 \text{ cm}} - \dfrac{1.333}{15 \text{ cm}} \right)}$$

$$= -12.6 \text{ cm}.$$

The negative answer indicates that we have a virtual image, as shown in Fig. 36-29b.

A special case of Equation 36-8 is a flat surface, for which $R = \infty$. The right-hand side is then zero, but the equation still relates object and image distances. Looking down into a swimming pool from above, for example, you see virtual images of objects in the pool. The objects appear closer than they actually are, as Fig. 36-30 and the following exercise illustrate.

EXERCISE The bottom of a swimming pool looks to be 1.5 m below the surface. What is the pool's actual depth?

Answer: 2.0 m

Some problems similar to Example 36-6: 29, 30, 32–34

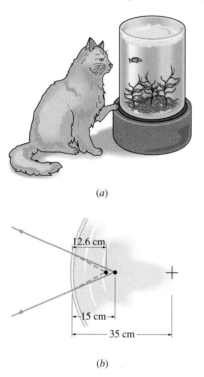

(a)

(b)

FIGURE 36-29 (a) A cylindrical aquarium. (b) Top view, showing formation of a virtual image of a fish that is actually 15 cm from the aquarium wall.

FIGURE 36-30 Refraction at a flat surface produces an image *I* that appears closer than the object *O*. Objects at the bottom of a swimming pool, for example, appear closer than they really are.

Lenses, Thick and Thin

Figure 36-31 shows a lens of arbitrary thickness t made from material with refractive index n. For simplicity we will consider the lens surrounded by air or vacuum, with refractive index equal to 1. An object O_1 lies a distance ℓ_1 to the left of the left-hand surface. This surface focuses light from O_1 to form image I_1. Light from this image impinges on the right-hand surface, forming a second image I_2. We want to relate the original object O_1 and final image I_2.

We developed Equation 36-8, our description of refraction at a single curved surface, using Fig. 36-28. In that figure we considered formation of a real image, but we quickly noted that Equation 36-8 applies to virtual and real images alike. In considering a two-surface lens it proves more convenient to place the object where its image in the first surface is a virtual one, as in Fig. 36-31. Again, our final result will apply whatever the object placement.

At the left-hand surface in Fig. 36-31, the quantities in Equation 36-8 are $\ell = \ell_1$, $\ell' = \ell'_1$, $n_1 = 1$, $n_2 = n$, and $R = R_1$, so Equation 36-8 becomes

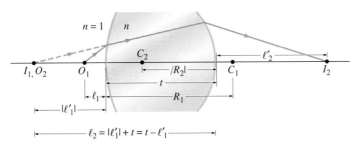

FIGURE 36-31 A thick lens. The left-hand surface forms the virtual image I_1, which is also labelled O_2 because it forms the object for the right-hand surface. Together, the two surfaces then produce the final real image I_2. Absolute value signs are used on distances whose algebraic signs in Equation 36-8 are negative—i.e., the distance ℓ'_1 to the virtual image and the curvature radius R_2 of the right-hand surface, which is concave toward the incident light.

$$\frac{1}{\ell_1} + \frac{n}{\ell'_1} = \frac{n-1}{R_1}. \qquad \text{(left-hand surface)}$$

The "object" O_2 for the right-hand surface is the image I_1 since light incident on that surface looks like it's coming from I_1. The distance ℓ in Equation 36-8 for this "object" is $\ell = \ell_2 = t - \ell'_1$, with t the lens thickness. Why minus? Because I_1 is a *virtual* image, so its image distance ℓ'_1 is a *negative* quantity. The physical distance between the image and the lens is the positive quantity $-\ell'_1$, and that quantity adds to the lens thickness to give the distance from this intermediate image to the right-hand surface. Also at the right-hand surface, $\ell' = \ell'_2, n_1 = n, n_2 = 1$, and $R = R_2$. Thus, at the right-hand surface Equation 36-8 reads

$$\frac{n}{t - \ell'_1} + \frac{1}{\ell'_2} = \frac{1-n}{R_2}. \qquad \text{(right-hand surface)}$$

Now we'll let the lens become arbitrarily thin, taking the limit $t \to 0$. Then the first term in our right-surface equation becomes $-n/\ell'_1$. But the term n/ℓ'_1 also occurs in the left-surface equation. So we can add the two equations to eliminate the intermediate-image distance ℓ'_1. Since we'll be left with only one object distance, ℓ_1, and one image distance, ℓ'_2, we can drop the subscripts 1 and 2 on these quantities. The result is

$$\frac{1}{\ell} + \frac{1}{\ell'} = (n-1)\left(\frac{1}{R_1} - \frac{1}{R_2}\right). \qquad (36\text{-}9)$$

The left-hand side of this equation is identical to the left-hand side of Equation 36-7. So we can equate the right-hand sides of the two equations to get an expression for the focal length of our lens:

$$\frac{1}{f} = (n-1)\left(\frac{1}{R_1} - \frac{1}{R_2}\right). \qquad \text{(lensmaker's formula)} \qquad (36\text{-}10)$$

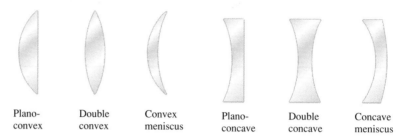

| Plano-convex | Double convex | Convex meniscus | Plano-concave | Double concave | Concave meniscus |

FIGURE 36-32 Common lens types.

We emphasize that the radii R_1 and R_2 in Equation 36-10 may be positive or negative, depending on the lens curvature. In Fig. 36-31, for example, R_1 is positive because the left-hand surface is convex toward the incident light but R_2 is negative because the right-hand surface is concave toward the light. One of the radii can also be infinite if the corresponding surface is flat.

Lenses are made in a variety of shapes, as shown in Fig. 36-32. Those that are thicker at the center are converging lenses, for which Equation 36-10 gives a positive focal length. Those that are thinner at the center are diverging lenses, with negative focal length. These characterizations of lens behavior reverse if the medium surrounding the lens has a higher refractive index than the lens itself (see Problem 71, which generalizes Equation 36-10 for an arbitrary refractive index in the surrounding medium).

● EXAMPLE 36-7   A PLANO-CONVEX LENS

The plano-convex lens in Fig. 36-32 is made from material with refractive index n. Given that the curved surface has curvature radius R, find an expression for the focal length of the lens. Show that the focal length is the same whether light is incident on the lens from right or left.

Solution
With light incident from the left in Fig. 36-32, $R_1 = R$. A flat surface has infinite curvature, so $R_2 = \infty$. Then Equation 36-10 gives

$$f = \left[(n - 1)\left(\frac{1}{R} - \frac{1}{\infty} \right) \right]^{-1} = \frac{R}{n - 1},$$

since $1/\infty = 0$. If light is incident from the right, then $R_1 = \infty$ and, since the curved surface is now concave toward the light, $R_2 = -R$. Then

$$f = \left[(n - 1)\left(\frac{1}{\infty} - \frac{1}{-R} \right) \right]^{-1} = \frac{R}{n - 1},$$

showing that the two focal points are indeed equidistant from the lens.

EXERCISE A double-convex lens (see Fig. 36-32) is made of glass with $n = 1.75$. If the curvature radii of the two sides are equal, what should that radius be in order to give the lens a 25-cm focal length?

Answer: 37.5 cm

Some problems similar to Example 36-7: 36–38 ●

Lens Aberrations

Lenses exhibit several types of optical defects. We've already described **spherical aberration** in connection with mirrors; this defect occurs because a spherical mirror only approximates the ideally focusing shape of a parabola. Lenses

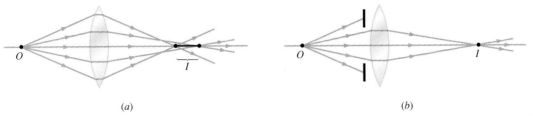

(a)   (b)

FIGURE 36-33 (a) Spherical aberration occurs because rays striking the lens farther from the axis and at greater angles focus closer to the lens than do near-axial rays. The result is a "smearing" of the image. (b) This defect can be minimized by using only the central portion of the lens, at the expense of a dimmer image.

FIGURE 36-34 Chromatic aberration occurs because the refractive index varies with wavelength, causing different colors to focus at different points.

with spherical surfaces also exhibit spherical aberration (Fig. 36-33a). In our analysis of image formation by a curved refracting surface we had to make the small-angle approximation; otherwise, we would not have reached the conclusion that all rays from an object come to a common focus. Spherical aberration can be minimized by ensuring that incident rays are as close and as parallel as possible to the lens axis. That's easy for distant objects but harder with objects close to the lens. Using only the central portion of the lens eliminates those rays at larger angles (Fig. 36-33b), leading to a sharper focus. That's why a camera focuses over a wider range when it is "stopped down," the outer part of its lens covered by an adjustable diaphragm. The tradeoff, of course, is that there is less light available.

Because the refractive index varies with wavelength, Equation 36-10 predicts different focal lengths for different colors. The result is **chromatic aberration** (Fig. 36-34). High-quality optical systems minimize this defect by using composite lenses made from different materials whose differing refractive indices allow several colors to focus at the same point. Chromatic aberration is unique to lenses since mirrors reflect light of all colors in exactly the same way. Figure 36-35 shows both spherical and chromatic aberration in a simple magnifying lens.

Figure 36-36 shows several other common lens defects. Modern lenses are often made with complicated nonspherical shapes to limit aberrations, giving undistorted images over wider ranges than the spherical lenses we have been discussing. Optical engineers still use ray tracing, but now they do so with computers.

FIGURE 36-35 This magnifying lens shows both spherical aberration (distortion of the lines) and chromatic aberration (colors).

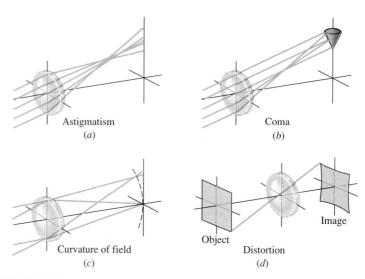

FIGURE 36-36 Lens aberrations. (*a*) Astigmatism results when the focusing quality of the lens varies with angular position, smearing images more in some directions. (*b*) Coma, caused by poor focusing of off-axis objects, produces images with comet-like tails. (*c*) Curvature of field describes the fact that extended objects focus on a curve rather than a plane. (*d*) Distortion changes straight lines into curves.

■ **APPLICATION** FRESNEL LENSES

Large-diameter lenses are impractically thick. Lighthouses, in particular, cannot use single lenses to concentrate their intense beams because too much light would be absorbed in the glass. To solve this problem the French scientist Augustin Fresnel, in 1822, devised a way of making large lenses thin. A Fresnel lens consists of concentric rings shaped to approximate segments of an ordinary lens surface, with step-like jumps to keep the overall structure thin (Fig. 36-37*a*). Fresnel lenses are still used in lighthouses (Fig. 36-37*b*), while Fresnel lenses made from plastic sheets appear in home, office, and car windows to provide magnification or wide-angle views (Fig. 36-37*c*). They are also used in some overhead projectors.

FIGURE 36-37 (*a*) A Fresnel lens approximates a normal curved lens by a series of short segments. (*b*) A giant Fresnel lens from a nineteenth-century lighthouse. (*c*) Fresnel lens made from sheet plastic mounted on the rear window of a van provides a wide-angle view of what's behind the van.

36-5 OPTICAL INSTRUMENTS

Numerous optical instruments make use of mirrors and/or lenses. We have already considered some simple applications using a single lens or mirror, including film and slide projectors, magnifying glasses, and vanity mirrors. More complicated systems use several optical elements (lenses or mirrors), but the principles we have developed still apply. In particular, we trace light through a sequence of optical elements by using the image formed by one element as the object for the next.

The Eye

Our eyes are our primary optical instruments. Each eye is a complex optical system with several refracting surfaces and mechanisms to vary both the focal length and the amount of light admitted. Figure 36-38 shows that the eye is essentially a fluid-filled ball about 2.3 cm in diameter. Light enters through the hard, transparent cornea and passes through a fluid called the aqueous humor before entering the lens. On exiting the lens, light traverses the vitreous humor that fills the main body of the eyeball and finally strikes the retina. The retina is covered with light-sensitive cells of two types. Cells called cones are sensitive to different colors but require moderate amounts of light, while rod cells function at low intensity but lack color discrimination. Both types of cells produce electrochemical signals that carry information to the brain.

A properly functioning eye produces a well-focused real image on the retina. Contrary to general belief, most of the refraction is provided not by the lens but by the cornea. The lens, however, is the adjustable element. Ciliary muscles pulling on the lens alter its focal length, allowing fine adjustment of the eye's focus—a process known as accommodation. Other muscles automatically adjust the iris, enlarging or contracting the pupil opening to compensate for different light levels.

When the ciliary muscles are relaxed, the lens is relatively flat, with focal length about 1.7 cm. But for nearsighted (myopic) people, the image still forms in front of the retina, causing distant objects to appear blurred. Diverging

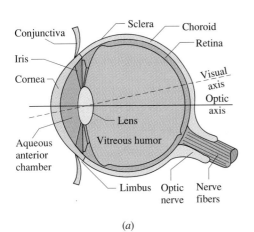

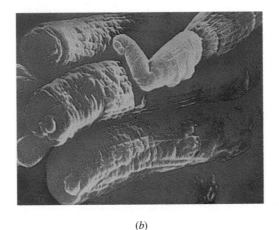

(a)   (b)

FIGURE 36-38   (a) The human eye. (b) Scanning-electron-microscope image showing rods and a cone (blue structure) in a human retina.

FIGURE 36-39 (*a*) A myopic eye focuses light from distant objects in front of the retina. (*b*) A diverging lens corrects the problem, creating a virtual image closer to the eye. The eye's image of this virtual image then falls on the retina.

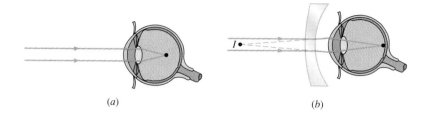

(*a*)

(*b*)

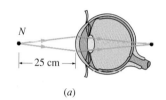

(*a*)

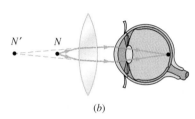

(*b*)

FIGURE 36-40 (*a*) A farsighted eye cannot focus light from the standard 25-cm near point at *N*. (*b*) A converging corrective lens makes light from an object at the near point appear to come from the point *N′* which is at the minimum distance where this eye can focus.

corrective lenses produce closer intermediate images that the myopic eye can then focus (Fig. 36-39). In farsighted (hyperopic) eyes, the image of nearby objects would focus behind the retina, and converging corrective lenses are used (Fig. 36-40). Actually, everyone is farsighted, in the sense that there is a minimum distance, called the **near point,** below which the eye cannot focus sharply. In the typical human the near point is at about 25 cm, and this value is taken as a standard. By the time one is 50, however, the typical near point is at about 40 cm, and it can easily be several meters.

Prescriptions for corrective lenses usually specify the corrective power, *P*, in **diopters,** where the diopter measure of a lens is the inverse of its focal length in meters. Thus, a 1-diopter lens has $f = 1$ m, while a 2-diopter lens has $f = 0.5$ m and is more powerful in that it refracts light more sharply.

The human eye is a complex biological system, in many ways a direct extension of the brain. Although we understand its optical behavior, we understand much less clearly how visual perception actually occurs. Recently biologists and computer scientists have teamed to produce an electronic "silicon retina" whose behavior mimics many subtle aspects of the natural eye and may therefore shed light on the process of perception.

● EXAMPLE 36-8 AN AUTHOR'S EYES

Writing this book has prematurely strained your author's eyes, which now cannot achieve clear focus at distances less than 1.2 m. What diopter lenses will correct this problem?

Solution

"Correct" in this context means to bring the range for clear focus to the standard 25-cm near point. According to Fig. 36-40*b*, that means we want a lens that will produce a virtual image at 1.2 m of an object at 25 cm. The required power follows directly from Equation 36-7:

$$P = \frac{1}{f} = \frac{1}{\ell} + \frac{1}{\ell'} = \frac{1}{0.25 \text{ m}} + \frac{1}{-1.2 \text{ m}} = 3.17 \text{ diopters},$$

where the image distance is negative because the image is virtual.

EXERCISE A nearsighted person cannot see clearly beyond 80 cm. Prescribe a lens power that will image the most distant objects at 80 cm, giving clear vision at all distances.

Answer: −1.25 diopters, the minus sign signifying a diverging lens

Some problems similar to Example 36-8: 46–48 ●

Cameras

A camera is much like the eye, except that it focuses its image on film or a light-sensitive electronic device (Fig. 36-41). Today most still cameras use the long-established technology of film based on light-sensitive chemicals, but the development of silicon chips known as CCDs (charge-coupled devices) has already made film obsolete in astronomical imaging and may someday have the

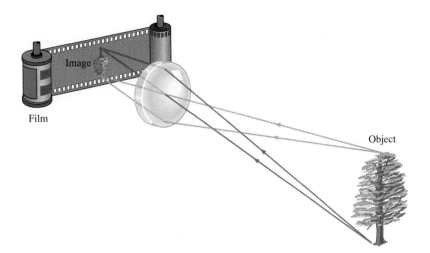

FIGURE 36-41 Optical system of a simple camera.

FIGURE 36-41 Optical system of a simple camera.

same effect on conventional photography. Video cameras also use CCDs.

Films and other light-sensitive media require that the amount of light admitted to a camera be regulated for optimum image quality. Too much light, and the image will be overexposed and "washed out"; too little and it will be too dark. Obviously, a larger lens admits more light. But the image brightness also depends on how much the lens concentrates the light. A lens with a short focal length bends light more, and will thus produce a brighter—but smaller—image than a lens of equal diameter that has a longer focal length. Suppose that light of a given intensity is incident on a lens of diameter d. Since intensity is power per unit area, the rate at which light energy enters the camera will be proportional to d^2. For distant objects, Equations 36-6 and 36-7 show that the image size scales with the focal length, and therefore, the image area is proportional to f^2. Thus, the intensity at the image will be proportional to $(d/f)^2$. The quantity d/f is called the **speed** of the lens. A fast lens—larger d/f—gives a brighter image and, therefore, can capture that image with a shorter exposure time.

Photographers and camera manufacturers usually use the inverse of the lens speed, called the focal ratio, **f-ratio,** or f-stop, equal to f/d. Increasing the f-ratio decreases the image intensity. Since that intensity scales as $(d/f)^2$, an increase in the f-ratio by a factor of $\sqrt{2}$ results in a halving of the image intensity. That's why adjustable cameras are commonly marked with f-ratios of 1.4, 2, 2.8, 4, 5.6, 8, 11, . . . ; these numbers are the rounded square roots of 2, 4, 8, 16, 32, 64, 128, ...

In a camera with a fixed-focus lens, changing the f-ratio is accomplished with an adjustable iris that covers part of the lens (Fig. 36-42). We've already seen that lenses focus best when only the central part of the lens is used. This is the cause of a photographer's common dilemma: to use a larger lens aperture to gather more light, or a smaller aperture to get a greater range of distances in focus.

Many of today's cameras automatically adjust both the focus and the f-ratio to optimize image quality, although professional photographers usually prefer to retain control over these settings. Increasingly, too, cameras incorporate zoom lenses whose focal lengths can be altered by physically moving the lens elements (Fig. 36-43). A lens of fixed diameter is necessarily slower at its telephoto (larger focal length) settings, as Problem 49 shows.

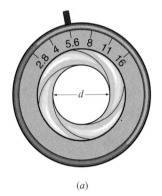

(a)

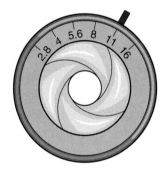

(b)

FIGURE 36-42 An adjustable diaphragm "stops down" the lens, decreasing its effective diameter and increasing the f-ratio.

FIGURE 36-43 Photos taken from the same point with (a) short and (b) long focal length settings of an adjustable zoom lens.

(a)　　　　　　　　　　　　　(b)

Magnifiers and Microscopes

We wouldn't need optical instruments to examine small objects if we could bring our eyes arbitrarily close to the objects. But we've seen that the average human eye cannot focus much closer than about 25 cm. We therefore use lenses to put enlarged images of small objects at distances where our eyes can focus.

What matters is not so much the actual size of the image, but how much of our field of view it occupies. Consequently we define the **angular magnification** m as the ratio of the angle subtended by an object seen through a lens to the angle subtended as seen by the naked eye when the object is at the standard 25-cm near point. Figure 36-44a shows that the former angle, α, is given by $\alpha = h/25$ cm, where h is the object height and α is in radians. The maximum magnification would occur with the image itself at the near point (see Problem 51), but it's more comfortable to view a very distant image. We therefore place the object at just inside the focal length, forming an enlarged virtual image at a great distance. Figure 36-44b shows this geometry, from which we see that the angle β subtended by the image is essentially h/f. Then the magnification is

$$m = \frac{\beta}{\alpha} = \frac{h/f}{h/25 \text{ cm}} = \frac{25 \text{ cm}}{f} \qquad \text{(simple magnifier).} \qquad (36\text{-}11)$$

This angular magnification is achieved only with the eye very close to the lens, which is not the way we normally hold an ordinary magnifying glass. But it is the way we place our eyes on many common instruments that use simple magnifying lenses for their eyepieces.

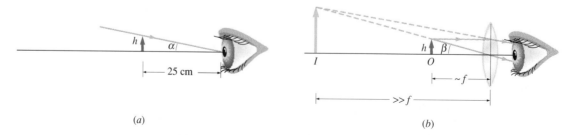

(a)　　　　　　　　　　　　　(b)

FIGURE 36-44 (a) An object of height h subtends a small angle $\alpha \simeq h/25$ cm at the standard 25-cm near point. (b) Putting the object near the focus of a converging lens gives an image that subtends an angle $\beta \simeq h/f$. The angular magnification is the ratio $m = \beta/\alpha$.

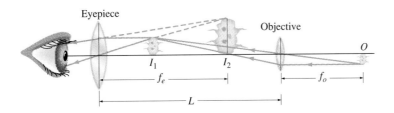

FIGURE 36-45 Image formation in a compound microscope. Figure is not to scale; distance L should be much greater than either focal length, and the image I_1 should be very near the eyepiece's focus, resulting in greater magnification.

Single lenses can produce angular magnifications of about 4 before aberrations compromise the image quality. Higher power magnification therefore requires more than one lens. A **compound microscope** is a two-lens system in which a lens of short focal length called the **objective** forms a magnified real image. This image is then viewed with a second lens, the **eyepiece,** positioned as a simple magnifier (Fig. 36-45). The object being viewed is positioned just beyond the focal length of the objective lens, and its image falls just inside the focal length of the eyepiece. If both focal lengths are small compared with the distance between the lenses, then the object distance for the objective lens is approximately the objective focal length f_o, and the resulting image distance is approximately the lens spacing L. The real image formed by the objective lens is larger than the object by the ratio of the image to object distance, or $-L/f_o$. The eyepiece enlarges the image further, by a factor of its angular magnification $25 \text{ cm}/f_e$. Then the overall magnification of the microscope is

$$M = M_o m_e = -\frac{L}{f_o}\left(\frac{25 \text{ cm}}{f_e}\right), \qquad \text{(compound microscope)} \quad (36\text{-}12)$$

where, as usual, the minus sign signifies an inverted image.

Optical microscopes work well as long as the approximation of geometrical optics holds—that is, when the object being viewed is much larger than the wavelength of light. Viewing smaller objects requires waves of shorter wavelength. In the widely used electron microscope, those "waves" are electrons, whose wavelike nature we will examine in Chapter 39.

Telescopes

A telescope collects light from distant objects, either forming an image or supplying light to instruments for analysis. Telescopes are classified as **refracting** or **reflecting,** depending on whether the main light-gathering element is a lens or mirror. Small hand-held telescopes and binoculars are refracting instruments, as are telephoto camera lenses and some older astronomical telescopes. Modern astronomical telescopes are almost invariably reflectors.

The simplest refracting telescope consists of a single lens that images distant objects at essentially its focal point. Film or a CCD placed at that point captures the image, or an eyepiece is used to view the image. The world's largest refracting telescope, at Yerkes Observatory in Wisconsin, has a 1-m-diameter lens with 12-m focal length. Photographers can think of it as a 12,000-mm telephoto lens. Figure 36-46 shows the imaging process in an astronomical refracting telescope. The focal points of the objective and eyepiece lenses are nearly coincident, so the real image of a distant object that forms at the objective's focus is then seen through the eyepiece as a greatly enlarged virtual image. The angular

FIGURE 36-46 Image formation in a refracting telescope. A distant object, in this case a galaxy, is imaged first at the focus of the objective lens (image I_1). An eyepiece with its focus at nearly the same point then gives an enlarged virtual image (I_2). The angles α and β are given by $\alpha = h_1/f_o$ and $\beta = h_1/f_e$ in the small-angle approximation, leading to Equation 36-13 for the angular magnification.

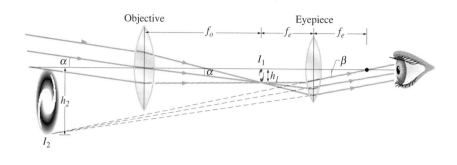

FIGURE 36-47 Hexagonal mirror of the Keck Telescope on Hawaii's Mauna Kea shows in this photo taken through the slit in the telescope's dome. The mirror is 10 m across.

magnification is the ratio of the angle β subtended by the final image to the angle α subtended by the actual object; Fig. 36-46 shows that this ratio is

$$ m = \frac{\beta}{\alpha} = \frac{f_o}{f_e}. \qquad \text{(refracting telescope)} \qquad (36\text{-}13) $$

Since a real image is inverted and a virtual image is upright, a two-lens refracting telescope gives an inverted image. This is fine for astronomical work, but telescopes designed for terrestrial use have an extra lens, a diverging eyepiece (see Problem 70) or a set of reflecting prisms (as in binoculars; see Fig. 35-14b) to produce an upright image.

Reflecting telescopes offer many advantages over refractors. Mirrors have reflective coatings on their front surfaces, eliminating chromatic aberration because light need not pass through glass. Having only one optically active surface allows much larger reflectors to be built since the mirrors can be supported across their entire back surfaces—unlike lenses, which can be supported only at the edges. Where the largest refracting telescope ever built has a 1-m-diameter lens, the newest reflectors boast diameters in the 10-m range (Fig. 36-47). These designs incorporate segmented and/or flexible mirrors whose shape can be adjusted under computer control for optimum focusing; with so-called adaptive optics, such systems may adjust rapidly enough to compensate for the atmospheric turbulence that has traditionally limited the resolution of ground-based telescopes.

The simplest reflecting telescope is a curved mirror with a CCD or film at its focus. Superb image quality results, in principle limited only by wave effects we will discuss in the next chapter. More often the telescope is used as a "light bucket," collecting light from stars or other distant sources too small to image even with today's large optical telescopes. Then a secondary mirror sends light to a focus at a convenient spot for telescope-mounted instrumentation. Optical fibers may also be used to bring light collected by the primary mirror to fixed instruments. Alternatively, an eyepiece can be mounted to examine the image visually. Figure 36-48 shows several common designs for reflecting telescopes.

Magnification is not a particularly important quantity in astronomical telescopes, which are used more for spectral and other analysis than for direct imaging. More important is the light-gathering power of the instrument, which is determined simply by the area of its objective lens or primary mirror. The 10-m Keck telescope, for instance, has 100 times the light-gathering power of the 1-m Yerkes refractor and more than 17 times the power of the 2.4-m Hubble Space Telescope.

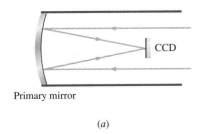

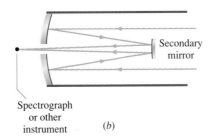

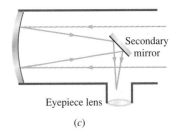

FIGURE 36-48 Common arrangements for reflecting telescopes. (*a*) Placing a CCD or film at the prime focus gives the best image quality. (*b*) In the Cassegrain design light reflects from a secondary mirror and passes through a hole in the primary. The secondary may be either plane or convex. This design is widely used in large telescopes, with the added option of placing detectors directly at the prime focus as in (*a*). (*c*) The Newtonian design, used primarily in small telescopes, has an angled secondary mirror to direct light to the eyepiece.

CHAPTER SYNOPSIS

Summary

1. Reflection and refraction can result in **images,** which are **real** or **virtual** depending on whether or not light actually comes from the image location.

2. Plane mirrors form only virtual images, equal in size to the object being imaged. Concave parabolic mirrors have a **focus** at which parallel rays meet; such mirrors form upright, magnified virtual images of objects located closer than the focus and inverted real images of objects beyond the focus. Concave spherical mirrors approximate the focusing behavior of parabolic mirrors. Convex curved mirrors form only virtual images, upright and reduced in size. The **mirror equation** quantifies the relation between object and image distances ℓ and ℓ':

$$\frac{1}{\ell} + \frac{1}{\ell'} = \frac{1}{f},$$

where the **focal length** f is the distance from the mirror's apex to its focus. The **magnification** is the ratio $M = -\ell'/\ell$, where a negative ℓ' corresponds to a virtual image and therefore to a positive M and so to an upright image.

3. **Lenses,** made from transparent materials, are shaped in such a way that they refract light to a focus. In the **thin-lens** approximation, the curvature radii of the lens surfaces are much greater than the thickness of the lens. Such a lens has two foci, located equal distances f on either side of the lens. The object and image distances and magnification then obey equations identical to those for mirrors. The focal length of a thin lens is given by the **lensmaker's equation:**

$$\frac{1}{f} = (n-1)\left(\frac{1}{R_1} - \frac{1}{R_2}\right),$$

where R_1 and R_2 are the curvature radii of the lens surfaces, taken as positive when the surface is convex toward the incident light. A positive focal length f corresponds to a **converging lens** and a negative f to a **diverging lens.** Lens **aberrations** are focusing defects resulting from imperfect lens shapes and the variation of refractive index with wavelength.

4. Optical instruments use lenses and/or mirrors to form useful images. The power of instrument lenses is often given in **diopters**—the inverse of the focal length in meters. Image brightness depends on the lens diameter d in relation to the focal length f and is characterized by the **f-ratio** f/d.

Terms You Should Understand

(Pairs are closely related terms whose distinction is important; number in parentheses is chapter section where term first appears.)

real image, virtual image (introduction)
focus, focal length (36-2)
magnification (36-2)
thin lens (36-3)
converging lens, diverging lens (36-3)
spherical aberration, chromatic aberration (36-2, 36-3)
near point (36-5)
diopter (36-5)
f-ratio (36-5)
angular magnification (36-5)
compound microscope (36-5)
reflecting telescope, refracting telescope (36-5)

Symbols You Should Recognize

ℓ, ℓ' (36-2, 36-3)
f (36-2, 36-3)
M, m (36-2–36-4)

Problems You Should be Able to Solve

finding mirror images and their sizes using ray tracing and algebraic techniques (36-2)
finding lens images and their sizes using ray tracing and algebraic techniques (36-3)
analyzing image formation at a single refracting surface (36-4)

designing lenses and analyzing their focal properties (36-4)
designing and analyzing simple optical instruments (36-5)

Limitations to Keep in Mind

Our description of imaging is based on the approximation of geometrical optics, valid only when objects and optical components are much larger than the wavelength of light.

Formulas for image formation in spherical mirrors and lenses are approximations, valid under the conditions that the curved surfaces used form only a small portion of a full sphere, and that all light rays are nearly parallel to the mirror or lens axis.

QUESTIONS

1. How can you see a virtual image, when it really "isn't there"?
2. You're trying to photograph yourself in a mirror, using an autofocus camera that sets its focus by bouncing ultrasound waves off the subject at which the camera is pointed. Why might the photo come out blurred?
3. You lay a magnifying glass (which is just a converging lens) on a printed page. Looking toward the glass, you move it away from the page. Explain the changes in what you see, especially as you move the lens beyond its focal length.
4. Under what circumstances will the image in a concave mirror be the same size as the object?
5. Describe the shapes of the mirrors making the images in Fig. 36-49.

FIGURE 36-49 Question 5.

6. If you're handed a converging lens, what can you do to estimate quickly its focal length?

7. What is the meaning of a negative object distance? Of a negative focal length?
8. A diverging lens always makes a reduced image. Could you use such a lens to start a fire by focusing sunlight? Explain.
9. Is there any limit to the temperature you can achieve by focusing sunlight? Think about the second law of thermodynamics.
10. Can a concave mirror make a reduced real image? A reduced virtual image? An enlarged real image? An enlarged virtual image? Give conditions for each image that is possible.
11. If you placed a screen at the location of a virtual image, would you see the image on the screen? Why or why not?
12. Where must the object be placed to form a reduced real image with a concave mirror or converging lens?
13. Where should you place a flashlight bulb in relation to the focus of its reflector?
14. Is the image on a movie screen real or virtual? How do you know?
15. Does a fish in a spherical bowl appear larger or smaller than it actually is?
16. A block of ice contains a hollow, air-filled space in the shape of a double-convex lens. Describe the optical behavior of this space.
17. The refractive index of the human cornea is about 1.38. If you can see clearly in air, why can't you see clearly underwater? Why do goggles help?
18. Why does "stopping down" a lens or mirror allow it to focus more sharply?
19. If you increase the f-stop setting on your camera from 4 to 5.6, what happens to the amount of light admitted through the lens?
20. The compound microscope and the refracting telescope are both two-lens optical systems. How do their designs reflect their different uses?
21. Cheap binoculars sometimes show blurred images with different colors evident across the blurred region. Why?

22. Do you want a long or short focal length for the objective lens of a telescope? Of a microscope?

23. Give several reasons reflecting telescopes are superior to refractors.

24. Given that star images cannot be resolved even with telescopes, why are large telescopes superior to ordinary cameras for studying stars?

PROBLEMS

Sections 36-1 and 36-2 Plane and Curved Mirrors

1. A shoe store uses small floor-level mirrors to let customers view prospective purchases. At what angle should such a mirror be inclined so that a person standing 50 cm from the mirror with eyes 140 cm off the floor can see her feet?

2. Two plane mirrors occupy the first four meters of the positive x and y axes, as shown in Fig. 36-50. Find the locations of all images of an object at $x = 2$ m, $y = 1$ m.

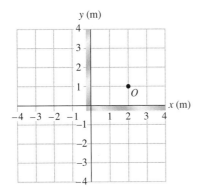

FIGURE 36-50 Problems 2, 3.

3. Suppose a stick were placed in Fig. 36-50 with its left end at O and extending parallel to the x axis. How many units long could it be for an observer at $x = 1$ m, $y = 1$ m to be able to see its entire length reflected in the x-axis mirror, which ends at $x = 4$ m.

4. A candle is 36 cm from a concave mirror with focal length 15 cm and on the mirror axis. (a) Where is its image? (b) How do the image and object sizes compare? (c) Is the image real or virtual?

5. An object is five focal lengths from a concave mirror. (a) How do the object and image heights compare? (b) Is the image upright or inverted?

6. The McMath-Pierce solar telescope at Kitt Peak National Observatory (Fig. 36-51) uses a single concave mirror to produce a real solar image 80 cm in diameter. (a) What is the focal length of the mirror? (b) How far is the image from the mirror? *Hint:* Consult Appendix E.

7. A virtual image is located 40 cm behind a concave mirror with focal length 18 cm. (a) Where is the object? (b) By how much is the image magnified?

FIGURE 36-51 The McMath-Pierce solar telescope (Problem 6).

8. (a) Where on the axis of a concave mirror would you place an object in order to produce a full-size image? (b) Will the image be real or virtual?

9. A 12-mm-high object is 10 cm from a concave mirror with focal length 17 cm. (a) Where, (b) how high, and (c) what type is its image?

10. An object's image in a 27-cm-focal-length concave mirror is upright and magnified by a factor of 3. Where is the object?

11. What is the curvature radius of a mirror that produces a 9.5-cm-high virtual image of a 5.7-cm-high object, when the object is located 22 cm from the mirror?

12. When viewed from Earth, the moon subtends an angle of 0.5° in the sky. How large an image of the moon will be formed by the 3.6-m-diameter mirror of the Canada-France-Hawaii telescope, which has a focal length of 8.5 m?

13. At what two distances could you place an object from a 45-cm-focal-length concave mirror in order to get an image 1.5 times the object's size?

14. Very distant objects are imaged close to the focus of a concave mirror. If a birdwatcher's telescope has a concave mirror with focal length 85 cm, what is the minimum distance at which a bird's image will be within 5 mm of the focal plane? (The focal plane is the plane through the focus and perpendicular to the mirror axis.)

FIGURE 36-52   Problem 15.

FIGURE 36-53   The Yerkes Observatory boasts the world's largest refracting telescope (Problem 22).

15. You look into a reflecting sphere 80 cm in diameter and see an image of your face at one-third its normal size (Fig. 36-52). How far are you from the sphere's surface?

Section 36-3 Lenses

16. By what factor is the image magnified when an object is placed 1.5 focal lengths from a converging lens? Is the image upright or inverted?
17. A lens with 50-cm focal length produces a real image the same size as the object. How far from the lens are image and object?
18. A real image is 4 times as far from a lens as is the object. What is the object distance, measured in focal lengths?
19. How far from a page should you hold a lens with 32-cm focal length in order to see the print magnified 1.6 times?
20. A converging lens has a focal length of 4.0 cm. A 1.0-cm-high arrow is located 7.0 cm from the lens with its lowest point 5.0 mm above the lens axis. Make a full-scale drawing of the situation, and use ray tracing to locate the image. Confirm using the lens equation.
21. A simple camera uses a single converging lens to focus an image on its film. If the focal length of the lens is 45 mm, what should be the lens-to-film distance for the camera to focus on an object 80 cm from the lens?
22. The largest refracting telescope in the world, at Yerkes Observatory, has a 1-m-diameter lens with focal length 12 m (Fig. 36-53). If an airplane flew 1 km above the telescope, where would its image occur in relation to the images of the very distant stars?
23. A small magnifying glass can focus sunlight sufficiently to ignite paper. Calculate the actual size of the solar image produced by a magnifying glass consisting of an optically perfect lens with 25-cm focal length. *Hint:* You'll need to consult Appendix E.
24. By holding a magnifying glass 25 cm from my desk lamp, I find I can focus an image of the lamp's bulb on a wall 1.6 m from the lamp. What is the focal length of my magnifying glass?

25. A lens has focal length $f = 35$ cm. Find the type and height of the image produced when a 2.2-cm-high object is placed at distances (a) $f + 10$ cm and (b) $f - 10$ cm.
26. How far apart are object and image produced by a converging lens with 35-cm focal length when the object is (a) 40 cm and (b) 30 cm from the lens?
27. An object and its lens-produced real image are 90 cm apart. If the lens has 20-cm focal length, what are the possible values for the object distance and magnification?
28. An object is placed two focal lengths from a diverging lens. (a) What type of image forms, (b) what is the magnification, and (c) where is the image?

Section 36-4 The Lensmaker's Formula

29. You're standing in a wading pool and your feet appear to be 30 cm below the surface. How deep is the pool?
30. A tiny insect is trapped 1.0 mm from the center of a spherical dew drop 4.0 mm in diameter. As you look straight into the drop at the insect, what is its apparent distance from the edge of the drop?
31. Use Equation 36-8 to show that an object at the center of a glass sphere will appear to be its actual distance—one radius—from the edge. Draw a ray diagram showing why this result makes sense.
32. You're underwater, looking through a spherical air bubble (Fig. 36-54). What is its actual diameter if it appears, along your line of sight, to be 1.5 cm in diameter?
33. Rework Example 36-6 for a fish 15 cm from the *far* wall of the tank.
34. Consider the inverse of Example 36-6: You're inside a 70-cm-diameter hollow tube containing air, and the tip of your nose is 15 cm from the wall of the tube. The tube is

FIGURE 36-54 Problem 32.

immersed in water, and a fish looks in. To the fish, what is the apparent distance from your nose to the tube wall?

35. Two specks of dirt are trapped in a crystal ball, one at the center and the other halfway to the surface. If you peer into the ball on a line joining the two specks, the outer one appears to be only one-third of the way to the other. What is the refractive index of the ball?

36. My magnifying glass is a double convex lens with equal curvature radii of 32 cm. If the lens glass has $n = 1.52$, what is its focal length?

37. Two lenses made of the same material have the same focal length. One is plano-convex, the other double convex with both curvatures the same. How do the curvature radii of the two lenses compare?

38. For what refractive index would the focal length of a plano-convex lens be equal to the curvature radius of its one curved surface?

39. An object is 28 cm from a double convex lens with $n = 1.5$ and curvature radii 35 cm and 55 cm. Where and what type is the image?

40. A double convex lens has equal curvature radii of 35 cm. An object placed 30 cm from the lens forms a real image at 128 cm. What is the refractive index of the lens?

41. A plano-convex lens has curvature radius 20 cm and is made from glass with $n = 1.5$. Use the generalized lens-maker's formula given in problem 71 to find the focal length when the lens is (a) in air, (b) submerged in water ($n = 1.333$) and (c) embedded in glass with $n = 1.7$. Comment on the sign of your answer to (c).

42. A slide projector has a double convex lens of focal length 104 mm. (a) What should be the slide-to-focal point distance to focus the image on a screen 4.5 m distant? What will be the magnification? (b) Repeat for the case when the lens is replaced by a 78-mm version.

43. Two plano-convex lenses are geometrically identical, but one is made from crown glass ($n = 1.52$), the other from flint glass. An object at 45 cm from the lens focuses to a real image at 85 cm with the crown-glass lens and at 53 cm with the flint-glass lens. Find (a) the curvature radius (common to both lenses) and (b) the refractive index of the flint glass.

44. A double convex lens with equal 38-cm curvature radii is made from glass with refractive indices $n_{red} = 1.51$, $n_{blue} = 1.54$ at the edges of the visible spectrum. If a point source of white light is placed on the lens axis at 95 cm from the lens, over what range will its visible image be smeared?

45. An object placed 15 cm from a plano-convex lens made of crown glass focuses to a virtual image twice the size of the object. If the lens is replaced with an identically shaped one made from diamond, what type of image will appear and what will be its magnification? See Table 35-1.

Section 36-5 Optical Instruments

46. You find that you have to hold a book 55 cm from your eyes for the print to be in sharp focus (Fig. 36-55). What power lens is needed to correct your farsightedness?

55 cm

FIGURE 36-55 Problem 46.

47. My grandmother's new reading glasses have 3.8-diopter lenses to provide full correction of her farsightedness. Her old glasses were 2.5 diopters. (a) Where is the near point for her unaided eyes? (b) Where will be the near point if she wears her old glasses?

48. A particular eye has a focal length of 2.0 cm instead of the 2.2 cm that would be required for a sharply focused image on the retina. (a) Is this eye nearsighted or farsighted? (b) What power of corrective lens is needed?

49. A camera's zoom lens covers the focal length range from 38 mm to 110 mm. (a) You point the camera at a distant object and photograph it first at 38 mm and then with the camera zoomed out to 110 mm. Compare the sizes of its images on the two photos. (b) If the camera's lowest f-ratio is 3.8 at 38 mm, what is it at 110 mm? Assume the effective lens area doesn't change.

50. A camera can normally focus as close as 60 cm, but it has provisions for mounting additional lenses at the outer end of the main lens to provide closeup capability. What type and power of auxiliary lens will allow the camera to focus as close as 20 cm? The distance between the two lenses is negligible.

51. The maximum magnification of a simple magnifier occurs with the image at the 25-cm near point. Show that the angular magnification is then given by $m = 1 + \dfrac{25 \text{ cm}}{f}$, where f is the focal length.

52. A compound microscope has objective and eyepiece focal lengths of 6.1 mm and 1.7 cm, respectively. If the lenses are 8.3 cm apart, what is the magnification of the instrument?

53. A 300-power compound microscope has a 4.5-mm-focal-length objective lens. If the distance from objective to eyepiece is 10 cm, what should be the focal length of the eyepiece?

54. To the unaided eye, the planet Jupiter has an angular diameter of 50 arc seconds. What will be its angular size when viewed through a 1-m-focal-length refracting telescope with an eyepiece whose focal length is 40 mm?

55. A Cassegrain telescope like that shown in Fig. 36-48b has 1.0-m focal length, and the convex secondary mirror is located 0.85 m from the primary. What should be the focal length of the secondary in order to put the final image 0.12 m behind the front surface of the primary mirror?

56. The Hubble Space Telescope is essentially a Cassegrain reflector like that shown in Fig. 36-48b. The focal lengths of the concave primary and convex secondary mirrors are 5520.00 mm and −679.00 mm, respectively. The secondary is located at 4906.071 mm from the apex of the primary. Draw a ray diagram showing the path of initially parallel incident rays through the telescope, and use appropriate equations to determine where in Fig. 36-48b such rays are finally focused.

Paired Problems

(Both problems in a pair involve the same principles and techniques. If you can get the first problem, you should be able to solve the second one.)

57. (a) How far from a 1.2-m-focal length concave mirror should you place an object in order to get an inverted image 1.5 times the size of the object? (b) Where will the image be?

58. (a) How far from a 48-cm-focal length concave mirror should you place an object in order to get an upright image 1.5 times the size of the object? (b) Where will the image be?

59. Find the focal length of a concave mirror if an object 15 cm from the mirror has a virtual image 2.5 times the object's actual size.

60. An object is held 6.0 cm from the surface of a reflecting ball, and its image appears three-quarters full size. What is the ball's diameter?

61. How far from a 1.6-m focal length concave mirror should you place an object to get an upright image magnified by a factor of 2.5?

62. How far from a 25-cm-focal-length lens should you place an object to get an upright image magnified by a factor of 1.8?

63. An object and its lens-produced real image are 2.4 m apart. If the lens has 55-cm focal length, what are the possible values for the object distance and magnification?

64. An object and its converging-lens-produced virtual image are 2.4 m apart. If the lens has 55-cm focal length, what are the possible values for the object distance and magnification?

65. An object is 68 cm from a plano-convex lens whose curved side has curvature radius 26 cm. The refractive index of the lens is 1.62. Where and of what type is the image?

66. Both surfaces of a double concave lens have curvature radii of 18 cm, and the refractive index is 1.5. If a virtual image appears 14 cm from the lens, where is the object?

Supplementary Problems

67. A distant object subtends an angle α at a lens of focal length f, as shown in Fig. 36-56. Take the object distance $\ell \gg f$, and assume the angle α is very small. In this approximation, find an expression for the size of the real image formed by the lens. *Hint:* For $\ell \gg f$, what is the approximate image distance?

68. Show that the powers of closely-spaced lenses add; that is, placing a 1-diopter lens in front of a 2-diopter lens gives the equivalent of a single 3-diopter lens.

69. Show that identical objects placed equal distances on either side of the focal point of a concave mirror or converging lens produce images of equal size. Are the images of the same type?

70. Galileo's first telescope used the arrangement shown in Fig 36-57, with a double convex eyepiece placed slightly before the focus of the objective lens. Use ray tracing to

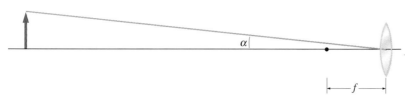

FIGURE 36-56 Problem 67.

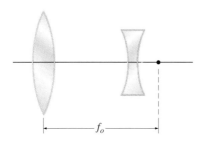

FIGURE 36-57   Problem 70.

show that this system gives an upright image, which makes the design useful for terrestrial observing.

71. Generalize the derivation of the lensmaker's formula (Equation 36-10) to show a lens of refractive index n_{lens} in an external medium with index n_{ext} has focal length given by

$$\frac{1}{f} = \left(\frac{n_{\text{lens}}}{n_{\text{ext}}} - 1\right)\left(\frac{1}{R_1} - \frac{1}{R_2}\right).$$

72. An object is located 40 cm from a lens made from glass with $n = 1.5$. In air, the resulting image is real and is located 32 cm from the lens. When the entire system is immersed in a liquid, the image becomes virtual and is reduced to half the object's size. Use the result given in the preceding problem to find the refractive index of the liquid.

73. A Newtonian telescope like that of Fig. 36-48c has a primary mirror with 20-cm diameter and 1.2-m focal length. (a) Where should the flat diagonal mirror be placed to put the focus at the edge of the telescope tube? (b) What shape should the flat mirror have to minimize blockage of light to the primary?

74. A parabola is described by the equation $y = x^2/4f$, where f is the distance from vertex to focus. (a) Find the equation for the circle (which would be a sphere in three dimensions) that closely approximates the parabola $y = x^2$. *Hint:* How are the focal length and curvature radius of a spherical mirror related? (b) Solve the circle's equation for y, and use the binomial theorem to show that the equation reduces to that of the parabola for $x \ll 1$.

75. Just before Equation 36-9 are two equations describing refraction at the two surfaces of a lens with thickness t. Combine these equations to show that the object distance ℓ and image distance ℓ' for such a lens are related by

$$\frac{1}{\ell} + \frac{1}{\ell'} - \frac{[(n-1)\ell - R_1]^2 t}{\ell R_1[t(\ell + R_1) + n\ell(R_1 - t)]}$$
$$= (n-1)\left(\frac{1}{R_1} - \frac{1}{R_2}\right).$$

76. Use the result of the preceding problem to find an expression for the focal length of a transparent sphere with radius R and refractive index n, as measured from the exit surface.

CHAPTER 37

INTERFERENCE AND DIFFRACTION

The colors seen in a soap bubble result from constructive interference of light waves reflecting off the inner and outer surfaces of the thin film that forms the bubble wall. Note that the far end of the bubble appears dark; this is because it is too thin for constructive interference at any visible wavelength.

The preceding chapters considered reflection and refraction, and their application in image formation, from the approximation of geometrical optics—an approximation valid when we can ignore the wave nature of light. We now turn to **physical optics,** which treats optical phenomena in which the wave nature of light plays an essential role. Two related phenomena, interference and diffraction, are central to much of physical optics.

37-1 INTERFERENCE

We considered wave **interference** in Chapter 16, showing how this phenomenon arises when two or more waves meet at the same place, their amplitudes adding in a manner that may be constructive or destructive depending on their relative phases. Interference of electromagnetic waves—including light—occurs in the same way; since both the electric and magnetic fields obey the superposition principle, the net fields at a point are the vector sums of the fields of individual waves. Depending on their directions, the superposition of those fields may result in an increase (constructive interference) or a decrease (destructive interference) in the overall field strength.

Coherence

Although interference occurs any time two waves interact, we only get steady interference patterns when the interacting waves are **coherent,** meaning that they maintain a constant phase relation (Fig. 37-1*a,b*). Coherence also requires that the two waves have exactly the same well-defined wavelength. (Waves of a single, sharply defined wavelength are called **monochromatic** because they correspond to a single color of light.) Very slight deviations from exactly equal wavelengths may still permit observable interference, the optical equivalent of beats. Even with lasers, whose light output is highly monochromatic and coherent, it is difficult to keep two different sources in exactly the same phase relation. But when light from a single source is split into two or more beams then, even if the source is itself incoherent, the beams will always be in the same phase relation and will therefore produce an interference pattern (Fig. 37-1*c*).

Interference in Thin Films

We know that light is partially reflected and partially transmitted at an interface between two transparent media. Figure 37-2*a* shows what happens when light propagating in air hits a thin film of transparent material. Partial reflection at

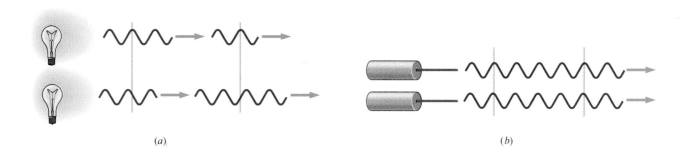

(*a*)   (*b*)

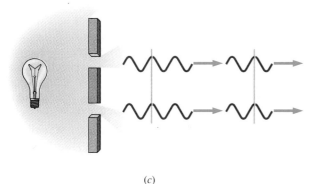

(*c*)

FIGURE 37-1   (*a*) Light bulbs emit incoherent light, consisting of wavetrains whose phases are related in a random way. Wavetrains from two different light bulbs would have random phase relations, and superposing them would not result in a steady interference pattern. (*b*) Laser light is coherent over much longer distances and therefore two lasers could be made to interfere if their wavelengths were exactly the same. (*c*) Two separate beams derived from a single point source maintain a constant phase relation even when the source itself is incoherent. Here light from a single bulb has been split by passing it through two holes in an opaque barrier. In this case the two beams are exactly in phase; making one beam traverse a longer path would introduce a constant phase difference. A single beam could also be split by partial reflection.

FIGURE 37-2   Reflection and refraction at a thin layer of transparent material. (*a*) Reflected beams 1 and 2 emerging at left are coherent and should therefore interfere. (*b*) The reflected and incident beams are 180° out of phase when they reflect at an interface that goes from a lower to a higher refractive index. (*c*) There is no phase difference on reflection when the incident medium has a higher refractive index.

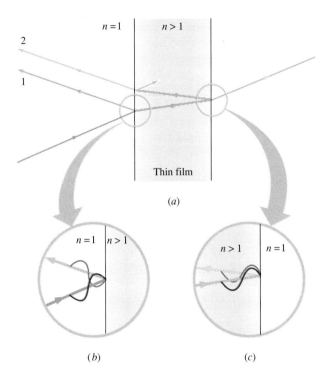

both surfaces results in two beams propagating back into the original medium. Since these beams derive from the same source they are coherent, and they should therefore interfere. (Additional reflections actually give rise to multiple beams, each successively fainter; here we consider only the first two reflected beams.)

Before we can analyze this interference, we need to understand exactly what happens when light reflects. In Section 17-5 we showed that waves on a string are partially reflected and partially transmitted at a junction between two strings with different mass per unit length. When the wave met a string with *greater* mass per unit length, it was reflected *with a 180° phase change.* When it met a string with *lower* mass per unit length, it was reflected *with no phase change.* The physical reason for these phase relations was shown in Fig. 17-7 for the extreme cases of infinite mass per unit length and zero mass per unit length. Although we will not prove it, Maxwell's equations show that light behaves in the same way, with the refractive index playing the role of mass per unit length. That is, light propagating in a material with refractive index n_1 reflects with a 180° phase change at the interface with a material of index n_2 if $n_1 < n_2$. If, on the other hand, $n_2 < n_1$, then there is no phase change. Figures 37-2*b* and *c* show these phase relations.

Now suppose the film in Fig. 37-2 were negligibly thin compared with the wavelength of light. Then the reflected beams 1 and 2 would be 180° out of phase since beam 1 reflects with a 180° phase change and beam 2 with no phase change. If the film has thickness d, then there will also be a phase difference because beam 2 travels an additional distance. For simplicity we assume that light is incident on the film at nearly normal incidence, so the extra distance traveled by beam 2 is essentially $2d$. What phase difference does this introduce? An extra half wavelength corresponds to a phase change of 180° (Fig. 37-3a), while a full wavelength separates the two waves by a full cycle, putting them back in phase (Fig. 37-3b). The 180° phase change would also occur if the path lengths differed by $\frac{3}{2}\lambda$, $\frac{5}{2}\lambda$, and so forth, and there would be no change if they differed by any integer number of wavelengths. There's also the 180° phase change introduced by the different reflections, which combines with the phase change due to the path-length difference to put beams 1 and 2 *in phase* if the path-length difference is $\frac{1}{2}\lambda$, $\frac{3}{2}\lambda$, $\frac{5}{2}\lambda$, and so forth. Conversely, beams 1 and 2 are out of phase by 180° if the path length is an integer multiple of the wavelength.

If beams 1 and 2 are in phase they interfere constructively, making the combined reflected beam brighter than it otherwise would be. We've just seen that the condition for this constructive interference is that the path-length difference $2d$ be an odd integer multiple of a half wavelength. We can express this mathematically by writing

$$2d = (m + \tfrac{1}{2})\lambda_n,$$

where m is a nonnegative integer and where λ_n is the wavelength *in the transparent film* since that's where the extra path length occurs. We saw in Chapter 35 that the wavelength in a material with refractive index n is reduced by a factor $1/n$ from its value in air or vacuum. If λ is the wavelength in air, then $\lambda_n = \lambda/n$, and our condition for constructive interference becomes

$$2nd = (m + \tfrac{1}{2})\lambda. \tag{37-1a}$$

Conversely, destructive interference occurs when

$$2nd = m\lambda. \tag{37-1b}$$

The destructive interference is not complete, however, since the two beams do not have the same intensity.

Suppose a film is illuminated with light of a single color—that is, a single wavelength λ. If the thickness of a film varies with position, then constructive interference described by Equation 37-1a will occur at different places. Thus there will be bright bands separated by darker areas in which destructive interference occurs. If, on the other hand, the film is illuminated with white light, then the interference conditions will be satisfied at different places for the different colors that comprise white light. The bands of color you see in a soap bubble or oil slick result from such thin-film interference (Fig. 37-4). If part of the film is too thin to meet the constructive interference condition of Equation 37-1a for *any* visible wavelength, then that part of the film will appear dark. You can see this at the top of the soap film in Fig. 37-4 and also in the photo at the beginning of this chapter, where one end of the soap bubble appears dark.

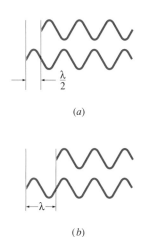

FIGURE 37-3 (a) Two waves begin to propagate at the same instant, but the lower one travels half a wavelength farther. When they come together they are 180° out of phase. (b) When the lower wave travels a full wavelength farther, they come together in phase.

FIGURE 37-4 Interference patterns formed in a soap film illuminated with white light. The film's thickness increases toward the bottom, resulting in constructive interference for a given wavelength occurring at several different vertical positions. The film is too thin at the top for constructive interference at any visible wavelength, so it appears dark.

● EXAMPLE 37-1 A SOAP FILM

A rectangular wire loop 20 cm high is dipped into a soap solution and then held vertically, producing a soap film whose thickness varies linearly from essentially zero at the top to 1.0 μm at the bottom (Fig. 37-5). (a) If the film is illuminated with 650-nm light, how many bright bands (as shown in Fig. 37-4) will appear? (b) What region of the film would be dark if it were illuminated with white light? The refractive index of the film is that of water, $n = 1.33$.

Solution

The film thickness at the position of a bright band is given by Equation 37-1a. To find the number of bright bands, we can solve this equation for the value of m at which the thickness d would equal the maximum film thickness:

$$m = \frac{2nd}{\lambda} - \frac{1}{2} = \frac{(2)(1.33)(1.0 \ \mu m)}{0.65 \ \mu m} - \frac{1}{2} = 3.6.$$

Since m must be an integer, the largest m corresponding to a thickness within the 1.0-μm maximum is 3. Since m ranges from zero to 3, there are a total of 4 bright bands.

At a given wavelength the minimum thickness for constructive interference occurs at $m = 0$. Taking the minimum wavelength for visible light at 400 nm = 0.40 μm, Equation 37-1a with $m = 0$ then gives

$$d = \frac{\lambda}{4n} = \frac{0.40 \ \mu m}{(4)(1.33)} = 0.0752 \ \mu m.$$

Since the film goes linearly from zero thickness to a maximum of 1.0 μm over its 20-cm height, this thickness occurs 0.0752 of the way from the top edge, or at $(0.0752)(20 \text{ cm}) = 1.5$ cm.

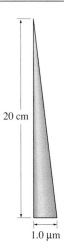

FIGURE 37-5 Cross section through a vertical soap film. The angle between the surfaces is greatly exaggerated.

Above this level the film is too thin for constructive interference in visible light, so the top 1.5 cm of film appears darker.

EXERCISE At what distances from the top of the film will bright bands appear if the film in Example 37-1 is illuminated with 450-nm light?

Answer: 1.69 cm, 5.08 cm, 8.46 cm, 11.8 cm, 15.2 cm, 18.6 cm

Some problems similar to Example 37-1: 1–3, 5 ●

We derived Equations 37-1 on the assumption that our thin film was surrounded by air. Frequently, though, a thin film is sandwiched between two different media—as when an oil film floats on water. If the intermediate medium has a higher refractive index than the medium at its rear (i.e., at the right in Fig. 37-2), then our results still apply. But if the rear medium has a greater refractive index than the thin film, then there will be an additional 180° phase change at the rear interface, and therefore, our interference conditions will reverse (see the application below, as well as Problems 63 and 64).

■ APPLICATION ANTIREFLECTION COATINGS

Partial reflection limits the amount of light that can be transmitted from one transparent material into another. That limitation, described quantitatively in Equation 35-7 and Figure 35-26, reduces the light-gathering power of cameras, binoculars, and other lens-based instruments and of nonimaging devices like windows and solar collectors.

Coating lenses, photovoltaic cells, and other critical light-gathering components with thin layers of appropriate materials can reduce reflection through the use of destructive interference. Since energy is conserved, the result is necessarily greater transmission into the light-gathering device. Normally such **antireflection coatings** have refractive indices between those of air and glass; consequently, there is a 180° phase change at *each* interface, and thus, Equation 37-1a rather than 37-1b gives the condition for *destructive* interference. Thus, the minimum thickness for an antireflection coating, given by Equation 37-1a with $m = 0$, is $d = \lambda/4n$. Since this result depends on wavelength, antireflection coatings are not equally effective for all colors. With composite lenses and multiple layers, however, reduced reflection is possible over the visible spectrum.

A perfect antireflection coating might seem impossible, even in principle, since the reflected rays 1 and 2 in Fig. 37-2 do not have the same intensity. But the situation is actually more complicated. There are additional rays resulting from multiple reflections inside the transparent material, and these also interfere. A full solution of the problem involving the application of Maxwell's equations at both interfaces shows that the reflection becomes exactly zero when the antireflection layer has not only the right thickness but also refractive index $n_2 = \sqrt{n_1 n_3}$, where n_1 and n_3 are the indices of the two media separated by the antireflection coating. A widely used antireflection coating, magnesium fluoride (MgF_2), has $n = 1.38$, a value that approximates the zero-reflection condition for high-index flint glasses in air.

Interference in thin layers is the basis of some very sensitive optical measuring techniques. The shape of a lens, for example, can be checked for accuracy to within a fraction of the wavelength of light using **Newton's rings**—interference patterns formed by interference of light reflected between the lens and a perfectly flat glass plate (Fig. 37-6; see also Problems 73-74).

The Michelson Interferometer

A number of optical instruments use interference for precise measurement of small distances. Among the simplest and most important of these is the **Michelson interferometer,** invented by the American physicist Albert Michelson and used in the 1880s by Michelson and his colleague Edward W. Morley in a famous experiment that paved the way for the theory of relativity. We discuss the Michelson-Morley experiment in the next chapter; here we describe the interferometer, which is still used for precision measurements.

Figure 37-7 shows the basic design of the Michelson interferometer. The key idea is that light from a monochromatic source is split into two beams by a half-silvered mirror called a **beam splitter.** The beam splitter is set at a 45° angle, so the reflected and transmitted beams travel perpendicular paths. Each then reflects off a flat mirror and returns to the beam splitter. The beam splitter again transmits and reflects half the light incident on it, with the result that some light from the originally separated beams is recombined. Since this light shares a common source, it is coherent and therefore interferes. The interference pattern is observed with a viewing lens located at the bottom of Fig. 37-7.

Suppose the path lengths in the two perpendicular arms of the interferometer were exactly the same or differed by a multiple of the wavelength. When the beams recombined they would then undergo fully constructive interference. If the path-length difference were a multiple of a half wavelength, on the other hand, the beams would interfere destructively. In reality, however, the mirrors are never exactly perpendicular, and therefore light reflecting from different parts of the mirrors experiences slightly different path lengths and thus recombines with different phase lags. To the observer, the result is a series of alternating light and dark fringes corresponding to constructive and destructive interference.

Now suppose one mirror is moved slightly. The path-length differences change and therefore, the interference pattern shifts. Moving the mirror a mere quarter wavelength, for example, adds an extra half wavelength to the round-trip

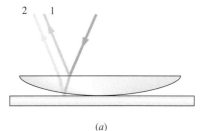

(a)

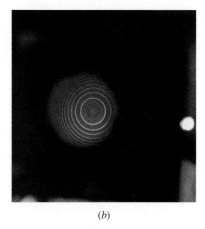

(b)

FIGURE 37-6 (a) Newton's rings arise from the difference in path lengths between rays like 1 and 2 shown here. (b) Ring pattern produced in a test of a telescope's optical system.

FIGURE 37-7 (*a*) Schematic diagram of a Michelson interferometer. The beam splitter splits incident light into reflected and transmitted beams of equal intensity. (*b*) The observer looking through the viewer sees interference fringes arising from differing optical path lengths.

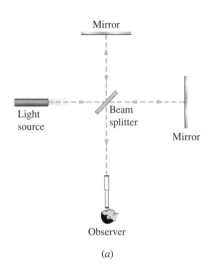

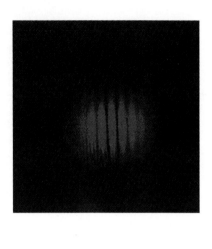

(*a*)

(*b*)

FIGURE 37-8 A modern interferometer, assembled from common laboratory optical components. The light source is a helium-neon laser producing 633-nm red light, and the lens at the right is used to project the interference fringes onto a screen. The fringes shown in Fig. 37-7*b* were produced with this interferometer.

path. That results in an additional 180° phase shift, moving dark fringes to where light ones were, and vice versa. An observer looking through the viewer as the mirror moves sees the fringes shift to their new positions. Shifts of a fraction of the distance between fringes are readily detected, allowing measurement of mirror displacements to within a small fraction of the wavelength.

A similar fringe shift occurs if a transparent material is placed in one path, retarding the beam because of its refractive index. This approach allows accurate measurements of the refractive indices of gases, which are so close to one that conventional methods would not be useful (see Problem 16).

Measurements with the Michelson interferometer depend not on the interference pattern itself but on *changes* in that pattern. Therefore, it doesn't really matter whether or not the path lengths are exactly equal. Whatever those lengths, there will be an interference pattern, and changes in that pattern will reflect changes in the relative optical paths. What is important is that components of the interferometer be mounted securely so their positions cannot change by even a fraction of the wavelength of light. The Michelson-Morley interferometer was constructed on a huge stone slab floated in mercury. Today, interferometers using lasers and precision optical components are easily assembled on optical tables designed for stability and isolation from vibration (Fig. 37-8).

● **EXAMPLE 37-2** SANDSTORM

A sandstorm has pitted the aluminum mirrors of a desert solar energy installation, and engineers want to know the depths of these pits so they can estimate the mirrors' useful lifetimes. They construct a Michelson interferometer with a sample from one of the pitted mirrors in place of one of the flat mirrors. With a 633-nm laser as the light source, the interference pattern shown in Fig. 37-9 results. What is the approximate depth of the pit?

Solution

The "bumps" in the interference pattern correspond to light that has traveled the extra distance into the bottom of the pit and back. A shift of one full cycle in the interference pattern would correspond to a full wavelength change in the path. As Fig. 37-9 shows, the pit causes a fringe shift corresponding to about 0.2 wavelengths. Since light makes a round trip through the pit, its actual depth is therefore 0.1λ, or 63 nm. Try measuring that with a meter stick!

EXERCISE A Michelson interferometer uses sodium light with wavelength 589.6 nm. As one mirror is moved, a photocell connected to a counter "watches" a fixed point through the viewer. If the counter records 4878 fringes passing that point, how far did the mirror move?

Answer: 1.438 mm

Some problems similar to Example 37-2: 14–16

$0.2\,\lambda$ λ

FIGURE 37-9 Fringe pattern resulting from a pitted mirror. The size of the bump is a measure of the pit depth.

37-2 HUYGENS' PRINCIPLE AND DIFFRACTION

The Dutch scientist Christian Huygens was the first to suggest that light might be a wave, and in 1678 he stated a principle that remains useful for describing how light waves propagate. Although Huygens worked nearly two centuries before Maxwell, his principle, like all results in classical optics, can be derived ultimately from Maxwell's equations of electromagnetism. **Huygens' principle** states the following:

> **All points on a wavefront act as point sources of spherically propagating "wavelets" that travel at the speed of light appropriate to the medium. At a short time Δt later, the new wavefront is the unique surface tangent to all the forward-propagating wavelets.**

Figure 37-10 shows how Huygens' principle accounts for the propagation of plane and spherical (or cylindrical) waves. It is also possible to derive the laws of reflection and refraction from Huygens' principle.

Diffraction

Diffraction is the bending of light or other waves as they pass by objects. Figure 37-11 shows plane waves incident on an opaque barrier containing a hole. Since the waves are blocked by the barrier, Huygens' wavelets produced near the barrier edge cause the wavefronts to bend at the barrier (see blowups in Fig. 37-11). When the width of the hole is much greater than the wavelength, as in Fig. 37-11a, this diffraction is of little consequence, and the waves effectively propagate straight through the hole in a beam defined by the hole size. But when the hole size and wavelength are comparable, then wavefronts emerging from the hole spread in a broad pattern (Fig. 37-11b). Here diffraction is the dominant effect on the propagation. Thus diffraction, although it always occurs when waves pass an object, is significant only on length scales comparable to or

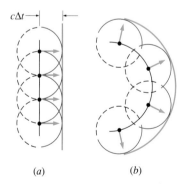

$c\Delta t$

(a)   (b)

FIGURE 37-10 Application of Huygens' principle to (a) plane and (b) spherical waves. In each case the wavefront acts like a set of point sources that emit circular waves. A short time Δt later the wavelets have expanded to radius $c\Delta t$, and the new wavefront is the surface tangent to the wavelets. To follow the propagation further one could draw new wavelets originating on the new wavefront.

FIGURE 37-11 Plane waves incident on an opaque barrier with a hole in it are diffracted at the edges of the hole, as shown by Huygens' wavelet constructions. (*a*) When the hole size is much larger than the wavelength this diffraction is negligible, and the waves essentially propagate in a straight beam defined by the hole. (*b*) When the hole size and wavelength are comparable, the emerging wavefronts spread in a broad beam, approaching circular wavefronts as the hole size becomes negligibly small.

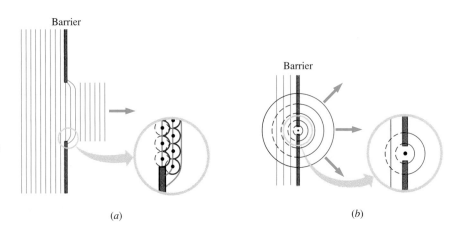

(*a*)

(*b*)

FIGURE 37-12 Diffraction of water waves as they pass through a gap in a breakwater results in circular wavefronts.

smaller than the wavelength. That's why we could ignore diffraction and assume that light always travels through a single medium in straight lines, when we considered optical systems with dimensions much larger than the wavelength of light.

Diffraction is not limited to light. When a stereo is playing in an adjacent room, you can hear the bass more clearly than the high notes. That's because the higher frequency sound does not diffract as much since its wavelength is shorter, and therefore, the situation is more like that of Fig. 37-11*a*. Lower notes, on the other hand, may have wavelengths comparable to the size of doorways or rooms, so they diffract readily. Similarly, diffraction of water waves occurs when their wavelength is comparable to the size of obstacles they encounter (Fig. 37-12).

In the remainder of this chapter we discuss interference and diffraction phenomena. Although we characterize phenomena by one or the other of these names, it's important to recognize that both interference and diffraction result from the wave nature of light and that both can play important roles in the same optical system. Often the wavefronts that result from diffraction subsequently interfere to produce characteristic patterns (Fig. 37-13). Ultimately this effect limits our ability to image small objects, as we will explore quantitatively later in this chapter.

37-3 DOUBLE-SLIT INTERFERENCE

In Chapter 16 we looked briefly at the interference patterns produced by a pair of coherent sources. With light, such a source pair can be made by passing light through a pair of narrow slits. In 1802 Thomas Young used this approach in an historic experiment that confirmed the wave nature of light. Young admitted sunlight to his laboratory through a small hole, then passed the light through a pair of closely spaced slits, after which it illuminated a screen. Waves diffract at each slit, resulting in cylindrical wavefronts that interfere in the region between slits and screen (Fig. 37-14*a*). Figure 37-14*b* shows how this interference should produce alternating regions of light and dark on the screen, while Fig. 37-14*c* shows that that is indeed what is observed. These light and dark regions are called interference **fringes.**

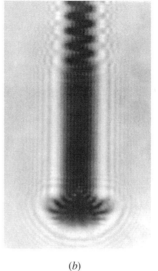

(a)

FIGURE 37-13 (*a*) Light waves diffract as they pass by the straight edge of an opaque barrier. Subsequent interference of the diffracted waves produces the interference fringes shown. (*b*) Diffraction of light passing by a screw again shows interference fringes and thus results in a fuzzy shadow of the screw. (*c*) Diffraction of light by a pair of crosshairs in a circular aperture. Diffraction effects from both the crosshairs (straight fringes) and the aperture itself (circular fringes) are evident.

(b)   **(c)**

The bright fringes in Fig. 37-14 occur at points where paths from the two slits are either the same length (for the central fringe) or differ by an integer number of wavelengths, causing the two waves to arrive in phase. Dark regions correspond to path lengths differing by an odd integer number of half wavelengths, making the waves 180° out of phase. We can relate the spacing of the interference fringes to the wavelength and slit spacing if we assume that the distance *L*

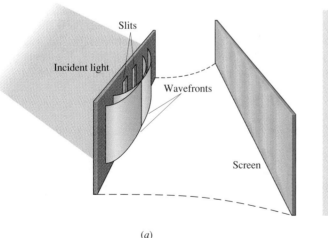

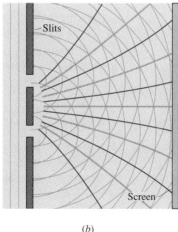

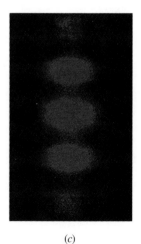

(a)   **(b)**   **(c)**

FIGURE 37-14 In Young's experiment, light passes through a pair of closely spaced slits. Light from the two slits subsequently interferes, producing a series of light and dark fringes on the screen. (*a*) Diagram showing the slits and one pair of wavefronts. Where the two wavefronts intersect the interference is constructive. Also shown is the screen, which in practice would be much farther from the slits. (*b*) Top view of the system, showing how interference between wavefronts should produce a pattern of light and dark fringes on the screen. (*c*) Photo produced by using film in place of the screen shows the expected fringe pattern.

FIGURE 37-15 (*a*) Geometry for finding locations of the interference fringes. A bright fringe occurs when the path-length difference $r_2 - r_1$ is an integer multiple of the wavelength. (*b*) For $L \gg d$ the paths to P are nearly parallel, and the difference in their lengths is $d \sin \theta$. Can you find similar triangles to prove that the angles labeled θ are indeed the same?

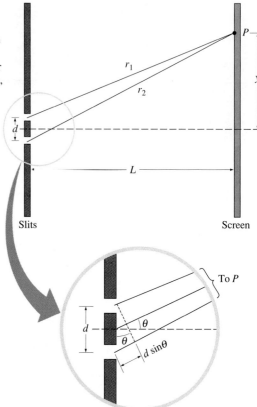

from slits to screen is much greater than the slit spacing d (Fig. 37-15). Then the two paths r_1 and r_2 from the slits to a point P on the screen are very nearly parallel, and the angle each makes with the horizontal in Fig. 37-15 is essentially the same as the angle θ made by a line to P from the point midway between the slits. Figure 37-15*b* then shows that the length difference between these paths is essentially $d \sin \theta$. The condition that a bright fringe be located at point P is that this difference be an integer number of wavelengths, or

$$d \sin \theta = m\lambda. \quad \text{(bright fringes)} \qquad (37\text{-}2a)$$

The integer m is the **order** of the fringe, with the central bright fringe being the zeroth order fringe and with higher order fringes on either side.

Waves interfere destructively when they arrive at the screen 180° out of phase, which occurs when their path lengths differ by an odd integer multiple of a half wavelength:

$$d \sin \theta = (m + \tfrac{1}{2})\lambda, \quad \text{(dark fringes)} \qquad (37\text{-}2b)$$

where m is any integer.

In a typical double-slit experiment, L may be on the order of 1 m, d a fraction of 1 mm, and λ the sub-μm wavelength of visible light. Then we have the additional condition that $\lambda \ll d$. This makes the fringes very closely spaced on the screen, so the angle θ in Fig. 37-15*b* is small even for large orders m. Then

$\sin\theta \simeq \tan\theta = y/L$, and Equations 37-2 and the geometry of Fig. 37-15 show that a fringe's position y on the screen, measured from the central maximum, is given by

$$y_{\text{bright}} = m\frac{\lambda L}{d} \quad \text{and} \quad y_{\text{dark}} = (m + \tfrac{1}{2})\frac{\lambda L}{d}. \quad (37\text{-}3a,b)$$

These equations show that the fringe spacing depends on wavelength, as confirmed experimentally in Fig. 37-16. Measurement of fringe spacing enabled Young to determine the wavelength of light.

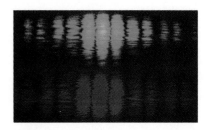

FIGURE 37-16 Interference patterns made with the same double-slit system, using two different wavelengths of light. The different fringe spacings reflect the wavelength dependence in Equations 37-3.

● **EXAMPLE 37-3** LASER WAVELENGTH

A pair of narrow slits are 0.075 mm apart and are located 1.5 m from a screen. Laser light shining through the slits produces an interference pattern whose third-order bright fringe is 3.8 cm from the screen center. Find the wavelength of the light.

Solution

Here we have $m = 3$, $L = 1.5$ m, and $d = 0.075$ mm, so we can solve Equation 37-3a for the wavelength λ:

$$\lambda = \frac{y_{\text{bright}}d}{mL} = \frac{(0.038\text{ m})(0.075\times10^{-3}\text{ m})}{(3)(1.5\text{ m})} = 633\text{ nm}.$$

This is in fact the wavelength of the red light from low-power helium-neon lasers commonly used in physics demonstrations.

EXERCISE What slit spacing will produce bright fringes 1.8 cm apart on a screen 85 cm from the slits, if the slits are illuminated with 589-nm light?

Answer: 27.8 μm

Some problems similar to Example 37-3: 20–23, 28
●

Intensity in the Interference Pattern

Geometric arguments allowed us to find the positions of the maxima and minima of a two-slit interference pattern. We can find the actual intensity variation by algebraically superposing the interfering waves. You might think this could be done by adding the wave intensities. But no! It's the electric and magnetic fields that obey the superposition principle, not the wave intensities. Intensity is proportional to the *square* of either field; if we added intensities we could never get the cancellation that occurs in destructive interference.

Consider again a point P on the screen of a double-slit apparatus (Fig. 37-17). Since light from the two slits reaches P over paths of different lengths, we expect that the expanding waves will have different intensities when they reach P. In the approximation $d \ll L$, however, the path-length difference $d\sin\theta$ is so small that we can neglect this effect. So we consider that the two waves arriving at P have electric fields of equal amplitude, given by expressions like Equation 34-10. Since we're considering a fixed point P, it is convenient to take the origin at P to eliminate the kx term in Equation 34-10. Then we can express the electric fields of the two waves in the form

$$E_1 = E_p \sin\omega t \quad \text{and} \quad E_2 = E_p \sin(\omega t + \phi),$$

where E_p is the common amplitude, ω the common frequency, and ϕ the phase difference that occurs because of the different path lengths. We have not bothered with vector notation because both waves, since they originate in the same source, are polarized in the same direction and thus their electric fields will simply add algebraically. Then net electric field at P is

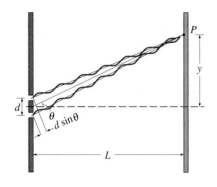

FIGURE 37-17 Waves leaving the slits with the same phase arrive at P displaced by the path-length difference $d\sin\theta$. This corresponds to phase difference $\phi = d\sin\theta(2\pi/\lambda)$.

$$E = E_1 + E_2 = E_p[\sin\omega t + \sin(\omega t + \phi)].$$

In Appendix A we find the trigonometric identity $\sin\alpha + \sin\beta = 2\sin[(\alpha + \beta)/2]\cos[(\alpha - \beta)/2]$, which, with $\alpha = \omega t$ and $\beta = \omega t + \phi$, transforms our expression for E into

$$E = 2E_p\sin\left(\omega t + \frac{\phi}{2}\right)\cos\left(\frac{\phi}{2}\right).$$

We also used $\cos(-x) = \cos x$ to eliminate a minus sign in the argument of the cosine. Thus, the electric field at P oscillates with the wave frequency ω, and its overall amplitude is scaled by the factor $\cos(\phi/2)$.

What is the phase difference ϕ? We've already seen that the path-length difference is $d\sin\theta$, with d the slit spacing and θ the angle subtended at the slits between P and the slit centerline. If this difference is one-half wavelength λ, then we have a phase difference of 180° or π radians. More generally, the phase difference in radians is whatever fraction of the full cycle (2π radians) the path difference $d\sin\theta$ is of the wavelength:

$$\phi = 2\pi\frac{d\sin\theta}{\lambda}.$$

Using this result in our expression for the electric field at P gives

$$E = 2E_p\sin\left(\omega t + \frac{\phi}{2}\right)\cos\left(\frac{\pi d\sin\theta}{\lambda}\right).$$

This equation describes a field whose peak amplitude is not E_p but $2E_p\cos(\pi d\sin\theta/\lambda)$. The average intensity then follows from Equation 34-20b:

$$\bar{S} = \frac{[2E_p\cos(\pi d\sin\theta/\lambda)]^2}{2\mu_0 c} = 4\bar{S}_0\cos^2\left(\frac{\pi d\sin\theta}{\lambda}\right), \qquad (37\text{-}4)$$

where $\bar{S}_0 = E_p^2/2\mu_0 c$ is the average intensity of either wave alone. Since the cosine function varies between -1 and 1, Equation 37-4 shows that the intensity varies between zero and $4\bar{S}_0$ as the angular position changes. We can also write Equation 37-4 in terms of position y on the screen. Under the approximation $d \gg \lambda$ even high-order fringes will occur at small angles θ, so we can write $\sin\theta \simeq \tan\theta \simeq y/L$, giving

$$\bar{S} = 4\bar{S}_0\cos^2\left(\frac{\pi d}{\lambda L}y\right). \qquad (37\text{-}5)$$

Now $\cos^2\alpha$ has its maximum value, 1, when its argument is an integer multiple of π. Thus, the maxima of Equation 37-5 occur when $dy/\lambda L$ is an integer m, or when $y = m\lambda L/d$. This is just the condition of Equation 37-3a, showing that our intensity calculation is fully consistent with the simpler geometrical analysis. But the intensity calculation tells us more: it gives not only the positions of bright and dark fringes, but also the intensity variation in between, as shown in Fig. 37-18.

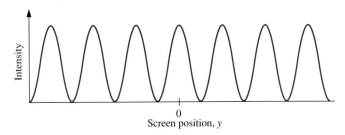

FIGURE 37-18   Intensity as a function of position y on the screen of a double-slit system.

37-4 MULTIPLE-SLIT INTERFERENCE AND DIFFRACTION GRATINGS

Systems with multiple slits play a crucial role in optical instrumentation and in the analysis of materials. As we will soon see, gratings manufactured with several thousand slits per centimeter make possible high-resolution spectroscopic analysis. At a much smaller scale the regularly spaced rows of atoms in a crystal act much like a multiple-slit system for x rays, and the resulting x-ray patterns reveal the crystal structure.

Figure 37-19 shows waves from three evenly spaced slits interfering at a screen. Maximum intensity requires that all three waves either be in phase, or differ in phase by an integer number of wavelengths. Our criterion for the maximum in a two-slit pattern, $d \sin \theta = m\lambda$, ensures that waves from two adjacent slits will add constructively. Since the slits are evenly spaced with distance d between each pair, waves coming through a third slit will be in phase with the other two if this criterion is met. So the criterion for a maximum in an N-slit system is still Equation 37-2a:

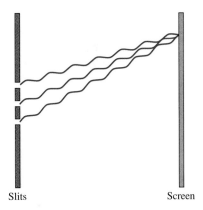

FIGURE 37-19   Waves from three evenly spaced slits interfere constructively when they arrive at the screen in phase.

$$d \sin \theta = m\lambda. \quad \text{(maxima in multi-slit interference)}$$

With more than two waves, however, the criterion for destructive interference is more complicated. Somehow all the waves need to sum to zero. Figure 37-20 shows that this happens for three waves when each is out of phase with the others by one-third of a cycle. Thus, the path-length difference $d \sin \theta$ must be either $(m + \frac{1}{3})\lambda$ or $(m + \frac{2}{3})\lambda$, where m is an integer. The case $(m + \frac{3}{3})\lambda$ is excluded because then the path lengths differ by a full wavelength, giving constructive interference and thus a maximum in the interference pattern. More generally we can write

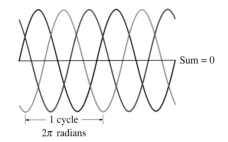

FIGURE 37-20   Waves from three slits must be out of phase by one-third of a cycle in order to interfere destructively. With this phase relationship the three waves sum to zero at every point.

$$d \sin \theta = \frac{m}{N}\lambda, \quad (37\text{-}6)$$

for destructive interference in an N-slit system, where m is an integer *but not an integer multiple of N*. The reason for the exclusion is that when m is an integer multiple of N then m/N is an integer; then the path-length difference is an integer number of wavelengths, resulting in constructive rather than destructive interference. Mathematically, when m/N is an integer, Equation 37-6 becomes equivalent to Equation 37-2a that gives the condition for constructive interference.

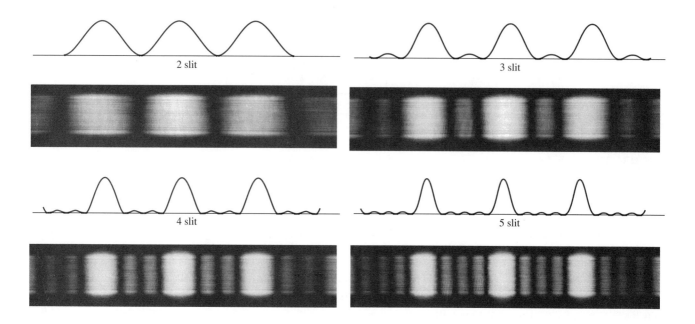

FIGURE 37-21 Interference patterns for multiple slit systems with the same slit spacing. The bright fringes stay in the same place, but become narrower and brighter as the number of slits increases. Intensity plots do not have the same vertical scale; peak intensity in fact scales as the *square* of the number of slits.

Figure 37-21 shows interference patterns and intensity plots from some multiple-slit systems. Note that the bright, or *primary,* maxima are separated by several minima and fainter, or *secondary,* maxima. Why this complex pattern? Our analysis of the three-slit system shows two minima between every pair of primary maxima; for example, we considered the minima at $d\sin\theta$ equal to $(m + \frac{1}{3})\lambda$ or $(m + \frac{2}{3})\lambda$, which lie between the maxima at $d\sin\theta$ equal to $m\lambda$ and $(m + 1)\lambda$. More generally, Equation 37-6 shows that there are $N - 1$ minima between each pair of primary maxima given by Equation 37-2a. The secondary maxima that lie between these minima result from interference that is neither fully destructive nor fully constructive. The figure shows that the primary maxima become much brighter and narrower as the number of slits increases, while the secondary maxima become relatively less bright. With a large number N of slits, then, we should expect a pattern of bright but narrow primary maxima, with broad, essentially dark regions in between.

Diffraction Gratings

A set of many very closely spaced slits is called a **diffraction grating** and proves very useful in the spectroscopic analysis of light. Diffraction gratings commonly measure several cm across and have several thousand slits—usually called lines—per cm. Gratings are made by photoreducing images of parallel lines or by ruling with a diamond stylus on aluminum-plated glass (Fig. 37-22). Gratings like the slit systems we have been discussing are **transmission gratings** since light passes through the slits. **Reflection gratings** produce similar interference effects by reflecting incident light.

We've seen that the maxima of the multi-slit interference pattern are given by the same criterion, $d\sin\theta = m\lambda$, that applies to a two-slit system. For $m = 0$ this equation implies that all wavelengths peak together at the central maximum, but for larger values of m the angular position of the maximum depends on wavelength. Thus, a diffraction grating can be used in place of a prism to

FIGURE 37-22 A diffraction grating disperses white light into its constituent colors. The enlarged spectrum at rear results from reflecting the diffracted beam off a concave mirror.

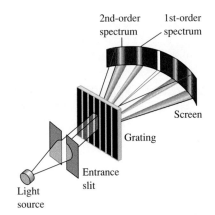

FIGURE 37-23 Essential elements of a grating spectrometer. The entrance slit regulates the amount of light entering the instrument. The grating disperses the light into its component wavelengths, which reach the screen to produce spectra of the different orders. An electronic detector would normally be used in place of the screen, and in most spectrometers the grating rotates to put different wavelengths on the detector.

disperse light into its component wavelengths, and the integer m is therefore called the **order** of the dispersion. Figure 37-23 shows a grating spectrometer working on this principle. Because the maxima in N-slit interference are very sharp for large N (recall Fig. 37-21), a grating with many slits diffracts individual wavelengths to very precise locations.

● **EXAMPLE 37-4** A GRATING SPECTROMETER

Light from glowing hydrogen contains discrete wavelengths ("spectral lines") called hydrogen-α and hydrogen-β, at 656.3 nm and 486.1 nm, respectively. Find the angular separation between these two wavelengths in a spectrometer using a grating with 6000 slits per cm. Consider both the first-order ($m = 1$) and second order ($m = 2$) dispersion.

Solution
With 6000 slits/cm, the slit spacing is $d = (1/6000)$ cm $= 1.667$ μm. Applying the criterion $d \sin \theta = m\lambda$ for the first-order spectrum, we then have

$$\theta_{1\alpha} = \sin^{-1}\left(\frac{\lambda}{d}\right) = \sin^{-1}\left(\frac{0.6563 \ \mu m}{1.667 \ \mu m}\right) = 23.2°$$

and

$$\theta_{1\beta} = \sin^{-1}\left(\frac{\lambda}{d}\right) = \sin^{-1}\left(\frac{0.4861 \ \mu m}{1.667 \ \mu m}\right) = 17.0°.$$

Thus, the angular separation is 6.2°. Repeating the same calculation with $m = 2$ gives $\theta_{2\alpha} = 51.9°$ and $\theta_{2\beta} = 35.7°$, for an angular spread of 16.2°. This wider spacing is characteristic of higher order dispersion.

EXERCISE The bright yellow light emitted by glowing sodium vapor actually consists of two spectral lines at 589.0 nm and 589.6 nm. Find the angular separation of these lines in a second-order spectrum taken with a 4800-slit/cm grating.

Answer: 0.04°

Some problems similar to Example 37-4: 35–37 ●

In Example 37-4 the two lines are near the ends of the visible spectrum, and the values calculated for the angular positions show that there is no overlap between the first- and second-order visible spectra. But higher order spectra do overlap, a fact that astronomers use in high-resolution spectroscopy when they wish to observe two different spectral lines simultaneously. Problem 39 explores this situation, while Fig. 37-24 shows the relative positions of the different orders.

Resolving Power

The detailed shapes and wavelengths of spectral lines contain a wealth of information about the systems in which light originates (Fig. 37-25). Studying

FIGURE 37-24 Positions of the different orders in a grating spectrum. The vertical separation between orders has been introduced for clarity; in actuality they would overlap. The central line of each spectrum is at 550 nm. Note the increased dispersion of the higher orders.

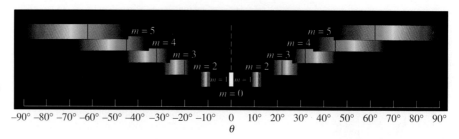

these details requires a high dispersion in order to separate nearby spectral lines or to analyze the intensity versus wavelength profile of a single line. Suppose we pass light containing two spectral lines of nearly equal wavelengths λ and λ' through a grating. Looking back at Fig. 37-21, you can see that the grating produces not just a single maximum for each wavelength in each order, but a

FIGURE 37-25 The interaction of a magnetic field with atomic electrons causes spectral lines to split. Plots show intensity versus wavelength for the 435.8-nm spectral line of mercury, with and without a 25-kG magnetic field. A high-resolution spectrometer is needed to resolve these lines, which are only 0.027 nm apart.

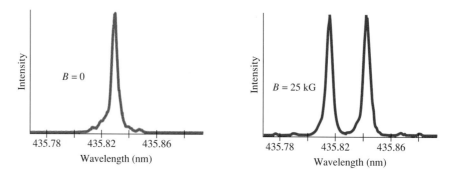

series of minima and lesser maxima in between. We will just be able to distinguish two spectral lines if the peak of one line corresponds to the first minimum of the other; any closer and the lines will blur together (Fig. 37-26). Suppose wavelength λ has its mth-order maximum at angular position θ. The criterion for this maximum is, as usual, $d \sin \theta_{max} = m\lambda$. We can equally well write this as

$d \sin \theta_{max} = \dfrac{mN}{N} \lambda$, with N the number of slits in the grating. Equation 37-6 then

shows that we get an adjacent minimum if we add 1 to the numerator mN. Thus, there is an adjacent minimum whose position satisfies

$$d \sin \theta_{min} = \frac{mN + 1}{N} \lambda.$$

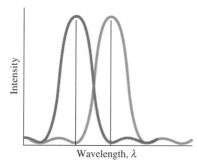

FIGURE 37-26 Two wavelengths are just distinguishable if the maximum of one falls on the first minimum of the other; any closer and they blur together.

Our criterion that the two wavelengths λ and λ' be just distinguishable is that the maximum for λ' fall at the location of this minimum for λ. But the maximum for λ' satisfies $d \sin \theta'_{max} = m\lambda' = \dfrac{mN}{N} \lambda'$, so for $\theta'_{max} = \theta_{min}$ we must have

$$(mN + 1)\lambda = mN\lambda'.$$

It's convenient to express this result in terms of the wavelength difference $\Delta\lambda = \lambda' - \lambda$. Solving our equation relating λ and λ' for λ' and using the result to write $\Delta\lambda$ in terms of λ alone then gives

$$\frac{\lambda}{\Delta\lambda} = mN. \qquad \text{(resolving power)} \qquad (37\text{-}7)$$

The quantity $\lambda/\Delta\lambda$ is the **resolving power** of the grating, a measure of its ability to distinguish closely spaced wavelengths. The higher the resolving power, the smaller the wavelength difference $\Delta\lambda$ that can be distinguished in the spectrum. That the resolving power increases with order is not surprising since we found that higher order spectra are more dispersed. Resolving power also depends on the number N of lines in the grating, although if the entire grating is not illuminated then N in Equation 37-7 becomes the number of lines actually illuminated.

● **EXAMPLE 37-5**   "SEEING" A DOUBLE STAR

A certain double star system consists of a massive star essentially at rest, with a smaller companion in circular orbit. The stars are far too close to each other, and the system far too distant from Earth, for the pair to appear as anything but a single point in even the largest telescopes. Yet astronomers can "see" the companion star through the Doppler shift in the wavelengths of its spectral lines that occurs because of the star's motion. In the system under observation the hydrogen-α line from the companion star, at 656.272 nm when the source is at rest, is shifted to 656.215 nm (this corresponds to an orbital speed of about 26 km/s) when the companion is moving toward Earth. If the telescope's spectrometer grating has 2000 lines/cm and measures 2.5 cm across, what order spectrum will resolve the hydrogen-α lines from the stationary star and its orbiting companion?

Solution

Here $N = (2000 \text{ lines/cm})(2.5 \text{ cm}) = 5000$ lines. Solving Equation 37-7 for m then gives

$$m = \frac{\lambda}{N\,\Delta\lambda} = \frac{656.272 \text{ nm}}{(5,000)(656.272 \text{ nm} - 656.215 \text{ nm})} = 2.3.$$

Since the order number must be an integer, the observation must be made in third order.

EXERCISE   To the nearest thousand, how many lines are needed to resolve the magnetically split spectral lines of Fig. 37-25 in first order, given a splitting of 0.027 nm at 435.8 nm?

Answer: 16,000

Some problems similar to Example 37-5:   44–47 ●

X-Ray Diffraction

The wavelengths of x rays, around 0.1 nm, are far too short to be dispersed with diffraction gratings produced mechanically or photographically. The regular spacing of atoms in a crystal, however, provides a "grating" of the appropriate scale. In the classical electromagnetic description, the reflection of electromagnetic waves occurs as electrons in each atom are set into oscillation by the electric field of the incident x-ray beam. The oscillating electrons then reradiate the energy they have absorbed, and the combined beam from all the atoms obeys the law of reflection. When the atomic spacing is regular, and comparable to the wavelength, however, interference enhances the reflected radiation at certain angles. Figure 37-27 shows an x-ray beam interacting with the layers of atoms in a crystal. In Fig. 37-27*b* we see that waves reflecting at one layer travel a total distance $2d\sin\theta$ farther than those reflecting at the layer above, where θ is the angle between the incident beams and the atomic planes. The outgoing beams will interfere constructively when this difference is an integer multiple of the wavelength:

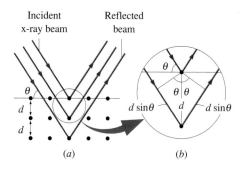

FIGURE 37-27   (a) X rays reflecting off the planes of atoms in a crystal. (b) X rays striking a lower plane travel an extra distance $2d\sin\theta$. The outgoing beam is enhanced by constructive interference when this distance is an integer multiple of the x-ray wavelength.

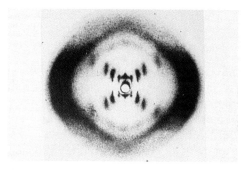

FIGURE 37-28   This x-ray diffraction image of DNA, made by the British scientist Rosalind Franklin in 1952, was crucial in establishing the structure of the DNA molecule.

$$2d\sin\theta = m\lambda. \quad \text{(Bragg condition)} \quad (37\text{-}8)$$

First derived by W. L. Bragg, this **Bragg condition** allows one to use a crystal with known spacing as a diffraction grating at x-ray wavelengths. More important is the converse: much of what we know about crystal structure comes from probing crystals with x rays and using the resulting diffraction patterns to deduce the positions of their atoms (Fig. 37-28).

■ APPLICATION   ACOUSTO-OPTIC MODULATION

Anything containing regularly spaced structures—like the lines in a diffraction grating or the atoms in a crystal—can act as a diffraction grating for waves of suitable wavelength. A recent and very useful technological development, the **acousto-optic modulator** (AOM), uses diffraction from the periodic variations in the structure of a quartz crystal caused by the presence of a sound wave in the crystal.

Figure 37-29 shows a diagram of an AOM. Attached to the crystal is a transducer—a loudspeaker-like device that converts an alternating voltage (usually in the ultrasound frequency range) into sound waves that propagate through the quartz. The regular spacing of the acoustic wavefronts constitutes a diffraction grating whose spacing may be varied by changing the driving frequency. Light incident on the quartz is diffracted at preferred angles given by Equation 37-8, just as are x rays by regularly spaced atomic planes. Changing the sound frequency allows the diffracted beam to be scanned in direction; turning the sound on and off switches the diffracted beam on and off. These effects are widely used in lightwave communication and other optical technologies; most laser printers, for example, use AOMs to control the laser beam that

"paints" a picture of the printed page on a light-sensitive surface. Setting up the AOM's "diffraction grating" requires propagating sound waves into the crystal. Since this takes time, the AOM is not as fast a lightwave modulator as is the electro-optic modulator discussed in Chapter 34. The AOM is, however, much less expensive and less demanding of the associated electronic circuitry.

Acoustic-optic modulators have another useful property. The diffraction grating established by the propagating sound waves is *moving* through the crystal, and the diffracted light is therefore Doppler shifted in frequency. A detailed analysis shows that the light frequency is shifted up or down by an amount equal to the sound frequency. Thus, the AOM may be used to encode information by altering slightly the frequency of light—an optical version of the process used to encode audio signals on radio waves in FM radio. Alternatively, AOMs can be used to produce light signals whose frequencies differ only slightly; bringing these signals together then produces the optical equivalent of the beat phenomenon we discussed in Section 16-5.

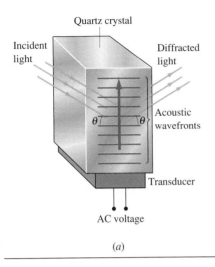

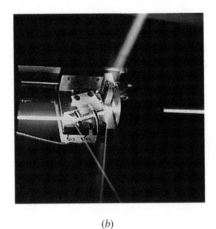

Quartz crystal

Incident
light

Diffracted
light

Acoustic
wavefronts

θ θ

Transducer

AC voltage

(a)

(b)

FIGURE 37-29 (a) Schematic diagram of an acousto-optic modulator. The transducer drives sound waves into the crystal, and the acoustic wavefronts act like a moving diffraction grating to incident light. (b) An acousto-optic modulator. A laser beam enters the transparent crystal while electrical signals supplied by the cable in the foreground generate sound waves that act as a diffraction grating to modify the beam's propagation direction and frequency. Such modulators are used in laser printers and many other applications.

37-5 SINGLE-SLIT DIFFRACTION

Why are we saving the single-slit case for last? Shouldn't that be the easiest? It would be—except that up to now we've neglected the slit width. According to Huygens' principle, each point in the slit acts as a source of cylindrical waves, and all these waves should interfere. Thus, a single slit is really like a system with infinitely many slits! Only when the slit width is very small can we make the approximation, implicit in our earlier analysis, that the slit behaves as a single source.

Figure 37-30 shows light incident on a slit of width a. Each point in the slit acts as a source of spherical wavelets propagating in all directions to the right of the slit. In Fig. 37-30 we focus on a particular direction described by the angle θ and will look at interference of light from the five points shown. Figure 37-30b concentrates on the points from which rays 1, 2, and 3 originate and shows that the path lengths for rays 1 and 3 differ by $\frac{1}{2}a\sin\theta$. These two beams will interfere destructively if this distance is half the wavelength; that is, if $\frac{1}{2}a\sin\theta = \frac{1}{2}\lambda$ or $a\sin\theta = \lambda$. But if rays 1 and 3 interfere destructively, so do rays 3 and 5, which have the same geometry, and so do rays 2 and 4, for the same reason. In fact, a ray leaving *any* point in the lower half of the slit will interfere destructively with the point located a distance $a/2$ above it. Therefore, an observer viewing the slit system at the angle θ satisfying $a\sin\theta = \lambda$ will see no light.

Similarly, the sources for rays 1 and 2 are $a/4$ apart and will therefore interfere destructively if $\frac{1}{4}a\sin\theta = \frac{1}{2}\lambda$, or $a\sin\theta = 2\lambda$. But then so will rays 2 and 3, and rays 3 and 4; in fact, any ray from a point in the lower three quarters of the slit will interfere destructively with a ray from the point $a/4$ above it, and therefore, an observer looking at an angle satisfying $a\sin\theta = 2\lambda$ will see no light.

We could equally well have divided the slit into six sections with seven evenly spaced points; we would then have found destructive interference if $\frac{1}{6}a\sin\theta = \frac{1}{2}\lambda$, or $a\sin\theta = 3\lambda$. We could obviously continue this process for any number of points in the slit, and therefore, we conclude that destructive interference occurs for all angles satisfying

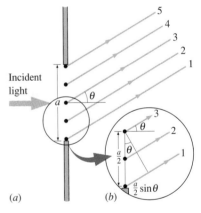

5
4
3
2
1

Incident
light

a θ

3
θ 2

$\frac{a}{2}$ θ

1

$\frac{a}{2}\sin\theta$

(a)   (b)

FIGURE 37-30 Each point in a slit acts as a source of Huygens' wavelets, which interfere in the region to the right of the slit. (a) Here we consider 5 points within the slit, and the relative phases of their light as viewed at an angle θ. (b) The path-length difference for two slits is their separation times $\sin\theta$.

$$a \sin \theta = m\lambda, \quad \text{(destructive interference, single-slit diffraction)} \quad (37\text{-}9)$$

with m any nonzero integer and a the slit width. Note that the case $m = 0$ is excluded; it produces not destructive interference but a central maximum in which all waves are in phase.

> **TIP** **Interference and Diffraction** Equation 37-9 for the *minima* of a single-slit diffraction pattern looks just like Equation 37-2a for the *maxima* of a multi-slit interference pattern, except that the slit width a replaces the slit spacing d. What's going on? Why does the same equation give the minima in one case and the maxima in another? Because we're dealing with two distinct but related phenomena. In the multi-slit case leading to Equation 37-2a, we considered each slit to be so narrow that it could be considered a single source, thus neglecting the interference of waves originating within the same slit. In the single-slit case leading to Equation 37-9, we do not neglect the slit width, and the resulting diffraction pattern occurs precisely because of the interference of waves from different points within the same slit.

● **EXAMPLE 37-6** DIFFRACTION: A NARROW SLIT

For what slit width, in terms of wavelength, will the first minimum lie at an angular position of 45°?

Solution
With $m = 1$, Equation 37-9 gives

$$a = \frac{\lambda}{\sin \theta} = \frac{\lambda}{\sin 45°} = \sqrt{2}\lambda.$$

Here, with the slit width nearly equal to the wavelength, diffraction dominates the wave propagation. The incident light is therefore spread over a wide angular range. The exercise below shows that the beam spreading decreases as the slit width grows.

EXERCISE Light is incident on a slit whose width is 20 times its wavelength. Find the angular spread of the beam, taken as the angle between the first minima on either side of the central maximum.

Answer: 5.7°

Some problems similar to Example 37-6: 49–51 ●

Intensity in Single-Slit Diffraction

We could try to find the intensity of the single-slit diffraction pattern by adding the electric fields of the individual beams, as we did for two-slit interference. Since there are infinitely many points along the slit, each constituting a point source of light, we would have to set up and evaluate a complicated integral. Instead we use the graphical method of phasors, introduced in Chapter 33 to deal with alternating currents and voltages of different phases. We have a similar situation here, as we want to combine waves with different phases that arise from different points on the slit. The phasor method is entirely equivalent to integrating over the electric field contributions from the individual sources in the slit.

We use phasors to represent the electric fields at the screen of the individual waves originating from points in the slit. The length of a phasor gives the field amplitude, and its direction gives its phase. Consider N such points, spaced evenly across a slit of width a, like the five points in Fig. 37-30. These divide the slit into $N - 1$ sections, each of width $a/(N - 1)$. At some angle θ, the phase difference in radians of light coming from two adjacent points will be the ratio of their path-length difference to the wavelength, times the 2π radians that comprise a complete cycle. As usual, the path-length difference is the point spacing times $\sin\theta$ (see Fig. 37-30b). Here the point spacing is $a/(N - 1)$, so we have

$$\Delta\phi = \left(\frac{a}{N - 1}\sin\theta\right)\left(\frac{2\pi}{\lambda}\right) \qquad (37\text{-}10)$$

for the phase difference between waves from adjacent points. All the waves have essentially the same amplitude, so we now need to add N equal-length phasors each of which differs in phase by $\Delta\phi$ from its nearest neighbors.

Consider first the case $\theta = 0$. Then Equation 37-10 gives $\Delta\phi = 0$ so, as Fig. 37-31a shows, the phasors are all in the same direction and sum to give a large total amplitude. This is the central maximum of the diffraction pattern. If, on the other hand, $\Delta\phi$ is such that the N phasors sum to a closed path, then the total amplitude—the net displacement from the beginning to the end of the phasor diagram—is zero (Fig. 37-31b). Here we have a point of zero intensity in the interference pattern. More generally, the phasors sum to give an amplitude E_θ that is neither zero nor as great as a simple sum of the individual amplitudes, as shown in Fig. 37-31c.

Now each point in the slit differs in phase by $\Delta\phi$ from its immediate neighbors. Thus, the second point differs by $\Delta\phi$ from the first point, the third point by $2\Delta\phi$ from the first point, and so forth until the Nth point differs in phase by $(N - 1)\Delta\phi$ from the first point. But this last quantity $(N - 1)\Delta\phi$ is just the total phase difference ϕ from one end of the slit to the other. Using Equation 37-10, we can write this phase difference as

$$\phi = (N - 1)\left(\frac{a}{N - 1}\sin\theta\right)\left(\frac{2\pi}{\lambda}\right) = \frac{2\pi}{\lambda}a\sin\theta. \qquad (37\text{-}11)$$

To consider all points in the slit, we take the limit as $N \to \infty$; then the end-to-end chain of phasors in Fig. 37-31c becomes a circular arc of radius R, as shown in Fig. 37-32. If we stretched this arc into a line, its length would be the amplitude E_0 of Fig. 37-31a; therefore, E_0 is also the length of the arc. From the geometry of Fig. 37-32 we see that

$$\sin(\phi/2) = \frac{E_\theta/2}{R}.$$

We can also write the angle $\phi/2$ using the definition of an angle in radians as the ratio of arc length to radius. Here $\phi/2$ subtends an arc of length $E_0/2$, so we have

$$\phi/2 = \frac{E_0/2}{R}.$$

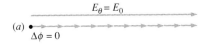
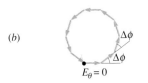
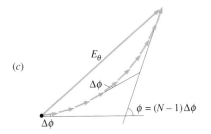

(a) $E_\theta = E_0$, $\Delta\phi = 0$

(b) $\Delta\phi$, $\Delta\phi$, $E_\theta = 0$

(c) E_θ, $\Delta\phi$, $\Delta\phi$, $\phi = (N - 1)\Delta\phi$

FIGURE 37-31 Phasor addition to find the amplitude in single-slit diffraction. (a) When $\theta = 0$ in Fig. 37-30 all waves have the same phase, and their amplitudes add to produce the central maximum. (b) When the phasor sum is a closed loop, the net phasor displacement is zero and so is the wave amplitude. (c) In general, the phasors add to produce an amplitude that is neither zero nor as great as in the central maximum. The angle between the first and last phasor is $N\Delta\phi$. Here $N = 10$ in all three frames, and in general we designate the resultant amplitude by E_θ. When the phasors all have the same phase, E_0 is the resultant amplitude.

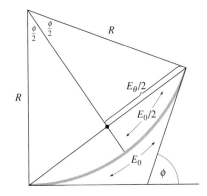

FIGURE 37-32 In the limit $N \to \infty$ the chain of phasors becomes a circular arc of length E_0.

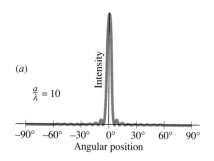

(a)

$\frac{a}{\lambda} = 10$

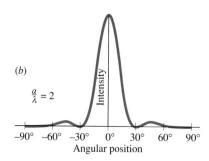

(b)

$\frac{a}{\lambda} = 2$

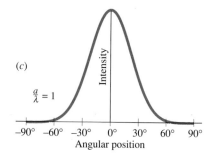

(c)

$\frac{a}{\lambda} = 1$

FIGURE 37-33 Intensity in single-slit diffraction, as a function of the angle θ from the centerline, for three values of of slit width a, expressed in units of wavelength λ. When the slit is much wider than the wavelength, most of the light is concentrated in a narrow central peak. But as the slit width approaches the wavelength, the diffracted beam becomes very wide.

Dividing these two equations gives the ratio of the amplitudes:

$$\frac{E_\theta}{E_0} = \frac{\sin(\phi/2)}{\phi/2}.$$

But the wave intensity $\bar{S}$ is proportional to the square of the amplitude, so

$$\bar{S}_\theta = \bar{S}_0 \left[\frac{\sin(\phi/2)}{\phi/2}\right]^2,$$

or, substituting ϕ from Equation 37-11,

$$\bar{S}_\theta = \bar{S}_0 \left[\frac{\sin\left(\frac{\pi a}{\lambda}\sin\theta\right)}{\frac{\pi a}{\lambda}\sin\theta}\right]^2 \qquad (37\text{-}12)$$

for the intensity in the single-slit diffraction pattern in terms of its value at the central maximum.

Figure 37-33 shows plots of Equation 37-12 for several values of the slit width a in relation to the wavelength λ. Note that for wide slits—large a/λ—the central peak is very narrow and the secondary peaks are much smaller. Here diffraction is negligible, and the beam essentially propagates through the slit in the ray approximation of geometrical optics. But as the slit narrows the diffracted beam spreads, until, with $a = \lambda$, it covers an angular width of some 120°.

The intensity given by Equation 37-12 will be zero when the numerator on the right-hand side is zero—that is, when the argument of the outermost sine function is an integer multiple of π. That occurs when $\frac{\pi a}{\lambda}\sin\theta = m\pi$, or when $a\sin\theta = m\lambda$. Thus, we recover our earlier result of Equation 37-9 for the angular positions where destructive interference gives zero intensity.

● **EXAMPLE 37-7** OTHER PEAKS

Figure 37-33 shows that secondary maxima lie approximately midway between the minima of the diffraction pattern. Use this approximate location to find the intensity at the first of these secondary maxima, in terms of the central-peak intensity $\bar{S}_0$.

Solution

The first and second minima are at positions given by $a\sin\theta_1 = \lambda$ and $a\sin\theta_2 = 2\lambda$, respectively. With the first of the secondary maxima midway between them, its angular position is given by $a\sin\theta = \frac{3}{2}\lambda$, or $\frac{a}{\lambda}\sin\theta = \frac{3}{2}$. (Here we assume θ is small, so $\sin\theta \simeq \theta$, and midway in angular position θ is therefore also midway in $\sin\theta$.) Using this result in Equation 37-12 then gives

$$\bar{S}_\theta = \bar{S}_0 \left[\frac{\sin\left(\dfrac{\pi a}{\lambda}\sin\theta\right)}{\dfrac{\pi a}{\lambda}\sin\theta} \right]^2 = \bar{S}_0 \left[\frac{\sin(3\pi/2)}{3\pi/2} \right]^2$$

$$= \frac{4\bar{S}_0}{9\pi^2} = 0.045\bar{S}_0.$$

Thus, the intensity at the first secondary maximum is only about 4.5% of the central-peak intensity. Note that this result is independent of the slit width, provided the secondary peak exists (which it may not; see Fig. 37-33c).

EXERCISE Show by direct substitution that the intensity in single-slit diffraction has half its maximum value when $\sin\theta = 1.3916\dfrac{\lambda}{\pi a}$.

Some problems similar to Example 37-7: 52–54 ●

In analyzing single-slit diffraction we considered only parallel rays in the diffraction region. To make the diffraction pattern actually appear on a screen, we would have to focus those rays with a lens between slit and screen (Fig. 37-34). Diffraction associated with parallel rays is called **Fraunhofer diffraction.** In general, Fraunhofer diffraction occurs when the distance from the diffracting system is large compared with the wavelength. Fraunhofer diffraction is an approximation to the more general case of **Fresnel diffraction,** whose analysis is more complicated because it also accounts for nonparallel rays that exist near a diffracting system.

Multiple Slits, Revisited

We neglected the width of slits in treating interference effects in multiple-slit systems. In effect, we were assuming the slits were so small that the central diffraction peak spread into the entire space beyond the slit system. When the slit width is not negligible, then the waves from each slit are not of uniform intensity, but instead exhibit a single-slit diffraction pattern. The superposition of waves from two or more slits then results in a pattern that combines single-slit diffraction with multiple-slit interference (Fig. 37-35).

37-6 THE DIFFRACTION LIMIT

Diffraction imposes a fundamental limit on the ability of optical systems to distinguish closely spaced objects. Consider two point sources of light illuminating a slit. The sources are so far from the slit that waves reaching the slit are essentially plane waves, but the different source positions mean the waves reach

FIGURE 37-34 Single-slit diffraction pattern produced by focusing light from a single slit onto a screen. Note the bright central peak and secondary maxima separated by points of destructive interference.

FIGURE 37-35 When the slit width is not negligible, the double-slit pattern shows the regular variations of double-slit interference confined to an "envelope" in the shape of a single-slit diffraction pattern. (Top) A plot of the intensity measured at each point on the screen. (Bottom) Interference/diffraction pattern as viewed on a screen. Compare with Fig. 37-33 for a single slit; note that the two patterns are essentially identical except that here a characteristic double-slit pattern of alternating light and dark bands appears within each bright band of the single-slit pattern.

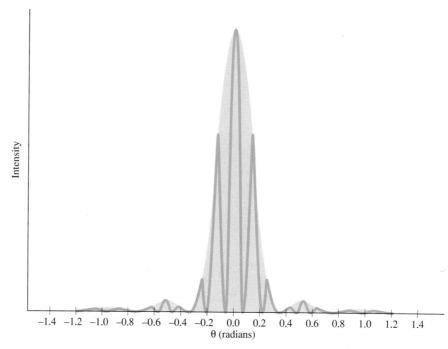

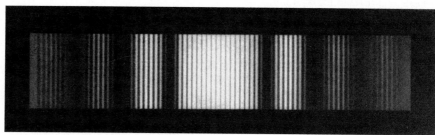

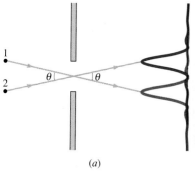

(a)

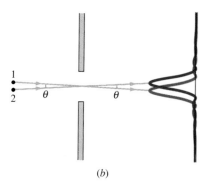

(b)

FIGURE 37-36 Two distant sources at different angular positions produce diffraction patterns whose central peaks have the same angular separation θ as the sources. (a) For sufficiently large θ the central peaks are clearly distinct, but (b) at small angular separations they overlap.

the slit at different angles. We assume the sources are incoherent, so there is no interference between the two. Then light diffracting at the slit produces two single-slit diffraction patterns, one for each source. Because the sources are at different angular positions, the central maxima of these patterns do not coincide, as shown in Fig. 37-36.

If the angular separation between the sources is great enough, then the central maxima of the two diffraction patterns will be entirely distinct. In that case the eye or any other optical system examining the diffraction pattern can clearly distinguish the two sources (Fig. 37-36a). But as the sources get closer the central maxima begin to overlap (Fig. 37-36b). The two sources remain distinguishable as long as the total intensity pattern shows two peaks. Since the sources are incoherent, they don't interfere, and the total intensity is just the sum of the individual intensities. Figure 37-37 shows how that sum loses its two-peak structure as the diffraction patterns merge. In general, two peaks are barely distinguishable if the central maximum of one coincides with the first minimum of the other. This condition is called the **Rayleigh criterion,** and when it is met we say that the two sources are just barely **resolved.**

What does all this have to do with optical instruments and images? Simply this: All optical systems are analogous to the single slit we've been considering. Every system has an aperture of finite size through which light enters the system. That aperture may be an actual slit or hole, like the diaphragm that "stops down" a camera lens, or it may be the full size of a lens or mirror. So all optical systems ultimately suffer loss of resolution if two sources—or two parts of the same extended object—have too small an angular separation. Thus, diffraction fundamentally limits our ability to probe the structure of objects that are either very small or very distant.

Figure 37-36 shows that the angular separation between the diffraction peaks is equal to the angular separation between the sources themselves. Then the Rayleigh criterion is just met if the angular separation between the two sources is equal to the angular separation between a central peak and the first minimum. We found earlier that the first minimum in single-slit diffraction occurs at angular position given by

$$\sin \theta = \frac{\lambda}{a},$$

with a the slit width and with θ measured from the central peak. In most optical systems the wavelength is much less than the size of any apertures, so we can use the small-angle approximation $\sin \theta \simeq \theta$ with θ in radians. Then the Rayleigh criterion—the condition that two sources be just resolvable—for single-slit diffraction becomes

$$\theta_{\text{min}} = \frac{\lambda}{a}. \qquad \text{(Rayleigh criterion, slit)} \qquad (37\text{-}13a)$$

Most optical systems have circular apertures rather than slits. The diffraction pattern from such an aperture is a series of concentric rings (Fig. 37-38). A more involved mathematical analysis shows that the angular position of the first ring relative to the central peak, and therefore, the minimum resolvable source separation for a circular aperture, is

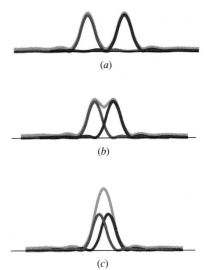

FIGURE 37-37 Since the two sources are incoherent, the total intensity is just the sum of the intensities of the two diffraction patterns. Here we see the intensities in two patterns and their sum (light blue). (*a*) Fully resolved; (*b*) barely resolved (Rayleigh criterion); (*c*) unresolved.

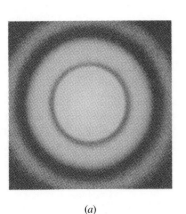

(*a*)

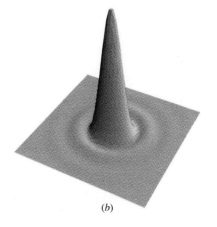

(*b*)

FIGURE 37-38 (*a*) Diffraction pattern produced by a circular aperture. (*b*) Three-dimensional plot of intensity versus position.

FIGURE 37-39 Diffraction patterns produced by a pair of point sources imaged through a circular hole. The angular separation of the sources decreases going from left to right, and when the sources are too close they cannot be resolved.

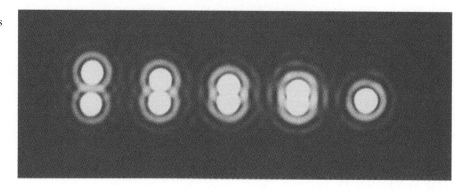

$$\theta_{min} = \frac{1.22\lambda}{D}, \qquad \text{(Rayleigh criterion, circular aperture)} \quad (37\text{-}13b)$$

with D the aperture diameter. In most optical systems the angle θ is very small, and Equations 37-13a and b hold for this case. For larger angles, θ_{min} in both equations should be replaced with $\sin\theta_{min}$. Figure 37-39 shows the loss of resolution as two nearby sources are imaged through a circular aperture.

Equations 37-13 show that increasing the aperture size allows smaller angular differences to be resolved. In optical instrument design, that means larger mirrors, lenses, and other components. An alternative is to decrease the wavelength used, which may or may not be an option depending on the source. In high-quality optical systems diffraction is often the limiting factor preventing perfectly sharp image formation; such systems are said to be **diffraction limited.** Among exceptions are large ground-based telescopes, whose imaging quality is limited by atmospheric turbulence. From its vantage point above the atmosphere, the Hubble Space Telescope is the first large diffraction-limited astronomical telescope.

● EXAMPLE 37-8 ASTEROID ALERT

An asteroid appears on a collision course with Earth, at a distance of 20×10^6 km. What is the minimum size asteroid that the 2.4-m-diameter diffraction-limited Hubble Space Telescope could resolve at this distance, using 550-nm reflected sunlight?

Solution

Resolving the asteroid means being able to distinguish its opposite edges in the telescope image. Suppose the asteroid's long dimension (it might not be spherical) is ℓ. Then at a distance $L \gg \ell$ it subtends an angle given very nearly by $\theta = \ell/L$. Using this result in Equation 37-13b with the mirror diameter and wavelength given, we have

$$\frac{\ell}{L} = \frac{1.22\lambda}{D},$$

or

$$\ell = \frac{1.22\lambda L}{D} = \frac{(1.22)(550\times10^{-9}\text{ m})(20\times10^9\text{ m})}{2.4\text{ m}} = 5.6\text{ km}.$$

This is a potentially dangerous object, comparable in size to the asteroid that some scientists believe caused the extinction of the dinosaurs, and somewhat larger than the comet fragments that slammed into Jupiter in 1994, causing Earth-sized disturbances on the giant planet.

●

● **EXAMPLE 37-9** STAR WARS: THE DIFFRACTION CHALLENGE

A system once proposed for defense against ballistic missiles calls for focusing high-power laser beams onto attacking missiles. A particular design specifies infrared laser light with 2.8-μm wavelength focused to a spot 50 cm in diameter on a missile 2500 km distant. What is the minimum diameter for a concave mirror that can achieve this spot size?

Solution
Figure 37-40a shows a parallel beam from the laser striking the mirror, then focusing on the distant missile. We assume the focal length is equal to the missile's distance, giving the best possible focus. Nevertheless, the mirror acts like a circular aperture that produces a central spot whose angular radius is given by the Rayleigh criterion of Equation 37-13b. We want the angular radius of that spot to correspond to a 25-cm-radius spot at a distance of 2500 km. Clearly the small-angle approximation applies, and Fig. 37-40b shows that we can write

$$\theta = \frac{25 \text{ cm}}{2500 \text{ km}} = 1.0 \times 10^{-7} \text{ radians}$$

for the angular radius of the spot. Solving Equation 37-13b for the minimum mirror diameter then gives

$$D = \frac{1.22\lambda}{\theta} = \frac{(1.22)(2.8 \times 10^{-6} \text{ m})}{1.0 \times 10^{-7}} = 34 \text{ m}.$$

This enormous mirror size is one reason many scientists have found "star wars" weapons designs implausible. A related problem is the long focal length; with $f = 2500$ km, a 34-m mirror is only about 40 μm from being perfectly flat, and it has to retain its exact shape despite the thermal stresses imposed by a 20-MW laser beam on a space-based mirror. An alternative approach is to use so-called phased arrays, in which a number of smaller sources with precisely related phases simulate the effect of the beam from a large mirror.

EXERCISE Two ants stand 1 cm apart. Assuming diffraction limited vision, at what distance could a human eye, with an iris aperture of 4 mm, tell that two are indeed separate creatures? Assume a wavelength of 550 nm, near the middle of the visible spectrum. (Other limitations of the human eye prevent its realizing this diffraction limit.)

Answer: 60 m

Some problems similar to Examples 37-8 and 37-9: 55–59

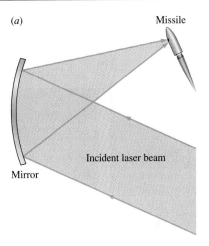

(a)

Missile

Incident laser beam

Mirror

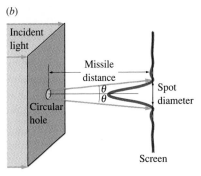

(b)

Incident light

Missile distance

θ
θ

Spot diameter

Circular hole

Screen

FIGURE 37-40 (a) In this laser-based antimissile system, the mirror acts like a circular aperture to limit the minimum size of the focused beam. (b) The equivalent single-hole diffraction system. The angle θ_{min} in the Rayleigh criterion is the ratio of spot radius to the missile distance.

■ **APPLICATION** LONG-PLAYING CDs

Standard compact discs have a maximum playing time of 74 minutes—a value that is in fact set by the optical diffraction limit. Information on a CD is encoded in a sequence of "pits" cut into the CD surface (Fig. 37-41). The pit depth is approximately one-quarter wavelength of the laser light used to "read" the information. The laser shines from below, where it "sees" each pit as a "bump" that introduces a round-trip change of one-half wavelength in the light path, causing destructive interference between incident and reflected beams (Fig. 37-42). As the disc rotates, the interference is alternately constructive and destructive, and the interfering light falling on a detector produces an electrical signal corresponding to the information encoded in the CD's pits.

The pits on a standard CD are about 0.5 μm wide, up to 3 μm long, and lie in spiral tracks spaced 1.6 μm apart. These are close to the minimum dimensions that can be used before diffraction effects prevent resolution of the individual pits. Thus, the diffraction limit determines the total number of pits, and therefore the total information stored on the disc. Each

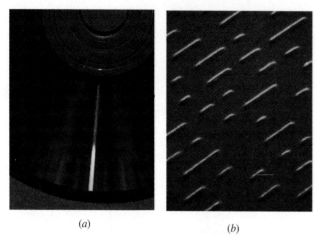

(a)

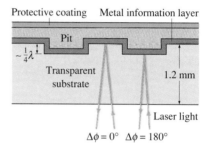

(b)

FIGURE 37-41 A compact disc encodes information in billions of microscopic pits arranged in spiral tracks 1.6 μm apart. (a) The pits are too small to see, but their spiral tracks act as a diffraction grating, producing the rainbow of colors seen here. (b) Electron micrograph showing individual pits in adjacent tracks. These CD pits are among the smallest manufactured structures. See also Fig. 1-11.

second of music recording requires on the order 10^6 pits, and the result is the 74-minute maximum recording time.

Today's CD players use 780-nm infrared laser light. The reason is economic: inexpensive semiconductor lasers operating at this wavelength are readily available. The diffraction limit on compact-disc playing time has spurred a vigorous research competition among electronics corporations, in the hope of making shorter wavelength semiconductor lasers. By the early 1990s several corporations had developed blue or blue-green lasers, with a Sony unit at 447 nm having the shortest wavelength. Expect longer playing CDs, thanks to the less stringent diffraction limit, when these new short-wavelength lasers become commercially viable.

FIGURE 37-42 Schematic diagram showing a laser beam "reading" from the underside of a compact disk. Destructive interference caused by the 180° phase difference occurs when the beam reflects from one of the pits. Diagram is not to scale; the metal layer containing the pit-encoded information is less than 100 nm thick, and the protective coating is 10–30 μm thick. See also Example 35-2.

■ APPLICATION INTERFEROMETRY

At a given wavelength, Equations 37-13 show that resolving small angular separations requires a large aperture. To see detail in distant astronomical objects, in particular, requires impractically large apertures. A technique called **interferometry** provides a way around this problem. Used most commonly at radio wavelengths, interferometry employs two or more individual apertures (i.e., radio telescopes), usually separated by a considerable distance. Signals from each are combined with their phases intact, so that they interfere. A two-telescope interferometer works like a two-slit system in reverse: the system

is most sensitive to radiation from angular positions that produce constructive interference (Fig. 37-43). The resolution of an interferometer is approximately that of a single slit-like aperture whose width is the spacing between the telescopes.

Today's interferometers include installations with many radio telescopes at a single site (Fig. 37-44), as well as coordinated instruments spread across the globe. The first interferometric arrays of optical telescopes began operation in the 1990s, and astronomers look forward to radio interferometers using the Earth-moon distance as their "aperture."

FIGURE 37-43 A radio interferometer consists of two or more radio telescopes at different locations. The interferometer can detect very slight changes in phase of the incoming wavefronts and can therefore resolve very small differences in angular position of objects in the sky.

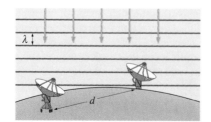

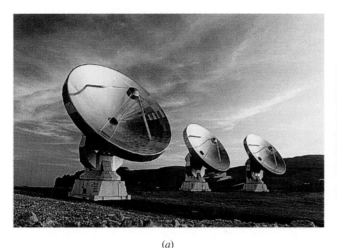

(a)

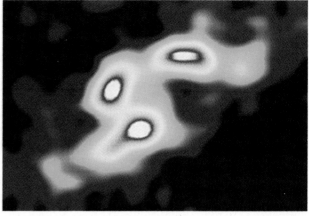

(b)

FIGURE 37-44   (a) Dish antennas of the IRAM interferometer in Spain, a joint French-German-Spanish instrument used extensively to study radio emission from interstellar molecules. (b) IRAM image of the spiral galaxy IC342 shows the distribution of interstellar hydrogen cyanide in the galaxy.

CHAPTER SYNOPSIS

Summary

1. **Interference** and **diffraction** are wave phenomena and constitute the subject of **physical optics.**
2. Interference effects require **coherent** beams of light, whose phases are related in a fixed way.
   a. Interference occurs in thin transparent films when light beams reflected from the front and back surfaces of the film recombine. For a film whose refractive index is larger than that of its surroundings, the interference is constructive when the film thickness d, refractive index n, and the light wavelength λ are related by $2nd = (m + \frac{1}{2})\lambda$, where m is an integer.
   b. The Michelson interferometer utilizes the interference of light traveling on two perpendicular paths to make precise distance measurements.
3. **Huygens' principle** treats each point on a wavefront as a source of expanding spherical wavelets. Superposition of the wavelets describes the propagation of the wavefront and shows that waves undergo **diffraction** when passing by an opaque edge.
4. When coherent light of wavelength λ diffracts on passing through a pair of narrow slits with spacing d, the resulting interference pattern shows bright fringes at angular positions given by

$$d \sin \theta = m\lambda,$$

where m is an integer. The intensity in the interference pattern is given by

$$\bar{S} = 4\bar{S}_0 \cos^2\left(\frac{\pi d \sin \theta}{\lambda}\right),$$

where $\bar{S}_0$ is the intensity incident on the slit system.
5. A multiple-slit system has primary interference maxima in the same position as a double-slit system, with multiple minima and secondary maxima in between. For large numbers of slits the primary maxima are narrow and bright and the region between them relatively dark. A system with many slits constitutes a **diffraction grating,** which disperses light into its component wavelengths. A grating produces spectra of many **orders,** corresponding to the integer m in the equation that locates the interference maxima. The **resolving power** of an N-slit grating is given by

$$\frac{\lambda}{\Delta \lambda} = mN,$$

where $\Delta \lambda$ is the minimum separation between wavelengths that can be distinguished in the mth-order spectrum. Other periodic systems, like the layers of atoms in a crystal, can serve as diffraction gratings for appropriate wavelengths.
6. When the width of a single slit is not negligible compared with the wavelength, interference of light passing through different parts of the slit produces a **diffraction pattern** with a central maximum surrounded by lesser maxima separated by points of zero intensity. The intensity as a function of angular position is given by

$$\bar{S}_\theta = \bar{S}_0 \left[\frac{\sin\left(\dfrac{\pi a}{\lambda}\sin\theta\right)}{\dfrac{\pi a}{\lambda}\sin\theta} \right]^2 .$$

7. The **diffraction limit** in optical systems arises because diffraction effects merge the images of two objects when their angular separation is sufficiently small. The **Rayleigh criterion** gives the minimum angular separation that can be resolved by a given aperture. For a circular aperture of diameter D, the Rayleigh criterion is

$$\theta_{min} = \frac{1.22\lambda}{D},$$

where θ_{min} is in radians. Large apertures are therefore necessary to resolve objects with small angular separation.

Terms You Should Understand

(Pairs are closely related terms whose distinction is important; number in parentheses is chapter section where term first appears.)
physical optics, geometrical optics (introduction and preceding chapters)
interference (37-1)
coherence (37-1)
monochromatic light (37-1)
antireflection coating (37-1)
Michelson interferometer (37-1)
beam splitter (37-1)
Huygens' principle (37-2)
diffraction (37-2)
order (37-3, 37-4)

diffraction grating (37-4)
resolving power (37-4)
Bragg condition (37-4)
diffraction limit (37-6)
Rayleigh criterion (37-6)

Symbols You Should Recognize

m (37-1, 37-3, 37-4)
θ_{min} (37-4, 37-6)

Problems You Should Be Able to Solve

determining conditions for constructive and destructive interference in thin films (37-1)
analyzing experiments using Michelson interferometers (37-1)
analyzing double-slit interference patterns (37-3)
finding locations of maxima and minima in multiple-slit interference (37-4)
analyzing dispersion of light with diffraction gratings (37-4)
finding minima and beam widths in single-slit diffraction (37-5)
finding the diffraction limit in optical systems (37-6)

Limitations to Keep in Mind

This chapter's analysis of double-slit interference assumes slit widths much smaller than the wavelength; otherwise the pattern combines diffraction and interference.
Our analysis of diffraction applies only in the Fraunhofer limit, in which the incident light is essentially parallel and the pattern is formed at large distances from the diffracting aperture.

QUESTIONS

1. A prism bends blue light more than red. Is the same true of a diffraction grating?
2. Why does an oil slick show colored bands?
3. Would a Michelson interferometer work even if its two arms were not exactly the same length?
4. Why does a soap bubble turn colorless just before it dries up and breaks?
5. Why don't you see interference effects between the front and back of your eyeglasses?
6. Hold two fingers very close together while looking through them at a source of light. Explain what you see.
7. Figure 37-45 shows the shadow cast by a ball bearing on the end of a needle, under monochromatic laser light. Explain the existence of the fringes shown and of the white spot at the center of the ball bearing's shadow.

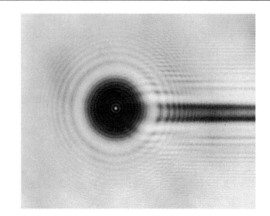

FIGURE 37-45 Question 7.

8. Why don't you see interference effects when you look out a window through venetian blinds?

9. You can hear around corners but you can't see around corners. Why the difference?

10. In deriving the intensity in double-slit interference, we could not simply add the intensities from the two slits. Why not?

11. In sketching the intensity pattern for two sources imaged through a single slit, we could simply add the intensities from the two sources. Why?

12. Explain the roles of diffraction and interference in the working of a diffraction grating.

13. The primary maxima in multiple-slit interference are in the same angular positions as those in double-slit interference. Why, then, do diffraction gratings have thousands of slits instead of just two?

14. In what way is a widely separated pair of small radio telescopes superior to a single large one? In what way is it inferior?

15. Describe the change in the diffraction pattern of a single slit as the slit is narrowed.

16. What pattern would result from passing red and blue light simultaneously through a double-slit system?

17. When the moon passes in front of a star, the intensity of the starlight fluctuates instead of dropping abruptly. Explain.

18. Why might it be desirable to observe a double star system in blue light rather than red?

19. In analyzing crystal structure using x-ray diffraction, it is necessary to take data with the crystal in several different orientations. Why?

20. Microwaves emerge from a rectangular horn, wider in the horizontal direction than in the vertical. Will the resulting beam be wider in the horizontal or in the vertical direction? Explain.

21. A double-slit system has one slit much narrower than the wavelength of the incident light, the other much wider. Describe the resulting intensity pattern.

22. Sketch roughly the diffraction pattern you would expect for light passing through a square hole a few wavelengths wide.

PROBLEMS

Section 37-1 Interference

1. Find the minimum thickness of a soap film ($n = 1.33$) in which 550-nm light will undergo constructive interference.

2. Light of unknown wavelength shines on a precisely machined wedge of glass with refractive index 1.52. The closest point to the apex of the wedge where reflection is enhanced occurs where the wedge is 98 nm thick. Find the wavelength.

3. Monochromatic light shines on a glass wedge with refractive index 1.65, and enhanced reflection occurs where the wedge is 450 nm thick. Find all possible values for the wavelength in the visible range.

4. White light shines on 100-nm-thick sliver of fluorite ($n = 1.43$). What wavelength is most strongly reflected?

5. As a soap bubble ($n = 1.33$) evaporates and thins, the reflected colors gradually disappear. (a) What is its thickness just as the last vestige of color vanishes? (b) What is the last color seen?

6. An oil film ($n = 1.25$) floats on water, and a soap film ($n = 1.33$) is suspended in air. Find the minimum thickness for each that will result in constructive interference with 500-nm green light.

7. Light reflected from a thin film of acetone ($n = 1.36$) on a glass plate ($n = 1.5$) shows maximum reflection at 500 nm and minimum at 400 nm. Find the minimum possible film thickness.

8. What minimum thickness of a coating with refractive index 1.35 should you use on a glass lens to minimize reflection at 500 nm, the approximate center of the visible spectrum?

9. An oil film with refractive index 1.25 floats on water. The film thickness varies from 0.80 μm to 2.1 μm. If 630-nm light is incident normally on the film, at how many locations will it undergo enhanced reflection?

10 Microwave ovens operate at a frequency of 2.4 GHz. What is the minimum thickness for a plastic tray with refractive index 1.45 that will cause enhanced reflection of microwaves incident normal to the plate?

11. Two perfectly flat glass plates are separated at one end by a piece of paper 0.065 mm thick. A source of 550-nm light illuminates the plates from above, as shown in Fig. 37-46. How many bright bands appear to an observer looking down on the plates?

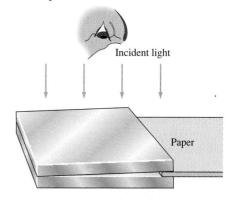

Incident light

Paper

FIGURE 37-46   An air wedge (Problems 11, 12, 13, 72).

12. An air wedge like that shown in Fig. 37-46 shows N bright bands when illuminated from above. Find an expression for the number of bands that will appear if the air is replaced by a liquid of refractive index n different from that of the glass.

13. You apply a slight pressure with your finger to the upper of a pair of glass plates forming an air wedge as in Fig. 37-46. The wedge is illuminated from above with 500-nm light, and you place your finger where, initially, there is a dark band. If you push gently so the band becomes light, then dark, then light again, by how much have you deflected the plate?

14. A Michelson interferometer uses light from glowing hydrogen at 486.1 nm. As you move one mirror, 530 bright fringes pass a fixed point in the viewer. How far did the mirror move?

15. What is the wavelength of light used in a Michelson interferometer if 550 bright fringes go by a fixed point when the mirror moves 0.150 mm?

16. One arm of a Michelson interferometer is 42.5 cm long and is enclosed in a box that can be evacuated. The box initially contains air, which is gradually pumped out to create a vacuum. In the process, 388 bright fringes pass a fixed point in the viewer. If the interferometer uses light with wavelength 641.6 nm, what is the refractive index of the air?

17. The evacuated box of the previous problem is filled with chlorine gas, whose refractive index is 1.000772. How many bright fringes pass a fixed point as the tube fills?

18. Roughly how many cycles should be in the wavetrains of Fig. 37-1a, given that the wavetrains last about 10^{-8} s? Assume a visible wavelength of about 600 nm.

19. Your personal stereo is in a dead spot caused by direct reception from an FM radio station at 89.5 MHz interfering with the signal reflecting off a wall behind you. How much farther from the wall should you move in order that the interference be fully constructive?

Section 37-3 Double-Slit Interference

20. A double-slit experiment with $d = 0.025$ mm and $L = 75$ cm uses 550-nm light. What is the spacing between adjacent bright fringes?

21. A double-slit experiment has slit spacing 0.12 mm. (a) What should be the slit-to-screen distance L if the bright fringes are to be 5.0 mm apart when the slits are illuminated with 633-nm laser light? (b) What will be the fringe spacing with 480-nm light?

22. With two slits separated by 0.37 mm, the interference pattern has bright fringes with angular spacing 0.065°. What is the wavelength of the light illuminating the slits?

23. The green line of gaseous mercury at 546 nm falls on a double-slit apparatus. If the fifth dark fringe is at 0.113° from the centerline, what is the slit separation?

24. What is the angular position θ of the second-order bright fringe in a double-slit system with 1.5-μm slit spacing if the light has wavelength (a) 400 nm or (b) 700 nm?

25. Light shines on a pair of slits whose spacing is three times the wavelength. Find the locations of the first- and second-order bright fringes on a screen 50 cm from the slits. *Hint:* Do Equations 37-2 apply?

26. A double-slit experiment has slit spacing 0.035 mm, slit-to-screen distance 1.5 m, and wavelength 500 nm. What is the phase difference between two waves arriving at a point 0.56 cm from the center line?

27. For a double-slit experiment with slit spacing 0.25 mm and wavelength 600 nm, at what angular position is the path difference equal to one-fourth of the wavelength?

28. A screen 1.0 m wide is located 2.0 m from a pair of slits illuminated by 633-nm laser light, with its center on the centerline of the slits. Find the highest order bright fringe that will appear on the screen if the slit spacing is (a) 0.10 mm; (b) 10 μm.

29. Laser light at 633 nm falls on a double-slit apparatus with slit separation 6.5 μm. Find the separation between (a) the first and second and (b) the third and fourth bright fringes, as seen on a screen 1.7 m from the slits.

30. A tube of glowing gas emits light at 550 nm and 400 nm. In a double-slit apparatus, what is the lowest order 550-nm bright fringe that will fall on a 400-nm dark fringe, and what are the corresponding orders?

Section 37-4 Multiple-Slit Interference and Diffraction Gratings

31. In a 5-slit system, how many minima lie between the zeroth-order and first-order maxima?

32. In a 3-slit system the first minimum occurs at an angular position of 5°. Where is the first maximum?

33. A 5-slit system with 7.5-μm slit spacing is illuminated with 633-nm light. Find the angular positions of (a) the first 2 maxima and (b) the 3rd and 6th minima.

34. On a screen 1.25 m from a multiple-slit system, the interference pattern shows bright maxima separated by 0.86° and 7 minima between each bright maximum. (a) How many slits are there? (b) What is the slit separation if the incident light has wavelength 656.3 nm?

35. Green light at 520 nm is diffracted by a grating with 3000 lines per cm. Through what angle is the light diffracted in (a) first and (b) fifth order?

36. Find the angular separation between the red hydrogen-α spectral line at 656 nm and the yellow sodium line at 589 nm if the two are observed in 3rd order with a 3500 line/cm grating spectrometer (Fig. 37-47).

37. Light is incident normally on a grating with 10,000 lines per cm. What is the maximum order in which (a) 450-nm and (b) 650-nm light will be visible?

FIGURE 37-47 Spectral lines of hydrogen-α and sodium. What would be their angular separation if they appeared on a single spectrum? (Problem 36)

38. Visible light has wavelengths between 400 nm and 700 nm. What is the lowest pair of consecutive orders for which there will be some overlap between the visible spectra as dispersed by a grating?

39. A solar astronomer is studying the Sun's 589-nm sodium spectral line with a 2500 line/cm grating spectrometer whose fourth order dispersion puts the wavelength range from 575 nm to 625 nm on a detector. The astronomer is interested in observing simultaneously the so-called calcium-K line, at 393 nm. What order dispersion will put this line also on the detector?

40. (a) What portions of the 4th and 5th order visible spectra overlap in a 3000 lines/cm grating spectrometer? (b) How would your answer change for a 1000 lines/cm grating? (c) For a 10,000 lines/cm grating?

41. Estimate the number of lines per cm in the grating used to produce Fig. 37-24.

42. A grating spectrometer's detector covers an angular range of 10° and can be swung at any angle within ±75° of the normal to the grating. If the grating has 1200 lines/cm, in what orders can the hydrogen-α spectral line at 656 nm and the calcium-K line at 393 nm both fall on the detector? (b) What will be the angular position of the K line under these conditions?

43. When viewed in 6th order, the 486.1-nm hydrogen-β spectral line is flanked by another line that appears at the position of 484.3 nm in the 6th order spectrum. Actually the line is from a different order of the spectrum. What are the possible visible wavelengths of this line?

44. (a) Find the resolving power of a grating needed to separate the sodium-D spectral lines, which are at 589.0 nm and 589.6 nm. (b) How many lines must the grating have to achieve this resolution in first order? (This is very low resolution by present-day spectroscopic standards.)

45. Echelle spectroscopy uses relatively course gratings in high order. Compare the resolving power of an 80 line/mm echelle grating used in 12th order with a 600 line/mm grating used in 1st order, assuming the two have the same width.

46. The International Ultraviolet Explorer satellite carries a spectrometer with a 2.0-cm-wide grating ruled at 102 lines/mm. What is the minimum wavelength difference it can resolve in 12th order when observing in the ultraviolet at 155 nm?

47. You wish to resolve the calcium-H line at 396.85 nm from the hydrogen-ε line at 397.05 nm in a 1st-order spectrum. To the nearest hundred, how many lines should your grating have?

48. X-ray diffraction in calcium chloride (KCl) results in a first-order maximum when the x rays graze the crystal plane at 8.5°. If the x-ray wavelength is 97 pm, what is the spacing between crystal planes?

Sections 37-5 and 37-6 Single-Slit Diffraction and the Diffraction Limit

49. For what ratio of slit width to wavelength will the first minima of a single-slit diffraction pattern occur at ±90°?

50. Light with wavelength 633 nm is incident on a 2.5 μm wide slit. Find the angular width of the central peak in the diffraction pattern, taken as the angular separation between the first minima.

51. A beam of parallel rays from a 29-MHz citizen's band radio transmitter passes between two electrically conducting (hence opaque to radio waves) buildings located 45 m apart. What is the angular width of the beam when it emerges from between the buildings?

52. Use trial-and-error with a calculator, or a more sophisticated root-finding method, to verify the number 1.3916 in the exercise following Example 37-7.

53. Find the intensity as a fraction of the central peak intensity for the second secondary maximum in single-slit diffraction, assuming the peak lies midway between the second and third minima.

54. The width of a peak is often given in terms of the *full width at half maximum* (FWHM), meaning the width measured where the peak has half its maximum value (Fig. 37-48). Use the number mentioned in Problem 52 to find an expression for the angular FWHM of the central peak in single-slit diffraction, in terms of the wavelength λ and slit width a.

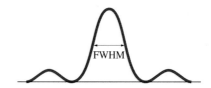

FIGURE 37-48 A peak's full width at half maximum is the width measured at half the maximum height (Problem 54).

FIGURE 37-49 CIA agents using satellite imaging to identify terrorists in the film *Patriot Games*. How big a mirror or lens must the satellite's optical system have? (Problem 55)

FIGURE 37-50 Jupiter, photographed with ground-based telescope (left) and from space (right). Atmospheric turbulence limits the ground-based image quality while diffraction limits the space-based image (Problem 61).

55. The movie *Patriot Games* has a scene in which CIA agents use spy satellites to identify individuals in a terrorist camp (Fig. 37-49). Suppose that a minimum resolution for distinguishing human features is about 5 cm. If the spy satellite's optical system is diffraction limited, what diameter mirror or lens is needed to achieve this resolution from an altitude of 100 km? Assume a wavelength of 550 nm.

56. Suppose the 10-m-diameter Keck telescope in Hawaii could be trained on San Francisco, 3400 km away. Would it be possible to read a (a) newspaper or (b) a billboard sign at this distance? Justify your answers by giving the minimum separation resolvable with Keck at 3400 km, assuming 550-nm light.

57. What is the minimum spot diameter to which a camera set at f-ratio of 16 can focus parallel light with 650-nm wavelength? *Hint:* Equation 37-13b gives the minimum angular spacing between the central maximum and first minimum; here you want the angular spread of the circle that marks the minimum.

58. The distance from the center of a circular diffraction pattern to the first minimum on a screen 0.85 m distant from the diffracting aperture is 15,000 wavelengths. What is the aperture diameter?

59. While driving at night, your eyes' irises have dilated to 3.1-mm diameter. If your vision were diffraction limited, what would be the greatest distance at which you could see as distinct the two headlights of an oncoming car, which are spaced 1.5 m apart? Take $\lambda = 550$ nm.

60. Two stars are 4.0 light-years apart, in a galaxy 20×10^6 light-years from Earth. What minimum separation of two radio telescopes, acting together as an interferometer, is needed to resolve them? The telescopes operate at 2.1 GHz and are pointing straight upward; the stars are directly overhead.

61. Under the best conditions, atmospheric turbulence limits the resolution of ground-based telescopes to about 1 arc second (1/3600 of a degree) as shown in Fig. 37-50. For what aperture sizes is this limitation more severe than that

of diffraction at 550 nm? Your answer shows why large ground-based telescopes do not produce better images than small ones, although they do gather more light.

62. Two objects are separated by approximately one wavelength of the light with which an observer is attempting to resolve them. Show that the Rayleigh criterion then requires that the distance to the objects be less than the diameter of the observing aperture, and thus the Fraunhofer diffraction approximation is violated. It is in fact impossible to resolve two objects as close as one wavelength.

Paired Problems

(Both problems in a pair involve the same principles and techniques. If you can get the first problem, you should be able to solve the second one.)

63. A thin film of toluene ($n = 1.49$) floats on water. What is the minimum film thickness if the most strongly reflected light has wavelength 460 nm?

64. Oil with refractive index 1.38 forms a 210-nm-thick film on a piece of glass. What color of visible light is most strongly reflected?

65. Find the total number of lines in a 2.5-cm-wide diffraction grating whose third-order spectrum has the 656-nm hydrogen-α spectral line at an angular position of 37°.

66. Light is diffracted by a 5000 line/cm grating, and a detector sensitive only to visible light finds a maximum intensity at 28° from the central maximum. (a) What is the wavelength of the light? (b) In what order is it seen at the 28° position?

67. A 400 line/mm diffraction grating is 3.5 cm wide. Two spectral lines whose wavelengths average to 560 nm are just barely resolved in the 4th-order spectrum of this grating. What is the difference between their wavelengths?

68. What order is necessary to resolve wavelengths of 647.98 nm and 648.07 nm using a 4500-line grating?

69. What diameter optical telescope would be needed to resolve a Sun-sized star 10 light-years from Earth? Take $\lambda = 550$ nm. Your answer shows why stars appear as point sources in optical astronomy.

70. Could the 305-m radio telescope at Arecibo in Puerto Rico resolve the star in the preceding problem, if it were observing at a wavelength of 1.0 cm?

Supplementary Problems

71. White light shines on a 250-nm-thick layer of diamond ($n = 2.42$). What wavelength *of visible light* is most strongly reflected?

72. An air wedge like that of Fig. 37-46 displays 10,003 bright bands when illuminated from above. If the region between the plates is then evacuated, the number of bands drops to 10,000. Calculate the refractive index of the air.

73. In Fig. 37-6 the *m*th Newton's ring appears a distance *r* from the center of the lens. Show that the curvature radius of the lens is given approximately by $R = r^2/(m + \frac{1}{2})\lambda$, where the approximation holds when the thickness of the air space is much less than the curvature radius.

74. Given the result of the preceding problem, how many bright Newton's rings would be seen with a 2.5-cm-diameter glass lens with curvature radius 7.5 cm, when illuminated with 500-nm light?

75. How many rings would be seen if the system of the preceding problem were immersed in water ($n = 1.33$)?

76. A thin-walled glass tube of length *L* containing a gas of unknown refractive index is placed in one arm of a Michelson interferometer using light of wavelength λ. The tube is then evacuated, and during the process *m* bright fringes pass a fixed point as seen in the viewer. Find an expression for the refractive index of the gas.

77. The signal from a 103.9-MHz FM radio station reflects off a building 400 m away, effectively producing two sources of the same signal. You're driving at 60 km/h along a road parallel to a line between the station's antenna and the building and located a perpendicular distance of 6.5 km from them. How often does the signal appear to fade when you're driving roughly opposite the transmitter and building?

78. A satellite dish antenna 1.5 m in diameter receives TV broadcasts at a frequency of about 6 GHz (Fig. 37-51). (a) What is the angular size of its beam, defined as the full width of the central diffraction peak? (b) How many communications satellites could fit in geosynchronous orbit above Earth's equator if antennas like this one were not to pick up signals from more than one satellite? Your answer shows why geosynchronous orbit is such valuable "real estate."

FIGURE 37-51   A satellite dish (Problem 78).

79. The component of a star's velocity in the radial direction relative to Earth is to be measured using the doppler shift in the hydrogen-β spectral line, which appears at 486.1 nm when the source is stationary relative to the observer. What is the minimum speed that can be detected by observing in 1st order with a 10,000 line/cm grating 5.0 cm across? *Hint:* See Equation 17-11.

80. Light is incident on a dispersion grating at an angle α to the normal. Show that the condition for maximum light intensity becomes $d(\sin\theta \pm \sin\alpha) = m\lambda$.

81. In a double-slit experiment, a thin glass plate with refractive index 1.56 is placed over one of the slits. The fifth bright fringe now appears where the second dark fringe previously appeared. How thick is the plate if the incident light has wavelength 480 nm?

82. An arrangement known as Lloyd's mirror (Fig. 37-52) allows interference between a direct and a reflected beam from the same source. Find an expression for the separation of bright fringes on the screen, given the distances *d* and *D* in Fig. 37-53, and the wavelength λ of the light. *Hint:* Think of two sources, one the virtual image of the other.

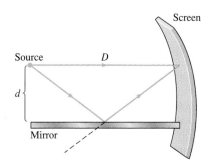

FIGURE 37-52   Lloyd's mirror (Problem 82).

PART 5   CUMULATIVE PROBLEMS

These problems combine material from chapters throughout the entire part or, in addition, from chapters in earlier parts, or they present special challenges.

1. A *grism* is a grating ruled onto a prism, as shown in Fig. 1. The grism is designed to transmit undeviated one wavelength of the spectrum in a given order, as refraction in the prism compensates for the deviation at the grating. Find an equation relating the separation d of the grooves that constitute the grating, the wedge angle α of the prism, the refractive index n, the undeviated wavelength λ_0, and order m_0.

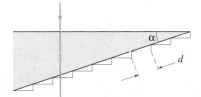

FIGURE 1   A grism (Cumulative Problem 1).

2. A double-slit system consists of two slits each of width a, with separation d between the slit centers ($d > a$). Light of intensity S_0 and wavelength λ is incident on the system, perpendicular to the plane containing the slits. Find an expression for the outgoing intensity as a function of angular position θ, taking into account both the slit width and the separation. Plot your result for the case $d = 4a$, and compare with Figure 37-35.

3. A closed cylindrical tube whose glass walls have negligible thickness measures 5.0 cm long by 5.0 mm in diameter. It is filled with water, initially at 15°C, and placed with its long dimension in one arm of a Michelson interferometer. The water is not perfectly transparent, and it absorbs 3.2% of the light energy incident on it. The laser power incident on the water is 50 mW, and the wavelength is 633 nm. The refractive index of water in the vicinity of 15°C is given approximately by $n = 1.335 - 8.4 \times 10^{-5}T$, where T is the temperature in °C. As the water absorbs light energy, how long does it take the interference pattern to shift by one whole fringe?

4. A radio antenna broadcasts a 6.5-MHz signal in all directions. The antenna is on top of a tower 300 m above sea level, and the tower is located 2.0 km from the shore. An airplane is flying toward the tower, at twice the altitude of the tower top, as shown in Fig. 2. Radio waves from the tower reflect off the ocean surface with a 180° phase change, but there is no significant reflection off the land. As the plane heads toward the tower, at what horizontal distance or distances from the tower will its radio equipment detect a minimum in the radio signal amplitude due to destructive interference of the direct and reflected radio waves?

5. In one type of optical fiber, called a *graded-index fiber*, the refractive index varies in a way that results in light rays

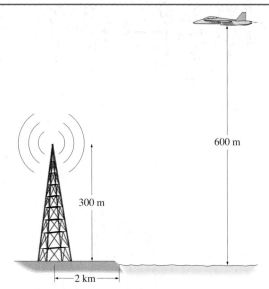

FIGURE 2   Cumulative Problem 4 (figure is not to scale).

being guided along the fiber on curved trajectories, rather than undergoing abrupt reflections. Figure 3 shows a simple model that demonstrates this effect; it also describes the basic optical effect in mirages. A slab of transparent material has refractive index $n(y)$ that varies with position y perpendicular to the slab face. A light ray enters the slab at $x = 0$, $y = 0$, making an angle θ_0 with the normal just inside the slab. The refractive index at this point is $n(y = 0) = n_0$. (a) By writing $\sin\theta$ in Snell's law in terms of the components dx and dy of the ray path, show that that path (written in the form of x as a function of y) is given by

$$ x = \int_0^y \frac{n_0 \sin\theta_0}{\sqrt{[n(y)]^2 - n_0^2 \sin^2\theta_0}} \, dy . $$

(b) Suppose $n(y) = n_0(1 - ay)$, where $n_0 = 1.5$ and $a = 1.0 \text{ mm}^{-1}$. If $\theta_0 = 60°$, find an explicit expression for x as a function of y, and plot your result to give the actual ray path. Explain the shape of your curve in terms of what happens when the ray reaches a point where $n(y) = n_0 \sin\theta_0$. What happens beyond this point?

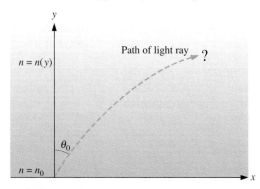

FIGURE 3   Cumulative Problem 5.

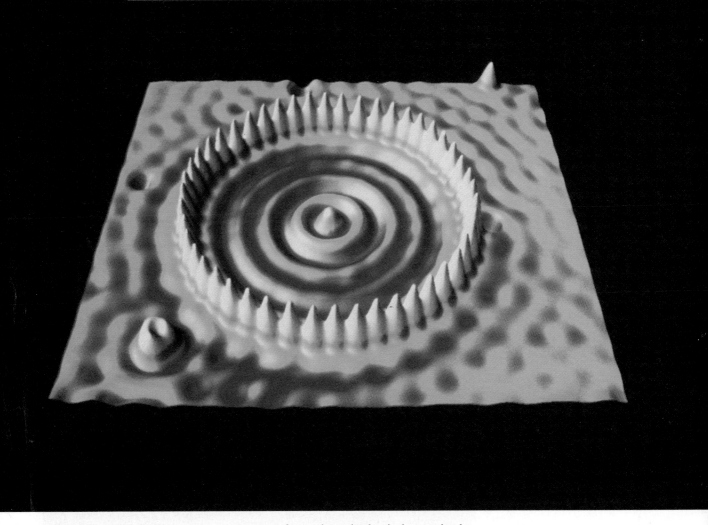

A "quantum corral" of 48 iron atoms on a copper surface encloses circular ripples associated with the wave-like nature of electrons on the surface. The image was taken with a scanning-tunneling microscope, which utilizes the principles of quantum mechanics. The ideas of modern physics allow scientists to not only understand but also, increasingly, to image and even manipulate matter at the atomic scale.

CHAPTER 38

THE THEORY OF RELATIVITY

Albert Einstein's theory of relativity revolutionized our understanding of space and time. Here Einstein explains his theory at a 1931 lecture in California.

By the last quarter of the nineteenth century, the basic laws of electromagnetism had been formulated. Maxwell's theory had demonstrated the electromagnetic nature of light, whose practical consequences we explored in the three preceding chapters. In the same era the work of Samuel Morse (telegraph), Alexander Graham Bell (telephone), Hertz and Marconi (radio), Thomas A. Edison (electric light, phonograph), and many others laid the foundation of electromagnetic technology. Yet at the same time the insights of Maxwell led to baffling questions and contradictions that shook the roots of physical understanding and even of common sense.

From the resolution of these contradictions arose the theory of relativity, a theory that radically altered the philosophical basis for our understanding of the

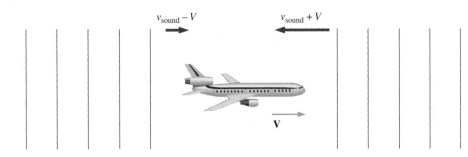

FIGURE 38-1 A plane moves at speed V through air, in which the sound speed is v_{sound}. Sound waves approaching the plane from the front move at $v_{sound} + V$ relative to the plane, while those approaching from behind move at $v_{sound} - V$ relative to the plane.

physical world, and whose influence spilled over into all areas of twentieth-century thought. The theory of relativity stands as a monument of human intellect and imagination. It transcends the everyday world of common sense, and shows us a universe whose richness is almost beyond imagination.

38-1 SPEED c RELATIVE TO WHAT?

Maxwell's equations show that electromagnetic waves can exist, and that all such waves travel with speed c. Speed c relative to what? When we described the mechanics of a taut string in Chapter 16, we encountered a wave equation showing that waves could propagate along the string with a certain speed. A certain speed relative to what? Clearly, to the string. Similarly, sound waves propagate through the air with a certain speed relative to the air. If you move through the air, the speed of sound *relative to you* will not be the same as its speed relative to the air (Fig. 38-1).

The Ether Concept

Each type of wave has a characteristic speed *relative to the medium in which it propagates.* What about light? What is the medium through which it propagates? Light reaches us from the most distant galaxies, traveling through seemingly empty space. Yet all our other experience with waves suggests that there should be a medium—and that the speed c in Maxwell's equations should be the speed relative to that medium. Nineteenth-century scientists, supposing that light waves were like mechanical waves, believed that light propagated through a tenuous substance called the **ether.** The ether permeated the entire universe, filling the smallest voids and permitting light to go anywhere. Electric and magnetic fields were visualized as stresses and strains in the ether. This mechanical view—that electromagnetic phenomena, including light, were disturbances of some substance—was deeply ingrained in the nineteenth-century scientists because of their previous experience that Newton's laws explained all known physical phenomena.

The ether had to have some unusual properties. First, it must be tenuous and without significant viscosity, or it could not creep into every corner of the universe. And it must offer no resistance to the motion of material bodies, or the planets would soon lose their energy and spiral into the Sun. At the same time the ether must be very stiff, for the speed of light is large. (If you make a spring stiffer, waves travel more quickly along it.) Indeed, the constants ε_0 and μ_0 must describe the mechanical properties of the ether that account for the high speed of light. These and other mechanical requirements make the ether a rather improbable substance, but without the ether it seemed there could be no waves, and the question "speed c relative to what?" would leave us floundering for an answer.

The existence of electromagnetic waves traveling at speed c follows from Maxwell's equations. But this result could be true only in a frame of reference fixed with respect to the ether, for if we move relative to the ether we should expect light to travel at a different speed relative to us. Therefore Maxwell's equations—that is, our description of electromagnetism—were presumably correct only in the ether's frame of reference.

This situation put electromagnetism in a rather different position from mechanics. In mechanics, the concept of absolute motion is meaningless. You can eat your dinner, or throw a ball, or do any mechanical experiment, as well on an airplane moving steadily at 1000 km/h as you can when the airplane is standing still on the ground. You need not take the uniform motion of the plane into account. This is the principle of **Galilean relativity,** which states that the laws of mechanics are valid in all frames of reference in uniform motion (see Section 3-8). But the laws of electromagnetism could only be valid in the ether's frame of reference, for it seemed that only in this frame could the prediction of electromagnetic waves moving at speed c be correct.

38-2 MATTER, MOTION, AND THE ETHER

Given the existence of the ether, it is natural to ask about Earth's motion relative to it. If Earth is moving through the ether, we should expect light to travel faster relative to us when it comes from the direction toward which Earth is moving. On the other hand, Earth might be at rest relative to the ether. Because other planets, stars, and galaxies move with respect to Earth, it is hard to imagine that ether is everywhere fixed with respect to Earth alone, for this violates the Copernican view that Earth does not occupy a privileged spot in the universe. But maybe Earth drags with it the ether in its immediate vicinity. If this "ether drag" occurs, then the speed of light must be independent of direction, but if ether drag does not occur then the speed of light measured on Earth must depend on direction. Through observation and experiment, nineteenth-century physicists sought to resolve the question of Earth's motion through the ether.

Aberration of Starlight

Imagine standing in a rainstorm with rain falling vertically. To keep dry, you hold your umbrella with its shaft straight up, as shown in Fig. 38-2a. But if you run, as in Fig. 38-2b, you will keep driest if you tilt your umbrella forward. Why? Because then the direction of rainfall *relative to you* is not straight down but at an angle, as shown in Fig. 38-2c. This argument presupposes that you do not drag with you a large volume of air. If such an "air drag" occurred, raindrops entering the region around you would be accelerated quickly in the horizontal direction by the air moving with you, so they would now fall vertically relative to you, as in Fig. 38-2d. No matter which way you ran, as long as you dragged air with you, you would point your umbrella vertically upward to stay dry.

FIGURE 38-2 (*a*) Standing still in vertically falling rain, you hold your umbrella overhead to keep dry. (*b*) Running, you tilt the umbrella to compensate for the rain's motion relative to you. (*c*) The situation of (*b*) seen from the runner's frame of reference. (*d*) If you dragged a volume of air with you, you would hold the umbrella overhead whatever your state of motion.

This umbrella example is exactly analogous to the observation of light from stars, with the rain being starlight and the umbrella a telescope. If Earth does not drag ether with it, the direction from which starlight comes will depend on the motion of Earth relative to the ether. But if "ether drag" occurs in analogy with Fig. 38-2*d*, then light from a particular star will always come from the same direction.

In fact we do observe a change in the direction of starlight. As Earth swings around in its orbit, we must first point a telescope one way to see a particular star. Then, six months later, Earth's orbital motion is in exactly the opposite direction, and we must point the telescope in a slightly different direction. This phenomenon is called **aberration of starlight** and shows that *Earth does not drag the ether.*

The Michelson-Morley Experiment

If we reject the pre-Copernican notion that Earth alone is at rest relative to the ether, then aberration of starlight forces us to conclude that Earth moves through the ether. Furthermore, the relative velocity of the motion must change throughout the year, as Earth orbits the Sun.

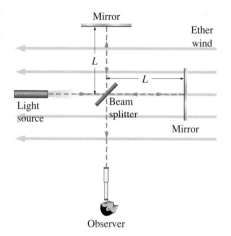

FIGURE 38-3 Simplified diagram of a Michelson interferometer. An "ether wind" blowing in the direction shown should result in a longer time for the light beam on the horizontal arm.

FIGURE 38-4 Vector diagram showing resultant velocity **u** of light moving at right angles to an ether wind with speed v. The speed of light relative to the ether is c.

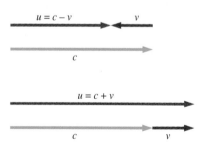

FIGURE 38-5 Vector diagrams showing the resultant velocities **u** for light moving with and against an ether wind whose speed is v.

In a series of experiments done in 1881–1887, the American scientists Albert A. Michelson and Edward W. Morley attempted to determine Earth's velocity relative to the ether. They used Michelson's interferometer (Fig. 38-3), whose operation we described in the preceding chapter. Recall that the interferometer produces a pattern of interference fringes that shifts if the round-trip travel time for light on one of its two perpendicular arms changes. The interference pattern reflects, among other things, possible differences in travel times that arise from differences in the speed of light in different directions—differences that should result from Earth's motion through the ether. Rotating the apparatus through 90° would interchange the directions of the arms and should therefore shift the interference pattern.

Now suppose that Earth moves at speed v relative to the ether. Then from the viewpoint of an observer on Earth, there is an "ether wind" blowing past Earth. Suppose that the Michelson-Morley apparatus is oriented with one light path parallel to the wind, the other perpendicular. Consider a light beam moving the distance L at right angles to the ether wind. The beam must be aimed slightly upwind, in order that it will actually move perpendicular to the wind. The light moves in this direction at speed c relative to the ether, but the ether wind sweeps it back so its path in the Michelson-Morley apparatus is at right angles to the wind. From Fig. 38-4, we see that its speed relative to the apparatus is

$$u = \sqrt{c^2 - v^2},$$

so the round-trip travel time is

$$t_{\text{perpendicular}} = \frac{2L}{u} = \frac{2L}{\sqrt{c^2 - v^2}}. \tag{38-1}$$

Light sent a distance L "upstream"—against the ether wind—travels at speed c relative to the ether but at speed $c - v$ relative to Earth. It therefore takes a time

$$t_{\text{upstream}} = \frac{L}{c - v}.$$

Returning, the light moves at $c + v$ relative to Earth, taking

$$t_{\text{downstream}} = \frac{L}{c + v}$$

(Fig. 38-5). So the round-trip time parallel to the wind is

$$t_{\text{parallel}} = \frac{L}{c - v} + \frac{L}{c + v} = \frac{2cL}{c^2 - v^2}. \tag{38-2}$$

The two round-trip travel times differ, with the trip parallel to the ether wind always taking longer (see Problems 1, 2, and 4). Light on the parallel trip is slowed when it moves against the ether wind, then speeds up when it moves with the wind. But slowing always dominates, because the light spends more time moving against the wind than with it.

The Michelson-Morley experiment of 1887 was sensitive enough to detect differences in the speed of light at least an order of magnitude smaller than Earth's orbital speed. The experiment was repeated with the apparatus oriented in different directions, and at different times throughout the year, and the same simple but striking result always emerged: there was never any difference in the travel times for the two light beams. In terms of the ether concept, the Michelson-Morley experiment showed that *Earth does not move relative to the ether*.

A Contradiction in Physics

Aberration of starlight shows that Earth does not drag ether with it. Earth must therefore move relative to the ether. But the Michelson-Morley experiment shows that it does not. This contradiction is a deep one, rooted in the fundamental laws of electromagnetism and in the analogy between mechanical waves and electromagnetic waves. The contradiction arises directly in trying to answer the simple question "with respect to what does light move at speed c?"

Physicists at the end of the nineteenth century made many ingenious attempts to resolve the dilemma of light and the ether, but their explanations were either inconsistent with experiment or lacked sound conceptual bases.

38-3   SPECIAL RELATIVITY

In 1905, at the age of 26, Albert Einstein (Fig. 38-6) proposed a theory that resolved the dilemma and at the same time altered the very foundation of physical thought. Einstein declared simply that the ether is a fiction. But then with respect to what does light move at speed c? With respect, Einstein declared, to anyone who cares to observe it. This statement is at once simple, radical, and conservative. Simple, because its meaning is clear and obvious. Anyone who measures the speed of light will get the value $c = 3.0 \times 10^8$ m/s. Radical, because it alters our commonsense notions of space and time. Conservative, because it asserts for electromagnetism what had long been true in mechanics: that the laws of physics do not depend on the motion of the observer. Einstein summarized his new ideas in the **special theory of relativity,** which is expressed in this simple sentence:

| **The laws of physics are the same in all inertial frames of reference.**

Recall that inertial frames are those which are not accelerated—i.e., those in which the laws of *mechanics* were already valid. Einstein's statement encompasses *all* laws of physics, including mechanics and electromagnetism. The prediction that electromagnetic waves move at speed c must, then, be a universal prediction that holds in *all* inertial frames of reference. The *special* theory of relativity is special because it is valid only for the special case of inertial frames. Later we will discuss the general theory of relativity, in which this restriction is removed.

Einstein's relativity readily explains the result of the Michelson-Morley experiment, for no matter what the speed of Earth relative to anything, an observer on Earth should measure the same speed for light in all directions. But at the same time, we will see that relativity flagrantly violates our common-sense notions of space and time.

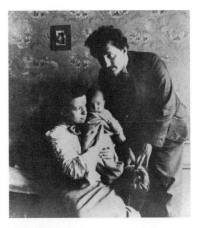

FIGURE 38-6   Albert Einstein was a young father when, at age 26, he formulated the special theory of relativity.

FIGURE 38-7   As the car passes the pedestrian, a light pulse goes by. Both driver and pedestrian measure the same speed c for the light, even though they are in relative motion.

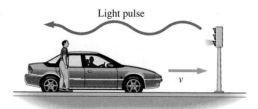

Light pulse

v

38-4 SPACE AND TIME IN RELATIVITY

Consider a car driving past a pedestrian standing by the roadside (Fig. 38-7). Driver and pedestrian each measure the speed of the light from a blinking traffic signal. Relativity says they will get the same answer, $c = 3.0 \times 10^8$ m/s, even though the car is moving toward the source of light. How can this be? Consider how each observer might make the measurement. Let each be equipped with a meter stick and an accurate, high-speed electronic stopwatch. Suppose that a light pulse passes the front ends of both meter sticks just as they coincide. Each observer measures the time it takes the light pulse to cross the meter stick, then divides the distance (one meter) by the measured time to get the speed of light. Since the stick on the car is moving toward the light source, common sense suggests that the light will pass the far end of this stick first, and therefore that the time on the car's stopwatch will be shorter. But this violates relativity, for if both observers use the same path length for their speed-of-light measurement, and if they get different times, then they will have measured different speeds for light. In fact, both stopwatches will read the same time, even though common sense tells us that the light passes the end of the "moving" meter stick "earlier."

How can this be? It follows logically from the statement of special relativity, which in turn is consistent with physical experiments. But how can it be? Something must be "wrong" with someone's meter stick or stopwatch. Maybe the motion of the car somehow affects the stopwatch on the car. But no: this suggestion violates the spirit of relativity, which says that steady, uniform motion is undetectable—that it makes no more sense to say that the car is moving and the pedestrian is at rest than to say the opposite. That is the whole point of relativity—any frame of reference in uniform motion is as good as any other for doing physics. So there can be nothing wrong with the clocks and meter sticks.

The only things left to go "wrong" are time and space. Time and space—the seemingly passive, universal backgrounds in which all physical events take place—must themselves depend on the observer. Two observers in different frames of reference, moving uniformly relative to each other, are measuring different quantities when they use clocks to record the passage of time and meter sticks to determine distances in space. In relativity it is the laws of physics—not measures of time and space—that must be the same for everyone. Time and space are altered in ways that allow the laws of physics, including Maxwell's theory and its prediction that light waves travel at speed c, to be the same for all observers in uniform motion.

In exploring space and time in relativity, we'll often speak of events. An **event** is an occurrence specified by giving its position—three spatial coordinates—and its time. Your birth, for example, is an event: it occurred at a certain place and a certain time, and both are needed to specify it fully.

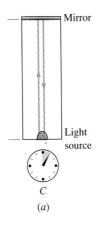

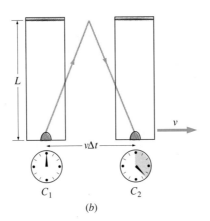

(a)

(b)

FIGURE 38-8 A "light-box clock" seen from two frames of reference. Light leaves the source, reflects off the mirror, and returns to the source. Clock C is attached to the box, while the box moves between clocks C_1 and C_2 at speed v. (a) In a frame of reference at rest with respect to the box, light travels a distance $2L$ at speed c. (b) In a frame of reference in which the box is moving, light travels a greater distance. But its speed c is the same in all frames of reference, so the time interval is longer in this frame. Part (b) shows the box both when the light is emitted and again when it returns to the source.

Time Dilation

To see how time is altered, consider the simple device shown in Fig. 38-8. It consists of a box of length L with a light source at one end and a mirror at the other. Let a flash of light leave the source, travel to the mirror, and return to the source. We want to know the time interval between two distinct events: the light flash leaving the source and the flash returning to the source.

In Fig. 38-8a we consider the experiment in a frame of reference S' at rest with respect to the box. The light travels a distance $2L$ in this frame, giving a round-trip travel time of

$$\Delta t' = \frac{2L}{c}.$$

Now consider the *same* experiment, viewed from a frame of reference S in which the box is moving to the right with speed v. In this frame there are two clocks along the path of the box. These clocks are synchronized, and are in the same frame of reference, so they both measure the same quantity. The box passes the first clock just as the light flash goes off, and at this instant the clock reads zero. Just as the light flash returns to the source, the box passes the second clock. The time this clock reads at the instant the box passes is the time, measured in frame S, between the emission and return of the light flash.

We can calculate this time interval Δt in frame S, just as we did in S', by figuring the total distance traveled by the light and dividing by its speed. Figure 38-8b shows the situation. In frame S, the box moves to the right a distance $v\,\Delta t$ in the time Δt between emission and return of the light flash. Meanwhile the light takes a diagonal path up to the mirror of the moving box, then back down to the source. The total length of this path is twice the diagonal from source to mirror, or, using the Pythagorean theorem, $2\sqrt{L^2 + (v\,\Delta t/2)^2}$. The time required for light to go this distance is just the distance divided by the speed of light, or

$$\Delta t = \frac{2\sqrt{L^2 + (v\,\Delta t/2)^2}}{c}. \qquad (38\text{-}3)$$

Notice that we explicitly used the theory of relativity in writing Equation 38-3. We did not vectorially add the horizontal speed of the box to the vertical speed of light to get a new speed of light in frame S, for relativity says that the speed of light is the same in all frames of reference. Had we altered the speed, we would have had an increased path length in S, but an increased speed of light as well, and would have found that the time intervals in both frames were the same. But no! Relativity requires that we use the same speed c in both frames, even though the path lengths differ. That is why we get different answers for the time.

The unknown time Δt appears on both sides of Equation 38-3. Multiplying through by c and squaring gives

$$c^2(\Delta t)^2 = 4L^2 + v^2(\Delta t)^2.$$

We then solve for $(\Delta t)^2$ to get

$$(\Delta t)^2 = \frac{4L^2}{c^2 - v^2} = \frac{4L^2}{c^2}\left(\frac{1}{1 - v^2/c^2}\right).$$

Taking the square root of both sides, and noting that $2L/c$ is just the time $\Delta t'$ measured in the frame S' at rest with respect to the box, we have

$$\Delta t = \frac{\Delta t'}{\sqrt{1 - v^2/c^2}} \qquad \text{or} \qquad \Delta t' = \Delta t \sqrt{1 - v^2/c^2}. \tag{38-4}$$

Equation 38-4 describes the phenomenon of **time dilation,** in which the time interval between two events is always shortest in a frame of reference in which the two events occur at the same place. (The time measured in this frame is called the **proper time,** although relativity precludes our considering it any "better" a measure of time than that made in any other frame of reference.) In our example, the two events are the emission and the return of the light flash, and they occur at the same place—the bottom of the box—in the box frame S', but at different places in S. Thus $\Delta t'$—the time interval measured in S', or the proper time—is shorter than Δt, as you can see from Equation 38-4.

Time dilation is sometimes characterized by saying that "moving clocks run slow," but this statement is not strictly correct because relativity rules out our saying that one clock is moving and another not. What the statement means is what we've just seen: that the time interval between two events is shortest in a frame of reference where the two events occur at the same place (Fig. 38-9).

We have illustrated time dilation with a very special device—a "light-box clock." But the phenomenon would occur with any other timing device, for it is not that something unusual happens to the clock, but to time itself. If we take away the light box in Fig. 38-8, giving Fig. 38-9, the clocks will still show the same discrepancy. There is no use searching for a physical mechanism that slows things down. All manifestations of time—the oscillations of the quartz crystal in a digital watch, the swing of a pendulum clock, the period of vibration of atoms in an atomic clock, biological rhythms, and human lifetimes—all are affected in the same way.

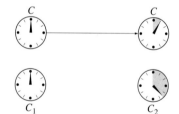

FIGURE 38-9 Clock C moves between clocks C_1 and C_2, which are at rest relative to each other and synchronized in their rest frame. Time between the event of C passing C_1 and the event of C passing C_2 is shorter in C's frame of reference.

● **EXAMPLE 38-1** STAR TREK

A spaceship leaves Earth on a one-way trip that Earthbound observers judge will take 25 years. If the ship travels at 0.95c relative to Earth, how long is the journey as measured by a clock on board the ship?

Solution

The spaceship is like the light box we used in deriving Equation 38-4 for time dilation, in that in its frame the beginning and end of the journey occur at the same place—namely, on the ship. So the time interval on the ship clock is the $\Delta t'$ of Equation 38-4. For the observers on Earth, the beginning and end of the journey occur at different places, so their 25-year time interval is Δt. Applying Equation 38-4 then gives

$$\Delta t' = \Delta t \sqrt{1 - v^2/c^2} = (25 \text{ y})(\sqrt{1 - (0.95c)^2/c^2})$$

$$= 7.8 \text{ y}.$$

Thus at this high relative speed, the ship's time is considerably shorter. We'll soon explore what happens if the ship turns around and returns to Earth.

EXERCISE An extraterrestrial spacecraft whizzes through the solar system at 0.80c. How long does it take to go the 8.3 light-minute distance from Earth to the Sun (a) according to an observer on Earth and (b) according to an alien aboard the ship?

Answers: (a) 10.4 min; (b) 6.23 min

Some problems similar to Example 38-1: 7, 8, 16

●

Why don't we notice time dilation as we travel about in our everyday lives? Because the factor v^2/c^2 in Equation 38-4 is so small for any velocities we have relative to Earth. Even in a jet airplane, we are moving at 1000 km/h or only about 10^{-6} c. Then time in the airplane is different from that on Earth by only about 1 part in $(10^6)^2$, which amounts to a few milliseconds per century. This illustrates an important point: any results predicted by relativity should agree with our common sense, Newtonian ideas when relative velocities are small compared with the speed of light. Only at substantial fractions of c do relativistic effects become obvious. Since our intuitions and common sense are built on experience at low relative velocities, it is not surprising that effects at high velocities seem counter to common sense.

■ **APPLICATION** MOUNTAINS AND MUONS: CONFIRMING TIME DILATION

Time dilation is clearly illustrated in experiments with subatomic particles moving, relative to us, at speeds near that of light. In a classic experiment, the "clocks" are the lifetimes of unstable particles called muons, which are created by the interaction of cosmic rays with Earth's upper atmosphere. The experiment consists in counting the number of muons incident each hour on the top of a mountain—Mt. Washington in New Hampshire, about 2000 m above sea level. The measurement is then repeated at sea level (Fig. 38-10).

Using a detector that records only those muons moving at about 0.994c at the mountaintop altitude, the experiment shows an average of about 560 muons with this speed are incident on the mountaintop each hour. If the mountain weren't there, the muons would travel from the mountaintop altitude to sea level in a time given by

$$\Delta t = \frac{2000 \text{ m}}{(0.994)(3.0 \times 10^8 \text{ m/s})} = 6.7 \ \mu s.$$

The muon's decay rate is such that one should expect only about 25 of the original 560 muons to remain after this 6.7-μs interval, so that's approximately the number we might expect to detect each hour at sea level. More muons would survive a shorter time interval; in particular, we should expect 414 muons surviving 0.73 μs after passing the mountaintop altitude.

Why the value 0.73 μs? Because the muons are moving at 0.994c, and the 6.7-μs interval from mountaintop to sea level is measured in Earth's frame of reference—not in the muons' frame. In the muons' frame, time dilation should reduce that interval to

$$\Delta t' = \Delta t \sqrt{1 - v^2/c^2} = (6.7 \ \mu s)(\sqrt{1 - 0.994^2})$$

$$= 0.73 \ \mu s.$$

Since the muons' decay is determined by their measure of time, we should therefore expect 414 muons surviving at sea level.

So what happens? When the experiment is done, muon counts of just over 400 per hour are observed at sea level. This is no subtle effect! The difference between 25 and 414 is dramatic. At 0.994c, the nonrelativistic description is hopelessly inadequate. At this speed, the factor $\sqrt{1 - v^2/c^2}$ is one-ninth, and time dilation is obvious.

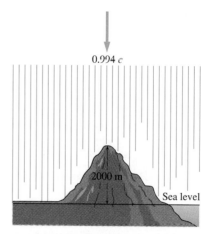

FIGURE 38-10   The rate at which high-speed muons are incident is measured on a mountaintop and again at sea level. The muons decay at such a rate that only a small fraction should survive to reach sea level. But the number measured at sea level is much larger, showing the effect of time dilation on muon lifetimes.

The Twin Paradox

FIGURE 38-11   At departure, the twins are the same age.

The phenomenon of time dilation has startling consequences, for it allows us to travel into the future! The famous "twin paradox" illustrates this possibility. One of two twins boards a fast spaceship for a journey to a distant star (Fig. 38-11). The other stays behind on Earth. Imagine that there are clocks at Earth and star, like the two clocks in frame S of our light-box experiment (Fig. 38-12a). There is a clock on the spaceship, like the one clock in our light-box frame S'. When the ship arrives at the distant star, less time will have elapsed on the ship clock than on the Earth and star clocks (Fig. 38-12b). Now the ship turns around and comes home. Again, the situation is identical to our light-box example, so again less time elapses on the ship clock, and the traveling twin arrives home younger than the earthbound twin (Fig. 38-13)! Depending on how far and how fast the traveling twin goes, the difference in ages could be arbitrarily large. The traveling twin could even return to Earth millions of years in the future, even though only hours had elapsed on the ship. But this is a one-way trip to the future! If the traveler doesn't like what he or she finds in the future, there is no going back!

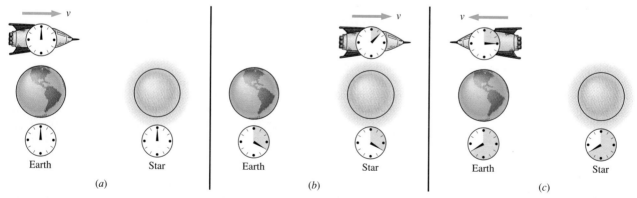

FIGURE 38-12   The traveling twin journeys to a distant star, then returns. Earth and star are at rest with respect to each other, and their clocks are synchronized. The figure is drawn from the Earth-star frame of reference. (a) Ship and Earth clocks agree as the ship leaves Earth. (b) When the ship reaches the star, less time has elapsed on its clock. (c) When the ship returns to Earth, its clock and Earth's no longer agree.

● EXAMPLE 38-2   THE TWIN PARADOX

Earth and a star are 10 light-years (ly) apart, measured in a frame at rest with respect to Earth and star. Twin A boards a space ship and travels at 0.80c to the star, then immediately turns around and returns to Earth at 0.80c. Twin B remains behind. Determine the round-trip travel time in the Earth-star frame of reference and in the ship frame. By how much will the twins' ages differ when they get back together?

Solution
At 0.80c, the time to go 10 ly in the Earth-star frame is just

$$\Delta t = \frac{10 \text{ ly}}{0.80 \text{ ly/y}} = 12.5 \text{ y}.$$

The round-trip time is then 25 y. Equation 38-4 for $\Delta t'$ then gives the one-way travel time in the ship frame:

$$\Delta t' = \Delta t \sqrt{1 - v^2/c^2} = (12.5 \text{ y})(\sqrt{1 - 0.80^2} = 7.5 \text{ y}.$$

Then the round-trip time in the ship frame is 15 y, so the twins' ages differ by 10 y when twin A returns.

TIP   Years, Light-Years, and the Speed of Light   A light-year (ly) is the distance light travels in one year. By definition, therefore, the speed of light is 1 ly/y. It's often easiest in relativity to work in units where the speed of light is 1, whether those units be light-years and years, light-seconds and seconds, or whatever.

EXERCISE   A spacecraft makes a round trip to a point a distance ℓ from Earth, as measured in Earth's frame of reference. The ship travels at 60% of the speed of light. If the round trip takes 1 hour by the ship's clock, (a) what is the distance ℓ and (b) how much time elapses on Earth between the ship's departure and its return?

Answers: (a) 0.375 light-hours (4×10^{11} m); (b) 1.25 h

Some problems similar to Example 38-2:   7, 9, 14, 15   ●

The paradox in the twin example is not just that something strange happens to time—we already expect that of special relativity. But now look at things from the spaceship's frame of reference. Doesn't the spaceship see Earth recede into the distance, turn around, and come back? And then shouldn't the earth-bound twin be younger? This is the paradox. It is resolved by considering what is *special* about the special theory of relativity. The special theory applies only to inertial—i.e., unaccelerated—frames of reference. The traveling twin must accelerate in order to return to Earth, and is therefore briefly in a noninertial frame. Absolute motion has no meaning in special relativity, but absolute acceleration does. The traveling twin feels inertial forces when the ship turns around, but the earthbound twin does not. Although we cannot say that one twin is moving and the other is not, we can say that one twin's motion changes and the other's does not. The situation is not symmetric, and that is why the traveling twin really does return younger.

Could it really happen? It could, and it has. Atomic clocks are now so accurate that experiments have been done to detect the minuscule time difference between a clock flown around the Earth and one left behind.

What if the traveling twin did not turn around? We could still argue that the ship clock runs slower than clocks on Earth and the star. But then isn't the situation symmetric? Couldn't the traveling twin argue that a clock on Earth should run slower than clocks on the ship? Yes—and we would find this to be true if we set up a series of clocks in the ship frame and measured time intervals on the Earth clock as Earth passed first one, then another of the ship clocks. But there is no contradiction. Unless the twins get together again, there is no way they can directly compare their clocks or their ages at the same place. Instead, they must compare one clock with a sequence of clocks that are all synchronized. And as we will soon see, clocks that are synchronized in one frame of reference are not synchronized in another frame! Only if the twins get back together can they compare just two clocks without having to worry about syn-

FIGURE 38-13   It is not only clock readings, but all manifestations of time that differ. The traveling twin really has aged less!

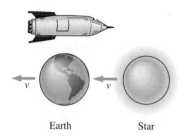

Earth   Star

FIGURE 38-14 Earth-star trip viewed from the spaceship's frame of reference. Note that Earth-star distance is contracted, and the ship lengthened, relative to Fig. 38-12, which is drawn from the Earth-star frame.

chronization of distant clocks. And they can get back together only if at least one of them accelerates.

The Lorentz Contraction

In Example 38-2 the spaceship moved 10 ly in 12.5 y at speed 0.80c. These quantities are related by the simple expression $\Delta x = v \Delta t$, where Δx is the distance between Earth and star *measured in the Earth-star frame of reference S*. Now an observer in the ship frame of reference S' sees Earth and star moving past at speed v. First Earth passes the ship, then the star passes (Fig. 38-14). We found that the time interval between these two events, measured in the ship frame, is $\Delta t' = \Delta t \sqrt{1 - v^2/c^2} = 7.5$ y. Since Earth and star are moving past at $v = 0.80\,c$, the distance between Earth and star as measured in the ship frame must be

$$\Delta x' = v \Delta t' = v \Delta t \sqrt{1 - v^2/c^2} = \Delta x \sqrt{1 - v^2/c^2}, \qquad (38\text{-}5)$$

or 6.0 ly in our example. This equation shows that the distance between two points is always greatest in a frame (the Earth-star frame S, in this example) fixed with respect to those points. In any other frame the distance is smaller. This phenomenon is called the **Lorentz contraction,** or the Lorentz-Fitzgerald contraction, after the Dutch physicist H. A. Lorentz and the Irish physicist George F. Fitzgerald, who, in the 1890s, independently proposed it as an ad hoc way of explaining the Michelson-Morley experiment. Only through Einstein's theory did the contraction acquire a solid conceptual basis.

Although we developed the Lorentz contraction using the distance between separate objects—Earth and a distant star—the effect occurs for any observer moving with respect to two points that are fixed with respect to each other. In particular, a rigid object like a meter stick or spaceship is shorter when measured by an observer with respect to whom it is moving. (An object's length in a frame in which it is at rest is called its **proper length,** although again relativity precludes our thinking of that frame as being in any way special.)

As with time dilation, do not look for some physical mechanism that squashes moving objects. Rather, it is space itself that is different for different observers. In order to accept the simple fact that absolute motion is meaningless, we must alter our common-sense notions of time and space. Lorentz contraction and time dilation are manifestations of that alteration.

● EXAMPLE 38-3 A STANFORD ELECTRON

At the Stanford Linear Accelerator Center (SLAC) (Fig. 38-15), subatomic particles are accelerated to high energies over a straight path whose proper length is 3.2 km. During a particular experiment, electrons are accelerated to 0.9999995 of the speed of light. In the SLAC frame, how long would it take electrons with this speed to travel the full length of the device?

How long would the trip take in the rest frame of the electrons? How long would the accelerator be in the rest frame of the electrons?

Solution

The electron speed is so close to that of light that the travel time is, to a very good approximation,

$$\Delta t = \frac{\Delta x}{c} = \frac{3.2 \times 10^3 \text{ m}}{3.0 \times 10^8 \text{ m/s}} = 1.1 \times 10^{-5} \text{ s}.$$

In the rest frame of the electrons, the time to traverse the accelerator is given by Equation 38-4, or

$$\Delta t' = \Delta t \sqrt{1 - v^2/c^2} = (1.1 \times 10^{-5} \text{ s})\sqrt{1 - 0.9999995^2}$$
$$= (1.1 \times 10^{-5} \text{ s})(1.0 \times 10^{-3})$$
$$= 1.1 \times 10^{-8} \text{ s}.$$

Our time calculation shows that the relativistic factor $\sqrt{1 - v^2/c^2}$ is 10^{-3}, so in the electron frame of reference the accelerator length is

$$\Delta x' = \Delta x \sqrt{1 - v^2/c^2} = 3.2 \times 10^{-3} \text{ km} = 3.2 \text{ m}.$$

FIGURE 38-15 The Stanford Linear Accelerator is 3.2 km (2 miles) long. But to electrons moving through it at 0.999 999 5c, the accelerator is only 3.2 m long.

● **EXAMPLE 38-4** A STANFORD STUDENT

A physics student from New York flies to San Francisco to do an experiment at SLAC. She travels a distance of 4800 km on a plane going at 1000 km/h. How long does the trip take in a frame at rest with respect to Earth? How long does it take according to the student's watch? How far is it from New York to San Francisco in the airplane's frame of reference?

Solution
At 1000 km/h, the 4800-km trip takes

$$\Delta t = \frac{\Delta x}{v} = \frac{4800 \text{ km}}{1000 \text{ km/h}} = 4.8 \text{ hours}.$$

In the frame of the moving airplane, time and distance are altered by the relativistic factor $\sqrt{1 - v^2/c^2}$. The speed of the plane is 1000 km/h, or 278 m/s, so

$$\sqrt{1 - v^2/c^2} = \left[1 - \frac{(278 \text{ m/s})^2}{(3.0 \times 10^8 \text{ m/s})^2}\right]^{1/2}$$
$$\simeq 1 - (\tfrac{1}{2})(8.6 \times 10^{-13}) = 0.99999999999957.$$

Here we used the binomial theorem, $(1 + x)^n \simeq 1 + nx$ for $|nx| \ll 1$ because most calculators do not carry enough significant figures to distinguish the result from 1 (see Appendix A). We really need not carry the calculation further. The time on the student's watch is the same as the time on the ground to about five parts in 10^{13}, and the distance in the plane's frame is the same as the ground distance to within this same factor. The student need take no account of time dilation and Lorentz contraction, except in a physics exam or high-energy experiment!

EXERCISE Our Milky Way galaxy is about 100,000 light-years in diameter. What is its diameter as measured by an intergalactic spacecraft traveling at 0.96c?

Answer: 28,000 ly

Some problems similar to Examples 38-3 and 38-4: 6, 11, 12 ●

Examples 38-3 and 38-4 show that relativistic effects are significant only at high relative velocities—so high that the quantity v^2/c^2 is comparable with 1. In our daily lives we have no experience with such velocities. It is for this reason that relativity so offends our common-sense notions of space and time. Those notions are built on a groundwork of limited experience that does not include high velocities. If we did move regularly with respect to our surroundings at

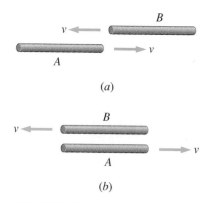

FIGURE 38-16 (a) In frame S, both sticks have the same speed v and both are contracted by the same amount. (b) Therefore their ends coincide at the same time.

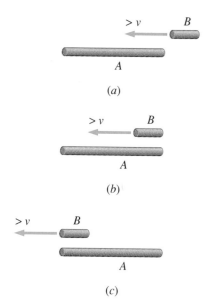

FIGURE 38-17 The passing sticks viewed in a frame S' at rest with respect to stick A. Relative speed is greater than the speed of either stick with respect to frame S. (a) First the left end of stick B passes the right end of A. (b) A while later, the right end of B passes the right end of A. This is event E_1. (c) Still later, the left end of B passes the left end of A. This is event E_2. Events E_1 and E_2 are not simultaneous in frame S'.

speeds near that of light, the relativity of space and time would be as obvious as our common-sense notions now seem.

For physicists working with high-energy elementary particles, relativistic effects *are* obvious. Unstable particles moving through the laboratory at high speeds live longer than they would at rest in the lab. And high-energy particle accelerators would not work if their design did not take relativity into account.

Events and Simultaneity

Consider two identical sticks A and B, each of proper length L. Suppose these sticks are moving toward each other. For a frame of reference S in which both sticks are moving at the same speed v, the situation is shown in Fig. 38-16. Both sticks are Lorentz contracted, but since both are moving at the same speed v relative to the frame S, both are contracted by the same amount and therefore have the same length. What happens as the sticks pass each other? First, the right end of stick A and the left end of stick B pass (Fig. 38-16a). A little while later, the right end of A passes the right end of B. At the same time, *because the sticks have the same length in S*, the left end of A passes the left end of B (Fig. 38-16b). The passing of the two right ends of the sticks is an event that we designate E_1. Similarly, the passing of the two left ends we designate E_2. We have shown that, in the frame S, the two events E_1 and E_2 are **simultaneous**—they occur at the same time.

Now look at the situation from a frame of reference S' in which stick A is at rest. In this frame, stick B moves toward stick A. Since we are in S', we are at rest relative to stick A, and it has its proper length L. But stick B is contracted more than in frame S because of its higher relative velocity. The situation is shown in Fig. 38-17. As the figure indicates, the event E_1 occurs before E_2; the two events are not simultaneous in the frame S'. What happens in a frame S'' at rest with stick B? As Fig. 38-18 shows, the events E_1 and E_2 are again not simultaneous, and this time event E_2 occurs first.

Isn't this all just an illusion arising from the apparent length differences due to motion of the sticks? Isn't the picture in frame S (Fig. 38-16) "really" the right one? No! Relativity theory assures us that all uniformly moving frames—including the frames S, S', and S'' of Figs. 38-16 through 38-18—are equally valid for describing physical reality. The length differences and the changes in time ordering of events E_1 and E_2 are not "apparent" and they are not "illusions." They arise from valid descriptions in different frames of reference, and each has equal claim to "reality." If you insist that one of the frames—say S—somehow has more validity, then you are reasserting the nineteenth-century notion that there is one favored reference frame in which alone the laws of physics are valid.

But how can observers disagree on the time order of events? Doesn't that violate causality? After all, if one event is a cause of another, we certainly expect the cause to precede the effect. It would be disturbing if some observer, with valid claim on "reality," found that cause and effect occurred in the reverse order. But there is no violation of causality. As we will soon show, the only events that can have their time order reversed are those that are so far apart in space, and so close in time, that not even light can travel fast enough to be at both events. There is no way that such events can influence each other, and therefore they cannot be causally related. In a very real sense it does not matter which event occurs first, and indeed different observers will disagree on their relative time order. For example, an event on Earth now and another occurring five minutes from now on the Sun cannot be causally related, for it takes light from the Sun eight minutes to reach Earth. For observers moving at high enough speeds relative to Earth and Sun, the solar event occurs first.

Only when events are close enough in space and separated enough in time so that light can travel from one to the other can the two be causally related. In that case, all observers will agree about their time order, although they may disagree about the actual time interval between the events. For example, an event on Earth now and another occurring fifteen minutes from now on the Sun could be causally related, and therefore all observers will agree that the terrestrial event occurs first. We will explore these notions more quantitatively in the next section.

The Lorentz Transformations

Our demonstration that the time order of events may be relative deals implicitly with the coordinates—position and time—of specific events, and suggests that those coordinates may differ for different observers. Similarly, time dilation and Lorentz contraction arise as specific instances of the way positions and times in one frame of reference are related to their values in another frame. We now seek more general expressions—called **Lorentz transformations**—relating the space and time coordinates of an event in two frames of reference in relative motion. Consider coordinate axes in a frame of reference S and in another frame S' moving in the x direction with speed v relative to S. Suppose that the origins of the two coordinate systems coincide at time $t = t' = 0$. If an event has coordinates x, y, z, and t in S, what are its coordinates x', y', z', and t' in S'? Were it not for relativity, we would expect the coordinates y, z, and the time t to remain unchanged from one coordinate system to the other. For an event occurring at time $t = 0$, when the two origins coincide, we would expect the x coordinates to be the same too. But as S' moves in the positive x direction relative to S, a given x value in S would correspond to a value $x' = x - vt$ in S' (Fig. 38-19).

How does relativity alter this coordinate transformation? First, there can be no change in the coordinates y and z at right angles to the relative motion, for if there were then in one frame distances along the y and z axes would be unambiguously shorter, and observers in both frames would agree about this (Fig. 38-20). But then there must be something special about the frame with the shorter y or z distances, and it is just such specialness that relativity prohibits.

Because of the Lorentz contraction of distances along the direction of relative motion, we expect that the simple expression $x' = x - vt$ will need

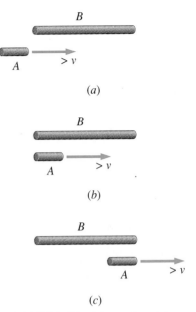

FIGURE 38-18 The passing sticks viewed in a frame of reference S'' at rest with respect to stick B. (a) First the right end of stick A passes the left end of B. (b) A while later, the left end of A passes the left end of B. This is event E_2. (c) Still later, the right end of A passes the right end of B. This is event E_1. The events E_1 and E_2 are not simultaneous in S'', and their time order is opposite what it was in S'.

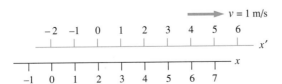

FIGURE 38-19 Nonrelativistic picture of two coordinate axes in relative motion. The x axis of frame S' moves to the right with speed v relative to S. At time $t = 0$ the two axes coincided. At a later time the x coordinates are related by $x' = x - vt$. Here $v = 1$ m/s, and the figure shows the two axes at $t = 2$ s.

FIGURE 38-20 A person holding a piece of chalk at $y' = 2$ in frame S' marks the y axis in S as the two axes pass. Observers in both frames agree unambiguously about the location of the mark. Unless it is at $y = 2$, there must be something special about one of the frames. Relativity precludes such "specialness." Thus coordinates perpendicular to the direction of relative motion are unchanged.

modification to be consistent with relativity. Any new expression we develop, however, must reduce to the nonrelativistic expression $x' = x - vt$ in the limit when $v \ll c$. A simple form that has this property is

$$x' = \gamma(x - vt), \qquad (38\text{-}6)$$

where γ depends on the relative speed v. We could also transform the other way. The only difference is the direction of relative motion; frame S' is moving in the positive x direction relative to S, while S is moving in the negative x direction relative to S'. Therefore, the transformation from x' and t' to x should look like Equation 38-6 except with v replaced by $-v$:

$$x = \gamma(x' + vt'). \qquad (38\text{-}7)$$

To see if we can make Equations 38-6 and 38-7 consistent with relativity, we impose the requirement that the speed of light be the same in both frames of reference. Suppose that a light flash goes off at the origin $x = 0$ at time $t = 0$. Since the origins of our frames coincide at this time, and since clocks at the origin in S and S' both read zero when the origins coincide, the light flash also occurs at $x' = 0$ and $t' = 0$ in frame S'. Let us call this event—the emission of the light flash—event E_1. Some time later, an observer at some position x in S observes the light flash. Let us call this event E_2. When does E_2 occur? Since light travels at speed c, we must have $x = ct$. Now in frame S' event E_2 has some coordinates x' and t'. But relativity requires that the speed of light in S' also be c, so we must have $x' = ct'$. Substituting $x = ct$ and $x' = ct'$ into Equations 38-6 and 38-7 gives

$$ct' = \gamma(ct - vt) = \gamma t(c - v)$$

and

$$ct = \gamma(ct' + vt') = \gamma t'(c + v).$$

Multiplying together the left-hand sides of these equations, and then the right-hand sides, and equating the results gives

$$c^2 t't = \gamma^2 tt'(c - v)(c + v),$$

so

$$c^2 = \gamma^2(c^2 - v^2),$$

or

$$\gamma = \frac{1}{\sqrt{1 - v^2/c^2}}. \qquad (38\text{-}8)$$

That we could find a value of γ depending on the relative velocity but not on the coordinates shows that our guess for the form of the transformation equations was correct. Equations 38-6 and 38-7, with γ given by Equation 38-8, are the relativistically correct transformations for the coordinates x and x'. Taking $v \ll c$ in Equation 38-8 shows that $\gamma \to 1$ in this limit, so our transformation equations correctly reduce to the nonrelativistic result at low relative velocities.

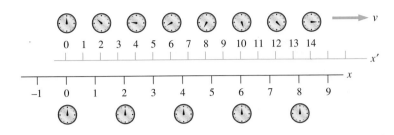

Solving Equations 38-6 and 38-7 simultaneously for t' and t (see Problem 25) gives the transformation equations for time:

$$t' = \gamma\left(t - \frac{vx}{c^2}\right) \tag{38-9}$$

and

$$t = \gamma\left(t' + \frac{vx'}{c^2}\right). \tag{38-10}$$

Because we know about time dilation, we should not be too surprised to find that measures of time differ between the two frames of reference. But why should the time in one frame depend not only on the time in the other frame but also on location in space? Because, as we have found, events that are simultaneous in one frame of reference are not simultaneous in a frame of reference moving relative to the first. In this case, our simultaneous events are the pointing of all clock hands to the same time. For the clocks in S', these events are simultaneous in S'. But they are not simultaneous in S. In fact, as Equation 38-9 shows, clocks that are farther to the right in S' read successively earlier times (Fig. 38-21). The term vx/c^2 in Equation 38-9 and its analog in Equation 38-10 account for this nonsynchronism of clocks in one frame as measured from the other frame.

Our earlier qualitative discussion of simultaneity can be made quantitative using the Lorentz transformations, as Example 38-5 and Problems 22 to 24 illustrate. Similarly, applying the Lorentz transformations to the coordinates describing the emission and return of the light flash in our light-box example results in a derivation of time dilation (see Problem 26). Table 38-1 summarizes the Lorentz transformations between coordinates in frames S and S', where S' is moving at speed v in the positive x direction relative to S.

▲ **TABLE 38-1** THE LORENTZ TRANSFORMATIONS

S TO S'	**S' TO S**
$y' = y$	$y = y'$
$z' = z$	$z = z'$
$x' = \gamma(x - vt)$	$x = \gamma(x' + vt')$
$t' = \gamma(t - vx/c^2)$	$t = \gamma(t' + vx'/c^2)$

where $\gamma = \dfrac{1}{\sqrt{1 - v^2/c^2}}$

● **EXAMPLE 38-5** GALACTIC FIREWORKS

Our Milky Way galaxy and the Andromeda galaxy are approximately at rest with respect to each other and are 2.0×10^6 light-years apart. At time $t = 0$ in the reference frame of these two galaxies, supernova explosions occur in both galaxies. Are these explosions simultaneous to the pilot of a spacecraft traveling at $0.80c$ from the Milky Way toward Andromeda? If not, what is the time interval between them, as measured in the spacecraft frame? Find also the spatial interval between the two explosions in the spacecraft frame.

Solution

The supernova explosions constitute two distinct events, and we're interested in their coordinates in the spaceship's frame of reference. Let the origin of the galaxy frame S be at the supernova in the Milky Way, and let the x axes of the galaxy frame S and spacecraft frame S' lie on the line connecting the two supernovae. Let the two frames coincide at time $t = t' = 0$. Then the space and time coordinates in S of the two supernova explosions are $x_1 = 0$, $t_1 = 0$, $x_2 = 2.0$ Mly, $t_2 = 0$. Similarly, the coordinates of the Milky Way explosion in the spacecraft frame S' are $x_1' = 0$, $t_1' = 0$. We seek the coordinates x_2' and t_2' of the Andromeda explosion in S'.

Referring to Table 38-1, we first calculate the relativistic factor γ:

$$\gamma = \frac{1}{\sqrt{1 - v^2/c^2}} = \frac{1}{\sqrt{1 - 0.80^2}} = 1.67.$$

Then using the Lorentz transformations, we have

$$t_2' = \gamma(t_2 - vx_2/c^2)$$

$$= (1.67)\left[0 \text{ y} - \frac{(0.80 \text{ ly/y})(2.0 \text{ Mly})}{(1 \text{ ly/y})^2}\right] = -2.7 \text{ My}.$$

and $x_2' = \gamma(x_2 - vt_2) = (1.67)[2.0 \text{ Mly} - (0.80 \text{ ly/y})(0 \text{ ly})]$

$$= 3.3 \text{ Mly}.$$

Do these results make sense? In the spacecraft frame, the Andromeda supernova occurs nearly three million years before the Milky Way supernova! (The minus sign tells us that t_2' is earlier than the time $t_1' = 0$ of the Milky Way supernova.) Here is an example of events that are simultaneous in one frame but not in another. To make matters worse, consider an observer moving at $0.80c$ from Andromeda toward the Milky Way. For an observer in this frame, we reverse the sign of v in the transformation equations, obtaining a time of $+2.7$ million years. For this observer, the Milky Way supernova occurs first! How can this be? This is no contradiction, and no violation of cause and effect. In the spaceship's frame S' the two supernova events occur 3.3 Mly apart in space but only 2.7 My apart in time. Light from the "earlier" event cannot travel to the "later" event, so there can be no causal influence between the two. It really doesn't matter which occurs first, and indeed which does depends on the observer.

Had we considered supernova events occurring not simultaneously but a long time apart— longer than 2 million years—in the galaxy frame of reference, we would find that all observers would agree on the time order of the two events, although not necessarily on the actual value of the time interval (see Problems 22 to 24).

EXERCISE Coordinate system S' is moving along the x axis of system S at $0.90c$. The two origins coincide when clocks in both systems read noon. An event occurs at $x = 5.0$ light-hours, $t = 3$ P.M. in frame S. Find the position and time of this event in S'.

Answers: 5.28 lh, 8:34 A.M.

Some problems similar to Example 38-5: 21–23, 53, 54 ●

Relativistic Velocity Addition

If you're in an airplane moving at 1000 km/h relative to the ground and you walk toward the front of the plane at 5 km/h, common sense suggests that you move at 1005 km/h relative to the ground. But relativity implies that measures of time and distance vary among frames of reference in relative motion. For this reason the velocity of an object with respect to one frame does not simply add to the relative velocity between frames to give the object's velocity with respect to another frame. In the airplane your speed with respect to the ground is actually a little less than 1005 km/h as you stroll down the aisle of the plane, though the difference is insignificant at such a low speed.

The correct expression for **relativistic velocity addition** follows from the Lorentz transformations. Consider a frame of reference S and another frame S' moving in the positive x direction with speed v relative to S. Let their origins coincide at time $t = t' = 0$, so the Lorentz transformations of Table 38-1 apply. Suppose an object moves with velocity u' along the x' axis in S'. In our airplane example, S' would be the airplane frame of reference, u' the velocity at which you walk through the plane, and v the velocity of the plane relative to the ground, or S frame. We seek the velocity u of the object relative to the frame S (that is, your velocity relative to the ground as you walk in the plane).

In either frame, velocity is the ratio of change in position to change in time, or

$$u = \frac{\Delta x}{\Delta t}.$$

Designating the beginning of the interval Δt by the subscript 1 and the end by 2, we can use Equations 38-7 and 38-10 to write

$$\Delta x = x_2 - x_1 = \gamma[(x_2' - x_1') + v(t_2' - t_1')] = \gamma(\Delta x' + v\Delta t')$$

and

$$\Delta t = t_2 - t_1 = \gamma[(t_2' - t_1') + v(x_2' - x_1')/c^2] = \gamma(\Delta t' + v\Delta x'/c^2).$$

Forming the ratio of these quantities, we have

$$\frac{\Delta x}{\Delta t} = \frac{\Delta x' + v\Delta t'}{\Delta t' + v\Delta x'/c^2} = \frac{(\Delta x'/\Delta t') + v}{1 + v(\Delta x'/\Delta t')/c^2}.$$

But $\Delta x'/\Delta t'$ is the velocity u' of the object in frame S', and $\Delta x/\Delta t$ is the velocity u, so

$$u = \frac{u' + v}{1 + u'v/c^2}. \tag{38-11}$$

The numerator of this expression is just what we would expect from common sense. But this simple sum of two velocities is altered by the second term in the denominator, which is significant only when both the object's velocity u' and the relative velocity v between frames are comparable with c. Solving Equation 38-11 for u' in terms of u, v, and c gives the inverse transformation:

$$u' = \frac{u - v}{1 - uv/c^2}. \tag{38-12}$$

● **EXAMPLE 38-6** COLLISION COURSE

Two spacecraft approach Earth from opposite directions, each moving at $0.80c$ relative to Earth, as shown in Fig. 38-22. How fast do the spacecraft move relative to each other?

Solution
Call the Earth frame of reference S', and let S be the frame of spacecraft A. Then S' is moving at speed $v = 0.80c$ relative to S, while spacecraft B is moving at $u' = 0.80c$ relative to S'. Then the velocity of B relative to A is given by Equation 38-11:

$$u = \frac{u' + v}{1 + u'v/c^2} = \frac{0.80c + 0.80c}{1 + (0.80c)(0.80c)/c^2} = \frac{1.6c}{1.64} = 0.98c.$$

The relative speed remains less than the speed of light. This result is quite general: Equations 38-11 and 38-12 imply that as long as an object moves at a speed $v < c$ relative to some frame of reference, its speed relative to any other frame of reference will also be less than c (see Problem 20).

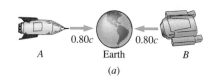

(a)

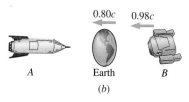

(b)

FIGURE 38-22 (a) Two spaceships approaching Earth at $0.80c$. (b) The situation in the frame of reference of the left-hand spaceship. Note changes in lengths of the ships and Earth. ●

● **EXAMPLE 38-7** CATCH THAT LIGHT

A light wave moves past Earth at the speed of light c. You try to chase the light wave by hopping a fast spacecraft, moving at $0.95c$ relative to Earth. What is the speed of the light relative to the spacecraft?

Solution
Call the Earth frame S, so $u = c$, and the spacecraft frame S', so $v = 0.95c$. Then u', the speed of light relative to the spacecraft, is given by Equation 38-12:

$$u' = \frac{u - v}{1 - uv/c^2} = \frac{c - 0.95c}{1 - 0.95c^2/c^2} = \frac{0.05c}{0.05} = c.$$

We really didn't need to calculate this result, since a fundamental premise of relativity is that the speed of light is the same for

all observers. The equations of relativistic velocity addition reflect this basic fact. No matter what the relative velocity v between two frames, light moving at c in one frame moves at c in any other frame. You cannot even begin to catch up with light!

EXERCISE A spacecraft whizzes by Planet X at $0.75c$, in excess of the Galactic Federation speed limit. A space cop takes off from Planet X at $0.90c$ relative to the planet. Find the relative speed of cop and speeder.

Answer: $0.46c$

Some problems similar to Examples 38-6 and 38-7: 17–19 ●

38-5 ENERGY AND MOMENTUM IN RELATIVITY

Conservation of momentum and conservation of energy are cornerstones of Newtonian mechanics, where they hold in any inertial—i.e., uniformly moving—frame of reference. But both the momentum and energy of a particle are functions of its velocity, and we've just seen that relativity alters the Newtonian picture of how velocities transform from one reference frame to another. How, then, can momentum and energy be conserved in all frames of reference?

Momentum

In Newtonian mechanics the momentum of a particle with mass m and velocity **u** is m**u**. (Here we use **u** for particle velocities, reserving **v** for the relative velocity between two reference frames.) But if this quantity is conserved in one frame of reference, then relativistic velocity addition suggests that it won't be conserved in another. The problem, however, lies not with momentum conservation but with our definition of momentum. The expression m**u** is an approximation valid only for speeds u much less than c. The measure of momentum valid at any speed is

$$\mathbf{p} = \frac{m\mathbf{u}}{\sqrt{1 - u^2/c^2}} = \gamma m\mathbf{u}, \qquad (38\text{-}13)$$

where γ is the familiar relativistic factor. The momentum given in expression 38-13 is conserved in all reference frames, and at low velocities it reduces to the Newtonian expression $\mathbf{p} = m\mathbf{u}$.

As $u \to c$ the factor γ grows arbitrarily large, and so does the relativistic momentum (Fig. 38-23). Since force is the rate of change of momentum, that means a very large force is required to produce even the slightest change in the velocity of a rapidly moving particle. This effect shows one answer to a common question about relativity: Why is it impossible to accelerate an object to the speed of light? The answer is that the object's momentum would approach infinity, and no matter how close to c it was moving it would still require infinite force to give it the last bit of speed needed to reach c.

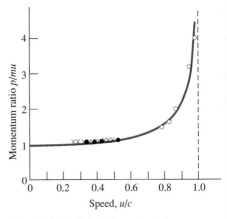

FIGURE 38-23 The ratio of the relativistic momentum to the Newtonian expression mu. Curve is a plot based on Equation 38-13, while data points correspond to actual measurements done on electrons. The relativistic momentum goes asymptotically to infinity as $u \to c$.

● **EXAMPLE 38-8** PUSHING TOWARD c

Compare the momentum changes required to accelerate an electron (a) from rest to $0.01c$ and (b) from $0.98c$ to $0.99c$.

Solution

The momentum at each speed is given by Equation 38-13:

$$p(0) = \frac{mu}{\sqrt{1 - u^2/c^2}} = 0,$$

$$p(0.01c) = \frac{m(0.01c)}{\sqrt{1 - 0.01^2}} = 0.010 \, mc,$$

$$p(0.98c) = \frac{m(0.98c)}{\sqrt{1 - 0.98^2}} = 4.92 \, mc,$$

and $\quad p(0.99c) = \dfrac{m(0.99c)}{\sqrt{1 - 0.99^2}} = 7.02 \, mc.$

In both cases the speed has increased by the same $0.01c$, but the momentum changes are very different: Going from rest to $0.01c$ gives $\Delta p = 0.010mc$, while from $0.98c$ to $0.99c$ the momentum change is $\Delta p = (7.02mc - 4.92mc) = 2.1mc$—a factor of 210 greater than the slow-speed case. To accomplish the second velocity change in the same time would therefore require a force 210 times greater. Note, incidentally, that we would have had sufficient accuracy using the Newtonian expression $p = mu$ in the low-speed case.

EXERCISE An electron is moving at $0.35c$. If its speed is doubled, by what factor does its momentum increase?

Answer: 2.62

Some problems similar to Example 38-8: 28, 30–32 ●

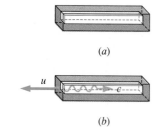

(a)

(b)

FIGURE 38-24 A light pulse is emitted at one end of a massive box. Since light carries momentum, the box recoils in the opposite direction.

Energy

The most widely known result of relativity is the famous equation $E = mc^2$. Here we show how that equation arises and what it means, then develop a general expression for relativistic energy and a relation between energy and momentum.

Einstein arrived at his equation through a simple "thought experiment." He imagined a closed, massive box of mass M and length L, initially at rest. A light flash is emitted from one end of the box. We found in Chapter 34 that light with energy E also carries momentum E/c; therefore, the box must recoil to conserve momentum (see Fig. 38-24). If the box is very massive its recoil speed u will be small compared with c, so we can express momentum conservation in the form $Mu = E/c$, or

$$u = \frac{E}{Mc}.$$

The light then moves down the box, taking a time

$$\Delta t = \frac{L}{c},$$

where again we assume that the box speed u is much less than c, so the distance traveled by the light is approximately L. In this time the box moves a very small distance Δx, given by

$$\Delta x = u\,\Delta t = \frac{EL}{Mc^2}.$$

Then the light flash hits the end of the box, transferring its momentum and bringing the box to a stop.

But now the box is in a new position. It looks as if its center of mass has moved, and yet the box is an isolated system whose center of mass cannot move! To escape this dilemma, Einstein assumed that light carries not only energy and momentum, but mass as well. If m is the mass carried by the light, we must have

$$mL = M\Delta x,$$

in order that center of mass of the system (box + light) will not move. Using our expression for Δx and solving for m gives

$$m = \frac{M\Delta x}{L} = \frac{M}{L}\frac{EL}{Mc^2} = \frac{E}{c^2},$$

or

$$E = mc^2,$$

where E is the energy of the light and m its equivalent mass.

Although we derived this expression for light energy, it is in fact a universal statement of the equivalence of mass and energy. Energy, like mass, exhibits

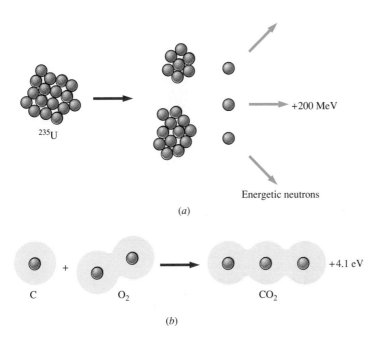

^{235}U

+200 MeV

Energetic neutrons

(a)

C + O_2 → CO_2 +4.1 eV

(b)

FIGURE 38-25 Energy release in (a) nuclear and (b) chemical reactions. A mass change occurs in both cases, but is approximately 10^7 times greater in the nuclear reaction.

inertia. A hot object is slightly harder to accelerate than a cold one because of the inertia of its thermal energy. A stretched spring is more massive than an unstretched one, because of its extra potential energy. When a system loses energy, it loses mass as well.

To the general public, $E = mc^2$ is synonymous with nuclear energy. This equation does describe mass changes that occur in nuclear reactions, but it applies equally well to chemical reactions and all other occurrences in which energy enters or leaves a system. If you weigh a nuclear power plant just after it has been refueled, then weigh it again a month later, you will find it weighs slightly less. If you weigh a coal-burning power plant and all the coal and oxygen that go into it for a month, and then weigh all the carbon dioxide and other combustion products that come out, you will find a discrepancy between the mass of what goes in and what comes out. If both plants produce the same amount of energy, the mass discrepancy will be the same for both. The only difference lies in the amount of mass released as energy in each individual reaction. The fissioning of a single uranium nucleus involves about 10 million times as much energy, and therefore mass, as the reaction of a single carbon atom with oxygen to make carbon dioxide (Fig. 38-25). That's why a coal-burning power plant consumes many hundred-car trainloads of coal each week, while a nuclear plant is refueled every 18 months with a few truckloads of uranium. Incidentally, neither process converts very much of the fuel mass to energy; if we could convert *all* the mass in a given object to energy, ordinary matter would be an almost limitless source of energy. Such conversion is in fact possible, but only in the annihilation of matter and antimatter. The opposite conversion also occurs, with a particle-antiparticle pair appearing where before there had been only energy (Fig. 38-26).

FIGURE 38-26 Pair creation events observed in a bubble chamber at Brookhaven National Laboratory. Each pair of curved paths represent the trajectories of a particle-antiparticle pair, bending in opposite directions in a magnetic field. The energy required to create the particles—mc^2 per particle—ultimately came from a single high-energy proton that collided with a stationary proton.

● **EXAMPLE 38-9**   ANNIHILATION

A positron is an antimatter particle with the same mass as the electron but the opposite electric charge. When an electron and a positron meet, they annihilate and produce a pair of gamma rays (bundles of electromagnetic wave energy) of equal energy. Find the energy of each gamma ray.

Solution
Here two electron masses annihilate to give two gamma rays. So the energy of each gamma ray is the energy equivalent of one electron mass, or

$$E = mc^2 = (9.11 \times 10^{-31} \text{ kg})(3.00 \times 10^8 \text{ m/s})^2$$

$$= 8.20 \times 10^{-14} \text{ J},$$

or 511 keV. The detection of 511-keV gamma rays from laboratory or astrophysical sources is a sure indication that electron-positron annihilation is occurring.

EXERCISE   The Sun radiates energy at the rate of 3.85×10^{26} W. Find the rate at which it loses mass.

Answer: 4.28×10^9 kg/s

Some problems similar to Example 38-9:   38, 39 ●

Rest Energy and Kinetic Energy

Einstein's result shows that a mass m is equivalent to an energy $E = mc^2$. This quantity is called the **rest energy** because it is associated with the mass itself and not with any bulk motion the object may have. As usual, the energy of bulk motion is **kinetic energy.** We can find an expression for relativistic kinetic energy much as we did in Chapter 7 for the Newtonian case, by considering the work done on an object as it is accelerated. Problem 63 covers the details; the result is

$$K = \frac{mc^2}{\sqrt{1 - u^2/c^2}} - mc^2 = (\gamma - 1)mc^2, \qquad (38\text{-}14)$$

where as usual $\gamma = 1/\sqrt{1 - u^2/c^2}$. This equation bears little resemblance to the Newtonian expression $K = \frac{1}{2}mu^2$. But Problem 43 uses the binomial approximation to show that the relativistic expression for kinetic energy really does reduce to $K = \frac{1}{2}mu^2$ for speeds u much less than c. Using Equation 38-14, we can now write an object's total energy as the sum of its kinetic energy K and rest energy mc^2:

$$E = \frac{mc^2}{\sqrt{1 - u^2/c^2}} = \gamma mc^2. \quad \text{(total energy)} \qquad (38\text{-}15)$$

● **EXAMPLE 38-10**   A RELATIVISTIC ELECTRON

An electron has a total energy of 2.5 Mev. Find (a) its kinetic energy and (b) its speed.

Solution
In Example 38-9 we found that the rest energy of an electron is 511 keV, or 0.511 MeV. The electron's kinetic energy is the difference between its total energy and its rest energy, or

$$K = E - mc^2 = 2.5 \text{ MeV} - 0.511 \text{ MeV} = 1.99 \text{ MeV}.$$

Equation 13-15 shows that the total energy is just γ times the rest energy. Thus we have

$$\gamma = \frac{1}{\sqrt{1 - u^2/c^2}} = \frac{E}{mc^2} = \frac{2.5 \text{ MeV}}{0.511 \text{ MeV}} = 4.89.$$

Solving for u then gives

$$u = c\sqrt{1 - \frac{1}{4.89^2}} = 0.979c = 2.94 \times 10^8 \text{ m/s}.$$

EXERCISE Find (a) the rest energy of a proton, and (b) its kinetic energy when moving at $0.98c$.

Answers: (a) 938 Mev; (b) 3.78 GeV

Some problems similar to Example 39-10: 40–42

●

The Energy-Momentum Relation

In Newtonian physics the equations $p = mu$ and $K = \frac{1}{2}mu^2$ yield the relation $p^2 = 2K/m$. Similarly, Problem 44 shows that in relativity we can combine the equations $p = \gamma mu$ and $E = \gamma mc^2$ to get

$$E^2 = p^2c^2 + (mc^2)^2, \tag{38-16}$$

where now the energy-momentum relation involves E rather than K because in relativity the energy includes both kinetic and rest energies. For a particle at rest, $p = 0$ and Equation 38-16 shows that the total energy is just the rest energy. For highly relativistic particles—those with speeds very near that of light—the rest energy is negligible and the total energy becomes very nearly $E = pc$. Some "particles"—like the photons that, in quantum physics, are "bundles" of electromagnetic energy, and possibly the neutral particles called neutrinos—have no rest mass. These particles exist only in motion at the speed of light, and for them Equation 38-16 gives the exact relation $E = pc$.

38-6 WHAT IS NOT RELATIVE

In relativity, space and time are not absolute but depend on the reference frame of the observer. But the speed of light c remains the same for all observers. Are there other such **relativistic invariants?** Yes—and these invariants, not the shifting measures of space and time, are at the basis of relativity's objective description of physical reality.

One invariant is electric charge. No matter what an electron's speed, all observers measure the same value for its charge. Other invariants may be formed from combinations of quantities that themselves are not invariant; for example, Equation 38-16 shows that the quantity $E^2 - p^2c^2$ for a particle is the same in all frames of reference and is in fact equal to the particle's rest energy.

In Newtonian physics the distance between two points is the same no matter who measures it. In relativity, distance depends on the observer. So does time. But there is a quantity, analogous to distance but incorporating time as well, that remains invariant. Called the **spacetime interval,** it is a kind of "four-dimensional distance," not between two points in space but between two events in space and time. The spacetime interval Δs is given by an expression that looks like a modified Pythagorean theorem:

$$(\Delta s)^2 = c^2(\Delta t)^2 - [(\Delta x)^2 + (\Delta y)^2 + (\Delta z)^2], \tag{38-17}$$

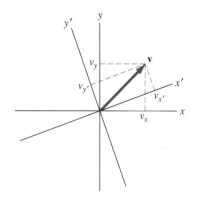

FIGURE 38-27 Although the x and y components of an ordinary vector depend on the choice of coordinate system, the length of the vector does not.

where the Δ quantities describe the differences between the space and time coordinates of two events. The invariance of Equation 38-17 follows directly from the Lorentz transformations (see Problem 45).

The invariance of the spacetime interval suggests that something absolute underlies the shifting sands of relativistic space and time. That absolute is **spacetime**—a four-dimensional framework linking space and time. The points in spacetime are events, specified by four coordinates. The time interval or space interval between two events depends on the particular frame of reference of the observer, but the spacetime interval—a four-dimensional "distance" that takes all four coordinates into account—is the same for all observers.

In more advanced treatments of relativity, it is convenient to consider four-dimensional vectors called **four-vectors.** The displacement between two events in spacetime—specified by the four quantities Δx, Δy, Δz, and Δt—is a four-vector, with "length" given by Equation 38-17. With two- and three-dimensional vectors in nonrelativistic physics, it is possible to break a vector into components in many different ways. Although the values of the individual components depend on your choice of coordinate system, the length of the vector does not (Fig. 38-27). Similarly, the individual space and time components of a four-vector depend on your choice of reference frame—that is, on your velocity. But the spacetime interval does not.

38-7 ELECTROMAGNETISM AND RELATIVITY

Historically, relativity arose from deep questions presented by Maxwell's equations with regard to electromagnetic waves. We have seen that relativity profoundly alters the basic concepts of space and time that stand at the foundation of Newtonian mechanics. As a result, fundamental ideas like momentum and energy must be altered for relativistic consistency. What analogous changes does relativity require of Maxwell's electromagnetic theory? The answer is simple: none. Maxwell's theory culminated in the prediction of light waves traveling through empty space at speed c. Relativity requires that the laws of physics be the same in all frames of reference in uniform motion. But that is exactly what Maxwell's equations suggest—that a light wave in one frame should be a light wave in any other frame, and that such a wave should have speed c with respect to any observer. Even the simple fact that electromagnetic induction occurs equally well when you move a magnet near a conductor, or a conductor near a magnet, suggests that only relative motion should be important in electromagnetism. Einstein thought a great deal about induction, and mentioned it at the beginning of his 1905 paper introducing special relativity. Even the title of that famous paper—"On the Electrodynamics of Moving Bodies"—shows how intimately related are electromagnetism and relativity. Maxwell's equations are relativistically correct, and require no modification.

Although electric and magnetic fields in any frame of reference obey the same Maxwell equations, this does not require the fields themselves to be independent of frame. If, for example, you sit in the rest frame of a point charge, you see a spherically symmetric point-charge field. If you move relative to the charge, you see a magnetic field as well, associated with the moving charge. Relativity accounts naturally for such a field transformation. In relativity, the electric and magnetic fields are not absolutes, but are manifestations of

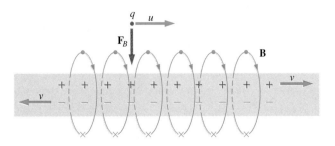

FIGURE 38-28 A current-carrying wire with equal densities of positive and negative charge moving in opposite directions. A magnetic field surrounds the wire, so a charge q moving parallel to the wire experiences a magnetic force toward the wire.

a more fundamental electromagnetic field. To one observer, this electromagnetic field breaks up in a certain way into electric and magnetic parts, while to another observer the individual electric and magnetic fields are different (see Problem 66).

We can illustrate the deep relationship that relativity imposes between electricity and magnetism by considering the force on a charged particle near a current-carrying wire. For simplicity, we consider a wire containing equal line charge densities of positive and negative charge, moving in opposite directions at the same speed v relative to the wire (Fig. 38-28).

Consider a particle of positive charge q a distance r from the wire. Since the densities of positive and negative charge are equal, the wire is neutral and has no electric field. If the particle is at rest with respect to the wire, it experiences no force. But there is a magnetic field associated with the current in the wire. The magnetic field lines encircle the wire, and the right-hand rule shows that they point out of the page at the location of the charged particle. Now suppose the particle is moving to the right with velocity **u** relative to the wire. It experiences a magnetic force $\mathbf{F} = q\mathbf{u} \times \mathbf{B}$; the right-hand rule shows that this force is toward the wire.

So the situation from the frame of reference of the wire is as follows: the wire is electrically neutral and therefore produces no electric field. But it does carry a current, and therefore produces a magnetic field. If a positively charged particle moves to the right, it experiences a magnetic force directed toward the wire. To describe the situation we needed to know about electric charges, about Ampère's law for magnetism, and about the magnetic force $q\mathbf{u} \times \mathbf{B}$—in short, about a variety of phenomena that were discovered independently during the nineteenth century.

Now let's look at the situation in the reference frame of the charged particle. Since the particle is moving to the right, the positive charges in the wire have a lower speed relative to the particle than do the negative charges. As measured by the particle, distances between the negative charges in the wire are therefore Lorentz contracted by *more* than the distances between the positive charges. But charge is invariant, so the charge density—the charge per unit length—is *greater* for the negative charges. So in the frame of the charged particle, *there is a net negative charge on the wire!* The negatively charged wire produces an electric field pointing toward the wire. As a result, our positively charged particle experiences an electric force toward the wire (Fig. 38-29). Of course there is still a magnetic field as well, but since the particle is at rest in its own frame of reference, it experiences no magnetic force. The force it does experience is entirely electric. What appeared as a magnetic phenomenon in the wire frame of reference—the existence of a force directed toward the wire—is explained entirely as an electric phenomenon in the particle's frame of reference.

FIGURE 38-29 The situation of Fig. 38-28 in the rest frame of the charged particle. The line of negative charges is Lorentz contracted more, and therefore the positively charged particle "sees" a net negative charge on the wire. Thus the particle experiences an electric force toward the wire.

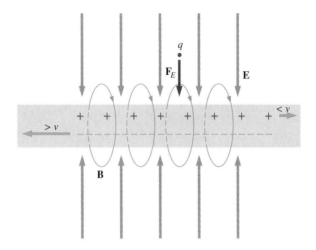

We have given two quite different descriptions of the force on a charged particle moving near a current-carrying wire. Our second description—from the frame of reference of the charged particle—required no knowledge of magnetism whatsoever. We only needed Coulomb's law for electric charge and the principle of relativity. And this illustrates a profound point: electricity and magnetism are not separate phenomena that happen to be related through Ampère's and Faraday's laws. Rather, they are two aspects of a single phenomenon—electromagnetism. In a universe obeying the principle of relativity, it is not even logically possible to have electricity without magnetism, or vice versa. Given Coulomb's law and the principle of relativity, all the rest of Maxwell's equations follow not as independent experimental results but as logically necessary consequences. Relativity provides for us the total unification of electricity and magnetism that we have hinted at since we began studying these phenomena.

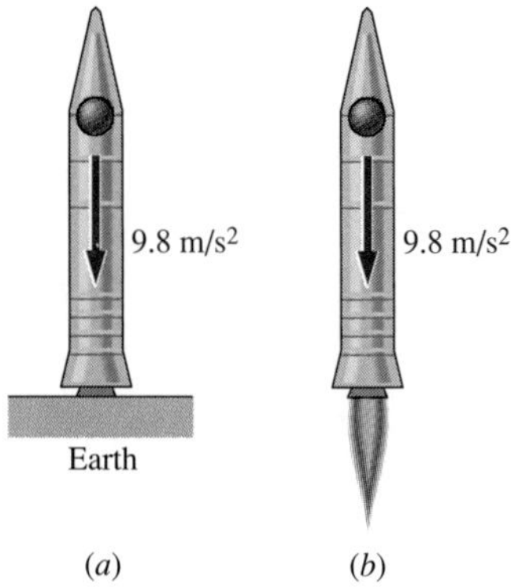

FIGURE 38-30 Why general relativity is a theory of gravity. (*a*) In a spaceship at rest on Earth, objects accelerate downward at *g* = 9.8 m/s². (*b*) Far from any gravitating body, a spaceship accelerates at 9.8 m/s². Relative to the accelerating reference frame of the spaceship, objects accelerate toward the back of the ship at 9.8 m/s². Situations (*a*) and (*b*) are impossible to distinguish, so a theory dealing with nonuniform motion must also consider gravity.

38-8 GENERAL RELATIVITY

The special theory of relativity is special because it is restricted to uniform motion. Following the development of special relativity, Einstein attempted to formulate a theory that would express the laws of physics in the same form in all frames of reference, including those in accelerated motion. But Einstein recognized that it is impossible to distinguish the effects of uniform acceleration from those of a uniform gravitational field (Fig. 38-30). Consequently, Einstein's general theory became a theory of gravity. Building on the notion of four-dimensional spacetime, Einstein introduced geometrical curvature of spacetime to account for gravity, and the result, published in 1916, is his **general theory of relativity.** The theory's predictions differ significantly from those of Newton's theory of gravity only in regions of very strong gravitational fields or when the overall structure of the universe is considered. By very strong fields we mean those of objects whose escape speed is comparable to that of light (see Problem 65). Because we have no direct laboratory experience of such fields, the general theory of relativity is not as solidly established as is the special theory. Nevertheless, general relativity is a cornerstone of modern astrophysics, playing a crucial role in the physics of such bizarre objects as neutron stars and black holes. General relativity also addresses cosmological questions of the origin and ultimate fate of the universe. Research in astrophysics and cosmology, in turn, is increasingly confirming the predictions of general relativity.

CHAPTER SYNOPSIS

Summary

1. The **ether** was a hypothetical medium whose properties were supposed to explain the propagation of electromagnetic waves. In particular, such waves were supposed to have speed c relative to the ether.
2. The **Michelson-Morley experiment** and the observation of the aberration of starlight led to a contradiction in physics: Earth's motion through the ether could not be detected, yet Earth did not drag ether with it.
3. Einstein's **special theory of relativity** (1905) resolved the contradiction by asserting that uniform motion is undetectable by any experiment, mechanical or electromagnetic. Einstein did away with the ether, declaring simply that **the laws of physics are the same in all inertial frames of reference.** Mechanics and electromagnetism alike are included in Einstein's theory, so Maxwell's prediction of electromagnetic waves moving at speed c is correct in all frames of reference.
4. The simple statement that the laws of physics are the same in all frames of reference requires profound changes in our common-sense notions of space and time. These changes are described by the **Lorentz transformations,** which relate space and time measurements made in different frames of reference:

$$y' = y$$
$$z' = z$$
$$x' = \gamma(x - vt)$$
$$t' = \gamma(t - vx/c^2),$$

where

$$\gamma = \frac{1}{\sqrt{1 - v^2/c^2}}.$$

Particular manifestations of these transformations include **time dilation, Lorentz contraction,** and the relativistic velocity addition formulas.

5. Relativistic transformations result in new expressions for momentum:

$$\mathbf{p} = \frac{m\mathbf{u}}{\sqrt{1 - u^2/c^2}} = \gamma m\mathbf{u}$$

and total energy:

$$E = \frac{mc^2}{\sqrt{1 - u^2/c^2}} = \gamma mc^2,$$

for a particle of mass m moving with velocity $\mathbf{u}$. Kinetic energy is the difference between the total energy E and the **rest energy** mc^2. The existence of rest energy shows that matter and energy may be converted into each other.

6. Relativity links space and time into a four-dimensional framework called **spacetime.** Although individual space and time measurements depend on one's frame of reference, the **spacetime interval** between two events does not.

7. Maxwell's equations of electromagnetism are fully consistent with relativity. But the fields themselves are different in different frames of reference; what appears as a magnetic field in one frame may be partially electric in another, and vice versa. Relativity imposes a logical relationship between electricity and magnetism, in that neither phenomenon is possible without the other.

8. The **general theory of relativity** is Einstein's generalization of relativity to include accelerated reference frames. Because the effects of acceleration mimic those of gravity, general relativity is a theory of gravity.

Terms You Should Understand

(Pairs are closely related terms whose distinction is important; number in parentheses is chapter section where term first appears.)

ether (38-1)
Michelson-Morley experiment (38-2)
special relativity, general relativity (38-2, 38-8)
time dilation, Lorentz contraction (38-4)
proper time, proper length (38-4)
Lorentz transformations (38-4)
rest energy (38-5)
spacetime (38-6)
spacetime interval (38-6)

Symbols You Should Recognize

γ (38-4)

Problems You Should Be Able to Solve

finding times and distances for different observers in relative motion (38-4)

transforming space and time coordinates among coordinate systems in relative motion (38-4)

transforming velocities (38-4)

applying mass-energy equivalence (38-5)

evaluating relativistic energy and momentum (38-5)

Limitations to Keep in Mind

The special theory of relativity applies only to inertial frames of reference—i.e., those that are not accelerated.

QUESTIONS

1. Why was the Michelson-Morley experiment a more sensitive test of motion through the ether than independent measurements of the speed of light in two perpendicular directions?

2. Why was it necessary to repeat the Michelson-Morley experiment at different times throughout the year?

3. Why do we reject the idea that the ether frame of reference is the Earth frame?

4. What is special about the special theory of relativity?

5. Does relativity require that the speed of sound be the same for all observers? Why or why not?

6. How would the world be different if the speed of light were 160 km/h (100 miles per hour)? Would our "common-sense" notions change?

7. A friend argues that the speed of light cannot be the same for all observers, for if one of them is moving toward a light source then that one will clearly measure a higher speed. How would you refute this argument?

8. Time dilation is sometimes described by saying that "moving clocks run slow." In what sense is this true? In what sense does the statement violate the spirit of relativity?

9. If you are in a spaceship moving at $0.95c$ relative to Earth, do you perceive time to be passing more slowly than it would on Earth? Think! Is your answer consistent with the theory of relativity?

10. In our light-box example for time dilation, we found that a time interval between two events measured in frame S' was shorter than in frame S. But you could equally well say that frame S is moving relative to frame S', so clocks in S should "run slow" compared with those in S'. An observer in each frame should judge the clocks in the other frame to "run slow." Is this a contradiction?

11. To try to circumvent the difficulty of accelerating an object to the speed of light, you build a series of conveyor belts, all running in the same direction, and each moving 10 m/s relative to the one next to it (Fig. 38-31). You step from the ground onto the first conveyor belt, then to the next, and so forth. By the time you reach the 3×10^7th conveyor belt, you should be moving at c relative to the ground. Why doesn't this scheme work?

FIGURE 38-31   A series of conveyer belts, each moving at 10 m/s relative to its neighbors (Question 11)

12. If you took your pulse while traveling in a high-speed spacecraft, would it be faster, slower, or the same as on Earth?

13. The Andromeda galaxy is 2 million light-years from our Milky Way. Although nothing can go faster than light, it would still be possible to travel to Andromeda in much less than 2 million years. How is this possible?

14. Is matter converted to energy in a nuclear reactor? In a burning candle? In your body?

15. An unstretched rubber band is weighed on an extraordinarily sensitive scale. It is then stretched and weighed again. Is there a difference in the weight? Why or why not?

16. The rest energy of an electron is 511 keV. What is the approximate speed of an electron whose total energy is 1 GeV (= 10^9 eV)? You need not do any calculations!

17. An atom in an excited state emits a burst of light. What happens to the mass of the atom?

18. In some of the hottest parts of the universe, the thermal energy of particles may be many millions of electron volts. At such temperatures, the number of particles within a closed volume may vary. How is this possible? *Hint:* The mass of the electron is equivalent to about 0.5 MeV.

19. The electric field is not invariant, but changes from one frame to another. Is this a violation of relativity? Relativity requires that the laws of physics be the same in all frames of reference. Does this mean that all physical quantities must be the same?

20. The quantity $\mathbf{E} \cdot \mathbf{B}$ is invariant. What does this say about how different observers will measure the angle between $\mathbf{E}$ and $\mathbf{B}$ in a light wave?

PROBLEMS

Section 38-2 Matter, Motion, and the Ether

1. Consider an airplane flying at 800 km/h airspeed between two points 1800 km apart. What is the round-trip travel time for the plane (a) if there is no wind? (b) if there is a wind blowing at 130 km/h perpendicular to a line joining the two points? (c) if there is a wind blowing at 130 km/h along a line joining the two points? Ignore relativistic effects. (Why are you justified in doing so?)

2. What would be the difference in light travel times on the two legs of the Michelson-Morley experiment if the ether existed and if Earth moved relative to it at (a) its orbital speed relative to the Sun (Appendix E)? (b) $10^{-2}c$? (c) $0.5c$? (d) $0.99c$? Assume each light path is exactly 11 m in length, and that the paths are oriented parallel and perpendicular to the ether wind.

3. Figure 38-32 shows a plot of James Bradley's data on the aberration of light from the star γ Draconis, taken in 1727–1728. (a) From the data, determine the magnitude of Earth's orbital velocity. (b) The data very nearly fit a perfect sine curve. What does this say about the shape of Earth's orbit?

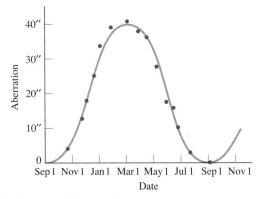

FIGURE 38-32 Bradley's data on the aberration of starlight (Problem 3). Aberration is measured in seconds of arc ($1'' = 1/3600°$).

4. Show that the time of Equation 38-2 is larger than that of Equation 38-1 as long as $0 < v < c$.

5. Suppose the speed of light differed by 100 m/s in two perpendicular directions. How long should the arms be in a Michelson interferometer if this difference is to cause the interference pattern to shift one-half cycle (i.e., a light fringe shifts to where a dark one was) relative to the pattern if there were no speed difference? Assume 550-nm light.

Section 38-4 Space and Time in Relativity

6. Two stars are 50 ly apart, measured in their common rest frame. How far apart are they to a spaceship moving between them at $0.75c$?

7. How long would it take a spacecraft traveling at 65% of the speed of light to make the 5.8×10^9 km journey from Earth to Pluto according to clocks (a) on Earth and (b) on the spacecraft?

8. A spaceship goes by at half the speed of light, and you determine that it is 35 m long. What is its length as measured in its rest frame?

9. Earth and Sun are 8.3 light-minutes apart, as measured in their rest frame. (a) What is the speed of a spacecraft that makes the trip in 5.0 min according to its on-board clocks? (b) What is the trip time as measured by clocks in the Earth-Sun frame?

10. The Andromeda galaxy is two million light-years from Earth, measured in the common rest frame of Earth and Andromeda (Fig. 38-33). Suppose you took a fast spaceship to Andromeda, so it got you there in 50 years as measured on the ship. If you sent a radio message home as soon as you reached Andromeda, how long after you left Earth would it arrive, according to timekeepers on Earth?

FIGURE 38-33 The Andromeda galaxy (Problem 10).

11. How fast would you have to move relative to a meter stick for its length to measure 99 cm in your frame of reference?

12. Electrons in a TV tube move through the tube at 30% of the speed of light. By what factor is the tube foreshortened in the electrons' frame of reference, relative to its length at rest?

13. You wish to travel to a star N light-years from Earth. How fast must you go if the one-way journey is to occupy N years of your life?

14. The nearest star beyond our solar system is about 4 light-years away. If a spaceship can get to the star in 5 years, as measured on Earth, (a) how long would the ship's pilot judge the journey to take? (b) How far from Earth would the pilot find the star to be?

15. Twins A and B live on Earth. On their 20th birthday, Twin B climbs into a spaceship and makes a round-trip journey

at 0.95c to a star 30 light-years distant, as measured in the Earth-star frame of reference. What are their ages when the twins are reunited?

16. Radioactive oxygen-15 decays at such a rate that half the atoms in a given sample decay every two minutes. If a tube containing 1000 O-15 atoms is moved at 0.80c relative to Earth for 6.67 minutes according to clocks at rest with respect to Earth, how many atoms will be left at the end of that time?

17. Two distant galaxies are receding from Earth at 0.75c, in opposite directions. How fast does an observer in one galaxy measure the other to be moving?

18. Two spaceships are having a race. The "slower" one moves past Earth at 0.70c, and the "faster" one moves at 0.40c relative to the slower one. How fast does the faster ship move relative to Earth?

19. Muons traveling vertically downward at 0.994c relative to Earth are observed from a rocket traveling upward at 0.25c. What speed does the rocket's crew measure for the muons?

20. Use relativistic velocity addition to show that if an object moves at speed $v < c$ relative to some uniformly moving frame of reference, then its speed relative to any other uniformly moving frame must also be less than c.

21. Earth and Sun are 8.33 light-minutes apart. Event A occurs on Earth at time $t = 0$, and event B on the Sun at time $t = 2.45$ min, as measured in the Earth-Sun frame. Find the time order and time difference between A and B for observers (a) moving on a line from Earth to Sun at 0.750c, (b) moving on a line from Sun to Earth at 0.750c, and (c) moving on a line from Earth to Sun at 0.294c.

22. Two civilizations are evolving on opposite sides of a galaxy, whose diameter is 10^5 light-years (Fig. 38-34). At time $t = 0$ in the galaxy frame of reference, civilization A launches its first interstellar spacecraft. Civilization B launches its first spacecraft 50,000 years later. A being from a more advanced civilization C is traveling through the galaxy at 0.99c, on a line from A to B. Which civilization does C judge to have first achieved interstellar travel, and how much in advance of the other?

23. Repeat the preceding problem, now assuming that civilization B lags A by 1 million years in the galaxy frame of reference.

24. Could there be observers who would judge the events in the two preceding problems to be simultaneous? If so, how must each be moving relative to the galaxy?

25. Derive the Lorentz transformations for time, Equations 38-9 and 38-10, from the transformations for space.

26. In the light box of Fig. 38-8, let event A be the emission of the light flash and event B its return to the source. Assign suitable space and time coordinates to these events in the frame in which the box moves with speed v. Apply the Lorentz transformations to show that the time $\Delta t'$ between the two events in the box frame is given by Equation 38-4.

27. Two spaceships are each 25 m long, as measured in their rest frames (Fig. 38-35). Ship A is approaching Earth at 0.65c. Ship B is approaching Earth from the opposite direction at 0.50c. Find the length of ship B as measured (a) in Earth's frame of reference and (b) in ship A's frame of reference.

FIGURE 38-35 Problem 27 (the drawing is in Earth's frame of reference).

Section 38-5 Energy and Momentum in Relativity

28. By what factor does the momentum of an object change if you double its speed when its original speed is (a) 25 m/s and (b) 1.0×10^8 m/s?

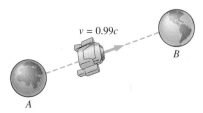

FIGURE 38-34 Civilizations A and B are on opposite sides of a galaxy, a distance of 10^5 light-years in the galaxy frame of reference (Problems 22, 23, 24).

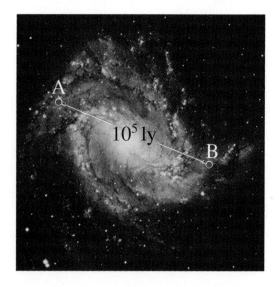

29. At what speed will the momentum of a proton (mass 1 u) equal that of an alpha particle (mass 4 u) moving at $0.5c$?

30. At what speed will the Newtonian expression for momentum be in error by 1%?

31. A particle is moving at $0.90c$. If its speed increases by 10%, by what factor does its momentum increase?

32. Compare the momentum changes needed to boost a spacecraft (a) from $0.1c$ to $0.2c$ and (b) from $0.8c$ to $0.9c$.

33. Find (a) the total energy and (b) the kinetic energy of an electron moving at $0.97c$.

34. A TV tube accelerates electrons from rest through a potential difference of 30 kV. Find their final speed.

35. At what speed will the relativistic and Newtonian expressions for kinetic energy differ by 10%?

36. Find (a) the speed and (b) the momentum of a proton whose kinetic energy is 500 MeV.

37. Among the most energetic cosmic rays ever detected are protons with energies around 10^{20} eV. Find the momentum of such a proton, and compare with that of a 25-mg insect crawling at 2 mm/s (Fig. 38-36).

(a)   (b)

FIGURE 38-36   (a) A cosmic ray proton slams into the upper atmosphere, creating a shower of elementary particles. How does the proton's energy compare with that of a crawling bug (b)? (Problem 37)

38. A large city consumes electrical energy at the rate of 10^9 W. If you converted all the rest mass in a 1-g raisin to electrical energy, for how long could it power the city?

39. In a nuclear fusion reaction, two deuterium nuclei (^{2}H) combine to give a helium nucleus (^{3}He) plus a neutron. The energy released in the process is 3.3 MeV. By how much do the combined masses of the helium nucleus and neutron differ from the combined masses of the original deuterium nuclei?

40. Find the kinetic energy of an electron moving at (a) $0.0010c$, (b) $0.60c$, and (c) $0.99c$. Use suitable approximations where possible.

41. Find the speed of an electron with kinetic energy (a) 100 eV, (b) 100 keV, (c) 1 MeV, (d) 1 GeV. Use suitable approximations where possible.

42. How much energy would it take to accelerate a proton (a) from rest to $0.10c$, (b) from $0.50c$ to $0.55c$, (c) from $0.95c$ to $0.99c$, and (d) from $0.99c$ to c.

43. Use the binomial approximation (Appendix A) to show that Equation 38-14 reduces to the Newtonian expression for kinetic energy in the limit $u \ll c$.

44. Show that Equation 38-16 follows from the expressions for relativistic momentum and total energy.

Section 38-6 What Is Not Relative

45. Show from the Lorentz transformations that the spacetime interval of Equation 38-17 has the same value in all frames of reference.

46. A spaceship travels at $0.80c$ from Earth to a star 10 light-years distant, as measured in the Earth-star frame (Fig. 38-37). Let event A be the ship's departure from Earth and event B its arrival at the star. (a) Find the distance and time between the two events in the Earth-star frame. (b) Find the distance and time between the two events in the ship frame. *Hint:* The distance in the ship frame is the distance an observer has to move *with respect to that frame* to be at both events—not the same as the Lorentz-contracted distance between Earth and star. (c) Compute the square of the spacetime interval in both frames to show explicitly that it is invariant.

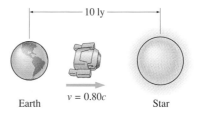

|← 10 ly →|

Earth $v = 0.80c$ Star

FIGURE 38-37   Problem 46.

47. Use Equation 38-17 to calculate the square of the spacetime interval between the events (a) of Problem 22 and (b) of Problem 23. Comment on the signs of your answers in relation to the possibility of a causal relation between the events.

48. A light beam is emitted at event A and arrives at event B. Show that the spacetime interval between the two events is zero.

Paired Problems

(Both problems in a pair involve the same principles and techniques. If you can get the first problem, you should be able to solve the second one.)

49. An extraterrestrial spacecraft passes Earth and 4.5 s later, according to its clocks, it passes the moon. Find its speed.

50. How fast would you have to go to reach a star 200 light-years distant in a 75-year human lifetime?

51. An electron moves down a 1.2-km-long particle accelerator at $0.999992c$. In the electron's frame, (a) how much time does the trip take and (b) how long is the accelerator?

52. An advanced civilization has developed a spaceship that goes, with respect to the galaxy, only 50 km/s slower than light. (a) How long, according to the ship's crew, does it take to cross the 100,000-ly diameter of the galaxy? (b) What is the galactic diameter as measured in the ship's frame of reference?

53. Event A occurs at $x = 0$ and $t = 0$ in a frame of reference S. Event B occurs at $x = 3.8$ light-years, $t = 1.6$ years in S. Find (a) the distance and (b) the time between A and B in a frame S' moving at $0.80c$ along the x axis of S.

54. Two nuclear reactions occur in a particle accelerator, at points 45 cm apart and separated by 6.8 ns in time, as measured in the accelerator's frame of reference. Find (a) the distance and (b) the time between the events as measured in the reference frame of a proton moving at $0.92c$ on a line between the two.

55. When a particle's speed doubles, its momentum increases by a factor of 3. What was the original speed?

56. Find the speed of an electron whose momentum is 50% greater than the value given by the Newtonian expression $p = mu$.

Supplementary Problems

57. How fast would you have to travel to reach the Crab Nebula, 6500 light-years from Earth, in 20 years? Give your answer to 7 significant figures.

58. At what speed are a particle's kinetic and rest energies equal?

59. A cosmic ray proton with energy 20 TeV is heading toward Earth. What is Earth's diameter measured in the proton's frame of reference?

60. Plot the ratio of relativistic to Newtonian kinetic energy as a function of speed, and from your graph find the speed

where the relativistic kinetic energy is 50% greater than the Newtonian value.

61. When the speed of an object increases by 5%, its momentum goes up by a factor of 5. What was the original speed?

62. Use the Lorentz transformations to show that if two events are separated in space and time so that a light signal leaving one event could not reach the other, then there is an observer for whom the two events are simultaneous. Show that the converse is true as well: that if a light signal could get from one event to the other, then no observer will find them simultaneous.

63. By writing force as the time derivative of relativistic momentum, formulate a relativistic version of Section 7-5's derivation of the expression for kinetic energy, and show that the result is Equation 38-14.

64. A source emitting light with frequency f moves toward you at speed u. By considering both time dilation and the effect of wavefronts "piling up" as shown in Fig. 17-20, show that you measure a Doppler shifted frequency given by

$$f' = f\sqrt{\frac{c + u}{c - u}}.$$

Use the binomial approximation to show that this result can be written in the form of Equation 17-12 for $u \ll c$.

65. Use Equation 9-7 to estimate the size to which you would have to squeeze each of the following before escape speed at its surface approximated the speed of light: (a) Earth; (b) the Sun; (c) the Milky Way galaxy, containing about 10^{11} solar masses. Your answers show why general relativity is not needed for most astronomical calculations.

66. Consider a line of positive charge with line charge density λ, as measured in a frame S at rest with respect to the charges. (a) Show that the electric field a distance r from this charged line has magnitude $E = \lambda/2\pi\varepsilon_0 r$, and that there is no magnetic field (no relativity needed here). Now consider the situation in a frame S' moving at speed v parallel to the line of charge. (b) Show that the line charge density as measured in S' is given by $\lambda' = \gamma\lambda$, with $\gamma = 1/\sqrt{1 - v^2/c^2}$. (c) Use the result of (b) to find the electric field in S'. Since the line of charge is moving with respect to S', there is a current in S'. (d) Find an expression for this current and (e) for the magnetic field it produces. Show that the quantities (f) $\mathbf{E} \cdot \mathbf{B}$ and (g) $E^2 - B^2$ are the same in both frames of reference. (In fact, these quantities are always invariant.)

CHAPTER 39

INSIDE ATOMS AND NUCLEI

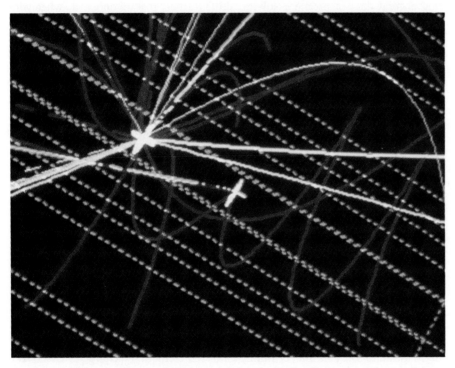

An electron-positron collision at the European Center for Nuclear Research produces a shower of subatomic particles whose trajectories reveal information about their mass and charge.

Newtonian mechanics and Maxwell's equations of electromagnetism constitute the core of **classical physics,** providing a rich and deep understanding of physical reality. Application of classical physics has led to many of the technological developments essential to modern civilization. Although these theories were firmly established by the middle of the nineteenth century, they remain central to the work of many scientists and engineers today. They may be classical, but they are also vitally contemporary.

Nevertheless, at the end of the nineteenth century a few seemingly minor phenomena defied classical explanation. Most physicists felt that it was only a matter of time before these, too, came under the classical umbrella. But that was not to be. We've already seen how questions about the speed of light led to a radical restructuring of our fundamental notions of space and time. Other questions, especially those concerning the behavior of matter at the atomic scale, were to bring an even more radical transformation of physical thought.

The essence of the new physics lies in the ideas of **quantum mechanics,** developed first to explain the workings of the atom and subsequently pushed into the nuclear realm and then to the subnuclear world of elementary particles. In this chapter we give a very brief introduction to quantum mechanics, to the structure of atoms and nuclei, and to current theories of the fundamental nature of matter. Today's physics has come full circle, linking questions about the smallest physical entities with those about the origin and evolution of the entire universe, and we conclude with a brief look at the connections between elementary particle physics and cosmology.

This chapter is not as comprehensive as those that precede it, nor will it give you a great deal of problem-solving skill. Rather, it is intended to whet your appetite for modern physics. Interested readers should consult the extended version of this text, whose seven new chapters explore further the ideas and applications of quantum physics.

39-1 TOWARD THE QUANTUM THEORY

Are matter and energy continuously dividable, or are there some minimum quantities of each from which all else is built? The essential difference between classical and quantum physics is that the former answers this question "in principle, yes," while the latter says definitively "no." Instead, most physical quantities in quantum physics are **quantized,** meaning that they come only in certain discrete values.

The idea that physical quantities might come in discrete "chunks" is not new to modern quantum theory. Some 2400 years ago the Greek philosopher Democritus proposed that all matter consists of indivisible atoms of different types. By the start of the twentieth century a more scientifically grounded atomic theory was widely accepted. J. J. Thomson's discovery of the electron in 1897 showed that atoms might be divisible after all, but at the same time it revealed a finer division of matter into discrete "chunks." Millikan's 1909 oil-drop experiment (Section 23-2) showed that electric charge is similarly quantized, seemingly available only in integer multiples of the elementary charge e. Discovery of the proton and later the neutron further solidified the notion that matter is comprised of fundamental building blocks with definite values for their various physical properties.

Quantization of matter into particles with discrete properties is not incompatible with classical physics as long as those particles behave according to classical laws—in particular, that they move continuously through space and can have *any* amount of energy. Add electromagnetism to the picture and the classical viewpoint requires that the fields be truly continuous, exerting forces on charged particles and perhaps changing, in a gradual and continuous way, the particles' energies.

The startling fact of quantum physics is that this classical behavior does not occur at the atomic scale; instead, energy itself is in many instances quantized. Reconciling the implications of that fact with our commonsense notions of matter and motion has proved impossible; instead, the quantum world speaks a different language, one in which deeply ingrained ideas about causality and the solid reality of matter seem no longer to apply. Here we look at three distinct phenomena that force on us the idea of energy quantization.

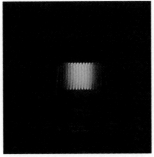

Blackbody Radiation

Heat a solid object hot enough and it begins to glow, emitting electromagnetic radiation in the form of light. As we saw in Section 19-6, the total power radiated is proportional to the fourth power of the temperature. There's also a change in wavelength with increasing temperature: The first visible glow is a dull red, changing with higher temperatures to orange and then yellow colors corresponding to ever shorter wavelengths. A familiar example is an incandescent lamp with a dimmer, which allows you to take the lamp filament from the faintest red glow through the yellow-white at full brightness (Fig. 39-1).

The radiation emitted by a dense, glowing substance that is a good absorber of radiation is **blackbody radiation,** so called because the substance appears black when viewed in reflected light only. Blackbody radiation covers a range of wavelengths, and experiment shows that the wavelength of greatest emission is inversely proportional to the temperature. Microscopically, blackbody radiation is associated with the thermal motions of atoms and molecules, so it's not surprising that the radiation increases with temperature. In the late 1800s physicists tried to apply the laws of electromagnetism and statistical mechanics to explain the experimental observations of blackbody radiation. They met with some success in describing the overall intensity and peak wavelength of the radiation, but could not reproduce completely the observed distribution of wavelengths.

In 1900 the German physicist Max Planck formulated an equation that fit the observed intensity-versus-wavelength curves of blackbody radiation at all wavelengths. Planck's equation, like its classical predecessors, describes blackbody radiation as arising from the thermally-induced vibration of individual molecules. But where classical physics allows a vibrating molecule to have *any* amount of energy, Planck's equation has a startling implication:

The energy of a vibrating molecule can take on only certain discrete values.

If f is a molecule's vibration frequency, Planck's equation requires that its energy be an integer multiple of the quantity hf, where h is a constant whose value Planck determined by fitting his equation to the observed blackbody radiation curves. Today we know h as one of the fundamental constants of nature, and call it **Planck's constant.** Its value is approximately 6.63×10^{-34} J·s, and it is precisely because h is so small that quantum phenomena usually

FIGURE 39-1 As the temperature of an electrically heated coil increases, it emits increasing amounts of radiation that peak at shorter and shorter wavelengths. The color, correspondingly, changes from a dull red to a bright yellow-white.

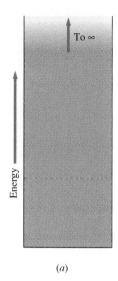

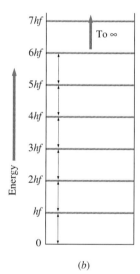

(b)

FIGURE 39-2 (a) In classical physics, a vibrating molecule can have any energy, and its energy can change through absorption or emission of any amount of energy. (b) Allowed energies in Planck's theory are integer multiples of hf and can change only through absorption or emission of discrete energy "bundles" of size hf. Two-headed arrows indicate these allowed transitions. Energy-level diagrams like this are used frequently in quantum physics. Usually, as here, the horizontal axis has no physical significance.

become obvious only when we delve into the atomic and molecular realm. Planck's quantization of the energy of vibrating molecules implies further that a molecule can absorb or emit energy only in discrete "bundles" of size hf, and that in doing so it jumps abruptly from one of its allowed energy levels to another (Fig. 39-2).

Planck's equation was simply a way of fitting observed data and had no deep theoretical justification. Thus, Planck could not say why energy should be quantized, nor could he answer questions such as, "What happens during the jump from one energy level to another?" Planck himself was reluctant to accept or elaborate on his theory's evident disagreement with classical physics; nevertheless, his revolutionary work won him the 1918 Nobel Prize. Other physicists emphasized the contrast between Planck's work and an earlier classical treatment of blackbody radiation. That treatment, based on the assumption that all vibrational energy levels are equally likely, led to the absurd conclusion that every glowing object should emit electromagnetic energy at an infinite rate, with that energy concentrated at the shortest wavelengths (Fig. 39-3). Since the shortest wavelength radiation known at the time was ultraviolet, this phenomenon was called the **ultraviolet catastrophe.**

The Photoelectric Effect

Experiments begun by Hertz in 1887 and carried on by other physicists in the early 1900s showed that some metals emit electrons when illuminated with light. This **photoelectric effect** is studied by putting a metal surface in an evacuated glass container that also contains a second conducting electrode (Fig. 39-4). Making this second electrode positive with respect to the emitting surface attracts the electrons, and the resulting current provides a measure of the number of electrons ejected. Making the second electrode sufficiently negative with respect to the emitting surface, on the other hand, produces an electric field so strong that the emitted electrons do not have enough energy to reach the elec-

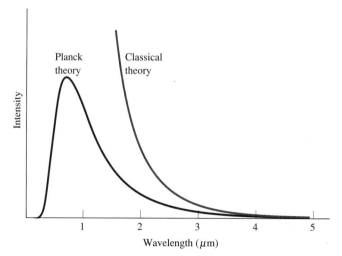

FIGURE 39-3 Intensity versus wavelength for blackbody radiation from an object at 4000 K. Classical and Planck theories agree at long wavelengths, but the classical theory predicts an excess of radiation at short wavelengths.

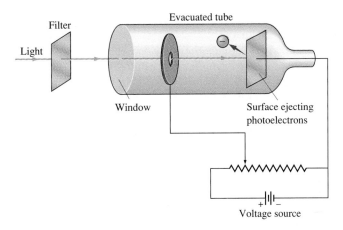

FIGURE 39-4 Apparatus for studying the photoelectric effect. An adjustable filter determines the wavelength of light incident on the apparatus. Light passes through a hole in a metal electrode and strikes a metal surface, where it ejects electrons. Varying the magnitude and polarity of the voltage between the electrode and the metal surface allows measurement of the number and energy of the ejected electrons.

trode. The value of this so-called stopping potential thus provides a measure of the ejected electrons' maximum kinetic energy.

Classical physics suggests that the photoelectric effect should occur because an electron in a metal experiences a force in the oscillating electric field of a light wave. As the wave continues past a given electron, the amplitude of the electron's oscillatory motion should grow until eventually it has enough energy to escape from the metal. Because the energy in a wave is spread throughout the entire wave, it might take a while for a single tiny electron to absorb enough energy to escape. Increasing the light intensity should increase the electric field in the light wave, so should result in the electron being ejected sooner and with more energy. Changing the wave frequency should have little effect on the behavior of the free electrons in the metal.

The photoelectric effect does occur, but not in the way classical physics suggests. The major disagreements are

1. Electrons are ejected almost immediately, even in dim light.
2. The maximum electron energy is independent of the light intensity.
3. Below a certain cutoff frequency *no* electrons are emitted, no matter how intense the light. Above the cutoff frequency electrons are emitted with a maximum energy that increases in proportion to the light-wave frequency.

In 1905, the same year in which he formulated the special theory of relativity, Albert Einstein offered an explanation for the photoelectric effect. Einstein suggested that the energy of an electromagnetic wave is not spread throughout the wave, but is instead concentrated in "bundles" called **photons.** Einstein applied to these photons the same energy-quantization condition Planck had already proposed for molecular vibrations: that photons in light with frequency *f* have energy *hf*, where again *h* is Planck's constant:

$$E = hf. \quad \text{(energy of a photon)} \quad (39\text{-}1)$$

The more intense the light, the more photons—but the energy of each photon is unrelated to the light intensity.

Einstein's idea readily explains the nonclassical aspects of the photoelectric effect. Since the minimum amount of energy in light of frequency f is the energy hf of a single photon, the photons in low-frequency light will not have enough energy to eject electrons—no matter how many photons there are. At the cutoff frequency the photons have just enough energy to eject electrons, and as the frequency increases still further the electrons emerge with kinetic energy equal to the difference between the photon energy and that needed for electron ejection. That excess energy depends only on the photon energy—that is, on the light frequency but not on its intensity. Finally, the immediate ejection of electrons follows because an individual photon delivers its entire bundle of energy to an electron all at once. Einstein received the 1921 Nobel Prize for his explanation of the photoelectric effect; his special and general theories of relativity were at that time sufficiently controversial that they bore no mention in the Nobel citation.

The work of Planck and Einstein applies quantization of energy to two quite different systems, in Planck's case vibrating molecules and in Einstein's case electromagnetic waves. We now turn to the energies of atoms themselves.

Atomic Spectra and the Bohr Atom

In Chapter 34 we found that accelerated charges are the source of electromagnetic radiation. By 1900 it was known that atoms contain negative electrons as well as regions of positive charge; by 1911 experiments by Ernest Rutherford and his colleague Hans Geiger and student Ernest Marsden had localized the positive charge in the tiny but massive nucleus. Electrons should orbit the nucleus under the influence of the electric force, radiating electromagnetic wave energy as they accelerated in their orbits. In fact, a classical calculation shows that atomic electrons should quickly radiate away all their energy and spiral into the nucleus. Thus, the very existence of atoms seems at odds with classical physics.

A more subtle problem involving radiation from atoms dates to 1804, when William Wollastan noticed apparent lines of demarcation between some of the colors dispersed by a prism. Ten years later the German optician Josef von Fraunhofer dispersed the solar spectrum sufficiently that he could see hundreds of narrow, dark lines against the otherwise continuous spectrum of colors (Fig. 39-5a). Studies of light emitted by diffuse gases excited by electric discharges show similar **spectral lines,** these being bright lines against an otherwise dark background (Fig. 39-5b). We now know that bright-line spectra—called **emission spectra**—are produced when atoms emit light of discrete frequencies. **Absorption spectra,** in contrast, arise when atoms in a diffuse gas absorb certain discrete frequencies of light from a continuous source. What Fraunhofer saw in sunlight were absorption lines arising from the diffuse gas overlying the bright visible surface of the Sun. We emphasize the word *diffuse:* Discrete spectra generally arise only when a gas is sufficiently diffuse that light from one atom stands a strong chance of escaping the gas before it interacts with other atoms. In dense gases, or in solids and liquids, multiple interactions result in the continuous spectrum characteristic of blackbody radiation.

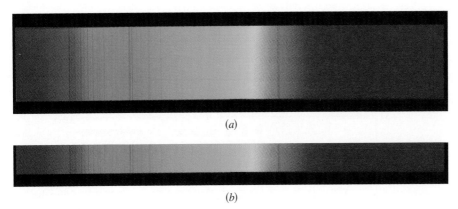

(a)

(b)

FIGURE 39-5 (a) The solar spectrum is a continuous band of colors overlain by dark lines where specific colors have been absorbed in the solar atmosphere. Prominent line in the red is hydrogen-alpha, and the pair in the yellow are the 589-nm sodium doublet. (b) Spectral lines emitted by a diffuse sample of hydrogen gas. This is the Balmer series of lines, with Hα the bright line in the red.

But why should atoms emit and absorb light of discrete frequencies? In 1884, several decades before a satisfactory answer was given, a Swiss schoolteacher named Johann Balmer realized that the wavelengths of the first four lines in the visible spectrum of hydrogen (Fig. 39-5b) were related by the equation

$$\frac{1}{\lambda} = R_H \left(\frac{1}{2^2} - \frac{1}{n^2} \right), \tag{39-2}$$

where $n = 3, 4, 5, 6 \ldots$ and R_H is a constant called the **Rydberg constant** for hydrogen, with value approximately $1.097 \times 10^7 \ \mathrm{m}^{-1}$.

In 1913 the great Danish physicist Niels Bohr (Fig. 39-6) proposed an atomic theory that accounted for the spectral lines of hydrogen. In the **Bohr atom** the electron moves in a circular orbit about the nucleus, held in orbit by the electric force. Classically, any orbital radius and, correspondingly, any energy and any angular momentum should be possible. But Bohr introduced quantization to the atom, stating that only certain orbits were possible—namely, those whose angular momentum is an integer multiple of Planck's constant divided by 2π. Angular momentum quantization implies energy quantization, and Bohr showed that the energy of an atomic electron should be proportional to $-1/n^2$, with n an integer (see Problem 42). The zero of potential energy is taken when the electron and proton are infinitely far apart, and Bohr's negative energy values thus imply that the electron in the atom is bound to the proton.

Bohr's theory asserted that an electron in any of the allowed orbits does not radiate energy, in direct contradiction to the predictions of classical electromagnetism. But an electron can jump from one orbit to another, emitting or absorbing a photon whose energy is equal to the energy difference between the two orbital levels. That difference, translated into wavelength through Equation 39-1 with $f = c/\lambda$ for the photon, leads directly to Balmer's formula for the spectral-line wavelengths. In fact, Bohr's model covers other series of spectral lines in which the 2 in Balmer's formula is replaced with other integers:

FIGURE 39-6 Niels Bohr (left) with Max Planck. Bohr was a powerful influence on many of the twentieth century's greatest physicists. In his later years he was also active in humanitarian causes, especially the effort to control nuclear weapons.

FIGURE 39-7 Energy-level diagram for the Bohr model of the hydrogen atom, showing transitions responsible for the first three series of spectral lines. Each series arises from jumps to a common final state; $n = 2$ gives the visible and ultraviolet lines of the Balmer series. The zero of energy corresponds to a widely separated electron-proton pair with no kinetic energy.

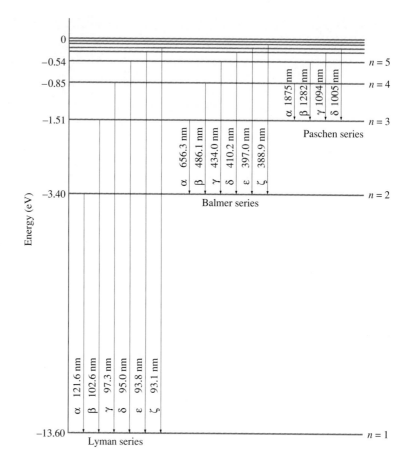

$$\frac{1}{\lambda} = R_H \left(\frac{1}{n_2^2} - \frac{1}{n_1^2} \right). \tag{39-3}$$

Here λ is the wavelength of light emitted when an atomic electron jumps from the n_1th energy level to the n_2th level. Thus, the Balmer series arises in jumps from the $n = 3, 4, 5, \ldots$ levels to the $n = 2$ level. Jumps to the lowest level, $n = 1$, give the ultraviolet lines of the Lyman series, while jumps to $n = 3$ give the infrared Paschen series. Figure 39-7 shows an energy level diagram for hydrogen in the Bohr theory.

● **EXAMPLE 39-1** BIG ATOMS

The Bohr theory predicts that the lowest energy electron orbit in the hydrogen atom—the state with $n = 1$—should have radius 0.0529 nm, consistent with measurements of the "size" of the atom. But in the diffuse hydrogen gas of interstellar space, atoms can be found in highly excited states with sizes approaching a fraction of a millimeter. Transitions between adjacent energy levels of these so-called Rydberg atoms result in emission of photons at radio frequencies, and therefore, radio astronomers can study the enormous atoms. One of the longest wavelengths observed is the so-called 272α, corresponding to a transition from $n = 273$ to $n = 272$. At what wavelength and frequency should a radio telescope be set to observe this transition?

Solution

We use Equation 39-3 with $n_1 = 273$ and $n_2 = 272$:

$$\lambda = \left[R_{\mathrm{H}} \left(\frac{1}{n_2^2} - \frac{1}{n_1^2} \right) \right]^{-1}$$

$$= \left[(1.097 \times 10^7 \text{ m}^{-1}) \left(\frac{1}{272^2} - \frac{1}{273^2} \right) \right]^{-1} = 92 \text{ cm}.$$

Using $f = c/\lambda$ gives the corresponding frequency of 325 MHz,
which happens to lie in a gap between the VHF and UHF bands used for television broadcasting.

EXERCISE Find the wavelength of the spectral line Hα, corresponding to the transition from $n = 3$ to $n = 2$.

Answer: 656.3 nm

Some problems similar to Example 39-1: 6, 7, 8, 9 ●

Bohr's theory was astoundingly successful in describing the hydrogen atom but proved less suitable for more complicated atoms. Furthermore, like Planck's original quantum hypothesis, Bohr's quantization of atomic energy levels lacked a convincing theoretical basis.

■ APPLICATION THE LASER

Scientists and engineers have developed an astonishing array of devices based on quantum principles, and we can hardly begin to do them justice here. But we will describe briefly one quantum system that surely ranks among the important inventions of the twentieth century: the laser.

Bohr knew that an atomic electron could absorb a photon and jump to one of its higher quantized energy levels. In 1917 Einstein suggested that it could be induced to jump downward by the nearby passage of a photon with energy equal to the energy difference between the two levels. This phenomenon of **stimulated emission** is the basis of the laser. In a laser, energy is supplied to a solid, liquid, or gaseous medium in a way that causes a great many atoms to jump into higher energy excited states (Fig. 39-8). When one of these atoms drops back to the lower state, the photon it emits stimulates others to drop as well. The result is a burst of photons, all with the same energy and phase, that gives the laser its bright, coherent light output. In some lasers the "pumping" of the atoms to their higher states is done first, followed by a burst of light; in others the pumping and emission occur all the time, and the laser shines continuously.

Today, lasers are employed wherever bright, coherent, monochromatic light is needed. Applications range from the commonplace, such as scanning bar codes at supermarket checkouts, to technological feats, such as focusing billions of watts of power on tiny pellets in an attempt to harness the energy of nuclear fusion (Fig. 39-9).

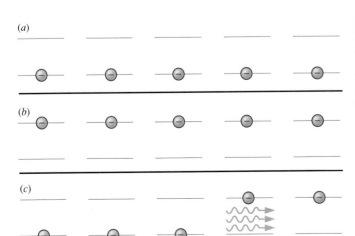

(a)

(b)

(c)

FIGURE 39-8 (a) Normally, atomic electrons are in their lowest possible energy states. Each pair of lines represents an atom with two energy levels shown; blue dots are electrons. (b) In a laser, an external energy source "pumps" the electrons into a higher state. (c) When one electron drops back it emits a photon that stimulates the other electrons to drop as well. The result is a burst of photons that comprise a beam of coherent, monochromatic light. Here, the fourth electron is about to drop as it's stimulated by passage of the photons.

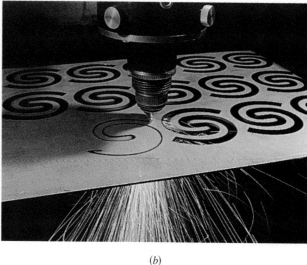

(a)

(b)

FIGURE 39-9 Laser applications. (a) Scanning bar codes in a supermarket checkout. (b) Cutting steel.

Matter Waves

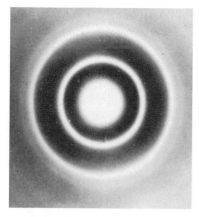

FIGURE 39-10 Diffraction pattern produced by passing an electron beam through a circular aperture shows that electrons have a wave-like character.

In classical physics, light is purely a wave phenomenon. Einstein's introduction of energy "bundles"—photons—gave light a particle-like quality as well. In 1923, ten years after Bohr published his atomic theory, the French prince Louis de Broglie (pronounced "de Broy") set forth a remarkable hypothesis in his doctoral thesis. If light has both wave-like and particle-like properties, he reasoned, why shouldn't matter also exhibit both properties?

We saw in Chapter 34 that light with energy E also carries momentum $p = E/c$. Combined with Equation 39-1, that means a photon of light with frequency f has momentum $p = hf/c$. Since $f\lambda = c$, the photon's momentum and wavelength are therefore related by

$$\lambda = \frac{h}{p}. \qquad \text{(de Broglie wavelength)} \qquad (39\text{-}4)$$

De Broglie proposed that this same relation should hold for particles of matter; at nonrelativistic speed, for example, an electron should have associated with it a "wavelength" given by h/mv. De Broglie's startling new idea was verified experimentally in 1927, with the demonstration that electrons passing through a narrow aperture exhibit a diffraction pattern that we know must be associated with waves (Fig. 39-10).

De Broglie used his matter-wave hypothesis to explain why atomic electron orbits should be quantized. He showed that Bohr's allowed orbits were those in which standing waves could exist (Fig. 39-11), in much the same way that a violin string can support only certain frequencies of standing waves.

If matter has wave-like properties, why don't we notice them all the time? As in optics, wave effects become important only in systems whose size is compara-

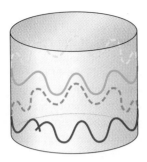

FIGURE 39-11 The allowed electron orbits in the Bohr atom are those that can fit an integral number of de Broglie wavelengths. Blue curve represents the de Broglie standing wave for the $n = 7$ level; red curve does not satisfy the Bohr condition.

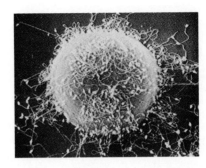

FIGURE 39-12 Diffraction limits the resolution of a microscope to approximately the wavelength of light. The electron microscope circumvents this limitation by using electrons whose wavelength, given by Equation 39-4, can be made much smaller than that of light. Here we see a scanning electron microscope photo of sperm fertilizing a sea-urchin egg.

ble to or smaller than a wavelength. For macroscopic objects, and even for electrons moving at substantial speed in, say, a TV tube, the electron wavelength is so small that wave properties are unnoticeable. In the electron microscope, we choose the electron energy—and, therefore, momentum and wavelength—so the wavelength will be smaller than the system we want to observe (Fig. 39-12). But when an electron is confined to the smaller orbits of the Bohr atom, its wave properties are dominant.

● **EXAMPLE 39-2** BASEBALLS AND ELECTRONS

Find the de Broglie wavelength of (a) a 150-g baseball pitched at 45 m/s and (b) an electron moving at 1.0×10^6 m/s. Compare with the sizes of home plate and an atom, respectively.

Solution
Equation 39-4 gives the wavelengths:

$$\lambda_{\text{baseball}} = \frac{h}{p} = \frac{h}{mv} = \frac{6.63 \times 10^{-34} \text{ J·s}}{(0.15 \text{ kg})(45 \text{ m/s})} = 9.8 \times 10^{-35} \text{ m}.$$

This is about 34 orders of magnitude smaller than home plate, showing that the baseball's wave properties are totally negligible in the systems with which it interacts. For the electron, on the other hand,

$$\lambda_{\text{electron}} = \frac{h}{mv} = \frac{6.63 \times 10^{-34} \text{ J·s}}{(9.11 \times 10^{-31} \text{ kg})(1.0 \times 10^6 \text{ m/s})} = 0.73 \text{ nm},$$

several times the size of an atom. Thus, we should expect wave effects to be prominent when electrons with this speed interact with atoms.

EXERCISE At what speed would a proton's de Broglie wavelength equal the approximately 0.1-nm diameter of the hydrogen atom?

Answer: 4 km/s

Some problems similar to Example 39-2: 10–13 ●

39-2 QUANTUM MECHANICS

Planck's quantized vibrations, Einstein's photons, Bohr's atom, de Broglie's matter waves, and several refinements of these ideas comprise what is known as the **old quantum theory.** Together they provided a reasonably accurate first try at modeling the strange world at the atomic scale. But the old quantum theory could not handle atoms with more than one electron, nor could it predict the rates at which jumps among energy levels should occur. It clung to a mixture of classical ideas—like well-defined electron orbits—constrained by the strange new ideas of energy quantization and matter waves. It left the physical interpretation of matter waves vague. Most importantly, the old quantum theory lacked the intellectual coherence of a single guiding principle.

In 1926, the Austrian physicist Erwin Schrödinger (Fig. 39-13) succeeded in giving a mathematical theory, called **wave mechanics,** that did for de Broglie's electron waves what Maxwell's electromagnetic theory had done for light. The **Schrödinger equation** describes the spatial and temporal behavior of a **wave function,** usually designated ψ, that embodies de Broglie's matter-wave idea.

The interpretation of the wave function is at the heart of the profound philosophical difference between classical and quantum physics. Where the equations of classical physics predict a definite outcome in a given physical situation, the wave function determines only the probabilities of different outcomes. This probabilistic interpretation means that even the language of our macroscopic world is ill suited to the quantum realm. An electron, for example, is neither here nor there; rather, it might have a greater probability of being here than there (Fig. 39-14). The discrete atomic orbits of the old Bohr atom make little sense in wave mechanics; they correspond instead to regions of higher probability. The hydrogen atom is not a miniature solar system, but a fuzzy, statistical object with its single electron spread out in a kind of "probability cloud."

Philosophical debate about the interpretation of quantum theory continues to this day, despite the theory's stunning success in describing the universe at the atomic scale. Einstein himself was among the most persistent critics of the probabilistic interpretation, writing: "The theory says a lot, but does not really bring us any closer to the secret of the Old One. I, in any case, am convinced that He does not throw dice." This quotation is often paraphrased as "I cannot believe that God plays dice with the universe." But neither Einstein nor anyone else has ever found a consistent theory of the atomic realm that does not require a probabilistic interpretation. A series of experiments done as recently as the 1980s appears to have ruled out even the possibility of theories in which a deterministic physics underlies the quantum description.

The Uncertainty Principle

Shortly before Schrödinger proposed his wave mechanics, the German physicist Werner Heisenberg developed a mathematical description of the atom based on the algebra of matrices. The two theories looked very different but were soon proved equivalent.

Heisenberg's formulation concentrated on what he called **observables**—quantities that could actually be measured in laboratory experiments. (Schrödinger's wave function, in contrast, could not be measured directly and, therefore, is not an observable). In 1927 Heisenberg presented his famous **uncertainty principle,** which states that certain pairs of observables cannot be measured simultaneously with arbitrary precision. Position and momentum constitute one such pair; if we measure a particle's position to within an uncertainty Δx, then its momentum must be uncertain by an amount Δp given by

$$\Delta x \, \Delta p \geq \frac{h}{2\pi},$$

(39-5)

FIGURE 39-13 Erwin Schrödinger.

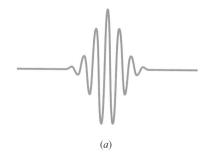

(a)

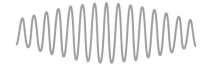

(b)

FIGURE 39-14 In quantum mechanics, the probability of finding a particle is proportional to the square of the wave function. (a) Here the wave function has significant amplitude only in a small region, and the particle is localized. (b) The particle described by this wave function has a nearly equal probability of being found anywhere in a broad region.

where h is Planck's constant. Similarly, if Δt is the time it takes to measure an energy to an accuracy ΔE, then

$$\Delta E \, \Delta t \geq \frac{h}{2\pi}. \qquad (39\text{-}6)$$

Why this limitation on the accuracy of measurements? The fundamental reason is quantization. To measure some property of a system requires interacting with that system—for example, shining light on it. That interaction invariably involves energy, and the interaction energy disturbs the system slightly. As a result of this disturbance, values inferred from the measurement are no longer quite the right ones. In classical physics the energy can be made as small as possible, resulting in an arbitrarily small disturbance. But in quantum theory the minimum energy is a single quantum, like a photon of light, and thus, the disturbance cannot be made arbitrarily small.

So why not use lower frequency light, whose photon energy hf is smaller? Because lower frequency means longer wavelength and, as we found in the preceding chapter, precise imaging requires light whose wavelength is much shorter than the system being imaged. Heisenberg summarized this dilemma with the "thought experiment" shown in Fig. 39-15. He imagined a "quantum microscope" observing a single photon interacting with an electron. Using a short-wavelength photon allows precise localization of the electron (Fig. 39-15a). But short wavelength means high frequency and, thus, through Equation 39-1, high photon energy. The high-energy photon imparts considerable momentum to the electron, and thus, any knowledge we had of the electron's momentum is seriously degraded by the very act that determines its position. We might instead choose a long-wavelength, low-energy photon, as in Fig. 39-15b. Then there is little disturbance, and we can get an accurate assessment of the electron's momentum. But now the photon wavelength is so long that diffraction effects severely limit our ability to tell just where the electron was. So we can measure the electron's position accurately, but only at the expense of very imprecise knowledge of its momentum. Or we can measure its momentum accurately, but then we'll know very little about its position. With a photon of

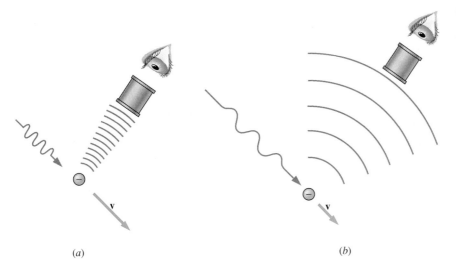

(a)   (b)

FIGURE 39-15 Heisenberg's "quantum microscope." (a) A short-wavelength, high-energy photon interacts with an electron. Diffraction is minimal, and the electron's position is accurately determined. But the high-energy photon disturbs the electron's trajectory, leaving its momentum very uncertain. (b) Using a long-wavelength, low-energy photon minimizes the disturbance and allows accurate determination of the electron's momentum. But now diffraction is significant, and the measurement does not give an accurate position.

intermediate wavelength we could measure both quantities, but neither with perfect accuracy. Equation 39-5 establishes the minimum uncertainties possible.

The uncertainty principle moves the observer from a passive onlooker to an active participant in physical events. To observe is necessarily to disturb, and quantum theory is therefore very much concerned with the role of the observer and the process of measurement. The uncertainty principle is fundamentally a statement about what can and cannot be learned through measurement.

So position and momentum cannot be measured simultaneously with arbitrary precision. Surely, though, an electron has well-defined values of both, even though we cannot know them? The answer is apparently no. The standard interpretation of quantum mechanics suggests that it makes no sense to talk about what cannot be measured, and recent experiments have ruled out "hidden variables" that might be active at a lower level to guide the electron in a deterministic path. We are left with uncertainty—or indeterminacy, as Heisenberg's word also translates—as a fundamental fact of our universe.

● EXAMPLE 39-3 MICROELECTRONICS

A beam of aluminum atoms is used to "dope" part of a semiconductor chip to give it the proper electrical characteristics. If the atoms' velocity is required to be $(4.5000 \pm 0.00001) \times 10^4$ m/s, how accurately can the atoms be localized?

Solution

With a velocity spread of $\pm 0.00001 \times 10^4$ m/s, the maximum tolerable uncertainty in the atoms' velocity is $\Delta v = 0.00002 \times 10^4$ m/s, or 0.20 m/s. Then corresponding momentum uncertainty is

$$\Delta p = m\Delta v = (26.98 \text{ u})(1.66 \times 10^{-27} \text{ kg/u})(0.20 \text{ m/s})$$

$$= 8.96 \times 10^{-27} \text{ kg·m/s},$$

where we found the atomic weight of aluminum in Appendix D. Equation 39-5 then gives the uncertainty in position:

$$\Delta x \geq \frac{h}{2\pi\Delta p} = \left(\frac{6.63 \times 10^{-34} \text{ J·s}}{(2\pi)(8.96 \times 10^{-27} \text{ kg·m/s})} \right)$$

$$= 1.2 \times 10^{-8} \text{ m}.$$

This is about 100 atomic diameters and shows that the uncertainty principle will ultimately limit our ability to manufacture very small microelectronic structures by the means now used.

EXERCISE You measure the velocity of a 3.0-g ping-pong ball to an accuracy of ± 0.01 mm/s. Find the minimum uncertainty in its position, and compare with the size of the ball.

Answer: 1.8×10^{-27} m, totally negligible compared with the ball

Some problems similar to Example 39-3: 14–20

FIGURE 39-16 Wolfgang Pauli and Niels Bohr watching a spinning top whose motion provides a classical analog of electron spin.

Electron Spin and the Exclusion Principle

We have seen how different energy states in the hydrogen atom are distinguished by values of the integer n, which is called a **quantum number.** The theories of Schrödinger and Heisenberg use two additional quantum numbers to describe the atom. But detailed observation of the hydrogen spectrum shows that spectral lines are actually "doublets" of two closely spaced lines. In 1925, Wolfgang Pauli (Fig. 39-16) proposed a fourth quantum number to account for the doublet lines. At the end of the same year, Samuel Goudsmit and George Uhlenbeck realized that the line splitting could be explained if the electron possessed an intrinsic angular momentum, called **spin.** Although it's tempting to think of the electron as a little spinning ball, that classical analogy is grossly inadequate. Spin is an entirely quantum mechanical property, and its behavior is far from classical; in particular, the spin angular momentum of an electron can take on only two possible orientations, loosely called "spin up" and "spin down."

Earlier, Pauli had stated his famous **exclusion principle,** saying that no two electrons in an atom can have the same set of quantum numbers. In the Schrödinger theory with electron spin added, the Pauli exclusion principle means that at most two electrons—one with spin "up" and the other with spin "down"—can share the first three quantum numbers. In multielectron atoms, the exclusion principle governs the distribution of electrons among the different quantum states, often called shells and subshells. It is thus the basis for the chemical properties of the elements and explains the groupings of the periodic table (inside back cover).

Relativistic Quantum Mechanics

Schrödinger's theory of quantum mechanics was based on the nonrelativistic relation between momentum and kinetic energy. In 1927, Paul A. M. Dirac developed a quantum-mechanical wave equation consistent with Equation 38-16 linking energy and momentum in special relativity: $E^2 = p^2c^2 + (mc^2)^2$. Dirac faced mathematical difficulties in dealing with the fact that energy appears squared in this equation. Two remarkable results followed from his handling of this problem: First, he found that consistency with relativity required that the electron have intrinsic angular momentum. Here was a fundamental justification for electron spin, proposed earlier on an *ad hoc* basis to account for spectral features that Schrödinger's theory left unexplained. Second, Dirac explored the fact that the relativistic equation for energy, like any quadratic equation, has two solutions. Rather than discarding the solution corresponding to negative energy, Dirac showed how it led to the concept of **antimatter**—in this case a particle with the mass of the electron but with opposite charge and other properties. The discovery of this particle—the **positron**—in 1932 was a remarkable vindication of Dirac's vision (Fig. 39-17). We know today that every particle has a corresponding antiparticle.

Quantum Electrodynamics

Another high point in the development of quantum mechanics was the theory of **quantum electrodynamics (QED),** set forth in 1948 by Sin-Itiro Tomonaga, Julian Schwinger, and Richard P. Feynman (Fig. 39-18). The theory of QED deals with the quantum-mechanical interaction between matter and electromagnetic radiation. In QED the electromagnetic interaction is visualized as an exchange of so-called **virtual photons** that exist only for the brief duration of the interaction and are never seen. Figure 39-19 shows a **Feynman diagram** for the quantum electrodynamical interaction of an electron and proton. Experiments have verified quantum electrodynamics to extremely high accuracy, and today it is our best-verified theory of physical reality.

The Wave-Particle Duality and Complementarity

One of the most disturbing aspects of quantum theory is the **wave-particle duality**—the seemingly contradictory fact that matter and light have both wave-like and particle-like properties. Bohr dealt with this conundrum through his **principle of complementarity.** The wave and particle pictures, he said, are complementary aspects of the same reality. If we do an experiment to measure a wave-like property—for example, the diffraction of electrons—then we find

FIGURE 39-17 Cloud chamber photograph on which Carl Anderson of the California Institute of Technology discovered the positron in 1932. Anderson's discovery resulted in his sharing the 1936 Nobel Prize in physics.

FIGURE 39-18 The American physicist Richard Feynman played a key role in the development of quantum electrodynamics. Some 40 years later, at the hearings into the explosion of the space shuttle Challenger, he argued that rubber O-rings in the rocket seals became brittle at low temperatures, allowing gases to leak from the booster rockets.

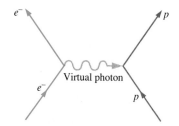

FIGURE 39-19 A Feynman diagram, showing electron-proton scattering as mediated by the exchange of a virtual photon. All electromagnetic interactions among particles can be described with such diagrams.

wave properties but not particle properties. If we do an experiment to measure a particle-like property—for example, localizing an electron—then we will not find wave properties. The two measurements require different experiments, and we can't perform both simultaneously on the same entity. So we will never catch wave and particle in an outright contradiction, and we can never definitively answer the question "Which is it, wave or particle?"

39-3 NUCLEAR PHYSICS

J. J. Thomson discovered the electron in 1897, but at that time it was not known how electrons and the equivalent positive charge were distributed in atoms. Thomson himself favored the **plum-pudding model,** with electrons embedded throughout a "pudding" of positive charge. In a 1909 to 1911 series of experiments suggested by Ernest Rutherford, Rutherford's colleague Hans Geiger and undergraduate student Ernest Marsden shot alpha particles (helium nuclei) from a radioactive source into thin gold foils. They expected the particles to pass straight through the gold or at most to be deflected very slightly. Most of them behaved as expected, but a few bounced back in nearly the direction they had come from (Fig. 39-20). It was, to paraphrase Rutherford, as though you shot a cannonball at a piece of tissue paper and the ball bounced back. Rutherford reasoned that nearly all of the atom's mass must be concentrated in a small, massive core: the **nucleus.** A new atomic model emerged, with the tiny, positively charged nucleus at the center, surrounded by electrons in distant orbits.

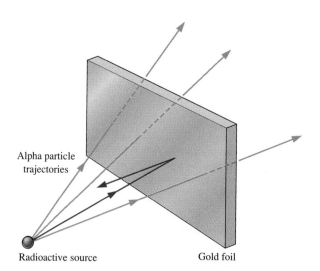

Alpha particle trajectories

Radioactive source   Gold foil

FIGURE 39-20 In the Rutherford-Geiger-Marsden experiment, alpha particles recoiling from a gold foil (red trajectory) revealed the existence of the massive atomic nucleus. Most alphas went right through the foil (green trajectories), showing that the nucleus, although it contains nearly all the mass of an atom, occupies only a tiny fraction of its volume.

But the mutual electrical repulsion of all the positive charges in the nucleus would be strong, so what holds the nucleus together? There must be another force—the **nuclear force**—acting attractively between the constituents of the nucleus. By 1920 Rutherford had also proposed the existence of a neutral particle in the nucleus to "dilute" the repulsive effect of the nuclear protons. In 1932 James Chadwick discovered such a particle, the **neutron.** Together the protons and neutrons making up a nucleus are called **nucleons.**

The number of electrons in an atom determines its chemical properties, and that number is in turn equal to the number of protons in the nucleus. But the number of neutrons can vary, giving rise to **isotopes** of a given element. Different isotopes have the same number of protons but different numbers of neutrons. Their chemical properties are essentially the same but their nuclear properties may be very different.

Fission, Fusion, and the Curve of Binding Energy

In the nucleus protons and neutrons alike experience the attractive nuclear force, but only the charged protons feel the electrical repulsion. The nuclear force is very strong, but, unlike the $1/r^2$ electrical force, it drops rapidly—exponentially, in fact—with distance. Thus, electrical repulsion is the dominant effect on distant protons, but the attractive nuclear force can nevertheless bind a cluster of closely spaced protons and neutrons.

The energy it would take to disassemble the nucleus—the so-called **binding energy**—is a measure of how tightly bound a given nucleus is. Comparing different nuclei makes more sense if we consider the binding energy per nucleon. As the number of nucleons grows, so does the binding energy per nucleon, because a given nucleon experiences attractive forces from more neighboring nucleons. But with larger nuclei the short-range nature of the nuclear force causes this effect to diminish. And with still larger nuclei the repulsion of relatively distant protons actually decreases the binding energy. The result is a **curve of binding energy** (Fig. 39-21) that has a broad peak in the vicinity of

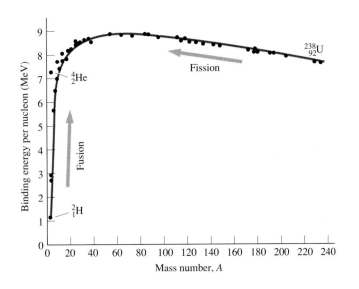

FIGURE 39-21 The curve of binding energy. Nuclei lighter than those at the peak of the curve can release energy by fusing to produce heavier nuclei. Nuclei heavier than those at the peak of the curve can release energy by fissioning into two smaller nuclei.

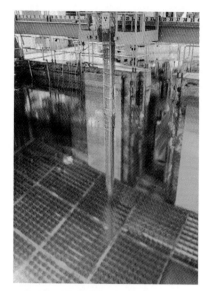

FIGURE 39-22 Nuclear energy is so concentrated that refueling a nuclear power plant may occur less than once a year. Here, a spent fuel rod is moved into a storage pool at the Vermont Yankee nuclear power plant during the plant's once-every-18-month refueling. The blue glow around the bottom of the fuel rod results from its intense radioactivity.

FIGURE 39-23 Lise Meitner and Otto Hahn. Meitner and her nephew Otto Frisch interpreted the experiments of Hahn and Strassman as evidence for uranium fission. In the 1920s Meitner, an Austrian, had braved sexist policies that denied her access to laboratories when men were present; by the 1930s she had become one of the world's most respected nuclear physicists.

60 nucleons. Among the most tightly bound nuclei near this peak is the common iron isotope $^{56}_{26}$Fe. (Here the lower number is the **atomic number,** signifying 26 protons, while the upper number is the **mass number** A, the total number of protons and nucleons.)

The binding energy is the energy it takes to disassemble the nucleus, so it's also the energy released when a nucleus is formed. Any reaction among nucleons that increases the binding energy therefore results in net energy release. As Fig. 39-21 shows, the two possibilities are **fusion** and **fission.** Fusion reactions release energy as lighter nuclei fuse to make heavier ones; fission reactions release energy as more massive nuclei break apart. The great strength of the nuclear force means that nuclear reactions are about 10^7 times more energetic than the electrical interactions among atoms that constitute chemical reactions, and that difference accounts for the concentrated power of nuclear energy sources (Fig. 39-22).

The Austrian physicist Lise Meitner and her nephew Otto Frisch first recognized the phenomenon of fission in 1939 (Fig. 39-23). From the safety of Norway, where Meitner had fled to avoid Hitler's persecution, they interpreted the results of experiments by their German colleagues Otto Hahn and Fritz Strassmann as indicating that uranium nuclei had indeed split, each releasing some 200 MeV of energy in the process. It was the eve of World War II, and a race for military applications of nuclear energy began at once. That race culminated in the devastation of the Japanese cities Hiroshima and Nagasaki in 1945 and continued for 45 more years largely as a nuclear competition between the United States and the former Soviet Union. Nuclear fission has also found a more benign use, in the nuclear reactors that produce 20% of the United States' electric power and more in many other countries.

Fusion of hydrogen powers the Sun and other stars and is the basis of thermonuclear weapons—"hydrogen bombs"—that afford the almost unlimited destructive power of today's nuclear arsenals. Harnessing fusion for electric power generation remains a more elusive goal. To overcome their electrical repulsion, the fusing nuclei must have energies corresponding to temperatures in the 100 million kelvin range. Achieving such temperatures and containing the hot material while it undergoes fusion have proved enormous challenges. After decades of research, it appears that fusion reactors might become commercially viable sometime in the first half of the twenty-first century. If they do, we will have in each gallon of ordinary water the fuel equivalent of about 400 gallons of gasoline.

Radioactivity

Not all combinations of nucleons are stable. If a nucleus has too many protons or gets too big, electrical repulsion can cause it to eject particles or even to fission spontaneously. It will also eject particles if it has too many neutrons. This phenomenon of **radioactivity** was first noticed by Henri Becquerel in 1896, when particles emitted by uranium compounds caused fogging of a nearby photographic plate. In the early 1900s his student Marie Curie and her husband Pierre Curie did pioneering experiments in the study of natural radioactivity (Fig. 39-24); later, in 1934, their daughter Irène and her husband Frédérick Joliot-Curie were the first to produce radioactive isotopes in the laboratory.

We distinguish three common types of radioactivity. In **alpha emission,** a nucleus with too many protons ejects an alpha particle—a helium-4 nucleus—and drops by two in atomic number and four in mass number. In **beta emission** a nucleus emits an electron or a positron, changing its atomic number by ±1, respectively, but not changing its mass number. Also emitted is a **neutrino,** an elusive, neutral particle whose mass is thought to be zero (in which case it travels at the speed of light) or possibly very small. Neutrinos interact with matter through the so-called **weak force** and, because this interaction is so weak, they can pass through solid bodies like Earth with very little probability of stopping. Finally nuclei, like atoms, can be excited to higher energy states. Dropping back, they undergo **gamma emission,** emitting a high-energy photon called a gamma ray.

Radioactive nuclei are characterized not only by the kind and energy of radiation they emit, but also by their **half-life,** meaning the time it takes half of the nuclei in a given sample to decay. Half-lives range from fractions of a second to billions of years.

In the byproducts of nuclear fission or in materials bombarded with high-energy particle beams we find a host of radioactive isotopes that are both dangerous and extremely useful. Common uses of radioactivity include medical diagnosis and cancer treatment, dating of ancient artifacts, geological dating, food preservation, tracing material flows in biological and other systems, and even detecting bombs in airline luggage. Figure 39-25 is a chart showing all the known nuclei, both stable and unstable.

FIGURE 39-24 Marie Curie and Pierre Curie in their laboratory.

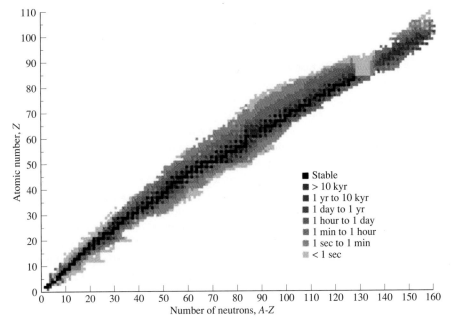

FIGURE 39-25 A chart of the nuclides, plotting all known nuclei according to the number of protons (atomic number Z) versus number of neutrons $(A - Z)$. The nuclei are color coded by half-life. Note that the more massive nuclei are richer in neutrons; this is necessary to counter the electrical repulsion of the protons.

- Stable
- > 10 kyr
- 1 yr to 10 kyr
- 1 day to 1 yr
- 1 hour to 1 day
- 1 min to 1 hour
- 1 sec to 1 min
- < 1 sec

Atomic number, Z

Number of neutrons, A-Z

● **EXAMPLE 39-4** FALLOUT FROM CHERNOBYL

The 1986 accident at the Chernobyl nuclear power plant in the then-Soviet Ukraine spread radioactive fallout over eastern Europe and Scandinavia. A particularly dangerous isotope released was iodine-131, which is absorbed by the thyroid gland and can cause thyroid cancer. The half-life of I-131 is 8.0 days. Following the accident, I-131 levels in milk in Romania rose to 15 times the Romanian government's safety standard. How long did Romanians have to wait until their milk was safe to drink?

Solution

In one half-life, the number of radioactive nuclei and, therefore, also the level of radioactivity drop to half their original value; in two half-lives they drop to $1/2^2$, or one-fourth of their original value. And in n half-lives they drop to $1/2^n$ of their original value. Thus, we want

$$\frac{1}{2^n} = \frac{1}{15}$$

or, inverting and taking logarithms of both sides,

$$\ln(2^n) = \ln 15 \, .$$

But $\ln(2^n) = n \ln 2$, so

$$n = \frac{\ln 15}{\ln 2} = 3.9 \text{ half-lives} \, .$$

Since I-131's half-life is 8 days, Romanians had to wait 31 days before their milk was considered safe.

EXERCISE A patient undergoing a PET scan is given a dose of radioactive oxygen-15 (half-life 2 minutes). How long will it take for the activity of the dose to decline to one one-millionth of its original level?

Answer: about 40 min (20 half-lives)

Some problems similar to Example 39-4: 21–24 ●

FIGURE 39-26 Maria Goeppert Mayer, shown with her daughter. She won the 1963 Nobel Prize for developing the shell model of nuclear structure.

Nuclear Structure

The proton and neutron both obey the Pauli exclusion principle, and thus, we might expect nucleons to arrange themselves into shells like the electrons in atoms. In the 1940s, Marie Goeppert Mayer (Fig. 39-26) and J. Hans Jensen independently developed such a **shell model,** which has proved reasonably successful in explaining nuclear properties. But the strength of the nuclear force and the close packing of nucleons makes it harder to model nuclei than it is to model atoms. As a result an alternative theory, the **liquid-drop model,** is also used to interpret nuclear experiments.

39-4 ELEMENTARY PARTICLES

By 1932 there were four "elementary" particles known: the electron, the proton, the neutron, and the neutrino. These, along with the photon of electromagnetic radiation, could explain the interactions of everyday matter. But as physicists began looking at higher energies, a host of new particles appeared, until there were soon more than 100 "elementary" particles. These fell into several categories. There were electrons and heavier electron-like particles called **muons.** There were three types of neutrinos. There were **mesons,** thought to be the "carriers" of the nuclear force in the way photons are the "carriers" of the electromagnetic force in quantum electrodynamics. There were a host of **baryons,** or heavy particles including protons and neutrons. And for each particle there was, as Dirac had foreseen, a corresponding antiparticle. Most of these particles—including the neutron when it's not in a stable nucleus—are unstable and quickly decay.

Quarks

In 1964 Murray Gell-Mann and, independently, George Zweig suggested a set of three particles that could be combined in different ways to form many of the known particles. Gell-Mann called these particles **quarks,** a word used in James Joyce's novel *Finnegan's Wake.* One surprising thing about quarks is that they carry fractional electric charges. The two least massive quarks, called the **up quark** (u) and the **down quark** (d) carry $+\frac{2}{3}e$ and $-\frac{1}{3}e$, respectively; their antiparticles have the opposite charges. Baryons, like the proton and the neutron, consist of three quarks (Fig. 39-27). Mesons, in constrast, contain a quark and an antiquark (Fig. 39-28).

The Pauli exclusion principle precludes three quarks having the same quantum numbers, and therefore, there must be an additional property distinguishing quarks. Called **color,** this property is a kind of "charge"—not to be confused with electric charge—that can take on any of three values called red, green, and blue. (The word "color" as used here has an entirely new meaning and has nothing to do with light.) The force binding quarks of different colors is called the **color force,** and the quark theory is therefore known as **quantum chromodynamics (QCD).** In QCD, particles called **gluons** play the role of photons in quantum electrodynamics, binding particles subject to the color force. The nuclear force, once thought to be fundamental, is actually a residual manifestation of the color force between quarks—in much the same way that the van der Waals force between gas molecules is a "residue" of the stronger electric force among the particles comprising the molecules.

A second surprising aspect of quarks is that the color force binding them into composite particles does not decrease with separation (Fig. 39-29). For that reason it appears impossible to isolate a single quark. As a result we never see individual particles with fractional electric charge.

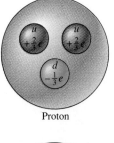

Proton

Neutron

FIGURE 39-27 Protons and neutrons consist of the quark combinations *uud* and *udd,* respectively.

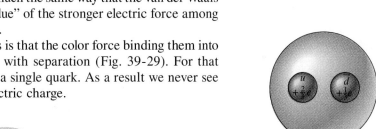

FIGURE 39-28 Mesons consist of quark-antiquark pairs. The meson shown here is the π^+, consisting of an up quark and a down antiquark. (The bar over the *d* signifies an antiparticle.)

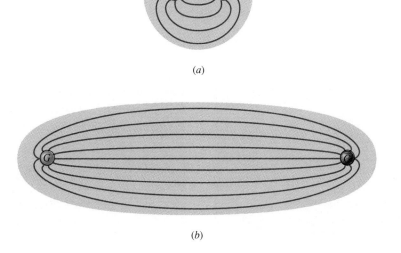

(a)

(b)

FIGURE 39-29 (*a*) A green and an antigreen quark form a meson. The quarks are bound by the color force, here depicted by field lines analogous to those of the electric field. (*b*) The field strength remains approximately constant even if the quarks move apart. Thus it would take infinite energy to separate them fully.

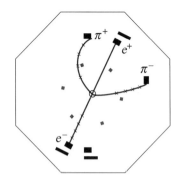

FIGURE 39-30 A computer reconstruction of one of the first ψ-particle events detected at Stanford. An electron-positron collision created the ψ particle, which decayed into four pions. Here we see the tracks of the four pions, which happen to make a shape that resembles the letter ψ.

Today the quark model is universally accepted, so we now know that many of the previously "elementary" particles—including protons and neutrons—are in fact composites of the simpler quarks. Electrons and their relatives—the so-called **leptons,** or light particles—seem, on the other hand, to be truly elementary.

The up and down quarks proved insufficient to account for all the observed particles. A third quark, the **strange quark,** was introduced to account for the "strange" properties—in particular, slower-than-expected decay rates—of some particles. But in 1964 the theorist Sheldon Glashow was arguing for a fourth quark, which he called the **charmed quark.** Ten years later, following intensive searches, experimental teams at Brookhaven National Laboratory and Stanford University announced the discovery of a particle that implies the existence of the charmed quark (Fig. 39-30). The Brookhaven group called the new particle J and the Stanford group called it ψ; to this day it is often called the J/ψ.

Since the J/ψ discovery yet another pair of quarks has been proposed. A 1977 experiment confirmed the existence of the **bottom quark,** and in 1994 the quark model was fully verified in Fermilab experiments that gave evidence for the **top quark.** The more exotic quarks are more massive and therefore, through mass-energy equivalence, it takes more energy to produce particles containing them. This need for higher energy is what drives the push for ever more powerful and expensive particle accelerators, such as the 82-km-circumference Superconducting Super Collider, whose construction had begun in Texas before Congress halted it in 1993 (Fig. 39-31).

The Standard Model

We now have six **flavors** of quarks—up, down, strange, charmed, top, bottom—which seem to be truly elementary constituents of matter. As we've seen, quarks join to form baryons or mesons; together, these quark-based particles are called **hadrons.** But there are other particles, namely, the neutrinos, and the electron and its relatives, that are not made from quarks. Most of these particles are less massive than hadrons and collectively are called **leptons.** They, like the quarks, are believed to be indivisible.

In this "zoo" of elementary particles, physicists recognize three distinct "families." The up and the down quarks comprise the neutron and proton; together with the electron and its related neutrino, these quarks account for the

FIGURE 39-31 (*a*) Photo showing the location of the Super Proton Synchrotron of CERN, the European Laboratory for Particle Physics, in Geneva, Switzerland. The accelerator's main ring, buried beneath farmland, is 7 km in circumference and produces proton-antiproton collisions with total energy 640 GeV. (*b*) Illustration showing magnets designed to guide proton beams around the 82-km circumference of the defunct Superconducting Super Collider.

(*a*)

(*b*)

▲ **TABLE 39-1** STANDARD MODEL OF QUARKS AND LEPTONS

QUARK NAME	SYMBOL	CHARGE	CORRESPONDING LEPTONS
Down	d	$-\frac{1}{3}e$	electron (e), electron neutrino (ν_e)
Up	u	$+\frac{2}{3}e$	
Strange	s	$-\frac{1}{3}e$	muon (μ), muon neutrino (ν_μ)
Charmed	c	$+\frac{2}{3}e$	
Bottom	b	$-\frac{1}{3}e$	tau (τ), tau neutrino (ν_τ)
Top	t	$+\frac{2}{3}e$	

properties of ordinary matter. A second family consists of the strange and charmed quarks, the electron-like muon, and the muon neutrino. The quarks of this family are more massive than the up and down quarks, and the muon is more massive than the electron. More massive still are the particles of the third family, consisting of the top and bottom quarks, the electron-like tau particle, and the tau neutrino. Table 39-1 shows these three particle families.

You may be expecting that future editions of this book will tell of additional quarks and, therefore, of additional families of matter. Whether or not such additional families exist was an open question until 1991, when physicists at the Large Electron Positron Collider in Geneva, Switzerland, examined over half a million particle decay events to conclude that the number of different types of neutrinos that can exist is 2.99 ± 0.06. Since there is presumably a neutrino type for each family, this result seems to preclude the existence of additional families.

The theory that currently describes elementary particles and their interactions is called the **standard model.** In addition to the constituent particles of matter shown in Table 39-1, the standard model includes the photons that mediate the electromagnetic interaction, the gluons of the color force, particles called the W and Z that mediate the weak force that couples leptons and hadrons, and the as yet undetected Higgs boson believed responsible for other particles' masses. The standard model is very successful in explaining the phenomena of particle physics, but it leaves many fundamental questions unanswered. Why, for example, do the quarks and leptons have the particular masses they do? Why are there only three families of elementary particles? Why are leptons and quarks distinct? Continuing theoretical work and experiments at ever higher energies may someday answer these questions.

39-5 UNIFICATION AND COSMOLOGY

We first introduced the fundamental forces of nature in Chapter 5, indicating that today physicists recognize just three such forces: gravity, the electroweak force, and the strong or color force. It was not always that simple, though. In Chapters 23 through 34 we studied the electric and magnetic forces, first separately but then with the realization that they fell under the single umbrella of electromagnetism. The unification of electricity and magnetism was a major step forward in our understanding of physical reality.

▲ **TABLE 39-2** NOBEL PRIZES FOR INTERPRETING THE STRUCTURE OF MATTER

PRIZE YEAR	PRIZE RECIPIENT	RESEARCH CONTRIBUTION[†]
1903	Antoine H. Becquerel	Discovery of radioactivity (1896)
	Marie and Pierre Curie	Studies of radioactivity (1896–1903)
1906	J. J. Thomson	Discovery of the electron and other studies (1897)
1908	Ernest Rutherford*	Disintegration of elements using alpha particles (1902)
1935	James Chadwick	Discovery of the neutron (1932)
	Irène and Frédéric Joliot-Curie	Synthesizing new radioactive elements (1934)
1936	Victor F. Hess	Discovery of cosmic rays (1911)
	Carl D. Anderson	Discovery of the positron (1932)
1938	Enrico Fermi	Production of new radioactive elements (1934–37)
1949	Hideki Yukawa	Theoretical prediction of mesons (1934)
1951	John D. Cockcroft and Ernest T. S. Walton	Transmutation of nuclei (1932)
1957	C. N. Yang and T. D. Lee	Theoretical prediction of parity violation (1956)
1959	Emilio G. Segrè and Owen Chamberlain	Production of the antiproton (1955)
1963	Maria Goeppert Mayer and J. Hans Jensen	Nuclear shell model (1949)
	Eugene P. Wigner	Application of symmetry to elementary particle theories

(a)

(b)

FIGURE 39-32 (a) Carlo Rubbia and Simon von der Meer headed the experimental efforts that led to the confirmation of electroweak unification. (b) This particle-track image led to the experimental discovery of the Z particle.

Electroweak Unification

In the 1960s and early 1970s, a century after Maxwell formalized the unification of electromagnetism, physicists Steven Weinberg, Abdus Salam, and Sheldon Glashow proposed that the electromagnetic force and the weak force are really aspects of the same thing. Their theory predicted the existence of a family of particles called W^+, Z^0, and W^-, which would join the photon as "carriers" of the unified electroweak force. In 1983 a huge international consortium headed by Carlo Rubbia discovered the W and Z particles, using advances in accelerator technology developed by Simon van der Meer (Fig. 39-32). That discovery confirmed the electroweak unification, and Rubbia and van der Meer joined a long list of physicists who had won the Nobel Prize for contributions to our understanding of the structure of matter (Table 39-2).

Further Unification

Electroweak unification led to the present situation in physics, which has gravity, electroweak, and strong forces as the fundamental interactions of matter. A

PRIZE YEAR	PRIZE RECIPIENT	RESEARCH CONTRIBUTION[†]
1968	Luis Alvarez	Contributions to particle accelerators (1946)
1969	Murray Gell-Mann	Theory classifying elementary particles (1964)
1975	Aage Bohr, Ben Mottelson, and James Rainwater	Studies of nuclear structure and dynamics (1950s)
1976	Burton Richter and Samuel C. C. Ting	Discovery of the J/ψ particle, the first charmed particle (1973–74)
1979	Sheldon L. Glashow, Abdus Salam, and Steven Weinberg	Unification of the electromagnetic and weak force (1961–72)
1980	Val L. Fitch and James W. Cronin	Experimental proof of violation of conservation of charge-parity (1964)
1984	Carlo Rubbia and Simon van der Meer	Discovery of W particles, predicted by the electroweak unification (1983); stochastic cooling for focusing particle beams (1968)
1988	Leon M. Lederman, Melvin Schwartz, and Jack Steinberger	Production of neutrino beams and discovery of the μ neutrino (1960–62)
1990	Jerome Friedman, Henry Kendall, and Richard Taylor	Using scattering of electrons from protons and neutrons to explore the quark model.
1992	Georges Charpak	Developing electronic detectors for detecting subatomic particles in accelerators.

† Note: Number in parentheses is the year discovery was made.
* Rutherford and Joliot-Curie prizes in chemistry; all others in physics.

further step, the **grand unification theories** (GUTs), attempts to merge the electroweak and strong forces. The simplest versions of this theory predict that the proton is not a stable particle but should decay on the very long timescale of some 10^{31} years. We can't wait that long, but we can put 10^{32} protons together (that's about 4000 tons of water) and watch for one year (Fig. 39-33). Experiments of this sort have not found the predicted decays, thus ruling out the simplest form of the GUTs. Two other GUTs predictions—that magnetic monopoles exist and that neutrinos have rest mass—are also under investigation. Despite the negative results from proton decay experiments, many physicists believe that grand unification will soon be achieved.

Even grand unification would still leave two forces, one of them gravity. Attempts to reconcile our current theory of gravity—Einstein's general relativity—with quantum mechanics have so far made little progress. Yet such a reconciliation is a necessary prerequisite for a final unification of all known forces. In the 1980s a flurry of interest developed in so-called **superstring theories,** which picture elementary particles as fundamental vibration modes on string-like structures that may be as short as 10^{-35} m. This model is set not in the four-dimensional spacetime to which we have become accustomed, but in a

FIGURE 39-33 The proton-decay experiment of the University of California at Irvine, The University of Michigan, and Brookhaven National Laboratory, known as IMB, contains 4000 tons of water in a salt mine in Ohio. Detectors around the room record light signals from particle reactions in the water. So far no proton decays have been detected, although the experiment did "see" neutrinos from the supernova 1987A.

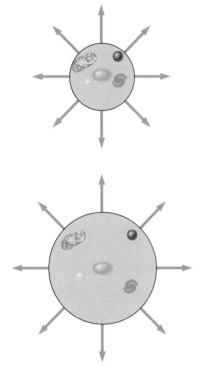

FIGURE 39-34 Expansion of a finite universe. Einstein's general relativity suggests that the universe could be a closed structure in four dimensions, analogous to the surface of a sphere in two dimensions. Such a surface has no boundary, and as it expands each point on the surface sees the other points moving away at speeds proportional to their distances.

ten-dimensional spacetime with six of its dimensions "compactified" in a way that makes them undetectable in normal interactions. To some physicists, string theories hold the promise of a "theory of everything," explaining all our observations about the behavior of the universe. To others they are another in a long list of unsuccessful attempts at a comprehensive explanation of physical phenomena. Time will tell.

Symmetry Breaking

Unification theories predict that phenomena that appear distinct under normal conditions will be clearly seen as one at sufficiently high energies. The observed unification represents a kind of symmetry that is "broken" as the energy level drops. A crude mechanical analog for such **symmetry breaking** is a ball rolling around the rim of a bowl. As long as the ball keeps rolling, there is no preferred angular direction. But if the ball stops, it drops into the bowl in a particular direction. The symmetry of the original situation has been broken. Analogously, at energies above 100 GeV, what we call the electromagnetic and weak forces are one and the same. But at lower energies the symmetry is broken, and we see two distinct forces. Particle accelerators now in construction will exceed the energy of electroweak symmetry breaking, allowing us to explore that interaction in its fundamental simplicity. But the energy at which symmetry breaking occurs increases to some 10^{15} GeV as we move from electroweak to grand unification, making it unlikely that we will achieve that energy in the foreseeable future. And the energy at which gravity, too, would join a single unified force is an even more remote 10^{19} GeV.

The Origin of the Universe

Although the energy above which the electroweak and strong forces would appear unified seems out of reach of our terrestrial laboratories and even of processes in the most bizarre astrophysical systems known, there is one "place" we can look for phenomena at such enormous energies. That "place" is, rather, a time—the very earliest instants after the universe began.

A host of evidence points to a universe that has evolved to its present state from a hot, explosive beginning called the **big bang.** In 1929 the American astronomer Edwin Hubble (after whom the Space Telescope was named) found that spectral lines from distant galaxies were shifted toward the red end of the spectrum, with the shift proportional to a galaxy's distance from Earth. Interpreting the shift as akin to the Doppler effect, astronomers concluded that the distant galaxies are rushing away from each other at speeds proportional to their distances. This is exactly what one would expect if a colossal explosion had flung them apart. Although Hubble's observation might suggest that Earth is at the center, that is not the case. In a universe infinite in extent, every observer would see the same universal expansion, and the same would be true in a finite universe subject to Einstein's general theory of relativity (Fig. 39-34).

In 1965 Arno Penzias and Robert Wilson at Bell Laboratories in New Jersey found a faint "noise" of microwave radiation in a satellite communications antenna they were testing. They could not eliminate the noise, which seemed to come from all directions. Theorists at Princeton concluded that Penzias and

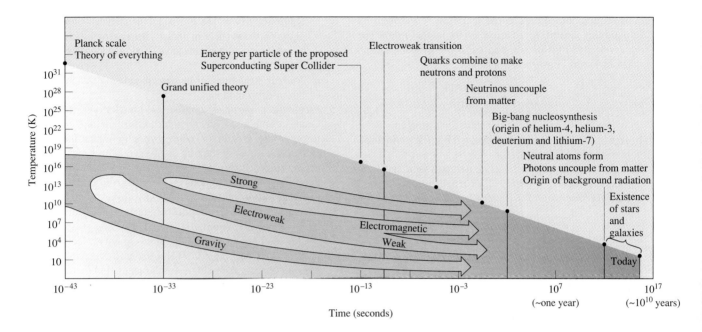

Time (seconds)

Wilson had detected radiation emitted during the formation of the first atoms, some 10^5 years after the universe began. This **cosmic microwave background radiation** is some of the strongest evidence for the big bang theory. That evidence was further strengthened in 1992 with the discovery that the microwave background contains very faint "ripples" that may have led to the formation of galaxies.

With the microwave background, we can "see" the universe as it was about 10^5 years after its beginning. We may soon be able to look further back by studying a similar "background" of neutrinos believed to date from about 1 second after the big bang began. Evidence for still earlier times comes indirectly from applying elementary particle physics to the behavior of matter at extreme temperatures and seeing if the result suggests an evolution to conditions we do observe. Today's accelerators achieve conditions that existed only 10^{-9} s from the beginning, and the Superconducting Super Collider would have pushed that back to 10^{-12} s. Thus, these early times provide a test of elementary particle physics, and particle physics in turn helps us to understand how the universe began and how it evolved to its present state. Figure 39-35 summarizes our understanding of the evolution of the universe, showing that the phenomena of particle physics and unification are inextricably tied with cosmic questions.

FIGURE 39-35 The evolution of the universe from the earliest times to the present. Note the highly logarithmic scale. As the universe has expanded, its average temperature and energy have fallen, giving rise to epochs in which different physical processes dominated. At first the fundamental forces were all one, but symmetry breaking caused them to separate as the energy dropped. Present-day physics cannot address times before 10^{-43} s, which requires a merging of quantum theory and general relativity.

CHAPTER SYNOPSIS

Summary

1. **Quantum physics** describes matter and energy on the small scale. The essential difference between quantum and classical physics is that some physical quantities in quantum physics come only in certain discrete values, or quanta.

   a. Studies of blackbody radiation led Planck to the quantization of molecular vibration energies.
   b. The photoelectric effect led Einstein to propose quantization of electromagnetic wave energy in "bundles" called **photons.**

c. Bohr was able to explain the spectrum of hydrogen by assuming quantized electron orbits in the atom. The wavelength of a photon emitted or absorbed when an atom jumps between two levels is given by

$$\frac{1}{\lambda} = R_\text{H}\left(\frac{1}{n_2^2} - \frac{1}{n_1^2}\right),$$

where R_H is the Rydberg constant for hydrogen. In all three cases quantization involved **Planck's constant,** $h = 6.63 \times 10^{-34}$ J·s.

2. Louis de Broglie proposed that matter, like light, has both particle and wave properties. If a particle has momentum p, then the associated wavelength is

$$\lambda = \frac{h}{p}.$$

3. **Quantum mechanics,** developed by Schrödinger, Heisenberg, and others, provides a nearly complete theory of atomic-scale phenomena. In Schrödinger's version, a particle's behavior is governed statistically by a **wave function** that satisfies Schrödinger's equation.

a. It is necessary to add the concept of electron **spin,** or intrinsic angular momentum, to Schrödinger's theory in order to account fully for atomic spectra. The **Pauli exclusion principle** says that no two electrons in an atom can be in the same state, where states with opposite spins count as distinct states.

b. Heisenberg's **uncertainty principle** says that the act of observing a quantum-mechanical system necessarily disturbs the system. As a result, it is impossible to measure simultaneously a particle's exact position and momentum; instead, the uncertainties Δx and Δp must satisfy the inequality

$$\Delta x \, \Delta p \ge \frac{h}{2\pi}.$$

c. Relativistic quantum mechanics, developed by Dirac, shows that spin arises naturally as a consequence of relativity. Dirac's theory also predicts the existence of antimatter.

d. **Quantum electrodynamics** describes the quantum-mechanical interaction of matter and electromagnetic radiation.

e. The principle of **complementarity** resolves the contradiction seemingly inherent in the **wave-particle duality** by showing that it is impossible simultaneously to observe both aspects of either matter or light in the same experiment.

4. The atomic nucleus consists of protons and neutrons bound by the **nuclear force.**

a. **Binding energy** measures how tightly bound a given nucleus is. Light nuclei undergoing **fusion** increase the binding energy per nucleon, releasing energy in the process. Heavier nuclei undergoing **fission** similarly release energy.

b. **Radioactivity** is the emission of particles—typically alpha particles (helium nuclei), beta particles (electrons or positrons), or gamma rays (high-energy photons)—by unstable nuclei.

c. Nuclear structure can be described using the shell and liquid-drop models.

5. Elementary particles consist of **quarks** and **leptons.** Quarks come in six flavors and three colors, and combine through the **strong** or **color force** to make **hadrons.** Hadrons with three quarks are **baryons** and include the proton and neutron. Hadrons with a quark and an antiquark are **mesons.** Leptons include electrons and the electron-like **muon** and **tau,** along with three types of **neutrino.** The **standard model** organizes the quarks and leptons into three families.

6. The fundamental forces of nature—gravity, electromagnetism, the weak force, and the strong force—may very well be aspects of a single fundamental interaction. Already **electroweak unification** has demonstrated that electromagnetism and the weak force are essentially one. **Grand unification** theories seek to unify the electroweak and strong interactions. Further off are theories that would unite gravity with the other forces.

7. Astronomical evidence, including the red shift of light from distant galaxies and the existence of the cosmic microwave background radiation, strongly suggests that the universe began with a hot, dense explosion called the **big bang.** Very early in the big bang event, particle energies would have been high enough that exotic phenomena predicted by elementary particle theories would have been commonplace. Today, cosmology and elementary-particle physics interact to push our understanding of the universe's evolution back toward the earliest times.

Terms You Should Understand

(Pairs are closely related terms whose distinction is important; number in parentheses is chapter section where term first appears.)

classical physics, quantum physics (introduction)
Planck's constant (39-1)
photon (39-1)
Bohr atom (39-1)
wave function (39-2)
exclusion principle (39-2)
spin (39-2)
uncertainty principle (39-2)
wave-particle duality (39-2)
principle of complementarity (39-2)
binding energy (39-3)
fission, fusion (39-3)
radioactivity (39-3)
quark (39-4)
hadron, baryon, meson, lepton (39-4)
unification (39-5)

Symbols You Should Recognize

h (39-1)

$^{56}_{26}$Fe and similar nuclear symbols (39-3)

Problems You Should Be Able to Solve

finding wavelengths emitted as electrons jump among energy levels in hydrogen (39-1)

evaluating de Broglie wavelengths (39-1)

applying the uncertainty principle (39-2)

working with radioactive half-lives (39-3)

distinguishing baryons and mesons given their quark composition (39-4)

Limitations to Keep in Mind

This chapter deals with the frontiers of physics, and therefore, especially in the later sections, the information given may well change.

QUESTIONS

1. Can you explain the workings of the telephone using classical physics? What about the laser? Why or why not?

2. Why does classical physics predict that atoms should collapse?

3. Imagine an atom that, unlike hydrogen, had only three energy levels. If these levels were evenly spaced, how many spectral lines would result? How would their wavelengths compare?

4. How many spectral lines are in the entire Balmer series?

5. Why are the lines of the Lyman series in the ultraviolet while some Balmer lines are visible (see Fig. 39-7)?

6. Why does the photoelectric effect suggest that light has particle-like properties?

7. Equation 39-4 suggests that a particle that's not moving should have infinite wavelength. Speculate on what this might mean.

8. If you measure a particle's position with perfect accuracy, what do you know about its momentum?

9. How might our everyday experience be different if Planck's constant had the value 1 J·s?

10. How could a neutrino have no mass yet still exist?

11. The electric force between charged particles drops as $1/r^2$. The stronger nuclear force between two nucleons drops exponentially with distance. Use these properties to explain why smaller nuclei are stable but very large ones, like uranium, are not.

12. Can you get energy from fissioning helium nuclei? From fusing lead nuclei? Explain.

13. Can you get energy from fission of iron-56 nuclei? From fusion? Explain.

14. Use Fig. 39-21 to tell which is greater: the energy release per nucleon in the fusion of deuterium ($^2_1$H) to helium ($^4_2$He) or in the fission of uranium into nuclei with mass number around 100.

15. Are electrons and protons both elementary particles?

16. Why are physicists always pushing for accelerators of ever higher energy?

17. If quarks carry fractional charges, why don't we ever detect particles with fractional charge?

PROBLEMS

Section 39-1 Toward the Quantum Theory

1. Find the energy in electron-volts of (a) 1.0-MHz radio photon, (b) a 5.0×10^{14}-Hz optical photon, and (c) a 3.0×10^{18}-Hz x-ray photon.

2. What is the wavelength of a 6.5-eV photon? In what spectral region is this?

3. A microwave oven uses electromagnetic radiation at 2.4 GHz (Fig. 39-36). (a) What is the energy of each microwave photon? (b) At what rate does a 625-W oven produce photons?

4. A red laser at 650 nm and a blue laser at 450 nm emit photons at the same rate. How do their total power outputs compare?

5. Find the rate of photon production by (a) a radio antenna broadcasting 1.0 kW at 89.5 MHz, (b) a laser producing 1.0 mW of 633-nm light, and (c) an x-ray machine producing 0.10-nm x rays with a total power of 2.5 kW.

FIGURE 39-36 How many photons does this microwave oven produce each second? (Problem 3)

6. Calculate the wavelengths of the first three lines in the Lyman series for hydrogen.
7. Which spectral line of the hydrogen Paschen series ($n_2 = 3$) has wavelength 1282 nm?
8. The wavelengths of a spectral line series tend to a limit as $n_1 \to \infty$. Evaluate the series limit for (a) the Lyman series and (b) the Balmer series in hydrogen.
9. A Rydberg hydrogen atom makes a downward transition to the $n = 225$ state. If the photon emitted has energy 9.32 μeV, what was the original state?
10. Find the de Broglie wavelength of (a) Earth, in its 30 km/s orbital motion, and (b) an electron moving at 10 km/s.
11. How slowly must an electron be moving for its de Broglie wavelength to equal 1 mm?
12. A proton and an electron have the same de Broglie wavelength. How do their speeds compare, assuming both are much less than that of light?
13. Electron microscopes can usually resolve smaller objects than optical microscopes because they illuminate their subjects with electrons whose de Broglie wavelengths are much smaller than that of light (Fig. 39-37). What is the minimum electron speed that would make an electron microscope superior to an optical microscope using 450-nm light?

FIGURE 39-37   An image of skin tissue shows on the screen of this electron microscope (Problem 13).

Section 39-2 Quantum Mechanics

14. A proton is confined to a space 1 fm wide (about the size of the atomic nucleus). What is the minimum uncertainty in its velocity?
15. Is it possible to measure an electron's velocity to an accuracy of ± 1.0 m/s while simultaneously finding its position to an accuracy of ± 1.0 μm? What about a proton?
16. A proton has velocity $\mathbf{v} = (1500 \pm 0.25)\hat{\imath}$ m/s. What is the minimum uncertainty in its position?
17. An electron is moving in the $+x$ direction with speed measured at 5.0×10^7 m/s, to an accuracy of $\pm 10\%$. What is the minimum uncertainty in its position?
18. A proton is moving along the x axis with speed $v = (1500 \pm 0.25)$ m/s, but its direction ($+$ or $-x$) is

unknown. Find the uncertainty in its position. *Hint:* What is the difference between the two extreme possibilities for the *velocity?*
19. An electron beam is accelerated at the back of a TV tube and then heads toward the center of the screen with a horizontal velocity of 2.2×10^7 m/s. As the electrons leave the acceleration region, their vertical position is known to within ± 45 nm. Find the minimum angular spread in the beam, as set by the uncertainty principle.
20. An electron has momentum of magnitude p, but the direction is unknown. (a) Argue that the uncertainty in its momentum is therefore $2p$. (b) Compare the minimum uncertainty in its position with its de Broglie wavelength. Why might the two be related?

Section 39-3 Nuclear Physics

21. What is the half-life of a radioactive material if 75% of it decays in 5.0 hours?
22. Iodine-131 has a half-life of 8 days. How many whole days after a nuclear power plant accidentally releases I-131 to the atmosphere will the number of I-131 nuclei be less than 1% of its original value?
23. Oxygen-15 is a radioactive isotope with a 2-minute half-life, widely used in medical studies because its short half-life ensures relative safety. Approximately how long will it take for 99.9% of the nuclei in a sample of O-15 to decay?
24. Carbon-14, widely used in archaeological dating, has a half-life of 5730 years (Fig. 39-38). How old is a bone sample if the amount of carbon-14 it contains has dropped to 15% of its value when the bone was in a living creature?

FIGURE 39-38   How old is the bone? Radiocarbon dating can tell (Problem 24).

25. The energy released in the fission of a single uranium-235 nucleus is about 200 MeV. Estimate the mass of uranium that fissioned in the Hiroshima bomb, shown in Fig. 39-39, whose explosive yield was 12.5 kilotons (see Appendix C for a useful conversion factor).
26. The dominant process powering the Sun is a sequence of fusion reactions that fuse four protons to make helium-4 along with neutrinos and positrons. The net energy re-

FIGURE 39-39 On August 6, 1945, the Japanese city of Hiroshima was devastated by a nuclear fission bomb. How much uranium fissioned to produce this blast? (Problem 25)

leased is about 27 MeV. At what rate must the Sun consume protons to produce its power output of about 4×10^{26} W?

Section 39-4 Elementary Particles

27. The lambda particle (Λ) consists of an up quark, a down quark, and a strange quark. What is its charge?
28. Find the charge of the omega minus particle (Ω^-), which consists of three strange quarks.
29. What is the quark composition of the antiproton?
30. Find all the baryons that can be made from the up and down quarks, and give the total charge of each.
31. One of two charged mesons called pions is formed from an up quark and a down antiquark. What is this meson's charge?
32. (a) Is the particle with quark composition $\bar{u}s$ a baryon or a meson? (b) What is its charge?

Paired Problems

(Both problems in a pair involve the same principles and techniques. If you can get the first problem, you should be able to solve the second one.)

33. (a) Find the energy of the highest energy photon that can be emitted as the electron jumps between two adjacent energy levels in the Bohr hydrogen atom. (b) Which energy levels are involved?
34. Find (a) the wavelength and (b) the energy in eV of the photon emitted as a Rydberg hydrogen atom drops from the $n = 180$ level to the $n = 179$ level.
35. In which of the following will wave properties be more evident: an electron moving at 10 Mm/s or a proton moving at 1 km/s?
36. Give the ratio of de Broglie wavelength to diameter for (a) a 100-g ball, 10 cm in diameter, moving at 5 m/s; (b) a 4-u

alpha particle, 3.4 fm in diameter, moving at 20 Mm/s; and (c) a 238-u uranium nucleus, 13 fm in diameter, moving at 1.4 km/s. (d) For which will wave properties be negligible, and for which will they be completely dominant?
37. An electron is known to be within ± 0.05 nm of the nucleus of an atom. What is the uncertainty in its velocity?
38. Typically, an atom remains in an excited state for about 10^{-8} s before it drops to a lower energy state, emitting a photon in the process. What is the uncertainty in the energy of this transition?

Supplementary Problems

39. An electron is accelerated from rest through a 4.5-kV potential difference. What is its de Broglie wavelength?
40. Consider an electron confined in one dimension to a region of width Δx. The uncertainty principle can still be satisfied if the magnitude p of the electron's momentum is known, provided we don't know which way it's moving; then the uncertainty in momentum is the magnitude of the difference of the two possible oppositely directed momentum vectors, or $2p$. Show under these conditions that the minimum possible value for p consistent with the uncertainty principle is $p = h/4\pi \Delta x$.
41. Use the result of the preceding problem to estimate the minimum possible kinetic energy for (a) an electron confined to a region of atomic dimensions, about 0.1 nm and (b) a proton confined to a region of nuclear dimensions, about 1 fm. Your answers show the order of magnitude of the energies to be expected in atomic and nuclear reactions, respectively.
42. Quantization in the Bohr model requires that the angular momentum mvr of an electron orbit be an integer multiple of Planck's constant divided by 2π. By adapting Section 9-4's discussion of energy in circular orbits to an electron orbiting a proton and applying Bohr's quantization condition, show that the total energies allowed are given by

$$E_n = -\frac{ke^2}{2a_0}\left(\frac{1}{n^2}\right),$$

where n is an integer, k is the constant in Coulomb's law and $a_0 = h^2/4\pi^2 m_e ke^2$. (Called the Bohr radius, a_0 is the radius of the lowest energy orbit in the Bohr model.)
43. The expansion of the universe is described by the so-called Hubble relation, $v = H_0 d$, where v is the recession speed of two galaxies a distance d apart, and where H_0 is called the Hubble constant. Estimates for the value of H_0 range from 50 km/s/megaparsec to 100 km/s/megaparsec. (See Appendix C for the distance unit parsec). If the expansion rate had been the same from the beginning, how long ago would two galaxies have been together? Give two answers, derived from the two extreme values for H_0. Your answers are estimates for the age of the universe.

APPENDIX A

MATHEMATICS

A-1 ALGEBRA AND TRIGONOMETRY

Quadratic Formula

If $ax^2 + bx + c = 0$ then $x = \dfrac{-b \pm \sqrt{b^2 - 4ac}}{2a}$

Circumference, Area, Volume

Where $\pi \approx 3.14159\ldots$

circumference of circle	$2\pi r$	
area of circle	πr^2	
surface area of sphere	$4\pi r^2$	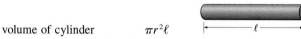
volume of sphere	$\frac{4}{3}\pi r^3$	
area of triangle	$\frac{1}{2}bh$	
volume of cylinder	$\pi r^2 \ell$	

Trigonometry

definition of angle (in radians): $\theta = \dfrac{s}{r}$

2π radians in complete circle
1 radian $\approx 57.3°$

Trigonometric Functions

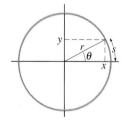

$\sin \theta = \dfrac{y}{r}$

$\cos \theta = \dfrac{x}{r}$

$\tan \theta = \dfrac{\sin \theta}{\cos \theta} = \dfrac{y}{x}$

Values at Selected Angles

$\theta \rightarrow$	0	$\dfrac{\pi}{6}$ (30°)	$\dfrac{\pi}{4}$ (45°)	$\dfrac{\pi}{3}$ (60°)	$\dfrac{\pi}{2}$ (90°)
$\sin\theta$	0	$\dfrac{1}{2}$	$\dfrac{\sqrt{2}}{2}$	$\dfrac{\sqrt{3}}{2}$	1
$\cos\theta$	1	$\dfrac{\sqrt{3}}{2}$	$\dfrac{\sqrt{2}}{2}$	$\dfrac{1}{2}$	0
$\tan\theta$	0	$\dfrac{\sqrt{3}}{3}$	1	$\sqrt{3}$	∞

Graphs of Trigonometric Functions

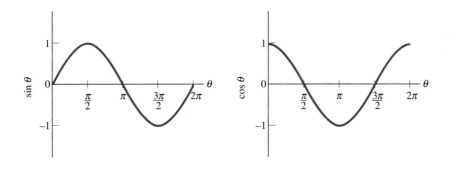

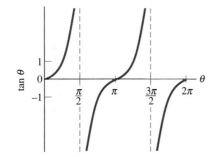

Trigonometric Identities

$$\sin(-\theta) = -\sin\theta$$

$$\cos(-\theta) = \cos\theta$$

$$\sin\left(\theta \pm \frac{\pi}{2}\right) = \pm\cos\theta$$

$$\cos\left(\theta \pm \frac{\pi}{2}\right) = \mp\sin\theta$$

$$\sin^2\theta + \cos^2\theta = 1$$

$$\sin 2\theta = 2\sin\theta\cos\theta$$

$$\cos 2\theta = \cos^2 \theta - \sin^2 \theta = 1 - 2\sin^2 \theta = 2\cos^2 \theta - 1$$

$$\sin(\alpha \pm \beta) = \sin \alpha \cos \beta \pm \cos \alpha \sin \beta$$

$$\cos(\alpha \pm \beta) = \cos \alpha \cos \beta \mp \sin \alpha \sin \beta$$

$$\sin \alpha \pm \sin \beta = 2\sin\left[\frac{1}{2}(\alpha \pm \beta)\right]\cos\left[\frac{1}{2}(\alpha \mp \beta)\right]$$

$$\cos \alpha + \cos \beta = 2\cos\left[\frac{1}{2}(\alpha + \beta)\right]\cos\left[\frac{1}{2}(\alpha - \beta)\right]$$

$$\cos \alpha - \cos \beta = -2\sin\left[\frac{1}{2}(\alpha + \beta)\right]\sin\left[\frac{1}{2}(\alpha - \beta)\right]$$

Laws of Cosines and Sines Where A, B, C are the sides of an arbitrary triangle and α, β, γ the angles opposite those sides:

Law of cosines

$$C^2 = A^2 + B^2 - 2AB\cos \gamma$$

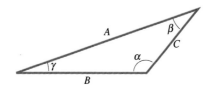

Law of sines

$$\frac{\sin \alpha}{A} = \frac{\sin \beta}{B} = \frac{\sin \gamma}{C}$$

Exponentials and Logarithms

Graphs

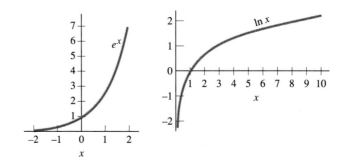

Exponential and Natural Logarithms Are Inverse Functions

$$e^{\ln x} = x, \quad \ln e^x = x \quad e = 2.71828 \ldots.$$

Exponential and Logarithmic Identities

$$a^x = e^{x \ln a} \qquad \ln(xy) = \ln x + \ln y$$

$$a^x a^y = a^{x+y} \qquad \ln\left(\frac{x}{y}\right) = \ln x - \ln y$$

$$(a^x)^y = a^{xy} \qquad \ln\left(\frac{1}{x}\right) = -\ln x$$

$$\log x \equiv \log_{10} x = \ln(10)\ln x \simeq 2.3\ln x$$

Expansions and Approximations

Series Expansions of Functions
Note: $n! = n(n-1)(n-2)(n-3)\cdots(3)(2)(1)$

$$e^x = 1 + x + \frac{x^2}{2!} + \frac{x^3}{3!} + \cdots \quad \text{(exponential)}$$

$$\sin x = x - \frac{x^3}{3!} + \frac{x^5}{5!} - \cdots \quad \text{(sine)}$$

$$\left. \right\} \ (x \text{ in radians})$$

$$\cos x = 1 - \frac{x^2}{2!} + \frac{x^4}{4!} - \cdots \quad \text{(cosine)}$$

$$\ln(1 + x) = x - \frac{x^2}{2} + \frac{x^3}{3} - \cdots \quad \text{(natural logarithm)}$$

$$(1 + x)^p = 1 + px + \frac{p(p-1)}{2!}x^2 + \frac{p(p-1)(p-2)}{3!}x^3 + \cdots$$

(binomial, valid for $|x| < 1$)

Approximations For $|x| \ll 1$, the first few terms in the series provide a good approximation; that is,

$$e^x \simeq 1 + x$$
$$\sin x \simeq x$$
$$\cos x \simeq 1 - \tfrac{1}{2}x^2 \qquad \text{for } |x| \ll 1$$
$$\ln(1 + x) \simeq x$$
$$(1 + x)^p \simeq 1 + px$$

Expressions that do not have the forms shown may often be put in the appropriate form. For example:

$$\frac{1}{\sqrt{a^2 + y^2}} = \frac{1}{a\sqrt{1 + \dfrac{y^2}{a^2}}} = \frac{1}{a}\left(1 + \frac{y^2}{a^2}\right)^{-1/2}.$$

For $y^2 \ll a^2$, this may be approximated using the binomial expansion $(1 + x)^p \simeq 1 + px$, with $p = -\frac{1}{2}$ and $x = y^2/a^2$:

$$\frac{1}{a}\left(1 + \frac{y^2}{a^2}\right)^{-1/2} \simeq \frac{1}{a}\left(1 - \frac{1}{2}\frac{y^2}{a^2}\right).$$

Vector Algebra

Vector Products

$$\mathbf{A} \cdot \mathbf{B} = AB\cos\theta$$

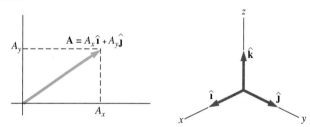

$|\mathbf{A} \times \mathbf{B}| = AB\sin\theta$, with direction of $\mathbf{A}\times\mathbf{B}$ given by right-hand rule:

Unit Vector Notation An arbitrary vector $\mathbf{A}$ may be written in terms of its components A_x, A_y, A_z and the unit vectors $\hat{\mathbf{i}}, \hat{\mathbf{j}}, \hat{\mathbf{k}}$ that have length 1 and lie along the x, y, z axes:

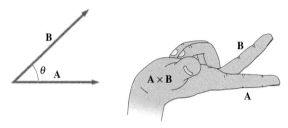

In unit vector notation, vector products become

$$\mathbf{A} \cdot \mathbf{B} = A_x B_x + A_y B_y + A_z B_z$$
$$\mathbf{A} \times \mathbf{B} = (A_y B_z - A_z B_y)\hat{\mathbf{i}} + (A_z B_x - A_x B_z)\hat{\mathbf{j}} + (A_x B_y - A_y B_x)\hat{\mathbf{k}}$$

Vector Identities

$$\mathbf{A} \cdot \mathbf{B} = \mathbf{B} \cdot \mathbf{A}$$

$$\mathbf{A} \times \mathbf{B} = -\mathbf{B} \times \mathbf{A}$$

$$\mathbf{A} \cdot (\mathbf{B} \times \mathbf{C}) = \mathbf{B} \cdot (\mathbf{C} \times \mathbf{A}) = \mathbf{C} \cdot (\mathbf{A} \times \mathbf{B})$$

$$\mathbf{A} \times (\mathbf{B} \times \mathbf{C}) = (\mathbf{A} \cdot \mathbf{C})\mathbf{B} - (\mathbf{A} \cdot \mathbf{B})\mathbf{C}$$

A-2 CALCULUS

Derivatives

Definition of the Derivative If y is a function of x $[y = f(x)]$, then the **derivative of y with respect to x** is the ratio of the change Δy in y to the corresponding change Δx in x, in the limit of arbitrarily small Δx:

$$\frac{dy}{dx} = \lim_{\Delta x \to 0} \frac{\Delta y}{\Delta x}.$$

Algebraically, the derivative is the rate of change of y with respect to x; geometrically, it is the slope of the y versus x graph—that is, of the tangent line to the graph at a given point:

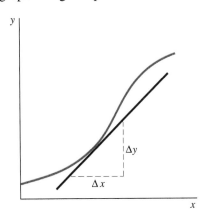

Derivatives of Common Functions Although the derivative of a function can be evaluated directly using the limiting process that defines the derivative, standard formulas are available for common functions:

$$\frac{da}{dx} = 0 \quad (a \text{ is a constant})$$

$$\frac{dx^n}{dx} = nx^{n-1} \quad (n \text{ need not be an integer})$$

$$\frac{d}{dx}\sin x = \cos x$$

$$\frac{d}{dx}\cos x = -\sin x$$

$$\frac{d}{dx}\tan x = \frac{1}{\cos^2 x}$$

$$\frac{de^x}{dx} = e^x$$

$$\frac{d}{dx}\ln x = \frac{1}{x}$$

Derivatives of Sums, Products, and Functions of Functions

1. Derivative of a constant times a function

$$\frac{d}{dx}[af(x)] = a\frac{df}{dx} \quad (a \text{ is a constant})$$

2. Derivative of a sum

$$\frac{d}{dx}[f(x) + g(x)] = \frac{df}{dx} + \frac{dg}{dx}$$

3. Derivative of a product

$$\frac{d}{dx}[f(x)g(x)] = g\frac{df}{dx} + f\frac{dg}{dx}$$

Examples

$$\frac{d}{dx}(x^2\cos x) = \cos x\frac{dx^2}{dx} + x^2\frac{d}{dx}\cos x = 2x\cos x - x^2\sin x$$

$$\frac{d}{dx}(x\ln x) = \ln x\frac{dx}{dx} + x\frac{d}{dx}\ln x = (\ln x)(1) + x\left(\frac{1}{x}\right) = \ln x + 1$$

4. Derivative of a quotient

$$\frac{d}{dx}\left[\frac{f(x)}{g(x)}\right] = \frac{1}{g^2}\left(g\frac{df}{dx} - f\frac{dg}{dx}\right)$$

Example

$$\frac{d}{dx}\left(\frac{\sin x}{x^2}\right) = \frac{1}{x^4}\left(x^2\frac{d}{dx}\sin x - \sin x\frac{dx^2}{dx}\right) = \frac{\cos x}{x^2} - \frac{2\sin x}{x^3}$$

5. Chain rule for derivatives
   If f is a function of u and u is a function of x, then

$$\frac{df}{dx} = \frac{df}{du}\frac{du}{dx}.$$

Examples

a. Evaluate $\dfrac{d}{dx}\sin(x^2)$. Here $u = x^2$ and $f(u) = \sin u$, so

$$\frac{d}{dx}\sin(x^2) = \frac{d}{du}\sin u\frac{du}{dx} = (\cos u)\frac{dx^2}{dx} = 2x\cos(x^2).$$

b. $\dfrac{d}{dt}\sin \omega t = \dfrac{d}{d\omega t}\sin \omega t\dfrac{d}{dt}\omega t = \omega\cos \omega t.$ (ω a constant)

c. Evaluate $\dfrac{d}{dx}\sin^2 5x$. Here $u = \sin 5x$ and $f(u) = u^2$, so

$$\frac{d}{dx}\sin^2 5x = \frac{d}{du}u^2\frac{du}{dx} = 2u\frac{du}{dx} = 2\sin 5x\frac{d}{dx}\sin 5x$$

$$= (2)(\sin 5x)(5)(\cos 5x) = 10\sin 5x\cos 5x = 5\sin 2x.$$

Second Derivative The second derivative of y with respect to x is defined as the derivative of the derivative:

$$\frac{d^2y}{dx^2} = \frac{d}{dx}\left(\frac{dy}{dx}\right).$$

Example

If $y = ax^3$, then $dy/dx = 3ax^2$, so

$$\frac{d^2y}{dx^2} = \frac{d}{dx}3ax^2 = 6ax.$$

Partial Derivatives When a function depends on more than one variable, then the partial derivatives of that function are the derivatives with respect to each variable, taken with all other variables held constant. If f is a function of x and y, then the partial derivatives are written

$$\frac{\partial f}{\partial x} \quad \text{and} \quad \frac{\partial f}{\partial y}.$$

Example

If $f(x, y) = x^3 \sin y$, then

$$\frac{\partial f}{\partial x} = 3x^2 \sin y \quad \text{and} \quad \frac{\partial f}{\partial y} = x^3 \cos y.$$

Integrals

Indefinite Integrals Integration is the inverse of differentiation. The **indefinite integral,** $\int f(x)\,dx$, is defined as a function whose derivative is $f(x)$:

$$\frac{d}{dx}\left[\int f(x)\,dx\right] = f(x).$$

If $A(x)$ is an indefinite integral of $f(x)$, then because the derivative of a constant is zero, the function $A(x) + C$ is also an indefinite integral of $f(x)$, where C is any constant. Inverting the derivatives of common functions listed in the preceding section gives some common integrals (a more extensive table appears at the end of this appendix).

$$\int a\,dx = ax + C$$

$$\int x^n\,dx = \frac{x^{n+1}}{n+1} + C, \quad n \neq -1$$

$$\int \sin x\,dx = -\cos x + C$$

$$\int \cos x\,dx = \sin x + C$$

$$\int e^x\,dx = e^x + C$$

$$\int x^{-1}\,dx = \ln x + C$$

Definite Integrals In physics we are most often interested in the **definite integral,** defined as the sum of a large number of very small quantities, in the limit as the number of quantities grows arbitrarily large and the size of each arbitrarily small:

$$\int_{x_1}^{x_2} f(x)\,dx \equiv \lim_{\substack{\Delta x \to 0 \\ N \to \infty}} \sum_{i=1}^{N} f(x_i)\,\Delta x,$$

where the terms in the sum are evaluated at values x_i between the limits of integration x_1 and x_2; in the limit $\Delta x \to 0$, the sum is over all values of x in the interval.

The definite integral is used whenever we need to sum over a quantity that is changing—for example, to calculate the work done by a variable force (Chapter 7), the entropy change in a system whose temperature varies (Chapter 22), or the flux of an electric field that varies with position (Chapter 24).

The key to evaluating the definite integral is provided by the **fundamental theorem of calculus.** The theorem states that, if $A(x)$ is an *indefinite* integral of $f(x)$, then the *definite integral* is given by

$$\int_{x_1}^{x_2} f(x)\,dx = A(x_2) - A(x_1) \equiv A(x)\Big|_{x_1}^{x_2}.$$

Geometrically, the definite integral is the area under the graph of $f(x)$ between the limits x_1 and x_2:

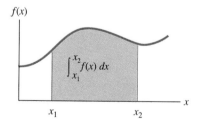

Evaluating Integrals The first step in evaluating an integral is to express all varying quantities within the integral in terms of a single variable. For example, in evaluating $\int E\,dr$ to calculate an electric potential (Chapter 25), it is necessary first to express E as a function of r. This procedure is illustrated in many examples throughout this text; Example 23-7 provides a typical case.

Once an integral is written in terms of a single variable, it is necessary to manipulate the integrand—the function being integrated—into a form whose integral you know or can look up in tables of integrals. Two common techniques are especially useful:

1. **Change of variables**

   An unfamiliar integral can often be put into familiar form by defining a new variable. For example, it is not obvious how to integrate the expression

   $$\int \frac{x\,dx}{\sqrt{a^2 + x^2}}.$$

   where a is a constant. But let $z = a^2 + x^2$. Then

   $$\frac{dz}{dx} = \frac{da^2}{dx} + \frac{dx^2}{dx} = 0 + 2x = 2x,$$

   so $dz = 2x\,dx$. Then the quantity $x\,dx$ in our unfamiliar integral is just $\frac{1}{2}dz$, while the quantity $\sqrt{a^2 + x^2}$ is just $z^{1/2}$. So the integral becomes

   $$\int \tfrac{1}{2} z^{-1/2}\,dz = \frac{\frac{1}{2}z^{1/2}}{(1/2)} = \sqrt{z},$$

   where we have used the standard form for the integral of a power of the independent variable. Substituting back $z = a^2 + x^2$ gives

   $$\int \frac{x\,dx}{\sqrt{a^2 + x^2}} = \sqrt{a^2 + x^2}.$$

2. **Integration by parts**

   The quantity $\int u\,dv$ is the area under the curve of u as a function of v between specified limits. In the figure below, that area can also be expressed as the area of the rectangle shown minus the area under the curve of v as a function of u. Mathematically, this relation among areas may be expressed as a relation among integrals:

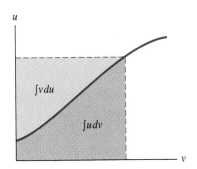

$$\int u\,dv = uv - \int v\,du. \qquad \text{(integration by parts)}$$

This expression may often be used to transform complicated integrals into simpler ones.

Example

Evaluate $\int x\cos x\, dx$. Here let $u = x$, so $du = dx$. Then $dv = \cos x\, dx$, so $v = \int dv = \int \cos x\, dx = \sin x$. Integrating by parts then gives

$$\int x\cos x\, dx = (x)(\sin x) - \int \sin x\, dx = x\sin x + \cos x,$$

where the $+$ sign arises because $\int \sin x\, dx = -\cos x$.

Table of Integrals [More extensive tables are available in many mathematical and scientific handbooks; see, for example, **Handbook of Chemistry and Physics** (Chemical Rubber Co.) or Dwight, **Tables of Integrals and Other Mathematical Data** (Macmillan).] Increasingly, sophisticated computer software is used instead of tables for the symbolic evaluation of integrals; among the most widely used are *Maple*, *Mathematica*, and *Derive*.

In the expressions below, a and b are constants. An arbitrary constant of integration may be added to the right-hand side.

$$\int e^{ax}\, dx = \frac{e^{ax}}{a}$$

$$\int \sin ax\, dx = -\frac{\cos ax}{a}$$

$$\int \cos ax\, dx = \frac{\sin ax}{a}$$

$$\int \tan ax\, dx = -\frac{1}{a}\ln(\cos ax)$$

$$\int \sin^2 ax\, dx = \frac{x}{2} - \frac{\sin 2ax}{4a}$$

$$\int \cos^2 ax\, dx = \frac{x}{2} + \frac{\sin 2ax}{4a}$$

$$\int x\sin ax\, dx = \frac{1}{a^2}\sin ax - \frac{1}{a}x\cos ax$$

$$\int x\cos ax\, dx = \frac{1}{a^2}\cos ax + \frac{1}{a}x\sin ax$$

$$\int \frac{dx}{\sqrt{a^2 - x^2}} = \sin^{-1}\left(\frac{x}{a}\right)$$

$$\int \frac{dx}{\sqrt{x^2 \pm a^2}} = \ln(x + \sqrt{x^2 \pm a^2})$$

$$\int \frac{dx}{x^2 + a^2} = \frac{1}{a}\tan^{-1}\left(\frac{x}{a}\right)$$

$$\int \frac{x\, dx}{\sqrt{a^2 - x^2}} = -\sqrt{a^2 - x^2}$$

$$\int \frac{x\, dx}{\sqrt{x^2 \pm a^2}} = \sqrt{x^2 \pm a^2}$$

$$\int \frac{dx}{(x^2 \pm a^2)^{3/2}} = \frac{\pm x}{a^2\sqrt{x^2 \pm a^2}}$$

$$\int xe^{ax}\, dx = \frac{e^{ax}}{a^2}(ax - 1)$$

$$\int x^2 e^{ax}\, dx = \frac{x^2 e^{ax}}{a} - \frac{2}{a}\left[\frac{e^{ax}}{a^2}(ax - 1)\right]$$

$$\int \frac{dx}{a + bx} = \frac{1}{b}\ln(a + bx)$$

$$\int \frac{dx}{(a + bx)^2} = -\frac{1}{b(a + bx)}$$

$$\int \ln ax\, dx = x\ln ax - x$$

THE INTERNATIONAL SYSTEM OF UNITS (SI)

This material is from the United States edition of the English translation of the sixth edition of "Le Système International d'Unités (SI)," the definitive publication in the French language issued in 1991 by the International Bureau of Weights and Measures (BIPM). The year the definition was adopted is given in parentheses.

unit of length (meter): The meter is the length of the path traveled by light in vacuum during a time interval of 1/299 792 458 of a second. (1983)

unit of mass (kilogram): The kilogram is the unit of mass; it is equal to the mass of the international prototype of the kilogram. (1889)

unit of time (second): The second is the duration of 9 192 631 770 periods of the radiation corresponding to the transition between the two hyperfine levels of the ground state of the cesium-133 atom. (1967)

unit of electric current (ampere): The ampere is that constant current which, if maintained in two straight parallel conductors of infinite length, of negligible circular cross section, and placed 1 meter apart in vacuum, would produce between these conductors a force equal to 2×10^{-7} newton per meter of length. (1948)

unit of thermodynamic temperature (kelvin): The kelvin, unit of thermodynamic temperature, is the fraction 1/273.16 of the thermodynamic temperature of the triple point of water. (1957) Also, the unit kelvin and its symbol K should be used to express an interval or a difference of temperature.

unit of amount of substance (mole): (1) The mole is the amount of substance of a system that contains as many elementary entities as there are atoms in 0.012 kilogram of carbon 12. (1971) (2) When the mole is used, the elementary entities must be specified and may be atoms, molecules, ions, electrons, other particles, or specified groups of such particles.

unit of luminous intensity (candela): The candela is the luminous intensity, in a given direction, of a source that emits monochromatic radiation of frequency 540×10^{12} hertz and that has a radiant intensity in that direction of (1/683) watt per steradian. (1979)

▲   SI BASE AND SUPPLEMENTARY UNITS

QUANTITY	SI UNIT	
	NAME	**SYMBOL**
Base Unit		
Length	meter	m
Mass	kilogram	kg
Time	second	s
Electric current	ampere	A
Thermodynamic temperature	kelvin	mol
Amount of substance	mole	mol
Luminous intensity	candela	cd
Supplementary Units		
Plane angle	radian	rad
Solid angle	steradian	sr

▲   SI PREFIXES

FACTOR	PREFIX	SYMBOL
10^{24}	yetta	Y
10^{21}	zetta	Z
10^{18}	exa	E
10^{15}	peta	P
10^{12}	tera	T
10^{9}	giga	G
10^{6}	mega	M
10^{3}	kilo	k
10^{2}	hecto	h
10^{1}	deka	da
10^{0}	—	—
10^{-1}	deci	d
10^{-2}	centi	c
10^{-3}	milli	m
10^{-6}	micro	μ
10^{-9}	nano	n
10^{-12}	pico	p
10^{-15}	femto	f
10^{-18}	atto	a
10^{-21}	zepto	z
10^{-24}	yocto	y

▲   SOME SI DERIVED UNITS WITH SPECIAL NAMES

QUANTITY	SI UNIT			
	NAME	**SYMBOL**	**EXPRESSION IN TERMS OF OTHER UNITS**	**EXPRESSION IN TERMS OF SI BASE UNITS**
Frequency	hertz	Hz		s^{-1}
Force	newton	N		$m \cdot kg \cdot s^{-2}$
Pressure, stress	pascal	Pa	N/m^2	$m^{-1} \cdot kg \cdot s^{-2}$
Energy, work, heat	joule	J	$N \cdot m$	$m^2 \cdot kg \cdot s^{-2}$
Power	watt	W	J/s	$m^2 \cdot kg \cdot s^{-3}$
Electric charge	coulomb	C		$s \cdot A$
Electric potential, potential difference, electromotive force	volt	V	J/C	$m^2 \cdot kg \cdot s^{-3} \cdot A^{-1}$
Capacitance	farad	F	C/V	$m^{-2} \cdot kg^{-1} \cdot s^4 \cdot A^2$
Electric resistance	ohm	Ω	V/A	$m^2 \cdot kg \cdot s^{-3} \cdot A^{-2}$
Magnetic flux	weber	Wb	$V \cdot s$	$m^2 \cdot kg \cdot s^{-2} \cdot A^{-1}$
Magnetic field	tesla	T	Wb/m^2	$kg \cdot s^{-2} \cdot A^{-1}$
Inductance	henry	H	Wb/A	$m^2 \cdot kg \cdot s^{-2} \cdot A^{-2}$
Radioactivity	becquerel	Bq	1 decay/s	s^{-1}
Absorbed radiation dose	gray	Gy	J/kg, 100 rad	$m^2 \cdot s^{-2}$
Radiation dose equivalent	sievert	Sv	J/kg, 100 rem	$m^2 \cdot s^{-2}$

APPENDIX C

CONVERSION FACTORS

The listings below give the SI equivalents of non-SI units. To convert from the units shown to SI, multiply by the factor given; to convert the other way, divide. For conversions within the SI system see table of SI prefixes in Appendix B, Chapter 1, or inside front cover. Conversions that are not exact by definition are given to, at most, 4 significant figures.

Length

1 inch (in.) = 0.0254 m
1 foot (ft) = 0.3048 m
1 yard (yd) = 0.9144 m
1 mile (mi) = 1609 m
1 nautical mile = 1852 m

1 angstrom (Å) = 10^{-10} m
1 light-year (ly) = 9.46×10^{15} m
1 astronomical unit (AU) = 1.5×10^{11} m
1 parsec = 3.09×10^{16} m
1 fermi = 10^{-15} m = 1 fm

Mass

1 slug = 14.59 kg
1 metric ton (tonne; T) = 1000 kg

1 unified mass unit (u) = 1.660×10^{-27} kg

Force units in the English system are sometimes used (incorrectly) for mass. The units given below are actually equal to the number of kilograms multiplied by g, the acceleration of gravity.

1 pound (lb) = weight of 0.454 kg
1 ton = 2000 lb = weight of 908 kg

1 ounce (oz) = weight of 0.02835 kg

Time

1 minute (min) = 60 s
1 hour (h) = 60 min = 3600 s

1 day (d) = 24 h = 86 400 s
1 year (y) = 365.2422 d* = 3.156×10^{7} s

---

* The length of the year changes very slowly with changes in Earth's orbital period.

Area

1 hectare (ha) $= 10^4$ m^2

1 square inch (in.^{2}) $= 6.452 \times 10^{-4}$ m^2

1 square foot (ft^2) $= 9.290 \times 10^{-2}$ m^2

1 acre $= 4047$ m^2

1 barn $= 10^{-28}$ m^2

1 shed $= 10^{-30}$ m^2

Volume

1 liter (L) $= 1000$ cm$^3 = 10^{-3}$ m^3

1 cubic foot (ft^3) $= 2.832 \times 10^{-2}$ m^3

1 cubic inch (in.^{3}) $= 1.639 \times 10^{-5}$ m^3

1 fluid ounce $= 1/128$ gal $= 2.957 \times 10^{-5}$ m^3

1 barrel $= 42$ gal $= 0.1590$ m^3

1 gallon (U.S.; gal) $= 3.785 \times 10^{-3}$ m^3

1 gallon (British) $= 4.546 \times 10^{-3}$ m^3

Angle, Phase

1 degree (°) $= \pi/180$ rad $= 1.745 \times 10^{-2}$ rad

1 revolution (rev) $= 360° = 2\pi$ rad

1 cycle $= 360° = 2\pi$ rad

Speed, Velocity

1 km/h $= (1/3.6)$ m/s $= 0.2778$ m/s

1 mi/h (mph) $= 0.4470$ m/s

1 ft/s $= 0.3048$ m/s

1 ly/y $= 3.00 \times 10^8$ m/s

Angular Speed, Angular Velocity, Frequency, and Angular Frequency

1 rev/s $= 2\pi$ rad/s $= 6.283$ rad/s (s^{-1})

1 Hz $= 1$ cycle/s $= 2\pi$s^{-1}

1 rev/min (rpm) $= 0.1047$ rad/s (s^{-1})

Force

1 dyne $= 10^{-5}$ N

1 pound (lb) $= 4.448$ N

Pressure

1 dyne/cm$^2 = 0.10$ Pa

1 atmosphere (atm) $= 1.013 \times 10^5$ Pa

1 torr $= 1$ mm Hg at 0°C $= 133.3$ Pa

1 lb/in.^{2} (psi) $= 6.895 \times 10^3$ Pa

1 in. H$_2$O (60°F) $= 248.8$ Pa

1 in. Hg (60°F) $= 3.377 \times 10^3$ Pa

Energy, Work, Heat

1 erg $= 10^{-7}$ J

1 calorie* (cal) $= 4.184$ J

1 electron-volt (eV) $= 1.602 \times 10^{-19}$ J

1 foot-pound (ft·lb) $= 1.356$ J

1 Btu* $= 1.054 \times 10^3$ J

1 kWh $= 3.6 \times 10^6$ J

1 megaton (explosive yield; Mt)
  $= 4.18 \times 10^{15}$ J

---

* Values based on the thermochemical calorie; other definitions vary slightly.

Power

$1 \text{ erg/s} = 10^{-7} \text{ W}$
$1 \text{ horsepower (hp)} = 746 \text{ W}$

$1 \text{ Btu/h (Btuh)} = 0.293 \text{ W}$
$1 \text{ ft·lb/s} = 1.356 \text{ W}$

Magnetic Field

$1 \text{ gauss (G)} = 10^{-4} \text{ T}$

$1 \text{ gamma } (\gamma) = 10^{-9} \text{ T}$

Radiation

$1 \text{ curie (ci)} = 3.7 \times 10^{10} \text{ Bq}$

$1 \text{ rad} = 10^{-2} \text{ Gy}$
$1 \text{ rem} = 10^{-2} \text{ Sv}$

▲ ENERGY CONTENT OF FUELS

ENERGY SOURCE	ENERGY CONTENT
Coal	2.9×10^7 J/kg $= 7300$ kWh/ton $= 25 \times 10^6$ Btu/ton
Oil	43×10^6 J/kg $= 39$ kWh/gal $= 1.3 \times 10^5$ Btu/gal
Gasoline	44×10^6 J/kg $= 36$ kWh/gal $= 1.2 \times 10^5$ Btu/gal
Natural gas	55×10^6 J/kg $= 30$ kWh/100 ft^3 $= 1000$ Btu/ft^3
Uranium (fission)	
Normal abundance	5.8×10^{11} J/kg $= 1.6 \times 10^5$ kWh/kg
Pure U-235	8.2×10^{13} J/kg $= 2.3 \times 10^7$ kWh/kg
Hydrogen (fusion)	
Normal abundance	7×10^{11} J/kg $= 3.0 \times 10^4$ kWh/kg
Pure deuterium	3.3×10^{14} J/kg $= 9.2 \times 10^7$ kWh/kg
Water	1.2×10^{10} J/kg $= 1.3 \times 10^4$ kWh/gal $= 340$ gal gasoline/gal
H_2O	
100% conversion, matter to energy	9.0×10^{16} J/kg $= 931$ MeV/u $= 2.5 \times 10^{10}$ kWh/kg

APPENDIX D

THE ELEMENTS

The atomic weights of stable elements reflect the abundances of different isotopes; values given here apply to elements as they exist naturally on Earth. For stable elements, parentheses express uncertainties in the last decimal place given. For elements with no stable isotopes (indicated in bold-face), sets of most important isotopes are given. (Exceptions are the unstable elements thorium, protactinium, and uranium, for which atomic weights reflect natural abundances of long-lived isotopes.) See also periodic table inside back cover.

ATOMIC NUMBER	NAMES	SYMBOL	ATOMIC WEIGHT
1	Hydrogen	H	1.00794 (7)
2	Helium	He	4.002602 (2)
3	Lithium	Li	6.941 (2)
4	Beryllium	Be	9.012182 (3)
5	Boron	B	10.811 (5)
6	Carbon	C	12.011 (1)
7	Nitrogen	N	14.00674 (7)
8	Oxygen	O	15.9994 (3)
9	Fluorine	F	18.9984032 (9)
10	Neon	Ne	20.1797 (6)
11	Sodium (Natrium)	Na	22.989768 (6)
12	Magnesium	Mg	24.3050 (6)
13	Aluminum	Al	26.981539 (5)
14	Silicon	Si	28.0855 (3)
15	Phosphorus	P	30.973762 (4)
16	Sulfur	S	32.066 (6)
17	Chlorine	Cl	35.4527 (9)
18	Argon	Ar	39.948 (1)
19	Potassium (Kalium)	K	39.0983 (1)
20	Calcium	Ca	40.078 (4)
21	Scandium	Sc	44.955910 (9)
22	Titanium	Ti	47.88 (3)
23	Vanadium	V	50.9415 (1)
24	Chromium	Cr	51.9961 (6)
25	Manganese	Mn	54.93805 (1)
26	Iron	Fe	55.847 (3)
27	Cobalt	Co	58.93320 (1)

ATOMIC NUMBER	NAMES	SYMBOL	ATOMIC WEIGHT
28	Nickel	Ni	58.69 (1)
29	Copper	Cu	63.546 (3)
30	Zinc	Zn	65.39 (2)
31	Gallium	Ga	69.723 (1)
32	Germanium	Ge	72.61 (2)
33	Arsenic	As	74.92159 (2)
34	Selenium	Se	78.96 (3)
35	Bromine	Br	79.904 (1)
36	Krypton	Kr	83.80 (1)
37	Rubidium	Rb	85.4678 (3)
38	Strontium	Sr	87.62 (1)
39	Yttrium	Y	88.90585 (2)
40	Zirconium	Zr	91.224 (2)
41	Niobium	Nb	92.90638 (2)
42	Molybdenum	Mo	95.94 (1)
43	**Technetium**	**Tc**	**97, 98, 99**
44	Ruthenium	Ru	101.07 (2)
45	Rhodium	Rh	102.90550 (3)
46	Palladium	Pd	106.42 (1)
47	Silver	Ag	107.8682 (2)
48	Cadmium	Cd	112.411 (8)
49	Indium	In	114.82 (1)
50	Tin	Sn	118.710 (7)
51	Antimony (Stibium)	Sb	121.75 (3)
52	Tellurium	Te	127.60 (3)
53	Iodine	I	126.90447 (3)
54	Xenon	Xe	131.29 (2)
55	Cesium	Cs	132.90543 (5)
56	Barium	Ba	137.327 (7)
57	Lanthanum	La	138.9055 (2)
58	Cerium	Ce	140.115 (4)
59	Praseodymium	Pr	140.90765 (3)
60	Neodymium	Nd	144.24 (3)
61	**Promethium**	**Pm**	**145, 147**
62	Samarium	Sm	150.36 (3)
63	Europium	Eu	151.965 (9)
64	Gadolinium	Gd	157.25 (3)
65	Terbium	Tb	158.92534 (3)
66	Dysprosium	Dy	162.50 (3)
67	Holmium	Ho	164.93032 (3)
68	Erbium	Er	167.26 (3)
69	Thulium	Tm	168.93421 (3)
70	Ytterbium	Yb	173.04 (3)
71	Lutetium	Lu	174.967 (1)
72	Hafnium	Hf	178.49 (2)
73	Tantalum	Ta	180.9479 (1)
74	Tungsten (Wolfram)	W	183.85 (3)
75	Rhenium	Re	186.207 (1)
76	Osmium	Os	190.2 (1)
77	Iridium	Ir	192.22 (3)
78	Platinum	Pt	195.08 (3)
79	Gold	Au	196.96654 (3)
80	Mercury	Hg	200.59 (3)
81	Thallium	Tl	204.3833 (2)
82	Lead	Pb	207.2 (1)
83	Bismuth	Bi	208.98037 (3)

ATOMIC NUMBER	NAMES	SYMBOL	ATOMIC WEIGHT
84	Polonium	Po	209, 210
85	Astatine	At	210, 211
86	Radon	Rn	211, 220, 222
87	Francium	Fr	223
88	Radium	Ra	223, 224, 226, 228
89	Actinium	Ac	227
90	Thorium	Th	232.0381 (1)
91	Protactinium	Pa	231.03588 (2)
92	Uranium	U	238.0289 (1)
93	Neptunium	Np	237, 239
94	Plutonium	Pu	238, 239, 240, 241, 242, 244
95	Americium	Am	241, 243
96	Curium	Cm	243, 244, 245, 246, 247, 248
97	Berkelium	Bk	247, 249
98	Californium	Cf	249, 250, 251, 252
99	Einsteinium	Es	252
100	Fermium	Fm	257
101	Mendelevium	Md	255, 256, 258, 260
102	Nobelium	No	253, 254, 255, 259
103	Lawrencium	Lr	256, 258, 259, 261
104	Rutherfordium	Rf	257, 259, 260, 261
105	Hahnium	Ha	260, 261, 262
106	Seaborgium	Sg	259, 260, 261, 263
107	Nielsbohrium	Ns	261, 262
108	Hassium	Hs	264, 265
109	Meitnerium	Mt	266

APPENDIX E

ASTROPHYSICAL DATA

SUN, PLANETS, PRINCIPAL SATELLITES

BODY	MASS (10^{24} kg)	MEAN RADIUS (10^6 m EXCEPT AS NOTED)	SURFACE GRAVITY (m/s^2)	ESCAPE SPEED (km/s)	SIDEREAL ROTATION PERIOD* (days)	MEAN DISTANCE FROM CENTRAL BODY† (10^6 km)	ORBITAL PERIOD	ORBITAL SPEED (km/s)
Sun	1.99×10^6	696	274	618	36 at poles 27 at equator	2.6×10^{11}	200 My	250
Mercury	0.330	2.44	3.70	4.25	58.6	57.6	88.0 d	48
Venus	4.87	6.05	8.87	10.4	−243	108	225 d	35
Earth	5.97	6.37	9.81	11.2	0.997	150	365.3 d	30
Moon	0.0735	1.74	1.62	2.38	27.3	0.385	27.3 d	1.0
Mars	0.642	3.38	3.74	5.03	1.03	228	1.88 y	24.1
Phobos	9.6×10^{-9}	9-13 km	0.001	0.008	0.32	9.4×10^{-3}	0.32 d	2.1
Deimos	2×10^{-9}	5-8 km	0.001	0.005	1.3	23×10^{-3}	1.3 d	1.3
Jupiter	1.90×10^3	69.1	26.5	60.6	0.414	778	11.9 y	13.0
Io	0.0889	1.82	1.8	2.6	1.77	0.422	1.77 d	17
Europa	0.478	1.57	1.3	2.0	3.55	0.671	3.55 d	14
Ganymede	0.148	2.63	1.4	2.7	7.15	1.07	7.15 d	11
Callisto	0.107	2.40	1.2	2.4	16.7	1.88	16.7 d	8.2
and 13 smaller satellites								
Saturn	569	56.8	11.8	36.6	0.438	1.43×10^3	29.5 y	9.65
Tethys	0.0007	0.53	0.2	0.4	1.89	0.294	1.89 d	11.3
Dione	0.00015	0.56	0.3	0.6	2.74	0.377	2.74 d	10.0
Rhea	0.0025	0.77	0.3	0.5	4.52	0.527	4.52 d	8.5
Titan	0.135	2.58	1.4	2.6	15.9	1.22	15.9 d	5.6
Iapetus	0.0019	0.73	0.2	0.6	79.3	3.56	79.3 d	3.3
and 12 smaller satellites								
Uranus	86.6	25.0	9.23	21.5	−0.65	2.87×10^3	84.1 y	6.79
Ariel	0.0013	0.58	0.3	0.4	2.52	0.19	2.52 d	5.5
Umbriel	0.0013	0.59	0.3	0.4	4.14	0.27	4.14 d	4.7
Titania	0.0018	0.81	0.2	0.5	8.70	0.44	8.70 d	3.7
Oberon	0.0017	0.78	0.2	0.5	13.5	0.58	13.5 d	3.1
and 11 smaller satellites								
Neptune	103	24.0	11.9	23.9	0.768	4.50×10^3	165 y	5.43
Triton	0.134	1.9	2.5	3.1	5.88	0.354	5.88 d	4.4
and 7 smaller satellites								
Pluto	0.015	1.2	0.4	1.2	−6.39	5.91×10^3	249 y	4.7
Charon	0.001	0.6			−6.39	0.02	6.39 d	0.2

*Negative rotation period indicates retrograde motion, in opposite sense from orbital motion. Periods are sidereal, meaning the time for the body to return to the same orientation relative to the distant stars rather than the Sun.

†Central body is galactic center for Sun, Sun for planets, and planet for satellites.

ANSWERS TO ODD-NUMBERED PROBLEMS

CHAPTER 1

3. 100,000 times bigger
5. 0.108 782 775 7 ns
7. 10^8
9. 0.79 rad
11. 28 g
13. 10^6
15. 7%
17. Yes, by 7 mi/h
19. 30 AU
21. (a) 0.0032 mpg; (b) 0.0014 km/L
23. Approximately 0.0175 rad
25. L/T^2
27. (c) Actually, the speed is given by $v = \sqrt{\lambda g/2\pi}$
29. 2.5×10^6 m
31. 7.4×10^6 m/s^2
33. (a) 2.5×10^{-4} mm^2; (b) 1.6×10^{-2} mm
35. 280 K
37. 41 m
39. 7
43. $\sim 1.3 \times 10^6$
45. 3600 (this assumes a 12-hour watch)
47. (a) $\sim 1.4 \times 10^{18}$ m^3; (b) $\sim 1.3 \times 10^{12}$
49. ~ 10 times as much in stars
51. 1
53. $M \cdot L/T^2$
55. $\sim 10^4$
57. $d_{Sun}/d_{moon} = 380$; $R_{Sun} \sim 7 \times 10^5$ km
   (a) 4 μm; (b) 7500
59. (a) $\sim 10^{28}$; (b) $\sim 10^{14}$
63. $\sqrt{F_0/\mu}$; this is in fact the equation for the wave speed

CHAPTER 2

1. 10.16 m/s
3. 5.431 m/s
5. (a) 24 km north; (b) 9.6 km/h; (c) 16 km/h; (d) 0; (e) 0
7. 26.7 km/h = 7.42 m/s
9. 48 mi/h
11. 1 m/s = 2.24 mi/h
13. 51 ft/s = 35 mi/h
15. (a) 2 d, 17 h; (b) 70 km/h
17. 2.6 h later; 1800 km from New York
21. (a) 2.0 m/s; (b) 0; (c) -5.0 m/s; (d) 1.2 m/s; (e) 0.17 m/s
23. (a) $b - 2ct$; (b) at 6.9 s after launch
25. (a) $t = 0$ s, 0.13 s, 2.5 s; (b) $v = 3bt^2 - 2ct + d$;
   (c) $v_0 = 1.0$ m/s; (d) $t = 0.065$ s, 1.7 s
27. 0
29. 11.9 m/s^2
31. 31 s
33. (a) 126 m/s; (b) 0.46 m/s^2
35. $a = 6bt - 2c$
37. 100 m by both methods
41. (a) 46 m/s^2; (b) 61 s
43. (a) 2.0 m/s^2; (b) 150 m
45. 27 ft/s^2
47. (a) $t = 2v_0/a$; (b) $v = v_0$
49. 22 m/s
51. (a) 0.42 m/s^2; (b) toward Chicago; (c) 1.1 km
53. $a = 125g = 1200$ m/s^2
55. Yes, $a = 370$ m/s^2
57. No collision; 10 m apart
59. 4.6×10^{-3} m/s^2
61. Collide at 12 km/h
63. 11.3 m/s
65. (a) 27 m; (b) 4.7 s
67. Venus
69. 273 m
71. 2.0 m/s
73. 2.4 s

75. (a) 25 km/h; (b) 13 km/h

77. 3.6 s; 8.3 m/s

79. 0.196 s

81. 5.0 s; 17 km/h

83. (a) 7.0 m/s; (b) in 2.3 s

85. (a) 10.45 m/s; (b) 8.98 m/s; (c) 8.88 m/s; (d) 1.83 m/s

87. (a) $\frac{1}{2}(v_1 + v_2)$; (b) $\dfrac{2v_1 v_2}{v_1 + v_2}$

89. (b) 3.8 s; (c) 19 m; (d) 100 m

91. (a) $v = \omega x_0 \cos \omega t$, $a = -\omega^2 x_0 \sin \omega t$; (b) $v_{max} = \omega x_0$, $a_{max} = \omega^2 x_0$

93. (a) 33 m; (b) 8%

95. 1.19 m

CHAPTER 3

1. 260 m, 7.9° N of W

3. 702 km, 21.3° east of north

5. 18.9 units long, 7.1° E of S

7. (a) 47.8 cm; (b) 30.4 cm

9. 409 km, 79.6° west of north

11. 130°

13. 1.24A at 234° to the horizontal

15. $-1.5, 2.5$

17. $\mathbf{A} + \mathbf{B} = 4.8\hat{\mathbf{i}} + 0.82\hat{\mathbf{j}}$; $\mathbf{A} - \mathbf{B} = 12\hat{\mathbf{i}} + 11\hat{\mathbf{j}}$;
    $\mathbf{A} + \mathbf{C} = 4.8\hat{\mathbf{i}} + 13\hat{\mathbf{j}}$; $\mathbf{A} + \mathbf{B} + \mathbf{C} = 1.4\hat{\mathbf{i}} + 8.1\hat{\mathbf{j}}$

19. (a) $A_x = 5.9$, $A_y = 8.1$; $A_x' = 5.4$, $A_y' = 8.4$ units

21. (a) $A_x = 8.7$, $A_y = 5.0$; (b) $A_x = 9.7$, $A_y = -2.6$;
    (c) $A_x = 10$, $A_y = 0$

23. $\mathbf{C} = -15\hat{\mathbf{i}} + 9\hat{\mathbf{j}} - 18\hat{\mathbf{k}}$

25. (a) in x-y system $\Delta\mathbf{r} = 5.0\hat{\mathbf{i}} + 3.5\hat{\mathbf{j}}$ km; in x'-y' system
    $\Delta\mathbf{r} = 6.1\hat{\mathbf{i}} + 0.50\hat{\mathbf{j}}$ km; (b) $\Delta r = 6.1$ km

27. (a) 5.4 mi at 32° E of N; (b) 15 mi/h at 32° E of N

29. (a) 264 km/h, 29° west of north; (b) $-128\hat{\mathbf{i}} + 231\hat{\mathbf{j}}$ km/h

31. $19\hat{\mathbf{i}} + 4.5\hat{\mathbf{j}} + 0.26\hat{\mathbf{k}}$ km/h

33. 5.12 m/s², 41° south of west

35. (a) 3.8×10^{-3} m/s²; (b) 5.4×10^{-3} km/s²; (c) 45°

37. $-4.9\hat{\mathbf{i}} - 2.8\hat{\mathbf{j}}$ m/s²

39. $\mathbf{v} = (3bt^2 + c)\hat{\mathbf{i}} + 2dt\hat{\mathbf{j}} + e\hat{\mathbf{k}}$
    $\mathbf{a} = 6bt\hat{\mathbf{i}} + 2d\hat{\mathbf{j}}$

41. (a) 18 km/h, 14 km/h, 10 km/h, 14 km/h;
    (b) 21 km/h, 17 km/h, 13 km/h, 17 km/h

43. $30\hat{\mathbf{i}} + 64\hat{\mathbf{j}}$ km/s

45. (a) 1.6 m, 2.8 m/s, both vertically downward; (b) 2.03 m, 3.57 m/s, both at 38.1° to the vertical; (c) 9.8 m/s², vertically downward in both frames of reference

47. 16.2 m

49. (a) $\mathbf{A} = 9.06\hat{\mathbf{i}} + 4.23\hat{\mathbf{j}}$ m; $\mathbf{B} = 10.0\hat{\mathbf{j}}$ m;
    $\mathbf{C} = -7.55\hat{\mathbf{i}} + 6.56\hat{\mathbf{j}}$ m; (b) $\mathbf{D} = -1.52\hat{\mathbf{i}} - 20.8\hat{\mathbf{j}}$ m;
    (c) $D = 20.9$ m

51. (a) 1.8 km/h/s = 0.50 m/s²; (b) 142°

53. (a) 0.32 cm/s; (b) 0.034 cm/s²; (c) 90°

55. 25° upstream

57. (a) 1.30×10^6 m; (b) 7.78 km/s; (c) 9.19 m/s², almost g

59. $-b\hat{\mathbf{i}} + a\hat{\mathbf{j}}$, $b\hat{\mathbf{i}} - a\hat{\mathbf{j}}$

61. $\frac{\sqrt{2}}{2}\hat{\mathbf{i}} + \frac{\sqrt{2}}{2}\hat{\mathbf{j}}$

65. 53.8°, 13.9 km/h

CHAPTER 4

1. 0°

3. $1.3\hat{\mathbf{i}} + 2.3\hat{\mathbf{j}}$ m/s

5. 4.49 m/s² at 58° below the x axis

7. (a) 23°; (b) 5.4×10^3 km

9. 1.09 m

11. (a) $t = 18$ s; (b) 300 m; (c) 22 m/s, at 120° to x axis

13. (a) 2.6×10^{17} cm/s², upward; (b) parabolic

15. (a) 1.4 s; (b) 10 m

17. 5.7 m/s

19. 1.27 m/s

21. 34 nm

23. 8.3 m/s at 61°

25. Yes

27. -14.6 m/s

29. (a) 6.64 km/s; (b) 16.0 min; (c) 8.28 km/s

31. 1.1 s

33. 1090 m

37. (a) 8.8 m; (b) 0.53 m

39. 11.2 m/s

43. 31.2° or 65.7°

45. 2.8×10^{-3} m/s²

47. 54 min

49. 20 cm

51. 0.344 ns

53. 14 s

55. (a) 148 m/s; (b) 0.974 m/s²

57. $t = \sqrt{r/a_t}$

59. 89 m/s

61. 32 m

63. 19 m

65. 300 m, 119 m

67. $\mathbf{v}_0 = 6.36\hat{\mathbf{i}} + 10.3\hat{\mathbf{j}}$ m/s, or 12.1 m/s at 58.3°

69. 83°

75. 7.2 m/s at 77° to horizontal

77. (a) $\tan\theta_0 = \dfrac{v_0^2}{gx} \pm \sqrt{\dfrac{v_0^4}{g^2 x^2} - 1}$;
    (b) taking $x \simeq 220$ m from the graph gives angles of 29.8° and 60.2°

79. 38°

81. v_x^2/g

83. 892 m/s²

87. (a) $\tan^{-1}\left(\tan\theta_0 - \dfrac{gt}{v_0 \cos\theta_0}\right)$;
    (b) $\tan^{-1}\left(\tan\theta_0 - \dfrac{gx}{v_0^2 \cos^2\theta_0}\right)$

CHAPTER 5

1. 3.8 MN

3. 1.5×10^3 kN

5. 10^{-4} N
7. 9.0×10^{22} m/s^2
9. (a) $m_B = 3.0 m_A$; (b) 2.0 m/s^2
11. Quadruples
13. (a) 11 m; (b) 24 m; (c) 43 m; (d) 53 m
15. $F_{driver} = 5.7$ kN; $F_{passenger} = 125$ kN, 22 times greater
17. 5.77 N at 72.2° to the x axis
19. Venus
21. (a) 3.3 N; (b) 12 oz
23. 9.1×10^3 kg
25. 8.3×10^{12} m/s^2
27. 6 m/s^2
29. 385 N
31. 2.9 m/s^2, downward
33. (a) 3.1×10^7 N; (b) 9.4×10^2 N
35. $a > g$
37. 0.53 s
39. 55 kN
41. 2.0 N
43. 1.3×10^{-21} cm
45. (a) 5.26 kN; (b) 1.08 kN; (c) 494 N; (d) 589 N
47. 130 N
49. 33 cm
51. 1.9 m/s^2
53. 830 g
55. (a) 132 cm; (b) 127 cm; (c) 120 cm; (d) 40 m/s^2
57. 30.7 kN
59. At least 2.9 cm
61. Apparent weight is 55% of actual weight
63. 240 N
65. 4.3 cm
67. 14 N
69. 7.2 m
71. (a) $0.40mg$; (b) $2.40mg$; (c) $1.40mg$
73. (a) 16 kN; (b) 1.5 kN
75. (a) $a = \dfrac{m_f - m_s}{m_s} g$;

(b) $y = \dfrac{m_f a_s h}{(m_f - m_s)(g + a_s)}$
77. $\ell + nm(a + g)/k$, where n is the spring number measured from the bottom
81. 900 N

CHAPTER 6

1. $2.7\hat{\imath} + 5.5\hat{\jmath}$ N, or 6.1 N at 64° to the x axis
3. (a) 2.0 kN; (b) 1.4 kN
7. 43 cm
9. 530 N; 3.6 times the weight
11. 98 N in horizontal string; $98\sqrt{2}$ N in vertical string
13. 230 N in short rope; 84 N in long rope
15. (a) 6.3 m/s^2; (b) 0.44 s
17. Right-hand mass 2.5 times left-hand mass
19. (a) 7.1 kg; (b) 3.9 kg
21. Left to right, 56.9 N, 34.4 N, 89.2 N

23. 8.18×10^{-8} N
27. (a) 13 m/s; (b) 4.4°
29. 132 m
31. (a) 310 N; (b) 0; (c) nothing
33. (a) 35°; (b) 22°
35. 37°
37. 17 m/s
39. 490 N
41. 45 kg
43. 0.18
45. 340 N; 1.6 times the weight
47. (a) 8.0°; (b) 0.50 m/s^2
49. 0.38 s
51. 4.1 m/s^2, accident deceleration is 21 times greater
53. 12 cm
55. (a) 1.6 m/s^2; (b) 3.3 N
57. 4.2 m/s^2
59. 95 km/h
61. 0.12
65. 0.72
67. 6.9 m/s^2
69. 3.45 rev/min
71. $\mu_s \geq 0.24$
73. (a) 10.0 N; (b) 10.4 N; (c) 10.4 N; (d) 90°, 73.5°, 90°
75. 0.70 m/s^2, upward
77. 7.75 m/s
79. 1.40 s
81. (a) 0.12; (b) toward the inside of the turn
83. 6.3 m/s

87. $\mu_k = \dfrac{v_0^2}{2gx_1} - \dfrac{x_2^2}{4x_1 h}$

93. 28 cm

CHAPTER 7

1. 1.3 kJ
3. 59 kJ
5. 9.6 MJ
7. (a) 400 J; (b) 31 kg
9. 5.9 MN
11. (a) 370 J; (b) 0.26
15. (a) $\hat{\imath} \cdot \hat{\imath} = \hat{\jmath} \cdot \hat{\jmath} = \hat{k} \cdot \hat{k} = 1$;
    (b) $\hat{\imath} \cdot \hat{\jmath} = \hat{\jmath} \cdot \hat{k} = \hat{k} \cdot \hat{\imath} = 0$
17. (a) 14; (b) -12; (c) -16
21. (a) 45°; (b) 111°; (c) 66°
23. (a) 630 J; (b) 0
25. 25°
27. (a) 60 kJ; (b) 20 kJ; (c) 80 kJ
29. (a) 360 J; (b) 350 J; (c) 357.5 J; (d) 359.375 J
31. $k_B = 8k_A$
33. 190 J
35. (a) 30 J; (b) 56 J; (c) 72 J
37. $F_0\left(x + \dfrac{x^2}{2\ell_0} + \dfrac{\ell_0^2}{\ell_0 + x} - \ell_0 \right)$
39. (a) 0; (b) $2FR$; (c) πFR

43. 90 J
45. (a) 14 GJ; (b) 3.3 MJ; (c) 28 J
47. $2.46{\times}10^7$ m/s $= 0.082c$
49. 2.3 kJ
51. 42 cm
53. (a) 24 J (b) 18 m/s
55. 4.1 m
57. (a) None
59. (a) 60 kW; (b) 1 kW; (c) 40 W
61. 9.4 MJ
63. (a) 36 MW; (b) 1.1 MW
67. 2.7 h
69. 300 million gal/day
71. 2.1 MJ
73. 7.7°
75. 1.6 m
77. $\frac{1}{2}F_0x_0$
79. 70.5°
81. (a) 28 kJ; (b) 18 kJ
83. (a) 450 W; (b) 8.0 kJ
85. (a) $W = \frac{1}{2}bt^2$; (b) $a = \sqrt{b/m}$
87. (a) 71 kW·y; (b) 93 kW·y
89. (a) $P = mgv\,\cos(vt/R)$
91. (a) 33 J; (b) 167 J
93. 6.3 J

CHAPTER 8

1. (a) $-2\mu mg\ell$; (b) $-\sqrt{2}\mu mg\ell$
3. (a) 0; (b) F_0a
5. (a) 170 MJ; (b) -7.6 MJ
9. (a) 1.07 J; (b) 1.12 J
11. 7.5 J
13. 54 cm
15. 0.53 J
17. $U = -\dfrac{1}{3}ax^3 - bx$
19. $U = F_0\left(x + \dfrac{x^2}{2\ell} + \dfrac{\ell^2}{\ell + x} - \ell\right)$
21. 50 m/s (180 km/h)
23. 96 m
27. 6.5 m/s
29. 26 MN/m
31. 15 km/h
35. 2.6 m/s
39. $x = \pm69$ cm
43. (a) $U = -\dfrac{a}{2}x^2 + \dfrac{b}{4}x^4$; (b) $x = 0.66$ m, x
    $= 2.1$ m
45. (a) -6.7 N; (b) 0; (c) 4.5 N
49. (a) 30 cm; (b) 10.4 N/m
51. 44%
53. 19%
55. 0.36
57. 2.6 m/s

59. 0.036
61. 62 cm from left end of frictional zone
63. 1000 MW, twice that of the coal plant
65. $x = 2(h_1 - h_2)$
69. (a) $v = \sqrt{\sqrt{2}g\ell}$; (b) $T = \dfrac{3}{\sqrt{2}}mg$
71. $x = \pm24$ cm
73. 14 m
75. (a) $2.53{\times}10^5$ m/s; (b) $2.91{\times}10^5$ m/s;
    (c) $2.93{\times}10^5$ m/s
77. 54.6 mJ
79. 75 cm
81. (a) 1.74 cm; (b) 0.78 cm; (c) $7.4{\times}10^7$ m/s
83. (a) $v = \left[\dfrac{x^2}{m}\left(a - \tfrac{1}{2}bx^2\right)\right]^{1/2}$; (c) $v_{\max} = a/\sqrt{2mb}$
85. $v = [2ax^{|b+1|}/m|b + 1|]^{1/2}$
87. (a) 11 m/s; (b) $x = \pm1.4$ m

CHAPTER 9

1. $R_E/\sqrt{2}$
3. 58%
5. 8.6 kg
7. 46 nN
9. 440 m
11. $1.2{\times}10^{-7}$
13. 3.1 km/s
15. 1.8 days
17. 1.0 hour
19. $2.6{\times}10^{41}$ kg, about 10^{11} solar masses
21. $6.3{\times}10^{10}$ m
23. $2.64{\times}10^{10}$ m
25. 2.47 times Earth's orbital radius
27. No
29. $3.2{\times}10^9$ J
31. 530 km
33. 58 MJ
35. $R_E/99 = 64$ km; underestimate
37. 3%
39. $\sqrt{2}$
41. $5.8{\times}10^6$ m
43. (a) 11.2 km/s; (b) 9.74 km/s; (c) no
47. $v = \sqrt{2GM\left(\dfrac{3}{R} + \dfrac{1}{r}\right)}$
49. $8.1{\times}10^{11}$ m, just beyond Jupiter
53. 109 min
55. 15 km/s, 23 km/s
57. 7.95 km/s
59. $8.85{\times}10^5$ m
61. (a) $9.0{\times}10^{10}$ m; (b) $5.3{\times}10^{11}$ J; (c) 38 km/s
63. $4.60{\times}10^{10}$ m
65. (b) 8.8 mm; (c) 2.9 km
67. 1400 km lower

CHAPTER 10

1. 0.75 m from the center
3. $X = 50$ cm; $Y = 69$ cm; with origin at lower left
5. $\ell/2\sqrt{3}$ along the perpendicular bisector of any side
7. $X = 44$ cm, $Y = 55$ cm, with origin at lower left
9. $0.115a$ above the vertex of the missing triangle
11. 6.5 pm from the oxygen
13. $\mathbf{R} = \left(t^2 + \dfrac{10}{3}t + \dfrac{7}{3}\right)\hat{\imath} + \left(\dfrac{2}{3}t + \dfrac{8}{3}\right)\hat{\jmath};$

$\mathbf{V} = \left(2t + \dfrac{10}{3}\right)\hat{\imath} + \dfrac{2}{3}\hat{\jmath}; \ \mathbf{A} = 2\hat{\imath}$

15. $\sim 10^{-10}$ m, the diameter of a hydrogen atom
17. $m_{\text{mouse}} = \frac{1}{4}m_{\text{bowl}}$
19. 3.0 m/s in the negative x direction
21. 21 g, $-x$ direction
23. 460 m/s
25. 8.4 km/h
27. (a) 0.14 N/m²; (b) 0.014 mm
29. (a) 0.99 m; (b) 3.9 m/s
31. 3.9 km/h
33. $26\hat{\imath} + 16\hat{\jmath}$ m/s
35. 1100 kg
37. (a) 3.3×10^6 N (b) 3.4×10^5 kg
39. 0.22
41. $K_{\text{cm}} = 1.1\times10^{-14}$ J before and after;
$K_{\text{int}} = 0$ before, $K_{\text{int}} = 1.3\times10^{-14}$ J after
43. before: $K_{\text{cm}} = 1.6$ MJ, $K_{\text{int}} = 21$ kJ; after: $K_{\text{cm}} = 1.6$ MJ, $K_{\text{int}} = 0$
45. (a) 1.7 cm above the bottom; (b) 2.7 cm above the bottom
47. 20 m
49. $-47\hat{\imath} - 68\hat{\jmath}$ m/s
51. (a) 3.5 m; (b) 1.3 m/s; (c) 0
53. (a) 0.096 m/s²; (b) 6.2 m/s
55. 9.3 m/s
57. (a) $m = \dfrac{\pi\rho h^2}{2a}$; (b) $Z = \dfrac{2}{3}h$
59. (a) 2×10^4 kg·m/s; (b) 2×10^8 J; (c) yes
61. (a) 37.7°; (b) 0.657 m/s
63. 5.8 s after explosion
65. (a) thrust $= m\dfrac{dv}{dt} = [(1 + f)V_{\text{ex}} - V]\dfrac{dM_{\text{in}}}{dt}$; (b) 1504 lb
67. $v_1 = \left(\dfrac{m_2 kx^2}{m_1^2 + m_1 m_2}\right)^{1/2}$; $v_2 = \left(\dfrac{m_1 kx^2}{m_2^2 + m_1 m_2}\right)^{1/2}$

CHAPTER 11

1. 95 N·s
3. 4.3×10^3 N, $7.1mg$
5. (a) $-2.6\hat{\imath} + 0.74\hat{\jmath}$ N·s; (b) $-51\hat{\imath} + 14\hat{\jmath}$ N
7. (a) 150 N·s, upward; (b) 3.0 kN, about 5 times your weight

9. (a) 7.3 MN·s; (b) 5.6 MN
11. (a) 6.8×10^{-3} N·s; (b) 2.3 N
13. $\Delta P/P = 2\%$
15. 12 ms
17. (a) 6.2 mi/h; (b) 12%
19. 19 kg
23. 10^{21} kg
25. $4.0\hat{\imath} + 21.5\hat{\jmath}$ Mm/s
29. 1.3 μJ
31. 120°
33. 46 m/s
35. $v_{1f} = -11$ Mm/s; $v_{2f} = 6.9$ Mm/s
37. $3 + 2\sqrt{2} \simeq 5.8$; it doesn't matter which is more massive
39. $v_A = -\frac{1}{3}v; \ v_B = \frac{2}{9}v; \ v_C = \frac{8}{9}v$
45. 22°
47. $(v_{1i} = 0.833$ m/s, $v_{2i} = 1.22$ m/s, $\theta_{2i} = 28.3°)$;
$(v_{1i} = 1.20$ m/s, $v_{2i} = 1.12$ m/s, $\theta_{2i} = 31.2°)$
49. 13 m/s at 27° to horizontal
51. $v_A = -\frac{1}{5}v_0\hat{\imath}; \ v_B = \frac{3}{5}v_0\hat{\imath} + \frac{1}{5}\sqrt{3}v_0\hat{\jmath} \ v_C = \frac{3}{5}v_0\hat{\imath} - \frac{1}{5}\sqrt{3}v_0\hat{\jmath}$
53. 350 N
55. 52 km/h at 33° north of east
57. $m_{\text{truck}} = 7.6m_{\text{car}}$
59. (a) $m_1 = 3m_2$; (b) $v_{2f} = 2v$
61. $v_{1f} = 1.66$ m/s; $v_{2f} = 0.703$ m/s; $\theta_{2f} = 67°$ clockwise from initial velocity of the first ball
63. (a) 12.0 m; (b) 15.4 m/s
65. 0.88
69. 44
71. $v_{1\text{kg}} = 4.0$ m/s; $v_{4\text{kg}} = 1.0$ m/s at 50° clockwise from the x axis
73. $v_{1200} = 2.2$ km/h, $v_{1800} = 18$ km/h
75. (a) $v_1 = 0.28v_0$, $v_2 = 0.48v_0$; (b) 3, $0.26v_0$, $0.31v_0$

CHAPTER 12

1. (a) 7.27×10^{-5} rad/s; (b) 1.75×10^{-3} rad/s;
(c) 1.45×10^{-4} rad/s; (d) 31.4 rad/s
3. (a) $v = (\pi/30)\omega r$; (b) $v = 2\pi\omega r$;
(c) $v = (\pi/180)\omega r$
5. (a) 66 rpm; (b) 3.7 s
7. (a) 21.7 rad/s, 207 rpm; (b) 34.7 rad/s, 331 rpm
9. (a) 0.068 rpm/s; (b) 7.1×10^{-3} rad/s²
11. (a) 12 min; (b) 2.2×10^4
13. 1.3 rad/s²
15. (a) 2.0 s; (b) 1.0 rev
17. 1.2 m
19. 0.079 N·m
21. 0.15 N·m
23. (a) 0.70 N·m, counterclockwise; (b) $\mathbf{F}_1$ and $\mathbf{F}_5$
25. 22 cm
27. (a) $\frac{2}{3}m\ell^2$; (b) $\frac{2}{3}m\ell^2$; (c) $\frac{4}{3}m\ell^2$
29. 45 kg·m²
33. (a) 9.7×10^{37} kg·m²; (b) 2.6×10^{19} N·m

35. (a) 1.29×10^{38} kg·m²; (b) 6.45×10^{33} N·m
39. (a) 430 min; (b) 1900 rev
41. 1900 N·m
43. 170 rpm
45. $m_{\text{pulley}} = 0.49$ kg; $m_1 = 0.41$ kg; $m_2 = 0.58$ kg
47. (a) 450 J; (b) 140 W
49. 0.089%
51. 12.2 rad/s
53. 7.0 m/s

55. $v = \sqrt{\dfrac{2gh}{\alpha + 1}}$

57. 17%
59. hollow
61. (a) 0.156; (b) 0.070 rad/s

63. $\dfrac{253}{512} MR^2 = 0.494 MR^2$

65. (a) 310 N; (b) 165 kg
67. (a) 3.5 m/s; (b) 24%

69. $\dfrac{27}{10} R$

71. $\omega = \sqrt{2A/I}$
73. $I = \frac{1}{2} Mb^2$
75. $2\sqrt{2g/3R}$
77. $\frac{1}{2} Mg\ell \cos \phi$

CHAPTER 13

1. 63 rad/s, west
3. 0.52 rad/s², −37°
5. 16.6 rad/s
7. (a) $-z$; (b) z; (c) in the x-y plane, 45° clockwise from the x axis
9. (a) $-12\hat{\mathbf{k}}$ N·m; (b) $36\hat{\mathbf{k}}$ N·m; (c) $12\hat{\mathbf{i}} - 36\hat{\mathbf{j}}$ N·m
11. $-17\hat{\mathbf{k}}$ N·m
13. Parallel to the x axis or 120° clockwise from x axis
17. $F_x = 1.33 F_y - 3.13$ N
19. 414 kg·m²/s
21. 7.6 rad/s
23. 0.017 kg·m²/s
25. 37 kg·m²/s
27. 2.7×10^5 kg·m²/s, out of page in Fig. 13-30
29. (a) $-4.2\hat{\mathbf{k}}$ kg·m²/s; (b) 0; (c) $6.4\hat{\mathbf{k}}$ kg·m²/s
31. (a) 0.17 rev/s; (b) 386 J
33. (a) 142 rpm; (b) 21%
35. 2.5 days
37. (a) 23.7 rpm; (b) 3.49 mJ

39. (a) $\dfrac{2M\omega_0}{2M + 3m}$; (b) $\dfrac{M\omega_0}{M + 6m}$; (c) same as (b)

41. (a) 0.537 rad/s; (b) 6.44 m/s; (c) 207 N
43. 6.0 cm

45. $I = \dfrac{mgd}{2\omega \Omega}$

47. 1.05×10^{-34} kg·m²/s $= \hbar$
49. $\tan^{-1}(\frac{1}{2}) = 26.6°$

51. 0.37 rev/s
53. 22 g

55. $\dfrac{I}{I + mR^2}$

57. (a) 1.61; (b) 2.22
61. Sun's rotation 2.8%; Jupiter's orbital motion 60%

63. $v = \left[\dfrac{8(m + M)g\ell}{m^2} \left(\dfrac{1}{4}m + \dfrac{1}{3}M \right) \right]^{1/2}$

65. Both wheels have stopped rotating about their axes, but the whole contraption is now rotating about the centerline between the two.

CHAPTER 14

1. (a) $\frac{1}{3}\ell F_3 - \frac{2}{3}\ell F_2 + \ell F_1 = 0$;
   (b) $\frac{1}{2}\ell F_1 - \frac{1}{6}\ell F_2 - \frac{1}{6}\ell F_3 + \frac{1}{2}\ell F_4 - \frac{1}{2}\ell F_5 = 0$
3. (b) $\boldsymbol{\tau}_{\text{origin}} = -7\hat{\mathbf{k}}$ N·m
5. (a) A vector of length $\sqrt{2}\,F$, oriented 45° clockwise from the negative y axis, applied at the point $x = 0$ m, $y = +1$ m (or anywhere on the line $y = (1 + x)$ m). (b) Not possible; the first two vectors sum to zero but produce a nonzero torque, so any other vector applied to balance torques will upset force balance.
7. Both sets have $-F_1 + F_2 \sin \phi + F_3 = 0$, $-F_2 \cos \phi + F_4 = 0$; torque equations are
   (a) $\ell_2 F_4 \cos \phi - \ell_2 F_3 \sin \phi - \ell_1 F_2 = 0$;
   (b) $(\ell_2 - \ell_1)F_2 - \ell_2 F_1 \sin \phi = 0$.
9. (a) $\tau_A = \frac{1}{2}\ell mg$; (b) $\tau_B = 0$; (c) $\tau_C = \frac{1}{2}\ell mg$
11. $m_2 = 0.384 m_1$
13. (a) 61 cm from left end; (b) 1.4 m from left end
15. 120 N
17. 11.7 kN
19. (a) 40 N·m; (b) 1300 N
21. Vertical forces both 73.5 N, downward, horizontal forces both 33.6 N, away from door jamb at top, toward jamb at bottom
23. 5.0 kN; tension
25. 0.87
27. 500 N
29. 50 kN
31. 6.05 kN
33. $\frac{1}{8}$
35. Maximum height of CM is at sphere center; lower for clown (b)
37. Two equilibria for $|a| > 2\sqrt{3}$; one metastable, other unstable
39. 1.2 m
41. 170 N
43. 74 kg
45. $\frac{1}{2}(\sqrt{2} - 1)mgs$

47. $\mu = \dfrac{\sin 2\theta}{3 + \cos 2\theta}$

49. $mg/2k$
51. 28°

53. Left scale 16.3 N; right scale 22.9 N
55. Tip
57. Slide
59. (a) $0.44mg$, at 12° to Earth's polar axis;
    (b) $0.036mgR_E$, out of the plane of Fig. 14-55

PART 1 CUMULATIVE PROBLEMS

1. 16.5 m from the post
3. $a = \dfrac{2g[(m_1 + m_2)\sin\theta - \mu m_1 \cos\theta]}{2m_1 + 3m_2}$
5. (a) $v = \frac{2}{7}\omega R$; (b) $\Delta x = \dfrac{2\omega^2 R^2}{49\mu g}$

CHAPTER 15

1. $T = 0.780$ s; $f = 1.28$ Hz
3. 11.5 fs
5. $A = 20$ cm, $\omega = \pi/2$ s^{-1}, $\phi = 0$;
   $A = 30$ cm, $\omega = 2.0$ s^{-1}, $\phi = -\pi/2$;
   $A = 40$ cm, $\omega = \pi/2$ s^{-1}, $\phi = \pi/4$
7. 63.3 kg
9. 1.7 kN/m
11. 0.69 s
13. (a) $\pi\sqrt{m/k}$; (b) $v_0\sqrt{m/k}$
17. (a) 1.0 cm; (b) 6.2 m; (c) 3.6 km
19. 0.11 N·m/rad
21. (a) $2\pi\sqrt{\ell/g}$; (b) $2\pi\sqrt{2\ell/3g}$; (c) $2\pi\sqrt{2\ell/g}$; (d) infinite
23. 0.34 s
25. $R = \sqrt{2\kappa/k}$
27. Within 1 μm
33. 5.0 g
35. $\omega^2 = \dfrac{k_1 k_2}{m(k_1 + k_2)}$
37. a and b are the amplitudes in the x and y directions, respectively
39. 400 J, or 1.4×10^{-3} of the total KE
41. $t = (0.14 + n)$ s, $t = (0.53 + n)$ s, n an integer;
    $x = \pm37$ cm
45. $\omega = \sqrt{2k/3M}$
47. 34
49. 77% at $0.90\,\omega$; 66% at 1.1ω
53. (a) 19 s^{-1}; (b) 0.33 s; (c) 92 m/s^2
55. (a) 6.5 cm; (b) 0.51 s
57. 1.64 s
59. 300 g
61. (a) $E_2 = \frac{1}{4}E_1$; (b) $a_{2\,\max} = \frac{1}{4}a_{1\,\max}$
63. 2.1 m/s^2
65. 0.44
67. $2\pi\sqrt{10ga/7}$
69. $f = 0.54$ Hz; $A = 22$ cm; $\phi = -0.11$ rad
71. $T = 2\pi\sqrt{R/g}$
73. $T = 2\pi\sqrt{mL/2F_0}$

CHAPTER 16

1. 3.4 s
3. 3.38 m
5. 1.81×10^8 m/s $= 0.604c$
7. (a) 400 nm; (b) 0.3 mm
9. 11 m
11. (a) 0.58 m^{-1}; (b) 1.53 s^{-1}
13. (a) 13.7 s^{-1}; (b) 0.393 cm^{-1};
    (c) $y = 2.5\cos(0.393x + 13.7t)$
15. (a) 25 cm; (b) 0.37 Hz; (c) 12 m; (d) 4.4 m/s
17. $y = \dfrac{2}{(x - 3t)^4 + 1}$
19. (a) 3.0 m; (b) 1.5 s; (c) 2.0 m/s; (d) $+x$
21. 250 m/s
23. (a) 7.6 N; (b) 1.7 m/s
25. 364 m/s
27. 94 N
29. 7.64 g/cm^3
31. 585 m/s
33. 9.9 W
35. 35 cm
37. $4\pi^2 A^2 F/\lambda$
39. 12 mW/m^2
41. (a) 9.1 kW/m^2; (b) 0.88 W/m^2
43. (a) 6.4 kW/m^2; (b) 4.9 W/m^2
45. 5.1 m
47. (a) 2 cm; (b) pulse 1 at $x = 0$, direction $+x$, pulse 2 at
    $x = 5$, direction $-x$; (c) $t = 2.5$ s
49. Every 6 s
51. 5.34 m
53. $\sqrt{gh}$
55. (a) 1.5 cm; (b) 63 cm; (c) 11 ms; (d) 56 m/s;
    (e) 18 W
57. $v = \sqrt{k\ell(\ell - \ell_0)/m}$
59. 10 m
61. Every 30 s
63. $u < 0.063v$
67. 5.2 km
69. 67 m

CHAPTER 17

3. $\lambda = 34$ cm; $T = 1.0$ ms; $\omega = 6.3\times10^3$ s^{-1};
   $k = 0.18$ cm^{-1}
5. 0.29 s
7. 0.14 kg/m^3
9. monatomic
11. 739 m/s
13. 190 m/s
15. 4.4 nm
17. (a) 3.8 mW/m^2; (b) 96 dB; (c) 1.8 N/m^2; (d) 1.6 μm
19. 1 kHz to 6.5 kHz
21. (a) 3.2 μW/m^2, 0.051 N/m^2; (b) 3.2×10^{-13} W/m^2;
    1.6×10^{-5} N/m^2

23. (a) 20 dB; (b) approximately 250 Hz
27. 6.3 m
29. 1
31. 3.2 km/s
33. 0.75 mm
35. 39 μs
37. (a) $\frac{2}{3}A$; (b) $\frac{1}{3}A$
39. (a) 0.12 s; (b) 1.1 cm; (c) 0.86 cm
41. 7
43. (a) 280 Hz; (b) 70 Hz; (c) 210 Hz
49. (a) 16.6 cm; (b) 424–457 Hz
51. 0.33 Hz
53. 91 Hz
55. 253 m/s
57. 43 m/s = 154 km/h
59. (a) 2800 Hz; (b) 933 Hz
61. 25 m/s
63. $u/v = 1/\sin 45° = 1.4$
65. (a) 5.5×10^{-4} W/m²; (b) 87 dB
67. (a) 112 m/s; (b) 4, 5
69. 960 m/s
71. 0.445 s
73. (a) $\lambda = 5.0$ m, $f = 0.56$ Hz; (b) $\lambda = 2.5$ m; $f = 0.79$ Hz
75. 1.36
79. 16 kHz
81. $\dfrac{4\sqrt{\mu_1 \mu_2}}{(\sqrt{\mu_1} + \sqrt{\mu_2})^2}$

CHAPTER 18

1. 1.2 kg
3. 10^{-14}
5. (a) 81 N; (b) 65 N
7. 1 in. $H_2O = 249$ Pa
9. 2×10^7 N, or 2000 tons
11. 21 N
13. 0.25 m²
15. No; $F = 2.3 \times 10^4$ N, or 2.5 tons
17. 1700 kg/m³
19. ~90 m
21. 890 Pa gauge
23. 8.1×10^{10} N
25. 3.6 mm
27. 93 cm higher in the eye
29. 8.11×10^3 kg
31. 44 kg
33. 0.75 %
35. 59 g
37. 27 m
39. (a) 49 kg; (b) 2500 kg
43. (a) 1.8×10^4 m³/s; (b) 1.5 m/s
45. 1.75 m/s
47. (a) $h_2 = h_1$; (b) $h_2 = h_1 - \dfrac{3v^2}{2g}$
49. 14.3 m

51. 7.2 cm³/s
53. (a) no; (b) yes
55. 13
57. 70%
59. $A\sqrt{2gh}$
61. (a) 14 m/s; (b) 2.2 m
65. (a) 25 L/s; (b) 55 m/s; (c) 1.8 kPa
67. $t = \dfrac{A_0}{A_1}\sqrt{\dfrac{2h}{g}}$
69. $P = P_a + \rho g h_0 + \dfrac{1}{2}\rho\omega^2 r^2$;   (b) $h = h_0 + \dfrac{\omega^2 r^2}{2g}$
71. (a) $\rho(h) = \dfrac{P_0}{h_0 g} e^{-h/h_0}$; (b) 5.7 km

PART 2   CUMULATIVE PROBLEMS

1. (a) $\ell = \dfrac{4M}{\pi d^2 \rho}$; (b) $T = 4\sqrt{\dfrac{\pi M}{d^2 \rho g}}$
5. 17.2 cm

CHAPTER 19

1. 720
3. 20°C
5. −40
7. −196°C, −321°F
9. (a) 138 kPa; (b) 33.4 kPa; (c) 233 kPa
11. 1.37 L
13. 586 mm
15. 240 kcal
17. 0.36 kg
19. 24 days
21. (a) 23 kJ; (b) 337 kJ; (c) 65 kJ
23. 7.5 kW
25. 2.4 kg
27. (a) 560 g; (b) 0.27 K/s
29. 1.8 kg
31. 0.70 K
33. 1.6 K/s
35. 197 g
37. 56.2°C
39. 3.7 kW
41. 0.293 W
43. 25 ft²·°F·h/Btu
45. 200 W
47. (a) 12.3 ft²·°F·h/Btu; (b) 715 Btu/h
49. Will save 10 gallons/month
51. 23°C
53. 80%
55. −25°C
57. Drop by 5.9 K
59. 24°C
61. 480 W
63. 1151 K

65. $-2.5°C$
67. (a) $87; (b) $10
69. 4.65%
71. $\Delta T_{copper} = 0.16$ K; $\Delta T_{iron} = 2.1$ K
73. 2.9 J
77. 10 hours

CHAPTER 20

1. 2.6 m^3
3. 1.8 MPa
5. (a) 27 L; (b) 330 K
7. 2.7×10^7
9. 11 L
11. 515 kPa
13. (a) 1.27 atm; (b) 0.0268 mol; (c) 0.786 atm
15. 2.88×10^3 K
17. 10^{10} K; gas molecules would dissociate first
19. 268 K, compared with ideal gas 292 K
21. 1.76 MPa
23. (a) 9.1×10^{20}; (b) 2.0×10^{20}
25. 22 kJ
27. 3.9 kg
29. 5.96 MJ
31. 1.3×10^{10} kg
33. 564 W
35. 44 minutes
37. 48 min
39. (a) 117 s
41. 3.55 MJ
43. 64°C
45. 177 g
47. 135 g solid in 865 g liquid, at 234 K
49. 5.0 kg
51. 4.9°C
53. 1.00021 cm
55. 3.9 km
57. 43.6 mm^3
59. 307 K
61. (a) Drop by 0.0115 km^3; (b) increase by 0.048 km^3
63. 120 mol/m^3
65. 19 kW
67. 79 g
69. 1.2 kg ice, 0.80 kg water, all at 0°C
71. 50 min
73. (a) 61 h; (b) 52 h
79. 34 km

CHAPTER 21

1. 29 kJ
3. Increases by 250 J
5. 140 kW
7. 0.02°C
9. $2P_1V_1$
11. 1.2 kJ

13. 4.3 kJ
15. 190 K
17. 1.99 kJ
19. (a) 399 J; (b) 264 kPa
21. (a) 571 kPa; (b) 438 J
23. 440°C
25. $V = 0.18V_0$
27. (b) Gas does 13 J of work; (c) 22 J heat lost from gas
29. (a) 300 K, 1.5 kJ; (b) 336 K, 0 J; (c) 326 K, 429 J
31. (a) 39.9 kPa; (b) 83.3 kPa; (c) 80.2 kJ
33. 928 J
35. (a) 211 J; (b) 12.9 L
37. 75°C
39. 128 J
41. (a) $\frac{9}{2}R$; (b) $\frac{11}{9}$
43. 57.7%
45. 79%
47. 20 mol
49. Drops 23.1 K
51. 343 K
53. 28 kPa
55. (a) 598 J; (b) 2500 J flows in
57. 25 m
59. $\frac{4}{3}P_1V_1$
63. (a) 2.5 kJ; (b) 447 K
69. (a) $M = \dfrac{P_0A}{g}\left[\dfrac{h_1}{h_2} - 1\right]$; (b) $M = \dfrac{P_0A}{g}\left[\left(\dfrac{h_1}{h_2}\right)^\gamma - 1\right]$

CHAPTER 22

1. (a) $12!/6!$, or 6.7×10^5 states; (b) about 1 in 1000
3. 5×10^{24} J, assuming oceans cover 75% of Earth to an average depth of 3 km; this is about 20,000 times annual use
5. (a) 27%; (b) 7.0%; (c) 77%
7. 0.95 K
9. 52% winter, 48% summer
11. (a) 1.75 GW; (b) 43%; (c) 505 K
13. 2×10^7 kg/s, slightly more than the Mississippi's flow
15. (a) 39%; (b) 550 J; (c) 190°C
17. 53.3 kJ
19. (a) 4.3; (b) maximum COP = 11
23. (a) COP$_{summer}$ = 13, COP$_{winter}$ = 3.5; (b) 0.076 J; (c) 0.22 J
25. (a) 561.7 J; (b) 464.1 J; (c) 97.66 J; (d) 17.4%; (e) $T_c = 403$ K, $T_h = 487$ K
29. 718 K
31. 1.22 kJ/K
33. 8.9°C
35. 1.36×10^8 J/K
39. (a) 53 J/K; (b) 74 J/K; (c) 0
41. (a) -109 J/K; (b) 122 J/K; (c) 13 J/K
43. $\Delta S_{AB} = 68.5$ J/K, $\Delta S_{BC} = 45.7$ J/K, $\Delta S_{CA} = -114.2$ J/K
45. 470 kPa
47. (a) 69%; (b) 967 K
49. Decrease in $T_{minimum}$
51. $58

53. (a) $W = Q = 345$ J; (b) $e = 24\%$
55. 598 J/K
57. About 30 km for 200 kg of water and a 60 kg bather
59. 166 MW
61. (a) $1 - 5^{1-\gamma}$; (b) $3T_{min}(5^{\gamma-1})$; (c) $e_{Carnot} = 1 - \frac{1}{3}(5^{1-\gamma})$
63. (a) 7.94; (b) 5.26; (c) $P_n = 2.96P_s$
65. (a) $T_h = T_{h0}e^{-P_0t/mc(T_{h0}-T_c)}$;

   (b) $P = 0$ at $t = \dfrac{mc(T_{h0} - T_c)}{P_0}\ln\left(\dfrac{T_{h0}}{T_c}\right)$

67.

	P	V	T	U − U_A	S − S_A
A					
B			$3.4\,T_0$	$6.0\,P_0V_0$	$3.1\,P_0V_0/T_0$
C	$1.5\,P_0$	$2.2\,V_0$	$3.4\,T_0$	$6.0\,P_0V_0$	$3.8\,P_0V_0/T_0$
D			$3.0\,T_0$	$5.0\,P_0V_0$	$3.8\,P_0V_0/T_0$

PART 3 CUMULATIVE PROBLEMS

1. $e = 1 - r^{1-\gamma}\left[\dfrac{r_c^\gamma - 1}{\gamma(r_c - 1)}\right]$

3. $W = an^2\left(\dfrac{1}{V_2} - \dfrac{1}{V_1}\right) + nRT\ln\left(\dfrac{V_2 - bn}{V_1 - bn}\right)$

5. (a) $t_1 = \dfrac{L_fMT_h}{P_hT_0}$; (b) $P = P_h\left[1 - \dfrac{T_0}{T_h}e^{P_h(t-t_1)/McT_h}\right]$,

   (c) $t_2 = t_1 + \dfrac{McT_h}{P_h}\ln\dfrac{T_h}{T_0}$, with L_f the heat of fusion of ice and c the specific heat of water.

CHAPTER 23

1. Several coulombs
3. (a) *uud*; (b) *udd*
5. About 10^9 N; about 10^6 times typical human weight
7. 8.2×10^{-8} N
9. 21.8 μC
11. -3.3 μC
13. $14\hat{\imath} - 7.4\hat{\jmath}$ N
17. 15 μC
19. $1.6\hat{\imath} - 0.33\hat{\jmath}$ N
21. $\dfrac{kq^2}{a^2}\left(\sqrt{2} + \dfrac{1}{2}\right)$
23. $q_2 = 143$ μC; $q_3 = 116$ μC
25. 3.8×10^9 N/C
27. (a) 2.2 MN/C; (b) 77 N
29. 5.15×10^{11} N/C
31. (a) 2.0 MN/C, upward; (b) 0.82 MN/C, downward; (c) 58 MN/C, downward
33. $-4e$
35. (a) $\mathbf{E} = \dfrac{2kqy}{(a^2 + y^2)^{3/2}}\hat{\jmath}$; (b) $y = \pm a/\sqrt{2}$
37. (a) $8.0\hat{\jmath}$ GN/C; (b) $190\hat{\jmath}$ MN/C; (c) $216\hat{\jmath}$ kN/C
39. 39 pm

43. 2.1 MN/C
45. 339 kN/C, upward
47. $-\dfrac{2k\lambda_0}{\pi\ell}\hat{\imath}$
49. -137 nC
53. 1.1 kN/C
55. (a) 2.5 μC/m; (b) 3.0×10^5 N/C; (c) 1.8 N/C
57. 3.3×10^{-12} kg
59. (a) 1.35 cm; (b) reverses direction, accelerates and exits field region at 3.8×10^5 m/s
61. $\ell\sqrt{eE/md}$
63. 2.8 Mm/s
65. -14 μC/m
67. (a) 3.0 mN·m; (b) 11 mJ
69. (b) Attractive
71. $x = -8.09$ nm
73. $\dfrac{k\lambda_0}{\ell}\hat{\imath}$
75. $2\sqrt{2}kQ/\pi a^2$
77. 1.4 cm
79. $-4q$, a distance $3a$ to the right of $-q$
81. 50.7 kN/C, downward
83. 7.0 cm; 0.54 μC
85. (a) 5.3×10^{-12} N, to right; (b) 5.3×10^{-12} N, to left

CHAPTER 24

1. $+3$ μC
5. (a) 1.7 kN·m^2/C; (b) 1.2 kN·m^2/C; (c) 0
7. 490 N·m^2/C
9. $\pi R^2 E$
11. (a) $-q/\varepsilon_0$; (b) $-2q/\varepsilon_0$; (c) 0; (d) 0
13. 4.9×10^4 N·m^2/C
15. (a) 0.69 MN·m^2/C (b) -0.69 MN·m^2/C (c) 0
17. 10 kN/C
19. 1.8×10^{12} N/C
21. (a) 2.2×10^5 N/C, outward; (b) 2.5×10^4 N/C, outward; (c) 4.0×10^3 N/C, inward
23. (a) $8kQ/R^2$, inward; (b) $kQ/4R^2$, inward; (c) (a) would not change, (b) would become 0
25. (a) 3.6 MN/C; (b) 3.8 MN/C; (c) (a) would not change, (b) would nearly double
27. as $1/r$
29. 6.3 μC/m^3
33. 3.6 mC/m^3
35. 58 nC/m^2
37. $E_1 = \sigma/2\varepsilon_0$, left; $E_2 = \sigma/2\varepsilon_0$, right; $E_3 = 3\sigma/2\varepsilon_0$, right; $E_4 = \sigma/2\varepsilon_0$, right
39. 18 N/C
41. (a) $x < 1.83$ cm; (b) $x > 54.5$ cm
43. 1.6×10^5 N/C
45. (a) $\rho = 0$; (b) $\sigma = 4.0$ mC/m^2; (c) other charges would destroy the symmetry, making σ nonuniform
47. (a) 0.50 μC/m^2; (b) 56 kN/C
49. (b) $-Q$

51. 1.8 MN/C
53. (a) 0; (b) 180 kN/C; (c) 0; (d) 20 kN/C
57. (a) $4kq/R^2$; (b) $3kq/4R^2$
59. (a) 0; (b) 1.3 MN/C; (c) 0
61. (a) 1.9×10^{11} N/C; (b) 3.6×10^{10} N/C
63. $\dfrac{\rho r}{3\varepsilon_0} - \dfrac{\rho a^3}{3\varepsilon_0 r^2}$
65. $\frac{1}{3} E_0 a^2$
67. 0.39 μs
69. $E_{\text{in}} = \rho_0 x^2/2\varepsilon_0 d$; $E_{\text{out}} = \rho_0 d/8\varepsilon_0$
73. (a) 0; (b) $(ac/\varepsilon_0 r^2)(e^{-1} - e^{-r/a})$;
   (c) $(ac/\varepsilon_0 r^2)(e^{-1} - e^{-b/a})$
77. $+10.6 \ \mu$C/m^2 on both outer faces, $\pm 36.9 \ \mu$C/m^2 on inner faces

CHAPTER 25

1. 600 μJ
3. 3.0 kV
5. 910 V
7. 5.6 kV/m
9. Proton and He$^+$ both gain 100 eV $= 1.6 \times 10^{-17}$ J; α gains 200 eV $= 3.2 \times 10^{-17}$ J
11. 4.5 V
13. 0.23 MC
17. 6.1 μC
19. 27.2 V
21. $Q = 5.4$ nC, $r = 17$ cm
23. (a) 442 kV; (b) 9.2 Mm/s
25. kQ/R.
27. $V(x) = -\frac{1}{2} ax^2$
29. 52 nC/m
31. $x = -a/2$, $x = a/4$
33. (a) 2.6 kV; (b) 1.8 kV; (c) 0
35. 12 μm
37. $2kQ/R$
39. $2\pi k\sigma(\sqrt{x^2 + b^2} - \sqrt{x^2 + a^2})$
41. (a) $V(x, y) = -E_0(x + y) = -150(x + y)$ V/m;
   (b) 150V
45. (a) $\mathbf{E} = -ay\hat{\mathbf{i}} - ax\hat{\mathbf{j}}$
47. $\mathbf{E} = 10\hat{\mathbf{i}} + 5.8 \, \hat{\mathbf{j}}$ V/m
49. (a) 4 V; (b) $E_x = 1$ V/m, $E_y = -12$ V/m, $E_z = 3$ V/m
51. $\mathbf{E} = \dfrac{kQx}{(x^2 + a^2)^{3/2}}\hat{\mathbf{i}}$
53. $E = V_0/R$, radially outward
55. 3 kV
57. (a) 34 kV, -9.0 kV; (b) 12.6 kV on each; (c) 24 nC
59. (a) 43 kV; (b) 1.7 MV/m; (c) 540 V; (d) 0
61. 1.55 keV $= 2.47 \times 10^{-16}$ J
63. (a) $\dfrac{2kqx}{x^2 - a^2}$; (b) $\dfrac{2kq}{x}$
65. (a) 27 kV; (b) no change
67. -7.5 V
69. (a) $x = -3$ m, 0 m, 1 m; (b) $\mathbf{E} = (3x^2 + 4x - 3)\hat{\mathbf{i}}$;
   (c) $x = -1.87$ m, 0.535 m

71. (a) 7.2 kV; (b) 14.4 kV
73. 14 cm, 1.7 nC
75. 23 nC/m
77. (a) $V(x) = \dfrac{k\lambda_0 x}{\ell^2}\left[x \ln\left(\dfrac{2x + \ell}{2x - \ell}\right) - \ell\right]$;
   (b) $\frac{1}{12}\lambda_0\ell$; (c) $\dfrac{k\lambda_0\ell}{12x}$
79. $y^2 + (x - \frac{5}{3}a)^2 = (\frac{4}{3}a)^2$; i.e., a circle
81. $\dfrac{kq}{2a}\ln\left(\dfrac{\sqrt{2} + 1}{\sqrt{2} - 1}\right) \simeq \dfrac{0.881 \ kq}{a}$

CHAPTER 26

1. $3kq^2/\ell$
3. 4.88 kJ
5. $v = 2q\sqrt{\dfrac{k}{m\ell}}$
7. $6kq^2/a$
9. (a) 2.0 MV/m; (b) 9.9 kV; (c) 5.5 mJ
11. (a) 0.74 μC; (b) 40 kV
13. $\frac{1}{2} kQ^2\left(\dfrac{1}{a} - \dfrac{1}{b}\right)$
15. (b) $dW = 2kq \ dq/a$; (c) $W = kQ^2/a$
17. 1 km^3
19. 9×10^{30} J/m^3
21. 24 μJ
25. $(kQ^2/R)(2^{2/3} - 1) = 0.60$ mJ
27. $U/\ell = \pi a^4 \rho^2/16\varepsilon_0$
29. ± 14 C
31. 6.5 mF
33. 0.74 nF
35. 55 pF
39. 70 nF
41. 1 μF stores 15 times as much energy
43. (a) 30 μF; (b) 0.1 μF; (c) 0.01 μF
45. (a) 5.0 kW; (b) 250 μF; (c) 0.50 W
47. (a) Increases by factor of 2.5; (b) drops to 40% of its original value
51. Equal
53. (a) 6.0 μF; (b) 0.55 μF; (c) 0.83 μF, 1.3 μF, 1.5 μF, 2.2 μF, 2.8 μF, 3.7 μF
55. (a) 2 series pairs in parallel or two parallel pairs in series; (b) 4 in series
57. 0.86 μF
61. $\pm 1\%$
63. (a) 64 V; (b) drops from 16 mJ to 14 mJ
65. (a) 3.5 mm; (b) 87 kV
67. 126 pF
69. (a) 50 nC, 170 nC; (b) 23 μJ, 77 μJ
71. 1.4 mm
73. (a) 0.90 J; (b) 1.8 J; energy comes from work done by agent separating the plates
75. (a) 1.2 μF; (b) 24 μC
77. 13 min

79. $\dfrac{kq^2}{a}\left(2\sqrt{3} - \dfrac{15}{2}\right) \simeq -4.04\dfrac{kq^2}{a}$

81. (a) $C = \frac{1}{2}(\kappa + 1)C_0$; (b) $U = \dfrac{C_0 V_0^2}{\kappa + 1}$;

   (c) $F = \dfrac{2C_0 V_0^2(\kappa - 1)}{L(\kappa + 1)^2}$, into capacitor

83. (a) 75 mF; (b) 1.4×10^{12} J
85. (a) $Q^2/2\varepsilon_0 a$; (b) $Q^2/\varepsilon_0 a$; (a) is right since (b) includes the fields of *both* plates, and a plate doesn't experience a force from its own field
87. $31d^2kp^2/1280\ell^5$
89. 7.2 pF/m

CHAPTER 27

1. 9.4×10^{18}
3. 2.9×10^5 C
5. (a) 480 m; (b) 50 mA; (c) 14 μA
7. 0.17 mm/s
9. 1.23 cm/s
11. 0.31 mA
13. (a) 37 A/m²; (b) 86%
15. (a) $2t^2 - t^3$; (b) $t = 2$ s
17. 0.17 V/m
19. 2.2×10^{-6} Ω·m
21. (a) 5.95×10^7 (Ω·m)$^{-1}$; (b) 4.5 (Ω·m)$^{-1}$
23. 6.4×10^{-15} s
25. 25 Ω
27. 2.34 mA
29. 25 mA
31. 1.5 kA
33. (a) 17 mΩ; (b) 86 A; (c) 18 MA/m²; (d) 1.7 V/m
35. $d_{Al} = 1.26\ d_{Cu}$
37. (a) 2.07 cm; (b) 2.60 cm; (c) aluminum
39. 34 mΩ
41. 1.38 kW
43. 160 μA
45. 240 Ω
47. 48 W
49. 960 W·h = 3.5 MJ
51. Resistor with more power has $\sqrt{2}$ times greater diameter
53. (a) 150 A; (b) 3.4 km
55. 0.54 mA
57. 0.94 Ω
59. (a) 8.7 kA; (b) 15%
61. 2.9 A
63. 2.5 A
65. 203 A
67. 17 J/K

CHAPTER 28

5. 1.4 hours
7. 6.0 V
9. 229 kΩ

11. 1.5 A
13. 0.02 Ω
15. 45 Ω
17. 30 A
19. R_1 for each
21. 24
23. (a) 162 Ω; (b) 125 mW
25. (a) $\dfrac{R_1 \mathscr{E}}{R + 2R_1}$; (c) $\frac{1}{2}\mathscr{E}$
27. 2.45 W
29. $(\mathscr{E}_2 R_1 - \mathscr{E}_1 R_2)/(R_1 + R_2)$
31. $\frac{7}{5}R$
33. $I_1 = 2.79$ A, $I_2 = 2.36$ A, $I_3 = 0.429$ A
35. $\mathscr{E}_2 > 5.49$ V
37. 1.6% low
39. 1.2 kW41.24.99 kΩ
41. 24.99 kΩ
43. (a) 20 V; (b) 3.0 mA
49. (a) 9.0 V; (b) 1.5 ms; (c) 0.32 μF
51. (a) 0.35 s; (b) 0.17 s
53. 3.4 μF
55. (a) $I_1 = 25$ mA, $I_2 = 0$, $V_C = 0$; (b) $I_1 = I_2 = 10$ mA, $V_C = 60$ V; (c) $I_1 = 0$, $I_2 = 10$ mA, $V_C = 60$ V; (d) $I_1 = I_2 = 0$, $V_C = 0$
57. (a) $3\mathscr{E}/4R$; (b) $2\mathscr{E}/3R$
59. 3.4 kΩ
61. (a) 4.51 V; (b) 35.2 Ω
63. 1.07 A, left to right
65. (a) 13.0 V; (b) 2.23 mA
67. 15.9 ms
69. 83 μs
71. (a) 0; (b) 1.0 A; (c) 0.75 A; (d) 0; (e) 1.0 A; (f) 3.0 A; (g) 3.0 A; (h) 1.0 A
73. (b) 22 hours
75. $I_1 = \dfrac{\mathscr{E}_1 R_2 + \mathscr{E}_1 R_3 - \mathscr{E}_2 R_3}{R_1 R_2 + R_1 R_3 + R_2 R_3}$;

   $I_2 = \dfrac{\mathscr{E}_1 R_3 - \mathscr{E}_2 R_1 - \mathscr{E}_2 R_3}{R_1 R_2 + R_1 R_3 + R_2 R_3}$;

   $I_3 = \dfrac{\mathscr{E}_1 R_2 + \mathscr{E}_2 R_1}{R_1 R_2 + R_1 R_3 + R_2 R_3}$;

CHAPTER 29

1. (a) 1.6 mT; (b) 2.3 mT
3. (a) 2.0×10^{-14} N; (b) 1.0×10^{-14} N; (c) 0
5. (a) $-1.1\hat{\imath} +1.5\hat{\jmath} + 1.7\hat{k}$ mN
7. $\mathbf{B} = 0.13\hat{k}$ T; $\mathbf{v}_2 = -14\hat{\imath}$ km/s
9. 7.86×10^{-14} N
11. 40.1° or 140°
13. (a) $1.2\hat{\imath} + 0.45\hat{\jmath}$ fN; (b) $-1.2\hat{\imath} + 35\hat{\jmath} - 15\hat{k}$ fN
15. 3.9 mm
17. 1.5 mT
19. $r_{proton} = 43\ r_{electron}$
21. 30 mT; yes
23. 1.3 μs

25. (a) 15 MHz; (b) 19 MeV; (c) 6467
29. 0.43%
31. 1.1 mm, 2.6 mm
33. 0.38 N
35. (a) 49 mT; (b) 0.73 N/m
37. 43 kN
39. 0.12 T
41. 21 mN, diagonally toward the upper right
45. 76 mT
47. (a) 1.1 mA·m²; (b) 1.0 mN·m
49. 0.15 A·m²
51. (a) 0.35 A·m²; (b) 4.2×10^{-2} N·m
53. 1.97×10^{-25} J $= 1.23$ μe V
55. $42\hat{\imath} + 88\hat{\jmath} - 25\hat{k}$ fN
57. 30 km
59. 0.27 N, to right
61. 0.010 T
63. 6.8 mm
65. (a) 12.8 N/m; (b) 24.9 cm
67. $\tan^{-1}(mg/2IaB)$
69. 77 mT
73. $\mu = 9.25 \times 10^{-24}$ A·m²

CHAPTER 30

1. 12 cm
3. 1.23 mT
5. 2.8×10^9 A
7. 0.875 cm²
9. 732
11. Between the wires, 2.0 cm from center of 5.0-A wire
13. 23° west of magnetic north
15. $\mu_0 I/4a$, into page
17. $\dfrac{\mu_0 I}{4ab}(b - a)$, out of the page
19. 5 μN
21. 3.8 mm
23. 13.2 μN/m, at 71.6° below right-pointing horizontal
25. 23.5 mN
27. 7.0 A
29. 24 A
31. 123 mA
33. (a) 0; (b) 0.36 mT; (c) 1.9 mT; (d) 0.50 mT
35. (a) $\mu_0 J_0 r^2/3R$; (b) $\mu_0 J_0 R^2/3r$
37. (a) 5.3 mm; (b) maximum; (c) 130 A
39. 0.83 mT
41. (a) $\sqrt{2}\mu_0 J_s/2$ both inside and outside (but inside and outside fields are mutually perpendicular)
43. (a) 0; (b) $\dfrac{\mu_0 I(r^2 - a^2)}{2\pi r(b^2 - a^2)}$; (c) $\dfrac{\mu_0 I}{2\pi r}$
45. 17 T
47. (a) 38 mT; (b) 5.9 μT
49. 3.3 m
51. $1/2\pi nR$
53. $\chi_M = -1.6 \times 10^{-3}$; diamagnetic

55. 7.2×10^3
57. $I_{\text{outer}} = 2I_{\text{inner}}$, in opposite direction
59. Out of page
61. 1 km
63. (a) 8.0 μT; (b) 4.0 μT; (c) 0
65. (a) 0.40 mT; (b) 1.0 nT
67. (a) $\dfrac{2\sqrt{2}\mu_0 Ia^2}{\pi(a^2 + 4x^2)\sqrt{a^2 + 2x^2}}$; (b) $\dfrac{\mu_0 Ia^2}{2\pi x^3} = \dfrac{\mu_0 \mu}{2\pi x^3}$
69. 55A

CHAPTER 31

3. 1.4×10^{-4} T·m²
5. 160 T/s
7. (a) 0.30 A; (b) 0.20 A
9. $B^2 \ell^3 v/R$
11. 39 V
13. 0.32 V
15. 9.0 μA, counterclockwise in Fig. 31-5
17. (a) $I = -25$ mA for $0 < t < 2$ s; $I = 0$ for 2 s $< t < 3$ s; $I = +25$ mA for 3 s $< t < 5$ s; (b) $P = 3.1$ mW for $0 < t < 2$ s and 3 s $< t < 5$ s; otherwise 0
19. (a) Left to right; (b) 140 μA; (c) 28 μA
21. (a) $I_R = -I_{\text{peak}}\cos \omega t$, where (b) $I_{\text{peak}} = 110$ mA; (c) 0
23. 0.16 s
25. 7.1 mV
27. (a) Downward; (b) $(B\ell v)^2/R$
29. $v_{\text{final}} = \mathcal{E}/B\ell$; R affects time to reach final speed
31. (a), (b) Both 6.7 mA, counterclockwise; (c) 0.44 mW in both cases
33. (a) 25 mA; (b) 1.25 mN; (c) 2.5 mW; (d) 2.5 mW
35. (a) 77 mV; (b) 94 mV
37. 1.08 T/ms
39. (a) 530 V/m; (b) counterclockwise (c) 4.0 keV
41. (a) 32.1 V/m; (b) 1.6 ms, 5.7 ms
43. $E = \frac{1}{2}bh$, to the left above the field region and to the right below
45. $\frac{16}{3}B_0 x_0^2$
47. 42 mA, clockwise
49. $mgR/B^2\ell^2$
51. $\frac{1}{2}BR^2\omega$
53. 58 T/ms
55. $I = at^2$, with $a = 0.81$ A/s²
59. (a) 2.4 km; (b) 2.89 kW; (c) $6.94; (d) their generators get harder to turn, so they use more fuel
61. (a) $(NAB\omega \sin \omega t)^2/R$; (b) $(NAB \sin \omega t)^2\omega/R$
65. $V = \frac{1}{2}B\ell\theta_0^2\sqrt{g\ell}\sin(2t\sqrt{g/\ell})$

CHAPTER 32

1. 120 V
3. 12 V
5. (a) $-2\pi fMI_p \cos 2\pi ft$; (b) 133 mH
7. $\mu_0 n^2 \pi R^2 \ell$
9. $\dfrac{\mu_0 \ell}{2\pi}\ln\left(\dfrac{a + w}{a}\right)$

11. 3.2 mH

13. 40 kV

15. 1350

17. (a) 2.1 A; (b) -1.5 A (minus indicates direction reversal)

19. 26 A/ms

21. 11 A

25. 130 Ω

27. 15 s

29. (a) 0.11 H; (b) 20 mA

31. (a) 76 mA; (b) 4.4 V (c) 7.6 V; (d) 2.2 A/s; (e) 0.58 W

33. 400 Ω

35. 50 Ω

37. (a) 2.0 A; (b) 6.0 A; (c) 60 V

39. 100 mA

41. (a) 52 J; (b) 1.8 s

43. (a) 2.5 kW; (b) no

45. 15 W

47. (a) 5.7 MJ; (b) 31 mΩ; (c) 39 s

49. 9.9×10^8 J/m^3

51. 10^{11} times that of gasoline; 2600 times that of U-235

53. Smaller by factor $1/4n^2R^2$

55. 10^{18} J

57. $|\mathcal{E}| = 2Mbt$

59. 48 H

61. (a) 4.4 A; (b) 0; (c) 70 V

63. $(\mu_0 I^2 \ln 100)/4\pi$

65. (a) $L = \dfrac{\mu_0 N^2 \ell}{2\pi} \ln\!\left(\dfrac{R + \ell}{R}\right)$

67. (a) $P = I_0^2 R e^{-2Rt/L}$

69. $E/B = 1/\sqrt{\mu_0\varepsilon_0} = 3.0\times10^8$ m/s $= c$

CHAPTER 33

1. $V = 325 \sin(314t)$, with V in volts and t in seconds.

3. 9.9 V

5. (a) 350 mA; (b) 1.50 kHz

9. $V_{\text{rms}} = V_p/\sqrt{3}$

11. $I_1 = I_p \cos\phi$, $I_2 = I_p \sin\phi$

13. 45 mA

15. (a) 804 Ω; (b) 48.2 Ω; (c) 2.41 Ω

17. 1.47 μF

19. 15.9 kHz

21. (a) 10 mA; (b) 14 V; (c) 7.7 mA

23. 2.7 mC

25. 0.32 A

27. 8.23 kHz or 5.17×10^4 s^{-1}

29. 4.9 pF to 42 pF

31. 3.65 F

33. (a) 1.63 A; (b) 4.58 μs

35. (a) 63 V; (b) 89 V; (c) 764 mA; (d) 99 μJ

37. (a) $1/\sqrt{2}$; (b) 1/2; (c) $-1/\sqrt{2}$; (d) 1/2

39. Close B, wait 0.70 s, close A and simultaneously open B, wait 0.35 s, open A

41. 0.22 μF

43. 16.2 V

45. (a) 0.43 μH; (b) $R > 16\ \Omega$

47. (a) $V_{Rp} = 23.86$ V, $V_{Cp} = 28.19$ V, $V_{Lp} = -28.05$ V; (b) $V_{R\,\text{rms}} = 16.9$ V, $V_{C\,\text{rms}} = 256$ V, $V_{L\,\text{rms}} = 255$ V

49. (a) above; (b) current lags by approximately $38°$

51. (a) I leads V by $63.6°$; (b) I lags V by $81.7°$

53. 500 W

55. (a) 80 Ω; (b) 97 Hz

57. (a) 5.5%; (b) 9.1%; (c) large

59. 5.0 mA

61. (a) 8.91 V; (b) 1.38 kΩ

63. $4f_1$

65. 13.2 V

67. 0.64 Ω

69. 8.25 V

71. 36.2 nF, 11.5 nF

73. $C = \dfrac{L[\ln(U_0/U_1)]^2}{4\pi^2 N^2 R^2}$

77. $R = 400\ \Omega$, $L = 67$ mH, $C = 0.094\ \mu$F

79. 1.3 kN/m

CHAPTER 34

1. 1.3 nA

5. (a) 0; (b) 8.33×10^{-13} T; (c) 9.26×10^{-13} T

7. $-x$

9. (a) $\sqrt{2}E$; (b) $(\hat{\jmath} - \hat{\imath})/\sqrt{2}$

13. 0.24 s

15. 1 ns

17. 600 m

19. 2.32

21. 5000 km

23. 27.3 MHz

25. 15 kV/m

27. 1.1 pT

29. $63°$

31. $30°$

33. $0.34S_0$

35. $0.304S_0$

37. $S = S_0 \cos^2 \omega t$, with $\omega = 20\pi$ s^{-1}

39. 12 GW/m^2

41. 2.7×10^{-10} W/m^2

43. (a) 157 W/m^2; (b) 344 V/m; (c) 1.15 μT

45. 1.1%

47. 3.9×10^{26} W

49. (a) 8.3×10^{-26} W/m^2; (b) 7.9×10^{-12} V/m

51. (a) 4.6 kW; (b) 52.5 mV/m

53. (a) as $1/r$; (b) as $1/r^2$

55. 6.0 mPa

57. 5.0 μN

59. 6.18×10^3 years

61. $E_p = 40$ V/m, $B_p = 0.13\ \mu$T

63. 1/4

65. 4.2 μN

67. 17.7 μPa
71. (a) 1.0 MV/m; (b) 4.3 mm; (c) 64 mJ;
    (d) 2.1×10^{-10} kg·m/s; (e) 64 W
73. Long, with cylindrical symmetry
75. 0.3 μm

PART 4 CUMULATIVE PROBLEMS

1. 6.0 MV
3. 15.7 mA, top to bottom
5. (b) $c/\sqrt{\kappa}$

CHAPTER 35

1. 15°
3. ±0.5°
5. (a) 2; (b) 210°
9. 126 nm
11. 1.57
13. Ethyl alcohol
15. 1.83
17. 5.1 m
19. (a) 3.20×10^{14} Hz; (b) 937 nm
21. (a) 49.8°; (b) 42.2°; (c) 22.4°
23. (a) 61.3°; (b) 80.9°; (c) there is none
27. $\sqrt{2}$
29. 2.62×10^8 m/s
31. 53.5°
35. 36.5° to 38.1°
37. About 3.4 cm
39. 98.2%
41. 1.76
43. 2.3
45. 1.522
47. 1.88
49. 1.96×10^8 m/s
51. 2.32 m
53. 28.8°
55. (a) base; (b) 63.2°
57. 35.2°
59. $n = 1.17$
65. (b) 42.1°

CHAPTER 36

1. Mirror surface 55° to horizontal
3. 5 m
5. (a) Image height is 1/4 object height; (b) inverted
7. (a) 12.4 cm in front of mirror; (b) 3.22 times
9. (a) 24.3 cm behind the mirror; (b) 29 mm; (c) virtual
11. 1.1 m
13. 75 cm, 15 cm
15. 40 cm
17. Each 1.0 m
19. 12 cm
21. 47.7 mm

23. 2.3 mm
25. (a) Real, inverted, 7.7 cm high; (b) virtual, upright, 7.7 cm high
27. ($\ell = 30$ cm, $M = -2$); ($\ell = 60$ cm, $M = -\frac{1}{2}$)
29. 40 cm
33. $\ell' = -67.9$ cm, closer
35. 2
37. $R_{\text{plano-convex}} = \frac{1}{2}R_{\text{double convex}}$
39. Virtual image, 81 cm from lens on same side as object
41. (a) 40 cm; (b) 160 cm; (c) −170 cm (the lens becomes a diverging lens)
43. (a) 15.3 cm; (b) 1.63
45. Real image, magnified 2.74 times
47. (a) 5.0 m; (b) 37 cm
49. (a) Enlarged 2.9 times at 110 mm; (b) 11
53. 1.85 cm
55. −17.7 cm
57. (a) 2.0 m; (b) 3.0 m in front of the mirror
59. 25 cm
61. 96 cm
63. ($\ell = 85.4$ cm, $m = -1.81$); ($\ell = 155$ cm, $m = -0.552$)
65. Real image, 109 cm on other side of lens
67. $h' = -f\alpha$
69. No, one is virtual, the other real
73. (a) 1.1 m from primary; (b) elliptical

CHAPTER 37

1. 103 nm
3. 424 nm, 594 nm
5. (a) 75.2 nm; (b) violet
7. 368 nm
9. 5
11. 236
13. 375 nm (i.e., $\frac{3}{4}\lambda$)
15. 545 nm
17. 1022
19. 83.8 cm
21. (a) 95 cm; (b) 3.8 mm
23. 1.25 mm
25. 17.7 cm, 44.7 cm
27. 0.034°
29. (a) 17.1 cm; (b) 20.0 cm
31. 4
33. (a) 4.8°, 9.7°; (b) 2.9°, 6.8°
35. (a) 8.97°; (b) 51.3°
37. (a) 2nd; (b) 1st
39. 6th order
41. 3200
43. 415 nm, 581 nm
45. Echelle grating has 60% greater resolving power
47. 2000
49. $a/\lambda = 1$
51. 26.6°

53. 0.0162
55. 1.3 m
57. 0.025 mm
59. 6.9 km
61. For diameters greater than 14 cm
63. 77.2 nm
65. 7645
67. 0.01 nm
69. 46 m
71. 484 nm
75. 5542
77. Every 2.8 s
79. 6.0 km/s
81. 3.0 μm
83. $m_0 = (n - 1)d \sin \alpha / \lambda_0$

PART 5 CUMULATIVE PROBLEMS

1. $m_0 = (n - 1)d (\sin\alpha)/\lambda_0$
3. 3.2 min
5. (b) $x = \dfrac{\sqrt{3}}{2a} \ln\left[2(1 - ay - \sqrt{\tfrac{1}{4} - 2ay + a^2y^2})\right]$

CHAPTER 38

1. (a) 4.50 h; (b) 4.56 h; (c) 4.62h
3. (a) $v/c = \tan\theta$, with θ the amplitude of the sine curve. Here $\theta = 20''$ (20/3600 degree), giving $v = 29$ km/s; (b) the orbit is nearly circular
5. 41 cm
7. (a) 8.26 h; (b) 6.28 h
9. (a) 0.857c; (b) 9.69 min
11. 0.14c
13. $c/\sqrt{2}$
15. Twin A: 83 years; twin B: 40 years
17. 0.96c
19. 0.996c
21. (a) B first, by 5.74 min; (b) A first, by 13.1 min; (c) essentially simultaneous
23. A, by 6.4 My
27. (a) 21.7 m; (b) 12.4 m

29. 0.918c
31. 3.4
33. (a) 2.1 MeV; (b) 1.6 MeV
35. 0.36c
37. 5.3×10^{-8} kg·m/s, about the same as the insect
39. 5.9×10^{-30} kg
41. (a) 0.020c; (b) 0.55c; (c) 0.94c; (d) essentially c
47. (a) -0.75×10^{10} ly^2; (b) 99×10^{10} ly^2; events can be causally related only if $(\Delta s)^2 > 0$
49. 0.274c
51. (a) 16 ns; (b) 4.8 m
53. (a) 4.2 ly; (b) -2.4 years (i.e., B occurs earlier)
55. 0.395c
57. 0.9999953c
59. 598 m
61. 0.95c
65. (a) 1 cm; (b) 3 km; (c) 0.03 ly

CHAPTER 39

1. (a) 4.1 neV; (b) 2.1 eV; (c) 12.4 keV
3. (a) 1.6×10^{-24} J; (b) 3.9×10^{26} photons/s
5. (a) 1.7×10^{28} s^{-1}; (b) 3.2×10^{15} s^{-1}; (c) 1.3×10^{18} s^{-1}
7. β ($n_1 = 5$)
9. $n = 229$
11. 73 cm/s
13. 1.6 km/s
15. Electron: no; proton: yes
17. 12 pm
19. 0.0034°
21. 2.5 h
23. 20 minutes
25. 0.64 kg
27. 0
29. $\bar{u}\,\bar{u}\,\bar{d}$
31. $+e$
33. (a) 10.2 eV; (b) $n_1 = 2$, $n_2 = 1$
35. The proton
37. 1.16×10^6 m/s
39. 18 pm
41. (a) 1 eV; (b) 5 MeV
43. 10–20 billion years

PHOTO CREDITS

Unless otherwise acknowledged, all photographs are the property of Scott, Foresman and Company. Page abbreviations are as follows: (T) top, (C) center, (B) bottom, (L) left, (R) right, (INS) inset.

435TR	Diane Schiumo/Fundamental Photographs	559B	Richard Megna/Fundamental Photographs
435B	NASA	572	Will & Deni McIntyre/Photo Researchers
437	Dan McCoy/Rainbow	575R	Leonard Lessin/Peter Arnold, Inc.
441R	Rod Planck/Tony Stone Images	575L	Courtesy, Panasonic
442T	Larry Mulvehill/Photo Researchers	577L	Object courtesy of SLAC and Sidney Drell, photo by Oscar Kapp, University of Chicago
442B	J. Toers/George Hall Photos		
443	Globus Brothers	577R	Stanford Linear Accelerator Center/SPL/Photo Researchers
444	Peter Menzel/Stock Boston		
445ALL	ONERA photo, H. Werle' 1974, from Milton Van Dyke, An Atlas of Fluid Motion	585	Chris Jones/The Stock Market
		587R	Harold M. Waage/Princeton University
449	Bruce Thomas/The Stock Market	593	Erik Borg
455	Soames Summerhays/Photo Researchers	615	Christopher Johnson, University of Utah
456	NASA/SS/Photo Researchers	618B	Hulon Forrester/Video Display Corporation
458T	Jay M. Pasachoff	618T	AP/Wide World
458B	Leonard Lessin/Peter Arnold, Inc.	633BL	EPRI
461	G. R. Roberts	633TR	c.1993 Environmental Elements Corporation, Baltimore, Maryland. All Rights Reserved
468	Manuel G. Velarde		
469T	Keith Wood/Tony Stone Images	635	Erik Borg
469B	Camerique/H. Armstrong Roberts	642	Sandia National Laboratory
471	NASA/Goddard Inst. for Space/SS/Photo Researchers	646	Adam Hart/SPL/Photo Researchers
		647	Cliff Holmes
480	Paolo Koch/Photo Researchers	651	Erik Borg
487	Jay M. Pasachoff	653T	Erik Borg
488TL	Courtesy of Central Scientific Company	653B	Erik Borg
488TR	Courtesy of Central Scientific Company	655TL	Lawrence Livermore National Laboratory
488BL	Paul Silverman/Fundamental Photographs	669	Steve Allen/Peter Arnold, Inc.
490	Alexander Tsiarps/Photo Researchers	671	Lisas/Lockheed/SPL/Photo Researchers
494BL	Hartford Steam Boiler and Insurance Company	676T	Kitt Peak National Observatory
494BR	AP/Wide World	676B	Tom Mcugh/Photo Researchers
500	Steve Allen/Peter Arnold, Inc.	677	Princeton Plasma Physics Laboratory
501	Jim Richardson/H. Armstrong Roberts	678T	Erik Borg
503	Jan Halaska/Photo Researchers	678B	Courtesy, Motorola
507	David R. Austed/Stock Boston	679R	John Greenleigh/Courtesy Apple Computer
511	Alan Bolesta/H. Armstrong Roberts	679L	Courtesy of the University of Pennsylvania
513B	Thomas Ives/The Stock Market	680	Hank Morgan/Rainbow
513T	Geoff Juckes/The Stock Market	682	Erik Borg
526	Sinclair Stammers/SPL/Photo Researchers	689	Peter Arnold, Inc.
528BR	Environmental Protection Agency	690	R. Krubner/H. Armstrong Roberts
533	Rick Browne/Stock Boston	695	Erik Borg
534	Martin Bond/SPL/Photo Researchers	702R	Courtesy, Tektronix
535B	Hank Morgan/Photo Researchers	702L	Falcon/Simpson Electric Company
535T	Lockheed	702C	Simpson Electric
539L	Consumers Power Company	705	Michael Dalton/Fundamental Photographs
555	Werner H. Muller/Peter Arnold, Inc.	713T	Erik Borg
556	Gary Ladd	713B	Erik Borg
559T	Lawrence Berkeley Laboratory/SPL/Photo Researchers	723	Karageorge/H. Armstrong Roberts
		724	Richard Megna/Fundamental Photographs

726	Fermi National Accelerator Laboratory
728TL	Richard Megna/Fundamental Photographs
728TR	Richard Megna/Fundamental Photographs
730T	Paul J. Sutton/Duomo Photography Inc.
730B	Ivan Massar
731	Peter Arnold, Inc.
732	Jay M. Pasachoff
733C	Lionel F. Stevenson/Photo Researchers
733T	Princeton Plasma Physics Laboratory
733B	Dr. L. A. Frank, Univ. of Iowa
736T	Japan National Railways
736C	Mitsuhiro Wada/Gamma-Liaison
736B	Mitsuhiro Wada/Gamma-Liaison
737	Walker Scientific Company
742L	Jay M. Pasachoff
742R	Jay M. Pasachoff
744	Lawrence Berkeley Laboratory/SPL/Photo Researchers
751T	Ray Pfortner/Peter Arnold, Inc.
751B	American Association of Physics Teachers
754	Richard Megna/Fundamental Photographs
756L	National Optical Astronomy Observatories
756R	National Optical Astronomy Observatories
757	Richard Megna/Fundamental Photographs
758	Boebinger, AT&T Bell Labs
760	National Center for Atmospheric Research
767	Richard Megna/Fundamental Photographs
768L	Courtesy of Automatic Switch Company
768R	Dan McCoy/Rainbow
770	Michael Seul, AT&T Bell Labs
771	Richard Megna/Fundamental Photographs
782	Daniel Zirinsky/Photo Researchers
791L	Milt & Joan Mann/Cameramann International, Ltd.
791R	T. J. Florian/Rainbow
792T	David R. Frazier Photolibrary
792BL	Erik Borg
792BR	Erik Borg
793	Robert Mathena/Fundamental Photographs
794	Erik Borg
797	NASA
800	Courtesy, GE Appliances
803	Courtesy of IBM
805	Jay M. Pasachoff
813	Milt & Joan Mann/Cameramann International, Ltd.

815L	Jay M. Pasachoff
823	NASA
833	"Copyright 1994 Tektronix, Inc., All Rights Reserved"
845	Erik Borg
853BR	Camerique/H. Armstrong Roberts
853TL	Chris Rogers/The Stock Market
853CL	H. Abernathy/H. Armstrong Roberts
853BL	G. Fritz/H. Armstrong Roberts
853TC	Coco McCoy/Rainbow
853TR	Coco McCoy/Rainbow
853CR	Michael Mathers/Peter Arnold, Inc.
862	Dale Boyer/Photo Researchers
875	P. Degginger/H. Armstrong Roberts
876T	J. Amos/H. Armstrong Roberts
876B	National Radio Astronomy Observatory
877R	NASA
877L	JPL/NASA
878	Diane Schiumo/Fundamental Photographs
879R	Sepp Geitz/Woodfin Camp & Associates
886L	Courtesy AT&T, Bell Laboratories
886R	Hughes Research Laboratories
891R	David R. Frazier Photolibrary
891L	NASA
895	Pete Saloutos/The Stock Market
896	Schott Corporation
898	Richard Megna/Fundamental Photographs
899	NASA
901	Richard Megna/Fundamental Photographs
904	John W. Dunay/Fundamental Photographs
905	Richard Megna/Fundamental Photographs
907R	Foto Forum
907L	Luettge/ZEFA/H. Armstrong Roberts
908B	R. Giovaneill & H. R. Gillett/ CSIRO National Measurement Laboratory, Australia
908T	Alfred Pasieka/Peter Arnold, Inc.
908C	Alfred Leitner/Rensselaer Polytechnic Institute
909	Simon Fraser/SPL/Photo Researchers
912	Bruno J. Zehnder/Peter Arnold, Inc.
913L	Leonard Lessin/Peter Arnold, Inc.
913R	Len Lessin/Peter Arnold, Inc.
915T	Library of Congress
915ALL	James E. Kettler/Ohio University, Belmont
920	Dan McCoy/Rainbow
923B	Jay M. Pasachoff
923T	The Exploratorium

INDEX

Note: Page numbers in *italics* refer to illustrations; page numbers followed by t refer to tables.